D1509314

Astrophysical Data

Kenneth R. Lang

Astrophysical Data
Planets and Stars

With 33 Illustrations

Springer-Verlag
New York Berlin Heidelberg London Paris
Tokyo Hong Kong Barcelona Budapest

Kenneth R. Lang
Department of Physics and Astronomy
Tufts University
Medford, MA 02155
USA

Library of Congress Cataloging-in-Publication Data
Lang, Kenneth R.
Astrophysical data / Kenneth R. Lang.
p. cm.
Includes bibliographical references and indexes.
Contents: Planets and stars.
ISBN 0-387-97104-2 (N.Y. : alk. paper). – ISBN
3-540-97104-2 (Berlin : alk. paper)
1. Astrophysics – Observations – Handbooks, manuals, etc.
I. Title.
QB461.L35 1991
523.01 – dc20 91-11236

Printed on acid-free paper.

Camera ready copy provided by the author using LaTex.
Printed and bound by Edwards Brothers, Inc., Ann Arbor, MI.
Printed in the United States of America.

9 8 7 6 5 4 3 2 (Corrected second printing)

ISBN 0-387-97109-2 Springer-Verlag New York Berlin Heidelberg
ISBN 3-540-97109-2 Springer-Verlag Berlin Heidelberg New York

Preface

This volume of Astrophysical Data deals with Planets and Stars; a second volume, Part II, will give data for Galaxies and the Universe. They both provide basic data for use by all scientists, from the amateur astronomer to the professional astrophysicist. In this first volume, we not only provide physical parameters of planets, stars and their environment, but we also provide the celestial coordinates required to observe them.

Here we use c.g.s. units, for they are the most commonly used in astronomy and astrophysics; but our volume begins with astronomical and physical constants and the conversion factors needed for other units.

The next section concerns the planets and their satellites; it singles out the Earth and Moon for special treatment. Spacecraft rendezvous with the planets and satellites have led to improved values for their atmospheric compositions, orbital parameters, magnetic fields, masses, radii, rotation periods, and surface pressures and temperatures. This section also contains data for the asteroids, comets and their debris.

We then discuss everyday stars, beginning with the Sun, and continuing with basic stellar data, the brightest stars and nearby stars. Special categories of stars, such as the Wolf-Rayet stars, magnetic stars, flare stars, and RS CVn binary stars, are included.

The section on star clusters includes globular clusters, open clusters and OB associations. The next section concerns regions of star formation; it includes molecular clouds, dust clouds, circumstellar disks or shells, T Tauri stars, and Herbig-Haro objects. Our following discussion of the stellar environment includes emission nebulae, at both optical and radio wavelengths, reflection nebulae, and planetary nebulae.

We next consider dying stars, such as white dwarf stars, pulsars, and candidate black holes, as well as interacting binary systems including classical novae, dwarf novae and symbiotic stars. Supernovae explosions and their remnants are also discussed; the supernova remnants are mainly visible in the radio and X-ray windows of the electromagnetic spectrum. Here we also include detailed information on objects of special current interest, such as polars with their mega-Gauss fields, Supernova 1987A, Cassiopeia A, and the Crab Nebula.

The high-energy X-ray and gamma ray windows on the Universe are highlighted in the next chapter that includes compact X-ray sources, like Cygnus

X-3 and SS 433, low mass X-ray binaries, and gamma ray sources. Our book concludes with an Appendix of the 2,241 nearest stars and an extensive bibliography.

And how was such a large collection of data compiled? It began in the fine library at the Harvard-Smithsonian Center for Astrophysics (CFA); many thanks go to Joyce M. Watson and her colleagues at the CFA for their pleasant hospitality over the years. After searching the astrophysical journals and supplements for relevant material, I combined existing catalogues or tables to create new collections of data that have usually not been previously published in exactly the form presented here. Of course, there were times when the existing data could not be improved, and in these cases the original data were usually extracted from computerized versions provided by either the original authors or by Wayne H. Warren Jr. at the National Space Science Data Center. The original sources for all of the data are referenced in the text, and gathered together in the complete references given near the end of the volume.

Although I checked each number in every table twice, and nearly ruined my eyesight, human error is inevitable. So, notification of any omissions, errors or corrections will be greatly appreciated by users of future editions.

But why go to all this trouble? It brings no respect from many practicing scientists who regard book-writing as an appropriate activity for the aging, senile, or unproductive. And money is not a factor. The author receives about 25 cents an hour from the publisher. The real reason is to share useful facts and ideas. My reward will come when current, or future, astronomers and astrophysicists turn to this volume for that constant or physical parameter they needed, or when they use it to find an object that they wanted to view.

A very special acknowledgment goes to William R. Leeson, who assiduously entered much of this data into the Tufts' computer using LaTeX software. Equally important acknowledgment and thanks go to those who either supplied unique and valuable data or checked and supplemented draft tables. They include Agnes Acker, Edward Anders, Pierre Bastien, Hilmar W. Duerbeck, Von R. Eshleman, Nicolas Grevesse, Kenneth A. Janes, James Liebert, Gösta Lynga, Jeffrey E. Mc Clintock, E. Myles Standish Jr., Bjorn Ragnvald Pettersen, Gary Schmidt, Fred Seward, Joseph H. Taylor, Yervant Terzian, Robert F. Willson and Robert J. Zinn.

Kenneth R. Lang
Tufts University
New Years Eve, 1991

Contents

Part IV. Star Clusters and Associations

Part V. The Stellar Environment

Part VI. Dying Stars

Part I
Constants and Conversion Factors

1
Constants and Units

1.1 Physical Constants

Universal Constants

Speed of light in vacuum	c	=	$2.99792458 \times 10^{10}$ cm s^{-1}
Newtonian constant of gravitation	G	=	$6.67259(85) \times 10^{-8}$ cm^3 g^{-1} s^{-2}
Planck constant	h	=	$6.6260755(40) \times 10^{-27}$ erg s
Planck constant	h	=	$4.1356692(12) \times 10^{-15}$ eV s
Atomic angular momentum, $\hbar = h/2\pi$	$\hbar$	=	$1.05457266(63) \times 10^{-27}$ g cm^2 s^{-1}
Planck mass, $(\hbar c/G)^{1/2}$	m_P	=	$2.17671(14) \times 10^{-5}$ g
Planck length, $\hbar/m_P c$	l_P	=	$1.61605(10) \times 10^{-33}$ cm
Planck time, l_P/c	t_P	=	$5.39056(34) \times 10^{-44}$ s
Permeability of vacuum	μ_0	=	$4\pi \times 10^{-7}$ henry m^{-1}
Permittivity of vacuum, $\epsilon_0 = 1/\mu_0 c^2$	ϵ_0	=	$8.854187817 \times 10^{-12}$ farad m^{-1}

*The numbers in parenthesis are the one-standard-deviation uncertainties in the last digits. Adapted from Kenneth R. Lang, Astrophysical Formulae, Springer Verlag, 1986, and E. Richard Cohen and Barry N. Taylor, "The 1986 adjustment of the fundamental physical constants", Reviews of Modern Physics 59, 1121-1148(1987).

Electron, Proton, Neutron

Electron charge	e	=	$4.803242(14) \times 10^{-10}$ esu
	e	=	$1.60217733(49) \times 10^{-19}$ coulomb
Mass of electron	m_e	=	$9.1093897(54) \times 10^{-28}$ g
Rest mass energy of electron	$m_e c^2$	=	$0.51099906(15)$ MeV
	$m_e c^2$	=	8.1871×10^{-7} erg
Classical electron radius, $r_e = e^2/m_e c^2$	r_e	=	$2.81794092(38) \times 10^{-13}$ cm
Electron Compton wavelength, $\lambda_c = h/m_e c$	λ_c	=	$2.42631058(22) \times 10^{-10}$ cm
Thomson cross section, $\sigma_e = 8\pi r_e^2/3$	σ_e	=	$0.66524616(18) \times 10^{-24}$ cm^2
Proton mass	m_p	=	$1.6726231(10) \times 10^{-24}$ g
Proton-electron mass ratio	m_p/m_e	=	$1836.152701(37)$
Neutron mass	m_n	=	$1.6749286(10) \times 10^{-24}$ g
Neutron-electron mass ratio	m_n/m_e	=	$1838.683662(40)$
Neutron-proton mass ratio	m_n/m_p	=	$1.001378404(9)$
Bohr magnetron, $\mu_B = e\hbar/2m_e$	μ_B	=	$9.2740154(31) \times 10^{-21}$ erg Gauss^{-1}
Nuclear magnetron, $\mu_N = e\hbar/2m_e$	μ_N	=	$5.0507866(17) \times 10^{-24}$ erg Gauss^{-1}
Electron magnetic moment	μ_e	=	$9.2847701(31) \times 10^{-21}$ erg Gauss^{-1}
	μ_e	=	$1.001159652193(10)\ \mu_B$
Proton magnetic moment	μ_p	=	$1.41060761(47) \times 10^{-23}$ erg Gauss^{-1}
	μ_p	=	$1.521032202(15) \times 10^{-3}\ \mu_B$

Atomic Constants

Fine structure constant, $\alpha = 2\pi e^2/hc$	α	=	$7.29735308(33) \times 10^{-3}$
Inverse fine structure constant	$1/\alpha$	=	137.0359895(61)
Rydberg constant,for infinite mass, $R_\infty = 2\pi^2 m_e e^4/ch^3$	R_∞	=	$109737.1534(13)$ cm^{-1}
	cR_∞	=	3.2898369×10^{15} s^{-1}
Inverse Rydberg constant	$1/R_\infty$	=	911.2684
Radius of first Bohr orbit, $a_0 = h^2/4\pi^2 m_e e^2$	a_0	=	$0.529177249(24) \times 10^{-8}$ cm
Atomic mass unit, $m(C^{12})/12$	amu	=	$1.6605402(10) \times 10^{-24}$ g
Mass of hydrogen atom	m_H	=	1.673534×10^{-24} g
	m_H	=	1.007825 amu
Hyperfine splitting, hydrogen atom ground state	ν_H	=	$1420.405751786(2) \times 10^6$ s^{-1}
Zeeman displacement (Magnetic field strength H in Gauss)	$eH/4\pi m_e c$	=	1.39961×10^6 H Hz

*The numbers in parenthesis are the one-standard-deviation uncertainties in the last digits. Adapted from Kenneth R. Lang, Astrophysical Formulae, Springer Verlag, 1986, and E. Richard Cohen and Barry N. Taylor, "The 1986 adjustment of the fundamental physical constants", Reviews of Modern Physics 59, 1121-1148 (1987).

Physicochemical Contants

Avogadro number	N_A	=	$6.0221367(36) \times 10^{23}$ mole^{-1}
Loschmidt constant	n_o	=	$2.686763(23) \times 10^{19}$ cm^{-3}
Molar volume, N_A/n_o	V_m	=	$22.41410(19) \times 10^{3}$ cm^3 mole^{-1}
Molar gas constant	R	=	$8.314510(70) \times 10^{7}$ erg °K^{-1} mole^{-1}
Boltzmann constant, R/N_A	k	=	$1.380658(12) \times 10^{-16}$ erg °K^{-1}
Atomic mass unit, m(C^{12})/12	amu	=	$1.6605402(10) \times 10^{-24}$ g
Atomic rest mass energy	(amu) c^2	=	931.49432(28) MeV
	(amu) c^2	=	1.492419×10^{-3} erg

Radiation Constants

Radiation density constant, $a = 8\pi^5k^4/15c^3h^3$	a	=	$7.56591(19) \times 10^{-15}$ erg cm^{-3} °K^{-4}
Stefan-Boltzmann constant, $\sigma = ac/4$	σ	=	$5.67051(19) \times 10^{-5}$ erg cm^{-2} °K^{-4} s^{-1}
First radiation constant (emittance), $c_1 = 2\pi hc^2$	c_1	=	$3.7417749(22) \times 10^{-5}$ erg cm^2 s^{-1}
Second radiation constant, $c_2 = hc/k$	c_2	=	1.438769(12) cm °K
Wien displacement law constant λ_{max} T = $c_2/4.96511423$	b	=	0.2897756(24) cm °K

Plasma Constants**

Frequencies

Electron gyrofrequency $\nu_{ce} = \omega_{ce}/2\pi = eH/2\pi m_e c$	ν_{ce}	=	2.80×10^6 H Hz
Ion gyrofrequency $\nu_{ci} = \omega_{ci}/2\pi = eH/2\pi m_i c$	ν_{ci}	=	1.52×10^3 Z μ^{-1} H Hz
Electron plasma frequency $\nu_{pe} = \omega_{pe}/2\pi = (e^2 N_e/\pi m_e)^{1/2}$	ν_{pe}	=	$8.98 \times 10^3\ N_e^{1/2}$ Hz
Ion plasma frequency $\nu_{pi} = \omega_{pi}/2\pi = (e^2 Z^2 N_i/\pi m_i)^{1/2}$	ν_{pi}	=	2.10×10^2 Z $\mu^{-1/2}\ N_i^{1/2}$ Hz
Index of refraction $n = [1 - (\nu_{pe}/\nu)^2]^{1/2}$	n	=	$[1 - 8.06 \times 10^7 N_e/\nu^2]^{1/2}$

Lengths

Electron de Broglie length $\lambda = \hbar/(m_e kT_e)^{1/2}$	λ	=	$2.97 \times 10^{-6}\ T_e^{-1/2}$ cm
Classical distance of minimum approach	e^2/kT	=	$1.67 \times 10^{-3}\ T^{-1}$ cm

**Here H is the magnetic field strength in Gauss, N_e is the electron density in cm^{-3}, N_i is the ion density in cm^{-3}, the μ is the ion mass m_i in units of the proton mass m_p, Z is the ion charge in units of the electron charge e, T_e and T_i are the electron and ion temperatures in °K, respectively, and γ is the adiabatic index. Adapted from Kenneth R. Lang, Astrophysical Formulae, Springer Verlag, 1986, and David L. Book's, NRL Plasma Formulary, ONR, 1983.

Plasma Constants**

Lengths

Electron gyroradius $r_e = v_{Te}/2\pi\nu_{ce}$	r_e	=	$2.21 \times 10^{-2}\ \mathrm{T}_e^{1/2}\ \mathrm{H}^{-1}$ cm
Ion gyroradius $r_i = v_{Ti}/2\pi\nu_{ci}$	r_i	=	$0.95\ \mu^{1/2}\ \mathrm{Z}^{-1}\ \mathrm{T}_i^{1/2}\ \mathrm{H}^{-1}$ cm
Plasma skin depth	c/ω_{pe}	=	$5.31 \times 10^5\ N_e^{-1/2}$ cm
Debye length $\lambda_D = (kT/4\pi e^2 N)^{1/2}$	λ_D	=	$6.90\ \mathrm{T}^{1/2}\ \mathrm{N}^{-1/2}$ cm

Velocities

Electron thermal velocity $V_{Te} = (kT_e/m_e)^{1/2}$	V_{Te}	=	$3.89 \times 10^5\ T_e^{1/2}$ cm s^{-1}
Ion thermal velocity $V_{Ti} = (kT_i/m_i)^{1/2}$	V_{Ti}	=	$9.09 \times 10^3\ \mu^{-1/2}\ T_i^{1/2}$ cm s^{-1}
Ion sound velocity $C_s = (\gamma Z k T_e/m_i)^{1/2}$	C_s	=	$9.09 \times 10^3\ (\gamma Z T_e/\mu)^{1/2}$ cm s^{-1}
Alfvén velocity $V_A = (H/4\pi N_i m_i)^{1/2}$	V_A	=	$2.18 \times 10^{11}\ \mu^{-1/2}\ N_i^{-1/2}$ H cm s^{-1}

**Here H is the magnetic field strength in Gauss, N_e is the electron density in cm^{-3}, N_i is the ion density in cm^{-3}, the μ is the ion mass m_i in units of the proton mass m_p, Z is the ion charge in units of the electron charge e, T_e and T_i are the electron and ion temperatures in °K, respectively, and γ is the adiabatic index. Adapted from Kenneth R. Lang, Astrophysical Formulae, Springer Verlag, 1986, and David L. Book's, NRL Plasma Formulary, ONR, 1983.

1.2 Astronomical Constants

General Astronomical Constants

Astonomical unit of distance = mean Sun-Earth distance
= semi-major axis of Earth orbit

Astronomical Unit	AU	=	$1.4959787061 \times 10^{13}$ cm
Parsec (= 206264.806 AU)	pc	=	3.085678×10^{18} cm
Parsec	pc	=	3.261633 light year
Light year	ly	=	9.460530×10^{17} cm
Light time for 1 AU		=	499.004782 s = 0.00577552 days
Solar mass	$M_\odot$	=	1.9891×10^{33} g
Solar radius	$R_\odot$	=	6.9599×10^{10} cm
Solar mean density	$\bar{\rho_\odot}$	=	1.409 g cm^{-3}
Solar absolute luminosity	$L_\odot$	=	3.8268×10^{33} erg s^{-1}
Solar effective temperature	T_{eff}	=	5780 °K
Earth mass	M_e	=	5.9742×10^{27} g
Earth equatorial radius	R_e	=	6.378140×10^{8} cm
Earth mean density	$\bar{\rho_e}$	=	5.52 g cm^{-3}
1 tropical year	y	=	$3.15569259747 \times 10^{7}$ s
1 tropical year	y	=	365.24219 days
1 day	d	=	86400 s
1 Julian century	cen	=	36525 days
1 radian	rad	=	$2.0626480625 \times 10^{5}$ ″
1 radian	rad	=	57.2957795131 degrees
Hubble's constant	H_o	=	100 h km s^{-1} Mpc^{-1}
			(h = 0.2 to 1.0)
Hubble time, $1/H_o$	t_0	=	0.9778×10^{10} h^{-1} years
Hubble distance, c/H_o	R_0	=	2.9979×10^{3} h^{-1} Mpc
Critical density, $3H_o^2/8\pi G$ (to close expanding universe)	ρ_c	=	1.8790×10^{-29} h^2 g cm^{-3}
Mass density of galaxies	ρ_G	=	2.0×10^{-31} h^2 g cm^{-3}
Mass density of galaxies	ρ_G	=	3.0×10^{9} $M_\odot$ h^2 Mpc^{-3}
Volume density of galaxies	V_G	=	0.02 h^2 Mpc^{-3}
Cosmic microwave background temperature	T_0	=	2.735 ± 0.060 °K

Numerical Constants

Radian	=	57.2957795131 deg
	=	$3.43774677078 \times 10^3$ arcmin
	=	2.0626480625×10^5 arcsec
Steradian	=	$32400/\pi^2$
	=	3.28280635×10^3 deg^2
	=	1.1818×10^7 arcmin^2
	=	4.2545×10^{10} arcsec^2
Degree	=	0.0174532925 rad
Arcmin	=	2.908882×10^{-4} rad
Arcsec	=	4.848137×10^{-6} rad
Deg^2	=	3.0462×10^{-4} steradian
Arcmin^2	=	8.4617×10^{-8} steradian
Arcsec^2	=	2.3504×10^{-11} steradian

$\pi = 3.1415926536$
$e = 2.7182818285$
$\ln 2 = 0.6931472$
$log_{10} 2 = 0.3010300$
$\ln 10 = 2.3025850930$
$log_{10} e = 0.4342944819$
$(2\pi)^{1/2} = 2.506628$
$\pi^2 = 9.8696044011$

Gaussian function $\exp(-\pi x^2)$:

a. Probable error	=	0.2691 =	0.6745 σ
b. Mean absolute error	=	0.3183 =	0.7979 σ
c. Standard deviation	=	0.3989 =	σ
d. Width to half-peak	=	0.9394 =	2.355 σ
e. Equivalent width	=	1.0000 =	2.5066 σ

Exponential interval exp = 0.4343 dex

Magnitude mag = – 0.4000 dex (in star brightness)

JPL Planetary Ephemerides*

Scale (km/AU)	149 597 870.61
Speed of light (km/sec)	299 792.458
Earth-Moon mass ratio	81.300587

Planetary Masses

	$(AU)^3/day^2$	GM(sun)/GM(i)	km^3/sec^2
Sun	.2959122083–003	1	132 712 439 800.910
Mercury	.4912547451–010	6 023 600	22 032.080
Venus	.7243427513–009	408 525.1	324 857.478
Mars	.9549528942–010	3 098 710	42 828.287
Jupiter	.2825344423–006	1 047.3492	126 712 700.983
Saturn	.8459675616–007	3 497.91	37 940 448.534
Uranus	.1292027173–007	22 902.94	5 794 559.117
Neptune	.1524358094–007	19 412.25	6 836 530.500
Pluto	.2276247752–011	130 000 000	1 020.865
Moon	.1093189230–010	27 068 708.7	4 902.799
Earth	.8887692608–009	332 946.043	398 600.442
Earth + Moon	.8997011531–009	328 900.555	403 503.241

*Courtesy of E. Myles Standish, Jr., private communication, 1989, and R. A. Jacobson (1991–Neptune system).

IAU System of Astronomical Constants*

[Square brackets indicate modified values used in constructing the ephemerides]

Defining constants:

1. Gaussian gravitational constant	k	=	17.202 098 95 $cm^3\ g^{-1}\ s^{-2}$
2. Speed of light	c	=	29 979 245 800 cm s^{-1}

Primary constants:

3. Light-time for unit time	τ_A	=	499.004 782 s [499.004 7837...]
4. Equatorial radius of Earth, [IUGG value]	a_e	=	637 814 000 cm [637 813 700 cm]
5. Dynamical form-factor for Earth	J_2	=	0.001 082 63
6. Geocentric gravitational constant	GE	=	$3.986\,005 \times 10^{20}$ $cm^3\ s^{-2}$ [$3.986\,004\,48... \times 10^{20}$]
7. Constant of gravitation	G	=	6.672×10^{-8} $cm^3\ g^{-1}\ s^{-2}$
8. Ratio of mass of Moon to that of Earth	μ	=	0.012 300 02 [0.012 300 034] [$1/\mu = 81.300\ 587$]
9. General precession in longitude per Julian century at standard epoch 2000	ρ	=	5029″.0966
10. Obliquity of the ecliptic at standard epoch 2000	ε	=	23° 26′ 21″.448 [23° 26′ 21″.4119]

IAU System of Astronomical Constants*

[Square brackets indicate modified values used in constructing the ephemerides]

Derived constants:

11. Constant of nutation at standard epoch 2000	N	=	$9''.2025$
12. Unit distance	AU	=	$1.495\,978\,70 \times 10^{13}$ cm
AU = $c\tau$			[$1.495\,978\,7066 \times 10^{13}$]
13. Solar parallax, $\arcsin(a_e/\mathrm{AU})$	$\pi_\odot$	=	$8''.794\,148$
14. Constant of aberration, for standard epoch 2000	κ	=	$20''.49\,552$
15. Flattening factor	f	=	0.003 352 81
for the Earth		=	1/298.257
16. Heliocentric gravitational	GS	=	$1.327\,124\,38 \times 10^{26}$ cm^3 s^{-2}
constant, GS = $(\mathrm{AU})^3k^2/D^2$			[$1.327\,124\,40... \times 10^{26}$]
17. Ratio of mass of the Sun	S/E	=	332 946.0
to that of the Earth			[332 946.038...]
S/E = (GS)/(GE)			
18. Ratio of mass of the Sun		=	328 900.5
to that of Earth + Moon			[328 900.55]
$(S/E)/(1+\mu)$			
19. Mass of the Sun	S	=	1.9891×10^{33} g
S = (GS)/G			

Adapted from the Astronomical Almanac for the Year 1989, Washington, U.S. Goverment Printing Office, and K. R. Lang's Astrophysical Formulae, Springer Verlag, 1986.

IAU System of Astronomical Constants*

Derived constants:

20. System of Planetary Masses

Ratios of mass of Sun to masses of the planets

Mercury	6 023 600	Jupiter	1 047.355	[1 047.350]
Venus	408 523.5	Saturn	3 498.5	[3 498.0]
Earth + Moon	328 900.5	Uranus	22 869	[22 960]
Mars	3 098 710	Neptune	19 412	
		Pluto	3 000 000	[130 000 000]

Other Quantities for Use in the Preparation of Ephemerides*

21. Masses of minor planets

Minor Planet	Mass in solar mass	
(1) Ceres	5.9×10^{-10}	
(2) Pallas	1.1×10^{-10}	[1.081×10^{-10}]
(4) Vesta	1.2×10^{-10}	[1.379×10^{-10}]

22. Masses of satellites

Planet	Satellite	Satellite/Planet
Jupiter	Io	4.70×10^{-5}
	Europa	2.56×10^{-5}
	Ganymede	7.84×10^{-5}
	Callisto	5.6×10^{-5}
Saturn	Titan	2.41×10^{-4}
Neptune	Triton	2.09×10^{-4}

23. Equatorial radii in km

Mercury	2 439	Jupiter	71 492
Venus	6 052	Saturn	60 268
Earth	6 378.140	Uranus	25 559
Moon	1 738	Neptune	24 764
Mars	3 397.2	Pluto	1 123
Sun	696 000		

Other Quantities for Use in the Preparation of Ephemerides*

24. Gravity fields of planets

Planet	J_2	J_3	J_4
Earth	+0.001 082 63	-0.254×10^{-5}	-0.161×10^{-5}
Mars	+0.001 964	$+0.36 \times 10^{-4}$	
Jupiter	+0.014 75		-0.58×10^{-3}
Saturn	+0.016 45		-0.10×10^{-2}
Uranus	+0.012		
Neptune	+0.004		

(Mars: $C_{22} = -0.000\,055$, $S_{22} = +0.000\,031$, $S_{31} = +0.000\,026$)

25. Gravity field of the Moon

$\gamma = (B-A)/C = 0.000\,2278$ $\quad C/MR^2 = 0.392$

$\beta = (C-A)/B = 0.000\,6313$ $\quad I = 5552''.7 = 1^\circ\ 32'\ 32''.7$

$C_{20} = -0.000\,2027$	$C_{30} = -0.000\,006$	$C_{32} = +0.000\,0048$
$C_{22} = +0.000\,0223$	$C_{31} = +0.000\,029$	$S_{32} = +0.000\,0017$
		$C_{33} = +0.000\,0018$
	$S_{31} = +0.000\,004$	$S_{33} = -0.000\,001$

*Adapted from the Astronomical Almanac for the Year 1989, Washington, U.S. Goverment Printing Office, and K. R. Lang's Astrophysical Formulae, Springer Verlag, 1986.

1.3 Conversion Factors

Length

Centimeter	cm	=	0.3937 in
Inch	in	=	2.5400 cm
Meter	m	=	100 cm = 3.2808 ft = 1.0936 yd
Foot	ft	=	30.4800 cm = 12 in
Yard	yd	=	91.44 cm = 3 ft
Fathom	fathom	=	6 ft = 182.88 cm
Kilometer	km	=	10^5 cm = 0.6214 mile (statute)
Mile (statute)	mile	=	1.609344 km = 5280 ft
Mile (nautical)	mile	=	1.8531 km = 6080 ft
Fermi	f	=	10^{-13} cm
Angstrom	Å	=	10^{-8} cm = 10^{-10} m
Wavelength associated with electron volt	λ_0	=	1.239854×10^{-4} cm
Micron	μ	=	10^{-4} cm = 10^{-6} m
Solar Radius	$R_\odot$	=	6.9599×10^{10} cm
Astronomical Unit	AU	=	$1.4959787061 \times 10^{13}$ cm
Light Year	ly	=	9.460530×10^{17} cm = 63239.74 AU
Parsec	pc	=	3.085678×10^{18} cm = 3.261633 ly
Megaparsec	Mpc	=	3.085678×10^{24} cm = 10^6 pc

Time

Second	s		
Hour	h	=	3600 s = 60 min
Day	d	=	86400 s = 24 h
Tropical year	y	=	$3.15569259747 \times 10^7$ s
Tropical year	y	=	365.24219 d
Sidereal second		=	0.9972696 s
Sidereal year		=	365.25636 d
Month:			
Draconic (node to node)		=	27.21222 d
Tropical (equinox to equinox)		=	27.32158 d
Sidereal (fixed star to fixed star)		=	27.32166 d
Anomalistic (perigee to perigee)		=	27.55455 d
Synodic (New Moon to New Moon)		=	29.53059 d
Year:			
Eclipse (lunar node to lunar node)		=	346.6201 d
Tropical (equinox to equinox)		=	365.24219 d
Average Gregorian		=	365.2425 d
Average Julian		=	365.2500 d
Sidereal (fixed star to fixed star)		=	365.25636 d
Anomalistic (perihelion to perihelion)		=	365.2596 d

Velocity

Meter per second	$m\ s^{-1}$	=	$100\ cm\ s^{-1}$
Mile per hour		=	$44.704\ cm\ s^{-1} = 17.60\ in\ s^{-1}$
Knot		=	$51.47\ cm\ s^{-1}$
Kilometer per hour	$km\ hr^{-1}$	=	$27.78\ cm\ s^{-1} = 0.6214$ mile per hour
Velocity of light	c	=	$2.9979245800 \times 10^{10}\ cm\ s^{-1}$
AU per year		=	$4.74057 \times 10^{5}\ cm\ s^{-1}$
Parsec per year		=	$9.7781 \times 10^{10}\ cm\ s^{-1}$
Speed of 1 eV electron		=	$5.93094 \times 10^{7}\ cm\ s^{-1}$

Area

Square centimeter	cm^2	=	$0.1550\ in^2$
Square inch	in^2	=	$6.452\ cm^2$
Square meter	m^2	=	$10^4\ cm^2 = 10.764\ ft^2$
Square foot	ft^2	=	$929.03\ cm^2$
Square kilometer	km^2	=	$10^{10}\ cm^2 = 0.3861$ square miles
Square mile	mi^2	=	$2.590\ km^2$
Acre		=	$4046.85\ m^2 = 4839.99$ square yards
Area First Bohr Orbit	πa_0^2	=	$8.797355 \times 10^{-17}\ cm^2$
Barn		=	$10^{-24}\ cm^2$

Volume

Cubic centimeter	cm^3	=	$0.0610\ in^3$
Cubic inch	in^3	=	$16.3872\ cm^3$
Cubic meter	m^3	=	$10^6\ cm^3 = 35.314\ ft^3$
Cubic foot	ft^3	=	$28316.8\ cm^3 = 1.3079\ yd^3$
Cubic yard	yd^3	=	$0.7646\ m^3$
Liter		=	$1000.027\ cm^3$
		=	1.0567 quarts (U.S. liquid)
		=	33.815 ounces (U.S. liquid)
Ounce (U.S. liquid)		=	$29.5735\ cm^3$
Gallon (U.S. liquid)		=	3.7853 liters
Solar volume	$4\pi R_{\odot}^3/3$	=	$1.4122 \times 10^{33}\ cm^3$
Cubic parsec	pc^3	=	$2.93800 \times 10^{55}\ cm^3$

Mass

Gram	g	=	0.03527 ounce (Avoirdupois)
Gram	g	=	0.03215 ounce (Troy)
Ounce (Avoirdupois)		=	23.3495 grams
Kilogram	kg	=	10^3 g = 2.20462 pounds (Avoirdupois)
Pound (Avoirdupois)	lb	=	453.59243 g = 7000 grains
Pound (Troy)		=	373.242 g = 5760 grains
Grain		=	0.0647989 g
Carat		=	0.2000 g
Slug		=	14.594 kg
Ton = tonne		=	2240 lb = 1.016047×10^6 g
Metric ton		=	10^6 g
Solar mass	$M_\odot$	=	1.9891×10^{33} g
Atomic mass unit	amu	=	$1.6605402(10) \times 10^{-24}$ g
Hydrogen mass	A_H	=	1.007825 amu
Deuterium mass	A_D	=	2.014102 amu
Helium mass	A_{He}	=	4.002603 amu

Density

Gram/cubic centimeter	g cm^{-3}	=
Kilogram/cubic meter	kg m^{-3}	= 10^{-3} g cm^{-3}
Density of water (4°C)		= 0.999972 g cm^{-3}
Density of mercury (0°C)		= 13.5951 g cm^{-3}
Solar mass/cubic parsec	$M_\odot$ pc^{-3}	= 6.770×10^{-23} g cm^{-3}
STP gas density (M_0 is molecular weight)		= 4.4616×10^{-5} M_0 g cm^{-3}

Temperature

Degree Kelvin	°K	=	°C + 273 = (5 °F/9) + 255.22
Degree Fahrenheit	°F	=	(9 °C/5) + 32 = (9 °K/5) - 459.4
Degree Centigrade	°C	=	5(°F - 32)/9 = °K - 273
Freezing Temperature of Water	273 °K	=	32 °F = 0 °C
Boiling Temperature of Water	373 °K	=	212 °F = 100 °C
Temperature associated with 1 eV		=	11604.8 °K

Energy

Erg	erg	=	10^{-7} joule
Joule	J	=	10^{7} erg
Erg	erg	=	2.39006×10^{-8} calories (gram)
Calorie (gram)	cal	=	4.1841×10^{7} erg
Erg	erg	=	9.48451×10^{-11} BTU
British Thermal Unit	BTU	=	1055 J = 252.0 cal
Kilowatt-hour		=	3.600×10^{6} J = 8.6013×10^{5} cal
Therm		=	10^{5} BTU
Foot-pound		=	1.35582×10^{7} erg
Explosion equivalent to 1000 tons of TNT		=	4.2×10^{19} erg
Electron volt	eV	=	$1.60217733(49) \times 10^{-12}$ erg
		=	10^{-6} MeV
Kilo-electron volt	keV	=	10^{3} eV = 1.6021917×10^{-9} erg
		=	10^{-3} MeV
Mega-electron volt	Mev	=	10^{6} eV = 1.6021917×10^{-6} erg
Energy equivalent of atomic mass unit	$(\text{amu})c^2$	=	1.492419×10^{-3} erg
		=	931.49432(28) MeV
Energy associated with 1°K	k	=	1.380622×10^{-16} erg
		=	8.61708×10^{-11} MeV
Photon energy associated with wavelength λ in cm	hc/λ	=	$1.98648 \times 10^{-16}/\lambda$ erg

Power or Luminosity

Erg/second	erg/s	=	10^{-7} J/s = 10^{-7} watt
Watt		=	10^{7} erg s^{-1}
Watt		=	0.001341 horsepower (U.S.)
Horsepower (U.S.)		=	745.7 watt
Force de cheval		=	735.5 watt
Watt		=	0.05688 BTU m^{-1}
Watt		=	0.73756 foot pound s^{-1}
Solar luminosity	$L_{\odot}$	=	3.826×10^{33} erg s^{-1}
Solar apparent visual magnitude	$m_{\odot}$	=	– 26.78
Solar absolute visual magnitude	$M_{V\odot}$	=	4.79
Star, $M_{bol} = 0$ radiation		=	2.97×10^{35} erg s^{-1}

Luminous Flux

Flux unit, Jansky	f.u. = Jy	=	10^{-23} erg cm^{-2} s^{-1} Hz^{-1}
		=	10^{-26} watt m^{-2} Hz^{-1}
Solar flux unit	s.f.u.	=	10^4 f.u.
Solar flux	$s_\odot$	=	1.368×10^6 erg s^{-1} cm^{-2}
		=	0.1368 watt cm^{-2}
Rayleigh		=	$(1/4\pi) \times 10^6$ photons cm^{-2} s^{-1} sr^{-1}
Lumen		=	flux from one candela into one steradian
		=	flux from $(1/60\pi)$ cm^2 of black body at the melting temperature of platinum (2044 °K)
Talbot		=	lumen second
Lambert		=	lumen cm^{-2}
Stilb	sb	=	lumen cm^{-2} sr^{-1}
Candle per square inch		=	0.487 lambert = 0.155 stilb
1 $m_V = 0$ star per square degree outside atmosphere		=	2.63×10^{-6} lambert
		=	0.84×10^{-6} stilb

Force

Dyne	dyn		
Newton	N	=	10^5 dyn
Poundal		=	1.3825×10^4 dyn
Pound weight		=	4.4482×10^5 dyn
Gram weight		=	980.665 dyn
Proton-electron attraction at Bohr radius a_0		=	8.238×10^{-3} dyn

Acceleration

Gravity (standard)	g	=	980.665 cm s^{-2}
Sun's surface		=	2.740×10^4 cm s^{-2}
At 1 A.U. from Sun		=	0.5931 cm s^{-2}

Pressure

Atmosphere (standard)	atm	=	1.013250×10^{6} dyn cm^{-2}
		=	760 torr = 1013.25 mb
Bar	bar	=	1.000×10^{6} dyn cm^{-2}
		=	10^{3} mb = 0.986923 atm
Millibar	mb	=	10^{-3} bar = 10^{3} μb
		=	10^{3} dyn cm^{-2}
Pascal		=	10 dyn cm^{-2} = 10 μb
Torr = millimeter of mercury (Hg)	Torr	=	1333.22 dyn cm^{-2}
		=	0.0013158 atm
Kilogram per square centimeter	kg cm^{-2}	=	0.96784 atm
		=	0.9806 bar = 14.2233 lb in^{-2}
Pounds per square inch	lb in^{-2}	=	6.8947×10^{4} dyn cm^{-2}
		=	0.068046 atm
Inch of mercury		=	3.38638×10^{4} dyn cm^{-2}
		=	0.033421 atm
Foot of water		=	0.03048 kg cm^{-2}
Kilometer of granite		$\sim$	265 kg cm^{-2}

Part II
Planets and their Satellites

2
Planet Earth

2.1 Physical Parameters and Motions of the Earth

Size

Equatorial Radius: $a_e = 6{,}378.140$ km
Polar Radius: $a_p = 6{,}356.755$ km
Flattening Factor: $f = (a_e - a_p)/a_e = 0.00335281 = 1/298.257$

Mean Radius $= R_E = (a_e^2 a_p)^{1/3} = 6{,}371.0$ km
Geodetic Equatorial Radius: 6,378.137 km

Surface Area: 5.10×10^{18} cm^2
Land Area: 1.49×10^{18} cm^2, Mean Land Elevation: 860 m
Ocean Area: 3.61×10^{18} cm^2, Mean Ocean Depth: 3,900 m
Volume: 1.0832×10^{27} cm^3

Mass

Inverse Mass of Earth: $M_\odot/M_E = 332{,}946.043$
Earth Mass: $M_E = 5.9742 \times 10^{27}$ grams
Surface Escape Velocity: $V_{esc} = 11.19$ km s^{-1}
Mean Density of Earth $= \bar{\rho_E} = 5.515$ grams cm^{-3}
Inverse Mass of Earth + Moon $= M_\odot/(M_E + M_m) = 328{,}900.555$

Mass of Atmosphere: 5.1×10^{21} grams
Mass of Ice: $25 - 30 \times 10^{21}$ grams
Mass of Oceans: 1.4×10^{24} grams
Mass of Crust: 2.5×10^{25} grams
Mass of Mantle: 4.05×10^{27} grams
Mass of Core: 1.90×10^{27} grams

Gravity

Geocentric Gravitational Constant: $GM_E = 398{,}600.442$ km^3 s^{-2}
Surface Gravity at Sea Level and Geographic Latitude, ϕ:
$g_E = 978.0327[1 + 0.0053024\ sin^2(\phi) - 0.0000058\ sin^2(2\phi)]$ cm s^{-2}
Gravity Field: $J_2 = 1.08263 \times 10^{-3}$
$J_3 = -0.254 \times 10^{-5}$
$J_4 = -0.161 \times 10^{-5}$

Age

Age of Earth: $4.55 \pm 0.05 \times 10^9$ years
Oldest Terrestrial Rock: 3.8×10^9 years
Oldest Fossil: 3.2×10^9 years

Orbital Motion

Mean Distance of Earth from Sun:
Astronomical unit = AU = 149,597,870.61 km
Solar Parallax = $a_e/\text{AU} = \pi_\odot = 8.''794148$
Perihelion Distance of Earth from Sun: $q = 147.1 \times 10^6$ km
Aphelion Distance of Earth from Sun: $Q = 152.1 \times 10^6$ km

Earth Orbital Eccentricity: e = 0.016722
Earth Orbital Period = 365.25636 days = Sidereal Year
Mean Orbital Speed of Earth: 29.78 km/s^{-1}
Obliquity of the Ecliptic: $\varepsilon = 23° \, 16' \, 21.''448 - 0.''46815$ T
ε = angle between planes of celestial equator and Earth's orbit
(T = Julian years from J2000.0)

Rotation

Rotation Period in mean solar time: 23h 56m 04.098904s = 86164.098904 sec
Earth Equatorial Rotational Velocity: 0.46512 km s^{-1}
Centrifugal Acceleration at Equator: – 3.3915 cm s^{-2}
Angular Velocity of Rotation: 7.292115×10^{-5} rad s^{-1} = 15.041067 $''s^{-1}$
Rotational Angular Momentum: 5.861×10^{40} cm^2 gram s^{-1}
Rotational Energy: 2.137×10^{36} erg

Lengthening of Day (l.o.d.)

Rotation Period = P = 8.6164×10^4 s
Angular Velocity = $\Omega = 7.292115 \times 10^{-5}$ rad s^{-1}
of Rotation
Deceleration = $\dot{\Omega} = 1100 \pm 100$ seconds of arc (century)$^{-2}$
of Earth Rotation

$\dot{\Omega} = (-5.5 \pm 0.5) \times 10^{-22}$ rad s^{-1}
(Lambeck, 1980)

Period Increase = $\dot{P} = P\dot{\Omega}/\Omega$

$\dot{P} = (2.0 \pm 0.2) \times 10^{-3}$ s (century)$^{-1}$
(Lambeck, 1980)

$\dot{P} = (2.40 \pm 0.04) \times 10^{-3}$ s (century)$^{-1}$
(Stephenson and Morrison, 1984)

Origin = Energy dissipated by tidal friction
Tidal Energy Dissipated = $dE/dt \approx 4 \times 10^{19}$ erg s^{-1}
(Earth-Moon system)

Lengthening of Day (l.o.d.)

Lunar Tidal Acceleration = $\dot{n}$ = $(-\ 1.35 \pm 0.10) \times 10^{-23}$ rad s^{-2}

$\dot{n}$ = $(-\ 28 \pm 2)$ seconds of arc $(\text{century})^{-2}$
(Lambeck, 1980)

$\dot{n}$ = $(-\ 25.3 \pm 1.2)$ seconds of arc $(\text{century})^{-2}$
(Stephenson and Lieske, 1988)

Increase in Mean = $\dot{a}_{Moon}$ = $(1.4 \pm 0.2) \times 10^{-7}$ cm s^{-1}
Earth-Moon Distance = $\dot{a}_{Moon}$ = (4.4 ± 0.6) cm yr^{-1}

Non-tidal Changes (decade)

$|\ \Delta l.o.d.\ | = 4$ ms in 7 years

$\dot{\Omega} \approx 1.4 \times 10^{-20}$ rad s^{-2}

Non-tidal Changes (long-term)

Δl.o.d. = − 8 ms between A.D.950 and present

$\dot{\Omega} = +2.7 \times 10^{-23}$ rad s^{-2} since A.D.950

Chandler Wobble = Polar Motion

Periodic Change of Orientation of Earth's Rotation Axis within the Earth. Free, or unforced, motion from west to east. Jeffreys (1968), Lambeck (1980), Lang (1985), Munk and Mac Donald (1975).

Period = 434.3 ± 2.2 sidereal days
≈ 1.20 years = 14 months
Frequency = 0.843 ± 0.004 cycles year^{-1}
Linear Radius ≈ 5.0 meters = 15 feet
(relative to reference pole)
Angular Displacement ≈ 0.15 seconds of arc
(of rotation axis)
Radius of Angular Displacement ≈ 0.15 seconds of arc
(of rotation axis, 0.01″= 1.01 ft)
Damping time = 10 to 20 years
Tidal-effective Love number = k = 0.29 ± 0.01
Excitation = elusive, earthquakes?

Annual Polar Motion

Annual Change of Orientation of Earth's Rotation Axis within Earth. **Forced motion by atmospheric changes. Lambeck (1980), Munk and** MacDonald (1975).

Period = 1.00 years = 12 months
Frequency = 1.00 cycles year^{-1}
Linear Radius ≈ 2.5 meters ≈ 7.5 feet
Angular Displacement ≈ 0.075 seconds of arc
Origin = seasonal variation in distribution
of air over the Earth.

Rapid Polar Motion

Changes in Orientation of Earth's Rotation Axis within the Earth. **Forced motion by atmospheric changes. Eubanks et al. (1988).**

Period = two weeks to several months
Linear Size ≈ 7 cm to 7 meters ≈ 0.2 to 2.0 feet
Size of Angular Displacement = 0.002 to 0.02 seconds of arc
Origin = changes in surface air pressure.

2.2 Timescales on Earth

Archeological Time Scale*

Age	Years Ago
Iron Age	3000 to 3200
Bronze Age	3740 to 6000
Neolithic (New Stone Age)	8000 to 10600
Paleolithic (Old Stone Age)	18800 to 2.4×10^6

Historical Time Scale*

Event (Reference Date AD 2000)	Years Ago
World War II	61
World War I	86
Victorian Age	129 to 163
Age of Revolution (French, Industrial)	186 to 240
European Empires	293 to 522
Middle Ages (Byzantine, Ming, Mongol)	547 to 1514
Classical Empires (Persian, Roman, Han)	1524 to 2550
Babylonian and Shang	2626 to 4030

*Adapted from Calder (1983).

Geological Time Scale*

Period	Millions of Years Ago	Key Events
Quaternary Period		
Pleistocene	0.10 to 1.80	Homo Erectus Breakout 0.75 Myr Ago
Cenozoic Era		
Tertiary Period		
Pliocene	1.8 to 5.3	Ape Man Fossils 4.0 Myr Ago
Miocene	5 to 25	Origin of Grass 24 Myr Ago
Oligocene	25 to 37	Rise of Cats, Dogs, Pigs and Bears 30 to 35 Myr Ago
Eocene	37 to 55	Debut of Hoofed Mammals, Rodents, Whales and Bats 55 Myr Ago
Paleocene	55 to 67	Earliest Primate Fossil 69 Myr Ago
Mesozoic Era		
Cretaceous	67 to 138	Demise of Dinosaurs 67 Myr Ago
Jurassic	138 to 208	First Birds 150 – 135 Myr Ago
Triassic	208 to 245	Dinosaurs Appeared 235 Myr Ago
Paleozoic Era		
Permian	245 to 290	Flowers and Insect Pollination Began 235 Myr Ago
Carboniferous	290 to 360	Conifers Appeared 350 Myr Ago
Devonian	360 to 410	First Trees and Vertebrates Ashore 370 Myr Ago
Silurian	410 to 435	Spore - Bearing Plants 425 Myr Ago
Ordovician	435 to 520	Animals Ashore (Invertebrates) 425 Myr Ago
Cambrian	520 to 570	Vertebrates 510 Myr Ago
Proterozoic Era	570 to 2500	Brains 620 Myr Ago, Jellyfish 670 Myr Ago, Sex 1000 Myr Ago, First Plants 1300 Myr Ago
Archean Era	2500 to 3800	Photosynthetic Bacteria 3500 Myr Ago
Hadean Era	3800 to 4450	Earth Formed 4600 Myr Ago

*Adapted from Calder (1983)

2.3 Terrestrial Atmosphere

Composition Of the Earth's Atmosphere*

Constituent	Formula	Abundance by Volume	Molecular Weight
Molecular Nitrogen	N_2	0.78084	28
Molecular Oxygen	O_2	0.20948	32
Argon	Ar	0.00934	40
Carbon Dioxide	CO_2	0.000333	44
Neon	Ne	18.18×10^{-6}	20.2
Helium	He	5.24×10^{-6}	4
Methane	CH_4	2.0×10^{-6}	16
Krypton	Kr	1.14×10^{-6}	83.8
Molecular Hydrogen	H_2	0.5×10^{-6}	2
Ozone	O_3	0.4×10^{-6}	48
Nitrous Oxide	N_2O	0.27×10^{-6}	44
Carbon Monoxide	CO	0.2×10^{-6}	28
Xenon	Xe	0.087×10^{-6}	131.3
Formaldehyde	H_2CO	1×10^{-8}	30
Ammonia	NH_3	4×10^{-9}	17
Hydrogen Peroxide	H_2O_2	1×10^{-9}	34
Sulphur Dioxide	SO_2	1×10^{-9}	64
Nitrogen Dioxide	NO_2	1×10^{-9}	46
Nitric Oxide	NO	5×10^{-10}	30
Dichlorofluoromethane	CF_2Cl_2	2×10^{-10}	121
Trichlorofluoromethane	$CFCl_3$	1×10^{-10}	137.5
Carbon Tetrachloride	CCl_4	1×10^{-10}	154
Hydrogen Sulfide	H_2S	5×10^{-11}	34
Hydroperoxyl	HO_2	3×10^{-11}	33
Hydrogen Bromide	HBr	$\sim 1 \times 10^{-11}$	81
Bromine Oxide	BrO	$\sim 1 \times 10^{-11}$	96
Hydroxyl	OH	$\sim 2 \times 10^{-13}$	17

*Adapted from Chamberlain and Hunten (1987). Dry air at sea level. Water vapor, H_2O, has a variable abundance of a few $\times 10^{-6}$. Apparent molecular weight of dry air is 28.964, and its total thickness is 2.15×10^{25} molecules/cm^2.

Standard Terrestrial Atmosphere*

Height z (km)	Temperature T (°K)	Pressure P (millibars)	Number density N (cm^{-3})	Mean molecular weight $<\mu>$ (gm/mole)	Pressure scale height H (km)
0	288	1.013×10^{3}	2.547×10^{19}	28.96	8.434
5	256	5.405×10^{2}	1.531×10^{19}	28.96	7.496
10	223	2.650×10^{2}	8.598×10^{18}	28.96	6.555
15	217	1.211×10^{2}	4.049×10^{18}	28.96	6.372
20	217	5.529×10^{1}	1.849×10^{18}	28.96	6.382
25	222	2.549×10^{1}	8.334×10^{17}	28.96	6.536
30	227	1.197×10^{1}	3.828×10^{17}	28.96	6.693
35	237	5.746	1.760×10^{17}	28.96	7.000
40	250	2.871	8.308×10^{16}	28.96	7.421
45	264	1.491	4.088×10^{16}	28.96	7.842
50	271	7.978×10^{-1}	2.135×10^{16}	28.96	8.047
55	261	4.253×10^{-1}	1.181×10^{16}	28.96	7.766
60	247	2.196×10^{-1}	6.439×10^{15}	28.96	7.368
65	233	1.093×10^{-1}	3.393×10^{15}	28.96	6.969
70	220	5.221×10^{-2}	1.722×10^{15}	28.96	6.570
75	208	2.388×10^{-2}	8.300×10^{14}	28.96	6.245
80	198	1.052×10^{-2}	3.838×10^{14}	28.96	5.962
85	189	4.457×10^{-3}	1.709×10^{14}	28.96	5.678
86	187	3.734×10^{-3}	1.447×10^{14}	28.95	5.621
90	187	1.836×10^{-3}	7.12×10^{13}	28.91	5.64

*U.S. Standard Atmosphere (1976).

Standard Terrestrial Atmosphere*

Height z (km)	Temperature T (°K)	Pressure P (millibars)	Number density N (cm^{-3})	Mean molecular weight $<\mu>$ (gm/mole)	Pressure scale height H (km)
95	189	7.597×10^{-4}	2.92×10^{13}	28.73	5.73
100	195	3.201×10^{-4}	1.19×10^{13}	28.40	6.01
110	240	7.104×10^{-5}	2.14×10^{12}	27.27	7.72
120	360	2.538×10^{-5}	5.11×10^{11}	26.20	12.09
130	469	1.250×10^{-5}	1.93×10^{11}	25.44	16.29
140	560	5.403×10^{-6}	9.32×10^{10}	24.75	20.03
150	634	4.542×10^{-6}	5.19×10^{10}	24.10	23.38
160	696	3.040×10^{-6}	3.16×10^{10}	23.49	26.41
180	790	1.527×10^{-6}	1.40×10^{10}	22.34	31.70
200	855	8.474×10^{-7}	7.189×10^{9}	21.30	36.18
220	899	5.015×10^{-7}	4.049×10^{9}	20.37	40.04
240	930	3.106×10^{-7}	2.429×10^{9}	19.56	43.41
260	951	1.989×10^{-7}	1.529×10^{9}	18.85	46.35
280	966	1.308×10^{-7}	9.818×10^{8}	18.24	48.93
300	976	8.770×10^{-8}	6.518×10^{8}	17.73	51.19
350	990	3.450×10^{-8}	2.528×10^{8}	16.64	55.83
400	996	1.452×10^{-8}	1.068×10^{8}	15.98	59.68
450	998	6.248×10^{-9}	4.687×10^{7}	15.25	63.64
500	999	3.024×10^{-9}	2.197×10^{7}	14.33	68.79
750	1000	2.260×10^{-10}	1.646×10^{6}	6.58	161.1
1000	1000	7.514×10^{-11}	5.445×10^{5}	3.94	288.2

*U.S. Standard Atmosphere (1976).

2.4 Terrestrial Impact Craters

	Name	Diameter (km)	Age (10^6 years)	Location
	Structures With Associated Meteorites			
1	Barringer, Arizona, USA	1.20	0.02 to 0.05	35°N 111°W
2	Boxhole, N.T. Australia	0.18	0.005	23°S 135°E
3	Campo del Cielo, Argentina	0.09	†	28°S 62°W
4	Dalgaranga, W.A., Australia	0.02	†	28°S 117°E
5	Haviland, Kansas, USA	0.01	†	38°N 99°W
6	Henbury, N.T. Australia	0.15	< 0.005	25°S 133°E
7	Kaalijärvi, Estonia SSR	0.11	0.005	58°N 23°E
8	Morasko, Poland	0.10	†	52°N 17°E
9	Odessa, Texas, USA	0.17	0.05	32°N 102°W
10	Sikhote Alin, Primorye Terr., Siberia, USSR	0.03	†	46°N 135°W
11	Sobolev, Siberia, USSR	0.05	†	46°N 138°E
12	Wabar, Saudi Arabia	0.10	< 0.005	21°N 50°E
13	Wolf Creek, W.A., Australia	0.85	†	19°S 128°E
	Structures With Shock Metamorphism			
1	Aouelloul, Mauritania	0.37	3.1 ± 0.3	20°N 13°W
2	Araguainha Dome, Brazil	40	< 250	17°S 53°W
3	Beyenchime Salaatin, USSR	8	< 65	72°N 123°E
4	Boltysh, Ukrainian SSR, USSR	25	100 ± 5	49°N 32°E
5	Bosumtwi, Ghana	10.5	1.3 ± 0.2	7°N 1°W
6	B.P. structure, Libya	2.8	< 120	25°N 24°E
7	Brent, Ontario, Canada	3.8	450 ± 30	46°N 78°W
8	Carswell, Saskatchewan, Canada	37	485 ± 50	58°N 109°W
9	Charlevoix, Quebec, Canada	46	360 ± 25	48°N 70°W
10	Clearwater L.East, Quebec, Canada	22	290 ± 20	56°N 74°W
11	Clearwater L.West, Quebec, Canada	32	290 ± 20	56°N 74°W
12	Conception Bay, Nfld., Canada		~ 500	47°N 53°W
13	Crooked Creek, Missouri, USA	5.6	320 ± 80	38°N 91°W
14	Decaturville, Missouri, USA	6	< 300	38°N 93°W
15	Deep Bay, Saskatchewan, Canada	12	100 ± 50	56°N 103°W
16	Dellen, Sweden	15	230	62°N 17°E
17	Flynn Creek, Tennessee, USA	3.8	360 ± 20	36°N 86°W
18	Gosses Bluff, N.T. Australia	22	130 ± 6	24°S 132°E
19	Gow L., Saskatchewan, Canada	5	< 200	56°N 104°W
20	Haughton Dome, N.W.T. Canada	20	15	75°N 90°W

†Ages < 10^6 years, and usually < 10^5. *After Grieve and Robertson (1979) and Wasson (1974).

Terrestrial Impact Craters*

	Name	Diameter (km)	Age (10^6 years)	Location
21	Holleford, Ontario, Canada	2	550 ± 100	44°N 77°W
22	He Rouleau, Quebec, Canada	4	< 300	51°N 74°W
23	Ilinsty, USSR	4.5	495 ± 5	49°N 28°E
24	Janisjärvi, USSR	14	700	62°N 31°E
25	Kaluga, USSR	15	360 ± 10	54°N 36°E
26	Kamensk, USSR	25	65	48°N 40°E
27	Kara, USSR	50	57	69°N 65°E
28	Karla, RSFSR	18	10	58°N 48°E
29	Kelley West, N.T. Australia	2.5	550	19°S 133°E
30	Kentland, Indiana, USA	13	300	41°N 87°W
31	Kjardla, Est.SSR	4	500 ± 50	57°N 23°E
32	Kursk, USRR	5	250 ± 80	52°N 36°E
33	Lac Couture, Quebec, Canada	8	420	60°N 75°W
34	Lac La Moinerie, Quebec, Canada	8	400	57°N 67°W
35	Lappajärvi, Finland	14	< 600	63°N 24°E
36	Liverpool, N.T. Australia	1.6	150 ± 70	12°S 134°E
37	Logoisk, Bel.SSR	17	100 ± 20	54°N 28°E
38	Lonar, India	1.83	0.05	20°N 77°E
39	Manicouagan, Quebec, Canada	70	210 ± 4	51°N 69°W
40	Manson, Iowa, USA	32	< 70	43°N 95°W
41	Mien L, Sweden	5	118 ± 2	56°N 15°E
42	Middlesboro, Kentucky, USA	6	300	37°N 84°W
43	Misarai, Lith.SSR	5	500 ± 80	54°N 24°E
44	Mishina Gora, USSR	2.5	< 360	59°N 28°E
45	Mistastin, Labrador, Canada	28	38 ± 4	56°N 63°W
46	Monturaqui, Chile	0.46	1	24°S 68°W
47	New Quebec Crater, New Quebec, Canada	3.2	5	61°N 74°W
48	Nicholson L., N.W.T., Canada	12.5	< 450	63°N 103°W
49	Oasis, Libya	11.5	< 120	25°N 24°E
50	Obolon, USSR	15	160	50°N 33°E
51	Pilot L., N.W.T., Canada	6	< 300	60°N 111°W
52	Popigai, USSR	100	38 ± 9	71°N 111°E
53	Puchezh-Katun, USRR	80	183 ± 3	57°N 44°E
54	Redwing CK., N. Dakota	9	200	48°N 102°W
55	Ries, Germany	24	14.8 ± 0.7	49°N 11°E

*After Grieve (1979) and Wasson (1974).

Terrestrial Impact Craters*

	Name	Diameter (km)	Age (10^6 years)	Location
56	Rochechouart, France	23	160 ± 5	46°N 1°E
57	Rotmistrovka, USSR	2.5	70	49°N 32°E
58	Sääksjärvi, Finland	5	490	61°N 22°E
59	St. Martin, Manitoba, Canada	23	225 ± 40	52°N 99°W
60	Serpent Mound, Ohio, USA	6.4	300	39°N 83°W
61	Serra da Canghala, Brazil	12	< 300	8°S 47°W
62	Shunak, Kaz, SSR	2.5	12	43°N 73°E
63	Sierra Madera, Texas, USA	13	100	31°N 103°W
64	Siljan, Sweden	52	365 ± 7	61°N 15°E
65	Slate Island, Ontario, Canada	30	350	49°N 87°W
66	Steen River, Alberta, Canada	25	95 ± 7	60°N 118°W
67	Steinheim, Germany	3.4	14.8 ± 0.7	49°N 10°E
68	Strangways, N.T.Australia	24	150 ± 70	15°S 134°E
69	Sudbury, Ontario, Canada	140	1840 ± 150	47°N 81°W
70	Tabun-Khara-Obo, Mongolia	1.3	< 30	44°N 110°E
71	Tenoumer, Mauritania	1.9	2.5 ± 0.5	23°N 10°W
72	Vepriaj, Lith. SSR	8	160 ± 30	55°N 25°E
73	Vredefort, South Africa	140	1970 ± 100	27°S 27°E
74	Wanapitei L., Ontario, Canada	8.5	37 ± 2	47°N 81°W
75	Wells Creek, Tennessee, USA	14	200 ± 100	36°N 88°W
76	West Hawk L., Manitoba, Canada	2.7	100 ± 50	50°N 95°W
77	Zeleny Gai, Ukr.SSR	1.4	120 ± 20	47°N 35°E
78	Zhamanshin, Aktyubinsk, USSR	10	4.5 ± 0.5	49°N 61°E

*After Grieve (1979) and Wasson (1974).

3
The Planets

3.1 The Planetary System

Total mass of planetary system	=	2.669×10^{30} grams = 447 M_E
	=	0.00134 $M_\odot$
Mass of Earth	=	$M_E = 5.9742 \times 10^{27}$ grams
Mass of the Sun	=	$M_\odot = 1.9891 \times 10^{33}$ grams
Total angular momentum of planetary system	=	3.148×10^{50} grams cm^2/sec
Total kinetic energy of planetary system	=	1.99×10^{42} erg
Total rotational energy of planets	=	0.7×10^{42} erg

Mean Orbital Elements Of The Planets*

Planet	Semi-major axis (AU)	Semi-major axis (10^6 km)	Sidereal Period (tropical years)	Sidereal Period (days)	Synodic Period (days)
Mercury	0.387099	57.9	0.24085	87.969	115.88
Venus	0.723332	108.2	0.61521	224.701	583.92
Earth	1.000000	149.6	1.00004	365.256	
Mars	1.523688	227.9	1.88089	686.980	779.94
Jupiter	5.202833481	778.3	11.86223	4332.589	398.88
Saturn	9.538762055	1427.0	29.4577	10759.22	378.09
Uranus**	19.19139128	2869.6	84.0139	30685.4	369.66
Neptune	30.06106906	4496.6	164.793	60189	367.49
Pluto	39.52940243	5900	247.7	90465	366.73

*These mean orbital elements are good for any epoch from 1950 to 2000 at their given accuracy. Greater accuracy for 1950 and any epoch thereafter may be obtained by using the formula in Seidelmann, Doggett and De Luccia (1974); they also give the relations for the mean longitudes of the planets.

**Mean solar distance for Uranus is now thought to be 19.28 AU.

Mean Orbital Elements Of The Planets

Planet	Mean Orbital Velocity V (km/sec)	Mean Daily Motion n (degrees)	Eccentricity e	Inclination to the Ecliptic i (degrees)	Longitude of Ascending Node Ω (degrees)	Longitude of Perihelion $\tilde{\omega}$ (degrees)
Mercury	47.89	4.0923	0.2056	7.00	47.7	76.7
Venus	35.03	1.6021	0.0068	3.39	76.2	130.9
Earth	29.79	0.9856	0.0167	0.01	174.4	102.1
Mars	24.13	0.5240	0.0933	1.85	49.2	335.1
Jupiter	13.06	0.0831	0.048	1.31	99.8	13.3
Saturn	9.64	0.0335	0.056	2.49	113.5	91.5
Uranus	6.81	0.0117	0.046	0.77	73.7	172.1
Neptune	5.43	0.0060	0.010	1.77	131.2	38
Pluto	4.74	0.0040	0.248	17.15	109.7	223

3.2 Physical Parameters of the Planets

Planet Names And The Titius - Bode Law

Planet	(genitive)	n	Distance by Titius - Bode Law (AU)	Actual Distance (AU)
Mercury		$-\infty$	0.4	0.39
Venus	Cytherean	0	0.7	0.72
Earth	Terrestrial	1	1.0	1.00
Mars	Martian	2	1.6	1.52
Asteroids	Asteroidal	3	2.8	2.2 to 3.2
Jupiter	Jovian	4	5.2	5.20
Saturn	Saturnian	5	10.0	9.54
Uranus	Uranian	6	19.6	19.2
Neptune	Neptunian	7	38.8	30.1
Pluto		8	77.2	39.5

Parameters of the Pluto-Charon System*

Semimajor Axis, a	19,130 km
Orbital Period, P	6.38718 days
Mass of System, M	1.36×10^{25} grams (0.0023 Earth masses)
Pluto's Radius	1,123 km ($\pm$ 20 km)
Charon's Radius	560 km ($\pm$ 20 km)
Mean Density of System	1.99 gm cm^{-3} ($\pm$ 0.09 gm cm^{-3})
Pluto's Density	1.8 gm cm^{-3} to 2.14 gm cm^{-3}
Charon's Density	1 to 3 gm cm^{-3}
Pluto's Mass	1.25×10^{25} grams (0.0021 Earth masses)
Orbit Tilt	120° (retrograde)

*Adapted from Mc Kinnon and Mueller (1988).

Physical Elements Of The Inner Planets*

	Mercury	Venus	Earth	Mars
Equatorial Radius, R_e (km)	2,439	6,051	6,378.140	3,397
Polar Radius, R_P (km)	2,439	6,051	6,356.775	3,372
Oblateness, $(R_e-R_P)/R_e$	0.0	0.0000	0.0033529	0.0074
Equatorial Radius, $(R_E=1.0)$	0.382	0.949	1.000	0.533
Angular Diameter at Closest Approach ($''$)	10.90	61.0		17.88
Angular Diameter at 1 AU ($''$)	6.74	16.92	17.60	9.36
Reciprocal Mass $(M_\odot/M_P)$	6023600	408525.1	332946.043	3098710
Mass, M_P (grams)	3.3022×10^{26}	4.8690×10^{27}	5.9742×10^{27}	6.4191×10^{26}
Mass, M_P $(M_E=1.0)$	0.05527	0.81499	1.00000	0.10745
Mean Density (grams cm^{-3})	5.43	5.25	5.52	3.93

*Equatorial radii and oblateness for Venus, Earth and Mars from Pettengill et al. (1980), Clemence (1965) and Lindal et al. (1979), respectively. Reciprocal masses $M_\odot/M_P$, are from the Jet Propulsion Laboratory ephemeris, courtesy of J. Myles Standish, Jr. (1988). Planetary masses, M_P, in grams are inferred from the reciprocal masses assuming a solar mass of $M_\odot = 1.9891 \times 10^{33}$ grams. Planetary masses, M_P, in units of the Earth's mass, M_E, are calculated from the ratio of the relevant reciprocal masses.

Physical Elements Of The Inner Planets*

	Mercury	Venus	Earth	Mars
Sidereal Rotation Period	58.6462 Earth days	243.01 Earth days (retrograde)	23 hours 56 minutes 4.099 seconds	24 hours 37 minutes 22.66 seconds
Inclination of Equator to Orbit (degrees)	7.0	177.4	23.45	23.98
Magnetic Moment (Gauss R_P^3)	0.0035	< 0.0003	0.31	≤0.0006
Tilt Angle of Magnetic Axis (degrees)	< 10		11.5	
Solar Wind Stagnation Point			11	
Equatorial Acceleration of Gravity (cm/s^2)	370	887	980	371
Equatorial Escape Velocity (km/sec)	4.25	10.36	11.18	5.02
Surface Temperature (°K)	100 to 700	730 ± 5	288 to 293	183 to 268
Surface Pressure (bars)		90 ± 2	1.0	0.007 to 0.010

*Rotation periods for Mercury and Venus are from Klassen (1976) and Shapiro, Campbell and De Campli (1979), respectively. Magnetic parameters are summarized by Lanzerotti and Krimigis (1985). Also see Whang (1977) and Russell, Elphic and Slavin (1980) for Mercury and Venus, respectively. The surface temperature and pressure of Venus are from Avduevsky et al. (1971), while those for Mars are summarized by Carr (1981).

Physical Elements Of The Outer Planets*

	Jupiter	Saturn	Uranus	Neptune	Pluto
Equatorial Radius, R_e (km)	71,492	60,268	25,559	24,764	1,123
Polar Radius, R_P (km)	66,854	54,364	24,973	24,341	
Oblateness, $(R_e\text{-}R_p)/R_e$	0.06487	0.09796	0.02293	0.0171	
Equatorial Radius, (R_E=1.0)	11.19	9.46	4.01	3.81	0.176
Angular Diameter at Closest Approach (″)	46.86	19.52	3.60	2.12	0.08
Angular Diameter at 1 AU (″)	196.74	165.6	65.8	33.9	2.9
Reciprocal Mass ($M_\odot/M_P$)	1047.3492	3497.91	22902.94	19434	13×10^7
Mass, M_P (grams)	1.8992×10^{30}	5.6865×10^{29}	8.6849×10^{28}	1.0235×10^{29}	1.36×10^{25}
Mass, M_P (M_E=1.0)	317.894	95.1843	14.5373	17.1321	0.02561
Mean Density (grams cm^{-3})	1.33	0.71	1.24	1.67	1.89 to 2.14

*Equatorial radii and polar radii are at the one-bar pressure level; from Lindal et al. (1987) for Jupiter, Saturn and Uranus. Pluto's radius and mass density are given by Mc Kinnon and Mueller (1988). Reciprocal masses, $M_\odot/M_P$, are from the Jet Propulsion Laboratory ephemeris, courtesy of J. Myles Standish, Jr. (1988). Planetary masses, M_P, in grams are inferred from the reciprocal masses assuming a solar mass of $M_\odot = 1.9891 \times 10^{33}$ grams. Planetary masses, M_P, in units of the Earth's mass, M_E, are calculated from the ratio of the relevant reciprocal masses. The equatorial radius and oblatness of Neptune are from Lindal et al. (1990) and Tyler et al. (1989). Eschleman (1989) gives Pluto's $R_e = 1,180 \pm 23$ km.

Physical Elements Of The Outer Planets*

	Jupiter	Saturn	Uranus	Neptune	Pluto
Sidereal Rotation Period	9 hours 55 minutes 29.7 seconds	10 hours 39 minutes 22.4 seconds	17.24 ± 0.01 hours	16.11 ± 0.05 hours	6.38718 days
Inclination of Equator to Orbit (degrees)	3.08	26.73	97.92	28.8	≥50
Magnetic Moment (Gauss R_P^3)	4.3	0.21	0.228	0.133	
Tilt Angle of Magnetic Axis (degrees)	9.6	0.8	58.6	47.0	
Solar Wind Stagnation Point (R_P)	70	22	18		
Equatorial Acceleration of Gravity (cm/s^2)	2,312	896	777	1,100	72
Equatorial Escape Velocity (km/sec)	59.54	35.49	21.29	23.71	1.27
Effective Temperature (°K)	124 ± 0.3	95.0 ± 0.4	58 ± 2	59.3 ± 1.0	40 to 60
Temperature at one-bar level (°K)	165 ± 5	134 ± 4	76 ± 2		
Heat Ratio**	1.9 ± 0.2	2.2 ± 0.7		2.1 ± 0.5	

*Rotation periods are for the magnetic field as inferred from periodic radio emission, Lindal et al. (1987), Warwick et al. (1987), and Warwick et al. (1989). The magnetic moments are in units of Gauss R_P^3 where the reference radii are 71372, 60330, 25600 and 24765 km for Jupiter, Saturn, Uranus and Neptune, respectively, Behannon et al. (1987), Connerney, Acuña and Ness (1987), Ness et al. (1979), Ness et al. (1982), Ness et al. (1986), and Ness et al. (1989). The tilted magnetic dipoles of Uranus and Neptune are displaced along the rotation axis from the center of the planet by 0.3 R_U and 0.55 R_N, respectively. The effective temperatures and thermal emission of Jupiter, Saturn, Uranus and Neptune are from Hanel et al. (1981), Hanel et al. (1983), Fazio et al. (1976) and Conrath et al. (1989), respectively. The atmospheric temperatures at the one-bar radius are from Lindal et al. (1987). **The heat ratio is the ratio of the thermal emission to solar energy absorbed.

Reference Size, Shape, And Rotational Elements For The Planets*

Planet	Reference Equatorial Radius (km)	Oblateness or Flattening	North Pole R.A.(2000) (degrees)	North Pole Dec.(2000) (degrees)	Prime Meridian, W (degrees)
Mercury	2,439	0	281.01 − 0.003 T	61.45 − 0.005 T	329.71 +6.1385025 d
Venus	6,051	0	272.69	67.17	160.39 − 1.4813291 d
Earth	6,378.140	0.00335281	0.0 (1950) − 0.640 T	90.0 (1950) − 0.557 T	99.87 +360.9856123 d
Mars	3,393.4	0.0051865	317.681 − 0.108 T	52.886 − 0.061 T	176.729 +350.8919830 d
Jupiter	71,398	0.0648088	268.05 − 0.009 T	64.49 +0.003 T	284.95 +870.5360000 d
Saturn	60,000	0.1076209	40.58 − 0.0036 T	83.54 − 0.004 T	38.90 +810.7939024 d
Uranus	25,400	0.030	257.43	− 15.10	360.00 − 501.1600928 d
Neptune	25,295	0.022	295.33	40.65	107.21 +468.7500000 d
Pluto	1,500	0	311.63	4.18	252.66 − 56.3640000 d

*From Davies et al. (1986, 1983). The time d is the interval in days from the standard epoch 2000 January 15, or JD 2451545.0, while T is the interval in Julian centuries (of 36525 days) from this standard epoch.

Gravitational Data For The Planets*

Planet	Mercury	Venus	Earth	Mars
Second Zonal Harmonic, J_2	$(8 \pm 6) \times 10^{-5}$	$(6 \pm 3) \times 10^{-6}$	1.0826×10^{-3}	1.959×10^{-3}
Centrifugal Potential $q = \omega^2 a^3/(GM_P)$	1.0×10^{-6}	6.1×10^{-8}	3.5×10^{-3}	4.6×10^{-3}

Planet	Jupiter	Saturn	Uranus	Neptune
Second Zonal Harmonic, J_2	1.4736×10^{-2}	1.6479×10^{-2}	3.3434×10^{-3}	3.4105×10^{-3}
Centrifugal Potential $q = \omega^2 a^3/(GM_P)$	0.089	0.153	0.035	0.04?

*Gravitational potentials of the planets are given by Esposito et al. (1977-Mercury), Aanda et al. (1980-Venus), Kaula (1966-Earth), Jordan and Lorell (1975-Mars), Christensen and Balmino (1979-Mars), Campbell and Synnott (1985-Jupiter), Null (1976-Jupiter), Null, Biller and Anderson (1981-Saturn), French et al. (1988-Uranus); Peale (1973-major planets), Jacobson (1991-Neptune).

3.3 Planetary Atmospheres

Basic Data*

Object	Surface Pressure (bar)	Surface Temperature (°K)	Effective Temperature (°K)	Major Gases	Acceleration of Gravity (cm s^{-2})	Scale Height (10^5 cm)
Mercury	$\sim 2 \times 10^{-15}$	440		He($\sim$ 0.98), H($\sim$ 0.02)	395	229
Venus	90	730	$\sim$ 230	CO_2(0.96), N_2($\sim$ 0.035)	888	15
Earth	1	288	$\sim$ 255	N_2(0.77), O_2(0.21)	978	8
Mars	0.007	218	$\sim$ 212	CO_2(0.95), N_2(0.027)	373	11
Jupiter			129	H_2(86.1%), He(13.8%)	2320	
Saturn			97	H_2(92.4%), He(7.4%)	877	
Uranus			58	H_2($\sim$ 0.89), He($\sim$ 0.11)	946	
Neptune			56	H_2($\sim$ 0.89), He($\sim$ 0.11)	1370	
Pluto	5 to 20×10^{-6}	50 to 70	50 to 70	CH_4	72	50

*Adapted from Henderson-Sellers (1983), Hubbard et al. (1988), Stern (1988) and Mc Kinnon and Mueller (1988). The numbers in parenthesis are the volume mixing ratios or percent if denoted by %. The scale height $H = kT/(\mu m_H g)$ where k is Boltzmann's constant, T is the surface temperature, μ is the molecular weight, m_H is the mass of the hydrogen atom, and g is the acceleration of gravity.

Surface Atmospheric Composition of the Terrestrial Planets*

Component	Venus Amount (% or ppm)	Venus Mass (grams)	Earth Amount (% or ppm)	Earth Mass (grams)	Mars Amount (% or ppm)	Mars Mass (grams)
Carbon Dioxide, CO_2	96.4%	4.6×10^{23}	0.03%	2.5×10^{18}	95.32%	3.6×10^{19}
Nitrogen, N_2	3.4%	9.7×10^{21}	78.08%	3.9×10^{21}	2.7%	4.0×10^{17}
Oxygen, O_2	69.3 ± 1.27 ppm		20.95%	1.0×10^{21}	0.13%	3.0×10^{16}
Water, H_2O	$\approx$ 0.1%	4.9×10^{20}	0.1 to 2.8%	3.0×10^{19}	0.03%	1.4×10^{15}
Carbon Monoxide, CO	20 ± 3 ppm		1 ppm		0.07%	
Argon, ^{40}Ar	4.3 ± 5 ppm	1.3×10^{20}	0.93%	6.3×10^{20}	1.6%	4.2×10^{18}
Neon, ^{20}Ne		4.9×10^{18}	18.2 ppm	5.7×10^{16}	2.5 ppm	3.1×10^{13}
Krypton ^{84}Kr		2.0×10^{18}	1.14 ppm	1.8×10^{16}	0.3 ppm	1.2×10^{13}
Xenon, ^{132}Xe			0.087 ppm	2.2×10^{16}	0.08 ppm	6.4×10^{12}
Total	100%	4.7×10^{23}	100%	5.6×10^{21}	100%	4.1×10^{19}

*Adapted from Ip (1981) and Baugher (1988). The term ppm means parts per million, and the masses of Venus, Earth and Mars are 4.8690×10^{27}, 5.9742×10^{27} and 6.4191×10^{26} grams, respectively.

Isotopic Abundance Ratios in the Atmospheres of the Terrestrial Planets*

Ratio	Venus	Earth	Mars
$^{12}C/^{13}C$	0.012	0.0112	0.0118
$^{14}N/^{15}N$	160	280	160
$^{17}O/^{16}O$	?	3.71×10^{-4}	3.91×10^{-4}
$^{18}O/^{16}O$	2.0×10^{-3}	2.05×10^{-3}	2.06×10^{-3}
$^{22}Ne/^{20}Ne$	?	0.097	0.10
$^{20}Ne/^{36}Ar$	0.5	0.6	0.5
$^{36}Ar/^{40}Ar$	0.80	3.4×10^{-3}	3.0×10^{-4}
$^{38}Ar/^{40}Ar$	0.14	6.4×10^{-4}	6.0×10^{-5}
$^{38}Ar/^{36}Ar$	0.18	0.19	0.20
$^{36}Ar/^{12}C$	1×10^{-5} to 6×10^{-5}	3×10^{-7}	6×10^{-6}
$^{84}Kr/^{36}Ar$	0.02	0.04	0.03
$^{129}Xe/^{132}Xe$		0.983	2.56
$^{36}Ar/^{132}Xe$		$> 3 \times 10^{3}$	1.3×10^{3}

*Adapted from Baugher (1988). Ratios of numbers of atoms. The isotopes ^{20}Ne, ^{36}Ar, ^{38}Ar, ^{84}Kr, ^{132}Xe are primordial; ^{40}Ar, ^{129}Xe are radiogenic.

Abundance Ratios in the Atmospheres of the Jovian Planets*

Ratio	Jupiter	Saturn	Uranus	Neptune
He/H**	0.05	0.03	0.06	?
C/H	0.0011	9.9×10^{-4}	0.0094	0.012
N/H	9.8×10^{-5}	2.4×10^{-4}	$< 9.8 \times 10^{-5}$	$< 9.8 \times 10^{-5}$
O/H	2.3×10^{-5}	?	?	?
P/H	2.4×10^{-7}	7.9×10^{-7}	?	?
D/H	3.2×10^{-5}	2.4×10^{-5}	4.8×10^{-5}	?
$^{13}C/^{12}C$	0.0062	0.011	?	?
$^{14}N/^{15}N$	0.0016	?	?	?

*Adapted from Baugher (1988). Ratios of the numbers of atoms.

**The mole fractions of helium, relative to hydrogen, for Jupiter, Saturn and Uranus are 0.117 ± 0.025, 0.062 ± 0.011 and 0.152 ± 0.033, respectively, Pollack (1984), Conrath et al. (1987).

3.4 Planetary Rings

Rings of Jupiter*

Ring	Distance (km)	Distance (R_J)	Width (km)	Thickness (km)	Optical Depth
Main Ring	122,800 (inner) 129,200 (outer)	1.72 1.81	6,400	≤ 30	3×10^{-5}
Inner Ring	71,398 (inner) 122,800 (outer)	1.00 1.72	51,402	$\sim 20,000$	$\leq 7 \times 10^{-6}$

*R_J = 71,398 km. Associated satellites J15 Adrastea (R = 10 km, D = 1.8064 R_J) and J16 Metis (R = 20 km, D = 1.7922 R_J). Adopted from Burns, Showalter and Morfill (1984).

Rings of Saturn (Main Rings)*

Ring	Distance (km)	Distance (R_S)	Width (km)	Thickness (meters)	Mass ($10^{-8} M_S$)
D Ring	69,970 (inner) 74,510 (outer)	1.110 1.235	4,540		
C Ring	74,510 (inner) 92,000 (outer)	1.235 1.525	17,500	≤ 10	~ 0.2
B Ring	92,000 (inner) 117,580 (outer)	1.525 1.949	25,500		~ 5.0
Cassini Division Cassini Gap	119,000 (center) 119,900 (center)	1.972 1.988	4,500 246		~ 0.1
A Ring	122,170 (inner) 136,780 (outer)	2.025 2.267	14,700	≈ 50	~ 1.1
Encke Gap Keeler Gap	135,706 (center) 136,526 (center)	2.214 2.263	325 ~ 35		

*R_S = 60,330 km, $M_S = 5.685 \times 10^{29}$ grams. The total mass of Saturn's rings is $M_r \sim 6 \times 10^{-8} M_S = 3.4 \times 10^{22}$ grams. A total ring mass of $\sim 3 \times 10^{-8} M_S$ is obtained from radio occultation studies assuming particles composed of solid water ice. The main rings are warped with a total vertical extent of ~ 1 km, but they are locally only 5 to 50 meters thick. Adapted from Tyler et al. (1982), Zebker and Tyler (1984) and Cuzzi et al. (1984).

Rings of Saturn (Tenuous Rings)*

Ring	Distance (km)	Distance (R_S)	Width (km)	Thickness (km)	Optical Depth
F Ring	140,300 (center)	2.326	30 to 500		10^{-2} to 1
G Ring	170,000 (center)	2.818	10,000	100 to 1,000	10^{-4} to 10^{-5}
E Ring	180,000 (inner) 480,000 (outer)	2.984 7.956	300,000	1,000	10^{-6} to 10^{-7}

*R_S = 60,330 km. Shepherd satellites for the F ring S16 1980S27 (D = 2.310 R_S) and S15 1980S26 (D = 2.349 R_S). Enceladus at D = 3.95 R_S may feed the E ring; other nearby satellites are Mimas (D = 3.09 R_S), Dione (D = 6.29 R_S) and Rhea (D = 8.78 R_S).Adopted from Burns, Showalter and Morfill (1984).

Rings of Uranus*

Ring	Distance (km)	Distance (R_U)	Width (km)	Optical Depth	Eccentricity (10^{-3})	Inclination (10^{-3} degrees)
1986 U2R	37,000 (inner) 39,500 (outer)	1.448 1.545	2,500	0.001 to 0.0001	0?	0?
6	41,850	1.637	1 to 3	0.2 to 0.3	1.0	63
5	42,240	1.653	2 to 3	0.5 to 0.6	1.9	52
4	42,580	1.666	2 to 3	0.3	1.1	32
α	44,730	1.750	7 to 12	0.3 to 0.4	0.8	14
β	45,670	1.787	7 to 12	0.2	0.4	5
η	47,180	1.856	0 to 2	0.1 to 0.4	(0)	(2)
γ	47,630	1.864	1 to 4	1.3 to 3.3	(0)	(11)
δ	48,310	1.890	1 to 9	0.3 to 0.4	(0)	4
1986U1R	50,040	1.956	1 to 2	0.1	0?	0?
ε	51,160	2.002	22 to 93	0.5 to 2.1	7.9	(1)

*R_U = 25,559 km. Adapted from Lane et al. (1986) and Stone and Miner (1986).

Rings of Neptune*

Ring	Distance (km)	Distance (R_N)	Width (km)	Optical Depth
1989 N3R	41,900	1.69	~ 1700	10^{-4}
1989 N2R	53,200	2.15	< 15	0.01–0.02
1989 N4R	53,200 to 59,000	2.15–2.4	5800	10^{-4}
1989 N1R**	62,930	2.53	~ 20	0.01–0.1

*R_N = 24,764 km. Courtesy of Joseph A. Burns and Philip D. Nicholson, Cornell University. All rings have low albedo, no detectable eccentricity or inclination, no measurable thickness and are quite dusty.

**Contains at least three optically thick arcs of length 4°, 4° and 10°.

4
The Moon

4.1 Physical Parameters and Motions of the Moon

Size

Mean Radius: R_m = 1738.0 km = 0.2725 R_E

Elevation with respect to R_m:
- Ringed maria = − 4.0 km
- Other maria = − 2.3 km
- Farside terrae = +1.8 km
- Nearside terrae = − 1.4 km

Surface Area = 3.788×10^{17} cm^2
Volume: 2.199×10^{25} cm^3

Mass

Inverse Mass of Moon: $M_\odot/M_m$ = 27,068,708.7
Earth-Moon Mass Ratio: $M_E/M_m = 1/\mu$ = 81.300587
Moon Mass: $M_m = 7.3483 \times 10^{25}$ grams
Surface Escape Velocity: V_{esc} = 2.37 km/sec
Mean Density of Moon: $\bar{\rho}_m$ = 3.344 grams/cm^3

Gravity

Gravitational Constant: GM_m = 4,902.799 km^3/sec^2
Surface Gravity: g_m = 162.0 cm/sec^2
Gravity Field: $J_2 = 202.7 \times 10^{-6}$

Age

Age of Moon: 4.6×10^9 years
Oldest Moon Rock: 4.6×10^9 years

Age and Diameter of Impact Basins*

Impact Basin	Impact Date (Billions of Years Ago)	Rim Diameter (km)
Orientale Basin	3.85	930
Imbrium Basin	3.95	1,500
Crisium-Humorum Basins	4.05 - 4.20	1,060 - 820
Nectaris Basin	4.25	860
Serenitatis Basin	4.28	880

*Adapted from Schaeffer (1977).

Rotation

Rotation Period: 27.321661 days = Sidereal Month
Moment of Inertia about Rotation Axis: $C = 0.392\ M_m R_m^2$
Moment of Inertia Differences: $\beta = (C-A)/B = 631.3 \times 10^{-6}$
$\gamma = (B-A)/C = 227.8 \times 10^{-6}$

Length of Mean Months

Sidereal Month: 27.321661 days = Fixed Star to Fixed Star
Synodic Month: 29.530589 days = New Moon to New Moon
Anomalistic Month (perigee to perigee): 27.554551 days
Tropical Month (equinox to equinox): 27.321582 days
Draconic Month (node to node): 27.212220 days

Orbital Motion

Mean Distance of Moon from Earth: a_m = 384,400 km
Mean Orbital Speed of Moon: 1.023 km/sec
Perigee Distance of Moon from Earth: q = 356,410 km
Apogee Distance of Moon from Earth: Q = 406,740 km
Eccentricity of Orbit: e = 0.0549
Inclination of Orbit to Ecliptic: 5° 8′ 43.″4
oscillating ± 9′ with period of 173 days
Inclination of Lunar Equator to Ecliptic = 1° 32′ 32.″7
Period of Rotation of Moon's Perigee: 3,232 days ≈ 8.85 tropical years
Period of Moon's Node (retrograde): 6,798 days ≈ 18.61 tropical years
Saros-period = 223 synodic months = 6,585.32 days
≈ 19 ecliptic years where Ecliptic year = 346.62 days

Observational Properties

Moon Semi-Diameter at Mean Distance: 15′ 32.″6 (geocentric)
15′ 48.″3 (topocentric, zenith)
Mean Equatorial Horizontal Parallax: π_m = 57′ 2.″61 = 3,422.″61
$\sin\pi_m = (a_e/a_m)$ = 57′ 2.″45 = 3,422.″45
Surface Area of Moon at Some Time Visible from Earth: 59%
Moon's Sidereal Mean Daily Motion: 13.176358 degrees per day

Acceleration

Acceleration of Moon's Orbit: $\dot{n} = -25.3 \pm 1.2$ arcsec century^{-2}
Increase in Moon's Mean Distance: $\Delta a_m \approx 5$ cm century^{-1}

Key Events in the History of the Moon

Feature Created	Time (Billions of Years Ago)	Process
Formation of Moon	4.6	Accretion
Magma Ocean	4.6 to 4.4	Accretion and Melting
Crust	4.4 to 4.2	Differentiation and Cooling
Highlands, Impact Basins	4.2 to 3.9	Intense Bombardment
Maria	3.9 to 3.1	Volcanism from Interior
Regolith, Smooth Surface	3.1 to now	Continued Bombardment

Libration of the Moon

Libration	In Longitude	In Latitude
Optical		
Selenocentric Displacement, maximum	$\pm$ 7°53′	$\pm$ 6°51′
Period	1 Sidereal Month	1 Sidereal Month
Physical		
Displacement, principal term	$\pm$ 66″	$\pm$ 105″
Period	1 year	1 month

4.2 Lunar Maria and Craters

Ages of Maria*

Maria	Duration of Volcanism (Billions of Years Ago)
Oceanus Procellarum	3.0 to 3.3
Mare Imbrium	3.1 to 3.4
Mare Crisium	3.2 to 3.3
Mare Tranquillitatis	3.5 to 3.9
Mare Serenitatis	3.7 to 3.9

*Adapted from Schaeffer (1977)

Ages and Diameters of Some Relatively Young Lunar Craters

Crater	Age (years)	Diameter (km)	Crater	Age (years)	Diameter (km)
South Ray	2 million	0.7	North Ray	50 million	0.9
Shorty	19 million	0.7	Camelot	90 million	0.7
Cone	24 million	0.4	Tycho	107 million	85.0
			Copernicus	900 million	93.0

Large Lunar Maria

Maria	Seas	Latitude (degrees)	Longitude (degrees)	Basin Diameter (km)
Oceanus Procellarum	Ocean Of Storms	10 N	47 W	3,200
Mare Imbrium	Sea Of Rain	36 N	16 W	1,500
Mare Crisium	Sea Of Crises	18 N	58 E	1,060
Mare Orientale	Eastern Sea	19 S	95 W	930
Mare Serenitatis	Sea Of Serenity	30 N	17 E	880
Mare Nectaris	Sea Of Nectar	14 S	34 E	860
Mare Smythii	Smyth's Sea	3 S	80 E	840
Mare Humorum	Sea Of Moisture	23 S	38 W	820
Mare Tranquillitatis	Sea Of Tranquillity	9 N	30 E	775
Mare Nubium	Sea Of Clouds	19 S	14 W	690
Mare Fecunditatis	Sea Of Fertility	4 S	51 E	690

Diameter And Location of Some Lunar Craters

Crater	Latitude (degrees)		Longitude (degrees)		Diameter (km)
Abulfeda	13.8°	S	13.9°	E	65
Aitken	16.5°	S	173.1°	E	131
Albategnius	11.2°	S	4.1°	E	136
Alphonsus	13.4°	S	2.8°	W	119
Alphonsus R	14.4°	S	1.9°	W	3
Anaxagoras	73.4°	N	10.1°	W	51
Antoniadi	69.8°	S	172.0°	W	135
Apollo	35.5°	S	149.6°	W	503
Apollonius	4.5°	N	61.1°	E	53
Arago	6.2°	N	21.4°	E	26
Archimedes	29.7°	N	4.0°	W	83
Ariadaeus	4.6°	N	17.3°	E	11
Aristarchus	23.7°	N	47.4°	W	40
Aristillus	33.9°	N	1.2°	E	55
Arzachel	18.2°	S	1.9°	W	97
Autolycus	30.7°	N	1.5°	E	39
Bailly	66.8°	S	69.4°	W	303
Beer	27.1°	N	9.1°	W	10
Berzelius	36.6°	N	50.9°	E	51
Bessel	21.8°	N	17.9°	E	16
Bode	6.7°	N	2.4°	W	19
Bonpland	8.3°	S	17.4°	W	60
Cauchy	9.6°	N	38.6°	E	12
Chaucer	3.7°	N	140.0°	W	45
Clavius	58.4°	S	14.4°	W	225
Compton	56.0°	N	105.0°	E	162
Copernicus	9.7°	N	20.0°	W	93
Copernicus H	6.9°	N	18.3°	W	5
Cyrillus	13.2°	S	24.0°	E	98
D'Alembert	51.3°	N	164.6°	E	225
Darwin	19.8°	S	69.1°	W	130
Davy	11.8°	S	8.1°	W	35
DeLisle	29.9°	N	34.6°	W	25
Descartes	11.7°	S	15.7°	E	48
Dionysius	2.8°	N	17.3°	E	18

Diameter And Location of Some Lunar Craters

Crater	Latitude (degrees)		Longitude (degrees)		Diameter (km)
Diophantus	27.6°	N	34.3°	W	18
Dollond B	7.7°	S	13.8°	E	37
Doppler	12.8°	S	159.9°	W	100
Eratosthenes	14.5°	N	11.3°	W	58
Euler	23.3°	N	29.2°	W	28
Fauth	6.3°	N	20.1°	W	12
Flammarion	3.4°	S	3.7°	W	75
Flamsteed	4.5°	S	44.3°	W	21
Fra Mauro	6.0°	S	17.0°	W	95
Fracastorius	21.2°	S	33.0°	E	124
Galilaei	10.5°	N	62.7°	W	16
Gassendi	17.5°	S	39.9°	W	110
Giordano Bruno	35.9°	N	102.8°	E	22
Grimaldi	5.2°	S	69.6°	W	140
Gruithuisen	32.9°	N	39.7°	W	16
H.G. Wells	41.0°	N	122.7°	E	103
Hausen	65.5°	S	88.4°	W	167
Hell	32.4°	S	7.8°	W	33
Herigonius	13.3°	S	33.9°	W	15
Herodotus	23.2°	N	49.7°	W	35
Herschel	5.7°	S	2.1°	W	41
Hevelius	2.2°	N	67.3°	W	106
Hippalus	24.8°	S	30.2°	W	58
Hommel	54.6°	S	33.0°	E	125
Humboldt	27.2°	S	80.9°	E	207
Hyginus	7.8°	N	6.3°	E	9
Icarus	5.3°	S	173.2°	W	96
Inghirami	47.5°	S	68.8°	W	91
Jansen	13.5°	N	28.7°	E	24
Janssen	44.9°	S	41.5°	E	190
Jules Verne	34.8°	S	146.9°	E	134
Julius Caesar	9.0°	N	15.4°	E	91
Kant	10.6°	S	20.1°	E	33
Kepler	8.1°	N	38.0°	W	32
King	5.0°	N	120.5°	E	77

Diameter And Location of Some Lunar Craters

Crater	Latitude (degrees)		Longitude (degrees)		Diameter (km)
Kopff	17.4°	S	89.6°	W	42
Korolev	4.4°	S	157.4°	W	453
Kramers	53.6°	N	127.6°	W	62
La Hire A	28.5°	N	23.4°	W	5
Lalande	4.4°	S	8.6°	W	24
Lambert	25.8°	N	21.0°	W	30
Langrenus	8.9°	S	60.9°	E	132
Lansberg	0.3°	S	26.6°	W	39
Lassell	15.5°	S	7.9°	W	23
Lavoisier	38.2°	N	81.2°	W	70
Le Monnier	26.6°	N	30.6°	E	61
Linné	27.7°	N	11.8°	E	2
Littrow	21.5°	N	31.4°	E	31
Longomontanus	49.5°	S	21.7°	W	145
Lyell	13.6°	N	40.6°	E	32
Macrobius	21.3°	N	46.0°	E	64
Marco Polo	15.4°	N	2.0°	W	28
Marius	11.9°	N	50.8°	W	41
Maskelyne	2.2°	N	30.1°	E	24
Maunder	14.6°	S	93.8°	W	55
Maurolycus	41.8°	S	14.0°	E	114
Mendeleev	5.6°	N	141.5°	E	330
Mutus	63.6°	S	30.1°	E	78
Nansen	81.3°	N	95.3°	E	122
Neper	8.8°	N	84.5°	E	137
Olbers	7.4°	N	75.9°	W	75
Omar Khayyam	58.0°	N	102.1°	W	70
Peirce	18.3°	N	53.5°	E	19
Petavius	25.3°	S	60.4°	E	177
Picard	14.6°	N	54.7°	E	23
Piccolomini	29.7°	S	32.2°	E	88
Planck	57.9°	S	135.8°	E	344
Plato	51.6°	N	9.3°	W	101
Playfair	23.5°	S	8.4°	E	48
Plutarch	24.1°	N	79.0°	E	68

Diameter And Location of Some Lunar Craters

Crater	Latitude (degrees)		Longitude (degrees)		Diameter (km)
Poincaré	56.7°	S	163.6°	E	319
Posidonius	31.8°	N	29.9°	E	95
Proclus	16.1°	N	46.8°	E	28
Ptolemaeus	9.2°	S	1.8°	W	153
Pythagoras	63.5°	N	62.8°	W	130
Regiomontanus	28.4°	S	1.0°	W	124
Regiomontanus A	28.0°	S	0.6°	W	6
Reiner	7.0°	N	54.9°	W	30
Rheita	37.1°	S	47.2°	E	70
Ritter	2.0°	N	19.2°	E	29
Ross	11.7°	N	21.7°	E	25
Rothmann G	28.4°	S	24.3°	E	92
Sabine	1.4°	N	20.1°	E	30
Schickard	44.4°	S	54.6°	W	227
Schroter	2.6°	N	7.0°	W	35
Sirsalis	12.5°	S	60.4°	W	42
Sulpicius Gallus	19.6°	N	11.6°	E	12
Taruntius	5.6°	N	46.5°	E	56
Taylor	5.3°	S	16.7°	E	42
Theophilus	11.4°	S	26.4°	E	100
Timocharis	26.7°	N	13.1°	W	34
Tsiolkovskiy	20.4°	S	129.1°	E	180
Tycho	43.3°	S	11.2°	W	85
Van de Graaff	27.0°	S	172.0°	E	234
Vasco da Gama	13.9°	N	83.8°	W	96
Vitello	30.4°	S	37.5°	W	42
Vitruvius	17.6°	N	31.3°	E	30
Wargentin	49.6°	S	60.2°	W	84
Widmanstatten	6.1°	S	85.5°	E	46
Wolf	22.7°	S	16.6°	W	25
Xenophon	22.8°	S	122.1°	E	25
Zwicky	15.9°	S	167.6°	E	135

Full Moon. This Earth-based telescopic view of the full Moon enhances the contrast between the dark maria and the bright rayed craters. The region near the Moon's south pole (top center) is dominated by the magnificent rays of ejecta emanating from the relatively young crater Tycho. The dark circular Mare Imbrium is prominent in the northwest (lower right), immediately below the bright rays of craters Copernicus and Kepler (middle right). The dark circular Mare Serenitatis lies to the east (left) of Imbrium. (Lick Observatory photograph.)

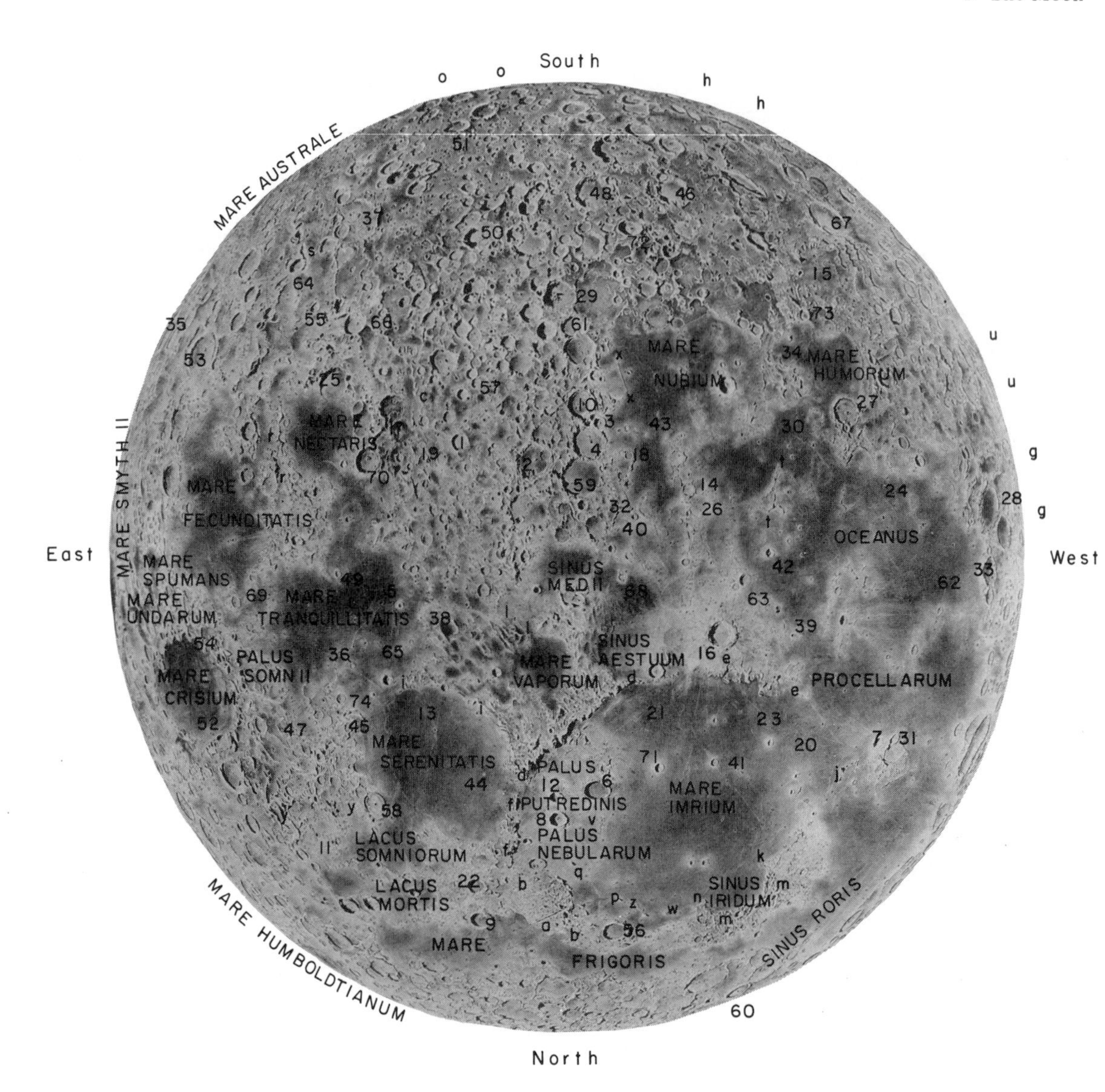
South
MARE AUSTRALE
MARE NUBIUM
MARE HUMORUM
MARE NECTARIS
MARE FECUNDITATIS
OCEANUS
East
MARE SMYTH II
MARE SPUMANS
MARE UNDARUM
MARE TRANQUILLITATIS
SINUS MEDII
West
SINUS AESTUUM
MARE VAPORUM
PALUS SOMNII
MARE CRISIUM
PROCELLARUM
MARE SERENITATIS
PALUS PUTREDINIS
MARE IMRIUM
LACUS SOMNIORUM
PALUS NEBULARUM
LACUS MORTIS
SINUS IRIDUM
MARE FRIGORIS
SINUS RORIS
MARE HUMBOLDTIANUM
North

MOUNTAINS AND VALLEYS

a. Alpine Valley
b. Alps Mts.
c. Altai Scarp
d. Apennine Mts.
e. Carpathian Mts.
f. Caucasus Mts.
g. Cordilleras Mts.
h. Doerfel Mts.
i. Haemus Mts.
j. Harbinger Mts.
k. Heraclides Prom.
l. Hyginus Cleft
m. Jura Mts.
n. Laplace Prom.
o. Leibnitz Mts.
p. Pico
q. Piton
r. Pyrenees Mts.
s. Rheita Mts.
t. Riphaeus Mts.
u. Rook Mts.
v. Spitzbergen Mts.
w. Straight Range
x. Straight Wall
y. Taurus Mts.
z. Teneriffe Mts.

PROMINENT LUNAR CRATERS

1. Abulfeda
2. Albategnius
3. Alpetragius
4. Alphonsus
5. Arago
6. Archimedes
7. Aristarchus
8. Aristillus
9. Aristoteles
10. Arzachel
11. Atlas
12. Autolycus
13. Bessel
14. Bonpland
15. Clausius
16. Copernicus
17. Cyrillus
18. Davy
19. Descartes
20. Diophantus
21. Eratosthenes
22. Eudoxus
23. Euler
24. Flamsteed
25. Fracastorius
26. Fra Mauro
27. Gassendi
28. Grimaldi
29. Hell
30. Herigonius
31. Herodotus
32. Herschel
33. Hevelius
34. Hippalus
35. Humboldt
36. Jansen
37. Janssen
38. Julius Caesar
39. Kepler
40. Lalande
41. Lambert
42. Lansberg
43. Lassell
44. Linné
45. Littrow
46. Longomontanus
47. Macrobius
48. Maginus
49. Maskelyne
50. Maurolycus
51. Mutus
52. Peirce
53. Petavius
54. Picard
55. Piccolomini
56. Plato
57. Playfair
58. Posidonius
59. Ptolemaeus
60. Pythagoras
61. Regiomontanus
62. Reiner
63. Reinhold
64. Rheita
65. Ross
66. Rothmann
67. Schickard
68. Schröter
69. Taruntius
70. Theophilus
71. Timocharis
72. Tycho
73. Vitello
74. Vitruvius

5
Satellites of the Planets

5.1 General Properties of the Satellites

Total measured mass of the satellites = 6.2×10^{26} grams = $0.10 M_E$
Mass of Earth = $M_E = 5.9742 \times 10^{27}$ grams

Planet	Number of Known Satellites	Planet Mass (10^{26} grams)	Total Mass of Satellites (10^{26} grams)	Planet Radius, R_P (km)	Satellite Distance (km)
Mercury	0	3.302		2,439	
Venus	0	48.690		6,051	
Earth	1	59.742	0.735	6,378	384.4
Mars	2	6.419		3,397	9.4 to 23.5
Jupiter	16	18,992	3.944	71,492	128 to 23,700
Saturn	17	5,686.5	1.409	60,268	138 to 12,954
Uranus	15	868.5	0.091	25,559	50 to 583
Neptune	8	1,024.3	0.214	24,764	354 to 5,510

Atmosphere of Titan*

Basic Data			
Surface Pressure, P	P	=	1.5 bar
Surface Temperature, T	T	=	94°K
Acceleration of Gravity, g	g	=	135 cm s^{-2}
Number Density, No	No	=	1.2×10^{20} cm^{-3}
Satellite Radius, R	R	=	2575 km
Atmospheric Composition			
Nitrogen, N_2	N_2	=	82–99%
Argon, Ar	Ar	=	0–12%
Methane, CH_4	CH_4	=	1–6%
Hydrogen, H_2	H_2	=	2000 ppm
Ethane, C_2H_6	C_2H_6	=	20 ppm
Propane, C_3H_8	C_3H_8	=	20 ppm
Ethylene, C_2H_4	C_2H_4	=	0.4 ppm
Diacetylene, C_4H_2	C_4H_2	=	0.1–0.01 ppm
Methylacetylene, C_3H_4	C_3H_4	=	0.03 ppm
Hydrogen Cyanide, HCN	HCN	=	0.2 ppm
Cyanogen, C_2N_2	C_2N_2	=	0.1–0.01 ppm
Cyanoacetylene, HC_3N	HC_3N	=	0.1–0.01 ppm
Carbon Monoxide, CO	CO	=	50–150 ppm
Carbon Dioxide, CO_2	CO_2	=	0.015 ppm

*Adapted from Lindal et al. (1983), Morrison and Owen (1988), and Yung, Allen and Pinto (1984). The ppm denotes parts per million.

Mean Orbital Elements For The Satellites*

Satellite	Distance from Planet Center (10^3 km)	Distance from Planet Center (R_P)	Orbital Period (days)	Eccentricity (degrees)	Inclination (degrees)
Earth					
Moon	384.4	60.2	27.3217	0.05490	18.2 to 28.6
Mars					
M1 Phobos	9.37	2.76	0.3189	0.0150	1.1
M2 Deimos	23.52	6.90	1.262	0.0008	0.9 to 2.7†
Jupiter					
J14 Adrastea	128	1.80	0.295	~ 0.0	~ 0.0
J16 Metis	128	1.80	0.295	~ 0.0	~ 0.0
J5 Amalthea	181	2.55	0.489	0.003	0.4
J15 Thebe	221	3.11	0.675	~ 0.0	~ 0.0
J1 Io	422	5.95	1.769	0.004†	0.0
J2 Europa	671	9.47	3.551	0.000†	0.5
J3 Ganymede	1070	15.1	7.155	0.001†	0.2
J4 Callisto	1880	26.6	16.69	0.010	0.2
J13 Leda	11110	156	240	0.146	26.7
J6 Himalia	11470	161	251	0.158	27.6
J10 Lysithea	11710	164	260	0.130	29.0
J7 Elara	11740	165	260	0.207	24.8
J12 Ananke	20700	291	617	0.17	147
J11 Carme	22350	314	692	0.21	164
J8 Pasiphae	23300	327	735	0.38	145
J9 Sinope	23700	333	758	0.28	153

Mean Orbital Elements For The Satellites*

Satellite	Distance from Planet Center (10^3 km)	Distance from Planet Center (R_P)	Orbital Period (days)	Eccentricity (degrees)	Inclination (degrees)
Saturn					
S17 Atlas	137.7	2.276	0.602	.002	0.3
S16 Prometheus	139.4	2.310	0.613	.004	0.0
S15 Pandora	141.7	2.349	0.629	.004	0.1
S10 Janus **	151.4	2.510	0.694	.009	0.3
S11 Epimetheus	151.5	2.511	0.695	.007	0.1
S1 Mimas	186	3.08	0.942	.020	1.5
S2 Enceladus	238	3.95	1.370	.004	0.0
S3 Tethys	295	4.88	1.888	.000	1.1
S13 Telesto	295	4.88	1.888		
S14 Calypso	295	4.88	1.888		
S4 Dione	377	6.26	2.737	.002	0.0
S12 Helene	377	6.26	2.737	.005	0.2
S5 Rhea	527	8.73	4.518	.001	0.4
S6 Titan	1222	20.3	15.95	.029	0.3
S7 Hyperion	1481	24.6	21.28	.104	0.4
S8 Iapetus	3561	59	79.33	.028	14.7†
S9 Phoebe	12954	215	550	.163	150

*Adopted from Masson (1984), Morrison (1982), Morrison, Cruikshank and Burns (1977), Owen and Synnott (1987), Stone and Miner (1982, 1986).
Inclinations for the satellites of Earth and Mars are relative to the planet's equator; those for all of the other planets are relative to the planet's orbital plane.
A † sign near a value means it is variable.

** The co-orbital satellites S11 (Epimetheus) and S10 (Janus) change orbits every four years.

Mean Orbital Elements For The Satellites*

Satellite	Distance from Planet Center (10^3 km)	Distance from Planet Center (R_P)	Orbital Period (days)	Eccentricity (degrees)	Inclination (degrees)
Uranus					
U13 Cordelia	49.7	1.94	0.333		
U14 Ophelia	53.8	2.10	0.375		
U15 Bianca	59.2	2.32	0.433		
U9 Cressida	61.8	2.42	0.463		
U12 Desdemona	62.7	2.45	0.475		
U8 Juliet	64.6	2.53	0.492		
U7 Portia	66.1	2.58	0.513		
U10 Rosalind	69.9	2.73	0.558		
U11 Belinda	75.3	2.94	0.621		
U6 Puck	86.0	3.36	0.763		
U5 Miranda	129.9	5.08	1.413	0.017	3.4
U1 Ariel	190.9	7.47	2.521	0.0028	
U2 Umbriel	266.0	10.41	4.146	0.0035	
U3 Titania	436.3	17.07	8.704	0.0024	
U4 Oberon	583.4	22.82	13.463	0.0007	
Neptune					
N1 Triton	354.8	14.33	5.877	0.00	160.0
N2 Nereid	5513.4	222.65	365.2	0.75	27.6
1989 N6	48.0	1.94			
1989 N5	50.0	2.01			
1989 N4	52.5	2.12			
1989 N3	62.0	2.50			
1989 N2	73.6	2.97			
1989 N1	117.6	4.75			

*Adopted from Masson (1984), Morrison (1982), Morrison, Cruikshank and Burns (1977), Owen and Synnott (1987), Stone and Miner (1982, 1986).
Inclinations for the satellites of Earth and Mars are relative to the planet's equator; those for all of the other planets are relative to the planet's orbital plane.
A † sign near a value means it is variable.

5.2 Physical Elements for the Principal Satellites*

Satellite	Radius (km)	Mass (10^{23} grams)	Density (grams/cm^3)	Visual Magnitude Opposition V_{opp}	Visual Magnitude Unit Distance V(1,0)	Geometrical Visual Albedo (ρ_v)
Earth						
Moon	1738	735	3.34		+0.21	0.12
Mars						
M1 Phobos	14 × 10	9.6×10^{-5}	≤ 2	11.6	+11.9	0.06
M2 Deimos	8 × 6	2.0×10^{-5}	≤ 2	12.7	+13.0	0.07
Jupiter						
J3 Ganymede	2631	1490	1.93	4.6	- 2.09	0.43
J4 Callisto	2400	1075	1.83	5.6	- 1.05	0.17
J1 Io	1815	892	3.55	5.0	- 1.68	0.63
J2 Europa	1569	487	3.04	5.3	- 1.41	0.64
Saturn						
S6 Titan	2575	1346	1.88	8.4	- 1.20	0.21
S5 Rhea	765	24.9	1.33	9.7	+0.16	0.60
S8 Iapetus	730	18.8	1.15	10.2 - 11.9	+1.6	0.12
S4 Dione	560	10.52	1.41	10.4	+0.88	0.60
S3 Tethys	530	7.55	1.20	10.3	+0.7	0.8
S2 Enceladus	250	0.74	1.13	11.8	+2.2	1.0
S1 Mimas	196	0.455	1.44	12.9	+3.3	0.7
Uranus						
U3 Titania	790	34.8	1.69	14.0	+1.3	0.28
U4 Oberon	762	29.2	1.64	14.2	+1.5	0.24
U2 Umbriel	586	12.7	1.58	15.3	+2.6	0.19
U1 Ariel	579	13.5	1.55	14.4	+1.7	0.40
U5 Miranda	236	0.8	1.25	16.5	+3.8	0.34
Neptune						
N1 Triton	1355	214	2.05	13.6	- 1.2	
N2 Nereid	150			18.7	+4.0	

*Adapted from Davies et al. (1987), Johnson, Brown and Pollack (1987), Masson (1984), Morrison, Cruikshank and Burns (1977), Smith et al. (1986), Stone and Miner (1982, 1986), Tyler et al. (1982, 1986), Anderson et al. (1987), and Tyler et al. (1989).

5.3 Physical Elements for the Small Satellites*

Satellite	Radius (km)	Geometric Albedo, P_V	Satellite	Radius (km)	Geometric Albedo, P_V
Jupiter			**Jupiter**		
J14 Adrastea	20 ± 5	≤ 0.1	J10 Lysithea	~ 10	
J16 Metis	20 ± 5	≤ 0.1	J7 Elara	40 ± 5	0.03
J5 Amalthea	$135 \times 85 \times 75$	0.056	J12 Ananke	~ 10	
J15 Thebe	40 ± 5	≤ 0.01	J11 Carme	~ 15	
J13 Leda	~ 5		J8 Pasiphae	~ 20	
J6 Himalia	90 ± 10	0.03	J9 Sinope	~ 15	
Saturn			**Saturn**		
S17 1980S28	20×10	0.4	S13 1980S13	$17 \times 14 \times 13$	0.6
S16 1980S27	$70 \times 50 \times 40$	0.6	S14 1980S25	$17 \times 11 \times 11$	0.8
S15 1980S26	$55 \times 45 \times 35$	0.6	S12 1980S6	$18 \times 16 \times 15$	0.5
S11 1980S3	$70 \times 60 \times 50$	0.5	S7 Hyperion	$205 \times 130 \times 110$	0.2
S10 1980S1	$110 \times 100 \times 80$	0.5	S9 Phoebe	110 ± 10	0.06
Uranus			**Uranus**		
U13 1986U7	~ 20	< 0.1	U8 1986U2	~ 40	< 0.1
U14 1986U8	~ 25	< 0.1	U7 1986U1	~ 40	< 0.1
U15 1986U9	~ 25	< 0.1	U10 1986U4	~ 30	< 0.1
U9 1986U3	~ 30	< 0.1	U11 1986U5	~ 30	< 0.1
U12 1986U6	~ 40	< 0.1	U6 1985U1	85 ± 5	< 0.1
Neptune			**Neptune**		
N3 1989N1	200 ± 10	0.06	N6 1989N4	90 ± 10	
N4 1989N2	95 ± 10	0.06	N7 1989N5	40 ± 8	
N5 1989N3	75 ± 15	0.06	N8 1989N6	27 ± 8	

*Adapted from Morrison (1982), Tyler et al. (1982), Veverka, Thomas and Synnott (1982), Smith et al. (1986), and Stone and Miner (1989).

6
Asteroids and their Debris

6.1 General Properties of the Asteroids

Number of minor planets with well-determined orbits (numbered asteroids) = 3445 in 1988
Total mass of asteroids $\approx 1.8 \times 10^{24}$ grams
Average orbital elements:
Orbital period = $\bar{P}$ = 4.5 years
Semi-major axis = $\bar{a}$ = 2.8 AU
Eccentricity = $\bar{e}$ = 0.15
Inclination = $\bar{i}$ = 9.4 degrees
Perihelion = $\bar{q}$ = 2.4 AU
Aphelion = $\bar{Q}$ = 3.2 AU
Kirkwood gaps = orbital periods of 1/3, 2/5, 3/7, 1/2, and 3/5 times Jupiter's orbital period of 11.86 years.

Mass, Radius and Density*

Asteroid	Mass ($10^{-10} M_{\odot}$)	Mass (10^{24} grams)	Radius (km)	Density (grams/cm^3)
1 Ceres	5.9 ± 0.3	1.180	466	2.7 ± 0.14
2 Pallas	1.08 ± 0.22	0.216	269	2.6 ± 0.9
4 Vesta	1.38 ± 0.12	0.276	261	3.3 ± 1.5
All others	0.65	0.130		

*Adapted from Shubert and Matson (1979), Cunningham (1988).

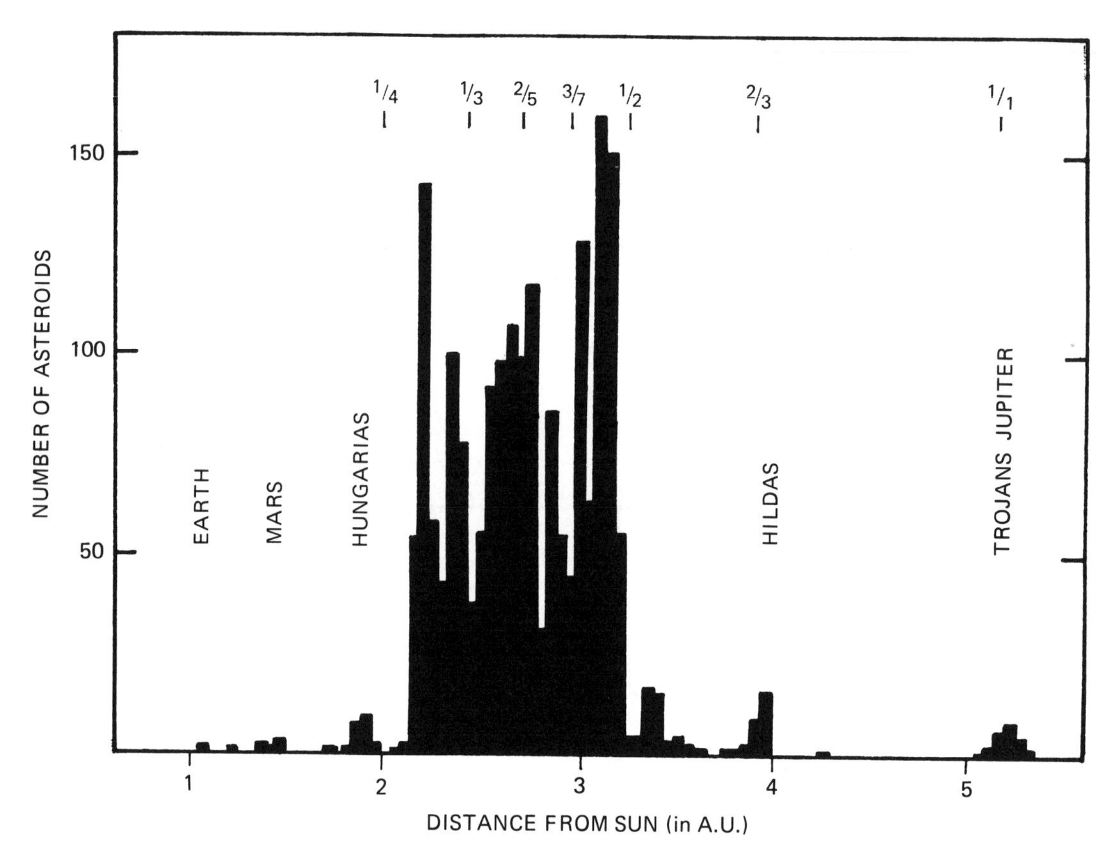

Kirkwood Gaps. The number of asteroids at different distances from the Sun. Most of the asteroids are found in the asteroid belt that lies between 2.2 and 3.3 A.U. from the Sun. Repeated gravitational interactions with Jupiter have tossed asteroids out of the Kirkwood gaps with orbital periods of 1/4, 1/3, 3/7 and 1/2 of Jupiter's orbital period. These periods are placed above the relevant gap in the figure.

Early Discoveries*

Number and Name	Discoverer	Date of Discovery	Number and Name	Discoverer	Date of Discovery
1 Ceres	G. Piazzi	1801 January 1	6 Hebe	K.L.Hencke	1847 July 1
2 Pallas	H.W.Olbers	1802 March 28	7 Iris	J.R.Hind	1847 August 13
3 Juno	K. Harding	1804 September 1	8 Flora	J.R.Hind	1847 October 18
4 Vesta	H.W.Olbers	1807 March 29	9 Metis	A. Graham	1848 April 25
5 Astrea	K.L.Hencke	1845 December 8	10 Hygeia	A. De Gasparia	1849 April 12

*Adapted from Pilcher (1979).

Asteroid Taxonomy*

Type	Visual Geometric Albedo	Color (B-V)	Meteorite Analogy	Type	Visual Geometric Albedo	Color (B-V)	Meteorite Analogy
C	0.03 to 0.08	0.63 to 0.80	Carbonaceous Chondrite	U	0.23	0.78	Basaltic Achondrite
M	0.09 to 0.11	0.70 to 0.72	Metal	S	0.10 to 0.16	0.82 to 0.92	
E	0.35	0.72	Enstatite Achondrite	R	0.26	0.97	Ordinary Chondrite

*Adapted from Chapman, Williams and Hartmann (1978).

6.2 Selected Asteroids: Physical and Orbital Elements*

Name	Diameter (km)	Spectral Class	Rotation Period (hours)	Mean Distance (AU)	Orbital Eccentricity	Orbital Inclination (degrees)
1 Ceres	907 × 959	C	9.078	2.768	0.077	10.60
2 Pallas	538 ± 12	U	7.881	2.773	0.233	34.80
3 Juno	267 ± 12	S	7.213	2.671	0.255	13.00
4 Vesta	470 × 530 × 580	U	5.342	2.362	0.090	7.14
6 Hebe	93 ± 5	S	7.274	2.424	0.203	14.79
7 Iris	211	S	7.135	2.386	0.230	5.50
9 Metis	110 × 190 × 245	S	5.064	2.387	0.122	5.59
10 Hygeia	414	C	18.	3.138	0.118	3.84
13 Egeria	233	C	7.045	2.575	0.088	16.49
15 Eunomia	248	S	6.081	2.642	0.188	11.76
16 Psyche	236	M	4.303	2.922	0.138	3.09
19 Fortuna	215	C	7.46	2.442	0.158	1.57
22 Kalliope	130 × 160 × 215	M	4.147	2.911	0.104	13.73
24 Themis	232 ± 12	C	8.369	3.129	0.133	0.76
31 Euphrosyne	257	C	5.54	3.148	0.278	26.33
39 Laetitia	85 × 150 × 255	S	5.138	2.768	0.112	10.38
45 Eugenia	237	U	5.70	2.720	0.084	6.60
51 Nemausa	67 × 70 × 85	U	7.785	2.365	0.066	9.97
52 Europa	277	C	11.258	3.095	0.109	7.46
87 Sylvia	238	CMEU	5.186	3.483	0.093	10.88

*Adapted from Bender (1979), Bowell et al. (1979), Cunningham (1988), Elliot (1979), and Tedesco (1979).

Selected Asteroids: Physical And Orbital Elements*

Name		Diameter (km)	Spectral Class	Rotation Period (hours)	Mean Distance (AU)	Orbital Eccentricity	Orbital Inclination (degrees)
88	Thisbe	232	C	6.042	2.769	0.163	5.24
107	Camilla	239	C	4.56	3.487	0.075	9.95
216	Kleopatra	224	CMEU	5.394	2.790	0.254	13.15
324	Bamberga	243	C	8.	2.685	0.337	11.17
433	Eros	14.1 × 14.5 × 40.5	S	5.270	1.458	0.223	10.83
451	Patientia	267	C	20.	3.065	0.068	15.20
511	Davida	318	C	5.17	3.181	0.172	15.90
532	Herculina	215 × 218 × 263	S	9.406	2.773	0.175	16.34
624	Hektor	150 × 300	U	6.923	5.153	0.026	18.26
1566	Icarus	1.4	U	2.273	1.078	0.827	22.95
1580	Betulia	6	U	6.130	2.196	0.490	52.04
1620	Geographos	1.5 × 4.0	S	5.223	1.245	0.335	13.33
1685	Toro	1.7 × 2.6	S	10.196	1.368	0.436	9.37
1862	Apollo	1.6		3.06	1.470	0.560	35.57
2060	Chiron	200			13.695	0.379	6.92

*Adapted from Bender (1979), Bowell et al. (1979), Cunningham (1988), Elliot (1979), and Tedesco (1979).

6.3 Concentrations of Asteroids

Zones of Asteroids*

Zone	Description	Number in Zone	Mean Distance from Sun, a (AU)	Limits for a
AAA	Apollo-Amor-Aten	36	1.831	
HU	Hungarias	30	1.900	$1.78 \leq a \leq 2.00$
MC	Mars crossers	29	2.285	
FL	Floras	421	2.230	$2.10 \leq a \leq 2.30$
PH	Phocaeas	62	2.368	$2.25 \leq a \leq 2.50$
NY	Nysas	44	2.448	$2.41 \leq a \leq 2.50$
I	Main belt	316	2.391	$2.30 \leq a \leq 2.50$
PAL	Pallas zone	4	2.755	$2.500 \leq a \leq 2.82$
IIa	Main belt	455	2.614	$2.500 \leq a \leq 2.706$
IIb	Main belt	298	2.761	$2.706 \leq a \leq 2.82$
KOR	Koronis zone	86	2.873	$2.83 \leq a \leq 2.91$
EOS	Eos zone	144	3.014	$2.99 \leq a \leq 3.03$
IIIa	Main belt	189	2.933	$2.82 \leq a \leq 3.03$
THE	Themis zone	165	3.145	$3.08 \leq a \leq 3.24$
GR	Griquas	3	3.243	$3.10 \leq a \leq 3.27$
IIIb	Main belt	480	3.140	$3.03 \leq a \leq 3.27$
CYB	Cybeles	51	3.431	$3.27 \leq a \leq 3.70$
HIL	Hildas	34	3.952	$3.70 \leq a \leq 4.20$
T	Trojans	35	5.203	$5.05 \leq a \leq 5.40$
Z	No zone	6		

*Adapted from Zellner, Tholen and Tedesco (1985).

Families of Asteroids*

Family Name	Minimum Number of Asteroids	Minimum Parent Diameter (km)	Largest Asteroid Diameter km (Name)	Parent Body Spectral Type	Mean Distance from Sun
1 Themis	62	300	249 (24 Themis)	~ C	3.13
2 Eos	74	189	98 (221 Eos)	S?	3.01
3 Koronis	42	90	51 (462 Eriphyla)	S	2.87
4 Maria	14	90	85 (695 Bella)	~ S	2.54
24 Nysa/Hertha	10	200	70 (135 Hertha)	E + M†	2.43
124 Budrosa	6	380	145 (349 Dembowska)	Vesta-like + M†	2.93
126 Leto	7	200	128 (68 Leto)	~ S	2.78
132 Concordia	8	250	190 (128 Nemesis)	~ C	2.75
138 Alexandra	12	270	177 (54 Alexandra)	C?	2.71
189 Flora	22	165	160 (8 Flora)	~ S	2.20

*Adapted from Gradie, Chapman and Williams (1979) and Williams (1979).
†Differentiated

6.4 Asteroid Debris

Meteorites

Number of Meteorites Cataloged in Museums ~ 2,600
Number of Meteorites Recovered in Antarctica $\geq$ 1,000
Mass Density of Meteorites:
Ordinary Chondrites (81%) 3.5 to 3.8 grams/cm^3
Carbonaceous Chondrites (5%) 2.2 to 2.9 grams/cm^3
Iron Meteorites (4%) 7.7 to 7.9 grams/cm^3

Atmospheric Entry Velocity: ~ 20 km/sec
Impact Velocity: 100 to 300 m/sec (Most Recoverable Meteorites)
: ~ 10 km/sec (Massive, Crater-Forming Meteorites)
Incoming Rate: Entire Earth ~ 10^7 per year ~ 10^9 grams per year (100 to 1,000 tons per year)
: Micrometeorites (Mass $\leq$ 1 gram): $\geq 10^{12}$ per year
: Possibly Recoverable Meteorites: $\approx 10^5$ per year
(1 gram $\leq$ Mass $\leq 10^8$ grams)
: Crater-Forming Meteorites: $\leq$ 5 per year
(Mass $\geq 10^8$ grams)

Impact Rate: 500 Meteorites per year entire Earth surface
(or about 50 tons per year entire Earth)
: 150 Meteorites per year on land
: 10 Meteorites per year actually recovered

Formation Age: (Time of Origin) ~ 4.6 billion years.
Cosmic-Ray Exposure Age: (Time in Space Since Parent Body Breakup)
: Stony Meteorites ~ 10 million years
: Iron Meteorites ~ 500 million years
Oldest Terrestrial Age: (Oldest Time on Earth)
: Stony Meteorites ~ few thousand years
: Iron Meteorites ~ 100 thousand years
But up to one million years
(older meteorites generally found in Antarctica)

Parent Bodies: Radius 50 to 200 kilometers.

Impact Craters

Terrestrial Impact Cratering Rate: $5.4 \pm 2.7 \times 10^{-15}$ km^{-2}year^{-1}
(for observed craters D > 20 km (Grieve, 1984))
Volume Ratio Crater/Meteorite = 10,000 to 60,000
Asteroid Cratering Rate: $6.0 \pm 3.0 \times 10^{-15}$ $km^{-2}year^{-1}$
(for Crater diameter D > 20 km (Shoemaker, 1979))

Catalogued Meteorites*

Class	Observed Falls	Fall Frequency (%)	Finds	Total
Stony-Chondrites	784	86.6	897	1681
Stony-Achondrites	69	7.7	63	132
Stony Irons	10	1.1	63	73
Irons	42	4.6	683	725
Total	905	100.0	1706	2611

*Well-classified specimens and some Antarctic meteorites, after Graham Bevan and Hutchinson (1985). The listed frequency only includes observed falls since weathered stony meteorites are more easily overlooked than weathered irons.

Asteroid-Earth Collision Rates*

Asteroid Type	Number to V(1,0)=18	Mean Collision Probability $10^{-9}yr^{-1}$	Collision Rate to V(1,0)=18 $10^{-6}yr^{-1}$
Atens	~ 100	9.1	~ 0.9
Apollos	700 ± 300	2.6	1.8 ± 0.8
Amors (Earth-crossing)	~ 500	~ 1	~ 0.5
Total	~ 1300		~ 3.5

*Adapted from Shoemaker et al. (1979).

Orbital Elements of Meteorites*

Meteorite Name	Mean Distance from Sun, a (AU)	Eccentricity e	Perihelion Distance q (AU)	Aphelion Distance Q (AU)	Inclination i (degrees)
Sikhote Alin	2.16	0.54	0.99	3.34	9.4
Pribram	2.42	0.67	0.79	4.05	10.4
Lost City	1.66	0.42	0.96	2.36	12.0
Innisfree	1.87	0.47	0.99	2.76	12.3
Dhajala	1.80	0.59	0.73	2.84	27.6

*Orbits inferred from trajectory in the terrestrial atmosphere. Adapted from Jessberger (1981).

6.5 Solar System Abundances of the Elements*

Element	Cameron (1982)	Anders & Grevesse (1989)	Element	Cameron (1982)	Anders & Grevesse (1989)
1 **H**	2.66×10^{10}	2.79×10^{10}	23 **V**	254	293
2 **He**	1.8×10^{9}	2.72×10^{9}	24 **Cr**	1.27×10^{4}	1.35×10^{4}
3 **Li**	60	57.1	25 **Mn**	9300	9550
4 **Be**	1.2	0.73	26 **Fe**	9.0×10^{5}	9.00×10^{5}
5 **B**	9	21.2	27 **Co**	2200	2250
6 **C**	1.11×10^{7}	1.01×10^{7}	28 **Ni**	4.78×10^{4}	4.93×10^{4}
7 **N**	2.31×10^{6}	3.13×10^{6}	29 **Cu**	540	522
8 **O**	1.84×10^{7}	2.38×10^{7}	30 **Zn**	1260	1260
9 **F**	780	843	31 **Ga**	38	37.8
10 **Ne**	2.6×10^{6}	3.44×10^{6}	32 **Ge**	117	119
11 **Na**	6.0×10^{4}	5.74×10^{4}	33 **As**	6.2	6.56
12 **Mg**	1.06×10^{6}	1.074×10^{6}	34 **Se**	67	62.1
13 **Al**	8.5×10^{4}	8.49×10^{4}	35 **Br**	9.2	11.8
14 **Si**	1.00×10^{6}	1.00×10^{6}	36 **Kr**	41.3	45
15 **P**	6500	1.04×10^{4}	37 **Rb**	6.1	7.09
16 **S**	5.0×10^{5}	5.15×10^{5}	38 **Sr**	22.9	23.5
17 **Cl**	4740	5240	39 **Y**	4.8	4.64
18 **Ar**	1.06×10^{5}	1.01×10^{5}	40 **Zr**	12	11.4
19 **K**	3500	3770	41 **Nb**	0.9	0.698
20 **Ca**	6.25×10^{4}	6.11×10^{4}	42 **Mo**	4.0	2.55
21 **Sc**	31	34.2	44 **Ru**	1.9	1.86
22 **Ti**	2400	2400	45 **Rh**	0.40	0.344

Solar System Abundances Of The Elements*

Element	Cameron (1982)	Anders & Grevesse (1989)	Element	Cameron (1982)	Anders & Grevesse (1989)
46 **Pd**	1.3	1.39	67 **Ho**	0.092	0.0889
47 **Ag**	0.46	0.486	68 **Er**	0.23	0.2508
48 **Cd**	1.55	1.61	69 **Tm**	0.035	0.0378
49 **In**	0.19	0.184	70 **Yb**	0.20	0.2479
50 **Sn**	3.7	3.82	71 **Lu**	0.035	0.0367
51 **Sb**	0.31	0.309	72 **Hf**	0.17	0.154
52 **Te**	6.5	4.81	73 **Ta**	0.020	0.0207
53 **I**	1.27	0.90	74 **W**	0.30	0.133
54 **Xe**	5.84	4.7	75 **Re**	0.051	0.0517
55 **Cs**	0.39	0.372	76 **Os**	0.69	0.675
56 **Ba**	4.8	4.49	77 **Ir**	0.72	0.661
57 **La**	0.37	0.4460	78 **Pt**	1.41	1.34
58 **Ce**	1.2	1.136	79 **Au**	0.21	0.187
59 **Pr**	0.18	0.1669	80 **Hg**	0.21	0.34
60 **Nd**	0.79	0.8279	81 **Tl**	0.19	0.184
62 **Sm**	0.24	0.2582	82 **Pb**	2.6	3.15
63 **Eu**	0.094	0.0973	83 **Bi**	0.14	0.144
64 **Gd**	0.42	0.3300	90 **Th**	0.045	0.0335
65 **Tb**	0.076	0.0603	92 **U**	0.027	0.0090
66 **Dy**	0.37	0.3942			

*Normalized to $N_{Si} = 10^6$, and based primarily on C1 carbonaceous meteorites. To convert from this meteoritic scale to the solar abundance scale ($\log N_H$=12.00) use log(Sun/meteorite)=1.554. Reprinted with permission from [Geochimica et Cosmochimica Acta, , Edward Anders and Nicolas Grevesse, Abundances of the Elements: Meteoritic and Solar], Copyright [1989], Pergamon Press plc.

7
Comets and their Debris

7.1 Structural Properties of Comets

Nucleus

Radius, R: 0.5 to 10 km.
(Halley's dimensions $8 \times 8 \times 16$ km).
Mass Density, ρ: 0.1 to 1 grams/cm^3.
(Halley $\rho = 0.1$ to 0.3 grams/cm^3).*
Mass, M: 10^{14} to 10^{18} grams.
(Halley M $\sim 10^{17}$ grams).*
Escape Velocity, $V_{esc} \sim 100$ cm/sec.
Water Loss, ΔH_2O : 10^{28} to 10^{30} molecules/sec at 1 AU
(Halley $\Delta H_2O \sim 10^{30}$ molecules/sec).*
Dust Production Rate : 10^4 to 10^7 grams/sec at 1 AU.
Dust/Gas Production Ratio, r: 0.1 to 0.5 by mass.
(Halley r $\sim$ 0.3).*
Mass Loss, ΔM: 10^5 to 10^8 gram/sec at 1 AU.
or, 0.1 to 0.7 percent per apparition.
(Halley $\Delta M \sim 5 \times 10^{14}$ grams last apparition)*
Geometric Visual Albedo, p_V: 0.01 to 1.0
mean $\bar{p_V} = 0.05 \pm 0.01$.
(Halley $p_V = 0.044$)*
Rotation Period, P: 4 to 70 hours, mean $\bar{P} = 23$ hours
(Halley P = 2.2 days, 7.4 days).*
*See Whipple (1987) for Comet Halley

Composition: Ices, Clathrates, Grains**

Parent Molecule	%	Parent Molecule	%	Parent Molecule	%
H_2O	83	C_3H_4	1	NH_3	0.5
CH_2O	6	CH_4	0.5	S_2	0.2
CO_2	4	HCN	1	H_2S	0.2
C_2H_2	1	N_2H_4	2	CS_2	0.2

**After Delsemme (1985).

Coma

Radius, R: 10^5 to 10^6 km
Apparent magnitude, m
(at solar and terrestrial distances r and Δ)
$m = H_o + 2.5\, n \log r + 5 \log \Delta$
where the absolute magnitude H_o is the magnitude at 1 AU from both Earth and Sun.
For reflected sunlight only, $n = 2$,
and for most comets one assumes $n \approx 4$.

Vaporization Onset Distance, r_o.*

Molecule	r_o (AU)	Molecule	r_o (AU)
Nitrogen, N_2	77.6	Carbon Dioxide, CO_2	8.3
Carbon Monoxide, CO	62.5	Hydrogen Cyanide, HCN	4.8
Methane, CH_4	38.0	Aqueous Ammonia	2.6
Formaldehyde, CH_2O	14.1	Clathrates	2.5
Ammonia, NH_3	9.7	Water, H_2O	2.5

*After Delsemme (1982). Sublimation from ice to gas becomes negligible for heliocentric distances greater than r_o.

Ion (Type I) Tail

Extent: up to 10^8 km
Width: 3,000 to 4,000 km
Color: Blue, $CO^+ \sim 10^3$ cm^{-3}
Direction: Anti-solar, nearly straight

Dust (Type II) Tail

Extent: up to 10^7 km
Color: Yellow
Direction: Anti-solar, curved
Anti-tail directed toward Sun.

Hydrogen Cloud

Diameter: up to 10^7 km
Color: Ultraviolet
Production Rate: 10^{29} to 10^{30} hydrogen atoms per second at 1 AU.

7.2 Origin and Statistics of Comets

Comet Cloud (Öpik-Oort Reservoir)

Distance from Sun, D: 10^4 to 10^5 AU
Total Mass, M_T: $\sim 10^{28}$ grams $\sim 2M_E$
Earth Mass = $M_E = 5.97 \times 10^{27}$ grams

Cometary Statistics

Total Number of Known Apparitions: 1187.
Total Number of Known Comets: 748.
Apparition Rate: 10 to 20 per year.
Discovery Rate of New Comets: 6 to 12 per year.

Statistics of Cometary Orbits*

Short Period Comets (P < 200 years).

Average Period, $\bar{P} = 14.044 \pm 21.772$ years.
Average Eccentricity, $\bar{e} = 0.600 \pm 0.191$.
Average Inclination, $\bar{i} = 18.668 \pm 25.781$ (°).
Average Perihelion Distance, $\bar{q} = 1.635 \pm 0.751$ AU
Average Absolute Magnitude, $H_o = 11.392 \pm 3.027$ magnitude

Long Period Comets (P > 200 years).

Eccentricity, $e \approx 1.000$
Inclination, i = random

Type of Cometary Orbit	Number of Comets	Percent of Total Comets
Short-Period Comets (P < 200 years) of More Than One Apparition	85	11.4
Short-Period Comets (P < 200 years) of Only One Apparition	50	6.7
Total Short-Period Comets (P < 200 years)	135	18.1
Long-Period Comets (P > 200 years) With Elliptical Orbits	179	23.9
Long-Period Comets (P > 200 years) With Parabolic Orbits	322	43.0
Long-Period Comets (P > 200 years) With Hyperbolic Orbits	112	15.0
Total Long-Period Comets (P > 200 years)	613	81.9
Total Number of Comets With Known Orbits	748	100.0

*Adapted from Marsden (1986).

7.3 Comets of Special Interest

Perihelion Passages of Halley's Comet*

Date			Perihelion Distance (AU)	Period (years)	Estimated Magnitude H_o
240	B.C.	May 25	0.5854	76.75	
164		November 13	0.5845	76.88	
87		August 6	0.58561	77.12	
12	B.C.	October 11	0.58720	76.33	
A.D. 66		January 26	0.58510	76.55	
141		March 22	0.58314	77.28	
218		May 18	0.58147	77.37	
295		April 20	0.57591	79.13	
374		February 16	0.57719	78.76	
451		June 28	0.57374	79.29	
530		September 27	0.57559	78.90	
607		March 15	0.58083	77.47	
684		October 3	0.57958	77.62	
760		May 21	0.58184	77.00	2.0
837		February 28	0.58232	76.90	2.0
912		July 19	0.58016	77.45	1.7
989		September 6	0.58191	77.14	3.5
1066		March 21	0.57450	79.26	– 1.0 or 0.0
1145		April 19	0.57479	79.02	2.0
1222		September 29	0.57421	79.12	
1301		October 26	0.57271	79.14	3.0
1378		November 11	0.57620	77.76	3.5
1456		June 10	0.57970	77.10	4.2
1531		August 26	0.58120	76.50	4.0
1607		October 28	0.583615	76.06	3.5
1682		September 15	0.582608	77.41	4.0
1759		March 13	0.584447	76.89	3.8
1835		November 16	0.586542	76.27	4.4
1910		April 20	0.587189	76.08	4.6
1986		February 9	0.587	76.80	
2061		July 28			
A.D. 2134		March 27			

*After Kronk (1984), Marsden (1986) and Seargent (1982) . The day of perihelion passage has been rounded, so a date like October 10.8 reads October 11.

Split Comets*

Comet	Name	Fragments	Distance from Sun (AU)	Maximum Separation (m/s)
1846 II	Biela	2	3.59	0.26
1852 III	Biela	2	3.59	
1860 I	Liais	2	2.49	5.48
1882 II	Great September comet	4	0.017	4.90
1888 I	Sawerthal	2	0.76	
1889 IV	Davidson	2	1.06	
1889 V	Brooks 2	5	4.25, 5.38	4.5
1896 V	Giacobini	2	2.36	
1899 I	Swift	3	0.48, 1.15	
1905 IV	Kopff	2	3.38	
1914 IV	Campbell	2	0.82	
1915 II	Mellish	5	2.09, 2.38	0.44
1916 I	Taylor	2	1.65	0.90
1943 I	Whipple-Fedtke-Tevzadze	2	1.43	
1947 XII	Southern comet	2	0.15	1.87
1955 V	Honda	2	8.2	
1957 VI	Wirtanen	2	9.25	0.24
1965 VIII	Ikeya-Seki	2	0.008	
1968 III	Wild	2	2.92	
1969 IX	Tago-Sato-Kosaka	2	1.20	
1970 III	Kohoutek	2	1.79	
1976 VI	West	4	0.22, 0.30, 0.41	1.72

*Adapted from Sekanina (1982).

Sun-Grazing Comets*

Comet Name	Perihelion Distance (AU)	Aphelion Distance (AU)	Period (years)	Eccentricity	Inclination (degrees)	Perihelion Longitude (degrees)	Perihelion Latitude (degrees)
1668	0.066604			1.0	144.375	248.61	+33.23
Great March (1843 I)	0.005527	128	517	0.999914	144.348	281.86	+35.31
Great southern (1880 I)	0.005494			1.0	144.660	281.68	+35.25
Great September (1882 II)	0.007751	166	759	0.999907	142.005	282.24	+35.24
Great southern (1887 I)	0.004834			1.0	144.377	281.85	+35.36
du Toit (1945 VII)	0.007516			1.0	141.867	282.87	+35.97
Pereyra (1963 V)	0.005065	187	903	0.999946	144.576	281.90	+35.37
Ikeya-Seki (1965 VIII)	0.007786	184	880	0.999915	141.858	282.24	+35.21
White-Ortiz-Bolelli (1970 VI)	0.008879			1.0	139.0652	282.26	+35.07

*Adapted from Marsden (1967, 1983).

7.4 Long Period and Short Period Comets

Selected Long Period Comets*

Comet	Name	a (AU)	P (years)	q (AU)	e	i (degrees)
1861 II	Great comet	55.08	409	0.822	0.985	85.4
1882 II	Great September comet	83.18	759	0.008	1.000	142.0
1965 VIII	Ikeya-Seki	91.82	880	0.008	1.000	141.9
1970 II	Bennett	141.2	1,678	0.538	0.996	90.0
1858 VI	Donati	156.1	1,951	0.578	0.996	117.0
1881 III	Great comet	181.2	2,439	0.735	0.996	63.4
1962 VIII	Humason	204.5	2,925	2.133	0.990	153.3
1811 I	Great comet	212.4	3,096	1.035	0.995	106.9
1844 III	Great comet	359.1	6,805	0.251	0.999	45.6
1680	Observed by Newton	430.0	8,917	0.006	1.000	60.7
1957 V	Mrkos	559		0.355	0.999	93.9
1976 VI	West	6,830		0.197	1.000	43.1
1902 III	Perrine	12,700		0.401	1.000	156.4
1910 I	Great January comet	25,100		0.129	1.000	138.8
1937 V	Finsler	58,800		0.863	1.000	146.4

*Adapted from Marsden (1983). Other well-known comets with hyperbolic orbits are 1973 XII (Kohoutek), 1957 III (Arend-Roland) and 1908 III (Morehouse) whose respective perihelion distances are q = 0.142, 0.316 and 0.945 AU.

a=mean distance from sun, P=orbital period, q=perihelion distance and e and i are respectively, the orbital excentricity and inclination.

Short Period Comets of Only One Apparition*

Comet	T (year)	P (years)	q (AU)	e	ω (°)	Ω (°)	i (°)	Q (AU)	H_o (mag)
Helfenzrieder	1766.32	4.35	0.406	0.848	178.7	75.6	7.9	4.92	6.8
Blanpain	1819.89	5.10	0.892	0.699	350.2	79.2	9.1	5.03	8.5
Barnard 1	1884.63	5.38	1.279	0.583	301.0	6.1	5.5	4.86	8.9
Brooks 1	1886.43	5.44	1.325	0.571	176.8	54.5	12.7	4.86	8.9
Lexell	1770.62	5.60	0.674	0.786	224.9	133.9	1.6	5.63	7.7
Hartley	1985.44	5.61	1.540	0.512	174.1	40.4	24.9	4.78	
Pigott	1783.89	5.89	1.459	0.552	354.6	58.0	45.1	5.06	6.9
Haneda-Campos	1978.77	5.97	1.101	0.665	240.4	131.6	6.0	5.48	12.6
Tritton	1977.82	6.35	1.439	0.580	147.7	300.0	7.0	5.42	16.5
Harrington-Wilson	1951.83	6.36	1.664	0.515	343.0	127.8	16.4	5.20	12.1
Spitaler	1890.82	6.37	1.818	0.471	13.4	45.9	12.8	5.06	9.0
Russell 4	1984.01	6.40	2.125	0.383	91.3	71.9	6.2	4.77	
Kowal 2	1979.04	6.51	1.521	0.564	189.4	247.2	15.8	5.45	14.5
Barnard 3	1892.95	6.52	1.432	0.590	170.0	207.3	31.3	5.55	9.8
Giacobini	1896.83	6.65	1.455	0.588	140.5	194.2	11.4	5.62	9.9
Schorr	1918.75	6.67	1.884	0.469	279.2	118.3	5.6	5.21	11.0
Swift	1895.64	7.20	1.298	0.652	167.8	171.1	3.0	6.16	11.4
Shoemaker 1	1984.71	7.23	1.977	0.471	18.7	339.3	26.3	5.50	
Takamizawa	1984.40	7.24	1.595	0.574	147.5	124.2	9.5	5.89	
Ciffréo	1985.83	7.27	1.703	0.546	357.7	53.1	13.1	5.80	

*After Marsden (1986), Meisel and Morris (1982) and Seargent (1982). The time of perihelion is T, the orbital period is P, the perihelion distance is q, the eccentricity e, the argument of perihelion ω (1950.0), the longitude of ascending node, Ω (1950.0), inclination i (1950.0), aphelion distance Q, and absolute magnitude at 1 AU from Earth and Sun is H_o.

Short Period Comets of Only One Apparition*

Comet	T (year)	P (years)	q (AU)	e	ω (°)	Ω (°)	i (°)	Q (AU)	H_o (mag)
Kowal-Mrkos	1984.43	7.32	1.951	0.483	338.1	248.5	3.0	5.59	
Denning	1894.11	7.42	1.147	0.698	46.3	85.1	5.5	6.46	10.4
Schuster	1978.02	7.47	1.628	0.574	353.9	50.8	20.4	6.01	13.2
Russell 3	1982.89	7.50	2.510	0.345	353.5	248.0	14.1	5.15	
Metcalf	1906.77	7.78	1.631	0.584	199.8	195.2	14.6	6.22	9.5
Shoemaker 2	1984.74	7.85	1.320	0.666	317.6	54.8	21.6	6.58	
Maury	1985.43	8.84	2.011	0.530	114.0	183.1	9.4	6.54	
Lovas 1	1980.67	9.06	1.676	0.615	72.6	342.3	12.3	7.02	
Sanguin	1977.71	12.5	1.811	0.664	162.1	182.3	18.6	8.96	13.5
IRAS	1983.64	13.2	1.697	0.696	356.9	357.2	46.2	9.45	
Kowal 1	1977.15	15.1	4.664	0.237	178.0	28.4	4.4	7.56	9.0
van Houten	1961.32	15.6	3.957	0.367	14.4	22.9	6.7	8.54	8.0
Bowell-Skiff	1983.20	15.7	1.945	0.689	169.0	345.6	3.8	10.6	
Chernykh	1978.12	15.9	2.568	0.594	266.7	134.1	5.7	10.1	6.5
Kowal-Vávrová	1983.25	15.9	2.609	0.588	19.5	201.8	4.3	10.1	
Hartley-IRAS	1984.02	21.5	1.282	0.834	47.1	0.8	95.7	14.2	
Pons-Gambart	1827.43	57.5	0.807	0.946	19.2	319.3	136.5	29.0	7.0
Dubiago	1921.34	62.3	1.115	0.929	97.4	66.5	22.3	30.3	10.5
de Vico	1846.18	76.3	0.664	0.963	12.9	79.0	85.1	35.3	7.2
Väisälä 2	1942.13	85.4	1.287	0.934	335.2	171.6	38.0	37.5	13.2
Swift-Tuttle	1862.64	120	0.963	0.960	152.8	138.7	113.6	47.7	4.0
Barnard 2	1889.47	145	1.105	0.960	60.2	271.9	31.2	54.2	9.0
Mellish	1917.27	145	0.190	0.993	121.3	88.0	32.7	55.1	7.4
Bradfield	1983.99	151	1.357	0.952	219.2	356.2	51.8	55.5	
Wilk	1937.14	187	0.619	0.981	31.5	57.6	26.0	64.9	

*After Marsden (1986), Meisel and Morris (1982) and Seargent (1982). The time of perihelion is T, the orbital period is P, the perihelion distance is q, the eccentricity e, the argument of perihelion ω (1950.0), the longitude of ascending node, Ω (1950.0), inclination i (1950.0), aphelion distance Q, and absolute magnitude at 1 AU from Earth and Sun is H_o.

Short Period Comets of More Than One Apparition*

Comet	T (year)	P (years)	q (AU)	e	ω (°)	Ω (°)	i (°)	Q (AU)	H_o (mag)
Encke	1984.24	3.31	0.341	0.846	186.0	334.2	11.9	4.10	11.0
Grigg-Skjellerup	1982.37	5.09	0.989	0.666	359.3	212.6	21.1	4.93	12.5
du Toit-Hartley	1982.24	5.21	1.195	0.602	251.7	308.6	2.9	4.81	
Tempel 2	1983.42	5.29	1.381	0.545	190.9	119.2	12.4	4.69	10.0
Honda-Mrkos-Pajdušáková	1985.39	5.30	0.542	0.822	325.6	88.7	4.2	5.54	10.6
Schwassmann-Wachmann 3	1979.67	5.36	0.941	0.693	198.7	69.3	11.4	5.19	11.5
Neujmin 2†	1927.04	5.43	1.338	0.567	193.7	328.0	10.6	4.84	11.3
Brorsen†† or †	1879.24	5.46	0.590	0.810	14.9	102.3	29.4	5.61	9.2
Tempel 1	1983.52	5.49	1.491	0.521	179.0	68.3	10.6	4.73	9.4
Wirtanen	1986.21	5.50	1.084	0.652	356.1	81.7	11.7	5.15	16.3
Clark	1984.41	5.50	1.551	0.502	209.0	59.1	9.5	4.68	12.0
Tuttle-Giacobini-Kresák	1978.98	5.58	1.124	0.643	49.4	153.3	9.9	5.17	13.0
Tempel-Swift†	1908.76	5.68	1.153	0.638	113.5	291.1	5.4	5.22	12.8
Howell	1981.34	5.94	1.615	0.508	214.6	75.4	5.6	4.94	
Russell 1	1985.51	6.10	1.612	0.517	0.4	230.1	22.7	5.06	15.0
West-Kohoutek-Ikemura	1981.28	6.12	1.401	0.581	358.1	84.6	30.1	5.29	9.6
Wild 2	1984.64	6.17	1.494	0.556	40.0	136.0	3.3	5.24	6.5
Kohoutek	1981.29	6.24	1.571	0.537	169.9	273.1	5.4	5.21	
Forbes	1980.73	6.27	1.479	0.565	262.5	23.0	4.7	5.32	10.0
de Vico-Swift	1965.31	6.31	1.624	0.524	325.4	24.4	3.6	5.21	14.5
Pons-Winnecke	1983.26	6.36	1.254	0.635	172.3	92.7	22.3	5.61	14.5
du Toit-Neujmin-Delporte	1983.42	6.37	1.708	0.503	115.2	188.4	2.9	5.16	14.0
d'Arrest	1982.70	6.38	1.291	0.625	177.0	138.9	19.4	5.59	6.5
Kopff	1983.61	6.44	1.576	0.545	162.8	120.3	4.7	5.35	13.4
Schwassmann-Wachmann 2	1981.21	6.50	2.135	0.387	357.5	125.9	3.7	4.83	11.4
Bus	1981.44	6.53	2.183	0.375	24.7	181.5	2.6	4.80	
Wolf-Harrington	1984.73	6.54	1.616	0.538	186.9	254.2	18.4	5.38	13.0

*After Marsden (1986), Meisel and Morris (1982) and Seargent (1982). The time of perihelion passage after 1990.0 is T, the orbital period is P, the perihelion distance is q, the eccentricity e, the argument of perihelion ω (1950.0), the longitude of ascending node, Ω (1950.0), inclination i (1950.0), aphelion distance Q, and absolute magnitude at 1 AU from Earth and Sun is H_o.
† Now lost. †† May have disappeared

Short Period Comets of More Than One Apparition*

Comet	T (year)	P (years)	q (AU)	e	ω (°)	Ω (°)	i (°)	Q (AU)	H_o (mag)
Giacobini-Zinner	1985.68	6.59	1.028	0.708	172.5	194.7	31.9	6.00	10.0
Churyumov-Gerasimenko	1982.86	6.61	1.306	0.629	11.3	50.4	7.1	5.74	10.0
Biela†	1852.73	6.62	0.861	0.756	223.2	247.3	12.6	6.19	8.1
Tsuchinshan 1	1985.01	6.67	1.508	0.575	22.8	96.2	10.5	5.58	14.0
Perrine-Mrkos††	1968.84	6.72	1.272	0.643	166.1	240.2	17.8	5.85	18.5
Reinmuth 2	1981.08	6.74	1.946	0.455	45.4	296.0	7.0	5.19	10.5
Borrelly	1981.14	6.77	1.319	0.631	352.8	75.1	30.2	5.84	13.0
Gunn	1982.90	6.82	2.459	0.316	197.0	67.9	10.4	4.74	10.0
Arend-Rigaux	1984.92	6.84	1.446	0.599	328.9	121.6	17.8	5.76	8.9
Tsuchinshan 2	1985.55	6.85	1.794	0.502	203.2	287.6	6.7	5.42	14.0
Harrington	1980.98	6.86	1.605	0.556	233.0	119.0	8.7	5.62	14.8
Wild 3	1980.76	6.89	2.288	0.368	179.3	72.0	15.5	4.95	
Brooks 2	1980.90	6.90	1.850	0.490	198.2	176.2	5.5	5.40	13.5
Giclas	1985.75	6.93	1.838	0.495	276.3	111.9	7.3	5.43	13.5
Johnson	1983.92	6.94	2.302	0.367	208.2	116.7	13.7	4.98	10.0
Finlay	1981.47	6.97	1.101	0.698	322.1	41.8	3.6	6.20	13.0
Longmore	1981.81	6.98	2.400	0.343	195.9	15.0	24.4	4.90	
Taylor	1984.02	6.99	1.961	0.464	355.6	108.2	20.5	5.35	12.0
Holmes	1979.14	7.06	2.160	0.413	23.6	327.4	19.2	5.20	13.5
Daniel	1985.59	7.07	1.651	0.552	10.8	68.5	20.1	5.72	11.5
Russell 2	1980.38	7.11	2.159	0.416	245.3	44.5	12.5	5.24	
Faye	1984.52	7.34	1.594	0.578	203.8	199.0	9.1	5.96	8.4
Shajn-Schaldach	1986.40	7.46	2.331	0.390	216.5	166.2	6.1	5.31	12.0
Ashbrook-Jackson	1986.07	7.47	2.307	0.396	348.8	2.0	12.5	5.34	7.1
Reinmuth 1	1980.83	7.59	1.982	0.487	9.5	121.1	8.3	5.75	14.0
Harrington-Abell	1983.92	7.62	1.785	0.539	138.6	336.7	10.2	5.96	15.0
Oterma	1958.44	7.88	3.388	0.144	354.9	155.1	4.0	4.53	9.5

*After Marsden (1986), Meisel and Morris (1982) and Seargent (1982). The time of perihelion passage after 1990.0 is T, the orbital period is P, the perihelion distance is q, the eccentricity e, the argument of perihelion ω (1950.0), the longitude of ascending node, Ω (1950.0), inclination i (1950.0), aphelion distance Q, and absolute magnitude at 1 AU from Earth and Sun is H_o.
†Now lost. ††May have disappeared

Short Period Comets of More Than One Apparition*

Comet	T (year)	P (years)	q (AU)	e	ω (°)	Ω (°)	i (°)	Q (AU)	H_o (mag)
Kojima	1986.26	7.89	2.414	0.391	348.4	154.2	0.9	5.51	
Gehrels 2	1981.88	7.98	2.362	0.409	183.5	215.5	6.7	5.62	
Arend	1983.39	8.02	1.857	0.536	46.9	355.6	19.9	6.15	14.5
Peters-Hartley	1982.35	8.12	1.627	0.598	338.8	259.3	29.8	6.46	8.0
Gehrels 3	1985.42	8.14	3.442	0.149	231.3	242.4	1.1	4.65	9.5
Wolf	1984.42	8.21	2.415	0.407	162.2	203.5	27.5	5.73	13.0
Schaumasse	1984.93	8.26	1.213	0.703	57.4	80.4	11.8	6.96	11.0
Jackson-Neujmin	1978.98	8.38	1.425	0.655	196.3	163.1	14.1	6.82	16.7
Whipple	1986.48	8.49	3.077	0.261	202.0	181.8	9.9	5.25	
Smirnova-Chernykh	1984.14	8.51	3.557	0.146	90.9	77.0	6.6	4.78	8.3
Comas Solá	1978.73	8.94	1.870	0.566	42.8	62.4	13.0	6.74	8.5
Kearns-Kwee	1981.91	8.99	2.224	0.486	131.4	315.3	9.0	6.42	11.2
Denning-Fujikawa	1978.75	9.01	0.779	0.820	334.0	41.0	8.7	7.88	11.5
Swift-Gehrels	1981.91	9.26	1.361	0.691	84.5	314.0	9.2	7.46	15.0
Neujmin 3	1972.37	10.6	1.976	0.590	146.9	150.2	3.9	7.66	14.5
Väisälä 1	1982.58	10.9	1.800	0.633	47.9	134.5	11.6	8.02	13.5
Klemola	1976.61	10.9	1.766	0.642	148.9	181.6	10.6	8.09	9.7
Gale[††]	1938.46	11.0	1.183	0.761	209.1	67.3	11.7	8.70	10.5
Boethin	1986.04	11.2	1.114	0.778	11.6	25.8	5.8	8.91	10.3
Slaughter-Burnham	1981.88	11.6	2.544	0.504	44.2	345.9	8.2	7.71	13.6
Van Biesbroeck	1978.92	12.4	2.395	0.553	134.3	148.6	6.6	8.31	7.5
Wild 1	1973.50	13.3	1.981	0.647	167.9	358.2	19.9	9.24	14.0
Tuttle	1980.95	13.7	1.015	0.823	206.9	269.9	54.5	10.4	8.0
Gehrels 1	1973.07	14.5	2.935	0.507	28.9	14.6	9.6	8.97	11.5
du Toit	1974.25	15.0	1.294	0.787	257.2	22.1	18.7	10.9	16.0
Schwassmann-Wachmann 1	1974.13	15.0	5.448	0.105	14.5	319.6	9.7	6.73	5.0
Neujmin 1	1984.77	18.2	1.553	0.776	346.8	346.3	14.2	12.3	10.2
Crommelin	1984.14	27.4	0.735	0.919	195.9	250.2	29.1	17.4	10.7
Tempel-Tuttle	1965.33	32.9	0.982	0.904	172.6	234.4	162.7	19.6	13.5
Stephan-Oterma	1980.93	37.7	1.574	0.860	358.2	78.5	18.0	20.9	5.3
Westphal[††]	1913.90	61.9	1.254	0.920	57.1	347.3	40.9	30.0	9.3
Olbers	1956.47	69.6	1.178	0.930	64.6	85.4	44.6	32.6	5.5
Pons-Brooks	1954.39	70.9	0.774	0.955	199.0	255.2	74.2	33.5	5.9
Brorsen-Metcalf	1919.79	72.0	0.485	0.972	129.5	311.2	19.2	34.1	8.7
Halley	1986.11	76.0	0.587	0.967	111.8	58.1	162.2	35.3	4.6
Herschel-Rigollet	1939.60	155	0.748	0.974	29.3	355.3	64.2	56.9	8.5

*After Marsden (1986), Meisel and Morris (1982) and Seargent (1982). The time of perihelion passage after 1990.0 is T, the orbital period is P, the perihelion distance is q, the eccentricity e, the argument of perihelion ω (1950.0), the longitude of ascending node, Ω (1950.0), inclination i (1950), aphelion distance Q, and absolute magnitude at 1 AU from Earth and Sun is H_o.
††May have disappeared.

7.5 Cometary Debris

One Comet

Water Loss: 10^{28} to 10^{30} molecules/sec at 1 AU.
: 10^5 to 10^7 grams/sec at 1 AU.
Dust Mass Loss: 10^4 to 10^7 grams/sec at 1 AU.
Mass Loss: 10^5 to 10^8 grams/sec at 1 AU.
Total Mass Loss: 10^{14} to 10^{15} grams per apparition

All Comets

Dust Mass Loss: 10^{14} to 10^{15} grams per year.
: 10^8 to 10^9 tons per year.
Total Mass Loss: 10^{15} to 10^{16} grams per year.
: 10^9 to 10^{10} tons per year.

Earth

Swept up Comet Dust: 10^{12} grams per year.
: 10^6 tons per year.
(mostly burnt up in atmosphere)

Zodiacal Light Material - Comet Dust Cloud

Loss to Sun: $\sim 10^{15}$ grams per year.
: $\sim 10^9$ tons per year
(replenished by short-period comets)

Comets or Asteroids Associated With Meteor Streams*

Meteor Stream	Comet or Asteroid	Meteor Stream	Comet or Asteroid
Lyrids	Thatcher (1861 I)	Epsilon Geminids	Ikeya (1964 VIII)
Eta Aquarids	Halley (1835 III)	Taurids-South	Encke (1971 II)
Daytime Arietids	Asteroid 1566 Icarus	Taurids-North	Encke (1971 II)
Scorpiids-Sagittariids	Encke (1971 II)	Andromedids	Biela (1852 III)
Bootids	Pons-Winnecke (1915 III)	Leonids	Temple-Tuttle (1965 IV)
Perseids	Swift-Tuttle (1862 III)	Geminids	Asteroid 3200 Phaethon
Aurigids	Kiess (1911 II)	Monocerotids	Mellish (1917 I)
Draconids	Giacobini-Zinner (1946 V)	Ursids	Tuttle (1939 X)
Orionids	Halley (1835 III)		

*Adapted from Drummond (1981), Kronk (1988), and Sekanina (1976).
Meteor Streams listed in order of month of maximum beginning in January.

Principal Meteor Streams*

Stream	Maximum Date	Radiant R.A. (°)	Radiant Dec. (°)	Visibility Dates	Meteors Per Hour
Quadrantids	Jan. 3-4	229	+49	Dec. 28 to Jan. 7	145
Virginids	Mar. 18	183	+3	Feb. 13 to Apr. 8	20
Lyrids	Apr. 20	272	+33	Apr. 16 to Apr. 25	45
Eta Aquarids	May 5	336	− 1	Apr. 21 to May 12	120
Daytime Arietids	June 8	45	+24	May 22 to July 2	66 (radar)
Daytime Zeta Perseids	June 13	63	+26	May 20 to July 5	42 (radar)
Scorpiids-Sagittariids	June 18	278	− 25	June 1 to July 15	20
Daytime Beta Taurids	June 30	79	+21	June 5 to July 18	27 (radar)
Delta Aquarids-South	July 29	339	− 17	July 14 to Aug. 18	60
Delta Aquarids-North	Aug. 13	344	+2	July 16 to Sept. 10	10
Perseids	Aug. 12-13	48	+57	July 23 to Aug. 22	200
Draconids	Oct. 9	262	+54	Oct. 6 to Oct. 10	
Orionids	Oct. 21	95	+16	Oct. 15 to Oct. 29	75
Taurids-South	Oct. 30 to Nov. 7	53	+12	Sept. 17 to Nov. 27	45
Taurids-North	Nov. 4 to 7	54	+21	Oct. 12 to Dec. 2	25
Andromedids	Nov. 14	26	+37	Sep. 25 to Dec. 6	5
Leonids	Nov. 17	153	+22	Nov. 14 to Nov. 20	45
Geminids	Dec. 13-14	112	+33	Dec. 6 to Dec. 19	150
Ursids	Dec. 22	217	+76	Dec. 17 to Dec. 24	45

*Adapted from Hoffmeister (1965) and Kronk (1988). Radiant positions are for the time of maximum. Radiants move at a rate of up to one degree per day. The hourly frequency of meteors is for the whole sky assuming the radiant is at the zenith. If the radiant is at the zenith distance, z, the rate is disminished by 0.3 cos(z) for one observer.

Meteor Stream Orbits*

Stream	Mean Distance from Sun (AU)	Eccentricity e	Inclination i (degrees)	Perihelion Distance q (AU)	Argument of Perihelion ω (degrees)	Longitude of Ascending Node, Ω (degrees)
Quadrantids	3.064	0.682	70.3	0.974	168.1	282.3
Virginids	1.7 to 1.9	0.7 to 0.8	3 to 11	0.3 to 0.5	282 to 304	~ 350
Lyrids	4.511	0.796	76.9	0.922	215.3	32.0
Eta Aquarids	2.823	0.834	161.2	0.468	79.5	44.9
Daytime Arietids	1.376	0.938	25.0	0.085	25.9	76.9
Daytime Zeta Perseids	1.492	0.755	6.5	0.365	60.5	80.8
Scorpiids-Sagittariids	1.908	0.799	2.5	0.384	113.8	270.4
Daytime Beta Taurids	1.653	0.834	0.3	0.274	52.3	102.0
Delta Aquarids-South	1.630	0.958	28.2	0.069	155.4	305.7
Delta Aquarids-North	1.259	0.866	19.2	0.169	323.2	141.7
Perseids	8.040	0.881	110.2	0.960	152.5	139.7
Draconids	2.221	0.553	49.2	0.994	187.6	194.5
Orionids	3.850	0.854	164.4	0.562	87.0	27.1
Taurids-South	1.910	0.806	5.0	0.370	114.3	27.3
Taurids-North	2.186	0.843	3.2	0.343	295.6	224.2
Andromedids	2.990	0.794	6.1	0.616	261.9	200.3
Leonids	9.956	0.901	161.0	0.983	173.1	235.4
Geminids	1.612	0.896	23.9	0.143	324.1	261.3
Ursids	6.112	0.848	52.6	0.929	208.4	267.3

*Adapted from Sekanina (1976) and Kronk (1988).

Part III
The Stars

8
The Sun

8.1 Basic Solar Data

Mass, size, age, gravity

Mass	$M_\odot$	=	1.989×10^{33} g
Radius	$R_\odot$	=	6.9599×10^{10} cm
Surface area	$A_\odot$	=	6.087×10^{22} cm^2
Volume	$V_\odot$	=	1.412×10^{33} cm^3
Mean density	$\bar{\rho}_\odot$	=	1.409 g cm^{-3}
Central density	$(\rho)_{central}$	=	148 g cm^{-3}
Age	$t_\odot$	=	4.6×10^9 years
Gravitational acceleration at surface	$g_\odot$	=	2.7398×10^4 cm s^{-2}
Escape velocity at surface	V_{esc}	=	6.177×10^7 cm s^{-1}

Radiation

Absolute luminosity	$L_\odot$	=	3.86×10^{33} erg s^{-1}
Surface flux	$F_\odot$	=	6.34×10^6 erg cm^{-2} s^{-1}
Effective temperature	T_{eff}	=	5780 °K
Central temperature	$T_{central}$	=	1.55×10^7 °K
Solar constant	S	=	1.372×10^6 erg cm^{-2} s^{-1}
(solar flux at 1 AU)	S	=	$1,372 \pm 4W\ m^{-2}$
Spectral type			G2V

Magnitudes

Apparent visual magnitude	V = $m_{v\odot}$	=	– 26.78
Apparent bolometric magnitude	$m_{bol\odot}$	=	– 26.85
Color index	B–V	=	0.68 ± 0.005
Color index	U–B	=	0.17 ± 0.01
Absolute visual magnitude	$M_{v\odot}$	=	4.82
Absolute bolometric magnitude	$M_{bol\odot}$	=	4.75

*Adapted from Abell (1976), Böhm-Vitense (1989), Taylor(1989).

Viewed from Earth

Mean distance (astronomical unit)	$D_{\odot} = AU$	=	$1.4959787061 \times 10^{13}$ cm
Mean equatorial horizontal parallax	$\pi_{\odot}$	=	8.794148 arcsec
Mean equatorial horizontal parallax	$\pi_{\odot}$	=	4.26354×10^{-5} radians
Minimum distance (perihelion)	$(D_{min})_{\odot}$	=	1.4710×10^{13} cm
Maximum distance (aphelion)	$(D_{max})_{\odot}$	=	1.5210×10^{13} cm
Angular radius (semi-diameter)	$R_{\odot}/D_{\odot}$	=	959.63 arc sec
Angular radius (semi-diameter)	$R_{\odot}/D_{\odot}$	=	0.0046526 radians
At mean distance $1''$= 1 arcsec	$1''$	=	7.253×10^{7} cm
At mean distance $1'$= 1 arcminute	$1'$	=	4.3518×10^{9} cm

Rotation

Period of synodic rotation (surface at latitude ϕ)	$P_{\odot syn}$	=	$26.75 + 5.7 \sin^2 \phi$ days
Period of synodic rotation (surface at latitude $\phi = 17°$)	$P_{\odot syn}$	=	27.275 days
Period of sidereal rotation (surface at latitude $\phi = 17°$)	$P_{\odot sid}$	=	25.38 days
Rotation velocity (surface at latitude $\phi = 17°$)		=	1.99×10^{5} cm s^{-1}
Angular rotation velocity (surface at latitude $\phi = 17°$)		=	2.865×10^{-6} rad s^{-1}
Angular momentum of rotation (based on surface rotation)		=	1.63×10^{48} g cm^{2} s^{-1}
Rotational energy (based on surface rotation)		=	2.4×10^{42} erg
Moment of inertia		=	5.7×10^{53} g cm^{2}

*Adapted from Abell (1976), Böhm-Vitense (1989), Taylor(1989).

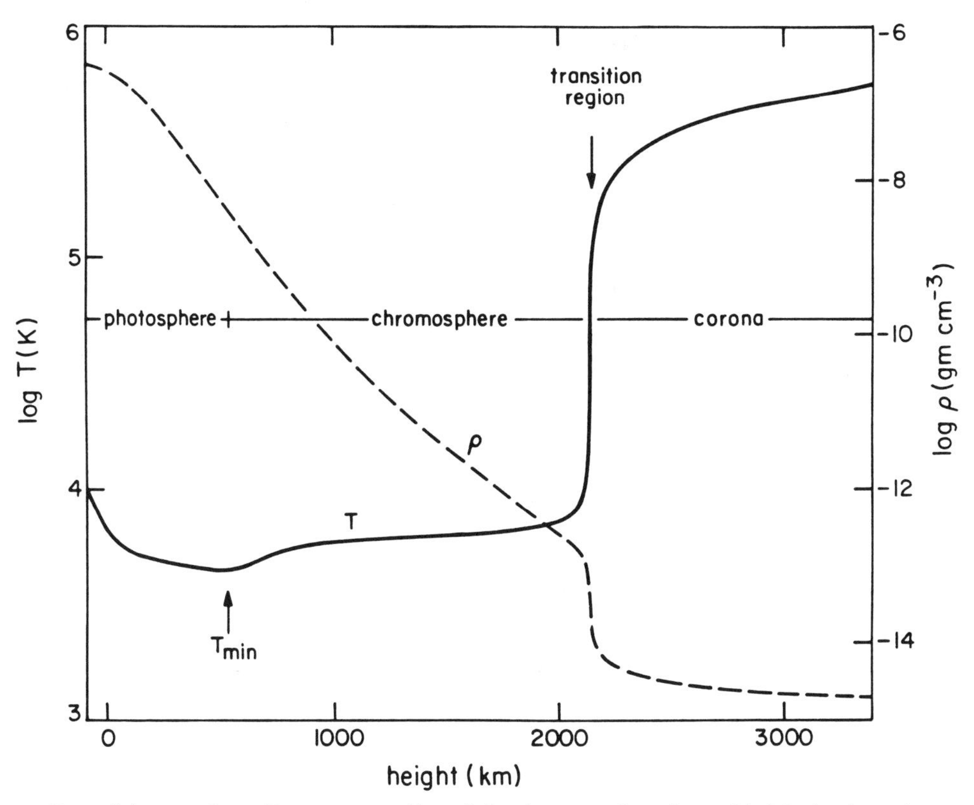

Transition region. Temperature, T, and density, ρ, as functions of height in the quiet, or non-active, solar atmosphere from the visible photosphere through the chromosphere, and into the million-degree corona. (Courtesy of Eugene H. Avrett, 1990)

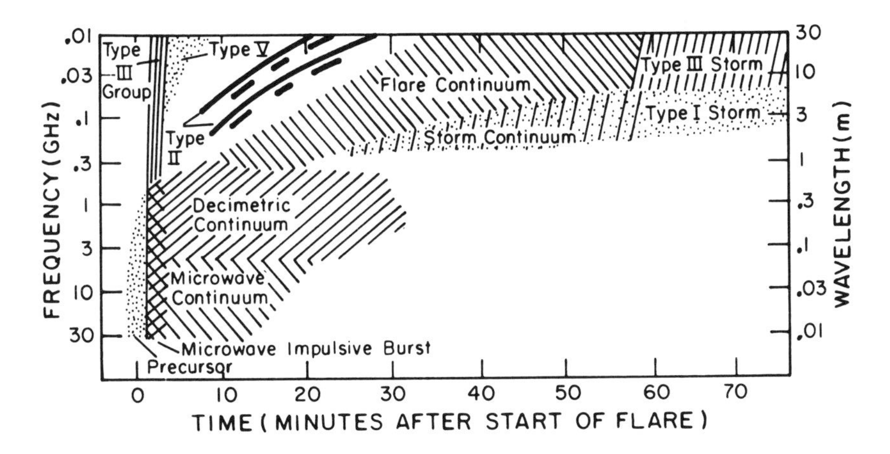

Radio bursts. Dynamic spectra of various types of solar radio bursts that might be associated with a powerful flare on the Sun. (Adapted from Dulk, 1985).

8.2 Standard Solar Model

Standard Model – Interior Structure*

Mass ($M_\odot$)	Radius ($R_\odot$)	Luminosity ($L_\odot$)	Temperature (10^6 °K)	Density (g cm^{-3})
0.0000	0.000	0.0000	15.513	147.74
0.0001	0.010	0.0009	15.48	146.66
0.001	0.022	0.009	15.36	142.73
0.020	0.061	0.154	14.404	116.10
0.057	0.090	0.365	13.37	93.35
0.115	0.120	0.594	12.25	72.73
0.235	0.166	0.845	10.53	48.19
0.341	0.202	0.940	9.30	34.28
0.470	0.246	0.985	8.035	21.958
0.562	0.281	0.997	7.214	15.157
0.647	0.317	0.992	6.461	10.157
0.748	0.370	0.9996	5.531	5.566
0.854	0.453	1.000	4.426	2.259
0.951	0.611	1.000	2.981	0.4483
0.9809	0.7304	1.0000	2.035	0.1528
0.9964	0.862	1.0000	0.884	0.042
0.9999	0.965	1.0000	0.1818	0.00361
1.0000	1.0000	1.0000	0.005770	1.99×10^{-7}

*Adapted from Turck-Chiéze et al. (1988).
Composition X = 0.7046, Y = 0.2757, Z = 0.0197

Standard Model – Evolution*

Time (10^9 years)	Luminosity ($L_\odot$)	Radius ($R_\odot$)	$T_{central}$ (10^6 °K)
Past			
0	0.7688	0.872	13.35
0.143	0.7248	0.885	13.46
0.856	0.7621	0.902	13.68
1.863	0.8156	0.924	14.08
2.193	0.8352	0.932	14.22
3.020	0.8855	0.953	14.60
3.977	0.9522	0.981	15.12
Now			
4.587	1.000	1.000	15.51
Future			
5.506	1.079	1.035	16.18
6.074	1.133	1.059	16.65
6.577	1.186	1.082	17.13
7.027	1.238	1.105	17.62
7.728	1.318	1.143	18.42
8.258	1.399	1.180	18.74
8.7566	1.494	1.224	18.81
9.805	1.760	1.361	19.25

*Adapted from Turck-Chiéze et al. (1988).
Composition X = 0.7046, Y = 0.2757, Z = 0.0197.
Present values are $R_\odot$ and $L_\odot$.
**For time t before the present age $t_\odot$=4.6 × 10^9 years,
$L/L_\odot \approx 1/[1+0.4(1-t/t_\odot)]$

8.3 Solar Neutrinos

Source	Reaction	Percent Terminations	Neutrino Energy (MeV)	Neutrino Flux (10^{10} cm^{-2} s^{-1})
pp	$p+p \to {}^2H+e^+ + \nu_e$	99.75	≤ 0.420	$6.0(1 \pm 0.02)$
pep	$p+e+p \to {}^2H+\nu_e$	0.5	1.442	$0.014(1 \pm 0.05)$
hep	${}^3He+p \to {}^4He+e^+ + \nu_e$	0.00002	≤ 18.77	$8 \times 10^{-7}(1 \pm ?)$
^{7}Be	${}^7Be + e^- \to {}^7Li+\nu_e$	15	0.861(90%) 0.383(10%)	$0.47(1 \pm 0.15)$
^{8}B	${}^8B \to {}^8Be^* + e^+ + \nu_e$		< 15	$5.8 \times 10^{-4}(1 \pm 0.37)$

*Adapted from Bahcall, Davis and Wolfenstein (1988).
**The standard solar model predicts a capture rate of 5.8 ± 1.3 SNU on ^{37}Cl, while the average neutrino capture rate observed from 1970–90 was 2.20 ± 0.24 SNU, (Davis,1991), where a SNU = solar neutrino unit = 10^{-36} interactions s^{-1} per atom. The predicted capture rate on ^{71}Ga is 125 ± 5 SNU (Turck-Chièze, et al., 1988).

8.4 The Outer Solar Atmosphere

Corona and Solar Wind*

Distance From Sun, D ($R_\odot$)	Electron Density, N_e (cm^{-3})	Electron Temperature, T_e (°K)	Velocity of Solar Wind, V ($km\ s^{-1}$)
1.02	4.0×10^8	2×10^6	0.2
1.4	2.3×10^7		2
2.0	2.8×10^6		8
3.0	4.0×10^5		26
6.0	3.1×10^4		80
15	2.5×10^3		160
50	1.0×10^2		220
215	5.0	3×10^5	400

*Adapted from Unsöld (1970). Velocity of the solar wind computed from the equation of continuity.

Solar Wind at 1 AU*

Parameter	Mean Value	Standard Deviation	Median
Density, N (cm^{-3})	8.7	6.6	6.9
Helium/Proton Density Ratio, N_α/N_p	0.047	0.019	0.048
Velocity, V ($km\ s^{-1}$)	468	116	442
Velocity variation, $< \delta V^2 >^{1/2}$ ($km\ s^{-1}$)	20.5	12.1	17.0
Flow direction, ϕ_V (degrees)	– 0.64	2.6	0.67
Magnetic Field Strength, H (10^{-6} Gauss)	6.2	2.9	5.6
Proton Temperature, T_p (°K)	1.2×10^5	0.9×10^5	0.95×10^5
Electron Temperature, T_e (°K)	1.4×10^5	0.4×10^5	1.33×10^5
Helium Temperature, T_α (°K)	5.8×10^5	5.0×10^5	4.5×10^5
Temperature Ratio, T_e/T_p	1.9	1.6	1.5
Temperature Ratio, T_α/T_p	4.9	1.8	4.7
$N_p(m_p V^2/2)$ (10^{-10} erg cm^{-3})	144	9	125
$H^2/8\pi$ (10^{-10} erg cm^{-3})	1.7	1.2	1.2
1.5 $N_p k T_p$ (10^{-10} erg cm^{-3})	1.9	2.3	1.3
1.5 $N_p k T_e$ (10^{-10} erg cm^{-3})	2.5	2.2	2.0
1.5 $N_\alpha k T_\alpha$ (10^{-10} erg cm^{-3})	0.4	0.5	0.28
1.5 $N_p m_p < \delta V >^2$ (10^{-10} erg cm^{-3})	1.1	3.3	0.37

*Adapted from Feldman et al. (1977).

Solar Wind at 1 AU*

Parameter	Mean Value	Standard Deviation	Median
Debye Length, $\lambda_D = 6.9 \times 10^{-2}\ (T_e/N_p)^2$ (meters)	9.9	3.0	9.8
Thermal Electron Gyroradius, R_e (km)	2.5		
Thermal Proton Gyroradius, R_p (km)	78	36	76
Thermal Alpha-Particle Gyroradius, R_α (km)	177		
Correlation Length Magnetic Field Variations, L_H (km)	1.0×10^7	0.4×10^7	0.94×10^7
Plasma Density Scale Height, L_N (km)	7.5×10^7		
Electron Plasma Frequency, ω_{pe} (s^{-1})	1.6×10^5	0.5×10^5	1.5×10^5
Proton Plasma Frequency, ω_{pp} (s^{-1})	3.7×10^3	1.2×10^3	3.5×10^3
Electron Gyrofrequency, ω_{He} (s^{-1})	1.1×10^3	0.5×10^3	0.95×10^3
Proton Gyrofrequency, ω_{Hp} (s^{-1})	0.57	0.25	0.52
Inverse Expansion Scale Time, $\tau_x^{-1} = V/L_N$ (s^{-1})	6.2×10^{-6}	1.5×10^{-6}	5.9×10^{-6}
Number of Particles in Debye Sphere, $N\lambda_D^3$	6.7×10^9	3.0×10^9	6.2×10^9
Collision Time/Expansion Time, $(\tau_c/\tau_x)_{ee}$	1.7	1.4	1.4
Collision Time/Expansion Time, $(\tau_c/\tau_x)_{pp}$	34	76	11
Collision Time/Expansion Time, $(\tau_c/\tau_x)_{\alpha p}$	53	89	18

*Adapted from Feldman et al. (1977).

Solar Wind Flows at 1 AU*

Parameter	Average Wind		Low Speed Wind		High Speed Wind	
	Mean Value	Standard Deviation	Mean Value	Standard Deviation	Mean Value	Standard Deviation
N (cm^{-3})	8.7	6.6	11.9	4.5	3.9	0.6
V ($km\ s^{-1}$)	468	116	327	15	702	32
NV ($cm^{-2}\ s^{-1}$)	3.8×10^8	2.4×10^8	3.9×10^8	1.5×10^8	2.7×10^8	0.4×10^8
$< \delta V^2 >^{1/2}$ ($km\ s^{-1}$)	20.5	12.1	9.6	2.9	34.9	6.2
ϕ_V (degrees)	– 0.6	2.6	1.6	1.5	– 1.3	0.4
T_p (°K)	1.2×10^5	0.9×10^5	0.34×10^5	0.15×10^5	2.3×10^5	0.3×10^5
T_e (°K)	1.4×10^5	0.4×10^5	1.3×10^5	0.3×10^5	1.0×10^5	0.1×10^5
T_α (°K)	5.8×10^5	5.0×10^5	1.1×10^5	0.8×10^5	14.2×10^5	3.0×10^5
T_e/T_p	1.9	1.6	4.4	1.9	0.45	0.07
T_α/T_p	4.9	1.8	3.2	0.9	6.2	1.3
N_α/N_p	0.047	0.019	0.038	0.018	0.048	0.005

*Adapted from Feldman et al. (1977). See the previous table for definition of symbols.

9
Basic Stellar Data

9.1 Physical Parameters of the Stars

M = mass

R = radius

$\bar{\rho}$ = $3M/(4\pi R^3)$ = mean density

g = GM/R^2 = surface gravitational acceleration
$G = 6.67259(85) \times 10^{-8}$ cm^3 g^{-1} s^{-2}

T_{eff} = effective temperature = T_e

L = $4\pi\sigma R^2 T_e^4$ = absolute luminosity
σ = Stefan Boltzmann constant
$\sigma = 5.67051(19) \times 10^{-5}$ erg cm^{-2} °K^{-4} s^{-1}
$\pi = 3.1415926536$

F = σT_e^4 = total radiant flux per unit stellar surface area

Sun = subscript $\odot$

$M_\odot$ = 1.989×10^{33} g

$R_\odot$ = 6.9599×10^{10} cm

$\bar{\rho}_\odot$ = 1.409 g cm^{-3}

$g_\odot$ = 2.7398×10^4 cm s^{-2}

$T_{e\odot}$ = 5780 °K

$L_\odot$ = 3.86×10^{33} erg s^{-1}

$F_\odot$ = 6.34×10^{10} erg cm^{-2} s^{-1}

m = apparent magnitude

Subscripts U, B, V, R and I for ultraviolet, blue, visual, red and infrared

U = m_U = ultraviolet apparent magnitude (at 3500 Å)

B = m_B = blue apparent magnitude (at 4350 Å)

V = m_V = visual apparent magnitude (at 5550 Å)

R = m_R = red apparent magnitude (at 6800 Å)

I = m_I = infrared apparent magnitude (at 8250 Å)

m_{bol} = bolometric apparent magnitude
= apparent magnitude integrated over all wavelengths

BC = bolometric correction = $m_{bol} - m_V$

M = absolute magnitude
= apparent magnitude standardized to a distance of 10 pc

M_V = absolute visual magnitude

M_{bol} = absolute bolometric magnitude

D = distance

π = parallax, in arcseconds = 1/D with D in pc

m - M = distance modulus = $5 \log D - 5 = -5(1 + \log \pi)$

μ = annual proper motion in arc seconds

v_t = transverse velocity, in km s^{-1} = $4.741\ \mu/\pi$

v_r = radial velocity away from observer along line of sight

P = pulsation period

Sp = spectral classification

LC = luminosity class

A = space absorption in magnitudes

E = color excess

E_{B-V} = $A_B - A_V$ = reddening

R = $A_V/E_{B-V} = 3.30 + 0.28\,(B\text{-}V)_o + 0.04\,E_{B-V}$

A_V = visual extinction

M_V = absolute visual magnitude corrected for interstellar reddening
= $(m_V - A_V) + 5 - 5 \log D$

$(B\text{–}V)_o$ = intrinsic color of unreddened star

(B–V) = observed color = $(B\text{–}V)_o + E_{B-V}$

C = color index = $m_{pg} - m_v = B - V - 0.11$

E_{U-B} = $A_U - A_B$

E_{U-B} = $(0.72 + 0.05\,E_{B-V})\,E_{B-V}$

$(U\text{–}B)_o$ = intrinsic color of unreddened star

(U–B) = observed color = $(U\text{–}B)_o + E_{U-B}$

9.2 Numerical Relations and Laws

Stefan - Boltzmann Law

$$L = 4\pi\sigma\, R^2\, T_e^4 = 7.1258 \times 10^{-4}\, R^2\, T_e^4 \text{ erg s}^{-1}$$

Magnitudes and Luminosities

$$\log(L_1/L_2) = 0.4\,(M_1 - M_2)$$

$$\log(F_1/F_2) = 0.4\,(m_1 - m_2)$$

Mass - Luminosity Relation
[Empirical, Mc Cluskey and Kondo (1972)]

$$\log(M/M_\odot) = 0.48 - 0.105\, M_{bol} \quad \text{for stars with } -8 < M_{bol} < 10.5$$

$$\log(L/L_\odot) = 3.8 \log(M/M_\odot) + 0.08 \text{ for stars with } M > 0.2\, M_\odot$$

Mass - Luminosity Relation
[Empirical, Harris, Strand and Worley (1963)]

$$\log(M/M_\odot) = 0.46 - 0.10\, M_{bol} \quad \text{for } 0 \le M_{bol} \le 7.5$$

$$\log(M/M_\odot) = 0.76 - 0.145\, M_{bol} \quad \text{for } 7.5 \le M_{bol} \le 11.0$$

$$\log(L/L_\odot) \approx 4 \log(M/M_\odot) \quad \text{for } L > L_\odot$$

$$\log(L/L_\odot) \approx 2.8 \log(M/M_\odot) \quad \text{for } L < L_\odot$$

Mass - Radius Relation
For the zero - age main sequence

$$\log(R/R_\odot) = 0.640 \log(M/M_\odot) + 0.011 \quad \text{for } 0.12 < \log(M/M_\odot) < 1.3$$

$$\log(R/R_\odot) = 0.917 \log(M/M_\odot) - 0.020 \quad \text{for } -1.0 < \log(M/M_\odot) < 0.12$$

Radius - Luminosity Relation
[Empirical, Barnes, Evans and Moffet (1978), Popper (1980)]

$$\log(R/R_\odot) = -0.2\, M_V - 2\, F_V + 8.451$$

$$\log(R/R_\odot) = -0.2\, M_{bol} - 2 \log(T_e) + 8.451$$

with M_{bol} in mag and T_e in °K, and $F_V = \log T_e + 0.1$ BC

Mass Loss rates

$$\log(-\dot{M}) = 1.769 \log (L/L_\odot) - 1.676 \log (T_{eff}) - 8.158$$

For population I, spectral types O through M, with M in units of solar masses, $M_\odot$, per year. De Jager, Nieuwenhuijzen and van der Hucht (1988).]

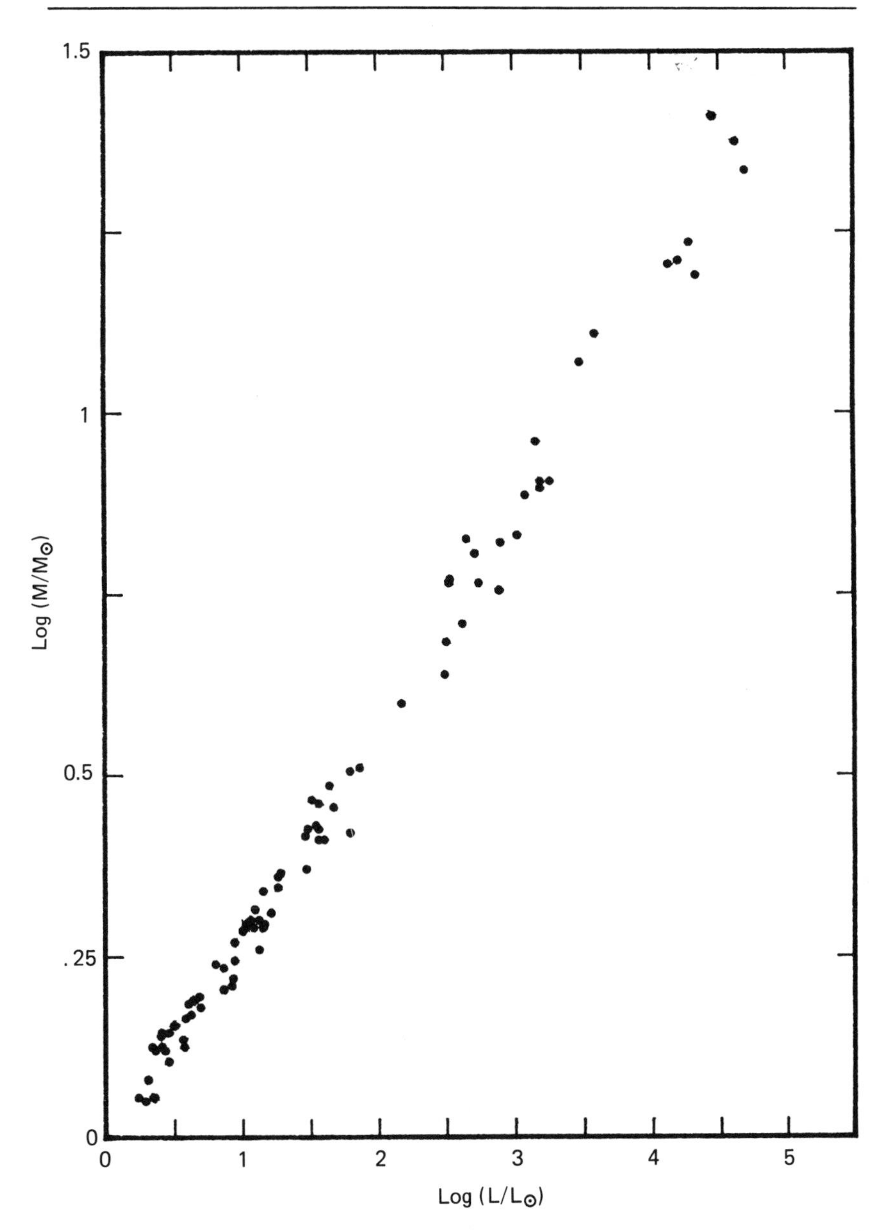

Mass-Luminosity Relation. Stellar masses, M, plotted as a function of their absolute luminosity, L, and normalized to the values, $M_\odot$ and $L_\odot$, for the Sun. [Using the data given by Popper (1980)].

9.3 Observed Angular Diameters of Stars

Angular diameters, θ, of stars measured interferometrically (Hanbury Brown et al., 1974), effective temperatures, T_e, and bolometric corrections, BC (Code et al., 1976).

Star	Name	Sp (MK)	θ (10^{-3} ″)	T_e °K	BC
HD 10144	α Eri	B3 Vp	1.92 ± 0.07	14510	- 1.32
HD 34085	β Ori	B8 Ia	2.55 ± 0.05	11550	- 0.60
HD 35468	γ Ori	B2 III	0.72 ± 0.04	21580	- 2.09
HD 37128	ε Ori	B0 Ia	0.69 ± 0.04	24820	- 2.46
HD 37742	ζ Ori	O9.5 Ib	0.48 ± 0.04	29910	- 2.93
HD 38771	κ Ori	B0.5 Ia	0.45 ± 0.03	26390	- 2.26
HD 44743	β CMa	B1 II–III	0.52 ± 0.03	25180	- 2.38
HD 45348	α Car	F0 Ib–II	6.60 ± 0.8	7460	+0.11
HD 47105	γ Gem	A0 IV	1.39 ± 0.09	9260	- 0.12
HD 48915	α CMa	A1 V	5.89 ± 0.16	9970	- 0.20
HD 52089	ε CMa	B2 II	0.80 ± 0.05	20990	- 2.06
HD 54605	δ CMa	F8 Ia	3.60 ± 0.50	6110	+0.07
HD 58350	η CMa	B5 Ia	0.75 ± 0.06	13310	- 0.87
HD 61421	α CMi	F5 IV–V	5.50 ± 0.17	6510	- 0.02
HD 66811	ζ Pup	O5	0.42 ± 0.03	32510	- 3.18
HD 68273	γ^2 Vel	WC8+O9 I	0.44 ± 0.05	32510	- 3.12
HD 80007	β Car	A1 IV	1.59 ± 0.07	9240	- 0.16
HD 87901	α Leo	B7 V	1.37 ± 0.06	12210	- 0.72
HD 102647	β Leo	A3 V	1.33 ± 0.10	8850	- 0.04
HD 106625	γ Crv	B8 III	0.75 ± 0.06	12450	- 0.72
HD 111123	β Cru	B0.5 III	0.722 ± 0.023	27600	- 2.75
HD 116658	α Vir	B1 IV	0.87 ± 0.04	23930	- 2.44
HD 118716	ε Cen	B1 III	0.48 ± 0.03	25740	- 2.54
HD 143275	δ Sco	B0.5 IV	0.46 ± 0.04	31460	- 3.07
HD 149757	ζ Oph	O9.5 V	0.51 ± 0.05	31910	- 2.96
HD 159561	α Oph	A5 III	1.63 ± 0.13	8020	+0.01
HD 169022	ε Sgr	A0 V	1.44 ± 0.06	9460	- 0.22
HD 172167	α Lyr	A0 V	3.24 ± 0.07	9660	- 0.25
HD 187642	α Aql	A7 IV,V	2.98 ± 0.14	8010	+0.02
HD 193924	α Pav	B2.5 V	0.80 ± 0.05	17880	- 1.80

Observed Angular Diameters of Stars

Angular diameters, θ, of stars measured interferometrically (Hanbury Brown et al., 1974), effective temperatures, T_e, and bolometric corrections, BC (Code et al., 1976), and other stars (Schmidt-Kaler, 1982).

Star	Name	Sp (MK)	θ (10^{-3} ″)	T_e °K	BC
HD 209952	α Gru	B7 IV	1.02 ± 0.07	14050	- 1.08
HD 216956	α PsA	A3 V	2.10 ± 0.14	8800	- 0.03
HD 18191	RZ Ari	M6III	10.18	3160	- 3.81
HD 29139	α Tau	K5III	24.6	3590	- 1.21
HD 44478	μ Gem	M3III	13.65	3650	- 2.02
HD 60179C	YY Gem	M0.5 V	0.458	3520	- 1.45
HD 86663	π Leo	M2III	5.9	3300	- 1.59
HD 87837	31 Leo	K4III	3.55	4000	- 0.99
HD 91232	46 Leo	gM2	5.6	2950	- 1.74
HD 99998	87 Leo	K4III	3.7	3690	- 1.13
HD 112142	ψ Vir	M3III	5.86	3530	- 1.96
HD 124897	α Boo	K2IIIp	23.1	4070	- 0.71
HD 139663	42 Lib	K4III	2.5	3920	- 0.74
HD 168574	HR 6861	gM5	3.6	4150	- 2.82
HD 169916	λ Sgr	K2III	4.4	4540	- 0.44
HD 175775	ε^2 Sgr	K1III	3.8	4210	- 0.50
HD 196777	υ Cap	M2III	4.72	3440	- 1.68
HD 207005	47 Cap	gM3	3.2	3760	- 2.23
HD 216386	λ Aqr	M2III	8.2	3590	- 1.76
HD 217906	β Peg	M2II–III	18	3450	- 1.96
HD 219215	ϕ Aqr	M2III	4.9	3940	- 1.47
HD 39801	α Ori	M2Iab	49	3450	- 1.50
HD 36389	119 Tau	M2Iab	10.9	3520	- 1.40
HD 148748	α Sco	M1Iab	42.5	3560	- 1.28
	U Ari	M4–6e	6.1 var		
	S Psc	M5–7e	3.8 var		
HD 84748	R Leo	M6.5–9e	60 var		
HD 14386	o Cet	M6IIIe	44 var		
HD 164875	W Sgr	M8e	5.2 var		
	AQ Sgr	C7(N3)	5.5 var		

9.4 Stellar Masses

Stellar Masses - Observed Parameters

The observed masses, radii, color indices, $(B-V)_o$, and extinction, E_{B-V}, for detached eclipsing binary stars are given in order of decreasing mass of the more massive component. Spectral types, Sp, were inferred from color indices as well as from the appearance of the spectra. Adapted from Popper (1980).

Eclipsing Binary System			Observed Parameters				
Name		Period (days)	Mass ($M_\odot$)	Radius ($R_\odot$)	$(B-V)_o$	E_{B-V}	Sp
Y Cyg	HD 198846	3.00	16.70	6.0	– 0.30	0.23	O9.8
			16.70	6.0	– 0.30		O9.8
V478 Cyg	HD 193611	2.88	15.60	7.3	– 0.29	0.84	O9.8
			15.60	7.3	– 0.29		O9.8
V453 Cyg	HD 227696	2.88	14.50	8.6	– 0.27	0.46	B0.4
			11.30	5.4	– 0.26		B0.7
CW Cep	HD 218066	2.73	11.80	5.4	– 0.28	0.67	B0.4
			11.10	5.0	– 0.28		B0.7
α Vir	HD 116658	4.01	10.80	8.1	– 0.25	0.00	B1.5
			6.80	4.4	– 0.18		B4
QX Car	HD 86118	4.48	9.20	4.3	– 0.23	0.04	B2
			8.50	4.0	– 0.22		B2
V539 Ara	HD 161783	3.17	6.13	4.4	– 0.19	0.07	B4
			5.25	3.7	– 0.18		B4
CV Vel	HD 77464	6.89	6.10	4.05	– 0.19	0.03	B3
			6.00	4.05	– 0.19		B3
BM Ori	HD 37021		5.90	2.90	– 0.21	0.30	B3
DI Her	HD 175227	10.55	5.20	2.55	– 0.175	0.19	B5
			4.60	2.55	– 0.16		B5
U Oph	HD 156247	1.68	5.16	3.43	– 0.175	0.22	B5
			4.60	3.11	– 0.165		B5
V760 Sco	HD 147683	1.73	5.00	3.00	– 0.165	0.34	B5
			4.65	2.70	– 0.155		B5
AG Per	HD 25833	2.03	4.53	3.00	– 0.185	0.19	B4
			4.12	2.60	– 0.17		B5

Stellar Masses - Observed Parameters

Eclipsing Binary System			Observed Parameters				
Name		Period (days)	Mass ($M_\odot$)	Radius ($R_\odot$)	$(B-V)_o$	E_{B-V}	Sp
ς Phe	HD 6882	1.67	3.92	2.85	− 0.145	0.00	B6
			2.54	1.85	− 0.05		B9
χ^2 Hya	HD 96314	2.27	3.61	4.39	− 0.09	0.02	B9
			2.64	2.16	− 0.05		A0
AS Cam	HD 35311	3.43	3.31	2.70	− 0.105	0.09	B8
			2.51	2.00	− 0.045		B9
V451 Oph	HD 170470	2.20	2.78	2.65	− 0.06	0.05	B9
			2.36	2.12	− 0.02		A0
RX Her	HD 170757	1.78	2.75	2.44	− 0.06	0.07	B9
			2.33	1.96	− 0.02		A0
AR Aur	HD 34364	4.13	2.48	1.83	− 0.08	0.00	B8
			2.29	1.83	− 0.04		B9
β Aur	HD 40183	3.96	2.35	2.49	+0.035	0.00	A1
			2.27	2.49	+0.035		A1
SZ Cen	HD 120359	4.11	2.28	3.62	+0.20	0.08	A7
			2.32	4.55	+0.22		A7
EE Peg	HD 206155	2.63	2.08	2.05	+0.10	0.00	A3
			1.32	1.29	+0.45		F5
V624 Her	HD 161321A	3.90	2.1	3.0	+0.19	0.00	Am
			1.8	2.2	+0.21		Am
V805 Aql	HD 177708	2.41	2.06	2.10	+0.13	0.13	A5
			1.60	1.75	+0.28		F0
RR Lyn	HD 44691	9.95	2.00	2.50	+0.22	0.00	Am
			1.55	1.93	+0.29		F0
WW Aur	HD 46052	2.52	1.98	1.89	+0.14	0.00	Am
			1.82	1.89	+0.19		Am
CM Lac	HD 209147	1.60	1.88	1.59	+0.08	0.03	A2
			1.47	1.42	+0.28		F0
RS Cha	HD 75747	1.67	1.86	2.28	+0.20	0.00	A8
			1.82	2.28	+0.26		A8
MY Cyg	HD 193637	4.00	1.81	2.20	+0.30	0.04	Am
			1.78	2.20	+0.31		Am
V477 Cyg	HD 190786	2.35	1.78	1.52	+0.07	0.00	A2
			1.34	1.20	+0.38		F2
EI Cep	HD 205234	8.44	1.68	2.54	+0.32	0.00	F0
			1.78	2.80	+0.36		F2

Stellar Masses - Observed Parameters

Eclipsing Binary System			Observed Parameters				
Name		Period (days)	Mass ($M_\odot$)	Radius ($R_\odot$)	$(B-V)_o$	E_{B-V}	Sp
XY Cet	HD 18597	2.78	1.76	1.88	+0.23	0.00	Am
			1.63	1.88	+0.27		Am
ZZ Boo	HD 121648	4.99	1.72	2.22	+0.37	0.00	F2
			1.72	2.22	+0.37		F2
TX Her	HD 156965	2.06	1.62	1.58	+0.26	0.00	A8
			1.45	1.48	+0.36		F2
CW Eri	HD 19115	2.73	1.52	2.11	+0.35	0.00	F2
			1.28	1.48	+0.39		F2
RZ Cha	HD 93486	2.83	1.51	2.26	+0.47	0.00	F5
			1.51	2.26	+0.47		F5
BK Peg	+25°5003	5.49	1.27	1.48	+0.54	0.00	F8
			1.43	2.03	+0.56		F8
DM Vir	HD 123423	4.67	1.46	1.76	+0.47	0.02	F7
			1.46	1.76	+0.47		F7
CD Tau	HD 34335	3.44	1.40	1.74	+0.48	0.00	F7
			1.31	1.61	+0.48		F7
TV Cet	HD 20173	9.10	1.39	1.50	+0.39	0.00	F2
			1.27	1.26	+0.46		F5
BS Dra	HD 190020	3.36	1.37	1.44	+0.44	0.00	F5
			1.37	1.44	+0.44		F5
HS Hya	HD 90242	1.57	1.34	1.36	+0.43	0.00	F5
			1.28	1.22	+0.46		F5
V1143 Cyg	HD 185912	7.64	1.33	1.31	+0.45	0.00	F5
			1.29	1.31	+0.45		F5
VZ Hya	HD 72257	2.90	1.23	1.35	+0.42	0.00	F5
			1.12	1.12	+0.48		F6
UX Men	HD 37513	4.18	1.17	1.28	+0.54	0.00	F8
			1.11	1.28	+0.55		F8
WZ Oph	HD 154676	4.18	1.12	1.34	+0.54	0.00	F8
			1.12	1.34	+0.54		F8
UV Leo	HD 92109	0.60	0.99	1.08	+0.62	0.00	G2
			0.92	1.08	+0.62		G2
YY Gem	+32°1582	0.81	0.59	0.62	+1.37	0.00	M1
			0.59	0.62	+1.37		M1
CM Dra	GL 630.1	1.27	0.24	0.25	+1.83	0.00	M4
			0.21	0.235	+1.83		M4

Stellar Masses - Derived Parameters

Derived quantities are the effective temperature, T_e, absolute luminosity, L, and absolute magnitude, M_V. These data were combined with the masses in the previous table to give the mass - luminosity plot shown on page 117. Adapted from Popper (1980).

Name	log T_e (°K)	log L ($L_\odot$)	M_V (mag)	Name	log T_e (°K)	log L ($L_\odot$)	M_V (mag)
Y Cyg	4.485	4.45	- 3.45	U Oph	4.22	2.90	- 1.00
	4.485	4.45	- 3.45		4.20	2.74	- 0.70
V478 Cyg	4.485	4.62	- 3.90	V760 Sco	4.20	2.71	- 0.65
	4.485	4.62	- 3.90		4.18	2.53	- 0.30
V453 Cyg	4.47	4.69	- 4.15	AG Per	4.245	2.89	- 0.85
	4.445	4.20	- 3.10		4.21	2.62	- 0.35
CW Cep	4.47	4.28	- 3.15	ζ Phe	4.160	2.50	- 0.37
	4.445	4.13	- 2.90		4.005	1.51	1.13
α Vir	4.39	4.33	- 3.5	χ^2Hya	4.063	2.49	- 0.98
	4.23	3.16	- 1.5		4.005	1.64	0.79
QX Car	4.34	3.59	- 2.10	AS Cam	4.088	2.17	0.00
	4.33	3.48	- 1.90		4.001	1.56	0.98
V539 Ara	4.255	3.26	- 1.70	V451 Oph	4.015	1.86	0.30
	4.23	3.02	- 1.25		3.985	1.54	0.95
CV Vel	4.255	3.19	- 1.55	RX Her	4.015	1.79	0.48
	4.255	3.19	- 1.55		3.985	1.48	1.12
BM Ori	4.30	3.08	- 1.05	AR Aur	4.048	1.67	0.99
DI Her	4.22	2.65	- 0.40		3.997	1.46	1.19
	4.19	2.52	- 0.25	β Aur	3.955	1.56	0.82
					3.955	1.56	0.82

Stellar Masses - Derived Parameters

Name	log T_e (°K)	log L ($L_\odot$)	M_V (mag)	Name	log T_e (°K)	log L ($L_\odot$)	M_V (mag)
SZ Cen	3.885	1.60	0.61	RZ Cha	3.800	0.86	2.51
	3.878	1.79	0.17		3.800	0.86	2.51
EE Peg	3.927	1.28	1.48	BK Peg	3.785	0.43	3.63
	3.806	0.40	3.68		3.780	0.69	3.01
V624 Her	3.890	1.47	0.98	DM Vir	3.800	0.64	3.06
	3.882	1.16	1.73		3.800	0.64	3.06
V805 Aql	3.915	1.26	1.52	CD Tau	3.798	0.625	3.11
	3.855	0.86	2.50		3.798	0.56	3.28
RR Lyn	3.880	1.26	1.49	TV Cet	3.820	0.58	3.20
	3.850	0.93	2.32		3.803	0.365	3.75
WW Aur	3.910	1.15	1.78	BS Dra	3.808	0.50	3.41
	3.890	1.065	1.98		3.808	0.50	3.41
CM Lac	3.935	1.095	1.97	HS Hya	3.810	0.46	3.52
	3.855	0.68	2.95		3.803	0.34	3.82
RS Cha	3.885	1.21	1.61	V1143 Cyg	3.806	0.41	3.64
	3.863	1.12	1.85		3.806	0.41	3.64
MY Cyg	3.848	1.03	2.07	VZ Hya	3.812	0.46	3.51
	3.845	1.02	2.11		3.798	0.24	4.07
V477 Cyg	3.940	1.08	2.03	UX Men	3.785	0.31	3.95
	3.820	0.41	3.66		3.782	0.29	3.98
EI Cep	3.840	1.12	1.84	WZ Oph	3.785	0.35	3.85
	3.827	1.15	1.77		3.785	0.35	3.85
XY Cet	3.875	1.00	2.15	UV Leo	3.768	0.09	4.52
	3.860	0.94	2.30		3.768	0.09	4.52
ZZ Boo	3.824	0.94	2.29	YY Gem	3.576	− 1.13	8.90
	3.824	0.94	2.29		3.576	− 1.13	8.90
TX Her	3.863	0.80	2.64	CM Dra	3.500	− 2.16	12.77
	3.827	0.60	3.15		3.500	− 2.16	12.92
CW Eri	3.830	0.92	2.35				
	3.820	0.57	3.23				

9.5 Spectral Classification of the Stars

The features of stellar line spectra permit a classification, Sp, in the scheme:

```
                     = C
                    ⏞
                   R – N
P                 /   S
|                    /
W – O – B – A – F – G – K – M    main series
|
Q

early   inter-   late types
        mediate
```

Stars With Predominant Absorption Lines*
Main Series O to M (99% of apparently bright stars)

Sp	Standards	Spectral characteristics
O		Intensive (blue) continuum, predominant absorption lines of He II, additionally of C III, N III, Si IV. If N III 4634/40/41 appears in emission the star is called Of, if He II 4686, it is Onfp.
O3	HDE 303308	He I 4471/He II 4541 = 0.1
O4f	HD 190429 A	= 0.2
O5(f)	HD 15558	= 0.3
O6fp	λ^1 Cep	= 0.6
O7	HD 190 864	= 0.9
O8	λ Ori	= 1.3
O9	ι Ori	= 1.8
B0	τ Sco, ε Ori	He I > He II; C III 4650 and Si IV 4089/4116:max; Hδ = 1.5 He I 4026.
B3	π^4 Ori	He I:max; H(Balmer lines) $\approx$ 0.5<A0>; O II and Si IV very weak
B5	φ Vel	Si II 4128/4131 > He I 4121
B8	β Per	He I 4471 = Mg II 4481; Hδ = 15 He I 4026; metal lines appear

Stars With Predominant Absorption Lines*
Main Series O to M (99% of apparently bright stars)

Sp	Standards	Spectral characteristics
A0	α CMa	Balmer series dominating (H:max); Mg II 4481 most conspicuous after Balmer lines; K(Ca II 3934) $\approx$ 0.1 Hδ; Si II:max; depression of UV continuum by continuous Balmer absorption.
A5	β Tri, α Pic	Ca II(K) = 0.9[CaII(H)+Hε] and >Hδ; Fe I 4299/4303 and Ti II 4303 strong
F0	δ Gem, α Car	Balmer series $\approx$ 0.5 <A0>; Ca II(K) = Ca II(H) + Hε = 3 Hδ; many metal lines; G-band 4307 (Fe, Ti, Ca) appearing.
F5	α CMi, ϱ Pup	Balmer lines $\approx$ 2$\odot$; Ca I 4227 = 0.5 Hγ; G-band = 0.6 Hγ.
G0	α Aur, β Hyi	Solar-type spectrum (Sun: G 2 or somewhat later); very intense metal lines; Ca I 4227 = Hδ; G-band = 2 Hγ = 3 Fe I 4325
K0	α Boo, α Phe	Metal lines further enhanced, Balmer lines further weakened; Ca I 4227 = 2 Fe II 4172 = 3 Fe I 4383; Fe I 4325 = 2 Hγ; Ca II(H,K):max.
K5	α Tau	Similar to sunspot spectrum; Ca I and II dominating; G-band dissolved in lines; green TiO-bands appearing.
M0···2(Ma)	β And (M0), α Ori (M2)	TiO-bands dominating (esp. 4762...4956,5168...5445); Ca I 4227 strongest line.
M3···5(Mb)	π Aur (M3)	Subclassification by increasing intensity of the green and red
M6···10(Mc)	ϱ Per (M6)	(6651,7054,7589) band systems.
M0e···M10e(Md)	o Cet (M6e)	Balmer lines (at least Hα) in emission.

*For any two lines a and b in the third column:

a > b	line a stronger than line b
a = b	lines a and b have the same intensity
a: max	line a has maximum intensity at this spectral type
a $\approx$ 0.5<A0>	line a has half the intensity of the same line in spectral type A0
a $\approx$ 0.1 Hδ	line a has one tenth of the intensity of the line Hδ
a $\approx$ 2$\odot$	line a has twice the intensity as the same line in the solar spectrum

Stars With Predominant Emission Lines

Sp	Standards	Spectral characteristics
P	NGC 7027 NGC 6720	Planetary Nebulae: many emission lines of very high excitation, including forbidden lines.
Q	GK Per 1901 DQ Her 1934	Novae
W	γ^2 Vel (WC7+O7)	Wolf-Rayet = WR stars : very broad, intensive emission lines of H, He I, He II are superposed on a continuum which is especially intensive in the blue-UV region
WN	HD 192163 (WN6)	a) nitrogen sequence WN5···8 with strong emission lines of N III, IV, V
WC	HD 192103 (WC7)	b) carbon sequence WC5···8 with strong emission lines of C II, III, IV, O III···O VI

Late-Type Stars

Sp	Standards	Spectral characteristics
S	π^1 Gru R Gem R And	Strong bands of ZrO and YO, LaO, TiO; subclassification by strength of TiO and ZrO bands; Ca II(H,K), Ca I 4227, Ba II 4554. Very rare type; transitions to class M
R	BD 10°5057(R0) HD 52432(R5)	Strong bands of CN and CO instead of TiO in class M; subclassification by strength of bands.
N	19 Psc (N0) (=Na)	Swan bands of C_2, Na I (D), Ca I 4227, for the rest similar to R; subclassification by strength of bands.

The "carbon series" C0···C7, corresponding to classes R, N, branches off the main series at G5, running parallel to the oxygen-rich G5···M4 main series;

Luminosity Class - MK System

For a given chemical composition, the line spectrum of a star is essentially determined by the degree of ionization, i.e. not only by the temperature but also by the electron pressure. This is taken into account in the Yerkes system by a second parameter, the luminosity class LC. Each luminosity class is subdivided into subclasses a, ab, b in the order of decreasing luminosity; often, transition classes are also used, e.g. Ib–II.

LC			Stars	Examples
0	0-Ia	Ia-0	super-supergiants	G0 Ia-0
Ia	Iab	Ib	supergiants	B1 Iab
IIa	IIab	IIb	bright giants	K0 II
IIIa	IIIab	IIIb	(normal) giants	K0 IIIa (gKO)
IVa	IVab	IVb	subgiants	K2 IV
Va	Vab	Vb	main-sequence stars (dwarfs)	G2V (dG2)
	(VI)		subdwarfs	(sdK3)
	(VII)		white dwarfs	(wA0,DB)

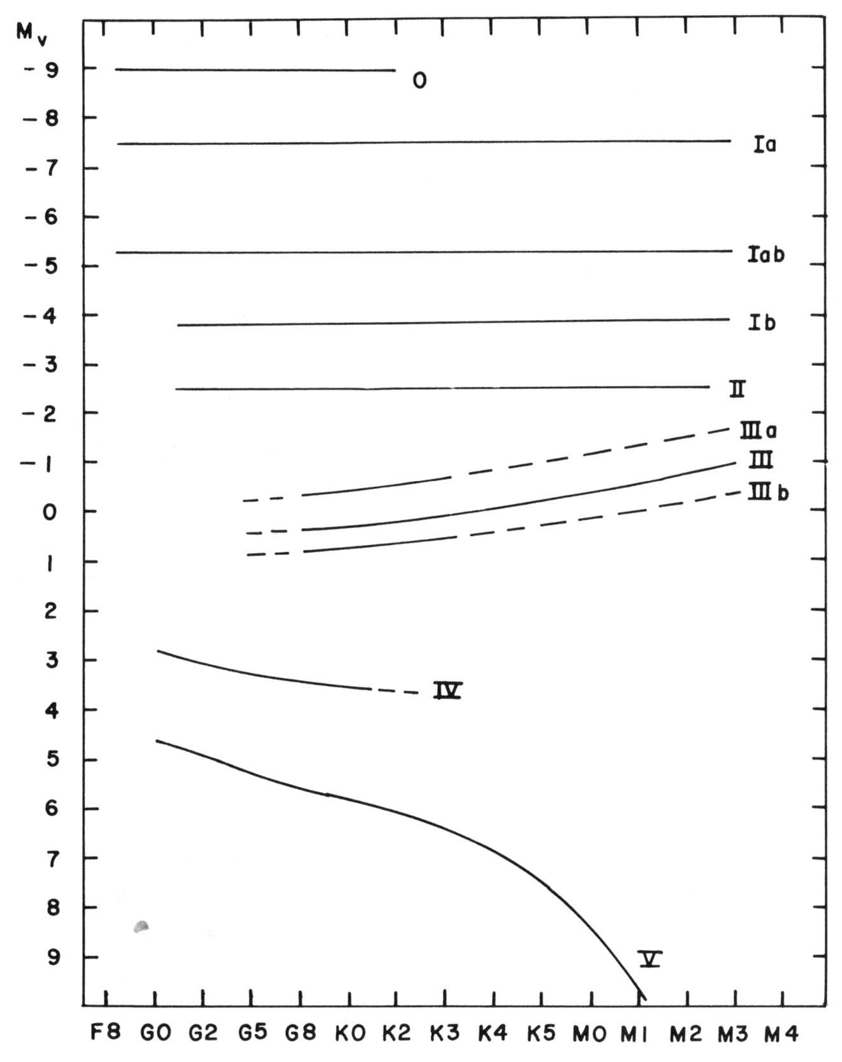

MK System. The two-dimensional MK system of stellar classification includes both spectral type and luminosity class. The luminosity class, LC, for stars is designated O, I, II, III, IV or V. Here their absolute visual magnitudes, M_V, are plotted as a function of spectral type ranging from F to G, K and M.

Criteria of the MK System

Sp	Criteria for Sp type	Sp	Criteria for luminosity
O3···O9.5	He I 4471/He II 4541	O stars	Si IV 4089, 4116/He I 4120, 4143
B0···B1	Si III 4552/Si IV 4089	O9···B3	(Si IV, He I 4116···21)/He I 4144
B2···B8	Si II 4128···30/He I 4121	B0···B3	N II 3995/He II 4009
B8···A2	He I 4471/Mg II 4481, He I 4026/Ca II 3934	B1···A5	wings of Balmer lines
		A3···F0	blend 4416/Mg II 4481
A2···F2	Mn I 4030···34/4128···32, 4300/4385	F0···F8	4172/Ca I 4227
F2···K	CH 4300 (G-Band)/Hγ 4341		
F5···G5	Fe I 4045/Hδ 4101, Ca I 4227/Hγ 4341	F2···K5	(Fe I 4045, Fe I 4063, Ca I 4227)/Sr II 4077
		G5···M	discontinuity at 4215
G5···K0	Fe I 4144/Hδ 4101		
K0···K5	Ca I 4227/4325,4290/4300	K3···M	4215/4260, Ca I increasing

Bright Standard Stars–MK Classification

Star	Sp	LC	Star	Sp	LC	Star	Sp	LC
ε Ori	B0	Ia	δ Ori	O9.5	II	γ Peg	B2	IV
χ^2 Ori	B2	Ia	ε CMa	B2	II	19 Tau	B6	IV
β Ori	B8	Ia	ο Sco	A5	II	γ Gem	A0	IV
α Cyg	A2	Ia	HR 292	F0	II	ε Cep	F0	IV
μ Cep	M2	Ia	ν Her	F2	II	ζ Her	G1	IV
χ Aur	B5	Iab	ε Leo	G1	II	η Boo	G0	IV
σ Cyg	B9	Iab	θ Lyr	$K0^+$	II	β Aql	G8	IV
44 Cyg	F5	Iab	γ Aql	K3	II	HD46223	O4	
o^1 CMa	K2,5	Iab	ι Ori	O9	III	ζ Pup	O5f	
α Ori	M1–2	Ia–Ib	ο Per	B1	III	10 Lac	O9	V
τ CMa	O9	Ib	η Tau	B7	III	τ Sco	B0	V
ζ Per	B1	Ib	θ^2 Tau	A7	III	η Aur	B3	V
α Lep	F0	Ib	ζ Leo	F0	III	18 Tau	B8	V
α Per	F5	Ib	ο Psc	G8	III	α Lyr	A0	V
β Cam	G0	Ib	β Gem	K0	IIIb	ϱ Gem	F0	V
ζ Cep	K1.5	Ib	α Ari	K2	IIIab	78 UMa	F2	V
119 Tau	M2	Iab–Ib	α Tau	K5	III	β Com	G0	V
			β And	M0	IIIa	σ Dra	K0	V
			χ Peg	$M2^+$	III	61 Cyg	K5	V
						HD 147379	M0	V

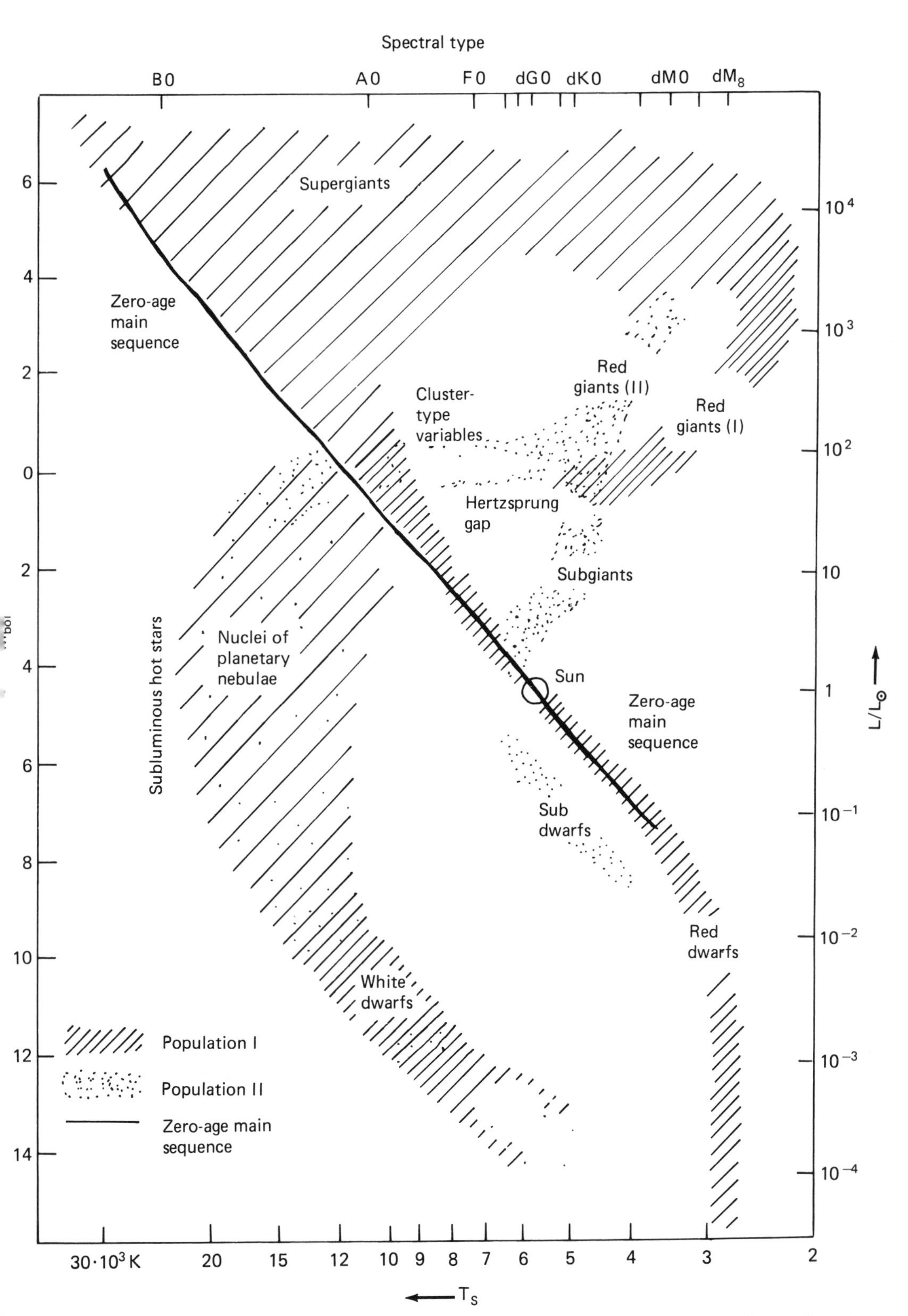

Composite Hertzsprung-Russell Diagram. Stars of different absolute luminosity, L - right axis, or bolometric absolute magnitude, M_{bol} - left axis, are plotted as a function of surface temperature, T_S - bottom axis, or spectral type - top axis. (Adapted from L. Goldberg and E.R. Dyer, Science in Space, eds. L.V. Berkner and H. Odishaw (1961).)

9.6 Stellar Mass, Radius, Density and Rotation

Mass and Radius - Main Sequence Stars

Stellar mass, M, and radius, R, in units of the Sun's values, $M_\odot$ and $R_\odot$, for the main-sequence stars (luminosity class LC = V) for different spectral types, Sp. Representative values of the surface gravity and mean density can be found in the following table under the V column. Schmidt-Kaler (1982).

Sp	M ($M_\odot$)	R ($R_\odot$)	Sp	M ($M_\odot$)	R ($R_\odot$)
O3	120	15	F0	1.6	1.5
O5	60	12	F5	1.3	1.3
O8	23	8.5	G0	1.05	1.1
O9	19	7.8	G5	0.92	0.92
B0	17.5	7.4	K0	0.79	0.85
B1	13	6.4	K5	0.67	0.72
B2	9.8	5.6	M0	0.51	0.60
B3	7.6	4.8	M3	0.33	0.45
B5	5.9	3.9	M5	0.21	0.27
B8	3.8	3.0	M7	0.12	0.18
A0	2.9	2.4	M8	0.06	0.1
A5	2.0	1.7			

Mass, Radius and Luminosity Class

Stellar mass, M, and radius, R, in units of the Sun's values, $M_\odot$ and $R_\odot$, for stars of different spectral type, Sp, and luminosity class, LC. Schmidt-Kaler (1982).

LC	V	III	I	V	III	I
Sp(MK)	M ($M_\odot$)			R ($R_\odot$)		
O3	120		140	15		
O5	60		70	12		30:
O6	37		40	10		25:
O8	23		28	8.5		20
B0	17.5	20	25	7.4	15	30
B3	7.6			4.8		
B5	5.9	7	20	3.9	8	50
B8	3.8			3.0		
A0	2.9	4	16	2.4	5	60
A5	2.0		13	1.7		60
F0	1.6		12	1.5		80
F5	1.4		10	1.3		100
G0	1.05	1.0	10	1.1	6	120
G5	0.92	1.1	12	0.92	10	150
K0	0.79	1.1	13	0.85	15	200
K5	0.67	1.2	13	0.72	25	400
M0	0.51	1.2	13	0.60	40	500
M2	0.40	1.3	19	0.50		800
M5	0.21		24	0.27		
M8	0.06			0.10		

Surface Gravity and Mean Density

Stellar surface gravity, g, and mean density, $\bar{\rho}$, in units of the Sun's values, $g_{\odot}$ and $\bar{\rho_{\odot}}$, for stars of different spectral type, Sp, and luminosity class, LC. Schmidt-Kaler (1982).

LC	V	III	I	V	III	I
Sp(MK)	$\log(g/g_{\odot})$			$\log(\rho/\rho_{\odot})$		
O3	- 0.3			- 1.5		
O5	- 0.4		- 1.1	- 1.5		- 2.6
O6	- 0.45		- 1.2	- 1.45		- 2.6
O8	- 0.5		- 1.2	- 1.4		- 2.5
B0	- 0.5	- 1.1	- 1.6	- 1.4	- 2.2	- 3.0
B3	- 0.5			- 1.15		
B5	- 0.4	- 0.95	- 2.0	- 1.00	- 1.8	- 3.8
B8	- 0.4			- 0.85		
A0	- 0.3		- 2.3	- 0.7	- 1.5	- 4.1
A5	- 0.15		- 2.4	- 0.4		- 4.2
F0	- 0.1		- 2.7	- 0.3		- 4.6
F5	- 0.1		- 3.0	- 0.2		- 5.0
G0	- 0.05	- 1.5	- 3.1	- 0.1	- 2.4	- 5.2
G5	+0.05	- 1.9	- 3.3	- 0.1	- 3.0	- 5.3
K0	+0.05	- 2.3	- 3.5	+0.1	- 3.5	- 5.8
K5	+0.1	- 2.7	- 4.1	+0.25	- 4.1	- 6.7
M0	+0.15	- 3.1	- 4.3	+0.35	- 4.7	- 7.0
M2	+0.2		- 4.5	+0.8		- 7.4
M5	+0.5			+1.0		
M8	+0.5			+1.2		

Rotational Velocities

Mean equatorial rotational velocities, $\bar{v}_{rot}$, determined from $v_{rot} \sin i$ assuming a random distribution with the mean $\sin i = \pi/4$. Rotational velocities less than 20 km s^{-1} are difficult to determine. Values of $\bar{v}_{rot}$ are given for stars of different spectral type, Sp, and luminosity class, LC. Schmidt-Kaler (1982). Main sequence stars are class V and giant stars are class III.

LC	V all	V normal	IV	III	II	Ib	Ia
Sp	$\bar{v}_{rot}$ [km s^{-1}]						
O8	210	200	180	145	140	130	120
9	215	200	175	130	135	125	105
B0	220	170	155	120	125	110	95
1	225	160	145	125	115	95	80
2	230	155	150	130	110	80	65
3	240	190	160	135	100	70	55
5	250	240	170	130	85	35	45
6	250	245	165	130	75	35	40
8	225	220	160	105	65	40	40
9	205	200	145	100	55	40	35
A0	185	180	125	100	50	45	35
1	180	160	130	115	45	50	30
2	175	160	160	140			
3	170	175	170	150			
5	160	170	175	155	45	45	<30
8	130	130	150	140			
F0	85	100	130	130	45	30	
2	70	70	90	110			
5	25	30	55	65	50	<20	
8	14		30	40			
G0	10		15	30	<20	<20	<30
2	<10		<15	25	<20	<20	<30
5	<10		<15	<20	<20	<20	<30
K0	<10		<15	<20	<20	<20	<30
2	<10		<15	<20	<20	<20	<30
5	<10		<15	<20	<20	<20	<30

9.7 Stellar Temperature and Luminosity

The Bolometric Correction

Population I

Luminosity Class		V	III	I
log T_{eff}	T_{eff} (°K)	BC (mag)	BC (mag)	BC (mag)
4.9	79000	− 6.15	− 6.15	− 6.15
4.8	63000	− 5.35	− 5.35	− 5.35
4.7	50000	− 4.60	− 4.60	− 4.60
4.6	40000	− 3.83	− 3.83	− 3.83
4.5	31600	− 3.23	− 3.09	− 3.09
4.4	25100	− 2.67	− 2.52	− 2.39
4.3	20000	− 2.16	− 1.97	− 1.77
4.2	15900	− 1.52	− 1.40	− 1.20
4.1	12600	− 0.92	− 0.85	− 0.82
4.0	10000	− 0.37	− 0.37	− 0.45
3.9	7940	− 0.13	− 0.13	− 0.00
3.8	6310	− 0.15	− 0.14	− 0.03
3.7	5010	− 0.38	− 0.36	− 0.27
3.6	3980	− 1.11	− 0.97	− 0.85
3.55	3550	− 1.97	− 1.82	− 1.45
3.5	3160	− 2.97	− 3.00	− 2.22
3.45	2820	− 3.71	− 3.71	− 3.38
3.4	2510	− 4.4	− 4.4	− 4.1
3.3	2000	− 6.1:	− 6.1:	− 6.1:
3.2	1600	− 8:	− 8:	− 8:

Main Sequence - Two Population

log T_{eff}	Pop I BC	Pop II BC
4.00	− 0.37	− 0.43
3.95	− 0.19	− 0.28
3.90	− 0.13	− 0.22
3.85	− 0.11	− 0.22
3.80	− 0.15	− 0.26
3.75	− 0.24	− 0.32
3.70	− 0.38	− 0.41
3.65	− 0.60	− 0.60

Effective Temperature, Bolometric Correction and Absolute Luminosity

Main Sequence Stars LC = V

Effective temperature, T_e, color index, $(CI)_o = (U-B)_o$, $(B-V)_o$ or $(R-I)_o$, absolute visual magnitude, M_V, bolometric correction, BC, absolute luminosity, L, in units of the solar value, $L_\odot$, for main sequence stars, or luminosity class LC = V. Schmidt-Kaler (1982).

Sp	log T_{eff}	T_{eff} (°K)	$(CI)_o$ (mag)	M_V (mag)	BC (mag)	M_{bol} (mag)	L ($L_\odot$)
			$(U-B)_o$				
O3	4.720	52500	- 1.22	- 6.0	- 4.75	- 10.7	1.4×10^6
4	4.680	48000	- 1.20	- 5.9	- 4.45	- 10.3	9.9×10^5
5	4.648	44500	- 1.19	- 5.7	- 4.40	- 10.1	7.9×10^5
6	4.613	41000	- 1.17	- 5.5	- 3.93	- 9.4	4.2×10^5
7	4.580	38000	- 1.15	- 5.2	- 3.68	- 8.9	2.6×10^5
8	4.555	35800	- 1.14	- 4.9	- 3.54	- 8.4	1.7×10^5
9	4.518	33000	- 1.12	- 4.5	- 3.33	- 7.8	9.7×10^4
B0	4.486	30000	- 1.08	- 4.0	- 3.16	- 7.1	5.2×10^4
1	4.405	25400	- 0.95	- 3.2	- 2.70	- 5.9	1.6×10^4
2	4.342	22000	- 0.84	- 2.4	- 2.35	- 4.7	5.7×10^3
3	4.271	18700	- 0.71	- 1.6	- 1.94	- 3.5	1.9×10^3
5	4.188	15400	- 0.58	- 1.2	- 1.46	- 2.7	8.3×10^2
6	4.146	14000	- 0.50	- 0.9	- 1.21	- 2.1	500
7	4.115	13000	- 0.43	- 0.6	- 1.02	- 1.6	320
8	4.077	11900	- 0.34	- 0.2	- 0.80	- 1.0	180
9	4.022	10500	- 0.20	+0.2	- 0.51	- 0.3	95
			$(B-V)_o$				
AO	3.978	9520	- 0.02	+0.6	- 0.30	+0.3	54
1	3.965	9230	+0.01	+1.0	- 0.23	+0.8	35
2	3.953	8970	+0.05	+1.3	- 0.20	+1.1	26
3	3.940	8720	+0.08	+1.5	- 0.17	+1.3	21

Effective Temperature, Bolometric Correction and Absolute Luminosity

Main Sequence Stars LC = V

Sp	log T_{eff}	T_{eff} (°K)	$(CI)_o$ (mag)	M_V (mag)	BC (mag)	M_{bol} (mag)	L ($L_\odot$)
			$(B-V)_o$				
A5	3.914	8200	+0.15	+1.9	− 0.15	+1.7	14
7	3.895	7850	+0.20	+2.2	− 0.12	+2.1	10.5
8	3.880	7580	+0.25	+2.4	− 0.10	+2.3	8.6
F0	3.857	7200	+0.30	+2.7	− 0.09	+2.6	6.5
2	3.838	6890	+0.35	+3.6	− 0.11	+3.5	2.9
5	3.809	6440	+0.44	+3.5	− 0.14	+3.4	3.2
8	3.792	6200	+0.52	+4.0	− 0.16	+3.8	2.1
G0	3.780	6030	+0.58	+4.4	− 0.18	+4.2	1.5
2	3.768	5860	+0.63	+4.7	− 0.20	+4.5	1.1
5	3.760	5770	+0.68	+5.1	− 0.21	+4.9	0.79
8	3.746	5570	+0.74	+5.5	− 0.40	+5.1	0.66
K0	3.720	5250	+0.81	+5.9	− 0.31	+5.6	0.42
1	3.706	5080	+0.86	+6.1	− 0.37	+5.7	0.37
2	3.690	4900	+0.91	+6.4	− 0.42	+6.0	0.29
3	3.675	4730	+0.96	+6.6	− 0.50	+6.1	0.26
4	3.662	4590	+1.05	+7.0	− 0.55	+6.4	0.19
5	3.638	4350	+1.15	+7.4	− 0.72	+6.7	0.15
7	3.609	4060	+1.33	+8.1	− 1.01	+7.1	0.10
			$(R-I)_o$				
M0	3.585	3850	+0.92	+8.8	− 1.38	+7.4	7.7×10^{-2}
1	3.570	3720	+1.03	+9.3	− 1.62	+7.7	6.1×10^{-2}
2	3.554	3580	+1.17	+9.9	− 1.89	+8.0	4.5×10^{-2}
3	3.540	3470	+1.30	+10.4	− 2.15	+8.2	3.6×10^{-2}
4	3.528	3370	+1.43	+11.3	− 2.38	+8.9	1.9×10^{-2}
5	3.510	3240	+1.61	+12.3	− 2.73	+9.6	1.1×10^{-2}
6	3.485	3050	+1.93	+13.5	− 3.21	+10.3	5.3×10^{-3}
7	3.468	2940	+2.1	+14.3	− 3.46	+10.8	3.4×10^{-3}
8	3.422	2640	+2.4	+16.0	− 4.1	+11.9	1.2×10^{-3}

Effective Temperature, Bolometric Correction and Absolute Luminosity

Giant Stars LC = III

Effective temperature, T_{eff}, color index, $(CI)_o = (U - B)_o$, $(B - V)_o$ or $(R - I)_o$, absolute visual magnitude, M_V, bolometric correction, BC, absolute luminosity, L, in units of the solar value, $L_\odot$, for giant stars, or luminosity class LC = III. Schmidt-Kaler (1982).

Sp	log T_{eff}	T_{eff} (°K)	$(CI)_o$ (mag)	M_V (mag)	BC (mag)	M_{bol} (mag)	L ($L_\odot$)
			$(U-B)_o$				
O3	4.698	50000	− 1.22	− 6.6	− 4.58	− 11.2	2.1×10^6
4	4.658	45500	− 1.20	− 6.5	− 4.28	− 10.8	1.5×10^6
5	4.628	42500	− 1.18	− 6.3	− 4.05	− 10.3	9.9×10^5
6	4.596	39500	− 1.17	− 6.1	− 3.80	− 9.9	6.5×10^5
7	4.568	37000	− 1.14	− 5.9	− 3.58	− 9.5	4.4×10^5
8	4.541	34700	− 1.13	− 5.8	− 3.39	− 9.2	3.4×10^5
9	4.505	32000	− 1.12	− 5.6	− 3.13	− 8.7	2.2×10^5
B0	4.463	29000	− 1.08	− 5.1	− 2.88	− 8.0	1.1×10^5
1	4.381	24000	− 0.97	− 4.4	− 2.43	− 6.8	3.9×10^4
2	4.308	20300	− 0.91	− 3.9	− 2.02	− 5.9	1.7×10^4
3	4.234	17100	− 0.74	− 3.0	− 1.60	− 4.6	5.0×10^3
5	4.177	15000	− 0.58	− 2.2	− 1.30	− 3.5	1.8×10^3
6	4.150	14100	− 0.51	− 1.8	− 1.13	− 2.9	1.1×10^3
7	4.120	13200	− 0.44	− 1.5	− 0.97	− 2.5	700
8	4.095	12400	− 0.37	− 1.2	− 0.82	− 2.0	460
9	4.042	11000	− 0.20	− 0.6	− 0.71	− 1.3	240
			$(B-V)_o$				
A0	4.005	10100	− 0.03	+0.0	− 0.42	− 0.4	106
1	3.977	9480	+0.01	+0.2	− 0.29	− 0.1	78
2	3.954	9000	+0.05	+0.3	− 0.20	+0.1	65
3	3.935	8600	+0.08	+0.5	− 0.17	+0.3	53

Effective Temperature, Bolometric Correction and Absolute Luminosity

Giant Stars LC = III

Sp	log T_{eff}	T_{eff} (°K)	$(CI)_o$ (mag)	M_V (mag)	BC (mag)	M_{bol} (mag)	L ($L_\odot$)
			$(B-V)_o$				
A5	3.908	8100	+0.15	+0.7	- 0.14	+0.6	43
7	3.884	7650	+0.22	+1.1	- 0.10	+1.0	29
8	3.873	7450	+0.25	+1.2	- 0.10	+1.1	26
F0	3.854	7150	+0.30	+1.5	- 0.11	+1.4	20
2	3.837	6870	+0.35	+1.7	- 0.11	+1.6	17
5	3.811	6470	+0.43	+1.6	- 0.14	+1.6	17
8	3.789	6150	+0.54		- 0.16		
G0	3.767	5850	+0.65	+1.0	- 0.20	+0.8	34
2	3.737	5450	+0.77	+0.9	- 0.27	+0.6	40
5	3.712	5150	+0.86	+0.9	- 0.34	+0.6	43
8	3.690	4900	+0.94	+0.8	- 0.42	+0.4	51
K0	3.676	4750	+1.00	+0.7	- 0.50	+0.2	60
1	3.663	4600	+1.07	+0.6	- 0.55	+0.1	69
2	3.646	4420	+1.16	+0.5	- 0.61	- 0.1	79
3	3.623	4200	+1.27	+0.3	- 0.76	- 0.5	110
4	3.602	4000	+1.38	+0.0	- 0.94	- 0.9	170
5	3.596	3950	+1.50	- 0.2	- 1.02	- 1.2	220
7	3.586	3850	+1.53	- 0.3	- 1.17	- 1.5	280
			$(R-I)_o$				
M0	3.580	3800	+0.90	- 0.4	- 1.25	- 1.6	330
1	3.570	3720	+0.96	- 0.5	- 1.44	- 1.9	430
2	3.559	3620	+1.08	- 0.6	- 1.62	- 2.2	550
3	3.548	3530	+1.30	- 0.6	- 1.87	- 2.5	700
4	3.535	3430	+1.60	- 0.5	- 2.22	- 2.7	880
5	3.522	3330	+1.91	- 0.3	- 2.48	- 2.8	930
6	3.510	3240	+2.20	- 0.2	- 2.73	- 2.9	1070

Effective Temperature, Bolometric Correction and Absolute Luminosity

Supergiant Stars LC = I

Effective temperature, T_{eff}, color index, $(CI)_o = (U-B)_o$, $(B-V)_o$ or $(R-I)_o$, absolute visual magnitude, M_V, bolometric correction, BC, absolute luminosity, L, in units of the solar value, $L_\odot$, for supergiant stars, or luminosity class approximately LC ≈ Iab. Schmidt-Kaler (1982).

Sp	log T_{eff}	T_{eff} (°K)	$(CI)_o$ (mag)	M_V (mag)	BC (mag)	M_{bol} (mag)	L ($L_\odot$)
			$(U-B)_o$				
O3	4.675	47300	- 1.21	- 6.8:	- 4.41	- 11.2:	2.2×10^6
4	4.644	44100	- 1.19	- 6.7:	- 4.17	- 10.9:	1.6×10^6
5	4.605	40300	- 1.17	- 6.6	- 3.87	- 10.5	1.1×10^6
6	4.591	39000	- 1.16	- 6.5	- 3.74	- 10.2	9.0×10^5
7	4.553	35700	- 1.14	- 6.5	- 3.48	- 10.0	7.1×10^5
8	4.535	34200	- 1.13	- 6.5	- 3.35	- 9.8	6.2×10^5
9	4.513	32600	- 1.13	- 6.5	- 3.18	- 9.7	5.3×10^5
B0	4.415	26000	- 1.06	- 6.4	- 2.49	- 8.9	2.6×10^5
1	4.318	20800	- 1.00	- 6.4	- 1.87	- 8.3	1.5×10^5
2	4.267	18500	- 0.94	- 6.4	- 1.58	- 8.0	1.1×10^5
3	4.209	16200	- 0.83	- 6.3	- 1.26	- 7.6	7.6×10^4
5	4.133	13600	- 0.72	- 6.2	- 0.95	- 7.2	5.2×10^4
6	4.114	13000	- 0.69	- 6.2	- 0.88	- 7.1	4.9×10^4
7	4.085	12200	- 0.64	- 6.2	- 0.78	- 7.0	4.4×10^4
8	4.048	11200	- 0.56	- 6.2	- 0.66	- 6.9	4.0×10^4
9	4.012	10300	- 0.50	- 6.2	- 0.52	- 6.7	3.5×10^4
			$(U-B)_o$				
A0	3.988	9730	- 0.38	- 6.3	- 0.41	- 6.7	3.5×10^4
1	3.965	9230	- 0.29	- 6.4	- 0.32	- 6.7	3.5×10^4
2	3.958	9080	- 0.25	- 6.5	- 0.28	- 6.7	3.6×10^4
3	3.943	8770	- 0.14	- 6.5	- 0.21	- 6.7	3.5×10^4

Effective Temperature, Bolometric Correction and Absolute Luminosity

Supergiant Stars LC = I

Sp	log T_{eff}	T_{eff} (°K)	$(CI)_o$ (mag)	M_V (mag)	BC (mag)	M_{bol} (mag)	L ($L_\odot$)
			$(U-B)_o$				
A5	3.930	8510	−0.07	−6.6	−0.13	−6.7	3.5×10^4
7	3.911	8150	+0.00	−6.6	−0.06	−6.7	3.3×10^4
8	3.900	7950	+0.11	−6.6	−0.03	−6.6	3.2×10^4
			$(B-V)_o$				
F0	3.886	7700	+0.17	−6.6	−0.01	−6.6	3.2×10^4
2	3.866	7350	+0.23	−6.6	−0.00	−6.6	3.1×10^4
5	3.839	6900	+0.32	−6.6	−0.03	−6.6	3.2×10^4
8	3.785	6100	+0.56	−6.5	−0.09	−6.6	3.1×10^4
G0	3.744	5550	+0.76	−6.4	−0.15	−6.6	3.0×10^4
2	3.716	5200	+0.87	−6.3	−0.21	−6.5	2.9×10^4
5	3.686	4850	+1.02	−6.2	−0.33	−6.5	2.9×10^4
8	3.663	4600	+1.15	−6.1	−0.42	−6.5	2.9×10^4
K0	3.645	4420	+1.24	−6.0	−0.50	−6.5	2.9×10^4
1	3.636	4330	+1.30	−6.0	−0.56	−6.6	3.0×10^4
2	3.628	4250	+1.35	−5.9	−0.61	−6.5	2.9×10^4
3	3.611	4080	+1.46	−5.9	−0.75	−6.6	3.3×10^4
4	3.597	3950	+1.53	−5.8	−0.90	−6.7	3.4×10^4
5	3.585	3850	+1.60	−5.8	−1.01	−6.8	3.8×10^4
7	3.568	3700	+1.63	−5.7	−1.20	−6.9	4.1×10^4
			$(R-I)_o$				
M0	3.562	3650	+0.96	−5.6	−1.29	−6.9	4.1×10^4
1	3.550	3550	+1.04	−5.6	−1.38	−7.0	4.4×10^4
2	3.538	3450	+1.15	−5.6	−1.62	−7.2	5.5×10^4
3	3.505	3200	+1.37	−5.6	−2.13	−7.7	5.6×10^4
4	3.474	2980	+1.59	−5.6	−2.75	−8.3	1.6×10^5
5	3.446	2800	+1.80	−5.6	−3.47	−9.1	3.0×10^5
6	3.415:	2600:	+2.02:	−5.6	−3.90	−9.5	4.5×10^5

Effective Temperature, Bolometric Correction and Absolute Luminosity

Wolf-Rayet Stars

Type	T_{eff} (°K)	BC (mag)	M_{bol} (mag)
WN3	50000	− 4.4	− 8.8
4	47000	− 4.0	− 7.8
5	43000	− 3.7	− 7.7
6	39000	− 3.6	− 8.0
7	32000	− 3.0	− 9.7
8	29000	− 2.8	− 8.4
WC6	60000	− 4.5	− 8.3
7	54000	− 3.9	− 8.3
8	46000	− 3.9	− 8.7
9	38000	− 3.1	− 8.1

9.8 Absolute Visual Magnitudes of the Stars

Absolute Visual Magnitudes in the MK System, Population I

LC	V	IV	III	II	Ib	Iab	Ia	Ia-0
Sp	Absolute Visual Magnitude, M_V							
O3	-6.0						-6.8	
4	-5.9	-6.1	-6.5				-6.8	
5	-5.7	-6.0	-6.3				-6.8	
6	-5.5	-5.8	-6.1			-6.5	-6.8	
7	-5.2	-5.5	-5.9	-6.0	-6.3	-6.5	-6.8	
8	-4.9	-5.4	-5.8	-6.0	-6.2	-6.5	-6.8	
9	-4.5	-5.2	-5.6	-5.9	-6.2	-6.5	-6.8	
B0	-4.0	-4.7	-5.1	-5.7	-6.1	-6.4	-6.9	-8.2
1	-3.2	-3.8	-4.4	-5.4	-5.8	-6.4	-6.9	-8.3
2	-2.45	-3.1	-3.9	-4.8	-5.7	-6.4	-6.9	-8.3
3	-1.6	-2.4	-3.0	-4.5	-5.5	-6.3	-7.0	-8.3
5	-1.2	-1.7	-2.2	-4.0	-5.4	-6.2	-7.0	-8.4
7	-0.6	-1.1	-1.5	-3.5	-5.3	-6.2	-7.1	-8.4
8	-0.25	-0.7	-1.2	-3.1	-5.2	-6.2	-7.1	-8.5
9	+0.2	-0.2	-0.6		-5.2	-6.2	-7.1	-8.5
A0	+0.65	+0.3	+0.0	-3.0	-5.2	-6.3	-7.1	-8.5
1	+1.0	+0.7	+0.2	-3.0	-5.2	-6.4	-7.2	-8.5
2	+1.3	+1.0	+0.3	-2.9	-5.2	-6.5	-7.2	-8.6
3	+1.5	+1.2	+0.5	-2.8	-5.2	-6.5	-7.2	-8.7
5	+1.95	+1.3	+0.7	-2.8	-5.1	-6.6	-7.4	-8.8
7	+2.2	+1.7	+1.1	-2.7	-5.1		-7.7	-8.9
8	+2.4	+2.0	+1.2	-2.6	-5.1		-7.8	-8.9

Absolute Visual Magnitudes in the MK System, Population I

LC	V	IV	III	II	Ib	Iab	Ia	Ia- 0
Sp	Absolute Visual Magnitude, M_V							
F0	+2.7	+2.2	+1.5	- 2.5	- 5.1	- 6.6	- 8.0	- 9.0
2	+3.6	+2.4	+1.7	- 2.4	- 5.1	- 6.6	- 8.0	- 9.0
5	+3.5	+2.5	+1.6	- 2.3	- 5.1	- 6.6	- 8.0	- 9.0
8	+4.0	+2.8		- 2.3	- 5.1	- 6.5	- 8.0	- 9.0
G0	+4.4	+3.0	+1.0	- 2.3	- 5.0	- 6.4	- 8.0	- 8.9
2	+4.7	+3.0	+0.9	- 2.3	- 5.0	- 6.3	- 8.0	- 8.8
5	+5.1	+3.1	+0.9	- 2.3	- 4.6	- 6.2	- 7.9	- 8.6
8	+5.5	+3.1	+0.8	- 2.3	- 4.4	- 6.1	- 7.8	- 8.5
K0	+5.9	+3.1	+0.7	- 2.3	- 4.3	- 6.0	- 7.7	- 8.5
1	+6.15	+3.1	+0.6	- 2.3	- 4.3	- 6.0	- 7.6	
2	+6.4		+0.5	- 2.3	- 4.3	- 5.9	- 7.6	
3	+6.65		+0.3	- 2.3	- 4.3	- 5.9	- 7.5	
4	+7.0		+0.0	- 2.3	- 4.3	- 5.8	- 7.5	
5	+7.35		- 0.2	- 2.3	- 4.4	- 5.8	- 7.5	
7	+8.1		- 0.3	- 2.3	- 4.4	- 5.7	- 7.4	
M0	+8.8		- 0.4	- 2.5	- 4.5	- 5.6	- 7.0	- 8.0
1	+9.3		- 0.5	- 2.5	- 4.6	- 5.6	- 7.0	- 8.0
2	+9.9		- 0.6	- 2.6	- 4.7	- 5.6	- 6.9	- 8.0
3	+10.4		- 0.6	- 2.6	- 4.8	- 5.6	- 6.9	- 8.0
4	+11.3		- 0.5	- 2.6	- 4.8	- 5.6	- 6.8	- 8.0
5	+12.3		- 0.3		- 4.8	- 5.6	- 6.8	

Absolute Visual Magnitudes, M_V, and Intrinsic Colors, $(B-V)_o$

Different Luminosity Classes, LC (Schmidt - Kaler, 1982).
V_o = Zero – Age Main Sequence (ZAMS)

LC	Ia	Ib	II	III	IV	V	V_o	VI	VII
$(B-V)_o$	Absolute Visual Magnitude, M_V								
- 0.3	- 6.8	- 6.2	- 5.8	- 5.3	- 4.9	- 4.0	- 3.25		+ 9.6
- 0.2	- 6.9	- 5.8	- 4.5	- 2.8	- 2.3	- 1.5	- 1.1		+10.2
- 0.1	- 7.0	- 5.4	- 3.1	- 1.0	- 0.5	- 0.1	+ 0.6		+10.6
0.0	- 7.1	- 5.2	- 3.0	+0.2	+0.6	+ 0.9	+ 1.5		+11.4
+0.1	- 7.5	- 5.1	- 2.8	+0.5	+1.2	+ 1.4	+ 1.9		+11.9
+0.2	- 8.0	- 5.1	- 2.6	+1.0	+1.7	+ 2.2	+ 2.4		+12.8
+0.3	- 8.0	- 5.1	- 2.4	+1.5	+2.2	+ 2.7	+ 2.8		+13.2
+0.4	- 8.0	- 5.1	- 2.3	+1.6	+2.5	+ 3.3	+ 3.4	+4.0	+13.8
+0.5	- 8.0	- 5.1	- 2.3		+2.7	+ 3.9	+ 4.1	+5.0	+14.1
+0.6	- 8.0	- 5.1	- 2.3		+3.0	+ 4.3	+ 4.7	+5.7	+14.5
+0.7	- 8.0	- 5.0	- 2.3	+1.0	+3.1	+ 5.2	+ 5.2	+6.4	+14.8
+0.8	- 8.0	- 5.0	- 2.3	+0.9	+3.1	+ 5.8	+ 5.8	+6.9	+15.1
+0.9	- 8.0	- 4.9	- 2.3	+0.8	+3.1	+ 6.3	+ 6.3	+7.4	+15.3
+1.0	- 7.8	- 4.6	- 2.3	+0.7	+3.1	+ 6.7	+ 6.7	+7.9	+15.7
+1.1	- 7.8	- 4.5	- 2.3	+0.5		+ 7.2	+ 7.2		
+1.2	- 7.8	- 4.3	- 2.3	+0.5		+ 7.5	+ 7.5		
+1.3	- 7.6	- 4.2	- 2.3	+0.3		+ 7.9	+ 8.0		
+1.4	- 7.6	- 4.3	- 2.3	0.0		+ 8.8	+ 8.8		
+1.5	- 7.5	- 4.3	- 2.3	- 0.2		+10.3	+10.3		
+1.6	- 7.5	- 4.4	- 2.6	- 0.6		+11.9	+12.0		
+1.7	- 7.0	- 4.8	- 2.6			+13.2	+13.2		
+1.8	- 6.8	- 4.8				+14.2	+14.2		
+1.9	- 6.8	- 4.8				+15.5:	+15.5:		
+2.0						+16.7:	+16.7:		

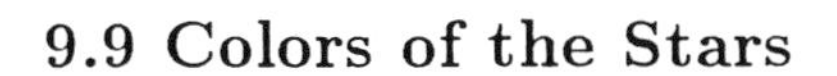

9.9 Colors of the Stars

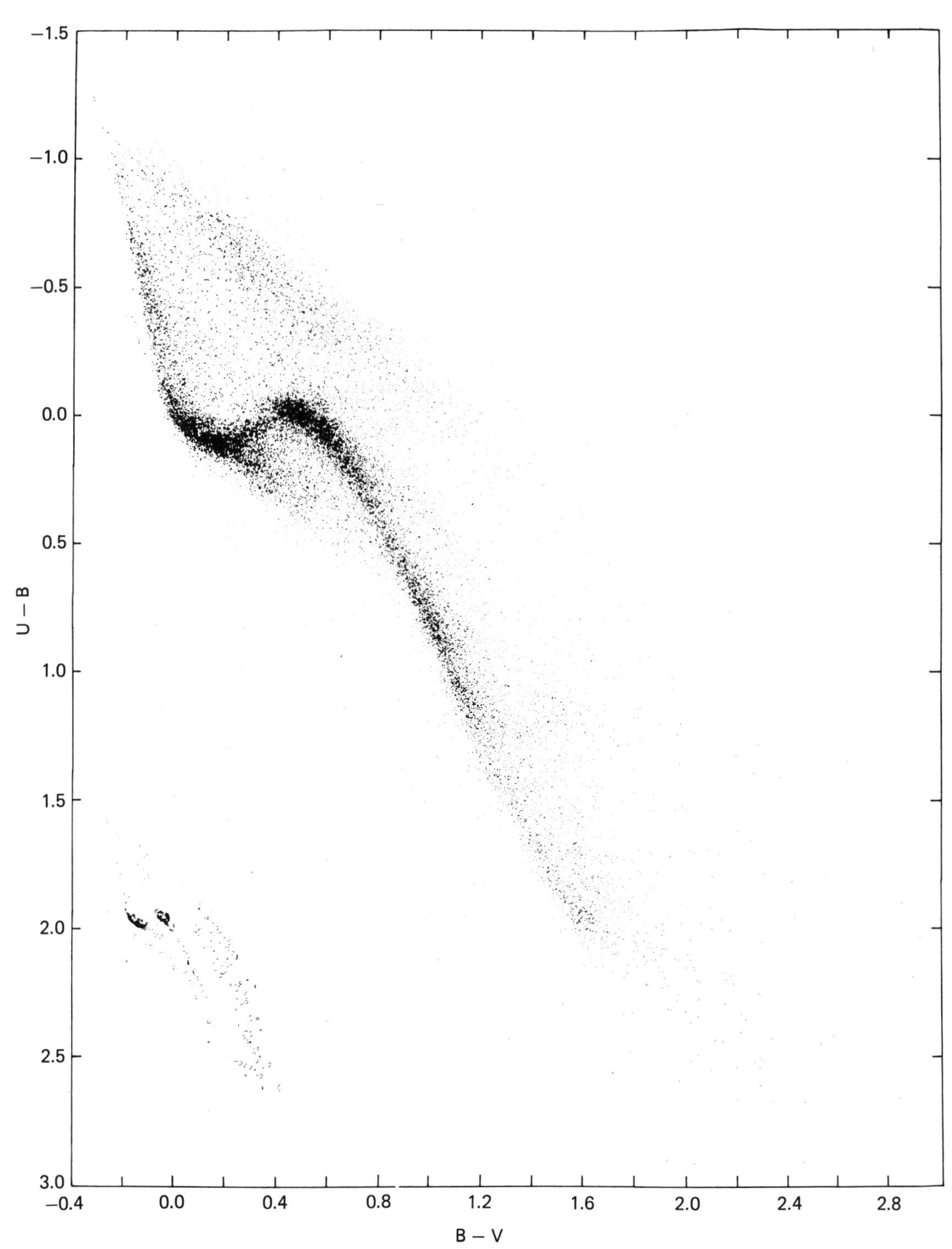

Colors of the Stars. The two-color (B-V, U-B) diagram for over 29,000 stars. (Courtesy of the Geneva Observatory).

Intrinsic Colors $(B-V)_o$ for Different Absolute Magnitudes, M_V

LC	Population I Ib	III	V	Population II (Globular Clusters) Blue branch	Red branch
M_V	$(B-V)_o$			$(B-V)_o$	
− 6	− 0.23	− 0.32	− 0.34		
− 5	+0.8	− 0.29	− 0.32		
− 4		− 0.24	− 0.30		
− 3		− 0.20	− 0.26	+1.6	+1.6
− 2		− 0.16	− 0.22	+1.2	+1.3
− 1		− 0.13	− 0.15	+0.9	+1.00
0		− 0.03	− 0.09	+0.55	+0.83
+1		+0.20	+0.01	− 0.05	+0.75
+2			+0.15	− 0.2	+0.65
+3			+0.35		+0.55
+4			+0.52		+0.45
+5			+0.67		+0.5
+6			+0.84		+0.7
+7			+1.05		

Intrinsic Colors Zero–Age Main Sequence, V_o

Locus of stars just starting hydrogen burning; distance modulus of the Hyades open cluster m_o–M = 3.28 (Schmidt–Kaler, 1982)

$(B-V)_o$	$(U-B)_o$	M_V	$Sp(V_o)$	$(B-V)_o$	$(U-B)_o$	M_V	$Sp(V_o)$
− 0.33	− 1.20	− 5.2	O4 Vo	+0.40	− 0.01	+ 3.4	F4 Vo
− 0.305	− 1.10	− 3.6	O9.5 Vo	+0.50	0.00	+ 4.1	F8 Vo
− 0.30	− 1.08	− 3.25	B0 Vo	+0.60	+0.08	+ 4.7	G0 Vo
− 0.28	− 1.00	− 2.6	B0.5 Vo	+0.70	+0.23	+ 5.2	G6 Vo
− 0.25	− 0.90	− 2.1	B1.5 Vo	+0.80	+0.42	+ 5.8	K0 Vo
− 0.22	− 0.80	− 1.5	B2.5 Vo	+0.90	+0.63	+ 6.3	K2 Vo
− 0.20	− 0.69	− 1.1	B3 Vo	+1.00	+0.86	+ 6.7	K3.5 Vo
− 0.15	− 0.50	− 0.2	B6 Vo	+1.10	+1.03	+ 7.1	K4.5 Vo
− 0.10	− 0.30	+0.6	B8 Vo	+1.20	+1.13	+ 7.5	K5.5 Vo
− 0.05	− 0.10	+1.1	B9.5 Vo	+1.30	+1.20	+ 8.0	K6.5 Vo
0.00	+0.01	+1.5	A0.5 Vo	+1.40	+1.22	+ 8.8	M0 Vo
+0.05	+0.05	+1.7	A2 Vo	+1.50	+1.17	+10.3	M2 Vo
+0.10	+0.08	+1.9	A4 Vo	+1.60	+1.20	+12.0	M4.5 Vo
+0.15	+0.09	+2.1	A5 Vo	+1.70	+1.32	+13.2	M5.5 Vo
+0.20	+0.10	+2.4	A7 Vo	+1.80	+1.43	+14.2	M7 Vo
+0.25	+0.07	+2.55	A8 Vo	+1.90	+1.53:	+15.5:	
+0.30	+0.03	+2.8	F0 Vo	+2.00	+1.64:	+16.7:	
+0.35	0.00	+3.1	F2 Vo				

Intrinsic Colors in the MK System, Population I

LC	V	III	II	Ib	Iab	Ia	w(VII)
Sp	Intrinsic Color, $(B-V)_o$						
O5	− 0.33	− 0.32	− 0.32	− 0.32	− 0.31	− 0.31	
6	− 0.33	− 0.32	− 0.32	− 0.32	− 0.31	− 0.31	
7	− 0.32	− 0.32	− 0.32	− 0.31	− 0.31	− 0.31	
8	− 0.32	− 0.31	− 0.31	− 0.29	− 0.29	− 0.29	
9	− 0.31	− 0.31	− 0.31	− 0.28	− 0.27	− 0.27	
B0	− 0.30	− 0.295	− 0.29	− 0.24	− 0.23	− 0.23	− 0.28
1	− 0.265	− 0.265	− 0.26	− 0.20	− 0.19	− 0.19	
2	− 0.24	− 0.24	− 0.23	− 0.18	− 0.17	− 0.16	
3	− 0.205	− 0.205	− 0.20	− 0.14	− 0.13	− 0.12	− 0.22
5	− 0.17	− 0.17	− 0.16	− 0.10	− 0.10	− 0.08	
6	− 0.15	− 0.15	− 0.14	− 0.08	− 0.08	− 0.06	
7	− 0.135	− 0.13	− 0.12	− 0.06	− 0.05	− 0.04	
8	− 0.11	− 0.11	− 0.10	− 0.04	− 0.03	− 0.02	− 0.15
9	− 0.075	− 0.075	− 0.07	− 0.03	− 0.02	0.00	− 0.06
A0	− 0.02	− 0.03	− 0.03	− 0.01	− 0.01	+0.02	− 0.00
1	+0.01	+0.01	+0.01	+0.03	+0.02	+0.02	+0.03
2	+0.05	+0.05	+0.03	+0.04	+0.03	+0.03	+0.07
3	+0.08	+0.08	+0.07	+0.06	+0.06	+0.05	+0.10
5	+0.15	+0.15	+0.11	+0.09	+0.09	+0.09	+0.16
7	+0.20	+0.22	+0.16	+0.13	+0.12	+0.12	+0.22
8	+0.25	+0.25	+0.18	+0.14	+0.14	+0.14	

Intrinsic Colors in the MK System, Population I

LC	V	III	II	Ib	Iab	Ia	w(VII)
Sp	Intrinsic Color, $(B-V)_o$						
F0	+0.30	+0.30	+0.25	+0.19	+0.17	+0.17	+0.29
2	+0.35	+0.35	+0.30	+0.23	+0.23	+0.22	+0.35
5	+0.44	+0.43	+0.38	+0.33	+0.32	+0.31	+0.42
8	+0.52	+0.54	+0.58	+0.56	+0.56	+0.56	+0.50
G0	+0.58	+0.65	+0.71	+0.76	+0.76	+0.75	+0.56
2	+0.63	+0.77	+0.81	+0.86	+0.87	+0.87	+0.60
5	+0.68	+0.86	+0.89	+1.00	+1.02	+1.03	+0.68
8	+0.74	+0.94	+0.99	+1.14	+1.14	+1.17	+0.73
K0	+0.81	+1.00	+1.08	+1.20	+1.25	+1.25	+0.81
1	+0.86	+1.07	+1.14	+1.24	+1.32	+1.32	+0.85
2	+0.91	+1.16	+1.29	+1.33	+1.36	+1.36	+0.89
3	+0.96	+1.27	+1.40	+1.46	+1.46	+1.46	+0.96
5	+1.15	+1.50	+1.49	+1.59	+1.60	+1.60	
7	+1.33	+1.53	+1.57	+1.61	+1.63	+1.63	
M0	+1.40	+1.56	+1.58	+1.64	+1.67	+1.67	
1	+1.46	+1.58	+1.59	+1.68	+1.69	+1.69	
2	+1.49	+1.60	+1.59	+1.68	+1.71	+1.71	
3	+1.51	+1.61	+1.60	+1.69	+1.69	+1.69	
4	+1.54	+1.62		+1.75	+1.76	+1.76	
5	+1.64	+1.63			+1.80		
6	+1.73	+1.52					
7	+1.80:	+1.50					
8	+1.93:	+1.50					

Intrinsic Colors in the MK System, Population I

LC	V	III	II	Ib	Iab	Ia	w(DA)
Sp	Intrinsic Color, $(U-B)_o$						
O5	− 1.19	− 1.18	− 1.17	− 1.17	− 1.17	− 1.17	
6	− 1.17	− 1.17	− 1.16	− 1.16	− 1.16	− 1.16	
7	− 1.15	− 1.14	− 1.14	− 1.14	− 1.14	− 1.14	
8	− 1.14	− 1.13	− 1.13	− 1.13	− 1.13	− 1.13	
9	− 1.12	− 1.12	− 1.12	− 1.13	− 1.13	− 1.13	
B0	− 1.08	− 1.08	− 1.08	− 1.07	− 1.06	− 1.05	
1	− 0.95	− 0.97	− 1.00	− 0.99	− 1.00	− 1.00	
2	− 0.84	− 0.91	− 0.92	− 0.92	− 0.93	− 0.96	
3	− 0.71	− 0.74	− 0.82	− 0.82	− 0.83	− 0.85	
5	− 0.58	− 0.58	− 0.69	− 0.70	− 0.72	− 0.76	
6	− 0.50	− 0.51	− 0.62	− 0.67	− 0.69	− 0.72	
7	− 0.43	− 0.44	− 0.54	− 0.60	− 0.63	− 0.68	
8	− 0.34	− 0.37	− 0.44	− 0.53	− 0.55	− 0.62	− 1.02
9	− 0.20	− 0.20	− 0.32	− 0.46	− 0.49	− 0.58	− 0.90
A0	− 0.02	− 0.07	− 0.20	− 0.33	− 0.38	− 0.44	− 0.79
1	+0.02	+0.07	− 0.12:	− 0.24	− 0.29	− 0.37	− 0.70
2	+0.05	+0.06	− 0.05:	− 0.20	− 0.25	− 0.30	− 0.63
3	+0.08	+0.10	+0.02:	− 0.10:	− 0.14	− 0.20:	− 0.60
5	+0.10	+0.11	+0.08	0.00:	− 0.08	− 0.10:	− 0.58
7	+0.10	+0.11	+0.10	+0.09:	0.00	+0.09:	− 0.58
8	+0.09	+0.10	+0.11	+0.11:	+0.11:	+0.12:	

Intrinsic Colors in the MK System, Population I

LC	V	III	II	Ib	Iab	Ia	w(DA)
Sp			Intrinsic Color, $(U-B)_o$				
F0	+0.03	+0.08	+0.12:	+0.15	+0.15	+0.15	− 0.56
2	+0.00	+0.08	+0.14:	+0.19	+0.18	+0.18	− 0.48
5	+0.02	+0.09	+0.16:	+0.27	+0.27	+0.27	
8	+0.02	+0.10	+0.24	+0.40	+0.41	+0.43	− 0.30
G0	+0.06	+0.21	+0.32	+0.50	+0.52	+0.52	− 0.20
2	+0.12	+0.39	+0.42	+0.62	+0.63	+0.65	− 0.12
5	+0.20	+0.56	+0.60	+0.81	+0.83	+0.82	
8	+0.30	+0.70	+0.78	+1.05	+1.07	+1.08	
K0	+0.45	+0.84	+0.95	+1.15	+1.17	+1.18	
1	+0.54	+1.01	+1.07	+1.21	+1.28	+1.28	
2	+0.64	+1.16	+1.33	+1.35	+1.32	+1.32	
3	+0.80	+1.39	+1.58	+1.56	+1.60	+1.60	(+0.37)
5	+0.98	+1.81	+1.74	+1.79	+1.80	+1.80	
7	+1.21	+1.83	+1.79	+1.84:	+1.84:	+1.84:	
M0	+1.22	+1.87	+1.91	+1.90	+1.90	+1.90	
1	+1.21	+1.88	+1.93	+1.92	+1.90	+1.90	
2	+1.18	+1.89	+1.94	+1.95	+1.95	+1.95	
3	+1.16	+1.88	+1.77	+1.98	+1.95	+1.95	
4	+1.15	+1.73		+2.12	+2.00:	+2.05:	
5	+1.24	+1.58		+1.60:	+1.60:	+1.60:	
6	+1.32:	+1.16					
7	+1.40:						
8	+1.53:						

10
Bright and Nearby Stars

10.1 Named Stars and Constellations

Data for named stars have been extracted from Hoffleit's (1982) catalogue of bright stars, (BS), including the right ascension, R.A., and declination, Dec., for epoch 2000.0, the BS number, the apparent visual magnitude, V, and spectral class, Sp.

The International Astronomical Union has legislated the boundaries of the constellations (Delaporte, 1930), from which the constellation areas in square degrees have been determined (Levin, 1935). These areas are given in the list of constellations together with the approximate position, names and three-letter abbreviations.

The ancients divided the ecliptic into only twelve constellations, the houses of the Zodiac. The well known rhyme by Isaac Watts enables us to remember them and the order in which they occur.

The Ram, the Bull, the Heavenly Twins,
And next the Crab, the Lion shines,
The Virgin and the Scales;
The Scorpion, Archer, and the Goat,
The Man that pours the Water out,
The Fish with glittering scales.

Named Stars

Name	R.A.(2000) h m s	Dec.(2000) ° ′	BS No	V	Sp	Other Designation
Acamar	02 58 15	− 40 18	897	3.42	A3V	θ Eridani
Achernar	01 37 42	− 57 15	472	0.47	B5IV	α Eridani
Acrux	12 26 36	− 63 06	4730	1.58	B1IV	α Crucis
Adara	06 58 38	− 28 58	2618	1.50	B2II	ϵ Canis Majoris
Al Na'ir	22 08 14	− 46 58	8425	1.73	B5V	α Gruis
Albireo	19 30 43	+27 58	7417	3.24	K5II?+B?	β Cygni
Alcor	13 25 13	+55 00	5062	4.01	A5V	80 Ursae Majoris
Alcyone	03 47 29	+24 07	1165	2.86	B7III	η Tauri
Aldebaran	04 35 55	+16 30	1457	0.86	K5III	α Tauri
Alderamin	21 18 35	+62 35	8162	2.41	A7IV-V	α Cephei
Alfard	09 27 35	− 08 40	3748	1.99	K4III	α Hydrae
Algeiba	10 19 59	+19 51	4057	2.61	KOIII	γ Leonis
Algenib	00 13 14	+15 11	39	2.83	B2IV	γ Pegasi
Algol	03 08 11	+40 57	936	2.2	B8V	β Persei
Alioth	12 54 02	+55 57	4905	1.76	AOpv	ϵ Ursae Majoris
Alkaid	13 47 32	+49 19	5191	1.86	B3V	η Ursae Majoris
Almach	02 03 53	+42 20	603	2.28	K3II	γ Andromedae
Alnilam	05 36 12	− 01 12	1903	1.70	BOIa	ϵ Orionis
Alphard	09 27 35	− 08 40	3748	1.99	K4III	α Hydrae
Alphecca	15 34 41	+26 43	5793	2.23	A0V	α Coronae Borealis
Alpheratz	00 08 23	+29 05	15	2.02	B9p	α Andromedae
Alruccabah	02 31 13	+89 15	424	2.5	F8Ib	α Ursae Minoris
Altair	19 50 47	+08 52	7557	0.77	A7V	α Aquilae
Ankaa	00 26 17	− 42 18	99	2.39	KOIII	α Phoenicis
Antares	16 29 25	− 26 26	6134	1.08	M1Ib	α Scorpii
Arcturus	14 15 40	+19 11	5340	0.06	K2IIIp	α Boötis
Atria	16 48 40	− 69 02	6217	1.91	K4III	α Trianguli Australis
Avior	08 22 31	− 59 30	3307	1.85	KOII+B	ϵ Carinae
Bellatrix	05 25 08	+06 21	1790	1.64	B2III	γ Orionis
Betelgeuse	05 55 10	+07 24	2061	0.80	M2Iab	α Orionis

Named Stars

Name	R.A.(2000) h m s	Dec.(2000) ° ′	BS No	V	Sp	Other Designation
Canopus	06 23 57	- 52 41	2326	- 0.73	F0Ib	α Carinae
Capella	05 16 41	+46 00	1708	0.09	G8III	α Aurigae
Caph	00 09 10	+59 09	21	2.25	F2IV	β Cassiopeiae
Castor	07 34 36	+31 53	2891	1.99	A1V	α Geminorum
Cor Caroli	12 56 02	+38 19	4915	2.89	B9.5pv	α Canum Venaticorum
Cynosura	02 31 13	+89 15	424	2.5	F8Ib	α Ursae Minoris
Deneb	20 41 26	+45 16	7924	1.26	A2Ia	α Cygni
Deneb Kaitos or Diphda	00 43 35	- 17 59	188	2.04	K1III	β Ceti
Denebola	11 49 04	+14 34	4534	2.14	A3V	β Leonis
Dubhe	11 03 44	+61 45	4301	1.79	K0II-III	α Ursae Majoris
Elnath	05 26 17	+28 36	1791	1.65	B7III	β Tauri
Eltanin	17 56 36	+51 29	6705	2.22	K5III	γ Draconis
Enif	21 44 11	+09 53	8308	2.42	K2Ib	ε Pegasi
Fomalhaut	22 57 39	- 29 37	8728	1.16	A3V	α Piscis Austrini
Gacrux	12 31 10	- 57 07	4763	1.62	M3II	γ Crucis
Gemma	15 34 41	+26 43	5793	2.23	AOV	α Coronae Borealis
Gienah	12 15 49	- 17 32	4662	2.60	B8III	γ Corvi
Hadar	14 03 50	- 60 22	5267	0.59	B1II	β Centauri
Hamal	02 07 10	+23 27	617	2.00	K2III	α Arietis
Kaus Australis	18 24 10	- 34 23	6879	1.84	B9IV	ε Sagittarii
Kochab	14 50 43	+74 09	5563	2.08	K4III	β Ursae Minoris
Markab	23 04 46	+15 12	8781	2.49	B9.5III	α Pegasi
Megrez	12 15 26	+57 02	4660	3.31	A3V	δ Ursae Majoris
Menkar	03 02 17	+04 06	911	2.52	M2III	α Ceti
Menkent	14 06 41	- 36 23	5288	2.05	KOIII-IV	θ Centauri
Merak	11 01 51	+56 23	4295	2.36	A1V	β Ursae Majoris
Merope	03 46 19	+23 57	1156	4.16	B6IVnn	23 Tauri
Miaplacidus	09 13 12	- 69 43	3685	1.67	A1IV	β Carinae
Mintaka	05 32 01	- 00 18	1852	2.20	O9.5II	δ Orionis

Named Stars

Name	R.A.(2000) h m s	Dec.(2000) ° ′	BS No	V	Sp	Other Designation
Mira	02 19 21	− 02 59	681	2.0	gM6e	o Ceti
Mirach	01 09 49	+35 37	337	2.03	MOIII	β Andromedae
Mirfak	03 24 20	+49 51	1017	1.79	F5Ib	α Persei
Mizar	13 23 56	+54 56	5054	2.40	A2V	ζ Ursae Majoris
Nunki	18 55 16	− 26 18	7121	2.10	B2V	σ Sagittarii
Peacock	20 25 38	− 56 44	7790	1.93	B3IV	α Pavonis
Phecda	11 53 49	+53 42	4554	2.44	A0V	γ Ursae Majoris
Polaris	02 31 13	+89 15	424	2.50	F8Ib	α Ursae Minoris
Pollux	07 45 19	+28 01	2990	1.15	KOIII	β Geminorum
Procyon	07 39 18	+05 14	2943	0.34	F5IV	α Canis Minoris
Pulcherrima	14 44 59	+27 05	5506	2.70	KOII-III	ϵ Boötis
Ras-Algethi	17 14 39	+14 23	6406	3.1	M5II	α Herculis
Rasalhague	17 34 56	+12 34	6556	2.08	A5III	α Ophiuchi
Regulus	10 08 22	+11 58	3982	1.36	B7V	α Leonis
Rigel	05 14 32	− 08 12	1713	0.08	B8Ia	β Orionis
Rigil Kentaurus	14 39 36	− 60 50	5459	0.33	G2V	α Centauri
Sabik	17 10 23	− 15 43	6378	2.44	A2.5V	η Ophiuchi
Scheat	23 03 47	+28 05	8775	2.56	M2II-III	β Pegasi
Schedar	00 40 31	+56 32	168	2.24	KOII-III	α Cassiopeiae
Shaula	17 33 36	− 37 06	6527	1.62	B1V	λ Scorpii
Sirius	06 45 09	− 16 43	2491	− 1.47	A1V	α Canis Majoris
Spica	13 25 11	− 11 09	5056	0.96	B1V	α Virginis
Suhail	09 08 00	− 43 26	3634	2.30?	K5Ib	λ Velorum
Thuban	14 04 24	+64 22	5291	3.64	A0III	α Draconis
Vega	18 36 56	+38 47	7001	0.04	A0V	α Lyrae
Zubenelgenubi	14 50 53	− 16 03	5531	2.75	Am	α Librae

The Constellations

Latin Name	R.A. h	Dec. °	Area °^{2}	Genitive	Abbrev.	Translation
Andromeda	1	+40	722	Andromedae	And	Princess of Ethiopia
Antlia	10	− 35	239	Antliae	Ant	Air Pump
Apus	16	− 75	206	Apodis	Aps	Bird of Paradise
Aquarius	23	− 15	980	Aquarii	Aqr	Water Bearer
Aquila	20	+5	653	Aquilae	Aql	Eagle
Ara	17	− 55	237	Arae	Ara	Altar
Aries	3	+20	441	Arietis	Ari	Ram
Auriga	6	+40	657	Aurigae	Aur	Charioteer
Boötes	15	+30	907	Boötis	Boo	Herdsman
Caelum	5	− 40	125	Caeli	Cae	Chisel
Camelopardalis	6	+70	757	Camelopardis	Cam	Giraffe
Cancer	9	+20	506	Cancri	Cnc	Crab
Canes Venatici	13	+40	465	Canum Venaticorum	CVn	Hunting Dogs
Canis Major	7	− 20	380	Canis Majoris	CMa	Big Dog
Canis Minor	8	+5	183	Canis Minoris	CMi	Little Dog
Capricornus	21	− 20	414	Capricorni	Cap	Goat
Carina	9	− 60	494	Carinae	Car	Ship's Keel**
Cassiopeia	1	+60	598	Cassiopeiae	Cas	Queen of Ethiopia
Centaurus	13	− 50	1060	Centauri	Cen	Centaur*
Cepheus	22	+70	588	Cephei	Cep	King of Ethiopia
Cetus	2	− 10	1231	Ceti	Cet	Whale, Sea Monster
Chamaeleon	11	− 80	132	Chamaeleonis	Cha	Chameleon
Circinus	15	− 60	93	Circini	Cir	Compass
Columba	6	− 35	270	Columbae	Col	Dove
Coma Berenices	13	+20	387	Comae Berenices	Com	Berenice's Hair*
Corona Australis	19	− 40	128	Coronae Australis	CrA	Southern Crown
Corona Borealis	16	+30	179	Coronae Borealis	CrB	Northern Crown
Corvus	12	− 20	184	Corvi	Crv	Crow
Crater	11	− 15	282	Crateris	Crt	Cup
Crux	12	− 60	68	Crucis	Cru	Southern Cross

The Constellations

Latin Name	R.A. h	Dec. °	Area °^{2}	Genitive	Abbrev.	Translation
Cygnus	21	+40	804	Cygni	Cyg	Swan
Delphinus	21	+10	189	Delphini	Del	Dolphin, Porpoise
Dorado	5	− 65	179	Doradus	Dor	Swordfish
Draco	17	+65	1083	Draconis	Dra	Dragon
Equuleus	21	+10	72	Equulei	Equ	Little Horse
Eridanus	3	− 20	1138	Eridani	Eri	River Eridanus*
Fornax	3	− 30	398	Fornacis	For	Furnace
Gemini	7	+20	514	Geminorum	Gem	Twins
Grus	22	− 45	366	Gruis	Gru	Crane
Hercules	17	+30	1225	Herculis	Her	Son of Zeus
Horologium	3	− 60	249	Horologii	Hor	Clock
Hydra	10	− 20	1303	Hydrae	Hya	Water Snake (female)
Hydrus	2	− 75	243	Hydri	Hyi	Water Snake (male)
Indus	21	− 55	294	Indi	Ind	Indian
Lacerta	22	+45	201	Lacertae	Lac	Lizard
Leo	11	+15	947	Leonis	Leo	Lion
Leo Minor	10	+35	232	Leonis Minoris	LMi	Little Lion
Lepus	6	− 20	290	Leporis	Lep	Hare
Libra	15	− 15	538	Librae	Lib	Balance, Scales
Lupus	15	− 45	334	Lupi	Lup	Wolf
Lynx	8	+45	545	Lyncis	Lyn	Lynx
Lyra	19	+40	287	Lyrae	Lyr	Lyre, Harp
Mensa	5	− 80	154	Mensae	Men	Table, Mountain
Microscopium	21	− 35	210	Microscopii	Mic	Microscope
Monoceros	7	− 5	482	Monocerotis	Mon	Unicorn
Musca	12	− 70	138	Muscae	Mus	Fly
Norma	16	− 50	165	Normae	Nor	Square, Level
Octans	22	− 85	291	Octantis	Oct	Octant
Ophiuchus	17	0	948	Ophiuchi	Oph	Serpent-bearer
Orion	5	+5	594	Orionis	Ori	Hunter

The Constellations

Latin Name	R.A. h	Dec. °	Area °²	Genitive	Abbrev.	Translation
Pavo	20	−65	378	Pavonis	Pav	Peacock
Pegasus	22	+20	1121	Pegasi	Peg	Winged Horse
Perseus	3	+45	615	Persei	Per	Rescuer of Andromeda
Phoenix	1	−50	469	Phoenicis	Phe	Phoenix
Pictor	6	−55	247	Pictoris	Pic	Painter, Easel
Pisces	1	+15	889	Piscium	Psc	Fish
Piscis Austrinus	22	−30	245	Piscis Austrini	PsA	Southern Fish
Puppis	8	−40	673	Puppis	Pup	Ship's Stern**
Pyxis	9	−30	221	Pyxidis	Pyx	Ship's Compass**
Reticulum	4	−60	114	Reticuli	Ret	Net
Sagitta	20	+10	80	Sagittae	Sge	Arrow
Sagittarius	19	−25	867	Sagittarii	Sgr	Archer
Scorpius	17	−40	497	Scorpii	Sco	Scorpion
Sculptor	0	−30	475	Sculptoris	Scl	Sculptor
Scutum	19	−10	109	Scuti	Sct	Shield
Serpens	17	0	637	Serpentis	Ser	Serpent
Sextans	10	0	314	Sextantis	Sex	Sextant
Taurus	4	+15	797	Tauri	Tau	Bull
Telescopium	19	−50	252	Telescopii	Tel	Telescope
Triangulum	2	+30	132	Trianguli	Tri	Triangle
Triangulum Australe	16	−65	110	Trianguli Australis	TrA	Southern Triangle
Tucana	0	−65	295	Tucanae	Tuc	Toucan
Ursa Major	11	+50	1280	Ursae Majoris	UMa	Big Bear
Ursa Minor	15	+70	256	Ursae Minoris	UMi	Little Bear
Vela	9	−50	500	Velorum	Vel	Ship's Sail**
Virgo	13	0	1294	Virginis	Vir	Maiden, Virgin
Volans	8	−70	141	Volantis	Vol	Flying Fish
Vulpecula	20	+25	268	Vulpeculae	Vul	Little Fox

*Proper names.

**Previously formed the constellation Argo Navis, the Argonauts' ship.

10.2 The Nearest and Brightest Stars

The sixty four stars nearer then five parsecs are given in the first accompanying table in order of increasing distance. These data have been adapted from Gliese (1982) with the distance, D, in light years, ly, computed from the trigonometric parallax, π_t, using the formula $D = 3.2616/\pi_t$. Gliese's values of annual proper motion, μ, and the position angle, θ, have been used to compute the components in right ascension, $\mu_{R.A.}$, and in declination, $\mu_{Dec.}$, from the relation $\mu_{R.A.} = \mu \sin\theta/[15\cos(Dec)]$ and $\mu_{Dec.} = \mu\cos\theta$. Here V_r denotes the radial velocity, Sp the spectral class, V the apparent visual magnitude, the color indices are B – V,U – B and R – I, the absolute visual magnitude is M_V, and the absolute luminosity is L in solar units $L_\odot = 3.8268 \times 10^{33} erg\ s^{-1}$. Also see van de Kamp (1971) for information on stars nearer than five parsecs. Data for the 2,241 stars within 22 parsecs of the Sun are listed in order of increasing right ascension within the last catalogue of this volume.

The table of stars nearer than five parsecs is followed by a table of the twenty brightest stars listed in order of decreasing brightness. This is followed by a more substantial compilation of the 446 stars brighter than apparent visual magnitude 4.00, listed in order of increasing right ascension, R.A., for epoch 2000.0. Here we also give the proper motions, μ, in both coordinates, the apparent visual magnitude, V, the color index B – V, the spectral class, Sp, the trigonometric parallax, π_t, the distance, D, the absolute visual magnitude, M_V, and the radial velocity, V_r. All the bright star data have been extracted from Hoffleit's (1982) bright star catalogue; the star's number in that catalogue, BS=HR, is given in the first column of our compilation.

Stars Nearer Than Five Parsecs*

Name	R.A.(1950) h m s	Dec.(1950) ° ′	$\mu_{R.A.}$ (s yr^{-1})	$\mu_{Dec.}$ (″ yr^{-1})	π_t (″)	D (ly)	V_r (km s^{-1})
Sun							
Proxima Cen	14 26 18	− 62 28	− 0.542	+0.800	0.772	4.2	− 16
α Cen A	14 36 12	− 60 38	− 0.491	+0.702	0.750	4.4	− 22
α Cen B							
Barnard's star	17 55 24	+04 33	− 0.048	+10.285	0.545	6.0	− 108
Wolf 359	10 54 06	+07 19	− 0.262	− 2.730	0.421	7.8	+13
BD +36°2147	11 00 36	+36 18	− 0.048	− 4.744	0.397	8.2	− 84
L 726–8A	01 36 24	− 18 13	+0.232	+0.583	0.387	8.4	+29
UV Ceti = B							+32
Sirius A	06 42 54	− 16 39	− 0.038	− 1.215	0.377	8.7	− 8
Sirius B							
Ross 154	18 46 42	− 23 53	+0.051	− 0.174	0.345	9.5	− 4
Ross 248	23 39 24	+43 55	+0.010	− 1.596	0.314	10.4	− 81
ε Eri	03 30 36	−09 38	− 0.066	+0.017	0.303	10.8	+16
Ross 128	11 45 06	+01 06	+0.043	− 1.218	0.298	10.9	− 13
61 Cyg A	21 04 42	+38 30	+0.350	+3.214	0.294	11.1	− 64
61 Cyg B							
ε Ind	21 59 36	− 57 00	+0.482	− 2.560	0.291	11.2	− 40
BD +43°44A	00 15 30	+43 44	+0.265	+0.404	0.290	11.3	+13
+43°44B							+20
L 789–6	22 35 42	− 15 36	+0.162	+2.265	0.290	11.3	− 60
Procyon A	07 36 42	+05 21	− 0.047	− 1.036	0.285	11.4	− 3
Procyon B							
BD +59°1915A	18 42 12	+59 33	− 0.173	+1.878	0.282	11.6	0
+59°1915B			− 0.091	+1.813			+10
CD −36°15693	23 02 36	− 36 09	+0.559	+1.317	0.279	11.7	+10
G 51–15	08 26 54	+26 57	− 0.084	− 0.596	0.278	11.7	
τ Cet	01 41 42	− 16 12	− 0.119	+0.872	0.277	11.8	− 16
BD +5°1668	07 24 42	+05 23	+0.039	− 3.724	0.266	12.3	+26
L 725–32	01 09 54	− 17 16	+0.081	+0.620	0.261	12.5	+28
CD −39°14192	21 14 18	− 39 04	− 0.281	− 1.126	0.260	12.6	+21
Kapteyn's star	05 09 42	− 45 00	+0.620	− 5.721	0.256	12.7	+245
Krüger 60 A	22 26 12	+57 27	− 0.097	− 0.350	0.253	12.9	− 26
Krüger 60 B							

*Adapted from Gliese (1982).

Stars Nearer Than Five Parsecs*

Name	Sp	V (mag)	B − V (mag)	U − B (mag)	R − I (mag)	M_V (mag)	L ($L_\odot$)
Sun	G2V	− 26.72	0.65	0.10		4.85	1.0
Proxima Cen	dM5e	11.05	1.97		1.65	15.49	0.00006
α Cen A	G2V	− 0.01	0.68		0.22	4.37	1.6
α Cen B	K0V	1.33	0.88		0.24	5.71	0.45
Barnard's star	M5V	9.54	1.74	1.29	1.25	13.22	0.00045
Wolf 359	dM8e	13.53	2.01	1.54	1.85	16.65	0.00002
BD +36°2147	M2V	7.50	1.51	1.12	0.91	10.50	0.0055
L 726–8A	dM6e	12.52	} 1.85	1.09 :	1.6	15.46	0.00006
UV Ceti = B	dM6e	13.02				15.96	0.00004
Sirius A	A1V	− 1.46	0.00	− 0.04	0.12	1.42	23.5
Sirius B	DA	8.3:	− 0.12:	− 1.03:		11.2	0.003
Ross 154	dM5e	10.45	1.70	1.17	1.30	13.14	0.00048
Ross 248	dM6e	12.29	1.91	1.48	1.56	14.78	0.00011
ε Eri	K2V	3.73	0.88	0.58	0.30	6.14	0.30
Ross 128	dM5	11.10	1.76	1.30	1.30	13.47	0.00036
61 Cyg A	K5V	5.22	1.17	1.11	0.47	7.56	0.082
61 Cyg B	K7V	6.03	1.37	1.23	0.60	8.37	0.039
ε Ind	K5V	4.68	1.05	1.00	0.40	7.00	0.14
BD +43°44A	M1V	8.08	1.56	1.24	0.88	10.39	0.0061
+43°44B	M6Ve	11.06	1.80	1.40	1.22	13.37	0.00039
L 789–6	dM7e	12.18	1.96	1.54	1.66	14.49	0.00014
Procyon A	F5IV-V	0.37	0.42	0.03	0.14	2.64	7.65
Procyon B	DF	10.7				13.0	0.00055
BD +59°1915A	dM4	8.90	1.54	1.11	1.07	11.15	0.0030
+59°1915B	dM5	9.69	1.59	1.14	1.14	11.94	0.0015
CD −36°15693	M2V	7.35	1.48	1.18	0.85	9.58	0.013
G 51–15		14.81	2.06		1.79	17.03	0.00001
τ Cet	G8V	3.50	0.72	0.22	0.26	5.72	0.45
BD +5°1668	dM5	9.82	1.56	1.12	1.19	11.94	0.0015
L 725–32	dM5e	12.04	1.83	1.46	1.44	14.12	0.00020
CD −39°14192	M0V	6.66	1.40	1.20	0.69	8.74	0.028
Kapteyn's star	sdM0pec	8.84	1.56	1.05	0.77	10.88	0.0039
Krüger 60 A	dM3	9.85	1.62	1.25	1.14	11.87	0.0016
Krüger 60 B	dM5e	11.3	1.8	1.3		13.3	0.0004

Stars Nearer Than Five Parsecs*

Name	R.A.(1950) h m s	Dec.(1950) ° ′	$\mu_{R.A.}$ (s yr^{-1})	$\mu_{Dec.}$ (″ yr^{-1})	π_t (″)	D (ly)	V_r (km s^{-1})
BD −12°4523	16 27 30	− 12 32	− 0.004	− 1.178	0.247	13.2	− 13
Ross 614 A	06 26 48	− 02 46	+0.049	− 0.682	0.246	13.3	+24
Ross 614 B							
Van Maanen's	00 46 30	+05 09	+0.085	− 2.710	0.232	14.1	+54:
Wolf 424 A	12 30 54	+09 18	− 0.117	+0.275	0.230	14.2	− 5
Wolf 424 B							
CD −37°15492	00 02 30	− 37 36	+0.477	− 2.289	0.225	14.5	+23
L 1159–16	01 57 30	+12 50	+0.074	− 1.791	0.224	14.6	
BD +50°1725	10 08 18	+49 42	− 0.140	− 0.496	0.222	14.7	− 26
CD −46°11540	17 24 54	− 46 51	+0.056	− 0.889	0.216	15.1	
G 158–27	00 04 12	− 07 48	− 0.056	− 1.864	0.214	15.2	
CD −49°13515	21 30 12	− 49 13	− 0.006	− 0.808	0.214	15.2	+8
CD −44°11909	17 33 30	− 44 17	− 0.065	− 0.926	0.213	15.3	
BD +68°946	17 36 42	+68 23	− 0.065	− 1.259	0.213	15.3	− 22
G 208–44 = A	19 52 18	+44 18	+0.041	− 0.591	0.211	15.5	
G 208–45 = B							
BD −15°6290	22 50 36	− 14 31	+0.065	− 0.637	0.209	15.6	+9
σ^2 (40)Eri A	04 13 00	− 07 44	− 0.149	− 3.422	0.207	15.8	− 42
40 Eri B	04 13 06	− 07 44	− 0.145	− 3.452			− 21
40 Eri C							− 45
BD +20°2465	10 16 54	+20 07	− 0.035	− 0.051	0.206	15.8	+11
L 145–141	11 43 00	− 64 33	+0.412	− 0.327	0.206	15.8	
70 Oph A	18 02 54	+02 31	+0.017	− 1.091	0.203	16.1	− 7
70 Oph B							
BD +43°4305	22 44 42	+44 05	− 0.064	− 0.464	0.200	16.3	− 2
Altair	19 48 18	+08 44	+0.036	+0.388	0.198	16.5	− 26
AC +79°3888	11 44 36	+78 58	+0.258	+0.485	0.193	16.9	− 119
G 9–38=A	08 55 24	+19 57	− 0.063	− 0.047	0.192	17.0	
LP 426–40=B							
BD +15°2620	13 43 12	+15 10	+0.123	− 1.447	0.192	17.0	+15

Stars Nearer Than Five Parsecs*

Name	Sp	V (mag)	B − V (mag)	U − B (mag)	R − I (mag)	M_V (mag)	L ($L_\odot$)
BD −12°4523	dM5	10.11	1.60	1.16	1.20	12.07	0.0013
Ross 614 A	dM7e	11.10	} 1.71	1.15	1.40	13.12	0.00049
Ross 614 B		14				16	0.00004
Van Maanen's	DG	12.37	0.56	0.02	0.16	14.20	0.00018
Wolf 424 A	dM6e	13.16	} 1.80	1.18	1.62	14.97	0.00009
Wolf 424 B	dM6e	13.4				15.2	0.00007
CD −37°15492	M4V	8.56	1.46	1.03	0.92	10.32	0.0065
L 1159–16	dM8e	12.26	1.82	1.35	1.35	14.01	0.00022
BD +50°1725	K7V	6.59	1.36	1.28	0.60	8.32	0.041
CD −46°11540	dM4	9.37	1.53	1.21	1.03	11.04	0.0033
G 158–27	dM	13.74	1.95		1.52	15.39	0.00006
CD −49°13515	M1V	8.67	1.46	1.05	0.93	10.32	0.0065
CD −44°11909	M5	10.96	1.65	1.20	1.26	12.60	0.00079
BD +68°946	M3.5V	9.15	1.50	1.08	1.10	10.79	0.0042
G 208–44 = A		13.41	1.90			15.03	0.00008
G 208–45 = B		13.99	1.98			15.61	0.00005
BD −15°6290	dM5	10.17	1.60	1.15	1.22	11.77	0.0017
σ^2 (40)Eri A	K1V	4.43	0.82	0.44	0.31	6.01	0.34
40 Eri B	DA	9.52	0.03	− 0.68	− 0.10	11.10	0.0032
40 Eri C	dM4e	11.17	1.66	0.83	1.31	12.75	0.00069
BD +20°2465	M4.5Ve	9.43	1.54	1.06	1.12	11.00	0.0035
L 145–141	DC	11.50	0.19	− 0.60	0.04	13.07	0.00052
70 Oph A	K0V	4.22	} 0.86	0.51	0.30	5.76	0.43
70 Oph B	K5V	6.00				7.54	0.084
BD +43°4305	dM5e	10.2	1.6	1.1	1.15	11.7	0.0018
Altair	A7IV,V	0.76	0.22	0.08	0.02	2.24	11.1
AC +79°3888	sdM4	10.80	1.60		1.18	12.23	0.0011
G 9–38=A	m	14.06	1.84			15.48	0.00006
LP 426–40=B	m	14.92	1.93			16.34	0.000025
BD +15°2620	M4V	8.49	1.44	1.10	0.86	9.91	0.0095

*Adapted from Gliese (1982).

The Twenty Brightest Stars

Star	R.A.(1950) h	m	Dec.(1950) °	′	D (pc)	μ ($''\text{yr}^{-1}$)	Spectral Class A	B	Visual Magnitude A	B	Absolute Magnitude A	B
Sirius	6	42.9	− 16	39	2.7	1.33	A1V	wd	− 1.46	+8.7	+1.4	+11.6
Canopus	6	22.8	− 52	40	30	0.02	F0Ib–II		− 0.72		− 3.1	
α Centauri	14	36.2	− 60	38	1.3	3.68	G2V	K0V	− 0.01	+1.3	+4.4	+5.7
Arcturus	14	13.4	+19	27	11	2.28	K2IIIp		− 0.06		− 0.3	
Vega	18	35.2	+38	44	8.0	0.34	A0V		+0.04		+0.5	
Capella	5	13.0	+45	57	14	0.44	GIII	M1V	+0.05	+10.2	− 0.7	+9.5
Rigel	5	12.1	− 08	15	250	0.00	B8 Ia	B9	+0.14	+6.6	− 6.8	− 0.4
Procyon	7	36.7	+05	21	3.5	1.25	F5IV–V	wd	+0.37	+10.7	+2.6	+13.0
Betelgeuse	5	52.5	+07	24	150	0.03	M2Iab		+0.41v		− 5.5	
Achernar	1	35.9	− 57	29	20	0.10	B5V		+0.51		− 1.0	
β Centauri	14	00.3	− 60	08	90	0.04	B1III	?	+0.63	+4	− 4.1	− 0.8
Altair	19	48.3	+08	44	5.1	0.66	A7IV–V		+0.77		+2.2	
α Crucis	12	23.8	− 62	49	120	0.04	B1IV	B3	+1.39	+1.9	− 4.0	− 3.5
Aldebaran	4	33.0	+16	25	16	0.20	K5III	M2V	+0.86	+13	− 0.2	+12
Spica	13	22.6	− 10	54	80	0.05	B1V		+0.91v		− 3.6	
Antares	16	26.3	− 26	19	120	0.03	MIIb	B4eV	+0.92v	+5.1	− 4.5	− 0.3
Pollux	7	42.3	+28	09	12	0.62	K0III		+1.16		+0.8	
Fomalhaut	22	54.9	− 29	53	7.0	0.37	A3V	K4V	+1.19	+6.5	+2.0	+7.3
Deneb	20	39.7	+45	06	430	0.00	A2Ia		+1.26		− 6.9	
β Crucis	12	44.8	− 59	24	150	0.05	B0.5IV		+1.28v		− 4.6	

*Capella has a third component C with a spectral class of M5V, a visual magnitude of +13.7 and an absolute visual magnitude of +13.

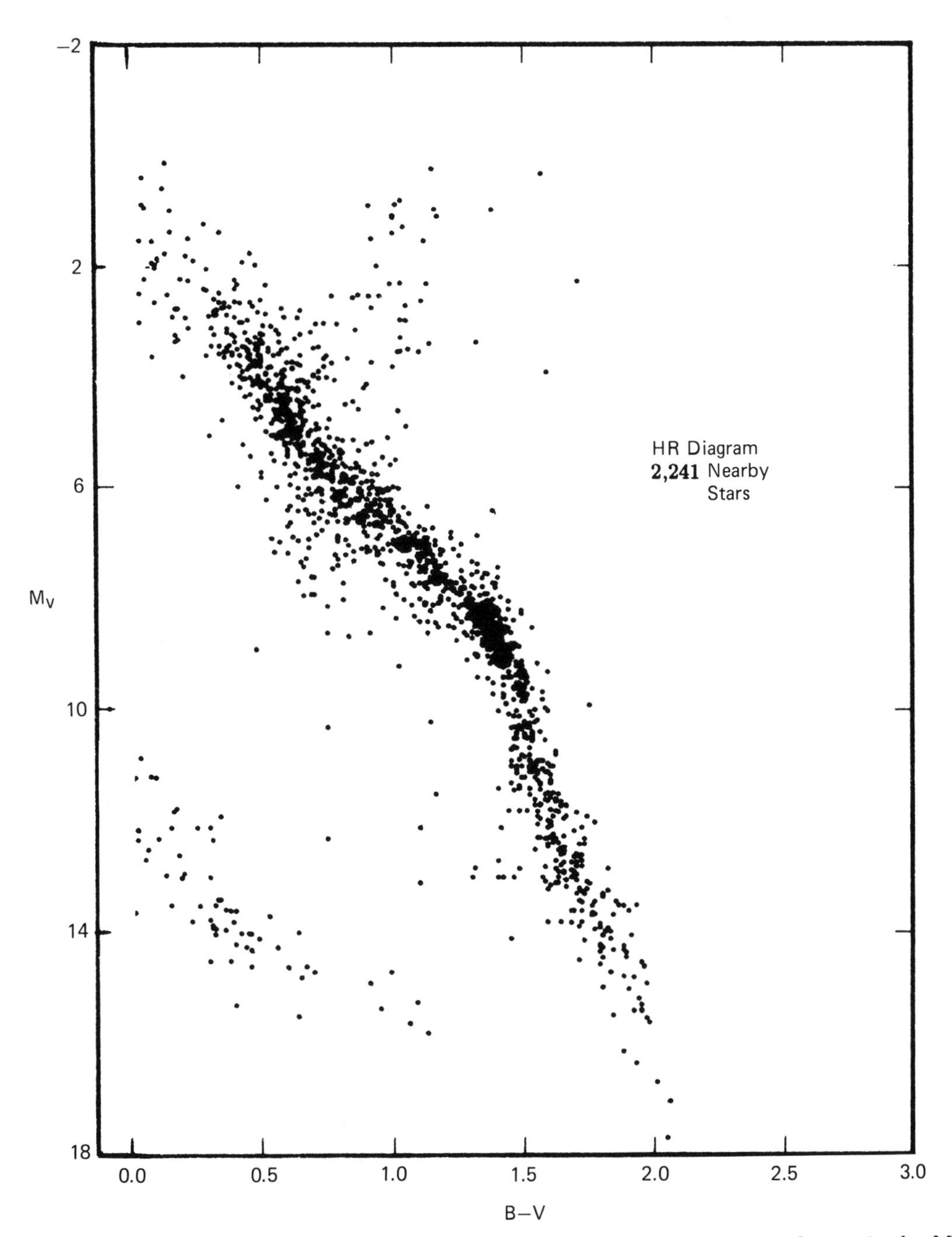

Hertzsprung-Russell Diagram for Nearby Stars. This plot of the absolute visual magnitude, M_V, against color index, B-V, illustrates the heavily populated main sequence running from the hot, luminous upper left to the cooler, less-luminous lower right, as well as the faint white dwarf stars in the lower left. It has been compiled from data for 2,241 nearby stars given in the Appendix of this book.

10.3 Catalogue of 446 Stars Brighter than Apparent Visual Magnitude 4.00

Brightest Stars

BS = HR	Name	HD	R.A.(2000) h m s	Dec.(2000) ° ′ ″	$\mu_{R.A.}$ (″ yr^{-1})	$\mu_{Dec.}$ (″ yr^{-1})
15	α And	358	00 08 23.2	+29 05 26	+0.137	− 0.158
21	β Cas	432	00 09 10.6	+59 08 59	+0.526	− 0.177
25	ε Phe	496	00 09 24.6	− 45 44 51	+0.129	− 0.177
39	γ Peg	886	00 13 14.1	+15 11 01	+0.003	− 0.007
74	ι Cet	1522	00 19 25.6	− 08 49 26	− 0.019	− 0.031
98	β Hyi	2151	00 25 45.3	− 77 15 16	+2.229	+0.327
99	α Phe	2261	00 26 17.0	− 42 18 22	+0.207	− 0.390
100	κ Phe	2262	00 26 12.1	− 43 40 48	+0.106	+0.035
153	ζ Cas	3360	00 36 58.2	+53 53 49	+0.019	− 0.005
165	δ And	3627	00 39 19.6	+30 51 40	+0.137	− 0.084
168	α Cas	3712	00 40 30.4	+56 32 15	+0.053	− 0.027
188	β Cet	4128	00 43 35.3	− 17 59 12	+0.232	+0.036
219	η Cas	4614	00 49 06.0	+57 48 58	+1.101	− 0.521
264	γ Cas	5394	00 56 42.4	+60 43 00	+0.025	+0.000
269	μ And	5448	00 56 45.1	+38 29 58	+0.152	+0.037
322	β Phe	6595	01 06 05.0	− 46 43 08	− 0.030	+0.005
334	η Cet	6805	01 08 35.3	− 10 10 56	+0.214	− 0.133
337	β And	6860	01 09 43.9	+35 37 14	+0.179	− 0.109
338	ζ Phe	6882	01 08 23.0	− 55 14 45	+0.020	+0.031
402	θ Cet	8512	01 24 01.3	− 08 11 01	− 0.083	− 0.218
403	δ Cas	8538	01 25 48.9	+60 14 07	+0.300	− 0.045
440	δ Phe	9362	01 31 15.0	− 49 04 22	+0.137	+0.159
437	η Psc	9270	01 31 28.9	+15 20 45	+0.028	− 0.002
424	α UMi	8890	02 31 50.5	+89 15 51	+0.046	− 0.004
464	51 And	9927	01 37 59.5	+48 37 42	+0.066	− 0.108
472	α Eri	10144	01 37 42.9	− 57 14 12	+0.104	− 0.028
509	τ Cet	10700	01 44 04.0	− 15 56 15	− 1.720	+0.858
539	ζ Cet	11353	01 51 27.5	− 10 20 06	+0.035	− 0.036
542	ε Cas	11415	01 54 23.6	+63 40 13	+0.033	− 0.015
544	α Tri	11443	01 53 04.8	+29 34 44	+0.010	− 0.229
553	β Ari	11636	01 54 38.3	+20 48 29	+0.097	− 0.108
566	χ Eri	11937	01 55 57.4	− 51 36 32	+0.679	+0.299
591	α Hyi	12311	01 58 46.2	− 61 34 12	+0.269	+0.033
603	γ^1 And	12533	02 03 53.9	+42 19 47	+0.046	− 0.048
617	α Ari	12929	02 07 10.3	+23 27 45	+0.190	− 0.144

Brightest Stars

BS = HR	V (mag)	B− V (mag)	S_p	π_t ($''$)	D (ly)	M_V (mag)	V_r (km s^{-1})
15	2.060	− 0.110	B9IV:pHgMn	+0.032	101.9	− 0.41	− 12
21	2.270	+0.340	F2III	+0.072	45.3	+1.56	+12
25	3.880	+1.030	K0III	+0.066	49.4	+2.98	− 9
74	3.560	+1.220	K1IIIb	+0.013	250.9	− 0.87	+19
98	2.800	+0.620	G1IV	+0.153	21.3	+3.72	+23
99	2.390	+1.090	K0IIIb	+0.035	93.2	+0.11	+75
100	3.940	+0.170	A7V	+0.072	45.3	+3.23	+11
153	3.660	− 0.200	B2IV	+0.004	815.4	− 3.33	+ 1
165	3.270	+1.280	K3III	+0.028	116.5	+0.51	− 7
168	2.230	+1.170	K0–IIIa	+0.016	203.8	− 1.75	− 4
188	2.040	+1.020	G9IIICH− 1CN	+0.061	53.5	+0.97	+13
219	3.440	+0.570	F9V+M0–V	+0.176	18.5	+4.67	+ 9
264	2.470	− 0.150	B0IVnpe (sh	+0.016	203.8	− 1.51	− 7
269	3.870	+0.130	A5V	+0.039	83.6	+1.83	+ 8
322	3.310	+0.890	G8III	+0.021	155.3	− 0.08	− 1
334	3.450	+1.160	K2–IIICN0.5	+0.041	79.6	+1.51	+12
337	2.060	+1.580	M0+IIIa	+0.049	66.6	+0.51	+ 3
338	3.920	− 0.080	B6V+B9V	+0.013	250.9	− 0.51	+15
402	3.600	+1.060	K0IIIb	+0.041	79.6	+1.66	+17
403	2.680	+0.130	A5IV	+0.037	88.2	+0.52	+ 7
424	2.020	+0.600	F7:Ib–II	+0.007	465.9	− 3.75	− 17
437	3.620	+0.970	G7IIIa	+0.015	217.4	− 0.50	+15
440	3.950	+0.990	G9III	+0.031	105.2	+1.41	− 7
464	3.570	+1.280	K3–III	+0.021	155.3	+0.18	+16
472	0.460	− 0.160	B3Vnp (shel	+0.026	125.4	− 2.47	+16
509	3.500	+0.720	G8V	+0.275	11.9	+5.70	− 16
539	3.730	+1.140	K0III	+0.031	105.2	+1.19	+ 9
542	3.380	− 0.150	B3IV:p (she	+0.010	326.2	− 1.62	− 8
544	3.410	+0.490	F6IV	+0.057	57.2	+2.19	− 13
553	2.640	+0.130	A5V	+0.074	44.1	+1.99	− 2
566	3.700	+0.850	G8III–IV*	+0.058	56.2	+2.52	− 6
585	4.000	+1.570	M0IIIb	+0.007	465.9	− 1.77	+18
591	2.860	+0.280	F0III–IVn	+0.048	67.9	+1.27	+ 1
603	2.260	+1.370	K3–IIb	+0.013	250.9	− 2.17	− 12
617	2.000	+1.150	K2–IIIabCa−	+0.049	66.6	+0.45	− 14

Brightest Stars

BS = HR	Name	HD	R.A.(2000) h m s	Dec.(2000) ° ′ ″	$\mu_{R.A.}$ (″ yr^{-1})	$\mu_{Dec.}$ (″ yr^{-1})
622	β Tri	13161	02 09 32.5	+34 59 14	+0.148	− 0.037
681	o Cet	14386	02 19 20.7	− 02 58 39	− 0.012	− 0.233
804	γ Cet	16970	02 43 18.0	+03 14 09	− 0.145	− 0.148
834	η Per	17506	02 50 41.8	+55 53 44	+0.019	− 0.011
838	41 Ari	17573	02 49 59.0	+27 15 38	+0.067	− 0.112
854	τ Per	17878	02 54 15.4	+52 45 45	+0.000	− 0.001
874	η Eri	18322	02 56 25.6	− 08 53 53	+0.074	− 0.217
897	θ^1 Eri	18622	02 58 15.6	− 40 18 17	− 0.051	+0.023
911	α Cet	18884	03 02 16.7	+04 05 23	− 0.012	− 0.074
915	γ Per	18925	03 04 47.7	+53 30 23	+0.000	− 0.002
921	ρ Per	19058	03 05 10.5	+38 50 25	+0.130	− 0.102
936	β Per	19356	03 08 10.1	+40 57 21	+0.003	+0.002
941	κ Per	19476	03 09 29.7	+44 51 27	+0.178	− 0.153
963	α For	20010	03 12 04.2	− 28 59 14	+0.333	+0.644
1003	τ^4 Eri	20720	03 19 30.9	− 21 45 28	+0.047	+0.036
1017	α Per	20902	03 24 19.3	+49 51 41	+0.025	− 0.022
1030	o Tau	21120	03 24 48.7	+09 01 44	− 0.068	− 0.075
1038	ξ Tau	21364	03 27 10.1	+09 43 58	+0.058	− 0.035
1084	ε Eri	22049	03 32 55.8	− 09 27 30	− 0.979	+0.019
1122	δ Per	22928	03 42 55.4	+47 47 15	+0.028	− 0.032
1131	o Per	23180	03 44 19.1	+32 17 18	+0.008	− 0.009
1135	ν Per	23230	03 45 11.6	+42 34 43	− 0.013	+0.000
1136	δ Eri	23249	03 43 14.8	− 09 45 48	− 0.099	+0.746
1142	17 Tau	23302	03 44 52.5	+24 06 48	+0.019	− 0.042
1165	η Tau	23630	03 47 29.0	+24 06 18	+0.019	− 0.044
1175	β Ret	23817	03 44 12.0	− 64 48 26	+0.314	+0.078
1178	27 Tau	23850	03 49 09.7	+24 03 12	+0.018	− 0.043
1203	ζ Per	24398	03 54 07.9	+31 53 01	+0.006	− 0.009
1208	γ Hyi	24512	03 47 14.5	− 74 14 21	+0.052	+0.117
1220	ε Per	24760	03 57 51.2	+40 00 37	+0.017	− 0.024
1231	γ Eri	25025	03 58 01.7	− 13 30 31	+0.057	− 0.110
1239	λ Tau	25204	04 00 40.8	+12 29 25	− 0.007	− 0.009
1251	ν Tau	25490	04 03 09.3	+05 59 22	+0.001	+0.000
1326	α Hor	26967	04 14 00.1	− 42 17 40	+0.040	− 0.208
1336	α Ret	27256	04 14 25.5	− 62 28 26	+0.046	+0.050

Brightest Stars

BS = HR	V (mag)	B–V (mag)	S_p	π_t ('')	D (ly)	M_V (mag)	V_r (km s^{-1})
622	3.00	+0.14	A5IV	0.022	148.3	- 0.29	+10
681	3.04	+1.42	M5.5-9IIIe+	0.024	135.9	- 0.06	+64
804	3.47	+0.09	A3V	0.052	62.7	+2.05	- 5
834	3.76	+1.68	K3-Ib-IIa	0.006	543.6	- 2.35	- 1
838	3.63	- 0.10	B8Vn	0.034	95.9	+1.29	+ 4
854	3.95	+0.74	G4III+A4V	0.019	171.7	+0.34	+ 2
874	3.89	+1.11	K1IIIb	0.033	98.8	+1.48	- 20
897	3.24	+0.14	A5IV	0.035	93.2	+0.96	+12
911	2.53	+1.64	M1.5IIIa	0.009	362.4	- 2.70	- 26
915	2.93	+0.70	G8III+A2V	0.016	203.8	- 1.05	+ 3
921	3.39	+1.65	M4II	0.011	296.5	- 1.40	+28
936	2.12	- 0.05	B8V	0.045	72.5	+0.39	+ 4
941	3.80	+0.98	K0III	0.033	98.8	+1.39	+29
963	3.87	+0.52	F8V	0.075	43.5	+3.25	- 21
1003	3.69	+1.62	M3.5IIICa–1				+42
1017	1.79	+0.48	F5Ib	0.016	203.8	- 2.19	- 2
1030	3.60	+0.89	G6IIIaFe–1	0.016	203.8	- 0.38	- 21
1038	3.74	- 0.09	B9Vn				- 2
1084	3.73	+0.88	K2V	0.303	10.8	+6.14	+15
1122	3.01	- 0.13	B5IIIe	0.016	203.8	- 0.97	+ 4
1131	3.83	+0.05	B1III	0.023	141.8	+0.64	+19
1135	3.77	+0.42	F5II	0.020	163.1	+0.28	- 13
1136	3.54	+0.92	K0+IV	0.109	29.9	+3.73	- 6
1142	3.70	- 0.11	B6III	0.020	163.1	+0.21	+12
1165	2.87	- 0.09	B7IIIe	0.008	407.7	- 2.61	+10
1175	3.85	+1.13	K2III	0.042	77.7	+1.97	+51
1178	3.63	- 0.09	B8III				+ 9
1203	2.85	+0.12	B1Ib	0.010	326.2	- 2.15	+20
1208	3.24	+1.62	M2III	0.005	652.3	- 3.27	+16
1220	2.89	- 0.20	B0.5V+A2V	0.009	362.4	- 2.34	+ 1
1231	2.95	+1.59	M0.5IIIbCa–	0.010	326.2	- 2.05	+62
1239	3.47	- 0.12	B3V+A4IV	0.002	1630.8	- 5.02	+18
1251	3.91	+0.03	A1V	0.030	108.7	+1.30	- 6
1326	3.86	+1.10	K1III	0.026	125.4	+0.93	+22
1336	3.35	+0.91	G8II–III	0.013	250.9	- 1.08	+36

Brightest Stars

BS = HR	Name	HD	R.A.(2000) h m s	Dec.(2000) ° ′ ″	$\mu_{R.A.}$ (″ yr^{-1})	$\mu_{Dec.}$ (″ yr^{-1})
1346	γ Tau	27371	04 19 47.5	+15 37 39	+0.116	− 0.024
1373	δ^1 Tau	27697	04 22 56.0	+17 32 33	+0.107	− 0.028
1393	43 Eri	28028	04 24 02.1	− 34 01 01	+0.062	+0.051
1409	ε Tau	28305	04 28 36.9	+19 10 49	+0.108	− 0.036
1411	θ^1 Tau	28307	04 28 34.4	+15 57 44	+0.102	− 0.026
1412	θ^2 Tau	28319	04 28 39.7	+15 52 15	+0.102	− 0.024
1457	α Tau	29139	04 35 55.2	+16 30 33	+0.065	− 0.189
1464	υ^2 Eri	29291	04 35 33.0	− 30 33 45	− 0.048	− 0.012
1465	α Dor	29305	04 33 59.8	− 55 02 42	+0.052	+0.001
1481	53 Eri	29503	04 38 10.7	− 14 18 15	− 0.080	− 0.161
1543	π^3 Ori	30652	04 49 50.3	+06 57 41	+0.463	+0.017
1552	π^4 Ori	30836	04 51 12.3	+05 36 18	− 0.004	+0.001
1567	π^5 Ori	31237	04 54 15.0	+02 26 26	− 0.004	+0.000
1577	ι Aur	31398	04 56 59.6	+33 09 58	+0.004	− 0.018
1605	ε Aur	31964	05 01 58.1	+43 49 24	+0.001	− 0.004
1612	ζ Aur	32068	05 02 28.6	+41 04 33	+0.009	− 0.022
1641	η Aur	32630	05 06 30.8	+41 14 04	+0.029	− 0.067
1654	ε Lep	32887	05 05 27.6	− 22 22 16	+0.018	− 0.071
1666	β Eri	33111	05 07 50.9	− 05 05 11	− 0.100	− 0.080
1702	μ Lep	33904	05 12 55.8	− 16 12 20	+0.033	− 0.027
1708	α Aur	34029	05 16 41.3	+45 59 53	+0.080	− 0.423
1713	β Ori	34085	05 14 32.2	− 08 12 06	− 0.003	− 0.002
1735	τ Ori	34503	05 17 36.3	− 06 50 40	− 0.021	− 0.008
1788	η Ori	35411	05 24 28.6	− 02 23 49	− 0.003	+0.001
1790	γ Ori	35468	05 25 07.8	+06 20 59	− 0.012	− 0.014
1791	β Tau	35497	05 26 17.5	+28 36 27	+0.025	− 0.175
1829	β Lep	36079	05 28 14.7	− 20 45 34	− 0.008	− 0.091
1852	δ Ori	36486	05 32 00.3	− 00 17 57	− 0.003	− 0.001
1862	ε Col	36597	05 31 12.7	− 35 28 14	+0.024	− 0.036
1865	α Lep	36673	05 32 43.7	− 17 49 20	− 0.006	+0.001
1879	λ Ori	36861	05 35 08.2	+09 56 03	− 0.001	− 0.005
1899	ι Ori	37043	05 35 25.9	− 05 54 36	− 0.004	+0.001
1903	ε Ori	37128	05 36 12.7	− 01 12 07	− 0.003	− 0.002
1910	ζ Tau	37202	05 37 38.6	+21 08 33	+0.001	− 0.022
1922	β Dor	37350	05 33 37.5	− 62 29 24	+0.001	+0.007

Brightest Stars

BS = HR	V (mag)	B–V (mag)	S_p	π_t ($''$)	D (ly)	M_V (mag)	V_r (km s^{-1})
1346	3.65	+0.99	G9.5IIIabCN	0.028	116.5	+0.89	+39
1373	3.76	+0.98	G9.5IIICN0.	0.021	155.3	+0.37	+39
1393	3.96	+1.49	K4III				+24
1409	3.53	+1.01	G9.5IIICN0.	0.020	163.1	+0.04	+39
1411	3.84	+0.95	G9IIIFe–0.5	0.038	85.8	+1.74	+40
1412	3.40	+0.18	A7III	0.029	112.5	+0.71	+40
1457	0.85	+1.54	K5+III	0.048	67.9	- 0.74	+54
1464	3.82	+0.98	G8IIIH				- 4
1465	3.27	- 0.10	A0IIISi	0.018	181.2	- 0.45	+26
1481	3.87	+1.09	K1.5IIIb	0.044	74.1	+2.09	+42
1543	3.19	+0.45	F6V	0.125	26.1	+3.67	+24
1552	3.69	- 0.17	B2III+B2IV	0.001	3261.6	- 6.31	+23
1567	3.72	- 0.18	B3III+B0V	0.003	1087.2	- 3.89	+23
1577	2.69	+1.53	K3II	0.021	155.3	- 0.70	+18
1605	2.99	+0.54	F0Iae+B	0.007	465.9	- 2.78	- 3
1612	3.75	+1.22	K4II+B8V	0.005	652.3	- 2.76	+13
1641	3.17	- 0.18	B3V	0.022	148.3	- 0.12	+ 7
1654	3.19	+1.46	K5III	0.011	296.5	- 1.60	+ 1
1666	2.79	+0.13	A3III	0.050	65.2	+1.28	- 9
1702	3.31	- 0.11	B9IIIpHgMn	0.023	141.8	+0.12	+28
1708	0.08	+0.80	G4:III:+G0I	0.073	44.7	- 0.60	+30
1713	0.12	- 0.03	B8Ia:	0.013	250.9	- 4.31	+21
1735	3.60	- 0.11	B5III	0.006	543.6	- 2.51	+20
1788	3.36	- 0.17	B1V+B2e	0.007	465.9	- 2.41	+20
1790	1.64	- 0.22	B2III	0.029	112.5	- 1.05	+18
1791	1.65	- 0.13	B7III	0.028	116.5	- 1.11	+ 9
1829	2.84	+0.82	G5II	0.020	163.1	- 0.65	- 14
1852	2.23	- 0.22	O9.5II	0.014	233.0	- 2.04	+16
1862	3.87	+1.14	K1IIIa	0.008	407.7	- 1.61	- 5
1865	2.58	+0.21	F0Ib	0.007	465.9	- 3.19	+24
1879	3.54	- 0.18	O8III((f))	0.007	465.9	- 2.23	+34
1899	2.77	- 0.24	O9III	0.025	130.5	- 0.24	+22
1903	1.70	- 0.19	BOIae				+26
1910	3.00	- 0.19	B4IIIpe	0.008	407.7	- 2.48	+20
1922	3.76	+0.82	F6Ia	0.012	271.8	- 0.84	+ 7

Brightest Stars

BS = HR	Name	HD	R.A.(2000) h m s	Dec.(2000) ° ′ ″	$\mu_{R.A.}$ (″ yr^{-1})	$\mu_{Dec.}$ (″ yr^{-1})
1931	σ Ori	37468	05 38 44.7	− 02 36 00	− 0.003	+0.001
1948	ζ Ori	37742	05 40 45.5	− 01 56 34	− 0.001	− 0.002
1956	α Col	37795	05 39 38.9	− 34 04 27	+0.001	− 0.027
1983	γ Lep	38393	05 44 27.8	− 22 26 54	− 0.294	− 0.373
1998	ζ Lep	38678	05 46 57.3	− 14 49 19	− 0.023	− 0.003
2004	κ Ori	38771	05 47 45.3	− 09 40 11	− 0.003	− 0.005
2012	ν Aur	39003	05 51 29.3	+39 08 55	− 0.005	+0.007
2020	β Pic	39060	05 47 17.1	− 51 03 59	+0.005	+0.087
2035	δ Lep	39364	05 51 19.2	− 20 52 45	+0.224	− 0.650
2040	β Col	39425	05 50 57.5	− 35 46 06	+0.050	+0.402
2061	α Ori	39801	05 55 10.3	+07 24 25	+0.025	+0.010
2077	δ Aur	40035	05 59 31.6	+54 17 05	+0.083	− 0.126
2085	η Lep	40136	05 56 24.2	− 14 10 04	− 0.049	+0.136
2088	β Aur	40183	05 59 31.7	+44 56 51	− 0.055	− 0.001
2095	θ Aur	40312	05 59 43.2	+37 12 45	+0.049	− 0.082
2120	η Col	40808	05 59 08.7	− 42 48 55	+0.015	− 0.016
2216	η Gem	42995	06 14 52.6	+22 30 24	− 0.067	− 0.013
2227	γ Mon	43232	06 14 51.3	− 06 16 29	− 0.009	− 0.019
2282	ζ CMa	44402	06 20 18.7	− 30 03 48	+0.005	+0.003
2286	μ Gem	44478	06 22 57.6	+22 30 49	+0.055	− 0.112
2294	β CMa	44743	06 22 41.9	− 17 57 22	− 0.013	− 0.004
2296	δ Col	44762	06 22 06.7	− 33 26 11	− 0.030	− 0.056
2326	α Car	45348	06 23 57.2	− 52 41 44	+0.026	+0.022
2421	γ Gem	47105	06 37 42.7	+16 23 57	+0.043	− 0.044
2429	ν^2 CMa	47205	06 36 41.0	− 19 15 22	+0.057	− 0.077
2473	ε Gem	48329	06 43 55.9	+25 07 52	− 0.004	− 0.015
2484	ξ Gem	48737	06 45 17.3	+12 53 44	− 0.115	− 0.194
2491	α CMa	48915	06 45 08.9	− 16 42 58	− 0.545	− 1.211
2540	θ Gem	50019	06 52 47.3	+33 57 40	− 0.001	− 0.051
2550	α Pic	50241	06 48 11.4	− 61 56 29	− 0.071	+0.266
2580	o^1 CMa	50877	06 54 07.8	− 24 11 02	− 0.012	+0.011
2618	ε CMa	52089	06 58 37.5	− 28 58 20	+0.001	+0.002
2646	σ CMa	52877	07 01 43.1	− 27 56 06	− 0.008	+0.002
2650	ζ Gem	52973	07 04 06.5	+20 34 13	− 0.008	− 0.003
2693	δ CMa	54605	07 08 23.4	− 26 23 35	− 0.008	+0.003

Brightest Stars

BS = HR	V (mag)	B–V (mag)	S_p	π_t ('')	D (ly)	M_V (mag)	V_r (km s^{-1})
1931	3.81	- 0.24	O9.5V	0.007	465.9	- 1.96	+29
1948	2.03	- 0.21	O9.7Ib	0.024	135.9	- 1.07	+18
1956	2.64	- 0.12	B7IVe	0.001	3261.6	- 7.36	+35
1983	3.60	+0.47	F6V	0.128	25.5	+4.14	- 10
1998	3.55	+0.10	A3Vn	0.049	66.6	+2.00	+20
2004	2.06	- 0.17	B0.5Ia	0.015	217.4	- 2.06	+21
2012	3.97	+1.13	K0IIICN1	0.017	191.9	+0.12	+10
2020	3.85	+0.17	A5V	0.061	53.5	+2.78	+20
2035	3.81	+0.99	K0IIIFe–1.5	0.022	148.3	+0.52	+99
2040	3.12	+1.16	K2III	0.023	141.8	- 0.07	+89
2061	0.50	+1.85	M1-M2Ia–Iab	0.005	652.3	- 6.01	+21
2077	3.72	+1.00	K0-III	0.022	148.3	+0.43	+ 8
2085	3.71	+0.33	F1III	0.066	49.4	+2.81	- 2
2088	1.90	+0.03	A2IV	0.041	79.6	- 0.04	- 18
2095	2.62	- 0.08	A0pSi	0.022	148.3	- 0.67	+30
2120	3.96	+1.14	K0III	0.019	171.7	+0.35	+17
2216	3.28	+1.60	M2.5III	0.014	233.0	- 0.99	+19
2227	3.98	+1.32	K1IIIBa0.5	0.013	250.9	- 0.45	- 5
2282	3.02	- 0.19	B2.5V	0.004	815.4	- 3.97	+32
2286	2.88	+1.64	M3IIIab	0.020	163.1	- 0.61	+55
2294	1.98	- 0.23	B1II–III	0.019	171.7	- 1.63	+34
2296	3.85	+0.88	G7II	0.019	171.7	+0.24	- 3
2326	-0.72	+0.15	F0II	0.028	116.5	- 3.48	+21
2421	1.93	0.00	A0IV	0.033	98.8	- 0.48	- 13
2429	3.95	+1.06	K1.5III–IVF	0.058	56.2	+2.77	+ 3
2473	2.98	+1.40	G8Ib	0.017	191.9	- 0.87	+10
2484	3.36	+0.43	F5III	0.055	59.3	+2.06	+25
2491	- 1.46	0.00	A1Vm	0.375	8.7	+1.41	- 8
2540	3.60	+0.10	A3III	0.021	155.3	+0.21	+21
2550	3.27	+0.21	A7IV	0.052	62.7	+1.85	+21
2580	3.87	+1.73	K2Iab	0.002	1630.8	- 4.62	+36
2618	1.50	- 0.21	B2II	0.001	3261.6	- 8.50	+27
2646	3.47	+1.73	K7Ib	0.024	135.9	+0.37	+22
2650	3.79	+0.79	F7-G3Ib				+ 7
2693	1.84	+0.68	F8Ia				+34

Brightest Stars

BS = HR	Name	HD	R.A.(2000) h m s	Dec.(2000) ° ′ ″	$\mu_{R.A.}$ (″ yr^{-1})	$\mu_{Dec.}$ (″ yr^{-1})
2736	γ^2 Vol	55865	07 08 45.0	− 70 29 57	+0.026	+0.102
2763	λ Gem	56537	07 18 05.5	+16 32 25	− 0.049	− 0.039
2773	π Pup	56855	07 17 08.5	− 37 05 51	− 0.012	+0.003
2777	δ Gem	56986	07 20 07.3	+21 58 56	− 0.025	− 0.014
2803	δ Vol	57623	07 16 49.8	− 67 57 27	− 0.006	− 0.004
2821	ι Gem	58207	07 25 43.5	+27 47 53	− 0.121	− 0.088
2845	β CMi	58715	07 27 09.0	+08 17 21	− 0.053	− 0.040
2878	σ Pup	59717	07 29 13.8	− 43 18 05	− 0.059	+0.186
2890	α Gem	60178	07 34 35.9	+31 53 18	− 0.170	− 0.102
2891	α Gem	60179	07 34 35.9	+31 53 18	− 0.170	− 0.102
2943	α CMi	61421	07 39 18.1	+05 13 30	− 0.706	− 1.029
2970	α Mon	61935	07 41 14.8	− 09 33 04	− 0.077	− 0.024
2985	κ Gem	62345	07 44 26.8	+24 23 53	− 0.030	− 0.055
2990	β Gem	62509	07 45 18.9	+28 01 34	− 0.627	− 0.051
3017		63032	07 45 15.2	− 37 58 07	− 0.014	− 0.001
3024	ζ Vol	63295	07 41 49.3	− 72 36 22	+0.031	+0.016
3045	ξ Pup	63700	07 49 17.6	− 24 51 35	− 0.007	− 0.004
3080		64440	07 52 13.0	− 40 34 33	− 0.016	+0.000
3117	χ Car	65575	07 56 46.7	− 52 58 56	− 0.034	+0.025
3185	ρ Pup	67523	08 07 32.6	− 24 18 15	− 0.088	+0.048
3207	γ^2 Vel	68273	08 09 31.9	− 47 20 12	− 0.006	+0.004
3249	β Cnc	69267	08 16 30.9	+09 11 08	− 0.044	− 0.052
3314		71155	08 25 39.6	− 03 54 23	− 0.070	− 0.027
3323	o UMa	71369	08 30 15.8	+60 43 05	− 0.131	− 0.110
3347	β Vol	71878	08 25 44.3	− 66 08 13	− 0.032	− 0.161
3438	β Pyx	74006	08 40 06.2	− 35 18 29	+0.015	− 0.018
3445		74180	08 40 37.6	− 46 38 55	− 0.004	+0.002
3461	δ Cnc	74442	08 44 41.0	+18 09 15	− 0.017	− 0.233
3482	ε Hya	74874	08 46 46.5	+06 25 08	− 0.191	− 0.055
3485	δ Vel	74956	08 44 42.2	− 54 42 30	+0.022	− 0.079
3487		75063	08 46 01.7	− 46 02 30	− 0.007	− 0.002
3547	ζ Hya	76294	08 55 23.6	+05 56 44	− 0.101	+0.010
3569	ι UMa	76644	08 59 12.4	+48 02 30	− 0.443	− 0.235
3579		76943	09 00 38.3	+41 46 58	− 0.439	− 0.253
3594	κ UMa	77327	09 03 37.5	+47 09 24	− 0.034	− 0.056

Brightest Stars

BS = HR	V (mag)	B–V (mag)	S_p	π_t ('')	D (ly)	M_V (mag)	V_r (km s^{-1})
2736	3.78	+1.04	K0III	0.016	203.8	- 0.20	+ 3
2763	3.58	+0.11	A3V	0.047	69.4	+1.94	- 9
2773	2.70	+1.62	K3Ib	0.032	101.9	+0.23	+16
2777	3.53	+0.34	F2IV	0.061	53.5	+2.46	+ 4
2803	3.98	+0.79	F6II	0.004	815.4	- 3.01	+23
2821	3.79	+1.03	G9IIIb	0.032	101.9	+1.32	+ 8
2845	2.90	- 0.09	B8Ve	0.019	171.7	- 0.71	+22
2878	3.25	+1.51	K5III	0.020	163.1	- 0.24	+88
2890	2.88	+0.04	A2Vm	0.067	48.7	+2.01	- 1
2891	1.98	+0.03	A1V	0.067	48.7	+1.11	+ 6
2943	0.38	+0.42	F5IV–V	0.288	11.3	+2.68	- 4
2970	3.93	+1.02	G9IIIFe–1	0.024	135.9	+0.83	+11
2985	3.57	+0.93	G8III	0.026	125.4	+0.64	+21
2990	1.14	+1.00	K0IIIb	0.094	34.7	+1.01	+ 3
3017	3.61	+1.73	K2.5Ib–II	0.006	543.6	- 2.50	+17
3024	3.95	+1.04	K0III	0.017	191.9	+0.10	+48
3045	3.34	+1.24	G6Iab–Ib	0.003	1087.2	- 4.27	+ 3
3080	3.73	+1.04	K1-2II+A0	0.031	105.2	+1.19	+24
3117	3.47	- 0.18	B3IVp	0.004	815.4	- 3.52	+19
3185	2.81	+0.43	F6IIpdel De	0.035	93.2	+0.53	+46
3207	1.78	- 0.22	WC8+O9I	0.017	191.9	- 2.07	+35
3249	3.52	+1.48	K4IIIBa0.5	0.012	271.8	- 1.08	+22
3314	3.90	- 0.02	A0V	0.026	125.4	+0.97	+10
3323	3.37	+0.85	G5III	0.009	362.4	- 1.86	+20
3347	3.77	+1.13	K1III	0.042	77.7	+1.89	+27
3438	3.97	+0.94	G7Ib–II	0.018	181.2	+0.25	- 13
3445	3.82	+0.70	F3Ia	0.031	105.2	+1.28	+25
3461	3.94	+1.08	K0IIIb	0.025	130.5	+0.93	+17
3482	3.38	+0.68	K1III:+F0V	0.027	120.8	+0.54	+36
3485	1.96	+0.04	A1V	0.051	64.0	+0.50	+ 2
3487	3.91	0.00	A1III	0.003	1087.2	- 3.70	+24
3547	3.11	+1.00	G9IIIa	0.035	93.2	+0.83	+23
3569	3.14	+0.19	A7IV	0.075	43.5	+2.52	+ 9
3579	3.97	+0.43	F5V	0.070	46.6	+3.20	+26
3594	3.60	0.00	A1Vn	0.016	203.8	- 0.38	+ 4

Brightest Stars

BS = HR	Name	HD	R.A.(2000) h m s	Dec.(2000) ° ′ ″	$\mu_{R.A.}$ (″ yr^{-1})	$\mu_{Dec.}$ (″ yr^{-1})
3614		78004	09 04 09.2	− 47 05 52	− 0.051	− 0.016
3634	λ Vel	78647	09 07 59.7	− 43 25 57	− 0.022	+0.012
3665	θ Hya	79469	09 14 21.8	+02 18 51	+0.129	− 0.313
3685	β Car	80007	09 13 12.1	− 69 43 02	− 0.151	+0.102
3690	38 Lyn	80081	09 18 50.6	+36 48 09	− 0.030	− 0.127
3699	ι Car	80404	09 17 05.4	− 59 16 31	− 0.019	+0.005
3705	α Lyn	80493	09 21 03.2	+34 23 33	− 0.223	+0.013
3734	κ Vel	81188	09 22 06.8	− 55 00 38	− 0.008	+0.008
3748	α Hya	81797	09 27 35.2	− 08 39 31	− 0.018	+0.028
3757	23 UMa	81937	09 31 31.7	+63 03 43	+0.109	+0.026
3775	θ UMa	82328	09 32 51.3	+51 40 38	− 0.952	− 0.540
3786	ψ Vel	82434	09 30 41.9	− 40 28 00	− 0.189	+0.069
3803		82668	09 31 13.3	− 57 02 04	− 0.034	− 0.001
3845	ι Hya	83618	09 39 51.3	− 01 08 34	+0.046	− 0.069
3852	o Leo	83808	09 41 09.0	+09 53 32	− 0.143	− 0.041
3873	ε Leo	84441	09 45 51.0	+23 46 27	− 0.045	− 0.015
3884		84810	09 45 14.8	− 62 30 28	− 0.016	+0.003
3888	υ UMa	84999	09 50 59.3	+59 02 19	− 0.293	− 0.156
3890	υ Car	85123	09 47 06.1	− 65 04 18	− 0.010	+0.007
3905	μ Leo	85503	09 52 45.8	+26 00 25	− 0.215	− 0.060
3975	η Leo	87737	10 07 19.9	+16 45 45	− 0.001	− 0.006
3982	α Leo	87901	10 08 22.3	+11 58 02	− 0.249	+0.003
3994	λ Hya	88284	10 10 35.2	− 12 21 15	− 0.207	− 0.095
4023		88955	10 14 44.1	− 42 07 19	− 0.150	+0.038
4031	ζ Leo	89025	10 16 41.3	+23 25 02	+0.018	− 0.012
4033	λ UMa	89021	10 17 05.7	+42 54 52	− 0.165	− 0.043
4050		89388	10 17 04.9	− 61 19 56	− 0.027	+0.003
4057	γ^1 Leo	89484	10 19 58.3	+19 50 30	+0.307	− 0.151
4058	γ^2 Leo	89485	10 19 58.6	+19 50 25	+0.313	− 0.173
4069	μ UMa	89758	10 22 19.7	+41 29 58	− 0.083	+0.030
4094	μ Hya	90432	10 26 05.4	− 16 50 11	− 0.132	− 0.083
4114		90853	10 27 52.7	− 58 44 22	− 0.014	− 0.006
4133	ρ Leo	91316	10 32 48.6	+09 18 24	− 0.009	− 0.006
4167		92139	10 37 18.0	− 48 13 33	− 0.159	− 0.020
4216	μ Vel	93497	10 46 46.1	− 49 25 12	+0.071	− 0.050

Brightest Stars

BS = HR	V (mag)	B–V (mag)	S_p	π_t ('')	D (ly)	M_V (mag)	V_r (km s^{-1})
3614	3.75	+1.20	K2III	0.020	163.1	+0.26	+24
3634	2.21	+1.66	K4.5Ib	0.022	148.3	- 1.08	+18
3665	3.88	- 0.06	B9.5V	0.027	120.8	+1.04	- 10
3685	1.68	0.00	A2IV	0.021	155.3	- 1.71	- 5
3690	3.82	+0.06	A3V	0.042	77.7	+1.94	+ 4
3699	2.25	+0.18	A8Ib	0.017	191.9	- 1.60	+13
3705	3.13	+1.55	K7IIIab	0.025	130.5	+0.12	+38
3734	2.50	- 0.18	B2IV–V	0.013	250.9	- 1.93	+22
3748	1.98	+1.44	K3II–III	0.022	148.3	- 1.31	- 4
3757	3.67	+0.33	F0IV	0.041	79.6	+1.73	- 10
3775	3.17	+0.46	F6IV	0.052	62.7	+1.75	+15
3786	3.60	+0.36	F3IV+F0IV	0.065	50.2	+2.66	+ 9
3803	3.13	+1.55	K5III	0.022	148.3	- 0.16	- 14
3845	3.91	+1.32	K2.5III	0.026	125.4	+0.98	+23
3852	3.52	+0.49	F6II+A1–5V	0.034	95.9	+1.18	+27
3873	2.98	+0.80	G1II	0.010	326.2	- 2.02	+ 4
3884	3.69	+1.22	G5Iab–Ib	0.027	120.8	+0.85	+ 3
3888	3.80	+0.29	F2IV	0.041	79.6	+1.86	+27
3890	3.01	+0.28	A6Ib	0.027	120.8	+0.17	+14
3905	3.88	+1.22	K2IIICN1Ca1	0.025	130.5	+0.87	+14
3975	3.52	- 0.03	A0Ib	0.003	1087.2	- 4.09	+ 3
3982	1.35	- 0.11	B7V	0.045	72.5	- 0.38	+ 6
3994	3.61	+1.01	K0IIICN0.5	0.027	120.8	+0.77	+19
4023	3.85	+0.05	A2V	0.034	95.9	+1.51	+ 7
4031	3.44	+0.31	F0III	0.017	191.9	- 0.41	- 16
4033	3.45	+0.03	A2IV	0.030	108.7	+0.84	+18
4050	3.40	+1.54	K2.5II	0.027	120.8	+0.56	+ 8
4057	2.61	+1.15	K1–IIIbFe-0	0.022	148.3	- 0.68	- 37
4058	3.47	0.00	G7IIIFe–1.5	0.022	148.3	+0.18	- 36
4069	3.05	+1.59	M0III	0.035	93.2	+0.77	- 21
4094	3.81	+1.48	K4+III	0.018	181.2	+0.09	+40
4114	3.82	+0.31	F2II	0.010	326.2	- 1.18	+ 9
4133	3.85	- 0.14	B1Ib	0.011	296.5	- 0.94	+42
4167	3.84	+0.30	F4IV+F3	0.040	81.5	+1.85	+19
4216	2.69	+0.90	G5III+G2V	0.022	148.3	- 0.60	+ 6

Brightest Stars

BS = HR	Name	HD	R.A.(2000) h m s	Dec.(2000) ° ′ ″	$\mu_{R.A.}$ (″ yr^{-1})	$\mu_{Dec.}$ (″ yr^{-1})
4232	ν Hya	93813	10 49 37.4	− 16 11 37	+0.089	+0.196
4247	46 LMi	94264	10 53 18.6	+34 12 53	+0.085	− 0.283
4257		94510	10 53 29.6	− 58 51 12	+0.069	+0.027
4295	β UMa	95418	11 01 50.4	+56 22 56	+0.081	+0.029
4301	α UMa	95689	11 03 43.6	+61 45 03	− 0.118	− 0.071
4337		96918	11 08 35.3	− 58 58 30	− 0.007	− 0.002
4357	δ Leo	97603	11 14 06.4	+20 31 25	+0.143	− 0.135
4359	θ Leo	97633	11 14 14.3	+15 25 46	− 0.061	− 0.083
4377	ν UMa	98262	11 18 28.7	+33 05 39	− 0.028	+0.023
4382	δ Crt	98430	11 19 20.4	− 14 46 43	− 0.128	+0.201
4390	π Cen	98718	11 21 00.4	− 54 29 27	− 0.031	− 0.006
4399	ι Leo	99028	11 23 55.4	+10 31 45	+0.166	− 0.079
4434	λ Dra	100029	11 31 24.2	+69 19 52	− 0.039	− 0.020
4450	ξ Hya	100407	11 33 00.1	− 31 51 27	− 0.207	− 0.042
4518	χ UMa	102224	11 46 03.0	+47 46 46	− 0.138	+0.024
4534	β Leo	102647	11 49 03.5	+14 34 19	− 0.497	− 0.119
4540	β Vir	102870	11 50 41.6	+01 45 53	+0.741	− 0.275
4554	γ UMa	103287	11 53 49.8	+53 41 41	+0.093	+0.007
4621	δ Cen	105435	12 08 21.5	− 50 43 20	− 0.032	− 0.012
4630	ε Crv	105707	12 10 07.4	− 22 37 11	− 0.072	+0.010
4638	ρ Cen	105937	12 11 39.1	− 52 22 06	− 0.035	− 0.019
4656	δ Cru	106490	12 15 08.6	− 58 44 56	− 0.038	− 0.010
4660	δ UMa	106591	12 15 25.5	+57 01 57	+0.102	+0.004
4689	η Vir	107259	12 19 54.3	− 00 40 00	− 0.064	− 0.022
4700	ε Cru	107446	12 21 21.5	− 60 24 04	− 0.176	+0.082
4730	α^1 Cru	108248	12 26 35.9	− 63 05 56	− 0.025	− 0.017
4731	α^2 Cru	108249	12 26 36.5	− 63 05 58	− 0.029	− 0.012
4757	δ Crv	108767	12 29 51.8	− 16 30 56	− 0.213	− 0.143
4786	β Crv	109379	12 34 23.2	− 23 23 48	+0.001	− 0.058
4787	κ Dra	109387	12 33 28.9	+69 47 17	− 0.059	+0.009
4802	τ Cen	109787	12 37 42.1	− 48 32 28	− 0.186	− 0.008
4819	γ Cen	110304	12 41 30.9	− 48 57 34	− 0.190	− 0.008
4825	γ Vir	110379	12 41 39.5	− 01 26 58	− 0.568	+0.008
4826	γ Vir	110380	12 41 39.5	− 01 26 58	− 0.568	+0.008
4844	β Mus	110879	12 46 16.9	− 68 06 29	− 0.032	− 0.024

Brightest Stars

BS = HR	V (mag)	B–V (mag)	S_p	π_t ($''$)	D (ly)	M_V (mag)	V_r (km s^{-1})
4232	3.11	+1.25	K1.5IIIbHde	0.028	116.5	+0.35	- 1
4247	3.83	+1.04	K0+III-IV	0.024	135.9	+0.73	+16
4257	3.78	+0.95	K1III	0.059	55.3	+2.63	+ 8
4295	2.37	- 0.02	A1V	0.053	61.5	+0.99	- 12
4301	1.80	+1.07	K0-IIIa	0.038	85.8	- 0.30	- 9
4337	3.91	+1.23	G40-Ia				+ 7
4357	2.56	+0.12	A4V	0.048	67.9	+0.97	- 20
4359	3.34	- 0.01	A2V	0.026	125.4	+0.41	+ 8
4377	3.48	+1.40	K3-III	0.020	163.1	- 0.01	- 9
4382	3.56	+1.12	G9IIIbCH0.2	0.024	135.9	+0.46	- 5
4390	3.89	- 0.15	B5Vn	0.011	296.5	- 0.90	+ 9
4399	3.94	+0.41	F4IV	0.052	62.7	+2.52	- 10
4434	3.84	+1.62	M0IIICa-1	0.026	125.4	+0.91	+ 7
4450	3.54	+0.94	G7III	0.027	120.8	+0.70	- 5
4518	3.71	+1.18	K0.5IIIb	0.019	171.7	+0.10	- 9
4534	2.14	+0.09	A3V	0.082	39.8	+1.71	- 0
4540	3.61	+0.55	F9V	0.104	31.4	+3.70	+ 5
4554	2.44	0.00	A0Ve	0.028	116.5	- 0.32	- 13
4621	2.60	- 0.12	B2IVne	0.026	125.4	- 0.33	+11
4630	3.00	+1.33	K2.5IIIa	0.027	120.8	+0.16	+ 5
4638	3.96	- 0.15	B3V	0.032	101.9	+1.49	+15
4656	2.80	- 0.23	B2IV	0.003	1087.2	- 4.81	+22
4660	3.31	+0.08	A3V	0.061	53.5	+2.24	- 13
4689	3.89	+0.02	A2IV+A4V	0.016	203.8	- 0.09	+ 2
4700	3.59	+1.42	K3-4III	0.026	125.4	+0.66	- 5
4730	1.33	- 0.24	B0.5IV	0.008	407.7	- 4.15	- 11
4731	1.73	- 0.26	B1V	0.008	407.7	- 3.75	- 1
4757	2.95	- 0.05	B9.5V	0.024	135.9	- 0.15	+ 9
4786	2.65	+0.89	G5IIb	0.034	95.9	+0.31	- 8
4787	3.87	- 0.13	B6IIIpe	0.013	250.9	- 0.56	- 11
4802	3.86	+0.05	A2V	0.024	135.9	+0.76	+ 5
4819	2.17	- 0.01	A1IV	0.016	203.8	- 1.81	- 6
4825	3.48	+0.36	F0V	0.099	32.9	+3.46	- 20
4826	3.50	0.00	F0V	0.099	32.9	+3.48	- 20
4844	3.05	- 0.18	B2.5V	0.015	217.4	- 1.07	+42

Brightest Stars

BS = HR	Name	HD	R.A.(2000) h m s	Dec.(2000) ° ′ ″	$\mu_{R.A.}$ (″ yr^{-1})	$\mu_{Dec.}$ (″ yr^{-1})
4905	ε UMa	112185	12 54 01.7	+55 57 35	+0.109	- 0.010
4910	δ Vir	112300	12 55 36.1	+03 23 51	- 0.470	- 0.058
4915	α^2 CVn	112413	12 56 01.6	+38 19 06	- 0.236	+0.052
4923	δ Mus	112985	13 02 16.3	- 71 32 56	+0.270	- 0.032
4932	ε Vir	113226	13 02 10.5	+10 57 33	- 0.275	+0.017
5020	γ Hya	115659	13 18 55.2	- 23 10 18	+0.065	- 0.049
5028	ι Cen	115892	13 20 35.7	- 36 42 44	- 0.340	- 0.089
5054	ζ UMa	116656	13 23 55.5	+54 55 31	+0.119	- 0.025
5055	ζ UMa	116657	13 23 56.3	+54 55 18	+0.115	- 0.033
5056	α Vir	116658	13 25 11.5	- 11 09 41	- 0.043	- 0.033
5089		117440	13 31 02.6	- 39 24 26	- 0.014	- 0.019
5107	ζ Vir	118098	13 34 41.5	- 00 35 46	- 0.286	+0.036
5191	η UMa	120315	13 47 32.3	+49 18 48	- 0.126	- 0.014
5235	η Boo	121370	13 54 41.0	+18 23 52	- 0.064	- 0.363
5267	β Cen	122451	14 03 49.4	- 60 22 22	- 0.020	- 0.023
5287	π Hya	123123	14 06 22.2	- 26 40 56	+0.043	- 0.144
5288	θ Cen	123139	14 06 40.8	- 36 22 12	- 0.520	- 0.523
5291	α Dra	123299	14 04 23.2	+64 22 33	- 0.058	+0.015
5340	α Boo	124897	14 15 39.6	+19 10 57	- 1.098	- 1.999
5429	ρ Boo	127665	14 31 49.7	+30 22 17	- 0.102	+0.117
5435	γ Boo	127762	14 32 04.6	+38 18 29	- 0.116	+0.149
5459	α^1 Cen	128620	14 39 36.2	- 60 50 07	- 3.608	+0.712
5460	α^2 Cen	128621	14 39 36.2	- 60 50 07	- 3.608	+0.712
5463	α Cir	128898	14 42 30.3	- 64 58 31	- 0.186	- 0.238
5470	α Aps	129078	14 47 51.6	- 79 02 41	- 0.003	- 0.018
5487	μ Vir	129502	14 43 03.5	- 05 39 30	+0.105	- 0.321
5506	ε Boo	129989	14 44 59.1	+27 04 27	- 0.051	+0.018
5511	109 Vir	130109	14 46 14.9	+01 53 34	- 0.114	- 0.031
5531	α^2 Lib	130841	14 50 52.6	- 16 02 31	- 0.108	- 0.071
5563	β UMi	131873	14 50 42.2	+74 09 20	- 0.035	+0.010
5602	β Boo	133208	15 01 56.6	+40 23 26	- 0.046	- 0.032
5603	σ Lib	133216	15 04 04.1	- 25 16 55	- 0.073	- 0.047
5649	ζ Lup	134505	15 12 17.0	- 52 05 57	- 0.107	- 0.070
5671	γ TrA	135382	15 18 54.6	- 68 40 46	- 0.060	- 0.031
5681	δ Boo	135722	15 15 30.1	+33 18 53	+0.083	- 0.116

Brightest Stars

BS = HR	V (mag)	B–V (mag)	S_p	π_t ($''$)	D (ly)	M_V (mag)	V_r (km s^{-1})
4905	1.77	- 0.02	A0pCr	0.009	362.4	- 3.46	- 9
4910	3.38	+1.58	M3+III	0.022	148.3	+0.09	- 18
4915	2.90	- 0.12	A0pSiEuHg	0.027	120.8	+0.06	- 3
4923	3.62	+1.18	K2III	0.030	108.7	+1.01	+37
4932	2.83	+0.94	G8IIIab	0.043	75.9	+1.00	- 14
5020	3.00	+0.92	G8IIIa	0.027	120.8	+0.16	- 5
5028	2.75	+0.04	A2V	0.062	52.6	+1.71	+ 0
5054	2.27	+0.02	A1VpSrSi	0.047	69.4	+0.63	- 6
5055	3.95	+0.13	A1m	0.047	69.4	+2.31	- 9
5056	0.98	- 0.23	B1III–IV+B2	0.023	141.8	- 2.21	+ 1
5089	3.88	+1.17	G9Ib	0.012	271.8	- 0.72	- 2
5107	3.37	+0.11	A3V	0.044	74.1	+1.59	- 13
5191	1.86	- 0.19	B3V	0.035	93.2	- 0.42	- 11
5235	2.68	+0.58	G0IV	0.108	30.2	+2.85	- 0
5267	0.61	- 0.23	B1III	0.009	362.4	- 4.62	+ 6
5287	3.27	+1.12	K2–IIIFe–0.5	0.049	66.6	+1.72	+27
5288	2.06	+1.01	K0–IIIb	0.065	50.2	+1.12	+ 1
5291	3.65	- 0.05	A0III	0.018	181.2	- 0.07	- 13
5340	- .04	+1.23	K1.5IIIFe–0	0.090	36.2	- 0.27	- 5
5429	3.58	+1.30	K3–III	0.029	112.5	+0.89	- 14
5435	3.03	+0.19	A7III	0.025	130.5	+0.02	- 37
5459	- .01	+0.71	G2V	0.751	4.3	+4.37	- 22
5460	1.33	+0.88	K1V	0.751	4.3	+5.71	- 21
5463	3.19	+0.24	ApSrEuCr:	0.056	58.2	+1.93	+ 7
5470	3.83	+1.43	K3IIICN0.5	0.029	112.5	+1.14	- 1
5487	3.88	+0.38	F2III	0.045	72.5	+2.15	+ 5
5506	2.70	+0.97	K0–II–III	0.016	203.8	- 1.28	- 17
5511	3.72	- 0.01	A0V	0.037	88.2	+1.56	- 6
5531	2.75	+0.15	A3IV	0.058	56.2	+1.57	- 10
5563	2.08	+1.47	K4–III	0.039	83.6	+0.04	+17
5602	3.50	+0.97	G8IIIaFe–0.4	0.037	88.2	+1.34	- 20
5603	3.29	+1.70	M2.5III	0.064	51.0	+2.32	- 4
5649	3.41	+0.92	G8III	0.043	75.9	+1.58	- 10
5671	2.89	0.00	A1V	0.010	326.2	- 2.11	- 3
5681	3.47	+0.95	G8IIIFe-1	0.030	108.7	+0.86	- 12

Brightest Stars

BS = HR	Name	HD	R.A.(2000) h m s	Dec.(2000) ° ′ ″	$\mu_{R.A.}$ (″ yr^{-1})	$\mu_{Dec.}$ (″ yr^{-1})
5685	β Lib	135742	15 17 00.3	− 09 22 59	− 0.098	− 0.023
5705	ϕ^1 Lup	136422	15 21 48.3	− 36 15 41	− 0.091	− 0.087
5708	ε Lup	136504	15 22 40.8	− 44 41 22	− 0.019	− 0.015
5735	γ UMi	137422	15 20 43.6	+71 50 02	− 0.024	+0.019
5744	ι Dra	137759	15 24 55.6	+58 57 57	− 0.015	+0.013
5747	β CrB	137909	15 27 49.7	+29 06 20	− 0.179	+0.083
5776	γ Lup	138690	15 35 08.3	− 41 10 00	− 0.016	− 0.031
5787	γ Lib	138905	15 35 31.5	− 14 47 22	+0.062	+0.003
5788	δ Ser	138917	15 34 48.0	+10 32 15	− 0.075	− 0.007
5789	δ Ser	138918	15 34 48.1	+10 32 21	− 0.075	+0.010
5793	α CrB	139006	15 34 41.2	+26 42 53	+0.120	− 0.091
5794	υ Lib	139063	15 37 01.4	− 28 08 06	− 0.008	− 0.002
5849	γ CrB	140436	15 42 44.5	+26 17 44	− 0.106	+0.043
5854	α Ser	140573	15 44 16.0	+06 25 32	+0.136	+0.044
5867	β Ser	141003	15 46 11.2	+15 25 18	+0.068	− 0.049
5881	μ Ser	141513	15 49 37.1	− 03 25 49	− 0.090	− 0.028
5892	ε Ser	141795	15 50 48.9	+04 28 40	+0.126	+0.061
5897	β TrA	141891	15 55 08.4	− 63 25 50	− 0.188	− 0.396
5933	γ Ser	142860	15 56 27.1	+15 39 42	+0.306	− 1.285
5944	π Sco	143018	15 58 51.0	− 26 06 51	− 0.009	− 0.027
5948	η Lup	143118	16 00 07.2	− 38 23 49	− 0.022	− 0.034
5984	β^1 Sco	144217	16 05 26.1	− 19 48 19	− 0.006	− 0.021
6030	δ TrA	145544	16 15 26.2	− 63 41 08	+0.010	− 0.014
6056	δ Oph	146051	16 14 20.6	− 03 41 40	− 0.048	− 0.145
6075	ε Oph	146791	16 18 19.2	− 04 41 33	+0.081	+0.039
6092	τ Her	147394	16 19 44.3	+46 18 48	− 0.017	+0.037
6095	γ Her	147547	16 21 55.1	+19 09 11	− 0.047	+0.042
6102	γ Aps	147675	16 33 27.1	− 78 53 49	− 0.118	− 0.069
6132	η Dra	148387	16 23 59.3	+61 30 51	− 0.024	+0.059
6134	α Sco	148478	16 29 24.4	− 26 25 55	− 0.007	− 0.023
6148	β Her	148856	16 30 13.1	+21 29 22	− 0.099	− 0.017
6149	λ Oph	148857	16 30 54.7	+01 59 02	− 0.030	− 0.075
6165	τ Sco	149438	16 35 52.9	− 28 12 58	− 0.008	− 0.025
6175	ζ Oph	149757	16 37 09.4	− 10 34 02	+0.012	+0.023
6212	ζ Her	150680	16 41 17.1	+31 36 10	− 0.471	+0.394

Brightest Stars

BS = HR	V (mag)	B–V (mag)	S_p	π_t ('')	D (ly)	M_V (mag)	V_r (km s^{-1})
5685	2.61	- 0.11	B8V				- 35
5705	3.56	+1.54	K4III	0.014	233.0	- 0.71	- 29
5708	3.37	- 0.18	B2IV-V	0.009	362.4	- 1.86	+ 8
5735	3.05	+0.05	A3II–III	0.003	1087.2	- 4.56	- 4
5744	3.29	+1.16	K2III	0.040	81.5	+1.30	- 11
5747	3.68	+0.28	F0p	0.032	101.9	+1.21	- 19
5776	2.78	- 0.20	B2IV	0.008	407.7	- 2.70	+ 2
5787	3.91	+1.01	G8.5III	0.041	79.6	+1.97	- 28
5788	3.80	+0.26	F0IV	0.021	155.3	+0.41	- 38
5789	3.80	+0.26	F0IV	0.021	155.3	+0.41	- 42
5793	2.23	- 0.02	A0V+G5V	0.045	72.5	+0.50	+ 2
5794	3.58	+1.38	K3.5III	0.044	74.1	+1.80	- 25
5849	3.84	0.00	B9IV+A3V	0.033	98.8	+1.43	- 11
5854	2.65	+1.17	K2IIIbCN1	0.053	61.5	+1.27	+ 3
5867	3.67	+0.06	A2IV	0.041	79.6	+1.73	- 1
5881	3.53	- 0.04	A0V	0.007	465.9	- 2.24	- 9
5892	3.71	+0.15	A2Vm	0.041	79.6	+1.77	- 9
5897	2.85	+0.29	F2III	0.083	39.3	+2.45	+ 0
5933	3.85	+0.48	F6V	0.069	47.3	+3.04	+ 7
5944	2.89	- 0.19	B1V+B2V	0.010	326.2	- 2.11	- 3
5948	3.41	- 0.22	B2.5IV	0.008	407.7	- 2.07	+ 8
5984	2.62	- 0.07	B1V	0.009	362.4	- 2.61	- 1
6030	3.85	+1.11	G2Ib–IIa	0.029	112.5	+1.16	- 5
6056	2.74	+1.58	M0.5III	0.034	95.9	+0.40	- 20
6075	3.24	+0.96	G9.5IIIbFe–1	0.043	75.9	+1.41	- 10
6092	3.89	- 0.15	B5IV	0.030	108.7	+1.28	- 14
6095	3.75	+0.27	A9III	0.024	135.9	+0.65	- 35
6102	3.89	+0.91	G8-K0III	0.056	58.2	+2.63	+ 5
6132	2.74	+0.91	G8–IIIab	0.051	64.0	+1.28	- 14
6134	0.96	+1.83	M1.5Iab–Ib	0.024	135.9	- 2.14	- 3
6148	2.77	+0.94	G7IIIaFe–0.2	0.024	135.9	- 0.33	- 26
6149	3.82	+0.01	A0V+A4V	0.023	141.8	+0.63	- 14
6165	2.82	- 0.25	B0V	0.020	163.1	- 0.67	+ 2
6175	2.56	+0.02	O9.5Vn	0.003	1087.2	- 5.05	- 15
6212	2.81	+0.65	G0IV	0.102	32.0	+2.85	- 70

Brightest Stars

BS = HR	Name	HD	R.A.(2000) h m s	Dec.(2000) ° ′ ″	$\mu_{R.A.}$ (″ yr^{-1})	$\mu_{Dec.}$ (″ yr^{-1})
6217	α TrA	150798	16 48 39.9	− 69 01 40	+0.028	− 0.034
6220	η Her	150997	16 42 53.7	+38 55 20	+0.033	− 0.083
6229	η Ara	151249	16 49 47.0	− 59 02 29	+0.041	− 0.031
6241	ε Sco	151680	16 50 09.7	− 34 17 36	− 0.610	− 0.255
6271	ζ^2 Sco	152334	16 54 34.9	− 42 21 41	− 0.125	− 0.236
6285	ζ Ara	152786	16 58 37.1	− 55 59 24	− 0.013	− 0.035
6299	κ Oph	153210	16 57 40.0	+09 22 30	− 0.294	− 0.010
6324	ε Her	153808	17 00 17.3	+30 55 35	− 0.049	+0.028
6378	η Oph	155125	17 10 22.6	− 15 43 29	+0.038	+0.095
6380	η Sco	155203	17 12 09.1	− 43 14 21	+0.023	− 0.285
6396	ζ Dra	155763	17 08 47.1	+65 42 53	− 0.025	+0.021
6406	α^1 Her	156014	17 14 38.8	+14 23 25	− 0.007	+0.034
6410	δ Her	156164	17 15 01.8	+24 50 21	− 0.023	− 0.157
6418	π Her	156283	17 15 02.7	+36 48 33	− 0.030	+0.003
6461	β Ara	157244	17 25 17.9	− 55 31 47	− 0.001	− 0.024
6510	α Ara	158427	17 31 50.4	− 49 52 34	− 0.024	− 0.071
6536	β Dra	159181	17 30 25.8	+52 18 05	− 0.022	+0.013
6553	θ Sco	159532	17 37 19.0	− 42 59 52	+0.016	+0.000
6556	α Oph	159561	17 34 56.0	+12 33 36	+0.117	− 0.227
6561	ξ Ser	159876	17 37 35.1	− 15 23 55	− 0.045	− 0.061
6582	η Pav	160635	17 45 43.9	− 64 43 26	− 0.003	− 0.052
6588	ι Her	160762	17 39 27.8	+46 00 23	− 0.010	+0.004
6603	β Oph	161096	17 43 28.3	+04 34 02	− 0.042	+0.159
6615	ι^1 Sco	161471	17 47 35.0	− 40 07 37	+0.001	− 0.006
6623	μ Her	161797	17 46 27.5	+27 43 15	− 0.309	− 0.747
6629	γ Oph	161868	17 47 53.5	+02 42 26	− 0.024	− 0.074
6630		161892	17 49 51.4	− 37 02 36	+0.054	+0.034
6688	ξ Dra	163588	17 53 31.6	+56 52 21	+0.087	+0.078
6695	θ Her	163770	17 56 15.1	+37 15 02	+0.001	+0.006
6698	ν Oph	163917	17 59 01.5	− 09 46 25	− 0.009	− 0.119
6703	ξ Her	163993	17 57 45.8	+29 14 52	+0.084	− 0.017
6705	γ Dra	164058	17 56 36.3	+51 29 20	− 0.013	− 0.020
6714	67 Oph	164353	18 00 38.6	+02 55 53	− 0.001	− 0.010
6746	γ^2 Sgr	165135	18 05 48.4	− 30 25 27	− 0.053	− 0.185
6771	72 Oph	165777	18 07 20.9	+09 33 50	− 0.062	+0.081

Brightest Stars

BS = HR	V (mag)	B–V (mag)	S_p	π_t ($''$)	D (ly)	M_V (mag)	V_r (km s^{-1})
6217	1.92	+1.44	K2IIb–IIIa	0.031	105.2	− 0.62	− 3
6220	3.53	+0.92	G7IIIFe-1	0.034	95.9	+1.19	+ 8
6229	3.76	+1.57	K5III	0.025	130.5	+0.75	+ 9
6241	2.29	+1.15	K2.5III	0.022	148.3	− 1.00	− 3
6271	3.62	+1.37	K4III	0.028	116.5	+0.86	− 19
6285	3.13	+1.60	K3III	0.044	74.1	+1.35	− 6
6299	3.20	+1.15	K2III	0.031	105.2	+0.66	− 56
6324	3.92	− 0.01	A0V	0.028	116.5	+1.16	− 25
6378	2.43	+0.06	A2V	0.052	62.7	+1.01	− 1
6380	3.33	+0.41	F3III–IVp	0.063	51.8	+2.33	− 28
6396	3.17	− 0.12	B6III	0.023	141.8	− 0.02	− 17
6406	3.48	+1.44	M5Ib–II				− 33
6410	3.14	+0.08	A3IV	0.044	74.1	+1.36	− 40
6418	3.16	+1.44	K3II	0.025	130.5	+0.15	− 26
6461	2.85	+1.46	K3Ib–IIa	0.034	95.9	+0.51	− 0
6510	2.95	− 0.17	B2Vne	0.007	465.9	− 2.82	+ 0
6536	2.79	+0.98	G2Ib–IIa	0.013	250.9	− 1.64	− 20
6553	1.87	+0.40	F1II	0.027	120.8	− 0.97	+ 1
6556	2.08	+0.15	A5III	0.065	50.2	+1.14	+13
6561	3.54	+0.26	F0IVdel Sct	0.030	108.7	+0.93	− 43
6582	3.62	+1.19	K2II	0.025	130.5	+0.61	− 8
6588	3.80	− 0.18	B3IV	0.005	652.3	− 2.71	− 20
6603	2.77	+1.16	K2IIICN0.5	0.033	98.8	+0.36	− 12
6615	3.03	+0.51	F2Iae	0.019	171.7	− 0.58	− 28
6623	3.42	+0.75	G5IV	0.108	30.2	+3.59	− 16
6629	3.75	+0.04	A0Vnp	0.039	83.6	+1.71	− 7
6630	3.21	+1.17	K2III	0.040	81.5	+1.22	+25
6688	3.75	+1.18	K2III	0.035	93.2	+1.47	− 26
6695	3.86	+1.35	K1IIaCN2	0.002	1630.8	− 4.63	− 27
6698	3.34	+0.99	G9IIIa	0.021	155.3	− 0.05	+13
6703	3.70	+0.94	G8.5III	0.021	155.3	+0.31	− 2
6705	2.23	+1.52	K5III	0.025	130.5	− 0.78	− 28
6714	3.97	+0.02	B5Ib				− 4
6746	2.99	+1.00	K0+III	0.025	130.5	− 0.02	+22
6771	3.73	+0.12	A4IV s	0.047	69.4	+2.09	− 24

Brightest Stars

BS = HR	Name	HD	R.A.(2000) h m s	Dec.(2000) ° ′ ″	$\mu_{R.A.}$ (″ yr^{-1})	$\mu_{Dec.}$ (″ yr^{-1})
6779	o Her	166014	18 07 32.5	+28 45 45	− 0.001	+0.010
6812	μ Sgr	166937	18 13 45.7	− 21 03 32	+0.003	+0.001
6832	η Sgr	167618	18 17 37.5	− 36 45 42	− 0.129	− 0.166
6859	δ Sgr	168454	18 20 59.6	− 29 49 41	+0.039	− 0.029
6869	η Ser	168723	18 21 18.5	− 02 53 56	− 0.554	− 0.697
6879	ε Sgr	169022	18 24 10.3	− 34 23 05	− 0.032	− 0.125
6895	109 Her	169414	18 23 41.8	+21 46 11	+0.194	− 0.244
6913	λ Sgr	169916	18 27 58.1	− 25 25 18	− 0.043	− 0.185
6927	χ Dra	170153	18 21 03.2	+72 43 58	+0.521	− 0.356
6973	α Sct	171443	18 35 12.3	− 08 14 39	− 0.019	− 0.312
7001	α Lyr	172167	18 36 56.2	+38 47 01	+0.200	+0.285
7106	β Lyr	174638	18 50 04.7	+33 21 46	+0.000	− 0.002
7150	ξ^2 Sgr	175775	18 57 43.7	− 21 06 24	+0.032	− 0.012
7178	γ Lyr	176437	18 58 56.5	+32 41 22	− 0.006	+0.002
7194	ζ Sgr	176687	19 02 36.6	− 29 52 49	− 0.014	− 0.001
7217	o Sgr	177241	19 04 40.9	− 21 44 30	+0.079	− 0.060
7234	τ Sgr	177716	19 06 56.3	− 27 40 14	− 0.053	− 0.249
7235	ζ Aql	177724	19 05 24.5	+13 51 48	− 0.007	− 0.095
7236	λ Aql	177756	19 06 14.8	− 04 52 57	− 0.021	− 0.088
7264	π Sgr	178524	19 09 45.7	− 21 01 25	+0.000	− 0.035
7310	δ Dra	180711	19 12 33.1	+67 39 42	+0.090	+0.093
7328	κ Cyg	181276	19 17 06.0	+53 22 07	+0.055	+0.125
7340	ρ^1 Sgr	181577	19 21 40.3	− 17 50 50	− 0.026	+0.024
7377	δ Aql	182640	19 25 29.8	+03 06 53	+0.253	+0.083
7417	β^1 Cyg	183912	19 30 43.2	+27 57 35	+0.001	− 0.002
7420	ι^2 Cyg	184006	19 29 42.2	+51 43 47	+0.018	+0.130
7525	γ Aql	186791	19 46 15.5	+10 36 48	+0.016	+0.002
7528	δ Cyg	186882	19 44 58.4	+45 07 51	+0.049	+0.049
7536	δ Sge	187076	19 47 23.2	+18 32 03	+0.007	+0.010
7557	α Aql	187642	19 50 46.9	+08 52 06	+0.537	+0.387
7570	η Aql	187929	19 52 28.3	+01 00 20	+0.007	− 0.006
7582	ε Dra	188119	19 48 10.3	+70 16 04	+0.078	+0.039
7590	ε Pav	188228	20 00 35.4	− 72 54 38	+0.082	− 0.131
7602	β Aql	188512	19 55 18.7	+06 24 24	+0.043	− 0.479
7615	η Cyg	188947	19 56 18.3	+35 05 00	− 0.033	− 0.026

Brightest Stars

BS = HR	V (mag)	B–V (mag)	S_p	π_t ($''$)	D (ly)	M_V (mag)	V_r (km s^{-1})
6779	3.83	- 0.03	B9.5V	0.005	652.3	- 2.68	- 30
6812	3.86	+0.23	B8Iap	0.012	271.8	- 0.74	- 6
6832	3.11	+1.56	M3.5III	0.045	72.5	+1.38	+ 1
6859	2.70	+1.38	K2.5IIIaCN0	0.047	69.4	+1.06	- 20
6869	3.26	+0.94	K0III–IV	0.058	56.2	+2.08	+ 9
6879	1.85	- 0.03	B9.5III	0.023	141.8	- 1.34	- 15
6895	3.84	+1.18	K2IIIab	0.025	130.5	+0.83	- 58
6913	2.81	+1.04	K1IIIb	0.053	61.5	+1.43	- 43
6927	3.57	+0.49	F7V	0.120	27.2	+3.97	+33
6973	3.85	+1.33	K3III	0.016	203.8	- 0.13	+36
7001	0.03	0.00	A0Va	0.123	26.5	+0.48	- 14
7106	3.45	+0.01	B7Ve+A8p				- 19
7150	3.51	+1.18	K1III	0.011	296.5	- 1.28	- 20
7178	3.24	- 0.05	B9III	0.021	155.3	- 0.15	- 21
7194	2.60	+0.08	A2III+A4IV	0.025	130.5	- 0.41	+22
7217	3.77	+1.01	G9IIIb	0.044	74.1	+1.99	+25
7234	3.32	+1.19	K1.5IIIb	0.044	74.1	+1.54	+45
7235	2.99	+0.01	A0Vn	0.045	72.5	+1.26	- 25
7236	3.44	- 0.09	B9Vn	0.032	101.9	+0.97	- 12
7264	2.89	+0.35	F2II	0.026	125.4	- 0.04	- 10
7310	3.07	+1.00	G9III	0.032	101.9	+0.60	+25
7328	3.77	+0.96	G9III	0.024	135.9	+0.67	- 29
7340	3.93	+0.22	F0IV–V	0.042	77.7	+2.05	+ 1
7377	3.36	+0.32	F3IV	0.072	45.3	+2.65	- 30
7417	3.08	+1.13	K3II+B9.5V	0.017	191.9	- 0.77	- 24
7420	3.79	+0.14	A5Vn	0.005	652.3	- 2.72	- 20
7525	2.72	+1.52	K3II	0.016	203.8	- 1.26	- 2
7528	2.87	- 0.03	B9.5IV+F1V	0.030	108.7	+0.26	- 20
7536	3.82	+1.41	M2II+A0V	0.001	3261.6	- 6.18	+ 3
7557	0.77	+0.22	A7V	0.198	16.5	+2.25	- 26
7570	3.90	+0.89	F6Ib	0.010	326.2	- 1.10	- 15
7582	3.83	+0.89	G7IIIbFe–1	0.016	203.8	- 0.15	+ 3
7590	3.96	- 0.03	A0V	0.016	203.8	- 0.02	- 2
7602	3.71	+0.86	G8IV	0.070	46.6	+2.94	- 40
7615	3.89	+1.02	K0III	0.015	217.4	- 0.23	- 27

Brightest Stars

BS = HR	Name	HD	R.A.(2000) h m s	Dec.(2000) ° ′ ″	$\mu_{R.A.}$ (″ yr^{-1})	$\mu_{Dec.}$ (″ yr^{-1})
7635	γ Sge	189319	19 58 45.3	+19 29 32	+0.065	+0.025
7665	δ Pav	190248	20 08 43.2	− 66 10 56	+1.198	− 1.141
7710	θ Aql	191692	20 11 18.2	− 00 49 17	+0.036	+0.007
7735	o^1 Cyg	192577	20 13 37.8	+46 44 29	+0.000	+0.005
7751	o^2 Cyg	192909	20 15 28.2	+47 42 52	− 0.003	+0.010
7754	α^2 Cap	192947	20 18 03.2	− 12 32 42	+0.061	+0.005
7776	β Cap	193495	20 21 00.6	− 14 46 53	+0.039	+0.003
7796	γ Cyg	194093	20 22 13.6	+40 15 24	+0.001	+0.002
7869	α Ind	196171	20 37 34.0	− 47 17 29	+0.056	+0.070
7882	β Del	196524	20 37 32.9	+14 35 43	+0.112	− 0.031
7906	α Del	196867	20 39 38.2	+15 54 43	+0.065	+0.000
7913	β Pav	197051	20 44 57.4	− 66 12 11	− 0.037	+0.017
7924	α Cyg	197345	20 41 25.8	+45 16 49	+0.001	+0.005
7949	ε Cyg	197989	20 46 12.6	+33 58 13	+0.355	+0.329
7950	ε Aqr	198001	20 47 40.5	− 09 29 45	+0.033	− 0.032
7957	η Cep	198149	20 45 17.3	+61 50 20	+0.091	+0.822
7986	β Ind	198700	20 54 48.5	− 58 27 15	+0.024	− 0.022
8028	ν Cyg	199629	20 57 10.3	+41 10 02	+0.008	− 0.012
8079	ξ Cyg	200905	21 04 55.8	+43 55 40	+0.005	+0.004
8115	ζ Cyg	202109	21 12 56.1	+30 13 37	− 0.001	− 0.052
8130	τ Cyg	202444	21 14 47.4	+38 02 44	+0.159	+0.437
8131	α Equ	202447	21 15 49.3	+05 14 52	+0.057	− 0.084
8162	α Cep	203280	21 18 34.7	+62 35 08	+0.150	+0.052
8204	ζ Cap	204075	21 26 39.9	− 22 24 41	+0.001	+0.027
8232	β Aqr	204867	21 31 33.4	− 05 34 16	+0.019	− 0.005
8238	β Cep	205021	21 28 39.5	+70 33 39	+0.010	+0.013
8254	ν Oct	205478	21 41 28.6	− 77 23 24	+0.057	− 0.239
8278	γ Cap	206088	21 40 05.4	− 16 39 45	+0.188	− 0.022
8308	ε Peg	206778	21 44 11.1	+09 52 30	+0.030	+0.005
8322	δ Cap	207098	21 47 02.3	− 16 07 38	+0.262	− 0.294
8353	γ Gru	207971	21 53 55.6	− 37 21 54	+0.103	− 0.017
8414	α Aqr	209750	22 05 46.9	− 00 19 11	+0.016	− 0.004
8425	α Gru	209952	22 08 13.9	− 46 57 40	+0.130	− 0.149
8430	ι Peg	210027	22 07 00.6	+25 20 42	+0.299	+0.028
8450	θ Peg	210418	22 10 11.9	+06 11 52	+0.275	+0.032

Brightest Stars

BS = HR	V (mag)	B–V (mag)	S_p	π_t ($''$)	D (ly)	M_V (mag)	V_r (km s^{-1})
7635	3.47	+1.57	M0–III	0.013	250.9	− 0.96	− 33
7665	3.56	+0.76	G6–8IV	0.170	19.2	+4.71	− 22
7710	3.23	− 0.07	B9.5III	0.012	271.8	− 1.37	− 27
7735	3.79	+1.28	K2II+B3V	0.007	465.9	− 1.98	− 8
7751	3.98	+1.52	K3Ib+B3V	0.014	233.0	− 0.29	− 14
7754	3.57	+0.94	G9III	0.034	95.9	+1.23	+ 0
7776	3.08	+0.79	F8V+A0	0.010	326.2	− 1.92	− 19
7796	2.20	+0.68	F8Ib	0.003	1087.2	− 5.41	− 8
7869	3.11	+1.00	K0IIICNIII-	0.046	70.9	+1.42	− 1
7882	3.63	+0.44	F5IV	0.028	116.5	+0.87	− 23
7906	3.77	− 0.06	B9IV	0.008	407.7	− 1.71	− 3
7913	3.42	+0.16	A7III	0.035	93.2	+1.14	+10
7924	1.25	+0.09	A2Iae				− 5
7949	2.46	+1.03	K0III	0.057	57.2	+1.24	− 11
7950	3.77	0.00	A1V	0.021	155.3	+0.38	− 16
7957	3.43	+0.92	K0IV	0.071	45.9	+2.69	− 87
7986	3.65	+1.25	KIII				− 5
8028	3.94	+0.02	A1Vn	0.010	326.2	− 1.06	− 28
8079	3.72	+1.65	K4.5Ib–II	0.007	465.9	− 2.05	− 20
8115	3.20	+0.99	G8+III–IIIa	0.027	120.8	+0.36	+17
8130	3.72	+0.39	F2IV	0.055	59.3	+2.42	− 21
8131	3.92	+++0.53	G0III+A5V	0.021	155.3	+0.53	− 16
8162	2.44	+0.22	A7V	0.068	48.0	+1.60	− 10
8204	3.74	+1.00	G4Ib				+ 3
8232	2.91	+0.83	G0Ib	0.006	543.6	− 3.20	+ 7
8238	3.23	− 0.22	B1IV	0.014	233.0	− 1.04	− 8
8254	3.76	+1.00	K0III	0.053	61.5	+2.38	+34
8278	3.68	+0.32	F0p	0.029	112.5	+0.99	− 31
8308	2.39	+1.53	K2Ib–II	0.006	543.6	− 3.72	+ 5
8322	2.87	+0.29	Am	0.087	37.5	+2.57	− 6
8353	3.01	− 0.12	B8III	0.013	250.9	− 1.42	− 2
8414	2.96	+0.98	G2Ib	0.012	271.8	− 1.64	+ 8
8425	1.74	− 0.13	B7IV	0.057	57.2	+0.52	+12
8430	3.76	+0.44	F5V	0.082	39.8	+3.33	− 4
8450	3.53	+0.08	A2Vp	0.049	66.6	+1.98	− 6

Brightest Stars

BS = HR	Name	HD	R.A.(2000) h m s	Dec.(2000) ° ′ ″	$\mu_{R.A.}$ (″ yr^{-1})	$\mu_{Dec.}$ (″ yr^{-1})
8465	ζ Cep	210745	22 10 51.2	+58 12 05	+0.013	+0.008
8502	α Tuc	211416	22 18 30.1	- 60 15 35	- 0.059	- 0.039
8518	γ Aqr	212061	22 21 39.3	- 01 23 14	+0.129	+0.012
8556	δ^1 Gru	213009	22 29 16.1	- 43 29 45	+0.028	- 0.003
8571	δ Cep	213306	22 29 10.2	+58 24 55	+0.012	+0.005
8585	α Lac	213558	22 31 17.4	+50 16 57	+0.135	+0.023
8634	ζ Peg	214923	22 41 27.6	+10 49 53	+0.080	- 0.008
8636	β Gru	214952	22 42 40.0	- 46 53 05	+0.138	- 0.006
8650	η Peg	215182	22 43 00.1	+30 13 17	+0.013	- 0.021
8667	λ Peg	215665	22 46 31.8	+23 33 56	+0.058	- 0.006
8675	ε Gru	215789	22 48 33.2	- 51 19 01	+0.109	- 0.064
8684	μ Peg	216131	22 50 00.1	+24 36 06	+0.148	- 0.036
8694	ι Cep	216228	22 49 40.7	+66 12 02	- 0.067	- 0.119
8698	λ Aqr	216386	22 52 36.8	- 07 34 47	+0.007	+0.039
8709	δ Aqr	216627	22 54 38.9	- 15 49 15	- 0.042	- 0.022
8728	α PsA	216956	22 57 39.0	- 29 37 20	+0.336	- 0.161
8762	ο And	217675	23 01 55.2	+42 19 34	+0.021	+0.000
8775	β Peg	217906	23 03 46.4	+28 04 58	+0.188	+0.142
8781	α Peg	218045	23 04 45.6	+15 12 19	+0.062	- 0.038
8812	88 Aqr	218594	23 09 26.7	- 21 10 21	+0.053	+0.036
8820	ι Gru	218670	23 10 21.5	- 45 14 48	+0.137	- 0.021
8848	γ Tuc	219571	23 17 25.7	- 58 14 08	- 0.025	+0.089
8852	γ Psc	219615	23 17 09.9	+03 16 56	+0.759	+0.022
8892	98 Aqr	220321	23 22 58.1	- 20 06 02	- 0.122	- 0.090
8961	λ And	222107	23 37 33.8	+46 27 30	+0.160	- 0.416
8974	γ Cep	222404	23 39 20.8	+77 37 57	- 0.065	+0.156

Brightest Stars

BS = HR	V (mag)	B–V (mag)	S_p	π_t ($''$)	D (ly)	M_V (mag)	V_r (km s^{-1})
8465	3.35	+1.57	K1.5Ib	0.017	191.9	- 0.50	- 18
8502	2.86	+1.39	K3III	0.026	125.4	- 0.07	+42
8518	3.84	- 0.05	A0V	0.046	70.9	+2.15	- 15
8556	3.97	+1.03	G6–8III	0.024	135.9	+0.87	+ 5
8571	3.75	+0.60	F5Ib–G2Ib	0.011	296.5	- 1.04	- 15
8585	3.77	+0.01	A1V	0.040	81.5	+1.78	- 4
8634	3.40	- 0.09	B8V	0.023	141.8	+0.21	+ 7
8636	2.10	+1.60	M4.5III	0.008	407.7	- 3.38	+ 2
8650	2.94	+0.86	G2II–III+F0	0.017	191.9	- 0.91	+ 4
8667	3.95	+1.07	G8IIIaCN0.5	0.042	77.7	+2.07	- 4
8675	3.49	+0.08	A3V	0.044	74.1	+1.71	+ 0
8684	3.48	+0.93	G8+III	0.040	81.5	+1.49	+14
8694	3.52	+1.05	K0–III	0.041	79.6	+1.58	- 12
8698	3.74	+1.64	M2.5IIIFe–0	0.017	191.9	- 0.11	- 9
8709	3.27	+0.05	A3V	0.038	85.8	+1.17	+18
8728	1.16	+0.09	A3V	0.149	21.9	+2.03	+ 7
8762	3.62	- 0.09	B6IIIpe+A2p	0.015	217.4	- 0.50	- 14
8775	2.42	+1.67	M2.5II–III	0.022	148.3	- 0.87	+ 9
8781	2.49	- 0.04	B9V	0.038	85.8	+0.39	- 4
8812	3.66	+1.22	K1.5III	0.010	326.2	- 1.34	+21
8820	3.90	+1.02	K1III	0.030	108.7	+1.29	- 4
8848	3.99	+0.40	F1III	0.043	75.9	+2.16	+18
8852	3.69	+0.92	G9III:Fe–2	0.036	90.6	+1.47	- 14
8892	3.97	+1.10	K1III	0.036	90.6	+1.75	- 7
8961	3.82	+1.01	G8III–IV	0.050	65.2	+2.31	+ 7
8974	3.21	+1.03	K1III–IVCN1	0.064	51.0	+2.24	- 42

11
Wolf-Rayet Stars

11.1 Basic Data for Wolf-Rayet Stars

The Wolf-Rayet (WR) stars are very rare, hot, luminous objects whose spectra are dominated by broad ($\sim 10^3$ km s^{-1}) emission lines that rise above a faint continuum in the optical region. Prominent emission lines from highly excited ions of helium, He, carbon, C, nitrogen, N, and oxygen, O, are used in their spectral classification (see accompanying table). There are two major subtypes of the WR stars; those dominated by lines of helium and nitrogen ions (the WN stars) and those that contain lines of carbon, oxygen and helium ions (the WC stars).

Wolf-Rayet stars are thought to be the remnants of initially massive stars that have lost their hydrogen-rich envelopes through mass transfer to a companion or by a strong stellar wind in the case of single objects. The WN stars are believed to be remnants at advanced stages of hydrogen burning, while the WC stars represent a more advanced evolutionary phase involving helium burning.

Only 159 galactic WR stars are listed by van der Hucht et al. (1981); their catalogue, as updated by Lundström and Stenholm (1984), is given in the following pages. It also includes a bibliography to 1981, a historical perspective, and finding charts that we do not reproduce here. Our listing of WR stars (Population I) contains the original catalogue number, or WR No., the star name or designation, the right ascension (R.A.) and declination (Dec.) for the epoch 1950.0, the corresponding galactic longitude and latitude, l and b, the spectral type, Sp, and the apparent visual magnitude, v, and color, b-v, in narrow-band photometry according to Smith's (1968) system; values in ( ) are V magnitudes in the B – V system, while those in [ ] are photographic.

Wolf-Rayet stars associated with ring nebulae and those found in open star clusters or stellar associations are listed in separate tables where distances, D, and absolute magnitudes, M_V, can also be found. The distances have been inferred from the true distance modulus, $y_o = m_o - M$, of the cluster or association, $D = 10^{[(y_o/5)-2]} kpc$; the adapted color excess E_{b-v}, and the true color $(b - v)_o$, are also given for these WR stars. Photometric distances for WR stars not found in clusters or associations are given by Hidayat, Supelli and van der Hucht (1982), but we did not reproduce this data.

Wolf-Rayet stars that are associated with emission nebulae, or H II regions, are given in a separate table, where r designates the sub-class of ring nebulae, as are those nearby WR stars with radio and X-ray emission.

The WR phase is also present in some central stars of planetary nebulae; these Population II objects are also given in a separate table.

General Properties*

Effective Temperature	T_{eff}	=	25,000 °K to 30,000 °K
Radius	R	=	8.6 $R_\odot$ to 35 $R_\odot$ (ten stars, photosphere)
Luminosity	L	=	2.4 to 5.8 $\times 10^5$ $L_\odot$ (ten stars)
Mass	M	=	7 $M_\odot$ to 50 $M_\odot$ (mean $\approx$ 17 $M_\odot$, twelve stars)
Mass-Luminosity Relation	$\log(L/L_\odot)$	$\approx$	$1.5 \log(M/M_\odot) + 3.8$
Mass Loss Rate	$\dot{M}$	=	$(0.8 \text{ to } 8.0) \times 10^{-5}$ $M_\odot$ yr^{-1} (mean $\approx 2 \times 10^{-5}$ $M_\odot$ yr^{-1}, 23 stars)
Terminal Wind Velocity	V_∞	=	1,100 km s^{-1} to 3,500 km s^{-1} (mean $\approx$ 2,300 km s^{-1}, 32 thermal wind)
Wind Kinetic Energy	K.E.	=	10^{38} erg s^{-1} kpc^{-2}
Initial Mass	M_i	$\geq$	30 $M_\odot$ to 50 $M_\odot$

*Adapted from Abbot et al. (1986), Abbot and Conti (1987), Beech (1987), Humphreys et al. (1985), and Underhill (1986).

Spectral Classification*

WN types	Nitrogen ions and (Other Criteria)
WN 9	N III present, N IV weak or absent (He I, lower Balmer series P Cyg)
WN 8	N III $\gg$ N IV (He I strong P Cyg, N III λ 4640 $\approx$ He II λ 4686)
WN 7	N III > N IV (He I weak P Cyg, N III λ 4640 < He II λ 4686)
WN 6	N III $\approx$ N IV, N V present but weak
WN 5	N III $\approx$ N IV $\approx$ N V
WN 4.5	N IV > N V, N III weak or absent
WN 4	N IV $\approx$ N V, N III weak or absent
WN 3	N IV $\ll$ N V, N III weak or absent
WN 2	N V weak or absent (Strong He II)

WC types	Carbon ions λ 5696 C III/λ 5805 C IV and (Other Criteria)	Carbon, oxygen ions λ 5696 C III/λ 5592 O V
WC 9	C III > C IV (C II present)	O V weak or absent
WC 8.5	C III > C IV (C II not present)	O V weak or absent
WC 8	C III $\approx$ C IV	O V weak or absent
WC 7	C III < C IV	C III $\gg$ O V
WC 6	C III $\ll$ C IV	C III > O V
WC 5	C III $\ll$ C IV	C III < O V
WC 4	C IV strong C III weak or absent	O V moderate

*Adapted from Smith (1968). Near infrared spectra with nitrogen lines (WN stars) and carbon lines (WC stars) are given by Vreux, Dennefeld and Andrillat (1983).

Absolute Magnitudes**

(Spectral type, Sp, Absolute visual magnitude, M_V)

Sp	WN 9	WN 8	WN 7	WN 6	WN 5
M_V	- 6.0:	- 6.0:	- 6.5	- 5.2	- 4.7

Sp	WN 3-4.5	WN 2	WC 7-9	WC 5-6	WC 4
M_V	- 3.9	- 2.5:	- 4.9	- 3.9	- 3.0:

**Adapted from Lundström and Stenholm (1984).

Radio and X-Ray Emission from Wolf-Rayet Stars*

WR No.	D (kpc)	E_{b-v} (mag)	M_V (mag)	S_R (mJy)	L_R 10^{18} erg s^{-1} Hz^{-1}	L_x 10^{32} erg s^{-1}
1	2.6		- 4.6	0.5	4.0	
2	2.5		- 2.4	< 0.2	< 1.5	
6	1.7	0.03	(- 4.8)	1.0	3.4	9.0
11	0.5	0.03	- 7.0/- 5.4	29.0	7.0	1.1
16	4.1	0.58				11
21	2.6	0.63				5.8
22	2.6	0.28				9.0
24	2.6	0.19				8.1
25	2.6	0.54				137
30	8.0	0.49				15

*Adapted from Abbot et al. (1986) and Pollock (1987). The radio data is at a frequency of 4.9×10^9 Hz.

Radio and X-Ray Emission from Wolf-Rayet Stars*

WR No.	D (kpc)	E_{b-v} (mag)	M_V (mag)	S_R (mJy)	L_R 10^{18} erg s^{-1} Hz^{-1}	L_x 10^{32} erg s^{-1}
38	2.0	1.19				1.7
40	4.1	0.44				0.7
46	8.7	0.21				35
47	3.8	1.05				19
48	2.4	0.15				20
54	7.4	0.67				27
67	3.6	0.99				8
78	1.9	0.46	− 6.7	1.3	6.5	2.6
79	2.0	0.34	− 6.0/− 5.0	1.1	4.8	4.4
81	1.9		(− 5.1)	0.3	1.3	
86	1.7		(− 4.7)	0.5	1.7	
87	2.9		− 6.4:	< 0.4	< 3.0	
89	2.9		− 7.1:	0.6	6.0	
90	1.8		(− 4.7)	< 0.3	< 1.2	
93	1.7		− 5.6/− 5.3:	0.9	3.1	
95	1.9		− 5.1:	< 0.4	< 1.7	
97	2.9	0.89	(− 4.0)	< 0.6	< 6.0	34
98	1.9	1.33	− 4.4:	0.9	3.9	
102	5.6	0.66				13
103	2.8		(− 5.1)	≤ 0.2	≤ 1.9	
104	1.6	1.86	− 4.6	< 0.4	< 1.2	3.8
105	1.6	2.16	− 6.1	3.6	11.0	8.2
108	2.6		(− 5.8)			
110	2.1		(− 5.3)	1.0	5.3	
111	1.6	0.16	− 3.5	0.3	0.9	0.4

Radio and X-Ray Emission from Wolf-Rayet Stars*

WR No.	D (kpc)	E_{b-v} (mag)	M_V (mag)	S_R (mJy)	L_R 10^{18} erg s^{-1} Hz^{-1}	L_x 10^{32} erg s^{-1}
113	2.0	0.80	– 5.4/– 4.7	≤ 0.4	≤ 1.9	2.0
114	2.2		– 3.5	< 0.3	< 1.2	
115	2.2		– 5.2	0.4	2.3	
121	2.3		(– 5.1)	< 0.3	< 1.9	
122	6.9	1.76				35
124	4.6	1.16				2
125	1.9	1.58	(– 4.7)	1.5	6.0	14
130	2.9		(– 5.8)	< 0.2	< 2.0	
133	2.1		– 6.2/– 3.6:	≤ 0.3	≤ 1.6	
134	2.1	0.50	– 5.3	0.8	4.2	4.6
135	2.1	0.44	– 4.3	0.6	3.2	0.3
136	1.8	0.50	– 5.6	1.6	6.2	0.6
137	1.8	0.39	– 5.0/– 3.6	0.4	1.5	1.6
138	1.8	0.47	– 5.2/– 3.8:	0.6	2.3	4.6
139	1.7	0.71	– 5.7/– 4.4:	0.3	1.0	7.7
140	1.3		(– 4.7)	1.5-25.0	3.0-50.0	
141	1.8		– 5.3	0.6	2.3	
142	1.0		– 2.4	< 0.6	< 0.7	
143	0.9		(– 3.9)	< 0.4	< 0.4	
145	2.0	1.84	– 6.7:	1.0	4.8	8.4
146	2.0	2.06	(– 3.0)			9.4
147	1.9	2.24	(– 5.8)	35.3	152.0	47
152	3.5	0.87				9.1
154	3.5	0.53				1.1
155	3.5	0.66				14
156	2.8		(– 5.8)			

*Adapted from Abbot et al. (1986) and Pollock (1987). The radio data is at a frequency of 4.9×10^9 Hz.

Masses of Binary Wolf-Rayet Stars

Double-lined binaries*

WR No.	Star	Period (days)	$M_{WR} \sin^3 i$ ($M_\odot$)	Mass ratio WR/O	a sin i ($R_\odot$)	Basis for i	M_{WR} ($M_\odot$)
11	γ^2 Vel	78.5002	17	0.54	285	$M_o \sim 35$	20
21	HD 90657	8.255	8.4	0.52	53	eclipse	11
30	HD 94305	18.82	15	0.47		$M_o \sim 35$	16
31	HD 94546	4.831	2.7	0.43	38	None	7
42	HD 97152	7.886	3.6	0.59	36	$M_o \sim 35$	20
47	HDE 311884	6.34	40	0.84	64	$M_o \sim 50$	50
79	HD 152270	8.893	1.8	0.36	34	$M_o \sim 60$	20
97	HDE 320102	8.83	1.8	0.33		$M_o \sim 35$	11
113	CV Ser	29.707	11	0.48	120	eclipse	13
127	HD 186943	9.5548	9-11	0.52	55	$M_o \sim 25$	13
133	HD 190918	112.8	0.7	0.26	130	$M_o \sim 35$	9
139	V444 Cyg	4.21238	9.3	0.40	35	eclipse	11
151	CX Cep	2.1267	5	0.43	18	eclipse	5-11
153	GP Cep	6.6884		> 0.22		eclipse	10-25
155	CQ Cep	1.6	23	1.19	20	eclipse	≥ 23

*Adapted from Abbot and Conti (1987), Massey (1981, 1982). Positions and spectral types are given in the main catalogue. These Wolf-Rayet (WR) binary stars have companion O stars whose spectral types are also given in the main catalogue; their estimated masses are given here in the basis for i column.

Single-lined binaries**

WR No.	Star	Period (days)	f(m) ($M_\odot$)	$M_{companion}$ if $M_{WR} = 15\ M_\odot$: $i = 90°$	$i = 60°$	$i = 25°.8$
6	EZ CMa	3.763	0.015	1.6	1.4	4.0
22	HD 92740	80.35	1.75	10	13	40
48	θ Mus	18.34	9.9	25	32	150
141	HD 193928	21.64	4.3	16	20	75
148	HD 197406	4.3174	0.025	4.6	5.4	14

**Adapted from Massey (1981).

Wolf-Rayet Ring Nebulae*

WR No	Nebula Name	and Type	D (kpc)	Angular Size (′)	N_e (cm^{-3})	M ($M_\odot$)	V_{ex} (km s^{-1})	Age (10^4 yr)
	A. Probable Galactic Wolf-Rayet Ring Nebulae							
6	S308	W	1.5	40	< 50	40	60	7
7	NGC 2359	W	5	4.5	1000	16	18	13
18	NGC 3199	W	3.3	16 × 20	< 100-300	400	20	25
23	Anon(MR 26)	W	2.5	15 × 35			15-30	18-36
40	RCW 58	E	3	7 × 9	< 100-500	5		
48	Anon(θ Mus)	R_s	0.9	80 × 45			< 7	> 100
52	Anon(MR 46)	R_s	4	20 × 30			< 12	> 100
55	RCW 78	R_a	7.2	25 × 35			NA	
75	RCW 104	W	3	4 × 9	190	147	25	8
102	G2.4+1.4	R_s	10-15?	8 × 9	1000		30	21?
124	M1-67	E	2.6	1.2	2500	0.5	42	0.6
128	Anon(MR 95)	R_s	5.5	20			< 10	> 100
131	L69.8+1.74	R_a	0.9	14	50-1500		NA	
134	Anon(MR 100)	W	2.1	17	50-500		50	9
136	NGC 6888	W	1.2	12 × 18	100-400	5	75	2.4
	B. Possible Galactic Wolf-Rayet Ring Nebulae							
22	Anon(MR 25)	W?	2.5	15				
85	RCW 118	R_s	2.4	27			< 10	> 57
93	NGC 6357	W R_s?	1.3	40 × 30				

*Adapted from Chu, Treffers and Kwitter (1983). The D denotes distance, N_e the electron density, M the nebula mass, and V_{ex} the expansion velocity. The names, positions and spectral types for these stars can be found in the main catalogue. The nebula type is coded as E for stellar ejecta, R for radiatively excited H II regions, with subscripts s and a for shell-structured and amorphous, and W for wind-blown.

Wolf-Rayet Stars Correlated with Emission Nebulae, or H II Regions*

WR No.	Star Name	H II Region
6	EZ Cma	S308 (r), RCW 11
7	HD 56925	NGC 2359 (r), S298
8	HD 62910	L247.26-3.88
9	HD 63099	Anon
11	γ^2 Vel	Gum Nebula
14	Ve 5-8	Gum 21
16	HD 86161	282.2-2.0
18	HD 89358	NGC 3199 (r), RCW 48, Gum 28
22	HD 92740	NGC 3372 (r?)
23	HD 92809	Anon (r)
24	HD 93131	NGC 3372
25	HD 93162	NGC 3372
36	LS6	Gum 34b
40	HD 96548	RCW 58 (r)
43	HD 97950	NGC 3603, Gum 38b
50	LSS 3013	RCW 75
51	LSS 3017	RCW 75
52	HD 115473	Anon (r)
53	HD 117297	RCW 78, BBW 27500
55	HD 117688	RCW 78 (r)
60	HD 121194	Anon
66	HD 134877	BBW 28802
67	LSS 3329	BBW 28802
75	HD 147419	RCW 104 (r)
76	LSS 3693	G338.9+0.6
77	LSS 3703	RCW 108, NGC 6193
78	HD 151932	RCW 113
79	HD 152270	RCW 113
85	HD 155603B	RCW 118 (r)
86	HD 156327	S 10, RCW 130
87	LSS 4064	RCW 122, 123
88	LSS 4068	S 5, RCW 130
89	LSS 4065	RCW 122, 123
90	HD 156385	RCW 114
91	St Sa 1	RCW 122

Wolf-Rayet Stars Correlated with Emission Nebulae, or H II Regions*

WR No.	Star Name	H II Region
92	HD 157451	RCW 114
93	HD 157504	NGC 6357 (r?), S 11, RCW 131
98	LSS 4282	Anon
102	LSS 4368	G2.4+1.4 (r)
104	Ve 2-45	S 28
111	HD 165763	S 34, RCW 149
113	CV Ser	S 54, RCW 167
115	LSS 5023	S 50
116	AS 306	Anon
121	AS 320	Anon
124	209 BAC	M1-67 (r), S 80
127	HD 186943	S 92
128	HD 187282	S 84 (r)
130	AS 374	S 98
131	MR 97	L69.80+1.74 (r)
132	HD 190002	Anon
133	HD 190918	S 109
134	HD 191765	S 109, Anon (r)
135	HD 192103	S 109
136	HD 192163	NGC 6888 (r), S 105
137	HD 192641	S 109
138	HD 193077	S 109
139	HD 193576	S 109
140	HD 193793	S 109
141	HD 193928	S 109
145	AS 422	S 108
146	HM 19-3	DWD 140
152	HD 211564	S 132 (r)
153	HD 211853	S 132 (r)
157	HD 219460	S 157 (r)
158	AS 513	S 165

*Adapted from van der Hucht et al. (1981), (r) denotes ring nebula

Population II - Central Stars of Planetary Nebulae*

Star	PN	R.A.(1950) h m s	Dec.(1950) ° ′ ″	Sp	V (mag)
HD 826	NGC 40	00 10 18.0	+72 15	WC8	11.64
	NGC 246	00 44 30.	- 12 09	OVI	11.95
HD 11758	IC 1747	01 54 00.	+63 05	WC6	14.9
	IC 351	03 44 18.	+34 54	cont.	15.
	IC 2003	03 53 10.5	+33 43 51	WC7–8	17.8
	NGC 1501	04 02 42.	+60 47	WC6	14.2
	M4–18	04 21 30.	+60 00		
HD 35914	IC 418	05 25 30.	- 12 44	WC7	9.57
	NGC 2371–2	07 22 26.2	+29 35 35	WC8	14.82
	NGC 2452	07 45 23.2	- 27 12 37	WC7–8	19.1
	A 30	08 44 03.5	+18 03 48	OVI	14.30
	NGC 2867	09 20 00.9	- 58 05 58	WC7–8	14.9
HD 117622	NGC 5189	13 30 10.	- 65 43 06	WC7–8	14.1
LSS 3169	PK 309–04°1	13 48 46.7	- 66 08 40	WC9	13.5
	NGC 5315	13 50 12.3	- 66 16 07	WC6	14.9
HD 125720	IC 4406	14 19 18.	- 43 55	W?	17.:
LSS 3299	HE3–1044	14 56 15.3	- 54 06 14		> 12.5
HD 151121	NGC 6210	16 42 24.	+23 54	WC7	11.3
CPD–56°8032	PK 333–09°1	17 04 48.0	- 56 51 01	WC11	12.2
HD 164963	NGC 6543	17 58 36.	+66 38	WN6	10.4
	M1–41	18 06 26.4	- 24 12 54	WN	
HD 166802	NGC 6572	18 09 42.	+06 50	WN6	> 11.0
HD 167362	VV 164	18 13 00.	- 30 53	WC10	11.
	VV 391	18 15 12.	+10 03	em.uncl.	13.
V348 SGR		18 37 24.	- 23 03	WC11	

Population II - Central Stars of Planetary Nebulae*

Star	PN	R.A.(1950) h m s	Dec.(1950) ° ′ ″	Sp	V (mag)
HD 173283	IC 4776	18 42 36.	- 33 24	WC6	15.
	VV 458	18 47 36.	+20 47	em.	13.
HD 177656	NGC 6751	19 03 15.4	- 06 04 10	WC6	12.87
	NGC 6790	19 20 42.	+01 25	cont.	10.5
HD 183889	NGC 6803	19 28 54.	+09 57	cont.	14.
HD 184738	VV503	19 32 48.	+30 24	WC9	10.10
HM SGE		19 39 41.4	+16 37 33.1	WC	12.
	VV 510	19 40 00.	+15 02	em.uncl.	
HD 186924	NGC 6826	19 43 30.	+50 24	WN6	10.2
	NGC 6879	20 08 12.	+16 46	Of-WR	(15.)
	NGC 6884	20 08 48.	+46 19	cont.	(19.:)
HD 192563	NGC 6891	20 12 48.	+12 33	Of-WR	11.09
HD 193538	IC 4997	20 17 54.	+16 35	WN7-8	11.4
HD 193949	NGC 6905	20 20 09.7	+19 56 38	WC6	13.9
HD 201192	NGC 7026	21 04 36.	+47 39	WC6	14.8
HD 205211	IC 5117	21 30 36.	+44 22	W II	
	A 78	21 33 20.4	+31 28 13	OVI	13.25
HD 212534	IC 5217	22 21 54.	+50 43	WN6	14.6
	VV 564	22 30 24.	+55 55	em.uncl.	
	VV 576	23 24 00.	+57 54	em.uncl.	(15.)

*Adapted from van der Hucht, Conti, Lundström and Stenholm (1981).

Wolf-Rayet Stars in Open Clusters*

WR No.	Cluster Name	IAU No.	m_o-M (mag)	D (kpc)	E_{b-v} (mag)	$(b-v)_o$ (mag)	M_V (mag)
6	Cr 121	C0652–245	9.8	0.91	0.05:	– 0.12:	– 3.1
10	Ru 44	C0757–284	13.3	4.57	0.47	– 0.29	– 4.1
12	Bo 7	C0843–458	13.8:	5.75	0.74:	– 0.26:	– 5.8
17	Ru 161	C1007–609			(0.30)	(– 0.26)	
23	Bo 10	C1040–588	11.6	2.09	0.26:	– 0.22	– 2.8
24	Cr 228	C1041–597	12.1	2.63	0.20	– 0.26	– 6.4
25	Tr 16	C1043–594	12.1	2.63	0.56:	– 0.27:	– 6.2
37	Ru 93	C1102–611	11.0	1.58			
43	NGC 3603	C1112–609	13.6:	5.25	1.38:	– 0.32:	– 7.8:
47	Ho 15	C1240–628	12.9	3.80	1.02:	– 0.30:	– 6.0
48A	Danks 1	C1309–624	12.2::	2.75	9.5		– 5
	Danks 2	C1310–624					
51	Sk 16	C1315–623	11.6	2.09	(1.31)	(– 0.26)	– 2.0
58	Cr 277	C1345–658			(0.68)	(– 0.26)	
62	Tr 22	C1427–609	11.1	1.66	(1.86:)	(– 0.26)	– 4.5:
67	Pi 20	C1511–588	12.8	3.63	1.00:	– 0.28:	– 4.7
78	NGC 6231	C1650–417	11.5	2.00	0.43:	– 0.22:	– 6.7
79					0.36	– 0.35	– 6.0
87	HM 1	C1715–387	12.3	2.88	1.64:	– 0.30:	– 6.4:
89					1.54	– 0.32	– 7.1:
93	Pi 24	C1722–343	11.2:	1.74	1.42:	– 0.27:	– 5.6:
95	Tr 27	C1732–334	11.4	1.91	1.90::	– 0.41::	– 5.1:
98					(1.34)	(– 0.26)	– 4.4
104	Bo 14	C1758–237	10.3:	1.15	(1.73)	(– 0.42)	– 3.9:
115	Do 28	C1822–146			(1.40)	(– 0.26)	
117	Do 29	C1828–066			(1.51)	(– 0.36)	
120	Do 33	C1838–044	12.7:	3.47	(1.29)	(– 0.26)	– 5.7:
133	NGC 6871	C2004+356	11.6	2.09	0.46	– 0.33	– 6.2
135	NGC 6883	C2009+357	10.7	1.38	0.26:	– 0.32:	– 3.3
137	Do 3	C2013+366			(0.45)	(– 0.30)	

Wolf-Rayet Stars in Open Clusters*

WR No.	Cluster Name	IAU No.	m_o-M (mag)	D (kpc)	E_{b-v} (mag)	$(b-v)_o$ (mag)	M_V (mag)
139	Be 86	C2018+385	11.2	1.74	(0.68)	(– 0.30)	– 5.7
142	Be 87	C2019+372	9.9	0.95	(1.94)	(– 0.22)	– 2.4
157	Ma 50	C2313+602	12.8	3.63	0.74	– 0.22	– 5.8

*Adapted from Lundström and Stenholm (1984). Positions, spectral types, Sp, the observed magnitudes, v, and the observed colors, b-v, can be found in the main catalogue. We have used the true cluster distance modulus $y_o = m_o$ - M, to derive the cluster distance, D, in kpc from the relation $D = 10^{[(y_o/5)-2]}$.

Wolf-Rayet Stars in Associations*

WR No.	Association	m_o-M (mag)	D (kpc)	E_{b-v} (mag)	$(b-v)_o$ (mag)	M_V (mag)
1	Cas OB7	12.1	2.63	0.74:	– 0.18:	– 4.6
2	Cas OB1	12.0	2.51	(0.44:)	(+0.26:)	– 2.4
8	Anon Pup a	12.7	3.47	0.68	+0.25	– 4.9
9	Anon Pup b	13.7:	5.50	(1.02)	(+0.26)	– 6.8:
11	Vel OB2	8.3	0.46	0.03	+0.35	– 7.0
14	Anon Vel a	11.5	2.00	(0.41)	(– 0.26)	– 3.8
15	Anon Vel b	12.0:	2.51	(1.01)	(– 0.26)	– 4.4:
18				(0.80)	(– 0.26)	– 4.2
21				(0.65)	(– 0.35)	– 5.0
22	Car OB1	12.1	2.63	0.23:	– 0.20:	– 6.6
23				(0.30)	(– 0.26)	– 3.6
42				(0.29)	(– 0.35)	– 5.0
48	Cen OB1	11.9	2.40	0.22:	– 0.33:	– 7.1
65				(1.77)	(– 0.42)	(– 5.6)
66	Anon Cir	(12.8)	3.63	1.03:	– 0.30:	(– 5.3)
68				(1.27)	(– 0.30)	(– 3.8)
74	Nor OB4	13.0	3.98	(1.78)	(– 0.26)	– 6.3
75				(0.89)	(– 0.26)	– 5.2

*Adapted from Lundström and Stenholm (1984). Positions, spectral types, Sp, the observed magnitudes, v, and the observed colors, b-v, can be found in the main catalogue. We have used the true association distance modulus $y_o = m_o$ - M, to derive the association distance, D, in kpc from the relation $D = 10^{[(y_o/5)-2]}$.

Wolf-Rayet Stars in Associations*

WR No.	Association	m_o-M (mag)	D (kpc)	E_{b-v} (mag)	$(b\text{-}v)_o$ (mag)	M_V (mag)
104				(1.73)	(- 0.42)	- 4.6
105				(1.97:)	(- 0.26)	- 6.1
108	Sgr OB1	11.0	1.58	(0.94)	(- 0.26)	- 4.7
110				(0.97)	(- 0.26)	- 4.8
111				(0.19)	(- 0.26)	- 3.5
113	Ser OB2	11.5	2.00	(0.82)	(- 0.35)	- 5.4
114	Ser OB1	11.7	2.19	(1.16)	(- 0.26)	- 3.5
115				(1.40)	(- 0.26)	- 5.2
121	Anon Sct	12.7:	3.47	(1.37)	(- 0.42)	- 6.0
127	Vul OB2	13.2	4.37	(0.56)	(- 0.35)	- 5.1
134				0.48:	- 0.23:	- 5.3
135	Cyg OB3	11.6	2.09	(0.30)	(- 0.36)	- 4.3
136				0.49:	- 0.24:	- 5.6
137				(0.45)	(- 0.30)	- 5.0
138	Cyg OB1	11.3	1.82	0.52:	- 0.26:	- 5.2
141				1.01	- 0.26	- 5.3
152				(0.61:)	(- 0.26)	- 3.6:
153	Cep OB1	12.7	3.47	0.67:	- 0.35:	- 6.2
154				(0.57)	(- 0.26)	- 3.4
155				0.63:	- 0.30:	- 6.3

*Adapted from Lundström and Stenholm (1984). Positions, spectral types, Sp, the observed magnitudes, v, and the observed colors, b-v, can be found in the main catalogue. We have used the true association distance modulus $y_o = m_o - M$, to derive the association distance, D, in kpc from the relation $D = 10^{[(y_o/5)-2]}$.

11.2 Catalogue of Wolf–Rayet Stars*

WR No.	Star	R.A.(1950) h m s	Dec.(1950) ° ′ ″	l °	b °	Sp	v (mag)	b-v (mag)
1	HD 4004	00 40 29.0	+64 29 19	122.08	+ 1.90	WN5	10.54	+0.56
2	HD 6327	01 02 17.1	+60 09 13	124.65	- 2.41	WN2	11.43	+0.18
3	HD 9974	01 35 37.7	+57 54 07	129.18	- 4.14	WN3 +abs	10.79	0.00
4	HD 16523	02 37 33.0	+56 31 00	137.59	- 2.98	WC5	10.61	+0.23
5	HD 17638	02 48 28.6	+56 43 51	138.87	- 2.15	WC6	11.12	+0.42
6	HD 50896 EZ Cma	06 52 08.1	–23 51 52	234.76	–10.08	WN5	6.94	- 0.07
7	HD 56925	07 16 10.1	–13 07 31	227.75	- 0.13	WN4	11.74	+0.33
8	HD 62910	07 43 01.7	–31 47 10	247.07	- 3.79	WN6 +WC4	10.56	+0.43
9	HD 63099	07 43 57.5	–34 12 30	249.27	- 4.84	WC5 +abs	11.04	+0.76
10	HD 65865 AS 193	07 57 44.3	–28 35 47	245.98	+ 0.58	WN4.5	11.08	+0.18
11	HD 68273 γ^2 Vel	08 07 59.5	–47 11 18	262.80	- 7.69	WC8 +O9I	1.74	- 0.32
12	Ve 5–5	08 43 05.6	–45 47 57	265.20	- 1.97	WN7	11.06	+0.48
13	Ve 6–15	08 48 08.6	–44 59 12	265.13	- 0.77	WC6	13.83	+0.82
14	HD 76536 Ve 5–8	08 53 18.2	–47 24 03	267.55	- 1.64	WC6	9.42	+0.15
15	HD 79573 Ve 5–10	09 11 31.0	–49 54 01	271.42	- 1.08	WC6	11.73	+0.75
16	HD 86161	09 53 14.4	–57 29 25	281.08	- 2.55	WN8	8.43	+0.25
17	HD 88500	10 08 52.8	–60 23 57	284.44	- 3.69	WC5	11.11	+0.04
18	HD 89358	10 15 14.9	–57 39 46	283.57	- 0.97	WN5	11.20	+0.54
19	LS 3	10 16 17.9	–58 01 23	283.89	- 1.19	WC4	13.85	+0.95
20	Th 35–24	10 17 32.5	–58 54 32	284.51	- 1.84	WN4.5	14.60	+0.74

***Adapted from van der Hucht, Conti, Lundström and Stenholm (1981) as corrected by Lundström and Stenholm (1984). Catalogued objects numbered WR 99 and WR 122 are not now thought to be Wolf-Rayet stars. The star W 48A was discovered by Danks et al. (1983).**

Catalogue of Wolf-Rayet Stars*

WR No.	Star	R.A.(1950) h m s	Dec.(1950) ° ′ ″	l °	b °	Sp	v (mag)	b-v (mag)
21	HD 90657	10 24 40.8	− 58 23 10	285.02	− 0.90	WN4 +O4–6	9.80	+0.30
22	HD 92740	10 39 22.6	− 59 24 55	287.17	− 0.85	WN7 +abs	6.44	+0.03
23	HD 92809	10 39 41.9	− 58 30 36	286.78	− 0.03	WC6	9.71	+0.04
24	HD 93131	10 41 56.7	− 59 51 18	287.67	− 1.08	WN7 +abs	6.49	− 0.06
25	HD 93162	10 42 14.2	− 59 27 24	287.51	− 0.71	WN7 +abs	8.17	+0.29
26	MS 1	10 42 33.0	− 57 34 37	286.68	+ 0.97	WN5	14.64	+0.72
27	LS 4	10 42 40.4	− 58 32 41	287.14	+ 0.12	WC6 +abs	14.73	+1.03
28	Th 35–112	10 46 59.3	− 58 47 44	287.75	+ 0.15	WN6	12.98	+0.72
29	Th 35–117	10 48 48.0	− 60 12 44	288.59	− 1.01	WN7	12.65	+0.64
29A	MS 4	10 49 41.0	− 60 40 38	288.90	− 1.38	WC4 +abs	13.70	+0.58
30	HD 94305	10 49 10.1	− 62 01 08	289.44	− 2.61	WC6 +abs	11.73	+0.27
31	HD 94546	10 51 43.3	− 59 14 48	288.50	+ 0.02	WN4 +O7	10.69	+0.28
32	MS 5	10 57 49.5	− 59 36 36	289.36	+ 0.02	WC5	15.5	
33	HD 95435	10 57 54.7	− 57 32 56	288.51	+ 1.90	WC5	12.34	+0.20
34	LS 5	10 58 04.9	− 61 10 22	290.03	− 1.39	WN4.5	14.50	+0.76
35	Th 35–159	10 58 20.0	− 60 57 44	289.97	− 1.18	WN6	13.83	+0.75
36	LS 6	11 00 27.7	− 59 10 11	289.48	+ 0.56	WN4	13.57	+0.76
37	MS 7	11 03 09.6	− 61 04 28	290.55	− 1.05	WN3	> 16.0	
38	MS 8	11 03 41.7	− 60 57 34	290.57	− 0.92	WC4	13.39	+0.93
39	MS 9	11 04 13.7	− 60 58 04	290.63	− 0.90	WC6	(13.2)	
40	HD 96548	11 04 18.0	− 65 14 21	292.31	− 4.83	WN8	7.85	+0.11

Catalogue of Wolf-Rayet Stars*

WR No.	Star	R.A.(1950) h m s	Dec.(1950) ° ′ ″	l °	b °	Sp	v (mag)	b-v (mag)
41	LS 7	11 05 48.6	- 61 11 25	290.89	- 1.03	WC5	(14.0)	
42	HD 97152	11 07 56.9	- 60 42 27	290.95	- 0.49	WC7 +O5-7	8.25	- 0.06
43	HD 97950 NGC 3603	11 12 57.6	- 60 59 13	291.62	- 0.52	WN7 +abs	(10.02:)	(+1.06:)
44	LSS 2289	11 14 45.3	- 59 10 09	291.18	+ 1.26	WN4	12.96	+0.37
45	LSS 2423	11 35 44.2	- 61 59 24	294.51	- 0.60	WC6	14.82	+0.85
46	HD 104994	12 02 42.8	- 61 46 26	297.56	+ 0.34	WN3	10.96	0.00
47	HDE 311884 Bh 16-31	12 40 52.9	- 62 48 51	302.07	- 0.23	WN6 +O5	11.09	+0.72
48	HD 113904 θ Mus	13 04 52.1	- 65 02 21	304.67	- 2.49	WC6 +O9.5I	5.69	- 0.11
48A		13 09 27.0	- 62 27 01	305.3	+ 0.1	WC9	~ 17	
49	Th 17-22	13 10 33.3	- 65 02 16	305.27	- 2.54	WN5	13.87	+0.56
50	Th 17-84	13 14 45.6	- 62 10 19	306.00	+ 0.28	WC6 +abs	12.49	+0.54
51	Th 17-85	13 15 07.1	- 62 12 34	306.04	+ 0.24	WN4	14.76	+1.05
52	HD 115473	13 15 18.8	- 57 52 29	306.50	+ 4.55	WC5	9.98	+0.15
53	HD 117297 Th 17-88	13 27 31.8	- 61 49 24	307.53	+ 0.44	WC8	11.06	+0.16
54	Th 17-89	13 29 14.1	- 64 46 00	307.27	- 2.50	WN4	12.99	+0.46
55	HD 117688 Th 17-90	13 30 07.1	- 62 03 36	307.80	+ 0.16	WN7	10.87	+0.40
56	LS 8	13 30 17.9	- 63 52 08	307.53	- 1.63	WC6	13.97	+0.26
57	HD 119078	13 39 34.0	- 67 08 57	307.89	- 5.03	WC7	10.11	- 0.18
58	LSS 3162	13 45 24.4	- 65 26 59	308.82	- 3.49	WN4	13.08	+0.42
59	LSS 3164	13 46 03.2	- 61 16 46	309.80	+ 0.57	WC8.5	13.90	+1.24
60	HD 121194	13 52 16.8	- 60 55 09	310.61	+ 0.74	WC8	13.25	+0.94

***Adapted from van der Hucht, Conti, Lundström and Stenholm (1981) as corrected by Lundström and Stenholm (1984). Catalogued objects numbered WR 99 and WR 122 are not now thought to be Wolf-Rayet stars. The star W 48A was discovered by Danks et al. (1983).**

Catalogue of Wolf-Rayet Stars*

WR No.	Star	R.A.(1950) h m s	Dec.(1950) ° ′ ″	l °	b °	Sp	v (mag)	b-v (mag)
61	LSS 3208	14 09 11.2	- 65 12 47	311.28	- 3.91	WN4.5	12.56	+0.28
62	NS 2	14 27 18.7	- 61 07 41	314.59	- 0.75	WN6	14.22	+1.6
63	LSS 3289	14 47 07.5	- 59 39 05	317.42	- 0.39	WN6	12.81	+1.28
64	BS 3	14 53 13.7	- 55 38 54	319.95	+ 2.82	WC7	15.08	+0.03
65	LSS 3319	15 09 45.3	- 59 00 28	320.27	- 1.20	WC9	14.45	+1.35
66	HD 134877	15 10 58.6	- 59 39 21	320.07	- 1.83	WN8	11.71	+0.73
67	MR 55	15 11 35.9	- 58 51 23	320.55	- 1.19	WN6	12.21	+0.74
68	BS 4	15 14 21.5	- 59 27 14	320.54	- 1.88	WC7	14.23	+0.97
69	HD 136488	15 19 58.2	- 62 30 00	319.48	- 4.82	WC9	9.43	+0.14
70	HD 137603	15 25 44.9	- 58 24 33	322.34	- 1.81	WC8 +abs	10.15	+0.91
71	HD 143414	15 59 23.5	- 62 33 20	323.08	- 7.61	WN6	10.22	+0.06
72	NS 1	16 03 12.6	- 35 37 10	341.55	+12.11	[WC3]	(14.24)	(+0.21)
73	NS 3	16 09 01.1	- 46 29 56	334.89	+ 3.38	WC8.5	(14.5:)	
74	BP 1	16 12 24.7	- 51 29 15	331.88	- 0.64	WN7	14.01	+1.52
75	HD 147419	16 20 35.6	- 51 25 11	332.84	- 1.48	WN4	11.42	+0.63
76	LSS 3693	16 36 27.5	- 45 35 20	338.88	+ 0.62	WC8.5	15.36:	+0.94:
77	LSS 3703	16 37 35.6	- 47 56 15	337.26	- 1.09	WC8.5	13.16	+0.60
78	HD 151932	16 48 48.4	- 41 46 17	343.22	+ 1.43	WN7	6.61	+0.21
79	HD 152270	16 50 48.7	- 41 44 21	343.49	+ 1.16	WC7 +O5–8	6.95	+0.01
80	LSS 3871	16 55 22.5	- 45 38 35	340.97	- 1.93	WC9	14.63	+1.11
81	MR 66	16 58 59.8	- 45 54 59	341.15	- 2.60	WC9	12.75	+1.14
82	LS 11	17 00 25.9	- 45 08 05	341.92	- 2.32	WN8	12.42	+0.81
83	MR 67	17 07 11.9	- 46 32 37	341.51	- 4.11	WN6	12.79	+0.65
84	LS 12	17 07 53.7	- 39 49 42	346.98	- 0.21	WN6	13.55	+1.18
85	HD 155603B	17 10 59.0	- 39 42 20	347.43	- 0.61	WN6	10.60	+0.56

Catalogue of Wolf-Rayet Stars*

WR No.	Star	R.A.(1950) h m s	Dec.(1950) ° ′ ″	l °	b °	Sp	v (mag)	b-v (mag)
86	HD 156327	17 15 04.4	- 34 21 2	352.25	+ 1.85	WC7 +abs	9.73	+0.44
87	LSS 4064	17 15 26.3	- 38 46 56	348.69	- 0.77	WN7	12.59	+1.34
88	LSS 4068	17 15 31.6	- 33 54 32	352.67	+ 2.04	WC9	13.38	+1.03
89	LSS 4065	17 15 34.0	- 38 45 44	348.72	- 0.78	WN7	11.53	+1.22
90	HD 156385	17 15 49.0	- 45 35 20	343.16	- 4.76	WC7	7.45	- 0.12
91	St Sa 1	17 16 55.2	- 38 53 47	348.76	- 1.07	WN7	(15.0)	
92	HD 157451	17 21 46.9	- 43 26 56	345.54	- 4.42	WC9	10.60	+0.06
93	HD 157504	17 21 50.0	- 34 08 34	353.23	+ 0.83	WC7	11.46	+1.15
94	HD 158860	17 29 49.9	- 33 36 20	354.60	- 0.25	WN6	12.27	+0.74
95	MR 74	17 33 02.3	- 33 24 18	355.13	- 0.70	WC9	14.10	+1.49
96	LSS 4265	17 33 07.5	- 32 52 39	355.58	- 0.43	WC8.5	14.14	+1.01
97	HDE 320102	17 33 34.9	- 34 00 45	354.68	- 1.12	WN3 +abs	11.15	+0.68
98	HDE 318016	17 33 56.0	- 33 26 08	355.21	- 0.87	WN7 +WC7	12.51	+1.08
99	DA 2	17 36 08.8	- 28 13 30	359.85	+ 1.54	em object	(16.0)	
100	HDE 318139	17 38 53.5	- 32 31 59	356.53	- 1.26	WN6	13.44	+1.17
101	DA 3	17 41 53.9	- 31 49 04	357.47	- 1.43	WC8	(14.9)	
102	LSS 4368	17 42 40.5	- 26 09 20	002.38	+ 1.41	WO1	14.64	+1.33
103	HD 164270	17 58 26.4	- 32 42 55	358.49	- 4.89	WC9	9.01	+0.03
104	MR 80	17 59 01.1	- 23 37 44	006.44	- 0.49	WC9	13.54	+1.31
105	AS 268	17 59 20.5	- 23 34 40	006.52	- 0.52	WN8	12.98	+1.71:
106	HDE 313643	18 01 44.0	- 21 09 44	008.90	+ 0.20	WC9	12.36	+0.72
107	DA 1	18 01 45.5	- 21 51 41	008.29	- 0.16	WN7–8	14.10	+1.32:
108	HDE 313846	18 02 23.5	- 23 00 38	007.36	- 0.85	WN9	10.16	+0.68
109	NS 5	18 04 30.7	- 35 10 47	356.94	- 7.19	WN3	(14.0)	
110	HD 165688	18 04 59.5	- 19 24 26	010.80	+ 0.39	WN6	10.23	+0.71

***Adapted from van der Hucht, Conti, Lundström and Stenholm (1981) as corrected by Lundström and Stenholm (1984). Catalogued objects numbered WR 99 and WR 122 are not now thought to be Wolf-Rayet stars. The star W 48A was discovered by Danks et al. (1983).**

Catalogue of Wolf-Rayet Stars*

WR No.	Star	R.A.(1950) h m s	Dec.(1950) ° ′ ″	l °	b °	Sp	v (mag)	b-v (mag)
111	HD 165763	18 05 28.7	− 21 15 41	9.24	− 0.61	WC5	8.25	− 0.07
112	CRL 2104	18 13 36.8	− 18 59 47	12.15	− 1.19	WC9	18.8	+1.30
113	HD 168206 CV Ser	18 16 19.8	− 11 39 16	18.91	+ 1.75	WC8 +O8–9	9.43	+0.47
114	HD 169010	18 20 26.3	− 13 45 01	17.54	− 0.13	WC5	12.92	+0.90
115	MR 87	18 22 38.8	− 14 40 23	16.98	− 1.03	WN6	12.26	+1.14
116	ST 1	18 24 15.8	− 12 24 40	19.16	− 0.32	WN8	(13.0)	
117	MR 88	18 28 21.1	− 06 37 59	24.74	+ 1.50	WC8	14.19	+1.15
118	CRL 2179	18 28 56.8	− 10 01 27	21.81	− 0.21	WC10	22.0:	+3.0:
119	LS 15	18 36 32.2	− 10 08 16	22.57	− 1.92	WC9	12.43	+0.64
120	MR 89	18 38 21.7	− 04 29 07	27.80	+ 0.29	WN7	12.28	+1.03
121	MR 80	18 41 35.0	− 03 51 04	28.73	− 0.13	WC9	12.34	+0.95
122	NaSt1	18 49 44.8	+00 56 03	33.92	+ 0.26	WN10	15.4	+1.50
123	HD 177230	19 01 20.4	− 04 23 31	30.51	− 4.75	WN8	11.27	+0.47
124	209 BAC	19 09 16.4	+16 46 35	50.20	+ 3.31	WN8	11.61	+0.87
125	MR 93	19 26 03.8	+19 27 09	54.44	+ 1.06	WC5	13.48	+1.35:
126	ST 2	19 37 52.7	+26 27 43	61.89	+ 2.11	WC5	13.3	+0.70
127	HD 186943	19 44 14.3	+28 08 56	64.06	+ 1.73	WN4 +O9	10.36	+0.21
128	HD 187282	19 46 18.0	+18 04 34	55.62	− 3.79	WN4	10.56	+0.02
129	MR 96	19 46 19.4	+30 19 20	66.16	+ 2.44	WN4	(13.3)	
130	AS 374	19 57 19.7	+31 19 15	68.25	+ 0.94	WN8	(12.28)	(+1.40)
131	MR 97	19 58 23.7	+33 07 30	69.90	+ 1.71	WN7 +abs	12.4	+0.73:
132	HD 190002	19 59 43.0	+32 26 02	69.46	+ 1.10	WC6	11.55	+1.13:
133	HD 190918	20 04 04.6	+35 38 39	72.65	+ 2.07	WN4.5 +O9.5Ib:	(6.81)	(+0.13)
134	HD 191765	20 08 21.6	+36 01 40	73.45	+ 1.55	WN6	8.31	+0.25
135	HD 192103 V1042 Cyg	20 10 00.8	+36 02 49	73.65	+ 1.28	WC8	8.51	− 0.06

Catalogue of Wolf-Rayet Stars*

WR No.	Star	R.A.(1950) h m s	Dec.(1950) ° ′ ″	l °	b °	Sp	v (mag)	b-v (mag)
136	HD 192163	20 10 17.1	+38 12 15	75.48	+ 2.43	WN6	7.73	+0.25
137	HD 192641	20 12 39.4	+36 30 28	74.33	+ 1.09	WC7 +abs	8.18	+0.15
138	HD 193077	20 15 08.6	+37 16 04	75.24	+ 1.11	WN5 +abs	8.21	+0.26
139	HD 193576 V444 Cyg	20 17 42.6	+38 34 24	76.60	+ 1.43	WN5 +O6	8.27	+0.38
140	HD 193793	20 18 46.7	+43 41 43	80.93	+ 4.18	WC7 +abs	7.19	+0.24
141	HD 193928	20 19 38.9	+36 45 37	75.33	+ 0.08	WN6	10.15	+0.75
142	Sand. 5	20 19 52.2	+37 12 53	75.73	+ 0.33	WO2	(13.56)	(+1.72)
143	HD 195177	20 26 31.5	+38 27 15	77.50	– 0.05	WC5	12.32	+1.19
144	MR 110	20 30 15.5	+41 05 06	80.04	+ 0.93	WC5	(14.3)	
145	MR 111	20 30 17.9	+40 38 16	79.69	+ 0.66	WN +WC	(12.3)	
146	MR 112	20 33 59.0	+41 12 45	80.57	+ 0.45	WC4	(13.2)	
147	AS 431	20 34 53.8	+40 10 38	79.85	– 0.32	WN7 or of	(13.8)	
148	HD 197406	20 39 54.1	+52 24 33	90.08	+ 6.47	WN7	10.50	+0.42
149	ST 4	21 05 29.5	+48 13 28	89.53	+ 0.65	WN6–7	14.9:	+1.25
150	ST 5	21 48 15.8	+50 28 22	96.13	– 2.48	WC6	(13.0:)	
151	CX Cep	22 07 48.7	+57 29 45	102.66	+ 1.39	WN4 +O8	12.40:	+0.70:
152	HD 211564	22 14 33.2	+55 22 36	102.23	– 0.89	WN3	11.62:	+0.35:
153	HD 211853 GP Cep	22 16 54.5	+55 52 30	102.78	– 0.65	WN6 +O	9.20	+0.32
154	HD 213049	22 25 23.5	+55 59 53	103.85	– 1.18	WC6	11.69	+0.31
155	HD 214419 CQ Cep	22 34 56.8	+56 38 46	105.32	– 1.29	WN7 +O	8.94	+0.33
156	MR 119	22 58 07.7	+60 39 29	109.82	+ 0.92	WN8	11.18	+0.88
157	HD 219460	23 13 02.0	+60 10 40	111.33	– 0.24	WN4.5	10.03	+0.52
158	AS 513	23 41 06.3	+61 39 09	115.03	+ 0.10	WN7	11.49	+0.81

*Adapted from van der Hucht, Conti, Lundström and Stenholm (1981) as corrected by Lundström and Stenholm (1984). Catalogued objects numbered WR 99 and WR 122 are not now thought to be Wolf-Rayet stars. The star W 48A was discovered by Danks et al. (1983).

12
Magnetic Stars

12.1 Basic Data for Magnetic Stars

Magnetic fields on nondegenerate stars have been detected using the Zeeman effect in absorption lines. Most of these magnetic stars are peculiar stars of A or B type, usually called Ap stars, that have certain sharp absorption lines with abnormal strengths, indicating a chemically peculiar situation. Bertaud and Floquet (1974) gave a catalogue of 1049 Ap stars, and Borra and Landstreet (1980) conducted a survey of the longitudinal magnetic field strengths among bright northern and southern Ap stars. Didelon (1983) gave a catalogue of stellar magnetic field measurements for both nondegenerate and degenerate stars, but here we only discuss the nondegenerate magnetic stars; magnetic fields of degenerate stars are given in the section on white dwarfs.

Techniques for measuring magnetic fields in nondegenerate stars have been discussed by Ledoux and Renson (1966), Landstreet (1980) and Borra and Landstreet (1982). The Zeeman effect leads to detectable broadening or splitting of absorption lines that become linearly and/or circularly polarized. Such effects are limited to very sharp absorption lines (observed rotation velocity $v \sin i \leq 10$ $km\ s^{-1}$) with large magnetic field strengths $H \geq 1000$ Gauss.

The most successful model for explaining the periodic magnetic variations of Ap stars is the oblique rotator model in which a rotating star presents, to terrestrial observers, different views of a frozen magnetic field that has, to the first approximation, a dipolar geometry. The inclination, i, of the axis of rotation to the line of sight for an oblique rotator is given by the relation $\sin i = Pv \sin i/(50.6R)$ for a rotation period, P, in days, observed rotation velocity $v \sin i$ in $km\ s^{-1}$, and a radius, R, in solar units, $R_{\odot}$. The radii of the Ap stars range between 2 and 5 $R_{\odot}$, with an average value of $R \sim 3R_{\odot}$, (Stift, 1974; Shore and Adelman, 1979). As examples, we give the HD number, name, hydrogen spectral type, Sp, the $v \sin i$, P, and extrema in the longitudinal magnetic field strength, $H_{\min}$ and $H_{\max}$, for the bright Ap stars, divided into short tables of slowly rotating ($v \sin i \leq 40$ $km\ s^{-1}$ and $P \geq 3$ days) and rapidly rotating Ap stars.

This is followed by the measured periods, P, in days for Ap stars whose periods are not questionable. This data has been adapted from Catalano and Renson (1984, 1988).

The main catalogue of nondegenerate Ap stars is adapted from Didelon (1983). Here we give the HD number, the star name, the right ascension, R.A., and the declination, Dec., at the epoch 1950.0, the apparent visual magnitude, V, the spectral type, Sp, and the lowest and highest longitudinal magnetic field strengths, $H_{\min}$ and $H_{\max}$. When more than one set of magnetic extremes was available, we chose the one with the greatest range. Stars with unmeasurable, questionable, or undetected magnetic fields have not been included. The coordinates for epoch 2000.0, proper motions, and distances for many of the included stars are given in the previous table of the brightest stars.

Bright Slowly–Rotating Ap Stars*

HD	Name	Sp	v sin i ($km\ s^{-1}$)	P (days)	H_{min} (Gauss)	H_{max} (Gauss)
32633	BD +33° 953	B5	23	6.43	- 5700	+3500
40312	θ Aur	B9	48	3.69	- 240	+360
65339	53 Cam	A2	14:	8.02	- 5400	+4200
112413	α^2 CVn	B7	24	5.47	- 1200	+1500
118022	78 Vir	A2	10	3.72	- 1400	- 400
125248	CS Vir	B8	$\leq$ 17	9.30	- 1800	+2200
133029	HR 5597	B9	20	2.88	+2905	+4065
137909	β CrB	F0	$\leq$ 3	18.50	- 700	+1000
152107	52 Her	A2	24	3.86	+900	+1400
215441	BD +54° 2846	B5	$\leq$ 5	9.49	+11000	+20500

*Adapted from Borra and Landstreet (1980). Positions and apparent visual magnitudes are given in the main catalogue.

Bright Rapidly–Rotating Ap Stars*

HD	Name	Sp	v sin i (km s^{-1})	P (days)	H_{min} (Gauss)	H_{max} (Gauss)
11503	γ Ari S	B9	69	1.61	- 900	+410
12447	α Psc A	B9	105	1.41	- 515	+433
19832	56 Ari	B7	200	0.73	- 346	+384
124224	CU Vir	B8	130	0.52	- 437	+811
148112	ω Her	B9	54	1.53	- 251	- 87
170000	ϕ Dra	B9	93	1.72	- 244	+696
196178	HR 7870	B8	55		- 1500	- 700

*Adapted from Borra and Landstreet (1980). Positions and apparent visual magnitudes are given in the main catalogue.

Magnetic Stars

Periods of the Ap Stars*

HD	P (days)	HD	P (days)	HD	P (days)
358	0.9636	10783	4.1	24769	1.49
2453	525:	10840	2.1	25267	3.84+1.21
3580	1.479	12767	1.9	25354	3.901
3980	3.952	15089	1.74	25823	7.2274
5601	1.11	19712	2.2	27309	1.569
6532	0.9	19832	0.72790	27463	2.83
8441	69.	21699	2.49	28843	1.3738
9531	0.67	22470	1.929	29009	3.8
9996	8500.	24155	2.53	29305	2.943
10221	3.2	24712	12.46	30849	15.9

Periods of the Ap Stars*

HD	P (days)	HD	P (days)	HD	P (days)
32549	4.6	42616	17.	71866	6.800
32633	6.430	42657	0.72	72968	11.3
32966	3.1	45530	1.58	73340	2.668
33331	1.14	45583	1.177	74521	4.24
33647	0.57	49333	2.18	74535	3.38
34364	4.1347	49976	2.98	74560	1.55
34452	2.466	51418	5.438	79416	2.91
34631	2.20	53116	11.98	81009	34.0
34797	2.287	54118	3.28	81289	1.86
35298	1.85	55892	0.94	83266	5.1
36313	0.589	56022	0.919	83368	2.852
36526	1.54	56350	1.9	83625	1.08
36540	2.17	56455	1.94	88385	14:
36668	2.12	57946	2.2	89822	11.6
36916	1.565	58292	2.9/1.48?	90044	4.4
37017	0.901	58448	0.83	90569	7.9
37058	14.6	61966	1.04	92385	0.549
37140	2.71	62140	4.3	92664	1.673
37151	0.804	63401	2.4	93030	1.78
37210	11.05	64740	1.330	96616	2.44
37479	1.191	64972	0.73	98088	5.905
37633	1.57	65339	8.027	98457	11.5
37642	1.08	66255	6.8	103192	2.34
37776	1.539	66295	2.45	108662	5.08
37808	1.1	66605	2.22	108945	2.2
38602	2.64	66624	2.01	110066	4900:
38823	8.64	66698	4.1	111133	16.3
39317	2.65	68074	1.18	112185	5.089
40312	3.62	68826	1.5846	112381	2.8
41089	1.38	71808	2.4	112413	5.4693

*Adapted from Catalano and Renson (1984, 1988).

Periods of the Ap Stars*

HD	P (days)	HD	P (days)	HD	P (days)
114365	1.27	148898	1.8/4.7?	197417	4.55
115599	0.675	149822	1.459	199728	2.2
116890	4.313	150549	3.8	203006	2.12
118022	3.722	151525	1.31	204411	0.73
119213	2.450	152107	3.86	206653	1.8
119419	2.6	153882	6.009	207188	2.7
120198	1.380	159376	9.74	207857	20.7
122532	1.8	162374	1.66	208217	8.4
123515	1.5	162588	1.95	208266	14.
124224	0.52068	164429	0.517	212385	2.5
125248	9.295	166469	2.9	213918	1.431
125630	2.2	166596	1.7	215038	2.038
125823	8.817	170000	1.7165	215441	9.487
126515	130.	170397	2.2	215661B	3.8
133029	2.89	170973	0.95	216533	17.2
133880	0.8775	171586	2.1	217833	5.4
134793	2.78	173650	9.98	219749	1.619
137193	4.9	175362	3.67	220825	0.585
137909	18.49	177410	1.1	221006	2.3
140160	1.596	179761	1.7	221394	2.84
140728	1.305	183806	2.9	221568	159.
142301	1.459	184905	1.845	221936	0.632
142884	0.803	184927	9.5	223640	3.7
143658	5.2	187473	4.72	224801	3.74
144231	4.4	187474	2450:	225289	6.43
145102	1.42	188041	225:	279021	2.8
146772	1.734	189832	19:	304842	2.3
147010	3.9	191287	1.623	343872	7.5
147890	4.34	192913	16.8	Cp D	
148330	2.14	196502	20.275	− 68°978	1.812

*Adapted from Catalano and Renson (1984, 1988).

12.2 Catalogue of Nondegenerate Magnetic Stars

HD	Name	R.A.(1950) h m s	Dec.(1950) ° ′	V (mag)	Sp	H_{min} (Gauss)	H_{max} (Gauss)
166		00 04 00	+28 45	6.1			+2125
358	α And	00 05 48	+28 49	2.1	A0 P	- 160	
886	γ Peg	00 10 42	+14 54	2.8	B2 IV	- 1100	+1700
2453		00 25 48	+32 10	6.9	A0 P	- 1030	- 250
3980	HR 183	00 39 30	- 56 47	5.7	A9 P	- 1600	+2000
4174	EG And	00 41 54	+40 24	7.5	M2 E,P	- 1200	+1100
5737	α Scl	00 56 12	- 29 38	4.3	B5	- 370	+300
5797		00 57 30	+60 11	8.5	A P		+1800
8441		01 21 24	+42 53	6.6	A2 P	- 750	+400
9996	HR 465	01 35 30	+45 09	6.4	A P	- 1700	+440
10221	43 Cas	01 38 36	+67 47	5.5	A0 P	- 1200	
10783		01 43 06	+08 19	6.6	A2 P	- 1170	+1850
11187		01 48 12	+54 41	7.1	A0 P	- 70	+1250
8890	α UMi	01 48 48	+89 02	2.0	Ceph	- 7	+16
11503	γ^2 Ari	01 50 48	+19 03	4.8	B9 CR	- 1100	+700
12288		01 59 24	+69 21	8.1	A P	+6100	+8600
12447	α Psc A/B	01 59 30	+02 31	3.8	B9 CR	- 500	+550
12767	ν For	02 02 18	- 29 32	4.7	A0 SI	- 300	+300
13480	6 Tri A	02 09 30	+30 04	4.9	G5 III		+160
15144	HR 710	02 23 36	- 15 34	5.8	A4 P,SB	- 1075	- 320
18078		02 52 48	+55 59	8.1	A P		+3800
18296	21 Per	02 54 12	+31 44	5.2	A0 P	- 1270	+1350
19445		03 05 30	+26 09	8.0	F6 VI		+415
19832	56 Ari	03 09 18	+27 04	5.8	B7 SI	- 400	+550
20210		03 12 54	+34 30	6.4	A2 A7	- 260	
20630		03 16 42	+03 11	4.8			+1000
22049		03 30 36	- 09 38	3.7			+1170
22374	9 Tau	03 34 00	+23 03	6.7	A P		+230
22649	HR 1105	03 37 48	+63 03	5.3	S		+450
22920	22 Eri	03 38 12	- 05 22	5.5	B5 HEW,SI	+200	+400
24712	HR 1217	03 52 54	- 12 15	5.9	F0 P	- 50	+1400
25267	τ^9 Eri	03 57 48	- 24 09	4.7	A0 SI	- 400	0
25354		03 59 54	+37 55	7.9	A0 P	- 380	0
25823	41 Tau	04 03 30	+27 28	5.3	A0 P,SB	- 530	+700
26965		04 13 00	- 07 44	4.4			+1850

*Adapted from Didelon (1983). Coordinates for epoch 2000.0, proper motions, and distances for many of these stars are given in the previous catalogue of the brightest stars.

Catalogue of Nondegenerate Magnetic Stars*

HD	Name	R.A.(1950) h m s	Dec.(1950) ° ′	V (mag)	Sp	H_{min}	H_{max} (Gauss)
27962	68 Tau	04 22 36	+17 49	4.3	A M,SB	- 1200	+770
29139	α Tau	04 33 00	+16 25	1.1	K5 III	- 140	+220
30466		04 46 06	+29 29	7.2	A0 P	+1890	+2320
32633	BD+33°953	05 02 48	+33 51	7.0	B9 P	- 3960	+2120
33254	16 Ori	05 06 36	+09 46	5.4	A2 F2 IV	- 420	+375
33904	μ Lep	05 10 42	- 16 16	3.3	B9 P	- 170	+325
34085	β Ori	05 12 06	- 08 15	0.1	B8 I,P		+130
34452	HR 1732	05 15 42	+33 42	5.4	B6 SI	- 600	+550
35298		05 21 12	+02 02	7.9	B3 HEW	- 2810	+2920
35456		05 22 12	- 02 32	6.9	B6 P	- 300	+1080
35502		05 22 30	- 02 52	7.4	B5 HEW	- 2250	- 95
36313		05 28 12	- 00 25	8.2	B9 P	- 1520	+1110
36526		05 29 42	- 01 38	8.3	B9 HEW	- 980	+3480
36629		05 30 30	- 04 36	7.7	B2 P		+1400
36916		05 32 24	- 04 09	6.7	B9 P,SI		+600
37017		05 32 54	- 04 32	6.6	B2 V,HE	- 2300	+400
37041	θ^2 Ori	05 32 54	- 05 27	5.1	O9.5 V	+100	+170
37058		05 33 06	- 04 52	7.3	B2 P,HEW		+2500
37479	σ Ori E	05 36 12	- 02 37	6.6	B2 HE	- 2200	+2800
37642		05 37 24	- 03 21	8.1	B9 P	- 2980	+2700
37776		05 38 24	- 01 32	7.0	B2 V,HE	- 2000	+250
40312	θ Aur	05 56 18	+37 13	2.6	B9 SI	- 260	+330
42474	WY Gem	06 08 54	+23 13	7.4	M3 E,P		+553
42616		06 10 12	+41 43	6.9	A2 P	- 840	- 430
45348	Canopus	06 22 48	- 52 40	- 0.7	F0 IB	- 103	+626
45677	MWC 142	06 26 00	- 13 01	7.5	B2 E	- 1600	
48915	α CMa	06 42 54	- 16 39	- 1.6	A0 V	- 25	+55
49976	HR 2534	06 48 18	- 07 59	6.3	A P	- 1940	+2220
50169		06 49 30	- 01 35	8.9	A4 P	+670	+2120
51418		06 55 48	+42 23	6.6	A P	- 220	+750
54118	HR 2683	07 03 24	- 56 40	5.3	A0 SI	- 1410	+1100
53791	R Gem	07 04 18	+22 47	7.4	S E	+370	+530
55719	HR 2727	07 10 36	- 40 25	5.3	A P	- 1090	+2010
56495		07 14 30	- 07 26	7.5	A3 P	+210	+570
58260		07 21 30	- 36 15	6.7	B2 V,HE S		+2200

Catalogue of Nondegenerate Magnetic Stars*

HD	Name	R.A.(1950) h m s	Dec.(1950) ° ′	V (mag)	Sp	H_{min} (Gauss)	H_{max} (Gauss)
61421	α CMi	07 36 42	+05 21	0.3	F5 IV	- 17	+7
62140	49 Cam	07 41 54	+62 57	6.3	F0 P	- 2230	+3223
62509	β Gem	07 42 18	+28 09	1.2	K0 III	- 130	
63700		07 47 12	- 24 44	3.3	G0 I,G3 IB	- 410	- 80
64740		07 51 42	- 49 29	4.6	B2 V,HE S	- 800	+400
64486	HR 3082	07 57 00	+79 37	5.3	A0 P	- 1360	+247
65339A	53 Cam	07 57 30	+60 28	6.0	A2 P	- 11400	+6800
71866		08 27 54	+40 24	6.7	A P	- 2440	+2850
72968	3 Hya	08 33 00	- 07 49	5.7	A2 P	- 480	+740
74521	49 Cnc	08 42 00	+10 16	5.6	A0 P	- 180	+1450
76151		08 51 48	- 05 15	6.0	G3		+750
77350	ν Cnc	08 59 48	+24 39	5.4	A0 P,B9 P	+105	+470
77581		09 00 12	- 40 21	6.9	B0 IA,SB	- 10000	+6300
78316	κ Cnc	09 05 00	+10 52	5.1	B8 P,SB	- 640	+460
79158	36 Lyn	09 10 30	+43 25	5.3	B7 HEW/SR	- 1300	- 300
81009	HR 3724	09 20 24	- 09 37	6.5	A P		+7900
84367		09 42 00	- 27 32	4.8	F7 V	- 210	
89069		10 16 18	+79 01	8.4	A P		+2300
89822	30 UMa	10 20 30	+65 49	4.9	A0 P,SB	- 290	+340
90569	45 Leo	10 25 00	+10 01	5.9	A2 P	- 80	+400
92664		10 38 30	- 64 50	5.5	B8 P,SI		+890
94660	HR 4263	10 52 42	- 41 59	6.1	A0 SI	- 3300	
96446		11 04 00	- 59 41	6.7	B2 V,HE	- 1400	
96707	HR 4330	11 06 30	+67 29	6.0	F0 P	- 3920	+830
98088		11 14 24	- 06 52	6.1	A P	- 8400	+7600
98230		11 15 30	+31 49	3.8			+1660
98231		11 15 30	+31 49	3.8			+2990
101065		11 35 12	- 46 26	8.0	A P	- 2500	- 2100
101501		11 38 24	+34 29	5.3			+915
102870	β Vir	11 48 06	+02 03	3.6	F8 V	- 270	
107168	8 Com	12 16 48	+23 19	6.3	A8 (M)	- 140	
108662	17 Com A	12 26 24	+26 11	5.4	A0 P	- 1050	+450
109026	γ Mus	12 29 30	- 71 51	3.9	B4 HEW	+140	+470
110066	HR 4816	12 36 54	+36 14	6.3	A4 P	- 55	+300
110073		12 37 12	- 39 43	4.8	B8 P		+580

*Adapted from Didelon (1983). Coordinates for epoch 2000.0, proper motions, and distances for many of these stars are given in the previous catalogue of the brightest stars.

Catalogue of Nondegenerate Magnetic Stars*

HD	Name	R.A.(1950) h m s	Dec.(1950) ° ′	V (mag)	Sp	H_{min} (Gauss)	H_{max} (Gauss)
110379	γ Vir N	12 39 06	− 01 11	3.7	F0 V	− 300	+250
111133	HR 4854	12 44 30	+06 13	6.3	A P	− 1400	− 280
112185	ε UMa	12 51 48	+56 14	1.8	A0 P	− 300	+600
112413	α^2 CVn	12 53 42	+38 35	2.9	A P	− 3500	+3500
115708		13 16 12	+26 38	8.3	A2 P	+680	+740
116458	HR 5049	13 22 12	− 70 22	5.7	A P	− 2400	− 1500
118022	78 Vir	13 31 36	+03 55	4.9	A P	− 1010	+135
119213		13 38 30	+57 28	6.3	A P	− 1750	+1500
	BD+46°1913	13 53 36	+46 00	9.7	A P/AM	− 500	
124224	CU Vir	14 09 42	+02 39	5.0	B8 SI	− 420	+780
125248	CS Vir	14 15 54	− 18 29	5.9	A0 P	− 1900	+2100
125823		14 19 54	− 39 17	4.4	B2.5 HEW/SI	− 450	+375
126515		14 23 24	+01 13	7.1	A P	− 2260	+1630
128898	α Cir	14 38 24	− 64 46	3.2	F0 P	− 1500	− 100
129174	π^1 Boo A	14 38 24	+16 38	4.9	A0 B8 P	− 75	+190
130559	μ Lib A	14 46 36	− 13 56	5.4	A0 P	− 1310	− 210
131156	ξ Boo A	14 49 06	+19 18	4.7	G8 V	− 46	+192
131511		14 51 06	+19 21	6.0			+1000
131977		14 54 30	− 21 10	5.6			+1000
133029	HR 5597	14 58 54	+47 28	6.2	A0 P	+1150	+3270
133880	HR 5624	15 04 54	− 40 24	5.8	A0 SI	− 2810	+3680
134793		15 09 06	+08 42	8.2	A3 P	− 530	+450
135297		15 11 48	+00 33	8.0	A0 P	− 1100	
137909	β CrB	15 25 48	+29 17	3.7	F0 P,SB	− 960	+1020
137949	33 Lib	15 26 42	− 17 16	6.7	F0 P	+980	+1780
142301	3 Sco	15 51 42	− 25 06	5.9	B HEW	− 4100	+1700
142884		15 54 48	− 23 23	6.8	B9 P,SI		+950
142990	HR 5942	15 55 36	− 24 41	5.4	B3 HEW	− 2500	+650
143807	ι CrB	15 59 24	+29 59	4.9	A0 P	− 340	+75
144334	HR 5988	16 03 06	− 23 28	5.9	B4 HEW/SI	− 1400	+500
145102		16 07 12	− 26 47	6.6	B9 P,SI		+450
145501	ν Sco CD	16 09 06	− 19 19	6.3	B4 HEW	− 1350	
147010		16 17 12	− 19 56	7.4	B9 P,SI,CR		+5050
148112	ω Her	16 23 06	+14 09	4.6	A P	− 1550	+200
149661		16 33 42	− 02 13	5.7			+665

Catalogue of Nondegenerate Magnetic Stars*

HD	Name	R.A.(1950) h m s	Dec.(1950) ° ′	V (mag)	Sp	H_{min} (Gauss)	H_{max} (Gauss)
149911	HR 6179	16 35 18	- 06 26	6.0	A0 P	- 2140	+450
152107	52 Her	16 47 48	+46 04	4.8	A P	- 3700	- 1000
153286		16 54 42	+47 27	6.9	A2 P,F	- 580	- 370
153882	BD+15°3095	16 59 18	+15 01	6.2	A0 P	- 2400	+2800
153919		17 00 30	- 37 46	6.5	O7	- 40	+900
155886		17 12 18	- 26 32	4.3			+2080
164975	W Sgr	18 01 48	- 29 35	5.1	Ceph	- 220	+270
165341	70 Oph A	18 02 54	+02 31	4.0	K0 V		+1800
165474	HR 6758 B	18 03 24	+12 00	7.4	A0 A7P		+900
168733	HR 6870	18 19 30	- 36 42	5.3	A P	- 1680	- 260
170000	ϕ Dra	18 21 30	+71 19	4.2	B9 SI	- 300	+720
170397	HR 6932	18 26 54	- 14 37	6.0	A0 CR	- 700	+900
234677	BY Dra	18 32 42	+51 41	8.6	M0 V,E		+40000
171586		18 33 06	+04 54	6.7	A2 P	- 740	
172167	α Lyr	18 35 12	+38 44	0.1	A0 V	- 60	+100
173524	46 Dra A	18 41 42	+55 29	5.0	A P	- 250	+500
173650		18 43 30	+21 56	6.4	A0 P	- 540	+700
174638	β Lyr	18 48 12	+33 18	3.4	B8 SB		+1900
175362	HR 7129	18 53 18	- 37 25	5.4	B HEW	- 5500	+7200
176232	10 Aql	18 56 30	+13 50	5.9	A4 P	- 315	+440
179761	21 Aql	19 11 12	+02 12	5.2	B8	- 590	+170
182989	RR Lyrae	19 23 54	+42 41	7.6	V	- 1580	+1170
185144		19 32 30	+69 35	4.7			+1160
184552	51 Sgr	19 33 00	- 24 50	5.7	A3	- 230	
187474	HR 7552	19 48 30	- 40 00	5.4	A0 P	0	+2880
188041	HR 7575	19 50 42	- 03 15	5.6	A P	+350	+1270
189849	15 Vul	19 59 00	+27 37	4.6	A M	- 630	+420
190073		20 00 36	+05 36	7.9	A0 E,P		+125
191742		20 08 06	+42 24	7.8	A7 P	- 910	- 175
192678		20 12 18	+53 30	7.4	A P		+4600
192913		20 14 24	+27 38	6.7	A0 P	- 670	+385
194093	γ Cyg	20 20 24	+40 06	2.2	F8 I,P	- 200	+200
196502	73 Dra	20 32 12	+74 47	5.2	A P	- 570	+850
196178	HR 7870	20 32 18	+46 31	5.8	B8 SI	- 1500	- 500
198149		20 44 18	+61 39	3.4			+915

*Adapted from Didelon (1983). Coordinates for epoch 2000.0, proper motions, and distances for many of these stars are given in the previous catalogue of the brightest stars.

Catalogue of Nondegenerate Magnetic Stars*

HD	Name	R.A.(1950) h m s	Dec.(1950) ° ′	V (mag)	Sp	H_{min} (Gauss)	H_{max} (Gauss)
335238	BD+29°4202	20 48 36	+29 37	9.0	A P	+9100	+11800
201091		21 04 42	+38 30	4.8			+2950
201601	γ Equ	21 07 54	+09 56	4.7	F0 P	- 300	+600
203006	θ^1 Mic	21 17 36	- 41 01	4.9	A2 P	- 650	
204075		21 23 48	- 22 38	3.7	G2 G5P	- 700	
204411	HR 8216	21 25 06	+48 37	5.3	A P		+500
205021	β Cep	21 28 00	+70 20	3.2	B	- 570	+1700
207260	ν Cep	21 44 00	+60 53	4.3	A2 IA	- 650	+2800
207757	BD+11°4673	21 48 36	+12 23	7.6	B E,P+M	- 1800	+520
207840		21 49 12	+19 35	5.8	B6 V,P		+1300
208816	VV Cep	21 55 12	+63 23	5.3	M2 I+B,SB	- 360	+850
213306	δ Cep	22 27 18	+58 10	3.8	Ceph	- 34	0
213918		22 31 54	+39 05	8.7	B6 P,SI,SR		+1730
215038		22 38 18	+75 24	8.0	A0 P	- 3000	
215441		22 42 06	+55 20	8.6	A0 P	+4140	+34400
216386		22 50 00	- 07 51	3.7	M2 III		+220
216533		22 50 42	+58 32	7.9	A2 P	- 1000	+95
219134		23 10 54	+56 54	5.6			+670
221507	β Scl	23 30 18	- 38 06	4.5	B9 P		+660
221568		23 30 24	+57 38	7.6	A P		+1800
222107	λ And	23 35 06	+46 11	3.8	G8 III–IV		+15
224085		23 52 30	+28 21	7.6	K2 IV,E	- 500	+100

*Adapted from Didelon (1983). Coordinates for epoch 2000.0, proper motions, and distances for many of these stars are given in the previous catalogue of the brightest stars.

13
Stellar Activity

13.1 General Properties of Flare Stars

Young stars of low luminosity apparently begin their life flaring, and continue flaring throughout most of their life. The term flare stars nevertheless designates a group of dwarf, or main sequence, red stars, generally of spectral type M, with transient (seconds to minutes) optical brightenings. About one fourth of all the stars in our Galaxy are such dwarf M flare stars. These stars are also characterized by strong chromospheric activity as indicated by the presence of emission lines of hydrogen (Hα) and other abundant elements; they are therefore usually classified as dMe stars.

Studies of flare stars in nearby clusters and associations indicate that younger stars emit more powerful flares, and that the proportion of flare stars is largest in younger stellar systems, but the flaring stage continues for 100 million years or longer.

About equal amounts of flare energy are recorded in the optical, UV and X-ray regions of the spectrum, with time averaged U-filter flare luminosities, $L_f(U) \approx 10^{-4} L_{bol}$, where L_{bol} is the bolometric absolute luminosity of the star. Typical flare luminosities are $L_f(U) \approx 10^{29} erg\ s^{-1}$, which is about one thousand times more powerful than similar flares on the Sun. Flare energies of up to 10^{37} erg have been observed, but weaker flares (about 10^{32} erg) occur more frequently. The weaker flares are detected optically at the rate of one every few hours from nearby flare stars. The radio emission from stellar flares is about one thousand times less energetic than that in the other spectral domains, with $L_f(R) \approx 10^{-7} L_{bol}$; the radio data is used to describe the energy release and acceleration mechanisms of flares.

Flare star properties are summarized in the first accompanying table; it is followed by the optical flare luminosities for some of them. Physical parameters of nearby flare stars are then provided (see Pettersen, 1980, 1983). They are followed by a short list of the nearest and brightest radio emitting flare stars (Lang, 1990). The main catalogue of nearby flare stars (within 22 pc of the Sun) was extracted from Kunkel's (1975) review, and supplemented by subsequent reports of flare emission (Gurzadyan, 1980; Pettersen, 1975, 1977, 1980, 1981, 1990-private communication). The positions, R.A. and Dec. for epoch 1950.0, proper motions, μ, apparent magnitude, V, spectral class, Sp, absolute magnitude, M_V, and distance, D, are from Gliese (1969), and Grenon and Rufener (1981). Bjorn Petterson (1990-private communication) has also supplied some average values of the color index B - V. Radial velocities, V_r, accurate to 1 $km\ s^{-1}$ and equivalent widths of Hα in absorption ($-$) or emission ($+$) are from Stauffer and Hartmann (1986). The X-ray luminosities, L_X, are extracted from the lists of Pallavicini (1990) and White, Jackson and Kundu (1989). This last reference provides a radio survey of nearby flare stars. About 40% of those visible with the VLA and nearer than 10 parsecs are detected with flux densities of several mJy at 6 cm wavelength and brightness temperatures of $T_B = 10^8$ to $10^{10} K$ for a radio source size equal to the stellar radius.

Flare Star Properties*

Effective Temperature	T_{eff}	=	2500 ° K to 4500 ° K
Absolute Luminosity	L	=	0.0003 $L_\odot$ to 0.1 $L_\odot$
Radius	R	=	0.13 $R_\odot$ to 0.80 $R_\odot$
Mass	M	=	0.06 $M_\odot$ to 0.80 $M_\odot$
Age	Age	=	10^8 to 10^9 years
Flare Luminosity	$L_f(U)$	=	10^{25} erg s^{-1} to 10^{29} erg s^{-1}
Flare Energy	E_f	=	10^{27} erg to 10^{35} erg

Optical Flare Luminosities*

G	Star	log $L_f(U)$ (erg s^{-1})	G	Star	log $L_f(U)$ (erg s^{-1})
15A	GX And	25.98	473AB	FL Vir	26.92
65AB	UV Cet	26.65	551	α Cen C	25.48
182	V1005 Ori	28.69	644AB	V1054 Oph	27.82
166C	DY Eri	26.91	719	BY Dra	28.26
229		25.57	752B	V1298 Aql	< 23.2
234AB	V577 Mon	26.97	799AB	AT Mic	28.20
278C	YY Gem	28.82	803	AU Mic	28.6
285	YZ CMi	27.35	867A	FK Aqr	27.74
388	AD Leo	27.82	873	EV Lac	27.33
406	CN Leo	26.00	896AB	EQ Peg	27.72
412B	WX UMa	< 23.3			

*Courtesy of Bjorn Ragnvald Pettersen, $L_f(U)$ is the time averaged U–filter flare luminosity.

Physical Parameters of Nearby Flare Stars*

Name	Gliese No.	log T_{eff}	$\log(L/L_\odot)$	$\log(R/R_\odot)$
Single Stars				
AC +71°532	48	3.525	− 1.53 ± 0.03	− 0.29 ± 0.07
V1005 Ori	182	3.553	− 1.42 ± 0.12	− 0.290 ± 0.11
V371 Ori	207.1	3.515	− 1.58 ± 0.13	− 0.30 ± 0.12
YZ CMi	285	3.498	− 1.909 ± 0.014	− 0.43 ± 0.06
AD Leo	388	3.538	− 1.6210 ± 0.0036	− 0.36 ± 0.05
CN Leo**	406	3.447	− 3.025 ± 0.016	− 0.88 ± 0.07
SZ Uma	424	3.553	− 1.451 ± 0.037	− 0.31 ± 0.07
Ross 128	447	3.488	− 2.431 ± 0.0006	− 0.67 ± 0.06
EQ Vir	517	3.610	− 1.10 ± 0.14	− 0.25 ± 0.11
CR Dra	616.2	3.550		
V1216 Sgr**	729	3.491	− 2.4164 ± 0.0054	− 0.67 ± 0.06
EV Lac	873	3.519	− 1.8547 ± 0.0032	− 0.44 ± 0.06
Components of Multiple Systems				
DM +43°44	15A	3.550	− 1.6176 ± 0.0041	− 0.39 ± 0.05
GQ And	15B	3.509	− 2.4379 ± 0.0038	− 0.72 ± 0.06
40 Eri C	166C	3.522	− 2.2051 ± 0.0017	− 0.62 ± 0.05
WX UMa	412B	3.459	− 2.957 ± 0.014	− 0.87 ± 0.07
Proxima Cen	551	3.484	− 2.7661 ± 0.0018	− 0.83 ± 0.06
Ross 868	669A	3.509	− 1.692 ± 0.064	− 0.34 ± 0.09
Ross 867	669B	3.491	− 2.088 ± 0.064	− 0.50 ± 0.09
HD 173739	725A	3.547	− 1.7885 ± 0.0002	− 0.47 ± 0.05
HD 173740	725B	3.532	− 2.0261 ± 0.0002	− 0.55 ± 0.05
VB 10	752B	3.342	− 3.3925 ± 0.0051	− 0.86 ± 0.08
L717–22	867B	3.481	− 2.019 ± 0.074	− 0.45 ± 0.09
Mean Components of Four Binaries				
FF And	29.1	3.550	− 1.24 ± 0.36	− 0.20 ± 0.14
L726–8A/B**	65A/B	3.470	− 2.835 ± 0.003	− 0.84 ± 0.06
YY Gem	278C	3.583	− 1.131 ± 0.039	− 0.21 ± 0.07
Wolf 424 A/B	473 A/B	3.481	− 2.686 ± 0.0033	− 0.78 ± 0.06

*Adapted from Pettersen (1980).

**CN Leo is Wolf 359, L 726B is UV Ceti; and V 1216 Sgr is Ross 652

Physical Parameters of Nearby Flare Stars*

Name	Gliese No.	log T_{eff}	$\log(L/L_{\odot})$	$\log(R/R_{\odot})$
Main Component of One Binary				
Ross 614A	234A	3.502	− 2.1685 ± 0.0085	− 0.56 ± 0.06
Assumed Single Stars				
Wolf 47	51	3.466	− 2.362 ± 0.073	− 0.59 ± 0.10
TZ Ari	83.1	3.495	− 2.531 ± 0.011	− 0.73 ± 0.06
AC +25°7918	109	3.522	− 1.713 ± 0.033	− 0.38 ± 0.07
BD − 21°1377	229	3.568	− 1.236 ± 0.032	− 0.23 ± 0.06
DT Vir	494	3.550	− 1.05 ± 0.14	− 0.10 ± 0.12
V 1285 Aql	735	3.515	− 1.398 ± 0.079	− 0.21 ± 0.09

*Adapted from Pettersen (1980).

Radio–Active Dwarf M Flare Stars*

Star	R.A.(1950.0) h m s	Dec.(1950.0) ° ′ ″	S_{Q6} (mJy)	S_{F20} (mJy)	D (pc)
L 726–8A**	01 36 33.314	− 18 12 23.20	1.0	20	2.7
UV Ceti**	01 36 33.404	− 18 12 21.56	3.2	100	2.7
YY Gem	07 31 25.691	+31 58 47.23	0.4	1	14.5
YZ CMi	07 42 02.962	+03 40 30.39	0.5	20	6.0
AD Leo	10 16 52.604	+20 07 17.59	1.1	100	4.9
Wolf 630A,B	16 52 46.455	− 08 15 13.715	0.9	3	6.2
AT Mic**	20 38 44.4	− 32 36 49.5	3.6	6	8.8
AU Mic	20 42 04.558	− 31 31 17.50	0.8	26	8.8
EQ Peg A**	23 29 20.910	+19 39 41.11	0.3	25	6.4

**Fully resolved binary radio stars.

*Accurate 6 cm positions, quiescent flux density, S_{Q6}, at 6 centimeters wavelength, peak flaring flux density, S_{F20}, at 20 centimeters wavelength, and distance, D, in parsecs. The 1950.0 radio coordinates for all sources except AT and AU Mic are for the epoch 1985.22; precession has been taken into account, but proper motion has not. The measured 1950.0 positions for AT Mic and AU Mic were on February 5 and March 22, 1985.0.

Conus Nebula. In this region active, young flare stars and T Tauri stars are found embedded within the dark gas and dust from which they formed. The bright nebulosity is excited by hot stars of the open star cluster NGC 2264, which is probably no more than a few million years old. This photograph was taken at the Karl-Schwarzschild Observatory using the 2 meter reflecting telescope of the JENA optical works. (Courtesy of the Central Institute of Astrophysics of the Academy of Sciences, DDR, and the Karl-Schwarzschild Observatory, Tautenburg.)

13.2 Catalogue of Nearby Flare Stars

G	Name	R.A.(1950) h m s	Dec.(1950) ° ′ ″	$\mu_{R.A.}$ (s yr^{-1})	$\mu_{Dec.}$ (″ yr^{-1})	V (mag)
15A	DM +43°44	00 15 31	+43 44 24	+0.2653	+0.404	8.090
15B	GQ And	00 15 31	+43 44 24	+0.2653	+0.404	11.063
22A	DM +66°34	00 29 20	+66 57 48	+0.298	- 0.21	10.51
29.1	FF And	00 40 05	+35 16 24	+0.020	+0.09	10.38
51	Wolf 47	01 00 08	+62 05 48	+0.109	+0.11	13.66
54.1	YZ Cet	01 09 54	- 17 16	+0.082	+0.62	11.6
65A	L726–8A	01 36 25	- 18 12 42	+0.232	+0.58	12.45
65B	UV Ceti	01 36 25	- 18 12 42	+0.232	+0.58	12.95
82	Ross 15	01 55 54	+58 16 54	+0.038	- 0.23	12.1
83.1	TZ Ari	01 57 28	+12 49 54	+0.073	- 1.79	12.284
103	CC Eri	02 32 28	- 44 00 36	+0.0053	- 0.301	8.909
109	G036–031	02 41 18	+25 19 00	+0.063	- 0.38	10.597
166C	40 Eri C	04 13 04	- 07 44 06	- 0.147	- 3.440	11.17
182	V 1005 Ori	04 56 59	+01 42 36	- 0.003	- 0.15	9.6
206	Ross 42	05 29 30	+09 47 18	- 0.014	- 0.21	11.50
207.1	V 371 Ori	05 31 09	+01 54 48	- 0.017	- 0.16	11.68
229	DM -21°1377	06 08 28	- 21 50 36	- 0.0094	- 0.725	8.149
234A	Ross 614A	06 26 48	- 02 46 12	+0.049	- 0.69	11.07
234B	Ross 614B	06 26 48	- 02 46 12	+0.049	- 0.69	14.4
	PZ Mon	06 45 46	+01 16 30			10.8
268	Ross 986	07 06 39	+38 37 30	- 0.043	- 0.95	11.536
277A	DM +36°1638A	07 28 40	+36 19 48	- 0.0285	- 0.274	10.529
277B	Ross 989	07 28 39	+36 20 24	- 0.0285	- 0.274	11.76
278C	YY Gem	07 31 26	+31 58 48	- 0.0154	- 0.108	9.07
285	YZ CMi	07 42 04	+03 40 48	- 0.026	- 0.47	11.20
388	AD Leo	10 16 54	+20 07 18	- 0.0346	- 0.049	9.420
398	L 1113–55	10 33 28	+05 22 42	- 0.043	+0.14	12.61
406	CN Leo	10 54 06	+07 19 12	- 0.259	- 2.70	13.53
412B	WX UMa	11 03 02	+43 46 42			14.53
424	SZ UMa	11 17 39	+66 07 00	- 0.4853	+0.148	9.32

Catalogue of Nearby Flare Stars

G	Sp	M_V (mag)	D (pc)	B − V (mag)	V_r (km s^{-1})	Hα (Å)	log L_X (erg s^{-1})
15A	dM2.5	10.32	3.48	1.56	+9.6	− 0.28	27.2
15B	dM4.5e	13.29	3.48	1.80	+9.7	0.00	
22A	dM2.5	10.42	10.42	1.50	− 5.0	− 0.14	
29.1	dM1e	8.7	21.28	1.38		2.27	29.47
51	dM5e	13.81	9.35	1.68			
54.1	dM5.5e	12.4	6.90	1.85			
65A	dM5.5e	15.27	2.73	1.85			}27.75
65B	dM6e	15.8	2.73				
82	dM4e	11.7	1.22		− 9.8	4.73	
83.1	dM8e	14.04	4.46	1.80			27.6
103	dK7e	8.63	11.36	1.39			29.69
109	dM3.5e	11.13	7.81	1.55	+29.6	− 0.31	
166C	dM4e	12.73	4.83	1.68			28.40
182	dM0.5e	8.8	14.71	1.37			29.28
206	dM4e	10.73	14.29	1.62	+31.2	3.17	29.1
207.1	dM2.5e	10.8	15.2	1.56	+21.0	4.42	
229	dM2.5	9.35	5.75	1.50		-1.1	26.9
234A	dM4.5e	13.08	3.97	1.74	+15.4	3.56	26.5
234B	dM7?	16.4	3.97				
PZ Mon	dK2e	7.1	16.				
268	dM5e	12.62	6.06	1.71	+65.0	1.86	27.5
277A	dM3.5e	10.10	12.20	1.46	+0.1	1.35	
277B	dM4.5e	11.48	12.20	1.79	− 0.9	2.15	
278C	dM0.5e	8.26	14.49	1.49		4.0	29.83
285	dM4.5e	12.29	6.06	1.61	26.4	7.48	28.61
388	dM3.5e	10.98	4.88	1.54	+12.0	3.43	28.73
398	dM4e	11.7	15.15	1.56			
406	dM6.5e	16.68	2.35	2.03	0.0	10.46	27.0
412B	dM5.5e	15.88	5.38	2.0	0.0	0.00	27.5
424	dM1.5	9.70	8.5	1.42	+58.2	− 0.40	

Catalogue of Nearby Flare Stars

G	Name	R.A.(1950) h m s	Dec.(1950) ° ′ ″	$\mu_{R.A.}$ (s yr^{-1})	$\mu_{Dec.}$ (″ yr^{-1})	V (mag)
447	Ross 128	11 45 09	+01 06 00	+0.043	− 1.22	11.128
473A	Wolf 424A	12 30 51	+09 17 36	− 0.118	+0.26	13.16
473B	Wolf 424B	12 30 51	+09 17 36	− 0.118	+0.26	13.4
490A	DM +36°2322	12 55 19	+35 29 48	− 0.021	− 0.15	10.60
490B	G 164–31	12 55 18	+35 29 36	− 0.021	− 0.15	13.16
493.1	Wolf 461	12 58 05	+05 57 06	− 0.064	+0.28	13.34
494	DT Vir	12 58 19	+12 38 42	− 0.046	0.00	9.79
516A	VW Com	13 30 18	+17 04 12	+0.020	− 0.22	12.00
516B	G 63–368B	13 30 18	+17 04 12	+0.020	− 0.22	12.3
517	EQ Vir	13 32 07	− 08 05 06	− 0.0186	− 0.098	9.252
540.2	Ross 845	14 10 26	− 11 47 12	− 0.045	− 0.44	13.5
551	Proxima Cen	14 26 19	− 62 28 06	− 0.544	+0.79	11.074
569	DM +16°2708	14 52 06	+16 18 18	+0.021	− 0.13	10.20
616.2	CR Dra	16 15 59	+55 23 48	+0.012	− 0.52	9.972
630.1	CM Dra	16 33 29	+57 14 48	− 0.133	+1.20	12.90
644A	Wolf 630A	16 52 48	− 08 14 42	− 0.0537	− 0.875	9.76
644B	Wolf 630B	16 52 48	− 08 14 42	− 0.0537	− 0.875	9.8
669A	Ross 868	17 17 54	+26 32 48	− 0.016	+0.39	11.36
669B	Ross 867	17 17 53	+26 32 48	− 0.016	+0.39	12.92
719	BY Dra	18 32 45	+51 41 00	+0.0197	− 0.311	8.198
729	V 1216 Sgr	18 46 45	− 23 53 30	+0.051	− 0.17	10.49
735	V 1285 Aql	18 53 03	+08 20 18	+0.007	− 0.08	10.07
752B	VB 10	19 14 30	+05 04 42	− 0.031	− 1.42	17.38
781	Wolf 1130	20 03 55	+54 18 12	− 0.141	− 0.90	12.015
791.2	HU Del	20 27 21	+09 31 12	+0.047	+0.10	13.06
799A	AT Mic A	20 38 44	− 32 36 36	+0.0216	− 0.331	10.83
799B	AT Mic B	20 38 44	− 32 36 36	+0.0216	− 0.331	10.9
803	AU Mic	20 42 04	− 31 31 06	+0.0212	− 0.348	8.701
815A	AC +39°57322	20 58 09	+39 52 42	+0.054	− 0.26	10.26
815B	G210–48B	20 58 09	+39 52 42	+0.054	− 0.26	12.7

Catalogue of Nearby Flare Stars

G	Sp	M_V (mag)	D (pc)	B − V (mag)	V_r (km s^{-1})	Hα (Å)	log L_X (erg s^{-1})
447	dM4.5	13.5	3.32	1.76	− 31.9	0.01	26.6
473A	dM5.5e	14.98	4.33	1.80	0.0	0.00	
473B	dM7?	15.2	4.33				
490A	dM1.5e	9.4	17.54	1.42	− 9.0	1.65	
490B	dM3.5e	11.9	17.54	1.61	− 3.8	3.65	
493.1	dM5e	13.3	10.1	1.73			27.9
494	dM1.5e	9.1	12.1	1.44	− 13.1	2.12	29.38
516A	dM3.5e	11.0	16.13	1.53	− 2.9	0.01	27.9
516B	dM4e	11.4	16.13		− 2.5	0.01	
517	dK5	7.95	18.18	1.18			29.30
540.2	dM5.5e	12.8	13.89	1.66			
551	dM5e	15.53	1.30	1.97			27.15
569	dM2e	10.1	10.42	1.48		3.58	28.71
616.2	dM1.5e	8.38	16.3	1.46	− 33.6	1.81	29.1
630.1	dM4e	12.90	15.05	1.60			28.51
644A	dM4.5e	10.79	6.21	1.62	+11.1	1.73	}29.03
644B	dM4.5	10.8	6.21				
669A	dM4e	11.25	10.53	1.55	− 34.7	1.63	
669B	dM4.5e	12.81	10.53	1.58	− 35.0	5.67	28.6
719	dM0e	7.23	15.63	1.19			29.75
729	dM4.5e	13.3	2.90	1.68			27.7
735	dM3e	9.9	10.87	1.75	− 12.6	1.69	28.83
752B	dM5e	18.57	5.78	2.12	0.0	0.00	27.0
781	dM3e	10.91	16.95	1.52	− 149.0	1.22	
791.2	dM6e	13.2	9.4	1.71			28.1
799A	dM4.5e	11.09	8.77	1.58			}29.25
799B	dM4.5e	11.2	8.77	1.58			
803	dM0e	8.87	9.26	1.44		8.70	29.33
815A	dM3e	9.8	14.29	1.49	17.7	1.28	
815B		11.8	14.29				

Catalogue of Nearby Flare Stars

G	Name	R.A.(1950) h m s	Dec.(1950) ° ′ ″	$\mu_{R.A.}$ (s yr^{-1})	$\mu_{Dec.}$ (″ yr^{-1})	V (mag)
841A	DM -51°13128	21 53 35	– 51 14 24	– 0.006	– 0.39	10.40
852A	Wolf 1561A	22 14 42	– 09 03 00	– 0.032	– 0.28	13.5
852B	Wolf 1561B	22 14 42	– 09 03 00	– 0.032	– 0.28	14.5
860A	DM +56°2783	22 26 13	+57 26 48	– 0.097	– 0.35	9.85
860B	DO Cep	22 26 13	+57 26 48	– 0.097	– 0.35	11.3
866	L 789–6	22 35 45	– 15 35 36	+0.162	+2.27	12.18
867A	DM -21°6267	22 36 01	– 20 52 48	+0.0325	– 0.056	9.062
867B	L 717–22	22 36 01	– 20 52 48	+0.0325	– 0.056	11.45
871.1A	L 574–62	22 42 18	– 33 31 00	+0.016	– 0.10	13.0
871.1B	L 574–61	22 42 18	– 33 31 00	+0.016	– 0.10	14.4
873	EV Lac	22 44 40	+44 04 36	– 0.064	– 0.46	10.048
896A	DM +19°5116	23 29 20	+19 39 42	+0.039	– 0.02	10.38
896B	EQ Peg B	23 29 20	+19 39 42	+0.039	– 0.02	12.4

Catalogue of Nearby Flare Stars

G	Sp	M_V (mag)	D (pc)	B − V (mag)	V_r (km s^{-1})	Hα (Å)	log L_X (erg s^{-1})
841A	dM0	9.6	14.49				
852A	dM4.5e	13.6	9.72	1.74			}28.57
852B	dM5e	14.6	9.71				
860A	dM3.5	11.87	3.94	1.62	− 34.2	0.01	
860B	dM4.5e	13.3	3.94	1.8			27.4
866	dM5.5e	14.60	3.3	1.96			27.0
867A	dM2e	9.25	9.17	1.49			
867B	dM3e	11.8	9.17	1.62			29.5
871.1A	dM3e	11.9					
871.1B	dM4e	13.3					
873	dM4.5e	11.56	4.98	1.60	− 1.4	3.36	28.67
896A	dM4e	11.33	6.45	1.56	+0.1	4.26	}29.15
896B	dM5e	13.4	6.45		+0.4	5.11	

13.3 Catalogue of RS CVn and BY Dra Binaries – Part I

The RS Canum Venaticorum, or RS CVn, and BY Draconis, or BY Dra, stars are late-type binary stars that show evidence of enhanced chromospheric or coronal activity when compared with other stars of the same spectral type. The enhanced activity is thought to be closely connected with strong magnetic fields on stars that are tidally locked into rapid synchronous rotation. As originally defined by Hall (1976), the RS CVn binaries have a main sequence dwarf, or subgiant, primary of spectral type F or G and a somewhat cooler, usually more massive and evolved, secondary star that is typically an early K subgiant. The BY Dra binaries contain K or M dwarfs. Both types of binary systems have relatively rapid orbital periods, P_{orb}, often a few days or less, photometric variability that is thought to be due to the rotation of large dark, cool starspots, and chromospheric activity signified by emission cores in the Ca II resonance lines, commonly called the H and K lines. Unusual coronal activity results in intense X-ray and radio emission. The X-ray luminosity, L_X, and radio luminosity, L_r, are about $L_X = 10^{-3}\ L_{bol}$ and $L_r = 10^{-7}\ L_{bol}$, where L_{bol} is the bolometric luminosity. These stars also often emit flares with a total energy of about 10^{35} *erg*. By way of comparison, flares on the Sun have about 10^{32} *erg*, and $L_X = 10^{-6} L_{bol}$ and $L_r = 10^{-12} L_{bol}$ for the Sun.

The accompanying catalogue of chromospherically active binary stars is in two parts. The first part contains right ascensions, R.A., and declinations, Dec., for the epoch 2000.0, proper motions μ, galactic longitude and latitude, l and b, maximum apparent visual magnitude, $V_{\max}$, spectral type, Sp, distance, D, in parsecs, absolute visual magnitude, M_V, orbital period, P_{orb}, semimajor axis, $a \sin i$, masses and radii in solar units, and eclipse data. These have been culled from the excellent catalogue of Strassmeier et al. (1988). In the second part, we have also given the logarithm of the X-ray luminosity, L_X, the average detected flux density, $< S_6 >$, measured at 6 cm wavelength in mJy, or an upper limit to this flux density, and the logarithm of the 6-cm monochromatic luminosity $L_6 = 1.20 \times 10^{12} < S_6 > D^2\ erg\ s^{-1} Hz^{-1}$. The X-ray and radio data come from the lists of Caillault, Drake, and Florkowski (1988), Drake, Simon, and Linsky (1989), and Morris and Mutel (1988).

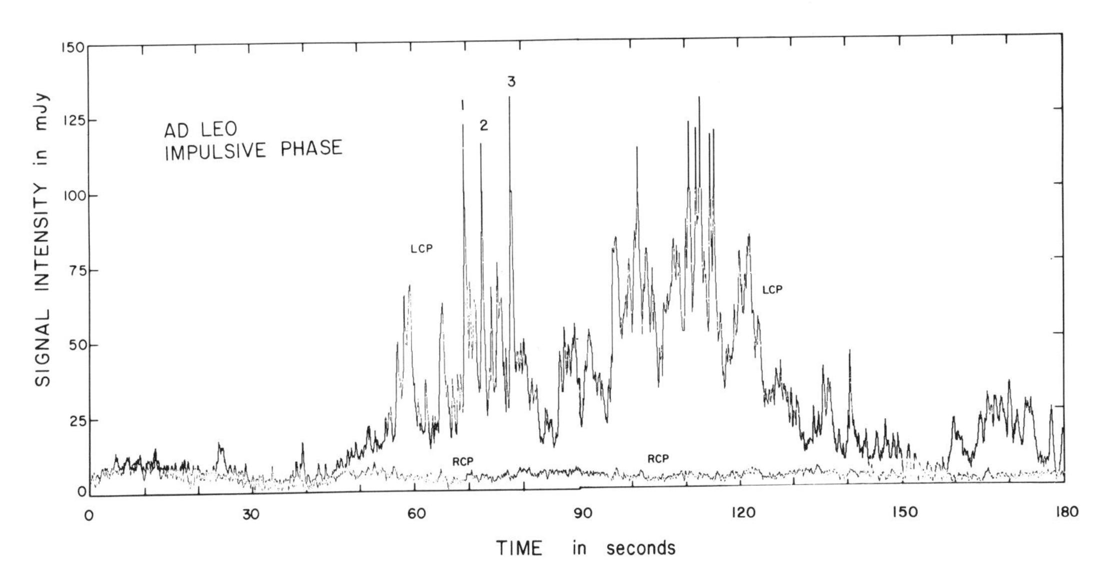

Flare Star AD Leonis. This time profile, or light curve, of the radio radiation from the flare star AD Leonis at 21 centimeters wavelength shows that the flaring emission is 100 percent left-hand circularly polarized (LCP), and that it consists of rapid spikes (1, 2, 3) with rise times of less than 20 milliseconds. An upper limit to the size of the emitter is provided by the distance that light travels in this rise time, thereby indicating that the emitting region is much smaller than the star in size. (Adapted from K.R. Lang et al. Astrophysical Journal Letters 272, L17 (1983).)

RS Canum Venaticorum stars. The RS CVn stars are binary systems that are tidally locked into rapid synchronous rotation. The fast rotation is thought to generate strong magnetic fields that may thread their way between the stars. A possible magnetic field configuration of a RS CVn system is shown here. (Adapted from Y. Uchida and T. Sakurai, Activity in Red-Dwarf Stars, eds. P.B. Byrne and M. Rodono (1983).)

RS CVn and BY Dra Binaries

Name	R.A.(2000) h m s	Dec.(2000) ° ′ ″	$\mu_{R.A.}$ (0.001 ″ yr^{-1})	$\mu_{Dec.}$	l (°)	b (°)	Vmax (mag)
33 Psc	00 05 20.1	- 05 42 27	- 13	+ 89	93.73	- 65.93	4.61
ADS 48A	00 05 36.7	+45 48 51	+878	- 127	114.64	- 16.32	8.23
5 Cet	00 08 12.0	- 02 26 52	+ 6	- 4	98.34	- 63.24	6.07
BD Cet	00 22 46.7	- 09 13 49	+ 3	- 51	100.84	- 70.86	8.2
13 Cet(A)	00 35 14.8	- 03 35 34	+410	- 21	112.87	- 66.15	5.20
FF And	00 42 47.3	+35 32 50	+250	+90	120.95	- 27.29	10.38
ζ And	00 47 20.3	+24 16 02	- 98	- 82	121.74	- 38.60	4.06
CF Tuc	00 52 58.3	- 74 39 07	+ 19	+28	302.81	- 42.48	7.47
η And	00 57 12.4	+23 25 04	- 34	- 37	124.65	- 39.43	4.42
BD +25°161	01 04 07.2	+26 35 14	- 10	- 23	126.44	- 36.20	8.4:
AY Cet	01 16 36.2	- 02 30 01	- 108	- 59	137.75	- 64.64	5.47
UV Psc	01 16 54.1	+06 48 43	+ 81	+ 28	134.15	- 55.50	9.22
CP -57°296	01 22 21.6	- 56 43 52	0	- 31	294.50	- 59.89	8.80
BD -00°210	01 22 50.7	+00 42 54	- 111	- 258	139.35	- 61.15	8.15
AR Psc	01 22 56.7	+07 25 11	+ 93	+246	136.49	- 54.62	7.28
CD -24°751	01 46 41.4	- 24 00 56	+171	+ 86	201.96	- 77.18	8.1
BD +34°363	02 03 47.3	+35 35 29	- 61	- 16	139.10	- 25.00	7.6:
6 Tri	02 12 22.2	+30 18 11	- 64	- 59	142.98	- 29.40	4.94
CP -71°166	02 18 23.9	- 71 28 26			293.53	- 44.03	8.19
CC Eri	02 34 21.8	- 44 47 54	+ 57	- 301	260.40	- 62.91	8.76
CD -38°899	02 43 25.5	- 37 55 40	+100	- 77	244.74	- 64.19	7.99
VY Ari	02 48 43.7	+31 06 55	+217	- 172	150.58	- 25.39	6.9
BD -05°592	03 10 39.3	- 05 23 38	+ 19	+ 16	185.96	- 50.40	8.22
LX Per	03 13 21.2	+48 06 40	+ 45	- 68	145.99	- 08.31	8.20
BD +43°657	03 13 51.0	+43 51 50	+ 58	- 83	148.33	- 11.88	7.30
UX Ari	03 26 35.2	+28 42 54	+ 38	- 107	159.55	- 22.91	6.5
V711 Tau*	03 36 47.2	+00 35 16	- 29	- 62	184.92	- 41.57	5.7
BD +25°580	03 37 10.2	+25 59 41	+249	- 283	163.39	- 23.58	8.1
V471 Tau	03 50 11.8	+17 15 17	+117	- 22	172.43	- 27.98	9.71
CP -52°497	04 07 32.1	- 52 34 05	+147	- 219	261.79	- 45.80	8.67
EI Eri	04 09 40.7	- 07 53 32	+ 15	+130	200.17	- 39.39	6.95
BD +23°635	04 12 26:	+23 40 30:			171.36	- 19.77	9.34
BD +16°577	04 17 38.3	+16 56 59	+118	- 2	177.62	- 23.37	8.33
BD +17°703	04 18 01.7	+18 15 24	+114	- 28	176.61	- 22.44	7.5
BD +36°903	04 31 56.9	+36 44 37	+ 10	- 63	164.50	- 07.77	6.72

*V711 Tauri = HR 1099

RS CVn and BY Dra Binaries

Name	Porb (days)	$a_h \sin i / a_c \sin i$ or $a \sin i$ (10^6 km)	Masses hot/cold ($M_\odot$)	Radii hot/cold ($R_\odot$)	Eclipse
33 Psc	72.93	15.8	f(m)=0.030		NONE
ADS 48A					
5 Cet	96.41	/31.3	$\approx 0.8/\geq 1.0$	[/$\approx$ 50]	POSSIBLE
BD Cet	35.100	15.0	f(m)=0.11	/$\geq$10	NONE
13 Cet(A)	2.08200	[1.27/]	f(m)=0.0189		NONE
FF And	2.170304	2.15/2.22	0.55/0.54		NONE
ζ And	17.7692	/6.4	0.78/2.70	$\approx$0.7/12.6	PARTIAL?
CF Tuc	2.79786	3.72/3.32	1.057/1.205	1.67/3.32	PARTIAL
η And	115.71	28.5/31.5	$\geq$0.338/$\geq$0.306		
BD +25°161					
AY Cet	56.815	/5.6	0.55/2.09	0.012/6.8	NONE
UV Psc	0.861046		$\geq$1.2/$\geq$0.9	1.219/0.929	PARTIAL
CP –57°296	0.65690	0.40/0.50	f(m)=0.0056		NONE
BD –00°210	0.515782	0.55/0.60	$\geq$0.9/$\geq$0.8	$\approx$0.9/$\approx$0.9	NONE
AR Psc	14.300	4.7/6.1	$\geq$0.110/$\geq$0.143	/$\geq$1.5	NONE
CD –24°751	15.05	1.9	f(m)=0.0013	$\geq$4	NONE
BD +34°363	23.9				
6 Tri	14.732	11.4/11.5	$\geq$1.12/$\geq$1.10		NONE
CP –71°166	18.379	10.4	f(m)=0.13		YES
CC Eri	1.56145	1.20/2.31	$\approx$0.47/$\approx$0.24	$\approx$0.7/	NONE
CD –38°899	0.95479	1.21/1.48	f(m)=0.077		NONE
VY Ari	13.198	5.67	f(m)=0.042		NONE
BD –05°592	48.263	11.2	f(m)=0.024	$\geq$8	NONE
LX Per	8.038207	8.37/8.27	1.23/1.32	1.6/2.8	TOTAL
BD +43°657					
UX Ari	6.43791	5.9/5.3	$\geq$0.63/$\geq$0.71	3/0.93	NONE
V711 Tau*	2.83774	1.9/2.4	1.1/1.4	1.3/3.9	NONE
BD +25°580	1.9299395	2.064/3.3	1.0/0.72	$\geq$1/$\geq$0.7	NONE
V471 Tau	0.521182993	/1.05	0.6/0.8:	0.012/$\geq$0.6	TOTAL
CP –52°497	2.562	2.04	f(m)=0.052		NONE
EI Eri	1.94722	0.729	$\geq$1.4/$\geq$0.53	$\geq$1.9/	NONE
BD +23°635	2.394357	2.178/	f(m)=0.0719		NONE
BD +16°577	5.609200	4.69/6.57	1.086/0.776		PARTIAL
BD +17°703	75.648	31.4/35.6	$\geq$1.06/$\geq$0.99		NONE
BD +36°903	21.295	8.6	f(m)=0.0565		

*V711 Tau = HR 1099

RS CVn and BY Dra Binaries

Name	R.A.(2000) h m s	Dec.(2000) ° ′ ″	$\mu_{R.A.}$ (0.001 ″ yr^{-1})	$\mu_{Dec.}$ (0.001 ″ yr^{-1})	l (°)	b (°)	Vmax (mag)
V833 Tau	04 36 47.6	+27 08 02	+251	− 138	172.51	− 13.37	8.42
3 Cam	04 39 54.7	+53 04 47	− 1	− 12	153.32	+04.26	5.05
RZ Eri	04 43 45.8	− 10 40 57	+ 21	− 6	208.00	− 33.17	7.70
BD +24°692	04 49 11.4	+24 48 43	+108	− 49	176.17	− 12.69	9.67
12 Cam	05 06 12.1	+59 01 16	+ 2	− 29	150.95	+10.82	6.1
CP −77°196	05 10 26.6	− 77 13 01	− 3	+ 10	289.30	− 31.92	7.69
α Aur	05 16 41.3	+45 59 53	+ 80	− 423	162.58	+04.57	0.08
HR 1908	05 37 04.3	+11 02 06	+ 44	− 12	194.33	− 11.02	5.94
TW Lep	05 40 39.6	− 20 17 56	+ 14	− 10	224.25	− 24.34	7.0
BD +03°1007	05 41 26.7	+03 46 40	+ 18	− 10	201.33	− 13.71	6.67
CD −28°2525	05 52 16.0	− 28 39 23	+ 34	− 22	233.87	− 24.73	9.05
SZ Pic	05 53 27.2	− 43 33 30	+ 2	− 75	250.04	− 28.41	7.9
CQ Aur	06 03 53.3	+31 19 48	− 9	− 10	179.93	+04.62	9.00
CP −54°973	06 07 56.8	− 54 26 22	− 7	+ 15	262.66	− 27.95	7.1
OU Gem	06 26 10.6	+18 45 33	− 119	− 183	193.41	+03.08	6.79
CP −58°718	06 31 05.8	− 59 00 16	+ 41	+ 47	268.29	− 25.57	7.61
W92/NGC2264	06 40 46.1	+09 49 22			202.98	+02.11	11.69
SV Cam	06 41 20.3	+82 16 36	+ 44	− 147	131.57	+26.52	8.4
VV Mon	07 03 18.3	− 05 44 05			219.39	+00.00	9.4
GL 268	07 10 03.7	+38 32 35	− 510	− 950	178.95	+19.90	11.48
SS Cam	07 16 21.6	+73 20 21			141.71	+27.68	10.1
AR Mon	07 20 48.0	− 05 15 31	+ 30	− 9	220.98	+04.08	8.62
YY Gem	07 34 38.0	+31 52 12	− 196	− 108	187.47	+22.47	9.07
CD −44°3573	07 36 17.2	− 44 57 26	+ 46	− 13	257.83	− 11.53	7.0
σ Gem	07 43 18.7	+28 53 01	+ 67	− 234	191.19	+23.26	4.14
54 Cam	08 02 35.8	+57 16 25	+ 30	− 62	160.34	+32.05	6.52
BD −06°2585	08 25 14.5	− 07 10 13	− 106	− 29	230.67	+17.11	8.0
GK Hya	08 30 37.9	+02 16 57	− 60	− 4	222.67	+23.01	9.4
HR 3385	08 32 59.8	− 34 39 48	− 15	+ 1	254.82	+03.14	6.36
RU Cnc	08 37 30.6	+23 33 46			201.28	+33.11	10.1
RZ Cnc	08 39 09.1	+31 47 51	+ 3	− 5	191.77	+35.65	8.69
TY Pyx	08 59 42.5	− 27 48 56	− 54	− 49	252.95	+11.83	6.87
WY Cnc	09 00 57.8	+26 40 48			199.42	+39.08	9.49
XY UMa	09 09 31.4	+54 23 51	− 48	− 175	162.86	+41.63	9.8
BD +40°2197	09 22 27.4	+40 12 22	− 353	− 363	182.05	+45.27	7.7

RS CVn and BY Dra Binaries

Name	Porb (days)	a_h sin i/a_c sin i or a sin i (10^6 km)	Masses hot/cold ($M_\odot$)	Radii hot/cold ($R_\odot$)	Eclipse
V833 Tau	~ 1.9		≈0.1/≈0.8		NONE
3 Cam	121	46.9	f(m)=0.282		NONE
RZ Eri	39.2826	12.3:	≥1.7/≥1.7	3.7/12.6	TOTAL
BD +24°692	11.9293	8.58/8.97	≥0.77/≥0.75	≈0.8/≈0.8	POSSIBLE
12 Cam	80.174469	23.5	f(m)=0.0801		NONE
CP –77°196	19.310	9.69	f(m)=0.11		NONE
α Aur	104.0214	44.0/37.3	2.80/3.31	7.8/12.7	NONE
HR 1908					
TW Lep	28.344	/16.3	f(m)=0.21	/≥9	NONE
BD +03°1007	53.580	18.6	f(m)=0.090	/≥11	NONE
CD –28°2525				1.12	
SZ Pic					NONE
CQ Aur	10.62148		≥1.6/≥2.0	1.9/8.7	TOTAL
CP –54°973	106.78	/46.8	f(m)=0.265		NONE
OU Gem	6.991868	5.38/6.35	≥0.71/≥0.59		NONE
CP –58°718	13.637	3.02	f(m)=0.0059		NONE
W92/NGC2264			1.7:	5.5:	NONE
SV Cam	0.593071	a=2.88	1.0/0.7	1.2/0.7	PARTIAL
VV Mon	6.05056	7.13/6.35	≥1.4/≥1.5	1.75/6.0	TOTAL
GL 268	10.428	4.47/5.37	≥0.191/≥0.159		POSSIBLE
SS Cam	4.8242459		1.7/2.2	2.0/7.0	TOTAL
AR Mon	21.20812	a=34	2.7/0.8	10.8/14.2	TOTAL
YY Gem	0.8142822	1.298/1.405	0.62/0.57	0.62/	PARTIAL
CD –44°3573	11.761	6.13	f(m)=0.066		NONE
σ Gem	19.60458	9.22	f(m)=0.0814		NONE
54 Cam	11.0764	9.2/9.4	1.64/1.61	3.14/2.64	NONE
BD –06°2585	16.537	2.44	f(m)=0.0021	≥5	NONE
GK Hya	3.587033		≥1.2:/≥1.3:	1.4/3.0	TOTAL
HR 3385	45.130	6.27	f(m)=0.0048	≥3	NONE
RU Cnc	10.172988	4.50/	≥1.5/≥1.5	1.9/5	TOTAL
RZ Cnc	21.643030	a=35.3	3.20/0.54	10.2/12.2	TOTAL
TY Pyx	3.198584	17.05	1.22/1.20	1.59/1.68	PARTIAL
WY Cnc	0.82937122		0.93/0.53	1/0.66	PARTIAL
XY UMa	0.4789944	a=2.12	0.95/0.70	0.98/0.73	PARTIAL
BD +40°2197	3.8025	7.47	0.76:/0.74:		NONE

RS CVn and BY Dra Binaries

Name	R.A.(2000) h m s	Dec.(2000) ° ′ ″	$\mu_{R.A.}$ (0.001 ″ yr^{-1})	$\mu_{Dec.}$	l (°)	b (°)	Vmax (mag)
IL Hya	09 24 48.8	- 23 49 35	+ 37	- 33	253.68	+18.71	7.4
CP -41°3888	09 37 13.2	- 42 01 16	- 86	+ 4	268.75	+07.60	9.0
DH Leo	10 00 02.4	+24 33 14	- 242	- 46	206.87	+51.53	7.75
BD -11°2916	10 36 02.2	- 11 54 34	+135	- 255	258.40	+38.98	7.58
DM UMa	10 55 44.2	+60 28 12	- 53	+ 7	145.37	+51.32	9.57
ξ UMa(B)	11 18 10.9	+31 31 45	- 432	- 591	195.08	+69.25	4.87
DF UMa	11 37 16.8	+47 27 11			153.57	+65.02	10.14
CD -38°7259	11 39 22.3	- 39 23 04	+ 17	- 36	288.05	+21.39	7.95
HR 4492	11 39 29.5	- 65 23 52	- 37	- 12	295.53	- 03.56	5.17
RW UMa	11 40 04.7	+51 59 54	- 36	- 6	146.44	+61.75	10.16
93 Leo	11 47 59.0	+20 13 08	- 150	- 8	235.01	+73.93	4.5
BD -08°3301	12 13 20.6	- 09 04 47	- 27	+ 9	287.30	+52.63	8.1
HR 4665	12 15 41.3	+72 33 03	- 21	- 38	126.67	+44.32	6.14
AS Dra	12 22 11.5	+73 14 52	- 31	+161	125.85	+43.71	8.00
BD +26°2347	12 25 02.5	+25 33 38	- 10	- 16	226.27	+83.87	8.16
BD +25°2511	12 29 38.3	+24 30 16	- 11	- 22	239.80	+84.43	9.7
UX Com	13 01 31.3	+28 37 41			67.63	+87.33	10.00
BD -04°3419	13 06 26.3	- 04 50 44	- 20	- 14	309.94	+57.82	8.27
RS CVn	13 10 37.3	+35 56 03	- 49	+ 4	99.26	+80.30	7.93
BD +39°2635	13 21 32.1	+38 52 51	- 69	- 5	96.74	+76.68	7.21
HR 5110	13 34 47.7	+37 10 56	+ 84	- 9	83.33	+76.41	4.95
CD -32°9477	13 36 08.0	- 33 28 45	- 6	- 14	313.52	+28.47	9.9
CP -60°4913	13 44 00.3	- 61 21 59	+ 35	+ 16	309.19	+00.87	7.86
BH Vir	13 58 26.6	- 01 40 15			334.84	+57.00	9.60
4 UMi	14 08 50.8	+77 32 51	- 34	+30	117.67	+38.78	4.82
CP -59°5631	14 34 15.7	- 60 24 27	- 99	- 40	315.30	- 00.03	8.63
RV Lib	14 35 47.9	- 18 02 15	- 22	- 26	335.09	+38.24	9.0
SS Boo	15 13 32.9	+38 34 03			63.17	+58.29	10.3
BD +26°2685	15 22 25.4	+25 37 27	+ 41	- 10	39.12	+56.28	7.21
GX Lib	15 23 25.8	- 06 36 37	- 15	- 120	356.01	+40.12	7.31
CP -62°4482	15 27 38.9	- 63 01 13	- 67	- 59	319.62	- 05.32	7.0
UZ Lib	15 33 14.2	- 08 32 07			356.40	+37.02	9.3
RT CrB	15 38 02.9	+29 29 10	+ 15	+ 5	46.68	+53.45	9.4
1E1548.7+1125	15 51 07.6	+11 16 21			21.29	+45.02	13.5
RS UMi	15 51 29.5	+72 07 31			106.76	+38.84	10.07

RS CVn and BY Dra Binaries

Name	Porb (days)	a_h sin i/a_c sin i or a sin i (10^6 km)	Masses hot/cold ($M_\odot$)	Radii hot/cold ($R_\odot$)	Eclipse
IL Hya	12.908	7.30	f(m)=0.073	/≥6	NONE
CP −41°3888	52.270	14.9	f(m)=0.048		NONE
DH Leo	1.070354	[3.26]	[0.83/0.58]	[0.97/0.67]	NONE
BD −11°2916	6.86569	5.43/5.46	≥0.547/≥0.545	≥0.8/≥0.8	NONE
DM UMa	7.492	2.8/	f(m)=0.016	≈4	NONE
ξ UMa(B)	3.9805	0.274	f(m)=0.0000517		NONE
DF UMa	1.033824	0.796/	≈0.56/≈0.3		NONE
CD −38°7259	11.710	6.27/6.14	0.28/0.29		NONE
HR 4492	61.360	/10.5	≈2.0/≈2.5		NONE
RW UMa	7.3282231	7.31/	1.50/1.45	1.25/3.17	TOTAL
93 Leo	71.6900	33.3/29.2	≥0.89/≥1.02	1.7/5.9	NONE
BD −08°3301	10.3880	6.64	f(m)=0.11	≥5	NONE
HR 4665	64.44	32.0/32.7	≥1.43/≥1.43	≥13/≥13	NONE
AS Dra	5.414905	4.99/5.20	≥0.736/≥0.706		NONE
BD +26°2347	0.9616		0.85/0.82	[1.1/1.1]	NONE
BD +25°2511					
UX Com	3.642386		≥0.95/≥1.12	1.0:/2.5:	TOTAL
BD −04°3419	≥ 20				
RS CVn	4.797851	a=13	1.34/1.40	1.88/4.10	TOTAL
BD +39°2635	20.625	3.33	f(m)=0.0035	≥6	NONE
HR 5110	2.6131738	0.40/0.74	1.5/0.8	3.10/2.85	NONE
CD −32°9477	22.740	4.60	f(m)=0.0075		NONE
CP −60°4913	11.987	1.62	f(m)=0.0012		NONE
BH Vir	0.81687099	1.55/1.52	0.87/0.90	1.05/1.15	PARTIAL
4 UMi	605.8	105	f(m)=0.124		
CP −59°5631	5.998	2.83	f(m)=0.025		NONE
RV Lib	10.722164		≥2.2/≥0.4		PARTIAL
SS Boo	7.606133	7.21/7.73	≥1.00/≥1.00	1.31/3.28	TOTAL
BD +26°2685	18.670	5.5	f(m)=0.020	≥16	
GX Lib	11.1345	/5.91	f(m)=0.066	/≥7	NONE
CP −62°4482	49.431	27.7/27.5	f(m)=2.89		NONE
UZ Lib	4.76783	2.16/	≈2/≈0.5	≥6/	NONE
RT CrB	5.1171590		≥1.27/≥1.34	2.6/2.6	PARTIAL
1E1548.7+1125	5–10				
RS UMi	6.16860				PARTIAL

RS CVn and BY Dra Binaries

Name	R.A.(2000) h m s	Dec.(2000) ° ′ ″	$\mu_{R.A.}$ (0.001 ″ yr^{-1})	$\mu_{Dec.}$	l (°)	b (°)	Vmax (mag)
BD +25°3003	15 58 44.2	+25 34 15	- 78	- 139	41.81	+48.27	8.36
σ^2 CrB	16 14 40.6	+33 51 30	- 275	- 80	54.67	+46.14	5.7
CM Dra	16 34 24.3	+57 08 58	- 1080	+1200	86.56	+40.91	12.90
WW Dra	16 39 03.6	+60 42 00	+12	- 61	90.87	+39.45	8.22
ε UMi	16 45 57.8	+82 02 14	+11	+3	115.00	+31.05	4.23
V792 Her	17 10 25.4	+48 57 55	- 12	- 25	75.41	+36.40	8.50
V824 Ara	17 17 25.7	- 66 56 56	- 8	- 133	324.91	- 16.29	6.72
HR 6469	17 21 43.4	+39 58 28	+3	- 65	64.69	+33.58	5.51
CD -33°12122	17 30 33.8	- 33 39 14	- 9	+17	354.30	+00.20	8.52
29 Dra	17 32 40.8	+74 13 39	- 78	+40	105.50	+31.35	6.55
BD +36°2975	17 55 25.2	+36 11 20	- 140	- 40	62.07	+26.31	8.0:
Z Her	17 58 06.8	+15 08 22	- 22	+70	40.87	+18.51	7.3
MM Her	17 58 38.5	+22 08 44	+2	- 34	47.78	+21.12	9.51
V772 Her	18 05 49.8	+21 26 48	- 28	- 53	47.76	+19.31	7.07
ADS 11060C	18 05 50v	+21 26 50	- 28	- 53	47.76	+19.31	10.62
CD -48°12280	18 07 00.2	- 48 14 49	+15	+51	345.06	- 13.05	7.11
V815 Her	18 08 16.0	+29 41 27	+115	- 27	56.19	+21.75	7.66
PW Her	18 10 23.7	+33 24 11			60.13	+22.54	9.9
CP -32°5229	18 15 10.7	- 32 47 19			359.78	- 07.40	10.4
AW Her	18 25 38.7	+18 17 32	- 17	0	46.68	+13.78	9.65
BY Dra	18 33 55.0	+51 43 25	+183	- 311	80.57	+23.59	8.07
o Dra	18 51 12.0	+59 23 18	+73	+27	89.31	+23.14	4.67
ν^2 Sgr	18 55 07.0	- 22 40 17	+102	- 26	12.91	- 10.89	4.99
V775 Her	18 55 52.4	+23 30 23	+125	- 288	54.54	+09.57	8.04
τ Sgr	19 06 56.3	- 27 40 14	- 53	- 249	9.34	- 15.37	3.32
V478 Lyr	19 07 32.3	+30 15 16	+110	+115	61.85	+10.13	7.72
HR 7275	19 08 25.6	+52 25 32	- 106	- 53	82.98	+18.73	5.81
HR 7333	19 20 32.8	- 05 24 57	+112	+48	31.39	- 08.93	5.01
BD -20°5516	19 22 38.4	- 20 38 29	- 3	- 94	17.51	- 15.88	6.76
CP -41°9096	19 28 01.8	- 40 50 04	+33	- 7	357.73	- 23.89	8.50
HR 7428	19 31 13.3	+55 43 55	- 7	- 10	87.52	+16.87	6.32
V1764 Cyg	19 36 42.4	+27 53 03	+13	0	62.68	+03.39	7.69
BD -06°5221	19 39 38.3	- 06 11 35	+15	- 20	32.85	- 13.51	8.50
HR 7578	19 54 17.6	- 23 56 28	- 133	- 412	17.18	- 23.91	6.16
V4091 Sgr	20 06 02.2	- 18 42 22	+26	- 9	23.61	- 24.56	8.35

RS CVn and BY Dra Binaries

Name	Porb (days)	a_h sin i/a_c sin i or a sin i (10^6 km)	Masses hot/cold ($M_\odot$)	Radii hot/cold ($R_\odot$)	Eclipse
BD +25°3003	9.01490	6.63/	≥0.86/≥0.71		NONE
σ^2 CrB	1.1397912	0.994/1.024	1.12/1.14	1.22/1.21	NONE
CM Dra	1.2683896	1.22/1.40	0.237/0.207	0.252/0.235	PARTIAL
WW Dra	4.6296166	5.6/5.7	≥1.4/≥1.4	2.3:/3.9:	PARTIAL
ε UMi	39.4809	/17.3	1.3/2.8	1.7/12	TOTAL
V792 Her	27.5384	a=47	1.40/1.46	2.58/12.28	TOTAL
V824 Ara	1.6817	1.96/2.10	≥0.491/≥0.459		NONE
HR 6469	2018.	/377	[1.65/1.15]—2.05	[≥1]/≥10	YES
CD -33°12122	30.969	20.7/20.5	≥1.7/≥1.7	=5.5:/=14	NONE
29 Dra	39.		0.55/	0.012/≥5	NONE
BD +36°2975					
Z Her	3.9928012	4.6/5.1	1.23/1.10	1.69/2.60	PARTIAL
MM Her	7.960322	7.72/7.91	1.18/1.27	1.58/2.83	PARTIAL
V772 Her	0.8794998	[a=3.2]	1.04/0.59	1.0:/0.6:	PARTIAL
ADS 11060C	≥ 5.				
CD -48°12280	> 600.				NONE
V815 Her	1.8098368	1.359/	f(m)=0.0306	0.93:/	NONE
PW Her	2.8810016		≥1.4/≥1.6		TOTAL
CP -32°5229	~P(phtm)				
AW Her	8.800760		≥1.4/≥1.4	1.12/3.00	TOTAL
BY Dra	5.975112	2.233/2.505	0.5–0.6/0.44	1.2–1.4/	NONE
o Dra	138.420	44.5/	f(m)=0.183		NONE
ν^2 Sgr					
V775 Her	2.879395	1.9585/	f(m)=0.0362	0.85/	NONE
τ Sgr					
V478 Lyr	2.130514	1.1/	f(m)=0.0118	≥0.9/	NONE
HR 7275	28.59	15.6/	f(m)=0.194	≥8	NONE
HR 7333	266.544	61.1	f(m)=0.128		NONE
BD -20°5516	13.048	2.40	f(m)=0.00083	≥9	NONE
CP -41°9096	45.180	20.75	f(m)=0.17		NONE
HR 7428	108.5707	/33.0	f(m)=0.121		NONE
V1764 Cyg	40.1425	22.65	f(m)=0.29	≥22	NONE
BD -06°5221	20.660	/3.70	f(m)=0.0047	/≥7.5	YES
HR 7578	46.817	22.7/22.6	≥0.85/≥0.85	≥0.8/≥0.8	POSSIBLE
V4091 Sgr	16.887	6.82	f(m)=0.045	/≥6	NONE

RS CVn and BY Dra Binaries

Name	R.A.(2000) h m s	Dec.(2000) ° ′ ″	$\mu_{R.A.}$ (0.001 ″ yr^{-1})	$\mu_{Dec.}$	l (°)	b (°)	Vmax (mag)
BD –21°5735	20 29 36.2	– 21 07 34	+28	– 14	23.30	– 30.59	8.98
CG Cyg	20 58 14.1	+35 10 29			78.47	– 06.86	10.0
V1396 Cyg	21 00 02.9	+40 04 28	+620	– 260	82.44	– 03.94	10.13
ER Vul	21 02 25.7	+27 48 26	+95	+7	73.34	– 12.29	7.27
CD –31°18145	21 14 45.2	– 31 11 00	– 88	– 91	14.41	– 42.97	7.72
BD +10°4514	21 18 34.6	+11 34 09	– 6	– 35	62.54	– 25.50	7.02
CP –53°10073	21 28 20.5	– 52 49 59			344.12	– 44.69	8.7
BD –00°4234	21 32 06.6	+00 13 14	+465	0	54.20	– 34.90	9.89
BD –14°6070	21 34 16.4	– 13 29 01	+16	+8	39.25	– 42.11	8.0
AD Cap	21 39 49.0	– 16 00 24	+46	– 24	336.81	– 44.35	9.8
42 Cap	21 41 32.7	– 14 02 51	– 123	– 307	39.55	– 43.95	5.18
FF Aqr	22 00 35.2	– 02 44 33			356.25	– 42.44	9.34
RT Lac	22 01 30.7	+43 53 22	+66	+37	93.41	– 09.02	8.84
HK Lac	22 04 56.5	+47 14 06	+61	+45	95.92	– 06.71	6.52
AR Lac	22 08 40.9	+45 44 31	– 48	+55	95.56	– 08.30	6.09
BD +29°4645	22 22 32.7	+30 21 27	+8	– 5	388.56	– 22.39	7.51
V350 Lac	22 30 06.3	+49 21 22	– 24	– 29	100.61	– 07.29	6.38
FK Aqr	22 38 42.8	– 20 37 22	+455	– 56	37.81	– 59.05	9.05
IM Peg	22 53 02.1	+16 50 28	– 11	– 29	86.36	– 37.48	5.60
BD –01°4364	22 58 52.7	– 00 18 58	+39	+16	73.71	– 51.46	7.3
CD –34°15853	23 00 27.7	– 33 44 34	– 44	– 132	10.64	– 65.25	8.42
KZ And	23 09 56.4	+47 57 29	+148	+2	105.90	– 11.53	7.98
RT And	23 11 10.2	+53 01 33	– 1	– 11	108.06	– 06.92	9.1
SZ Psc	23 13 24.0	+02 40 30	– 1	– 35	80.66	– 51.96	7.2
EZ Peg	23 16 53.4	+25 43 09	– 70	+13	97.58	– 32.45	9.53
λ And	23 37 33.7	+46 27 30	+162	– 420	109.90	– 14.53	3.7
BD +27°4588	23 39 30.8	+28 14 49	+297	+241	104.22	– 32.00	7.04
II Peg	23 55 03.9	+28 38 01	+546	+31	108.22	– 32.62	7.2

RS CVn and BY Dra Binaries

Name	Porb (days)	a_h sin i/a_c sin i or a sin i (10^6 km)	Masses hot/cold ($M_\odot$)	Radii hot/cold ($R_\odot$)	Eclipse
BD −21°5735	23.206	9.37	f(m)=0.061	≥10	NONE
CG Cyg	0.63114347	1.08/1.08	0.52/0.52	0.88/0.87	PARTIAL
V1396 Cyg	3.275631	1.78/2.57	≥0.181/≥0.125		NONE
ER Vul	0.69809510	1.33/1.43	1.07/0.98	1.23/1.23	PARTIAL
CD −31°18145	63.089	23.7	f(m)=0.13		NONE
BD +10°4514	3.9660306	[3.61/3.77]	≥0.52/≥0.50		NONE
CP −53°10073	22.349	4.21	f(m)=0.0060		NONE
BD −00°4234	3.7569	2.23/2.77	0.69/0.55	≈0.55/≈0.45	NONE
BD −14°6070	49.137	14.5	f(m)=0.050	/≥13	NONE
AD Cap	2.96		≥0.5:/≥1.1:		POSSIBLE
42 Cap	13.1740	4.13	f(m)=0.0160		NONE
FF Aqr	9.207755	a=27	0.5/2.00	0.1/6.12	TOTAL
RT Lac	5.074015	4.36/8.09	0.78/1.66	4.2/3.4	TOTAL
HK Lac	24.4284	11.6	f(m)=0.105		NONE
AR Lac	1.98322195	3.17/3.15	≥1.30/≥1.30	1.8/3.1	TOTAL
BD +29°4645					
V350 Lac	17.755	9.8	f(m)=0.12	≥11	NONE
FK Aqr	4.08322	2.63/3.26	≥0.27/≥0.22		NONE
IM Peg	24.649	11.3	f(m)=0.0937	/≥12	NONE
BD −01°4364				≥10	NONE
CD −34°15853	1.64250	1.37/1.98	f(m)=0.054		NONE
KZ And	3.032867	2.82/2.97	≥0.43/≥0.41	≥0.74/	NONE
RT And	0.62892984	1.14/1.77	1.50/0.99	1.1/0.97	TOTAL
SZ Psc	3.9658663	6.02/4.44	≥1.3/≥1.6	1.38/5.08	PARTIAL
EZ Peg	11.6598	3.93/3.89	≥0.070/≥0.071		NONE
λ And	20.5212	1.86/	f(m)=0.000611		NONE
BD +27°4588	6.20197	3.52/5.2	≥0.40/≥0.27		NONE
II Peg	6.724183	3.40	f(m)=0.035	2.2	NONE

13.4 Catalogue of RS CVn and BY Dra Binaries — Part II

Name	Sp hot/cold	log L_X (erg s^{-1})	log L_6 (erg s^{-1} Hz^{-1})	$<S_6>$ (mJy)	D (pc)	M_V hot/cold (mag)
33 Psc	K0III	28.67	<15.11	< 0.22	27	+2.4
ADS 48A	dK6				11	+8.0
5 Cet	~F/K1III	29.50	<15.66	< 0.16	140	− 0.1
BD Cet	K1III				71	+3.9
13 Cet(A)	{F7V}/G4V	29.16	<13.77	< 0.15	17	+4.0
FF And	dMle/dMle				21.3	+8.5
ζ And	/K1II	30.35	<14.96	< 0.26	31	+1.9
CF Tuc	G0V/K4IV	31.15			54	[+4.4/+3.1:]
η And	G8IV–III/G8IV–III	29.99	<14.82	< 0.22	111	− 0.8
BD +25°161	G2V				[55:]	[+4.7]
AY Cet	WD/G5III	31.32	16.15	8.63	66.7	/+1.32
UV Psc	G4–6V/K0–2V	31.25	16.46	0.89	125	+4.6/+6.1
CP −57°296	G6–8IV–IIIe				100:	+3.5:
BD −00°210	G5V:/G5V:		16.2	3.23	60	+5
AR Psc	/G8IV		15.4	2.91	70:	+3.2
CD −24°751	K0IV				[100]	[+3.1]
BD +34°363	G5IV				77	+3.2
6 Tri	G5III/[G5III]	30.83			75	+0.3
CP −71°166	G1:Vp				77	+3.8
CC Eri	K7Ve/	29.67	14.0	0.62	11.4	+8.47
CD −38°899	G5–8V/[G]				[35]	[+5.3:]
VY Ari	G9V				21	+5.2
BD −5°592	G8IV–III				[146]	[+2.4]
LX Per	G0IV/K0IV	30.71	16.02	0.51	145	+4.6/+3.3
BD +43°657	G5IV				65	+3.2
UX Ari	G5V/K0IV	31.32	17.08	21.5	50	+2.5
V711 Tau*	G5IV/K1IV	30.96	16.17	12.7	36	+4.2/+3.7
BD +25°580	G2V/KV		16.0	0.80	55	+4.38
V471 Tau	WD/K2V	30.21	15.28	1.00	59	+9.75/+5.35
CP −52°497	K1Vp				[32]	[+6.15]
EI Eri	G5IV		16.25	4.3	75	+2.75
BD +23°635	dK0/dM0:				40:	+6.4:
BD +16°577	G6V/K6V				44.7	+5.35
BD +17°703	G4V/G8V				46.3	+4.95
BD +36°903	K1III				50	+3.2

*V711 = HR 1099

RS CVn and BY Dra Binaries

Name	Sp hot/cold	log L_X (erg s^{-1})	log L_6 (erg s^{-1} Hz^{-1})	<S_6> (mJy)	D (pc)	M_V hot/cold (mag)
V833 Tau	dK5e				16.7	+6.8
3 Cam	K0III	30.07	<15.43	< 0.25	85	+0.2
RZ Eri	Am/K0IV	31.45	<16.21	< 0.14	143	+0.6
BD +24°692	K3V/K3V				42	+7.5
12 Cam	K0III	31.18			134	+0.2
CP −77°196	K1IIIp				82	+3.2
α Aur	F9III/G6III	30.55	13.76	0.21	14.5	+0.16/-0.1
HR 1908	K4III				160	− 0.3
TW Lep	F/G8III		17.3	4.70	220	+0.3
BD +3°1007	F:/K1III				[164]	[+0.6]
CD −28°2525	G1V	31.35			85	+4.5
SZ Pic	G8V		14.74	0.37	[30]	[+5.5]
CQ Aur	G2/K0	30.94	16.36	0.43	220	+2.8/+3.0
CP −54°973	F/G8–K0III				110	+0.2
OU Gem	K3V/K5V				12	+6.4
CP −58°718	K1IV–IIIp				60:	− 0.1
W92/NGC2264	K0:IVp	31.25	<16.98	< 0.19	900	+2.0
SV Cam	G2–3V/K4V	30.79	<15.54	< 0.40	74	+5.0/+6.74
VV Mon	G2IV/K0IV	31.29	17.54	1.37	380	+3.0/+2.5
GL 268	dM5e/dM5e				5.9	+12.62
SS Cam	F5V– IV/K0IV–III	30.89	<16.49	< 0.40	255	+3.8/+3.4
AR Mon	G8III/K2–3III	32.01	17.34	0.92	525	+0.85/+1.25
YY Gem	dM1e/dM1e	29.60			13.7	+9.14
CD −44°3573	K1III				[190]	[+0.6]
σ Gem	K1III	31.06	15.75	5.8	59	+0.22
54 Cam	F9IV/F9IV	29.94	15.30	3.65	38	+2.50/+2.88
BD −6°2585	K1IV				[95]	[+3.1]
GK Hya	F8/G8IV	30.90	16.57	1.04	220	+2.7
HR 3385	K0III				[135]	[+0.7]
RU Cnc	F5IV/K1IV	30.82	16.64	0.56	190	+3.2/+4.5
RZ Cnc	K1III/K3–4III	31.56	<16.55	< 0.27	395	+1.35/+1.90
TY Pyx	G5IV/G5IV	30.71	15.48	0.70	55	+3.9/+3.8
WY Cnc	G5V/[∼ M2]	< 31.01	15.39	0.40	160	+3.5/+3.2
XY UMa	G2–5V/K5V	< 31.32	15.56	0.30	100	+4.8
BD +40°2197	K2V/[dK]				29	+5.4

RS CVn and BY Dra Binaries

Name	Sp hot/cold	log L_X (erg s^{-1})	log L_6 (erg s^{-1} Hz^{-1})	$\langle S_6 \rangle$ (mJy)	D (pc)	M_V hot/cold (mag)
IL Hya	K1III		17.36	5.0	138	+3:
CP −41°3888	K2IIIp				[500:]	[+0.5:]
DH Leo	{K0V/K7V}K5V	< 30.67	14.92	0.77	32	[+5.9/+8.1/+7.35]
BD −11°2916	K3–4V/K3–4V				34	[+6.8/+6.8]
DM UMa	K0–1IV–III	31.08	16.32	3.0	130	+4.0
ξ UMa(B)	G5V	29.35	<13.49	< 0.40	7.9	+4.9
DF UMa	dM0e/[dM5]				22	+8.4
CD −38°7259	G5V/K1IV		16.0	0.91	62	+4.0
HR 4492	A0/K2–4III		17.9	65.00	140	+0.3
RW UMa	F8IV/K1IV	32.21	<16.23	< 0.17	150	+3.6/+4.7
93 Leo	A6:V/G5IV–III	30.05	<14.51	< 0.17	36	+1.24
BD −8°3301	K0III		16.7	0.51	[302]	[+0.7]
HR 4665	K1III/K1III	30.53			130	[+0.6/+0.6]
AS Dra	G4V/G9V	29.82	<14.65	< 0.15	29.4	+4.8/+5.9
BD +26°2347	F8V/F8V				86	+4.4
BD +25°2511	G9V				55	+6.0
UX Com	G2/K1[IV]	31.22	16.69	1.08	350	+3.6
BD −4°3419	K2IV–III	32.36			165	+2.06
RS CVn	F5IV/K0IV	31.69	17.13	5.72	145	+3.0/+2.7
BD +39°2635	K1III				250	+0.2
HR 5110	F2IV/K2IV	31.11	16.3	5.00	53	+1.36
CD −32°9477	K2IIIp				[760:]	[+0.5:]
CP −60°4913	K2IV–III				80:	+3.5:
BH Vir	F8V–IV/G2V		<15.63	< 0.18	[166:]	[+3.4/+4.7]
4 UMi	K3III				100	− 0.2
CP −59°5631	K1IV				63	+4.6
RV Lib	G5IV/K3IV	30.49	16.50	1.4	270	+2.1/+2.8
SS Boo	G0V/K1IV	31.10	16.40	0.22	220	+3.9/+4.0
BD +26°2685	K1III				230	0.0
GX Lib	[G–KV]/K1III		<15.94	< 0.18	[219]	[+0.6]
CP −62°4482	K2IV/K2IV		16.9	7.70	54	[+3.1/+3.1]
UZ Lib	K2III/[dM]	32.03	<17.28	< 0.40	[575:]	[+0.5/]
RT CrB	G0/K0–2	30.55	<16.23	< 0.17	360	+1.6
1E1548.7+1125	K5V–IV				[500:]	[+5:]
RS UMi	F8/	30.95	<16.77	< 0.40	350	+2.4

RS CVn and BY Dra Binaries

Name	Sp hot/cold	log L_X (erg s^{-1})	log L_6 (erg s^{-1} Hz^{-1})	$\langle S_6 \rangle$ (mJy)	D (pc)	M_V hot/cold (mag)
BD +25°3003	K2V/K6V				30	+6.2/+7.5
σ^2 CrB	F6V/G0V	30.59	15.7	8.50	21	+4.1
CM Dra	M4Ve/M4Ve				14.5	+12.77/+12.92
WW Dra	G2IV/K0IV	30.84	16.87	3.6	180	+2.8/+2.9
ε UMi	A8–F0V/G5III	30.61	15.26	0.42	71	0.0
V792 Her	F3V/K0III	31.00	16.3	2.70	310	[+3.6/+0.7]
V824 Ara	G5IV/K0V–IV	30.99			30	[+3.1/+4.5:]
HR 6469	{F2V/[G0V]?}G5IV		<14.69	< 0.23	50	+4.0
CD −33°12122	F2IV/K1III				400:	+0.5:
29 Dra	WD/K0–2III		17.50	15.82	87.9	− 0.1
BD +36°2975	F/G5IV				83	+3.2
Z Her	F4V–IV/K0IV	30.48	15.72	1.6	75	+3.3/+3.6
MM Her	G2IV/G8IV	30.76	16.62	2.24	190	+3.9/+3.0
V772 Her	{G0V/[M1V]}G5V	30.56	15.32	0.98	41.7	+4.0
ADS 11060C	K7:V				41.7	+7.5
CD −48°12280	WD/G8III				170	+0.2
V815 Her	G5V/[M1–2V]		15.38	1.56	31	+5.2
PW Her	G2/K0	30.68			285	+3.7/+3.0
CP −32°5229	K5Ve				40	+7.4
AW Her	G0/K1[IV]	30.90	16.80	0.86	315	+2.6/+3.1
BY Dra	K4V/K7.5V	29.5	14.3	0.61	15.6	+7.4/+8.6
o Dra	G9III				67	+0.5
ν^2 Sgr	K3III				29.4	[+0.3]
V775 Her	K0V/[K5–M2V]		14.91	0.55	24	+6.1
τ Sgr	K1III				26.3	+1.0
V478 Lyr	G8V/[dK–dM]		15.50	1.48	26	+5.5
HR 7275	K1IV–III	31.10	15.96	1.49	48	[+2.4]
HR 7333	G8IV–III				33	+1.8
BD −20°5516	K1III				31	+4.3
CP −41°9096	K2–3III				[417]	[+0.4]
HR 7428	A0V/K2III–II	31.15	16.48	0.40	302	− 1.1
V1764 Cyg	K1III:		<16.74	< 0.40	340	+0.0
BD −6°5221	sdB/K0IV–III				[209]	[/+1.9]
HR 7578	K2–3V/K2–3V				15	+6.0/+6.0
V4091 Sgr	K0III				[340]	[+0.7]

RS CVn and BY Dra Binaries

Name	Sp hot/cold	log L_X (erg s^{-1})	log L_6 (erg s^{-1} Hz^{-1})	<S_6> (mJy)	D (pc)	M_V hot/cold (mag)
BD −21°5735	K2III				99	+4.0
CG Cyg	G9.5V/K3V		<15.35	< 0.23	[~63]	[+5.8/+6.65]
V1396 Cyg	dM3e				15.9	+9.7
ER Vul	G0V/G5V	30.38	16.08	4.97	45	+4.8/+4.6
CD −31°18145	K1IIIp				[265]	[+0.6]
BD +10°4514	{F9V/G0V}GIV				50	+4.4
CP −53°10073	K1IIICNIVp				[417:]	[+0.6:]
BD −00°4234	K3Ve/K7Ve		<14.83	< 0.28	50	+5.5
BD −14°6070	K1III		17.8	7.17	93	+3.2
AD Cap	G5/G5	31.31	16.53	0.78	250	+2.7/+4.4
42 Cap	G2IV		<14.25	< 0.19	34	+3.0
FF Aqr	sd0−B/G8IV−III		17.32	2.94	300	+4.7/+2.1
RT Lac	G9IV/K1IV	31.38	16.87	1.78	205	+2.9/+3.4
HK Lac	F1V/K0III	31.69	17.12	4.57	150	+3.0/+0.8
AR Lac	G2IV/K0IV	31.28	16.30	8.34	47	+3.5/+3.3
BD +29°4645	F/G5IV				[80:]	[/+3.0]
V350 Lac	K2III	30.66	16.8	10.00	69	+2.2
FK Aqr	dM2e/dM3e				8.3	+9.5
IM Peg	K2III−II	31.44	17.02	8.65	50	+2:
BD −01°4364	K0III				260	+0.2
CD −34°15853	G5Vp				[46:]	[+5.1:]
KZ And	dK2/dK2				[~23]	[+6.4/+6.4]
RT And	F8V/G5V	< 31.69	<15.33	< 0.16	95	+5.5/+3.9
SZ Psc	F8IV/K1IV	31.43	17.27	21.75	100	+2.8/+2.8
EZ Peg	G5V−IV/K0IV:		15.58	0.22	83	+5.12
λ And	G8IV−III	30.48	15.18	6.00	23	+2.2:
BD +27°4588	G5V/KV				[25]	[+5.1/]
II Peg	K2−3V−IV	30.95	15.77	12.6	29.4	+5.4

Part IV
Star Clusters and Associations

14
Globular Clusters

14.1 Basic Data for Globular Clusters

The statistical properties of globular cluster parameters are given in the first accompanying table, including their ages that apparently range between 10 and 22 billion years. These ages are rather uncertain and somewhat controversial. Petersen (1987) has tabulated the ages of many clusters in a self-consistent fashion, but the uncertainties of specific age estimates are large. Different authors obtain age estimates for the same globular cluster that can vary as much as 3 billion years (see Alcaino and Liller, 1988, Sandage 1982, 1983, and Vandenberg 1988), and there is evidence for an age spread of 5 billion years for galactic globular clusters (Sarajedini and King, 1989).

The Hertzsprung-Russell diagram for the globular cluster M 5 is given in the first diagram, together with some of its descriptive features. It is followed by the distribution of globular clusters in galactic coordinates.

Tables of millisecond pulsars and compact X-ray sources in globular clusters are then provided.

The main catalogue of globular clusters is in two parts. The first part begins with the cluster name, continues with accurate right ascensions (R.A) and declinations (Dec.) for the cluster centers in epoch 1950.0 from Shawl and White (1986) together with the corresponding galactic longitude and latitude, l and b, and ends with the cluster's IAU number.

Each page of the second part of our catalogue begins with the IAU number. The integrated visual magnitudes, V_T, are from Webbink (1985), and the angular diameters, θ, in minutes of arc, are from Alcaino (1979). The apparent magnitudes of the horizontal branch, V_{HB}, come from Armanoff (1988), Harris (1980), Zinn (1985) and Zinn (1989). The heliocentric radial velocities, V_r, and the velocities with respect to the local standard of rest, V_{LSR}, are from Peterson (1985) and Zinn (1985). In these tables a colon(:) or double colon(::) indicate sources of poor or very poor quality.

The integrated absolute magnitudes, M_V, were computed from the relation $M_V = V_T - (m-M)_V$, where the distance modulii, $(m-M)_V$, are from Zinn (1989); the approximate number of stars, N, has been estimated from $N = 10^{0.4(4.79-M_V)}$. The heliocentric distances, D_o, and the galactocentric coordinates, Z, and distances, D_{gc}, are also from Zinn (1989) who assumed that the distance from the Sun to the galactic center is 8.0 *kpc*.

The cluster metallicities, [Fe/H], are from Armanoff and Zinn (1988), Da Costa and Seitzer (1989), Zinn (1985), and Zinn and West (1984). The color excesses, E_{B-V}, are also from Zinn (1989), while the core radius, R_c, and the tidal, or limiting radius, R_t, are from Webbink (1985). Here they are compared with the observed radius $R_o = \theta D_o/2$.

Parameters for individual clusters have been supplemented by recent data for NGC 288 (Pound et al. 1987), NGC 1904 (Heasley et al. 1986), NGC 4590 (McClure et al. 1987), NGC 6362 (Alcaino and Liller 1986), and NGC 7099 (Richer et al. 1988).

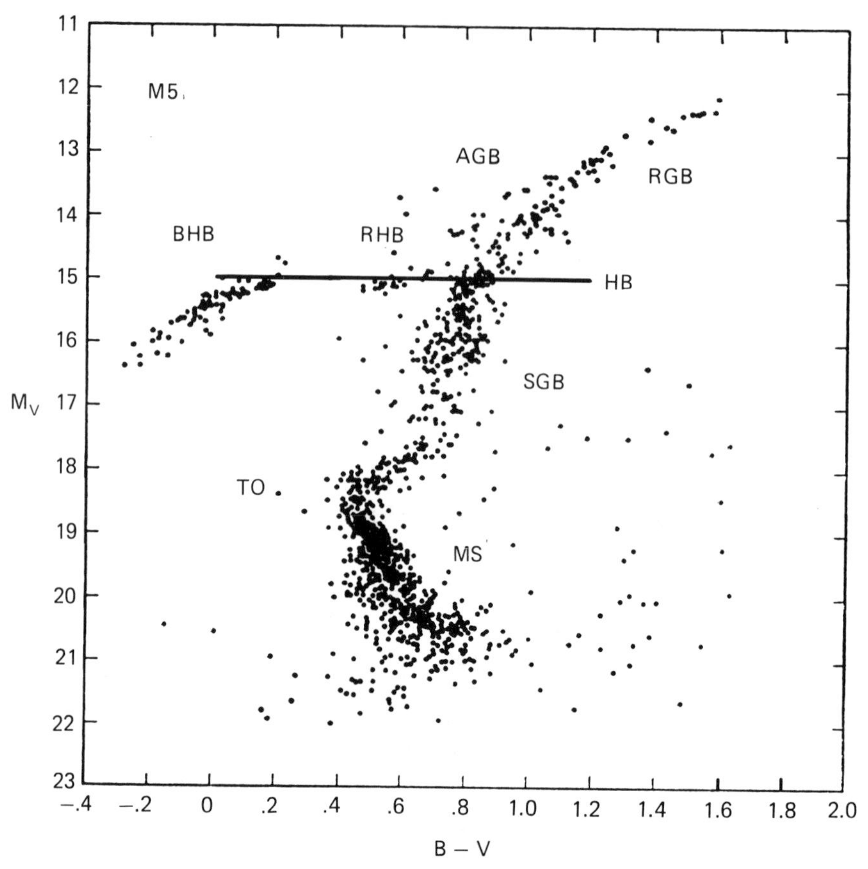

Hertzsprung-Russell Diagram for the Globular Cluster M 5. Halton Arp's plot of the absolute visual magnitudes, M_V, and colors, B-V, of stars in the globular cluster M 5 [Astrophysical Journal 135, 311-322 (1962)] with the addition of typical evolutionary phases. They are the main sequence, MS, the turnoff, TO, point from the main sequence, the subgiant branch, SGB, the horizontal branch, HB, often subdivided into the red, RHB, and blue, BHB, parts that are separated by an instability strip, and the giant branch, GB, often subdivided into the red, RGB, or first, FGB, and asymptotic, AGB parts. Allan Sandage provided a similar pioneering color-magnitude diagram for the stars in the globular cluster, M 3 [Astronomical Journal 58, 61-75 (1953)], and subsequently compared those of M 3, M 13, M 15 and M 92 with theoretical models of stellar evolution [Astrophysical Journal 162, 841-870 (1970)]. Gonzalo Alcaino has compiled the *Atlas of Globular Clusters with Color-Magnitude Diagrams* (Santiago: Universidad Catholica de Chile, 1973). A.G. Davis Philip has provided a bibliography of globular cluster data and 165 color-magnitude plots [Vistas in Astronomy 21, 407-445 (1977)]. A bibliography of globular-cluster color-magnitude diagram studies has also been given by Charles J. Petersen [Publications of the Astronomical Society of the Pacific 98, 1258-1272 (1986)], and Don A. Vandenberg has provided synthetic color-magnitude diagrams for the oldest clusters [Astrophysical Journal Supplement 51, 29-66 (1983)].

Globular Cluster Statistics

Parameter	Mean	Max.	Min.	Disp., σ
Integrated Visual Magnitude, V_T (mag)	9.83	20.10	3.52	3.06
Angular Diameter, θ (′)	8.55	36.30	1.20	5.98
Apparent Magnitude Horizontal Branch, V_{HB} (mag)	17.41	26.20	12.90	2.25
Heliocentric Velocity, V_r (km/s)	- 3.22	494.00	- 384.80	98.41
Integrated Absolute Magnitude, M_V (mag)	- 7.12	-1.43	- 10.30	1.46
Number of Stars, N ($\times 10^5$)	1.20	10.86	0.02	1.59
Heliocentric Distance, D_o (kpc)	16.6	112.3	1.9	19.5
Galactocentric Coordinate, Z (kpc)	0.4	85.4	- 83.4	14.9
Galactocentric Distance, D_{gc} (kpc)	13.7	113.7	0.9	20.4
Metallicity, $[F_e/H]$	- 1.32	0.24	- 2.24	0.56
Distance Modulus, $(m-M)_V$ (mag)	15.91	20.25	12.26	1.49
Color-Excess, E_{B-V}	0.50	3.80	0.01	0.64
Observed Radius, R_o (pc)	13.00	56.20	3.10	8.17
Core Radius, R_c (pc)	2.75	27.35	0.10	4.51
Tidal, or Limiting, Radius, R_t (pc)	42.87	284.80	5.50	37.57
Age ($\times 10^9$ years)	15.51	21.88	10.47	2.48

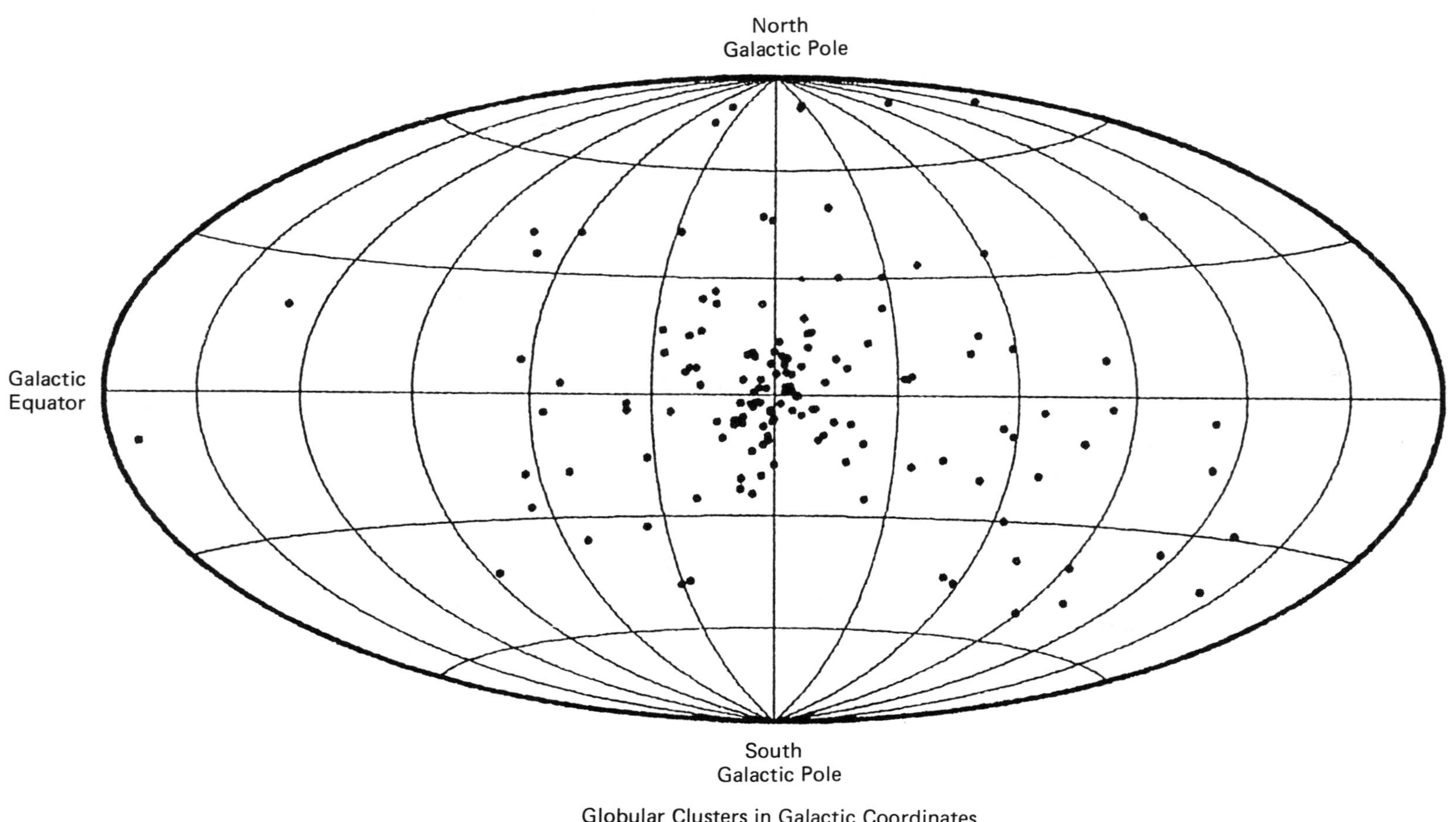

Globular Clusters in Galactic Coordinates

Millisecond Pulsars In Globular Clusters*

PSR	Globular Cluster	R.A.(1950) h m s	Dec.(1950) ° ′ ″	Period (seconds)	$\dot{P}$ (10^{-15})	DM (cm^{-3} pc)	D (kpc)
0021-71A+	NGC 104, 47 Tuc	00 21 53	– 72 21 30	0.004478953		65+	2.50
0021-72B+	NGC 104, 47 Tuc	00 21 53	– 72 21 30	0.006127		65+	2.50
1620–26	NGC 6121, M 4	16 20 34.14	– 26 24 58.0	0.01107575081	0.00080	62.87	2.20
1821–24	NGC 6626, M 28	18 21 27.382	– 24 53 50.95	0.00305431449	0.00155	120.0	3.90
2127+11++	NGC 7078, M 15	21 27 33.22	+11 56 49.4	0.11066470954	– 0.02	67.25	2.20

* Adapted from Lyne et al (1987), Taylor (1989), and Wolszczan et al. (1989). Since this table was prepared more than twenty millisecond pulsars have been found in twelve globular clusters, including three in M 15 and eleven in 47 Tuc.
+ PSR 0021-71C and ten other millisecond pulsars, all with dispersion measures close to 24.5 $cm^{-3}pc$ have been found in 47 Tuc [Nature 352, 219 (1991)], now suggesting that PSR 0021-71A and B are not within the cluster.
++ M 15 contains at least three pulsars 2127+11 A, B, and C with respective periods of 110.665, 56.133 and 30.529 milliseconds [Nature 346, 42 (1990)]

Compact X-Ray Sources In Globular Clusters*

Globular Cluster	X-Ray Source	R.A.(1950) h m s	Dec.(1950) ° ′ ″	Position Error (″)
NGC 104, 47 Tuc	1E 0021.8–7221	00 21 51.58	– 72 21 34.6	±1
NGC 1851	4U 0513–40	05 12 27.88	– 40 05 59.7	±1
NGC 1904, M 79	1E 0522.1–2433	05 22 07.7	– 24 33 55	±60
NGC 5139, ω Cen	1E 1323.8–4713	13 23 49.8	– 47 13 13	±30
NGC 5272, M 3	1E 1339.8+2837	13 39 51.4	+28 37 52	±30
NGC 5824	1E 1500.7–3251	15 00 49.4	– 32 52 07	±10
Terzan 2	4U 1722–30	17 24 20.09	– 30 45 39.4	±1
Grindlay 1	4U 1728–34	17 28 39.2	– 33 47 55	±5
Liller 1	MXB 1730–335	17 30 06.63	– 33 21 13.2	±1
NGC 6440	MX 1746–20	17 45 55.0	– 20 21 07	±60
NGC 6441	4U 1746–37	17 46 48.49	– 37 02 17.8	±1
NGC 6541	1E 1804.4–4343	18 04 24.8	– 43 43 36	±60
NGC 6624	4U 1820–30†	18 20 27.84	– 30 23 17.0	±1
NGC 6656, M 22	1E 1833.3–2357	18 33 20.4	– 23 56 56	±20
NGC 6712	4U 1850–08	18 50 21.18	– 08 46 04.4	±1
NGC 7078, M 15	4U 2127+11†	21 27 33.14	+11 56 51.0	±1

Compact X-Ray Sources In Globular Clusters*

X-Ray Source	Assumed Distance (kpc)	X-Ray Luminosity (erg/sec)	X-Ray Source	Assumed Distance (kpc)	X-Ray Luminosity (erg/sec)
1E 0021.8–7221	4.6	34.59	MXB 1730–335	10.0	36.81
4U 0513–40	10.8	36.12	MX 1746–20	3.5	32.82
1E 0522.1–2433	13.3	33.89	4U 1746–37	10.1	36.81
1E 1323.8–4713	5.1	32.93	1E 1804.4–4343	6.8	33.33
1E 1339.8+2837	9.8	33.56	4U 1820–30†	8.7	37.95
1E 1500.7–3251	23.5	34.29	1E 1833.3–2357	3.0	32.39
4U 1722–30	10.5	36.71	4U 1850–08	7.4	36.41
4U 1728–34	10.0	37.61	4U 2127+11†	9.4	36.75

*Adapted from Hertz and Grindlay (1983), Grindlay et al. (1984), Grindlay (1988). The positions are those of the X-ray sources. The optical counterpart of the X-ray source in NGC 7078 is AC 211 (Aurière et al. 1985). For NGC 5139 and NGC 6656 the positions of the cluster centers are given, these two clusters contain multiple X-ray sources and the luminosity of the brightest one is given. †X-ray sources 4U 1820-30 and 4U 2127+11 are periodic, with respective periods of 11.4 minutes and 8.5 hours.

14.2 Catalogue of Globular Clusters — Part I

NAME	R.A.(1950) h m s	Dec.(1950) ° ′ ″	l °	b °	IAU Number
47 Tuc, NGC 104	00 21 52.28	– 72 21 27.5	305.90	– 44.89	C0021–723
NGC 288	00 50 21.20	– 26 51 40.8	152.33	– 89.38	C0050–268
NGC 362	01 01 32.64	– 71 06 59.3	301.53	– 46.25	C0100–711
NGC 1261	03 10 53.23	– 55 24 12.8	270.54	– 52.13	C0310–554
Pal 1	03 26 03.72	+79 24 38.8	130.07	+19.03	C0325+794
NGC 1466, SL 1	03 44 49	– 71 49 42	286.70	– 39.54	C0344–718
ESO 1, AM 1	03 53 35	– 49 45 36	258.36	– 48.47	C0354–498
Eridanus, SW 2	04 22 35	– 21 18 06	218.11	– 41.33	C0422–213
Reticulum, Ser	04 35 21	– 58 57 48	268.66	– 40.27	C0435–590
Pal 2	04 42 53.67	+31 17 27.0	170.53	– 09.07	C0443+313
NGC 1841	04 52 45	– 84 04 48	297.02	– 30.15	C0444–840
NGC 1851	05 12 27.63	– 40 06 13.2	244.51	– 35.04	C0512–400
M 79, NGC 1904	05 22 06.98	– 24 34 07.8	227.23	– 29.35	C0522–245
NGC 2298	06 47 12.91	– 35 56 50.4	245.63	– 16.01	C0647–359
NGC 2419	07 34 45.50	+38 59 44.3	180.37	+25.24	C0734+390
AM 2	07 36 53	– 33 43 42	248.13	– 05.88	C0737–337
NGC 2808	09 11 03.90	– 64 39 22.5	282.19	– 11.25	C0911–646
ESO 3	09 21 31	– 77 04 12	292.27	– 19.02	C0921–770
UKS 2	09 23 43	– 54 30 06	276.00	– 03.01	C0923–545
Pal 3, Sex C	10 02 57.46	+00 18 53.7	240.14	+41.86	C1003+003
NGC 3201	10 15 33.50	– 46 09 38.3	277.23	+08.64	C1015–461
ESO 093– 08	11 17 34	– 64 56 48	293.51	– 04.04	C1117–649
Pal 4, UMa	11 26 37.89	+29 14 57.4	202.31	+71.80	C1126+292
NGC 4147	12 07 33.31	+18 49 11.6	252.85	+77.19	C1207+188
NGC 4372	12 22 49.23	– 72 22 55.8	300.99	– 09.88	C1223–724
Rup 106	12 35 53	– 50 52 36	300.89	+11.67	C1235–509
M 68, NGC 4590	12 36 48.71	– 26 28 06.4	299.63	+36.05	C1236–264
NGC 4833	12 56 13.42	– 70 36 17.8	303.61	– 08.01	C1256–706
M 53, NGC 5024	13 10 28.27	+18 26 02.3	332.97	+79.77	C1310+184
NGC 5053	13 14 00.15	+17 57 40.5	335.70	+78.94	C1313+179
ω Cen, NGC 5139	13 23 45.76	– 47 13 02.8	309.10	+14.97	C1323–472
M 3, NGC 5272	13 39 52.94	+28 37 37.9	042.21	+78.71	C1339+286
NGC 5286	13 43 16.04	– 51 07 25.1	311.61	+10.57	C1343–511
AM 4	13 53 31	– 26 55 36	320.28	+33.51	C1353–269
NGC 5466	14 03 12.44	+28 46 22.4	042.15	+73.59	C1403+287

Catalogue of Globular Clusters

NAME	R.A.(1950) h m s	Dec.(1950) ° ′ ″	l °	b °	IAU Number
NGC 5634	14 26 59.37	− 05 45 15.5	342.21	+49.26	C1427–057
NGC 5694	14 36 41.60	− 26 19 24.9	331.06	+30.36	C1436–263
IC 4499	14 52 09.28	− 82 00 49.0	307.35	− 20.47	C1452–820
NGC 5824	15 00 53.75	− 32 52 22.6	332.55	+22.07	C1500–328
Pal 5, Ser	15 13 31.51	+00 04 19.7	000.85	+45.86	C1513+000
NGC 5897	15 14 31.35	− 20 49 38.7	342.95	+30.29	C1514–208
M 5, NGC 5904	15 16 01.90	+02 15 50.9	003.86	+46.80	C1516+022
NGC 5927	15 24 22.98	− 50 29 58.6	326.60	+04.86	C1524–505
NGC 5946	15 31 49.53	− 50 29 37.4	327.58	+04.19	C1531–504
BH 176	15 35 28	− 49 52 54	328.42	+04.34	C1535–499
NGC 5986	15 42 46.78	− 37 37 51.0	337.02	+13.27	C1542–376
Pal 14, AvdB	16 08 47	+15 05 12	028.76	+42.18	C1608+150
M 80, NGC 6093	16 14 03.37	− 22 51 09.2	352.67	+19.46	C1614–228
NGC 6101	16 20 05.69	− 72 05 13.6	317.75	− 15.82	C1620–720
M 4, NGC 6121	16 20 31.35	− 26 24 36.4	350.97	+15.97	C1620–264
NGC 6144	16 24 10.53	− 25 54 48.1	351.93	+15.70	C1624–259
NGC 6139	16 24 17.39	− 38 44 15.8	342.37	+06.94	C1624–387
Ter 3	16 25 24	− 35 14 12	345.08	+09.19	C1625–352
M 107, NGC 6171	16 29 43.85	− 12 56 54.1	003.37	+23.01	C1629–129
ESO 452- 11	16 36 18	− 28 18 00	351.91	+12.10	C1636–283
M 13, NGC 6205	16 39 54.19	+36 33 16.3	059.01	+40.91	C1639+365
M 12, NGC 6218	16 44 38.63	− 01 51 33.9	015.72	+26.31	C1644–018
NGC 6229	16 45 34.42	+47 36 56.8	073.64	+40.31	C1645+476
NGC 6235	16 50 25.59	− 22 05 45.4	358.92	+13.52	C1650–220
M 10, NGC 6254	16 54 30.65	− 04 01 20.6	015.14	+23.08	C1654–040
NGC 6256, Ter 12	16 56 10.24	− 37 02 48.6	347.79	+03.31	C1656–370
Pal 15	16 57 28	− 00 28 06	018.87	+24.29	C1657–004
M 62, NGC 6266	16 58 01.46	− 30 02 23.4	353.57	+07.32	C1658–300
M 19, NGC 6273	16 59 32.06	− 26 11 49.7	356.87	+09.38	C1659–262
NGC 6284	17 01 25.22	− 24 41 46.3	358.35	+09.94	C1701–246
NGC 6287	17 02 08.48	− 22 38 24.8	000.13	+11.02	C1702–226
NGC 6293	17 07 04.08	− 26 31 11.4	357.62	+07.83	C1707–265
NGC 6304	17 11 21.82	− 29 24 19.9	355.83	+05.37	C1711–294
NGC 6316	17 13 28.62	− 28 05 08.7	357.18	+05.76	C1713–280
NGC 6325	17 14 56.63	− 23 42 48.4	000.97	+08.00	C1714–237

Catalogue of Globular Clusters

NAME	R.A.(1950) h m s	Dec.(1950) ° ′ ″	l °	b °	IAU Number
M 92, NGC 6341	17 15 35.01	+43 11 21.1	068.34	+34.86	C1715+432
M 9, NGC 6333	17 16 16.03	– 18 27 54.7	005.54	+10.71	C1716–184
NGC 6342	17 18 12.98	– 19 32 19.3	004.90	+09.72	C1718–195
NGC 6356	17 20 40.06	– 17 46 02.0	006.72	+10.22	C1720–177
NGC 6355	17 20 52.24	– 26 18 28.7	359.59	+05.43	C1720–263
NGC 6352	17 21 41.06	– 48 22 43.1	341.42	– 07.17	C1721–484
Ter 2, HP 3	17 24 20	– 30 45 36	356.32	+02.30	C1724–307
NGC 6366	17 25 04.70	– 05 02 09.5	018.41	+16.04	C1725–050
NGC 6362	17 26 44.83	– 67 00 38.5	325.55	– 17.57	C1726–670
Ter 4, HP 4	17 27 24	– 31 33 30	356.02	+01.31	C1727–315
HP 1	17 27 53	– 29 56 54	357.42	+02.11	C1727–299
Gri 1	17 28 40	– 33 47 54	354.30	– 00.15	C1728–338
Lil 1	17 30 07	– 33 21 18	354.84	– 00.16	C1730–333
NGC 6380, Ton 1	17 30 59	– 39 01 54	350.18	– 03.41	C1731–390
Ter 1, HP 2	17 32 34	– 30 27 00	357.56	+00.99	C1732–304
NGC 6388	17 32 37.50	– 44 42 14.3	345.56	– 06.74	C1732–447
Ton 2, Pis 26	17 32 43	– 38 31 12	350.80	– 03.42	C1733–390
M 14, NGC 6402	17 34 58.62	– 03 13 01.8	021.32	+14.80	C1735–032
NGC 6401	17 35 33.69	– 23 52 51.5	003.45	+03.98	C1735–238
NGC 6397	17 36 37.13	– 53 38 52.0	338.16	– 11.96	C1736–536
Pal 6	17 40 36	– 26 12 06	002.09	+01.78	C1740–262
NGC 6426	17 42 24.63	+03 11 23.9	028.09	+16.23	C1742+031
Ter 5	17 45 00	– 24 45 48	003.84	+01.69	C1745–247
NGC 6440	17 45 54.14	– 20 20 39.4	007.73	+03.80	C1746–203
NGC 6441	17 46 48.68	– 37 02 13.7	353.53	– 05.01	C1746–370
NGC 6453	17 47 31.95	– 34 35 07.9	355.72	– 03.87	C1748–346
Ter 6, HP 5	17 47 32	– 31 15 42	358.57	– 02.16	C1747–312
UKS 1	17 51 24	– 24 08 12	005.13	+00.76	C1751–241
NGC 6496	17 55 23.06	– 44 15 42.3	348.03	– 10.01	C1755–442
Ter 9	17 58 31	– 26 50 24	003.60	– 01.99	C1758–268
NGC 6517	17 59 06.26	– 08 57 33.7	019.23	+06.76	C1759–089
Ter 10	17 59 51	– 26 04 06	004.42	– 01.86	C1800–260
NGC 6522	18 00 21.66	– 30 02 10.9	001.03	– 03.93	C1800–300
NGC 6535	18 01 16.56	– 00 18 00.1	027.18	+10.44	C1801–003
NGC 6528	18 01 37.16	– 30 03 34.9	001.14	– 04.17	C1801–300

Catalogue of Globular Clusters

NAME	R.A.(1950) h m s	Dec.(1950) ° ′ ″	l °	b °	IAU Number
NGC 6539	18 02 07.06	− 07 35 24.3	020.80	+06.78	C1802–075
NGC 6544	18 04 15.71	− 25 00 15.8	005.84	− 02.20	C1804–250
NGC 6541	18 04 24.56	− 43 42 46.9	349.29	− 11.18	C1804–437
NGC 6553	18 06 09.45	− 25 55 01.6	005.25	− 03.02	C1806–259
NGC 6558	18 07 03.25	− 31 46 26.5	000.20	− 06.03	C1807–317
Pal 7, IC 1276	18 08 02.03	− 07 13 08.4	021.83	+05.67	C1808–072
Ter 11	18 09 14	− 22 45 18	008.36	− 02.10	C1809–227
NGC 6569	18 10 23.67	− 31 50 27.7	000.48	− 06.68	C1810–318
Kod 1	18 12 19	− 12 04 12	018.07	+02.42	C1812–121
NGC 6584	18 14 37.86	− 52 14 07.2	342.14	− 16.41	C1814–522
NGC 6624	18 20 27.92	− 30 23 15.2	002.79	− 07.91	C1820–303
M 28, NGC 6626	18 21 28.27	− 24 53 51.9	007.80	− 05.58	C1821–249
NGC 6638	18 27 50.84	− 25 31 55.3	007.90	− 07.15	C1827–255
M 69, NGC 6637	18 28 07.45	− 32 23 02.5	001.72	− 10.27	C1828–323
NGC 6642	18 28 51.65	− 23 30 46.6	009.81	− 06.44	C1828–235
NGC 6652	18 32 29.03	− 33 01 53.8	001.53	− 11.38	C1832–330
M 22, NGC 6656	18 33 21.12	− 23 56 44.2	009.89	− 07.55	C1833–239
Pal 8	18 38 32	− 19 52 30	014.10	− 06.80	C1838–198
M 70, NGC 6681	18 39 57.28	− 32 20 31.7	002.85	− 12.51	C1840–323
NGC 6712	18 50 20.53	− 08 46 05.7	025.35	− 04.32	C1850–087
M 54, NGC 6715	18 51 51.18	− 30 32 34.3	005.61	− 14.09	C1851–305
NGC 6717, Pal 9	18 52 05.18	− 22 45 55.1	012.88	− 10.90	C1852–227
NGC 6723	18 56 11.19	− 36 42 03.7	000.07	− 17.30	C1856–367
NGC 6749	19 02 43	+01 49 30	036.20	− 02.20	C1902+017
NGC 6752	19 06 27.30	− 60 03 50.5	336.50	− 25.63	C1906–600
NGC 6760	19 08 39.38	+00 56 48.6	036.11	− 03.92	C1908+009
Ter 7	19 14 25	− 34 44 48	003.39	− 20.06	C1914–347
M 56, NGC 6779	19 14 38.42	+30 05 39.5	062.66	+08.34	C1914+300
Pal 10	19 15 49	+18 28 54	052.44	+02.73	C1916+184
Arp 2	19 25 34	− 30 27 42	008.54	− 20.79	C1925–304
M 55, NGC 6809	19 36 49.08	− 31 04 40.8	008.79	− 23.27	C1936–310
Ter 8	19 38 29	− 34 07 06	005.76	− 24.56	C1938–341
Pal 11	19 42 32	− 08 07 42	031.81	− 15.58	C1942–081
M 71, NGC 6838	19 51 32.28	+18 38 49.2	056.74	− 04.56	C1951+186
M 75, NGC 6864	20 03 07.87	− 22 03 54.7	020.30	− 25.75	C2003–220

Catalogue of Globular Clusters

NAME	R.A.(1950) h m s	Dec.(1950) ° ′ ″	l °	b °	IAU Number
NGC 6934	20 31 44.45	+07 13 54.5	052.10	– 18.89	C2031+072
M 72, NGC 6981	20 50 43.18	– 12 43 37.3	035.16	– 32.68	C2050–127
NGC 7006	20 59 09.26	+15 59 24.7	063.77	– 19.41	C2059+160
M 15, NGC 7078	21 27 33.35	+11 56 48.8	065.01	– 27.31	C2127+119
M 2, NGC 7089	21 30 54.88	– 01 02 44.2	053.38	– 35.78	C2130–010
M 30, NGC 7099	21 37 31.64	– 23 24 23.1	027.18	– 46.84	C2137–234
Pal 12, Cap	21 43 50	– 21 29 00	030.51	– 47.68	C2143–214
Pal 13, Peg	23 04 14.24	+12 30 05.2	087.10	– 42.70	C2304+124
NGC 7492	23 05 48.60	– 15 52 57.1	053.39	– 63.48	C2305–159

14.3 Catalogue of Globular Clusters — Part II

IAU Number	V_T (mag)	θ (′)	V_{HB} (mag)	V_r (km/s)	V_{LSR} (km/s)	M_V (mag)	N/10^5	D_o (kpc)
C0021–723	3.83	30.9	14.06	– 14.1	– 24.1	– 9.36	4.57	4.1
C0050–268	8.11	13.8	15.30	– 48.2	– 55.7	– 6.46	0.32	7.8
C0100–711	6.46	12.9	15.43	+232.4	+221.5	– 8.21	1.58	7.9
C0310–554	8.28	6.9	16.63	+55	+40	– 7.60	0.90	15.0
C0325+794*	13.62	1.8	16.76			– 2.54	0.01	13.7
C0344–718*	10.66:		18.8			–7.54	0.86	39.4
C0354–498	15.84		20.93	+116	+98.9	– 4.41	0.05	112.3
C0422–213	14.70:		20.24	– 21	– 39	– 4.80:	0.07	76.0
C0435–590*	12.7		19.17			– 5.87	0.18	50.4
C0443+313*	13.04	1.9	20.9			– 7.26	0.66	13.6
C0444–840*	10.90:		18.88			–7.38	0.74	40.9
C0512–400	7.15	11.0	16.05	+318.6	+299.2	– 8.15	1.50	11.1
C0522–245	7.81	8.7	16.20	+185.4	+165.8	– 7.71	1.00	12.5
C0647–359	9.20	6.8	16.10	+44	+24	– 6.24	0.26	9.5
C0734+390	10.32	4.1	20.50	– 20	– 26	– 9.58	5.60	91.4
C0737–337*	14.0		21.1:			– 6.50	0.33	57.7
C0911–646	6.13	13.8	16.20	+104.1	+89.8	– 9.33	4.45	9.0
C0921–770	11.35		15.93			– 3.88	0.03	7.3
C0923–545*	13.0		17.75			– 4.15	0.04	9.0
C1003+003	13.92:	2.8	20.48	+89	+80	– 5.90:	0.19	88.1
C1015–461	6.68	18.2	14.75	+494.0	+481.1	– 7.37	0.73	4.8
C1117–649*	14.5:		22.0::			– 6.90	0.47	59.5
C1126+292	14.20	2.1	20.45	+75	+77	– 5.57	0.14	89.9
C1207+188	10.22	4.0	16.85	+182	+185	– 5.97	0.20	16.8
C1223–724	7.29:	18.6	15.50	+83	+74	– 7.61:	0.91	5.0
C1235–509*	10.9		18.5			– 7.00	0.52	26.7
C1236–264	7.74	12.0	15.60	– 59	– 62	– 7.26	0.66	9.6
C1256–706	7.03	13.5	15.45	+216	+208	– 7.77	1.06	5.8
C1310+184	7.48	12.6	16.94	– 78.9	– 71.2	– 8.85	2.86	18.4
C1313+179	9.94	10.5	16.63	+43.5	+51.4	– 6.10	0.23	15.9
C1323–472	3.52:	36.3	14.52	+228.3	+224.7	– 10.30:	10.86	5.0
C1339+286	5.92	16.2	15.65	– 147.1	– 136.2	– 9.05	3.44	9.7
C1343–511	7.18	9.1	16.20	+48.6	+45.1	– 8.36	1.82	8.7
C1353–269	15.9		18.03			– 1.43	0.01	26.9
C1403+287	8.95	11.0	16.58	+119.9	+132.1	– 7.05	0.54	15.9

*M_V, inferred N and D_o from Webbink (1985)

Catalogue of Globular Clusters

IAU Number	Z (kpc)	D_{gc} (kpc)	$[F_e/H]$	$(m-M)_V$ (mag)	E_{B-V}	R_o (pc)	R_c (pc)	R_t (pc)
C0021–723	− 2.9	7.3	− 0.71	13.19	0.04	19.0	0.52	60.3
C0050–268	− 7.7	11.2	− 1.40	14.57	0.04	16.0	3.96	37.0
C0100–711	− 5.7	9.0	− 1.28	14.67	0.06	15.3	0.52	25.9
C0310–554	− 11.8	16.9	− 1.29	15.88	0.00	15.5	1.86	37.1
C0325+794*	+4.5	20.3	− 1.01		0.12		0.63	19.9
C0344–718*	− 25.1	38.4	− 2.15		0.07		2.81	73.9
C0354–498	− 83.4	113.7	− 1.69	20.25	0.00		9.54	81.2
C0422–213	− 50.1	81.0	− 1.35	19.50	0.03		6.19	85.5
C0435–590*	− 32.6	51.3	− 2.01		0.02		14.66	146.6
C0443+313*	− 2.1	22.2	− 1.68		1.2:		0.31	12.5
C0444–840*	− 20.5	38.3	− 1.56		0.07		7.85	94.4
C0512–400	− 6.4	15.8	− 1.29	15.30	0.02	18.4	0.23	28.3
C0522–245	− 6.1	18.4	− 1.68	15.52	0.01	16.3	0.66	40.5
C0647–359	− 2.6	14.6	− 1.81	15.44	0.18	9.7	1.34	24.4
C0734+390	+39.1	98.8	− 2.10	19.90	0.03	56.2	11.08	284.8
C0737–337*	− 5.9	61.5	− 1.47		0.53		25.97	98.7
C0911–646	− 1.8	10.8	− 1.37	15.46	0.22	18.7	0.70	39.2
C0921–770	− 2.4	8.7	− 1.60	15.23	0.30		4.48	25.2
C0923–545*	− 0.5	11.9	− 0.29		0.74		2.04	9.3
C1003+003	+58.8	91.4	− 1.78	19.82	0.03	37.0	11.96	109.1
C1015–461	+0.7	8.8	− 1.56	14.05	0.21	13.0	1.58	52.4
C1117–649*	− 4.2	56.6			0.79		1.07	75.5
C1126+292	+85.4	92.6	− 1.68	19.77	0.00	28.3	14.58	83.9
C1207+188	+16.4	19.1	− 1.80	16.19	0.02	10.1	0.65	36.4
C1223–724	− 0.9	7.0	− 2.08	14.90	0.45	14.0	3.76	45.3
C1235–509*	+5.4	23.5			0.24		7.42	89.2
C1236–264	+5.6	9.7	− 2.09	15.00	0.03	17.2	1.72	82.1
C1256–706	− 0.8	6.8	− 1.86	14.80	0.32	11.7	1.95	33.8
C1310+184	+18.1	18.9	− 2.04	16.33	0.00	34.8	2.20	118.0
C1313+179	+15.6	16.5	− 2.17	16.04	0.01	25.1	11.31	63.6
C1323–472	+1.3	6.3	− 1.59	13.82	0.11	27.1	3.96	65.7
C1339+286	+9.5	11.6	− 1.66	14.97	0.01	23.6	1.26	117.4
C1343–511	+1.6	7.0	− 1.79	15.54	0.27	11.9	0.60	33.8
C1353–269	+14.8	22.6	− 1.60	17.33	0.06		3.43	35.1
C1403+287	+15.2	16.2	− 2.22	16.00	0.00	26.2	7.31	96.3

*Z and D_{gc} from Webbink (1985)

Catalogue of Globular Clusters

IAU Number	V_T (mag)	θ (′)	V_{HB} (mag)	V_r (km/s)	V_{LSR} (km/s)	M_V (mag)	N/10^5	D_o (kpc)
C1427–057	9.38	4.9	17.75	– 63	– 54	– 7.72	1.01	24.8
C1436–263	9.17:	3.6	18.40	– 118	– 113	– 8.59:	2.25	30.9
C1452–820	9.42:	7.6	17.80			– 7.66:	0.95	17.8
C1500–328	7.84:	6.2	18.00	– 28	– 24	– 9.51:	5.25	24.2
C1513+000	11.75	6.9	17.35	– 55.4	– 42.8	– 4.88	0.07	20.3
C1514–208	8.64	12.6	16.25	+10	+18	– 6.93	0.49	11.9
C1516+022	5.69	17.4	15.15	+51.9	+65.0	– 8.73	2.56	7.3
C1524–505	8.02	12.0	16.70	– 89	– 89	– 7.74	1.03	7.5
C1531–504	9.48	7.1	17.14	+115	+116	– 6.92	0.48	9.0
C1535–499*	14.0:		22.6::			– 8.00	1.30	85.7
C1542–376	7.48	9.8	16.50	– 35	– 30	– 8.34	1.79	9.6
C1608+150	14.68	2.1	20.08	+72	+89	– 4.68	0.06	71.3
C1614–228	7.33	8.9	15.82	+12.9	+23.0	– 7.80	1.09	8.5
C1620–720	9.16:	10.7	16.72	+163	+158	– 6.90:	0.47	15.4
C1620–264	5.76	26.3	13.35	+64.3	+73.6	– 6.83	0.44	1.9
C1624–259	9.01	9.3	16.60	+131	+141	– 6.92	0.48	9.2
C1624–387	8.91	5.5	17.50	+7.6	+13.4	– 7.90	1.19	7.6
C1625–352*	12.0		18.8			– 6.93	0.49	27.2
C1629–129	8.10	10.0	15.63	– 34	– 21	– 6.72	0.40	5.9
C1636–283	12.0		16.66			– 3.96	0.03	10.0
C1639+365	5.68	16.6	14.95	– 247.8	– 228.9	– 8.58	2.23	6.9
C1644–018	6.77	14.5	14.90	– 43.5	– 27.7	– 7.44	0.78	5.4
C1645+476	9.36	4.5	18.05	– 154.2	– 135.8	– 7.98	1.28	29.4
C1650–220	9.99	5.0	16.66	+89	+100	– 5.94	0.20	9.4
C1654–040	6.55	15.1	14.65	+70.1	+85.7	– 7.40	0.75	4.1
C1656–370	11.29:		18.20			– 6.02:	0.21	9.3
C1657–004	14.2		20.20:	+69	+85	– 5.30	0.11	64.2
C1658–300	6.68	14.1	16.30	– 60.9	– 51.7	– 8.87	2.91	6.5
C1659–262	6.73	13.5	16.95	+121	+131	– 9.54	5.40	10.7
C1701–246	8.85	5.6	16.85	+22	+33	– 7.24	0.65	10.9
C1702–226	9.31	5.1	16.52	– 211	– 200	– 6.60	0.36	6.3
C1707–265	8.15	7.9	16.16	– 94	– 84	– 7.37	0.73	7.5
C1711–294	8.36	6.8	16.10	– 110	– 100	– 6.85	0.45	5.2
C1713–280	8.75	4.9	17.78	+69	+79	– 8.13	1.47	12.0
C1714–237	10.56	4.3	17.30	– 6	+5	– 6.01	0.21	5.8

*M_V, inferred N and D_o from Webbink (1985)

Catalogue of Globular Clusters

IAU Number	Z (kpc)	D_{gc} (kpc)	$[F_e/H]$	$(m-M)_V$ (mag)	E_{B-V}	R_o (pc)	R_c (pc)	R_t (pc)
C1427–057	+18.8	20.8	- 1.82	17.10	0.04	18.2	1.71	55.2
C1436–263	+15.7	25.5	- 1.91	17.76	0.10	16.7	0.46	138.0
C1452–820	- 6.2	14.8	- 1.50	17.08	0.27	20.3	6.61	83.2
C1500–328	+9.1	18.2	- 1.85	17.35	0.14	22.5	0.39	142.6
C1513+000	+14.6	15.8	- 1.47	16.63	0.03	21.0	17.97	113.4
C1514–208	+6.0	7.0	- 1.68	15.57	0.06	22.5	4.43	39.5
C1516+022	+5.3	6.1	- 1.40	14.42	0.03	19.1	0.97	64.1
C1524–505	+0.6	4.5	- 0.31	15.76	0.45	13.4	1.23	35.4
C1531–504	+0.7	4.9	- 1.39	16.40	0.53	9.5	0.44	26.6
C1535–499*	+6.5	78.3			0.73		11.13	153.6
C1542–376	+2.2	4.3	- 1.67	15.82	0.29	14.2	1.46	38.3
C1608+150	+47.9	66.4	- 1.47	19.36	0.03	22.5	16.62	102.5
C1614–228	+2.8	3.0	- 1.64	15.13	0.16	11.3	0.24	21.7
C1620–720	- 4.2	11.2	- 1.81	16.06	0.04	24.7	3.72	74.2
C1620–264	+0.5	6.3	- 1.28	12.59	0.40	7.3	0.75	26.5
C1624–259	+2.5	2.9	- 1.75	15.93	0.36	12.8	3.42	32.6
C1624–387	+0.9	2.6	- 1.65	16.81	0.78	6.2	0.52	29.5
C1625–352*	+4.4	19.0			0.32		6.59	60.1
C1629–129	+2.3	3.5	- 0.99	14.82	0.31	8.9	1.31	41.5
C1636–283	- 1.4	3.0	- 1.60	15.96	0.31		1.17	10.4
C1639+365	+4.5	8.3	- 1.65	14.26	0.02	17.2	1.72	55.5
C1644–018	+2.4	4.3	- 1.61	14.21	0.17	11.8	1.82	28.1
C1645+476	+19.0	28.8	- 1.54	17.34	0.00	19.9	1.60	51.7
C1650–220	+2.2	2.5	- 1.40	15.93	0.34	7.1	0.73	21.9
C1654–040	+1.6	4.7	- 1.60	13.95	0.28	9.4	0.92	31.2
C1656–370	+0.5	2.3	- 0.60	17.31	0.80		1.00	17.8
C1657–004	+24.4	57.5	- 1.60	19.50	0.15		27.35	108.9
C1658–300	+0.8	2.0	- 1.29	15.55	0.48	13.7	0.43	18.4
C1659–262	+1.8	3.2	- 1.68	16.27	0.36	21.7	1.55	53.8
C1701–246	+1.9	3.3	- 1.24	16.09	0.29	9.2	0.48	29.8
C1702–226	+1.2	2.2	- 2.05	15.91	0.62	4.8	0.69	20.0
C1707–265	+1.0	1.2	- 1.92	15.52	0.37	8.9	0.37	32.9
C1711–294	+0.5	2.9	- 0.59	15.21	0.53	5.3	0.46	21.9
C1713–280	+1.2	4.2	- 0.55	16.88	0.48	8.8	0.65	53.8
C1714–237	+0.8	2.4	- 1.44	16.57	0.89	3.7	0.44	17.9

*Z and D_{gc} from Webbink (1985)

Catalogue of Globular Clusters

IAU Number	V_T (mag)	θ (′)	V_{HB} (mag)	V_r (km/s)	V_{LSR} (km/s)	M_V (mag)	N/10^5	D_o (kpc)
C1715+432	6.39	11.2	15.05	− 120.5	− 101.3	− 8.09	1.42	7.6
C1716–184	7.61	9.3	15.89	+224.7	+237.4	− 7.62	0.92	6.8
C1718–195	9.84	3.0	16.90	+86	+99	− 6.19	0.25	8.7
C1720–177	8.18	7.2	17.67	+36	+49	− 8.59	2.25	15.4
C1720–263	9.68	5.0	17.20	− 165	− 154	− 6.81	0.44	6.5
C1721–484	8.13	7.1	15.15	− 114	− 110	− 6.12	0.23	5.0
C1724–307*	14.29:	1.5	19.8:	+109	+118	− 4.91	0.08	10.0
C1725–050	8.88:	8.3	15.70	− 123.2	− 107.0	− 5.98:	0.20	3.7
C1726–670	7.52	10.7	15.34	− 6.2	− 8.9	− 7.03	0.53	6.9
C1727–315*	16.0:		21.6::	− 50	− 41	− 5.00	0.08	16.1
C1727–299*	12.49	2.9	20.0:	+44	+54	− 6.91	0.48	9.5
C1728–338*	17.66:		26.2::			− 7.94	1.24	11.8
C1730–333*	15.84:		24.4	+52	+61	− 7.96	1.26	7.9
C1731–390*	11.12:	3.9	18.0			− 6.28	0.27	4.0
C1732–304*	15.9	2.8	20.6:	+35	+45	− 4.10	0.04	10.6
C1732–447	6.73	8.7	17.24	+80	+85	− 9.62	5.81	11.3
C1733–390*	12.24	3.4	18.2			− 5.36	0.11	8.7
C1735–032	7.57	11.7	17.20	− 123.4	− 106.9	− 8.89	2.96	8.4
C1735–238	9.45:	5.6	17.30	− 70	− 59	− 7.06:	0.55	6.1
C1736–536	5.75	25.7	12.90	+19.0	+21.0	− 6.51	0.33	2.2
C1740–262	11.55:	7.2	19.1			− 6.43:	0.31	5.9
C1742+031	11.13	3.2	17.30	− 155	− 137	− 5.59	0.14	11.9
C1745–247*	13.85:	2.1	21.7	− 94	− 83	− 7.25	0.65	7.1
C1746–203	9.05	5.4	18.40	− 70	− 57	− 8.41	1.91	6.5
C1746–370	7.19	7.8	17.10	+19	+27	− 9.01	3.31	8.9
C1748–346	9.78	3.5	17.70	− 78	− 70	− 7.21	0.63	10.6
C1747–312*	13.85:	1.2	20.8::	+126	+135	− 6.35	0.29	3.9
C1751–241*	17.29:		25.5			− 7.61	0.91	10.4
C1755–442	8.48:	6.9	16.36	− 79	− 74	− 6.97:	0.51	9.7
C1758–268*	16.0:		20.3:			− 3.70	0.02	7.0
C1759–089	10.30	4.3	18.00	− 40	− 25	− 6.95	0.50	6.0
C1800–260*	14.9:		21.9::			− 6.40	0.30	5.8
C1800–300	8.35	5.6	16.97	+8	+18	− 7.89	1.18	9.3
C1801–003	10.48	3.6	15.80	− 126	− 109	− 4.65	0.06	6.7
C1801–300	9.56	3.7	16.75	+143	+153	− 6.28	0.27	6.5

*M_V, inferred N and D_o from Webbink (1985)

Catalogue of Globular Clusters

IAU Number	Z (kpc)	D_{gc} (kpc)	$[F_e/H]$	$(m-M)_V$ (mag)	E_{B-V}	R_o (pc)	R_c (pc)	R_t (pc)
C1715+432	+4.4	9.2	- 2.24	14.48	0.02	12.8	0.69	37.2
C1716–184	+1.3	1.9	- 1.78	15.23	0.34	9.5	0.91	33.7
C1718–195	+1.5	1.7	- 0.66	16.03	0.43	3.9	0.45	29.5
C1720–177	+2.7	7.8	- 0.54	16.77	0.27	16.6	1.43	55.7
C1720–263	+0.6	1.6	- 1.50	16.49	0.78	4.9	0.39	15.7
C1721–484	- 0.6	3.7	- 0.51	14.25	0.24	5.3	1.30	23.1
C1724–307*	+0.4	1.4	- 0.47		1.31		0.31	9.0
C1725–050	+1.0	4.9	- 0.85	14.86	0.65	4.6	2.56	23.3
C1726–670	- 2.1	5.0	- 1.08	14.55	0.11	11.1	3.57	37.4
C1727–315	+0.4	7.4	- 0.21		1.55		$<$ 0.47	7.8
C1727–299*	+0.4	0.9	- 1.68		1.41		0.58	12.9
C1728–338*	+0.0	3.2			3.2:		1.84	29.9
C1730–333*	+0.0	1.2	- 0.21		2.91		0.14	7.6
C1731–390*	- 0.2	5.0	- 1.00		1.38		0.40	6.5
C1732–304*	+0.2	1.9	+0.24		1.52		0.30	12.0
C1732–447	- 1.3	4.2	- 0.60	16.35	0.35	14.8	0.58	32.6
C1733–390*	- 0.5	1.5			0.91		1.83	12.9
C1735–032	+2.2	3.7	- 1.39	16.46	0.59	14.8	2.36	29.7
C1735–238	+0.4	2.0	- 1.13	16.51	0.84	5.1	0.49	27.4
C1736–536	- 0.5	6.1	- 1.91	12.26	0.18	8.4	0.50	24.7
C1740–262	+0.2	2.9	- 0.74	17.98:	1.46	6.4	1.16	12.8
C1742+031	+3.3	6.7	- 2.20	16.72	0.43	5.7	1.81	37.8
C1745–247*	- 0.2	1.8	+0.24		2.14		0.10	8.4
C1746–203	+0.4	1.9	- 0.34	17.46	1.10	5.2	0.24	24.7
C1746–370	- 0.8	1.5	- 0.53	16.20	0.47	10.4	0.47	25.9
C1748–346	- 0.7	2.8	- 1.53	16.99	0.60	5.6	0.51	15.6
C1747–312*	- 0.5	4.0			1.46		0.31	5.6
C1751–241*	+0.1	1.8	- 1.18		3.07		0.66	17.7
C1755–442	- 1.7	2.9	- 0.48	15.45	0.17	10.0	3.23	18.2
C1758–268*	- 0.2	1.9	- 0.38		1.71		0.25	5.5
C1759–089	+0.7	3.1	- 1.34	17.25	1.08	3.9	0.18	18.5
C1800–260*	- 0.5	5.9			1.71		0.49	6.4
C1800–300	- 0.6	1.5	- 1.44	16.24	0.45	7.8	0.34	19.1
C1801–003	+1.2	3.9	- 1.75	15.13	0.32	3.6	0.40	20.1
C1801–300	- 0.5	1.6	- 0.23	15.79	0.56	3.6	0.26	9.3

*Z and D_{gc} from Webbink (1985)

Catalogue of Globular Clusters

IAU Number	V_T (mag)	θ (′)	V_{HB} (mag)	V_r (km/s)	V_{LSR} (km/s)	M_V (mag)	N/10^5	D_o (kpc)
C1802–075	9.77	6.9	18.33	– 37	– 21	– 7.69	0.98	8.4
C1804–250	8.07	8.9	15.00	– 39	– 28	– 6.23	0.26	2.3
C1804–437	6.08	13.1	15.10	– 152.8	– 147.4	– 8.37	1.84	6.0
C1806–259	8.06	8.1	16.95	– 12	– 1	– 7.95	1.25	5.2
C1807–317	9.82	3.7	16.70	– 135	– 126	– 6.15	0.24	8.7
C1808–072*	10.34:	7.1	18.5			– 7.56	0.87	9.8
C1809–227*	16.4:		22.5::			– 5.50	0.13	23.7
C1810–318	8.71	5.8	17.05	– 26	– 17	– 7.50	0.82	8.0
C1812–121	20.1::							
C1814–522	8.63	7.9	15.90	+252	+255	– 6.56	0.35	9.3
C1820–303	7.99	5.9	16.05	+39	+49	– 7.13	0.59	7.4
C1821–249	6.83	11.2	15.60	+1.8	+13.1	– 8.04	1.36	5.3
C1827–255	9.05	5.0	16.50	– 1.5	+20	– 6.67	0.38	7.8
C1828–323	7.56	7.1	16.20	+44	+53	– 7.75	1.04	8.9
C1828–235	9.36	4.5	15.50	– 67	– 55	– 5.39	0.12	5.3
C1832–330	8.76	3.5	16.70	– 86	– 77	– 7.11	0.58	13.3
C1833–239	5.07	24.0	14.15	– 152.5	– 141.0	– 8.41	1.91	3.2
C1838–198	11.15	4.7	17.27	– 38	– 25	– 5.21	0.10	11.9
C1840–323	7.95	7.8	15.40	+225	+234	– 6.74	0.41	7.7
C1850–087	8.17	7.2	16.11	– 107.5	– 108.7	– 7.13:	0.59	5.8
C1851–305	7.57	9.1	17.71	+131.0	+140.4	– 9.41	4.79	20.1
C1852–227	9.18	3.9	15.73	+20	+32	– 5.80	0.17	7.2
C1856–367	7.17	11.0	15.48	– 87	– 80	– 7.52	0.84	8.7
C1902+017*	12.44	6.31	19.2			– 6.16	0.24	12.8
C1906–600	5.48	20.4	13.75	– 32.2	– 32.7	– 7.56	0.87	3.8
C1908+009	9.08	6.6	17.50	– 21	– 4	– 7.52	0.84	7.6
C1914–347*	12.0		18.6			– 6.00	0.21	36.4
C1914+300	8.26	7.1	16.20	– 138.1	– 118.8	– 7.31	0.69	9.6
C1916+184*	13.22	3.5	19.4			– 5.58	0.14	10.6
C1925–304	12.3	3.7	18.21			– 5.21	0.10	27.2
C1936–310	6.36	19.0	14.40	+166.6	+175.0	– 7.39	0.74	5.2
C1938–341*	12.4		19.4			– 6.40	0.30	48.2
C1942–081	9.80::	3.2	17.30	– 68	– 54	– 6.63:	0.37	11.9
C1951+186	8.00	7.2	14.41	– 19.3	– 1.4	– 5.52	0.13	3.4
C2003–220	8.53	6.0	17.45	– 195.2	– 185.1	– 8.17	1.53	17.2

*M_V, inferred N and D_o from Webbink (1985)

Catalogue of Globular Clusters

IAU Number	Z (kpc)	D_{gc} (kpc)	$[F_e/H]$	$(m-M)_V$ (mag)	E_{B-V}	R_o (pc)	R_c (pc)	R_t (pc)
C1802–075	+1.0	3.2	-0.66	17.46	0.91	8.7	0.48	12.0
C1804–250	-0.1	5.7	-1.56	14.30	0.79	3.1	0.35	8.8
C1804–437	-1.2	2.7	-1.83	14.45	0.18	11.8	0.57	64.1
C1806–259	-0.3	2.9	-0.29	16.01	0.78	6.3	1.13	17.4
C1807–317	-0.9	1.1	-1.44	15.97	0.41	4.8	0.40	16.2
C1808–072*	+1.0	3.8	-0.84		0.92		3.59	35.9
C1809–227*	-0.9	15.1			1.57		1.35	13.8
C1810–318	-0.9	0.9	-0.86	16.21	0.55	6.9	0.89	18.3
C1812–121					3.8:			
C1814–522	-2.6	3.8	-1.54	15.19	0.11	11.1	1.74	38.0
C1820–303	-1.0	1.3	-0.37	15.12	0.25	6.5	0.20	26.8
C1821–249	-0.5	2.9	-1.44	14.87	0.40	9.0	0.64	25.5
C1827–255	-1.0	1.5	-1.15	15.72	0.41	5.8	0.46	11.5
C1828–323	-1.6	1.8	-0.59	15.31	0.18	9.5	0.99	31.3
C1828–235	-0.6	3.0	-1.29	14.75	0.36	3.6	0.12	12.8
C1832–330	-2.6	5.7	-0.89	15.87	0.08	7.0	0.63	32.3
C1833–239	-0.4	5.0	-1.75	13.48	0.32	11.3	1.10	30.2
C1838–198	-1.4	4.7	-0.48	16.36	0.32	8.4	3.25	18.3
C1840–323	-1.7	1.8	-1.51	14.69	0.08	9.0	0.25	29.6
C1850–087	-0.4	3.7	-1.01	15.30	0.48	6.3	1.23	19.9
C1851–305	-4.9	12.6	-1.43	16.98	0.15	27.5	0.65	46.4
C1852–227	-1.4	2.3	-1.32	14.98	0.22	4.2	0.13	18.0
C1856–367	-2.6	2.6	-1.09	14.69	0.00	14.3	2.22	33.6
C1902+017*	-0.5	7.7	-0.37		0.96:		1.78	10.0
C1906–600	-1.7	5.3	-1.54	13.04	0.04	11.7	0.59	36.5
C1908+009	-0.5	4.9	-0.52	16.60	0.71	7.5	0.53	14.5
C1914–347*	-12.5	28.4			0.06		3.85	36.8
C1914+300	+1.4	9.3	-1.94	15.57	0.21	10.3	1.25	34.3
C1916+184*	+0.5	8.7			1.15		0.99	17.3
C1925–304	-9.7	20.0	-1.60	17.51	0.11	15.1	16.42	46.3
C1936–310	-2.0	4.0	-1.82	13.75	0.06	14.7	2.59	27.7
C1938–341*	-20.0	40.4			0.12		10.89	122.2
C1942–081	-3.2	7.1	-0.70	16.43	0.34	5.7	5.27	27.9
C1951+186	-0.3	6.8	-0.58	13.52	0.27	3.7	1.06	17.6
C2003–220	-7.5	11.3	-1.32	16.70	0.17	15.5	0.51	32.5

*Z and D_{gc} from Webbink (1985)

Catalogue of Globular Clusters

IAU Number	V_T (mag)	θ (′)	V_{HB} (mag)	V_r (km/s)	V_{LSR} (km/s)	M_V (mag)	N/10^5	D_o (kpc)
C2031+072	8.72	5.9	16.82	− 379	− 364	− 7.39	0.74	13.5
C2050–127	9.32	5.9	16.85	− 278	− 268	− 6.82	0.44	16.0
C2059+160	10.46	2.8	18.80	− 384.8	− 369.9	− 7.64	0.94	38.8
C2127+119	6.02	12.3	15.86	− 112.1	− 99.2	− 9.25	4.13	9.8
C2130–010	6.36	12.9	16.05	− 6.1	+4.5	− 8.99	3.25	11.4
C2137–234	7.32	11.0	15.11	− 172.3	− 166.7	− 7.19	0.62	7.6
C2143–214	11.71	2.9	17.10	+9	+15	− 4.61	0.06	17.8
C2304+124	13.80:	1.8	17.70	− 28	− 21	− 3.24:	0.02	23.8
C2305–159	11.43	6.2	17.63	− 188.5	− 186.9	− 5.49	0.13	24.2

Catalogue of Globular Clusters

IAU Number	Z (kpc)	D_{gc} (kpc)	$[F_e/H]$	$(m - M)_V$ (mag)	E_{B-V}	R_o (pc)	R_c (pc)	R_t (pc)
C2031+072	- 4.4	11.0	- 1.54	16.11	0.15	11.9	0.89	38.7
C2050–127	- 8.6	12.0	- 1.54	16.14	0.04	14.2	1.80	43.1
C2059+160	- 12.9	36.3	- 1.59	18.10	0.05	16.3	2.02	71.7
C2127+119	- 4.5	10.1	- 2.17	15.27	0.10	18.1	0.26	59.1
C2130–010	- 6.7	10.3	- 1.58	15.35	0.02	22.1	1.18	56.4
C2137–234	- 5.5	6.9	- 2.13	14.51	0.04	12.5	0.16	33.4
C2143–214	- 13.2	14.7	- 1.14	16.32	0.02	7.8	2.70	60.4
C2304+124	- 16.2	24.9	- 1.79	17.04	0.05	6.4	2.70	27.0
C2305–159	- 21.6	23.4	- 1.51	16.92	0.00	22.5	4.51	42.0

15
Open Clusters

15.1 Basic Data for Open Clusters

We have used the Lund catalogue of Open Cluster Data, kindly supplied by Gösta Lynga (1987) and updated by Kenneth Janes (1989). This data has been used by Lynga and Palous (1987) to describe the local kinematics of open star clusters, and by Janes, Tilley, and Lynga (1988) to describe the general properties of the open cluster system. The galactic distribution of open clusters was discussed by Janes and Adler (1982).

The statistical properties of some of the catalogue data are given in the first table, including age estimates between a million and seven billion years. Mermilliod (1981) has provided the ages for some of the best-observed clusters such as the Pleiades (77.6 million years) and the Hyades (0.66 billion years). A composite Hertzsprung-Russell diagram for these well-known open clusters is provided on the next page. It is followed by a plot of the location of open clusters in galactic coordinates.

Because the metallicity, [Fe/H], is known for relatively few open clusters, it is provided with the cluster name and IAU number in the second table. This is followed by a table of cluster masses in units of the Sun's mass, $M_{\odot}$ (Pandey, Bhatt, and Mahra, 1987; also see Bruch and Sanders, 1983).

The main catalogue of open cluster parameters includes the cluster name, right ascension, R.A., and declination, Dec., for epoch 1950.0, the galactic coordinates, l and b, the total cluster magnitude, V_T, its IAU number, the angular diameter, θ, in minutes of arc, the heliocentric radial velocity, V_r, the magnitude of the brightest star, MBS, the integrated color, B – V, the turn off color, TOC, from the main sequence in the Hertzsprung-Russell diagram, the color excess, E_{B-V}, the radius R, the heliocentric distance, D_o, in parsecs, and the age in millions of years.

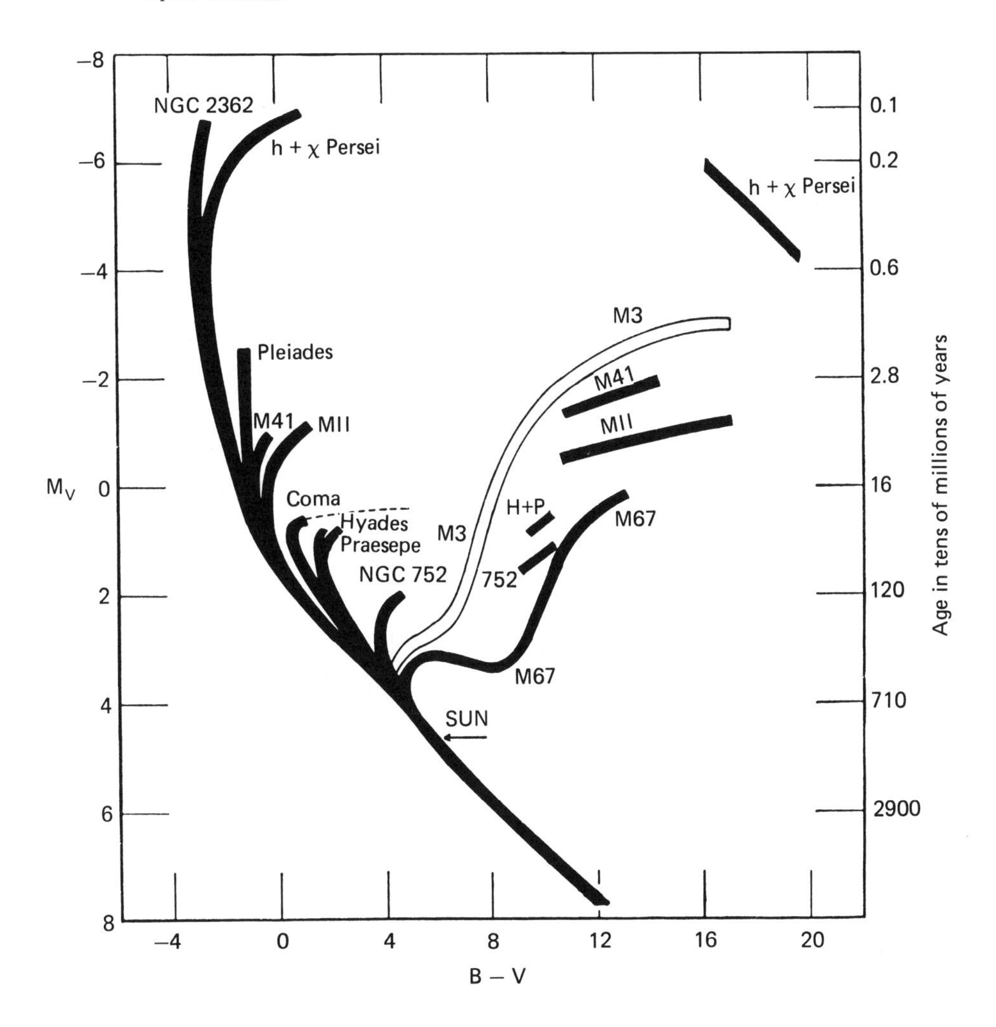

Hertzsprung-Russell Diagram for Open Clusters. The color-magnitude diagrams for the stars in ten open, or galactic, clusters are compared with that of one globular cluster M 3. It is Allan Sandage's plot of absolute visual magnitudes, M_V, and colors, B-V [Astrophysical Journal 125, 422-443 (1957)]. The ordinate at the right gives Sandage's estimates of the ages of the open clusters derived from the magnitude of the upper end of the main sequence. For comparison, J.C. Mermilliod [Astronomy and Astrophysics 97, 235-244 (1981)], obtains ages of 77.6 million years for the Pleiades and 0.66 billion years for the Hyades. Mermilliod has also provided an atlas of color-magnitude diagrams for 75 young open clusters [Astronomy and Astrophysics Supplement 44, 467-500 (1981)].

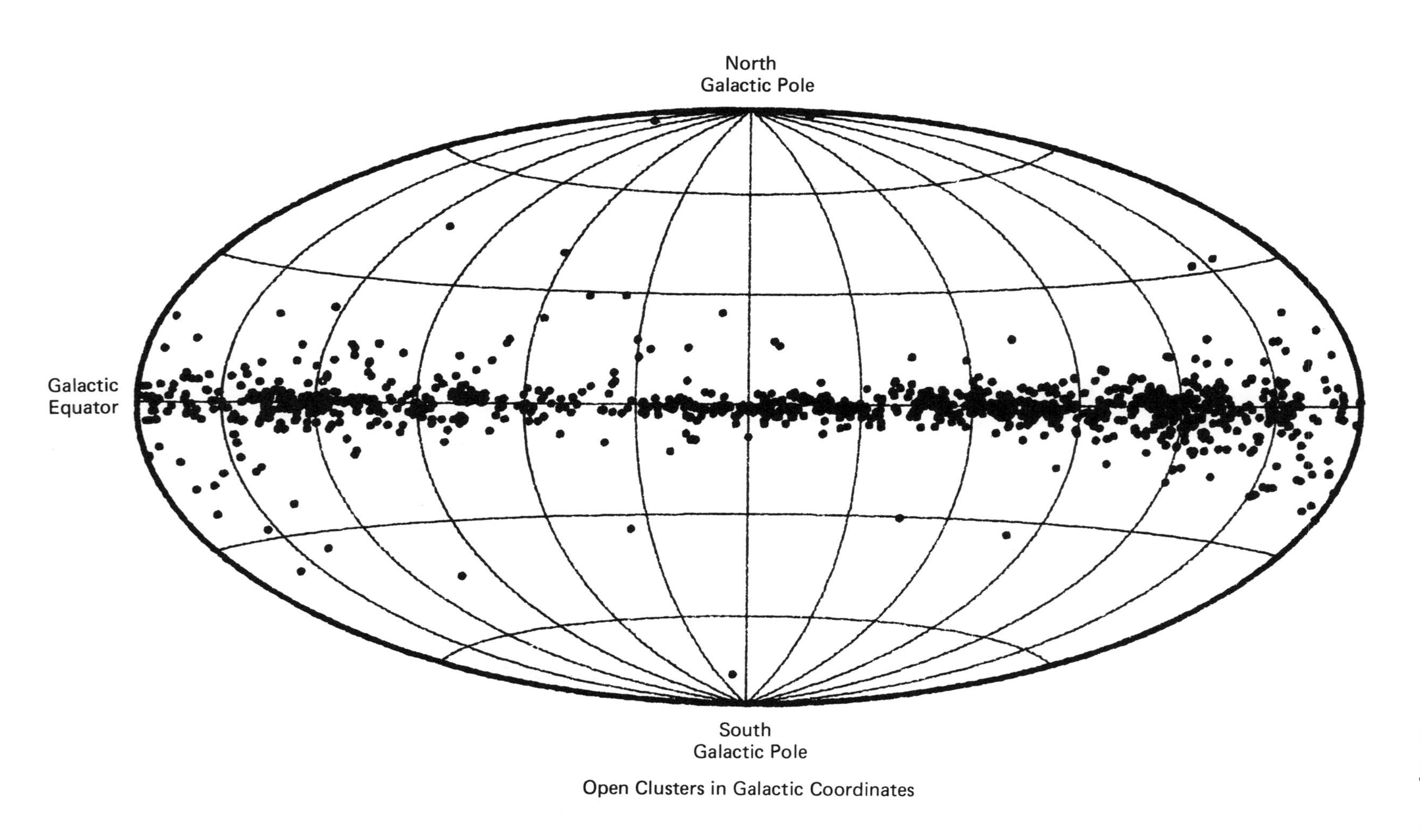

Open Clusters in Galactic Coordinates

Open Cluster Statistics

Parameter	Mean	Max.	Min.	Disp., σ
Angular Diameter, θ (′)	2.44	50.50	0.50	2.98
Magnitude of Brightest Cluster Star		18.00	2.00	
Metallicity, [Fe/H]	− 0.11	0.30	− 0.76	0.22
Integrated B–V (mag)	0.53	1.54	0.01	0.30
Turn-off Colour	− 0.12	0.70	− 0.40	0.19
Color Excess, E_{B-V}	0.45	1.72	0.01	0.31
Linear Radius (pc)	2.56	18.10	0.30	2.31
Heliocentric Distance, D_o (pc)		15000	20	
Age ($\times 10^6$ years)	346	7079	1	890.55

[Fe/H]

Name	[Fe/H]	IAU Number	Name	[Fe/H]	IAU Number
Sc 1, Blanco 1	+0.03	C0001− 302	Hyades 1	+0.12	C0424+157
NGC 188	− 0.06	C0039+850	NGC 1662	− 0.20	C0445+108
NGC 559	− 0.76	C0126+630	NGC 1817	− 0.26	C0509+166
NGC 752	− 0.21	C0154+374	Berkeley 19	− 0.50	C0520+295
M 34, NGC 1039	− 0.26	C0238+425	NGC 1907	− 0.10	C0524+352
NGC 1245	+0.14	C0311+470	M 38, NGC 1912	− 0.11	C0525+358
Melotte 20	+0.10	C0318+484	King 8	− 0.50	C0546+336
NGC 1342	− 0.13	C0328+371	M 37, NGC 2099	+0.01	C0549+325
M 45, Pleiades 1	+0.11	C0344+239	NGC 2141	− 0.46	C0600+104
NGC 1528	− 0.10	C0411+511	NGC 2158	− 0.60	C0604+241

[Fe/H]

Name	[Fe/H]	IAU Number	Name	[Fe/H]	IAU Number
NGC 2204	− 0.58	C0613− 186	M 67, NGC 2682	− 0.10	C0847+120
NGC 2232	− 0.20	C0624− 047	NGC 3114	+0.01	C1001− 598
NGC 2243	− 0.55	C0627− 312	θ Car, IC 2602	− 0.20	C1041− 641
S Mon, NGC 2264	− 0.15	C0638+099	NGC 3680	− 0.07	C1123− 429
M 41, NGC 2287	+0.09	C0644− 206	NGC 3960	− 0.30	C1148− 554
NGC 2281	− 0.04	C0645+411	Melotte 111	− 0.03	C1222+263
NGC 2301	+0.04	C0649+005	NGC 5316	+0.01	C1350− 616
NGC 2324	− 0.13	C0701+011	NGC 5617	− 0.51	C1426− 605
NGC 2335	+0.20	C0704− 100	NGC 5822	− 0.07	C1501− 541
NGC 2343	− 0.20	C0705− 105	NGC 5823	− 0.13	C1502− 554
NGC 2360	− 0.13	C0715− 155	NGC 6025	+0.23	C1559− 603
Collinder 140	+0.05	C0722− 321	NGC 6067	− 0.05	C1609− 540
Ruprecht 18	+0.00	C0722− 261	NGC 6134	+0.25	C1624− 490
Melotte 66	− 0.36	C0724− 476	NGC 6259	+0.29	C1657− 446
NGC 2423	− 0.04	C0734− 137	NGC 6281	+0.10	C1701− 378
Melotte 71	− 0.29	C0735− 119	IC 4651	− 0.16	C1720− 499
NGC 2420	− 0.61	C0735+216	M 61, NGC 6405	+0.10	C1736− 321
NGC 2447	+0.00	C0742− 237	NGC 6633	− 0.11	C1825+065
NGC 2451	− 0.45	C0743− 378	M 25, IC 4725	− 0.06	C1828− 192
NGC 2477	+0.04	C0750− 384	IC 4756	+0.04	C1836+054
NGC 2482	+0.12	C0752− 241	M 11, NGC 6705	+0.10	C1848− 063
NGC 2516	− 0.23	C0757− 607	NGC 6716	− 0.28	C1851− 199
NGC 2506	− 0.55	C0757− 106	Stephenson 1	− 0.10	C1851+368
Ruprecht 46	− 0.20	C0759− 193	NGC 6791	+0.04	C1919+377
NGC 2527	+0.00	C0803− 280	NGC 6819	− 0.11	C1939+400
NGC 2539	− 0.20	C0808− 126	NGC 6885	− 0.16	C2009+263
NGC 2547	− 0.13	C0809− 491	NGC 6939	− 0.11	C2030+604
NGC 2546	+0.30	C0810− 374	NGC 6940	+0.04	C2032+281
M 48, NGC 2548	+0.10	C0811− 056	NGC 7082	+0.03	C2127+468
NGC 2567	+0.20	C0816− 304	NGC 7142	− 0.11	C2144+655
NGC 2571	+0.08	C0816− 295	NGC 7261	− 0.46	C2218+578
M 44, NGC 2632	+0.07	C0837+201	NGC 7789	− 0.10	C2354+564
o Vel, IC 2391	− 0.04	C0838− 528			
NGC 2660	+0.06	C0840− 469			
Trumpler 10	-0.08	C0846-423			

Cluster Masses*

Cluster	Mass ($M_\odot$)	Cluster	Mass ($M_\odot$)	Cluster	Mass ($M_\odot$)
NGC 129	186.21	NGC 3680	93.33	NGC 7092	42.66
NGC 188	208.93	NGC 5460	112.20	NGC 7209	63.10
NGC 654	295.12	NGC 5617	389.05	NGC 7243	89.13
NGC 1039	97.72	NGC 6025	154.88	NGC 7380	478.63
NGC 1778	120.23	NGC 6134	204.17	NGC 7788	89.13
NGC 1912	204.17	NGC 6208	181.97	NGC 7790	112.20
NGC 1960	151.36	NGC 6530	467.74	IC 1369	138.04
NGC 2099	489.78	NGC 6611	524.81	IC 1805	645.65
NGC 2169	79.43	NGC 6613	112.20	IC 2391	72.44
NGC 2264	199.53	NGC 6633	147.91	IC 2581	501.19
NGC 2516	169.82	NGC 6705	398.11	IC 2602	97.72
NGC 2527	51.29	NGC 6716	39.81	IC 4651	147.91
NGC 2539	275.42	NGC 6811	302.00	IC 4665	72.44
NGC 2546	257.04	NGC 6823	537.03	Pleiades	100.00
NGC 2567	95.50	NGC 6866	549.54	Hyades	77.62
NGC 2571	208.93	NGC 6913	380.19	Cr 185	50.12
NGC 2632	112.20	NGC 6939	338.84	Hogg 17	72.44
NGC 2682	467.74	NGC 7062	93.33	Pismis-1	44.67
				TR-1	109.65

*Adapted from Pandey, Bhatt and Mahra (1987).

15.2 Catalogue of Open Clusters

Open Cluster Parameters

Name	R.A.(1950) h m s	Dec.(1950) ° ′	l °	b °	V_T (mag)	IAU Number
Berkeley 59	00 00 00	+67 06	118.25	+04.95		C0000+671
Berkeley 104	00 00 54	+63 19	117.63	+01.22		C0000+633
Blanco 1, ζ Scl	00 01 42	- 30 13	014.97	- 79.26	4.5	C0001- 302
Stock 19	00 01 48	+55 45	116.35	- 06.24		C0001+557
Czernik 1	00 05 06	+61 08	117.73	- 01.02		C0005+611
Berkeley 1	00 07 00	+60 09	117.79	- 02.03		C0007+601
King 13	00 07 30	+60 56	117.98	- 01.28		C0007+609
Berkeley 60	00 15 00	+60 41	118.85	- 01.64		C0015+606
King 1	00 19 12	+64 07	119.75	+01.69		C0019+641
Mayer 1	00 19 12	+61 28	119.45	- 00.93		C0019+614
Stock 20	00 22 06	+62 22	119.89	- 00.07		C0022+623
Berkeley 2	00 22 30	+60 07	119.70	- 02.31		C0022+601
NGC 103	00 22 30	+61 04	119.80	- 01.38	9.8	C0022+610
NGC 110	00 24 24	+71 07	121.00	+08.61		C0024+711
NGC 129	00 27 06	+59 57	120.25	- 02.54	6.5	C0027+599
Stock 21	00 27 18	+57 42	120.09	- 04.78		C0027+577
NGC 133	00 28 24	+63 05	120.67	+00.57	9.4	C0028+630
NGC 136	00 28 42	+61 15	120.56	- 01.26		C0028+612
King 14	00 29 00	+62 53	120.72	+00.36	8.5	C0029+628
King 15	00 30 06	+61 35	120.75	- 00.95		C0030+615
NGC 146	00 30 12	+63 01	120.87	+00.49	9.1	C0030+630
NGC 189	00 36 42	+60 48	121.51	- 01.77	8.8	C0036+608
Stock 24	00 36 48	+61 41	121.55	- 00.88	8.8	C0036+616
NGC 188	00 39 24	+85 04	122.78	+22.46	8.1	C0039+850
NGC 225	00 40 30	+61 31	121.99	- 01.08	7.0	C0040+615
King 16	00 40 42	+63 55	122.09	+01.33	10.3	C0040+639
Czernik 2	00 40 48	+59 53	121.97	- 02.70		C0040+598
Berkeley 4	00 42 30	+64 08	122.29	+01.54	10.6	C0042+641
Berkeley 61	00 45 24	+66 58	122.65	+04.37		C0045+669
Dolidze 13	00 47 00	+63 52	122.78	+01.27		C0047+638
King 2	00 48 06	+57 55	122.88	- 04.67		C0048+579
NGC 281	00 49 54	+56 21	123.13	- 06.24	7.4	C0049+563
Berkeley 62	00 57 54	+63 41	123.99	+01.10	9.3	C0057+636
Czernik 3	00 60 00	+62 32	124.27	- 00.04	9.9	C0100+625
NGC 366	01 03 18	+61 58	124.68	- 00.59		C0103+619

Open Cluster Parameters

IAU Number	θ (′)	V_r (km s^{-1})	MBS	B–V (mag)	TOC	E_{B-V}	R (pc)	D_o (pc)	Age ($\times 10^6$)
C0000+671	10	− 14.5	11.0						
C0000+633	4.0		16.0						
C0001− 302	89	+5	8.0		− 0.15	0.02	2.5	190	50
C0001+557	3.0		8.0						
C0005+611	9								
C0007+601	5								
C0007+609	7		12.0						
C0015+606	4.0		14.0						
C0019+641	7		13.0						
C0019+614									
C0022+623	1.0		13.0						
C0022+601	4.0		15.0						
C0022+610	5		11.0	0.48	− 0.22	0.38	2.4	3000	38
C0024+711									
C0027+599	21	− 38.5	11.0	0.83	− 0.19	0.58v	4.9	1600	151
C0027+577	5.0		12.0						
C0028+630	7								
C0028+612	1.2		13.0		− 0.05	0.56	0.8	4400	200
C0029+628	7		10.0	0.50	− 0.20	0.47	2.8	2600	16
C0030+615	1.5		18.0						
C0030+630	6			0.49	− 0.20	0.58	3.0	2900	13
C0036+608	3.7			0.27	− 0.33	0.42	0.6	1080	20
C0036+616	4.0		13.0	0.22	− 0.38	0.41	1.1	1900	3
C0039+850	13	− 50	10.0	0.81	+0.60	0.08	3.2	1550	5012
C0040+615	12			0.43	− 0.10	0.29	1.1	630	141
C0040+639	3.0		11.0	0.38			1.0	2300	
C0040+598	10								
C0042+641	5.0		18.0	0.19					
C0045+669	4.0		14.0						
C0047+638	12								
C0048+579	5		17.0						
C0049+563	4.0		9.0						
C0057+636	10		13.0	0.68	− 0.31	0.86	3.0	2050	10
C0100+625	3.0			0.14					
C0103+619	3.0		10.0						

Open Cluster Parameters

Name	R.A.(1950) h m s	Dec.(1950) ° ′	l °	b °	V_T (mag)	IAU Number
NGC 381	01 05 12	+61 19	124.94	- 01.22	9.3	C0105+613
Stock 3	01 09 06	+62 04	125.35	- 00.44		C0109+620
NGC 433	01 12 06	+59 52	125.90	- 02.60		C0112+598
NGC 436	01 12 30	+58 33	126.07	- 03.91	8.8	C0112+585
NGC 457	01 15 54	+58 04	126.56	- 04.35	6.4	C0115+580
NGC 559	01 26 06	+63 03	127.19	+00.75	9.5	C0126+630
M 103, NGC 581	01 29 54	+60 27	128.02	- 01.76	7.4	C0129+604
Czernik 4	01 32 00	+61 11	128.16	- 00.99		C0132+611
Trumpler 1	01 32 18	+61 02	128.22	- 01.14	8.1	C0132+610
NGC 609	01 33 42	+64 18	127.83	+02.11	11.0	C0133+643
NGC 637	01 39 24	+63 45	128.55	+01.70	8.2	C0139+637
NGC 657	01 40 30	+55 37	130.28	- 06.25		C0140+556
NGC 654	01 40 36	+61 38	129.09	- 00.35	6.5	C0140+616
NGC 659	01 40 48	+60 27	129.34	- 01.51	7.9	C0140+604
NGC 663	01 42 36	+61 00	129.46	- 00.94	7.1	C0142+610
Berkeley 5	01 44 18	+62 41	129.28	+00.76		C0144+626
Collinder 463	01 44 18	+71 42	127.36	+09.57	5.7	C0144+717
Collinder 21	01 47 18	+27 00	138.66	- 33.83	8.2	C0147+270
Berkeley 6	01 47 42	+60 50	130.09	- 00.96		C0147+608
IC 166	01 49 00	+61 35	130.08	- 00.19	11.7	C0149+615
Stock 4	01 49 24	+56 49	131.21	- 04.82		C0149+568
Berkeley 7	01 50 36	+62 07	130.13	+00.37		C0150+621
Czernik 5	01 51 36	+61 05	130.48	- 00.61		C0151+610
NGC 752	01 54 48	+37 26	137.17	- 23.36	5.7	C0154+374
NGC 744	01 55 06	+55 14	132.39	- 06.16	7.9	C0155+552
NGC 743	01 55 12	+59 56	131.23	- 01.61		C0155+599
Berkeley 8	01 56 24	+75 20	127.33	+13.30		C0156+753
Czernik 6	01 58 24	+62 38	130.88	+01.10		C0158+626
Czernik 7	01 58 48	+62 01	131.08	+00.51		C0158+620
Stock 5	02 00 48	+64 12	130.74	+02.65		C0200+642
Stock 2	02 11 24	+59 02	133.44	- 01.93	4.4	C0211+590
Basel 10	02 15 12	+58 05	134.21	- 02.64	9.9	C0215+580
h Per, NGC 869	02 15 30	+56 55	134.63	- 03.72	5.3	C0215+569
Berkeley 63	02 15 48	+63 31	132.49	+02.52		C0215+635
Per, Berkeley 64	02 17 06	+65 40	131.92	+04.60		C0217+656

Open Cluster Parameters

IAU Number	θ (′)	V_r (km s^{-1})	MBS	B–V (mag)	TOC	E_{B-V}	R (pc)	D_o (pc)	Age ($\times 10^6$)
C0105+613	6		10.0						
C0109+620	2.0		11.0						
C0112+598	2.5		9.0						
C0112+585	5		10.0	0.01	– 0.09	0.19	1.9	2200	79
C0115+580	13	– 25.1	6.0	0.57	– 0.28	0.46	5.4	2800	25
C0126+630	4.4		9.0	0.35	+0.05	0.54	0.6	900	1259
C0129+604	6	– 42.2	9.0	0.32	– 0.22	0.40v	2.7	2700	40
C0132+611	3.0								
C0132+610	4.5		10.0	0.25	– 0.20	0.53	1.5	2200	26
C0133+643	3.0		14.0	1.31	– 0.15	0.95	1.4	3100	200
C0139+637	3.5	– 46.2	8.0	0.36	– 0.04	0.38	1.0	2100	40
C0140+556									
C0140+616	5.0	– 33.8	10.0	0.84	– 0.27	0.96v	1.1	1600	15
C0140+604	5.0		10.0	0.50	– 0.14	0.55	1.5	2100	20
C0142+610	16	– 33.1	9.0	0.68	– 0.32	0.87v	5.3	2200	22
C0144+626	3.0		17.0						
C0144+717	36			0.59	– 0.15	0.00v	3.2	600	151
C0147+270	5								
C0147+608	5		14.0						
C0149+615	4.5	– 36	17.0	1.30	+0.35	0.80	2.2	3300	1585
C0149+568	20		11.0						
C0150+621	4.0		14.0						
C0151+610	7								
C0154+374	50	– 3	8.0	0.77	+0.35	0.03	2.9	400	1096
C0155+552	11		10.0	0.48	– 0.05	0.41	2.4	1500	39
C0155+599	5		10.0						
C0156+753	7		14.0				1.8	1490	
C0158+626	3.0								
C0158+620	5.0								
C0200+642	14	– 18	7.0						
C0211+590	60	+2		0.49	– 0.06	0.38v	2.8	320	100
C0215+580				1.29	– 0.31	0.98		2600	16
C0215+569	29	– 44.8	7.0	0.44	– 0.24	0.54v	9.6	2200	6
C0215+635	4.0		15.0						
C0217+656	4.0		14.0						

Open Cluster Parameters

Name	R.A.(1950) h m s	Dec.(1950) ° ′	l °	b °	V_T (mag)	IAU Number
NGC 884, χ Per	02 18 54	+56 53	135.08	- 03.60	6.1	C0218+568
Stock 6	02 19 48	+63 38	132.88	+02.78		C0219+636
Tombaugh 4	02 25 06	+61 34	134.19	+01.06		C0225+615
Markarian 6	02 25 54	+60 26	134.68	+00.04	7.1	C0225+604
IC 1805	02 28 54	+61 14	134.74	+00.92	6.5	C0228+612
NGC 956	02 29 12	+44 26	141.21	- 14.61	8.9	C0229+444
Czernik 8	02 29 18	+58 31	135.79	- 01.58	9.7	C0229+585
Czernik 9	02 29 54	+59 40	135.43	- 00.48		C0229+596
NGC 957	02 30 00	+57 19	136.34	- 02.66	7.6	C0230+573
Czernik 10	02 30 06	+59 58	135.34	- 00.20		C0230+599
King 4	02 32 00	+58 47	136.02	- 01.20	10.5	C0232+587
Czernik 11	02 32 06	+59 25	135.78	- 00.61		C0232+594
Trumpler 2	02 33 42	+55 46	137.41	- 03.89	5.9	C0233+557
Berkeley 65	02 35 12	+60 12	135.84	+00.27	10.2	C0235+602
Czernik 12	02 35 36	+54 43	138.07	- 04.74		C0235+547
NGC 1027	02 38 48	+61 20	135.78	+01.48	6.7	C0238+613
M 34, NGC 1039	02 38 48	+42 34	143.64	- 15.60	5.2	C0238+425
Czernik 13	02 40 48	+62 08	135.67	+02.31	10.4	C0240+621
Dol–Dzim 1	02 44 36	+16 59	158.64	- 37.51		C0244+169
IC 1848	02 47 18	+60 14	137.19	+00.92	6.5	C0247+602
Latysev 1	02 49 30	+27 16	153.39	- 28.15		C0249+272
Collinder 33	02 55 24	+60 12	138.10	+01.35	5.9	C0255+602
Collinder 34	02 57 00	+60 13	138.28	+01.45	6.8	C0257+602
Berkeley 66	03 00 24	+58 34	139.42	+00.23		C0300+585
NGC 1193	03 02 30	+44 11	146.81	- 12.18	12.6	C0302+441
Trumpler 3	03 07 36	+63 04	137.98	+04.56	7.0	C0307+630
NGC 1220	03 08 00	+53 09	143.05	- 03.97	11.8	C0308+531
King 5	03 11 00	+52 32	143.75	- 04.27		C0311+525
NGC 1245	03 11 12	+47 04	146.64	- 08.93	8.4	C0311+470
Stock 23	03 12 18	+59 51	140.13	+02.10		C0312+598
Czernik 14	03 13 00	+58 25	140.93	+00.91		C0313+584
Melotte 20	03 18 30	+48 26	146.95	- 07.11	1.2	C0318+484
Czernik 15	03 19 30	+52 04	145.09	- 03.96		C0319+520
King 6	03 24 12	+56 17	143.35	- 00.07		C0324+562
Czernik 16	03 27 06	+52 29	145.84	- 02.98		C0327+524

Open Cluster Parameters

IAU Number	θ (′)	V_r (km s^{-1})	MBS	B–V (mag)	TOC	E_{B-V}	R (pc)	D_o (pc)	Age ($\times 10^6$)
C0218+568	29	− 42.5	7.0	0.48	− 0.27	0.60v	10.1	2300	3
C0219+636	20		11.0						
C0225+615	2.5		16.0						
C0225+604	4.5		8.8	0.43	− 0.26	0.58	0.3	485	25
C0228+612	21		9.0	0.53	− 0.35	0.81v	6.7	2100	1
C0229+444	7		9.0						
C0229+585	6			1.19	− 0.35	1.06	2.5	2400	1
C0229+596	5								
C0230+573	11	− 28.6	11.0	0.74	− 0.25	0.80v	3.5	2200	15
C0230+599	3.0								
C0232+587	3.0		13.0	0.92	− 0.26	0.86	0.9	2200	32
C0232+594	7								
C0233+557	20			0.56	− 0.05	0.32	1.8	600	78
C0235+602	5.0		13.0	0.97	− 0.25	1.13	2.4	3300	6
C0235+547	3.0								
C0238+613	20		9.0	0.58	− 0.15	0.40	2.9	1000	347
C0238+425	35		9.0	0.17	− 0.11	0.11	2.3	440	195
C0240+621	5			0.70	− 0.30	0.77			1
C0244+169	12								
C0247+602	12	− 44.2		0.37	− 0.35	0.72v	3.8	2200	1
C0249+272									
C0255+602	39								
C0257+602	25								
C0300+585	6		16.0						
C0302+441	1.5	− 82	14.0						
C0307+630	23								
C0308+531	1.6		13.0						
C0311+525	7		13.0						
C0311+470	10		12.0	0.76	+0.20	0.27	3.3	2300	1096
C0312+598	14								
C0313+584	6								
C0318+484	185	− 2	3.0		− 0.20	0.09	4.6	170	51
C0319+520	3.0								
C0324+562	6		10.0						
C0327+524	9								

Open Cluster Parameters

Name	R.A.(1950) h m s	Dec.(1950) ° ′	l °	b °	V_T (mag)	IAU Number
NGC 1342	03 28 24	+37 10	154.99	- 15.38	6.7	C0328+371
Berkeley 9	03 29 00	+52 32	146.04	- 02.76		C0329+525
NGC 1348	03 30 12	+51 16	146.94	- 03.71		C0330+512
Berkeley 10	03 34 48	+66 22	138.58	+08.91		C0334+663
IC 348	03 41 24	+32 08	160.43	- 17.74	7.3	C0341+321
Tombaugh 5	03 43 42	+58 54	143.94	+03.58	8.4	C0343+589
M 45, Pleiades 1	03 44 00	+23 58	166.56	- 23.53	1.2	C0344+239
NGC 1444	03 45 36	+52 31	148.16	- 01.29	6.6	C0345+525
Czernik 17	03 48 06	+61 51	142.53	+06.22		C0348+618
King 7	03 55 12	+51 39	149.76	- 01.04		C0355+516
NGC 1496	04 00 36	+52 29	149.87	+00.14	9.6	C0400+524
NGC 1502	04 03 18	+62 12	143.65	+07.62	6.9	C0403+622
Dolidze 14	04 03 30	+27 18	167.54	- 18.08		C0403+273
NGC 1513	04 06 18	+49 23	152.60	- 01.57	8.4	C0406+493
NGC 1528	04 11 36	+51 07	152.04	+00.27	6.4	C0411+511
IC 361	04 14 48	+58 11	147.49	+05.71	11.7	C0414+581
Mayer 2	04 15 36	+53 05	151.12	+02.13		C0415+530
Berkeley 11	04 17 00	+44 48	157.08	- 03.65	10.4	C0417+448
NGC 1545	04 17 06	+50 08	153.36	+00.17	6.2	C0417+501
Hyades 1	04 24 00	+15 45	180.05	- 22.40	.5	C0424+157
Czernik 18	04 24 48	+30 49	168.28	- 12.28		C0424+308
NGC 1582	04 28 30	+43 45	159.28	- 02.91	7.0	C0428+437
NGC 1605	04 31 24	+45 09	158.61	- 01.58	10.7	C0431+451
Berkeley 67	04 34 18	+50 39	154.88	+02.51		C0434+506
NGC 1624	04 36 36	+50 21	155.35	+02.58	11.8	C0436+503
Berkeley 68	04 41 00	+41 58	162.13	- 02.41	9.8	C0441+419
Berkeley 12	04 41 06	+42 35	161.68	- 01.99		C0441+425
Ruprecht 148	04 42 54	+44 39	160.33	- 00.39	9.5	C0442+446
NGC 1647	04 43 06	+18 59	180.39	- 16.78	6.4	C0443+189
NGC 1662	04 45 42	+10 51	187.70	- 21.12	6.4	C0445+108
NGC 1663	04 45 48	+13 04	185.79	- 19.81		C0445+130
NGC 1664	04 47 30	+43 37	161.66	- 00.44	7.6	C0447+436
Berkeley 13	04 51 36	+52 40	155.09	+05.87		C0451+526
Czernik 19	04 53 54	+28 42	174.09	- 08.85		C0453+287
Berkeley 14	04 56 36	+43 24	162.86	+00.71		C0456+434

Open Cluster Parameters

IAU Number	θ (′)	V_r (km s^{-1})	MBS	B–V (mag)	TOC	E_{B-V}	R (pc)	D_o (pc)	Age ($\times 10^6$)
C0328+371	14		8.0	0.65	+0.10	0.28	1.1	550	302
C0329+525	5.0		15.0						
C0330+512									
C0334+663	12		14.0						
C0341+321	7	+18	10.0	0.73	– 0.24	0.62v	0.4	390	126
C0343+589	17		14.0	0.61			4.4	1800	
C0344+239	110	+7	3.0		– 0.10	0.04	2.0	125	78
C0345+525	4.0	+2		0.44	– 0.20	0.70v	0.6	1000	25
C0348+618	5.0								
C0355+516	5		16.0						
C0400+524	6		12.0						
C0403+622	7	– 18	7.0	0.57	– 0.30	0.77v	1.1	950	20
C0403+273	12								
C0406+493	9		11.0	0.31	– 0.10	0.53	1.0	820	427
C0411+511	23		10.0	0.43	– 0.03	0.30	2.8	800	269
C0414+581	6		14.0	1.22	+0.25	0.55			1259
C0415+530									
C0417+448	5		15.0	0.74	– 0.31	0.96	1.9	2200	30
C0417+501	18	– 15	9.0	1.00	– 0.05	0.36	2.1	800	195
C0424+157	330	+43	4.0	0.40	+0.12	0.00	2.3	48	661
C0424+308	10								
C0428+437	37		9.0						
C0431+451	5.0			0.93	– 0.20	0.97	2.0	2800	126
C0434+506	7		15.0						
C0436+503	1.9								
C0441+419	12			0.82			5.6	3200	
C0441+425	6		16.0						
C0442+446	3.0		14.0	0.85			2.4	5500	
C0443+189	45		9.0	0.42	– 0.10	0.39	3.6	550	214
C0445+108	20		9.0	0.60	– 0.10	0.34	1.1	400	302
C0445+130									
C0447+436	18		10.0	0.42	– 0.02	0.20	3.2	1200	302
C0451+526	7		15.0						
C0453+287	18								
C0456+434	9		16.0						

Open Cluster Parameters

Name	R.A.(1950) h m s	Dec.(1950) ° ′	l °	b °	V_T (mag)	IAU Number
Berkeley 15	04 58 42	+44 23	162.33	+01.61		C0458+443
NGC 1724	04 59 42	+49 26	158.45	+04.85		C0459+494
NGC 1746	05 00 36	+23 45	179.02	- 10.65	6.1	C0500+237
NGC 1758	05 01 24	+23 43	179.15	- 10.52		C0501+237
NGC 1778	05 04 42	+36 59	168.88	- 02.00	7.7	C0504+369
King 17	05 05 00	+39 01	167.30	- 00.73		C0505+390
NGC 1807	05 07 48	+16 28	186.08	- 13.50	7.0	C0507+164
NGC 1798	05 08 06	+47 34	160.76	+04.85		C0508+475
NGC 1817	05 09 12	+16 38	186.13	- 13.13	7.7	C0509+166
Dolidze 16	05 11 18	+32 40	173.17	- 03.48		C0511+326
Collinder 464	05 16 12	+73 14	139.38	+19.85	4.2	C0516+732
Czernik 20	05 16 36	+39 25	168.30	+01.32		C0516+394
NGC 1857	05 16 42	+39 18	168.41	+01.26	7.0	C0516+393
Berkeley 17	05 17 24	+30 33	175.65	- 03.65		C0517+305
NGC 1901	05 18 00	- 68 30	279.05	- 33.64		C0518- 685
Berkeley 18	05 18 30	+45 21	163.63	+05.01		C0518+453
Collinder 62	05 19 00	+40 57	167.26	+02.50	4.2	C0519+409
NGC 1893	05 19 24	+33 21	173.59	- 01.70	7.5	C0519+333
Dol–Dzim 2	05 20 24	+11 25	192.15	- 13.72		C0521+114
Dolidze 18	05 20 48	+33 15	173.83	- 01.50		C0520+332
Berkeley 19	05 20 54	+29 33	176.90	- 03.59	11.4	C0520+295
Dolidze 19	05 21 00	+08 08	195.11	- 15.32		C0521+081
Berkeley 69	05 21 18	+32 36	174.42	- 01.79		C0521+326
Berkeley 70	05 22 12	+41 51	166.90	+03.58		C0522+418
NGC 1883	05 22 12	+46 30	163.04	+06.19	12.0	C0522+465
Collinder 65	05 23 00	+16 03	188.48	- 10.71	3.0	C0523+160
Czernik 21	05 23 06	+35 57	171.87	+00.41		C0523+359
Stock 8	05 24 18	+34 23	173.35	- 00.23		C0524+343
Dolidze 21	05 24 42	+07 02	196.58	- 15.10		C0524+070
NGC 1907	05 24 42	+35 17	172.62	+00.30	8.2	C0524+352
M 38, NGC 1912	05 25 18	+35 48	172.27	+00.70	6.4	C0525+358
Dolidze 20	05 25 18	+33 45	173.96	- 00.46		C0525+337
NGC 1931	05 28 06	+34 13	173.90	+00.28	10.1	C0528+342
Berkeley 20	05 30 24	+00 11	203.50	- 17.28		C0530+001
Dol–Dzim 3	05 30 36	+26 27	180.68	- 03.54		C0530+264

Open Cluster Parameters

IAU Number	θ (′)	V_r (km s^{-1})	MBS	B–V (mag)	TOC	E_{B-V}	R (pc)	D_o (pc)	Age ($\times 10^6$)
C0458+443	9		15.0						
C0459+494									
C0500+237	42	+2	8.0				2.5	420	
C0501+237									
C0504+369	6	+6.3		0.50	– 0.10	0.34	1.4	1350	159
C0505+390	1.5		14.0						
C0507+164	17		9.0	0.86					
C0508+475	5		13.0						
C0509+166	15	+69	9.0	0.74	+0.15	0.28	4.1	1750	794
C0511+326	12								
C0516+732	120								
C0516+394	18								
C0516+393	5		11.0	1.08			1.6	1900	
C0517+305	13		16.0						
C0518– 685					+0.15	0.06		300	501
C0518+453	20		16.0						
C0519+409	28								
C0519+333	11	– 8		0.40	– 0.34	0.54v	6.4	4000	1
C0521+114	12								
C0520+332	12								
C0520+295	6		15.0	0.93	+0.40	0.40	4.1	4000	3020
C0521+081	23								
C0521+326	5		14.0						
C0522+418	12		15.0						
C0522+465	2.5		14.0						
C0523+160	220								
C0523+359	9								
C0524+343	5.0		9.0						
C0524+070	12								
C0524+352	6		11.0	0.67	– 0.10	0.42v	1.4	1380	437
C0525+358	21		8.0	0.29	– 0.08	0.24	4.1	1320	224
C0525+337	12								
C0528+342	1.0								
C0530+001	3.0		15.0						
C0530+264	14								

Open Cluster Parameters

Name	R.A.(1950) h m s	Dec.(1950) ° ′	l °	b °	V_T (mag)	IAU Number
Collinder 69	05 32 18	+09 54	195.05	- 12.00	2.8	C0532+099
NGC 1981	05 32 42	- 04 28	208.09	- 18.98	4.2	C0532- 044
M 36, NGC 1960	05 32 48	+34 06	174.52	+01.04	6.0	C0532+341
Dol–Dzim 4	05 32 48	+25 55	181.39	- 03.41		C0532+259
Trapezium 1	05 32 54	- 05 25	209.01	- 19.37		C0532- 054
NGC 1977	05 32 54	- 04 52	208.50	- 19.11		C0532- 048
NGC 1980	05 32 54	- 05 58	209.54	- 19.60	2.5	C0532- 059
Collinder 70	05 33 00	- 01 08	204.98	- 17.44	0.4	C0533- 011
Stock 10	05 35 36	+37 54	171.63	+03.56		C0535+379
Sigma Ori 0	05 36 12	- 02 38	206.82	- 17.34		C0536- 026
Berkeley 71	05 38 00	+32 22	176.57	+01.02		C0538+323
Basel 4	05 45 18	+30 12	179.23	+01.20	9.1	C0545+302
Collinder 74	05 45 48	+07 23	198.98	- 10.40	14.4	C0545+073
King 8	05 46 06	+33 37	176.40	+03.12	11.2	C0546+336
Czernik 23	05 46 30	+28 55	180.46	+00.76		C0546+289
Berkeley 72	05 47 18	+22 11	186.32	- 02.57		C0547+221
Berkeley 21	05 48 42	+21 46	186.83	- 02.50	1.1	C548+217
M 37, NGC 2099	05 49 06	+32 32	177.65	+03.09	5.6	C0549+325
NGC 2112	05 51 18	+00 22	205.90	- 12.61	9.1	C0551+003
Czernik 24	05 52 12	+20 52	188.04	- 02.27		C0552+208
Berkeley 22	05 55 42	+07 50	199.80	- 08.05		C0555+078
NGC 2129	05 58 00	+23 18	186.61	+00.13	6.7	C0558+233
NGC 2126	05 59 06	+49 54	163.24	+13.21	10.2	C0559+499
NGC 2141	06 00 18	+10 26	198.07	- 05.79	9.4	C0600+104
IC 2157	06 01 54	+24 00	186.45	+01.25	8.4	C0601+240
NGC 2158	06 04 24	+24 06	186.64	+01.76	8.6	C0604+241
NGC 2169	06 05 36	+13 58	195.63	- 02.92	5.9	C0605+139
M 35, NGC 2168	06 05 48	+24 21	186.58	+02.18	5.1	C0605+243
NGC 2175	06 06 48	+20 20	190.20	+00.42	6.8	C0606+203
NGC 2175	06 07 54	+20 37	190.08	+00.79		C0607+206
NGC 2186	06 09 30	+05 28	203.56	- 06.20	8.7	C0609+054
Czernik 25	06 10 24	+07 00	202.31	- 05.26		C0610+070
NGC 2194	06 11 00	+12 49	197.26	- 02.33	8.5	C0611+128
NGC 2192	06 11 42	+39 52	173.41	+10.64	10.9	C0611+398
NGC 2204	06 13 30	- 18 38	226.01	- 16.07	8.6	C0613- 186

Open Cluster Parameters

IAU Number	θ (′)	V_r (km s^{-1})	MBS	B–V (mag)	TOC	E_{B-V}	R (pc)	D_o (pc)	Age ($\times 10^6$)
C0532+099	64	+35					4.8	500	
C0532− 044	25	+28	10.0				1.5	400	
C0532+341	12	− 4	9.0	0.09	− 0.20	0.24	2.2	1270	25
C0532+259	28								
C0532− 054	47	+23			− 0.20	0.06v	3.2	450	25
C0532− 048									
C0532− 059	13	+21							
C0533− 011	150						9.4	430	
C0535+379	25								
C0536− 026									
C0538+323	5		15.0						
C0545+302	7			0.46	− 0.34	0.53	6.8	5900	13
C0545+073	3.0								
C0546+336	7	− 1	15.0	0.70	− 0.06	0.68	4.1	3500	794
C0546+289	5								
C0547+221	5		15.0						
C0548+217	6	+5	6.0	0.16	− 0.30	1.00	15.3	15000	30
C0549+325	23		11.0	0.59	− 0.05	0.31	4.7	1350	302
C0551+003	11		10.0		+0.70	0.50			
C0552+208	5								
C0555+078	2.0		15.0						
C0558+233	6	+17.6	10.0	0.61	− 0.20	0.67v	2.0	2000	16
C0559+499	6		13.0				1.4	1400	
C0600+104	10		15.0	1.03	+0.45	0.30	6.4	4400	3981
C0601+240	6		12.0	0.51	− 0.16	0.50	2.0	2000	100
C0604+241	5.0		15.0	0.89	+0.31	0.49	3.5	4900	1096
C0605+139	6	+20.8		0.09	− 0.22	0.16	1.1	1100	50
C0605+243	28	− 10	8.0	0.21	− 0.10	0.16	3.5	870	107
C0606+203	18		8.0	0.20	− 0.35	0.63v	5.1	1950	1
C0607+206									
C0609+054	4.0		12.0	0.50	− 0.05	0.31	1.0	1800	200
C0610+070	6								
C0611+128	10		13.0	0.53	+0.02	0.42	2.3	1600	794
C0611+398	5		14.0						
C0613− 186	12		13.0	0.75	+0.31	0.08	5.8	3100	3020

Open Cluster Parameters

Name	R.A.(1950) h m s	Dec.(1950) ° ′	l °	b °	V_T (mag)	IAU Number
Collinder 89	06 15 00	+23 39	188.21	+03.70	5.7	C0615+236
NGC 2215	06 18 36	- 07 16	216.01	- 10.10	8.4	C0618- 072
Collinder 91	06 19 06	+02 23	207.42	- 05.52	6.4	C0619+023
Berkeley 73	06 19 36	- 06 19	215.28	- 09.41		C0619- 063
Collinder 92	06 20 12	+05 09	205.10	- 03.99	8.6	C0620+051
Dolidze 22	06 20 36	+04 41	205.56	- 04.12		C0620+046
Bochum 1	06 22 30	+19 48	192.43	+03.41	7.9	C0622+198
NGC 2232	06 24 06	- 04 43	214.36	- 07.65	3.9	C0624- 047
NGC 2236	06 27 00	+06 52	204.37	- 01.69	8.5	C0627+068
Collinder 96	06 27 42	+02 54	207.97	- 03.38	7.3	C0627+029
Collinder 95	06 27 42	+09 58	201.72	- 00.07		C0627+099
NGC 2243	06 27 54	- 31 15	239.50	- 17.97	9.4	C0627- 312
NGC 2239	06 28 18	+04 59	206.18	- 02.29		C0628+049
Czernik 26	06 28 18	- 04 11	214.35	- 06.51		C0628- 041
vdBergh 80	06 28 24	- 09 38	219.27	- 08.95		C0628- 096
Collinder 97	06 28 36	+05 57	205.37	- 01.76	5.4	C0628+059
Rosette, NGC 2244	06 29 42	+04 54	206.42	- 02.02	4.8	C0629+049
NGC 2250	06 30 18	- 05 00	215.30	- 06.45	8.9	C0630- 050
Berkeley 23	06 30 30	+20 35	192.61	+05.44		C0630+205
Basel 8	06 31 30	+08 07	203.79	- 00.12		C0631+081
NGC 2251	06 32 00	+08 24	203.60	+00.13	7.3	C0632+084
NGC 2252	06 32 18	+05 25	206.27	- 01.21	7.7	C0632+054
NGC 2254	06 33 18	+07 43	204.35	+00.09	9.1	C0633+077
Collinder 104	06 33 48	+04 52	206.93	- 01.13	9.6	C0633+048
Basel 7	06 33 54	+08 24	203.81	+00.54	8.5	C0633+084
Trumpler 5	06 34 00	+09 29	202.86	+01.05	10.9	C0634+094
Ruprecht 1	06 34 06	- 14 08	223.98	- 09.69		C0634- 141
Collinder 106	06 34 24	+06 00	206.02	- 00.42	4.6	C0634+060
vdBergh 1	06 34 24	+03 07	208.56	- 01.80	9.5	C0634+031
Collinder 107	06 35 00	+04 47	207.14	- 00.91	5.1	C0635+047
Berkeley 24	06 35 12	- 00 52	212.18	- 03.46		C0635- 008
NGC 2259	06 35 48	+10 56	201.79	+02.12	10.8	C0635+109
Collinder 110	06 35 48	+02 03	209.65	- 01.98	10.5	C0635+020
NGC 2262	06 35 48	+01 13	210.39	- 02.37	11.3	C0635+012
Collinder 111	06 36 00	+06 57	205.34	+00.33	7.0	C0636+069

Open Cluster Parameters

IAU Number	θ (′)	V_r (km s^{-1})	MBS	B–V (mag)	TOC	E_{B-V}	R (pc)	D_o (pc)	Age ($\times 10^6$)
C0615+236	35						6.6	1300	
C0618– 072	11		11.0	0.45	+0.10	0.10	1.6	1000	355
C0619+023	17								
C0619– 063	3.0		16.0						
C0620+051	11								
C0620+046	18								
C0622+198				0.18	– 0.35	0.55		4060	1
C0624– 047	29				– 0.20	0.01	1.8	400	22
C0627+068	6		12.0	0.58	+0.13	0.37	3.5	3400	1259
C0627+029	7			0.33	– 0.20	0.48	1.3	1100	25
C0627+099	19								
C0627– 312	5.0			0.72	+0.40	0.03	2.6	3580	6026
C0628+049	15								
C0628– 041	5.0								
C0628– 096									
C0628+059	21								
C0629+049	23	+27.3	7.0	0.46	– 0.31	0.47	5.9	1700	3
C0630– 050	7		12.0						
C0630+205	4.0		15.0						
C0631+081								1300	
C0632+084	10			0.39	+0.05	0.20	2.3	1550	302
C0632+054	20		9.0						
C0633+077	4.0		12.0	0.65	– 0.01	0.33	1.3	2200	631
C0633+048	21								
C0633+084				0.36	– 0.01	0.28			631
C0634+094	7		17.0		+0.57	0.64	2.8	2400	1259
C0634– 141	11		11.0						
C0634+060	45								
C0634+031				0.92				1700	25
C0635+047	35			0.44	– 0.30	0.54v	8.6	1700	
C0635– 008	10		17.0						
C0635+109	4.5		14.0						
C0635+020	12								
C0635+012	3.5								
C0636+069	3.2								

Open Cluster Parameters

Name	R.A.(1950) h m s	Dec.(1950) ° ′	l °	b °	V_T (mag)	IAU Number
vdBergh 83	06 38 00	- 27 16	236.55	- 14.40		C0638- 272
S Mon, NGC 2264	06 38 18	+09 56	202.94	+02.20	3.9	C0638+099
Berkeley 25	06 38 48	- 16 28	226.61	- 09.69		C0638- 164
Ruprecht 2	06 39 06	- 29 30	238.78	- 15.05		C0639- 295
NGC 2266	06 40 06	+27 01	187.78	+10.28	9.5	C0640+270
Ruprecht 3	06 40 12	- 29 24	238.78	- 14.81		C0640- 294
Dolidze 23	06 40 36	+00 03	211.99	- 01.84		C0640+000
NGC 2269	06 41 18	+04 37	208.01	+00.41	10.0	C0641+046
Dolidze 24	06 41 36	+01 38	210.68	- 00.88		C0641+016
Dolidze 25	06 42 30	+00 21	211.94	- 01.29	7.6	C0642+003
Collinder 115	06 43 54	+01 49	210.80	- 00.30	9.2	C0643+018
vdBergh 85	06 44 18	+01 23	211.23	- 00.41		C0644+013
M 41, NGC 2287	06 44 54	- 20 41	231.10	- 10.23	4.5	C0644- 206
NGC 2286	06 45 06	- 03 07	215.32	- 02.30	7.5	C0645- 031
NGC 2281	06 45 48	+41 07	175.00	+17.06	5.4	C0645+411
Bochum 2	06 46 18	+00 26	212.30	- 00.40	9.7	C0646+004
Ruprecht 4	06 46 30	- 10 29	222.05	- 05.35		C0646- 104
Berkeley 75	06 47 12	- 23 54	234.30	- 11.12		C0647- 239
Biurakan 12	06 47 36	+05 49	207.66	+02.37		C0647+058
Biurakan 11	06 48 36	+05 50	207.76	+02.60		C0648+058
NGC 2301	06 49 12	+00 31	212.55	+00.28	6.0	C0649+005
NGC 2302	06 49 30	- 07 00	219.29	- 03.08	8.9	C0649- 070
Biurakan 10	06 49 36	+03 00	210.39	+01.52	10.4	C0649+030
Ruprecht 149	06 50 06	- 23 34	234.28	- 10.38		C0650- 235
Berkeley 29	06 50 24	+16 59	197.98	+08.03		C0650+169
Collinder 121	06 52 06	- 24 34	235.39	- 10.40	2.6	C0652- 245
NGC 2304	06 52 06	+18 05	197.17	+08.87	10.0	C0652+180
Ruprecht 5	06 53 12	- 18 40	230.14	- 07.58		C0653- 186
Ruprecht 6	06 53 42	- 13 13	225.30	- 05.02		C0653- 132
NGC 2309	06 53 48	- 07 08	219.86	- 02.25	10.5	C0653- 071
Biurakan 7	06 54 54	+08 20	206.26	+05.12		C0654+083
Biurakan 9	06 55 06	+03 17	210.80	+02.89		C0655+032
NGC 2311	06 55 18	- 04 31	217.73	- 00.69	9.6	C0655- 045
Biurakan 8	06 55 24	+06 30	207.95	+04.40		C0655+065
Ruprecht 7	06 55 24	- 13 09	225.43	- 04.62		C0655- 131

Open Cluster Parameters

IAU Number	θ (′)	V_r (km s^{-1})	MBS	B–V (mag)	TOC	E_{B-V}	R (pc)	D_o (pc)	Age ($\times 10^6$)
C0638− 272									
C0638+099	20	+24.1	5.0		− 0.30	0.06v	2.2	750	20
C0638− 164	6		16.0						
C0639− 295	6		12.0						
C0640+270	6		11.0				3.5	3400	
C0640− 294	2.8		11.0						
C0640+000	12								
C0641+046	4.0		10.0	0.42	− 0.10	0.44	0.8	1400	126
C0641+016	18								
C0642+003	23			0.59	− 0.35	0.81	18.1	5200	1
C0643+018	7								
C0644+013									
C0644− 206	38	+25.9	8.0	0.39	− 0.05	0.01	4.1	740	100
C0645− 031	14		9.0	0.66	+0.00	0.41	2.8	1300	316
C0645+411	14	+18	8.0	0.65	+0.07	0.08	1.1	500	302
C0646+004				0.54	− 0.35	0.89		5500	1
C0646− 104	5		14.0						
C0647− 239	4.0		16.0						
C0647+058	4.0		17.0						
C0648+058	3.0		15.0				3.5	8000	
C0649+005	12		8.0	0.30	− 0.10	0.03	1.4	750	107
C0649− 070	2.5		12.0	0.16	− 0.15	0.25	0.4	1100	63
C0649+030	4.0		15.0	0.25	− 0.22	0.95	2.9	5000	
C0650− 235	5		13.0						
C0650+169	4.0		15.0						
C0652− 245	50	+36		0.42	− 0.20	0.03	8.5	1170	2
C0652+180	5								
C0653− 186	2.0		13.0						
C0653− 132	2.0								
C0653− 071	3.0		13.0						
C0654+083	5		14.0				3.5	4000	
C0655+032	4.0		15.0						
C0655− 045	6		12.0						
C0655+065	5		14.0				7.8	9000	
C0655− 131	4.0		14.0						

Open Cluster Parameters

Name	R.A.(1950) h m s	Dec.(1950) ° ′	l °	b °	V_T (mag)	IAU Number
Biurakan 13	06 57 48	+00 11	214.18	+01.87		C0657–001
Tombaugh 1	06 58 12	– 20 24	232.22	– 07.32		C0658–204
Ruprecht 8	06 59 24	– 13 31	226.20	– 03.92		C0659–135
Czernik 27	07 00 30	+06 29	208.55	+05.52		C0700+064
M 50, NGC 2323	07 00 48	– 08 16	221.67	– 01.24	5.9	C0700–082
Bochum 3	07 00 54	– 05 00	218.81	+00.33	9.9	C0700–050
Tombaugh 2	07 01 12	– 20 47	232.90	– 06.84		C0701–207
NGC 2324	07 01 36	+01 08	213.45	+03.31	8.4	C0701+011
Haffner 3	07 01 36	– 06 03	219.83	+00.02		C0701–060
vdBergh 92	07 01 36	– 11 27	224.60	– 02.50		C0701–114
Auner 1	07 02 06	– 19 40	231.99	– 06.15		C0702–196
Ruprecht 150	07 03 54	– 28 22	240.04	– 09.67		C0703–283
Haffner 4	07 03 54	– 14 54	227.93	– 03.59		C0703–149
NGC 2331	07 04 06	+27 26	189.67	+15.30	8.5	C0704+274
Berkeley 76	07 04 06	– 11 32	224.98	– 01.97		C0704–115
NGC 2335	07 04 12	– 10 00	223.62	– 01.27	7.2	C0704–100
Ruprecht 10	07 04 12	– 20 02	232.55	– 05.88		C0704–200
Collinder 466	07 04 54	– 10 44	224.35	– 01.45	11.1	C0704–107
Ruprecht 11	07 05 12	– 20 43	233.26	– 05.98		C0705–207
Ruprecht 12	07 05 12	– 28 08	239.95	– 09.31		C0705–281
Ruprecht 13	07 05 48	– 25 47	237.88	– 08.14		C0705–257
NGC 2343	07 05 54	– 10 34	224.32	– 01.16	6.7	C0705–105
NGC 2345	07 06 00	– 13 05	226.57	– 02.30	7.7	C0706–130
Haffner 23	07 07 12	– 16 52	230.05	– 03.79		C0707–168
Berkeley 35	07 08 12	+02 49	212.70	+05.56		C0708+028
NGC 2353	07 12 12	– 10 13	224.72	+00.38	7.1	C0712–102
NGC 2354	07 12 12	– 25 39	238.42	– 06.80	6.5	C0712–256
Collinder 132	07 12 30	– 31 05	243.34	– 09.21	3.6	C0712–310
Ruprecht 14	07 13 06	– 31 16	243.58	– 09.17		C0713–312
Berkeley 36	07 13 48	– 13 01	227.37	– 00.58		C0713–130
NGC 2355	07 14 06	+13 52	203.36	+11.80	9.7	C0714+138
Basel 11	07 14 48	– 13 53	228.26	– 00.78	8.2	C0714–138
Collinder 135	07 15 12	– 36 45	248.76	– 11.20	2.1	C0715–367
NGC 2360	07 15 30	– 15 32	229.80	– 01.42	7.2	C0715–155
Haffner 5	07 16 00	– 22 34	236.08	– 04.58		C0716–225

Open Cluster Parameters

IAU Number	θ (′)	V_r (km s^{-1})	MBS	B–V (mag)	TOC	E_{B-V}	R (pc)	D_o (pc)	Age ($\times 10^6$)
C0657– 001	5		17.0						
C0658– 204	5		14.0		+0.18	0.27			
C0659– 135	4.0		12.0						
C0700+064	5.0								
C0700– 082	16	+9	9.0	0.37	– 0.15	0.24	2.2	910	78
C0700– 050				0.14	– 0.15	0.24			63
C0701– 207	3.0		16.0		+0.19	0.08			
C0701+011	7		12.0	0.43	– 0.25	0.28	3.3	2900	661
C0701– 060	2.7		14.0						
C0701– 114									
C0702– 196	2.5								
C0703– 283	2.5		13.0						
C0703– 149	2.4		14.0						
C0704+274	18		9.0						
C0704– 115	6		16.0						
C0704– 100	12		10.0	0.56	– 0.11	0.38	1.8	1000	159
C0704– 200	4.5		12.0						
C0704– 107	4.0								
C0705– 207	2.9		11.0						
C0705– 281	5.0		14.0						
C0705– 257	4.2		13.0						
C0705– 105	6		8.0	0.35	– 0.14	0.16	1.0	1000	100
C0706– 130	12		9.0	0.88	– 0.15	0.70v	3.2	1800	79
C0707– 168	11		13.0						
C0708+028	7		16.0						
C0712– 102	20	+27	9.0	0.20	– 0.15	0.12	3.2	1100	13
C0712– 256	20			0.78	+0.15	0.14	5.4	1850	182
C0712– 310	95	+26			– 0.15	0.03			25
C0713– 312	2.3		13.0						
C0713– 130	5		17.0						
C0714+138	9		13.0						
C0714– 138				0.46	+0.00	0.00			316
C0715– 367	50	+17		0.59					
C0715– 155	12			0.62	+0.30	0.07	3.1	1630	1288
C0716– 225	5		15.0						

Open Cluster Parameters

Name	R.A.(1950) h m s	Dec.(1950) ° ′	l °	b °	V_T (mag)	IAU Number
τ CMa, NGC 2362	07 16 42	- 24 51	238.18	- 05.53	4.1	C0716- 248
Ruprecht 15	07 17 30	- 19 34	233.57	- 02.88		C0717- 195
Haffner 6	07 17 48	- 13 02	227.85	+00.25	9.2	C0717- 130
Berkeley 37	07 17 54	- 01 00	217.23	+05.95		C0717- 010
NGC 2367	07 18 00	- 21 50	235.65	- 03.84	7.9	C0718- 218
NGC 2368	07 18 36	- 10 17	225.53	+01.73	11.8	C0718- 102
Berkeley 77	07 19 06	- 03 14	219.36	+05.17		C0719- 032
King 23	07 19 18	+00 53	217.29	+06.31		C0719- 008
Ruprecht 16	07 21 00	- 19 21	233.78	- 02.06		C0721- 193
Haffner 8	07 21 06	- 12 14	227.52	+01.34	9.1	C0721- 122
Haffner 7	07 21 12	- 29 24	242.68	- 06.79		C0721- 294
Ruprecht 17	07 21 30	- 23 07	237.16	- 03.74		C0721- 231
NGC 2374	07 21 42	- 13 10	228.42	+01.04	8.0	C0721- 131
Berkeley 78	07 21 48	+05 27	211.90	+09.77		C0721+054
Collinder 140	07 22 00	- 32 06	245.18	- 07.87	3.5	C0722- 321
Haffner 9	07 22 18	- 16 54	231.78	- 00.62		C0722- 169
NGC 2383	07 22 36	- 20 50	235.26	- 02.44	8.4	C0722- 208
Ruprecht 18	07 22 42	- 26 07	239.94	- 04.92	9.4	C0722- 261
NGC 2384	07 22 54	- 20 56	235.39	- 02.41	7.4	C0722- 209
Ruprecht 19	07 23 30	- 21 25	235.89	- 02.51		C0723- 214
Trumpler 6	07 24 00	- 24 12	238.39	- 03.75	10.0	C0724- 242
NGC 2395	07 24 18	+13 41	204.62	+13.96	8.0	C0724+136
Ruprecht 20	07 24 42	- 28 47	242.51	- 05.79	9.5	C0724- 287
Melotte 66	07 24 54	- 47 38	259.61	- 14.29	7.8	C0724- 476
Trumpler 7	07 25 12	- 23 56	238.28	- 03.38	7.9	C0725- 239
Ruprecht 21	07 25 12	- 31 05	244.60	- 06.78		C0725- 310
NGC 2396	07 25 48	- 11 38	227.56	+02.66	7.4	C0725- 116
Czernik 29	07 26 00	- 15 18	230.80	+00.93	10.3	C0726- 153
Haffner 10	07 26 18	- 15 17	230.82	+01.00	11.5	C0726- 152
NGC 2401	07 27 06	- 13 52	229.67	+01.85	12.6	C0727- 138
Ruprecht 22	07 27 12	- 29 07	243.07	- 05.47		C0727- 291
Mayer 3	07 27 12	- 18 25	233.67	- 00.33		C0727- 184
Dolidze 26	07 27 18	+12 00	206.51	+13.89		C0727+120
Ruprecht 23	07 28 18	- 23 13	238.01	- 02.43		C0728- 232
Bochum 5	07 28 42	- 16 58	232.57	+00.69	7.0	C0728- 169

Open Cluster Parameters

IAU Number	θ (′)	V_r (km s^{-1})	MBS	B–V (mag)	TOC	E_{B-V}	R (pc)	D_o (pc)	Age ($\times 10^6$)
C0716– 248	8	+34	8.0		– 0.30	0.11	2.0	1550	25
C0717– 195	1.7		12.0						
C0717– 130	4.0		16.0	0.77	+0.20	0.00	0.6	1100	794
C0717– 010	7		15.0						
C0718– 218	3.5			0.08	– 0.30	0.35	1.0	2000	1
C0718– 102	5.0								
C0719– 032	6		13.0						
C0719– 008	4.0								
C0721– 193	11		13.0						
C0721– 122	4.2		12.0	0.51	+0.04	0.28	1.0	1700	501
C0721– 294	2.9		14.0						
C0721– 231	5.0		12.0						
C0721– 131	19			0.46	– 0.03	0.15	3.7	1300	316
C0721+054	7		16.0						
C0722– 321	42	+17		0.15	– 0.17	0.03	1.9	300	22
C0722– 169	2.1		14.0						
C0722– 208	5			0.48	– 0.20	0.27	1.8	2000	25
C0722– 261	4.0		12.0	0.50	– 0.10	0.67			126
C0722– 209	2.5	+67.3		0.07	– 0.30	0.29	0.8	2000	1
C0723– 214	8		13.0						
C0724– 242	6								
C0724+136	12			0.81	– 0.20	0.72	2.2	1200	50
C0724– 287	10		11.0	0.72	+0.00	0.10			316
C0724– 476	10			0.93	+0.50	0.17	5.8	3950	6026
C0725– 239	5		10.0	0.12	– 0.20	0.29	1.3	1600	25
C0725– 310	11		11.0						
C0725– 116	10		11.0						
C0726– 153	7			0.51	– 0.15	0.55			63
C0726– 152	1.6		15.0	1.10	– 0.25	0.55			6
C0727– 138	2.0								
C0727– 291	2.8		12.0						
C0727– 184									
C0727+120	23								
C0728– 232	3.0		13.0						
C0728– 169	11			0.74	– 0.30	0.63	3.7	2290	1

Open Cluster Parameters

Name	R.A.(1950) h m s	Dec.(1950) ° ′	l °	b °	V_T (mag)	IAU Number
Bochum 4	07 28 48	- 16 51	232.48	+00.77	7.3	C0728- 168
Czernik 30	07 28 54	- 09 52	226.36	+04.17		C0728- 098
Ruprecht 24	07 29 36	- 12 39	228.91	+02.99		C0729- 126
Bochum 6	07 29 48	- 19 20	234.77	- 00.23	9.9	C0729- 193
NGC 2414	07 31 00	- 15 20	231.41	+01.96	7.9	C0731- 153
Ruprecht 40	07 31 12	- 20 23	235.84	- 00.45		C0731- 203
Haffner 11	07 33 24	- 27 37	242.41	- 03.55		C0733- 276
NGC 2421	07 34 06	- 20 30	236.24	+00.08	8.3	C0734- 205
M 47, NGC 2422	07 34 18	- 14 23	230.97	+03.13	4.4	C0734- 143
Ruprecht 25	07 34 24	- 23 15	238.71	- 01.21		C0734- 232
Czernik 31	07 34 36	- 20 23	236.23	+00.24		C0734- 203
NGC 2423	07 34 48	- 13 45	230.47	+03.55	6.7	C0734- 137
Ruprecht 26	07 34 54	- 15 32	232.03	+02.70		C0734- 155
Melotte 71	07 35 12	- 11 57	228.95	+04.51	7.1	C0735- 119
Ruprecht 27	07 35 24	- 26 29	241.65	- 02.58		C0735- 264
NGC 2420	07 35 30	+21 41	198.11	+19.65	8.3	C0735+216
vdB-Hagen 4	07 36 00	- 35 55	249.96	- 07.10		C0736- 359
NGC 2425	07 36 00	- 14 45	231.47	+03.30		C0736- 147
Melotte 72	07 36 00	- 10 34	227.85	+05.38	10.1	C0736- 105
Collinder 467	07 36 54	- 10 26	227.83	+05.61	11.8	C0736- 104
Ruprecht 28	07 37 30	- 30 49	245.64	- 04.34		C0737- 308
Bochum 15	07 38 12	- 33 26	248.00	- 05.50	6.3	C0738- 334
Haffner 13	07 38 30	- 30 00	245.04	- 03.75		C0738- 300
NGC 2432	07 38 42	- 18 58	235.48	+01.79	10.2	C0738- 189
NGC 2439	07 38 54	- 31 32	246.42	- 04.42	6.9	C0738- 315
Ruprecht 151	07 39 00	- 16 08	233.06	+03.25		C0739- 161
Ruprecht 29	07 39 00	- 24 13	240.07	- 00.77		C0739- 242
M 46, NGC 2437	07 39 30	- 14 42	231.87	+04.07	6.1	C0739- 147
Ruprecht 30	07 40 18	- 31 21	246.40	- 04.08		C0740- 313
Ruprecht 31	07 40 54	- 35 28	250.06	- 06.01		C0740- 354
M 93, NGC 2447	07 42 30	- 23 45	240.07	+00.15	6.2	C0742- 237
Haffner 14	07 42 48	- 28 15	243.99	- 02.06		C0742- 282
Ruprecht 32	07 42 54	- 25 24	241.54	- 00.60	8.4	C0742- 254
Haffner 15	07 43 24	- 32 40	247.88	- 04.17	9.4	C0743- 326
Ruprecht 33	07 43 36	- 21 49	238.51	+01.33		C0743- 218

Open Cluster Parameters

IAU Number	θ (′)	V_r (km s^{-1})	MBS	B–V (mag)	TOC	E_{B-V}	R (pc)	D_o (pc)	Age ($\times 10^6$)
C0728– 168	23			0.18	– 0.20	0.19			25
C0728– 098	3.0								
C0729– 126	2.0		11.0						
C0729– 193				0.45	– 0.35	0.70		4000	1
C0731– 153	4.0	+75		0.43	– 0.25	0.55	1.5	2500	6
C0731– 203	3.5		13.0						
C0733– 276	4.4		16.0						
C0734– 205	10		11.0	0.39	– 0.20	0.47	2.8	1900	25
C0734– 143	29	+29	5.0	0.02	– 0.15	0.08	2.1	480	78
C0734– 232	0.7		14.0						
C0734– 203	9								
C0734– 137	19			0.45	+0.04	0.21	2.4	870	355
C0734– 155	10		12.0						
C0735– 119	9			0.34	+0.03	0.01	3.8	2800	417
C0735– 264	18		12.0						
C0735+216	10	+70.0	11.0	0.70	+0.39	0.02	3.5	2300	3981
C0736– 359	2.5								
C0736– 147	3.3		14.0						
C0736– 105	9								
C0736– 104	2.0								
C0737– 308	4.0		14.0						
C0738– 334				0.44	– 0.30	0.55			1
C0738– 300	14		8.0						
C0738– 189	7								
C0738– 315	10	+73	9.0	0.41	– 0.19	0.35	2.3	1610	66
C0739– 161	14		12.0						
C0739– 242	2.5		13.0						
C0739– 147	27	+41	10.0	0.23	+0.00	0.14	5.6	1410	302
C0740– 313	4.0		11.0						
C0740– 354	2.0		11.0						
C0742– 237	22		9.0	0.37	+0.00	0.06	3.6	1100	98
C0742– 282	3.8		14.0						
C0742– 254	5		10.0	0.04	– 0.25	0.50	3.6	4100	1
C0743– 326	3.5		12.0	1.28	– 0.25	1.16	1.3	2500	6
C0743– 218	6		13.0						

Open Cluster Parameters

Name	R.A.(1950) h m s	Dec.(1950) ° ′	l °	b °	V_T (mag)	IAU Number
NGC 2451	07 43 36	- 37 51	252.41	- 06.72	2.8	C0743- 378
Ruprecht 34	07 43 42	- 20 16	237.20	+02.15	9.5	C0743- 202
Ruprecht 35	07 44 12	- 31 08	246.63	- 03.25		C0744- 311
Berkeley 39	07 44 12	- 04 29	223.47	+10.09		C0744- 044
NGC 2453	07 45 42	- 27 07	243.33	- 00.94	8.3	C0745- 271
Ruprecht 36	07 46 24	- 26 10	242.60	- 00.31	9.6	C0746- 261
Haffner 25	07 46 36	- 25 46	242.28	- 00.07		C0746- 257
NGC 2455	07 46 48	- 21 10	238.34	+02.31	10.2	C0746- 211
Ruprecht 37	07 47 36	- 17 09	234.96	+04.53		C0747- 171
Haffner 16	07 48 12	- 25 19	242.08	+00.48	10.0	C0748- 253
Ruprecht 38	07 48 24	- 20 04	237.59	+03.20		C0748- 200
Czernik 32	07 48 24	- 29 42	245.86	- 01.74		C0748- 297
Haffner 17	07 49 48	- 31 43	247.72	- 02.53		C0749- 317
Ruprecht 39	07 50 00	- 22 20	239.71	+02.37		C0750- 223
Haffner 18	07 50 24	- 26 14	243.11	+00.44	9.3	C0750- 262
NGC 2477	07 50 30	- 38 25	253.58	- 05.83	5.8	C0750- 384
NGC 2467	07 50 30	- 26 15	243.14	+00.41	7.1	C0750- 263
Haffner 19	07 50 36	- 26 07	243.04	+00.52	9.4	C0750- 261
Ruprecht 41	07 51 48	- 26 51	243.80	+00.37		C0751- 268
NGC 2479	07 52 48	- 17 35	235.97	+05.36	9.6	C0752- 175
NGC 2482	07 52 48	- 24 10	241.63	+01.97	7.3	C0752- 241
Ruprecht 152	07 52 48	- 38 08	253.58	- 05.29		C0752- 381
Trumpler 9	07 53 12	- 25 48	243.07	+01.19	8.7	C0753- 258
NGC 2483	07 53 54	- 27 48	244.86	+00.28	7.6	C0753- 278
NGC 2489	07 54 12	- 29 56	246.71	- 00.78	7.9	C0754- 299
Haffner 20	07 54 18	- 30 16	247.01	- 00.94	11.0	C0754- 302
Ruprecht 42	07 55 30	- 25 45	243.30	+01.65		C0755- 257
Ruprecht 43	07 56 42	- 28 47	246.02	+00.28		C0756- 287
Ruprecht 44	07 57 00	- 28 27	245.77	+00.52	7.2	C0757- 284
Ruprecht 45	07 57 18	- 16 10	235.32	+07.02		C0757- 161
NGC 2516	07 57 30	- 60 44	273.94	- 15.88	3.8	C0757- 607
NGC 2506	07 57 48	- 10 39	230.57	+09.91	7.6	C0757- 106
Ruprecht 153	07 58 12	- 30 07	247.32	- 00.15		C0758- 301
NGC 2509	07 58 30	- 18 56	237.86	+05.83	9.3	C0758- 189
Haffner 21	07 59 06	- 27 02	244.81	+01.66	10.3	C0759- 270

Open Cluster Parameters

IAU Number	θ (′)	V_r (km s^{-1})	MBS	B–V (mag)	TOC	E_{B-V}	R (pc)	D_o (pc)	Age ($\times 10^6$)
C0743– 378	45	+27	6.0	0.56	– 0.20	0.04	1.5	220	36
C0743– 202	4.0		12.0	0.54					
C0744– 311	0.8		14.0						
C0744– 044	12		16.0						
C0745– 271	5.0			0.39	– 0.28	0.47	1.1	1500	40
C0746– 261	4.0		12.0	0.04	– 0.04	0.14			200
C0746– 257	1.0		14.0						
C0746– 211	7		12.0						
C0747– 171	0.8		14.0						
C0748– 253	1.1		12.0	0.14	– 0.20	0.19	0.6	3900	25
C0748– 200	6		13.0						
C0748– 297	3.0								
C0749– 317	1.6		15.0						
C0750– 223	1.5		14.0						
C0750– 262	1.0		11.0	0.79	– 0.30	0.50	1.0	6900	1
C0750– 384	27		12.0	0.84	+0.23	0.27	5.1	1300	708
C0750– 263	15				– 0.35	0.54	7.9	3400	1
C0750– 261	1.8		14.0	0.20	– 0.25	0.45v	1.8	6900	6
C0751– 268	1.1		14.0						
C0752– 175	7								
C0752– 241	12			0.41	+0.05	0.04	1.5	800	398
C0752– 381	1.7		16.0						
C0753– 258	5			0.26	– 0.19	0.24	1.9	2200	63
C0753– 278	10			0.29	– 0.30	0.45	4.2	2900	10
C0754– 299	8		11.0	0.68	+0.00	0.36	1.5	1200	240
C0754– 302	1.8		14.0	0.66	– 0.05	0.55			200
C0755– 257	1.5		14.0						
C0756– 287	13		12.0						
C0757– 284	5.0		12.0	0.46	– 0.30	0.64	3.3	4600	1
C0757– 161	11		13.0						
C0757– 607	29	+19	7.0	0.04	– 0.16	0.13	1.9	440	107
C0757– 106	6		11.0	0.68	+0.34	0.05	2.8	2750	3388
C0758– 301	2.8		14.0						
C0758– 189	8								
C0759– 270	1.1		15.0	0.39	+0.00	0.20			316

Open Cluster Parameters

Name	R.A.(1950) h m s	Dec.(1950) ° ′	l °	b °	V_T (mag)	IAU Number
Ruprecht 46	07 59 54	- 19 20	238.37	+05.89	9.1	C0759- 193
Ruprecht 154	08 00 06	- 44 16	259.57	- 07.30		C0800- 442
Ruprecht 47	08 00 18	- 30 58	248.28	- 00.20	9.6	C0800- 309
Ruprecht 48	08 00 54	- 31 53	249.13	- 00.60		C0800- 318
Ruprecht 49	08 01 00	- 26 39	244.72	+02.22	9.6	C0801- 266
Ruprecht 50	08 01 24	- 30 45	248.23	+00.10		C0801- 307
Ruprecht 51	08 01 36	- 30 30	248.04	+00.27		C0801- 305
Collinder 173	08 02 00	- 46 08	261.32	- 08.06	0.6	C0802- 461
Ruprecht 155	08 03 00	- 31 39	249.17	- 00.10		C0803- 316
Ruprecht 52	08 03 06	- 31 49	249.32	- 00.17		C0803- 318
NGC 2527	08 03 12	- 28 01	246.13	+01.89	6.5	C0803- 280
NGC 2533	08 05 00	- 29 45	247.81	+01.29	7.6	C0805- 297
vdB–Hagen 19	08 05 06	- 32 14	249.90	- 00.04		C0805- 322
NGC 2539	08 08 24	- 12 41	233.73	+11.12	6.5	C0808- 126
Ruprecht 53	08 08 42	- 26 52	245.83	+03.55		C0808- 268
NGC 2547	08 09 12	- 49 07	264.55	- 08.55	4.7	C0809- 491
Ruprecht 54	08 09 18	- 31 48	250.03	+00.95		C0809- 318
Ruprecht 55	08 10 18	- 32 27	250.69	+00.76	7.8	C0810- 324
Haffner 22	08 10 30	- 27 44	246.78	+03.41		C0810- 277
NGC 2546	08 10 36	- 37 29	254.90	- 01.98	6.3	C0810- 374
Ruprecht 56	08 10 48	- 40 19	257.29	- 03.51		C0810- 403
M 48, NGC 2548	08 11 18	- 05 39	227.92	+15.37	5.8	C0811- 056
vdB–Hagen 23	08 12 30	- 36 15	254.10	- 00.97		C0812- 362
Ruprecht 57	08 12 42	- 26 50	246.30	+04.31		C0812- 268
Ruprecht 58	08 12 42	- 31 48	250.43	+01.55		C0812- 318
Haffner 26	08 13 24	- 30 41	249.58	+02.29		C0813- 306
Pismis 1	08 15 54	- 36 56	255.08	- 00.74	10.7	C0815- 369
Pismis 2	08 16 18	- 41 28	258.83	- 03.29		C0816- 414
NGC 2567	08 16 36	- 30 29	249.81	+02.98	7.4	C0816- 304
NGC 2571	08 16 54	- 29 35	249.11	+03.54	7.0	C0816- 295
Ruprecht 59	08 17 12	- 34 18	253.03	+00.92	9.0	C0817- 343
NGC 2580	08 19 36	- 30 09	249.91	+03.70	9.7	C0819- 301
NGC 2579	08 20 18	- 36 15	254.67	+00.27	7.5	C0819- 360
NGC 2588	08 21 12	- 32 49	252.28	+02.45	11.8	C0821- 328
NGC 2587	08 21 24	- 29 20	249.44	+04.48	9.2	C0821- 293

Open Cluster Parameters

IAU Number	θ (′)	V_r (km s^{-1})	MBS	B–V (mag)	TOC	E_{B-V}	R (pc)	D_o (pc)	Age ($\times 10^6$)
C0759– 193	2.0		12.0	0.95	+0.45	0.07			3162
C0800– 442	0.6		13.0						
C0800– 309	5.0		13.0	0.22	– 0.15	0.28			63
C0800– 318	2.0		12.0						
C0801– 266	2.5		13.0	0.21	+0.00	0.26			
C0801– 307	3.5		12.0						
C0801– 305	4.5		14.0						
C0802– 461	370						17.8	330	
C0803– 316	2.0		13.0						
C0803– 318	2.5		13.0						
C0803– 280	22			0.31	+0.09	0.10	2.0	600	1000
C0805– 297	3.5			0.59	– 0.14	0.20	0.9	1700	182
C0805– 322	2.5								
C0808– 126	21		9.0	0.55	+0.05	0.10	4.1	1280	661
C0808– 268	18		10.0						
C0809– 491	20	+15	7.0	0.05	– 0.19	0.05	1.1	400	58
C0809– 318	2.5		15.0						
C0810– 324	17		11.0	0.34	– 0.30	0.56	10.9	4400	1
C0810– 277	4.8		15.0						
C0810– 374	40	+16	7.0	0.16	– 0.10	0.16	5.9	1000	42
C0810– 403	42		9.0						
C0811– 056	54		8.0	0.31	+0.00	0.04	4.8	610	302
C0812– 362	12								
C0812– 268	5.0		12.0						
C0812– 318	10		12.0						
C0813– 306	6		14.0						
C0815– 369	4.6		11.0	0.63	– 0.17	0.58	1.6	2400	85
C0816– 414	4.3		15.0						
C0816– 304	10		11.0	0.37	+0.00	0.13	2.5	1700	68
C0816– 295	13			0.16	– 0.25	0.10	4.1	2100	22
C0817– 343	5.0		12.0	0.37					
C0819– 301	7								
C0819– 360	10			0.36	– 0.20	0.15	1.5	1000	13
C0821– 328	2.0								
C0821– 293	9								

Open Cluster Parameters

Name	R.A.(1950) h m s	Dec.(1950) ° ′	l °	b °	V_T (mag)	IAU Number
Collinder 185	08 21 30	- 36 11	254.76	+00.45	7.8	C0820- 360
Collinder 187	08 22 06	- 28 59	249.24	+04.82	9.6	C0822- 289
Ruprecht 60	08 22 54	- 47 02	264.09	- 05.49		C0822- 470
Ruprecht 61	08 23 00	- 34 00	253.47	+02.07		C0823- 340
Ruprecht 157	08 27 30	- 18 56	241.63	+11.57		C0827- 189
Pismis 3	08 29 06	- 38 30	257.86	+00.46		C0829- 385
vdB–Hagen 34	08 29 36	- 44 19	262.57	- 02.95		C0829- 443
Ruprecht 62	08 30 18	- 19 29	242.48	+11.80		C0830- 194
Ruprecht 63	08 31 06	- 48 08	265.80	- 05.01		C0831- 481
Pismis 4	08 32 48	- 44 06	262.74	- 02.37	5.9	C0832- 441
vdB–Hagen 37	08 34 00	- 43 25	262.32	- 01.78		C0834- 434
NGC 2627	08 35 12	- 29 46	251.58	+06.65	8.4	C0835- 297
Ruprecht 64	08 35 36	- 39 55	259.72	+00.58		C0835- 399
Pismis 5	08 35 54	- 39 24	259.39	+00.86	9.9	C0835- 394
NGC 2635	08 36 30	- 34 35	255.60	+03.96	11.2	C0836- 345
M 44, NGC 2632	08 37 12	+20 10	205.54	+32.52	3.1	C0837+201
Ruprecht 65	08 37 24	- 43 51	263.04	- 01.56		C0837- 438
Pismis 6	08 37 36	- 46 02	264.81	- 02.87	7.0	C0837- 460
Ruprecht 66	08 38 36	- 37 53	258.47	+02.29		C0838- 378
Waterloo 6	08 38 42	- 45 58	264.86	- 02.67	8.4	C0838- 459
o Vel,IC 2391	08 38 48	- 52 53	270.36	- 06.88	2.5	C0838- 528
Pismis 7	08 39 12	- 38 30	259.03	+02.00		C0839- 385
IC 2395	08 39 30	- 48 01	266.57	- 03.81	4.6	C0839- 480
Pismis 8	08 39 48	- 46 06	265.08	- 02.63	9.5	C0839- 461
Ruprecht 67	08 40 00	- 43 12	262.82	- 00.78	9.1	C0840- 432
NGC 2660	08 40 36	- 46 58	265.86	- 03.03	8.8	C0840- 469
NGC 2659	08 40 54	- 44 46	264.16	- 01.63	8.6	C0840- 447
vdB–Hagen 47	08 41 00	- 47 56	266.65	- 03.58		C0841- 479
NGC 2658	08 41 24	- 32 28	254.56	+06.07	9.2	C0841- 324
Ruprecht 68	08 42 30	- 35 45	257.28	+04.22		C0842- 357
Collinder 197	08 42 54	- 41 11	261.56	+00.88	6.7	C0842- 411
Collinder 196	08 43 00	- 31 27	253.95	+06.97	10.5	C0843- 314
Ruprecht 69	08 43 00	- 47 25	266.45	- 02.99		C0843- 474
Bochum 7	08 43 06	- 45 48	265.20	- 01.97	6.8	C0843- 458
Collinder 198	08 43 18	- 31 35	254.09	+06.93	11.2	C0843- 315

Open Cluster Parameters

IAU Number	θ (′)	V_r (km s^{-1})	MBS	B–V (mag)	TOC	E_{B-V}	R (pc)	D_o (pc)	Age ($\times 10^6$)
C0820– 360	9			0.41	– 0.10	0.21	2.0	1500	79
C0822– 289	7								
C0822– 470	6		13.0						
C0823– 340	2.5		14.0						
C0827– 189	17		11.0						
C0829– 385	6		13.0						
C0829– 443	12								
C0830– 194	6		11.0						
C0831– 481	5.0		13.0						
C0832– 441	18		8.0	0.26	– 0.20	0.03	1.6	600	25
C0834– 434	2.5								
C0835– 297	11		11.0						
C0835– 399	67		9.0						
C0835– 394	2.0		11.0	0.46					
C0836– 345	3.0		13.0	0.86					
C0837+201	95	+33	6.0	0.36	+0.15	0.00	2.5	180	661
C0837– 438	11		13.0						
C0837– 460	1.5		9.0	0.21	– 0.25	0.41	0.3	1600	32
C0838– 378	2.0		15.0						
C0838– 459	2.2			0.19	– 0.20	0.48	0.6	1900	32
C0838– 528	50	+15	4.0		– 0.17	0.01	1.0	140	36
C0839– 385	2.5		13.0						
C0839– 480	7	+8			– 0.30	0.11	1.0	850	16
C0839– 461	2.0		10.0	0.60	– 0.20	0.75	0.4	1400	32
C0840– 432	5		12.0	0.40	– 0.15	0.47			63
C0840– 469	4.0		13.0	1.05	+0.28	0.38	1.2	2100	1585
C0840– 447	2.7		10.0	0.15			0.6	1450	
C0841– 479	12								
C0841– 324	12		12.0						
C0842– 357	10		14.0						
C0842– 411	17			0.31	– 0.25	0.58	2.5	1000	6
C0843– 314	5.0								
C0843– 474	2.2		14.0						
C0843– 458				0.14	– 0.36	0.86		5800	
C0843– 315	5								

Open Cluster Parameters

Name	R.A.(1950) h m s	Dec.(1950) ° ′	l °	b °	V_T (mag)	IAU Number
NGC 2669	08 43 24	- 52 47	270.71	- 06.27	6.1	C0843- 527
NGC 2670	08 43 54	- 48 36	267.47	- 03.61	7.8	C0843- 486
NGC 2671	08 44 24	- 41 42	262.15	+00.79	11.6	C0844- 417
vdB–Hagen 52	08 45 00	- 52 44	270.82	- 06.06		C0845- 527
Trumpler 10	08 46 00	- 42 18	262.81	+00.64	4.6	C0846- 423
Ruprecht 70	08 47 24	- 46 39	266.33	- 01.93		C0847- 466
M 67, NGC 2682	08 47 42	+12 00	215.58	+31.72	6.9	C0847+120
Ruprecht 71	08 47 48	- 46 37	266.35	- 01.85		C0847- 465
vdB–Hagen 54	08 48 00	- 44 14	264.53	- 00.31		C0848- 442
Ruprecht 72	08 50 12	- 37 25	259.55	+04.37		C0850- 374
Ruprecht 158	08 50 24	- 37 23	259.55	+04.42		C0850- 373
vdB–Hagen 55	08 54 12	- 39 20	261.52	+03.73		C0854- 393
vdB–Hagen 56	08 55 30	- 43 01	264.48	+01.51		C0855- 430
Muzzio 1	08 55 30	- 47 35	267.93	- 01.47		C0855- 475
Markarian 18	08 58 54	- 48 47	269.21	- 01.85	7.8	C0858- 487
Ruprecht 73	08 59 42	- 50 44	270.76	- 03.03		C0859- 507
vdB–Hagen 58	09 08 48	- 56 05	275.64	- 05.63		C0908- 560
NGC 2818	09 14 00	- 36 24	261.98	+08.59	8.2	C0914- 364
vdB–Hagen 60	09 14 06	- 49 48	271.64	- 00.71		C0914- 498
Pismis 11	09 14 06	- 49 48	271.53	- 00.40		C0915- 495
NGC 2849	09 17 24	- 40 20	265.27	+06.33	12.5	C0917- 403
Pismis 12	09 18 06	- 44 55	268.64	+03.20	9.7	C0918- 449
vdB–Hagen 63	09 18 24	- 49 02	271.59	+00.32		C0918- 490
Ruprecht 74	09 19 00	- 36 51	263.00	+09.00		C0919- 368
Ruprecht 159	09 19 00	- 60 11	279.55	- 07.50		C0919- 601
Pismis 13	09 20 36	- 50 55	273.15	- 00.78	10.2	C0920- 509
Ruprecht 75	09 20 42	- 56 04	276.79	- 04.43		C0920- 560
Ruprecht 76	09 22 30	- 51 31	273.80	- 00.99	10.8	C0922- 515
vdB–Hagen 66	09 23 36	- 54 34	276.04	- 03.07		C0923- 545
vdB–Hagen 67	09 25 00	- 51 06	273.79	- 00.42		C0925- 511
Ruprecht 77	09 25 30	- 54 54	276.47	- 03.12	10.4	C0925- 549
IC 2488	09 26 06	- 56 46	277.81	- 04.40	7.4	C0926- 567
Pismis 14	09 27 36	- 52 30	275.03	- 01.17		C0927- 525
Ruprecht 78	09 27 36	- 53 27	275.69	- 01.85		C0927- 534
NGC 2910	09 28 42	- 52 41	275.29	- 01.18	7.2	C0928- 526

Open Cluster Parameters

IAU Number	θ (′)	V_r (km s^{-1})	MBS	B–V (mag)	TOC	E_{B-V}	R (pc)	D_o (pc)	Age ($\times 10^6$)
C0843–527	12			0.25	– 0.20	0.19	1.8	1000	63
C0843–486	9		13.0	0.51	– 0.20	0.48	1.3	1000	96
C0844–417	4.0								
C0845–527	6								
C0846–423	14	+19			– 0.20	0.06	0.9	380	47
C0847–466	5		13.0						
C0847+120	29	+32	9.0	0.70	+0.50	0.08	3.2	720	3981
C0847–465	6		11.0						
C0848–442	3.0								
C0850–374	1.4		13.0						
C0850–373	2.5		12.0						
C0854–393	4.0								
C0855–430	12								
C0855–475								1700	5
C0858–487	2.0			0.50	– 0.30	0.77	0.4	1600	1
C0859–507	3.6		14.0						
C0908–560	3.0								
C0914–364	9			0.61	+0.10	0.22	4.2	3200	1000
C0914–498	2.5								
C0915–495	2.0		12.0						
C0917–403	2.3								
C0918–449	4.5		12.0	0.79					
C0918–490	1.5								
C0919–368	2.2		13.0						
C0919–601	0.7		15.0						
C0920–509	2.0		12.0	0.57	– 0.08	0.67	0.8	2600	63
C0920–560	4.0		12.0						
C0922–515	5		13.0	0.39	– 0.05	0.38			200
C0923–545	3.5								
C0925–511	4.0								
C0925–549	2.0		14.0	0.52	– 0.21	0.65	1.5	5200	
C0926–567	14		10.0						
C0927–525	0.6		14.0						
C0927–534	0.6		15.0		– 0.19	0.61	0.3	3500	
C0928–526	5.0			0.29	– 0.02	0.07	0.9	1320	40

Open Cluster Parameters

Name	R.A.(1950) h m s	Dec.(1950) ° ′	l °	b °	V_T (mag)	IAU Number
vdB–Hagen 72	09 29 18	– 52 46	275.41	– 01.18		C0929– 527
vdB–Hagen 73	09 30 24	– 49 59	273.65	+00.98		C0930– 499
Basel 11	09 55 12	+21 58	187.44	– 01.11	8.9	C0555+219
Basel 20	09 31 00	– 56 06	277.86	– 03.46		C0931– 561
NGC 2925	09 32 00	– 53 13	276.02	– 01.24	8.3	C0932– 532
Pismis 15	09 32 54	– 47 54	272.55	+02.80		C0932– 479
vdB–Hagen 75	09 33 12	– 54 30	277.01	– 02.07		C0933– 545
NGC 2972	09 38 30	– 50 06	274.69	+01.77	9.9	C0938– 501
Ruprecht 79	09 39 18	– 53 36	277.09	– 00.80	9.2	C0939– 536
Ruprecht 80	09 40 18	– 43 48	270.80	+06.73		C0940– 438
vdB–Hagen 78	09 42 06	– 56 20	279.17	– 02.62		C0942– 563
vdB–Hagen 79	09 42 24	– 53 05	277.10	– 00.11		C0942– 530
Ruprecht 81	09 43 30	– 43 54	271.30	+07.02		C0943– 439
Ruprecht 82	09 43 54	– 53 45	277.70	– 00.47	8.1	C0943– 537
NGC 3033	09 47 06	– 56 11	279.60	– 02.06	8.8	C0947– 561
Ruprecht 83	09 47 30	– 54 20	278.48	– 00.59	9.8	C0947– 543
Ruprecht 84	09 47 54	– 65 02	285.33	– 08.83		C0947– 650
Pismis 16	09 49 18	– 52 57	277.82	+00.66	8.0	C0949– 529
Hogg 2	09 49 54	– 56 04	279.83	– 01.72		C0949– 560
Collinder 213	09 52 48	– 50 29	276.72	+02.93	9.2	C0952– 504
Ruprecht 160	09 53 36	– 46 47	274.53	+05.92		C0953– 467
Hogg 3	09 56 06	– 54 25	279.51	+00.12		C0956– 544
Hogg 4	09 56 12	– 54 21	279.48	+00.18		C0956– 543
NGC 3105	09 59 00	– 54 32	279.92	+00.28	9.7	C0959– 545
vdB–Hagen 84	09 59 54	– 57 56	282.06	– 02.36		C0959– 579
Ruprecht 85	09 59 54	– 54 48	280.19	+00.15		C0959– 548
vdB–Hagen 85	09 60 00	– 49 24	276.97	+04.49		C1000– 494
Ruprecht 86	09 60 00	– 59 14	282.85	– 03.40		C1000– 592
NGC 3114	10 01 06	– 59 52	283.34	– 03.83	4.2	C1001– 598
vdB–Hagen 87	10 02 30	– 55 12	280.72	+00.05		C1002– 552
Schuster 1	10 02 48	– 55 36	280.99	– 00.25		C1002– 556
Trumpler 11	10 03 24	– 61 22	284.46	– 04.85	8.1	C1003– 613
vdB–Hagen 88	10 04 30	– 51 20	278.69	+03.35		C1004– 513
Hogg 5	10 04 36	– 60 08	283.85	– 03.78		C1004– 601
Hogg 6	10 04 48	– 60 15	283.94	– 03.86		C1004– 602

Open Cluster Parameters

IAU Number	θ (′)	V_r (km s^{-1})	MBS	B–V (mag)	TOC	E_{B-V}	R (pc)	D_o (pc)	Age ($\times 10^6$)
C0929– 527	3.0								
C0930– 499	1.5								
C0555+219	9			0.56			2.0	1500	200
C0931– 561	10				– 0.11	0.56	5.7	3870	
C0932– 532	12				– 0.15	0.08	1.4	810	79
C0932– 479	4.5		14.0						
C0933– 545	5.0								
C0938– 501	4.0			0.82	+0.05	0.35	0.7	1200	398
C0939– 536	11		11.0	0.78	– 0.20	0.72	2.2	1370	13
C0940– 438	12		13.0						
C0942– 563	1.5								
C0942– 530	2.5								
C0943– 439	5.0		12.0						
C0943– 537	3.6		12.0	0.45	+0.00	0.29	1.3	2500	
C0947– 561	5.0			0.44	+0.05	0.33	0.8	1100	398
C0947– 543	3.4		12.0	0.46	– 0.12	0.47	1.1	2240	
C0947– 650	3.6		11.0						
C0949– 529	1.5		9.0	0.35	– 0.26	0.61	0.8	3700	6
C0949– 560	5								
C0952– 504	17								
C0953– 467	1.7		14.0						
C0956– 544	4.0								
C0956– 543	4.0								
C0959– 545	2.0			1.03	– 0.23	1.09	2.3	8000	13
C0959– 579	4.5								
C0959– 548	1.7		15.0						
C1000– 494	5.0								
C1000– 592	12		12.0						
C1001– 598	35	0	9.0	0.27	– 0.15	0.06	4.6	900	107
C1002– 552	3.0								
C1002– 556	1.2								
C1003– 613	5			0.11					
C1004– 513	3.0								
C1004– 601	3.0								
C1004– 602	3.0								

Open Cluster Parameters

Name	R.A.(1950) h m s	Dec.(1950) ° ′	l °	b °	V_T (mag)	IAU Number
Trumpler 12	10 04 48	- 60 04	283.83	- 03.70	8.8	C1004- 600
Ruprecht 161	10 07 12	- 60 57	284.59	- 04.26		C1007- 609
Loden 27	10 09 18	- 56 21	282.17	- 00.34		C1009- 563
vdB-Hagen 90	10 10 12	- 57 50	283.11	- 01.48	10.3	C1010- 578
Ruprecht 87	10 13 30	- 50 28	279.36	+04.88		C1013- 504
vdB-Hagen 91	10 15 30	- 58 28	284.05	- 01.62		C1015- 584
vdB-Hagen 92	10 17 06	- 56 11	282.97	+00.41		C1017- 561
Ruprecht 88	10 17 48	- 62 50	286.68	- 05.11		C1017- 628
NGC 3228	10 19 48	- 51 28	280.74	+04.58	6.0	C1019- 514
Trumpler 13	10 22 00	- 59 50	285.50	- 02.31	11.3	C1022- 598
Westerlund 2	10 22 06	- 57 30	284.26	- 00.33	10.5	C1022- 575
NGC 3247	10 24 06	- 57 41	284.59	- 00.35	7.6	C1024- 576
NGC 3255	10 24 42	- 60 25	286.08	- 02.64	11.0	C1024- 604
IC 2581	10 25 30	- 57 23	284.60	+00.01	4.3	C1025- 573
Ruprecht 89	10 26 42	- 57 55	285.00	- 00.36		C1026- 579
Loden 143	10 27 06	- 58 32	285.37	- 00.87		C1027- 585
Hogg 7	10 27 18	- 60 28	286.38	- 02.51		C1027- 604
Loden 59	10 28 18	- 53 53	283.11	+03.21		C1028- 538
Collinder 223	10 28 42	- 59 34	286.19	- 01.87	9.4	C1028- 595
Ruprecht 90	10 29 06	- 57 59	285.31	- 00.26		C1029- 579
Loden 112	10 30 24	- 56 25	284.67	+01.18		C1030- 564
Bochum 9	10 33 54	- 59 52	286.79	- 01.58	6.3	C1033- 598
NGC 3293	10 33 54	- 57 58	285.86	+00.07	4.7	C1033- 579
Loden 165	10 35 00	- 58 29	285.93	- 00.36		C1035- 584
NGC 3324	10 35 24	- 58 22	286.22	- 00.17	6.7	C1035- 583
vdB-Hagen 99	10 36 00	- 58 56	286.56	- 00.63		C1036- 589
NGC 3330	10 36 36	- 53 53	284.18	+03.83	7.4	C1036- 538
Bochum 10	10 40 18	- 58 53	287.03	- 00.32	6.2	C1040- 588
Melotte 101	10 40 18	- 64 50	289.85	- 05.56	8.0	C1040- 648
Collinder 228	10 41 06	- 59 45	287.52	- 01.03	4.4	C1041- 597
θ Car, IC 2602	10 41 24	- 64 08	289.60	- 04.90	1.9	C1041- 641
Trumpler 14	10 42 00	- 59 18	287.42	- 00.58	5.5	C1041- 593
Trumpler 15	10 42 48	- 59 06	287.40	- 00.36	7.0	C1042- 591
Collinder 232	10 42 54	- 59 18	287.51	- 00.53	6.8	C1042- 593
η Car, Trumpler 16	10 43 12	- 59 27	287.61	- 00.65	5.0	C1043- 594

Open Cluster Parameters

IAU Number	θ (′)	V_r (km s^{-1})	MBS	B–V (mag)	TOC	E_{B-V}	R (pc)	D_o (pc)	Age ($\times 10^6$)
C1004– 600	4.0			0.30					
C1007– 609	32		11.0						
C1009– 563									
C1010– 578	4.0			0.51	– 0.10	0.51			126
C1013– 504	2.2		11.0						
C1015– 584	5.0								
C1017– 561	1.5								
C1017– 628	5		12.0						
C1019– 514	18			0.15	– 0.10	0.03	1.3	500	42
C1022– 598	5.0								
C1022– 575	1.5			1.09	– 0.35	1.68	1.1	5000	1
C1024– 576	6			0.39	– 0.08	0.25	1.5	1400	50
C1024– 604	2.0			0.54	– 0.05	0.25			200
C1025– 573	7	– 6		0.49	– 0.25	0.41	2.0	1660	10
C1026– 579	2.0		13.0						
C1027– 585									
C1027– 604	4.0								
C1028– 538									
C1028– 595	9								
C1029– 579	9		12.0						
C1030– 564									
C1033– 598				0.23					
C1033– 579	5	– 13	8.0	0.14	– 0.29	0.32v	2.3	2600	25
C1035– 584									
C1035– 583	5	– 12		0.21	– 0.35	0.46v	2.9	3300	2
C1036– 589	14								
C1036– 538	6			0.15	– 0.15	0.18	1.4	1390	50
C1040– 588				0.20	– 0.32	0.35		3600	7
C1040– 648	13			0.50	– 0.15	0.48v	4.3	2100	63
C1041– 597	14			0.18	– 0.36	0.37v	5.7	2600	1
C1041– 641	50	+22	3.0		– 0.22	0.04	1.1	155	10
C1041– 593	5.0	– 14		0.20	– 0.34	0.56v	2.1	2900	10
C1042– 591	3.0			0.41	– 0.26	0.46	1.1	2600	6
C1042– 593	4.0								
C1043– 594	10			0.35	– 0.33	0.49v	4.2	2900	10

Open Cluster Parameters

Name	R.A.(1950) h m s	Dec.(1950) ° ′	l °	b °	V_T (mag)	IAU Number
Bochum 11	10 45 18	− 59 50	288.03	− 00.87	7.9	C1045− 598
Ruprecht 91	10 45 42	− 57 13	286.89	+01.48		C1045− 572
vdB–Hagen 106	10 50 36	− 53 59	286.06	+04.69		C1050− 539
Ruprecht 162	10 51 06	− 62 01	289.64	− 02.51		C1051− 620
Ruprecht 92	10 51 54	− 61 28	289.49	− 01.97	8.6	C1051− 614
Graham 1	10 54 12	− 62 45	290.28	− 03.01		C1054− 627
Trumpler 17	10 54 12	− 58 57	288.66	+00.43	8.4	C1054− 589
Collinder 236	10 55 00	− 60 46	289.52	− 01.18	7.7	C1055− 607
Bochum 12	10 55 24	− 61 28	289.86	− 01.79	9.7	C1055− 614
Hogg 9	10 56 18	− 58 47	288.84	+00.69	10.6	C1056− 587
NGC 3496	10 57 48	− 60 04	289.55	− 00.40	8.2	C1057− 600
Sher 1	10 58 00	− 60 06	289.58	− 00.42	8.8	C1058− 601
Pismis 17	10 59 00	− 59 33	289.48	+00.14	9.4	C1059− 595
Ruprecht 93	11 02 18	− 61 06	290.47	− 01.12	7.7	C1102− 611
Ruprecht 163	11 02 54	− 67 40	293.15	− 07.11		C1102− 676
Feinstein 1	11 03 54	− 59 33	290.04	+00.38	4.7	C1103− 595
NGC 3532	11 04 18	− 58 24	289.64	+01.46	3.0	C1104− 584
vdB–Hagen 110	11 05 12	− 61 12	290.83	− 01.07		C1105− 612
Loden 309	11 07 24	− 60 16	290.72	− 00.11		C1107− 602
Loden 280	11 07 24	− 58 50	290.18	+01.22		C1107− 588
vdB–Hagen 111	11 07 24	− 63 35	291.98	− 03.17		C1107− 635
Loden 282	11 08 12	− 58 46	290.24	+01.32	7.7	C1108− 587
NGC 3572	11 08 18	− 59 58	290.71	+00.21	6.6	C1108− 599
Hogg 10	11 08 36	− 60 06	290.80	+00.10	6.9	C1108− 601
Collinder 240	11 09 06	− 60 01	290.82	+00.20	3.9	C1109− 600
Trumpler 18	11 09 18	− 60 24	290.99	− 00.14	6.9	C1109− 604
Hogg 11	11 09 24	− 60 06	290.89	+00.14	8.1	C1109− 601
Hogg 12	11 10 12	− 60 29	291.12	− 00.18	8.8	C1110− 604
NGC 3590	11 10 48	− 60 31	291.21	− 00.18	8.2	C1110− 605
Stock 13	11 10 54	− 58 39	290.52	+01.55	7.0	C1110− 586
Trumpler 19	11 12 06	− 57 19	290.20	+02.86	9.6	C1112− 573
NGC 3603	11 12 54	− 60 59	291.62	− 00.52	9.1	C1112− 609
Hogg 13	11 14 06	− 60 00	291.39	+00.45		C1114− 600
IC 2714	11 15 42	− 62 26	292.44	− 01.76	8.2	C1115− 624
Melotte 105	11 17 18	− 63 14	292.89	− 02.45	8.5	C1117− 632

Open Cluster Parameters

IAU Number	θ (′)	V_r (km s^{-1})	MBS	B–V (mag)	TOC	E_{B-V}	R (pc)	D_o (pc)	Age ($\times 10^6$)
C1045– 598				0.26	– 0.30	0.54		3600	1
C1045– 572	1.7		10.0						
C1050– 539	5.0								
C1051– 620	4.5		12.0						
C1051– 614	2.2		12.0	0.68	– 0.21	0.44	0.9	2700	126
C1054– 627									
C1054– 589	5.0			0.53	– 0.28	0.67	1.0	1400	35
C1055– 607	7								
C1055– 614				0.42	– 0.20	0.24			25
C1056– 587	1.5			0.34					
C1057– 600	9			0.82	+0.00	0.50	1.5	1100	229
C1058– 601				0.34	– 0.35	1.45		3100	1
C1059– 595	0.6		9.0	0.23	– 0.30	0.51	0.3	4200	1
C1102– 611	4.0		11.0	0.35	– 0.10	0.27			200
C1102– 676	1.7		12.0						
C1103– 595		+1		0.14	– 0.30	0.40			1
C1104– 584	55	+7	8.0	0.28	– 0.01	0.04	4.0	500	347
C1105– 612	2.0								
C1107– 602									
C1107– 588									
C1107– 635	5								
C1108– 587				0.28	– 0.09	0.16			
C1108– 599	6	– 8	7.0	0.22	– 0.29	0.45v	2.3	2300	13
C1108– 601	3.0			0.21	– 0.24	0.48	0.9	2200	2
C1109– 600	25			0.33	– 0.35	0.41			1
C1109– 604	12			0.23	– 0.15	0.28v	4.3	2500	25
C1109– 601	1.5			0.11	– 0.25	0.32	0.5	2300	6
C1110– 604	3.0			0.30					
C1110– 605	4.0			0.30	– 0.20	0.49v	1.1	1900	50
C1110– 586	3.0		10.0	0.02	– 0.30	0.24	1.2	2700	25
C1112– 573	10								
C1112– 609	2.5			0.95	– 0.34	1.38v	2.0	5300	1
C1114– 600	3.0								
C1115– 624	12		10.0		– 0.05	0.46	2.1	1200	200
C1117– 632	4.0			0.55	+0.00	0.38	1.2	2100	59

Open Cluster Parameters

Name	R.A.(1950) h m s	Dec.(1950) ° ′	l °	b °	V_T (mag)	IAU Number
Loden 336	11 17 36	- 58 36	291.34	+01.93		C1117- 586
vdB-Hagen 118	11 20 36	- 58 16	291.58	+02.36		C1120- 582
NGC 3680	11 23 18	- 42 58	286.77	+16.93	7.6	C1123- 429
Ruprecht 94	11 28 06	- 63 10	294.02	- 01.98		C1128- 631
Loden 372	11 28 24	- 58 13	292.53	+02.74		C1128- 582
Ruprecht 164	11 28 48	- 60 32	293.29	+00.55		C1128- 605
NGC 3766	11 33 48	- 61 20	294.11	- 00.03	5.3	C1133- 613
IC 2944	11 34 18	- 62 45	294.57	- 01.37	4.5	C1134- 627
vdB-Hagen 121	11 35 48	- 63 04	294.82	- 01.63		C1135- 630
Lynga 15	11 40 00	- 62 13	295.05	- 00.68		C1140- 622
Ruprecht 95	11 41 06	- 60 52	294.83	+00.65		C1141- 608
Stock 14	11 41 36	- 62 13	295.23	- 00.63	6.3	C1141- 622
NGC 3960	11 48 24	- 55 25	294.41	+06.18	8.3	C1148- 554
Ruprecht 96	11 48 42	- 61 48	295.94	- 00.03		C1148- 618
Loden 480	11 53 30	- 58 09	295.72	+03.66		C1153- 581
Ruprecht 97	11 54 48	- 62 22	296.76	- 00.42	9.1	C1154- 623
Ruprecht 98	11 55 30	- 64 12	297.22	- 02.20	7.0	C1155- 642
NGC 4052	11 59 18	- 62 55	297.38	- 00.85	8.8	C1159- 629
Ruprecht 99	12 00 30	- 63 32	297.63	- 01.44		C1200- 635
Ruprecht 100	12 03 06	- 62 16	297.69	- 00.14		C1203- 622
NGC 4103	12 04 06	- 60 58	297.57	+01.17	7.4	C1204- 609
Stock 15	12 04 18	- 59 13	297.30	+02.91		C1204- 592
Loden 565	12 06 00	- 60 38	297.78	+01.75		C1206- 606
Ruprecht 101	12 06 18	- 62 36	298.11	- 00.41		C1206- 626
Ruprecht 102	12 10 24	- 62 21	298.54	- 00.08		C1210- 623
Ruprecht 103	12 13 12	- 58 08	298.28	+04.14		C1213- 581
NGC 4230	12 14 36	- 54 51	298.03	+07.42	9.4	C1214- 548
Loden 615	12 17 54	- 64 30	299.66	- 02.10		C1217- 645
NGC 4337	12 21 12	- 57 51	299.30	+04.56	8.9	C1221- 578
NGC 4349	12 21 42	- 61 37	299.77	+00.82	7.4	C1221- 616
Ruprecht 104	12 22 00	- 60 10	299.65	+02.26		C1222- 601
Melotte 111	12 22 36	+26 23	221.20	+84.01	1.8	C1222+263
vdB-Hagen 131	12 23 48	- 63 09	300.16	- 00.69		C1223- 631
vdB-Hagen 132	12 24 12	- 63 48	300.27	- 01.33		C1224- 638
vdB-Hagen 133	12 24 24	- 60 30	299.97	+01.96		C1224- 605

Open Cluster Parameters

IAU Number	θ (′)	V_r (km s^{-1})	MBS	B–V (mag)	TOC	E_{B-V}	R (pc)	D_o (pc)	Age ($\times10^6$)
C1117– 586									
C1120– 582	2.5								
C1123– 429	12		10.0	0.80	+0.45	0.04	1.4	800	1820
C1128– 631	21		10.0						
C1128– 582									
C1128– 605	2.0		14.0						
C1133– 613	12	– 19	8.0	0.36	– 0.24	0.19	3.0	1700	22
C1134– 627	14			0.12	– 0.31	0.33	4.6	2100	10
C1135– 630	5								
C1140– 622									
C1141– 608	5		12.0						
C1141– 622	4.0	– 4	10.0	0.03	– 0.27	0.26	1.5	2600	6
C1148– 554	6			0.77	+0.20	0.29	1.7	1660	1000
C1148– 618	5		13.0						
C1153– 581									
C1154– 623	3.5		12.0	0.67	+0.07	0.23	2.0	4000	1000
C1155– 642	10		9.0	0.29	+0.00	0.17	0.6	400	316
C1159– 629	7						2.2	1900	138
C1200– 635	4.0		12.0						
C1203– 622	6		12.0						
C1204– 609	6		10.0		– 0.20	0.32	1.9	1860	22
C1204– 592	12		10.0						
C1206– 606									
C1206– 626	10		13.0						
C1210– 623	5.0		13.0						
C1213– 581	2.7		13.0						
C1214– 548	5								
C1217– 645									
C1221– 578	3.5			0.54					
C1221– 616	15		11.0	0.61	+0.00	0.33	4.0	1700	224
C1222– 601	3.6		13.0						
C1222+263	275	0	5.0	0.46	+0.05	0.00	3.5	86	398
C1223– 631	4.5								
C1224– 638	3.0								
C1224– 605	5								

Open Cluster Parameters

Name	R.A.(1950) h m s	Dec.(1950) ° ′	l °	b °	V_T (mag)	IAU Number
NGC 4439	12 25 36	− 59 49	300.07	+02.66	8.4	C1225− 598
Hogg 14	12 25 48	− 59 32	300.07	+02.94	9.5	C1225− 595
Ruprecht 165	12 25 54	− 56 11	299.77	+06.27		C1225− 561
Harvard 5	12 26 12	− 60 29	299.96	+01.97	7.1	C1226− 604
Hogg 23	12 26 18	− 60 39	300.22	+01.83		C1226− 606
NGC 4463	12 27 06	− 64 31	300.65	− 02.01	7.2	C1227− 645
Ruprecht 105	12 31 12	− 61 17	300.86	+01.24		C1231− 612
Upgren 1	12 32 36	+36 35	142.68	+80.18		C1232+365
Harvard 6	12 34 54	− 68 12	301.69	− 05.64	10.7	C1234− 682
Ruprecht 106	12 36 00	− 50 53	300.91	+11.66		C1236− 508
Trumpler 20	12 36 48	− 60 20	301.48	+02.24	10.1	C1236− 603
Coal–Sa., NGC 4609	12 39 24	− 62 42	301.90	− 00.11	6.9	C1239− 627
Hogg 15	12 40 36	− 62 50	302.04	− 00.24	10.3	C1240− 628
vdB–Hagen 140	12 50 00	− 66 55	303.10	− 04.32		C1250− 669
κ Cru, NGC 4755	12 50 36	− 60 04	303.21	+02.53	4.2	C1250− 600
NGC 4815	12 54 54	− 64 41	303.63	− 02.09	8.6	C1254− 646
NGC 4852	12 57 06	− 59 20	304.03	+03.25	8.9	C1257− 593
Loden 757	13 09 06	− 65 02	305.12	− 02.51		C1309− 650
Danks 1	13 09 18	− 62 26	305.35	+00.07		C1309− 624
Danks 2	13 09 42	− 62 25	305.39	+00.09		C1310− 624
vdB–Hagen 144	13 11 48	− 65 40	305.35	− 03.17		C1311− 656
Harvard 8	13 15 24	− 66 56	305.59	− 04.46	9.5	C1315− 669
Stock 16	13 15 48	− 62 18	306.11	+00.14	9.1	C1315− 623
Ruprecht 107	13 17 12	− 64 41	306.01	− 02.25	9.7	C1317− 646
Loden 807	13 21 36	− 62 10	306.79	+00.20	7.9	C1321− 621
Ruprecht 166	13 22 42	− 63 09	306.80	− 00.80	10.8	C1322− 631
NGC 5138	13 24 06	− 58 45	307.55	+03.55	7.6	C1324− 587
Basel 18	13 25 00	− 62 06	307.20	+00.20	8.2	C1325− 621
Hogg 16	13 26 00	− 60 57	307.47	+01.34	8.4	C1326− 609
Collinder 271	13 26 18	− 63 56	307.09	− 01.62	8.7	C1326− 639
Collinder 272	13 27 18	− 61 01	307.62	+01.25	7.7	C1327− 610
NGC 5168	13 27 54	− 60 41	307.74	+01.57	9.1	C1327− 606
Trumpler 21	13 28 48	− 62 32	307.57	− 00.30	7.7	C1328− 625
Ruprecht 108	13 28 54	− 58 14	308.25	+03.96	7.5	C1328− 582
Loden 848	13 30 36	− 64 16	307.50	− 02.03		C1330− 642

Open Cluster Parameters

IAU Number	θ (′)	V_r (km s^{-1})	MBS	B–V (mag)	TOC	E_{B-V}	R (pc)	D_o (pc)	Age ($\times 10^6$)
C1225−598	4.0			0.33	−0.16	0.34	0.9	1600	63
C1225−595	3.0			0.37	+0.00	0.28			316
C1225−561	21		7.0						
C1226−604	5			0.29	−0.17	0.18			40
C1226−606									
C1227−645	5.0			0.42	−0.20	0.44			25
C1231−612	12		12.0						
C1232+365	14						0.3	140	
C1234−682	5.0								
C1236−508	3.0		15.0						
C1236−603	7								
C1239−627	5.0		10.0	0.42	−0.20	0.36	1.1	1510	36
C1240−628	2.0			0.91	−0.35	1.16	1.2	4200	8
C1250−669	4.5								
C1250−600	10	−18	7.0	0.31	−0.23	0.39	3.4	2340	7
C1254−646	3.0			1.01					
C1257−593	11								
C1309−650									
C1309−624	0.8								
C1310−624	1.2								
C1311−656	1.5								
C1315−669	4.0			0.51	−0.10	0.36			126
C1315−623	3.0		10.0	0.54	−0.29	0.49	1.0	2300	28
C1317−646	5		12.0	0.69	−0.10	0.48			126
C1321−621				0.24					
C1322−631	2.8		13.0	0.88					
C1324−587	7			0.60	−0.15	0.27	2.1	1800	151
C1325−621	4.0			0.30	−0.13	0.15	0.9	1556	159
C1326−609	4.0			0.29	−0.17	0.41	0.3	603	25
C1326−639	6			0.40	−0.15	0.32	1.6	1600	63
C1327−610	9			0.55	−0.29	0.72	3.8	2900	2
C1327−606	4.0			0.45	−0.15	0.32	0.8	1400	159
C1328−625	4.0			0.08	−0.15	0.21	0.6	1100	63
C1328−582	12		9.0	0.23	−0.10	0.18	1.3	700	126
C1330−642									

Open Cluster Parameters

Name	R.A.(1950) h m s	Dec.(1950) ° ′	l °	b °	V_T (mag)	IAU Number
Collinder 275	13 31 18	– 59 52	308.29	+02.29	10.2	C1331– 598
Pismis 18	13 33 12	– 61 54	308.18	+00.26	9.7	C1333– 619
vdB–Hagen 150	13 34 24	– 63 05	308.11	– 00.93		C1334– 630
Loden 894	13 35 54	– 64 03	308.11	– 01.91		C1335– 640
vdB–Hagen 151	13 36 54	– 61 28	308.69	+00.60		C1336– 614
Loden 991	13 41 54	– 61 46	309.21	+00.20		C1341– 617
Loden 1010	13 42 00	– 60 01	309.58	+01.91		C1342– 600
NGC 5281	13 43 06	– 62 39	309.17	– 00.70	5.9	C1343– 626
Collinder 277	13 45 00	– 65 50	308.71	– 03.86	9.2	C1345– 658
NGC 5288	13 45 06	– 64 26	309.01	– 02.49	11.8	C1345– 644
Loden 1095	13 50 00	– 59 30	310.68	+02.19		C1350– 595
NGC 5316	13 50 24	– 61 37	310.23	+00.12	6.0	C1350– 616
Loden 1002	13 50 36	– 65 04	309.44	– 03.23		C1350– 650
vdB–Hagen 155	13 54 00	– 59 20	311.21	+02.23		C1354– 593
Loden 1101	13 54 42	– 61 32	310.74	+00.07		C1354– 615
Loden 1177	13 55 54	– 57 36	311.89	+03.84		C1355– 576
Loden 1152	13 56 00	– 59 08	311.51	+02.36		C1356– 591
Loden 1171	13 56 00	– 58 08	311.77	+03.32		C1356– 581
Lynga 1	13 56 36	– 61 57	310.86	– 00.38		C1356– 619
NGC 5381	13 57 06	– 59 19	311.60	+02.14		C1357– 593
Loden 1202	14 00 54	– 58 27	312.30	+02.84		C1400– 584
Loden 1194	14 02 12	– 59 28	312.18	+01.82		C1402– 594
Ruprecht 110	14 02 30	– 67 20	310.01	– 05.74		C1402– 673
NGC 5460	14 04 24	– 48 05	315.78	+12.65	5.6	C1404– 480
Loden 1225	14 05 36	– 59 29	312.59	+01.68		C1405– 594
Loden 1289	14 10 36	– 57 38	313.77	+03.25		C1410– 576
Loden 1282	14 12 18	– 58 55	313.58	+01.96		C1412– 589
Loden 1256	14 14 30	– 61 12	313.12	– 00.28		C1414– 612
Ruprecht 167	14 14 36	– 58 44	313.92	+02.04		C1414– 587
Lynga 2	14 20 18	– 61 10	313.78	– 00.49	6.4	C1420– 611
NGC 5593	14 22 24	– 54 35	316.35	+05.59		C1422– 545
NGC 5606	14 24 06	– 59 25	314.84	+00.99	7.7	C1424– 594
NGC 5617	14 26 00	– 60 30	314.67	– 00.10	6.3	C1426– 605
Pismis 19	14 26 54	– 60 46	314.68	– 00.38		C1426– 607
Trumpler 22	14 27 24	– 60 57	314.67	– 00.59	7.9	C1427– 609

Open Cluster Parameters

IAU Number	θ (′)	V_r (km s^{-1})	MBS	B–V (mag)	TOC	E_{B-V}	R (pc)	D_o (pc)	Age ($\times 10^6$)
C1331– 598	6								
C1333– 619	4.0		10.0	0.60					
C1334– 630	2.5								
C1335– 640									
C1336– 614	3.0								
C1341– 617									
C1342– 600									
C1343– 626	5.0		10.0	0.25	– 0.20	0.26	0.9	1300	51
C1345– 658	15								
C1345– 644	4.0								
C1350– 595									
C1350– 616	13		11.0	0.08	– 0.05	0.18	2.3	1120	195
C1350– 650									
C1354– 593	11								
C1354– 615									
C1355– 576									
C1356– 591									
C1356– 581									
C1356– 619	3.0								
C1357– 593	13		12.0						
C1400– 584									
C1402– 594									
C1402– 673	28		10.0						
C1404– 480	25		9.0	0.25	– 0.10	0.14	1.8	500	107
C1405– 594									
C1410– 576									
C1412– 589									
C1414– 612									
C1414– 587	13		11.0						
C1420– 611	12			0.25	– 0.20	0.19	1.9	1100	32
C1422– 545	7								
C1424– 594	3.0			0.25	– 0.24	0.49	0.8	1700	13
C1426– 605	10		10.0	0.70	– 0.13	0.53v	1.8	1200	46
C1426– 607	2.2		12.0						
C1427– 609	6		12.0	0.58	– 0.19	0.56	1.8	1700	110

Open Cluster Parameters

Name	R.A.(1950) h m s	Dec.(1950) ° ′	l °	b °	V_T (mag)	IAU Number
Loden 1339	14 29 36	- 61 47	314.60	- 01.46		C1429- 617
Hogg 17	14 29 54	- 61 10	314.87	- 00.90	8.3	C1429- 611
NGC 5662	14 31 36	- 56 20	316.90	+03.47	5.5	C1431- 563
Ruprecht 111	14 32 06	- 59 44	315.66	+00.31		C1432- 597
Loden 1373	14 37 48	- 62 39	315.14	- 02.63		C1437- 626
NGC 5715	14 39 42	- 57 20	317.53	+02.12	9.8	C1439- 573
vdB- Hagen 164	14 39 54	- 66 11	313.89	- 05.95		C1439- 661
Loden 1375	14 40 18	- 63 10	315.18	- 03.23		C1440- 631
Collinder 285	14 40 18	+69 47	109.87	+44.67		C1440+697
Loden 1409	14 40 48	- 61 30	315.93	- 01.74		C1440- 615
NGC 5749	14 45 18	- 54 19	319.50	+04.53	8.8	C1445- 543
Hogg 18	14 47 12	- 52 03	320.75	+06.43	8.0	C1447- 520
NGC 5764	14 50 00	- 52 29	320.97	+05.86	12.6	C1450- 524
Ruprecht 112	14 53 00	- 62 21	316.85	- 03.12		C1453- 623
NGC 5822	15 01 30	- 54 09	321.71	+03.58	6.5	C1501- 541
NGC 5823	15 02 00	- 55 24	321.16	+02.46	7.9	C1502- 554
Pismis 20	15 11 30	- 58 53	320.52	- 01.21	7.8	C1511- 588
Lynga 3	15 12 36	- 58 08	321.04	- 00.64		C1512- 581
Pismis 21	15 12 48	- 59 28	320.36	- 01.79		C1512- 594
NGC 5925	15 23 54	- 54 21	324.40	+01.70	8.4	C1523- 543
Lynga 4	15 29 30	- 55 03	324.67	+00.67	11.4	C1529- 550
Harvard 9	15 30 00	- 53 26	325.66	+01.97		C1530- 534
vdB- Hagen 176	15 35 48	- 49 54	328.45	+04.30		C1535- 499
Lynga 5	15 38 06	- 56 28	324.82	- 01.19		C1538- 564
Collinder 292	15 46 42	- 57 31	325.12	- 02.72	7.9	C1546- 575
NGC 5999	15 48 12	- 56 19	326.02	- 01.93	9.0	C1548- 563
NGC 6005	15 51 48	- 57 17	325.79	- 02.99	10.7	C1551- 572
Ruprecht 113	15 53 00	- 59 19	324.60	- 04.66		C1553- 593
Trumpler 23	15 56 36	- 53 23	328.82	- 00.43	11.2	C1556- 533
Moffat 1	15 57 36	- 54 00	328.53	- 01.00		C1557- 540
NGC 6025	15 59 24	- 60 22	324.54	- 05.97	5.1	C1559- 603
Lynga 6	16 01 00	- 51 47	330.37	+00.34	9.5	C1601- 517
Ruprecht 114	16 02 36	- 56 47	327.23	- 03.56		C1602- 567
NGC 6031	16 03 42	- 53 56	329.25	- 01.53	8.5	C1603- 539
Lynga 7	16 07 00	- 55 10	328.78	- 02.77		C1607- 551

Open Cluster Parameters

IAU Number	θ (′)	V_r (km s^{-1})	MBS	B–V (mag)	TOC	E_{B-V}	R (pc)	D_o (pc)	Age (×10^6)
C1429– 617									
C1429– 611	6			0.57	– 0.20	0.54	1.8	1700	182
C1431– 563	12		10.0	0.68	– 0.14	0.31			63
C1432– 597	8		13.0						
C1437– 626									
C1439– 573	5		11.0						
C1439– 661	29								
C1440– 631									
C1440+697			2.0		– 0.05	0.00		20	162
C1440– 615									
C1445– 543	7				– 0.15	0.45	1.0	900	91
C1447– 520	3.0			0.52	– 0.15	0.51	0.5	1100	50
C1450– 524	2.0						0.3	1000	
C1453– 623	8		14.0						
C1501– 541	39		10.0		+0.23	0.19	4.4	760	891
C1502– 554	10		13.0	0.71	+0.11	0.39	1.9	1260	200
C1511– 588	4.5			0.89	– 0.35	1.18	2.9	4400	2
C1512– 581	5.0								
C1512– 594	2.0		13.0						
C1523– 543	14								
C1529– 550	3.0			1.17					
C1530– 534	3.0								
C1535– 499	3.0								
C1538– 564	5								
C1546– 575	15								
C1548– 563	5.0		12.0						
C1551– 572	3.0			1.20					
C1553– 593	45		9.0						
C1556– 533	5.0								
C1557– 540									
C1559– 603	12	– 3	7.0	0.16	– 0.16	0.16	1.5	840	107
C1601– 517	5.0	– 59.4		0.78	– 0.21	1.34	1.1	1600	40
C1602– 567	8		12.0						
C1603– 539	2.0			0.42	– 0.10	0.38	0.4	1590	22
C1607– 551	2.2								

Open Cluster Parameters

Name	R.A.(1950) h m s	Dec.(1950) ° ′	l °	b °	V_T (mag)	IAU Number
Ruprecht 115	16 09 00	- 52 15	330.97	- 00.83		C1609- 522
NGC 6067	16 09 18	- 54 05	329.76	- 02.20	5.6	C1609- 540
Pismis 22	16 09 36	- 51 47	331.36	- 00.55		C1609- 517
Ruprecht 176	16 11 00	- 51 12	331.92	- 00.28		C1611- 512
NGC 6087	16 14 42	- 57 47	327.76	- 05.40	5.4	C1614- 577
Harvard 10	16 15 54	- 54 52	329.91	- 03.42		C1615- 548
Lynga 8	16 15 54	- 50 06	333.24	- 00.02		C1615- 501
Lynga 9	16 16 48	- 48 26	334.51	+01.07		C1616- 484
Ruprecht 116	16 19 42	- 51 53	332.41	- 01.71		C1619- 518
Ruprecht 117	16 19 48	- 51 46	332.50	- 01.63		C1619- 517
Pismis 23	16 20 06	- 48 48	334.64	+00.43		C1620- 488
Ruprecht 118	16 20 42	- 51 51	332.54	- 01.79	9.8	C1620- 518
NGC 6124	16 22 12	- 40 33	340.77	+05.96	5.8	C1622- 405
Grasdalen 1	16 22 54	- 24 20	352.96	+16.96		C1622- 243
Collinder 302	16 23 00	- 26 07	351.59	+15.76	1.0	C1623- 261
NGC 6134	16 24 00	- 49 02	334.92	- 00.19	7.2	C1624- 490
Ruprecht 119	16 24 30	- 51 24	333.28	- 01.90	8.8	C1624- 514
Hogg 19	16 25 06	- 49 01	335.06	- 00.31		C1625- 490
NGC 6152	16 28 48	- 52 31	332.93	- 03.14	8.1	C1628- 525
NGC 6169	16 30 30	- 43 57	339.38	+02.51	6.6	C1630- 439
NGC 6167	16 30 36	- 49 30	335.32	- 01.28	6.7	C1630- 495
Collinder 307	16 31 24	- 50 52	334.42	- 02.31	9.2	C1631- 508
Ruprecht 120	16 31 42	- 48 12	336.40	- 00.54		C1631- 482
NGC 6178	16 32 06	- 45 32	338.40	+01.23	7.2	C1632- 455
Dolidze 27	16 33 48	- 08 51	007.62	+24.66		C1633- 088
Lynga 11	16 34 24	- 46 13	338.17	+00.47		C1634- 462
NGC 6192	16 36 48	- 43 16	340.65	+02.12	8.5	C1636- 432
NGC 6193	16 37 30	- 48 40	336.70	- 01.57	5.2	C1637- 486
Ruprecht 121	16 38 24	- 46 05	338.73	+00.04		C1638- 460
NGC 6200	16 40 30	- 47 23	338.00	- 01.09	7.4	C1640- 473
Hogg 20	16 40 42	- 47 28	337.96	- 01.17		C1640- 474
Lynga 12	16 42 06	- 50 42	335.67	- 03.46		C1642- 507
Hogg 21	16 42 12	- 47 40	337.97	- 01.49		C1642- 476
NGC 6204	16 42 48	- 46 56	338.59	- 01.08	8.2	C1642- 469
Hogg 22	16 43 00	- 47 01	338.55	- 01.16	6.7	C1643- 470

Open Cluster Parameters

IAU Number	θ (′)	V_r (km s^{-1})	MBS	B–V (mag)	TOC	E_{B-V}	R (pc)	D_o (pc)	Age (×10^6)
C1609– 522	5		13.0						
C1609– 540	12	– 39.9	10.0	0.61	– 0.09	0.32	4.0	2100	78
C1609– 517	4.0		13.0						
C1611– 512									
C1614– 577	12	+1	8.0	0.45	– 0.16	0.18	1.5	900	55
C1615– 548	29				– 0.04	0.36			
C1615– 501	1.0								
C1616– 484	5								
C1619– 518	5.0		9.0						
C1619– 517	1.7		12.0						
C1620– 488	1.0		15.0						
C1620– 518	3.4		11.0	0.49	– 0.10	0.41			126
C1622– 405	29		9.0	0.90	– 0.05	0.68	2.1	490	51
C1622– 243									
C1623– 261	505								
C1624– 490	6		11.0	0.69	+0.00	0.45	0.8	790	631
C1624– 514	8		10.0	0.44					
C1625– 490	4.0								
C1628– 525	29		11.0				4.5	1030	
C1630– 439	6	+6					1.1	1100	
C1630– 495	7			0.88	– 0.15	0.89v	1.4	1200	40
C1631– 508	5			1.02					
C1631– 482	3.4		12.0						
C1632– 455	4.0			0.08	– 0.25	0.24			6
C1633– 088	23								
C1634– 462	4.0								
C1636– 432	7		11.0						
C1637– 486	14			0.14	– 0.32	0.46v	3.0	1350	1
C1638– 460	8		13.0						
C1640– 473	12			0.34	– 0.30	0.63	4.2	2400	1
C1640– 474	4.0								
C1642– 507	5								
C1642– 476	4.0								
C1642– 469	5.0	– 39		0.31	– 0.18	0.47v	1.9	2600	13
C1643– 470	1.5			0.36	– 0.30	0.66	0.6	2800	1

Open Cluster Parameters

Name	R.A.(1950) h m s	Dec.(1950) ° ′	l °	b °	V_T (mag)	IAU Number
vdB–Hagen 197	16 43 12	– 45 44	339.55	– 00.36		C1643– 457
Dol–Dzim 6	16 43 36	+38 22	061.49	+40.38		C1643+383
Westerlund 1	16 44 36	– 45 45	339.56	– 00.39		C1644– 457
NGC 6208	16 45 30	– 53 44	333.69	– 05.82	7.2	C1645– 537
Lynga 13	16 45 42	– 43 21	341.65	+00.85		C1645– 433
NGC 6216	16 45 48	– 44 39	340.67	+00.01	10.1	C1645– 446
NGC 6222	16 46 06	– 44 39	340.70	– 00.04		C1646– 446
vdB–Hagen 200	16 46 24	– 44 08	341.13	+00.25		C1646– 441
NGC 6231	16 50 30	– 41 43	343.47	+01.22	2.6	C1650– 417
Lynga 14	16 51 36	– 45 14	340.88	– 01.16	9.7	C1651– 452
vdB–Hagen 202	16 51 48	– 40 52	344.28	+01.57		C1651– 408
Collinder 316	16 52 00	– 40 45	344.39	+01.62	3.4	C1652– 407
NGC 6242	16 52 12	– 39 25	345.46	+02.43	6.4	C1652– 394
vdB–Hagen 205	16 52 36	– 40 34	344.61	+01.65		C1652– 405
Trumpler 24	16 53 30	– 40 35	344.70	+01.51	8.6	C1653– 405
NGC 6249	16 54 00	– 44 42	341.56	– 01.16	8.2	C1654– 447
NGC 6250	16 54 18	– 45 52	340.79	– 01.83	5.9	C1654– 457
NGC 6253	16 55 06	– 52 38	335.46	– 06.25	10.2	C1655– 526
vdB–Hagen 208	16 56 12	– 37 02	347.81	+03.31		C1656– 370
NGC 6259	16 57 06	– 44 36	341.98	– 01.52	8.0	C1657– 446
vdB–Hagen 211	16 58 42	– 41 00	344.99	+00.47		C1658– 410
NGC 6268	16 58 54	– 39 40	346.04	+01.31	9.5	C1658– 396
NGC 6281	17 01 24	– 37 50	347.82	+02.01	5.4	C1701– 378
Harvard 13	17 01 36	– 48 07	339.69	– 04.30		C1701– 481
vdB–Hagen 214	17 02 18	– 36 36	348.90	+02.61		C1702– 366
Dol–Dzim 7	17 08 18	+15 36	036.30	+29.18		C1708+156
vdB–Hagen 217	17 12 36	– 40 46	346.76	– 01.49		C1712– 407
vdB–Hagen 218	17 12 42	– 39 22	347.90	– 00.68		C1712– 393
Bochum 13	17 14 00	– 35 30	351.19	+01.37	7.2	C1714– 355
NGC 6318	17 14 18	– 39 24	348.06	– 00.96	11.8	C1714– 394
NGC 6322	17 14 54	– 42 54	345.27	– 03.07	6.0	C1714– 429
vdB–Hagen 221	17 15 12	– 32 18	353.94	+03.02		C1715– 323
vdB–Hagen 222	17 15 24	– 38 14	349.13	– 00.45		C1715– 382
Hav–Moffat 1	17 15 30	– 38 47	348.69	– 00.78		C1715– 387
vdB–Hagen 223	17 17 12	– 35 50	351.30	+00.64		C1717– 358

Open Cluster Parameters

IAU Number	θ (′)	V_r (km s^{-1})	MBS	B–V (mag)	TOC	E_{B-V}	R (pc)	D_o (pc)	Age ($\times10^6$)
C1643– 457	3.0								
C1643+383	17								
C1644– 457	2.0						0.4	1400	
C1645– 537	15			0.86	+0.25	0.18	2.3	1000	1000
C1645– 433	7								
C1645– 446	4.0		12.0						
C1646– 446	4.0								
C1646– 441	3.5								
C1650– 417	14	– 23	6.0	0.23	– 0.30	0.46	4.3	2000	3
C1651– 452	2.0			1.15	– 0.35	1.48	0.6	2300	1
C1651– 408	3.0								
C1652– 407	105		14.0						
C1652– 394	9			0.69	– 0.17	0.39	1.5	1200	51
C1652– 405	4.0								
C1653– 405	60				– 0.25	0.37	13.9	1600	1
C1654– 447	6			0.52	– 0.20	0.45			25
C1654– 457	7			0.24	– 0.23	0.38	1.2	1020	14
C1655– 526	5.0		13.0						
C1656– 370	1.5								
C1657– 446	10		11.0	0.85	– 0.05	0.65	1.1	770	224
C1658– 410	4.0								
C1658– 396	6				– 0.20	0.41	1.0	1100	25
C1701– 378	8		9.0	0.38	– 0.05	0.15	0.8	600	224
C1701– 481	14								
C1702– 366	3.0								
C1708+156	20								
C1712– 407	4.0								
C1712– 393	5.0								
C1714– 355				0.64	– 0.25	0.88		1700	6
C1714– 394	4.0		12.0						
C1714– 429	10			0.32	– 0.35	0.65	1.8	1200	10
C1715– 323	10								
C1715– 382	2.5								
C1715– 387									
C1717– 358	5.0								

Open Cluster Parameters

Name	R.A.(1950) h m s	Dec.(1950) ° ′	l °	b °	V_T (mag)	IAU Number
NGC 6334	17 18 00	- 36 02	351.23	+00.40		C1718- 360
Ruprecht 123	17 20 00	- 37 53	349.94	- 00.99		C1720- 378
IC 4651	17 20 48	- 49 54	340.07	- 07.88	6.9	C1720- 499
Trumpler 25	17 21 18	- 38 57	349.20	- 01.80	11.7	C1721- 389
Pismis 24	17 22 06	- 34 23	353.12	+00.71	9.6	C1722- 343
Dol–Dzim 8	17 24 06	+24 14	047.02	+28.81		C1724+242
vdB–Hagen 228	17 24 18	- 30 45	356.32	+02.31		C1724- 307
IC 1257	17 24 24	- 07 03	016.54	+15.13		C1724- 070
Ruprecht 124	17 24 30	- 40 44	348.07	- 03.32		C1724- 407
Trumpler 26	17 25 18	- 29 27	357.53	+02.86	9.5	C1725- 294
Antalova 1	17 25 36	- 31 32	355.83	+01.64		C1725- 315
Ruprecht 125	17 26 12	- 40 28	348.48	- 03.44		C1726- 404
Antalova 2	17 26 24	- 32 28	355.15	+00.98	8.8	C1726- 324
Collinder 332	17 27 24	- 37 03	351.46	- 01.73	8.9	C1727- 370
Collinder 333	17 28 00	- 34 03	354.02	- 00.17	9.8	C1728- 340
Harvard 16	17 28 00	- 36 49	351.73	- 01.72		C1728- 368
vdB–Hagen 231	17 28 36	- 31 52	355.91	+00.93		C1728- 318
NGC 6383	17 31 30	- 32 32	355.68	+00.05	5.5	C1731- 325
Ruprecht 126	17 31 48	- 34 16	354.27	- 00.95		C1731- 342
Trumpler 27	17 32 54	- 33 27	355.07	- 00.70	6.7	C1732- 334
Trumpler 28	17 33 30	- 32 27	355.98	- 00.26	7.7	C1733- 324
Ruprecht 127	17 34 18	- 36 14	352.89	- 02.45	8.8	C1734- 362
Collinder 338	17 34 48	- 37 32	351.86	- 03.25	8.0	C1734- 375
NGC 6396	17 34 48	- 34 58	354.02	- 01.86	8.5	C1734- 349
NGC 6404	17 36 18	- 33 13	355.66	- 01.17	10.6	C1736- 332
M 6, NGC 6405	17 36 48	- 32 11	356.58	- 00.70	4.2	C1736- 321
NGC 6400	17 36 54	- 36 57	352.65	- 03.34	8.8	C1737- 369
Trumpler 29	17 38 06	- 40 05	350.04	- 05.15	7.5	C1738- 400
Ruprecht 128	17 40 54	- 34 51	354.78	- 02.85		C1740- 348
NGC 6416	17 41 06	- 32 20	356.94	- 01.54	5.7	C1741- 323
Collinder 345	17 41 18	- 33 44	355.78	- 02.32	10.9	C1741- 337
vdB–Hagen 245	17 43 12	- 29 41	359.43	- 00.54		C1743- 296
Collinder 347	17 43 12	- 29 17	359.78	- 00.32	8.8	C1743- 292
NGC 6425	17 43 42	- 31 31	357.97	- 01.60	7.2	C1743- 315
IC 4665	17 43 48	+05 44	030.61	+17.08	4.2	C1743+057

Open Cluster Parameters

IAU Number	θ (′)	V_r (km s^{-1})	MBS	B–V (mag)	TOC	E_{B-V}	R (pc)	D_o (pc)	Age ($\times 10^6$)
C1718− 360	20						6.7	2300	
C1720− 378	9		10.0						
C1720− 499	12		10.0	0.86	+0.36	0.14	1.3	710	2399
C1721− 389	5.0								
C1722− 343	4.0		10.0	1.40	− 0.35	1.72	1.2	2100	1
C1724+242	13								
C1724− 307	1.5								
C1724− 070									
C1724− 407	2.0		12.0						
C1725− 294	17						6.9	2800	
C1725− 315									
C1726− 404	13		11.0						
C1726− 324	3.0			0.21					
C1727− 370	10								
C1728− 340	5.0								
C1728− 368	14								
C1728− 318	4.0								
C1731− 325	5.0	− 2		0.06	− 0.21	0.34	1.0	1380	4
C1731− 342	6		13.0						
C1732− 334	6			1.54	− 0.15	1.30v	1.7	1650	10
C1733− 324	7			0.71	− 0.05	0.74	1.8	1500	200
C1734− 362	8		11.0	1.00	− 0.25	1.03	1.9	1500	6
C1734− 375	25								
C1734− 349	3.0			0.91	− 0.20	0.96			25
C1736− 332	5.0								
C1736− 321	14	− 6	7.0	0.28	− 0.15	0.15	1.3	600	51
C1737− 369	7		9.0						
C1738− 400	9								
C1740− 348	3.4		13.0						
C1741− 323	18			0.62	+0.00	0.33	2.1	800	316
C1741− 337	5								
C1743− 296	2.0								
C1743− 292	4.0			0.62	− 0.25	1.16	0.9	1500	6
C1743− 315	7			0.70	− 0.15	0.50	0.9	800	63
C1743+057	40	− 15	6.0	0.29	− 0.15	0.17	2.5	430	36

Open Cluster Parameters

Name	R.A.(1950) h m s	Dec.(1950) ° ′	l °	b °	V_T (mag)	IAU Number
Ruprecht 129	17 44 06	- 29 35	359.62	- 00.66		C1744- 295
Ruprecht 130	17 44 24	- 30 06	359.22	- 00.98		C1744- 301
Collinder 350	17 45 36	+01 19	026.74	+14.67	6.1	C1745+013
Ruprecht 131	17 45 48	- 29 12	000.14	- 00.77		C1745- 292
NGC 6444	17 46 12	- 34 48	355.39	- 03.75		C1746- 348
Collinder 351	17 46 12	- 28 43	000.60	- 00.60	9.3	C1746- 287
vdB–Hagen 249	17 47 24	- 31 17	358.54	- 02.15		C1747- 312
NGC 6451	17 47 30	- 30 12	359.48	- 01.61	8.2	C1747- 302
Basel 5	17 49 00	- 30 04	359.76	- 01.82		C1749- 300
Ruprecht 133	17 49 18	- 28 42	000.97	- 01.17		C1749- 287
Ruprecht 134	17 49 36	- 29 33	000.27	- 01.66		C1749- 295
Ruprecht 168	17 49 42	- 28 27	001.21	- 01.13		C1749- 284
NGC 6469	17 49 54	- 22 20	006.50	+01.98	8.2	C1749- 223
Czernik 37	17 50 06	- 27 21	002.20	- 00.64		C1750- 273
M 7, NGC 6475	17 50 36	- 34 48	355.86	- 04.53	3.3	C1750- 348
Trumpler 30	17 53 06	- 35 19	355.67	- 05.23	8.8	C1753- 353
M 23, NGC 6494	17 53 54	- 19 01	009.85	+02.85	5.5	C1753- 190
Ruprecht 135	17 55 24	- 11 41	016.39	+06.22		C1755- 116
Ruprecht 136	17 56 12	- 24 42	005.19	- 00.47		C1756- 247
Ruprecht 169	17 56 24	- 24 47	005.14	- 00.55		C1756- 248
NGC 6507	17 56 42	- 17 24	011.58	+03.10	9.6	C1756- 174
Trumpler 31	17 56 42	- 28 11	002.25	- 02.32	9.8	C1756- 281
Ruprecht 137	17 56 48	- 25 11	004.84	- 00.83		C1756- 251
Ruprecht 138	17 56 54	- 24 41	005.29	- 00.59		C1756- 246
Ruprecht 139	17 58 06	- 23 33	006.41	- 00.26		C1758- 235
Melotte 186	17 58 36	+02 54	029.75	+12.54	3.0	C1758+029
Bochum 14	17 58 54	- 23 42	006.37	- 00.50	9.3	C1758- 237
NGC 6514	17 59 18	- 23 02	006.99	- 00.25	6.3	C1759- 230
NGC 6520	18 00 12	- 27 54	002.88	- 02.86	7.6	C1800- 279
M 21, NGC 6531	18 01 36	- 22 30	007.72	- 00.44	5.9	C1801- 225
NGC 6530	18 01 42	- 24 20	006.13	- 01.36	4.6	C1801- 243
NGC 6540	18 03 12	- 27 49	003.26	- 03.36	14.6	C1803- 278
Collinder 468	18 03 30	- 27 28	003.61	- 03.26	13.4	C1803- 274
NGC 6546	18 04 12	- 23 20	007.31	- 01.37	8.0	C1804- 233
vdBergh 113	18 05 36	- 21 28	009.07	- 00.74		C1805- 214

Open Cluster Parameters

IAU Number	θ (′)	V_r (km s^{-1})	MBS	B–V (mag)	TOC	E_{B-V}	R (pc)	D_o (pc)	Age ($\times 10^6$)
C1744– 295	8		12.0						
C1744– 301	3.9		14.0						
C1745+013	45								
C1745– 292	10		11.0						
C1746– 348	12		11.0						
C1746– 287	9								
C1747– 312	0.5								
C1747– 302	7		12.0		+0.46	0.07	0.6	570	6310
C1749– 300	9				+0.13	0.32	1.1	850	1259
C1749– 287	5		12.0						
C1749– 295	5		12.0						
C1749– 284	3.4		12.0						
C1749– 223	12						2.9	1600	
C1750– 273	3.0								
C1750– 348	80	– 12	7.0	0.15	– 0.10	0.06	2.8	240	224
C1753– 353	10								
C1753– 190	27		10.0	0.53	– 0.05	0.38v	2.6	660	224
C1755– 116	11								
C1756– 247	2.2		13.0						
C1756– 248	2.5		14.0						
C1756– 174	6		12.0						
C1756– 281	8				+0.13	0.35	1.3	1000	1259
C1756– 251	3.0		13.0						
C1756– 246	4.0		13.0						
C1758– 235	10		12.0						
C1758+029	240						7.0	200	
C1758– 237				1.42	– 0.35	1.62		1150	1
C1759– 230	28	0	6.0	0.45			6.5	1600	
C1800– 279	6	– 26	9.0		+0.05	0.25	1.6	1650	1000
C1801– 225	13	– 12	8.0	0.12	– 0.20	0.27	2.5	1300	5
C1801– 243	14	– 11	6.0	0.14	– 0.31	0.35v	3.5	1600	2
C1803– 278	0.8								
C1803– 274	1.5								
C1804– 233	13			0.65	– 0.25	0.72	1.6	830	40
C1805– 214									

Open Cluster Parameters

Name	R.A.(1950) h m s	Dec.(1950) ° ′	l °	b °	V_T (mag)	IAU Number
Collinder 367	18 06 30	− 24 00	006.97	− 02.17	6.4	C1806− 240
Dol–Dzim 9	18 06 54	+31 31	058.09	+22.26		C1806+315
NGC 6568	18 09 48	− 21 37	009.43	− 01.66	8.6	C1809− 216
vdB–Hagen 261	18 11 00	− 28 40	003.35	− 05.29		C1811− 286
Markarian 38	18 12 18	− 19 01	011.99	− 00.94		C1812− 190
NGC 6583	18 12 48	− 22 09	009.28	− 02.54	10.0	C1812− 221
Collinder 469	18 13 30	− 18 14	012.80	− 00.80	9.1	C1813− 182
NGC 6595	18 14 00	− 19 54	011.39	− 01.69	7.0	C1814− 199
NGC 6605	18 14 12	− 14 59	015.73	+00.62	6.0	C1814− 149
NGC 6596	18 14 36	− 16 41	014.29	− 00.28		C1814− 166
Trumpler 32	18 14 42	− 13 22	017.22	+01.26	12.2	C1814− 133
NGC 6604	18 15 18	− 12 15	018.26	+01.69	6.5	C1815− 122
NGC 6603	18 15 30	− 18 26	012.86	− 01.32	11.1	C1815− 184
M 16, NGC 6611	18 16 00	− 13 48	016.99	+00.79	6.0	C1816− 138
NGC 6613	18 17 00	− 17 09	014.15	− 01.01	6.9	C1817− 171
NGC 6618	18 17 54	− 16 12	015.09	− 00.74	6.0	C1817− 162
Ruprecht 140	18 18 30	− 33 12	000.01	− 08.86		C1818− 332
NGC 6625	18 20 24	− 12 05	019.01	+00.67	9.0	C1820− 120
Trumpler 33	18 21 48	− 19 43	012.43	− 03.22	7.8	C1821− 197
Ruprecht 170	18 22 24	− 10 05	021.01	+01.16		C1822− 100
Dolidze 28	18 22 30	− 14 41	016.96	− 01.01		C1822− 146
NGC 6631	18 24 24	− 12 04	019.47	− 00.19	11.7	C1824− 120
NGC 6633	18 25 18	+06 32	036.09	+08.28	4.6	C1825+065
Ruprecht 141	18 28 30	− 12 21	019.70	− 01.20		C1828− 123
NGC 6647	18 28 36	− 17 23	015.25	− 03.57	8.0	C1828− 173
Dolidze 29	18 28 42	− 06 40	024.75	+01.41		C1828− 066
M 25, IC 4725	18 28 42	− 19 17	013.58	− 04.48	4.6	C1828− 192
Ruprecht 171	18 29 18	− 16 05	016.48	− 03.11		C1829− 160
Ruprecht 142	18 29 18	− 12 17	019.85	− 01.34		C1829− 122
NGC 6645	18 29 42	− 16 56	015.77	− 03.59	8.5	C1829− 169
Ruprecht 143	18 29 48	− 12 10	020.01	− 01.40		C1829− 121
Ruprecht 144	18 30 36	− 11 28	020.70	− 01.25		C1830− 114
NGC 6649	18 30 42	− 10 26	021.64	− 00.78	8.9	C1830− 104
Graff 1	18 32 24	+05 07	035.65	+06.03		C1832+051
NGC 6664	18 34 00	− 08 16	023.95	− 00.50	7.8	C1834− 082

Open Cluster Parameters

IAU Number	θ (′)	V_r (km s^{-1})	MBS	B–V (mag)	TOC	E_{B-V}	R (pc)	D_o (pc)	Age ($\times 10^6$)
C1806– 240	37								
C1806+315	34								
C1809– 216	12								
C1811– 286	1.5								
C1812– 190	2.0	0					0.3	1000	
C1812– 221	2.8								
C1813– 182	5.0			0.51	– 0.25	0.60	1.5	1970	20
C1814– 199	11								
C1814– 149									
C1814– 166									
C1814– 133	4.0								
C1815– 122	2.0			0.56	– 0.32	0.97	0.5	1640	4
C1815– 184	5.0		14.0				2.1	2880	
C1816– 138	6	+23	11.0	0.58	– 0.32	0.80v	2.5	2500	5
C1817– 171	9	– 14		0.34	– 0.20	0.47	1.7	1200	32
C1817– 162	11			0.69			2.4	1500	
C1818– 332	3.5		11.0						
C1820– 120									
C1821– 197	6			0.20	– 0.24	0.42	1.3	1300	13
C1822– 100	2.5		13.0						
C1822– 146	12								
C1824– 120	5.0								
C1825+065	27	– 28	8.0	0.41	+0.10	0.17	1.3	320	661
C1828– 123	6		12.0						
C1828– 173									
C1828– 066	18								
C1828– 192	32	+2.4	8.0	0.69	– 0.20	0.50	2.7	560	89
C1829– 160	6		14.0						
C1829– 122	4.0		13.0						
C1829– 169	10		12.0						
C1829– 121	5		14.0						
C1830– 114	5.0	– 8.8	12.0						
C1830– 104	5		13.2	1.45	– 0.12	1.34	1.4	1630	50
C1832+051	5								
C1834– 082	16	+17.8	9.0	1.01	– 0.10	0.60v	3.3	1370	37

Open Cluster Parameters

Name	R.A.(1950) h m s	Dec.(1950) ° ′	l °	b °	V_T (mag)	IAU Number
IC 4756	18 36 30	+05 24	036.37	+05.26	4.6	C1836+054
Trumpler 34	18 37 06	- 08 32	024.07	- 01.30	8.6	C1837- 085
Dolidze 32	18 37 48	- 04 09	028.03	+00.57		C1837- 041
Dolidze 33	18 38 00	- 04 27	027.80	+00.37		C1838- 044
Dolidze 34	18 39 18	- 04 38	027.78	- 00.01		C1839- 046
NGC 6683	18 39 30	- 06 20	026.28	- 00.81	9.4	C1839- 063
Trumpler 35	18 40 18	- 04 11	028.29	- 00.01	9.2	C1840- 041
M 26, NGC 6694	18 42 30	- 09 27	023.86	- 02.92	8.0	C1842- 094
Berkeley 79	18 42 36	- 01 16	031.14	+00.84		C1842- 012
Basel 1	18 45 30	- 05 54	027.36	- 01.93	8.9	C1845- 059
Iskudarian 1	18 46 12	+36 48	066.37	+16.52		C1846+368
Czernik 38	18 47 12	+04 52	037.12	+02.64		C1847+048
Ruprecht 145	18 47 36	- 18 09	016.62	- 07.94		C1847- 181
NGC 6704	18 48 12	- 05 16	028.23	- 02.23	9.2	C1848- 052
M 11, NGC 6705	18 48 24	- 06 20	027.31	- 02.77	5.8	C1848- 063
NGC 6709	18 49 06	+10 17	042.16	+04.70	6.7	C1849+102
Ruprecht 146	18 49 36	- 21 11	014.08	- 09.70		C1849- 211
Collinder 394	18 50 30	- 20 27	014.83	- 09.55	6.3	C1850- 204
NGC 6716	18 51 36	- 19 57	015.39	- 09.59	7.5	C1851- 199
Stephenson 1	18 51 48	+36 51	066.85	+15.51	3.8	C1851+368
Berkeley 80	18 51 54	- 01 19	032.16	- 01.25		C1851- 013
Berkeley 81	18 59 00	+00 35	033.64	- 02.51		C1859- 005
NGC 6738	18 59 06	+11 32	044.40	+03.09	8.3	C1859+115
Berkeley 42	19 02 36	+01 48	036.17	- 02.19		C1902+018
Czernik 39	19 05 06	+04 13	038.60	- 01.62		C1905+042
NGC 6755	19 05 18	+04 09	038.55	- 01.70	7.5	C1905+041
NGC 6756	19 06 12	+04 36	039.06	- 01.69	10.6	C1906+046
Berkeley 82	19 09 06	+12 59	046.82	+01.58		C1909+129
Berkeley 43	19 13 12	+11 08	045.66	- 00.19		C1913+111
Ruprecht 147	19 13 48	- 16 22	021.02	- 12.81		C1913- 163
Berkeley 44	19 15 00	+19 28	053.20	+03.35		C1915+194
Berkeley 45	19 16 54	+15 37	050.03	+01.15		C1916+156
NGC 6791	19 19 00	+37 45	070.01	+10.96	9.5	C1919+377
NGC 6793	19 21 00	+22 05	056.19	+03.34		C1921+220
King 25	19 22 12	+13 36	048.86	- 00.94		C1922+136

Open Cluster Parameters

IAU Number	θ (′)	V_r (km s^{-1})	MBS	B–V (mag)	TOC	E_{B-V}	R (pc)	D_o (pc)	Age ($\times 10^6$)
C1836+054	52	− 18	8.0	0.36	+0.15	0.20	3.0	400	575
C1837− 085	7			0.72					
C1837− 041	12								
C1838− 044	6								
C1839− 046	4.0								
C1839− 063	11			0.53	− 0.26	0.54	2.0	1250	100
C1840− 041	9			1.24	− 0.15	1.19	2.1	1610	42
C1842− 094	14	+4	11.0	0.87	− 0.14	0.57	3.4	1550	89
C1842− 012	10		15.0						
C1845− 059	9			0.44	− 0.29	0.57	1.9	1460	58
C1846+368	110						4.0	250	
C1847+048	13								
C1847− 181	35		10.0						
C1848− 052	5		12.0	0.80	− 0.21	0.72	1.6	1810	20
C1848− 063	13		11.0	0.52	− 0.05	0.42	3.5	1720	224
C1849+102	13	− 13	9.0	0.48	− 0.15	0.30	1.9	950	78
C1849− 211	3.0		12.0						
C1850− 204	22				− 0.15	0.25			
C1851− 199	6			0.20	− 0.12	0.12	0.6	600	159
C1851+368	20	− 26		0.78	− 0.20	0.05			25
C1851− 013	4.0		15.0						
C1859− 005	7		15.0						
C1859+115	15								
C1902+018	5		18.0						
C1905+042	6								
C1905+041	14		11.0	1.10	− 0.25	0.93	3.3	1500	35
C1906+046	4.0		13.0		− 0.17	1.18	0.9	1650	47
C1909+129	4.0		14.0						
C1913+111	5		15.0						
C1913− 163	47		9.0						
C1915+194	5		16.0						
C1916+156	4.0		15.0						
C1919+377	15	− 68	15.0	1.02	+0.65	0.18	12.1	5200	7079
C1921+220									
C1922+136	5.0								

Open Cluster Parameters

Name	R.A.(1950) h m s	Dec.(1950) ° ′	l °	b °	V_T (mag)	IAU Number
Collinder 399	19 23 12	+20 05	054.68	+01.96	3.6	C1923+200
Dolidze 35	19 24 00	+11 30	047.24	- 02.35		C1924+115
NGC 6800	19 25 06	+25 02	059.24	+03.93		C1925+250
Berkeley 47	19 26 24	+17 18	052.60	- 00.05		C1926+173
King 26	19 26 42	+14 46	050.41	- 01.33		C1926+147
NGC 6802	19 28 24	+20 10	055.34	+00.93	8.8	C1928+201
Stock 1	19 33 42	+25 06	060.24	+02.26	5.3	C1933+251
Collinder 401	19 35 48	+00 12	038.64	- 10.27	7.0	C1935+002
NGC 6811	19 36 42	+46 27	079.44	+11.95	6.8	C1936+464
NGC 6819	19 39 36	+40 04	073.98	+08.47	7.3	C1939+400
Czernik 40	19 40 24	+21 04	057.50	- 01.08		C1940+210
NGC 6823	19 41 00	+23 11	059.41	- 00.15	7.1	C1941+231
Roslund 1	19 42 48	+17 24	054.64	- 03.42		C1942+174
Roslund 2	19 43 18	+23 48	060.21	- 00.29		C1943+238
Berkeley 48	19 46 42	+21 04	058.24	- 02.35		C1946+210
Czernik 41	19 48 30	+25 02	061.87	- 00.67		C1948+250
NGC 6830	19 48 54	+22 56	060.14	- 01.83	7.9	C1948+229
NGC 6834	19 50 12	+29 17	065.70	+01.18	7.8	C1950+292
Harvard 20	19 50 54	+18 12	056.30	- 04.67	7.7	C1950+182
NGC 6837	19 51 06	+11 33	050.55	- 08.08		C1951+115
NGC 6846	19 54 24	+30 13	066.98	+00.88	14.2	C1954+302
Roslund 3	19 56 30	+20 21	058.81	- 04.68		C1956+203
Berkeley 49	19 57 54	+34 28	070.98	+02.50		C1957+344
Berkeley 83	19 59 18	+28 29	066.07	- 00.93		C1959+284
NGC 6866	20 02 06	+43 51	079.54	+06.87	7.6	C2002+438
Roslund 4	20 02 54	+29 04	066.96	- 01.26	10.0	C2002+290
Berkeley 84	20 02 54	+33 43	070.90	+01.23		C2002+337
NGC 6871	20 04 00	+35 38	072.64	+02.08	5.2	C2004+356
Dol–Dzim 10	20 04 00	+40 23	076.64	+04.63		C2004+403
Melotte 227	20 04 30	- 79 28	314.54	- 30.43	5.3	C2004- 794
Basel 6	20 05 00	+38 12	074.91	+03.29	7.7	C2005+382
Biurakan 1	20 05 36	+35 32	072.73	+01.74		C2005+355
Dolidze 1	20 06 18	+36 24	073.53	+02.10		C2006+364
Biurakan 2	20 07 18	+35 20	072.76	+01.35	6.3	C2007+353
Roslund 5	20 08 06	+33 37	071.40	+00.25		C2008+336

Open Cluster Parameters

IAU Number	θ (′)	V_r (km s^{-1})	MBS	B–V (mag)	TOC	E_{B-V}	R (pc)	D_o (pc)	Age ($\times 10^6$)
C1923+200	60	− 18		0.33			1.1	130	200
C1924+115	7								
C1925+250			10.0						
C1926+173	5		16.0						
C1926+147	1.0								
C1928+201	3.2		14.0	1.27	+0.20	0.81v	0.4	990	1660
C1933+251	60								
C1935+002	1.0								
C1936+464	12		11.0	0.67	+0.15	0.16	1.7	900	537
C1939+400	5	− 7	11.0	0.91	+0.34	0.27	1.6	2200	3467
C1940+210	5.0								
C1941+231	12	+11		0.61	− 0.34	0.76v	6.1	3470	5
C1942+174									
C1943+238									
C1946+210	4.0		15.0						
C1948+250	9								
C1948+229	12		10.0	0.51	− 0.18	0.56v	2.5	1470	100
C1950+292	5.0		11.0	0.66	− 0.19	0.71v	1.6	2300	79
C1950+182	6			0.70	− 0.10	0.26	1.8	1750	60
C1951+115									
C1954+302	0.5		15.0						
C1956+203									
C1957+344	4.0		16.0						
C1959+284	4.0		17.0						
C2002+438	6		10.0	0.44	+0.08	0.16	1.2	1200	229
C2002+290				0.73	− 0.25	0.91			6
C2002+337	4.0		16.0						
C2004+356	20	− 15		0.23	− 0.25	0.46v	4.8	1650	10
C2004+403	20								
C2004− 794	50								
C2005+382	13			0.69			4.3	2100	
C2005+355	14								
C2006+364	5								
C2007+353	12		16.0	0.31	− 0.30	0.41			1
C2008+336	45						2.0	300	

Open Cluster Parameters

Name	R.A.(1950) h m s	Dec.(1950) ° ′	l °	b °	V_T (mag)	IAU Number
Dolidze 2	20 08 12	+41 13	077.78	+04.42		C2008+412
Berkeley 50	20 08 30	+34 47	072.42	+00.84		C2008+347
IC 1311	20 08 36	+41 04	077.70	+04.25	13.1	C2008+410
NGC 6883	20 09 24	+35 42	073.29	+01.19	8.0	C2009+357
Ruprecht 172	20 09 48	+35 29	073.15	+01.00		C2009+354
NGC 6885	20 09 54	+26 20	065.53	- 04.07	8.1	C2009+263
Berkeley 51	20 10 06	+34 12	072.12	+00.24		C2010+342
Berkeley 52	20 12 18	+28 49	067.90	- 03.14		C2012+288
Dolidze 3	20 13 48	+36 38	074.56	+00.98		C2013+366
Dolidze 39	20 14 36	+37 43	075.56	+01.43		C2014+377
IC 4996	20 14 36	+37 29	075.36	+01.31	7.3	C2014+374
Dolidze 4	20 15 54	+36 31	074.71	+00.57		C2015+365
vdBergh 130	20 15 54	+39 10	076.89	+02.06	9.3	C2015+391
Collinder 419	20 16 18	+40 34	078.09	+02.79	5.4	C2016+405
Dolidze 40	20 16 24	+37 41	075.72	+01.14		C2016+376
Berkeley 85	20 17 00	+37 33	075.68	+00.96		C2016+375
Dolidze 41	20 17 24	+37 35	075.75	+00.92		C2017+375
Dolidze 42	20 17 54	+37 58	076.12	+01.05		C2017+379
Berkeley 86	20 18 36	+38 32	076.66	+01.26	7.9	C2018+385
Dolidze 5	20 18 42	+39 13	077.24	+01.64		C2018+392
Dolidze 6	20 19 00	+41 13	078.91	+02.74		C2019+412
Berkeley 87	20 19 48	+37 12	075.71	+00.31		C2019+372
Berkeley 88	20 20 00	+47 55	084.55	+06.38		C2020+479
NGC 6910	20 21 18	+40 37	078.66	+02.03	7.4	C2021+406
Collinder 421	20 21 30	+41 32	079.43	+02.52	10.1	C2021+415
M 29, NGC 6913	20 22 06	+38 22	076.92	+00.60	6.6	C2022+383
Dolidze 8	20 22 42	+42 06	080.04	+02.66		C2022+421
Berkeley 89	20 23 00	+45 49	083.11	+04.78		C2023+458
Dolidze 9	20 23 54	+41 46	079.89	+02.30		C2023+417
Dolidze 10	20 24 30	+39 57	078.48	+01.15		C2024+399
Dolidze 11	20 24 42	+41 17	079.59	+01.90		C2024+412
Roslund 6	20 27 00	+39 13	078.12	+00.27		C2027+392
Dolidze 44	20 27 54	+41 33	080.15	+01.57		C2027+415
NGC 6939	20 30 24	+60 28	095.88	+12.30	7.8	C2030+604
NGC 6940	20 32 30	+28 08	069.90	- 07.16	6.3	C2032+281

Open Cluster Parameters

IAU Number	θ (′)	V_r (km s^{-1})	MBS	B–V (mag)	TOC	E_{B-V}	R (pc)	D_o (pc)	Age ($\times 10^6$)
C2008+412	10								
C2008+347	4.0		14.0						
C2008+410	9		17.0						
C2009+357	14	– 8					3.0	1380	15
C2009+354	4.0		12.0						
C2009+263	7		6.0	0.61	+0.25	0.08	0.6	590	
C2010+342	4.0		15.0						
C2012+288	4.0		18.0						
C2013+366	14								
C2014+377	12								
C2014+374	5		8.0	0.57	– 0.40	0.64	1.4	1620	10
C2015+365	6								
C2015+391				0.57					
C2016+405	4.5	– 7							
C2016+376	12								
C2016+375	7		15.0						
C2017+375	11								
C2017+379	11								
C2018+385	7		13.0	0.76	– 0.20	0.99	2.0	1720	6
C2018+392	6								
C2019+412	10								
C2019+372	12		13.0		– 0.30	1.35			
C2020+479	3.0		15.0						
C2021+406	7	– 30		0.73	– 0.30	1.05v	1.9	1650	10
C2021+415	5								
C2022+383	6	– 16.4	9.0	0.70	– 0.28	0.78v	1.3	1250	10
C2022+421	5								
C2023+458	5.0		15.0						
C2023+417	7								
C2024+399	15								
C2024+412	7								
C2027+392									
C2027+415	12								
C2030+604	7			1.05	+0.27	0.50	1.5	1250	1820
C2032+281	31	+4	11.0	0.68	+0.30	0.26v	3.7	800	1096

Open Cluster Parameters

Name	R.A.(1950) h m s	Dec.(1950) ° ′	l °	b °	V_T (mag)	IAU Number
Berkeley 90	20 33 42	+46 38	084.86	+03.76		C2033+466
Dolidze 47	20 39 42	+36 28	077.49	- 03.33		C2039+364
Ruprecht 173	20 39 48	+35 22	076.64	- 04.04		C2039+353
Ruprecht 174	20 41 36	+36 52	078.04	- 03.38		C2041+368
Ruprecht 175	20 43 12	+35 19	077.02	- 04.60		C2043+353
Dol–Dzim 11	20 49 00	+35 46	078.12	- 05.23		C2049+357
Roslund 7	20 50 30	+37 44	079.83	- 04.21		C2050+377
Barkhatova 1	20 52 00	+45 51	086.21	+00.79		C2052+458
NGC 6996	20 54 42	+44 26	085.46	- 00.47	10.0	C2054+444
Berkeley 53	20 55 00	+50 50	090.34	+03.66		C2055+508
NGC 6991	20 55 00	+47 13	087.53	+01.39		C2055+472
NGC 6994	20 56 18	- 12 50	035.73	- 33.95	8.9	C2056- 128
Collinder 427	20 58 54	+67 58	103.93	+14.34	13.8	C2058+679
NGC 7023	20 59 54	+67 58	103.99	+14.27	7.1	C2059+679
Berkeley 54	21 01 18	+40 16	083.13	- 04.14		C2101+402
Collinder 428	21 01 24	+44 23	086.21	- 01.41	8.7	C2101+443
NGC 7031	21 05 42	+50 38	091.32	+02.26	9.1	C2105+506
Dolidze 45	21 07 00	+37 24	081.74	- 06.89		C2107+374
Basel 12	21 08 42	+46 02	088.30	- 01.24		C2108+460
NGC 7039	21 09 24	+45 27	087.96	- 01.73	7.6	C2109+454
Berkeley 91	21 10 00	+48 16	090.07	+00.13		C2110+482
IC 1369	21 10 24	+47 32	089.59	- 00.42	8.8	C2110+475
Basel 13	21 10 30	+46 22	088.75	- 01.24		C2110+463
NGC 7044	21 11 06	+42 17	085.87	- 04.13	12.0	C2111+422
Basel 15	21 14 12	+48 38	090.82	- 00.11		C2114+486
Berkeley 55	21 15 06	+51 34	093.01	+01.84		C2115+515
Berkeley 56	21 15 48	+41 41	086.04	- 05.18		C2115+416
Basel 14	21 19 24	+44 36	088.58	- 03.59		C2119+446
NGC 7062	21 21 24	+46 10	089.93	- 02.72	8.3	C2121+461
NGC 7067	21 22 24	+47 48	091.19	- 01.67	9.7	C2122+478
NGC 7063	21 22 24	+36 17	083.09	- 09.90	7.0	C2122+362
Berkeley 92	21 23 42	+57 17	097.94	+05.01		C2123+572
NGC 7082	21 27 36	+46 52	091.19	- 02.95	7.2	C2127+468
NGC 7086	21 28 48	+51 22	094.41	+00.20	8.4	C2128+513
M 39, NGC 7092	21 30 24	+48 13	092.46	- 02.28	4.6	C2130+482

Open Cluster Parameters

IAU Number	θ (′)	V_r (km s^{-1})	MBS	B–V (mag)	TOC	E_{B-V}	R (pc)	D_o (pc)	Age (×10^6)
C2033+466	5		14.0						
C2039+364									
C2039+353	50		8.0						
C2041+368	5.0		14.0						
C2043+353	9		11.0						
C2049+357	12								
C2050+377									
C2052+458	20	− 9							
C2054+444	6						0.5	500	
C2055+508	12		16.0						
C2055+472									
C2056− 128	2.8		10.0						
C2058+679	4.0								
C2059+679	5.0								
C2101+402	5		17.0						
C2101+443	13						1.0	480	
C2105+506	5.0			0.91	− 0.11	0.83	0.8	1000	56
C2107+374	18								
C2108+460	7				+0.00	0.57	1.8	1540	
C2109+454	25			0.46	− 0.20	0.14	2.5	700	126
C2110+482	5		16.0						
C2110+475	4.0			0.99	+0.00	0.52	0.9	1500	1259
C2110+463	10				+0.10	0.34	1.9	1270	
C2111+422	3.5		15.0						
C2114+486	6				+0.00	0.67	1.5	1440	
C2115+515	5		14.0						
C2115+416	7		16.0						
C2119+446	12				− 0.05	0.65	1.8	1030	
C2121+461	6			0.69	+0.06	0.41	2.0	1900	631
C2122+478	3.0			0.94	− 0.20	0.84v	1.5	3500	13
C2122+362	7			0.32	− 0.10	0.08	0.8	660	141
C2123+572	4.0		15.0						
C2127+468	25			0.50	− 0.15	0.28	5.1	1400	1585
C2128+513	9			0.85	− 0.10	0.70	1.6	1200	85
C2130+482	31		7.0	0.06	− 0.05	0.02	1.3	270	269

Open Cluster Parameters

Name	R.A.(1950) h m s	Dec.(1950) ° ′	l °	b °	V_T (mag)	IAU Number
IC 1396	21 37 30	+57 16	099.29	+03.73	3.5	C2137+572
NGC 7129	21 40 12	+65 52	105.26	+09.98	11.5	C2140+658
NGC 7127	21 42 12	+54 23	097.91	+01.12		C2142+543
NGC 7128	21 42 18	+53 29	097.35	+00.42	9.7	C2142+534
Barkhatova 2	21 42 30	+50 59	095.76	- 01.51		C2142+509
NGC 7142	21 44 42	+65 34	105.42	+09.45	9.3	C2144+655
IC 5146	21 51 30	+47 02	094.39	- 05.50	7.2	C2151+470
NGC 7160	21 52 18	+62 22	104.02	+06.45	6.1	C2152+623
Berkeley 93	21 54 48	+63 42	105.07	+07.32		C2154+637
NGC 7209	22 03 12	+46 15	095.51	- 07.34	7.7	C2203+462
Collinder 471	22 06 12	+71 45	110.88	+13.08		C2206+717
IC 1434	22 08 36	+52 35	099.93	- 02.70	9.0	C2208+525
NGC 7226	22 08 42	+55 10	101.42	- 00.60	9.6	C2208+551
NGC 7235	22 10 48	+57 02	102.72	+00.78	7.7	C2210+570
vdBergh 150	22 11 54	+73 05	112.06	+13.90		C2211+730
vdBergh 152	22 12 18	+70 00	110.27	+11.37		C2212+700
NGC 7243	22 13 18	+49 38	098.86	- 05.55	6.4	C2213+496
NGC 7245	22 13 24	+54 05	101.37	- 01.87	9.2	C2213+540
King 9	22 13 36	+54 09	101.45	- 01.84		C2213+541
IC 1442	22 14 36	+53 48	101.36	- 02.20	9.1	C2214+538
NGC 7261	22 18 36	+57 50	104.04	+00.86	8.4	C2218+578
Berkeley 94	22 20 48	+55 36	103.10	- 01.18	8.7	C2220+556
NGC 7281	22 22 54	+57 35	104.40	+00.34		C2222+575
NGC 7296	22 26 12	+52 02	101.89	- 04.63	9.7	C2226+520
Berkeley 95	22 26 30	+58 52	105.47	+01.19		C2226+588
Berkeley 96	22 27 30	+55 09	103.67	- 02.06		C2227+551
Berkeley 97	22 37 36	+58 45	106.66	+00.37		C2237+587
Czernik 42	22 37 54	+59 37	107.11	+01.11		C2237+596
Berkeley 98	22 41 06	+52 09	103.94	- 05.67		C2241+521
NGC 7380	22 45 00	+57 50	107.08	- 00.90	7.2	C2245+578
King 18	22 50 06	+58 01	107.78	- 01.04		C2250+580
NGC 7419	22 52 18	+60 34	109.13	+01.14	13.0	C2252+605
King 10	22 52 54	+58 54	108.49	- 00.40		C2252+589
Berkeley 57	22 53 12	+56 52	107.65	- 02.25		C2253+568
NGC 7429	22 53 54	+59 43	108.95	+00.28		C2253+597

Open Cluster Parameters

IAU Number	θ (′)	V_r (km s^{-1})	MBS	B–V (mag)	TOC	E_{B-V}	R (pc)	D_o (pc)	Age (×10^6)
C2137+572	50	− 10		1.37	− 0.35	0.50	5.8	800	1
C2140+658	2.7								
C2142+543	2.8						0.6	1300	
C2142+534	3.1			0.84	− 0.20	0.92v	1.1	2600	10
C2142+509									
C2144+655	4.3	− 44	11.0	1.06	+0.40	0.41	0.6	1000	3981
C2151+470	9	− 6.5		0.62	− 0.05	0.45v	1.4	1000	229
C2152+623	7	− 23.6		0.24	− 0.20	0.30	1.0	900	10
C2154+637	4.0		16.0						
C2203+462	25		9.0	0.54	+0.00	0.15	3.3	900	302
C2206+717	130								
C2208+525	7		12.0						
C2208+551	1.8			0.51	− 0.04	0.49	0.6	2200	501
C2210+570	4.0	− 52		0.88	− 0.30	0.96v	2.2	3800	2
C2211+730									
C2212+700									
C2213+496	21	− 9	8.0	0.14	− 0.12	0.24	2.7	880	107
C2213+540	5.0			0.61	− 0.08	0.49	1.4	1900	398
C2213+541	2.5		18.0						
C2214+538			12.0	0.43	− 0.21	0.43		1800	126
C2218+578	5			0.80	− 0.22	0.94	0.8	900	40
C2220+556	4.0		13.0	0.40	− 0.22	0.61	0.9	1600	63
C2222+575									
C2226+520	4.0		10.0						
C2226+588	5		15.0						
C2227+551	2.0		13.0		− 0.33	0.68			
C2237+587	5		11.0						
C2237+596	3.0								
C2241+521	6		15.0						
C2245+578	12	− 38	10.0	0.42	− 0.29	0.56v	6.3	3600	4
C2250+580	4.0		12.0						
C2252+605	2.0		10.0				0.6	1920	
C2252+589	3.0		11.0						
C2253+568	5		15.0						
C2253+597			11.0					1920	

Open Cluster Parameters

Name	R.A.(1950) h m s	Dec.(1950) ° ′	l °	b °	V_T (mag)	IAU Number
1	23 06 00	+60 36	110.69	+00.48		C2306+606
King 19	23 06 12	+60 15	110.57	+00.15	9.2	C2306+602
NGC 7510	23 09 24	+60 18	110.96	+00.05	7.9	C2309+603
Markarian 50	23 13 06	+60 12	111.36	- 00.20	8.5	C2313+602
Berkeley 99	23 19 36	+71 29	115.95	+10.11		C2319+714
M 52, NGC 7654	23 22 00	+61 19	112.76	+00.46	6.9	C2322+613
Czernik 43	23 23 30	+61 02	112.83	+00.13		C2323+610
Berkeley 100	23 24 24	+63 28	113.72	+02.40		C2324+634
NGC 7686	23 27 48	+48 51	109.52	- 11.63	5.6	C2327+488
King 20	23 31 00	+58 14	112.86	- 02.84		C2331+582
Berkeley 101	23 31 12	+63 56	114.58	+02.61		C2331+639
Czernik 44	23 31 12	+61 38	113.90	+00.41		C2331+616
Stock 12	23 34 48	+52 09	111.61	- 08.81		C2334+521
Av–Hunter 1	23 34 48	+48 08	110.43	- 12.65		C2334+481
Berkeley 102	23 36 18	+56 22	113.02	- 04.82		C2336+563
Berkeley 103	23 42 48	+59 02	114.56	- 02.48		C2342+590
Stock 17	23 43 36	+61 54	115.38	+00.27		C2343+619
King 11	23 45 24	+68 21	117.16	+06.47		C2345+683
NGC 7762	23 47 24	+67 45	117.20	+05.84		C2347+677
King 21	23 47 24	+62 26	115.94	+00.68		C2347+624
NGC 7772	23 49 12	+15 59	102.75	- 44.27		C2349+159
King 12	23 50 30	+61 41	116.12	- 00.13		C2350+616
Harvard 21	23 51 36	+61 29	116.21	- 00.37	9.0	C2351+614
Czernik 45	23 53 48	+64 16	117.05	+02.30		C2353+642
NGC 7788	23 54 12	+61 07	116.43	- 00.79	9.4	C2354+611
NGC 7789	23 54 30	+56 27	115.49	- 05.36	6.7	C2354+564
Frolov 1	23 54 54	+61 21	116.56	- 00.57	9.2	C2354+613
NGC 7790	23 55 54	+60 56	116.59	- 01.01	8.5	C2355+609
Berkeley 58	23 57 36	+60 41	116.75	- 01.29	9.7	C2357+606
Stock 18	23 59 00	+64 22	117.63	+02.28		C2359+643
Harvard 21	23 51 36	+61 29	116.21	- 00.37	9.0	C2351+614
Czernik 45	23 53 48	+64 16	117.05	+02.30		C2353+642

Open Cluster Parameters

IAU Number	θ (′)	V_r (km s^{-1})	MBS	B–V (mag)	TOC	E_{B-V}	R (pc)	D_o (pc)	Age ($\times 10^6$)
C2306+606	3.0						1.8	4000	
C2306+602	6		12.0	1.07	– 0.20	0.82	1.4	1350	40
C2309+603	4.0		10.0	0.87	– 0.26	1.06v	1.9	3160	10
C2313+602	5.0	– 81		0.59	– 0.27	0.90	1.6	2250	10
C2319+714	6		14.0						
C2322+613	12	– 35	11.0	0.70	– 0.06	0.57v	2.8	1470	35
C2323+610	13								
C2324+634	4.0		16.0						
C2327+488	14	+6	7.0	1.19			2.2	1000	
C2331+582	4.0		13.0						
C2331+639	4.0		16.0						
C2331+616	5								
C2334+521	20		8.0						
C2334+481									
C2336+563	5		18.0						
C2342+590	4.0		15.0						
C2343+619	1.0								
C2345+683	3.5		17.0						
C2347+677	11		11.0		+0.15	1.02	1.7	1020	794
C2347+624	2.5		10.0						
C2349+159									
C2350+616	2.0		10.0		– 0.27	0.60v			
C2351+614	4.0								
C2353+642	3.0								
C2354+611	9				– 0.15	0.28	3.2	2410	16
C2354+564	15		10.0	0.98	+0.30	0.25	4.4	1900	1585
C2354+613				0.54					
C2355+609	17		10.0	0.68	– 0.22	0.62	9.4	3680	78
C2357+606	7		15.0	0.69					
C2359+643	5.0								
C2351+614	4.0								
C2353+642	3.0								

16
OB Associations

16.1 Basic Data for OB Associations

The bright, massive O and B stars are often found in loose groups called OB associations. Because these luminous stars deplete their thermonuclear fuel relatively quickly, they appear in relatively young groups with ages on the order of a million years.

The OB associations typically contain between 10 to 100 stars within a radius between 8 and 160 parsecs. Their total mass lies between a few hundred and a few thousand solar masses, $M_{\odot}$, with mass densities of between 0.001 and 0.1 $M_{\odot}$ pc^{-3}. Because of their weak internal gravitation, the associations tend to disperse with time, often expanding with velocities between 5 and 10 $km\ s^{-1}$. Such expansions originated about a million years ago; these kinematic ages are comparable to the thermonuclear ages of the bright O and B stars that make up the associations.

Masses, ages and sizes for various associations are given in relatively short tables, followed by our main catalogue. It contains the IAU designation and spread in galactic longitude, l, and latitude, b, given by Ruprecht (1966), together with right ascensions, R.A., and declinations, Dec., for epoch 2000.0, and radial velocities, V_r, when provided by Hirshfeld and Sinnott (1985). The distances, D1, given by Ruprecht (1966) are compared with those calculated, D2, from the distance moduli adopted by Humphreys (1978). In many cases, the central part of the association contains one or more early-type open star clusters that are designated in the last column of our catalogue.

Masses of Eleven OB Associations*

Name	Mass ($M_{\odot}$)	Name	Mass ($M_{\odot}$)	Name	Mass ($M_{\odot}$)
Cas OB4	237	Sgr OB5	8614	Cas OB5	3379
Per OB2	288	Cyg OB2	8077	Cep OB4(i)	1581
Gem OB1	2834	Lac OB1	947	Cep OB4(ii)	608
Ori OB1	1939	Cep OB3	1216		

* Adapted from Bruch and Sanders (1983).

Sizes of OB Associations*

Name	θ (′)	D (kpc)	R (pc)	Name	θ (′)	D (kpc)	R (pc)
Cas OB1	120	2.51	45	Cen OB2	66	2.50	25
Per OB1	360	2.29	124	Cen OB1	360	2.51	136
Cas OB6	480	2.19	158	Ara OB1	225	1.38	
Per OB2	390	0.40	23	Sco OB1	81	1.91	23
Aur OB1	330	1.32	65	Sgr OB1	405	1.58	96
Ori OB1	960	0.46	66	Ser OB1	240	2.19	79
Gem OB1	300	1.51	68	Ser OB2	500:	2.00:	150:
Mon OB1	570	0.55	47	Cyg OB1	330	1.82	90
Mon OB2	305	1.51	69	Cyg OB2	30	1.82	8
CMa OB1	240	1.32	47	Lac OB1	720	0.60	65
Pup OB1	210	2.51	79	Cep OB2	460	0.83	54
Vel OB1	300	1.40	63	Cep OB1	210	3.47	109
Car OB1	93	2.51	35	Cas OB5	150	2.51	56
Car OB2	240	2.00	72	Mean:	321	1.76	70

*Adapted from Hirshfeld and Sinnott (1985). The mean angular diameter, θ, is given when more than one value was available. Values of distance, D, are used with the θ to infer the association radius, R.

The Nearest Associations with Subgroups**

Name	D (pc)	R (pc)	N	Age (10^6 years)
Scorpio–Centaurus				
Upper Scorpius	170	22.5	16	14
Upper Centaurus –Lupus	170	40	18	10–20
Lower Centaurus –Crux	160			
Per OB2	330	25	14	1.5–4
Ori OB1a†	460	25	20	12
Ori OB1b	460	10	15	8
Ori OB1c	460	7.5	18	6
Ori OB1d	460	1		2
Lac OB1a	600	50	10	16
Lac OB1b	600	15	11	12
Mon OB2	715	4	4	1
Cep OB3a	730	8.5	16	8
Cep OB3b	730	5	15	4

**Adapted from Blauw (1964). Positions of the associations are given in the main catalogue, except for the nearest OB association, Sco–Cen, extending from Scorpius to Crux, or galactic longitude 292°to 2°and galactic latitude – 10° to +30°, with right ascension ≈ 16*h* and declination ≈ – 25°. The radius, R, is half the largest projected dimension, N is the number of stars B3 and earlier, and the age estimates are from color–magnitude diagrams or kinematic data, including runaway stars.

†The Orion OB1 association has been subdivided into subgroups 1a (northwest region), 1b (Belt region), 1c (outer Sword) and 1d (Orion–nebula cluster) by Warren and Hesser (1977, 1978).

16.2 Catalogue of OB Associations

Catalogue of OB Associations*

Name	R.A.(1950) h m	Dec.(1950) ° ′	l (° to °)	b (° to °)	V_r (km s^{-1})	D1 (pc)	D2 (pc)	Open Clusters
Cas OB4	00 28.4	+62 42	119.0 to 121.6	− 2.1 to + 2.0	− 43	2650	2884	NGC 103
Cas OB14	00 28.8	+63 22	119.7 to 121.1	− 1.3 to + 2.5	− 8	1180	1096	
Cas OB7			121.7 to 125.2	− 0.9 to + 2.6		2340	2512	
Cas OB1	01 00.8	+61 30	122.3 to 125.8	− 2.3 to − 0.4	− 38	2630	2512	
Cas OB8	01 46.2	+61 19	129.2 to 129.7	− 1.5 to − 0.2	− 30	2940	2884	NGC 581, 683
Per OB1	02 14.5	+57 19	132 to 136	− 2.5 to − 5	− 41	23	2291	NGC 869, 884 (h, χ Per)
Cas OB6	02 43.2	+61 23	133.8 to 138.0	− 0.3 to + 3.0	− 47	2420	2188	IC 1805
Cam OB1	03 31.6	+58 38	134 to 151	− 3 to + 7	− 6	9	1 0	
Per OB3	03 27.8	+49 54	142 to 152	+ 2 to + 4		170	170	α Per cl.
Cam OB3			146.3 to 147.7	+ 2.0 to + 3.9		35	3311	
Per OB2	03 42.2	+33 26	156.4 to 162.1	− 13.0 to − 21.3		4	398	
Aur OB1	05 21.7	+33 52	168.1 to 178.1	− 7.4 to + 4.2	− 3	1340	1318	NGC 1912, 1960
Aur OB2	05 28.3	+34 54	172 to 174	− 1.8 to + 2.0	− 13	36	3162	NGC 1893, 36 , IC 410
Ori OB1	05 31.4	− 02 41	198 to 214	− 13 to − 25		5	501	NGC 1976; Trapezium
Gem OB1	06 09.8	+21 35	187.4 to 190.8	− 2.1 to + 4.2	+13	15	1514	

*Adapted from Ruprecht (1966), Humphreys (1978), and Hirshfeld and Sinnott (1985).

Catalogue of OB Associations*

Name	R.A.(1950) h m	Dec.(1950) ° ′	l (° to °)	b (° to °)	V_r (km s^{-1})	D1 (pc)	D2 (pc)	Open Clusters
Mon OB1	06 33.1	+08 50	196 to 210	− 2.5 to + 2.5	+22	715	714	NGC 2264
Mon OB2	06 37.2	+04 50	205 to 209	− 2.7 to + 0.8	+28	14	1514	NGC 2244
CMa OB1	07 07.0	− 10 28	222 to 226	− 3.4 to + 0.7	+27	1315	1318	NGC 2335, 2353
Pup OB1**	07 54.8	− 27 05	242 to 246	− 1 to + 2	+43	25	2512	NGC 2467
Vela OB1	08 49.9	− 45	262 to 268	− 2.7 to + 1.4		1450	1819	NGC 2650
Car OB1	10 46.7	− 59 05	283 to 292	− 2 to + 2		26	2512	NGC 3293
Car OB2(?)	11 06.0	− 59 51	289.9 to 290.2	+ 0.3 to + 0.4		19	1995	NGC 3572, Tr 18
Cen OB2	11 35.3	− 62 36	294.3	− 1.0		21		IC 2944
Cen OB1	13 04.8	− 62 04	301 to 308	− 2.5 to + 4		15	2512	NGC 4755
Ara OB1	16 39.5	− 46 46	335 to 341	− 3 to + 3		13	1380	NGC 6169, 6193
Sco OB1	16 53.5	− 41 57	343.3	+ 1.2	− 18	14	1905	NGC 6231
Sco OB2	16 14.9	− 25 55	347.1 to 353.0	+ 12.3 to + 23.3		160	158	
Sgr OB5			358.8 to 1.5	− 3.9 to + 1.4		26	3020	
Sgr OB1	18 07.9	− 21 28	4 to 14	− 2.6 to + 1.4	− 4	1560	1585	NGC 6514, 6523, 6530
Sgr OB7(?)			10.5 to 10.8	− 1.7 to − 1.3		1860	1738	

** The Puppis cluster has been divided by Havlen (1972) into Pup OB1 and Pup OB2 located at mean distances of 2.5 kpc and 4.2 kpc, respectively, with estimated ages of 4 and 2 million years.

Catalogue of OB Associations*

Name	R.A.(1950) h m	Dec.(1950) ° ′	l (° to °)	b (° to °)	V_r (km s^{-1})	D1 (pc)	D2 (pc)	Open Clusters
Sgr OB4	18 14.4	− 19 03	11.2 to 12.7	− 1.4 to − 0.5	+3	2130	2399	Burakan 5, NGC 6603
Sgr OB6			13.4	+ 1.0		2 0	1995	NGC 6561
Ser OB1	18 20.8	− 14 35	14 to 19	− 1.5 to + 1.5	− 23	17	2188	NGC 6611
Sct OB3			16.7 to 18.0	− 1.6 to − 0.1		16	1659	
Ser OB2	18 18.6	− 11 58	18.0 to 19.1	+ 1.1 to + 2.3	+6	2 0	1995	NGC 6604
Sct OB2			20 to 26	− 2.8 to + 1.5		730	1 0	
Vul OB4			59 to 62	− 3 to + 4		1020	1019	
Vul OB1	19 44.0	+24 13	59.1 to 61.5	− 1.2 to + 1.5	+8	2050	1995	NGC 6823
Cyg OB3	20 04.7	+35 50	71.3 to 73.8	+ 1.2 to + 3.4	0	23	2291	NGC 6871
Cyg OB1	20 17.8	+37 38	74 to 77	− 0.6 to + 2.8	− 7	17	1819	NGC 6913, IC 4996
Cyg OB8			76.3 to 79.2	+ 2.1 to + 5.4		2190	2291	
Cyg OB9	20 23.3	+39 56	77 to 79	+ 0.8 to + 2.2	− 20	1170	1202	NGC 6910
Cyg OB2	20 32.4	+41 17	80.1	+ 0.9		15	1819	
Cyg OB4	21 13.1	+37 52	81 to 84	− 8.3 to − 6.3		1 0	1 0	
Cyg OB7	21 02.7	+49 43	84 to 96	− 4.9 to + 9.0	− 10	740	794	

*Adapted from Ruprecht (1966), Humphreys (1978), and Hirshfeld and Sinnott (1985).

Catalogue of OB Associations*

Name	R.A.(1950) h m	Dec.(1950) ° ′	l (° to °)	b (° to °)	V_r (km s^{-1})	D1 (pc)	D2 (pc)	Open Clusters
Cep OB2	21 47.9	+61 04	97 to 108	− 0.9 to + 12.3	− 20	7	832	NGC 7160, IC 1396
Lac OB1	22 41.2	+39 05	96 to 98	− 18.7 to − 15.6		6	6	
Cep OB1	22 24.6	+55 14	98 to 108	− 0.7 to − 3.0	− 51	36	3467	NGC 7380
Cep OB5			108.3 to 108.6	− 3.2 to − 2.3		2090	2089	
Cep OB3	23 00.4	+64 03	109.4 to 112.9	+ 2.3 to + 5.2	− 21	960	871	
Cas OB2			110.1 to 114.0	− 1.3 to + 1.8		2680	2630	
Cas OB5	23 58.7	+60 22	114.9 to 118.0	− 2.4 to − 1.3	− 46	2450	2512	NGC 7788
Cep OB4	23 59.5	+67 35	117.4 to 118.6	+ 3.9 to + 6.5			843	

*Adapted from Ruprecht (1966), Humphreys (1978), and Hirshfeld and Sinnott (1985).

Part V
The Stellar Environment

17
Regions of Star Formation

17.1 Molecular Clouds and Dust Clouds

Present-day star formation takes place within molecular clouds whose interiors are hidden from view at optical wavelengths by an obscuring veil of interstellar dust. Star-forming regions can nevertheless be observed in considerable detail at infrared, millimeter and submillimeter wavelengths.

Observations of line emission from carbon monoxide, CO, at 2.6 millimeters wavelength indicate that star formation in our Galaxy takes place within giant molecular clouds with typical radii of R = 10 parsecs and masses, M, that are 100,000 times as massive as the Sun, or $M = 10^5\ M_\odot$. As indicated in the first accompanying table [Goldsmith (1987)], the largest molecular cloud complex has a radius of about 50 parsecs with a mass $M \geq 10^6\ M_\odot$. Our table also shows that the giant molecular clouds contain denser, hotter cores and clumps.

The molecular gas that gives birth to young stars is extremely cold (10 K to 20 K) and mainly composed of molecular hydrogen, or H_2. However, molecular hydrogen is generally not directly observable from the ground, and its presence and properties are inferred from observations of less-abundant molecules like CO. These observations indicate that the total mass of molecular hydrogen in our Galaxy exceeds a billion times that of the Sun, or $M \geq 10^9\ M_\odot$. The mean density, N, of molecular hydrogen in a giant molecular cloud is roughly N = 200 cm^{-3}. Between 1% and 10% of the mass in a giant molecular cloud exists in relatively dense cores with $N \geq 10^4\ cm^{-3}$. Star formation is probably located within these dust-enshrouded cores.

Giant molecular clouds are associated with high-mass star formation. The massive O and B stars illuminate the surrounding gas, creating the colorful regions of ionized hydrogen, or H II regions. Such high-mass stars arrive at the main sequence and begin burning hydrogen while still gaining mass from the collapsing molecular core, and they therefore do not reveal much information about their origins.

Stars of low mass $M \leq M_\odot$ can be observed in a state of collapse before reaching the main sequence and becoming hot enough to sustain fusion reactions. Such pre-main-sequence objects are found within nearby dark clouds that contain a wide variety of molecules. As indicated in the second accompanying table, these dark nebulae have typical radii of R = 1 parsec with masses of $M = 100\ M_\odot$, but they often exist in a larger, more-massive complex like the well-known one in Taurus.

The dark nebulae have long been known to optical astronomers who noticed the way they block the light of distant stars, and gave them the fascinating names provided in the next table. It is followed by a table of the right ascensions, R.A., and the declinations, Dec., of selected dense molecular clouds for epoch 1950.0 [Klingsmith and Hollis (1987)]. The masses and distances of molecular clouds within one kiloparsec of the Sun are given by Dame et al. [Astrophysical Journal 322, 706-720 (1987)].

The molecular clouds in Orion and Monoceros are portrayed as a contour map of integrated intensity of CO emission; the lowest contour level is 1.28 K $km\ s^{-1}$ with subsequent levels at 3, 5, 7 ... times this value [Maddalena et al. (1986)]. The location of NGC objects, the Orion Nebula, and Barnard's Loop (shaded arc) are shown in the diagrams that proceed and follow the CO contour diagram.

A CO survey of the molecular clouds in Perseus, Taurus and Auriga is next shown in a diagram of velocity-integrated contours; the lower contour is 0.5 K $km\ s^{-1}$, and the separation between contours is 1.5 K $km\ s^{-1}$ [Ungerechts and Thaddeus (1987)]. It is accompanied by a diagram that shows the location of NGC and IC objects as well as dark clouds in the Lynds, L, and Barnard, B, catalogues.

During evolution to the main sequence, most, if not all, stars experience vigorous episodes of mass loss, often characterized by massive bipolar flows of cold molecular gas with enormous kinetic energies of 10^{43} to 10^{47} *erg*. The next table provides a catalogue of high-velocity molecular flow sources with CO emission line widths of $\Delta V \geq 10\ km\ s^{-1}$. It has been adapted from Lada (1985), who provided celestial coordinates and distance, D, and supplemented with mass-loss rates, $\dot{M}$, and ages given by Levreault (1988) for pre-main-sequence objects. The molecular outflow data indicate typical extents of about 1 parsec, outflow masses between 0.1 and 100 $M_\odot$, outflow velocities of about 10 $km\ s^{-1}$, outflow ages of about 10^5 years, and a mean $\dot{M}$ of 3 x 10^{-7} $M_\odot$ per year for stellar wind velocities of about 300 km s^{-1}. Detailed information for the classical bipolar molecular outflow source L 1551 is given by Moneti et al. (1988) and Moriarty-Schieven and Snell (1988).

As indicated in the next table, an elongated concentration of gas is also detected near the central star of L 1551 [Sargent et al. (1988)]. One circumstellar gaseous disk is in Keplerian rotation about another star, HL Tauri; it contains 0.1 $M_\odot$ of gas distributed in a disklike structure of about 4000 AU in extent.

Circumstellar dust shells have been inferred from the unexpectedly-intense, far-infrared radiation of nearby stars such as α Lyrae, ϵ Eridani, α Piscis Austrini and β Pictoris. Direct imaging has proved in the case of β Pictoris that this infrared excess is caused by a disk of dust surrounding the star. Our final table of such stars is adapted from Aumann (1985); it includes the celestial coordinates of the star, together with its proper motion, spectral type, luminosity class, and visual magnitude, as well as the calculated temperature and radius of the circumstellar dust shell. Shell radii of between 3 AU and 100 AU are obtained, leading to speculation that these disks could be related to the formation of planetary systems.

Giant Molecular Clouds*

Category	R (pc)	N (cm^{-3})	Mass ($M_\odot$)	T (°K)	ΔV ($km\ s^{-1}$)	Examples
Giant molecular Cloud Complex	10 – 40	100 – 300	$10^5 - 10^6$	7 – 15	6 – 15	W 51, M 17, W 3
Giant Molecular Cloud	1 – 10	$10^3 - 10^4$	$10^3 - 10^5$	15 – 40	4 – 12	Orion OMC1 W 33, W 3A
Giant Molecular Cloud Core	0.2 – 1.5	$10^4 - 10^6$	$10 - 10^3$	30 – 100	1 – 3	Orion (Ridge)
Giant Molecular Cloud Clump	< 0.2	$> 10^6$	$30 - 10^3$	30 – 200	4 – 15	M 17 (Kleinmann–Weight) Orion (Hot Core), W3 (Mrt)

*Associated with high–mass star formation, adapted from Goldsmith (1987).

Dark Nebulae*

Category	R (pc)	N (cm^{-3})	Mass ($M_\odot$)	T (°K)	ΔV ($km\ s^{-1}$)	Examples
Dark Nebula Complex	3 – 10	$10^2 - 10^3$	$10^3 - 10^4$	≈ 10	1 – 3	Taurus, Sco–Oph
Dark Nebula	0.1 – 2	$10^2 - 10^4$	5 – 500	8 – 15	0.5 – 1.5	B 227, B 335, B 5, B 18
Dark Nebula Core	0.05 – 0.2	$10^4 - 10^5$	0.3 – 10	≈ 10	0.2 – 0.4	B 335, L 1535

*Sometimes associated with low–mass star formation, adapted from Goldsmith (1987).

Named Dark Nebulae

Name	R.A.(2000) h m	Dec.(2000) ° ′
Black Hole (in Sagittarius), B 92	18 15.5	− 18 11
Bok's Valentine, ESO 210–6A	08 25.5	− 51 01
Coalsack (7°× 5°)	12 53	− 63
Dragon (in Lagoon nebula)	18 04.8	− 24 30
Fish on a Platter, B 144	19 59	+35
Fish's Mouth (in Orion nebula)	05 35.5	− 05 23
Hoffmeister's cloud (∼ 20 square degrees)	20 47	− 42
Horsehead Nebula, B 33	05 40.9	− 02 28
Keyhole Nebula (in η Carinae nebula)	10 43.8	− 59 52
Northern Coalsack	20 40	+42
Parrot's head, B 87	18 04.3	− 32 30
Pipe Nebula, B 59, 65–7, 78	17 33	− 26
Rho, ρ, Ophiuchi, B 39	16 25.6	− 24
Snake Nebula, B 72	17 23.5	− 23 38
Taurus Dark Cloud	04 30	+27

Selected Molecular Clouds

Source	R.A.(1950) h m s	Dec.(1950) ° ′ ″	V_{LSR} km s^{-1}
W3M	02 21 57.0	+61 52 45	− 40.2
NGC 1333	03 25 55.6	+31 10 10	+7.9
B5–NH_3	03 44 28.7	+32 44 30	+10.0
S206	03 59 30.0	+51 10 34	− 26.2
S209	04 07 19.8	+51 01 46	− 49.5
L1551	04 28 40.0	+18 01 52	+6.5
TMC 1	04 38 38.6	+25 36 20	+5.8
Orion KL (OMC1)	05 32 47.0	− 05 24 21	+8.8
OMC 2	05 32 59.1	− 05 12 10	+11.1
NGC 1999	05 34 00.0	− 06 47 01	+8.8
NGC 2023	05 39 10.0	− 02 17 49	+9.8
NGC 2024	05 39 13.0	− 01 57 04	+10.4
L1622–NH_3	05 51 54.0	+01 47 45	+1.0
S247	06 05 51.4	+21 38 40	+6.0
NGC 2264	06 38 24.4	+09 32 12	+7.3
IRC +10216	09 45 14.8	+13 30 41	− 26.0
L134N	15 51 34.0	− 02 43 31	+2.6
ρ Oph–CO	16 23 12.0	− 24 17 40	+4.0
ρ Oph–NH_3	16 23 55.0	− 24 27 00	+4.0
L63	16 47 17.0	− 18 00 00	+5.6
B255	17 17 40.9	− 23 25 30	+5.0

Selected Molecular Clouds

Source	R.A.(1950) h m s	Dec.(1950) ° ′ ″	V_{LSR} km s^{-1}
B68	17 19 36.0	− 23 47 30	+3.5
Sgr A–NH_3	17 42 28.0	− 29 01 30	+23.8
Sgr B2	17 44 11.0	− 28 22 30	+60.0
M8	18 00 36.8	− 24 22 52	+9.8
M16	18 16 07.0	− 13 49 48	+25.0
M17 SW	18 17 26.5	− 16 14 54	+18.8
B133–NH_3	19 03 32.0	− 06 58 00	+12.0
B134–CO	19 04 15.6	− 06 19 20	+11.0
W49	19 07 53.0	+09 01 00	+6.0
W51	19 21 27.0	+14 24 30	+58.3
B335–NH_3	19 34 34.0	+07 27 00	+8.0
G75.84+0.40	20 19 46.5	+37 21 34	− 10.0
S106–IRS	20 25 33.7	+37 12 50	− 1.7
W75N	20 36 50.0	+42 26 58	+11.8
DR 21	20 37 14.0	+42 09 00	− 2.2
B361–NH_3	21 10 28.0	+47 11 00	+2.4
IC 1396	21 35 16.0	+57 17 27	− 8.0
IC 5146	21 43 59.6	+47 20 43	+3.6
S140	22 17 41.0	+63 03 45	− 9.0
NGC 7538	23 11 37.0	+61 11 45	− 58.3
Cas A	23 21 11.0	+58 32 48	− 20.0
NGC 7822	23 58 35.0	+67 05 00	− 10.0

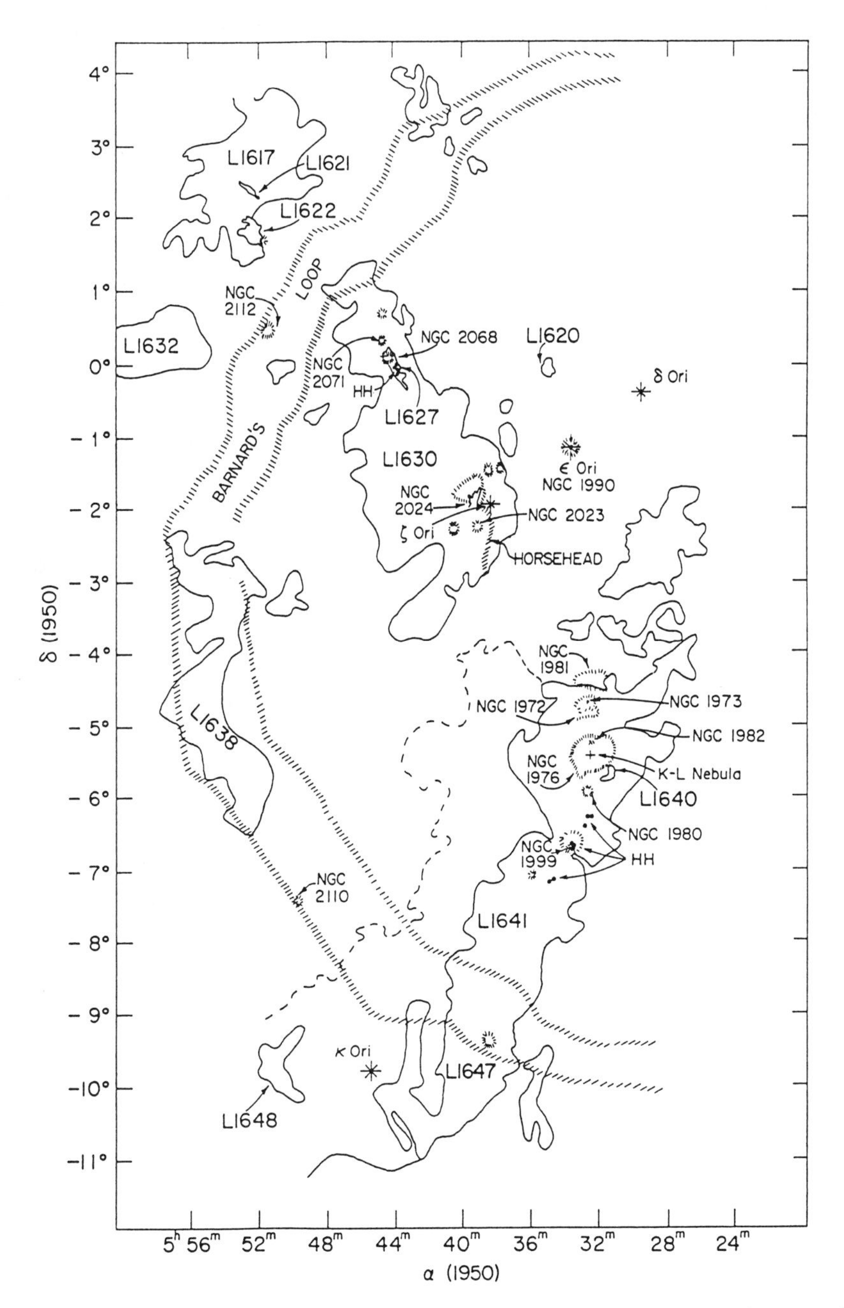

Dust Clouds in Orion. Solid lines outline dust clouds in the Orion region; they are designated by their L numbers in the Lynds catalogue of dark nebulae. The hatched boundaries indicate the boundaries of optical emission or reflection nebulosity.

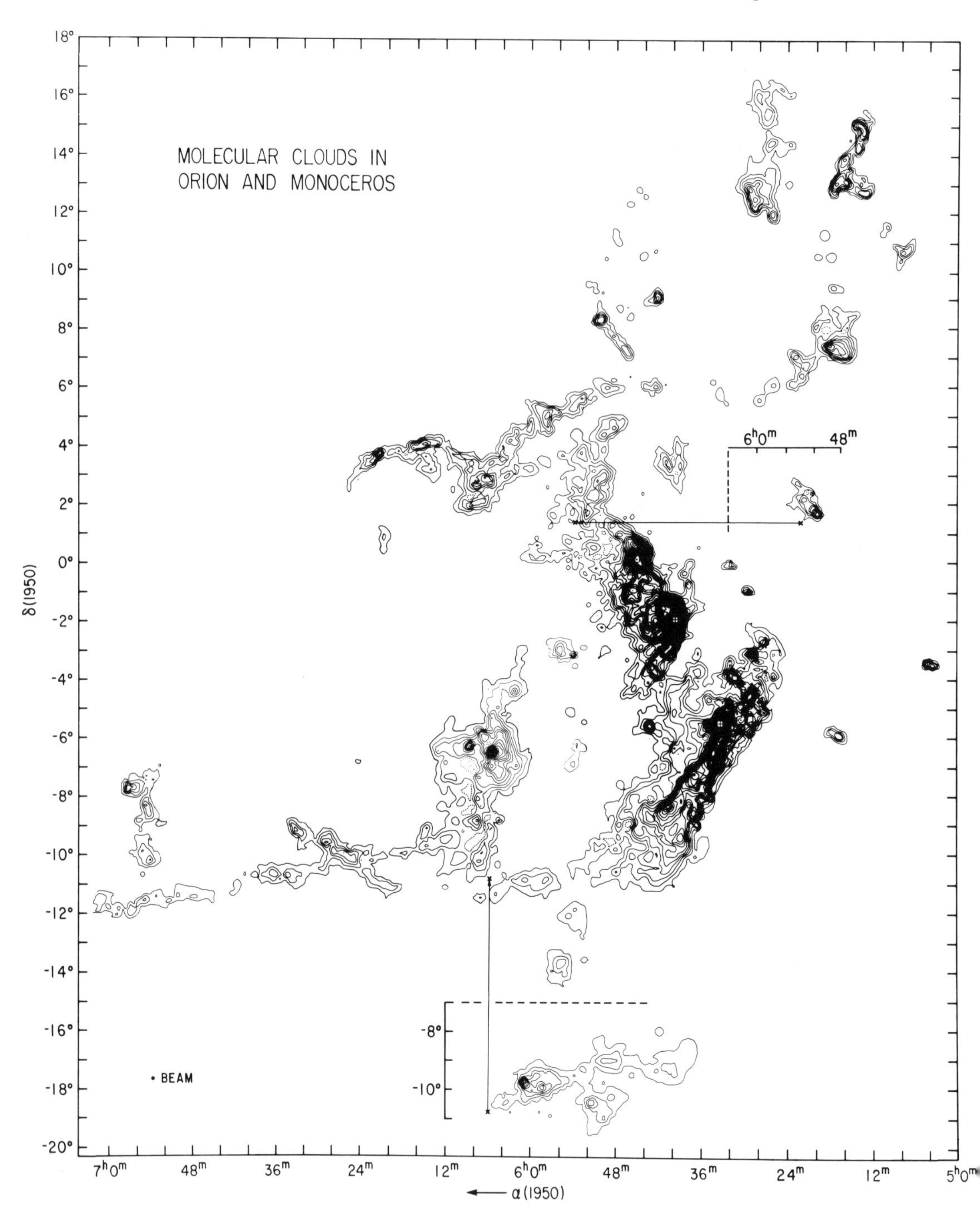

Molecular Clouds in Orion and Monoceros. This contour map of the velocity-integrated intensity of CO emission shows giant molecular clouds in the regions near the Orion Nebula. The lowest contour level is 1.28 K km s^{-1} with subsequent levels at 3, 5, 7 ... times this value. (Adapted from Maddalena et al. (1986).)

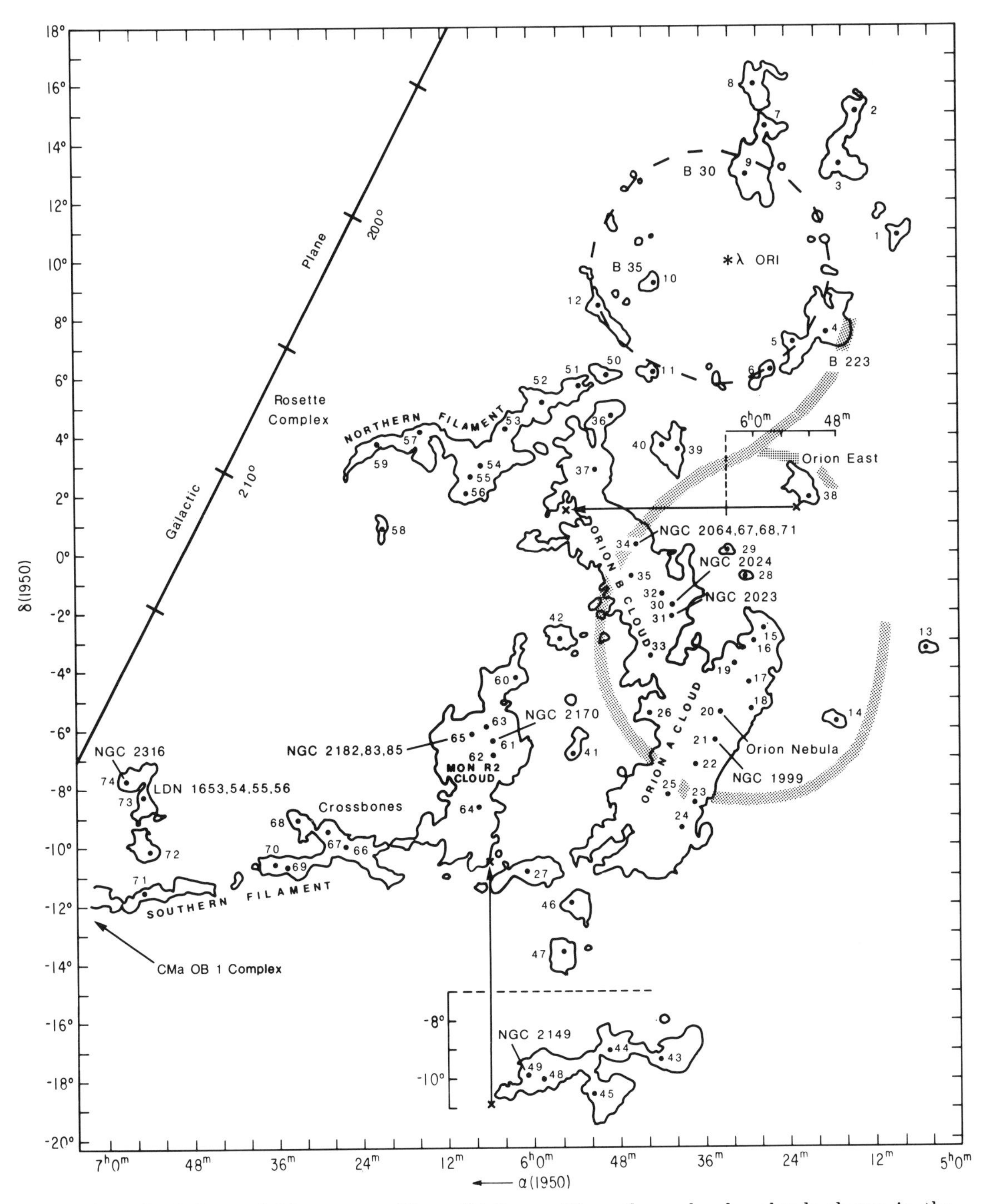

NGC Objects in Orion and Monoceros. The solid line outlines the molecular clouds shown in the previous figure. Here the location of NGC objects, the Orion Nebula, and Barnard's Loop (shaded arc) are shown. (Adapted from Maddalena et al. (1986).)

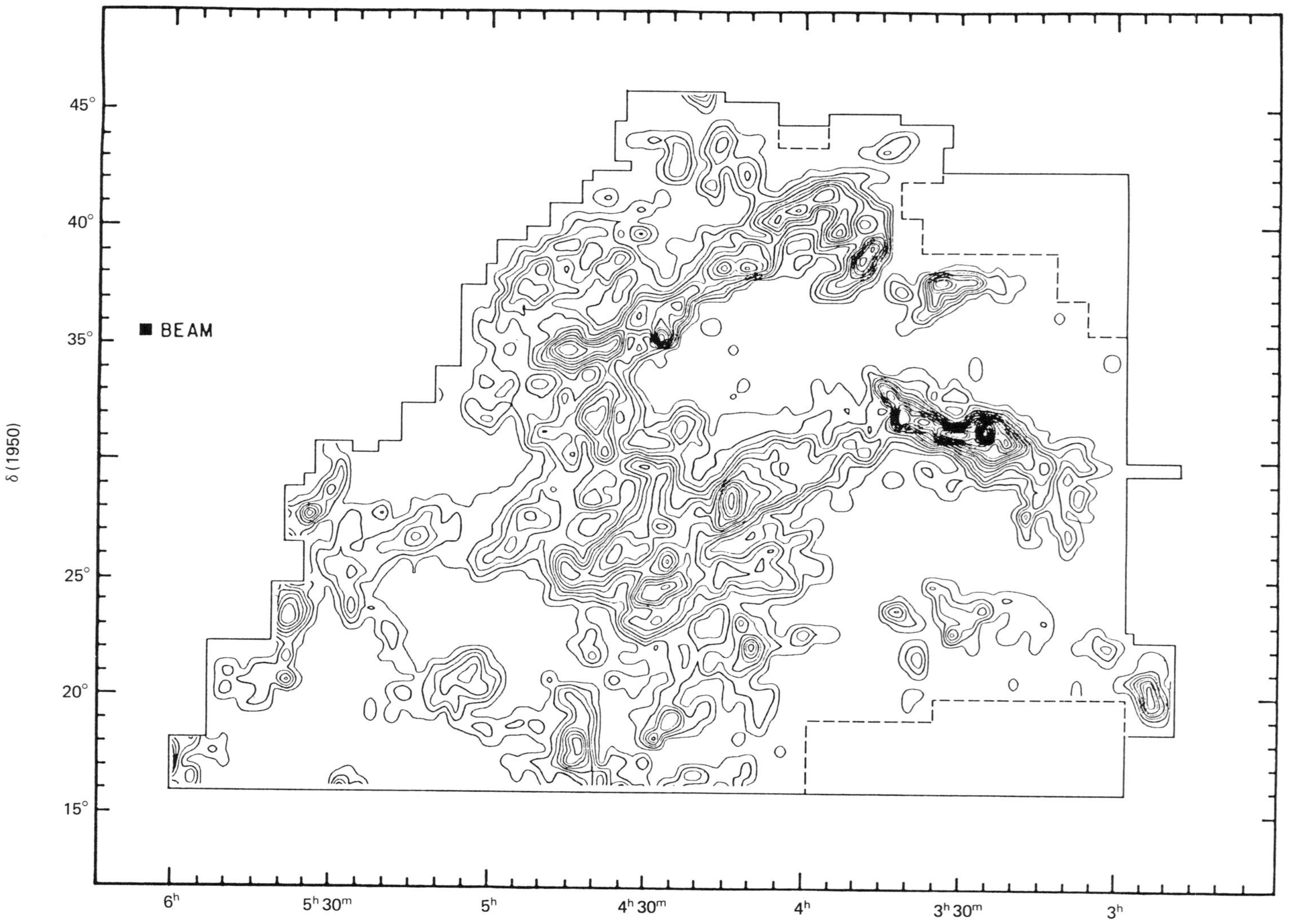

Molecular Clouds in Perseus, Taurus and Auriga. A contour map of the velocity-integrated intensity of CO emission in Perseus, Taurus and Auriga. The lower contour is 0.5 K km s^{-1}, and the separation between contours is 1.5 K km s^{-1}. (Adapted from Ungerechts and Thaddeus (1987).)

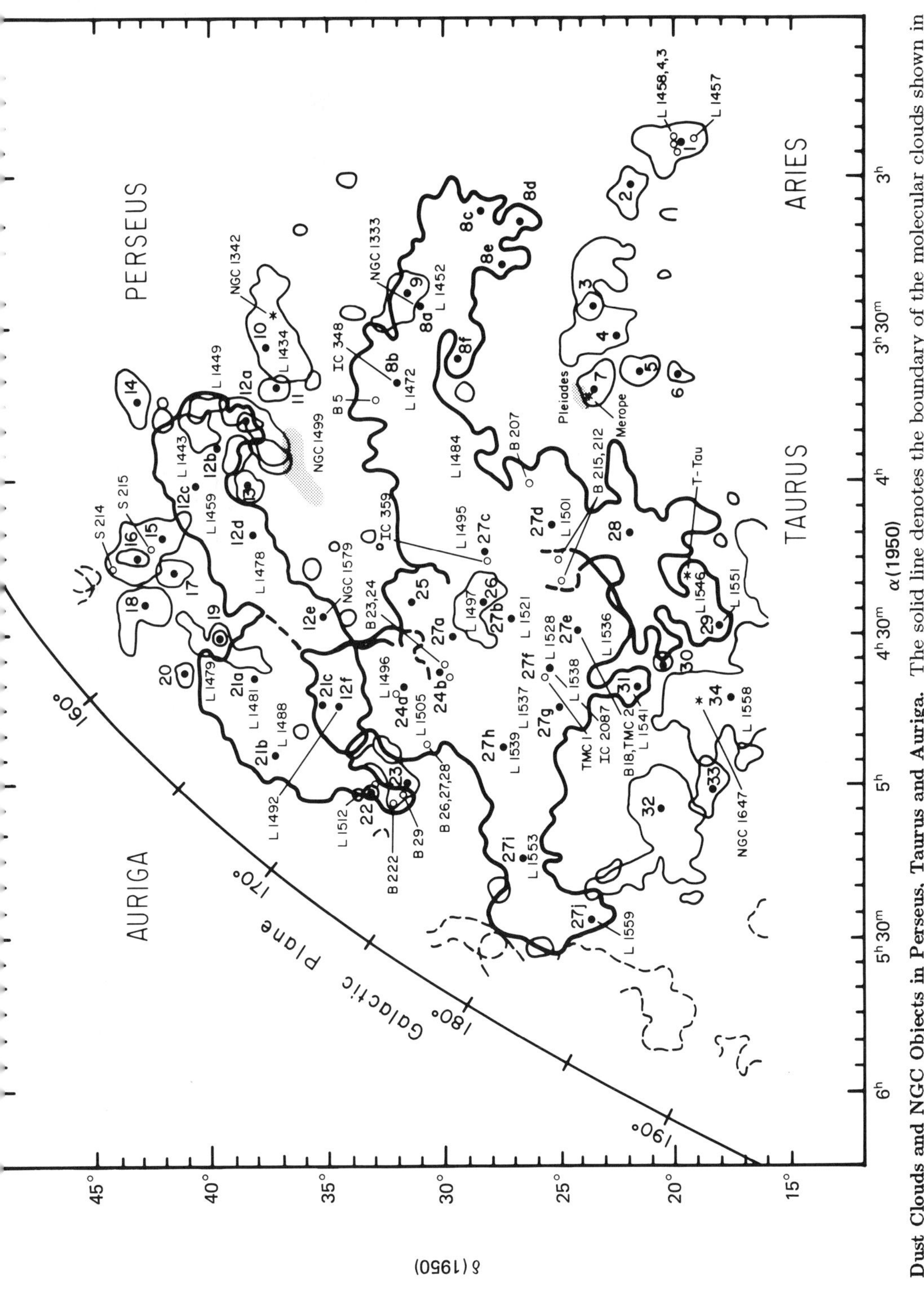

Dust Clouds and NGC Objects in Perseus, Taurus and Auriga. The solid line denotes the boundary of the molecular clouds shown in the previous figure. Here dust clouds are denoted by the L number in the Lynds catalogue of dark nebulae or by the B number in Barnard's description of dark nebulae. NGC or IC objects are also designated by the appropriate number in these catalogues. (Adapted from Ungerechts and Thaddeus (1987).)

Molecular Outflows from Young Stellar Objects*

Name†	R.A.(1950) h m s	Dec.(1950) ° ′ ″	D (kpc)	ΔV (km s^{-1})	Age ($\times 10^4$ yr)	$\dot{M}$ ($M_\odot$ yr^{-1})
LKHα 198†	00 08 47	+58 33 06	1.0	13	25	3.2×10^{-7}
S187–IRS	01 19 58	+61 33 08	2.0	30		
W3–IRS5†	02 21 53	+61 52 21	2.3	52		
W3 OH	02 23 17	+61 39 00	2.3	26		
AFGL 437†	03 03 32	+58 19 37		24		
RNO 13	03 22 04.8	+30 35 50	0.2	12	2.3	1.5×10^{-7}
AFGL 490†	03 23 38.8	+58 36 39	0.9	65	8	5.1×10^{-6}
RNO 15 NW†	03 24 24	+30 05	0.35	7	6	1.6×10^{-7}
RNO 15 FIR	03 24 36	+30 02 40	0.35	15	2.2	9.2×10^{-8}
RNO 15 L1455†	03 24 43.5	+30 01 43	0.35	12	4	3.6×10^{-7}
LKHα 325	03 25 48	+30 33 60	0.5	13		
HH 12	03 25 57	+31 10 00	0.5	20		
HH 7–11† SVS 13	03 25 58.2	+31 05 46	0.5	40	7	3.0×10^{-6}
T Tau	04 19 04	+19 25 05	0.14	13	4	1.5×10^{-7}
LKHα 101	04 26 57	+35 09 56	0.8	15		
L1551–IRS5†	04 28 40.2	+18 01 42	0.16	30	2.3	7.0×10^{-7}
HL Tau	04 28 44.4	+18 07 37	0.16	11	0.7	4.1×10^{-8}
L1529	04 29 43.4	+24 16 54	0.14	30		
L1527	04 36 00	+26 10 00	0.16	6		
RNO 43S†	05 29 35.0	+12 50	0.4	11		
RNO 43N†	05 29 38.0	+12 55	0.4	11		
Orion A†	05 32 47	− 05 24 14	0.5	127	0.1	8.8×10^{-4}
CS–Star	05 33 55.4	− 06 47 24	0.5	8		
V380 Ori	05 33 59.5	− 06 44 26	0.5	10	3	3.2×10^{-7}
V380 Ori, NE†	05 34 00	− 06 41	0.5	12	4	2.4×10^{-7}

Molecular Outflows from Young Stellar Objects*

Name†	R.A. 1950 h m s	Dec. 1950 ° ′ ″	D (kpc)	ΔV (km s^{-1})	Age ($\times 10^4$ yr)	$\dot{M}$ ($M_\odot$ yr^{-1})
NGC 1999	05 34 00.0	− 06 45 00	0.5	15		
V380 Ori S	05 34 00	− 06 56	0.5	13	4	2.1×10^{-6}
Haro 4–255†	05 36 55	+07 27 40	0.5	10	5	4.2×10^{-7}
S235 B	05 37 31	+35 39 55	0.8	22		
NGC 2023	05 39 06	− 02 17 24	0.5	13		
NGC 2024	05 39 14	− 01 55 59	0.5	35		
HH 26 IR†	05 43 31.6	− 00 15 23	0.5	30		
HH 24†	05 43 38	− 00 13 23	0.5	15		
NGC 2068, H_2O	05 43 58	− 00 04	0.5	11		
NGC 2071 IR†	05 44 30	+00 20 40	0.5	75		
HD 250550	05 59 07	+16 13 06	1.0	7		
Mon R2†	06 05 22	− 06 22 25	0.8	31		
GGD 12–15†	06 08 25	− 06 10 45	1.0	25		
S255–IRS1	06 09 59	+18 00 15	0.81	26		
AFGL 961†	06 31 59	+04 15 09	1.6	30		
NGC 2264	06 38 25	+09 32 12	0.76	28		
ρ Oph 2	16 23 59	− 24 19 49	0.16	16		
IRAS 16293–2422	16 29 20.9	− 24 22 13	0.16	40	> 1	$> 10^{-5}$
RNO 91, L43	16 31 37.9	− 15 40 50	0.16	5	2.5	8.9×10^{-9}
NGC 6334	17 16 37	− 35 54 49	1.7	60		
M8E–IR	18 01 48.8	− 24 26 56	1.5	16		
S68†	18 27 25	+01 12 40	0.5	28		
R CrA†	18 58 31.6	− 37 01 30	0.7	17	0.85	3.3×10^{-7}
L723†	19 15 42	+19 06 49	0.30	20		
AS 353†	19 19 09.4	+10 56 30	0.20	11	~ 3	1.9×10^{-8}

†Objects with a name designated by a † mark are established bipolar outflows. *Positions are for the central object, D denotes the distance, and ΔV is the CO line width. Adapted from Lada (1985) as updated from Cabrit, Goldsmith and Snell (1988), Hirano et al. (1988), Lizano et al. (1988), Margulis, Lada and Snell (1988), Little et al. (1988), Mathieu et al. (1988), Mitchell et al. (1988), Moneti et al. (1988), Moriarity-Schieven and Snell (1988), Sandell, Magnani and Lada (1988) and Walker et al. (1988). Flow lifetimes, or ages,and mass loss rates, $\dot{M}$, are from Levreault (1988) with a few additions from the aforementioned references.

Molecular Outflows from Young Stellar Objects*

Name†	R.A. 1950 h m s	Dec. 1950 ° ′ ″	D (kpc)	ΔV ($km\ s^{-1}$)	Age ($\times 10^4$ yr)	$\dot{M}$ ($M_\odot\ yr^{-1}$)
B335† IRAS 1934+0727	19 34 34.7	+07 27 20	0.25	8	25	1.0×10^{-8}
S87	19 44 14	+24 28 00	2.7	20		
S88 B	19 44 41	+25 05 11	2.0	17		
IRAS 20188+3928†	20 18 50.7	+39 28 18	0.4 to 4.0	14		
S106	20 25 34	+37 12 52	1.3	18		
AFGL 2591†	20 27 35.8	+40 01 14	1.2	42		
DR 21	20 37 13	+42 08 50	3.0	53		
PV Ceph†	20 45 23.3	+67 46 36	0.5	5	18	5.9×10^{-8}
V1057 Cyg	20 57 06.2	+44 03 47	0.7	10	6	2.4×10^{-8}
V1331 Cyg	20 59 32.1	+50 09 56	0.7	7	7	1.0×10^{-7}
V645 Cyg†	21 38 11	+50 00 43	6.0	28	50	1.4×10^{-6}
NGC 7129, FIR†	21 41 52	+65 49 50	1.0	13		
LKHα 234	21 41 57	+65 53 05	1.0	15		
Elias 1–12	21 45 26.8	+47 18 08	0.9	10	7	1.1×10^{-7}
BD +46°3471	21 50 38.5	+46 59 34	0.9	10	7	3.2×10^{-7}
S140	22 17 41	+63 03 40	1.0	42		
Cep A†	22 54 19	+61 45 44	0.7	40	10	2.0×10^{-6}
NGC 7538	23 11 36	+61 11 49	2.8	35		
MWC 1080	23 15 16.3	+60 33 52	2.5	60	22	3.1×10^{-6}

†Objects with a name designated by a † mark are established bipolar outflows. *Positions are for the central object, D denotes the distance, and ΔV is the CO line width. Adapted from Lada (1985) as updated from Cabrit, Goldsmith and Snell (1988), Hirano et al. (1988), Lizano et al. (1988), Margulis, Lada and Snell (1988), Little et al. (1988), Mathieu et al. (1988), Mitchell et al. (1988), Moneti et al. (1988), Moriarity-Schieven and Snell (1988), Sandell, Magnani and Lada (1988) and Walker et al. (1988). Flow lifetimes, or ages,and mass loss rates, $\dot{M}$, are from Levreault (1988) with a few additions from the aforementioned references.

Circumstellar Gaseous (CO) Disks**

Star	R.A.(1950) h m s	Dec.(1950) ° ′ ″	D (kpc)	θ (″)	M ($M_\odot$)	R (AU)
AFGL 490	03 23 38.8	+58 36 39	0.9	3	6–50	1300
L1551–IRS5	04 28 40.2	+18 01 42	0.16	10	0.1	700
HL Tau	04 28 44.4	+18 07 37	0.16	30	0.1	2000

**Source distance, D, angular extent, θ, mass, M, and radius, R. Adapted from Mundy and Adelman (1988), Sargent and Beckwith (1987), Sargent et al. (1988).

Circumstellar Dust Shells with Infrared (60-micron) Excess*

Name	R.A.(1950) h m s	Dec.(1950) ° ′ ″	$\mu_{R.A.}$ (s yr^{-1})	$\mu_{Dec.}$ (″ yr^{-1})	Sp LC	V (mag)	T (°K)	R (AU)
τ(1) Eri	02 42 46	- 18 47 00	+0.0229	+0.042	F6V	4.46	71	22
ε Eri	03 30 34	- 09 37 36	- 0.0662	- 0.979	K2V	3.73	106	3.5
γ Dor	04 14 43	- 51 36 42	+0.0113	+0.105	F0V	4.24	77	32
DM+79°169	05 14 17	+79 10 42	- 0.0285	- 0.080	F6V	5.04	81	17
β Pic	05 46 06	- 51 05 00	+0.0005	+0.005	A5V	3.85	106	31
β UMa	10 58 50	+56 39 06	+0.0055	+0.029	A1V	2.40	120	48
β Leo	11 46 31	+14 51 06	- 0.0343	- 0.497	A3V	2.14	111	35
α CrB	15 32 34	+26 52 54	+0.0079	- 0.098	A0V	2.2	141	35
γ Ser	15 54 08	+15 49 24	+0.0212	+0.306	F6V	3.86	76	19
α Lyr	18 35 15	+38 44 12	+0.0171	+0.200	A0V	0.03	84	99
DM−47°13928	21 45 01	- 47 31 54	+0.0162	+0.164	G2V	5.58	105	7.1
α PsA	22 54 54	- 29 53 18	+0.0258	+0.336	A3V	1.16	75	76

*Adapted from Aumann (1985) for nearby stars associated with infrared sources that show large excess fluxes at 60 microns. These positions are for the associated stars; the infrared positions, which are between 0.05′and 0.5′away, are listed by Aumann (1985).

17.2 T Tauri Stars and Related Objects

T Tauri stars are young pre-main-sequence objects of low mass M = 0.2 to 2 $M_\odot$ with ages typically from 10^5 to 10^6 years. They are characterized by emission lines of hydrogen (Balmer Hα) and the H and K lines of Ca II. These lines are similar to those observed in the solar chromosphere and in solar active regions, but the intensity and width of the T Tauri emission lines are considerably larger than those for the Sun.

The emission lines probably arise in a hot circumstellar envelope. They are usually superposed upon a bright continuous spectrum, which often displays the absorption line spectrum of a star with late spectral type, usually K. The emission line profiles have been interpreted in terms of mass outflow with mass-loss rates on the order of 10^{-8} $M_\odot$ per year. Some of the T Tauri stars are also associated with bipolar outflows of cool molecular material.

The majority of T Tauri stars are found embedded within dark clouds of dust and gas, such as those found in Ophiuchus and Taurus-Auriga; others are found in regions of bright nebulosity, such as Orion, or within young star clusters like NGC 2264. Groups of T Tauri stars associated with dark cloud complexes are called T Associations. They typically subtend between 1 and 10 degrees on the sky [Kholopov (1959)].

The T Tauri stars exhibit irregular and unpredictable changes of light. A few have shown dramatic increases in brightness, dubbed the FU Orionis phenomenon; an example is V 1057 Cyg which brightened by about 6 magnitudes in 200 days.

Infrared observations suggest that many T Tauri stars are surrounded by extended, flattened disks of gas and dust. Unexpectedly intense infrared radiation has been attributed to thermal emission by the circumstellar dust grains. The gaseous disks can occasionally be resolved, as was the case with HL Tauri.

Some T Tauri stars also exhibit detectable soft X-ray radiation with luminosities of 10^{30} to $10^{31} erg\, s^{-1}$, which is about a thousandth of the star's bolometric luminosity [Fiegelson and De Campli (1981)]. Other pre-main-sequence stars exhibit relatively intense X-ray radiation, but they lack the strong emission lines and infrared excesses which characterize the classical T Tauri stars. They have been dubbed naked T Tauri stars because they are found in regions of active star formation and are probably related to the classical objects.

The first accompanying table is a catalogue of naked T Tauri stars in the Taurus-Auriga complex [Walter et al. (1988)]. It provides the names, right ascensions, R.A., and declinations, Dec., for epoch 1950.0, spectral type, Sp, radial velocity, V_r, maximum apparent visual magnitude, V_{max}, mean visual extinction, A_V, and the logarithm of the stellar luminosity, L, in units of the Sun's luminosity, $L_\odot$, for an assumed distance of 140 *pc* to the complex.

It is followed by our main catalogue of T Tauri stars and related objects. The names, R.A., Dec., and galactic coordinates, l and b, are from Herbig and Rao (1972), while the V, Sp, A_V, and bolometric luminosity, L, are from Cohen and Kuhi (1979). If a single zero appears for the extinction, then the color is bluer than would be expected for an unreddened star. The assumed distances for the well-known groups of T Tauri stars were 160 *pc* for the Taurus-Auriga complex, 170 *pc* for the Ophiuchus dark clouds, 460 *pc* for the Orion nebula, 500 *pc* for NGC 1333, 700 *pc* for NGC 7000/IC 5070, 800 *pc* for NGC 2264, and 845 *pc* for Cepheus IV. These distances were used to compute $\log(L/L_\odot)$; when no distance was available, the luminosity column contains $(L/L_\odot D^2)$ with the power of ten given in parenthesis. The last entries in our main catalogue provide the locations and numbers in the Herbig and Rao (1972) catalogue.

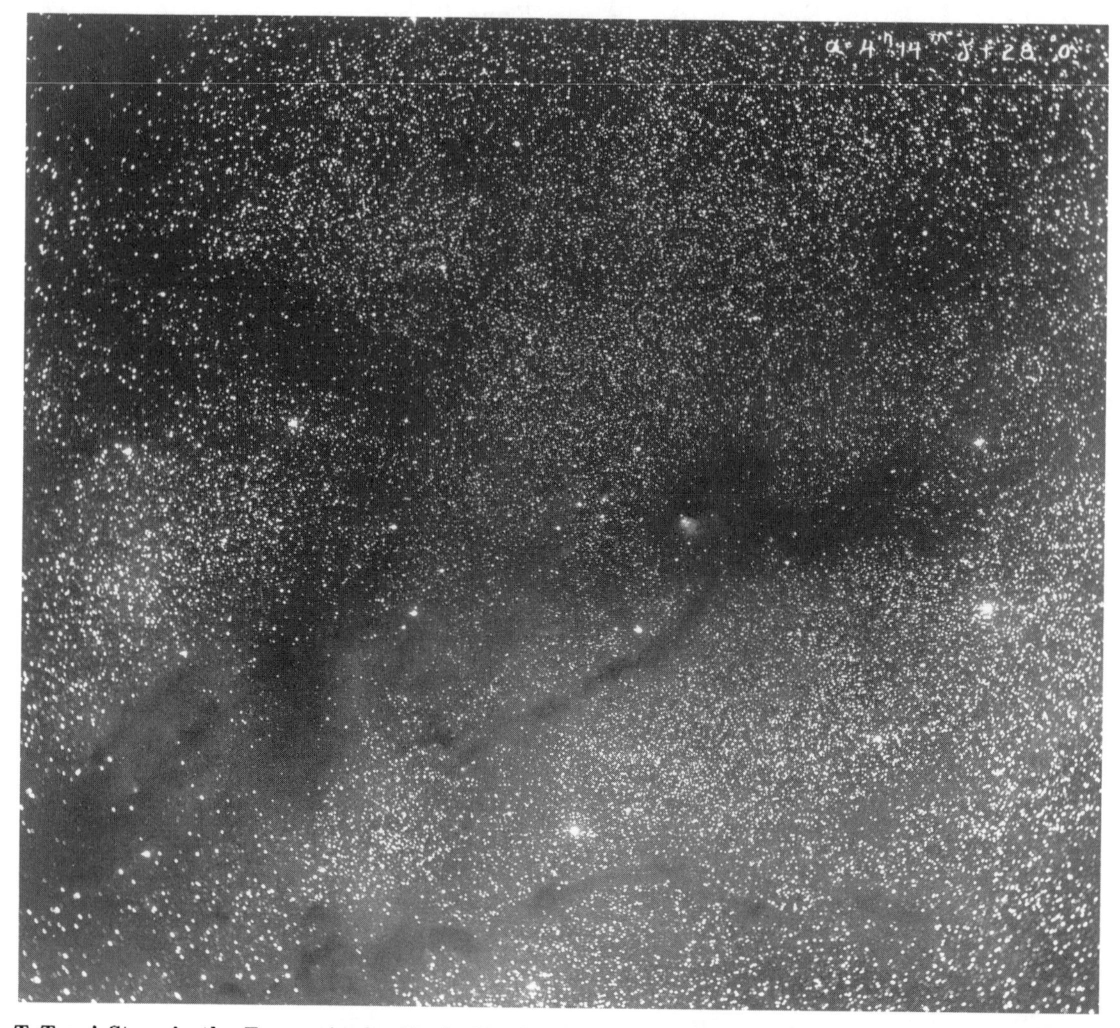

T Tauri Stars in the Taurus-Auriga Dark Clouds. Interstellar dust blocks the light of distant stars, marking the location of dark filaments that contain new-born T Tauri stars. This photograph, published by Edward Emerson Barnard in 1927, is 9.6 degrees in angular width, which corresponds to a linear width of 78 light years at the distance of the Taurus clouds.

Naked T Tauri Stars*

Name	R.A.(1950) h m s	Dec.(1950) ° ′ ″	Sp	V_r (km s^{-1})	V_{max} (mag)	A_V (mag)	log ($L/L_\odot$)
03261+2420	03 26 40.5	+24 20 24	K1	15:	12.00	0.5	- .30
034903+2431	03 49 02.7	+24 30 55	K5	9.8	12.17	0.0	- .30
035120+3154SW	03 51 20.2	+31 54 14	G0	16:	11.80	0.9	- .11
035120+3154NE	03 51 20.8	+31 54 18	G5	15.1	12.29	1.0	- .24
035135+2528NW	03 51 34.7	+25 28 25	K3	9.9	13.72	1.2	
035135+2528SE	03 51 35.0	+25 28 22	K2	11.3	12.63	0.5	- .51
SAO 76411A	03 59 55.1	+21 59 59	G1	19.0	8.85	0.0	+.68
040012+2545N+S	04 00 12.2	+25 44 45	K2		12.83	0.7	- .78
040047+2603W	04 00 46.9	+26 02 42	M2		14.34	0.5	- .47
040047+2603E	04 00 48.6	+26 02 43	M2	14.4	14.13	0.2	- .56
SAO 76428	04 01 31.3	+21 47 56	F8	15.3	9.46	0.1	+.46
040142+2150W	04 01 42.1	+21 50 11	M3		14.93	0.3	- .60
040142+2150NE	04 01 42.6	+21 50 13	M3	16.2	15.08	0.3	- .76
040234+2143	04 02 33.8	+21 43 05	M2	16.7	14.61	0.3	- .70
041529+1652	04 15 29.4	+16 51 30	K5	15.8	13.23	0.0	- .75
041559+1716	04 15 59.1	+17 16 01	K7	18:	12.21	0.0	- .32
V819 Tau	04 16 19.9	+28 19 02	K7	14.4	13.14	1.2	- .03
041636+2743	04 16 35.8	+27 42 38	M0	16.0	12.17	0.3	- .01
HDE 283572	04 18 52.5	+28 11 07	G6	15:	8.99	0.4	+.84
042417+1744	04 24 17.2	+17 44 03	K1	15.4	10.30	0.1	+.22
042835+1700	04 28 34.5	+17 00 02	K5	16.7	12.48	0.2	- .37
042916+1751	04 29 15.7	+17 51 04	K7	19.0	11.98	0.0	- .18
V827 Tau	04 29 20.5	+18 13 54	K7	16.6	12.05	0.3	+.04
V826 Tau	04 29 22.1	+17 55 21	K7	17.9	12.06	0.3	- .30
042950+1757	04 29 50.0	+17 56 40	K7	18.1	13.18	0.7	- .36
V830 Tau	04 30 08.0	+27 27 26	K7	17.7	12.08	0.4	- .05
043124+1824	04 31 23.7	+18 23 55	G8	18.5	12.65	1.1	- .30
043220+1815	04 32 19.8	+18 15 29	F8	22.1	10.94	0.8	+.16
043230+1746	04 32 40.4	+17 45 34	M2	20.7	14.12	0.7	- .42
045226+3013	04 52 25.9	+30 13 12	K5	14.6	10.76	0.0	+.20
045251+3016	04 52 51.0	+30 16 20	K7	15:	11.54	0.0	+.01
V836 Tau	05 00 02.2	+25 19 09	K7	20.5	13.02	0.9	- .15

*Adapted from Walter et al. (1988). A distance of 140 pc is assumed for the Taurus-Auriga complex.

Catalogue of T Tauri Stars and Related Objects

Name	R.A.(1900) h m s	Dec.(1900) ° ′ ″	l °	b °	V (mag)	Sp
MacC H12	00 01 50	+65 05 13	118.4	+ 3.2	16.1	A5–F:
LkH 197	00 05 20.5	+58 16 43	117.7	– 3.6	15.3	K7–M0
LkH 198	00 06 10.3	+58 16 24	117.8	– 3.6	14.7	B3
MacC H10	00 07 34	+65 00 46	119.0	+ 3.0	14.5	K4
MacC H9	00 08 08	+65 02 37	119.0	+ 3.0	14.8	K4
LkH 200	00 36 42	+61 23	121.9	– 0.9	13.7	K1
LkH 201	00 37 30	+61 05	122.0	– 1.2	13.7	B3
LkH 262	02 50 27.4	+19 39 06	158.8	– 34.0	14.1	M0
Lkh 263	02 50 27.8	+19 39 20	158.8	– 34.0	14.7	M2
LkH 264	02 50 56.8	+19 41 22	158.9	– 39.9	12.7	K3:(C)
LkH 325	03 22 42	+30 23	158.7	– 21.0	13.5	K7–M0
LkH 270	03 23 06.8	+31 01 59	158.3	– 20.4	14.5	K7–M0
LkH 271	03 23 11.3	+30 54 51	158.4	– 20.5	14.7	K3
LkH 326	03 24 48	+30 13	159.2	– 20.9	16.0	M0
LkH 327	03 27 18	+30 50	159.2	– 20.1	15.0	K2
LkH 328	03 27 48	+30 54	159.3	– 19.9	16.8	M1
LH 92	03 38 10.2	+31 45 32	160.5	– 17.9	16.5	K?e
LH 97	03 38 28.8	+31 45 06	160.6	– 17.8	18.0	–e
LkH 329	03 39 19.8	+32 07 07	160.5	– 17.4	14.9	K3
LkH 330	03 39 31.2	+32 05 23	160.5	– 17.4	12.2	G3:
LkH 272	03 42 28.2	+30 37 50	156.8	– 12.0	14.5:	K2:e
LkH 273	03 42 38.8	+38 38 12	156.8	– 12.0	14.0:	–e
FM Tau	04 08 02.7	+27 57 34	168.2	– 16.3	14.0	M0
FN Tau	04 08 03.1	+28 12 43	168.0	– 16.2	15.7	M5
CW Tau	04 08 06.2	+27 55 43	168.2	– 16.3	15.1	K3
FP Tau	04 08 40.1	+26 31 16	169.4	– 17.2	13.8	M5
CX Tau	04 08 40.6	+26 33 00	169.4	– 17.2	14.8	M2.5
CY Tau	04 11 22.0	+28 05 57	168.6	– 15.7	14.5	M1
V410 Tau	04 12 17	+28 13 12	168.7	– 15.5	10.6	K7
DD Tau	04 12 19.5	+28 01 47	168.8	– 15.6	14.0	M1
CZ Tau	04 12 19.9	+28 02 17	168.8	– 15.6	15.3	M1.5
BP Tau	04 13 01.8	+28 51 51	168.3	– 14.9	12.1	K7
DE Tau	04 15 44.5	+27 40 51	169.6	– 15.3	12.9	M1
RY Tau	04 15 44.8	+28 12 21	169.2	– 14.9	10.4	K1
T Tau	04 16 09.4	+19 17 52	176.2	– 20.9	10.3	K1

Catalogue of T Tauri Stars and Related Objects

Name	Av (mag)	$\log(L/L_\odot)$	Location	No.
MacC H12	5.45	+2.16	Cep IV	1n
LkH 197	1.32	5.34(-6)	Anon dark neb	2
LkH 198	4.53	3.55(-4)	Anon dark neb	3n
MacC H10	2.10	+0.92	Cep IV	4
MacC H9	2.02	+0.79	Cep IV	5
LkH 200	0.72	7.93(-6)	NGC 225	6
LkH 201	4.24	1.24(-4)	NGC 225	7
LkH 262	0.08	1.09(-5)	Anon dark neb	8
Lkh 263	0.22	1.04(-5)	Anon dark neb	9
LkH 264	0.52	1.77(-4)	Anon dark neb	10
LkH 325	0.46	+0.92	B 205	11n
LkH 270	1.96	+1.13	NGC 1333, B 205	12
LkH 271	3.07	+0.41	NGC 1333, B 205	13
LkH 326		+0.90	Near NGC 1333, B 205	14n
LkH 327	3.85	+0.40	Tau-Aur, B1	15
LkH 328	2.40	-0.62	Tau-Aur, B1	16
LH 92			IC 348, B4	17
LH 97			IC 348, B4	18
LkH 329	0.69	2.67(-5)	IC 348	19
LkH 330	1.78	1.51(-4)	IC 348	20
LkH 272			Near XY Per	21
LkH 273			Near XY Per	22
FM Tau	0.90	-0.28	Tau-Aur, B 209	23n
FN Tau	1.40	-0.10	Tau-Aur, B 209	24
CW Tau	2.34	+0.35	Tau-Aur, B 209	25n
FP Tau	0.26	-0.21	Tau-Aur	26
CX Tau	0.85	-0.21	Tau-Aur	27
CY Tau	0.92	-0.15	Tau-Aur, B7	28
V410 Tau	0.03	+0.38	Tau-Aur, B7	29
DD Tau	3.50	+0.29	Tau-Aur, B10, B7	30n
CZ Tau	1.35	-0.03	Tau-Aur, B10, B7	31n
BP Tau	0.55	+0.21	Tau-Aur, near B7	32
DE Tau	0.20	+0.24	Tau-Aur	33
RY Tau	1.88	+1.24	Tau-Aur, B 214	34n
T Tau	1.44	+1.45	Tau-Aur	35n

Catalogue of T Tauri Stars and Related Objects

Name	R.A.(1900) h m s	Dec.(1900) ° ′ ″	l °	b °	V (mag)	Sp
DF Tau	04 20 56.9	+25 28 49	172.1	- 16.0	11.7	M0.5
DG Tau	04 20 57.7	+25 52 42	171.8	- 15.7	12.7	C
DH Tau	04 23 33.1	+26 19 46	171.9	- 14.9	13.5	M0
DI Tau	04 23 34.0	+26 19 37	171.9	- 14.9	12.6	M0
LkH 101	04 23 42	+35 04	165.4	- 9.0	15.9	C
IQ Tau	04 23 44.2	+25 53 34	172.2	- 15.2	14.1	M0.5
UX Tau B	04 24 16.0	+18 00 41	178.6	- 20.2	13.5	M1
UX Tau A	04 24 16.3	+18 00 41	178.6	- 20.2	11.3	K2
FX Tau	04 24 26.6	+24 13 38	173.7	- 16.2	13.1	M1
DK Tau	04 24 37.1	+25 48 20	172.5	- 15.1	12.2	K7
ZZ Tau	04 24 47.7	+24 29 19	173.5	- 16.0		
LkH 331	04 25 21.4	+23 57 54	174.0	- 16.2	14.1	M5.5
HK Tau	04 25 47.5	+24 11 22	173.9	- 16.0	14.7	M0.5
HL Tau	04 25 50.7	+18 01 03	178.9	- 20.0	14.9	(K7?)
XZ Tau	04 25 52.3	+18 01 01	178.9	- 20.0	14.6	M3
LkH 266	04 26 09.7	+18 08 43	178.8	- 19.8	14.5	M1
UZ Tau f	04 26 35.9	+25 39 43	172.9	- 14.9	14.1	M1.5
UZ Tau p	04 26 35.9	+25 39 43	172.9	- 14.9	14.6	M0.5
GG Tau	04 26 44.2	+17 18 51	179.6	- 20.2	12.4	K7–M0
GH Tau	04 27 03.7	+23 56 49	174.3	- 15.9	13.3	M2
GI Tau	04 27 31.0	+24 08 36	174.2	- 15.7	13.1	K7
GK Tau	04 27 31.4	+24 08 25	174.2	- 15.7	12.5	K7
DL Tau	04 27 33.4	+25 07 57	173.4	- 15.0	12.2	C
IS Tau	04 27 46.6	+26 00 48	172.8	- 14.4	15.1	K2
HN Tau	04 27 52.3	+17 39 12	179.5	- 19.8	13.4	K5
CI Tau	04 27 52.8	+22 37 51	175.5	- 16.6	13.0	K7
DM Tau	04 28 00.9	+17 57 31	179.2	- 19.6	14.1	M0.5
AA Tau	04 28 51.8	+24 16 22	174.3	- 15.4	12.6	K7–M0
HO Tau	04 29 21.7	+22 19 46	175.9	- 16.6	15.8	M0.5
DN Tau	04 29 24.3	+24 02 33	174.6	- 15.4	12.0	M0
HP Tau	04 29 53.3	+22 42 00	175.7	- 16.2	13.4	K3
DO Tau	04 32 20.2	+25 58 48	173.5	- 13.7	13.6	K7–M0
VY Tau	04 33 17.8	+22 35 58	176.3	- 15.7	13.5	M0
LkH 332	04 36 01.1	+25 11 40	174.7	- 13.6	13.9	K7–M0
DP TAU	04 36 31.3	+25 04 10	174.8	- 13.6	14.3	M0.5

Catalogue of T Tauri Stars and Related Objects

Name	Av (mag)	$\log(L/L_\odot)$	Location	No.
DF Tau	1.90	+0.73	Tau–Aur	36
DG Tau		>0.88	Tau–Aur, B 217	37n
DH Tau	1.01	– 0.12	Tau–Aur, B 19	38
DI Tau	0.82	0.00	Tau–Aur, B 19	39
LkH 101		>1.07(–2)	NGC 1579	40n
IQ Tau	0.77	– 0.05	Tau–Aur, B 218	41
UX Tau B	0.16	– 0.28	Tau–Aur	42
UX Tau A	0.20	+0.28	Tau–Aur	43
FX Tau	1.57	+0.13	Tau–Aur, B 18	44
DK Tau	1.18	+0.60	Tau–Aur, near B 19	45
ZZ Tau			Tau–Aur, B 18	46
LkH 331	0.40	– 0.29	Tau–Aur, B 18	47
HK Tau	3.02	+0.05	Tau–Aur, B 18	48
HL Tau		>0.64	Tau–Aur	49n
XZ Tau	3:	+1.03	Tau–Aur	50n
LkH 266	0.88	– 0.13	Tau–Aur	51
UZ Tau f	0.85	+0.23	Tau–Aur, B 19	52
UZ Tau p	1.53	+0.03	Tau–Aur, B 19	53
GG Tau	0.78	+0.45	Tau–Aur	54
GH Tau	0.53	+0.09	Tau–Aur, B 18	55
GI Tau	1.66	+0.28	Tau–Aur, B 18	56n
GK Tau	1.12	+0.54	Tau–Aur, B 18	57n
DL Tau		>0.17	Tau–Aur, near B 19	58
IS Tau	3.86	+0.26	Tau–Aur, B 19	59
HN Tau	0.54	+0.09	Tau–Aur	60
CI Tau	1.20	+0.22	Tau–Aur	61
DM Tau	0.00	– 0.43	Tau–Aur	62
AA Tau	0.93	+0.14	Tau–Aur, B 18	63
HO Tau	1.15	– 0.64	Tau–Aur	64
DN Tau	0.42	+0.21	Tau–Aur, B 18	65
HP Tau	2.32	+0.48	Tau–Aur	66n
DO Tau	1.35	+0.54	Tau–Aur, B 22	67n
VY Tau	0.85	– 0.12	Tau–Aur	68
LkH 332	2.74	+0.31	Tau–Aur, B 22	69n
DP TAU	0.94	– 0.07	Tau–Aur, B 22	70

Catalogue of T Tauri Stars and Related Objects

Name	R.A.(1900) h m s	Dec.(1900) ° ′ ″	l °	b °	V (mag)	Sp
GO Tau	04 36 57	+25 08 48	174.8	− 13.4	14.5	M0
DQ Tau	04 41 07.3	+16 49 08	182.2	− 17.9	14.1	K7–M0
Haro 6–37	04 41 13.1	+16 51 47	182.2	− 17.8	13.7	K6
DR Tau	04 41 20.5	+16 47 52	182.3	− 17.8	13.0	C
DS Tau	04 41 30.0	+29 14 26	172.3	− 10.0	12.3	K3
UY Tau	04 45 24.4	+30 37 00	171.8	− 8.5	12.6	K7
GM Aur	04 48 48.9	+30 12 15	172.6	− 8.2	12.0	K7–M0
AB Aur	04 49 23.1	+30 23 25	172.5	− 8.0	7.1	B9–A0
SU Aur	04 49 36.6	+30 24 24	172.5	− 7.9	9.0	G2
RW Aur	05 01 26.2	+30 16 07	174.2	− 6.0	10.2	C
RW Aur B	05 01 26.2	+30 16 07	174.2	− 6.0	12.7	K3:e
Lk H 333	05 02 00	− 03 28	203.5	− 24.7	14.3	K5
V534 Ori	05 15 30	− 05 53	207.6	− 22.9		
CO Ori	05 22 04.3	+11 20 33	192.8	− 12.8	10.8	G5
GW Ori	05 23 33.3	+11 47 20	192.6	− 12.2	11.1	G5:
V649 Ori	05 23 48.9	+11 47 06	192.6	− 12.2	13.2	K4
V442 Ori	05 23 56.6	+12 51 32	191.7	− 11.6	14.3	K7–M0
V370 Ori	05 24 26.0	+12 08 24	192.4	− 11.9	12.5	K7
GX Ori	05 24 26.0	+12 08 50	192.4	− 11.9	13.2	K1
V447 Ori	05 24 59.3	+12 32 12	192.1	− 11.6	14.9	K7
V448 Ori	05 25 16.0	+12 03 59	192.6	− 11.7	14.9	K6
V449 Ori	05 25 42.1	+11 20 57	193.2	− 12.0		
HI Ori	05 25 47.8	+12 05 10	192.6	− 11.6	14.0	K4
HK Ori	05 25 52.2	+12 04 38	192.6	− 11.6	12.2	A4
V452 Ori	05 26 06.8	+12 25 51	192.4	− 11.4	14.0	K7
V453 Ori	05 26 12.4	+12 27 30	192.4	− 11.4	15.3	K2
San 1	05 27 12	− 03 10	206.5	− 19.0	12.6	M1.5–2
HS Ori	05 28 05.7	− 04 53 36	208.2	− 19.6	16.3	C
HT Ori	05 28 06.7	− 06 11 19	209.4	− 20.2	13.9	M8 III
V466 Ori	05 28 07.6	− 05 30 42	208.8	− 19.9	13.4	K1
V384 Ori	05 28 16	− 05 46 30	209.1	− 20.0	15.6	M1.5
P1207	05 28 37.9	− 05 37 45	209.0	− 19.8	13.3	K5
VX Ori	05 28 38	− 04 47 30	208.2	− 19.5	15.6	M0
VY Ori	05 28 40.3	− 05 05 45	208.5	− 19.6	16.3	M2.5
San 2	05 28 48	− 01 13	204.9	− 17.7	13.6	K5

Catalogue of T Tauri Stars and Related Objects

Name	Av (mag)	$\log(L/L_\odot)$	Location	No.
GO Tau	3.14	− 0.06	Tau–Aur, B 22	71
DQ Tau	0.52	− 0.03	Tau–Aur	72
Haro 6–37	2.18	+0.29	Tau–Aur	73
DR Tau		>0.69	Tau–Aur	74
DS Tau	0.33	+0.14	Tau–Aur	75
UY Tau	1.05	+0.63	Tau–Aur	76
GM Aur	0.14	− 0.03	Tau–Aur	77
AB Aur	0.65	+2.00	Tau–Aur	78n
SU Aur	0.93	+1.25	Tau–Aur	79n
RW Aur		>0.74		80
RW Aur B		>0.74		81
Lk H 333	1.94	9.07(− 6)	NGC 1788	82
V534 Ori			Anon dark neb	83
CO Ori	1.08	+1.68	B 225	84
GW Ori	0.82	+1.82	B 225	85
V649 Ori	0.27	+0.91	B 225	86
V442 Ori	1.15	+0.45	B 30	87
V370 Ori	0.28	+0.49	B 30	88
GX Ori	0.43	+0.63	B 30	89
V447 Ori	0.38	+0.11	B 30	90
V448 Ori	0.32	+0.06	B 30	91
V449 Ori				92
HI Ori	1.10	+0.57	B 30	93
HK Ori	0.66	+1.45	B 30	94n
V452 Ori	0.14	+0.04	B 30	95
V453 Ori	1.98	+0.52	B 30	96
San 1	0.07	+1.26	Orion neb	97
HS Ori		≥1.12	Orion neb	98
HT Ori	0.00		Orion neb	99
V466 Ori	0.20	+0.53	Orion neb	100
V384 Ori	0.48	− 0.13	Orion neb	101
P1207	0.04	+0.21	Orion neb	102
VX Ori	1.00	− 0.35	Orion neb	103
VY Ori	0.64	− 0.13	Orion neb	104
San 2	0	+0.41	Orion neb	105

Catalogue of T Tauri Stars and Related Objects

Name	R.A.(1900) h m s	Dec.(1900) ° ′ ″	l °	b °	V (mag)	Sp
VZ Ori	05 28 50.8	− 05 35 00	208.9	− 19.8	13.5	K4(C)
V386 Ori	05 28 51	− 05 35 18	208.9	− 19.8	15.1	K7
P1267	05 28 52.2	− 06 08 36	209.5	− 20.0	15.3	M0
P1270	05 28 53.0	− 06 01 23	209.4	− 20.0	12.3	K1
SU Ori	05 28 59.7	− 04 51 59	208.3	− 19.4	16.5	M1
WX Ori	05 29 12.2	− 05 17 53	208.7	− 19.6	14.1	K3
P1397	05 29 20.8	− 05 50 52	209.2	− 19.8	15.6	M6 III
P1404	05 29 22.2	− 05 40 52	209.1	− 19.7	12.4	G5
EZ Ori	05 29 23.1	− 05 08 54	208.6	− 19.4	12.2	F8
SW Ori	05 29 23	− 06 40 18	210.0	− 20.2	13.6	K4
IU Ori	05 29 40.7	− 05 45 40	209.2	− 19.7		K2 III
XX Ori	05 29 43.0	− 06 09 38	209.6	− 19.8	16.5	M2.5
IX Ori	05 29 46.1	− 05 26 45	208.9	− 19.5		
YY Ori	05 29 54.1	− 06 01 59	209.5	− 19.8	12.8	K5
YZ Ori	05 29 57.5	− 05 07 28	208.6	− 19.3	15.0	K5
P1649	05 29 57.8	− 05 37 10	209.1	− 19.6	10.5	K1
KM Ori	05 30 01.2	− 05 27 14	209.0	− 19.5	13.5	K1
KN Ori	05 30 01.6	− 05 15 34	208.8	− 19.4	13.6	K6
KP Ori	05 30 01.7	− 05 45 27	209.2	− 19.6	14.8	M0
KR Ori	05 30 05.3	− 05 27 02	209.0	− 19.4	14.9	K6
LL Ori	05 30 10.9	− 05 29 19	209.0	− 19.4	13.7	K3
SY Ori	05 30 13.2	− 04 31 39	208.1	− 19.0	14.2	K3
LN Ori	05 30 13.7	− 05 36 43	209.1	− 19.5		
V356 Ori	05 30 14.6	− 05 33 58	209.1	− 19.5	15.3	K3
AA Ori	05 30 16.5	− 05 50 32	209.4	− 19.6	13.0	K4
V486 Ori	05 30 18.4	− 05 47 17	209.3	− 19.6	14.2	K7
P1817	05 30 18.9	− 04 48 41	208.4	− 19.1	12.6	K2
LX Ori	05 30 18.9	− 05 43 33	209.2	− 19.5	12.3	K3
V488 Ori	05 30 19.1	− 05 34 56	209.1	− 19.4		–e
AB Ori	05 30 20.2	− 05 47 15	209.3	− 19.6	14.0	K7
TT Ori	05 30 20.5	− 04 49 42	208.4	− 19.1	14.7	K1
AD Ori	05 30 22.6	− 05 26 33	209.0	− 19.4	14.6	C
P1931	05 30 24.3	− 06 02 24	209.6	− 19.6	13.9	K7
P1946	05 30 25.9	− 04 32 48	208.2	− 19.0	15.7	M0.5
AG Ori	05 30 27.4	− 05 38 44	209.2	− 19.4	15.4	K6

Catalogue of T Tauri Stars and Related Objects

Name	Av (mag)	$\log(L/L_\odot)$	Location	No.
VZ Ori	1.66	+0.92	Orion neb	106
V386 Ori	2.39	+0.15	Orion neb	107
P1267	0.40	- 0.47	Orion neb	108
P1270	0.12	+0.68	Orion neb	109
SU Ori	0.75	- 0.73	Orion neb	110
WX Ori	0.44	+0.18	Orion neb	111
P1397	2.76		Orion neb	112
P1404	0.36	+0.84	Orion neb	113
EZ Ori	0.40	+0.91	Orion neb	114
SW Ori	0.52	+0.60	Orion neb	115
IU Ori	0:	+1.78	Orion neb	116
XX Ori	0.47	- 0.80	Orion neb	117
IX Ori			Orion neb	118
YY Ori	0	+0.45	Orion neb	119
YZ Ori	0.24	- 0.12	Orion neb	120
P1649	0.32	+1.34	Orion neb	121
KM Ori	0.08	- 0.03	Orion neb	122
KN Ori	0	+0.90	Orion neb	123
KP Ori	1.85	+0.39	Orion neb	124
KR Ori	0.16	- 0.52	Orion neb	125
LL Ori	0.41	+0.03	Orion neb	126
SY Ori	0.00	+0.21	Orion neb	127
LN Ori			Orion neb	128
V356 Ori		>0.70	Orion neb	129
AA Ori	1.46	+1.05	Orion neb	130
V486 Ori	0.60	+0.23	Orion neb	131
P1817	0.32	+0.72	Orion neb	132
LX Ori	0.69	+1.04	Orion neb	133
V488 Ori			Orion neb	134
AB Ori	1.08	+0.67	Orion neb	135
TT Ori	1.08	+0.44	Orion neb	136
AD Ori		>0.30	Orion neb	137n
P1931	0.76	+0.56	Orion neb	138
P1946	1.00	- 0.11	Orion neb	139
AG Ori	0.86	- 0.38	Orion neb	140

Catalogue of T Tauri Stars and Related Objects

Name	R.A.(1900) h m s	Dec.(1900) ° ′ ″	l °	b °	V (mag)	Sp
AI Ori	05 30 31.6	− 05 15 03	208.8	− 19.2		K0:e
NS Ori	05 30 32	− 06 05 24	209.6	− 19.6	17.2	M1
AL Ori	05 30 35.0	− 04 59 13	208.6	− 19.1	15.1	K7–M0
V360 Ori	05 30 36.6	− 05 13 23	208.8	− 19.2	12.5	K6
TW Ori	05 30 39	− 06 49 18	210.3	− 19.9	15.6	M0.5
AM Ori	05 30 40.4	− 05 25 22	209.2	− 19.3	14.3	K4
TV Ori	05 30 40.8	− 05 05 10	208.7	− 19.1	14.2	K6
NY Ori	05 30 41	− 05 16 24	208.9	− 19.2		cont+e
BO Ori	05 30 46.1	− 04 28 50	208.1	− 18.8	13.9	K4
AN Ori	05 30 47.4	− 05 32 07	209.1	− 19.3		K0–1(e)IV
V573 Ori	05 30 51	− 06 38 42	210.2	− 19.8	16.2	M0
CE Ori	05 30 52	− 05 05 30	208.7	− 19.1	16.2	K5
OT Ori	05 30 55.8	− 05 20 22	209.0	− 19.2	15.2	K5
T Ori	05 30 55.9	− 05 32 28	209.2	− 19.3	13.0	A5
AR Ori	05 30 58.6	− 05 08 07	208.8	− 19.1	14.3	K7–M0
V390 Ori	05 31 04	− 05 02 36	208.7	− 19.0	15.9	M1
AU Ori	05 31 04.2	− 06 01 10	209.6	− 19.5	15.6	K7–M0
V577 Ori	05 31 06	− 06 46 42	210.3	− 19.8	14.0	K7
AV Ori	05 31 08	− 06 46 24	210.3	− 19.8	13.7	K4
PQ Ori	05 31 19	− 02 14 48	206.1	− 17.7	13.2	F5:
PU Ori	05 31 19	− 00 46 00	203.4	− 16.2	14.6	K3:
AZ Ori	05 31 21.8	− 05 15 32	209.0	− 19.1	13.1	K6
Haro 4–125	05 31 32	− 05 34 54	209.3	− 19.2	16.4	M1
V380 Ori	05 31 33.7	− 06 46 25	210.4	− 19.7	10.7	B9(C)
BD Ori	05 31 39.8	− 06 23 06	210.0	− 19.5	13.2	K3
BC Ori	05 31 42.6	− 05 30 11	209.2	− 19.1	15.6	M0
P2441	05 31 54.3	− 04 29 24	208.3	− 18.6	10.7	G5
BE Ori	05 32 07.8	− 06 37 08	210.3	− 19.5	15.6	K0
BF Ori	05 32 21.3	− 06 38 41	210.4	− 19.5	12.2	A0:
RR Tau	05 33 17	+26 19 00	181.5	− 02.5	11.5	A6
LkH 206	05 33 24	+26 16	181.5	− 02.5	13.3	K3
TX Ori	05 33 32	− 02 48 06	206.9	− 17.4	12.9	K4
TY Ori	05 33 34	− 02 47 30	206.9	− 17.4	13.6	K3
Haro 7–1	05 33 44	− 07 53 48	211.7	− 19.8	13.3	K5
Haro 7–5	05 34 22	− 08 36 30	212.5	− 19.9	13.8	M4.5 III

Catalogue of T Tauri Stars and Related Objects

Name	Av (mag)	$\log(L/L_\odot)$	Location	No.
AI Ori			Orion neb	141
NS Ori	0.40	– 1.17	Orion neb	142
AL Ori	0.58	+0.01	Orion neb	143
V360 Ori	0.97	+1.09	Orion neb	144
TW Ori	0.47	+0.02	Orion neb	145
AM Ori	0.10	– 0.28	Orion neb	146
TV Ori	0.54	+0.46	Orion neb	147
NY Ori			Orion neb	148
BO Ori	0.12	+0.35	Orion neb	149
AN Ori			Orion neb	150
V573 Ori	1.40	– 0.23	Orion neb	151
CE Ori	1.22	– 0.12	Orion neb	152
OT Ori	0.84	+0.16	Orion neb	153
T Ori	1.24	+1.63	Orion neb	154n
AR Ori	0.00	+0.33	Orion neb	155
V390 Ori	0.11	– 0.28	Orion neb	156
AU Ori	0.95	– 0.43	Orion neb	157
V577 Ori	0.12	+0.34	Orion neb	158
AV Ori	0.18	+0.35	Orion neb	159
PQ Ori	0.00	+0.03	Orion east	160
PU Ori	0.18	– 0.43	Orion east	161
AZ Ori	0.44	+0.62	Orion neb	162
Haro 4–125	0.94	– 0.64	Orion neb	163
V380 Ori	1.86	+1.93	NGC 1999	164n
BD Ori	0.04	+0.31	Orion neb	165
BC Ori	1.19	– 0.28	Orion neb	166
P2441	0	+1.12	Orion neb	167
BE Ori	3.28	+0.96	Orion neb	168
BF Ori	0.25	+1.28	Orion neb	169
RR Tau	0.80	2.06(– 4)	Anon dark	170n
LkH 206	0.40	8.16(– 5)	Anon dark	171
TX Ori	0.20	+0.34	Orion east	172
TY Ori	0.20	+0.06	Orion east	173
Haro 7–1	0.12	+0.15	Orion neb	174
Haro 7–5	2.01		Orion neb	175

Catalogue of T Tauri Stars and Related Objects

Name	R.A.(1900) h m s	Dec.(1900) ° ′ ″	l °	b °	V (mag)	Sp
Haro 4–255	05 34 30	– 07 29	211.4	– 19.4	15.7	K7
V510 Ori	05 34 38.4	– 02 34 42	206.8	– 17.1	14.1	C
BH Ori	05 35 14	– 06 19 24	210.4	– 18.7		K3
Haro 7–4	05 35 39	– 08 10 24	212.2	– 19.4	14.2	K3
V614 Ori	05 36 06.9	+09 04 46	196.6	– 11.0	15.1	K6
DL Ori	05 36 37	– 08 08 42	212.2	– 19.2	12.9	K1
Haro 7–2	05 37 02	– 08 03 54	212.3	– 19.1	15.1	G5:
V625 Ori	05 37 52.2	+09 03 18	196.8	– 10.7	13.8	K6
V630 Ori	05 38 48.4	+09 08 18	196.9	– 10.4	15.6	K5
V631 Ori	05 39 13.0	+08 54 21	197.1	– 10.5	15.0	K1
FU Ori	05 39 54.0	+09 01 40	197.1	– 10.2	10.2	F2I
LkH 314	05 42 06.4	+00 06 53	205.3	– 14.2	14.7	C
LkH 334	05 48 30	+01 36 54	204.8	– 12.0	13.0	K4
LkH 335	05 48 47	+01 42 48	204.7	– 11.9	14.1	K4
LkH 336	05 49 08	+01 41 42	204.8	– 11.9	14.3	K7–M0
LkH 337	05 49 26	+01 28 18	205.0	– 11.9	14.2	K7
HD 250550	05 56 13.0	+16 30 56	192.6	– 03.0	9.7	B6
LkH 208	06 01 57.0	+18 40 10	191.4	– 00.8	12.4	F0
LkH 209	06 02 15.9	+18 39 13	191.5	– 00.7	12.2	G3
Bretz 3	06 02 36	+18 09	192.0	– 00.9	14.5	K0:
LkH 338	06 05 54.2	– 06 11 35	213.9	– 11.8	15.1	F2
LkH 339	06 06 01.6	– 06 13 25	213.9	– 11.8	13.6	F2
Bretz 4	06 07 54	– 06 12	214.1	– 11.4	15.0	K4
MWC 137	06 13 00	+15 19	195.6	– 00.1	13.5	C
LkH 340	06 24 48.5	+10 35 49	201.2	+00.2	13.3	B3
LkH 341	06 25 18.0	+10 37 15	201.2	+00.3	14.9	O9
VY Mon	06 25 35.1	+10 30 13	201.3	+00.3	12.5	K4
LkH 274	06 25 38.2	+10 30 12	201.3	+00.3	14.9	M0
LkH 342	06 25 58.0	+10 36 44	201.3	+00.4	14.0	K3
LkH 343	06 26 33.5	+10 29 34	201.5	+00.5	15.7	K7
LkH 344	06 26 33.9	+10 40 43	201.3	+00.6	13.1	B0
R Mon	06 33 42	+08 49 30	203.8	+01.3		
KV Mon	06 34 12	+09 51 42	202.9	+01.9	15.0	K3
G–G 405	06 34 12	+09 05	203.6	+01.5	13	G8e
LL Mon	06 34 56	+09 56 24	202.9	+02.0	16.0	K2

Catalogue of T Tauri Stars and Related Objects

Name	Av (mag)	$\log(L/L_\odot)$	Location	No.
Haro 4–255	3.68	+1.02	Orion neb	176n
V510 Ori		≥0.25	Orion east	177
BH Ori			Orion neb	178
Haro 7–4	1.54	+0.35	Orion neb	179
V614 Ori	0.10	+0.20	B 35	180
DL Ori	0.28	+0.93	Orion neb	181
Haro 7–2	0	+1.02	Orion neb	182n
V625 Ori	0	+0.83	B 35	183
V630 Ori	0.20	+0.31	B 35	184
V631 Ori	0.58	+0.03	B 35	185
FU Ori	2.50	+2.43	B 35	186n
LkH 314		>1.99	NGC 2068	187
LkH 334	0.20	+0.31	Orion east	188
LkH 335	0.99	+0.15	Orion east	189
LkH 336	0.78	+0.05	Orion east	190n
LkH 337	0.49	– 0.08	Orion east	191
HD 250550	0.02	+4.16	Anon dark neb	192n
LkH 208	0.36	1.03(– 4)	Ced 62	193n
LkH 209	0.26	2.22(– 5)	Anon dark neb	194
Bretz 3	0.86	3.36(– 6)	Anon dark neb	195
LkH 338	3.22	1.64(– 5)	NGC 2183/85	196n
LkH 339	1.42	2.40(– 5)	NGC 2183/85	197n
Bretz 4	1.24	3.68(– 6)	NGC 2183/85	198n
MWC 137		>2.94(– 5)	Anon	199n
LkH 340	3.24	5.76(– 5)	IC 446	200n
LkH 341	7.38	2.21(– 3)	IC 446	201
VY Mon	1.04	4.84(– 5)	IC 446	202n
LkH 274	1.45	1.05(– 5)	IC 446	203
LkH 342	0.12	5.55(– 6)	IC 446	204
LkH 343	3.68	1.57(– 5)	IC 446	205
LkH 344	4.24	1.21(– 3)	IC 446	206
R Mon			NGC 2261	207n
KV Mon	0.12	– 0.08	NGC 2264	208
G–G 405			Anon dark neb	209
LL Mon	0	– 0.29	NGC 2264	210

Catalogue of T Tauri Stars and Related Objects

Name	R.A.(1900) h m s	Dec.(1900) ° ′ ″	l °	b °	V (mag)	Sp
PT Mon	06 35 02	+09 55 06	202.9	+02.1	17.7	K5:
LM Mon	06 35 07	+09 56 36	202.9	+02.1	16.2	K3
NW Mon	06 35 10	+09 40 36	203.2	+02.0	14.9	K4
LP Mon	06 35 11	+09 56 36	202.9	+02.1	16.1	M0
LH 21	06 35 11	+09 59 48	202.9	+02.1	14.0	G8–K0
NX Mon	06 35 11	+09 39 30	203.2	+02.0	15.1	C
W 84	06 35 12	+09 39 12	203.2	+02.0	13.3	F8
LQ Mon	06 35 13	+09 52 36	203.0	+02.1	17.2	K6
LH 25	06 35 15	+09 53 30	203.0	+02.1	13.3	B9–A0
LR Mon	06 35 17	+09 55 06	203.0	+02.1	14.5	K3
LT Mon	06 35 20	+09 42 24	203.2	+02.0	15.6	K1
W 108	06 35 23	+09 50 24	203.1	+02.1	12.3	F9
V419 Mon	06 35 24	+09 39 00	203.2	+02.0	15.0	K3
V347 Mon	06 35 24	+09 54 54	203.0	+02.1	17.0	M1
LU Mon	06 35 27	+09 43 24	203.2	+02.0	15.0	K3
IO Mon	06 35 29	+09 36 24	203.3	+02.0	14.3	K3
IP Mon	06 35 31	+09 38 12	203.2	+02.0	13.7	K3
LY Mon	06 35 35	+09 30 30	203.4	+02.0	15.8	K5
LX Mon	06 35 35	+09 53 54	203.0	+02.2	15.2	K7
SS Mon	06 35 35.3	+10 32 15	202.5	+02.5	12.7	K3
V360 Mon	06 35 36	+09 42 00	203.2	+02.1	15.1	F8
V426 Mon	06 35 37	+09 33 06	203.3	+02.0		G–Ke
V363 Mon	06 35 43	+09 32 06	203.4	+02.0		
MM Mon	06 35 43	+09 58 18	203.0	+02.2	14.0	K3
LH 61	06 35 43	+09 31 48	203.4	+02.0	14.1	K4
V365 Mon	06 35 44	+09 31 42	203.4	+02.0	14.9	K4
V432 Mon	06 35 49	+09 39 30	203.3	+02.1	15.0	K5
MO Mon	06 36 02	+09 32 36	203.4	+02.1	14.1	K2
OW Mon	06 36 05	+10 14 42	202.8	+02.4	14.8	K2
OY Mon	06 36 10	+09 46 06	203.2	+02.2	14.1	K3
MQ Mon	06 36 38.7	+09 47 04	203.2	+02.3	15.2	K3
PY Mon	06 37 01.8	+09 15 12	203.8	+02.2	14.2	K1
Z CMa	06 59 02.0	– 11 24 16	224.6	– 02.6		eq
LH 332–20	10 56 29	– 76 29 30	296.6	– 15.6		K0e
CPD–76 652	11 04 34	– 77 05 48	297.3	– 15.9	10.6	G1Iep

Catalogue of T Tauri Stars and Related Objects

Name	Av (mag)	log(L/L$_\odot$)	Location	No.
PT Mon	2.11	- 0.14	NGC 2264	211
LM Mon	1.46	+0.43	NGC 2264	212
NW Mon	0.14	0.00	NGC 2264	213
LP Mon	0.10	- 0.52	NGC 2264	214
LH 21	0.00	+0.24	NGC 2264	215
NX Mon		>0.31	NGC 2264	216
W 84	0	+1.03	NGC 2264	217
LQ Mon	0.70	- 0.66	NGC 2264	218
LH 25	0.64	+1.81	NGC 2264	219
LR Mon	0	+0.31	NGC 2264	220
LT Mon	0.50	- 0.20	NGC 2264	221
W 108	0.00	+1.02	NGC 2264	222
V419 Mon	0.00	- 0.15	NGC 2264	223
V347 Mon	1.57	- 0.21	NGC 2264	224
LU Mon	0.79	+0.33	NGC 2264	225
IO Mon	0.12	+0.48	NGC 2264	226
IP Mon	0.98	+1.04	NGC 2264	227n
LY Mon	0.10	- 0.40	NGC 2264	228
LX Mon	0.23	+0.32	NGC 2264	229
SS Mon	0.16	- 0.12	NGC 2264	230
V360 Mon	1.02	+0.45	NGC 2264	231
V426 Mon			NGC 2264	232
V363 Mon			NGC 2264	233
MM Mon	0.23	+0.70	NGC 2264	234
LH 61	0.06	+0.85	NGC 2264	235
V365 Mon	0.44	+0.50	NGC 2264	236
V432 Mon	0	+0.08	NGC 2264	237
MO Mon	0.04	+0.77	NGC 2264	238
OW Mon	0.54	+0.18	NGC 2264	239
OY Mon	0	+0.30	NGC 2264	240
MQ Mon	0.00	- 0.17	NGC 2264	241
PY Mon	0	+0.14	NGC 2264	242
Z CMa			Anon dark neb	243n
LH 332–20			Anon dark neb	244
CPD–76 652			Anon dark neb	245

Catalogue of T Tauri Stars and Related Objects

Name	R.A.(1900) h m s	Dec.(1900) ° ′ ″	l °	b °	V (mag)	Sp
HD 97048	11 05 15.1	- 77 06 43	297.4	- 15.9	8.4	A0e
LH 332-21	11 09 17	- 76 11 30	297.2	- 15.0		G8e
CoD-33 10685	15 38 48	- 33 58	339.2	+16.1	10.5	Ge
LH 450-6	15 40 18	- 34 12	339.3	+15.7		M0-1e
CoD-35 10525	15 42 42	- 35 21	338.9	+14.5	10.0	cont+e
RU Lup	15 50 07	- 37 31 48	338.6	+11.9		cont+e
RY Lup	15 52 43	- 40 04 54	337.3	+09.6		G0e V
EX Lup	15 56 20	- 40 01 42	337.9	+09.2		cont+e
AS 205	16 05 48	- 18 23 48	355.2	+23.3	12.3	K0
Do-Ar 9	16 12 59.8	- 23 01 52	352.8	+18.9	13	G
Haro 1-1	16 15 28.1	- 25 58 07	350.9	+16.5	13.2	K5
Haro 1-4	16 19 10.9	- 23 05 23	353.7	+17.8	13.1	K6
V852 Oph	16 19 21.7	- 24 15 55	352.8	+17.0	14.7	M0
S-R 4	16 19 53.9	- 24 07 03	353.0	+17.0	12.6	K7
Do-Ar 22	16 20 18.5	- 23 29 39	353.6	+17.4		
Haro 1-8	16 20 47.8	- 23 01 14	354.0	+17.6	13.6	K5
S-R 24	16 20 55.0	- 24 31 59	352.9	+16.6	15.9	K2:
S-R 12	16 21 16.2	- 24 28 06	353.0	+16.6	13.0	M1
S-R 9	16 21 37.8	- 24 08 33	353.3	+16.7	11.8	K7
S-R 10	16 21 52.8	- 24 12 49	353.3	+16.6	14.4	M1.5
V853 Oph	16 22 42.4	- 24 14 56	353.4	+16.5	13.4	M1.5
Haro 1-14	16 25 02.3	- 23 51 29	354.0	+16.3	14.0	M0
Haro 1-16	16 25 30.3	- 24 14 37	353.8	+16.0	12.4	K3
Do-Ar 58	16 28 23.8	- 24 01 08	354.4	+15.6		
V1121 Oph	16 43 36	- 14 13	004.8	+19.0	12.1	K5
AK Sco	16 48 02	- 36 43 24	347.4	+04.2		
IX Oph	17 03 33	- 27 09 12	357.0	+07.5	11.1	G
KK Oph	17 03 54	- 27 07 42	357.1	+07.4	11.6	B-A
LkH 345	17 04 36	- 27 32	356.8	+07.1	12.7	K7
LkH 346	17 04 54	- 27 17	357.1	+07.2	14.3	K7
LkH 347	17 06 00	- 27 13	357.3	+07.0	14.1	M0
LkH 122	17 55 54	- 22 52	007.1	- 00.1	14.2	G4
V1752 Sgr	17 56 46	- 24 20 54	005.9	- 01.0	14.8	K5:
LkH 102	17 56 48	- 24 18	006.0	- 01.0	14.5	K5
SV Sgr	17 57 49.7	- 24 25 42	006.0	- 01.2	14.2	K5

Catalogue of T Tauri Stars and Related Objects

Name	Av (mag)	$\log(L/L_\odot)$	Location	No.
HD 97048			Anon dark neb	246n
LH 332–21			Anon dark neb	247
CoD–33 1068			B 228	248n
LH 450–6			B 228	249
CoD–35 1052			B 228	250
RU Lup			Anon dark neb	251
RY Lup			Anon dark neb	252
EX Lup			Anon dark neb	253
AS 205	3.02	5.96(– 4)	Anon dark neb	254
Do–Ar 9			B 42	255
Haro 1–1	1.67	+0.05	Near B 42	256
Haro 1–4	3.36	+0.70	B 42	257
V852 Oph	1.23	– 0.22	B 42	258
S–R 4	2.10	+0.71	B 42	259
Do–Ar 22			B 42	260
Haro 1–8	2.70	+0.18	B 42	261
S–R 24	4.49	+0.60	B 42	262n
S–R 12	0.35	+0.02	B 42	263
S–R 9	0.90	+0.59	B 42	264
S–R 10	0.14	– 0.24	B 42	265
V853 Oph	0	+0.18	B 42, 44	266
Haro 1–14	1.96	+0.18	B 42	267n
Haro 1–16	1.16	+0.53	B 44	268
Do–Ar 58			B 44	269
V1121 Oph	1.80	2.60(– 4)	Anon dark neb	270n
AK Sco				271
IX Oph	2.60	4.27(– 4)	B 59	272
KK Oph	4.49	2.99(– 3)	B 59	273
LkH 345	2.07	6.45(– 5)	B 59	274
LkH 346	1.16	7.49(– 6)	B 59	275
LkH 347	1.39	1.08(– 5)	B 59	276
LkH 122	2.20	1.61(– 5)	NGC 6514	277
V1752 Sgr	1.04	3.58(– 6)	NGC 6523	278
LkH 102	0.91	4.08(– 6)	NGC 6523	279
SV Sgr	0.79	5.32(– 6)	NGC 6523	280

Catalogue of T Tauri Stars and Related Objects

Name	R.A.(1900) h m s	Dec.(1900) ° ′ ″	l °	b °	V (mag)	Sp
LkH 118	17 59 42	− 24 16	006.3	− 01.5	11.0	
VV Ser	18 23 40.8	+00 04 51	030.5	+05.1	11.8	A:
MWC 300	18 24 04.1	− 06 08 31	025.0	+02.1	11.7	O9:
AS 310	18 28 02.2	− 05 02 34	026.4	+01.8	12.8	B0
LkH 348	18 28 06	− 00 31	031.4	+04.3	14.2	C
S CrA	18 54 25	− 37 05 18	359.9	− 17.7		
TY CrA	18 54 56	− 37 00 54	000.0	− 17.8		
R CrA	18 55 09	− 37 05 36	359.9	− 17.8		
DG CrA	18 55 09	− 37 32 00	359.5	− 18.0		
T CrA	18 55 14	− 37 06 24	359.9	− 17.9		
VV CrA	18 56 22	− 37 21 12	359.8	− 18.2		
AS 353	19 15 48	+10 51	046.0	− 01.3	12.7	C
LH 483–41	19 22 24	+23 42	058.1	+03.5	11.4	F5
V536 Aql	19 34 12	+10 16 36	047.8	− 05.6	14.9	K7
LkH 228	20 21 24	+42 10	080.3	+02.7	13.2	K1–2
LkH 149	20 47 33	+44 01 18	084.7	− 00.1	15.1	G8–K0
V751 Cyg	20 48 41	+43 56 48	084.7	+00.1	13.9	A5:
LkH 172	20 48 55	+43 54 30	084.7	+00.1	14.5	K4
V521 Cyg	20 54 49	+43 30 00	085.1	− 01.2	13.6	K5
V1057 Cyg	20 55 19	+43 52 12	085.4	− 01.0		
LkH 191	20 55 30	+43 33 42	085.2	− 01.3	12.8	K6
V1331 Cyg	20 57 54	+49 58	090.3	+02.7	11.9	(B0.5)
LkH 321	20 58 36	+49 28	090.0	+02.2	12.2	G1
FU Cep	21 00 30	+67 45 12	104.0	+14.2		
LkH 324	21 00 42	+49 51	090.5	+02.3	13.4	B5:
FV Cep	21 01 04	+68 02 42	104.3	+14.3	14.3	K7
EH Cep	21 02 13	+67 35 36	104.0	+13.9	13.0	K2
LkH 349	21 34 12	+56 49 36	099.0	+03.7	13.0	F8:
LkH 234	21 40 48	+65 39	105.4	+09.9	12.7	O8
BD+46 3471	21 48 44	+46 45 18	094.2	− 05.4		
LkH 245	21 49 39	+46 44 36	094.4	− 05.5	15.3	K4
LkH 257	21 50 27	+46 43 18	094.5	− 05.6	13.0	A2:
LkH 233	22 30 18	+40 08	096.7	− 15.2		
LkH 350	22 45 18.0	+61 39 25	109.2	+02.6	13.5	B2–5
DI Cep	22 52 06	+58 08 00	108.4	− 00.9	11.2	G8

Catalogue of T Tauri Stars and Related Objects

Name	Av (mag)	log(L/L$_\odot$)	Location	No.
LkH 118			NGC 6523	281
VV Ser			Anon dark neb	282n
MWC 300	3.86	1.10(- 3)	Anon dark neb	283
AS 310	3.34	1.67(- 3)		284n
LkH 348		>4.69(- 6)	Anon dark neb	285
S CrA			Anon dark neb	286
TY CrA			NGC 6726–7	287n
R CrA			NGC 6729	288n
DG CrA			Anon dark neb	289
T CrA			NGC 6729	290n
VV CrA			Anon dark neb	291
AS 353		>6.79(- 5)	Anon dark neb	292n
LH 483–41	1.00	9.28(- 5)	Anon dark neb	293n
V536 Aql	3.59	7.66(- 5)	B 142	294
LkH 228	0.18	1.21(- 5)	NGC 6914b	295
LkH 149	2.74	+0.77	IC 5070	296
V751 Cyg	0.34	+0.39	IC 5070	297
LkH 172	2.29	+1.00	IC 5070	298
V521 Cyg	1.38	+1.22	NGC 7000	299
V1057 Cyg			NGC 7000	300
LkH 191	0.48	+1.08	NGC 7000	301
V1331 Cyg		>6.86	Anon dark neb	302n
LkH 321	2.23	1.72(- 4)	Anon dark neb	303n
FU Cep			NGC 7023	304
LkH 324	3.94	3.00(- 4)	Anon dark neb	305n
FV Cep	0.16	3.33(- 6)	NGC 7023	306
EH Cep	0.41	3.35(- 5)	NGC 7023	307
LkH 349	3.65	1.50(- 4)	IC 1396	308n
LkH 234	3.64	1.12(- 2)	NGC 7129	309n
BD+46 3471			IC 5146	310n
LkH 245	2.36	1.39(- 5)	IC 5146	311n
LkH 257	1.50	2.38(- 5)	IC 5146	312
LkH 233			Anon dark neb	313n
LkH 350	6.56	5.12(- 3)	Cep III	314
DI Cep	0	6.48(- 5)	Anon dark neb	315

Catalogue of T Tauri Stars and Related Objects

Name	R.A.(1900) h m s	Dec.(1900) ° ′ ″	l °	b °	V (mag)	Sp
AS 501	22 53 36	+58 15	108.6	- 00.9		
MWC 1080	23 12 54	+60 18	111.7	+00.0	11.5	B0(C)
BM And	23 32 48	+47 51	110.5	- 12.7	13.1	K1:
MacC H3	23 49 30	+66 20 54	117.4	+04.6	13.8	K2
MacC H19	23 52 33	+64 13 24	117.2	+02.5	14.1	K3
LkH 259	23 53 39	+65 52 48	117.7	+04.1	15.0	A9
MacC H5	23 54 17	+65 49 48	117.7	+04.0	15.7	K5
MacC H18	23 57 06	+64 20 54	117.7	+02.5	14.5	K7

Catalogue of T Tauri Stars and Related Objects

Name	Av (mag)	$\log(L/L_\odot)$	Location	No.
AS 501			Anon dark neb	316
MWC 1080	5.24	2.79(- 3)	Anon dark neb	317n
BM And	0.64	2.69(- 5)	Anon dark neb	318
MacC H3	1.25	+1.08	Cep IV	319
MacC H19	1.65	+1.15	Cep IV	320
LkH 259	4.84	+2.05	Cep IV	321n
MacC H5	4.63	+1.51	Cep IV	322
MacC H18	2.25	+1.26	Cep IV	323

17.3 Herbig-Haro Objects

Herbig-Haro, or HH, objects are small, bright nebulae discovered in the early 1950s by George Herbig and Guillermo Haro. They consist of tightly grouped, semi-stellar knots located within regions of star formation. Their enigmatic forbidden emission lines are now thought to be shock-excited by a strong stellar wind that escapes supersonically from newborn stars and interacts with the interstellar material.

The HH objects owe their semi-stellar appearance to their small size, with radii, R, of R = 300 to 2000 AU. They have densities of about $10^4 cm^{-3}$, and visible masses, M, of about ten Earth masses, M_E, with a range of M = 0.5 to 30 M_E. That places the Herbig-Haro objects among the lowest mass objects detected outside the solar system.

Radial velocity and proper motion measurements indicate that HH objects have high velocities of 100 to 400 $km\ s^{-1}$. When the proper motions are extrapolated backwards, a T Tauri star or infrared object is found. In some cases, the HH objects are linearly aligned with the young stellar object and moving in opposite directions from it. High-velocity H_2O masers and bipolar molecular outflows are also found in the vicinity, suggesting that they are all manifestations of strong winds from a central newborn star, and that the HH objects may have been ejected from its surface.

Our catalogue of Herbig-Haro objects has been adapted from that of von Hippel, Burnell and Williams (1988). The HH name is followed by the right ascension, R.A., and declination, Dec., for epoch 1950.0, the proper motions, $\mu_{R.A.}$ and $\mu_{Dec.}$, in arcseconds per century, and the radial velocity, V_r, in $km\ s^{-1}$.

Catalogue of Herbig–Haro Objects*

Name	R.A.(1950) h m s	Dec.(1950) ° ′ ″	$\mu_{R.A.}$ (″cen^{-1})	$\mu_{Dec.}$ (″cen^{-1})	V_r (km s^{-1})
14C	03 25 44.1	+30 50 29			
14E	03 25 44.4	+30 50 56			
14D	03 25 44.8	+30 51 05			
14B	03 25 45.0	+30 50 50			
12D	03 25 52.39	+31 09 51.1	0.5 ± 3.8	5.5 ± 2.0	- 52 ± 3
12C	03 25 52.4	+31 10 04			- 52 ± 3
12G	03 25 53.1	+31 10 26			- 52 ± 3
15	03 25 53.5	+30 57 43			
12B	03 25 53.52	+31 10 13.0	5.0 ± 0.7	14.4 ± 0.6	- 52 ± 3
12E	03 25 53.69	+31 09 48.6	1.5 ± 1.7	3.2 ± 1.7	- 52 ± 3
12F	03 25 53.72	+31 09 31.0	- 3.1 ± 1.2	10.2 ± 1.1	- 52 ± 3
11B	03 25 58.99	+31 05 33.1	1.4 ± 2.1	- 4.3 ± 2.1	- 133 ± 17
11A	03 25 59.05	+31 05 34.7	3.0 ± 0.5	- 1.8 ± 0.8	- 133 ± 17
10	03 25 59.83	+31 05 29.2	- 1.1 ± 2.0	- 1.0 ± 2.0	- 16
8A	03 26 00.68	+31 05 18.7	- 2.2 ± 2.0	1.7 ± 2.0	- 31
9	03 26 00.9	+31 05 35			
7C	03 26 02.3	+31 05 08			- 45 ± 14
7B	03 26 02.56	+31 05 10.1	3.0 ± 2.1	1.9 ± 2.1	- 45 ± 14
7A	03 26 02.78	+31 05 10.8	1.4 ± 2.0	2.9 ± 2.0	- 45 ± 14
6D	03 26 05.8	+31 08 10			
6C	03 26 06.5	+31 08 15			
6E	03 26 06.6	+31 08 24			
6F	03 26 07.0	+31 08 23			
6B	03 26 07.2	+31 08 28			
5	03 26 14.78	+31 02 32.3	1.8 ± 2.0	- 2.1 ± 2.5	
N1555/HH	04 18 34	+19 25 05			- 34 ± 14
DCTau/HH	04 23 55.5	+25 58 28			- 125 ± 10
31D	04 24 53.3	+26 12 41			+131 ± 55
31B	04 25 13.8	+26 11 35			
31C	04 25 14.3	+26 10 24			
31A	04 25 15.2	+26 11 33			
28	04 28 13.5	+17 57 02	- 16. ± 3.	- 13. ± 3.	- 21 ± 14
102sb	04 28 25.3	+18 00 57			- 44
29	04 28 33.7	+18 00 01	- 13. ± 1.	- 19. ± 1.	- 28 ± 2
RNO40	05 17 13.80	- 05 55 44.8	- 3.6 ± 0.3	- 1.9 ± 1.1	

*Adapted from von Hippel, Burnell and Williams (1988).

Catalogue of Herbig–Haro Objects*

Name	R.A.(1950) h m s	Dec.(1950) ° ′ ″	$\mu_{R.A.}$ (″cen^{-1})	$\mu_{Dec.}$ (″cen^{-1})	V_r (km s^{-1})
RNO43A	05 29 38.97	+12 51 12.1	3.9 ± 0.4	14.9 ± 0.4	
RNO43D	05 29 43.98	+12 57 44.0	− 0.6	5.5	
RNO43B	05 29 44.64	+12 55 46.7	0.4	2.3	
M42/HH2	05 32 44	− 05 24 40			+25 ± 3
M42/HH1A	05 32 44.01	− 05 23 48.6	− 6.3 ± 0.6	4.3 ± 0.4	+23 ± 3
M42/HH1B	05 32 44.01	− 05 23 48.6	− 6.3 ± 0.6	4.3 ± 0.4	− 254 ± 3
M42/HH5	05 32 44.29	− 05 22 22.9	− 2.9 ± 0.8	10.5 ± 0.6	
M42/HH6	05 32 45.0	− 05 22 32			
M42/HH7	05 32 45.2	− 05 22 43			
M42/HH8	05 32 46.3	− 05 24 15			
M42/HH9	05 32 46.4	− 05 23 40			
M42/HH10	05 32 47.93	− 05 22 37.8	2.8 ± 0.5	9.5 ± 0.7	
44	05 32 48.5	− 05 12 19			
33	05 32 51.5	− 06 19 35			
40	05 32 54.5	− 06 20 16			
M42/HH3	05 32 54.8	− 05 26 51			+18 ± 3
M42/HH4	05 32 55.2	− 05 27 06			+20 ± 3
34 sb	05 33 05.4	− 06 30 28			− 97 ± 10
45	05 33 06.3	− 04 52 43			
41	05 33 34.1	− 05 04 40			
42A	05 33 37.3	− 05 06 31			
42B	05 33 40.9	− 05 06 31			
3	05 33 45.77	− 06 44 53.4	− 0.4 ± 0.5	0.8 ± 0.3	+7 ± 6
1F	05 33 54.54	− 06 46 57.0	− 9.9 ± 2.6	12.7 ± 2.0	− 6
1D	05 33 54.70	− 06 46 59.0	− 6.2 ± 0.9	9.4 ± 0.6	− 6
1C	05 33 54.83	− 06 46 59.2	− 5.7 ± 0.6	7.4 ± 1.0	− 6
1A	05 33 54.85	− 06 47 01.2	− 2.7 ± 0.8	6.6 ± 1.3	− 6
35	05 33 56.6	− 06 43 40			
2D	05 33 59.35	− 06 49 04.0	− 0.9 ± 0.6	− 3.8 ± 1.0	
2A	05 33 59.44	− 06 48 59.2	3.0	− 6.3	+21 ± 4
2I	05 33 59.65	− 06 49 08.5	2.8 ± 0.5	− 10.5 ± 2.1	
2C	05 33 59.67	− 06 48 55.5	6.9 ± 0.8	13.5 ± 1.1	+31 ± 5
2H	05 33 59.69	− 06 49 03.8	5.6 ± 0.4	− 9.6 ± 0.6	+13 ± 2
2B	05 33 59.89	− 06 48 56.3	1.9 ± 0.5	− 4.2 ± 0.6	+13 ± 6
2G	05 34 00.11	− 06 48 57.1	3.9 ± 0.5	− 5.8 ± 0.4	− 2 ± 8

Catalogue of Herbig–Haro Objects*

Name	R.A.(1950) h m s	Dec.(1950) ° ′ ″	$\mu_{R.A.}$ (″cen^{-1})	$\mu_{Dec.}$ (″cen^{-1})	V_r (km s^{-1})
2E	05 34 00.70	- 06 49 00.4	- 0.3 ± 0.6	- 2.8 ± 1.0	+39 ± 6
2L	05 34 00.8	- 06 49 15			+11 ± 3
43A	05 35 45.4	- 07 11 04			+44 ± 28
43B	05 35 45.4	- 07 11 04			- 34 ± 19
43C	05 35 45.4	- 07 11 04			- 143 ± 21
38	05 35 56.5	- 07 13 18			
N2023/HH2A	05 38 55.58	- 02 24 30.0	- 5.50	- 3.20	
N2023/HH2B	05 38 56.33	- 02 24 34.8	- 7.69	12.0	
N2023/HH1C	05 39 00.35	- 02 18 54.6	- 3.07	- 5.94	- 8 ± 17
N2023/HH1B	05 39 00.37	- 02 18 49.8	- 3.07	- 5.94	- 8 ± 17
N2023/HH1A	05 39 00.86	- 02 18 52.2	- 1.96	- 0.84	- 100 ± 22
N2023/HH3A	05 39 01.31	- 02 25 11.0	2.93	2.24	- 30 ± 21
N2023/HH3B	05 39 01.90	- 02 25 07.8	- 2.61	- 7.49	- 30 ± 21
N2023/HH3C	05 39 02.21	- 02 25 02.7			
N2023/HH5	05 39 02.85	- 02 18 43.6			
N2023/HH4	05 39 03.9	- 02 18 24			
19C	05 43 15.28	- 00 06 12.0	- 2.5	- 2.9	- 4
19A	05 43 15.98	- 00 06 19.7	- 1.6 ± 0.5	1.9 ± 0.5	- 33
20	05 43 21.43	- 00 04 11.3	- 2.3 ± 0.5	6.5 ± 0.2	- 109
21	05 43 22.0	- 00 05 36			
26	05 43 31.14	- 00 15 42.8	- 0.5	- 0.3	- 130
25	05 43 33.4	- 00 14 31			+143
24C2	05 43 34.15	- 00 10 46.5	- 6.6 ± 0.7	13 ± 1.3	- 184
24Csb	05 43 34.4	- 00 10 53			- 88
24B	05 43 34.44	- 00 11 11.7	- 0.2 ± 0.2	0.8 ± 0.2	- 72
24E	05 43 35.0	- 00 10 34			+60
24A	05 43 35.66	- 00 11 31.4	- 1.4 ± 0.5	- 0.8 ± 1.0	+48
24G1	05 43 37.5	- 00 10 30			- 133
24G2	05 43 38.3	- 00 10 17			- 129
24G3	05 43 39.3	- 00 10 10			- 140
23	05 43 40.86	- 00 04 36.5	- 1.3	1.1	
27	05 43 49.4	- 00 14 45			+60
GM10A	05 59 53.5	- 09 06 25			- 53 ± 31
GGD18	06 31 57.0	+04 15 12			
39C	06 36 21.08	+08 53 50.0	1.7 ± 0.5	8.0 ± 0.5	+20 ± 20

*Adapted from von Hippel, Burnell and Williams (1988).

Catalogue of Herbig–Haro Objects*

Name	R.A.(1950) h m s	Dec.(1950) ° ′ ″	$\mu_{R.A.}$ (″cen^{-1})	$\mu_{Dec.}$ (″cen^{-1})	V_r (km s^{-1})
39B	06 36 21.30	+08 54 07.0			
39D	06 36 21.43	+08 53 40.8	-1.0 ± 0.6	1.2 ± 1.4	
39E	06 36 21.54	+08 53 48.1	-3.6 ± 0.6	6.3 ± 3.4	
39A	06 36 21.60	+08 54 13.2	-1.5 ± 1.1	7.1 ± 0.5	$+20 \pm 20$
39F	06 36 22.8	+08 53 35			
N2264HH11–6	06 38 16.9	+09 42 28			
N2264HH11–7	06 38 17.2	+09 42 42			$+34 \pm .7$
N2264HH14–6	06 38 17.7	+09 47 09			
N2264HH14–4	06 38 18.4	+09 47 24			
N2264HH14–5	06 38 18.7	+09 47 19			
120	08 07 40.0	− 35 56 02			-42 ± 12
Re 4 head	08 19 28.9	− 49 25 12			-29 ± 2
47CS	08 24 05.64	− 50 51 48.5	-11.0	-5.2	$+91 \pm 15$
47CN	08 24 05.68	− 50 51 45.3	-9.6	-7.1	$+124 \pm 15$
46	08 24 16.77	− 50 50 41.1	1.7 ± 10	10.2 ± 23	-189 ± 15
47BS	08 24 18.19	− 50 50 30.1	-4.6 ± 6.2	1.9 ± 23	-138 ± 15
47BN	08 24 19.26	− 50 50 23.3	1.8 ± 2.7	7.7 ± 8.8	-144 ± 15
47AN	08 24 22.83	− 50 49 59.3	4.0 ± 1.4	7.4 ± 2.4	-123 ± 15
47AS	08 24 22.91	− 50 50 03.4	0.6 ± 4.5	9.0 ± 5.4	-111 ± 15
49	11 04 37.13	− 77 17 21.8	4.5 ± 1.2	-15.0 ± 3.2	$+25 \pm 3$
50A	11 04 39.33	− 77 16 53.3	-12.9 ± 2.4	-44.9 ± 0.3	
50D	11 04 39.37	− 77 16 45.1	-6.3	-22.4	$+30 \pm 5$
50C	11 04 39.57	− 77 16 47.2	-6.3	-20.1	
50B	11 04 39.70	− 77 16 51.1	-9.0 ± 2.0	-34.0 ± 1.0	
50E	11 04 43.06	− 77 16 29.8	-1.1	-23.7	
51	11 08 21.5	− 76 08 01			
52	12 51 28.03	− 76 41 35.5	3.7 ± 4.2	2.4 ± 5.5	-99 ± 10
53B	12 51 35.21	− 76 41 12.4	-1.6 ± 2.7	-0.6 ± 0.8	-101 ± 23
53A	12 51 36.21	− 76 41 12.0	-3.8 ± 2.7	-2.0 ± 4.9	-101 ± 23
53C	12 51 36.7	− 76 41 12			-101 ± 23
54X	12 52 00.31	− 76 41 20.2	-6.6	-8.8	
54G	12 52 05.3	− 76 39 55			-78 ± 6
54A1	12 52 08.6	− 76 40 16			
54F	12 52 08.8	− 76 40 01			-32 ± 7
54A2	12 52 09.2	− 76 40 13			

Catalogue of Herbig–Haro Objects*

Name	R.A.(1950) h m s	Dec.(1950) ° ′ ″	$\mu_{R.A.}$ (″cen^{-1})	$\mu_{Dec.}$ (″cen^{-1})	V_r (km s^{-1})
54C1	12 52 10.1	- 76 39 45			
54B	12 52 10.59	- 76 40 04.1	4.0 ± 1.5	- 1.1 ± 0.9	- 41 ± 4
54C2	12 52 10.9	- 76 39 47			
54J	12 52 11.8	- 76 40 00			
54H	12 52 12.1	- 76 39 49			
54K	12 52 13.1	- 76 40 07			
54E	12 52 13.2	- 76 40 05			- 67 ± 5
54I	12 52 13.7	- 76 39 51			
55	15 53 18.7	- 37 42 12			- 24 ± 21
Th28/HHW	16 05 00	- 38 56 30			+64
Th28/HHE	16 05 38	- 38 54 00	(48 ± 3)		- 26
56	16 28 54.02	- 44 48 36.6	4.2 ± 3.2	3.5 ± 1.7	+36 ± 5
57	16 28 56.63	- 44 49 16.5	- 1.5 ± 1.1	1.1 ± 0.8	- 59
M16/HH1	18 16 05.1	- 13 53 03			+10 ± 5
101S	18 58 12.35	- 37 07 20.2	- 3.3	- 3.0	- 51 ± 9
101N	18 58 12.51	- 37 07 14.4	- 1.0 ± 2.6	- 5.8 ± 0.7	- 89 ± 14
96	18 58 18.80	- 37 05 10.5			- 52 ± 15
100	18 58 26.7	- 37 02 36			- 139
98	18 58 30.42	- 37 01 56.5			+3 ± 15
104B	18 58 36.01	- 37 01 42.0	2.6	8.2	- 56 ± 15
104A	18 58 36.56	- 37 01 37.8	2.3 ± 2.1	2.2 ± 4.5	- 46 ± 15
99A	18 58 43.10	- 36 59 00.8	10.8 ± 2.5	7.1 ± 3.6	+5 ± 15
99B	18 58 43.31	- 36 58 56.2			+14 ± 15
32A	19 18 07.91	+10 56 21.7	- 3.8 ± 2.1	0.6 ± 1.5	+20 ± 4
32B	19 18 08.19	+10 56 17.2	- 13.2 ± 2.5	5.1 ± 1.5	

*Adapted from von Hippel, Burnell and Williams (1988).

Catalogue of Herbig–Haro Objects*

Name	R.A.(1950) h m s	Dec.(1950) ° ′ ″	$\mu_{R.A.}$ (″cen^{-1})	$\mu_{Dec.}$ (″cen^{-1})	V_r (km s^{-1})
32C	19 18 09.7	+10 56 11			– 320
32D	19 18 09.71	+10 56 10.4	– 0.3 ± 1.5	– 1.8 ± 1.5	
GN20183/HH1	20 18 21.27	+37 00 43.2			– 143
V645Cyg/HH	21 38 14	+50 00 38			
103	21 41 15.8	+65 49 55			– 52 ± 15
GGD32	21 41 17.99	+65 50 42.4			– 38 ± 34
105	21 42 12.92	+65 53 59.6			
GGD34	21 42 20.76	+65 54 50.3			– 175 ± 37
GGD35	21 42 33.47	+65 54 37.9			– 162 ± 18
GGD37A	22 54 03.01	+61 46 15.1			
GGD37S	22 54 03.03	+61 45 57.7			
GGD37H	22 54 03.61	+61 45 54.3			
GGD37B	22 54 03.98	+61 46 07.3			
GGD37C	22 54 04.35	+61 46 04.7			
GGD37D	22 54 04.83	+61 45 59.5			
GGD37G	22 54 05.68	+61 46 11.6			
GGD37E	22 54 07.39	+61 45 56.0			
GGD37F	22 54 07.75	+61 46 04.7			
GGD37HW	22 54 09.95	+61 45 42.2			

*Adapted from von Hippel, Burnell and Williams (1988).

18
Diffuse Emission Nebulae or H II Regions

18.1 Basic Data for Emission Nebulae

Energetic ultraviolet light from hot, luminous O- or early B- type stars ionizes nearby interstellar gas, producing bright, extended objects called diffuse emission nebulae or H II regions, for ionized hydrogen. The hot stars ionize the surrounding hydrogen and heat it up to a temperature of about 10,000 degrees. The size of these H II regions depends upon the temperature of the exciting star and the density of the hydrogen. Representative radii, call the Strömgren radii R after the man who first calculated them, are given in the first accompanying table for exciting stars of different spectral class, Sp, absolute visual magnitude, M_V, effective temperature, T_{eff}, and a typical hydrogen density of $N_H = 1\ cm^{-3}$.

The sizes of the H II regions (R = 10 to 100 *pc*) are much smaller than that of our Galaxy ($\sim$ 30,000 *pc*), and only the hot, luminous stars can ionize nearby hydrogen. The vast majority of interstellar matter therefore consists of cold clouds of neutral, or unionized, hydrogen called H I regions.

The rarefied, ionized H II regions have an emission-line spectrum dominated at optical wavelengths by the emission lines of common elements such as the red permitted line of hydrogen, Hα λ 6563, and the red forbidden transitions of nitrogen [N II] $\lambda\lambda$ 6548, 6583. Permitted lines of hydrogen also appear in the blue, Hβ λ 4861, and violet, Hγ λ 4360, parts of the optical spectrum; and forbidden transitions of ionized oxygen are detected at ultraviolet, [O II] $\lambda\lambda$ 3726, 3729, and green, [O III] $\lambda\lambda$ 4959, 5007, wavelengths. The permitted emission lines of excited hydrogen atoms are radiated when the free electrons recombine with the protons. The forbidden transitions of ionized elements are not really forbidden, but just rarely occur in the higher-density laboratory situation on Earth.

The net result at optical wavelengths is a colorful display that is easily visible on photographs or even at the eyepiece of a telescope. The best-known visible objects have received the names given in the first table together with their approximate right ascension, R.A., and declination, Dec., for the epoch 2000.0. This is followed by a table of selected optically-visible, evolved H II regions together with the equatorial coordinates, R.A. and Dec., for epoch 1950.0, galactic coordinates, 1 and b, maximum angular extent, θ, distance, D, maximum linear extent, R, in Hα, electron temperature, T_e, emission measure, E, electron density, N_e, and mass of ionized gas, M_{HII}.

We next provide a more comprehensive catalogue of emission nebulae at optical wavelengths. It is based on the Hα surveys of Gum (1955), designated G, Sharpless (1959), designated S, Rogers, Campbell, and Whiteoak

(1960), designated RCW and Lynds (1965). The distances, D, are those of the exciting star; when the stellar distances are unavailable we give the kinematic distances (see Goergelin and Goergelin (1970) for both distances).

When the free electrons in H II regions recombine with the free protons to make a hydrogen atom, they do not always fall into the lowest energy state, but instead cascade through a series of intermediate levels, producing the recombination emission lines detected at both optical and radio wavelengths. The bound-bound, recombination transitions at high energy levels, or remote distances from the atomic nucleus, are detected at radio wavelengths such as the 6 cm H109α transition. Our catalogue of emission nebulae at radio wavelengths is based on an all-sky survey at this transition by Reifenstein et al. (1970) and Wilson et al. (1970). It includes the galactic coordinates, or G number, the R.A. and Dec. for epoch 1950.0., the angular extents, $\theta_{R.A}$ and $\theta_{Dec.}$, the flux density, S_{6cm}, at 6 cm wavelength, the velocity with respect to the local standard of rest, V_{LSR}, the electron temperature, T_e, emission measure, E, kinematic distance, D, radius, R, electron density, N_e, and mass of ionized gas, M_{HII}. Both values of the kinematic distance are given when this ambiguity exists, and the listed radius is based on the nearest distance. Here the angular sizes are Gaussian; to obtain the linear size multiply by 1.47. When the H II region was not resolved, the angular extent was given as 0.0. The mean diameter, 2R, is obtained from the geometric mean of the measured angular diameters. Here we do not give the errors in T_e, which are on the order of 1,000° K, or in V_{LSR}, which are about 1.0 $km\ s^{-1}$.

Strömgren Radii of H II Regions*

Sp	M_V (mag)	T_{eff} (°K)	R (pc)
O5	- 5.6	48,000	108
O6	- 5.5	40,000	74
O7	- 5.4	35,000	56
O8	- 5.2	33,500	51
O9	- 4.8	32,000	34
O9.5	- 4.6	31,000	29
B0	- 4.4	30,000	23
B0.5	- 4.2	26,200	12

*Adapted from Osterbrock (1974), for a hydrogen density of $N_H = 1\ cm^{-3}$.

Named Bright Nebulae*

Name	NGC, other	R.A.(2000) h m	Dec.(2000) ° ′
America Nebula (Also North America)	NGC 7000	20 58.8	+44 20
Bubble Nebula	NGC 7635	23 20.7	+61 12
Bug Nebula	NGC 6302	17 13.7	- 37 06
California Nebula	NGC 1499	04 00.7	+36 37
Carina Nebula (Also η Carinae)	NGC 3372	10 43.8	- 59 52
Checkmark Nebula (Also Omega or Swan)	NGC 6618, M 17	18 20.8	- 16 11
Cocoon Nebula	IC 5146	21 53.5	+47 16
Cone Nebula	NGC 2264	06 40.9	+09 54
Crescent Nebula	NGC 6888	20 12.0	+38 21
Cygnus Nebula (Also γ Cygni)	IC 1318?	20 28.5	+39 57
Eagle Nebula (Also Star Queen)	NGC 6611, M 16	18 18.8	- 13 47
Eta, η, Carinae Nebula (Also Carina)	NGC 3372	10 43.8	- 59 52
Flaming Star Nebula	IC 405	05 16.2	+34 16
Gamma, γ, Cassiopeiae Nebula	IC 59-63	00 58.1	+60 56
Gamma, γ, Cygni Nebula (Also Cygnus)	IC 1318?	20 28.5	+39 57
Great Looped Nebula (Also Tarantula or 30 Doradus)	NGC 2070	05 38.7	- 69 06
Great Nebula in Orion (Also Orion)	NGC 1976, M 42	05 35.4	- 05 27
Homunculus Nebula (Core of η Carinae)	NGC 3372	10 43.8	- 59 52
Hourglass Nebula (Bright Part of Lagoon)	NGC 6523, M 8	18 03.8	- 24 23
Lace–Work Nebula	NGC 6960	20 45.7	+30 43
Lagoon Nebula	NGC 6523, M 8	18 03.8	- 24 23

*Adapted from Hirshfeld and Sinnott (1985).

Named Bright Nebulae*

Name	NGC, other	R.A.(2000) h m	Dec.(2000) ° ′
Lamda, λ, Centauri Nebula	IC 2944,48	11 38.3	− 63 22
Loop Nebula	NGC 2070	05 38.7	− 69 06
(Also Tarantula or 30 Doradus)			
North America Nebula	NGC 7000	20 58.8	+44 20
(Also America)			
Omega Nebula	NGC 6618, M 17	18 20.8	− 16 11
(Also Checkmark or Swan)			
Orion Nebula	NGC 1976, M 42	05 35.4	− 05 27
(Also Great Nebula in Orion)			
Pelican Nebula	IC 5067,70	20 50.8	+44 21
Rho, ρ, Ophiuchi Nebula	IC 4604	16 25.6	− 23 26
Rosette Nebula	NGC 2237–9	06 32.3	+05 03
Running Chicken Nebula	IC 2944	11 35.8	− 63 01
Star Queen Nebula	NGC 6611, M 16	18 18.8	− 13 47
(Also Eagle)			
Struve's Lost Nebula	NGC 1554–5	04 21.8	+19 32
(Also Hind's Variable Nebula)			
Swan Nebula	NGC 6618, M 17	18 20.8	− 16 11
(Also Checkmark or Omega)			
Tarantula Nebula	NGC 2070	05 38.7	− 69 06
(Also Great Looped, Loop, 30 Doradus)			
Thirty, 30, Doradus	NGC 2070	05 38.7	− 69 06
Trifid Nebula	NGC 6514, M 20	18 02.6	− 23 02

*Adapted from Hirshfeld and Sinnott (1985).

Selected Optically Visible Evolved H II Regions*

NGC, other	Name	R.A.(1950.0) h m	Dec.(1950.0) ° ′	l °	b °
NGC 7822, S 171, W 1		00 01	+68 20	118.6	+06.2
NGC 1491, S 206		03 59	+51 11	150.6	- 01.0
S 209		04 07	+51 02	151.6	- 00.3
NGC 1976, M 42, W 10	Orion Nebula	05 33	- 05 30	209.1	- 19.4
IC 434, W 12		05 39	- 02 20	206.9	- 16.6
NGC 2237–46, W 16	Rosette Nebula	06 30	+05 00	206.4	- 01.9
NGC 3372, RCW 53	Eta, η, Carinae Nebula	10 42	- 59 36	287.5	- 00.9
NGC 6514, M 20, W 28	Trifid Nebula	17 59	- 23 00	7.0	- 00.2
NGC 6523, M 8, W 29	Lagoon Nebula	18 01	- 24 20	6.1	- 01.2
NGC 6618, M 17, W 38	Omega Nebula	18 18	- 16 00	15.7	+01.7
NGC 6611, M 16, W 37	Eagle Nebula	18 16	- 13 50	17.0	+00.8
IC 1318, W 67		20 26	+39 50	78.6	+00.9
IC 5067–70, W 80	Pelican Nebula	20 49	+44 10	84.6	+00.1
NGC 7000, W 80	North America Nebula	21 00	+44 00	85.8	- 01.5

NGC, other	θ (′)	D (kpc)	2R (pc)	Te (°K)	E (pc cm^{-6})	Ne (cm^{-3})	M_{HII} ($M_\odot$)
NGC 7822, S 171, W 1	20 x 4	0.9	30		$< 10^4$	< 20	> 1600
NGC 1491, S 206	6 x 9	3.0	90	7800	$< 5 \times 10^3$	< 14	> 2800
S 209	12	5.4	50	8400	$< 3 \times 10^3$	< 8	> 9600
NGC 1976, M 42, W 10	90 x 60	0.5	13	8000	6×10^6	5000	10
IC 434, W 12	90 x 30	0.5	16	$\leq$ 3500	6×10^2	6	300
NGC 2237–46, W 16	80 x 60	1.6	37	8000	3×10^4	16	11000
NGC 3372, RCW 53	180 x 120	2.5	175	7000	3×10^5	200	2000
NGC 6514, M 20, W 28	20 x 20	2.1	12	8000	5×10^4	100	200
NGC 6523, M 8, W 29	45 x 30	1.4	18	8000	4×10^5	600	200
NGC 6618, M 17, W 38	20 x 15	2.2	12	7700	3×10^6	500	600
NGC 6611, M 16, W 37	120 x 25	2.2	77	8000	4×10^5	200	700
IC 1318, W 67	45 x 20	1.5	10	7000	2×10^4	40	2000
IC 5067–70, W 80	25 x 10	1.2 }					
NGC 7000, W 80	120 x 30	1.2 }	42	7000	4×10^3	10	18000

*Adapted from Scheffler (1982), and Scheffler and Elsässer (1987).

18.2 Emission Nebulae at Optical Wavelengths

Name	l b ° °	R.A.[††] h m	Dec.[††] ° ′	θ ′	V_{LSR} (km s^{-1})	D[†] (kpc)
NGC 7822, S 171	118.6+06.1	00 01	+68 20	20	- 11.0	0.88
S 173	119.6- 00.9	00 20	+61 30	25	- 37.1	3.33
S 175	120.3+01.9	00 24	+64 20	1		
S 176	120.3- 05.5	00 29	+57 00	12		
S 181	122.9+02.4	00 48	+65 00	15		
NGC 281, S 184	123.2- 06.3	00 50	+56 20	35	- 27.7	2.53
S 185, γ Cas, IC 63	123.9- 01.9	00 56	+60 40	120	+4.1	0.22
S 178	123.2+23.4	01 00	+86 00	450		
S 187	126.7- 00.8	01 20	+61 35	3		
S 188	128.0- 04.1	01 27	+58 07	10	- 14.1	1.01[†]
IC 1795, S 190	133.8+01.1	02 22	+61 45	12	- 38.1	2.78[†]
IC 1805, S 190	135.0+00.8	02 30	+61 00	60	- 45.0	2.42
S 191	136.3- 00.5	02 36	+59 22	1		
S 195	136.5- 00.4	02 38	+59 21	2		
S 192, S 193	136.2+02.1	02 44	+61 43	1		
IC 1848, S 199	137.2+00.8	02 47	+60 05	40	- 35.7	2.30
S 196	136.4+02.5	02 47	+62 00	6	- 38.7	2.88[†]
IC 1871	138.5+01.6	02 59	+60 14	4		
S 200	138.2+04.1	03 07	+62 35	8		
S 202	140.1+01.6	03 10	+59 30	240	- 17.5	1.30[†]
S 203	143.4- 02.2	03 16	+54 30	90		
IC 348	160.5- 17.9	03 41	+31 59	10		
S 204	145.8+02.8	03 51	+57 10	30		
S 205	149.0- 00.1	03 55	+52 50	100	- 16.6	1.50[†]
NGC 1499, S 220	160.2- 12.1	03 58	+36 30	160	+2.6	0.50
NGC 1491, S 206	150.6- 01.0	03 59	+51 11	6	- 23.0	3.6
S 245	187.4- 34.9	04 00	+03 00	600		
S 206	150.9- 01.1	04 00	+50 50	45	- 27.5	2.50
IC 360	170.0- 18.3	04 10	+25 30	180		
IC 359	169.0- 15.5	04 16	+28 05	15		
S 207	151.3+02.1	04 16	+53 00	7		
S 213	157.1- 03.6	04 17	+44 50	1		
S 214	157.6- 03.9	04 18	+44 15	4		
NGC 1554	176.2- 20.8	04 19	+19 30	7		
C 33	171.9- 15.7	04 24	+25 59	5		

Emission Nebulae at Optical Wavelengths*

Name	l b ° °	R.A.†† h m	Dec.†† ° ′	θ ′	V_{LSR} (km s^{-1})	D† (kpc)
S 210	152.7+02.8	04 26	+52 30	20	- 15.2	1.27†
S 239	178.9- 20.2	04 28	+18 00	6		
S 251	189.8- 27.1	04 30	+05 50	10		
S 211	154.7+02.4	04 33	+50 45	2		
S 250	189.7- 25.0	04 37	+07 10	10		
IC 2087	174.1- 13.7	04 37	+25 39	4		
NGC 1624, S 212	155.4+02.6	04 37	+50 20	5	- 37.5	4.36†
S 216	158.7+00.8	04 42	+46 40	50		
S 221	160.1+01.9	04 52	+46 20	60		
S 219	159.3+02.5	04 52	+47 19	2	- 22.6	2.76†
S 260	193.7- 22.7	04 53	+05 20	40		
S 217	159.2+03.3	04 55	+47 56	9	- 14.5	1.66†
S 246	187.0- 16.6	04 59	+14 00	90		
S 262	194.7- 20.0	05 04	+06 00	30		
S 226	168.5- 00.9	05 08	+37 58	2	- 19.4	
S 228	169.2- 00.9	05 10	+37 20	2	- 9.4	
IC 405, S 229	172.2- 02.0	05 14	+34 20	50	+14.0	0.64
S 223	165.9+02.4	05 14	+42 00	70		
S 265	194.8- 16.5	05 16	+07 45	40		
S 227	168.7+00.9	05 16	+38 50	30	- 20.8	
S 278	207.3- 22.7	05 18	- 05 30	50		
S 263	194.7- 15.6	05 19	+08 20	20		
IC 410, S 236	173.6- 01.8	05 19	+33 20	40	+2.7	3.80
S 225	168.1+03.0	05 23	+40 30	10		
S 224	166.2+04.5	05 24	+43 00	20		
IC 417	173.5- 00.2	05 25	+34 20	13		
NGC 1931, S 237	173.9+00.2	05 28	+34 10	4	+3.9	
C 54	194.7- 12.4	05 30	+10 00	270		
NGC 1952	184.6- 05.7	05 32	+22 00	8		
NGC 1976	209.1- 19.4	05 33	- 05 30	90		
NGC 1980	209.8- 19.1	05 35	- 06 00	240		
S 233	173.3+02.4	05 35	+35 50	2	- 6.2	
S 243	184.0- 04.2	05 36	+23 20	8		
S 240	180.1- 01.7	05 36	+28 00	200		
S 231	173.6+02.9	05 38	+35 50	10	- 12.7	

*Adapted from Gum (1955), Sharpless (1959), Rogers, Campbell and Whiteoak (1960) and Lynds(1965).
†Distances of the exciting star except when designated by † for kinematic distance.
††R.A. and Dec. are for the epoch 1950.0.

Emission Nebulae at Optical Wavelengths*

Name	l b ° °	R.A.†† h m	Dec.†† ° ′	θ ′	V_{LSR} (km s^{-1})	D† (kpc)
IC 434, S 277	206.9– 16.6	05 39	– 02 20	90	+17.6	0.41
S 232	173.5+03.2	05 39	+36 10	40	– 10.9	1.05
C 59	197.2– 10.1	05 43	+09 02	3		
NGC 2064	205.4– 14.4	05 44	+00 00	10		
NGC 2071	205.3– 14.0	05 45	+00 12	7		
S 218	159.8+12.5	05 45	+52 30	90		
S 242	182.3+00.1	05 48	+27 00	7	+4.0	
S 261	194.1– 01.9	06 06	+15 47	42	+9.3	1.63
S 247	189.0+00.9	06 06	+21 36	10		
S 270	196.8– 03.2	06 07	+12 50	1		
S 268	195.6– 02.5	06 07	+14 10	30		
NGC 2175	190.1+00.6	06 07	+20 30	40		
S 268	196.1– 02.5	06 08	+13 45	30		
S 254	192.5– 00.2	06 09	+18 02	10	+9.8	2.83
IC 2162, S 255	192.6– 00.0	06 10	+18 00	4	+8.5	1.53
S 269	196.5– 01.6	06 12	+13 51	3	+16.9	2.50†
S 267	196.2– 01.2	06 13	+14 18	8		
IC 443	189.1+03.0	06 14	+22 30	50		
S 249	188.7+04.3	06 18	+23 30	50		
NGC 2238	206.4– 01.9	06 30	+05 00	80		
NGC 2245	201.8+00.6	06 30	+10 12	2		
NGC 2247	201.6+00.6	06 30	+10 23	2		
S 280	208.7– 02.5	06 32	+02 40	40	+14.4	1.90
S 282	210.0– 02.1	06 36	+01 40	40	+17.9	1.62
NGC 2261	203.7+01.2	06 36	+08 46	2		
NGC 2264, S 273	202.9+02.2	06 38	+09 57	10	+5.3	1.0
S 284	211.6– 01.5	06 41	+00 30	60	+30.4	2.62†
S 304	234.0– 12.0	06 43	– 24 00	180		
RCW 9	234.4– 12.2	06 43	– 24 20	90		
S 289	218.9– 04.4	06 44	– 07 17	8		
S 308, RCW 11	234.6– 10.0	06 52	– 23 40	20		
S 291	220.4– 02.7	06 53	– 07 50	11		
S 303	233.8– 09.4	06 53	– 22 40	60		
S 287	218.1– 00.3	06 57	– 04 40	12		
S 293	224.2– 02.7	07 00	– 11 10	20		

Emission Nebulae at Optical Wavelengths*

Name	l b ° °	R.A.†† h m	Dec.†† ° ′	θ ′	V_{LSR} (km s^{-1})	D† (kpc)
S 295	224.5– 02.6	07 01	– 11 25	15		
IC 2177, S 292, RCW 2	223.7– 01.9	07 02	– 10 20	20	+14.4	0.67
S 297	225.4– 02.5	07 03	– 12 11	10	+11.2	0.79†
S 296, RCW 1	224.5– 01.7	07 04	– 11 00	150	+9.8	0.84
S 288	218.7+01.8	07 06	– 04 10	1		
RCW 6, S 301, G 5	231.4– 04.4	07 07	– 18 23		+54.3	4.02†
RCW 15	237.5– 07.3	07 08	– 25 00	300		
S 294	224.2+01.2	07 14	– 09 25	7	+38.5	2.84†
NGC 2359, IC 468, S 298, RCW 5	227.7– 00.1	07 16	– 13 02	10	+59.6	6.70
RCW 14, NGC 2367	235.6– 04.1	07 17	– 21 50	6		
RCW 4	224.4+03.2	07 21	– 08 30	60		
S 310	237.9– 04.2	07 21	– 24 00	300	+16.8	1.55
S 305, RCW 8	233.8– 00.2	07 28	– 18 29	3	+32.2	2.32†
S 306	234.3– 00.4	07 28	– 19 03	30		
RCW 10	234.4– 00.2	07 29	– 18 54	18		
S 302, RCW 7	232.7+01.0	07 30	– 16 54	20	+18.2	1.30†
S 309, RCW 13	234.8– 00.2	07 30	– 19 22	10	+35.0	2.33†
S 307, RCW 12, G 7	234.6+00.8	07 33	– 18 40	4	+38.1	2.76†
NGC 2467, RCW 16, G 9	243.2+00.3	07 50	– 26 20	8	+49.5	4.80
S 311	243.3+00.5	07 51	– 26 20	50	+46.6	4.1
RCW 19, G 10	253.7– 00.5	08 13	– 35 42	48	+28.9	3.0
RCW 20, NGC 2579, G 11	254.5+00.0	08 17	– 36 00	10		
RCW 27, G 14	260.1+00.5	08 36	– 40 12	100		
RCW 32, G 15	261.6+00.9	08 43	– 41 09	27	+1.2	
RCW 33, G 17	263.0+01.4	08 49	– 41 54	95		
RCW 35, G 18	264.6+00.1	08 50	– 43 55		+10.0	
S 312	250.5+13.1	08 55	– 25 00	840	– 1.1	
RCW 36, G 20	265.2+01.4	08 57	– 43 33	12		
RCW 38, G 22– 23– 24	268.0– 01.0	08 57	– 47 16		+2.3	1.2
RCW 37, NGC 2736	267.0+00.1	08 58	– 45 45	13		
RCW 40, G 25	269.3– 01.4	09 01	– 48 27		– 6.3	
G 12a	269.3– 01.3	09 02	– 48 30		+9.7	0.25
RCW 22	258.1+12.1	09 13	– 31 10	45		
RCW 41	270.3+00.8	09 15	– 47 45		– 1.0	
RCW 42, G 26	274.1– 01.3	09 22	– 51 54		+24.5	3.54

*Adapted from Gum (1955), Sharpless (1959), Rogers, Campbell and Whiteoak (1960) and Lynds(1965).
†Distances of the exciting star except when designated by † for kinematic distance.
††R.A. and Dec. are for the epoch 1950.0.

Emission Nebulae at Optical Wavelengths*

Name	l b ° °	R.A.[††] h m	Dec.[††] ° ′	θ ′	V_{LSR} (km s^{-1})	D[†] (kpc)
RCW 47	283.0- 02.7	10 03	- 58 42	25		
RCW 46	282.4- 01.3	10 06	- 57 15		+0.0	
RCW 45	282.2- 00.1	10 10	- 56 09	16		
NGC 3199, RCW 48, G 28	283.5- 01.0	10 14	- 57 36	15	- 8.5	3.58
RCW 49, NGC 3247, G 29	284.3- 00.3	10 22	- 57 27		- 3.0	4.7
RCW 50	284.3+00.4	10 24	- 56 54	12		
RCW 51	286.0+00.5	10 36	- 57 42	12		
RCW 53, η Car	287.4- 00.9	10 40	- 59 30		- 20.0	2.50
RCW 52, G 32	287.2+00.4	10 43	- 58 18	15		
RCW 55	290.4- 03.0	10 54	- 62 45	8		
G 35, RCW 54	289.7- 01.3	10 56	- 60 59	210	+22.2	7.90
G 34b	289.6+00.5	11 01	- 59 16		+0.0	
RCW 58	292.4- 04.9	11 04	- 65 18	7		
G 37	290.6+00.3	11 08	- 59 50		- 20.7	
G 38a	291.3- 00.6	11 10	- 60 56		- 18.8	
RCW 57	291.6- 00.5	11 12	- 60 56	170	- 25.5	3.50
IC 2872, RCW 60, G 39,40	293.7- 01.4	11 26	- 62 30		- 23.8	2.92
RCW 61, G 41	294.2- 02.3	11 28	- 63 30	15	- 24.9	1.75
RCW 59	293.0+04.5	11 35	- 56 40		- 21.0	2.93
IC 2944, RCW 62, G 42	294.6- 01.1	11 35	- 62 30		- 20.8	2.00
RCW 63	264.6+06.9	12 05	- 55 00	250		
RCW 65, G 43	301.0+01.2	12 31	- 61 18	11		
RCW 68	301.7- 00.7	12 37	- 61 36	15		
RCW 69, G 45	302.2+00.3	12 41	- 62 18		- 14.2	1.20[†]
RCW 74	305.2+00.0	13 07	- 62 33		- 30.8	2.40[†]
RCW 75, G 48a	306.3+00.2	13 16	- 62 15	18	- 23.8	1.90
RCW 78, G 48b	307.9+00.2	13 30	- 62 00	45		
RCW 79, G 48c	308.7+00.6	13 36	- 61 30	9	- 41.2	3.20[†]
RCW 80, G 48d	309.3- 00.5	13 43	- 62 24	21	- 50.0	4.10[†]
RCW 82	311.0+00.4	13 56	- 61 12	5	- 45.2	3.61[†]
RCW 83	311.9- 00.5	14 05	- 61 50	60	- 17.0	1.25[†]
RCW 85	313.5- 00.4	14 17	- 61 10	25	- 17.6	1.30[†]
RCW 87	320.2+00.8	15 01	- 57 19		- 41.0	2.90[†]
RCW 88	320.2+00.5	15 03	- 57 36		- 19.0	1.40[†]
RCW 91	321.2- 00.5	15 12	- 58 01	11	- 60.4	4.30[†]

Emission Nebulae at Optical Wavelengths*

Name	l b ° °	R.A.†† h m	Dec.†† ° ′	θ ′	V_{LSR} (km s^{-1})	D† (kpc)
RCW 92	322.2+00.6	15 14	- 56 30	8	- 57.4	4.10†
RCW 94	326.2+00.9	15 37	- 54 00	20	- 46.3	3.40†
RCW 95	326.7+00.8	15 40	- 53 47		- 45.4	3.30†
RCW 97	327.3- 00.5	15 49	- 54 25	6	- 49.2	3.38
RCW 98, G 49	327.5- 00.8	15 51	- 54 30	6	- 49.3	3.15
RCW 99	328.6- 00.5	15 56	- 53 35		- 49.6	3.70†
S 33	8.4+36.2	15 58	- 01 30	30		
S 7	349.8+22.0	15 58	- 23 00	130		
S 36	11.5+36.1	16 04	+00 30	40		
RCW 105, G 51	332.8+01.9	16 06	- 49 00	45	- 21.1	1.86
S 24	5.0+30.9	16 08	- 06 58	30		
S 73	37.6+44.8	16 08	+22 00	60		
S 14	4.5+29.9	16 10	- 07 55	35		
RCW 103	332.4- 00.4	16 13	- 51 00	5		
RCW 102	331.9- 01.0	16 14	- 51 48			
S 9	351.1+17.2	16 17	- 25 30	17		
RCW 104, RCW 106	332.9- 00.6	16 17	- 50 48		- 49.0	4.0
RCW 129, S 9, σ Sco	351.3+17.0	16 18	- 25 28	180	+1.3	0.16
IC 4606	351.8+15.0	16 26	- 26 30	60		
S 27, ζ Oph	6.0+24.7	16 30	- 10 00	480	+3.0	0.15
NGC 6165, RCW 107,G 52	336.4- 00.2	16 30	- 48 00	8	- 50.5	4.20†
RCW 108, NGC 6193	336.5- 01.3	16 35	- 48 40	210	- 20.9	1.30
IC 4628, RCW 116	344.3+02.1	16 50	- 40 30		- 13.7	1.60
RCW 110, RCW 111, G 54	341.0- 01.0	16 51	- 45 00	7	- 25.4	2.59†
RCW 113, NGC 6231	342.8+00.0	16 53	- 43 00	360	- 21.7	2.00
RCW 119, S 1, NGC 6281	347.7+01.9	17 01	- 38 00	180	- 11.3	1.60†
RCW 120, G 58, S 2	348.3+00.5	17 09	- 38 24	6	- 12.4	
RCW 126	350.6+01.0	17 13	- 36 18	16		
RCW 130, S 10	352.4+02.1	17 14	- 34 06	30	- 12.9	2.60
RCW 125	350.0+00.2	17 15	- 37 09	8		
NGC 6334, RCW 127, S 8	351.4+00.7	17 17	- 35 48	50	- 6.0	1.51
RCW 123, NGC 6337, G 59	349.5- 00.8	17 18	- 38 09	75	- 14.8	
RCW 114	343.9- 04.7	17 18	- 45 00	330		
RCW 128	351.4- 00.1	17 20	- 36 15	10		
NGC 6357, RCW 131, S 11	353.2+00.7	17 22	- 34 18	170	- 5.6	1.58

*Adapted from Gum (1955), Sharpless (1959), Rogers, Campbell and Whiteoak (1960) and Lynds(1965).

†Distances of the exciting star except when designated by † for kinematic distance.

††R.A. and Dec. are for the epoch 1950.0.

Emission Nebulae at Optical Wavelengths*

Name	l b (° °)	R.A.[††] (h m)	Dec.[††] (° ′)	θ (′)	V_{LSR} (km s^{-1})	D[†] (kpc)
RCW 133, G 68, S 10	355.9+01.5	17 26	- 31 36	45	+5.9	
S 12, RCW 132,NGC6383	355.8+00.5	17 30	- 32 10	150	+3.9	1.38
RCW 138, S 17	0.1+00.2	17 41	- 28 49	8	- 6.3	
S 16, RCW 140	359.7- 00.3	17 43	- 29 20	12		
S 20, RCW 137, S 6, RCW 141	0.3- 00.2	17 44	- 28 45	2	+9.2	
S 18	359.9- 00.4	17 44	- 29 12	2		
S 19	0.1- 00.6	17 45	- 29 10	14		
S 15, RCW 134, G 69	358.5- 02.1	17 47	- 31 20	45	- 4.1	2.0
S 21	0.6- 01.0	17 48	- 28 55	3		
RCW 144, G 71	4.4+00.5	17 50	- 25 00	85		
S 22	4.6+00.3	17 52	- 24 50	60		
RCW 145, G 47a,b	6.6+00.1	17 57	- 23 15	90		
NGC 6514, RCW 147, G 76, M 20	7.0- 00.2	17 59	- 23 00	20	+8.1	2.47
NGC 6523, RCW 146, M 8, S 25	6.1- 01.2	18 01	- 24 20	45	+2.6	1.60
S 46, RCW 158, G 80	15.1+03.4	18 03	- 14 10	30		
RCW 154	12.7+02.0	18 03	- 17 00	40		
RCW 149, G 77a, S 24	8.7- 00.6	18 04	- 21 45	120		
S 34	8.6- 00.9	18 05	- 22 00	60		
S 38	12.0+00.8	18 06	- 18 10	3		
NGC 6559, S 29	7.0- 02.3	18 07	- 24 00	15		
IC 1274	7.3- 02.3	18 08	- 23 45	20		
RCW 155, S 31	12.9+00.3	18 09	- 17 41	10		
S 40	12.7+00.4	18 09	- 17 45	15		
S 70	35.1+11.4	18 12	+07 03	8		
IC 4701, RCW 157, S 42	14.1+00.1	18 13	- 16 40	60		
S 41, RCW 151	12.9- 00.6	18 13	- 18 00	210	+14.8	2.58
S 35, NGC 6526, G 77b	10.9- 01.7	18 13	- 20 20	10	+5.5	1.82
S 47	15.3+00.4	18 14	- 15 30	5		
IC 1283, RCW 153, S 37	11.8- 01.5	18 14	- 19 30	15	+8.5	
S 54	18.5+01.9	18 15	- 12 00	60	+31.1	2.2 †
RCW 156, S 34	13.7- 00.8	18 15	- 17 30	50		
NGC 6611, RCW 165, M 16, S 49	17.0+00.8	18 16	- 13 50	120	+26.5	3.30
RCW 167, NGC 6604, G 84	19.0+01.3	18 18	- 11 54	180	+26.1	2.33
S 50	16.3- 00.1	18 18	- 14 50	20		
RCW 161	16.1- 00.3	18 18	- 15 09	80		

Emission Nebulae at Optical Wavelengths*

Name	l b (° °)	R.A.†† (h m)	Dec.†† (° ′)	θ (′)	V_{LSR} (km s^{-1})	D† (kpc)
NGC 6618, RCW 160, M 17, S 45	15.3− 00.7	18 18	− 16 00	40	+19.8	3.00
S 48, RCW 162, G 82	16.7− 00.5	18 20	− 14 40	15		
S 68	30.6+06.4	18 22	+00 50	8		
RCW 166, S 45	18.4− 00.3	18 22	− 13 09	15		
RCW 159	15.3− 01.8	18 22	− 16 36	15		
RCW 164, S 46	16.9− 01.2	18 23	− 14 51	8		
S 62	26.8+03.5	18 25	− 03 50	3		
RCW 171	23.2+00.6	18 28	− 08 27	5		
S 57	22.9+00.6	18 28	− 08 40	2		
RCW 170	22.6+00.3	18 28	− 09 09	7		
S 56	22.0+00.2	18 28	− 09 40	10		
RCW 169	22.0+00.1	18 28	− 09 48	7		
S 64, RCW 174	28.8+03.4	18 29	− 02 10	20		
S 58	23.2+00.5	18 29	− 08 30	8		
IC 1287	21.1− 00.6	18 29	− 10 50	20		
S 55	20.4− 01.3	18 30	− 11 48	20		
S 61	26.5+02.6	18 31	− 05 00	2	+30.6	2.48†
S 60, RCW 173	25.4+00.2	18 34	− 06 40	15	+30.6	2.54†
S 59	24.5− 00.2	18 34	− 07 40	30		
S 51	16.8− 05.3	18 38	− 16 50	40		
S 69, RCW 177	32.4+01.9	18 41	− 00 20	15	+46.0	3.38†
S 66, RCW 176	30.5+00.4	18 43	− 02 00	10	+43.4	3.04†
S 65, RCW 175	29.1− 00.6	18 44	− 03 47	10		
S 67	30.7− 00.6	18 47	− 02 23	15		
RCW 181	38.8+02.0	18 52	+06 00	5		
S 76	40.5+02.5	18 54	+07 45	8		
S 71	36.0− 01.3	18 59	+02 05	2		
RCW 179	36.4− 01.7	19 01	+02 09	20		
S 74	40.0− 01.4	19 07	+05 30	3		
S 80	50.1+03.4	19 09	+16 45	1		
S 94	64.9+06.7	19 26	+31 20	50		
S 96	66.1+07.1	19 27	+32 35	20		
S 82	53.5+00.0	19 28	+18 09	10	+11.9	0.90†
S 91	64.2+04.2	19 35	+29 30	120		
NGC 6820, S 86	59.4− 00.2	19 41	+23 10	40	+32.8	2.58

*Adapted from Gum (1955), Sharpless (1959), Rogers, Campbell and Whiteoak (1960) and Lynds(1965).
†Distances of the exciting star except when designated by † for kinematic distance.
††R.A. and Dec. are for the epoch 1950.0.

Emission Nebulae at Optical Wavelengths*

Name	l b ° °	R.A.[††] h m	Dec.[††] ° ′	θ ′	V_{LSR} (km s^{-1})	D[†] (kpc)
S 87	61.0– 00.0	19 44	+24 35	17		
S 88	61.5+00.3	19 44	+25 13	18	+16.5	1.70[†]
S 92	64.0+01.8	19 44	+28 10	45	+14.6	1.33[†]
S 84	55.7– 03.6	19 46	+18 15	15		
S 90	63.1+00.5	19 47	+26 41	8		
S 93	64.2– 00.4	19 53	+27 09	1		
S 95	65.9+00.6	19 53	+29 10	1		
S 97	66.7+00.9	19 54	+30 00	10		
S 63	27.1– 21.1	19 55	– 14 30	70		
S 98	68.1+00.9	19 57	+31 10	10		
S 101, Ced 173	71.6+02.9	19 58	+35 10	20	+7.1	2.71
S 99	70.1+01.7	19 59	+33 20	3	– 23.4	8.8
S 100	70.2+01.5	20 00	+33 20	3		
NGC 6888	75.5+02.3	20 11	+38 10	20		
IC 1310	72.9+00.3	20 12	+34 50	15		
IC 1318A	78.8+03.6	20 15	+41 40	45		
S 104	74.8+00.6	20 16	+36 40	7		
S 108, S 109	77.6+01.6	20 20	+39 30	110	+3.8	2.10
IC 1318B	78.6+00.9	20 26	+39 50	45		
S 102	71.6– 05.5	20 31	+30 30	70		
C 181	83.8+03.3	20 32	+45 30	13		
S 115	84.8+03.9	20 33	+46 40	30	+2.1	2.0
S 107	77.4– 03.7	20 41	+36 10	8		
IC 5068, S 109	83.2– 01.0	20 49	+42 20	40	– 2.5	
IC 5070	84.5+00.0	20 49	+44 00	60		
IC 5067	84.6+00.1	20 49	+44 10	25		
NGC 6960	74.5– 08.4	20 50	+31 00	210		
NGC 7000, S 117	85.8– 01.5	21 00	+44 00	120	+1.5	1.17
S 129, Ced 190	98.3+07.8	21 10	+59 30	150	– 11.0	0.40
S 119	87.2– 03.8	21 15	+43 30	150	+0.5	0.72
S 110	79.2– 12.1	21 17	+32 00	100		
S 113	83.7– 08.3	21 19	+37 50	20		
S 114	84.2– 07.8	21 19	+38 35	12		
S 127	96.2+02.5	21 27	+54 20	1		
S 133	103.1+09.7	21 27	+64 10	80		

Emission Nebulae at Optical Wavelengths*

Name	l b ° °	R.A.†† h m	Dec.†† ° ′	θ ′	V_{LSR} (km s^{-1})	D† (kpc)
S 128	97.6+03.1	21 31	+55 40	1		
S 111	80.5- 16.5	21 36	+29 50	70		
IC 1396, S 131	99.3+03.8	21 37	+57 15	14	- 3.5	0.82
S 124	94.8- 01.8	21 39	+50 10	90		
S 118	89.6- 08.0	21 40	+42 00	270		
S 123	91.0- 06.4	21 40	+44 12	15		
NGC 7129	105.3+09.9	21 41	+65 50	2		
IC 5146, S 125	94.4- 05.6	21 52	+47 00	10	+3.4	1.00
S 137	105.5+07.8	21 55	+64 20	120		
S 134	104.3+02.1	22 15	+59 00	190		
S 140	106.6+05.2	22 17	+62 50	35	- 11.2	0.98
S 132	102.8- 00.9	22 18	+55 40	80	- 50.9	3.68
S 135	104.4+01.3	22 19	+58 25	22		
S 141	106.8+03.3	22 27	+61 20	5		
S 126	95.0- 17.0	22 30	+38 00	110		
S 150	108.9+06.1	22 30	+64 50	35		
S 139	106.3- 00.3	22 38	+57 58	10		
S 144	107.6+00.8	22 43	+59 34	10		
NGC 7380, S 142	107.0- 01.0	22 45	+57 45	25	- 36.7	3.07
S 146	108.2+00.5	22 48	+59 38	3		
S 154	109.0+01.7	22 49	+61 00	55	- 12.2	1.22†
S 149	108.3- 01.0	22 54	+58 15	2		
S 155	110.1+02.4	22 55	+62 10	50	- 15.8	0.73
S 153	108.8- 01.0	22 57	+58 29	5	- 48.3	4.17†
S 152	108.8- 01.0	22 57	+58 31	2	- 35.7	3.19†
S 151	108.6- 02.7	23 01	+56 50	30		
S 160	112.5+03.8	23 10	+64 20	100		
NGC 7538	111.6+00.8	23 12	+61 13	8		
S 157	111.1- 00.7	23 13	+59 40	3	- 38.3	3.50
S 158	111.7+00.3	23 14	+60 50	7	- 60.1	2.8
S 161	112.0+00.9	23 15	+61 30	90		
NGC 7635, S 162	112.2+00.2	23 18	+60 55	15	- 42.8	3.20
S 163	113.5- 00.7	23 31	+60 30	18	- 43.1	3.49†
S 164	113.1- 04.5	23 36	+56 41	5		
S 165	114.6+00.2	23 37	+61 39	8	- 27.3	2.23†

*Adapted from Gum (1955), Sharpless (1959), Rogers, Campbell and Whiteoak (1960) and Lynds(1965).
†Distances of the exciting star except when designated by † for kinematic distance.
††R.A. and Dec. are for the epoch 1950.0.

Emission Nebulae at Optical Wavelengths*

Name	l ° b °	R.A.†† h m	Dec.†† ° ′	θ ′	V_{LSR} (km s^{-1})	D† (kpc)
S 166	114.6– 00.9	23 40	+60 38	10		
S 174	120.3+18.4	23 45	+80 40	15		
S 168	115.7– 01.6	23 50	+60 10	5	– 43.9	3.44†
S 170	117.6+02.3	23 59	+64 23	20	– 53.6	4.10†
C 214	118.1+04.8	23 59	+66 55	20		

*Adapted from Gum (1955), Sharpless (1959), Rogers, Campbell and Whiteoak (1960) and Lynds(1965).
†Distances of the exciting star except when designated by † for kinematic distance.
††R.A. and Dec. are for the epoch 1950.0.

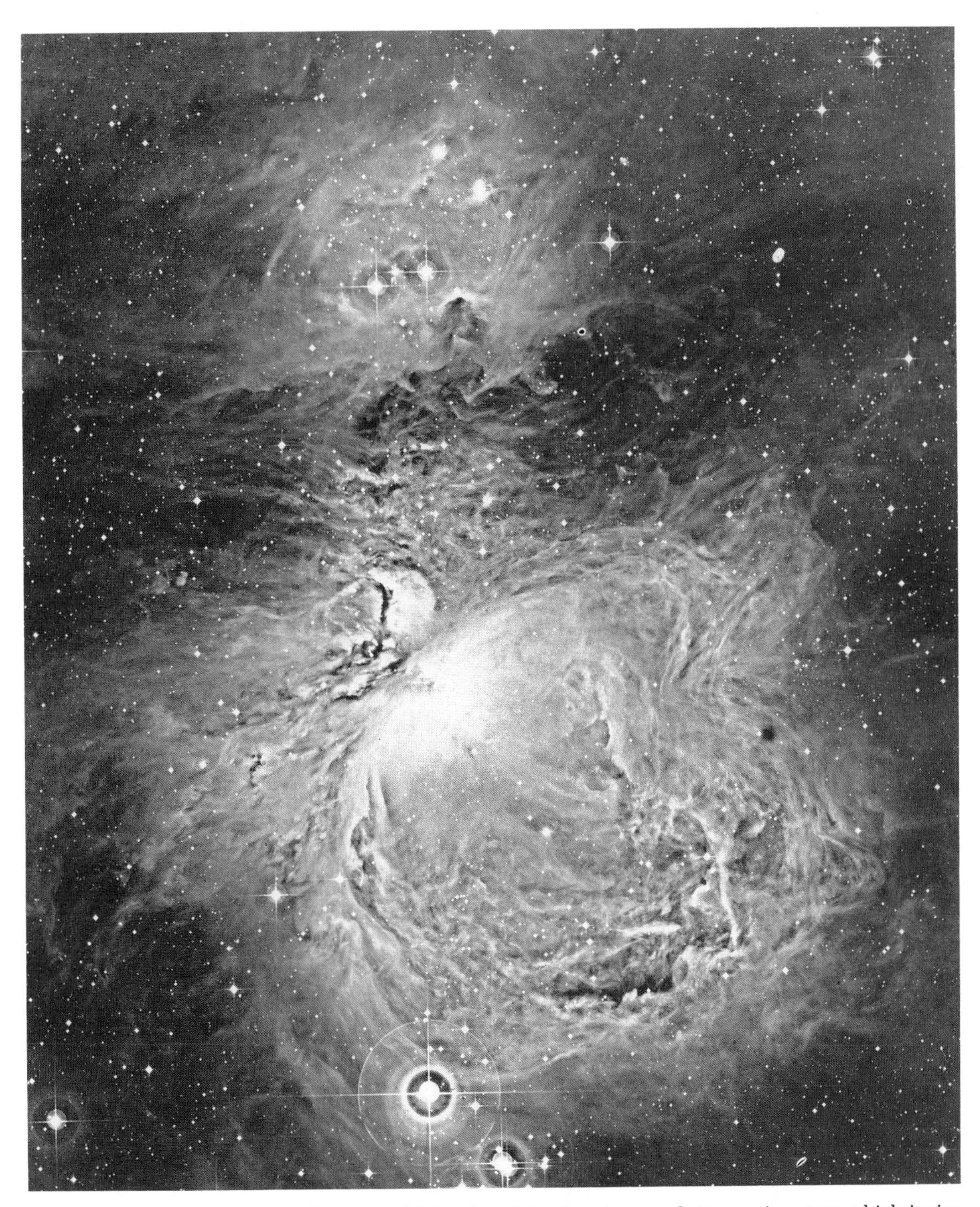

Orion Nebula. Interstellar gas becomes visible when it is close to very hot, massive stars which ionize it to fluorescence, creating an emission nebula or H II region. Such stars are found in the bright central region of the Orion nebula (M 42 or NGC 1976). The bright region is about 6 light years across. This photograph was taken in red light during a 90 minute exposure on the 1.2 meter UK Schmidt Telescope. The original photographic plate has been copied using the unsharp masking technique for controlling contrast, thereby permitting fine detail to be resolved in both the bright central region and the fainter outlying filaments. (Royal Observatory Edinburgh ©1978, prepared by David F. Malin.)

18.3 Emission Nebulae at Radio Wavelengths

Galactic Coordinate	R.A. h m s	Dec. ° ′ ″	θ_{RA} (′)	θ_{Dec} (′)	S_{6cm} (Jy)	V_{LSR} (km s^{-1})
133.7+ 1.2	02 21 55	+61 51 59	2.9	4.7	68.1	− 42.3
209.0− 19.4	05 32 51	− 05 25 39	3.8	4.3	382.2	− 2.8
208.9− 19.3	05 33 03	− 05 18 18	2.5	2.5	16.7	+7.7
279.4− 31.7	05 39 00	− 69 06 42	4.0	5.0	33.7	
206.5− 16.4	05 39 12	− 01 55 42	3.6	3.0	53.8	+7.0
189.2+ 3.0	06 14 12	+22 26 48				
189.0+ 3.2	06 14 48	+22 40 48				
260.7− 3.2	08 22 37	− 42 56 48				
265.1+ 1.5	08 57 38	− 43 33 36	2.3	2.5	23.8	+2.8
267.8− 0.9	08 57 40	− 47 07 12	8.1	7.5	25.9	+7.8
268.0− 1.1	08 58 05	− 47 20 12	1.3	2.1	172.3	+1.8
282.0− 1.2	10 04 53	− 56 57 30	1.8	2.1	25.9	+22.4
283.3− 1.0	10 13 17	− 57 38 18	3.9	5.1	4.8	
284.0− 0.9	10 18 03	− 57 49 36	6.3	9.2	7.7	+4.5
284.1− 0.4	10 22 44	− 57 28 48	4.3	8.4	8.0	− 16.0
284.3− 0.3	10 22 19	− 57 32 00	4.7	6.2	178.8	− 0.7
284.6− 0.2	10 24 32	− 57 32 42	5.5	7.2	10.7	
285.3+ 0.0	10 29 36	− 57 46 42	1.7	1.7	16.5	− 1.6
287.2− 0.7	10 39 48	− 59 18 41	7.1	11.7	42.3	− 22.4
287.3− 0.9	10 39 50	− 59 29 11	6.3	4.3	15.1	− 17.0
287.4− 0.6	10 41 36	− 59 19 11	6.2	6.1	114.5	− 18.1
287.5− 0.6	10 42 50	− 59 22 59	5.7	5.3	86.0	− 23.6
287.6− 0.9	10 42 09	− 59 42 17	6.7	7.0	28.0	− 24.1
287.7− 0.6	10 43 54	− 59 29 29	4.2	2.2	10.4	− 26.0
287.8− 0.8	10 43 43	− 59 40 29	6.0	2.0	7.3	− 21.3
287.9− 0.9	10 44 09	− 59 50 17	7.9	5.4	23.9	− 20.7
287.9− 0.8	10 44 53	− 59 44 41	7.5	13.2	57.4	− 25.8
289.1− 0.4	10 54 29	− 59 49 48	3.2	3.1	7.8	+27.0
289.8− 1.1	10 56 51	− 60 50 48	7.4	6.5	16.9	+21.9
291.3− 0.7	11 09 45	− 61 02 36	0.9	0.9	97.4	− 23.4
291.0− 0.1	11 09 47	− 60 22 18	4.2	4.5	8.4	
291.6− 0.5	11 12 53	− 60 59 24	3.8	6.5	172.1	+9.4
291.9− 0.7	11 14 24	− 61 13 18	0.0	1.8	3.0	+25.5
292.0+ 1.8	11 22 02	− 58 59 24	2.5	2.8	8.8	
295.1− 1.6	11 37 47	− 63 08 42	5.7	7.2	6.8	− 20.9

Emission Nebulae at Radio Wavelengths*

Galactic Coordinate	Te (°K)	E (pc cm^{-6})	D (kpc)	2R (pc)	Ne (cm^{-3})	M_{HII} ($M_\odot$)
133.7+ 1.2	6,800	5.4×10^2	3.1 ± 0.8	3.3	330	450
209.0− 19.4	7,000	2.5×10^3	0.5	0.6	1700	13
208.9− 19.3	6,300	2.8×10^5	0.7 ± 0.7	0.7	613	3.1
279.4− 31.7	9,400	2.0×10^5	55.0	105.2	44	6.2×10^5
206.5− 16.4	7,200	5.5×10^2	0.5 ± 0.5	0.6	790	6.7
189.2+ 3.0						
189.0+ 3.2						
260.7− 3.2						
265.1+ 1.5	7,400	4.6×10^5	0.6 ± 0.6	0.6	863	2.4
267.8− 0.9	6,000	4.3×10^4	1.8 ± 1.8	6.0	85	220
268.0− 1.1	7,900	7.1×10^6	0.6 ± 0.6	0.4	4106	3.7
282.0− 1.2	7,300	7.6×10^5	6.6 ± 1.0	5.5	371	730
283.3− 1.0						
284.0− 0.9	3,500	1.1×10^4	5.4 ± 4.4	17.6	25	1700
284.1− 0.4	10,700	2.8×10^4	2.8 ± 1.2	7.2	62	280
284.3− 0.3	6,800	6.6×10^5	6.0 ± 1.2	13.8	218	6900
284.6− 0.2						
285.3+ 0.0	9,100	6.8×10^5	$0.2, 5.1 \pm 0.9$	0.1	2157	0.08
287.2− 0.7	7,200	5.5×10^4	2.7	10.5	72	1000
287.3− 0.9	5,900	5.7×10^4	2.7	6.0	97	250
287.4− 0.6	7,200	3.3×10^5	2.7	7.1	216	920
287.5− 0.6	6,100	2.9×10^5	2.7	6.3	215	660
287.6− 0.9	5,100	5.8×10^4	2.7	7.9	86	510
287.7− 0.6	6,700	1.2×10^5	2.7	3.5	185	96
287.8− 0.8	4,900	5.9×10^4	2.7	4.0	121	93
287.9− 0.9	4,400	5.2×10^4	2.7	7.5	83	430
287.9− 0.8	8,200	6.6×10^4	2.7	11.5	76	1400
289.1− 0.4	13,700	1.1×10^5	8.9 ± 0.8	12.0	94	1900
289.8− 1.1	9,100	4.1×10^4	8.7 ± 0.8	25.8	40	8200
291.3− 0.7	7,700	1.4×10^7	3.6 ± 1.9	1.4	3122	99
291.0− 0.1						
291.6− 0.5	6,900	7.5×10^5	8.2 ± 0.9	17.4	208	13000
291.9− 0.7	7,000	$> 3.6 \times 10^5$	9.6 ± 0.8	<3.9	>305	<220
292.0+ 1.8						
295.1− 1.6	3,800	1.5×10^4	2.4 ± 1.3	6.6	47	160

*Adapted from Reifenstein et al. (1970)., and Wilson et al. (1970).

Emission Nebulae at Radio Wavelengths*

Galactic Coordinate	R.A. h m s	Dec. ° ′ ″	θ_{RA} (′)	θ_{Dec} (′)	S_{6cm} (Jy)	V_{LSR} (km s^{-1})
295.2− 0.6	11 41 05	− 62 09 12	4.9	4.2	9.9	+51.1
298.2− 0.3	12 07 24	− 62 33 18	1.8	2.1	30.6	+30.6
298.8− 0.3	12 12 38	− 62 39 12	4.7	4.7	16.0	+25.0
298.9− 0.4	12 12 44	− 62 44 48	3.8	2.8	32.7	+24.2
301.0+1.2	12 32 03	− 61 22 54	0.0	0.0	2.8	− 46.6
301.1+1.0	12 33 11	− 61 34 48	0.0	0.0	3.5	− 48.1
305.1+0.1	13 07 06	− 62 22 36	5.0	5.7	14.2	− 37.9
305.2+0.0	13 08 03	− 62 29 12	3.6	3.9	28.0	− 40.0
305.2+0.2	13 08 23	− 62 17 30	4.9	6.8	52.0	− 39.1
305.4+0.2	13 09 20	− 62 18 48	3.0	2.5	39.4	− 39.1
305.6+0.0	13 11 06	− 62 28 54	4.6	5.0	17.8	− 44.7
307.1+1.2	13 23 01	− 61 07 24	1.3	0.0	4.0	
308.6+0.6	13 36 37	− 61 29 00	3.5	6.3	9.1	− 51.8
308.7+0.6	13 37 18	− 61 29 48	3.0	4.5	5.1	− 46.4
309.8+1.7	13 42 43	− 60 14 12	6.0	4.9	20.7	
311.5+0.4	14 00 01	− 61 02 12	3.0	4.5	5.1	− 64.8
311.6+0.3	14 01 16	− 61 05 36	0.0	0.0	2.8	− 63.5
311.9+0.1	14 03 52	− 61 13 06	3.0	1.8	7.1	− 47.3
311.9+0.2	14 03 49	− 61 05 18	3.0	3.5	7.8	− 45.5
314.2+0.4	14 21 16	− 60 09 12	2.7	2.8	5.2	− 50.5
316.8− 0.1	14 41 31	− 59 36 54	2.3	1.3	28.7	− 36.1
317.0+0.3	14 41 45	− 59 13 36	6.1	8.2	14.7	− 47.4
317.3+0.2	14 44 14	− 59 08 18	6.2	8.6	6.7	− 46.8
319.2− 0.4	14 59 14	− 58 51 42	5.1	4.6	8.2	− 22.8
319.4+0.0	14 59 23	− 58 24 30	5.0	2.3	8.2	− 14.1
320.2+0.8	15 01 34	− 57 19 24	1.3	1.3	7.7	− 36.0
320.3− 0.3	15 06 17	− 58 14 36	2.7	3.0	7.2	− 67.7
320.3− 0.2	15 06 14	− 58 06 12	1.3	1.6	5.8	− 7.7
320.4− 1.0	15 09 45	− 58 50 24	5.8	5.5	12.2	
321.0− 0.5	15 12 06	− 58 00 30	3.2	2.3	7.8	− 61.6
321.1− 0.5	15 12 45	− 57 59 36	3.8	4.1	7.8	− 55.5
322.2+0.6	15 14 49	− 56 27 54	0.9	0.9	12.2	− 51.8
324.2+0.1	15 29 02	− 55 46 18	0.0	0.0	3.3	− 86.6
326.5+0.9	15 38 33	− 53 48 54	0.9	0.9	7.2	− 39.0
326.7+0.6	15 40 56	− 53 57 12	2.8	3.8	35.7	− 44.5

Emission Nebulae at Radio Wavelengths*

Galactic Coordinate	Te (°K)	E (pc cm^{-6})	D (kpc)	2R (pc)	Ne (cm^{-3})	M_{HII} ($M_\odot$)
295.2– 0.6	7,100	5.1×10^4	12.5 ± 0.8	24.2	46	7900
298.2– 0.3	8,100	9.2×10^5	11.7 ± 0.7	9.7	308	3400
298.8– 0.3	11,900	9.2×10^4	11.5 ± 0.7	23.1	63	9400
298.9– 0.4	6,000	3.1×10^5	11.5 ± 0.8	16.0	140	6900
301.0+ 1.2	6,100	$> 1.2 \times 10^6$	5.2 ± 1.2	<1.1	>1022	<17
301.1+ 1.0	9,100	$> 1.7 \times 10^6$	5.2 ± 1.2	<1.1	>1221	<20
305.1+ 0.1	5,200	4.8×10^4	$3.4, 8.1 \pm 1.2$	7.8	79	440
305.2+ 0.0	5,200	2.0×10^5	$3.6, 7.9 \pm 1.2$	5.8	184	420
305.2+ 0.2	5,700	1.6×10^5	$3.5, 8.0 \pm 1.1$	8.6	135	1000
305.4+ 0.2	5,200	5.1×10^5	8.1 ± 1.3	9.5	233	2400
305.6+ 0.0	4,600	7.3×10^4	$4.3, 7.3 \pm 1.4$	8.8	91	750
307.1+ 1.2						
308.6+ 0.6	10,200	5.0×10^4	$4.7, 7.8 \pm 1.3$	9.4	73	740
308.7+ 0.6	7,600	4.2×10^4	$3.9, 8.6 \pm 1.1$	6.1	83	230
309.8+ 1.7						
311.5+ 0.4	5,100	3.6×10^4	6.6 ± 1.2	10.4	59	800
311.6+ 0.6	3,300	9.6×10^5	6.6 ± 1.2	<1.4	>824	<28
311.9+ 0.1	4,700	1.2×10^5	9.8 ± 1.3	9.7	113	1300
311.9+ 0.2	4,300	6.9×10^4	$3.5, 9.9 \pm 0.9$	4.8	119	160
314.2+ 0.4	3,000	5.7×10^4	$3.8, 10.1 \pm 0.9$	4.5	113	120
316.8– 0.1	5,900	9.8×10^5	12.1 ± 0.9	8.9	332	2800
317.0+ 0.3	4,200	2.7×10^4	11.2 ± 0.9	34.0	28	13000
317.3+ 0.2	2,700	9.9×10^3	$3.3, 11.4 \pm 0.9$	10.3	31	410
319.2– 0.4	11,200	4.4×10^4	$1.5, 13.6 \pm 0.8$	3.2	117	46
319.4+ 0.0	7,400	7.9×10^4	$1.0, 14.2 \pm 0.7$	1.5	233	8.5
320.2+ 0.8	8,400	5.2×10^5	2.5 ± 0.7	1.4	614	20
320.3– 0.3	8,400	1.0×10^5	$5.0, 10.4 \pm 0.8$	6.1	129	350
320.3– 0.2	5,100	2.7×10^5	$0.6, 14.8 \pm 0.4$	0.4	860	0.52
320.4– 1.0						
321.0– 0.5	4,900	1.0×10^5	$4.4, 11.1 \pm 0.8$	5.1	141	230
321.1– 0.5	4,400	4.6×10^4	$4.0, 11.6 \pm 0.8$	6.7	83	300
322.2+ 0.6	5,400	1.5×10^6	$3.7, 12.1 \pm 0.8$	1.4	1029	35
324.2+ 0.1	4,300	1.2×10^6	$6.9, 9.3 \pm 1.2$	<1.5	>911	<35
326.5+ 0.9	7,000	9.7×10^5	$2.1, 14.6 \pm 0.7$	0.8	1094	6.9
326.7+ 0.6	6,400	3.5×10^5	3.2 ± 0.7	4.5	281	300

***Adapted from Reifenstein et al. (1970)., and Wilson et al. (1970).**

Emission Nebulae at Radio Wavelengths*

Galactic Coordinate	R.A. h m s	Dec. ° ′ ″	θ_{RA} (′)	θ_{Dec} (′)	S_{6cm} (Jy)	V_{LSR} (km s^{-1})
327.3− 0.5	15 49 12	− 54 26 30	0.9	0.9	44.9	− 48.8
327.6− 0.4	15 50 02	− 54 05 54	3.5	4.3	5.7	− 69.8
328.0− 0.1	15 50 54	− 53 39 24	1.6	3.3	4.6	− 44.7
328.3+0.4	15 50 18	− 53 02 30	2.1	2.3	10.0	− 96.2
328.4+0.2	15 51 46	− 53 08 36	2.5	2.1	10.5	
330.9− 0.4	16 06 27	− 51 58 36	1.8	1.6	9.5	− 56.1
331.0− 0.2	16 06 20	− 51 42 30	4.6	4.7	8.3	− 89.2
331.4+0.0	16 07 18	− 51 23 06	1.8	3.3	6.6	− 79.0
331.3− 0.2	16 07 36	− 51 34 54	1.3	0.0	4.2	− 84.4
331.5− 0.1	16 08 22	− 51 19 24	2.5	2.1	31.1	− 88.7
332.1− 0.5	16 08 23	− 51 55 30	0.0	1.8	3.2	
331.3− 0.3	16 08 33	− 51 39 24	2.5	2.3	7.8	− 64.4
332.2− 0.5	16 12 52	− 51 10 06	1.3	0.9	13.2	− 55.0
332.5− 0.1	16 13 17	− 50 40 18	1.8	0.0	2.2	− 55.9
332.4− 0.4	16 13 43	− 50 56 54	5.5	6.8	13.0	
333.2− 0.1	16 15 56	− 50 12 18	4.2	6.7	12.3	− 90.8
332.7− 0.6	16 15 58	− 50 56 00	3.6	3.8	20.1	− 47.0
332.8− 0.6	16 16 25	− 50 47 18	3.2	6.7	18.7	− 57.2
333.0− 0.4	16 16 51	− 50 33 12	4.9	3.8	39.7	− 53.8
333.1− 0.4	16 17 16	− 50 29 12	3.2	4.9	42.1	− 55.8
333.3− 0.4	16 17 45	− 50 19 18	0.0	3.2	34.2	− 50.1
333.6− 0.1	16 17 53	− 49 53 54	4.1	5.0	15.0	− 53.7
333.6− 0.2	16 18 26	− 49 58 54	0.9	0.9	84.4	− 48.3
333.7− 0.5	16 19 58	− 50 06 12	3.5	4.9	3.2	− 49.9
335.8− 0.2	16 27 27	− 48 24 06	4.5	3.6	7.6	− 52.1
336.5+0.0	16 29 33	− 47 46 06	3.3	4.7	3.9	− 63.4
336.4− 0.2	16 29 52	− 47 56 54	5.0	4.5	7.6	− 68.5
336.4− 0.3	16 30 33	− 47 59 06	5.1	4.5	10.0	− 93.1
336.5− 0.2	16 30 37	− 47 51 06	5.9	8.1	14.6	− 88.6
336.8+0.0	16 30 49	− 47 30 24	10.7	6.1	57.5	− 78.9
336.9− 0.1	16 32 09	− 47 32 30	3.3	6.6	14.4	− 73.1
337.1− 0.2	16 33 01	− 47 25 18	0.9	1.8	15.1	− 72.7
337.3− 0.1	16 33 27	− 47 16 24	3.5	4.6	7.4	− 53.5
338.0− 0.1	16 34 14	− 46 45 18	6.0	5.7	11.1	− 52.5
337.6+0.0	16 34 30	− 46 58 06	4.3	5.0	6.1	− 54.8

Emission Nebulae at Radio Wavelengths*

Galactic Coordinate	Te (°K)	E (pc cm^{-6})	D (kpc)	2R (pc)	Ne (cm^{-3})	M_{HII} ($M_\odot$)
327.3− 0.5	6,100	5.8×10^6	3.5 ± 0.7	1.3	2069	60
327.6− 0.4	3,300	3.1×10^4	$5.1, 11.8 \pm 0.8$	8.5	61	450
328.0− 0.1	5,600	8.7×10^4	3.3 ± 0.6	3.2	165	65
328.3+ 0.4	4,800	2.0×10^5	9.4 ± 0.6	8.8	150	1200
328.4+ 0.2						
330.9− 0.4	4,700	3.1×10^5	$4.2, 13.3 \pm 0.7$	3.0	322	110
331.0− 0.2	3,200	3.2×10^4	$6.6, 10.9 \pm 0.8$	13.1	49	1300
331.4+ 0.0	5,400	1.1×10^5	$5.8, 11.8 \pm 0.7$	6.0	135	360
331.3− 0.2	5,900	$> 6.6 \times 10^5$	$6.2, 11.3 \pm 0.8$	<2.1	>557	<65
331.5− 0.1	5,200	5.8×10^5	11.1 ± 0.9	10.9	232	3500
332.1− 0.5	4,500	$> 3.3 \times 10^5$	$5.0, 12.5 \pm 0.7$	>2.0	>406	<40
331.3− 0.3	3,400	1.2×10^5	$4.7, 12.8 \pm 0.7$	4.9	155	210
332.2− 0.5	5,400	1.1×10^6	4.1 ± 0.7	1.9	771	63
332.5− 0.1	2,200	$> 1.8 \times 10^5$	$4.2, 13.5 \pm 0.7$	<1.7	>329	<19
332.4− 0.4						
333.2− 0.1	4,900	4.1×10^4	$6.7, 11.1 \pm 0.7$	15.2	52	2200
332.7− 0.6	4,100	1.3×10^5	$3.6, 14.2 \pm 0.7$	5.7	153	340
332.8− 0.6	4,900	8.3×10^4	4.3 ± 0.7	8.5	99	730
333.0− 0.4	3,700	1.9×10^5	$4.2, 13.6 \pm 0.7$	7.7	156	870
333.1− 0.4	4,700	2.5×10^5	13.4 ± 0.8	22.7	106	15000
333.3− 0.4	5,000	2.1×10^6	3.9 ± 0.7	2.1	994	110
333.6− 0.1	4,200	6.7×10^4	$4.1, 13.8 \pm 0.7$	7.9	92	550
333.6− 0.2	7,200	1.1×10^7	$3.7, 14.1 \pm 0.8$	1.4	2835	98
333.7− 0.5	2,400	1.5×10^4	$3.9, 14.0 \pm 0.8$	6.9	46	180
335.8− 0.2	4,800	4.4×10^4	$4.2, 14.2 \pm 0.8$	7.2	78	360
336.5+ 0.0	3,300	2.1×10^4	$5.0, 11.7 \pm 0.7$	8.4	50	360
336.4− 0.2	2,200	2.5×10^4	$5.3, 13.0 \pm 0.6$	10.8	48	720
336.4− 0.3	3,600	3.8×10^4	$6.9, 11.4 \pm 1.6$	14.1	52	1800
336.5− 0.2	3,200	2.5×10^4	$6.6, 11.7 \pm 0.6$	19.5	36	3200
336.8+ 0.0	6,000	9.0×10^4	12.4 ± 0.7	42.8	46	43000
336.9− 0.1	6,000	6.7×10^4	5.7 ± 0.6	11.4	77	1400
337.1− 0.2	5,100	9.1×10^5	$5.7, 12.7 \pm 0.6$	3.1	542	190
337.3− 0.1	3,900	4.1×10^4	4.4 ± 0.7	7.5	74	380
338.0− 0.1	3,500	2.8×10^4	$4.4, 14.1 \pm 0.7$	11.0	50	810
337.6+ 0.0	4,700	2.7×10^4	$4.5, 14.0 \pm 0.7$	8.9	55	470

***Adapted from Reifenstein et al. (1970)., and Wilson et al. (1970).**

Emission Nebulae at Radio Wavelengths*

Galactic Coordinate	R.A. h m s	Dec. ° ′ ″	θ_{RA} (′)	θ_{Dec} (′)	S_{6cm} (Jy)	V_{LSR} (km s^{-1})
337.8− 0.1	16 35 22	− 46 52 06	4.3	4.0	7.2	
338.1− 0.1	16 36 14	− 46 45 18	2.3	4.1	7.3	− 41.7
336.5− 1.5	16 36 22	− 48 46 18	0.0	0.0	8.5	− 24.9
338.9+0.6	16 36 42	− 45 34 24	3.2	3.5	9.9	− 63.0
338.1− 0.2	16 36 58	− 46 41 54	3.5	5.4	6.3	− 47.7
338.4+0.0	16 37 10	− 46 18 18	5.0	8.8	46.6	− 36.9
337.9− 0.5	16 37 27	− 47 01 36	1.3	1.3	17.3	− 40.4
338.4− 0.2	16 38 00	− 46 29 36	1.6	1.6	3.9	− 4.3
338.9− 0.1	16 39 36	− 46 00 48	1.8	1.8	3.4	− 40.0
340.3− 0.2	16 45 19	− 45 04 06	1.8	2.3	4.9	− 43.3
340.8− 1.0	16 50 39	− 45 12 12	1.8	2.7	12.1	− 24.8
345.0+1.5	16 54 22	− 40 19 36	6.9	11.3	14.8	− 17.4
345.3+1.5	16 55 37	− 40 09 24	3.8	3.3	12.6	− 15.5
343.5+0.0	16 55 48	− 42 30 12	3.5	3.6	12.6	− 30.3
345.4+1.4	16 56 10	− 40 07 00	1.3	1.3	13.1	− 14.6
345.2+1.0	16 57 09	− 40 29 18	1.3	1.6	10.4	− 9.4
347.6+0.2	17 08 07	− 39 04 42	8.9	5.4	23.4	− 96.2
348.2+0.5	17 08 56	− 38 26 01	4.2	5.2	7.6	
349.5+1.1	17 10 22	− 37 02 51	1.8	2.9	3.2	
348.7+0.3	17 10 54	− 38 08 00	6.4	6.4	12.7	
348.4+0.1	17 11 05	− 38 28 25	8.3	6.6	23.7	
348.5+0.1	17 11 13	− 38 26 48	0.0	0.0	0.0	
349.7+0.2	17 14 37	− 37 23 16	1.9	2.6	8.4	
350.1+0.1	17 16 04	− 37 07 39	2.4	3.4	6.5	− 69.8
348.7− 1.0	17 16 37	− 38 54 19	3.7	5.0	39.0	− 12.8
348.7− 1.0	17 16 39	− 38 54 36	1.6	2.1	33.4	− 12.8
351.4+0.7	17 17 12	− 35 45 58	7.0	10.2	171.4	− 2.5
349.8− 0.6	17 17 53	− 37 44 04	5.1	5.2	9.2	− 25.8
351.6+0.2	17 19 54	− 35 51 28	5.9	3.0	15.7	− 42.5
353.2+0.9	17 21 30	− 34 08 38	8.3	5.4	128.4	− 3.8
353.1+0.7	17 22 17	− 34 18 19	11.7	13.4	259.4	− 2.9
353.1+0.6	17 22 18	− 34 20 06	5.5	5.7	111.3	− 4.1
353.1+0.3	17 23 19	− 34 33 26	8.5	4.9	31.3	− 3.8
351.6− 1.2	17 25 52	− 36 37 35	4.0	6.6	36.8	− 12.2
353.5− 0.0	17 26 07	− 34 23 24	2.0	2.0	3.3	− 51.0

Emission Nebulae at Radio Wavelengths*

Galactic Coordinate	Te (°K)	E (pc cm^{-6})	D (kpc)	2R (pc)	Ne (cm^{-3})	M_{HII} ($M_\odot$)
337.8− 0.1						
338.1− 0.1	3,500	6.7×10^4	3.6 ± 0.8	4.7	119	150
336.5− 1.5	9,700	4.1×10^6	2.1 ± 0.9	0.4	3030	3.3
338.9+ 0.6	5,000	8.5×10^4	$5.2, 13.5 \pm 8.9$	7.4	107	530
338.1− 0.2	3,200	2.8×10^4	$4.1, 14.5 \pm 0.8$	7.6	61	320
338.4+ 0.0	4,900	1.0×10^5	$3.3, 15.3 \pm 0.8$	9.4	104	1000
337.9− 0.5	5,400	1.0×10^6	$3.5, 15.0 \pm 0.8$	1.9	725	64
338.4− 0.2	4,700	1.5×10^5	$0.4, 18.2 \pm 1.0$	0.3	730	0.18
338.9− 0.1	5,700	1.1×10^5	$3.6, 15.1 \pm 0.8$	2.8	196	50
340.3− 0.2	5,400	1.2×10^5	$4.0, 14.8 \pm 0.8$	3.5	184	93
340.8− 1.0	5,000	2.4×10^5	$2.5, 16.4 \pm 1.0$	2.4	320	50
345.0+ 1.5	5,700	1.9×10^4	$2.1, 17.2 \pm 1.1$	7.9	49	290
345.3+ 1.5	5,800	1.0×10^5	$2.0, 17.3 \pm 1.3$	3.0	184	61
343.5+ 0.0	7,800	1.1×10^5	$3.3, 15.9 \pm 1.0$	5.0	150	220
345.4+ 1.4	4,500	7.3×10^5	$1.9, 17.4 \pm 1.3$	1.1	833	12
345.2+ 1.0	4,800	4.8×10^5	$1.2, 18.1 \pm 1.5$	0.7	806	4
347.6+ 0.2	3,800	4.3×10^4	$7.9, 11.6 \pm 1.0$	23.4	43	6600
348.2+ 0.5						
349.5+ 1.1						
348.7+ 0.3						
348.7+ 0.3						
348.4+ 0.1						
348.5+ 0.1						
349.7+ 0.2						
350.1+ 0.1	8,200	9.1×10^4	$7.4, 12.3 \pm 0.5$	6.1	100	870
348.7− 1.0	5,300	2.1×10^5	$2.0, 17.6 \pm 0.5$	2.5	240	140
351.4+ 0.7	6,400	2.5×10^5	0.7 ± 1.9	1.6	320	54
349.8− 0.6	4,900	3.4×10^4	$4.1, 15.6 \pm 1.3$	6.1	61	530
351.6+ 0.2	6,600	9.4×10^4	$6.2, 13.5 \pm 1.0$	7.6	92	1500
353.2+ 0.9	6,100	3.0×10^5	1.0 ± 2.3	1.8	330	77
353.1+ 0.7	6,000	1.7×10^5	1.0 ± 2.4	3.1	200	210
353.1+ 0.6	6,000	1.7×10^5		3.1	200	200
353.1+ 0.3	7,600	8.4×10^4	1.0 ± 3.0	1.9	170	47
351.6− 1.2	6,000	1.4×10^5	$2.6, 17.2 \pm 2.0$	3.9	160	360
353.5− 0.0	3,300	7.0×10^4	$7.3, 12.5 \pm 0.6$	4.3	100	310

*Adapted from Reifenstein et al. (1970)., and Wilson et al. (1970).

Emission Nebulae at Radio Wavelengths*

Galactic Coordinate	R.A. h m s	Dec. ° ′ ″	θ_{RA} (′)	θ_{Dec} (′)	S_{6cm} (Jy)	V_{LSR} (km s^{-1})
353.4− 0.4	17 27 08	− 34 39 56	2.0	2.5	8.8	− 12.8
4.5+6.8	17 27 43	− 21 26 41				
355.2+0.1	17 30 09	− 32 52 43	7.5	9.3	13.6	
357.7− 0.1	17 37 05	− 30 56 55	4.4	3.4	15.7	
359.5− 0.1	17 41 24	− 29 23 57	5.7	6.7	15.4	
0.0− 0.0	17 42 28	− 28 58 40	3.7	3.7	190.0	
0.1+0.0	17 42 34	− 28 51 00	7.0	5.0	80.0	
0.2+0.0	17 42 48	− 28 45 53	16.9	5.0	180.0	− 12.7
0.2− 0.0	17 42 59	− 28 47 00	17.0	5.0	157.0	
0.5− 0.0	17 43 49	− 28 29 06	4.4	5.2	35.5	+47.2
0.5+0.0	17 43 56	− 28 30 35	8.1	3.0	41.0	+47.1
0.7− 0.0	17 44 07	− 28 21 36	2.9	3.6	47.8	+61.6
0.9+0.1	17 44 07	− 28 07 36	6.4	1.6	7.9	
0.7− 0.1	17 44 14	− 28 23 17	5.0	2.1	59.8	+64.8
1.1− 0.1	17 45 25	− 27 58 12	2.6	3.8	10.4	− 21.7
3.3− 0.1	17 50 28	− 26 10 23	8.2	8.0	11.0	+4.0
4.4+0.1	17 52 18	− 25 06 07	3.8	4.2	6.2	+9.1
6.1− 0.1	17 56 52	− 23 45 29	10.2	6.1	13.0	+11.5
6.5+0.1	17 57 05	− 23 16 29	8.1	6.9	15.2	+14.1
5.9− 0.4	17 57 34	− 24 04 42	5.9	5.9	23.3	+10.5
6.6− 0.1	17 57 45	− 23 18 59	3.9	5.8	11.4	+14.7
6.7− 0.2	17 58 39	− 23 18 59	6.9	7.7	20.4	+11.5
6.6− 0.3	17 58 39	− 23 21 00	7.0	7.0	12.0	+14.2
5.3− 1.1	17 58 46	− 24 52 49	12.2	18.0	20.2	
7.0− 0.2	17 59 19	− 23 02 09	5.4	5.8	13.3	+16.4
8.1+0.2	18 00 00	− 21 48 06	1.0	2.1	5.8	+19.3
6.0− 1.2	18 00 41	− 24 23 20	8.5	7.6	85.1	+3.0
8.5− 0.3	18 02 47	− 21 46 20	8.7	2.4	7.7	
10.3− 0.1	18 05 58	− 20 06 05	3.2	1.9	13.6	+9.7
10.2− 0.3	18 06 24	− 20 19 53	3.3	3.6	51.8	+13.9
10.6− 0.4	18 07 33	− 19 56 32	5.7	4.2	10.2	+0.3
11.2− 0.4	18 08 32	− 19 26 52	2.7	3.7	8.9	
13.2+0.0	18 11 10	− 17 29 28	4.0	1.0	4.7	+57.0
12.8− 0.2	18 11 15	− 17 57 02	4.9	3.6	44.9	+36.3
14.6+0.1	18 13 57	− 16 14 38	11.0	9.0	24.3	+37.2

Emission Nebulae at Radio Wavelengths*

Galactic Coordinate	Te (°K)	E (pc cm^{-6})	D (kpc)	2R (pc)	Ne (cm^{-3})	M_{HII} ($M_\odot$)
353.4– 0.4	8,000	2.0×10^5	$3.2, 16.6 \pm 2.4$	2.1	250	90
4.5+ 6.8						
355.2+ 0.1						
357.7– 0.1						
359.5– 0.1						
0.0– 0.0						
0.1+ 0.0						
0.2+ 0.0	7,600	2.3×10^5	10.0	39.3	77	56000
0.2– 0.0						
0.5– 0.0	6,300	1.6×10^5	10.0	13.9	90	9100
0.5+ 0.0	5,400	1.7×10^5	10.0	21.1	89	10000
0.7– 0.0	6,900	5.0×10^5	10.0	9.4	190	5900
0.9+ 0.1						
0.7– 0.1	7,900	6.4×10^5	10.0	13.9	215	6900
1.1– 0.1	6,300	1.1×10^5	10.0	9.1	91	2600
3.3– 0.1	4,200	1.6×10^4	$2.2, 17.8 \pm 3.4$	5.1	45	230
4.4+ 0.1	5,200	3.8×10^4	$3.4, 16.5 \pm 3.4$	4.0	80	200
6.1– 0.1	2,800	1.7×10^4	3.3 ± 1.7	11.1	39	640
6.5+ 0.1	7,500	3.0×10^4	3.7 ± 1.5	11.8	50	1000
5.9– 0.4	6,400	7.1×10^4	$3.0, 16.9 \pm 2.7$	5.1	97	490
6.6– 0.1	5,900	5.1×10^4	3.8 ± 1.4	7.7	81	450
6.7– 0.2	10,000	4.6×10^4	3.1 ± 1.6	9.7	69	750
6.6– 0.3	5,800	2.5×10^4	3.6 ± 2.2	7.3	48	710
5.3– 1.1						
7.0– 0.2	7,300	4.7×10^4	3.8 ± 2.0	6.2	72	650
8.1+ 0.2	4,000	2.5×10^5	$3.8, 16.0 \pm 1.8$	1.6	320	53
6.0– 1.2	7,300	1.4×10^5	1.0 ± 3.7	2.2	210	88
8.5– 0.3						
10.3– 0.1	7,000	2.4×10^5	$1.8, 17.8 \pm 1.9$	1.3	360	30
10.2– 0.3	5,900	4.5×10^5	$2.5, 17.2 \pm 1.6$	2.5	350	210
10.6– 0.4	5,400	4.3×10^4	$0.1, 19.6 \pm 2.1$	0.1	450	<1
11.2– 0.4						
13.2+ 0.0	4,700	1.1×10^5	$6.0, 13.5 \pm 0.7$	3.5	150	240
12.8– 0.2	7,800	2.9×10^5	$4.6, 14.9 \pm 1.0$	5.6	190	1200
14.6+ 0.1	4,500	2.3×10^4	$4.2, 15.1 \pm 1.0$	12.2	36	2500

***Adapted from Reifenstein et al. (1970)., and Wilson et al. (1970).**

Emission Nebulae at Radio Wavelengths*

Galactic Coordinate	R.A. h m s	Dec. ° ′ ″	θ_{RA} (′)	θ_{Dec} (′)	S_{6cm} (Jy)	V_{LSR} (km s^{-1})
18.5+1.9	18 14 50	- 11 57 20	27.6	29.1	102.6	+32.9
17.0+0.8	18 15 54	- 13 47 32	14.1	18.2	107.8	+24.5
15.1- 0.7	18 17 36	- 16 12 17	4.9	7.3	534.6	+17.2
15.0- 0.7	18 17 37	- 16 13 06	4.1	6.7	478.3	+18.4
18.8+0.3	18 21 10	- 12 26 22	7.0	15.4	12.8	
19.1- 0.3	18 23 53	- 12 29 02	12.4	4.2	17.0	+67.8
19.7- 0.2	18 24 46	- 11 53 42	5.8	5.9	12.4	+43.4
20.7- 0.1	18 26 29	- 10 54 47	5.6	7.2	14.5	+57.4
28.8+3.5	18 28 51	- 02 07 29	5.2	5.8	35.1	+0.7
21.8- 0.6	18 30 16	- 10 13 00	10.9	7.0	20.3	
21.5- 0.9	18 30 47	- 10 36 35	1.0	1.3	6.5	
22.8- 0.3	18 31 01	- 09 11 39	19.8	18.1	45.5	+82.5
23.1- 0.3	18 31 36	- 08 57 10	13.0	17.4	18.9	
23.4- 0.2	18 31 58	- 08 34 55	7.0	2.8	13.2	+101.5
24.8+0.1	18 33 30	- 07 13 24	6.6	4.4	11.5	+114.1
24.6- 0.2	18 34 09	- 07 30 09	7.1	6.2	8.2	+109.2
25.8+0.2	18 34 55	- 06 18 43	4.3	7.9	9.5	+110.0
25.4- 0.2	18 35 32	- 06 50 09	4.2	3.7	23.8	+60.4
27.3- 0.2	18 39 01	- 05 08 29	6.3	13.7	7.7	+97.6
28.6+0.0	18 40 52	- 03 52 38	8.3	14.3	14.4	+96.2
29.9- 0.0	18 43 29	- 02 44 46	3.2	5.4	21.1	+96.4
30.2- 0.2	18 44 26	- 02 32 02	17.8	3.1	9.2	+101.0
30.8- 0.0	18 45 01	- 02 00 03	5.4	5.3	97.4	+92.3
31.1+0.0	18 45 18	- 01 42 06	4.0	4.0	8.0	+99.2
31.9+0.0	18 46 48	- 00 58 59	3.7	2.4	9.4	
32.8+0.2	18 47 57	- 00 05 33	2.9	2.2	4.4	
33.7+0.0	18 50 04	+00 35 56	7.3	5.6	5.4	
34.3+0.1	18 50 48	+01 11 07	1.5	2.5	15.0	+53.9
34.8- 0.3	18 53 18	+01 25 21				
34.7- 0.5	18 54 00	+01 18 24				
34.7- 0.6	18 54 04	+01 14 17	8.9	18.9	50.6	
34.7- 0.8	18 54 58	+01 07 24				
37.6- 0.1	18 57 48	+03 59 36	6.0	4.0	7.9	+55.8
34.7- 0.5	18 54 00	+01 18 24				
34.7- 0.8	18 54 58	+01 07 24				

Emission Nebulae at Radio Wavelengths*

Galactic Coordinate	Te (°K)	E ($pc\ cm^{-6}$)	D (kpc)	2R (pc)	Ne (cm^{-3})	M_{HII} ($M_\odot$)
18.5+ 1.9	3,000	1.1×10^4	3.2 ± 0.9	26.5	17	12000
17.0+ 0.8	6,100	4.4×10^4	2.7 ± 1.0	12.8	48	2.3×10^8
15.1− 0.7	6,400	1.6×10^6	2.1 ± 1.1	3.7	540	1000
15.0− 0.7	6,200	1.8×10^6	2.3 ± 1.2	5.2	594	970
18.8− 0.3						
19.1− 0.3	4,000	3.0×10^4	$5.7, 13.2 \pm 0.6$	12.0	41	2700
19.7− 0.2	5,300	3.6×10^4	$4.0, 14.8 \pm 0.9$	6.9	60	730
20.7− 0.1	4,400	3.4×10^4	$4.9, 13.8 \pm 0.7$	9.1	50	1400
28.8+ 3.5	8,500	1.4×10^5	0.1 ± 0.8	0.2	760	<1
21.8− 0.6						
21.5− 0.9						
22.8− 0.3	6,400	1.3×10^4	$6.2, 12.2 \pm 0.6$	34.4	16	25000
23.1− 0.3						
23.4− 0.2	6,300	7.1×10^4	$7.5, 10.9 \pm 1.0$	9.6	71	2400
24.8+ 0.1	8,000	4.5×10^4	9.0 ± 1.4	14.2	46	5100
24.6− 0.2	4,500	1.8×10^4	9.1 ± 1.0	16.0	27	4200
25.8+ 0.2	4,700	2.7×10^4	9.0 ± 1.4	15.3	35	4600
25.4− 0.2	9,100	1.8×10^5	$4.7, 13.4 \pm 0.7$	5.3	150	880
27.3− 0.2	5,000	8.8×10^3	$7.2, 10.6 \pm 1.6$	19.5	18	4900
28.6+ 0.0	6,700	1.3×10^4	$7.2, 10.4 \pm 1.6$	22.6	20	8700
29.9− 0.0	6,900	1.3×10^5	$7.3, 10.0 \pm 1.3$	8.8	100	2600
30.2− 0.2	3,800	1.5×10^4	8.1 ± 1.2	17.6	24	4900
30.8− 0.0	5,600	3.4×10^5	$7.0, 10.2 \pm 1.6$	10.9	150	7200
31.1+ 0.0	6,700	5.4×10^4	8.5 ± 1.3	9.9	61	2200
31.9+ 0.0						
32.8+ 0.2						
33.7+ 0.0						
34.3+ 0.1	7,500	4.5×10^5	$3.8, 12.7 \pm 0.8$	2.1	380	140
34.8− 0.3						
34.7− 0.5						
34.7− 0.6						
34.7− 0.8						
37.6− 0.1	5,600	3.4×10^4	$4.0, 11.9 \pm 0.7$	5.6	64	430
34.7− 0.5						
34.7− 0.8						

*Adapted from Reifenstein et al. (1970)., and Wilson et al. (1970).

Emission Nebulae at Radio Wavelengths*

Galactic Coordinate	R.A. h m s	Dec. ° ′ ″	θ_{RA} (′)	θ_{Dec} (′)	S_{6cm} (Jy)	V_{LSR} (km s^{-1})
35.2− 1.7	18 59 15	+01 09 04	1.0	1.8	15.3	+46.5
37.9− 0.4	18 59 19	+04 09 26	19.3	4.4	24.4	+60.2
39.2− 0.3	19 01 38	+05 22 31	4.5	4.3	8.7	
41.1− 0.3	19 05 05	+07 03 23	9.8	9.8	16.9	
43.2− 0.0	19 07 54	+09 01 01	1.5	2.0	49.8	+8.6
45.5+0.1	19 11 59	+11 04 17	3.9	2.9	12.4	
48.6+0.0	19 18 07	+13 49 18	4.6	4.8	10.6	+17.0
49.0− 0.3	19 20 08	+14 02 49	12.2	13.0	111.3	+63.2
49.1− 0.4	19 20 32	+14 03 30	4.0	5.6	15.0	+72.4
49.2− 0.3	19 20 42	+14 13 57	2.0	2.0	12.0	+67.2
49.4− 0.3	19 20 53	+14 18 36	3.5	3.5	37.2	+52.8
49.0− 0.6	19 21 02	+13 55 42	22.6	12.4	87.9	
49.5− 0.4	19 21 23	+14 24 29	4.7	3.0	117.4	+58.2
49.2− 0.7	19 21 56	+13 60 12	11.3	14.5	50.3	
51.2− 0.1	19 23 10	+16 08 14	20.9	22.4	37.0	+55.3
61.5+0.1	19 44 42	+25 05 04	1.2	1.0	6.1	+20.2
63.2+0.4	19 47 12	+26 43 08	2.7	2.5	4.8	+15.3
69.9+1.5	19 59 17	+33 03 52	2.0	1.0	4.4	
70.3+1.6	19 59 51	+33 24 27	3.4	2.2	14.5	− 24.4
74.9+1.2	20 14 04	+37 03 46	9.3	4.7	7.2	
78.9+3.7	20 14 37	+41 45 05	37.2	37.9	136.2	
74.8+0.6	20 15 49	+36 36 07	4.7	3.9	4.6	
78.3+2.5	20 18 16	+40 36 29	13.3	9.5	13.3	
75.8+0.4	20 19 47	+37 19 53	1.7	6.2	14.1	− 4.8
78.1+1.8	20 20 44	+40 02 20	6.9	14.0	25.6	+5.1
78.5+1.2	20 24 08	+40 00 20	8.4	26.2	40.2	− 7.1
78.0+0.6	20 25 22	+39 15 56	3.4	3.4	9.8	+2.9
78.6+1.0	20 25 25	+39 56 04	12.2	14.7	22.5	
76.4− 0.6	20 25 33	+37 12 55	1.8	2.2	13.5	+5.6
79.3+1.3	20 26 22	+40 41 39	3.7	3.6	15.6	− 41.4
78.0+0.0	20 27 44	+38 52 12	3.9	2.8	5.6	− 3.0
80.0+1.5	20 27 53	+41 24 34	21.1	35.5	73.4	− 15.4
79.8+1.2	20 28 32	+41 04 17	13.2	29.0	28.5	
78.5− 0.1	20 29 31	+39 13 08	9.5	4.6	5.6	
78.2− 0.4	20 29 52	+38 48 00	8.6	8.8	24.3	+1.1

Emission Nebulae at Radio Wavelengths*

Galactic Coordinate	Te (°K)	E (pc cm^{-6})	D (kpc)	2R (pc)	Ne (cm^{-3})	M_{HII} ($M_\odot$)
35.2– 1.7	6,500	9.0×10^5	3.4 ± 0.8	1.3	680	60
37.9– 0.4	6,500	3.1×10^4	$4.2, 11.5 \pm 1.0$	11.4	43	2400
39.2– 0.3						
41.1– 0.3						
43.2– 0.0	7,700	1.9×10^6	13.8 ± 0.8	7.0	430	5400
45.5+ 0.1						
48.6+ 0.0	5,900	4.9×10^4	$1.3, 11.9 \pm 0.7$	1.8	140	29
49.0– 0.3	7,400	7.8×10^4	6.6 ± 1.7	24.0	47	25000
49.1– 0.4	4,300	6.2×10^4	6.6 ± 1.7	13.4	68	1900
49.2– 0.3	4,700	2.9×10^5	6.6 ± 1.7	3.8	230	470
49.4– 0.3	7,000	3.3×10^5	8.5 ± 1.0	12.7	161	4000
49.0– 0.6						
49.5– 0.4	7,600	9.3×10^5	7.3 ± 1.7	7.1	300	4000
49.2– 0.7						
51.2– 0.1	4,200	7.3×10^3	6.3 ± 2.0	39.6	11	26000
61.5+ 0.1	3,400	4.4×10^5	$2.0, 7.5 \pm 2.0$	0.7	670	7
63.2+ 0.4	3,700	6.3×10^4	$1.8, 7.2 \pm 1.1$	1.4	180	17
69.9+ 1.5						
70.3+ 1.6	7,200	2.1×10^5	8.9 ± 0.8	7.0	140	1900
74.9+ 1.2						
78.9+ 3.7						
74.8+ 0.6						
78.3+ 2.5						
75.8+ 0.4	6,200	1.4×10^5	5.5 ± 2.2	5.2	130	730
78.1+ 1.8	9,700	3.2×10^4	2.1 ± 2.8	5.9	61	470
78.5+ 1.2	5,800	1.9×10^4	5.0 ± 2.0	21.6	24	9200
78.0+ 0.6	7,200	9.3×10^4	$0.9, 3.2 \pm 1.9$	0.9	270	7
78.6+ 1.0						
76.4– 0.6	6,800	3.7×10^5	2.3 ± 3.0	1.4	430	41
79.3+ 1.3	7,300	1.3×10^5	8.0 ± 0.9	8.5	100	2300
78.0+ 0.0	6,100	5.3×10^4	4.6 ± 2.6	4.5	90	300
80.0+ 1.5	6,300	1.0×10^4	5.5 ± 1.1	43.6	13	40000
79.8+ 1.2						
78.5– 0.1						
78.2– 0.4	8,100	3.7×10^4	$0.3, 3.8 \pm 1.8$	0.7	180	3

*Adapted from Reifenstein et al. (1970)., and Wilson et al. (1970).

Emission Nebulae at Radio Wavelengths*

Galactic Coordinate	R.A. h m s	Dec. ° ′ ″	θ_{RA} (′)	θ_{Dec} (′)	S_{6cm} (Jy)	V_{LSR} (km s^{-1})
79.2+0.3	20 30 22	+40 03 00	5.0	5.0	7.5	− 1.6
82.4+2.5	20 30 53	+43 55 39	5.7	23.8	17.2	− 4.6
80.4+0.5	20 33 15	+41 05 00	4.4	32.2	29.0	+7.5
81.3+1.2	20 33 23	+42 15 17	9.1	15.5	41.8	+11.3
80.1− 0.1	20 34 48	+40 31 26	23.8	8.1	30.0	+1.3
80.9+0.4	20 35 10	+41 26 02	3.5	4.6	7.6	+1.3
81.7+0.5	20 37 13	+42 08 51	0.3	0.4	19.0	+1.7
80.9− 0.2	20 37 42	+41 07 35	7.4	6.4	21.7	− 1.0
81.5+0.0	20 39 01	+41 42 49	16.4	21.8	64.5	+0.2
84.8− 0.4	20 52 00	+44 00 00	65.0	95.0	340.0	
84.8− 3.5	21 04 37	+41 57 51	¡0.1	¡0.1	6.4	+5.2
111.6+0.9	23 11 39	+61 21 42	2.3	1.9	25.0	− 60.6

Emission Nebulae at Radio Wavelengths*

Galactic Coordinate	Te (°K)	E (pc cm^{-6})	D (kpc)	2R (pc)	Ne (cm^{-3})	M_{HII} ($M_\odot$)
79.2+ 0.3	7,000	3.3×10^4	4.0 ± 2.0	5.8	62	470
82.4+ 2.5	6,300	1.3×10^4	3.6 ± 2.2	12.0	27	1800
80.4+ 0.5	9,100	2.4×10^4	1.7 ± 2.1	5.8	53	390
81.3+ 1.2	6,500	3.2×10^4	1.5 ± 1.0	5.2	64	340
80.1– 0.1	8,900	1.8×10^4	$0.4, 3.0 \pm 1.7$	1.7	85	16
80.9+ 0.4	5,600	4.8×10^4	$0.5, 2.7 \pm 1.9$	0.6	240	2
81.7+ 0.5	7,200	1.7×10^7	1.5	0.2	8800	1.2×10^8
80.9– 0.2	6,000	4.7×10^4	3.4 ± 1.8	6.7	69	800
81.5+ 0.0	7,000	2.0×10^4	1.5	8.3	40	850
84.8– 0.4						
84.8– 3.5	15,000	9.0×10^7	0.9 ± 2.1		48000	<1
111.6+ 0.9	6,700	6.1×10^2	4.9 ± 0.8	3.0	370	380

*Adapted from Reifenstein et al. (1970)., and Wilson et al. (1970).

19
Reflection Nebulae

19.1 Basic Data for Reflection Nebulae

Reflection nebulae are diffuse, extended, irregularly-shaped sources that shine by the reflected light of bright luminous nearby stars. The reflection nebulae contain interstellar dust particles that reflect starlight, and reproduce the stellar absorption lines. Selected bright reflection nebulae and their illuminating stars are listed in the first two accompanying tables.

The right ascension, R.A., and declination, Dec., of the illuminating star for the epoch 1950.0 are followed by the largest angular extent, θ, of the reflection nebula in blue light. This extent increases with the brightness of the star, following the relation B = 11.0 − 5 log(θ/2) for the apparent blue magnitude of stars illuminating northern reflection nebulae (van den Bergh, 1966). The distance, D, and surface brightness, V_{sur}, of the bright reflection nebulae are followed by the color difference Δ(B − V) between the nebula and the illuminating star, the star's HD or BD number, and its spectral type, Sp, apparent visual magnitude, V, and color B − V. These data are adapted from Scheffler (1982) and Witt and Schild (1986), with the addition of Hubble's variable nebula, NGC 2261 (Johnson, 1966).

Recently attention has been directed toward a special class of bipolar reflection nebulae. They are symmetrically placed about a central object, with two fan-shaped projections that resemble a butterfly in appearance. Their colorful names include the boomerang nebula, the egg nebula, Minkowski's footprint, and the red rectangle. Here we only give the names and positions of these intriguing objects; further details are contained in the references listed at the bottom of the table. The bipolar reflection nebulae may represent an evolutionary stage that precedes planetary nebulae.

Associations of reflection nebulae, called R-Associations, can be used as tracers of nearby spiral arms that contain concentrations of interstellar dust. Our table of R-Associations is adapted from Herbst (1975b). It includes the name, galactic coordinates, l and b, the true distance modulus, $V_0 - M_V$, distance, D, color excess, E(B − V), and the number, No., of reflection nebulae for 37 R-Associations.

Our main catalogue of reflection nebulae contains the number, N, from catalogues of van den Bergh (1966) and van den Bergh and Herbst (1975), followed by the letters N or S for the northern and southern objects. They are listed in order of increasing right ascension, R.A.. The positions for the illuminating stars of the northern reflection nebulae were taken from the SAO star catalogue or supplied by Donna Colletti and Joyce Watson using the SIMBAD data base at the Harvard-Smithsonian Center for Astrophysics. These positions are followed by the galactic coordinates and a type designation. Type I reflection nebulae are those in which the illuminating star is imbedded in the nebulosity; in Type II reflection nebulae the illuminating star is situated outside the nebula. The maximum angular radius, $\theta/2$, of the reflection nebula is measured on blue plates, and the distance, D, of the illuminating star is inferred from the relation $V_0 - M_V = 5 \log D - 5$ = the true distance modulus. The HD or BD numbers, absolute visual magnitudes, M_V, colors, B − V, and spectral types, Sp, of these stars are taken from Racine (1968) and Herbst (1975a); uncertain numerical values are denoted by a colon (:).

Bright Reflection Nebulae*

Nebula	R.A.(1950) h m s	Dec.(1950) ° ′ ″	θ (′)	D (pc)	V_{Sur} **	Δ(B−V)*** (mag)
NGC 1333	03 26 18	+31 16 00	4	500		
IC 348	03 41 25.8	+32 00 23	10	390	23.3	− 0.71
Electra nebula†	03 41 54.055	+23 57 27.82	20	125	21.2	− 0.19
Maia nebula, NGC 1432†	03 42 50.76	+24 12 47.0	30	125	21.4	− 0.45
Merope nebula, NGC 1435†	03 43 21.20	+23 47 39.0	30	125	21.0	− 0.11
NGC 1788	05 04 24	− 03 23 00	7	420		
Cederblad 44	05 19 00	+08 22 51	22		23.6	− 0.59
IC 426	05 33 45.41	− 00 20 02.0		575		
IC 435	05 40 29.5	− 02 20 05	5	575	22.6	
NGC 2068, M 78	05 44 09.58	+00 03 33.6	8	500	20.6	− 0.37
NGC 2071	05 44 33.82	+00 16 54.4	7	500		
NGC 2261††	06 05 00	+18 42 00		750		
IC 2169	06 28 00	+10 06 00	25			
NGC 2245	06 28 29.67	+09 49 35.8	3	900		
NGC 2247	06 30 19.4	+10 21 38	4	800	21.8	
Antares nebula	16 26 20.206	− 26 19 21.94	126		23.7	− 1.19
IC 1287	18 28 39.23	− 10 49 55.2	20		23.2	− 0.49
NGC 6914	20 22 30	+42 08 00	3	1100		
NGC 7023	21 00 59.70	+67 57 55.5	10	390	22.0	− 0.24
NGC 7129	21 41 42	+65 53 00	8		20.8	
Cederblad 201	22 12 14.0	+70 00 12	7	400		

*Adapted from H. Scheffler (1982), and Witt and Schild (1986).

†Nebula around the Pleiades stars.

††NGC 2261 is Hubble's variable nebula.

**The visual surface brightness, V_{Sur}, is in units of magnitude per square second of arc.

***The mean color difference, Δ(B−V), between the nebula, n, and its illuminating star, s, is $\Delta(B-V) = (B-V)_n - (B-V)_s$.

Bright Reflection Nebulae – Illuminating Stars*

Nebula	Star	Sp	V (mag)	B–V (mag)
NGC 1333	DM +30°549	B8 V	10.47	+0.48
IC 348	HD 281159	B5 V	8.53	+0.69
Electra nebula†	HD 23302, 17 Tau	B6 III	3.71	– 0.11
Maia nebula, NGC 1432†	HD 23408, 20 Tau	B7 III	3.88	– 0.07
Merope nebula, NGC 1435†	HD 23480, 23 Tau	B6 IVnn	4.18	– 0.06
NGC 1788	HD 293815	B9 V	10.12	+0.24
Cederblad 44	HD 34989	B1 V	5.77	– 0.13
IC 426	HD 37140	B8p	8.55	+0.09
IC 435	HD 38087	B3n	8.30	+0.13
NGC 2068, M 78	HD 38563	B5	10.56	+1.18
NGC 2071	HD 290861	F8	10.14	+1.03
NGC 2261††	R Mon, LkHα 208			
IC 2169	HD 258686	B7 IIIp	9.12	– 0.06
NGC 2245	HD 258853	B4nn	8.83	+0.02
NGC 2247	HD 259431	B3pe	8.74	+0.28
Antares nebula	HD 148478/9	M1 Ib(+B2.5V)	0.92	+1.84
IC 1287	HD 170740	B2 V	5.72	+0.24
NGC 6914	BD +41°3731	B3n	9.90	+0.10
NGC 7023	HD 200775	B5e	7.39	+0.38
NGC 7129	BD +65°1637/8	B2nne+B3	10.15	+0.41
Cederblad 201	BD +69°1231	B9.5 V	9.29	+0.14

*Adapted from H. Scheffler (1982), and Witt and Schild (1986).
†Nebula around the Pleiades Stars.
††Hubble's variable nebula.

Bipolar Reflection Nebulae*

Name	Illuminating Star	R.A.(1950) h m s	Dec.(1950) ° ′ ″
CRL 618	CRL 618	04 39 33.8	+36 01 15
The Red Rectangle	HD 44179	06 17 37.0	− 10 36 52
Toby Jug Nebula, IC 2220 (Also Butterfly Nebula)	HD 65750	07 55 54.5	− 58 59 00
The Boomerang Nebula		12 41 55.0	− 54 14 54
The Footprint Nebula (Also Minkowki's Footprint)	M 1–92	19 34 20	+29 26 00
The Egg Nebula	CRL 2688	21 00 19.9	+36 29 45

*Adapted from Cohen et al. (1975), Crampton, Cowley and Humphreys (1975), Dachs and Isserstedt (1973), Dachs, Isserstedt and Rahe (1978), Herbig (1975), Perkins, King and Scarrott (1981), Taylor and Scarrott (1980), Wegner and Glass (1979), and Westbrook et al. (1975).

R–Associations**

Name	l °	b °	V_0-M_V	D (pc)	E(B–V)	No.
Sgr R1	13	− 1	10.3	1150	0.60	11
Ros 4	66	− 1	12.3	2880	1.06	12
Cyg R1/ vdB 130	77	+2	10.0	1000	0.58	17
Cep R1	109	+4	9.1	660	0.92	3
Cep R2	111	+12	8.0	400	0.35	5
Cas R1	118	− 3	8.6	530	0.39	5
Cam R1	142	+2	9.7	870	0.51	5
Per R1	158	− 18	7.6	330	0.53	4
Tau R1	166	− 24	5.2	110	0.06	4
Tau R2	171	− 17	5.7	135	0.22	7

**Adapted from Herbst (1975b).

R–Associations**

Name	l °	b °	V_0–M_V	D (pc)	E(B–V)	No.
Tau-Ori R1	201	− 17	7.8	360	0.17	8
Mon R1	201	− 1	10.1	1050	0.20	5
Ori R1/R2	208	− 17	8.9	600	0.41	8
Mon R2	216	− 12	9.6	830	0.57	27
CMa R1	224	− 3	9.2	690	0.35	28
CMa R2	237	− 14	9.5	790	0.35	6
Pup R3	252	− 1	11.2	1740	0.52	4
Pup R2	256	− 2	10.0	1000	0.43	10
Pup R1	260	− 3	12.9	3800	0.88	4
Vela R1	263	− 4	8.3	460	0.40	10
Vela R2	265	+2	9.7	870	0.61	9
Vela R3	268	− 3	11.7	2190	0.73	5
Car R1	286	− 1	12.3	2880	0.24	31
Car R2	291	− 2	13.8	5750	0.73	7
Cen R2	295	− 3	12.7	3470	0.61	8
Cen R3	296	0	10.5	1260	0.42	5
Cen R1	305	0	11.7	2190	0.64	7
Cir R1	317	+1	12.2	2750	0.91	3
Cir R2	317	− 4	10.8	1450	1.11	4
Ara R1	336	− 2	10.1	1050	0.12	8
Sco R2	337	− 6	9.4	760	0.27	5
Sco R3	342	− 4	11.4	1900	1.02	3
Sco R4	344	+2	11.4	1900	0.77	8
Sco R5	348	0	9.7	870	0.75	4
Sco R6	352	+1	12.6	3310	1.53	7
Sco R7	359	− 2	10.7	1380	0.66	3
Sco R1	354	+20	5.8	145	0.38	12

**Adapted from Herbst (1975b).

Horsehead Nebula. Dusty clouds obscure the light of bright stars that lie behind them, producing the horsehead and reflection nebulae. The horsehead, NGC 2024, is illuminated by IC 434, a long, bright strip of ionized hydrogen. This photograph was taken at the Karl-Schwarzschild Observatory using the 2 meter reflecting telescope of the JENA optical works. (Courtesy of the Central Institute of Astrophysics of the Academy of Sciences, DDR, Karl-Schwarzschild Observatory, Tautenburg.)

19.2 Catalogue of Reflection Nebulae*

No.		R.A.(1950) h m s	Dec.(1950) ° ′	l b ° °	Type	θ/2 (°)	D (pc)
1	N	00 08 08	+58 29.5	117.7 − 03.7	I	4.3	457
2	N	00 10 42	+65 20.0	119.0 +03.0	I	1.7	631
3	N	00 31 38	+69 09.5	121.4 +06.6	I	0.9	182
4	N	00 40 22	+61 38.2	121.9 − 01.0	I– II	4.2	
5	N	00 53 40	+60 26.8	123.6 − 02.1	II	60.0	229
6	N	01 40 16	+61 35.0	129.1 − 00.4	II	0.9	1820
7	N	02 44 23	+69 25.5	132.9 +09.1	II	3.1	79
8	N	02 47 09	+67 36.6	133.9 +07.6	I	2.3	209
9	N	02 47 29	+68 41.0	133.5 +08.5	II	3.7	302
10	N	03 11 57	+56 57.4	141.6 − 00.4	I	8.0	912
11	N	03 20 05	+61 21.7	140.1 +03.9	II	3.1	832
12	N	03 22 18	+31 33.4	157.4 − 20.6	I	2.3	166
13	N	03 22 45	+30 45.4	158.0 − 21.3	I	2.3	288
14	N	03 25 00	+59 46.1	141.5 +02.9	II	23.0	759
16	N	03 25 17	+29 37.6	159.2 − 21.9	II	4.5	138
15	N	03 25 54	+58 42.4	142.2 +02.1	II	27.0	1000
17	N	03 26 18	+31 16.0	158.3 − 20.4	I	4.0	502
18	N	03 31 56	+37 51.0	155.1 − 14.4	II	7.1	288
19	N	03 41 26	+32 00.4	160.5 − 17.8	I	2.8:	302
20	N	03 41 54	+23 57.5	166.2 − 23.9	I	11.0:	115
21	N	03 42 51	+24 12.8	166.2 − 23.5	I	26.0:	115
22	N	03 43 21	+23 47.7	166.6 − 23.8	I	26.0:	115
23	N	03 44 30	+23 57.1	166.6 − 23.5	I	17.0:	96
24	N	03 46 17	+38 49.8	156.8 − 11.9	II	4.5	1738:
25	N	04 09 26	+23 26.8	171.4 − 19.8	II	4.3	115
26	N	04 10 50	+10 05.2	182.7 − 28.4	I	5.7	126
27	N	04 18 57	+28 19.6	169.3 − 14.9	I	2.8	
28	N	04 19 04	+19 25.1	176.2 − 20.9	II pec	0.9	
29	N·	04 45 13	+29 41.1	172.1 − 09.7	II	6.8	151
30	N	04 49 04	+66 15.6	144.1 +14.0	II	20.0	794
31	N	04 52 34	+30 28.4	172.5 − 08.0	I	4.5	120
32	N	04 58 45	+44 11.9	162.5 +01.5	I– II	0.9	302
33	N	05 04 24	− 03 23.0	203.5 − 24.7	I	0.9:	417
36	N	05 12 08	− 08 15.5	209.2 − 25.3	II	410.0	275
35	N	05 12 11	+12 57.5	189.7 − 14.6	I	0.9	457

Catalogue of Reflection Nebulae*

No.		Star Name HD or BD	M_V (mag)	V_0-M_V (mag)	V (mag)	B–V (mag)	Sp LC
1	N	627, +57 22	- 0.9	8.3	8.24	+0.08	B5 V
2	N	+64 13	- 1.2	9.0	10.29	+0.57	B2.5 V
3	N	3037, +68 34	+1.2	6.3	8.39	+1.29	K0 III
4	N	+61 154			10.98	+1.02	Beq
5	N	5394, +59 144	- 4.6	6.8	2.6v	- 0.19	B0 IV:e
6	N	+61 315	- 4.7	11.3	9.58	+0.85	A0 Ib
7	N	17138, +69 179	+1.7	4.5	6.6v	+0.15	A2 V
8	N	17443, +67 230	+1.1	6.6	8.74	+0.29	B9 V
9	N	17463, +68 200	- 2.5	7.4	5.9v	+0.79v	F6 Ib–IIv
10	N	20041, +56 798	- 6.1	9.8	5.81	+0.70	A0 Ia
11	N	20798, +61 570	- 2.8	9.6	8.37	+0.25	B2 III
12	N	21110, +31 597	+0.6	6.1	7.29	+1.51	K4 III–IV
13	N	+30 540	+0.4	7.3	9.21	+0.34	B8 V
14	N	21291, +59 660	- 6.6	9.4	4.23	+0.41	B9 Ia
16	N	+29 565	+2.6	5.7	9.16	+0.53	F0 V
15	N	21389, +58 607	- 7.2	10.0	4.58	+0.57	A0 Ia
17	N	+30 549	+0.2	8.5	10.47	+0.48	B8 V
18	N	22114, +37 794	- 0.1	7.3	7.58	+0.07	B8 Vp
19	N	281159, +31 643	- 1.2	7.4	8.53	+0.69	B5 V
20	N	23302, +23 507	- 1.7	5.3	3.71	- 0.11	B6 III
21	N	23408, +23 516	- 1.6	5.3	3.88	- 0.07	B7 III
22	N	23480, +23 522	- 1.4	5.3	4.18	- 0.06	B6 IVnn
23	N	23630, +23 541	- 2.1	4.9	2.87	- 0.09	B7 III
24	N	275877, +38 811	- 3.0:	11.2:	9.36	+0.52	A2 II+B6
25	N	26514, +23 642	+1.1	5.3	7.19	+1.06	G6 III
26	N	26676, +09 549	+0.2	5.5	6.25	+0.04	B7 V
27	N	283571, +28 645			11.0v	+1.04	dF8e–dG2e
28	N	284419, +19 706			10.27	+1.22	K0 IIIe
29	N	30378, +29 741	+1.1	5.9	7.41	+0.03	B9.5 V
30	N	30614, +66 358	- 6.1	9.5	4.30	+0.03	O9.5 Ia
31	N	31293, +30 741	+1.3	5.4	7.08	+0.12	A0ep
32	N	+44 1080	- 1.2	7.4	9.16	+0.77	B5 IV
33	N	293815, - 03 1013	+1.0	8.1	10.12	+0.24	B9 V
36	N	34085, - 08 1063	- 7.0	7.2	0.15	- 0.02	B8 Ia
35	N	34033, +12 754	+0.0	8.3	8.66	+1.09	G8 II–III

*Adapted from van den Bergh (1966), van den Bergh and Herbst(1975), Herbst (1975) and Racine (1968).

Catalogue of Reflection Nebulae*

No.	R.A.(1950) h m s	Dec.(1950) ° ′	l b ° °	Type	θ/2 (°)	D (pc)
34 N	05 13 00	+34 15.4	172.1 − 02.3	I	10.0	316
37 N	05 15 15	+13 21.9	189.8 − 13.7	I	2.0	
38 N	05 19 00	+08 22.8	194.6 − 15.6	I	14.0	316
39 N	05 21 13	+32 46.5	174.3 − 01.7	II	0.3	724
40 N	05 22 57	+06 32.7	196.8 − 15.7	I:	0.3	240
41 N	05 26 06	+23 39.0	182.5 − 05.9	I	0.6	159
42 N	05 28 51	− 05 44.4	208.8 − 20.4	II	2.8	166
43 N	05 29 27	+06 00.7	198.1 − 14.6	I	1.1	380
44 N	05 29 46	− 04 33.2	207.8 − 19.7	II	4.3	550
45 N	05 33 24	+31 49.0	176.5 − 00.1	I	0.3	871:
46 N	05 34 00	− 06 44.4	210.4 − 19.7	I:	0.9	724:
48 N	05 35 33	− 00 12.8	204.5 − 16.3	II	14.0	363
47 N	05 36 12	+23 17.8	184.0 − 04.2	II	2.3	871
49 N	05 36 33	+04 05.7	200.7 − 14.0	I	21.0	316
50 N	05 37 42	− 01 29.3	205.9 − 16.5	I	3.1	794
51 N	05 38 24	− 01 31.9	206.0 − 16.3	I	5.1	661
53 N	05 39 00	− 10 21.0	214.4 − 20.2	I	0.9	
52 N	05 39 07	− 02 17.0	206.8 − 16.5	I	4.3	550
54 N	05 39 24	− 06 16.0	210.6 − 18.3	I	0.6	437:
55 N	05 39 57	− 08 09.4	212.4 − 19.0	I	0.9	380
57 N	05 40 29	− 02 20.1	207.1 − 16.3	I	2.6	575
56 N	05 40 50	+16 20.2	190.5 − 06.9	I	0.6	263
58 N	05 42 00	− 08 44.0	213.2 − 18.8	I	0.3	
59 N	05 44 10	+00 03.5	205.3 − 14.3	I	6.8	550:
60 N	05 44 34	+00 16.9	205.2 − 14.1	I	4.3	
61 N	05 49 57	+05 09.2	201.5 − 10.6	II	0.6	479
62 N	05 51 30	+01 40.0	204.8 − 12.0	I	1.4	
63 N	05 53 42	+01 51.5	204.9 − 11.4	I	0.3	
64 N	05 55 30	− 14 04.0	219.8 − 18.1	I	0.6	794:
66 N	06 00 48	− 09 43.0	216.3 − 15.1	I	0.9	832:
65 N	06 01 18	+30 31.0	180.7 +04.3	I	2.3	1148:
67 N	06 05 06	− 06 23.5	213.7 − 12.7	I	1.1	724:
68 N	06 05 37	− 06 13.1	213.6 − 12.5	I	3.4	912
69 N	06 05 39	− 06 21.2	213.6 − 12.6	I	3.1	759
70 N	06 05 59	− 05 19.9	212.8 − 12.0	I– II	6.0	832

Catalogue of Reflection Nebulae*

No.		Star Name HD or BD	M_V (mag)	V_0-M_V (mag)	V (mag)	B–V (mag)	Sp LC
34	N	34078, +34 980	- 3.0	7.5	5.97	+0.23	O9.5 V
37	N	34454, +13 852			7.84	+1.75	M2: II- III
38	N	34989, +08 933	- 2.2	7.5	5.77	- 0.13	B1 V
39	N	243202, +32 970	+0.6	9.3	11.09	+0.33	B5 Vp
40	N	243588, +06 921	+1.6	6.9	9.08	+1.15	G8 III
41	N	244068, +23 921	+3.3	6.0	10.06	+0.72	F4 V
42	N	36412, - 05 1281	+2.1	6.1	9.46	+0.75	A7 V
43	N	+05 951	+0.3	7.9	8.69	+0.06	B8 V
44	N	36540, - 04 1162	- 1.2	8.7	8.11	+0.05	B7 III- IV
45	N	245259, +31 1022	- 1.4:	9.7:	10.51	+0.50	B8
46	N	- 06 1253	- 1.2:	9.3:	10.2v	+0.47	
48	N	37370, - 00 1034	- 0.7	7.8	7.46	- 0.04	B6 V
47	N	37387, +23 982	- 3.8	9.7	7.47	+1.97	K2 Ib
49	N	37490, +04 1002	- 3.4	7.5	4.50	- 0.08	B3 IIIe
50	N	37674, - 01 1001	- 2.3	9.5	7.67	- 0.08	B3n
51	N	37776, - 01 1005	- 2.5	9.1	6.98	- 0.14	B2 V
53	N	- 10 1261			10.04	+0.44	
52	N	37903, - 02 1345	- 2.0	8.7	7.82	+0.09	B1.5 V
54	N	- 06 1287	+0.4:	8.2:	10.60	+0.54	
55	N	38023, - 08 1199	- 0.7	7.9	8.87	+0.32	B4 V
57	N	38087, - 02 1350	- 1.5	8.8	8.30	+0.13	B3n
56	N	38065, +16 852	+1.8	7.1	9.15	+0.15	A2 V
58	N	- 08 1208			10.22	+1.45	
59	N	38563, +00 1177	- 0.6:	8.7:	10.49	+0.62	B5
60	N	290861, +00 1181			9.90	+1.05	F8
61	N	39398, +05 1035	- 0.5	8.4	8.44	+1.06	G6 II- III
62	N	288313, +01 1156			9.93	+1.07	K2
63	N	288309, +01 1163			9.17	+1.19	K
64	N	- 14 1294			10.41	+1.03	
66	N	- 09 1310	- 0.7:	9.6:	10.12	+0.19	
65	N	+30 1096	- 2.3:	10.3:	10.40	+0.54	
67	N	- 06 1415	- 1.7:	9.3:	10.27	+0.65	B1
68	N	42004, - 06 1417	- 1.9	9.8	9.65	+0.32	B1.5 V
69	N	- 06 1418	- 2.0	9.4	9.21	+0.31	B2.5 V
70	N	42050, - 05 1515	- 2.5	9.6	8.09	+0.06	B1 V

*Adapted from van den Bergh (1966), van den Bergh and Herbst(1975), Herbst (1975) and Racine (1968).

Catalogue of Reflection Nebulae*

No.		R.A.(1950) h m s	Dec.(1950) ° ′	l b ° °	Type	θ/2 (°)	D (pc)
72	N	06 07 05	− 06 18.9	213.8 − 12.2	I	2.3	692
71	N	06 07 11	+14 05.1	195.7 − 02.5	II	1.7	1047
73	N			213.9 − 11.8	I	1.4	
74	N	06 09 24	− 06 09.0	213.9 − 11.6	I	0.6	832:
75	N	06 16 20	+23 17.8	188.7 +03.8	I	3.4	550
76	N	06 28 00	+10 06.0	201.6 +00.0	II		1047
77	N	06 28 12	+09 51.0	201.9 − 00.0	I– II	0.9:	1445
80	N	06 28 27	− 09 37.1	219.3 − 09.0	I	5.7	759
78	N	06 28 30	+09 49.6	201.9 +00.0	I– II	2.8:	912
79	N	06 28 58	+10 22.6	201.5 +00.4	II	2.6	
81	N	06 30 12	+07 22.3	204.3 − 00.8	I	18.0	832
82	N	06 30 19	+10 21.6	201.6 +00.7	I	3.1	
83	N	06 37 42	− 27 12.0	236.5 − 14.4	I	0.3	
84	N			236.6 − 14.4	I	1.4	
85	N	06 44 16	+01 22.2	211.2 − 00.4	I– II	1.1	1738
86	N	06 54 46	− 10 12.8	222.7 − 03.4	II	2.0	417
87	N	06 58 05	− 08 47.7	221.8 − 02.0	I	1.4	501
88	N	06 59 29	− 11 13.7	224.2 − 02.8	I	7.1	871:
89	N	07 00 18	− 12 10.0	225.1 − 03.1	I	1.1	661:
90	N	07 00 22	− 11 22.8	224.4 − 02.7	I	4.0	525
91	N	07 01 00	− 10 38.0	223.8 − 02.2	II	0.9	794:
92	N	07 01 36	− 10 30.0	224.7 − 02.5	I	2.0	759:
93	N	07 02 04	− 10 22.7	223.7 − 01.9	I	10.0	724
94	N	07 02 57	− 12 15.0	225.5 − 02.6	I	5.6	832
95	N	07 04 20	− 11 12.9	224.7 − 01.8	II	2.8	794
96	N	07 17 32	− 23 55.8	237.4 − 04.9	I	2.8	1047:
97	N	07 30 12	− 15 48.0	232.6 +01.1	I– II	1.1	1380:
98	N	07 34 27	− 25 13.2	240.4 − 02.2	I	8.0	380
1	S	07 59 06	− 45 19	260.4 − 08.0	I	1.3	1820
3a	S	08 02 54	− 39 11	255.5 − 04.2	I– II	1.3	1148
3b	S	08 03 00	− 39 12	255.5 − 04.2	I– II	0.8	759
3c	S	08 03 06	− 39 10	255.5 − 04.1	I– II		955
2	S	08 03 12	− 31 22	249.0 +00.1	I	1.3	1820
5	S	08 05 30	− 35 48	252.9 − 01.9	I	0.6	1318
4	S	08 07 36	− 38 43	255.6 − 03.1	II	3.5	912

Catalogue of Reflection Nebulae*

No.		Star Name HD or BD	M_V (mag)	V_0-M_V (mag)	V (mag)	B–V (mag)	Sp LC
72	N	42261, – 06 1431	– 1.3	9.2	9.18	+0.17	B3 V
71	N	252680, +14 1171	– 1.9	10.1	9.10	+0.04	B2 V
73	N	– 06 1440					
74	N	– 06 1444	– 0.8	9.6:	10.82:	+0.43	B6
75	N	43836, +23 1301	– 3.2	8.7	6.96	+0.48	B9 II
76	N	258686, +10 1158	– 1.3	10.1	9.12	– 0.06	B7 IIIp
77	N	258749, +09 1266	– 1.9	10.8	9.80	+0.10	B5 III
80	N	46060, – 09 1498	– 2.0	9.4	9.0	+0.31	B3n
78	N	258853, +09 1269	– 1.8	9.8	8.83	+0.02	B4nn
79	N	258973, +10 1163			10.03	+0.25	A2
81	N	46300, +07 1337	– 5.2	9.6	4.48	+0.02	A0 Ib
82	N	259431, +10 1172			8.74	+0.28	B3pe
83	N	– 27 3174					
84	N	– 27 1408					
85	N	289120, +01 1503	– 2.4	11.2	10.18	+0.18	B2n(e)
86	N	51479, – 10 1773	– 0.1	8.1	8.42	– 0.02	B7 V
87	N	52329, – 08 1664	– 0.1	8.5	8.62	– 0.08	B6 V
88	N	52721, – 11 1747	– 4.0:	9.7:	6.58	+0.06	B3e
89	N	– 12 1748	– 0.2:	9.1:	10.23	+0.25	
90	N	52942, – 11 1755	– 1.6	8.6	8.14	+0.14	B3n
91	N	– 10 1839	– 1.5:	9.5:	9.76	+0.36	
92	N	– 11 1763	– 1.0:	9.4:	9.25	+0.07	
93	N	53367, – 10 1848	– 4.5	9.3	6.97	+0.44	B0 IV:e
94	N	53623, – 12 1771	– 2.4	9.6	7.93	– 0.06	B2 V
95	N	53974, – 11 1790	– 5.2	9.5	5.38	+0.05	B0.5 IV
96	N	57281, – 23 5277	– 1.0:	10.1:	8.97	– 0.06	B9
97	N	– 16 2003	– 1.4:	10.7:	10.42	+0.12	
98	N	61071, – 25 4775	– 1.0	7.9	6.84	– 0.07	B6 V
1	S			11.3	12.45	+0.30	
3a	S		+0.25	10.3	11.43	+0.14	B6 V
3b	S		+0.90	9.4	11.14	+0.18	B9 V
3c	S		+1.45	9.9	12.27	+0.27	A0 V
2	S			11.3	13.40	+0.51	
5	S			10.6	13.47V?	+0.62	
4	S		– 2.35	9.8	9.77	+0.45	B1 V

*Adapted from van den Bergh (1966), van den Bergh and Herbst(1975), Herbst (1975) and Racine (1968).

Catalogue of Reflection Nebulae*

No.		R.A.(1950) h m s	Dec.(1950) ° ′	l b ° °	Type	θ/2 (°)	D (pc)
6	S	08 11 06	− 34 26	252.4 − 00.2	II	1.1	1820
7	S	08 11 30	− 38 17	255.7 − 02.3	I	7.2	832
8	S	08 12 00	− 35 59	253.8 − 00.9	II	1.1	1318
9a	S	08 14 00	− 35 59	254.0 − 00.6	I− II	0.3	3802
9b	S	08 14 06	− 36 00	254.1 − 00.6	I− II	0.3	9120
10	S	08 15 05	− 42 05	259.3 − 03.8	II	0.8	479
11	S	08 16 18	− 41 57	259.2 − 03.6	I− II	2.4	550
12b	S	08 18 30	− 49 58	266.1 − 07.8	I	0.3	631
13a	S	08 19 06	− 36 05	254.7 +00.2	I	1.1	417:
13b	S	08 19 06	− 36 05	254.7 +00.2	II	0.3	6607:
12a	S	08 19 06	− 50 09	266.3 − 07.8	I	2.4	347
15d	S	08 20 24	− 41 55	259.6 − 02.9	II		3162
14a	S	08 20 30	− 40 17	258.3 − 02.0	I	0.5	263
14b	S	08 20 42	− 40 15	258.3 − 01.9	I		457:
15c	S	08 20 54	− 41 56	259.7 − 02.9	II		2630
15b	S	08 21 12	− 41 55	259.7 − 02.8	II		2512
15a	S	08 21 24	− 41 56	259.8 − 02.8	II	2.1	5754
16	S	08 26 12	− 50 59	267.7 − 07.3	II	0.8	6310::
17a	S	08 34 00	− 40 30	260.0 − 00.0	I	5.3	955
18	S	08 35 12	− 39 14	259.1 +00.9	I− II	0.6	1202::
19	S	08 35 36	− 47 01	265.4 − 03.7	II	2.4	575
20	S	08 37 36	− 40 17	260.3 +00.7	II	1.6	912
21a	S	08 44 18	− 43 29	263.5 − 00.4	I	1.1	1995
21b	S	08 45 24	− 43 32	263.7 − 00.2	I	0.3	1995
22e	S	08 49 30	− 48 54	268.3 − 03.1	II		2630
22d	S	08 49 42	− 48 54	268.3 − 03.1	II		2884
22c	S	08 49 42	− 48 56	268.3 − 03.2	II		1259
22a	S	08 50 48	− 48 33	268.2 − 02.7	I		1738
22b	S	08 51 06	− 48 35	268.2 − 02.7	II		2089
23	S	08 52 54	− 46 44	267.0 − 01.3			457
24	S	08 53 18	− 43 16	264.5 +01.0	I− II	1.6	550
25a	S	08 54 42	− 42 54	264.4 +01.4	I− II	0.5	6310:
25b	S	08 54 48	− 43 01	264.4 +01.4	II	0.8	603
27b	S	08 56 12	− 42 30	264.3 +01.9	II		794
27a	S	08 56 12	− 42 38	264.3 +01.9	I− II	0.8	1047

Catalogue of Reflection Nebulae*

No.		Star Name HD or BD	M_V (mag)	V_0-M_V (mag)	V (mag)	B–V (mag)	Sp LC
6	S	RS Pup, 68860		11.3	7.01	+1.49	F8– K5
7	S	68982, – 38 2054		9.6	7.59	+0.09	B2 IV
8	S		– 0.03	10.6	11.03	+0.02	B6 V
9a	S			12.9	13.17	+0.52	
9b	S			14.8	14.66	+0.84	
10	S		+1.47	8.4	11.01	+0.33	A0 V
11	S	70024, – 41 235	– 0.25	8.7	9.07	+0.55	B5 V
12b	S			9.0	11.83	+0.75	
13a	S	NGC 2580		8.1:	10.18	+1.39	K1 III
13b	S			14.1:	13.11	+0.81	
12a	S	70584, – 49 1602	+0.10	7.7	9.17	+0.29	B7 V
15d	S			12.5	13.57	+0.56	
14a	S		+1.59	7.1	9.63	+0.34	A2 V
14b	S			8.3:	10.36	+1.24	K0 III:
15c	S			12.1	12.23	+0.46	
15b	S			12.0	13.00	+0.58	
15a	S			13.8	11.83	+0.78	
16	S			14.0::	15.60	+1.88	
17a	S	NGC 2626	– 2.30	9.9	10.02	+0.46	B1 V
18	S			10.4::	15.59	+1.22	
19	S	73589, – 46 2738	– 0.55	8.8	8.93	+0.04	B5 V
20	S		– 0.02	9.8	10.86	+0.19	B6 V
21a	S			11.5	12.22	+0.43	
21b	S			11.5	13.13	+0.52	
22e	S			12.1	14.85	+0.72	
22d	S			12.3	12.83	+0.49	
22c	S		– 0.50	10.5	11.58	+0.30	B5 V
22a	S			11.2	12.14	+0.38	
22b	S			11.6	11.79	+0.70	
23	S		+1.53	8.3	10.89	+0.32	A0 V
24	S	76534, – 42 3114	– 1.85	8.7	8.04	+0.11	B2 Vpe
25a	S	RCW 34		14.0:	11.81	+0.92	O9: I:pe
25b	S	76764, – 42 3140	– 0.50	8.9	9.27	+0.10	B5 V
27b	S			9.5	13.00	+1.07	
27a	S		– 0.90	10.1	10.11	+0.09	B3 V

*Adapted from van den Bergh (1966), van den Bergh and Herbst(1975), Herbst (1975) and Racine (1968).

Catalogue of Reflection Nebulae*

No.		R.A.(1950) h m s	Dec.(1950) ° ′	l b ° °	Type	θ/2 (°)	D (pc)
26	S	08 56 24	− 47 11	267.7 − 01.1	I	1.7	6309:
28	S	08 56 36	− 43 14	264.9 +01.6	I	1.0	502
29c	S	09 09 30	− 45 21	267.9 +01.8	II	1.3	794
29b	S	09 09 30	− 45 23	267.9 +01.8		1.6	955
29a	S	09 10 00	− 45 26	268.1 +01.8	I	0.3	479
30	S	09 16 54	− 48 11	270.8 +00.8			955
31a	S	09 21 30	− 48 00	271.2 +01.4			
32a	S	10 00 24	− 57 26	281.8 − 01.9	I	0.6	3802
33d	S	10 01 12	− 59 02	282.9 − 03.1	I	0.5	1738
34	S	10 14 00	− 60 03	284.8 − 03.1	I	2.4	3020
35	S	10 16 36	− 60 39	285.4 − 03.4	I	1.1	3311:
36	S	10 22 30	− 57 19	284.1 − 00.3	II	1.0	2089
37	S	10 30 54	− 58 24	285.7 − 00.5	II	3.2	2089:
39	S	10 31 36	− 59 25	286.3 − 01.3	I	0.5	4365
38	S	10 31 54	− 61 27	287.2 − 03.0	I	0.3	2512
40c	S	10 33 06	− 58 45	285.9 − 00.8	II	1.6	2512:
40b	S	10 33 54	− 58 59	286.0 − 00.8	I– II	1.7	3162:
40a	S	10 34 30	− 58 42	286.3 − 00.5	I– II	1.6	2630
41b	S	10 35 12	− 58 24	286.2 − 00.2	I	0.3	8318
41c	S	10 35 24	− 58 13	286.1 − 00.0	I	0.3	2754
41a	S	10 35 30	− 58 28	286.3 − 00.3	I	2.4	1585:
42b	S	10 35 42	− 57 45	286.0 +00.4	II	1.6	2630
42a	S	10 36 42	− 57 47	286.1 +00.4	I	1.6	2754
43	S	10 44 24	− 59 45	287.9 − 00.9	I	0.8	2291
44	S	10 50 06	− 55 54	286.8 +02.9	I	0.8	2188:
45a	S	10 54 24	− 62 46	290.3 − 03.0	II		6607
45b	S	10 54 24	− 62 46	290.3 − 03.0	I		2884:
46	S	10 59 24	− 59 35	289.5 +00.1	I	2.4	3631
47b	S	11 08 54	− 61 05	291.2 − 00.8	I– II	0.6	4571
47a	S	11 09 06	− 61 02	291.2 − 00.7	I	0.8	7586
47c	S	11 09 18	− 61 06	291.3 − 00.8	I		10471:
48a	S	11 21 48	− 58 38	291.9 +02.1	II:	1.6	1514
49	S	11 26 36	− 62 49	293.8 − 01.7	II	1.1	3981
50a	S	11 28 06	− 63 33	294.1 − 02.3	I– II		2884
50b	S	11 28 06	− 63 33	294.1 − 02.3	I– II		871

Catalogue of Reflection Nebulae*

No.		Star Name HD or BD	M_V (mag)	V_0-M_V (mag)	V (mag)	B–V (mag)	Sp LC
26	S			14.0:	14.20	+1.21	
28	S		- 0.10	8.5	11.28	+0.72	B5 Vp
29c	S			9.5	12.42	+0.62	
29b	S			9.9	11.97	+0.45	
29a	S		+0.90	8.4	10.96	+0.43	B5 V:
30	S			10.1?	14.16	+1.36	
31a	S						
32a	S			12.9	13.49	+0.54	
33d	S			11.2	14.34	+0.46	
34	S			12.4	12.65	+0.15	
35	S			12.6:	14.92	+0.63:	
36	S		- 2.30	11.6	10.17	+0.01	B1 V
37	S	91533, - 58 2285		11.6:	6.03	+0.31	A2 Ia
39	S			13.2	14.48	+0.60	
38	S			12.0	13.04	+0.57	
40c	S			12.0:	9.97	+0.67	B1.5 II
40b	S			12.5:	10.39	+0.50	B2 II
40a	S	92061, - 58 2372		12.1	9.00	+0.07	B1.5 III
41b	S			14.6	13.78	+0.57	
41c	S			12.2	11.86	+0.33	
41a	S	92207, - 58 2411		11.0:	5.52	+0.49	A1 Ia+
42b	S		- 3.65	12.1	9.70	+0.08	B0 V
42a	S	92383, - 57 3621		12.2	9.34	+0.07	B1 IVn
43	S		- 2.35	11.8	11.30	+0.29	B1 Ve
44	S			11.7:	9.46	+1.61	M0 II:
45a	S			14.1	12.23	+0.41	
45b	S		- 3.35	12.3:	11.36	+0.44	B1 Vp?
46	S	NGC 3503	- 4.00	12.8	10.46	+0.21	B0 Ve
47b	S			13.3	12.99	+0.47	
47a	S			14.4	12.07	+0.43	
47c	S			15.1:	13.79V?	+0.97	
48a	S		- 1.55	10.9	10.48	+0.11	B2 V
49	S			13.0	14.12	+0.81	
50a	S	100099, - 63 1904		12.3	8.08	+0.11	O9 III
50b	S			9.7	12.52	+0.73	

*Adapted from van den Bergh (1966), van den Bergh and Herbst(1975), Herbst (1975) and Racine (1968).

Catalogue of Reflection Nebulae*

No.		R.A.(1950) h m s	Dec.(1950) ° ′	l b ° °	Type	θ/2 (°)	D (pc)
51	S	11 30 18	− 63 11	294.3 − 01.9		1.6	2630
52	S	11 38 12	− 64 15	295.4 − 02.7	I	0.5	3467
53b	S	11 43 54	− 65 17	296.3 − 03.5	I		3467:
53a	S	11 44 06	− 65 18	296.3 − 03.6	I	0.5	4169
54	S	11 46 06	− 61 13	295.5 +00.5	II	1.4	1259
55	S	11 46 12	− 62 03	295.7 − 00.3	I	0.8	955
56	S	11 47 42	− 64 36	296.5 − 02.8	I	2.6	3467
57a	S	12 19 30	− 63 01	299.7 − 00.6	I	0.5	1202
58	S	12 55 42	− 66 02	303.7 − 03.4	mB	0.5	1738
59	S	13 05 36	− 62 02	305.0 +00.5	I	0.8	1738
60c	S	13 15 18	− 62 26	306.0 +00.0	II		2089
60a	S	13 15 42	− 62 18	306.1 +00.1	II		1738
60b	S	13 15 42	− 62 19	306.1 +00.1	II		2399
61	S	13 16 42	− 62 45	306.2 − 00.3	I	0.8	2630
62	S	14 39 42	− 58 32	317.0 +01.0	II	1.6	3162
63	S	14 45 12	− 65 02	314.9 − 05.1	I	1.3	417
64b	S	14 54 24	− 59 34	318.3 − 00.7	I	0.5	525
64a	S	14 54 36	− 59 28	318.4 − 00.7	I	0.5	912
65a	S	14 57 00	− 63 05	316.9 − 04.0	pec		1259
65b	S	14 59 12	− 63 11	317.1 − 04.2	I		1445
66	S	15 10 54	− 62 35	318.6 − 04.3	I	1.6	1047
67	S	15 43 06	− 56 53	325.1 − 01.9	I	0.5	525
68	S	15 53 34	− 53 50	328.2 − 00.5	I	1.1	1995
99	N	15 55 50	− 25 59	347.2 +20.2	II	100.0	138
69	S	16 08 00	− 51 08	331.6 +00.1	II		912:
100	N	16 09 05	− 19 19.9	354.6 +22.7	II	5.7:	151
70	S	16 09 18	− 50 16	332.4 +00.6	I	0.5	3802
101	N	16 16 12	− 20 05.9	355.2 +21.0	I	4.3	66
102	N	16 17 08	− 19 55.5	355.5 +20.9	I	7.1	126
103	N	16 17 34	− 19 59.9	355.5 +20.8	I	4.5	132
104	N	16 18 09	− 25 28.0	351.3 +17.0	I	28.0:	145
105	N	16 20 07	− 24 16.9	352.9 +17.0	I	14.0:	165
106	N	16 22 35	− 23 27.0	353.7 +17.7	I	31.0	145
107	N	16 26 21	− 26 19.5	351.9 +15.1	I	41.0	100:
108	N	16 27 11	− 26 00.5	353.1 +15.8	I	14.0	145

Catalogue of Reflection Nebulae*

No.	Star Name HD or BD	M_V (mag)	V_0-M_V (mag)	V (mag)	B–V (mag)	Sp LC
51 S		- 3.05	12.1	10.21	+0.08	B0.5 V
52 S			12.7	13.11	+0.30	
53b S			12.7:	15.39	+1.20	
53a S			13.1	13.07	+0.45	
54 S		- 1.20	10.5	9.96	- 0.01	B3 V
55 S		+0.35	9.9	11.15	+0.17	B8 V
56 S	IC 2966	- 3.60	12.7	11.47	+0.44	B0.5 V
57a S			10.4	13.08	+0.80	
58 S		- 0.68	11.2	12.36	+0.40	B4 V
59 S			11.2	12.57	+0.82	
60c S			11.6	12.02	+0.49	
60a S		- 2.30	11.2	10.79	+0.30	B2 V
60b S			11.9	12.70	+0.35	
61 S		- 2.60	12.1	10.73	+0.09	B1 V
62 S			12.5	14.12	+0.75	
63 S	130079, - 64 3016	+1.15	8.1	10.49	+0.32	B9 V:p?
64b S		+1.05	8.6	11.25	+0.41	B9 V
64a S		- 0.80	9.8	10.63	+0.32	B3 V
65a S			10.5	13.86V?	+1.84	
65b S			10.8	12.75	+1.13	
66 S		- 1.20	10.1	11.64	+0.62	B3 V
67 S			8.6	11.89	+0.70	
68 S		- 3.50	11.5	11.00	+0.49	B0.5 V
99 N	143018, - 25 11228	- 2.60	5.7	2.89	- 0.19	B1 V+B2V
69 S			9.8	11.27v	+2.56:	M6: III:
100 N	145502, - 19 4333	- 2.60	5.9	4.03	+0.07	B2 IV- V
70 S			12.9	12.82	+0.95	
101 N	146834, - 19 4357	+1.30	4.1	6.35	+1.14	G5 III
102 N	147009, - 19 4358	+1.40	5.5	8.05	+0.27	B9.5 V
103 N	147103- 4, - 19 4361	+0.50	5.6	7.56	+0.50	A0 V
104 N	147165, - 25 11485	- 4.10	5.8	2.89	+0.14	B1 III
105 N	147889, - 24 12684	- 1.60	6.1	7.86	+0.85	B2 V
106 N	147933- 4, - 23 1286	1 - 2.70	5.8	4.59	+0.24	B2 IV+B2V
107 N	148478- 9, - 26 11359	- 4.80:	5.0:	0.92	+1.84	M1 Ib+B2.5V
108 N	148605, - 24 12695	- 1.40	5.8	4.78	- 0.12	B2 V

*Adapted from van den Bergh (1966), van den Bergh and Herbst(1975), Herbst (1975) and Racine (1968).

Catalogue of Reflection Nebulae*

No.		R.A.(1950) h m s	Dec.(1950) ° ′	l b ° °	Type	θ/2 (°)	D (pc)
71	S	16 27 48	- 38 17	343.2 +06.8	II	1.6	174
72b	S	16 35 00	- 48 47	336.3 - 01.3	II		1445
72c	S	16 35 24	- 48 50	336.3 - 01.4	I- II		1820
72a	S	16 36 18	- 48 42	336.4 - 01.4	I	0.5	1148
109	N	16 38 41	- 17 38.8	000.8 +18.4	II	18.0	229:
73e	S	16 42 18	- 41 08	342.9 +02.8	I		4169
73c	S	16 42 30	- 41 07	343.0 +02.8	I	0.5	1514
73d	S	16 42 48	- 41 06	343.0 +02.7	I		5495:
73a	S	16 42 54	- 41 09	343.0 +02.7	I	1.6	1905:
73b	S	16 43 00	- 41 08	343.0 +02.7	I	0.2	1148
74	S	16 45 00	- 48 00	338.0 - 02.1	II	1.0	912
75	S	16 53 06	- 40 11	345.0 +01.8	I	1.3	2884
76	S	16 54 42	- 45 39	340.9 - 01.9	II	1.3	7586::
79a	S	16 55 18	- 46 03	340.7 - 02.2	II	4.8	1000
77	S	16 55 30	- 40 32	345.0 +01.2	I	0.5	871
80	S	16 55 30	- 42 38	343.3 - 00.1	pec	1.0	
78	S	16 55 48	- 40 05	345.4 +01.5		1.4	5495:
81	S	17 00 06	- 51 01	337.2 - 05.9	II	6.4	692
82	S	17 02 12	- 39 55	346.3 +00.6	II	0.5	1096
88c	S	17 07 30	- 45 51	342.1 - 03.7	I		2089
88b	S	17 07 30	- 45 56	342.0 - 03.8	I:	0.8	1586
88a	S	17 07 30	- 46 04	341.9 - 03.9	I:	0.5	1380
83	S	17 09 18	- 42 40	344.9 - 02.1	I	0.8	631
84	S	17 09 42	- 38 44	348.1 +00.2	I	0.5	661
85a	S	17 15 36	- 35 38	351.3 +01.0			2399
110	N			003.0 +09.9	I	0.6	
86	S	17 16 24	- 36 02	351.0 +00.6			1820
111	N	17 16 26	+06 08.2	027.7 +23.3	II	6.2	
85b	S	17 16 42	- 35 37	351.2 +00.6			4169
85c	S	17 17 06	- 35 47	351.3 +00.7			3020
87	S	17 18 12	- 44 05	344.7 - 04.3	I	1.0	19055::
89	S	17 21 30	- 34 07	353.2 +00.9		0.5	2884
90	S	17 23 24	- 35 51	352.0 - 00.4	I- II	0.8	871
91	S	17 30 42	- 39 21	349.9 - 03.5	II	0.5	
92	S	17 45 54	- 31 24	358.3 - 01.9	I	0.3	1514

Catalogue of Reflection Nebulae*

No.		Star Name HD or BD	M_V (mag)	V_0-M_V (mag)	V (mag)	B–V (mag)	Sp LC
71	S	148657, - 38 6420	+1.45	6.2	8.56	+0.28	A0 V
72b	S		- 2.15	10.8	10.79	+0.41	B2 V
72c	S		- 2.85	11.3	11.50	+0.66	B1 Vp?
72a	S		- 1.20	10.3	11.03	+0.38	B3 V
109	N	150416, - 17 4618	- 2.00	6.8:	4.96	+1.11	G8 II
73e	S			13.1:	14.45	+0.80	
73c	S		- 1.75	10.9	11.36	+0.43	
73d	S			13.7:	14.28	+0.92	
73a	S	- 41 7613	- 2.70	11.4:	10.13	+0.17	B2 V
73b	S			10.3	13.57	+0.87	
74	S		+0.70	9.8	11.21	+0.15	B8 V
75	S		- 2.95	12.3	11.43v?	+0.35	B1 Vp?
76	S			14.4::	13.95	+1.04	
79a	S	152979, - 45 8234		10.0	8.19	+0.15	B2 IV
77	S			9.7:	10.19	+0.68	B5 III:
80	S	- 42 11721			11.43v	+1.28	Bep
78	S			13.7:	13.13	+1.20	
81	S	153772, - 50 9813	- 1.80	9.2	8.32	+0.06	B2 V
82	S		- 1.30	10.2	11.60	+0.62	B3 V
88c	S			11.6	13.97	+0.93	
88b	S			11.0	13.41	+0.71	
88a	S			10.7	12.75	+0.85	
83	S		+0.55	9.0	10.92	+0.33	B8 V
84	S		- 0.45	9.1	12.44	+0.99	B2 V:
85a	S		- 4.00	11.9	11.86	+0.90	B0 V
110	N	- 20 4696			10.52	+0.73	
86	S		- 5.40	11.3	11.09	+1.24	O7
111	N	156697, +06 3386			6.50	+0.41	F0n
85b	S			13.1	13.23	+1.06	
85c	S			12.4	12.25	+1.34	
87	S			16.4:	15.16	+0.99:	
89	S			12.3	12.77	+1.53	
90	S		- 1.05	9.7	10.62	+0.41	B3 V
91	S				11.02	+0.31	Bep
92	S			10.9	12.58	+0.71	

*Adapted from van den Bergh (1966), van den Bergh and Herbst(1975), Herbst (1975) and Racine (1968).

Catalogue of Reflection Nebulae*

No.		R.A.(1950) h m s	Dec.(1950) ° ′	l b ° °	Type	θ/2 (°)	D (pc)
93a	S	17 48 12	− 31 15	358.7 − 02.3	I− II	1.4	1097
93b	S	17 48 18	− 31 17	358.6 − 02.3	I:		1259
112	N	17 50 48	− 05 36.0	021.2 +10.2	II	0.6	
114	N	18 05 36	− 18 24.0	011.8 +00.8	I	0.6	660
113	N	18 05 39	− 21 27.5	009.1 − 00.7	II	8.5	1660
115	N	18 06 06	− 23 26.7	007.4 − 01.8	I	0.9	912
116	N	18 07 42	− 17 45.0	012.6 +00.6	I	0.3	603:
117	N	18 11 43	− 17 22.7	013.3 − 00.0	I	1.1	759
118	N	18 13 55	− 19 47.8	011.5 − 01.6	I	2.6	1259
119	N	18 14 08	− 19 53.0	011.4 − 01.7	I	2.8	1584
120	N	18 14 23	− 16 56.9	014.0 − 00.4	I	0.6	
121	N	18 16 18	− 17 00.9	014.3 − 01.0	I− II	0.3	1660
122	N	18 21 43	− 13 41.3	017.8 − 00.4	I	0.6	
123	N	18 27 53	+01 11.3	031.6 +05.2	I	1.7	437
124	N	18 28 39	− 10 49.9	021.0 − 00.5	I	18.0	166
125	N	19 23 44	+15 27.4	050.6 − 00.4	I	0.6	
126	N	19 24 04	+22 39.4	057.0 +03.0	I	4.0	479
127	N	19 45 09	+18 24.6	055.8 − 03.4	II pec	28.0	182:
128	N	20 02 38	+32 04.5	069.5 +00.4	I	4.0	832
129	N	20 08 43	− 00 58.3	041.6 − 18.1	II	48.0:	40
130	N	20 15 54	+39 12.0	076.9 +02.1	I	0.3:	871:
131	N	20 22 30	+42 08.0	080.0 +02.7	I	2.8	1097
132	N	20 23 02	+42 13.8	080.2 +02.7	I	2.8:	1000
134	N	20 28 31	+48 47.0	086.1 +05.7	I− II	11.0	229
133	N	20 29 05	+36 46.0	076.4 − 01.4	I	5.7	1514:
135	N	20 34 45	+32 16.8	073.5 − 05.1	II	0.9	346
136	N	20 36 30	+41 53.8	081.4 +00.5	II	2.0	398
137	N	20 54 08	+47 13.5	087.5 +01.4	II	6.8	1820
138	N	20 55 36	+48 06.1	088.3 +01.8	I	3.7	3311
139	N	21 01 00	+67 57.9	104.1 +14.2	I	8.5	
141	N	21 15 42	+68 03.0	105.0 +13.2	I	1.7	
140	N	21 15 56	+58 24.1	098.0 +06.5	I− II	10.0	501
142	N	21 35 09	+57 16.6	099.1 +03.9	I	0.6	724
143	N	21 36 01	+67 57.6	106.4 +11.8	I	4.8	794
144	N	21 39 04	+54 38.6	097.7 +01.6	I	1.7	66:

Catalogue of Reflection Nebulae*

No.		Star Name HD or BD	M_V (mag)	V_0-M_V (mag)	V (mag)	B–V (mag)	Sp LC
93a	S		- 0.2	10.2	11.80	+0.42	B6 V
93b	S			10.5	13.68	+0.83	
112	N	- 05 4524			10.23	+1.48	
114	N	165811, - 11 4800	- 0.5:	9.1:	10.26	+0.38	A
113	N	165784, - 21 4866	- 7.3	11.1	6.53	+0.85	A2 Ia
115	N	165872, - 23 13974	- 1.5	9.8	9.66	+0.23	B3 V
116	N	166288, - 17 5049	- 0.6:	8.9	9.97	+0.37	B8
117	N	167143, - 17 5080	- 1.6	9.4	9.10	+0.22	B3 V
118	N	167638, - 19 4940	- 2.3:	10.5:	9.85	+0.26	
119	N	313095, - 19 4946	- 2.0	11.0:	11.11	+0.41	B5
120	N	167746, - 16 4790			8.63	+1.35	G5
121	N	168418, - 17 5141	- 3.5	11.1	9.43	+0.31	B2 III
122	N	- 13 4965			10.01	+0.74	B3n+B0
123	N	170634, +01 3694	- 0.5	8.2	9.85	+0.58	B7 V
124	N	170740, - 10 4713	- 1.8	6.1	5.72	+0.24	B2 V
125	N	182830, +15 3811			8.05	+0.82	
126	N	182918, +22 3693	- 0.8	8.4	8.61	+0.16	B6 V
127	N	187076- 7, +18 4240	- 2.5:	6.3:	3.82	+1.41	M2 II+B6
128	N	190603, +31 3925	- 6.2	9.6	5.63	+0.55	B1.5 Ia
129	N	191692, - 01 3911	+0.2	3.0	3.24	- 0.07	B9.5 III
130	N	228789, +38 3993	- 1.8:	9.7:	10.21	+0.50	B
131	N	+41 3731	- 1.3	10.2	9.90	+0.10	B3n
132	N	+41 3737	- 1.7	10.0	9.26	+0.13	B3 Vn
134	N	195556, +48 3142	- 2.2	6.8	4.95	- 0.09	B2 V
133	N	195593, +36 4105	- 6.5:	10.9:	6.19	+1.01	F5 Iab
135	N	+31 4152	- 0.4	7.7	8.43	+1.95	M1 IIIe
136	N	196819, +41 3836	- 2.0	8.0	7.50	+1.89	K3 II
137	N	199478, +46 3111	- 7.0	11.3	5.69	+0.46	B8 Ia
138	N	199714, +47 3237	- 5.4	12.6	8.29	+0.28	B8 Ib
139	N	206775, +67 1283			7.39	+0.38	B5e
141	N	+67 1300			10.73	+1.36	
140	N	203025, +57 2309	- 3.5	8.5	6.44	+0.21	B2 IIIe
142	N	239710, +56 2604	- 1.5	9.3	9.50	+0.36	B3 V
143	N	206135, +67 1332	- 2.3	9.5	8.40	+0.18	B3 V
144	N	206509, +54 2595	+1.4	4.1:	6.22	+1.18	K0 III

*Adapted from van den Bergh (1966), van den Bergh and Herbst(1975), Herbst (1975) and Racine (1968).

Catalogue of Reflection Nebulae*

No.		R.A.(1950) h m s	Dec.(1950) ° ′	l °	b °	Type	$\theta/2$ (°)	D (pc)
146	N	21 41 42	+65 53.0	105.4	+09.9	I	4.5	2754:
145	N	21 41 46	+48 39.2	094.2	− 03.2	I	4.3	275:
147	N	21 50 39	+46 59.6	094.3	− 05.4	I	0.6	436
148	N	22 05 31	+55 59.4	101.5	+00.3	II	1.7	
149	N	22 08 12	+72 38.0	111.6	+13.7	I	2.6	251
150	N	22 08 50	+73 08.7	111.9	+14.1	I	5.7	501
151	N	22 11 44	+39 28.0	092.8	− 13.8	II	5.4	52:
152	N	22 12 14	+70 00.2	110.3	+11.4	I	6.8	398
153	N			106.8	+04.6	I	1.4	603
154	N	22 29 35	+65 12.5	109.0	+06.4	II	5.1:	871
155	N	22 51 18	+61 52.8	109.6	+02.4	I	3.1	575
156	N	22 59 37	+42 03.4	102.2	− 16.1	II	94.0	138
157	N	23 00 36	+72 27.6	114.9	+11.6	pec	2.3	263
158	N	23 35 26	+48 13.2	110.6	− 12.6	I	1.7:	479

*Adapted from van den Bergh (1966), van den Bergh and Herbst(1975), Herbst (1975) and Recine (1968).

Catalogue of Reflection Nebulae*

No.		Star Name HD or BD	M_V (mag)	V_0-M_V (mag)	V (mag)	B–V (mag)	Sp LC
146	N	+65 1637	- 4.1	12.2:	10.15	+0.41	B2nne
145	N	206887, +48 3485	+0.0	7.2:	8.11	+0.58	F2 II- III
147	N	+46 3471	+0.4	8.2	10.15	+0.41	B9.5 Ve
148	N	239856, +55 2682			8.7	+1.07	F2 Iab
149	N	+72 1018	+0.7	7.0	9.80	+0.55	B8 V
150	N	210806, +72 1020	- 0.9	8.5	8.38	+0.09	B8 IV
151	N	211073, +38 4711	+0.6	3.6:	4.49	+1.38	K3 III
152	N	+69 1231	+0.7	8.0	9.29	+0.14	B9.5 V
153	N	+61 2292	- 1.5	8.9	10.06	+0.67	B2 Vn
154	N	+64 1677	- 3.2	9.7	8.96	+0.64	B2 IV- III
155	N	216629, +61 2361	- 2.4	8.8	9.29	+0.72	B2pe
156	N	217675- 6, +41 4664	- 2.4	5.7	3.62	- 0.09	B6 IIIp
157	N	217903, +71 1181	+0.2	7.1	8.09	+0.11	B9 V
158	N	222142, +47 4220	+0.4	8.4	9.58	+0.15	B8 V

*Adapted from van den Bergh (1966), van den Bergh and Herbst(1975), Herbst (1975) and Racine (1968).

20
Planetary Nebulae

20.1 Average Basic Data for Planetary Nebulae

The table of planetary nebulae parameters contains the name, Perek-Kohoutek (PK) designation (galactic longitude and latitude, l and b, followed by discovery order) and accurate optical right ascensions (R.A.) and declinations (Dec.) in the epoch 1950.0 from Blackwell and Purton (1981) and Milne (1973, 1976). When accurate positions were not available, those with less precision were obtained from the Perek-Kohoutek (1967) catalogue and its supplements. The optical angular diameter, θ, heliocentric radial velocity, V_r, expansion velocity, V_e, flux density, S_{6cm} at 6 cm wavelength, absolute flux in the Hβ line, F(Hβ), electron density, N_e, electron temperature, T_e, the B and V magnitudes of the central star, and the extinction constant, C, for the brightest planetary nebulae were obtained from Lang (1980) and Pottasch (1984).

The F(Hβ), N_e, T_e, B, V and C for numerous northern planetary nebulae come from Shaw and Kaler (1985) while those for southern planetary nebulae are due to Acker, Stenholm, and Tylenda (1989) or Tylenda, Acker, Gleizes and Stenholm (1989). Expansion velocities, V_e, are from Robinson, Reay, and Atherton (1982), Sabbadin and Hamzaoglu (1982), Sabbadin (1984), Sabbadin, Strafella, and Bianchini (1986) and Metheringham, Wood, and Faulkner (1988). Radial velocities, V_r, are from Schneider, Terzian, Purgathofer, and Perinotto (1983), Aller and Keyes (1987), and Metheringham, Wood, and Faulkner (1988). Angular diameters, θ, are from Acker, Stenholm and Tylenda (1989) and Maciel (1984), heliocentric distances, D, are from Maciel (1984), and linear radii, R, were computed from $R = \theta D/2$.

Finally the entire table was checked and extended by comparison with the Strasbourg-ESO Catalogue of Galactic Planetary Nebulae compiled by Agnes Acker with the collaboration of Franҫois Ochsenbein, Björn Stenholm, Romuald Tylenda, James Marcout, and Constant Schon (1989). This catalogue also contains line intensities, radio dimensions, infrared magnitudes, IRAS fluxes, notes, numerous bibliographical references, and finding charts.

Planetary Nebulae Statistics

Parameter	Mean	Max.	Min.	Disp., σ
Magnitude, m_V (mag)	12.9	20.9	3.9	2.3
Angular Diameter, θ ($''$)	34.7	972.0	1.0	96.1
Expansion Velocity, V_e (km s^{-1})	21.8	92.0	4.0	14.5
Flux Density, S_{6cm} (mJy)	350.8	6370	3	730.2
Heliocentric Distance, D (kpc)	2.9	12.7	0.2	1.9
Radius, R ($\times 10^{10}$ km)	438.2	3734.0	14.9	595.8
Age, T (years)	8420	266755	355	22258

Named Planetary Nebulae

Name	NGC, other	PK l	PK b	R.A. (1950) h m	Dec. (1950) ° ′
Ant Nebula	Dipolar	331	−01	16 13.5	−51 52
Blinking Planetary	NGC 6826	83	+12	19 43.5	+50 24
Blue Planetary	NGC 3918	294	+04	11 47.8	−56 54
Blue Snowball	NGC 7662	106	−17	23 23.5	+42 16
Box Nebula	NGC 6309	09	+14	17 11.3	−12 51
Butterfly Nebula (also Little Dumb-bell, M 76)	NGC 650-1	130	−10	01 39.2	+51 19
Clown Face Neb. (also Eskimo Nebula)	NGC 2392	197	+17	07 26.2	+21 01
Cork Nebula (also Little Dumb-bell, M76)	NGC 650-1	130	−10	01 39.2	+51 19
Cygnus Egg* (also Egg Nebula)	CRL 2688			21 00.3	+36 30
Diabolo Nebula (also Dumbbell Neb-ula, M 87)	NGC 6853	60	−03	19 57.4	+22 35
Double-Headed Shot (Also Dumb-bell Nebula, M 87)	NGC 6853	60	−03	19 57.4	+22 35
Dumbbell Nebula (also M 87)	NGC 6853	60	−03	19 57.4	+22 35
Egg Nebula* (also Cygnus Egg)	CRL 2688			21 00.3	+36 30
Eight-Burst Planetary	NGC 3132	272	+12	10 04.9	−40 12
Eskimo Nebula (also Clown Face)	NGC 2392	197	+17	07 26.2	+21 01
Ghost of Jupiter (also Hydra and Jupiter)	NGC 3242	261	+32	10 22.4	−18 23
Helix Nebula	NGC 7293	36	−57	22 26.9	−21 03
Hydra Nebula	NGC 3242	261	+32	10 22.4	−18 23
Jupiter Nebula	NGC 3242	261	+32	10 22.4	−18 23
Little Dumbbell (also Butterfly and Cork, M76)	NGC 650-1	130	−10	01 39.2	+51 19
Little Gem	NGC 6445	8	+03	17 46.3	−20 00
Medusa Nebula M 56	Abell 21			07 26.2	+13 21
Owl Nebula, M 97	NGC 3587	148	+57	11 11.9	+55 17
Red Rectangle*	HD 44179			06 17.6	−10 37
Ring Nebula (in Lyra, M 57)	NGC 6720	63	+13	18 51.7	+32 58
Saturn Nebula	NGC 7009	37	−34	21 01.5	−11 34

*See table of bipolar reflection nebulae for accurate positions.

20.2 Catalogue of Planetary Nebulae — Part I

Planetary Nebulae Parameters

Name	PK Number l b	R.A.(1950) h m s	Dec.(1950) ° ′ ″	m_V (mag)	S_{6cm} (mJy)
NGC 40	120 +09 1	00 10 18	+72 14 35	10.7	460
VY 1–1	118 − 08 1	00 16 02	+53 35 41		28
BV 1	119 +00 1	00 17 15.0	+62 42 22		
HU 1–1	119 − 06 1	00 25 30	+55 41 20	13.3	
BoBn 1	108 − 76 1	00 34 46.8	− 13 58.7		
BV 2	121 +00 1	00 37 26.0	+62 34 49		
A 2	122 − 04 1	00 42 42.0	+57 41 00		
NGC 246	118 − 74 1	00 44 35.3	− 12 09 03	8.0	247
PHL 932	125 − 47 1	00 57 19.6	+15 28		
M 1–1	130 − 11 1	01 34 12.9	+50 12 57		
NGC 650–1	130 − 10 1	01 39 10	+51 19 23	12.2	125
BV 3	131 − 05 1	01 49 42.0	+56 09 36		
IC 1747	130 +01 1	01 53 58	+63 04 42	13.6	124
M 1–2	133 − 08 1	01 55 32.6	+52 39 15		
A 6	136 +04 1	02 54 31	− 64 18 10	>15.5	
IC 289	138 +02 1	03 06 16.4	+61 07 39	12.3	170
NGC 1360	220 − 53 1	03 31 06.6	− 26 02 12		600
M 1–4	147 − 02 1	03 37 59.1	+52 07 26		
IC 351	159 − 15 1	03 44 20.2	+34 53 35	12.4	33
BA 1	171 − 25 1	03 50 44.7	+19 20 36	13.9	
IC 2003	161 − 14 1	03 53 10.5	+33 43 51	12.6	43
NGC 1501	144 +06 1	04 02 42	+60 47 12	13.3	210
NGC 1514	165 − 15 1	04 06 08.3	+30 38 43	10.	300
M 2–2	147 +04 1	04 09 10.2	+56 49 20		
NGC 1535	206 − 40 1	04 11 57.0	− 12 51 42	9.6	160
H 3–29	174 − 14 1	04 34 20.2	+24 57 00		
K3–67	165 − 06 1	04 36 27.5	+36 39 52		
A 7	215 − 30 1	05 00 52.4	− 15 40 24	13.2	305
J 320	190 − 17 1	05 02 48.6	+10 38 25	12.9	31
A 8	167 − 00 1	05 03 12.4	+39 04 09		
K 2–1	173 − 05 1	05 04 54	+30 44.		
IC 2120	169 − 00 1	05 14 48	+37 33.		
IC 418	215 − 24 1	05 25 09.5	− 12 44 15	10.7	1550
K 3–68	178 − 02 1	05 28 25.0	+28 56 31		
K 1–7	197 − 14 1	05 29 03.8	+06 54 00		

Planetary Nebulae Parameters

Name	θ ($''$)	V_r (km s^{-1})	V_{LSR} (km s^{-1})	V_e (km s^{-1})	D (kpc)	R/10^{10} (km)	T (years)
NGC 40	36	- 20	- 12	29	0.8	215	2354
VY 1–1	6	- 50	- 43	11			
BV 1		- 72	- 64				
HU 1–1	5	- 59	- 47	13	1.9	71	1732
BoBn 1		+196					
BV 2		- 39	- 40				
A 2	31	- 42	- 36	68	3.0	695	3242
NGC 246	245	- 46	- 51	38	0.4	733	6115
PHL 932	270	+15			1.1	2221	
M 1–1	6	- 38	- 35	12	6.1	273	7232
NGC 650–1	128	- 23		39	0.7	670	5447
BV 3		- 59	- 56				
IC 1747	13	- 67	- 62	25	0.9	87	1109
M 1–2		- 12	- 10	16			
A 6	182				1.5	2042	
IC 289	45	- 13	- 11	27	1.3	437	5137
NGC 1360	380	+42	+26	28	0.5	1421	16090
M 1–4	4	- 33	- 34	13	0.5	14	364
IC 351	7	- 9	- 15	15	3.0	157	3319
BA 1	40	- 20	- 30	71	3.6	1077	4809
IC 2003	7	- 16	- 22	21	2.4	125	1896
NGC 1501	54	+36	+37	39	1.1	444	3611
NGC 1514	100	+60	+52	25	0.8	598	7587
M 2–2	12	- 7	- 7	14	1.4	125	2845
NGC 1535	21	- 3	- 20	20	1.6	251	3983
H 3–29	18	- 20	- 30				
K3–67	15	- 77	- 84				
A 7	750	+18	- 1		0.5	2805	
J 320	6	- 23	- 38	12	4.1	184	4860
A 8	60	+58	+51		2.6	1166	
K 2–1	132	+19	+10		1.4	1382	
IC 2120	47	- 8	- 16	15	1.5	527	11144
IC 418	12	+61	+43	12	0.8	71	1896
K 3–68	11	+65	+55				
K 1–7	34				3.0	762	

Planetary Nebulae Parameters

Name	PK Number l b	R.A.(1950) h m s	Dec.(1950) ° ′ ″	m_V (mag)	S_{6cm} (mJy)
H 3–75	193 – 09 1	05 37 56.1	+12 19 47		
K 3–69	170 +04 1	05 37 54.1	+39 13 39		
NGC 2022	196 – 10 1	05 39 22.0	+09 03 54	12.4	91
M 1–5	184 – 02 1	05 43 48.0	+24 21 00		80
Pu 1	181 +01 1	05 49 39.0	+28 05 48		
IC 2149	166 +10 1	05 52 40.8	+46 05 53	11.2	280
K 3–70	184 +00 1	05 55 40.2	+25 18 30		
A 12	198 – 06 1	05 59 37.7	+09 39 07		
A 13	204 – 08 1	06 02 08.0	+03 57 00	>16.0	7
A 14	197 – 03 1	06 08 21.1	+11 47 30	>18.2	
K 3–71	184 +04 1	06 10 47.4	+26 53 48		
VV 1–4	197 – 02 1	06 12 05.0	+12 22 22		
IC 2165	221 – 12 1	06 19 24.2	– 12 57 40	12.9	188
J 900	194 +02 1	06 23 01.8	+17 49 15	12.4	110
A 15	233 – 16 1	06 25 00.0	– 25 21 00	16.3	
M 1–6	211 – 03 1	06 33 11.0	– 00 03 11		
M 1–7	189 +07 1	06 34 17.8	+24 03 12		
A 17	221 – 04 1	06 46 12.0	– 09 29 00	>18.5	
K 2–2	204 +04 1	06 49 45.2	+10 01 30		54
M 1–8	210 +01 1	06 50 56.5	+03 12 11		
A 18	216 – 00 1	06 53 43.8	– 02 49 10		
A 19	200 +08 1	06 57 06.0	+14 41 00	17.0	
PB 1	226 – 03 1	07 00 28.6	– 13 38 24		
M 3–1	242 – 11 1	07 00 55.5	– 31 31 14		
M 1–9	212 +04 1	07 02 42.5	+02 51 35		
K 1–9	219 +01 1	07 04 47.6	– 05 05 13		
K 2–3	234 – 06 1	07 04 49.3	– 21 57 30		
NGC 2346	215 +03 1	07 06 49.2	– 00 43 24		86:
M 1–11	232 – 04 1	07 09 05.4	– 19 45 55		
K 2–10	229 – 02 1	07 10 20.0	– 16 00 36		
M 3–2	240 – 07 1	07 12 49.2	– 27 45 01		
M 1–12	235 – 03 1	07 17 12.0	– 21 38 17		
M 1–13	232 – 01 1	07 19 00.8	– 18 02 46		
A 20	214 +07 1	07 20 22.0	+01 51 26	16.3	
NGC 2371–2	189 +19 1	07 22 25	+29 35 26	13.0	87

Planetary Nebulae Parameters

Name	θ ($''$)	V_r (km s^{-1})	V_{LSR} (km s^{-1})	V_e (km s^{-1})	D (kpc)	$R/10^{10}$ (km)	T (years)
H 3–75	24	+23	+8				
K 3–69		+24					
NGC 2022	19	+14	− 1	29	2.2	312	3417
M 1–5	2	+37	+26	5			
Pu 1		+39					
IC 2149	10	− 31	− 36	10	0.7	52	1659
K 3-70	2	+25					
A 12	37				2.6	719	
A 13	156			40	1.8	2100	16645
A 14	34				2.8	712	
K 3–71	3	+16					
VV 1–4	127	+31			0.8	759	
IC 2165	8	+54	+35	20	1.9	113	1802
J 900	11	+47	+34	18	2.1	172	3042
A 15	34				3.1	788	
M 1–6	5				3.0	112	
M 1–7	9	+2	− 9	18	5.7	383	6757
A 17	43				3.7	1190	
K 2–2	414				1.0	3096	
M 1–8	22	+52	+36	11	3.5	575	16598
A 18	73	+52			1.7	928	
A 19	69				3.2	1651	
PB 1							
M 3–1	11	+70	+51		4.3	353	
M 1–9	2	+137	+121	10			
K 1–9	37				2.6	719	
K 2–3	63				2.4	1130	
NGC 2346	55	+20	+5		1.5	617	
M 1–11	2	+29	+10				
K 2–10	62				2.1	973	
M 3–2	8	+84	+65		3.2	191	
M 1–12	2						
M 1–13	10	+46	+28		4.5	336	
A 20	64				3.0	1436	
NGC 2371–2	44	+21	+11	45	1.5	493	3477

Planetary Nebulae Parameters

Name	PK Number l b	R.A.(1950) h m s	Dec.(1950) ° ′ ″	m_V (mag)	S_{6cm} (mJy)
M 3–3	221 +05 1	07 24 06.3	− 05 16 00		
M 1–14	235 − 01 1	07 25 46.0	− 20 06 58		
NGC 2392	197 +17 1	07 26 13.2	+21 00 51	9.9	251
A 21	205 +14 1	07 26 15.7	+13 20 44		387
M 4–2	248 − 08 1	07 27 06	− 35 39		
K 1–11	215 +11 1	07 33 30.0	+02 49 00	15.4	
M 1–16	226 +05 1	07 34 54.9	− 09 31 55		
M 1–17	228 +05 1	07 38 01.0	− 11 25 02		
NGC 2438	231 +04 2	07 39 32.5	− 14 36 56	10.1	83
NGC 2440	234 +02 1	07 39 41.5	− 18 05 26	10.8	410
M 1–18	231 +04 1	07 39 45.2	− 14 14 06		
A 23	249 − 05 1	07 41 26.9	− 34 38 00		
NGC 2452	243 − 01 1	07 45 24.7	− 27 12 43	12.6	57
HE 2–5	264 − 12 1	07 46 01.1	− 51 07 41		
K 1–12	236 +03 1	07 47 58.9	− 19 10 30		80
A 24	217 +14 1	07 49 01.6	+03 08 11	13.6	
M 3–4	241 +02 1	07 53 03.1	− 23 29 47		
NGC 2474	164 +31 1	07 54 00	+53 33 24	14.	
M 3–5	245 +01 1	08 00 25.2	− 27 33 31		
K 1–13	224 +15 1	08 04 14.3	− 02 44 01	15.4	
A 26	250 +00 1	08 07 03.8	− 32 31 24	18.1	
HE 2–7	264 − 08 1	08 10 02.1	− 48 34 13	12.4	
PB 2	263 − 05 1	08 19 03.3	− 46 10 39		
HE 2–9	258 − 00 1	08 26 38.0	− 39 13 41		
K 1–1	252 +04 1	08 29 51.9	− 31 55 54	>17.9	
NGC 2610	239 +13 1	08 31 05.0	− 15 58 39	13.6	
HE 2–11	259 +00 1	08 35 17	− 39 15 54		
M 3–6	254 +05 1	08 38 42.9	− 32 11 26		
A 30	208 +33 1	08 44 04	+18 03 35	15.6	
A 31	219 +31 1	08 51 30.0	+09 06 00	12.2	
HE 2–15	261 +02 1	08 51 37.7	− 39 52 09		
PB 3	269 − 03 1	08 52 43.0	− 50 20 51		
K 1–2	253 +10 1	08 55 42	− 28 45		
IC 2448	285 − 14 1	09 06 37.3	− 69 44 07	11.5	73
HE 2–18	273 − 03 1	09 07 08.0	− 53 06 55		

Planetary Nebulae Parameters

Name	θ (″)	V_r (km s^{-1})	V_{LSR} (km s^{-1})	V_e (km s^{-1})	D (kpc)	R/10^{10} (km)	T (years)
M 3–3	13	+95	+78	10	2.5	243	7706
M 1–14	5	+131	+113				
NGC 2392	45	+75	+63	54	1.1	370	2173
A 21	614	+29	+15	32	0.5	2296	22748
M 4–2	8				3.6	215	
K 1–11	200				2.1	3141	
M 1–16	3	+49	+32	20	5.5	123	1956
M 1–17	3	+100	+33	10	5.8	130	4125
NGC 2438	70	+74	+59	46	1.5	785	5412
NGC 2440	33	+63	+45	22	1.1	271	3912
M 1–18	32	+18	+1	26	2.1	502	6128
A 23	54				2.0	807	
NGC 2452	20	+65	+50		2.7	403	
HE 2–5	3				5.4	121	
K 1–12	37				3.9	1079	
A 24	356	+12	− 2	14	0.9	2396	54265
M 3–4	14	+74	+56		4.6	481	
NGC 2474	388	− 84	− 86	35	0.9	2612	23657
M 3–5	7	+64	+46		4.9	256	
K 1–13	180				2.1	2827	
A 26	40				2.0	598	
HE 2–7	13	+88			3.2	311	
PB 2	3				4.7	105	
HE 2–9	4				0.8	23	
K 1–1	43				1.9	611	
NGC 2610	39	+89	+72	67	2.6	758	3588
HE 2–11	65				1.2	583	
M 3–6	10	+50	+34		2.6	194	
A 30	128			40	3.1	2968	23521
A 31	972				0.5	3635	
HE 2–15	20				2.1	314	
PB 3	7				3.2	167	
K 1–2	64	+66	+51		3.3	1579	
IC 2448	9	− 24	− 37		2.9	195	
HE 2–18	10	+40	+25		3.1	231	

Planetary Nebulae Parameters

Name	PK Number l b	R.A.(1950) h m s	Dec.(1950) ° ′ ″	m_V (mag)	S_{6cm} (mJy)
NGC 2792	265 +04 1	09 10 33.7	- 42 13 08	13.5	114
HE 2-21	275 - 04 2	09 12 22.9	- 55 15 53		
PB 4	275 - 04 1	09 13 36.5	- 54 40 07		
NGC 2818	261 +08 1	09 13 59.4	- 36 24 58	13.0	33
PB 5	268 +02 1	09 14 21.0	- 45 16 12		
HE 2-25	275 - 03 1	09 16 30	- 54 27		
HE 2-26	278 - 06 1	09 18 06.4	- 58 59 23		
NGC 2867	278 - 05 1	09 20 00.8	- 58 05 57	9.7	252
HE 2-28	275 - 02 1	09 20 31.5	- 53 56 55		
HE 2-29	275 - 02 2	09 23 10.0	- 54 23 21		
NGC 2899	277 - 03 1	09 25 31.3	- 55 53 13		
HE 2-32	278 - 04 1	09 29 24	- 57 23		
A 33	238 +34 1	09 36 38.6	- 02 35 05	13.4	35
IC 2501	281 - 05 1	09 37 20.9	- 59 51 52	11.3	
HE 2-34	274 +02 1	09 39 24	- 49 09		
HE 2-35	274 +02 2	09 39 47.9	- 49 44 02		
HE 2-36	279 - 03 1	09 41 50.7	- 57 03 12		
A 34	248 +29 1	09 43 10	- 12 56 22	14.5	12
HE 2-37	274 +03 1	09 45 32.6	- 48 44 22		
HE 2-39	283 - 04 1	10 02 14	- 60 29 12		
NGC 3132	272 +12 1	10 04 55.1	- 40 11 29	8.2	228
HE 2-41	286 - 06 1	10 05 54.0	- 63 39 50		
IC 2553	285 - 05 1	10 07 47.9	- 62 21 55	13.0	
NGC 3195	296 - 20 1	10 10 06	- 80 37		
PB 6	278 +05 1	10 11 18.8	- 50 05 07		
HF 4	283 - 01 1	10 13 54	- 58 36		
NGC 3211	286 - 04 1	10 16 12.5	- 62 25 06	11.8	85
HE 2-47	285 - 02 1	10 21 24.0	- 60 17 22		
NGC 3242	261 +32 1	10 22 21.3	- 18 23 23	8.6	860
MY 60	283 +02 1	10 29 36.0	- 55 05 27		
HE 2-50	283 +03 1	10 32 18.0	- 53 25 27		
HE 2-51	288 - 05 1	10 34 02.3	- 64 03 30	14.2	
PE 1-1	285 +01 1	10 36 36	- 56 31		
PE 1-3	288 - 02 1	10 42 38.1	- 61 23 54		
HE 2-55	286 +02 1	10 46 42	- 55 47		

Planetary Nebulae Parameters

Name	θ ($''$)	V_r (km s^{-1})	V_{LSR} (km s^{-1})	V_e (km s^{-1})	D (kpc)	R/10^{10} (km)	T (years)
NGC 2792	13	+14	- 1		1.8	175	
HE 2-21	2				7.2	107	
PB 4	11				2.9	238	
NGC 2818	49	- 1	- 17		2.3	842	
PB 5							
HE 2-25	5						
HE 2-26	3				4.7	105	
NGC 2867	17	+12			1.6	203	
HE 2-28	10				4.7	351	
HE 2-29	14	+25			4.0	418	
NGC 2899	90	+3	- 10	15	1.5	1009	21340
HE 2-32	35				2.2	575	
A 33	268	+60	+50	31	1.6	3207	32798
IC 2501	2	+33	+20		2.0	29	
HE 2-34							
HE 2-35	3				5.9	132	
HE 2-36	10	- 7	- 20		2.7	201	
A 34	274				1.4	2869	
HE 2-37	23	+12			2.3	395	
HE 2-39	10	- 23	- 35		3.4	254	
NGC 3132	54	- 10	- 29	14	1.1	444	10060
HE 2-41	10				3.6	269	
IC 2553	9	+37	+25		2.8	188	
NGC 3195	39	- 7	- 20		2.4	700	
PB 6	12				4.0	359	
HF 4	21				1.9	298	
NGC 3211	16	- 22	- 34		2.5	299	
HE 2-47	3				2.0	44	
NGC 3242	40	+5	- 5	20	0.8	239	3793
MY 60	7				3.4	178	
HE 2-50	12	+79	+68		2.8	251	
HE 2-51	12	+8		10	3.0	269	8536
PE 1-1	3				2.2	49	
PE 1-3	8				4.7	281	
HE 2-55	18				2.6	350	

Planetary Nebulae Parameters

Name	PK Number l b	R.A.(1950) h m s	Dec.(1950) ° ′ ″	m_V (mag)	S_{6cm} (mJy)
HE 2–57	289 – 01 1	10 54 00	– 61 11		
IC 2621	291 – 04 1	10 58 23.5	– 64 58 47		
NGC 3587	148 +57 1	11 11 54	+55 17	12.0	180
HE 2–63	289 +07 1	11 21 40.8	– 52 34 52		
K 1–22	283 +25 1	11 24 18	– 34 05 44		
HE 2–64	291 +03 1	11 25 18	– 57 01 24		
NGC 3699	292 +01 1	11 25 42.0	– 59 41 00		
FG 1	290 +07 1	11 26 14.6	– 52 39 34		
HE 2–67	292 +01 2	11 26 30.5	– 59 50 00		
HE 2–68	294 – 04 1	11 29 31.8	– 65 41 40		
PB 8	292 +04 1	11 30 57.5	– 56 49 43		
HE 2–71	296 – 06 1	11 36 54.1	– 68 35 30		
HE 2–73	296 – 03 1	11 46 12.8	– 64 51 53		
NGC 3918	294 +04 1	11 47 50.1	– 56 54 10	8.4	850
BLDZ 1	293 +10 1	11 50 33.0	– 50 34.		
NGC 4071	298 – 04 1	12 01 39.5	– 67 01 53		
HE 2–76	298 – 01 2	12 05 48	– 63 55		
HE 2–77	298 – 00 1	12 06 23.8	– 62 59 20		
HE 2–78	297 +03 1	12 06 36	– 58 26		
K 2–4	275 +72 1	12 15 45.7	+11 19 47		
HE 2–81	299 – 01 1	12 20 12	– 63 45		
HE 2–82	299 +02 1	12 21 06	– 59 57		
NGC 4361	294 +43 1	12 21 55.0	– 18 30 32	10.3	205
HE 2–84	300 – 00 1	12 26 00	– 63 27		
HE 2–86	300 – 02 1	12 27 38.7	– 64 34 35		
IC 3568	123 +34 1	12 31 47	+82 50 22	11.6	95
A 35	303 +40 1	12 51 01.3	– 22 35 26	12.0	255
H 4–1	049 +88 1	12 57 02.7	+27 54 24	16.0	
HE 2–88	304 +05 1	13 02 42	– 57 23		
IC 4191	304 – 04 1	13 05 28.0	– 67 22 33	12.0	
HE 2–90	305 +01 1	13 06 30	– 61 04		
TH 2–A	306 – 00 1	13 19 15.3	– 63 05 15		
NGC 5189	307 – 03 1	13 29 59.5	– 65 43 00	10.3	455
MY CN 18	307 – 04 1	13 35 54.4	– 67 07 33	12.2	
A 36	318 +41 1	13 37 57.8	– 19 37 33	13.0	215

Planetary Nebulae Parameters

Name	θ ($''$)	V_r (km s^{-1})	V_{LSR} (km s^{-1})	V_e (km s^{-1})	D (kpc)	R/10^{10} (km)	T (years)
HE 2–57	20				2.2	329	
IC 2621	5	+14	+3		2.1	78	
NGC 3587	202	+6	+12	44	0.7	1057	7620
HE 2–63	3				7.8	175	
K 1–22	182	– 22		28			
HE 2–64	8	+72	+63		4.6	275	
NGC 3699	45	– 22	– 16	29	2.0	673	7358
FG 1	27	+29	+19		2.4	484	
HE 2–67	5				4.3	160	
HE 2–68	3				5.1	114	
PB 8	5				5.1	190	
HE 2–71	3				7.8	175	
HE 2–73	4				3.4	101	
NGC 3918	14	– 17	– 25	20	0.9	94	1493
BLDZ 1	82	– 17					
NGC 4071	63	+11	+2	19	2.4	1130	18869
HE 2–76	18				1.6	215	
HE 2–77	25				0.6	112	
HE 2–78	3				3.8	85	
K 2–4	690				0.6	3096	
HE 2–81	6				0.9	40	
HE 2–82	24				2.0	359	
NGC 4361	114	+9	+8	38	0.9	767	6402
HE 2–84	28	– 44	– 51		1.6	335	
HE 2–86	4				2.6	77	
IC 3568	18	– 41	– 32	8	2.1	282	11203
A 35	900	– 7	– 7	4	0.5	3366	266755
H 4–1	6	– 141	– 133	22	12.7	569	8212
HE 2–88							
IC 4191	14	– 13	– 20	14	2.0	209	4742
HE 2–90							
TH 2–A	24	– 45			2.5	448	
NGC 5189	140	– 9	– 15	36	0.7	733	6454
MY CN 18	6	– 55	– 61	10	2.8	125	3983
A 36	478	+37	+40	39	0.6	2145	17437

Planetary Nebulae Parameters

Name	PK Number l b	R.A.(1950) h m s	Dec.(1950) ° ′ ″	m_V (mag)	S_{6cm} (mJy)
HE 2–97	307 – 09 1	13 41 24.0	– 71 13 47		
NGC 5307	312 +10 1	13 47 51.6	– 50 57 26	12.1	
HE 2–99	309 – 04 1	13 48 46.3	– 66 08 37		
NGC 5315	309 – 04 2	13 50 12.7	– 66 16 06	13.0	442
HE 2–101	311 +03 1	13 51 36	– 58 11		
HE 2–102	311 +02 1	13 54 45.9	– 58 39 54		
IC 972	326 +42 1	14 01 41.8	– 16 59 13	14.9	
HE 2–103	310 – 02 1	14 01 50.9	– 64 26 37		
HE 2–104	315 +09 1	14 08 33.5	– 51 12 19		
HE 2–105	308 – 12 1	14 10 43.4	– 73 58 52		
HE 2–108	316 +08 1	14 14 47.5	– 51 56 50		
HE 2–107	312 – 01 1	14 14 55.1	– 62 53 22		
IC 4406	319 +15 1	14 19 15.5	– 43 55 27	10.6	
HE 2–111	315 – 00 1	14 29 31.4	– 60 36 33		
HE 2–112	319 +06 1	14 37 00.7	– 52 22 00		
HE 2–114	318 – 02 1	15 00 09.7	– 60 41 39		
HE 2–115	321 +02 1	15 01 34.2	– 54 59 34		
HE 2–116	318 – 02 2	15 01 59.4	– 61 09 48		
HE 2–117	321 +02 2	15 02 14.5	– 55 47 45		
HE 2–118	327 +13 1	15 02 55.2	– 42 48 24		
HE 2–119	317 – 05 1	15 06 23.1	– 64 28 57		
HE 2–120	321 +01 1	15 08 10.4	– 55 28 34		
NGC 5873	331 +16 1	15 09 38.0	– 37 56 16	13.3	
NGC 5882	327 +10 1	15 13 24.9	– 45 27 56	10.5	334
HE 2–123	323 +02 1	15 18 35.2	– 53 57 34		
ME 2–1	342 +27 1	15 19 23.0	– 23 27 05		38
PE 2–8	322 – 00 1	15 19 48.9	– 56 58 39		
HE 2–125	324 +02 1	15 19 51.7	– 53 40 46		
HE 2–127	325 +04 2	15 21 18.3	– 51 39 10		
HE 2–128	325 +04 1	15 21 29.7	– 51 09 08		
HE 2–129	325 +03 1	15 21 50.6	– 52 40 11		
MZ 1	322 – 02 1	15 30 13.8	– 58 58 57	12.5	
HE 2–131	315 – 13 1	15 31 54.0	– 71 45 00		325
HE 2–132	323 – 02 1	15 33 57.9	– 58 35 02		
HE 2–133	324 – 01 1	15 38 00.9	– 56 27 11		

Planetary Nebulae Parameters

Name	θ ('')	V_r (km s^{-1})	V_{LSR} (km s^{-1})	V_e (km s^{-1})	D (kpc)	R/10^{10} (km)	T (years)
HE 2–97	5				4.9	183	
NGC 5307	13	+40	+37	15	2.3	223	4726
HE 2–99	17	- 2	- 8		4.2	534	
NGC 5315	5	- 34	- 39	40	1.2	44	355
HE 2–101							
HE 2–102	9				4.0	269	
IC 972	44	- 27	- 21	16	3.8	1250	24778
HE 2–103	20	- 30	- 26		4.0	598	
HE 2–104	5	- 105	- 107		6.3	235	
HE 2–105	31				3.6	834	
HE 2–108	11	- 8	- 10	12	3.9	320	8476
HE 2–107	10				3.1	231	
IC 4406	35	- 41	- 40	6	1.7	445	23514
HE 2–111	12	- 11	- 13	22	2.8	251	3621
HE 2–112	14				2.5	261	
HE 2–114	30	- 37	- 39		4.1	920	
HE 2–115	3	- 63	- 64		2.0	44	
HE 2–116	51				1.1	419	
HE 2–117	5	- 29	- 30		1.7	63	
HE 2–118	3	- 164	- 162		8.4	188	
HE 2–119	50	- 11	- 14		1.7	635	
HE 2–120	27	- 20	- 15		3.2	646	
NGC 5873	5	- 128	- 124		3.8	142	
NGC 5882	14	+10	+12		1.6	167	
HE 2–123	5	- 12	- 12		2.9	108	
ME 2–1	7	+46	+54	13	4.0	209	5107
PE 2–8	2				2.9	43	
HE 2–125	3	- 27	- 27		1.8	40	
HE 2–127	3	- 57	- 56		9.1	204	
HE 2–128	3	- 79	- 78		4.7	105	
HE 2–129	1				10.3	77	
MZ 1	25	- 33	- 32		2.4	448	
HE 2–131	6	- 1	- 6		1.6	71	
HE 2–132	18	- 131	- 132		3.7	498	
HE 2–133	4				1.7	50	

Planetary Nebulae Parameters

Name	PK Number l b	R.A.(1950) h m s	Dec.(1950) ° ′ ″	m_V (mag)	S_{6cm} (mJy)
NGC 5979	322 − 05 1	15 43 26.0	− 61 03 48		
CN 1–1	330 +04 1	15 47 38.5	− 48 36 00		
HE 2–136	322 − 06 1	15 47 47.9	− 62 21 54		
SP 1	329 +02 1	15 47 56.8	− 51 22 24	13.6	
HE 2–138	320 − 09 1	15 51 19.2	− 66 00 26		
HE 2–140	327 − 01 2	15 54 11.4	− 55 33 17		
HE 2–141	325 − 04 1	15 55 02.3	− 58 15 19		
HE 2–142	327 − 02 1	15 55 59.5	− 55 46 57		
HE 2–143	327 − 01 1	15 57 03.3	− 54 57 14		
NGC 6026	341 +13 1	15 58 07.4	− 34 24 16		
NGC 6058	064 +48 1	16 02 42	+40 49	13.3	10
HE 2–145	331 +00 1	16 05 13.0	− 50 54 08		
HE 2–146	328 − 02 1	16 06 43.1	− 54 49 44		
IC 4593	025 +40 1	16 09 23.3	+12 12 08	10.9	98
NGC 6072	342 +10 1	16 09 41.6	− 36 06 01	14.1	164
HE 2–147	327 − 04 1	16 09 54	− 56 52		
H 1–1	343 +11 1	16 10 12	− 34 28		
HE 2–149	329 − 02 1	16 10 26.6	− 54 40 06		
MZ 2	329 − 02 2	16 10 33.5	− 54 49 31	12.6	
HE 2–151	326 − 06 1	16 11 25.4	− 59 46 34		
HE 2–152	333 +01 1	16 11 37.4	− 49 05 56		
HE 2–153	330 − 02 1	16 13 19.3	− 53 24 41		
MZ 3	331 − 01 1	16 13 23.3	− 51 51 44		
HE 2–155	338 +05 1	16 15 54.7	− 42 08 23		
HE 2–157	331 − 02 1	16 18 17.1	− 53 33 53		
SN 1	013 +32 1	16 18 30.2	− 00 09 13		
HE 2–158	327 − 06 1	16 19 18.7	− 58 12 26		
K 1–3	346 +12 1	16 20 06.0	− 31 38 00	>15.7	
PE 1–6	336 +01 1	16 20 16.5	− 46 35 17		
HE 2–159	330 − 03 1	16 20 21.9	− 54 29 09		
HE 2–161	331 − 02 2	16 20 41.7	− 53 15 41		
HE 2–162	331 − 03 1	16 23 53.6	− 53 54 47		
HE 2–163	327 − 07 1	16 25 13.9	− 59 02 48		
HE 2–164	332 − 03 1	16 25 56.8	− 53 16 32		
HE 2–165	331 − 03 2	16 26 00.8	− 54 03 05		

Planetary Nebulae Parameters

Name	θ ($''$)	V_r (km s^{-1})	V_{LSR} (km s^{-1})	V_e (km s^{-1})	D (kpc)	$R/10^{10}$ (km)	T (years)
NGC 5979	8	+23	+21		2.6	155	
CN 1–1							
HE 2–136	10	- 135	- 137		4.5	336	
SP 1	72	- 31	- 31		1.6	861	
HE 2–138	7	- 47	- 49		3.1	162	
HE 2–140	3	- 60	- 60		2.9	65	
HE 2–141	14	- 46	- 46		2.8	293	
HE 2–142	4	- 73	- 73		3.7	110	
HE 2–143	5	- 35	- 34		2.7	100	
NGC 6026	54	- 103	- 105		1.9	767	
NGC 6058	26	+3	+20	28	4.1	797	9027
HE 2–145	10				1.2	89	
HE 2–146	22	+62	+58		1.7	279	
IC 4593	12	+22	+39	13	2.4	215	5253
NGC 6072	70	+7	+13		1.3	680	
HE 2–147							
H 1–1	2				8.8	131	
HE 2–149	3	- 113	- 112		8.4	188	
MZ 2	23	- 30	- 28	19	2.3	395	6601
HE 2–151	3	- 128	- 129		8.4	188	
HE 2–152	11	- 64	- 61		2.0	164	
HE 2–153	13	- 40	- 39		1.2	116	
MZ 3	15	- 21	- 19		1.0	112	
HE 2–155	14	- 27	- 22		2.7	282	
HE 2–157	3	- 70	- 68		5.4	121	
SN 1		- 87	- 72				
HE 2–158	2	- 39	- 39		10.0	149	
K 1–3	100	- 12	- 4	17	3.0	2244	41844
PE 1–6	7	- 76	- 72		4.0	209	
HE 2–159	10	- 89	- 88		4.3	321	
HE 2–161	10	- 98	- 96		4.0	299	
HE 2–162	2	+88	+85		5.6	83	
HE 2–163	20	- 45	- 45		3.1	463	
HE 2–164	16	- 77	- 75		2.3	275	
HE 2–165	50	- 18	- 16		1.6	598	

Planetary Nebulae Parameters

Name	PK Number l b	R.A.(1950) h m s	Dec.(1950) ° ′ ″	m_V (mag)	S_{6cm} (mJy)
PE 1–7	337 +01 1	16 26 48.1	– 45 56 22		
NGC 6153	341 +05 1	16 28 05.0	– 40 08 58	11.5	477
NGC 6164	336 – 00 1	16 30 09.5	– 48 00 24		
HE 2–169	335 – 01 1	16 30 29.0	– 49 15 06		
HE 2–171	346 +08 1	16 30 48	– 34 59		
HE 2–170	332 – 04 1	16 31 22.9	– 53 43 59		
PC 11	331 – 05 1	16 33 37.1	– 55 36 25		
HE 2–175	345 +06 1	16 36 06	– 36 29		
VD 1–1	344 +04 1	16 39 12	– 38 49		
PC 12	000 +17 1	16 40 54	– 18 51.		
NGC 6210	043 +37 1	16 42 23.5	+23 53 17	9.3	260
VD 1–2	345 +04 1	16 43 18	– 38 32		
H 1–2	347 +05 1	16 45 36	– 35 42		
A40	359 +15 1	16 45 35.8	– 20 55 26	16.8	
VD 1–3	344 +03 1	16 46 06	– 39 15		
VD 1–4	345 +03 1	16 47 00	– 39 03		
VD 1–5	344 +02 1	16 47 12	– 39 58		
HE 2–182	325 – 12 1	16 49 49.3	– 64 09 39		
H 1–3	342 +00 1	16 50 00	– 42 34		
VD 1–6	345 +03 2	16 50 48	– 38 40		
H 1–5	344 +00 1	16 53 54	– 41 33		
HE 2–186	336 – 05 1	16 55 40.5	– 51 37 36		
HE 2–185	321 – 16 1	16 55 45.4	– 70 01 40		
M 2–4	349 +04 1	16 57 47.4	– 34 45 18		
HE 2–187	337 – 05 1	16 57 45.9	– 50 18 36		
IC 4634	000 +12 1	16 58 34.6	– 21 45 28	10.7	140
M 2–5	351 +05 1	16 59 03.0	– 33 05 48		
M 1–19	351 +04 1	17 00 30.6	– 33 25 36		
M 2–6	353 +06 2	17 01 06.0	– 30 49 21		
H 2–1	350 +04 1	17 01 19.4	– 33 55 05		
IC 4637	345 +00 1	17 01 39.2	– 40 48 52	13.6	
VD 1–8	347 +01 1	17 01 48	– 37 49		
M 2–7	353 +06 1	17 02 01.8	– 30 28 12		
M 2–8	352 +05 1	17 02 15.6	– 32 28 06		
M 2–9	010 +18 2	17 02 52.8	– 10 04 24		

Planetary Nebulae Parameters

Name	θ ($''$)	V_r (km s^{-1})	V_{LSR} (km s^{-1})	V_e (km s^{-1})	D (kpc)	R/10^{10} (km)	T (years)
PE 1–7	2	- 33	- 29		2.6	38	
NGC 6153	23	+37	+45		1.0	172	
NGC 6164	340	- 52			0.2	508	
HE 2–169	8				2.5	149	
HE 2–171		- 88	- 76				
HE 2–170	3	+64	+65		7.1	159	
PC 11	3				6.2	139	
HE 2–175	7	- 25			6.4	335	
VD 1–1		- 142	- 136				
PC 12	5	- 44	- 32		6.1	228	
NGC 6210	16	- 36	- 18	21	1.3	155	2348
VD 1–2							
H 1–2		- 102	- 94				
A 40	32				4.7	1124	
VD 1–3							
VD 1–4	5						
VD 1–5							
HE 2–182	3	- 91	- 92		3.8	85	
H 1–3	15				1.4	157	
VD 1–6	30						
H 1–5	5				0.9	33	
HE 2–186	3	- 67	- 64		6.2	139	
HE 2–185	10	- 6	- 9		4.9	366	
M 2–4	2	- 184	- 176				
HE 2–187	6	+58			3.2	143	
IC 4634	8	- 33	- 21	15	2.5	149	3161
M 2–5	5	- 98	- 89		1.5	56	
M 1–19	3	- 40	- 31				
M 2–6	2	- 89	- 79				
H 2–1	5	- 20	- 12		3.4	127	
IC 4637	14	- 15	+18		1.4	146	
VD 1–8							
M 2–7	8	- 56	-47		4.4	263	
M 2–8	4	+27	+36		3.1	92	
M 2–9	17	+88	+103		3.3	419	

Planetary Nebulae Parameters

Name	PK Number l b	R.A.(1950) h m s	Dec.(1950) ° ′ ″	m_V (mag)	S_{6cm} (mJy)
PC 14	336 − 06 1	17 02 16.0	− 52 25 55		
H 1–6	344 − 01 1	17 03 24	− 42 37		
H 1–7	345 − 01 1	17 06 54	− 41 49		
M 4–3	357 +07 1	17 07 34.7	− 27 05 03		
IC 4642	334 − 09 1	17 07 37.2	− 55 20 24	13.4	
M 3–36	358 +07 1	17 09 33.6	− 25 40 06		
NA 1	018 +20 1	17 10 13.8	− 03 12 27	13.4	
NGC 6302	349 +01 1	17 10 21.1	− 37 02 38	12.8	3488
M 2–10	354 +04 1	17 10 53.5	− 31 16 16		
NGC 6309	009 +14 1	17 11 14.9	− 12 51 11	10.8	151
H 1–8	352 +03 2	17 11 26.0	− 33 21 23		
TH 3–3	356 +05 1	17 14 06	− 28 56		
TH 3–4	354 +03 1	17 15 36	− 31 36		
HE 2–207	342 − 04 1	17 15 48	− 45 50		
TH 3–6	355 +03 3	17 16 06	− 31 08		
M 3–37	359 +06 1	17 16 08.4	− 25 14 12		
NGC 6326	338 − 08 1	17 16 49.1	− 51 42 20	12.2	70
M 2–11	356 +04 1	17 17 23.1	− 28 57 40		
M 3–38	356 +04 2	17 17 54.0	− 29 00 03		
M 3–39	358 +05 1	17 18 04.1	− 27 08 32		
H 1–9	355 +03 2	17 18 18	− 30 18		
H 1–11	002 +08 1	17 18 16.8	− 22 15 36		
NGC 6337	349 − 01 1	17 18 50.0	− 38 26 06		
M 3–40	358 +05 2	17 19 20.8	− 27 05 45		
M 2–12	359 +05 1	17 20 55.6	− 25 56 40		
M 3–7	357 +03 1	17 21 23.6	− 29 21 33		
M 3–8	358 +04 1	17 21 43.2	− 28 03 15		
TH 3–12	356 +03 1	17 21 54	− 29 43		
M 3–41	357 +03 2	17 22 48.6	− 29 19 18		
M 3–9	359 +05 2	17 22 37.2	− 26 09 18		
CN 1–3	345 − 04 1	17 22 42	− 44 09		
H 1–12	352 +00 1	17 23 06	− 34 59		
M 3–42	357 +03 4	17 23 49.2	− 29 13 00		
M 3–10	358 +03 1	17 24 11.0	− 28 25 22		
CN 1–4	342 − 06 1	17 24 12	− 46 53		

Planetary Nebulae Parameters

Name	θ ($''$)	V_r (km s^{-1})	V_{LSR} (km s^{-1})	V_e (km s^{-1})	D (kpc)	R/10^{10} (km)	T (years)
PC 14	7	- 49	- 47		4.5	235	
H 1-6	12				1.4	125	
H 1-7	9						
M 4-3	2	+156	+166				
IC 4642	15	+44	+46		2.7	302	
M 3-36	4	+12	+23		5.9	176	
NA 1	6	+22	+38				
NGC 6302	45	- 39	- 31	10	0.4	134	4268
M 2-10	4	- 75	- 66		3.9	116	
NGC 6309	14	- 48	- 33	34	2.1	219	2050
H 1-8	8				1.4	83	
TH 3-3	17						
TH 3-4							
HE 2-207		- 33	- 33				
TH 3-6							
M 3-37	10	- 67	- 56		9.8	733	
NGC 6326	12	+9	+13	17	2.5	224	4184
M 2-11	10	+78	+88		5.1	381	
M 3-38	25	- 156	- 146		3.6	673	
M 3-39	10	+4	+13		1.5	112	
H 1-9	1	- 143	- 138				
H 1-11	5	+2	+14		6.7	250	
NGC 6337	47	- 71	- 67		1.7	597	
M 3-40	10	+37	+48		4.1	306	
M 2-12	5	+73	+84		5.7	213	
M 3-7	7	- 191	- 181		4.6	240	
M 3-8	6	+95	+105		4.9	219	
TH 3-12	1						
M 3-41	4	- 69	- 59		3.4	101	
M 3-9	18	- 82	- 71		3.4	457	
CN 1-3		- 78	- 72				
H 1-12	7				0.9	47	
M 3-42	5	- 250	- 240		4.0	149	
M 3-10	3	- 96	- 86		6.1	136	
CN 1-4		- 88	- 83				

Planetary Nebulae Parameters

Name	PK Number l b	R.A.(1950) h m s	Dec.(1950) ° ′ ″	m_V (mag)	S_{6cm} (mJy)
H 2–10	358 +03 2	17 24 23.4	– 28 28 42		
H 1–14	001 +05 1	17 25 00	– 24 22 54		
H 1–13	352 – 00 1	17 25 06	– 35 05		
H 1–15	001 +05 2	17 25 36	– 24 49		
M 4–4	357 +02 5	17 25 38.4	– 30 05 24		
M 2–13	011 +11 1	17 25 44.6	– 13 23 49		
M 1–20	006 +08 1	17 26 00.7	– 19 13 31		
A 41	009 +10 1	17 26 12	– 15 12		
NGC 6369	002 +05 1	17 26 17.9	– 23 43 12	12.9	1950
H 2–11	000 +04 1	17 26 18	– 25 47		
H 1–16	000 +04 2	17 26 18	– 26 24		
H 1–17	358 +03 7	17 26 30.6	– 28 38 06		
H 1–18	357 +02 4	17 26 32.0	– 29 30 30		
H 1–19	358 +03 4	17 26 53.4	– 27 57 00		
H 1–20	358 +03 6	17 27 34.8	– 28 01 51		
TH 3–24	357 +02 7	17 27 36	– 30 15		
TH 3–25	359 +03 2	17 27 42	– 27 04		
TH 3–55	356 +01 2	17 27 48	– 30 59		
H 2–13	357 +02 6	17 27 54	– 30 08		
TH 3–26	358 +03 8	17 28 00	– 28 13		
H 1–22	350 – 02 1	17 28 54	– 37 55		
H 2–14	349 – 03 1	17 29 24	– 39 50		
H 1–23	357 +01 1	17 29 35	– 29 58 09		
H 1–24	004 +06 2	17 30 37.5	– 21 44 16		
M 1–21	006 +07 1	17 31 20.5	– 19 07 23		
H 2–15	003 +05 1	17 31 25.0	– 22 51 22		
HE 2–250	000 +03 1	17 31 48.0	– 26 34 06		
PC 17	343 – 07 1	17 31 56.3	– 46 57 57		
M 4–6	358 +01 1	17 32 03.6	– 29 01 18		
HE 2–248	341 – 09 1	17 32 16.3	– 49 23 43		
M 3–11	005 +06 1	17 32 22.2	– 20 55 24		
TH 3–27	002 +04 1	17 32 54.6	– 24 23 30		
H 1–26	350 – 03 1	17 33 00	– 39 20		
M 3–12	005 +05 1	17 33 22.8	– 21 29 24		
M 1–23	007 +06 1	17 34 25.8	– 18 45 00		

Planetary Nebulae Parameters

Name	θ ($''$)	V_r (km s^{-1})	V_{LSR} (km s^{-1})	V_e (km s^{-1})	D (kpc)	$R/10^{10}$ (km)	T (years)
H 2–10	2	+50	+60				
H 1–14	7	+34	+46		6.6	345	
H 1–13	10				0.8	59	
H 1–15	5				5.6	209	
M 4–4	7	+29	+39		3.0	157	
M 2–13	2	+86	+100				
M 1–20		+92	+105				
A 41	18				5.4	727	
NGC 6369	29	- 101	- 90		0.6	130	
H 2–11	3						
H 1–16	2	+54	+65				
H 1–17	1	+1	+11				
H 1–18	1	- 204	- 194				
H 1–19	1	+65	+76				
H 1–20	4	+186	+197		4.1	122	
TH 3–24							
TH 3–25	2	- 93	- 82				
TH 3–55							
H 2–13							
TH 3–26	6	+204	+215		3.2	143	
H 1–22		- 213	- 205				
H 2–14	20						
H 1–23	3	- 72	- 62		0.8	17	
H 1–24	8	+160	+172		3.5	209	
M 1–21		+141	+154				
H 2–15	5	- 59	- 47		7.5	280	
HE 2–250	5	- 200	- 189		2.7	100	
PC 17	5				4.8	179	
M 4–6	2	- 292	- 282				
HE 2–248	5				6.7	250	
M 3–11	7				3.1	162	
TH 3–27	10						
H 1-26	18	- 37	- 30		2.5	336	
M 3–12	6	+19	+30		3.1	139	
M 1–23	7	- 65	- 52		3.7	193	

Planetary Nebulae Parameters

Name	PK Number l b	R.A.(1950) h m s	Dec.(1950) ° ′ ″	m_V (mag)	S_{6cm} (mJy)
M 1–24	007 +06 2	17 35 13.8	− 19 36 00		
M 1–25	004 +04 1	17 35 30.2	− 22 07 05		
FG 2	346 − 06 1	17 35 41.4	− 44 08 00		
HE 2–260	008 +06 1	17 36 01.5	− 18 15 57		
H 2–16	005 +05 2	17 36 55.0	− 21 12 32		
H 2–17	003 +03 1	17 37 03.0	− 24 24 11		
HE 2–262	001 +02 1	17 37 06	− 26 43		
H 1–27	005 +04 1	17 37 17.0	− 22 17 45		
M 3–13	005 +04 2	17 38 36	− 22 12		
HB 4	003 +02 1	17 38 48.4	− 24 40 34		
M 2–14	003 +03 2	17 38 53.4	− 24 09 48		
H 1–28	350 − 05 1	17 39 24	− 39 35		
K 1–14	045 +24 1	17 40 28.9	+21 28 17		
H 2–18	006 +04 1	17 40 29.3	− 21 08 33		
H 1–29	355 − 02 2	17 40 54	− 34 16		
M 3–14	355 − 02 1	17 41 01.6	− 34 05 25		
H 1–30	352 − 04 1	17 41 42	− 38 08		
IC 4663	346 − 08 1	17 41 48.4	− 44 53 00	13.1	
TC 1	345 − 08 1	17 41 52.6	− 46 04 10		
H 1–31	355 − 02 4	17 42 12.6	− 34 32 42		
M 3–15	006 +04 2	17 42 32.4	− 20 56 52		
H 2–20	002 +01 1	17 42 35.6	− 25 36 42		
H 1–32	355 − 02 3	17 42 47.4	− 34 02 36		
M 1–26	358 − 00 2	17 42 45.0	− 30 11 02		
K 1–15	051 +25 1	17 42 57.8	+27 21 17		
TH 4–1	007 +04 1	17 43 18	− 20 12		
TH 4–2	008 +05 1	17 43 18	− 18 39		
M 1–27	356 − 02 2	17 43 28.2	− 33 07 30		
M 2–15	011 +06 1	17 44 01.0	− 16 16 20		
H 1–33	355 − 03 1	17 44 30	− 34 07		
H 2–22	006 +03 1	17 44 36	− 21 46		
M 1–28	006 +03 2	17 44 37.4	− 22 05 19		
HB 5	359 − 00 1	17 44 44.5	− 29 58 53	13.6	
H 1–34	005 +02 1	17 45 06	− 22 46		
NGC 6439	011 +05 1	17 45 26.0	− 16 27 44	13.8	53

Planetary Nebulae Parameters

Name	θ ($''$)	V_r (km s^{-1})	V_{LSR} (km s^{-1})	V_e (km s^{-1})	D (kpc)	R/10^{10} (km)	T (years)
M 1–24	6	- 7	+6		3.8	170	
M 1–25	5	+26	+38		3.6	134	
FG 2		+35	+41				
HE 2–260	1						
H 2–16	16	- 56	- 44		5.4	646	
H 2–17	4	+87	+94		7.2	215	
HE 2–262	4	- 172	- 161		1.5	44	
H 1–27	7	+17	+29		4.7	246	
M 3–13							
HB 4	6	- 63	- 51	23	2.2	98	1360
M 2–14		- 35	- 23				
H 1–28	8	- 45	- 38		1.7	101	
K 1–14	47				5.3	1863	
H 2–18	4	- 116	- 104		3.9	116	
H 1–29		- 1	- 9				
M 3–14	7	- 76	- 67		1.6	83	
H 1–30	5	- 8	- 0		2.2	82	
IC 4663	14	- 49	- 48	15	3.2	335	7081
TC 1	10	- 88	- 78		1.0	74	
H 1–31	1	+86	+95				
M 3–15	4	+100	+113		1.5	44	
H 2–20	4				1.0	29	
H 1–32		- 18	- 179				
M 1–26	4	- 5	+5		1.4	41	
K 1–15	43				7.2	2315	
TH 4–1	10	- 108	- 95				
TH 4–2	19	+44	+57		2.5	355	
M 1–27	5	- 54					
M 2–15	6	+4	+17		5.4	242	
H 1–33		- 117	- 108				
H 2–22	6				2.5	112	
M 1–28	26	+18	+30		3.5	680	
HB 5	20	- 28	- 18		1.2	179	
H 1–34							
NGC 6439	5	- 94	- 80	24	3.8	142	1877

Planetary Nebulae Parameters

Name	PK Number l b	R.A.(1950) h m s	Dec.(1950) ° ′ ″	m_V (mag)	S_{6cm} (mJy)
H 2–24	004 +01 1	17 45 32.5	– 24 15 39		
TH 4–3	006 +02 1	17 45 36.0	– 22 15 53		
H 2–23	355 – 03 2	17 45 36	– 34 21		
H 1–35	355 – 03 3	17 45 54.6	– 34 21 59		
H 2–25	004 +02 1	17 45 57.5	– 23 42 00		
NGC 6445	008 +03 1	17 46 17.2	– 19 59 41	13.2	368
H 1–36	353 – 04 1	17 46 24	– 37 01		
H 2–26	356 – 03 1	17 46 36	– 34 01		
M 1–29	359 – 01 1	17 47 04.8	– 30 34 06		
M 3–43	000 – 01 1	17 47 12	– 29 24.		
H 1–38	353 – 05 1	17 47 18	– 37 23		
H 1–37	351 – 06 1	17 47 18	– 39 17		
TH 4–4	008 +03 2	17 47 30	– 19 52		
TH 4–5	009 +04 1	17 47 31.6	– 19 02 24		
HF 2–1	355 – 04 1	17 47 51.9	– 34 54 40		
TH 4–6	009 +04 2	17 48 00.8	– 18 46 00		
M 3–44	359 – 01 2	17 48 06.0	– 30 23 12		
H 2–27	356 – 03 2	17 48 30	– 33 47		
M 3–45	359 – 01 3	17 48 53.4	– 30 04 36		
M 2–17	010 +04 1	17 49 09.6	– 17 35 34		
M 2–16	357 – 03 2	17 49 17.7	– 32 45 12		
TH 4–7	006 +02 3	17 49 22.0	– 21 50 33		
B1 3–15	000 – 01 2	17 49 24	– 29 06		
M 3–16	359 – 02 2	17 49 32.4	– 30 49 00		
M 1–30	355 – 04 2	17 49 39.0	– 34 37 48		
M 1–31	006 +02 5	17 49 40.2	– 22 21 18		
TH 4–8	007 +02 1	17 49 42.0	– 21 14 00		
H 2–29	357 – 03 3	17 50 00	– 32 40		
H 1–39	356 – 03 3	17 50 02.4	– 33 55 24		
M 2–18	357 – 03 4	17 50 21.0	– 32 58 12		
M 2–19	000 – 01 5	17 50 33.6	– 29 43 12		
BLM	001 – 01 1	17 50 36	– 28 27		
B1 O	000 – 01 3	17 50 36	– 28 54		
A 43	036 +17 1	17 51 11.1	+10 37 57	14.7	
M 2–20	000 – 01 6	17 51 13.2	– 29 35 42		

Planetary Nebulae Parameters

Name	θ ('')	V_r (km s^{-1})	V_{LSR} (km s^{-1})	V_e (km s^{-1})	D (kpc)	R/10^{10} (km)	T (years)
H 2–24	6				6.3	282	
TH 4–3							
H 2–23							
H 1–35	2	+160	+169		2.7	40	
H 2–25	5				6.3	235	
NGC 6445	33	+16	+29	38	1.0	246	2059
H 1–36		- 121	- 113				
H 2–26	5				2.2	82	
M 1–29	7	- 62	- 52		0.7	36	
M 3–43	4	+20	+30		0.9	26	
H 1–38	7				5.7	298	
H 1–37	9	- 16	- 9				
TH 4–4							
TH 4–5	7	- 18	- 5		4.4	230	
HF 2–1	10				2.4	179	
TH 4–6							
M 3–44	4	- 89	- 79		1.0	29	
H 2–27							
M 3–45	6	+ 32	+42		1.0	44	
M 2–17	7	- 21	- 8		2.7	141	
M 2–16	5	+106	+115		1.8	67	
TH 4–7							
B1 3–15							
M 3–16	8	+81	+91		1.4	83	
M 1–30	4	- 98	- 89				
M 1–31	7	+73	+85				
TH 4–8							
H 2–29	5				1.2	44	
H 1–39	2	- 50	- 41				
M 2–18	2	- 3	+6				
M 2–19	7	- 45	- 34		1.0	52	
BLM	4				1.0	29	
B1 O							
A 43	80	- 42	- 24		2.7	1615	
M 2–20		+75	+85				

Planetary Nebulae Parameters

Name	PK Number l b	R.A.(1950) h m s	Dec.(1950) ° ′ ″	m_V (mag)	S_{6cm} (mJy)
CN 2–1	356 – 04 1	17 51 13.6	– 34 21 50	13.9	
B1 Q	001 – 01 2	17 51 24	– 28 12		
K 2–5	014 +06 1	17 51 36.0	– 12 47 47		
M 3–46	359 – 02 4	17 51 54	– 31 12		
HB 6	007 +01 1	17 52 06.8	– 21 44 10		
H 1–40	359 – 02 3	17 52 23.0	– 30 33 05		
VY 1–2	053 +24 1	17 52 24	+28 00		
B1 3–13	000 – 02 1	17 52 50.7	– 29 10 53		
H 2–31	001 – 01 3	17 52 54	– 28 13		
TH 4–9	009 +02 1	17 53 00.0	– 19 28 00		
HE 2–306	348 – 09 1	17 53 00	– 43 04		
M 3–17	359 – 03 1	17 53 12.0	– 31 04 00		
M 1–32	011 +04 1	17 53 26.0	– 16 28 39		
H 1–41	356 – 04 2	17 54 00.1	– 34 09 30		
H 1–42	357 – 04 1	17 54 06	– 33 35		
TH 4–10	010 +03 1	17 54 18	– 18 06		
M 3–47	000 – 02 5	17 54 36	– 30 02		
H 1–44	358 – 03 1	17 54 54	– 31 43		
H 1–43	357 – 04 3	17 54 55.8	– 33 47 24		
M 2–21	000 – 02 4	17 54 57.8	– 29 44 06		
H 2–33	359 – 03 2	17 55 00	– 31 08		
M 3–19	000 – 02 6	17 55 06.6	– 30 00 24		
H 1–45	002 – 02 1	17 55 12.0	– 28 14 36		
M 2–22	357 – 04 2	17 55 14.7	– 33 28 23		
PE 2–11	002 – 01 1	17 55 24	– 27 37		
H 1–46	358 – 04 1	17 55 46.3	– 32 21 33		
M 3–20	002 – 02 2	17 56 09.7	– 28 13 38		
M 1–33	013 +04 1	17 56 06.6	– 15 32 06		
M 3–48	359 – 04 1	17 56 42	– 31 54		
FG 3	352 – 07 1	17 56 44.4	– 38 49 45	11.4	
H 2–35	356 – 05 1	17 57 00	– 34 28		
TH 4–11	011 +02 1	17 57 14.0	– 17 40 24		
H 1–47	001 – 03 1	17 57 26.4	– 29 21 42		
M 3–20	002 – 02 3	17 58 01.2	– 27 38 24		
M 1–34	357 – 05 1	17 58 04.7	– 33 17 43		

Planetary Nebulae Parameters

Name	θ ($''$)	V_r (km s^{-1})	V_{LSR} (km s^{-1})	V_e (km s^{-1})	D (kpc)	$R/10^{10}$ (km)	T (years)
CN 2–1	2	- 271	- 262		3.6	53	
B1 Q	5						
K 2–5	25				5.7	1065	
M 3–46	4				2.6	77	
HB 6	6	+9	+22	20	1.9	85	1351
H 1–40		+108	+118				
VY 1–2	5	- 102	- 32	15	4.7	175	3714
B1 3–13	5						
H 2–31							
TH 4–9							
HE 2–306	3				7.1	159	
M 3–17	3	- 28	- 18				
M 1–32	8	- 73	- 59		0.8	47	
H 1–41	12	+76	+85		4.1	368	
H 1–42	6	- 79	- 70		4.2	188	
TH 4–10	2	+29	+43				
M 3–47		- 16	- 6				
H 1–44	4	+99	+109		3.2	95	
H 1–43		+76	+85				
M 2–21		- 139	- 123				
H 2–33	7				3.5	183	
M 3–19	6	+158	+168		2.2	98	
H 1–45	6	+4	+15				
M 2–22	5	- 92	- 83		2.2	82	
PE 2–11	5				1.3	48	
H 1–46	5	- 13	- 4		4.3	160	
M 3–20	10	+55	+66		3.7	276	
M 1–33	5	- 37	- 23	13	1.2	44	1094
M 3–48	5	- 11	- 1		4.4	164	
FG 3	25	+5	+12		2.0	374	
H 2–35	11	- 195	- 186		3.9	320	
TH 4–11							
H 1–47	3	+109	+120				
M 3–20	5				3.7	138	
M 1–34	11	+2	+11		3.5	287	

Planetary Nebulae Parameters

Name	PK Number l b	R.A.(1950) h m s	Dec.(1950) ° ′ ″	m_V (mag)	S_{6cm} (mJy)
PE 1–11	358 − 05 1	17 58 24	− 33 15		
M 2–23	002 − 02 4	17 58 32.7	− 28 25 46		
NGC 6543	096 +29 1	17 58 34	+66 38 05	8.8	850
M 2–24	356 − 05 2	17 58 43.3	− 34 27 49		
M 3–22	000 − 03 1	17 59 06	− 30 14		
M 3–21	355 − 06 1	17 59 08.0	− 36 38 55		
M 3–49	356 − 06 1	17 59 12	− 35 13		
M 2–25	359 − 04 3	17 59 30.6	− 32 09 36		
M 2–26	003 − 02 2	18 00 06	− 26 59		
IC 4673	003 − 02 3	18 00 10.0	− 27 06 24		
M 1–35	003 − 02 1	18 00 31.9	− 26 43 34		
H 1–50	358 − 05 3	18 00 36.6	− 32 41 48	14.5	
M 2–27	359 − 04 2	18 00 38.1	− 31 17 55		
M 3–50	357 − 06 1	18 00 45.6	− 34 28 48		
H 2–36	359 − 04 4	18 00 48	− 31 40		
H 1–51	356 − 06 2	18 01 06	− 34 58		
H 2–37	002 − 03 2	18 01 18	− 28 38		
M 3–51	358 − 05 4	18 01 39.6	− 32 54 12		
M 2–28	000 − 04 1	18 01 48.6	− 30 58 36		
NGC 6537	010 +00 1	18 02 15.5	− 19 50 30	12.5	640
M 1–37	002 − 03 3	18 02 16.2	− 28 22 24		
H 1–53	004 − 02 1	18 02 50.0	− 26 30 01		
M 1–38	002 − 03 5	18 02 55.6	− 28 40 54		
SP 3	342 − 14 1	18 03 22.8	− 51 01 35	11.9	
M 2–29	004 − 03 1	18 03 34.7	− 26 55 43		
M 3–23	000 − 04 2	18 03 54	− 30 35		
H 1–54	002 − 04 1	18 03 56.1	− 29 13 20		
H 1–55	001 − 04 1	18 04 02.7	− 29 41 49		
M 1–39	015 +03 1	18 04 40.8	− 13 29 18		
H 1–56	001 − 04 2	18 04 42	− 29 45.		
M 3–24	005 − 02 1	18 04 48.0	− 25 24 30		
H 2–39	002 − 03 6	18 04 54	− 28 24		
H 2–40	000 − 05 1	18 05 18	− 31 37		
M 1–40	008 − 01 1	18 05 24.2	− 22 17 23		
H 1–58	005 − 03 1	18 06 06	− 26 03		

Planetary Nebulae Parameters

Name	θ ($''$)	V_r (km s^{-1})	V_{LSR} (km s^{-1})	V_e (km s^{-1})	D (kpc)	$R/10^{10}$ (km)	T (years)
PE 1–11	9				2.3	154	
M 2–23	9	+224	+235				
NGC 6543	19	- 66	- 51	24	0.7	99	1314
M 2–24	7	+173	+182		7.2	376	
M 3–22	6	- 54	- 44		3.0	134	
M 3–21	5	- 69	- 61		4.9	183	
M 3–49	10	- 47	- 38		4.8	359	
M 2–25	14	+20	+30		2.0	209	
M 2–26	9	- 52	- 41		1.3	87	
IC 4673	15	- 15	- 5		3.4	381	
M 1–35	5	+82	+93		0.7	26	
H 1–50	1	+28	+37				
M 2–27	8	+170	+180				
M 3–50	4	+24	+33		8.3	248	
H 2–36	10						
H 1–51	13				5.1	495	
H 2–37	5				2.0	74	
M 3–51	9	+34	+43		4.6	309	
M 2–28	5	- 17	- 7		3.7	138	
NGC 6537	7	- 167	- 4	10	0.9	47	1493
M 1–37	3	+241	+252				
H 1–53		+75	+86				
M 1–38	3	- 70	- 60				
SP 3	35	+46	+49	22	2.2	575	8299
M 2–29	5	- 114	- 103		1.2	44	
M 3–23	11	- 156	- 146		2.6	213	
H 1–54	8	- 116	- 106				
H 1–55	3	- 52	- 42				
M 1–39	4	+128	+143				
H 1–56	3	- 106	- 96		3.6	80	
M 3–24	10	+128	+140		1.1	82	
H 2–39	4						
H 2–40	13				6.4	622	
M 1–40	5	- 32	- 20		1.9	71	
H 1–58	6						

Planetary Nebulae Parameters

Name	PK Number l b	R.A.(1950) h m s	Dec.(1950) ° ′ ″	m_V (mag)	S_{6cm} (mJy)
M 1–41	006 − 02 1	18 06 26.4	− 24 12 54		
H 1–57	356 − 07 2	18 06 30	− 35 45		
M 3–52	018 +04 1	18 07 42	− 10 30		
M 1–42	002 − 04 2	18 07 54.0	− 28 59 42		
H 1–59	003 − 04 3	18 08 20.3	− 27 46 59		
AP 1–12	003 − 04 7	18 08 25.3	− 28 23 21		
NGC 6565	003 − 04 5	18 08 43.3	− 28 11 23	13.2	37
NGC 6563	358 − 07 1	18 08 44.6	− 33 52 46	13.8	69
M 1–43	011 − 00 1	18 08 54	− 18 47		
H 2–41	003 − 04 4	18 09 12	− 27 53		
H 1–60	004 − 04 1	18 09 16.2	− 27 29 42		
H 2–42	005 − 03 2	18 09 18	− 26 34		
M 4–8	018 +03 1	18 09 23.0	− 10 43 38		
M 2–30	003 − 04 8	18 09 24.9	− 27 59 01		
H 1–61	006 − 03 1	18 09 30	− 24 51		
H 2–43	003 − 04 9	18 09 36	− 28 21		
NGC 6572	034 +11 1	18 09 41.7	+06 50 37	9.0	1307
H 1–62	359 − 06 1	18 10 00	− 32 21		
M 2–31	006 − 03 3	18 10 10.2	− 25 30 54		
NGC 6567	011 − 00 2	18 10 48.2	− 19 05 13	11.7	166
M 2–32	359 − 07 1	18 11 34.2	− 32 37 48		
M 4–9	024 +05 1	18 11 39.0	− 05 00 18		
M 2–33	002 − 06 1	18 11 53.8	− 30 16 32		
M 3–25	019 +03 1	18 12 31.2	− 10 11 12		
SWST 1	001 − 06 2	18 12 58.8	− 30 53 10	12.0	
M 3–26	004 − 05 1	18 13 03.3	− 27 16 01		
H 1–63	002 − 06 2	18 13 06.1	− 30 08 40		
M 1–44	004 − 04 2	18 13 09.5	− 27 05 37		
NGC 6578	010 − 01 1	18 13 18.6	− 20 28 04	13.1	170
M 2–34	007 − 03 1	18 14 12.6	− 24 00 00		
M 2–35	000 − 07 1	18 14 21.6	− 31 57 48		
M 2–36	003 − 06 1	18 14 30.0	− 29 09 30	13.0	
PE 1–12	004 − 05 3	18 14 32.4	− 28 18 29		
IC 4699	348 − 13 1	18 14 48.5	− 46 00 16	11.9	
PE 2–13	006 − 04 1	18 15 06	− 25 39		

Planetary Nebulae Parameters

Name	θ ($''$)	V_r (km s^{-1})	V_{LSR} (km s^{-1})	V_e (km s^{-1})	D (kpc)	R/10^{10} (km)	T (years)
M 1–41	8	- 5	+7		1.4	83	
H 1–57	13				4.4	427	
M 3–52	12	- 16	- 1		5.2	466	
M 1–42	8	- 92	- 33		4.3	257	
H 1–59	6				2.6	116	
AP 1–12	12	+152	+163		2.2	197	
NGC 6565	9	- 2	+6		3.5	235	
NGC 6563	45	- 30	- 21	11	1.9	639	18430
M 1–43	5				0.7	26	
H 2–41	8				2.5	149	
H 1–60	4				4.1	122	
H 2–42	10				3.1	231	
M 4–8		+28	+43				
M 2–30	9	+174	+185				
H 1–61		+53	+65				
H 2–43	9	- 21	- 10		2.6	175	
NGC 6572	13	- 9	+10	16	0.8	77	1541
H 1–62		- 84	- 75				
M 2–31		+157	+169				
NGC 6567	9	+119	+133	18	2.1	141	2489
M 2–32		- 49	- 40				
M 4–9	44	- 21	- 5		1.5	493	
M 2–33	4	- 112	- 102		4.9	146	
M 3–25	4	+175	+190	13			
SWST 1	3	- 19	- 4	13	2.1	47	1149
M 3–26	7	- 10	+1		3.1	162	
H 1–63	5	- 1	+9		5.7	213	
M 1–44	4	- 75	- 64		1.8	53	
NGC 6578	10	+5	+17		2.1	157	
M 2–34	8	+70	+82		2.0	119	
M 2–35	5	- 21	- 11				
M 2–36	14	+100	+111		4.0	418	
PE 1–12	12				5.9	529	
IC 4699	7	- 124	- 119	10	5.2	272	8631
PE 2–13	6				6.0	269	

Planetary Nebulae Parameters

Name	PK Number l b	R.A.(1950) h m s	Dec.(1950) ° ′ ″	m_V (mag)	S_{6cm} (mJy)
CN 3–1	038 +12 1	18 15 10.7	+10 08 02	12.4	75
H 1–64	008 − 03 1	18 15 24	− 23 26		
H 2–46	000 − 07 2	18 15 24	− 31 56		
M 2–37	004 − 05 5	18 15 29.3	− 28 09 20		
M 2–38	005 − 05 1	18 16 18.0	− 26 36 37		
H 1–65	007 − 04 1	18 17 05.0	− 24 16 27		
M 2–40	024 +03 1	18 18 43.0	− 06 03 26		
M 2–39	008 − 04 1	18 18 57.5	− 24 12 09		
M 2–41	002 − 07 1	18 19 21.2	− 30 45 01		
M 2–42	008 − 04 2	18 19 28.1	− 24 11 00		
NGC 6620	005 − 06 1	18 19 46.8	− 26 50 50	15.0	
M 1–45	012 − 02 1	18 20 11.0	− 19 18 41		
K 3–1	032 +07 1	18 20 52.7	+03 34 56		
M 3–53	019 +00 1	18 21 21.0	− 11 08 24		
H 1–66	007 − 06 1	18 21 51.6	− 25 43 36		
H 1–67	009 − 04 1	18 22 06	− 22 37		
PC 19	032 +07 2	18 22 13.6	+02 27 48		
K 3–2	028 +05 1	18 22 24.6	− 01 32 36		
NGC 6629	009 − 05 1	18 22 41.2	− 23 13 45	11.6	275
M 2–43	027 +04 1	18 24 03.0	− 02 44 47		
VY 2–1	007 − 06 2	18 24 53.2	− 26 08 36	13.3	
M 1–46	016 − 01 1	18 25 04.5	− 15 34 53		
M 3–27	043 +11 1	18 25 31.6	+14 27 11		
CN 1–5	002 − 09 1	18 25 57.0	− 31 32 00	12.6	
M 1–47	011 − 05 1	18 26 10.8	− 21 48 54		
M 1–48	013 − 03 1	18 26 33.0	− 19 07 47		
PE 2–14	013 − 04 1	18 27 02.4	− 19 42 42		
A 44	015 − 03 1	18 27 17.5	− 16 47 32	17.4	
K 3–4	032 +05 1	18 28 29.5	+02 23 26		
HE 2–406	008 − 07 1	18 28 48	− 24 49		
A 46	055 +16 1	18 29 18.0	+26 54 05		
HF 2–2	005 − 08 1	18 29 18	− 28 46		
NGC 6644	008 − 07 2	18 29 30.0	− 25 10 08	12.2	
M 3–28	021 − 00 1	18 29 55.6	− 10 08 05		
M 3–54	018 − 02 1	18 30 13.7	− 13 46 34		

Planetary Nebulae Parameters

Name	θ ($''$)	V_r (km s^{-1})	V_{LSR} (km s^{-1})	V_e (km s^{-1})	D (kpc)	$R/10^{10}$ (km)	T (years)
CN 3-1	6	+4	+23	10	2.9	130	4125
H 1-64	8	+84	+96		2.1	125	
H 2-46	4						
M 2-37	7	+57	+68		3.8	198	
M 2-38	8	- 72	- 61		3.7	221	
H 1-65	3	+161	+173				
M 2-40	5	+89	+105		1.6	59	
M 2-39	3	+71	+83		8.8	197	
M 2-41	14	- 82	- 72		3.9	408	
M 2-42	4	+157	+169		2.3	68	
NGC 6620	6	+73	+84		4.0	179	
M 1-45		+126	+139				
K 3-1							
M 3-53	5	+35	+50		0.9	33	
H 1-66	8	+42	+53		4.0	239	
H 1-67	6	- 13	- 1		2.5	112	
PC 19		+20	+37				
K 3-2	3	+42	+59				
NGC 6629	15	+14	+27	6	1.6	179	9484
M 2-43	2	+95	+112				
VY 2-1	8	+115	+126		3.9	233	
M 1-46	15	+30			2.4	269	
M 3-27		- 6	+12	14			
CN 1-5	7	- 30	- 20				
M 1-47	5	- 72	- 60		2.8	104	
M 1-48	5	+140	+153		7.3	273	
PE 2-14	5	+32			3.1	115	
A 44	56	+4			1.8	753	
K 3-4	12				6.5	583	
HE 2-406	3						
A 46	63				3.1	1460	
HF 2-2	19				4.5	639	
NGC 6644	3	+194	+205		1.7	38	
M 3-28	9	+21	+36		0.9	60	
M 3-54	6	+157	+171				

Planetary Nebulae Parameters

Name	PK Number l b	R.A.(1950) h m s	Dec.(1950) ° ′ ″	m_V (mag)	S_{6cm} (mJy)
M 1–50	014 – 04 1	18 30 25.2	– 18 18 49		
M 3–55	021 – 00 2	18 30 28.9	– 10 17 26		
M 1–51	021 – 01 1	18 30 42.0	– 11 09 42		
K 3–6	030 +04 1	18 30 43.8	+00 09 32		
IC 4732	010 – 06 1	18 30 53.3	– 22 40 57	13.3	49
M 1–52	017 – 02 1	18 31 07.0	– 14 54 48		
M 4–10	019 – 02 1	18 31 24.1	– 13 14 43		
K 3–7	028 +02 1	18 31 37.0	– 02 29 59		
PE 1–13	010 – 06 2	18 31 48	– 22 46		
A 47	030 +03 1	18 32 48.0	– 00 16 00	>20.9	
M 1–53	015 – 04 1	18 32 53.5	– 17 38 38		
M 1–54	016 – 04 1	18 33 14.4	– 17 02 30		
M 1–55	011 – 06 1	18 33 36	– 21 52		
M 2–44	028 +01 1	18 34 59.4	– 03 08 36		
M 1–56	016 – 04 2	18 34 52.2	– 17 08 25		
M 3–29	004 – 11 1	18 36 12.9	– 30 43 21	13.1	
M 2–45	027 +00 1	18 36 43.0	– 04 22 36		
M 1–57	022 – 02 1	18 37 34.0	– 10 42 37		
M 3–30	017 – 04 1	18 38 23.0	– 15 36 40		
K 3–11	023 – 01 2	18 38 23.6	– 08 58 51		
PE 1–14	025 – 00 1	18 39 24	– 06 44		
M 1–58	022 – 03 1	18 40 10.3	– 11 09 54		
PC 20	031 +01 1	18 40 29.3	– 00 19 37		
M 2–36	003 – 06 1	18 40 30.5	– 29 09 31		
M 1–59	023 – 02 1	18 40 35.3	– 09 07 44		
M 1–60	019 – 04 1	18 40 48.0	– 13 47 51		
HE 2–418	004 – 11 2	18 41 00	– 30 22		
M 3–31	014 – 07 1	18 41 12	– 19 57		
M 3–32	009 – 09 1	18 41 36	– 25 23.		
IC 4776	002 – 13 1	18 42 34.1	– 33 23 52	11.7	71
PC 21	013 – 07 1	18 42 36	– 20 38 12		
PE 2–15	026 – 01 2	18 42 45.4	– 07 00 01		
K 3–13	034 +02 1	18 42 53.0	+01 58 15		
K 4–5	026 – 01 1	18 42 54.1	– 06 21 42		
M 1–61	019 – 05 1	18 43 03.6	– 14 31 01		

Planetary Nebulae Parameters

Name	θ ($''$)	V_r (km s^{-1})	V_{LSR} (km s^{-1})	V_e (km s^{-1})	D (kpc)	$R/10^{10}$ (km)	T (years)
M 1–50	6	+27	+40		3.9	175	
M 3–55	7	+26	+41		1.4	73	
M 1–51	9	+3	+18		1.2	80	
K 3–6							
IC 4732	5	- 145	- 133		3.4	127	
M 1–52	7	+133	+147		2.1	109	
M 4–10		+50	+64				
K 3–7	7						
PE 1–13	8				6.6	394	
A 47	16				5.5	658	
M 1–53	7	+63	+76	13	3.6	188	4596
M 1–54	17	- 41	- 27		3.2	406	
M 1–55		- 22	+10				
M 2–44	8	+106	+122	12	0.8	47	1264
M 1–56	10	+88	+102		4.5	336	
M 3–29	8	+50	+60	7	3.0	179	8129
M 2–45	7				2.5	130	
M 1–57	9	+92	+107		3.1	208	
M 3–30	17	+71	+85		4.3	546	
K 3–11	3						
PE 1–14	5				0.9	33	
M 1–58	7	+60	+75		0.4	20	
PC 20							
M 2–36	14				4.0	418	
M 1–59	5	+99	+114	13	2.5	93	2279
M 1–60	3	+76	+90				
HE 2–418	13				5.8	563	
M 3–31		- 89	- 76				
M 3–32	6	+46	+57		4.6	206	
IC 4776	7	+19	+28		3.3	172	
PC 21	13				5.5	534	
PE 2–15	3				1.4	31	
K 3–13							
K 4–5							
M 1–61		+41	+55				

Planetary Nebulae Parameters

Name	PK Number l b	R.A.(1950) h m s	Dec.(1950) ° ′ ″	m_V (mag)	S_{6cm} (mJy)
H 2–48	011 − 09 1	18 43 30	− 23 30		
PE 1–15	025 − 02 1	18 43 42	− 07 18		
M 2–46	024 − 02 1	18 43 51.0	− 08 31 18		
PE 1–16	026 − 02 1	18 44 50.6	− 06 57 17		
PE 1–17	024 − 03 1	18 45 06	− 09 12		
M 3–33	009 − 10 1	18 45 07.2	− 25 32 18		
K 4–6	034 +01 1	18 45 46.5	+01 39 41		
PE 1–18	027 − 02 1	18 46 06	− 05 59		
K 3–14	042 +05 1	18 46 11.4	+10 32 38		
PE 1–19	026 − 02 3	18 47 03.0	− 07 05 06		
M 1–62	012 − 09 1	18 47 25.2	− 22 37 54		
HU 2–1	051 +09 1	18 47 39.2	+20 47 12	12.2	106
M 1–64	064 +15 1	18 48 12	+35 11		
M 1–63	021 − 05 1	18 48 41.6	− 13 14 14		
K 3–15	041 +04 1	18 49 19.0	+09 51 13		
K 3–16	044 +05 1	18 50 42.0	+12 12 17		
A 49	027 − 03 1	18 50 48	− 06 33		
VY 1–4	027 − 03 2	18 51 20.6	− 06 30 07	13.4	
M 4–11	024 − 05 1	18 51 32.4	− 10 09 00		
K 4–8	025 − 04 1	18 51 36.1	− 08 57 22		
IC 1295	025 − 04 2	18 51 53.5	− 08 54 28	15.0	
NGC 6720	063 +13 1	18 51 44	+32 57 52	9.7	360
HB 7	003 − 14 1	18 52 23.8	− 32 19 49	10.9	
YM 16	038 +02 1	18 52 29.9	+05 58 48		
K 3–17	039 +02 1	18 53 52.5	+07 03 24		
M 1–65	043 +03 1	18 54 11.9	+10 48 14		
PE 1–20	028 − 04 1	18 54 36.6	− 06 03 43		
PE 1–21	028 − 03 1	18 55 12	− 05 32		
AP 2–1	035 − 00 1	18 55 38.4	+01 32 54		
M 1–66	032 − 02 1	18 55 50.3	− 01 07 39		
K 4–10	052 +07 1	18 56 53.7	+20 32 52		
K 3–18	032 − 03 1	18 57 58.0	− 02 16 13		
A 50	078 +18 1	18 58 00	+48 24		
A 51	017 − 10 1	18 58 06.0	− 18 16 20	15.4	
K 3–19	032 − 02 2	18 59 01.3	− 01 23 20		

Planetary Nebulae Parameters

Name	θ ($''$)	V_r (km s^{-1})	V_{LSR} (km s^{-1})	V_e (km s^{-1})	D (kpc)	R/10^{10} (km)	T (years)
H 2–48	2						
PE 1–15	13				1.9	184	
M 2–46	4	+83	+98		1.4	41	
PE 1–16	8				1.2	71	
PE 1–17	5				3.2	119	
M 3–33	8	+174	+185		5.6	335	
K 4–6							
PE 1–18	3				1.5	33	
K 3–14							
PE 1–19	4				1.4	41	
M 1–62	4	+34			4.4	131	
HU 2–1	3	+14	+33	10	1.9	42	1351
M 1–64	17	- 25	- 6		3.1	394	
M 1–63	4	+26	+40	8	3.2	95	3793
K 3–15							
K 3–16							
A 49	44				2.1	691	
VY 1–4		+110	+126				
M 4–11		+19	+34				
K 4–8							
IC 1295	108	- 36	- 21		0.7	565	
NGC 6720	70	- 19	- 0	30	0.7	366	3872
HB 7	4	- 65	- 56		4.8	143	
YM 16	304				0.9	2046	
K 3–17	15				2.4	269	
M 1–65	4	+20	+38		2.3	68	
PE 1–20	6				3.4	152	
PE 1–21	9				2.5	168	
AP 2–1	33				1.6	394	
M 1–66		+43	+59				
K 4–10							
K 3–18	4				6.2	185	
A 50	47	- 159	- 141		2.8	984	
A 51	67	+23			2.7	1353	
K 3–19							

Planetary Nebulae Parameters

Name	PK Number l b	R.A.(1950) h m s	Dec.(1950) ° ′ ″	m_V (mag)	S_{6cm} (mJy)
SH 2–71	036 – 01 1	18 59 28.0	+02 04 56		
K 3–20	032 – 03 2	18 59 34.1	– 01 53 03		
NGC 6741	033 – 02 1	19 00 02.0	– 00 31 12	10.8	230
K 3–21	047 +04 1	19 00 24	+14 24		
K 1–17	051 +06 1	19 01 26.0	+19 16 53		
A 52	050 +05 1	19 02 19.2	+17 52 36	16.5	
HB 8	003 – 17 1	19 02 20.5	– 33 16 15	13.4	
K 4–16	048 +04 2	19 02 36.0	+15 43 03		
NGC 6751	029 – 05 1	19 03 15.0	– 06 04 07	12.5	63
A 53	040 – 00 1	19 04 19.3	+06 19 15	16.9	
A 54	055 +06 1	19 06 32.6	+22 54 00	17.1	
K 4–17	033 – 04 1	19 06 49.4	– 01 13 52		
K 3–22	045 +01 1	19 07 06.3	+11 55 54		
NGC 6765	062 +09 1	19 09 12	+30 28		
M 1–67	050 +03 1	19 09 16.7	+16 46 29		
K 3–24	048 +02 1	19 09 49.5	+15 03 56		
A 56	037 – 03 2	19 10 36.0	+02 47 40	>15.5	
HE 2–428	049 +02 1	19 10 49.5	+15 41 32		
M 2–47	039 – 02 1	19 11 05.8	+04 32 55		
K 4–20	042 – 01 1	19 11 07.0	+07 21 19		
HE 2–429	048 +01 1	19 11 21.2	+14 54 18		
M 1–69	038 – 03 2	19 11 24.0	+03 32 33		
K 2–11	038 – 03 3	19 11 47.1	+03 29 33		
HE 2–430	051 +03 1	19 11 50.9	+17 26 20		
NGC 6772	033 – 06 1	19 12 00.5	– 02 47 35	14.2	95
K 3–26	035 – 05 1	19 12 06.0	+00 08 22		
K 3–29	048 +01 2	19 13 12.4	+13 58 33		
IC 4846	027 – 09 1	19 13 44.9	– 09 08 06	12.7	47
IC 1297	358 – 21 1	19 13 57.6	– 39 42 12		
K 3–30	040 – 03 1	19 13 59.4	+05 07 58		
NA 2	026 – 11 1	19 15 33.1	– 11 11 42	13.3	
A 58	037 – 05 1	19 15 48.5	+01 41 20	>18.8	
NGC 6778	034 – 06 1	19 15 49.4	– 01 41 24	13.3	55
NGC 6781	041 – 02 1	19 16 01.9	+06 26 46	11.8	350
A 59	053 +03 1	19 16 29.5	+19 28 23	17.2	

Planetary Nebulae Parameters

Name	θ ($''$)	V_r (km s^{-1})	V_{LSR} (km s^{-1})	V_e (km s^{-1})	D (kpc)	R/10^{10} (km)	T (years)
SH 2–71	100	+25	+42	16	1.0	748	14819
K 3–20							
NGC 6741	8	- 41	+58	21	1.7	101	1535
K 3–21	7				10.5	549	
K 1–17	45				3.9	1312	
A 52	37				3.1	857	
HB 8	5	- 172	- 163		5.3	198	
K 4–16							
NGC 6751	21	- 39	- 23	40	2.8	439	3485
A 53	41				1.6	490	
A 54	56				4.1	1717	
K 4–17							
K 3–22							
NGC 6765	38	- 72	- 53	35	4.1	1165	10555
M 1–67	100	+202			1.0	748	
K 3–24	6				3.6	161	
A 56	181				1.7	2301	
HE 2–428	10				1.7	127	
M 2–47	9	+36	+53	13	2.9	195	4760
K 4–20							
HE 2–429	4	+14	+32		0.8	23	
M 1–69		+7	+24				
K 2–11	13				4.4	427	
HE 2–430							
NGC 6772	112		+16	12	1.4	1172	30983
K 3–26							
K 3–29							
IC 4846	2	+151	+166	19	3.3	49	823
IC 1297	7	+14	+20	13	3.0	157	3830
K 3–30							
NA 2							
A 58	40				3.8	1136	
NGC 6778	16	+91	+107	21	2.2	263	3974
NGC 6781	70	+4	+22	12	0.9	471	12448
A 59	87				1.8	1171	

Planetary Nebulae Parameters

Name	PK Number l b	R.A.(1950) h m s	Dec.(1950) ° ′ ″	m_V (mag)	S_{6cm} (mJy)
A 60	025 − 11 1	19 16 30.0	− 12 20 26	15.7	
K 3–31	052 +02 1	19 16 50.6	+18 56 51		
M 4–14	043 − 03 1	19 18 35.7	+07 31 17		
ANON	032 − 08 1	19 19 23.0	− 04 17 36		
K 3–33	045 − 01 1	19 20 04.2	+10 35 36		
NGC 6790	037 − 06 1	19 20 24.5	+01 25 02	10.2	275
HE 2–432	055 +02 1	19 21 14.8	+21 02 21		
HE 1–1	055 +02 2	19 21 37.1	+21 00 46		
VY 2–2	045 − 02 1	19 21 59.0	+09 47 59	12.7	220*
M 3–34	031 − 10 1	19 24 20.7	− 06 41 00		
HE 1–2	055 +02 3	19 24 28.0	+21 03 30		
PB 9	046 − 03 1	19 25 22.3	+10 18 15		
K 3–35	056 +02 1	19 25 35.0	+21 23 55		
PB 10	048 − 02 1	19 25 54.4	+12 13 35		
HE 2–434	320 − 28 1	19 27 33.2	− 74 39 25		
K 4–28	050 − 01 1	19 27 58.7	+14 40 57		
HE 2–436	004 − 22 1	19 28 51.5	− 34 18 59		
NGC 6803	046 − 04 1	19 28 53.5	+09 57 00	11.3	114
NGC 6804	045 − 04 1	19 29 12.0	+09 07 13	12.2	135
K 3–36	044 − 05 1	19 30 13.1	+07 21 29		
HE 2–437	061 +03 1	19 30 54.9	+26 46 13		
A 62	047 − 04 1	19 30 56.2	+10 30 43	14.8	
K 4–30	057 +01 1	19 31 00.7	+22 52 02		
K 3–37	059 +02 1	19 31 41.0	+24 25 54		
NGC 6807	042 − 06 1	19 32 03.4	+05 34 30.3	13.8	
BD+30°3639	064 +05 1	19 32 47	+30 24 18	9.6	600
K 3–39	059 +02 2	19 33 49.0	+24 48 10		
M 1–71	055 − 00 1	19 34 14.6	+19 35 45		
K 3–40	058 +01 1	19 34 14.7	+23 33 05		
K 4–31	050 − 03 1	19 34 45.6	+13 34 42		
HE 2–440	060 +01 1	19 36 03.5	+25 09 00		
ME 1–1	052 − 02 2	19 36 53.5	+15 49 52	12.6	
K 3–41	052 − 02 1	19 37 00.0	+16 13 49		
K 3–42	056 − 00 1	19 37 24.5	+20 12 07		
K 3–43	055 − 01 1	19 38 12.0	+18 42 18		

Planetary Nebulae Parameters

Name	θ ($''$)	V_r (km s^{-1})	V_{LSR} (km s^{-1})	V_e (km s^{-1})	D (kpc)	$R/10^{10}$ (km)	T (years)
A 60	78				3.1	1808	
K 3–31							
M 4–14	7	+48	+65		1.6	83	
ANON							
K 3–33							
NGC 6790	7	+40	+57	15	1.5	78	1659
HE 2–432	5						
HE 1–1	6				3.6	161	
VY 2–2	1	− 71	− 54	15			
M 3–34	6	+37	+52	14	4.4	197	4471
HE 1–2	5						
PB 9	12						
K 3–35							
PB 10							
HE 2–434		+38	+33				
K 4–28							
HE 2–436	10				4.5	336	
NGC 6803	6	+13	+30	15	2.5	112	2371
NGC 6804	40	− 12	+5	27	1.6	478	5620
K 3–36							
HE 2–437	7				7.0	366	
A 62	161				1.1	1324	
K 4–30							
K 3–37							
NGC 6807	2	− 68	− 51	13	5.5	82	2006
B+30 3639	11	− 31	− 13	23	0.6	49	680
K 3–39							
M 1–71	4	+51	+69				
K 3–40							
K 4–31							
HE 2–440							
ME 1–1	10	− 6	+12	9	3.8	284	10011
K 3–41							
K 3–42	3				0.8	17	
K 3–43	3				1.5	33	

Planetary Nebulae Parameters

Name	PK Number l b	R.A.(1950) h m s	Dec.(1950) ° ′ ″	m_V (mag)	S_{6cm} (mJy)
M 1-73	051 − 03 1	19 38 51.4	+14 49 50		
M 1-72	054 − 02 1	19 39 19.3	+17 38 14		
PC 22	051 − 04 1	19 39 44.2	+13 43 32		
A 63	053 − 03 1	19 39 55.2	+16 58 00	17.1	
M 1-74	052 − 04 1	19 40 01.3	+15 01 57		
K 4-32	060 +01 1	19 40 01.6	+24 23 06		
NGC 6818	025 − 17 1	19 41 09.0	− 14 16 21	9.9	300
A 64	044 − 09 1	19 43 07.1	+05 26 35	15.3	
HE 2-447	057 − 01 1	19 43 11.7	+21 12 46		
NGC 6826	083 +12 1	19 43 27	+50 24 10	9.8	385
A 65	017 − 21 1	19 43 33.2	− 23 15 19	15.2	
HE 1-3	059 − 01 1	19 46 15.5	+22 02 28		
NGC 6833	082 +11 1	19 48 20.9	+48 50 01	13.8	19
M 2-48	062 − 00 1	19 48 23.1	+25 46 41		
K 4-37	066 +02 1	19 49 02.4	+30 54 48		
PC 23	068 +03 1	19 49 57.2	+32 51 33		
K 3-48	063 +00 1	19 50 05.6	+27 10 49		
K 3-49	069 +02 1	19 52 05.9	+33 14 20		
NGC 6842	065 +00 1	19 53 00	+29 09	13.6	
A 66	019 − 23 1	19 54 34.8	− 21 44 43	14.9	
K 4-41	068 +01 1	19 54 37.0	+32 14 13		
A 67	043 − 13 1	19 55 58.5	+02 54 12	16.0	
HE 1-4	068 +01 2	19 57 18	+31 47		
NGC 6853	060 − 03 1	19 57 26.6	+22 34 45	7.6	1325
A 68	060 − 04 1	19 58 00.0	+21 34 40	16.6	
NGC 6852	042 − 14 1	19 58 07.6	+01 35 33		
K 3-51	056 − 06 1	20 00 20.6	+17 28 23		
K 3-52	067 − 00 1	20 01 11.3	+30 24 09		
K 3-53	064 − 02 1	20 01 18.0	+26 52 28		
M 1-75	068 − 00 1	20 02 48	+31 19		
K 3-54	063 − 03 1	20 02 52.0	+25 18 04		
K 3-55	069 +00 1	20 04 58.3	+32 07 52		
M 4-17	079 +05 1	20 07 24	+43 35		
NGC 6879	057 − 08 1	20 08 09.9	+16 46 24	13.0	
NGC 6884	082 +07 1	20 08 49	+46 18 44	12.6	200

Planetary Nebulae Parameters

Name	θ ($''$)	V_r (km s^{-1})	V_{LSR} (km s^{-1})	V_e (km s^{-1})	D (kpc)	$R/10^{10}$ (km)	T (years)
M 1–73	5	+7	+25	11	1.9	71	2047
M 1–72							
PC 22							
A 63	40				2.7	807	
M 1–74		+10	+28	8			
K 4–32							
NGC 6818	19	- 13	- 1	28	1.5	213	2413
A 64	37				3.6	996	
HE 2–447							
NGC 6826	25	- 6	+11	8	0.7	130	5186
A 65	38	+13			1.5	426	
HE 1–3	8						
NGC 6833	2	- 109	- 92	13	2.6	38	948
M 2–48	8				1.6	95	
K 4–37							
PC 23							
K 3–48							
K 3–49							
NGC 6842	48	- 5	+13	35	1.7	610	5528
A 66	268				1.2	2405	
K 4–41							
A 67	67				3.2	1603	
HE 1–4	22	+12	+29		1.9	312	
NGC 6853	200	- 41	- 24	15	0.4	598	12646
A 68	38				3.2	909	
NGC 6852	28	- 11			3.4	712	
K 3–51							
K 3–52							
K 3–53							
M 1–75	14	- 9	+9		3.1	324	
K 3–54							
K 3–55							
M 4–17	15	- 26	- 9		1.8	201	
NGC 6879	5	+9	+26	21	3.7	138	2088
NGC 6884	8	- 36	- 19	23	1.7	101	1402

Planetary Nebulae Parameters

Name	PK Number l b	R.A.(1950) h m s	Dec.(1950) ° ′ ″	m_V (mag)	S_{6cm} (mJy)
NGC 6881	074 +02 1	20 09 01.6	+37 15 44	14.3	180
HE 1–5	060 − 07 1	20 09 42.9	+20 11 04		
NGC 6886	060 − 07 2	20 10 29.6	+19 50 16	12.2	108
K 3–57	072 +00 1	20 10 52.2	+34 11 22		
HE 2–459	068 − 02 1	20 11 55.1	+29 24 52		
NGC 6891	054 − 12 1	20 12 47.7	+12 32 59	11.7	105
NGC 6894	069 − 02 1	20 14 22.8	+30 24 36	14.4	59
HE 1–6	065 − 05 1	20 15 13.9	+25 12 22		
IC 4997	058 − 10 1	20 17 51.0	+16 34 27	11.6	127
M 3–35	071 − 02 1	20 19 04.8	+32 19 49		
NGC 6905	061 − 09 1	20 20 08.5	+19 56 39	11.9	52
A 70	038 − 25 1	20 28 53.2	− 07 15 30	14.3	
A 71	085 +04 1	20 30 47	+47 10 48	15.2	
A 72	059 − 18 1	20 47 40.4	+13 22 13	14.6	
NGC 7008	093 +05 2	20 59 05	+54 20 44	13.3	250
NGC 7009	037 − 34 1	21 01 27.6	− 11 33 54	8.3	750
NGC 7026	089 +00 1	21 04 35	+47 39 00	12.7	260
NGC 7027	084 − 03 1	21 05 09	+42 02 00	10.4	6370
NGC 7048	088 − 01 1	21 12 24	+46 04	11.3	
A 74	072 − 17 1	21 14 39.0	+23 59 42	>12.2	
M 1–77	089 − 02 1	21 17 21	+46 06		
M 1–78	093 +01 1	21 19 05.5	+51 40 41		
K 3–60	098 +04 1	21 25 57.9	+57 26 05		
PS 1	065 − 27 1	21 27 33.0	+11 57 12	14.9	3
K 3–61	096 +02 1	21 28 23.8	+54 14 17		
K 3–62	095 +00 1	21 30 08.8	+52 20 37		
A 77	097 +03 1	21 30 36	+55 40		
IC 5117	089 − 05 1	21 30 37	+44 22 29	13.3	230
ANON, HU 1–2	086 − 08 1	21 31 07	+39 24 40	12.7	155
A 78	081 − 14 1	21 33 20	+31 28 18	16.0	
K 4–45	096 +01 1	21 33 41.0	+53 33 46		
NGC 7094	066 − 28 1	21 34 27.9	+12 33 49		
M 1–79	093 − 02 1	21 35 12	+48 43		
M 2–49	095 − 02 1	21 41 29.9	+50 11 29		
NGC 7139	104 +07 1	21 44 36	+63 25		

Planetary Nebulae Parameters

Name	θ ($''$)	V_r (km s^{-1})	V_{LSR} (km s^{-1})	V_e (km s^{-1})	D (kpc)	R/10^{10} (km)	T (years)
NGC 6881	5	- 14	+3	18	1.7	63	1119
HE 1–5	30	+39	+56	34	2.8	628	5858
NGC 6886	6	- 36	- 19	16	2.8	125	2489
K 3–57							
HE 2–459							
NGC 6891	13	+42	+59	13	2.1	204	4979
NGC 6894	44	- 58	- 40	43	1.5	493	3639
HE 1–6	15	- 22	- 5		6.9	774	
IC 4997	2	- 66	- 50	15	2.3	34	727
M 3–35		- 176	- 159				
NGC 6905	47	- 8	+8	48	1.8	632	4179
A 70	42	- 79			3.5	1099	
A 71	157				0.9	1056	
A 72	128	- 59	- 44		1.1	1053	
NGC 7008	86	- 76	- 61	37	0.9	578	4960
NGC 7009	27	- 44	- 37	21	0.9	181	2743
NGC 7026	19	- 41	- 26	42	0.9	127	965
NGC 7027	12	+9	+24	18	0.7	62	1106
NGC 7048	55	- 50	- 35	15	1.2	493	10433
A 74	832				0.6	3734	
M 1–77	7			13	0.7	36	893
M 1–78	6	- 88	- 74		3.1	139	
K 3–60							
PS 1	1	- 141	- 128		9.3	69	
K 3–61	62				1.1	510	
K 3–62		- 53	- 39	16			
A 77	37	- 113	- 100	15	1.5	415	8773
IC 5117	2	- 26	- 12	17	1.7	25	474
ANON HU 1–2	5	+10	+24	28	2.2	82	931
A 78	107	+17	+31		4.0	3201	
K 4–45							
NGC 7094	88	- 87	- 75	90	2.9	1908	6723
M 1–79	33	- 24	- 10	19	1.0	246	4118
M 2–49		- 134	- 121				
NGC 7139	76	- 54	- 42	19	1.6	909	15175

Planetary Nebulae Parameters

Name	PK Number l b	R.A.(1950) h m s	Dec.(1950) ° ′ ″	m_V (mag)	S_{6cm} (mJy)
M 2–50	097 – 02 1	21 55 54	+51 27		
IC 5148–50	002 – 52 1	21 56 30	– 39 37		
M 2–51	103 +00 1	22 14 18	+57 14		
BL 2–1	104 +00 1	22 18 28.0	+57 59 01		
M 2–52	103 +00 2	22 18 42	+57 21		
IC 5217	100 – 05 1	22 21 56	+50 42 52	12.6	163
NGC 7293	036 – 57 1	22 26 54.6	– 21 05 50		1292
ME 2–2	100 – 08 1	22 29 37.7	+47 32 37.5		40
M 2–53	104 – 01 1	22 30 24	+55 55		
NGC 7354	107 +02 1	22 38 26.2	+61 01 17	12.9	579
IC 1454	117 +18 1	22 42 12	+80 11		
M 2–54	104 – 06 1	22 49 29.0	+51 34 44		
M 1–80	107 – 02 1	22 54 14.6	+56 53 18	14.0	
VY 2–3	107 – 13 1	23 20 36	+46 38	13.9	
NGC 7662	106 – 17 1	23 23 29	+42 15 36	9.2	600
HB 12	111 – 02 1	23 23 57	+57 54 24	14.0	300
M 2–55	116 +08 1	23 29 42	+70 06		
ANON.	114 +03 1	23 33 11.0	+64 35 58		
JN 1	104 – 29 1	23 33 24	+30 11 26	15.1	
A 82	114 – 04 1	23 43 24	+56 47		
A 84	112 – 10 1	23 45 16	+51 07 17	14.4	
M 2–56	118 +08 1	23 54 06.6	+70 31 31		

Planetary Nebulae Parameters

Name	θ ($''$)	V_r (km s^{-1})	V_{LSR} (km s^{-1})	V_e (km s^{-1})	D (kpc)	R/10^{10} (km)	T (years)
M 2–50	4	- 136		15	1.5	44	948
IC 5148–50	113	- 28	- 27				
M 2–51	39	- 11	+1	11	1.5	437	12610
BL 2–1							
M 2–52	14	- 92	- 80	8	1.5	157	6224
IC 5217	7	- 99	- 87	21	2.8	146	2213
NGC 7293	780	- 28	- 25	14	0.2	1166	26421
ME 2–2	1	- 152	- 141	5			
M 2–53	15	- 62	- 51	11	0.9	100	2910
NGC 7354	20	- 4	- 31	25	0.8	119	1517
IC 1454	34			92	3.3	839	2891
M 2–54							
M 1–80	8	- 58	- 47	9	1.6	95	3372
VY 2–3	5	- 63		16	7.4	276	5483
NGC 7662	15	- 13	- 5	26	0.8	89	1094
HB 12	10	- 5	+5	13	0.5	37	911
M 2–55	39	- 23	- 13		1.9	554	
ANON.							
JN 1	314				0.9	2113	
A 82	72	- 31	- 22	50	2.0	1077	6828
A 84	130			32	1.1	1069	10596
M 2–56							

20.3 Catalogue of Planetary Nebulae — Part II

Planetary Nebulae Parameters

Name	Log F(Hβ) (erg cm^{-2} s^{-1})	$N_e/10^3$ (cm^{-3})	$T_e/10^4$ (°K)	B (mag)	V (mag)	C
NGC 40	- 10.64	1.3	0.85		11.6	0.82
VY 1-1	- 11.54	2.0	1.12	14.11	14.45	0.96
BV 1						
HU 1-1	- 11.73		1.14			0.32
BoBn 1						
BV 2						
A 2	- 12.50	0.10	1.20	16.07	15.85	0.37
NGC 246	- 10.9	0.09	1.0		11.92	0.00
PHL 932						
M 1-1						
NGC 650-1	- 11.36	5.62	1.18	16.10	15.87	0.15
BV 3						
IC 1747	- 11.45	3.8	1.0		15.8	0.94
M 1-2	- 12.43				18.3	
A 6	- 12.05	1000.0	1.50	>12.8	>13.13	0.98
IC 289	- 11.78	0.50	1.70	16.8	>15.9	1.15
NGC 1360	- 10.84	0.05	1.80	10.86	11.35	0.01
M 1-4	- 12.21	10.0	1.00		16.7	1.36
IC 351	- 11.40	8.7	1.27		15.	0.57
BA 1					17.2	
IC 2003	- 11.18	3.7	1.07	16.2	16.5	0.41
NGC 1501	- 11.52	1.91	1.08	15.11	14.45	1.10
NGC 1514	- 11.24	0.25	1.42	9.91	9.40	0.92
M 2-2						
NGC 1535	- 10.36	8.91	1.3		11.6	0.00
H 3-29	- 12.51	1.10	1.60	>17.0	>18.1	1.17
K3-67						
A 7					15.5	
J 320	- 11.37	15.	1.26	14.09	14.31	0.22
A 8						
K 2-1					18.2	
IC 2120					16.2	
IC 418	- 9.62	10.0	0.94	10.13	10.33	0.32
K 3- 68						
K 1-7						

Planetary Nebulae Parameters

Name	Log F(Hβ) (erg cm^{-2} s^{-1})	$N_e/10^3$ (cm^{-3})	$T_e/10^4$ (°K)	B (mag)	V (mag)	C
H 3–75						
K 3–69						
NGC 2022	– 11.10	1.4	1.4		14.9	0.45
M 1–5	– 12.05	28.0	1.1			
Pu 1				>18.		
IC 2149	– 10.50	1.8	1.10	11.28	11.59	0.28
K 3–70						
A 12						
A 13	– 12.62				18.8	
A 14					15.0	
K 3–71						
VV 1–4					11.4	
IC 2165	– 10.87	4.4	1.34	>15.8	>16.4	0.48
J 900	– 11.23	4.7	1.21	16.46	16.26	0.63
A66 15	– 12.9	0.1	1.60	15.9	16.4	0.04
M 1–6	– 12.28	6.9	0.93	16.7	15.8	2.01
M 1–7						
A 17					19.9	
K 2–2	– 11.12				18.2	
M 1–8						
A 18						
A 19						
PB 1				16.2	16.0	0.74
M 3–1	– 11.3	4.0	1.24	15.1	15.4	0.16
M 1–9	– 11.65	10.		15.7	15.5	0.80
K 1–9					18.6	
K 2–3						
NGC 2346	– 12.39	0.4	1.3		11.1	1.67
M 1–11	– 11.76	60.	2.00	14.73	13.95	1.60
K 2–10						
M 3–2	– 13.1					
M 1–12	– 11.60	13.	0.90	14.31	14.08	0.85
M 1–13						
A 20					15.7	
NGC 2371–2	– 10.96	1.12	1.95		14.7	0.00

Planetary Nebulae Parameters

Name	Log F(Hβ) (erg cm^{-2} s^{-1})	$N_e/10^3$ (cm^{-3})	$T_e/10^4$ (°K)	B (mag)	V (mag)	C
M 3–3						
M 1–14	- 11.51	4.0		14.37	14.09	0.84
NGC 2392	- 10.9	2.69	1.42	10.35	10.53	0.16
A 21	- 10.4					
M 4–2	- 11.7	1.0		16.9	17.0	0.20
K 1–11					19.1	
M 1–16						
M 1–17						
NGC 2438	- 10.97	0.28	1.46		17.5	0.45
NGC 2440	- 10.52	1.29	1.30		14.3	0.17
M 1–18					19.1	
A 23						
NGC 2452	- 11.54	2.0	1.2		19.1	
HE 2–5	- 11.24	8.0	1.11	15.1:	16.4:	0.32
K 1–12					>21.	
A 24	- 11.35				16.5	
M 3–4						
NGC 2474	- 11.26				16.5	
M 3–5						
K 1–13					17.8	
A 26						
HE 2–7	- 10.9	4.0		16.6	16.6	0.35
PB 2						
HE 2–9	- 12.30	10.	1.00	17.9	16.8	2.08
K 1–1					>21.	
NGC 2610					15.5	
HE 2–11	- 12.2					
M 3–6	- 11.0	6.5	0.72	13.5	13.9	0.64
A 30	- 12.19				14.3	
A 31	- 10.54				15.5	
HE 2–15						
PB 3						
K 1–2						
IC 2448	- 00.83	3.0	1.32	13.63	13.94	0.12
HE 2–18						

Planetary Nebulae Parameters

Name	Log F(Hβ) (erg cm^{-2} s^{-1})	$N_e/10^3$ (cm^{-3})	$T_e/10^4$ (°K)	B (mag)	V (mag)	C
NGC 2792	- 11.23	2.95	1.8			0.66
HE 2–21						
PB 4	- 11.6	10.	1.09	15.9	15.9	0.73
NGC 2818	- 10.9	0.8	1.4			
PB 5						
HE 2–25	- 12.60					
HE 2–26	- 11.3	10.		17.2:	16.7:	0.53
NGC 2867	- 10.7	4.0	1.17	17.4:	17.3:	0.50
HE 2–28						
HE 2–29						
NGC 2899						
HE 2–32	- 12.9					
A 33	- 12.54	0.10	1.00	14.32	14.74	0.00
IC 2501						
HE 2–34	- 13.17			18.1	16.5	2.76
HE 2–35	- 12.03		0.90	16.3	16.2	0.91
HE 2–36			1.06	11.75	11.22	1.39
A 34	- 11.84				16.2	
HE 2–37	- 12.3					
HE 2–39						
NGC 3132	- 10.43	1.0	0.95		8.8	0.27
HE 2–41						
IC 2553	- 10.8	7.0	1.02	15.8	15.6	0.38
NGC 3195						
PB 6	- 11.9					
HF 4	- 13.0					
NGC 3211	- 10.99	0.9	1.30			0.27
HE 2–47	- 11.17	30.	0.80	13.18	12.96	0.90
NGC 3242	- 10.51	3.3	2.95		11.5	0.18
MY 60						
HE 2–50						
HE 2–51	- 12.6	0.9	1.07	16.7	15.60	0.96
PE 1–1	- 12.27					
PE 1–3						
HE 2–55	- 12.6					

Planetary Nebulae Parameters

Name	Log F(Hβ) (erg cm^{-2} s^{-1})	$N_e/10^3$ (cm^{-3})	$T_e/10^4$ ($^\circ$K)	B (mag)	V (mag)	C
HE 2–57	– 12.8					
IC 2621	– 11.1					
NGC 3587	– 10.33	6.31	1.0		13.2	0.00
HE 2–63						
K 1–22	– 11.42					
HE 2–64	– 12.7					
NGC 3699						
FG 1						
HE 2–67						
HE 2–68	– 11.7					
PB 8	– 11.41	5.1	1.00	14.09	14.05	0.43
HE 2–71	– 11.67					
HE 2–73	– 11.95					
NGC 3918	– 10.04	3.2	1.1			0.36
BLDZ 1	– 11.3					
NGC 4071	– 11.6					
HE 2–76	– 12.8					
HE 2–77	– 12.9					
HE 2–78	– 12.60					
K 2–4					17.7	
HE 2–81	– 13.5					
HE 2–82	– 12.5					
NGC 4361	– 10.48	0.3	2.3		12.9	0.00
HE 2–84						
HE 2–86	– 12.27	17.	0.85	19.2:	18.4:	2.08
IC 3568	– 10.77	5.5	1.08	12.47	12.31	0.18
A 35	– 11.3		1.5			
H 4–1	– 12.70	1.0	1.2			
HE 2–88	– 11.9		1.25	15.56	15.26	0.96
IC 4191	– 11.00					
HE 2–90	– 11.2	5.0	1.00	16.1	15.3	1.54
TH 2–A	– 12.7			15.39	14.48	2.06
NGC 5189	– 10.87			14.1		
MY CN 18	– 11.2	4.7	0.90	14.7	13.9	1.83
A 36	– 10.86				11.5	

Planetary Nebulae Parameters

Name	Log F(Hβ) (erg cm^{-2} s^{-1})	N$_e$/10^3 (cm^{-3})	T$_e$/10^4 (°K)	B (mag)	V (mag)	C
HE 2–97	– 11.3	8.0	0.58	15.4	15.4	0.60
NGC 5307	– 11.2	2.0	1.28	14.58	14.69	0.45
HE 2–99		0.7	1.33	14.13	13.84	1.12
NGC 5315	– 10.42	11.	0.92	14.7	14.3	0.27
HE 2–101	– 13.0					
HE 2–102						
IC 972					17.7	
HE 2–103	– 12.0					
HE 2–104	– 11.6	1.1	1.09	16.8	16.3	1.44
HE 2–105	– 12.0	0.3		15.45	15.61	0.42
HE 2–108	– 11.2	1.6	0.98	12.68	12.63	0.54
HE 2–107	– 12.2	3.8	1.00	16.8	15.9	1.70
IC 4406	– 10.3	0.62			14.7	0.26
HE 2–111	– 11.6					
HE 2–112	– 11.4	1.9	1.28	17.8:	17.1:	1.59
HE 2–114	– 12.0					
HE 2–115	– 12.2	46.	0.88	17.0	15.6	2.41
HE 2–116	– 12.0					
HE 2–117	– 12.3					
HE 2–118	– 11.1	50.	0.98	16.0	16.1	0.17
HE 2–119	– 11.6					
HE 2–120	– 12.3					
NGC 5873	– 11.2	5.0	1.22	16.1	16.4	0.14
NGC 5882	– 10.38	17.	0.90	13.66	13.62	0.45
HE 2–123	– 12.3					
ME 2–1	– 11.6	2.1	1.26	17.0	17.4	0.11
PE 2–8	– 13.6					
HE 2–125	– 12.9					
HE 2–127	– 12.31					
HE 2–128	– 12.1					
HE 2–129	– 12.9					
MZ 1	– 11.5					
HE 2–131	– 10.17	22.	0.90	10.92	11.08	0.30
HE 2–132	– 12.3					
HE 2–133	– 13.4					

Planetary Nebulae Parameters

Name	Log F(Hβ) (erg cm^{-2} s^{-1})	N$_e$/10^3 (cm^{-3})	T$_e$/10^4 (°K)	B (mag)	V (mag)	C
NGC 5979	- 10.9		1.00	16.7	16.7	0.74
CN 1–1	- 11.74	100.	1.50	11.44	10.83	1.27
HE 2–136	- 11.7	4.0		16.5	16.6	0.60
SP 1						
HE 2–138	- 10.6	12.	1.10	10.69	10.83	0.40
HE 2–140	- 12.42	7.3	0.73	18.4	17.2	1.90
HE 2–141	- 11.8					
HE 2–142	- 12.2	44.	0.71	17.2	15.7	1.73
HE 2–143	- 13.20					
NGC 6026						
NGC 6058	- 11.80	5.5	1.34	13.55	13.78	0.05
HE 2–145						
HE 2–146						
IC 4593	- 10.55	5.5	0.88	10.92	11.13	0.12
NGC 6072	- 11.37	1.7	0.9		17.5:	1.05
HE 2–147	- 13.4	5.4			16.9:	1.27
H 1–1	- 12.1					
HE 2–149	- 12.3	0.6	1.49	17.1	16.1	2.72
MZ 2	- 11.2					
HE 2–151	- 11.95	49.	0.66	13.19	13.07	0.65
HE 2–152	- 11.6					
HE 2–153	- 11.8					
MZ 3	- 10.8	5.8	0.65	16.8:	16.0	1.93
HE 2–155						
HE 2–157	- 12.7					
SN 1	- 11.78	3.0	1.01	14.54	14.73	0.22
HE 2–158	- 12.2					
K 1–3					20.	
PE 1–6	- 13.2					
HE 2–159	- 12.5					
HE 2–161	- 12.4					
HE 2–162	- 12.07	1.0	1.33	13.53	13.43	0.91
HE 2–163	- 12.8					
HE 2–164	- 12.2					
HE 2–165	- 11.7					

Planetary Nebulae Parameters

Name	Log F(Hβ) (erg cm^{-2} s^{-1})	$N_e/10^3$ (cm^{-3})	$T_e/10^4$ (°K)	B (mag)	V (mag)	C
PE 1–7	– 12.39	3.6	1.72	18.3	17.1	2.33
NGC 6153	– 10.85	3.84	1.56			0.93
NGC 6164					6.8	
HE 2–169	– 13.2					
HE 2–171	– 12.2					
HE 2–170	– 12.04					
PC 11	– 11.45	7.7	1.34	13.0	12.67	0.61
HE 2–175	– 13.0					
VD 1–1	– 12.0					
PC 12						
NGC 6210	– 10.08	3.0	1.07	13.6	13.7	0.29
VD 1–2	– 13.25					
H 1–2						
A40					19.6	
VD 1–3	– 13.39					
VD 1–4	– 12.92					
VD 1–5	– 13.51					
HE 2–182	– 11.08	100.	0.73	13.35	13.33	0.40
H 1–3	– 13.1					
VD 1–6	– 12.4					
H 1–5	– 13.8					
HE 2–186						
HE 2–185						
M 2–4	– 11.84	3.6	1.11	17.3	16.9	1.40
HE 2–187	– 12.8			13.0	12.78	1.41
IC 4634	– 10.92	3.39	1.06			0.53
M 2–5	– 12.03	1.5	0.56	16.4	16.0	1.10
M 1–19	– 12.0					
M 2–6	– 12.16					
H 2–1	– 11.45	8.0	1.08	13.50	13.21	0.91
IC 4637	– 11.2	2.5	1.09	12.88	12.54	1.12
VD 1–8	– 13.74					
M 2–7	– 12.5	0.9		14.61	13.92	1.28
M 2–8	– 12.5					
M 2–9	– 11.6					

Planetary Nebulae Parameters

Name	Log F(Hβ) (erg cm^{-2} s^{-1})	$N_e/10^3$ (cm^{-3})	$T_e/10^4$ (°K)	B (mag)	V (mag)	C
PC 14	− 11.5					
H 1–6	− 13.2					
H 1–7	− 11.8					
M 4–3	− 12.5					
IC 4642	− 11.0	0.8	2.10	16.5:	16.6:	0.33
M 3–36	− 12.45					
NA 1	− 11.96	1.00	1.06	15.8	>16.0	0.09
NGC 6302	− 9.9	20.	1.65	16.7	16.0	1.32
M 2–10	− 12.62					
NGC 6309	− 11.16	4.27	1.35	14.79	13.74	0.07
H 1–8						
TH 3–3	− 12.8					
TH 3–4	− 13.4					
HE 2–207						
TH 3–6	− 13.67					
M 3–37						
NGC 6326	− 11.06	1.70	1.9			0.24
M 2–11	− 12.60					
M 3–38	− 13.46					
M 3–39						
H 1–9	− 12.32			16.6	15.8	1.70
H 1–11	− 12.3					
NGC 6337					17.	
M 3–40	− 13.58					
M 2–12	− 12.25	10.		15.14	14.74	1.18
M 3–7	− 12.36	13.	1.13	17.2	16.5	1.55
M 3–8	− 13.03	6.			16.7	1.96
TH 3–12	− 13.75					
M 3–41	− 12.96					
M 3–9	− 12.8				18.	
CN 1–3						
H 1–12	− 14.2					
M 3–42	− 13.3					
M 3–10	− 12.51					
CN 1–4						

Planetary Nebulae Parameters

Name	Log F(Hβ) (erg cm^{-2} s^{-1})	$N_e/10^3$ (cm^{-3})	$T_e/10^4$ (°K)	B (mag)	V (mag)	C
H 2–10	– 13.17					
H 1–14						
H 1–13	– 13.3					
H 1–15	– 12.7					
M 4–4	– 13.8				>21.	
M 2–13	– 12.25					
M 1–20						
A 41	– 12.3	0.5	1.00	16.1	16.1	0.50
NGC 6369	– 10.9	4.4	1.12	17.3	16.2	2.09
H 2–11	– 13.9					
H 1–16	– 13.1					
H 1–17	– 13.5					
H 1–18						
H 1–19	– 13.27					
H 1–20	– 13.03					
TH 3–24	– 14.2					
TH 3–25	– 14.1					
TH 3–55	– 14.1					
H 2–13	– 13.50					
TH 3–26	– 13.4					
H 1–22						
H 2–14	– 12.6					
H 1–23						
H 1–24	– 12.7					
M 1–21						
H 2–15	– 13.35					
HE 2–250	– 13.33					
PC 17						
M 4–6	– 14.3				>21.	
HE 2–248						
M 3–11	– 12.7					
TH 3–27	– 13.0					
H 1–26	– 12.4					
M 3–12	– 12.3					
M 1–23						

Planetary Nebulae Parameters

Name	Log F(Hβ) (erg cm^{-2} s^{-1})	N$_e$/10^3 (cm^{-3})	T$_e$/10^4 ($^\circ$K)	B (mag)	V (mag)	C
M 1–24	– 11.9					
M 1–25	– 11.90	5.2	0.79	16.5	15.9	1.38
FG 2	– 11.80					
HE 2–260	– 12.13	93.	0.65	14.43	14.25	0.63
H 2–16	– 12.9					
H 2–17	– 13.39					
HE 2–262	– 13.73					
H 1–27	– 12.7					
M 3–13	– 13.4					
HB 4	– 12.4					
M 2–14						
H 1–28						
K 1–14						
H 2–18	– 13.15					
H 1–29	– 12.4					
M 3–14	– 12.3					
H 1–30						
IC 4663						
TC 1	– 10.8	1.8	0.81	11.71	11.85	0.30
H 1–31	– 12.63					
M 3–15						
H 2–20	– 13.48					
H 1–32	– 13.5					
M 1–26	– 11.11	80.	1.00	13.57	12.88	1.65
K 1–15						
TH 4–1	– 12.8					
TH 4–2	– 13.0					
M 1–27	– 12.25	7.0	1.00	15.70	14.70	2.12
M 2–15						
H 1–33	– 12.74					
H 2–22	– 13.5					
M 1–28	– 13.9					
HB 5	– 11.2	6.3	1.24	18.4	17.4	1.84
H 1–34	– 13.26					
NGC 6439	– 11.59					

Planetary Nebulae Parameters

Name	Log F(Hβ) (erg cm^{-2} s^{-1})	N$_e$/10^3 (cm^{-3})	T$_e$/10^4 (°K)	B (mag)	V (mag)	C
H 2–24	– 13.58					
TH 4–3	– 13.4					
H 2–23	– 12.78					
H 1–35	– 11.38	19.	1.18	15.7	15.3	1.31
H 2–25	– 13.35					
NGC 6445	– 11.5	1.10	1.5		19.0:	0.80
H 1–36						
H 2–26	– 13.7					
M 1–29	– 12.2					
M 3–43					>21.	
H 1–38	– 13.6					
H 1–37	– 12.2					
TH 4–4	– 13.05					
TH 4–5	– 12.9					
HF 2–1	– 12.2					
TH 4–6						
M 3–44	– 14.0					
H 2–27	– 14.1					
M 3–45	– 13.3					
M 2–17	– 11.5					
M 2–16						
TH 4–7						
B1 3–15	– 14.6					
M 3–16	– 12.6					
M 1–30	– 11.77	5.9	0.68	16.5	16.3	0.91
M 1–31	– 11.9	5.4	0.89	19.2:	18.2:	1.75
TH 4–8						
H 2–29	– 13.7					
H 1–39	– 12.58					
M 2–18	– 12.48					
M 2–19	– 12.8					
B1 M	– 14.1					
B1 O	– 13.5					
A 43	– 12.93	0.10	1.40	14.70	14.79	0.38
M 2–20						

Planetary Nebulae Parameters

Name	Log F(Hβ) (erg cm^{-2} s^{-1})	N$_e$/10^3 (cm^{-3})	T$_e$/10^4 ($^\circ$K)	B (mag)	V (mag)	C
CN 2–1						
B1 Q	– 14.0					
K 2–5						
M 3–46	– 14.0					
HB 6						
H 1–40						
VY 1–2	– 11.53	7.94	0.93	17.3	17.6	0.06
B1 3–13	– 13.7					
H 2–31						
TH 4–9						
HE 2–306	– 11.8	1.2		15.8	16.0	0.33
M 3–17	– 13.0					
M 1–32						
H 1–41	– 11.6	0.6		16.1	16.1	0.62
H 1–42	– 11.5	3.3	1.00	17.2:	17.2:	0.85
TH 4–10	– 12.93					
M 3–47	– 14.3					
H 1–44	– 13.35					
H 1–43	– 12.56	3.1		15.95	15.53	1.22
M 2–21	– 12.0	2.7	1.25	15.49	14.64	1.60
H 2–33	– 13.9					
M 3–19	– 12.9			18.2	16.9	1.52
H 1–45	– 12.7			18.6	16.8	2.40
M 2–22	– 12.5	2.2		17.0	16.4	0.95
PE 2–11	– 13.9					
H 1–46						
M 3–20	– 12.2					
M 1–33						
M 3–48	– 13.56					
FG 3	– 10.9	15.	0.93	14.3	14.3	0.27
H 2–35	– 13.0					
TH 4–11	– 13.11					
H 1–47	– 12.61					
M 3–20						
M 1–34	– 13.2					

Planetary Nebulae Parameters

Name	Log F(Hβ) (erg cm^{-2} s^{-1})	N$_e$/10^3 (cm^{-3})	T$_e$/10^4 (°K)	B (mag)	V (mag)	C
PE 1–11	– 13.0					
M 2–23	– 11.1	13.	1.07	16.6	16.7	0.65
NGC 6543	– 9.56	7.24	0.81	10.87	11.31	0.12
M 2–24	– 11.8					
M 3–22	– 12.6					
M 3–21	– 11.12				14.0	
M 3–49	– 13.1					
M 2–25	– 12.5					
M 2–26	– 13.1					
IC 4673	– 11.6					
M 1–35	– 12.37					
H 1–50	– 11.73	6.0		17.0	16.2	0.72
M 2–27	– 12.0					
M 3–50	– 12.96					
H 2–36	– 13.1					
H 1–51	– 13.2					
H 2–37	– 12.81					
M 3–51	– 13.4					
M 2–28	– 12.59					
NGC 6537	– 11.63	3.98	1.60	>16.1		2.02
M 1–37	– 12.0	4.5		15.32	14.98	1.15
H 1–53	– 12.59					
M 1–38	– 12.01	11.		14.92	14.42	1.20
SP 3						
M 2–29	– 12.18					
M 3–23	– 12.2					
H 1–54	– 11.4	24.	1.05	15.7	15.4	0.91
H 1–55	– 12.66					
M 1–39	– 13.07					
H 1–56	– 12.10					
M 3–24	– 12.3					
H 2–39	– 12.91					
H 2–40	– 13.2					
M 1–40	– 12.84					
H 1–58	– 12.5	5.8	1.50	18.6	17.4	2.26

Planetary Nebulae Parameters

Name	Log F(Hβ) (erg cm^{-2} s^{-1})	N$_e$/10^3 (cm^{-3})	T$_e$/10^4 (°K)	B (mag)	V (mag)	C
M 1–41	– 13.6					
H 1–57	– 12.8					
M 3–52						
M 1–42	– 11.9	1.7	0.94	18.4	17.3	0.72
H 1–59						
AP 1–12	– 11.5	5.0		13.42	13.27	0.72
NGC 6565	– 11.23	9.12	1.15			0.31
NGC 6563	– 10.96	0.62			18.	0.34
M 1–43	– 13.4			16.9	16.0	1.49
H 2–41	– 12.8					
H 1–60	– 12.4					
H 2–42	– 13.2					
M 4–8						
M 2–30	– 11.8	1.1	1.09	17.7	17.1	0.87
H 1–61	– 13.48					
H 2–43	– 12.2					
NGC 6572	– 9.75	10.0	1.05		9.06	0.36
H 1–62	– 11.98	6.		15.03	14.89	0.62
M 2–31						
NGC 6567	– 10.95	6.61	1.09			0.68
M 2–32	– 11.6		1.17	15.4	15.7	0.17
M 4–9	– 12.6				19.7	
M 2–33						
M 3–25						
SWST 1	– 10.3	100.		11.6	11.8	0.0
M 3–26						
H 1–63						
M 1–44						
NGC 6578	– 11.75	4.68				1.72
M 2–34	– 13.0					
M 2–35	– 12.50					
M 2–36						
PE 1–12	– 12.9					
IC 4699						
PE 2–13						

Planetary Nebulae Parameters

Name	Log F(Hβ) (erg cm^{-2} s^{-1})	N$_e$/10^3 (cm^{-3})	T$_e$/10^4 (°K)	B (mag)	V (mag)	C
CN 3–1	– 10.95	7.2	0.8			
H 1–64	– 13.2					
H 2–46	– 12.84					
M 2–37	– 13.0					
M 2–38	– 12.7					
H 1–65	– 12.21	10.		14.99	14.66	0.83
M 2–40						
M 2–39	– 12.13	2.8		16.1	15.8	1.01
M 2–41	– 12.5					
M 2–42	– 12.12					
NGC 6620	– 12.31					0.92
M 1–45						
K 3–1						
M 3–53	– 14.3					
H 1–66	– 12.4					
H 1–67	– 12.1					
PC 19						
K 3–2						
NGC 6629	– 10.94	1.82	0.85	13.03	12.77	1.04
M 2–43	– 13.1					
VY 2–1	– 11.50	3.4	0.86	17.5:	17.6:	0.59
M 1–46	– 11.5	3.9		13.14	12.83	1.11
M 3–27						
CN 1–5	– 11.3	4.3	0.84	16.4	16.6	0.34
M 1–47						
M 1–48						
PE 2–14						
A 44						
K 3–4						
HE 2–406						
A 46	– 12.33	0.10	1.00	14.62	14.87	0.00
HF 2–2	– 12.3					
NGC 6644	– 11.00	7.24	1.22		>15.9	0.41
M 3–28	– 13.8					
M 3–54	– 13.3					

Planetary Nebulae Parameters

Name	Log F(Hβ) (erg cm^{-2} s^{-1})	$N_e/10^3$ (cm^{-3})	$T_e/10^4$ (°K)	B (mag)	V (mag)	C
M 1–50	− 11.9					
M 3–55	− 14.2					
M 1–51	− 12.9					
K 3–6						
IC 4732	− 11.51	2.69	1.44	>16.5	>16.2	0.47
M 1–52	− 12.8					
M 4–10						
K 3–7	− 14.0					
PE 1–13	− 12.7					
A 47						
M 1–53	− 12.1	0.01	1.07	15.05	14.50	1.09
M 1–54	− 11.7					
M 1–55	− 11.86	9.9		13.96	13.99	0.55
M 2–44						
M 1–56	− 12.18					
M 3–29	− 11.7	0.6	0.95	15.30	15.55	0.07
M 2–45	− 13.6					
M 1–57	− 12.4	5.3	1.24	17.3	16.3	1.77
M 3–30	− 12.7			16.9	16.0	1.32
K 3–11	− 14.1					
PE 1–14	− 14.2					
M 1–58						
PC 20						
M 2–36						
M 1–59						
M 1–60	− 12.28					
HE 2–418	− 12.6			15.6	16.0	0.35
M 3–31						
M 3–32						
IC 4776	− 10.5	7.1	1.18	14.2	13.9	0.50
PC 21	− 12.1					
PE 2–15	− 13.70					
K 3–13						
K 4–5						
M 1–61	− 11.43	6.4	0.98	17.1:	16.8:	1.09

Planetary Nebulae Parameters

Name	Log F(Hβ) (erg cm^{-2} s^{-1})	$N_e/10^3$ (cm^{-3})	$T_e/10^4$ (°K)	B (mag)	V (mag)	C
H 2–48	– 11.30	10.		13.55	13.37	0.81
PE 1–15	– 13.09					
M 2–46	– 13.67					
PE 1–16						
PE 1–17	– 13.37					
M 3–33	– 11.88		1.01	15.6	15.9	0.26
K 4–6						
PE 1–18	– 13.23					
K 3–14	– 12.81					
PE 1–19	– 12.56					
M 1–62	– 11.95					
HU 2–1	– 10.76	9.5	0.91	13.45	13.63	0.49
M 1–64						
M 1–63						
K 3–15						
K 3–16	– 13.55					
A66 49	– 12.9					
VY 1–4						
M 4–11	– 12.3					
K 4–8	– 12.35					
IC 1295	– 11.4				15.	
NGC 6720	– 10.06	0.6	1.0		14.7:	0.03
HB 7	– 11.25	2.7	0.86	13.76	13.97	0.28
YM 16						
K 3–17						
M 1–65						
PE 1–20						
PE 1–21	– 13.3					
AP 2–1						
M 1–66						
K 4–10						
K 3–18	– 14.2					
A 50						
A 51						
K 3–19						

Planetary Nebulae Parameters

Name	Log F(Hβ) (erg cm^{-2} s^{-1})	$N_e/10^3$ (cm^{-3})	$T_e/10^4$ (°K)	B (mag)	V (mag)	C
SH 2–71	– 12.1					
K 3–20	– 13.23					
NGC 6741	– 11.5	11.	1.16	18.2:	17.6:	1.25
K 3–21	– 13.7					
K 1–17						
A 52						
HB 8	– 11.92					
K 4–16						
NGC 6751	– 11.7	1.1	1.00	15.67	15.44	0.86
A 53						
A 54						
K 4–17						
K 3–22						
NGC 6765						
M 1– 67						
K 3–24	– 14.0					
A 56						
HE 2–428	– 13.6					
M 2–47						
K 4–20						
HE 2–429						
M 1–69						
K 2–11						
HE 2–430						
NGC 6772	– 11.65	6.31			18.7:	1.07
K 3–26						
K 3–29						
IC 4846	– 11.33	10.0	1.0			0.53
IC 1297	– 11.3	2.9	0.99	16.7	16.6	0.28
K 3–30						
NA 2						
A 58						
NGC 6778	– 11.14	1.7	1.0		14.8:	0.34
NGC 6781	– 11.19	0.49			16.2:	1.26
A 59						

Planetary Nebulae Parameters

Name	Log F(Hβ) (erg cm^{-2} s^{-1})	$N_e/10^3$ (cm^{-3})	$T_e/10^4$ (°K)	B (mag)	V (mag)	C
A 60						
K 3–31	− 13.78					
M 4–14						
ANON						
K 3–33						
NGC 6790	− 11.0	11.	1.13	16.3	15.5	1.25
HE 2–432	− 13.32					
HE 1–1	− 13.4					
VY 2–2	− 11.56	1.	1.49	15.94	14.84	1.48
M 3–34	− 11.7			15.7	15.9	0.47
HE 1–2						
PB 9	− 12.4					
K 3–35						
PB 10						
HE 2–434		1.5		14.46	14.45	0.34
K 4–28						
HE 2–436	− 12.4					
NGC 6803	− 11.15	13.49	1.02		14.:	0.71
NGC 6804	− 11.28	1.20			14.1	0.85
K 3–36						
HE 2–437						
A 62	− 12.01					
K 4–30						
K 3–37						
NGC 6807	− 11.41	3.98	1.17	>15.5		0.33
BD+30°3639	− 10.01	10.	0.8	10.1		1.72
K 3–39						
M 1–71						
K 3–40						
K 4–31						
HE 2–440						
ME 1–1						
K 3–41						
K 3–42						
K 3–43						

Planetary Nebulae Parameters

Name	Log F(Hβ) (erg cm^{-2} s^{-1})	$N_e/10^3$ (cm^{-3})	$T_e/10^4$ (°K)	B (mag)	V (mag)	C
M 1–73						
M 1–72	– 12.54	3.5	1.43	18.9	17.6	2.43
PC 22						
A 63						
M 1–74	– 11.89	7.7	0.84	18.8	18.1	1.21
K 4–32						
NGC 6818	– 10.45	3.80	1.27	>15.0		0.32
A 64						
HE 2–447						
NGC 6826	– 9.97	1.62	0.97	10.22	10.69	0.04
A 65						
HE 1–3						
NGC 6833	– 11.22	10.:	1.3			0.10
M 2–48						
K 4–37						
PC 23						
K 3–48						
K 3–49						
NGC 6842						
A 66						
K 4–41						
A 67						
HE 1–4						
NGC 6853	– 9.44	2.95	1.3		13.9	0.00
A 68						
NGC 6852						
K 3–51	– 13.13					
K 3–52						
K 3–53						
M 1–75						
K 3–54						
K 3–55						
M 4–17						
NGC 6879						
NGC 6884	– 11.11	7.94	0.92	16.1	>15.6	0.81

Planetary Nebulae Parameters

Name	Log F(Hβ) (erg cm^{-2} s^{-1})	$N_e/10^3$ (cm^{-3})	$T_e/10^4$ (°K)	B (mag)	V (mag)	C
NGC 6881	– 12.25	10.	1.41	16.8	16.7	1.77
HE 1–5						
NGC 6886	– 11.50	851	1.3			0.94
K 3–57						
HE 2–459						
NGC 6891	– 10.68	2.5	0.96	12.30	12.44	0.32
NGC 6894	– 11.46	0.35			17.5:	0.56
HE 1–6						
IC 4997	– 10.8	14.	1.70	16.0:	15.4:	1.52
M 3–35						
NGC 6905	– 11.2	0.9	0.97	15.8	15.5	0.51
A 70	– 12.3					
A 71	– 11.75					
A 72						
NGC 7008	– 11.51	3.98	1.10	13.79	13.21	0.84
NGC 7009	– 9.78	3.48	1.05		>10.9	0.15
NGC 7026	– 10.93	3.39	0.88	15.33	14.20	0.65
NGC 7027	– 10.17	10.0	1.4			1.35
NGC 7048						
A 74						
M 1–77						
M 1–78						
K 3–60						
PS 1	– 12.15	2.0	1.2			
K 3–61						
K 3–62						
A 77	– 12.43	0.10	1.00	>16.0	15.68	2.34
IC 5117	– 11.40	10.0	1.12	17.5	16.7	1.38
ANON, HU 1–2	– 11.18	7.76	1.76		16.1	0.59
A 78	– 12.04					
K 4–45						
NGC 7094	– 11.8			13.62	13.73	0.08
M 1–79						
M 2–49						
NGC 7139						

Planetary Nebulae Parameters

Name	Log F(Hβ) (erg cm^{-2} s^{-1})	N$_e$/10^3 (cm^{-3})	T$_e$/10^4 (°K)	B (mag)	V (mag)	C
M 2–50						
IC 5148–50	− 11.3					
M 2–51	− 12.24	0.40	1.00	14.76	13.45	1.39
BL 2–1						
M 2–52						
IC 5217	− 11.18	6.03	1.2		14.6	0.78
NGC 7293	− 8.8	0.1		13.18	13.62	0.04
ME 2–2	− 11.17	18.	1.3			
M 2–53						
NGC 7354	− 11.60	6.31	1.20		>16.2	1.76
IC 1454						
M 2–54	− 11.98	19.95	1.00	12.29	12.10	0.85
M 1–80						
VY 2–3	− 11.96		1.30	14.56	14.90	0.19
NGC 7662	− 10.93	4.90	1.30	13.6	13.2	0.15
HB 12	− 11.01	39.81	1.40	15.2	14.3	0.98
M 2–55						
ANON.						
JN 1	− 11.48			15.6		
A 82						
A 84	− 11.74				18.0	
M 2–56					>21.	

Part VI
Dying Stars

21
White Dwarf Stars

21.1 Basic Data for White Dwarf Stars

White dwarf stars represent the most common endpoint of stellar evolution. In fact, about 90% of all stars will end up as white dwarfs. Their high temperatures and low luminosities imply that they are small - only about the size of the Earth ($R_e = 0.009 R_\odot$). The mean radius for white dwarfs is R = 0.01 $R_\odot$ (see the first table). Their rather ordinary masses of $M \approx 0.6 M_\odot$ indicate that they have been gravitationally compressed to enormously high densities of about 100,000 $g\ cm^{-3}$. Degenerate electron pressure stabilizes these stars against further collapse.

The second table gives the spectral classification scheme for white dwarf, WD, stars. The capital letter D denotes a degenerate star. It is followed by a capital letter that denotes the primary spectroscopic type in the optical spectrum. White dwarfs usually consist of a pure hydrogen, DA, or pure helium, DB, atmosphere; about two-thirds are DA and one-third DB. The second capital letter is usually followed by a number from 0 to 9 that indexes the temperature, T, defined by 50,400/T. Effective temperatures of white dwarf stars can range from 70,000 °K down to the present limit of observation at about 4,000 °K. The photospheric emission from the hottest (T > 25,000 °K) white dwarfs can sometimes be detected at X-ray wavelengths. The detected stars are listed in the third table. Interacting binary stars with a white dwarf component can emit X-ray radiation of an entirely different sort; it is discussed in the section on cataclysmic binary systems.

White dwarf components of widely-spaced, non-interacting, or detached, binary systems have been investigated by Eggen and Greenstein (1965, 1967), Greenstein (1986), and Wegener (1973). Most of these binary systems consist of a main-sequence star and a white dwarf star; the smaller set of detached pairs of white dwarf stars, or double degenerates, is discussed by Greenstein (1986), Saffer, Liebert, and Olszewski (1988), and Ruiz and Maza (1988).

On cooling through the narrow instability strip, extending from about 13,500 °K to about 10,500 °K, the hydrogen-rich DA white dwarfs become pulsationally unstable, causing variations in their luminous output with periods between 100 and 1200 seconds and amplitudes between 0.003 and 0.30 magnitudes. These variations are due to nonradial gravity mode (g mode) pulsations (for reviews see McGraw, 1980; Robinson, 1979; and Winget, 1986). They are called ZZ Ceti stars after their stellar prototype.

The table of ZZ Ceti stars includes the WD number, alternative names, peak-to-peak amplitude variations, Amp., the dominant periods, P, of variation, and the upper limits to the rate of change, $\dot{P}$, of the period. These stars are usually multi-periodic, with several periodicities simultaneously present in their light curves. Specific periods are nevertheless extremely stable, with accuracies as great as one part in 10^{14}; this makes the ZZ Ceti stars amongst the most accurate astronomical clocks known. We also include a table of pulsating, helium-rich DB white dwarfs whose existence was predicted by the theory of nonradial, g mode pulsations.

The high gravity of the compact white dwarf stars leads to large gravitational redshifts of 20 to 90 $km\ s^{-1}$. Accurate masses have also been inferred for a few white dwarfs that are members of well-studied binary systems. These masses can be combined with the gravitational redshifts to infer radii (see the sixth accompanying table). A longer table of measured gravitational redshifts, V_{grs}, is also included. They can be used to infer masses, M, or radii, R, from the relation $V_{grs} = 0.635(M/M_{\odot})(R_{\odot}/R)\ km\ s^{-1}$, provided one assumes a mean value for one of them.

The radii of white dwarf stars can also be computed from their color and flux (photometry), distances (parallaxes), and model atmospheres. The models are used to convert a measured stellar color into the radiation flux at the stellar surface, and this is combined with the distance and the flux measured at Earth to infer the radius (Shipman, 1972; 1979; Koester, Schulz, and Weidemann, 1979). When this technique is applied to 110 hydrogen-rich white dwarfs, for example, a mean radius of R = 0.01 $R_{\odot}$ is obtained with an estimated uncertainty of 5%. This radius can be used with the theoretical mass-radius relation of Hamada and Salpeter (1961) to infer a mean mass of $M \approx 0.6\ M_{\odot}$. These results are not significantly different form those obtained by others (also see Moffet, Barnes, and Evans, 1978).

Because magnetic flux is conserved in gravitational collapse, the stellar magnetic fields can be amplified to more than a million Gauss when a star contracts to the white dwarf stage. These intense magnetic fields are measured by either the Zeeman displacement of their spectral lines or from their wavelength-dependent, circularly-polarized light (Angel, 1980; Kemp et al., 1970; Landstreet, 1980; Liebert, 1980). The following table of the mega-Gauss (10^6 G) magnetic field strengths, H, of isolated, or single, white dwarf stars is adapted from Schmidt (1987, 1988, 1990). Our table does not include the measured mega-Gauss magnetic fields of the white dwarf components of interacting binary systems; they are discussed in the section on cataclysmic binaries. Didelon (1983) and Angel, Borra, and Landstreet (1981) list about forty additional white dwarf stars with magnetic field strengths between 40 and 1,000 kilo-Gauss (10^3 Gauss), but the uncertainties of these measurements are so large that they are listed as questionable or unmeasurable.

The catalogue of white dwarf stars has been adapted from McCook and Sion (1987). Our list is less extensive, being limited to those with known values for both the apparent visual magnitude, V, and the absolute visual magnitude, M_V. The white dwarf number, WD, is followed by the right ascension, R.A., and declination, Dec., for the epoch 1950.0, the spectral classification, Sp, the V, wide-band colors B − V and U − B, the M_V, the proper motion, μ, and position angle, θ, and the trigonometric parallax, π_t. The distances, D, in parsecs can be computed from $D = \pi_t^{-1}$; the white dwarfs are so faint that they must be within about 200 pc to be detected. Components of proper motion in R.A. and Dec. can be calculated from the relations $\mu_{R.A.} = \mu \sin\theta/[15\cos(Dec.)]$ and $\mu_{Dec} = \mu\cos\theta$.

The accompanying diagram gives the Hertzsprung-Russell plot of M_V versus B − V for all the available data. It can be compared with the similar diagram for nearby stars given on pages 167 and 179 or with the diagram given by Jahreiss (1987).

The final table of this section gives a cross-reference for white dwarf star names; it is also adopted from McCook and Sion (1987).

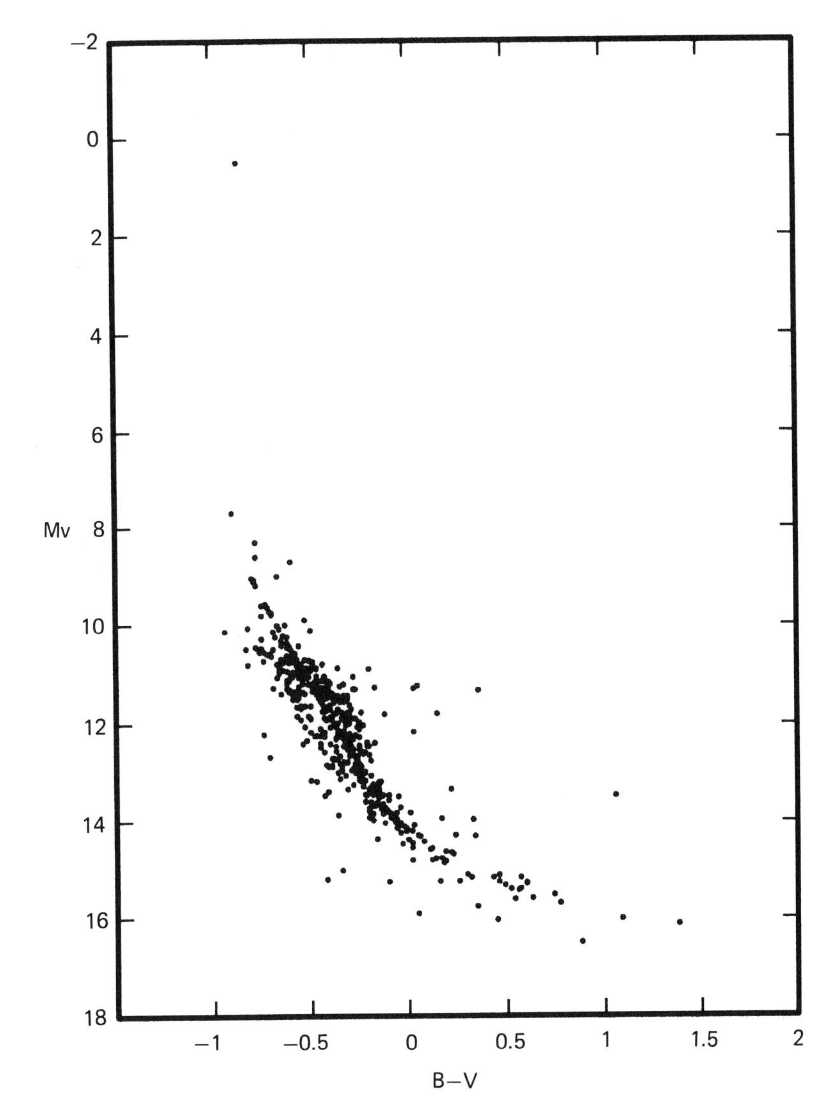

Hertzsprung-Russell Diagram for White Dwarf Stars. Plot of the absolute visual magnitude, M_V, and color index, B-V, for white dwarf stars. They are relatively faint objects with initially high temperatures.

Mean Parameters of White Dwarf Stars*

Mass	$\bar{M}$	=	$0.60 \pm 0.10\ M_\odot$
	$\bar{M}$	=	1.2×10^{33} grams
Radius	$\bar{R}$	=	$0.010 \pm 0.002\ R_\odot$
	$\bar{R}$	=	7×10^{8} cm
Mass Density	$\bar{\rho}$	=	10^{5} g cm^{-3}
Gravity	$\bar{g}$	=	10^{8} cm s^{-2}
Absolute Luminosity	$\bar{L}$	=	$10^{-2.9\pm0.6}\ L_\odot$
Effective Temperature	T_{eff}	$\leq$	4,000 to 70,000 °K
Magnetic Field Strength	H	$\leq$	10^{5} to 10^{9} Gauss
Apparent Visual Magnitude	$\bar{V}$	=	15.14 ± 1.30 mag
Color Index	$\overline{B-V}$	=	0.17 ± 0.31 mag
Color Index	$\overline{U-B}$	=	-0.66 ± 0.33 mag
Absolute Visual Magnitude	$\bar{M}_V$	=	12.13 ± 1.54 mag
Local Space Density†		=	0.003 stars pc^{-3}

*Adapted from Koester, Shulz and Weidemann (1979), Shipman (1972, 1979), Shipman and Sass (1980) and the following catalogues.
†From Liebert, Dahn and Monet (1988).

Spectral Classification of White Dwarf Stars

Sp	Characteristics
DA*	Only Balmer lines; no He I or metals present
DB	He I lines; no H or metals present
DC	Continuous spectrum, no lines deeper then 5% in any part of the electromagnetic spectrum
DO	He II strong; He I or H present
DZ	Metal lines only; no H or He lines
DQ	Carbon features, either atomic or molecular in any part of the electromagnetic spectrum

*The uppercase D denotes degenerate, and the second uppercase letter denotes the primary spectral type in the optical spectrum.

Soft X–ray Emission from Hot White Dwarfs*

WD	Name	Sp	T_{eff} (10^3 °K)	n(He)/n(H)
0232+035	Feige 24	DA1	55.0 ± 5.4	$> 10^{-3}$
0302+027	Feige 31	DAwk		$< 10^{-4}$
0346–011	GR 288	DA2	47.5 ± 2.5	$30 - 200 \times 10^{-5}$
0501+527	G191 2B2	DA1	62.3 ± 3.5	$> 10^{-3}$
0548+000	GR 289	DA1	55.0 ± 5.0	$80 - 600 \times 10^{-5}$
0642–166	Sirius B	DA2	25.5 ± 0.5	$< 2 \times 10^{-5}$
1052+273	GD 125	DA3	26.0	< 1 to 5×10^{-5}
1254+223	EG 187	DA1	42.0 ± 2.0	2 to 40×10^{-5}
1314+293	HZ 43A	DA1	50.0 ± 5.0	$< 1 \times 10^{-5}$
1634–573	HD 149499B	DOZ1	70.0	$> 10^{-3}$
2028+390	GD 391	DA2		$< 10^{-4}$
2309+105	EG 233 (BPM 97859)	DA1	55.0 ± 5.0	1.5 to 15×10^{-5}

*The values of effective temperature, T_{eff}, are those consistent with model fitting to the X–ray data, or those inferred from Lyman alpha profiles (Holberg and Wesemael, 1987). The relative abundance of helium, He, to hydrogen, H, is also inferred from models for the X–ray data. Positions, magnitudes and parallaxes for these objects can be found in the following catalogue of white dwarf stars (also see Mc Cook and Sion, 1987). The first degenerate dwarf detected in soft X–rays was Sirius B using the ANS satellite (Mewe et al., 1975), followed by HZ 43 using SAS–3 (Hearn et al., 1976). These bright X–ray objects have subsequently been observed using the transmission grating spectrograph of EXOSAT (Heise et al., 1988; Paerels et al., 1988). Instruments aboard the Einstein Observatory have detected EG 187, GR 288, GR 289 and LB 1663 (Kahn et al., 1984 – we do not include LB 1663 because is not listed in Mc Cook and Sion's (1987) catalogue), and EG 233 and GD 125 (Petre, Shipman and Canizares, 1986). The white dwarfs detected by the EXOSAT satellite (Heise, 1985) are, in order of decreasing count rate, HZ 43, G191 2B2, Sirius B, Feige 24, EG 187, EG 233, GR 289, GD 391, HD 149499B, and Feige 31, ranging from 28 c/s (HZ 43) and 19 c/s (G191 2B2) to a barely detectable 0.03 c/s (Feige 31).

ZZ Ceti Variable Stars*

WD	Name	Amp. (mag)	Dominant Periods, P	$\dot{P}$ (s s^{-1})
0104− 464	BPM 30551	0.18	607, 745, 823	
0133− 116	ZZ Ceti = Ross 548	0.02	212.77, 213.13, 274.25, 274.77	$< 2 \times 10^{-13}$
0341− 459	BPM 31594	0.21	311, 404, 617	
0416+272	HL Tau 76	0.34	494, 626, 661, 746	
0417+361	G038–029	0.21	925, 1020	
0455+553	G191–016	0.30	~ 865	
0517+307	GD 066 = GR 572	0.31	197, 256, 273, 304	
0625− 253†	GR 903	0.003	555	
0858+363	GD 099	0.13	260, 480, 590	
0921+354	G117–B15A	0.022	107.6, 119.8, 126.2, 215.2, 271.0, 304.4	$< 6.5 \times 10^{-14}$
1159+803	G255–002	0.30	380?, 450?, 685, 830	
1307+354	GD 154	0.10	780, 1186	$< 10^{-8}$
1350+656	G238–053	0.009	206	
1425− 811	L0019–002	0.006	113.8, 118.5, 143.4, 192.6, 350.1	$< 4 \times 10^{-13}$
1559+369	Ross 808	0.15	513, 830	
1647+591	G226–029	0.006	109.086, 109.279, 109.472	
1855+338	G207–009	0.06	292, 318, 557, 739	
1935+276	G185–032	0.03	100 to 1000	
1950+250	GD 385	0.05	128.115, 256.127, 256.332	
2336+049	G029–038	0.28	694, 820, 930	

*Adapted from Dolez, Vauclair and Chevreton (1983), Fontaine and Wesemael (1984), Fontaine et al. (1982), Kepler (1984), Kepler, Robinson and Nather (1983), Kepler et al. (1982), McGraw (1980), McGraw et al. (1981), O'Donahue and Warner (1982), Robinson (1979), Robinson and Kepler (1980), Robinson et al. (1978), Stover et al. (1980), Vauclair, Dolez and Chevreton (1981), and Winget (1986).

†WD 0625–253 = GR 903 is the hot DA component of an eclipsing and spectroscopic binary star V471 Tauri = BD +16°0516 with an orbital period of 0.521 days. Both the X–ray and optical radiation vary with a 555 second period (Jensen et al., 1986; Robinson et al., 1988); this periodicity could be due to rotation of the white dwarf star if it is magnetized and accreting material from its K2 V companion, or it could be caused by the g mode pulsations of the white dwarf.

Pulsating DB White Dwarfs*

WD	R.A.(1950.0) h m s	Dec.(1950.0) ° ′ ″	Period (seconds)
1115+158	11 15 45	+15 49 54	100 to 8000
1351+489	13 51 12	+48 55 06	489.5
1456+103	14 56 07.4	+10 20 15	423 to 854
1645+325†	16 45 25	+32 33 42	140 to 950
1654+160	16 54 42	+16 01 00	148 to 851

*Adapted from Grauer et al. (1988), Mc Cook and Sion (1987), Winget et al. (1982, 1984, 1987).
†WD 1645+325 = GD 358 = GW Vir, g–modes split by rotation with a rotational period of 5400 seconds.

Astrometric Masses and Gravitational Redshifts*

WD	Name	Mass ($M_\odot$)	V_{grs} (km s^{-1})	Radius ($R_\odot$)	T_{eff} (°K)
0413–077	40 Eri B†	0.43 ± 0.02	23.9 ± 1.3	0.013 ± 0.003	16850 to 16950
0642–166	Sirius B†	1.053 ± 0.028	89.0 ± 16.0	0.0078 ± 0.0002	26000 to 28000
1213+528	Case 1	0.38 ± 0.07	39.2 ± 6.0	0.006 ± 0.001	
0429+176	HZ 9	0.51 ± 0.10	37.3 ± 7.2	0.008 ± 0.001	
0727+482	G107–70A,B	0.46 ± 0.10			

†40 Eri B = o^2 Eri B, Sirius B = α CMa B.
*The astrometric solutions for the masses, M, are, from top to bottom, due to Heintz (1974), Gatewood and Gatewood (1978), Stauffer (1987), and Liebert (1980). The measured gravitational redshifts, V_{grs}, are from Wegener (1980), Greenstein, Oke and Shipman (1971), and Stauffer (1987). The radii, R, for 40 Eri B and Sirius B are from Wegener (1980) and Greenstein Oke and Shipman (1971), while their respective effective temperatures are due to Wegener (1980) and Holberg et al. (1984). The other two radii have been inferred from the relation $V_{grs} = 0.635\ (M/M_\odot)(R_\odot/R)$ km s^{-1}.

Gravitational Redshifts of White Dwarf Stars*

WD	Name	V_{grs} (km s^{-1})	R ($R_\odot$)	M ($M_\odot$)
0214+568	EG 017	74	(0.010)	1.17
0326− 273(?)	L587–077B(or A?)	36.1	0.0135	0.77
0349+247	EG 025	26	0.0081	0.33
0352+096	EG 026	69	0.0120	1.30
0401+250	EG 028	52	0.0109	0.98
0407+179	EG 030	43	(0.010)	0.68
0410+117	EG 031	27	(0.010)	0.43
0413− 077	40 Eri B	23.9	0.013†	0.43†
0421+162	EG 036	44	0.0133	0.92
0425+168	EG 037	68	0.0110	1.18
0429+176	HZ 09	37.3	0.008†	0.51†
0431+125	EG 039	79	0.0108	1.34
0438+108	EG 042	65	0.0098	1.00
0501+527	EG 247	34	0.0164	0.88
0642− 166	Sirius B	89.0	0.0078†	1.053†
0727+482	G107–070A,B			0.46†
1105− 048	L0970–030	20.1	0.0132	0.42
1121+216	EG 079	62	0.0123	1.20
1143+321	EG 185	78	0.0124	1.52
1213+528	Case 1	39.2	0.006†	0.38†

Gravitational Redshifts of White Dwarf Stars*

WD	Name	V_{grs} (km s^{-1})	R ($R_\odot$)	M ($M_\odot$)
1327– 083	G014–058	24.9	0.0141	0.56
1334– 160	LDS 455B	51.2	0.0085	0.69
1348– 273	L0619–050B	23.3		
1544+008	EG 113	52	(0.010)	0.82
1544– 377	CD –37°6571B	27.9	0.0145	0.64
1620– 391	CD –38°10980	37.9	0.0118	0.70
1659– 531	L0268–092	30.2	0.0132	0.63
1716+020(?)	Wolf 672B (or A?)	22.0	0.0151	0.53
2032+248	EG 139	38	(0.010)	0.60
	Mean	43.8 ±5.7	0.0123 ±0.0026	0.778 ±0.245

†Sources with astrometric masses are denoted by †; in this case the radii are infered from the formula $V_{grs} = 0.635\ (M/M_\odot)(R_\odot/R)$ km s^{-1} (also see the previous table).

*Gravitational redshifts are from Greenstein et. al. (1977), Greenstein, Oke and Shipman (1971), Greenstein and Trimble (1967), Koester (1987), Stauffer (1987), Trimble and Greenstein (1972), and Wegener (1980). Radii are from Koester (1987), Koester, Schulz and Weidmann (1979) and Shipman (1979), except those in parenthesis which have the mean value of $\bar{R} = 0.010$. Masses are calculated from

$$V_{grs} = 0.635\ (M/M_\odot)(R_\odot/R)\ \text{km s}^{-1}$$

except for the astrometric masses denoted by †.

Isolated Magnetic White Dwarf Stars With Mega–Gauss Fields*

WD	Name	R.A.(1950.0) h m s	Dec.(1950) ° ′ ″	μ (″ yr^{-1})	θ (°)
0041– 102	Feige 7	00 41 15	– 10 16 48	0.167	230
0136+251	PG 0136+251	01 36 05	+25 08 12		
0253+508	KPD 0253+5052	02 53 44	+50 52 12		
	KUV 03292+0035	03 29 11.2	+00 35 09		
0548– 001	G 099–037	05 48 46	– 00 11 12	0.27	21
0553+053	G 099–047	05 53 47	+05 22 00	1.06	207
0816+376	GD 090	08 16 31	+37 40 54		
0912+536	G 195–019	09 12 27	+53 38 36	1.56	222
1015+014	PG 1015+015	10 15 30	+01 26 30		
	GD 116	10 17 59.5	+36 41 54		
1031+234	PG 1031+234	10 31 05	+23 24 42		
1036– 204	LP 790–029	10 36 32	– 20 25 48	0.628	330.3
	ESO 439–162	11 27 24.0	– 31 06 19		
	LBQS 1136–0132	11 36 05.9	– 01 32 26		
1312+098	PG 1313+095	13 12 38	+09 53 06		
1533– 057	PG 1533–057	15 33 55	– 05 42 42		
1639+537	GD 356	16 39 49	+53 46 54	0.236	212
1658+440	PG 1658+441	16 58 17	+44 05 30		
1743– 521	BPM 25114	17 43 48	– 52 06 00	0.112	148
1748+708	G 240–072	17 48 58	+70 52 42	1.66	312
1818+126	G 141–002	18 18 13	+12 37 24	0.28	19
1829+547	G 227–035	18 29 21	+54 45 12	0.385	318.3
1900+705	GRW +70°8247	19 00 40	+70 35 12	0.536	9.6
2010+310	GD 229	20 10 23	+31 04 24	0.102	18.4
2316+123	KUV 813–14	23 16 14	+12 19 36		

*Adapted from Schmidt (1988, 1990 private communication) and Mc Cook and Sion (1987). Individual objects are discussed by Latter et al. (1987), Liebert et al. (1985), Schmidt et al. (1986), Wickramasinghe and Cooper (1988), Wickramasinghe and Ferrario (1988).

Isolated Magnetic White Dwarf Stars With Mega–Gauss Fields*

WD	Name	Sp	P_{rot}	T_{eff} (10^3 °K)	H (10^6 G)
0041−102	Feige 7	DBAP3	2.2 h	22.	35.
0136+251	PG 0136+251	DAP1		45.	2.
0253+508	KPD 0253+5052	DAP	4.1 h	20.	18.
	KUV 03292+0035			15.	12.
0548−001	G 099−037	DQP8		6.2	10.
0553+053	G 099−047	DAP9		5.7	25.
0816+376	GD 090	DAH5		12.	9.
0912+536	G 195−019	DCP7	1.3 d	8.	~100.
1015+014	PG 1015+015	DCX5	1.6 h	10.	120.
	GD 116			15.	65.
1031+234	PG 1031+234	DAP2	3.4 h	15.	500.
1036−204	LP 790−029	DQP9		8.6	200.
	ESO 439−162			6.3	~100.
	LBQS 1136−0132				24.
1312+098	PG 1313+095	DC	5.4 h		~50.
1533−057	PG 1533−057	DAP3		20.	31.
1639+537	GD 356	DCP7		7.5	15.
1658+440	PG 1658+441	DAP2		30.	3.5
1743−521	BPM 25114	DAP6	2.8 d	20.	36.
1748+708	G 240−072	DXP9		6.	~200.
1818+126	G 141−002	DCP8		5.6	5.
1829+547	G 227−035	DXP8		7.	~150.
1900+705	GRW +70°8247	DXP5		14.	320.
2010+310	GD 229	DXH2		16.	>200
2316+123	KUV 813−14	DAP4		10.4	29.

*Adapted from Schmidt (1988, 1990 private communication) and Mc Cook and Sion (1987). HS 1254+3430 has been verbally reported to have a magnetic field of 1.5 × 10^7 Gauss but the result is as yet unpublished. Polarized and unpolarized magnetic stars are denoted by P and H, respectively, and X denotes peculiar or unclassifiable spectra.

21.2 Catalogue of White Dwarf Stars*

Name	R.A.(1950) h m s	Dec.(1950) ° ′	Sp	V (mag)	M_V (mag)
0000- 170	00 00 58	- 17 00.8	DB4	14.69	11.69
0000- 345	00 00 06	- 34 30.0	DC9	14.90	14.06
0004+330	00 04 58	+33 00.8	DA1	13.82	9.11
0007+308	00 07 01	+30 52.2	DC6	16.85	13.15
0009+501	00 09 39	+50 09.0	DA8	14.36	13.92
0009- 058	00 09 58	- 05 50.2	DA6	16.00	12.63
0011+000	00 11 04	+00 02.8	DA6	15.31	12.72
0011- 134	00 11 41	- 13 27.3	DC8	15.82	14.58
0016- 220	00 16 57	- 22 05.8	DA6	15.31	11.05
0017+136	00 17 26	+13 36.1	DB3	15.22	11.12
0019+423	00 19 36	+42 20.2	DC9	16.44	14.64
0023- 109	00 23 30	- 10 54.4	DA7	16.22	13.48
0024- 556	00 24 17	- 55 41.6	DA5	15.14	12.46
0026+136	00 26 16	+13 38.1	DA1	15.94	9.79
0031+150	00 31 51	+15 01.6	DA7	16.87	13.48
0031- 274	00 31 25	- 27 24.9	DA1	14.18	10.56
0033+016	00 33 02	+01 36.9	DA5	15.52	12.06
0034- 211	00 34 56	- 21 10.1	DA7	14.53	13.49
0037+312	00 37 15	+31 16.1	DA1	14.66	9.64
0038+555	00 38 30	+55 33.7	DQ5	14.10	12.17
0038- 226	00 38 57	- 22 37.3	DC9	14.53	14.80
0046+051	00 46 31	+05 09.0	DZ8	12.36	14.30
0047- 524	00 47 48	- 52 25.	DA2	14.20	10.95
0048- 544	00 48 54	- 54 29.	DA3	15.29	11.07
0050- 332	00 50 54	- 33 16.2	DA1	13.36	10.29
0052+226	00 52 05	+22 40.0	DA5	16.16	12.41
0053- 117	00 53 20	- 11 44.1	DA7	15.26	13.42
0058- 044	00 58 24	- 04 27.5	DA3	15.38	11.04
0100- 068	01 00 53	- 06 48.3	DB3	13.95	10.94
0101+048	01 01 14	+04 48.3	DA6	14.10	13.00
0104- 331	01 04 24	- 33 09.	DA1	13.57	9.07
0104- 464	01 04 42	- 46 26.	DA6	15.26	12.38
0106+372	01 06 34	+37 16.8	DA2	15.25	9.57
0107+172	01 07 45	+17 12.6	DA3	16.46	10.51
0107+267	01 07 27	+26 45.1	DA4	14.93	11.14

Pages 546 to 577, Section 21.2 Catalogue of White Dwarf Stars. The absolute magnitudes, M_V, appear to be inconsistent with the trigonometric parallaxes, π_t, but this is because M_V was only inferred from π_t when π_t was greater than 0.1. Usually M_V was inferred from the measured colors and a color magnitude relation given in the original reference, for there are large errors in the small parallax values.

Catalogue of White Dwarf Stars*

Name	B–V (mag)	U–B (mag)	μ ($'' \; yr^{-1}$)	θ (°)	π_t^{-1} ($''$)
0000– 170	– 0.05	– 0.94	0.241	86.	
0000– 345	+0.465	– 0.435	0.76	168.	0.0715
0004+330	– 0.29	– 1.21			
0007+308	+0.25	– 0.70	0.55	220.	
0009+501	+0.42	– 0.42	0.718	216.0	0.0908
0009– 058	+0.22	– 0.50	0.27	130.	
0011+000	+0.24	– 0.58	0.55	114.	0.031
0011– 134	+0.61	– 0.27	0.97	217.	0.0549
0016– 220	+0.22	– 0.58	0.09	220.	
0017+136	– 0.12	– 0.97	0.04	77.	
0019+423	+0.72	– 0.16	0.35	218.	0.044
0023– 109	+0.36	– 0.45	0.27	160.	
0024– 556	+0.17	– 0.66	0.59	214.	
0026+136	– 0.20	– 1.14			
0031+150	+0.36	– 0.48	0.27	192.	
0031– 274	– 0.26	– 1.20			
0033+016	+0.24	– 0.68	0.45	201.	0.026
0034– 211	+0.45	– 0.57	0.32	230.	0.062
0037+312	– 0.22	– 1.20			
0038+555	+0.00		0.32	103.	0.046
0038– 226	+0.62	– 0.14	0.60	232.	0.0969
0046+051	+0.56	+0.04	2.97	154.	0.2291
0047– 524	– 0.015	– 0.79	0.11	84.	
0048– 544	– 0.02	– 0.78	0.186	66.	
0050– 332	– 0.25	– 1.07			
0052+226	+0.20	– 0.67	0.56	109.	
0053– 117	+0.35	– 0.50	0.47	350.	0.043
0058– 044	+0.03	– 0.75			
0100– 068	– 0.12	– 0.37	0.21	166.	
0101+048	+0.14	– 0.50	0.45	53.	0.0478
0104– 331	– 0.29	– 1.11	0.182	77.	
0104– 464	+0.29	– 0.58	0.234	70.	
0106+372	– 0.23	– 1.05			
0107+172	– 0.10	– 0.81			
0107+267	+0.09	– 0.64	0.25	124.	

*Adapted from McCook and Sion (1987).

Catalogue of White Dwarf Stars*

Name	R.A.(1950) h m s	Dec.(1950) ° ′	Sp	V (mag)	M_V (mag)
0109− 264	01 09 48	− 26 29.3	DA1	13.15	10.74
0115+159	01 15 20	+15 55.0	DQ6	13.82	12.71
0121+401	01 21 25	+40 08.3	DC9	17.10	15.16
0126+101	01 26 46	+10 07.9	DA6	14.38	13.06
0126+422	01 26 46	+42 12.8	DA3	14.0 pg	11.06
0126− 532	01 26 06	− 53 17.	DA4	14.48	11.27
0127+270	01 27 19	+27 00.9	DA2	15.5 pg	11.2
0133− 116	01 33 42	− 11 36.	DAV4	14.10	11.77
0135− 052	01 35 27	− 05 14.8	DA7	12.84	13.62
0136+768	01 36 45	+76 53.8	DA3	14.85pg	11.33
0138− 559	01 38 42	− 55 58.	DB3	14.86	11.29
0141− 675	01 41 36	− 67 32.	DA7	13.90	13.95
0142+230	01 42 32	+23 02.9	DC4	17.41	11.64
0142+312	01 42 19	+31 17.9	DA6	14.80	12.78
0143+216	01 43 56	+21 39.8	DA5	15.05	12.27
0145− 174	01 45 45	− 17 26.3	DC7	17.57	13.64
0148+467	01 48 56	+46 45.2	DA3	12.44	11.35
0148+641	01 48 12	+64 11.2	DA6 *	14.8 pg	12.51
0151+017	01 51 38	+01 46.7	DA4	15.00	11.47
0203+207	02 03 03	+20 43.0	DABH3	17.36	11.12
0204− 233	02 04 26	− 23 21.6	DA6	15.60	12.44
0205+250	02 05 56	+25 00.1	DA4	13.22	11.08
0208+396	02 08 13	+39 41.5	DAZ7	14.51	13.46
0213+396	02 13 13	+39 37.6	DA6	14.54	12.81
0213+427	02 13 47	+42 44.5	DA9	16.23	14.68
0214+568	02 14 00	+56 53.	DA2	13.68	10.71
0220+222	02 20 44	+22 14.0	DA5	15.83	11.67
0227+050	02 27 42	+05 03.	DA3	12.65	11.04
0231− 054	02 31 37	− 05 24.9	DA6	14.24	11.89
0232+035	02 32 31	+03 30.1	DA1	12.25	12.22
0235+064	02 35 52	+06 24.8	DA8	15.09	13.95
0236+745	02 36 22	+74 30.0	DA6	15.93	12.50
0239+109	02 39 26	+10 59.9	DA6	16.18	12.90
0245+541	02 45 03	+54 11.1	DA9	15.32	15.16
0255− 705	02 55 48	− 70 34.	DA6	14.08	12.23

Catalogue of White Dwarf Stars*

Name	B–V (mag)	U–B (mag)	μ ($''$ yr^{-1})	θ (°)	π_t^{-1} ($''$)
0109– 264	- 0.24	- 1.04	0.00		
0115+159	+0.11	- 0.81	0.6391	182.3	0.066
0121+401	+0.82	+0.21	0.74	103.	0.031
0126+101	+0.26	- 0.54	0.42	189.	0.027
0126+422					
0126– 532	+0.01	- 0.725	0.204	90.	
0127+270					
0133– 116	+0.20	- 0.54	0.51	119.	0.014
0135– 052	+0.34	- 0.50	0.670	121.	0.0821
0136+768	+0.06	- 0.63			
0138– 559	- 0.12	- 1.06	0.16	100.	
0141– 675	+0.43	- 0.44	1.05	198.	0.102
0142+230	- 0.05	- 0.62	0.31	112.	
0142+312	+0.19	- 0.56	0.34	115.	
0143+216	+0.25	- 0.65	0.27	240.	
0145– 174	+0.37		1.186	187.6	0.0157
0148+467	+0.06	- 0.53	0.07	5.	0.061
0148+641			0.29	129.	
0151+017	+0.16	- 0.59	0.32	95.	
0203+207	- 0.12	- 1.02	0.34	216.	0.0070
0204– 233	+0.19		0.228	115.	
0205+250	- 0.04	- 0.85	0.42	105.	0.029
0208+396	+0.37	- 0.43	1.15	116.	0.0593
0213+396	+0.23	- 0.61			0.045
0213+427	+0.73	+0.02	1.07	127.	0.036
0214+568	- 0.12	- 0.92	0.160	94.	
0220+222	+0.07	- 0.71	0.15	113.	
0227+050	- 0.06	- 0.83	0.10	101.	0.048
0231– 054	+0.21	- 0.68			
0232+035	- 0.24	- 1.24	0.078	88.6	0.0105
0235+064	+0.44	- 0.51			
0236+745	+0.20	- 0.52	0.34	156.	
0239+109	+0.23	- 0.62	0.36	152.	
0245+541	+0.93	+0.35	0.613	224.1	0.0969
0255– 705	+0.23	- 0.59	0.67	98.	0.043

*Adapted from McCook and Sion (1987).

Catalogue of White Dwarf Stars*

Name	R.A.(1950) h m s	Dec.(1950) ° ′	Sp	V (mag)	M_V (mag)
0257+080	02 57 17	+08 00.0	DA8	15.90	13.95
0300−013	03 00 21	−01 20.2	DBZ4	15.56	11.08
0302+621	03 02 11	+62 10.8	DA6	14.95	12.63
0308−565	03 08 30	−56 34.	DB3	14.07	11.35
0310−688	03 10 00	−68 48.	DA3	11.40	11.27
0311−543	03 11 18	−54 18.	DZ7	14.75	14.54
0314+648	03 14 13	+64 49.2	DA3	16.54	10.75
0316+345	03 16 34	+34 31.6	DA3	14.16	11.29
0319+054	03 19 00	+05 28.0	DA3	15.04	10.86
0322−019	03 22 39	−01 59.2	DZ9	16.12	15.10
0324+738	03 24 35	+73 51.7	DC7	16.46	13.48
0326−273	03 26 42	−27 18.6	DA5	14.0	12.86
0332+320	03 32 09	+32 02.2	DA5	15.49	12.31
0339+523	03 39 18	+52 21.	DA3	15.75	11.43
0339−035	03 39 24	−03 32.3	DA3	15.20	11.55
0341+182	03 41 42	+18 17.5	DQ8	15.20	13.92
0341−459	03 41 48	−45 58.	DA4	15.03	11.70
0342−673	03 42 00	−67 19.	DA3	15.74	11.25
0346−011	03 46 17	−01 07.5	DA2	13.98	10.08
0348+339	03 48 48	+33 58.6	DA4	15.20	11.70
0352+096	03 52 06	+09 37.	DA4	14.47	11.24
0354+463	03 54 48	+46 20.	DA6	15.52	13.16
0354+556	03 54 12	+55 38.	DA6	16.72	13.34
0357+081	03 57 46	+08 06.0	DC9	15.87	14.62
0401+250	04 01 33	+25 00.9	DA4	13.80	11.62
0402+543	04 02 12	+54 19.	DA3	15.39	10.86
0406+169	04 06 18	+16 59.	DA4	15.35	11.43
0407+179	04 07 18	+17 54.	DA4	14.14	11.43
0407+197	04 07 18	+19 46.9	DC9	17.80	15.16
0408−041	04 08 33	−04 06.1	DA5	15.50	10.88
0410+117	04 10 06	+11 45.	DA3	13.86	10.80
0413−077	04 13 00	−07 44.	DA4	9.52	11.05
0416+272	04 16 48	+27 13.4	DAV5	15.2	11.99
0416+701	04 16 06	+70 07.4	DA4	14.74	11.47
0417+361	04 17 00	+36 09.5	DA5	15.63	11.88

Catalogue of White Dwarf Stars*

Name	B–V (mag)	U–B (mag)	μ ($''$ yr^{-1})	θ (°)	π_t^{-1} ($''$)
0257+080	+0.43	- 0.41	0.491	199.4	0.035
0300- 013	- 0.05	- 0.92			
0302+621	+0.22	- 0.57			
0308- 565	- 0.11	- 1.06	0.113	76.	
0310- 688	+0.05	- 0.55	0.09	147.	0.0859
0311- 543	+0.52	- 0.42	0.081	163.	0.091
0314+648	+0.01	- 0.67			
0316+345	+0.08	- 0.68			
0319+054	- 0.05	- 0.91			
0322- 019	+0.80	- 0.20	0.88		0.066
0324+738	+1.56	+1.01:	0.43	137.	0.026
0326- 273			0.840	64.	0.0588
0332+320	+0.13	- 0.51	0.3752	132.7	0.010
0339+523	+0.19	- 0.40	0.15	130.	
0339- 035	+0.15	- 0.59	0.219	80.	
0341+182	+0.30	- 0.52	1.25	162.	0.0557
0341- 459	+0.21	- 0.66	0.17	33.	
0342- 673	- 0.06	- 0.77	0.111	30.	
0346- 011	- 0.16	- 1.16			
0348+339	+0.10	- 0.63			
0352+096	+0.15	- 0.67	0.16	95.	0.022
0354+463	+0.28	- 0.57	0.11	300.	
0354+556	+0.19	- 0.82	0.12	140.	
0357+081	+0.69	+0.00	0.54	223.	0.0569
0401+250	+0.10	- 0.58	0.28	146.	0.0367
0402+543	- 0.05	- 0.74	0.10	310.	
0406+169	+0.09	- 0.65	0.13	81.	0.023
0407+179	+0.17	- 0.58	0.12	137.	
0407+197	+1.07	+0.36	0.55	121.	0.0243
0408- 041	+0.14	- 0.54			
0410+117	- 0.05	- 0.88	0.11	150.	
0413- 077	+0.03	- 0.68	4.08	213.	0.2084
0416+272	+0.20	- 0.50	0.13		
0416+701	+0.11	- 0.49			
0417+361	+0.16	- 0.53	0.29	172.	0.013

*Adapted from McCook and Sion (1987).

Catalogue of White Dwarf Stars*

Name	R.A.(1950) h m s	Dec.(1950) ° ′	Sp	V (mag)	M_V (mag)
0418+153	04 18 54	+15 22.	DA5	16.62	12.12
0418−539	04 18 06	−53 58.	DB3	15.32	11.34
0419+500	04 19.24	+50 01.0	DC5	16.28	12.34
0419−487	04 19 36	−48 46.	DA8	14.36	14.80
0421+162	04 21 00	+16 14.	DA3	14.29	11.05
0423+044	04 23 40	+04 26.3	DC9	17.11	15.29
0423+120	04 23 07	+12 05.2	DC8	15.7 pg	15.05
0425+168	04 25 42	+16 52.	DA2	14.17	10.63
0426+588	04 26 48	+58 53.	DC5	12.45	13.68
0429+176	04 29 24	+17 38.	DA2	13.93	14.36
0431+125	04 31 00	+12 35.	DA2	14.18	10.71
0433+270	04 33 39	+27 03.9	DC8	15.94	11.80
0433+406	04 33 00	+40 40.	DA6	16.55	12.31
0435−088	04 35 24	−08 53.9	DQ7	14.10	13.87
0438+108	04 38 12	+10 53.	DA2	13.83	10.33
0446−789	04 46 30	−78 57.	DA3	13.47	10.65
0452+103	04 52 09	+10 23.3	DA6	16.6 pg	12.81
0453+418	04 53 50	+41 51.5	DA3	13.77	11.30
0501+527	05 01 31	+52 44.8	DA1	11.8	10.07
0503−174	05 03 39	−17 27.3	DZ9?	15.97	14.28
0511+079	05 11 21	+07 57.1	DA8	15.89	13.92
0517+307	05 17 25	+30 45.5	DA6	15.56	12.63
0518+005	05 18 02	+00 34.6	DA3	16.64	11.20
0518+333	05 18 26	+33 19.2	DAE	16.20	13.41
0532+414	05 32 48	+41 28.2	DA7	14.75	13.37
0533+322	05 33 36	+32 13.5	DA4	16.43	11.81
0543+436	05 43 08	+43 37.9	DA5	17.09	12.12
0543+579	05 43 35	+57 58.3	DA5	16.0 pg	12.12
0548+000	05 48 04	+00 05.2	DA1	15.10	9.04
0548−001	05 48 46	−00 11.2	DQP8	14.58	14.03
0549+158	05 49 34	+15 52.7	DA1	13.06	9.81
0551+123	05 51 05	+12 23.8	DC5	15.86	12.48
0552−041	05 52 40	−04 09.2	DZ9	14.52	15.42
0553+053	05 53 47	+05 22.0	DAP9	14.12	14.55
0557+237	05 57 24	+23 43.3	DA6	16.90	13.19

Catalogue of White Dwarf Stars*

Name	B–V (mag)	U–B (mag)	μ ($'' \ \mathrm{yr}^{-1}$)	θ (°)	π_t^{-1} ($''$)
0418+153	+0.14	- 0.57	0.04	259.	
0418- 539	- 0.105	- 1.08	0.110	200.	
0419+500	- 0.02	- 0.77	0.130	158.	
0419- 487	+0.52	- 0.45	0.556	175.6	0.123
0421+162	- 0.02	- 0.84	0.11	92.	
0423+044	1.10		0.856	132.3	0.043
0423+120			0.26	210.	0.081
0425+168	- 0.12	- 0.93	0.10	99.	
0426+588	+0.33	- 0.49	2.37	146.	0.1810
0429+176	+0.34	- 0.72	0.11	109.	0.023
0431+125	- 0.03	- 0.89	0.10	93.	0.023
0433+270	+0.65	- 0.06	0.36	116.	0.0613
0433+406	+0.17	- 0.59	0.06	110.	
0435- 088	+0.14	- 0.65	1.49	171.	0.1054
0438+108	- 0.15	- 1.04	0.09	87.	0.023
0446- 789	- 0.105	- 1.025	0.04	257.	
0452+103	+0.25	- 0.59	0.36	131.	
0453+418	+0.23	- 0.61			0.0228
0501+527	- 0.32	- 1.20	0.10	160.	0.021
0503- 174	+0.74	+0.00	0.687	15.6	0.0464
0511+079	+0.41	- 0.38	0.36	219.	0.040
0517+307	+0.22	- 0.59			
0518+005	+0.17	- 0.63			
0518+333	+0.31	- 0.64	0.18	113.	
0532+414	+0.32	- 0.54			0.053
0533+322	+0.38	- 0.59	0.44	141.	0.003
0543+436	+0.14	- 0.63	0.27	159.	
0543+579					
0548+000	- 0.30	- 1.21			
0548- 001	+0.46	- 0.53	0.27	21.	0.099
0549+158	- 0.25	- 1.16			
0551+123	+0.05	- 0.76	0.29	185.	0.014
0552- 041	+1.06	+0.81	2.3784	166.6	0.155
0553+053	+0.62	- 0.16	1.06	207.	0.124
0557+237	+0.35	- 0.47	0.27	170.	

*Adapted from McCook and Sion (1987).

Catalogue of White Dwarf Stars*

Name	R.A.(1950) h m s	Dec.(1950) ° ′	Sp	V (mag)	M_V (mag)
0559+158	05 59 37	+15 53.3	DA7	16.80	13.42
0606+282	06 06 51	+28 15.1	DA4	14.63	11.20
0612+177	06 12 24	+17 44.8	DA2	13.40	10.33
0615- 591	06 15 30	- 59 11.	DB4	14.09	11.42
0618+067	06 18 05	+06 46.7	DA9	16.37	15.91
0624- 756	06 24 24	- 75 40.	DA6	15.38	11.77
0625+100	06 25 01	+10 02.5	DZ6	16.58	12.95
0625+415	06 25 30	+41 32.8	DA3	14.99	11.00
0632+409	06 32 59	+40 57.1	DA6	16.75	12.87
0637+477	06 37 26	+47 47.3	DA5	14.80	11.79
0639+447	06 39 36	+44 43.	DA3	16.70	10.86
0642- 166	06 42 57	- 16 38.9	DA2	8.30	11.18
0644+025	06 44 47	+02 34.4	DA8	15.68	13.84
0648+368	06 48 51	+36 49.4	DA5	16.16	12.17
0648+641	06 48 40	+64 07.6	DA9	16.64	14.44
0651- 020	06 51 42	- 02 04.6	DA1	14.82	9.72
0654+027	06 54 46	+02 45.1	DC5	16.15	12.46
0657+320	06 57 36	+32 02.5	DC9	16.57	15.25
0658+624	06 58 49	+62 27.3	DA4	15.55	11.60
0701- 587	07 01 30	- 58 46.	DA3	14.46	11.30
0706+377	07 06 52	+37 45.4	DQ8	15.64	13.79
0710+216	07 10 24	+21 39.2	DA5	15.29	12.24
0710+741	07 10 54	+74 06.	DA3	14.97	10.79
0713+584	07 13 20	+58 29.8	DA4	12.0 pg	11.58
0714+458	07 14 23	+45 53.4	DBA6	15.19	12.84
0716+404	07 16 32	+40 27.1	DB3	14.94	11.13
0727+482	07 27 06	+48 17.4	DC	14.64	15.32
0730+487	07 30 44	+48 47.9	DA2	14.96	10.58
0732- 427	07 32 06	- 42 47.	DAE	14.16	11.41
0738- 172	07 38 02	- 17 17.4	DZQ6	13.00	13.42
0740- 570	07 40 48	- 57 02.	DA2	15.06	10.98
0743+442	07 43 54	+44 16.3	DA5	14.92	11.88
0747+073.1	07 47 32	+07 20.9	DC9	16.98	15.69
0752+365	07 52 09	+36 30.1	DA7	16.11	13.34
0752- 146	07 52 48	- 14 38.	DA3	13.56	10.96

Catalogue of White Dwarf Stars*

Name	B–V (mag)	U–B (mag)	μ ($'' \text{ yr}^{-1}$)	θ (°)	π_t^{-1} ($''$)
0559+158	+0.35	- 0.56	0.31	110.	
0606+282	0.00	- 0.82			
0612+177	- 0.15	- 0.98	0.43	188.	0.022
0615- 591	- 0.09	- 0.95	0.301	184.	
0618+067	+0.55	- 0.17	0.58	93.	0.0445
0624- 756	+0.26	- 0.73	0.086	17.	
0625+100	+0.21	- 0.72	0.10	151.	
0625+415	- 0.03	- 0.81			
0632+409	+0.26	- 0.62	0.29	208.	
0637+477	+0.13	- 0.64			
0639+447	- 0.05	- 0.95			
0642- 166	- 0.12	- 1.03	1.323	203.	0.367
0644+025	+0.32	- 0.53	0.49	275.	0.0546
0648+368	+0.53	- 0.04			
0648+641	+0.52	- 0.17	0.41	199.	0.0292
0651- 020	- 0.21	- 1.19			
0654+027	+0.13	- 0.69	0.27	161.	0.0260
0657+320	+0.96	+0.42	0.684	150.6	0.051
0658+624	+0.06	- 0.71			
0701- 587	+0.22	- 0.72	0.217	342.	
0706+377	+0.30	- 0.55	0.44	222.	0.035
0710+216	+0.19	- 0.52			
0710+741	- 0.06	- 0.75			
0713+584					0.082
0714+458	+0.08	- 0.79	0.195	202.	
0716+404	- 0.11	- 0.99			
0727+482	+0.99	+0.37	1.285	189.8	0.0869
0730+487	- 0.09	- 0.74	0.240	228.	
0732- 427	+0.11	- 0.66	0.659	5.	0.0091
0738- 172	+0.30	- 0.62	1.252	116.6	0.1120
0740- 570	- 0.16	- 0.79	0.07	336.	
0743+442	+0.07	- 0.73			
0747+073.1	+1.27		1.778	173.2	0.058
0752+365	+0.31	- 0.58	0.49	196.	0.029
0752- 146	- 0.02	- 0.79	0.32	204.	

*Adapted from McCook and Sion (1987).

Catalogue of White Dwarf Stars*

Name	R.A.(1950) h m s	Dec.(1950) ° ′	Sp	V (mag)	M_V (mag)
0752−676	07 52 48	−67 38.	DQ9	14.09	15.24
0756+437	07 56 32	+43 43.4	DC?6	16.28	12.88
0800−533	08 00 42	−53 19.	DAE	15.76	11.37
0802+386	08 02 18	+38 41.0	DZ5	15.51	12.17
0802+387	08 02 41	+38 42.8	DC9	16.88	15.2
0806−661	08 06 24	−66 09.	DQ5	13.92	12.10
0810+234	08 10 44	+23 29.7	DA3	15.0 pg	11.08
0813+217	08 13 46	+21 47.2	DA8	17.02	14.10
0816+376	08 16 31	+37 40.9	DAH5	15.74	12.31
0816+387	08 16 44	+38 44.2	DA7	16.58	13.18
0817+386	08 17 14	+38 38.3	DA2	15.89	10.8
0823+316	08 23 58	+31 40.0	DA1	15.82	7.7
0827+328	08 27 32	+32 52.3	DA7	15.69	13.24
0830+371	08 30 47	+37 09.8	DA5	16.01	12.45
0836+199	08 36 54	+19 57.	DA5	18.23	11.54
0836+201	08 36 54	+20 11.	DA5	17.85	12.25
0837+199	08 37 36	+19 54.	DA5	17.66	11.47
0839−327	08 39 34	−32 46.7	DA6	11.90	12.11
0840+262	08 40 59	+26 14.1	DB3	14.78	11.36
0845−188	08 45 18	−18 48.	DB4	15.55	11.47
0846+346	08 46 02	+34 40.9	DA7	15.71	13.45
0846+557	08 46 04	+55 46.3	DA1	16.42	10.01
0848−730	08 48 06	−73 02.	DA2	15.30	10.96
0850−617	08 50 00	−61 43.	DA2	14.73	10.88
0852+602	08 52 57	+60 17.5	DA2	16.47	10.37
0856+331	08 56 11	+33 08.7	DQ6	15.18	13.16
0858+363	08 58 41	+36 19.0	DAV4	14.55	11.81
0900+734	09 00 25	+73 27.0	DC9	17.02	15.01
0902+092	09 02 33	+09 16.7	DC6	16.8	12.75
0912+536	09 12 27	+53 38.6	DCP7	13.79	13.79
0913+442	09 13 26	+44 12.5	DA6	15.32	12.66
0913−010	09 13 35	−01 04.4	DA7	15.33	13.24
0921+354	09 21 13	+35 29.8	DAV4	15.52	11.62
0928−713	09 28 36	−71 20.	DA5	15.44	12.79
0934−587	09 34 18	−58 43.	DA4	15.46	10.12

Catalogue of White Dwarf Stars*

Name	B–V (mag)	U–B (mag)	μ ($'' \; yr^{-1}$)	θ ($^\circ$)	π_t^{-1} ($''$)
0752−676	+0.66	−0.17	2.05	135.	0.1336
0756+437	+0.11	−0.44	0.34	294.	
0800−533	−0.05	−0.89	0.124	78.	
0802+386	+0.07	−0.91	0.2912	187.2	0.023
0802+387	+0.08		0.830	226.2	0.045
0806−661	+0.05	−0.90	0.47	128.	0.043
0810+234					
0813+217	+0.48	−0.29	0.44	186.	
0816+376	+0.22	−0.63			
0816+387	+0.36	−0.58	0.38	212.	0.019
0817+386	−0.17	−0.95			
0823+316	−0.40	−1.14			
0827+328	+0.35	−0.54	0.64	209.	0.044
0830+371	+0.21	−0.59	0.28	216.	
0836+199	+0.05	−0.49	0.04	250.	
0836+201	+0.16	−0.68	0.04	250.	
0837+199	+0.04	−0.60	0.04	250.	
0839−327	+0.25	+0.59	1.69	321.	0.1134
0840+262	−0.10	−1.01			
0845−188	−0.06	−0.93	0.13	247.	
0846+346	+0.28	−0.35			
0846+557	−0.17	−1.05			
0848−730	−0.09	−0.72	0.111	291.	
0850−617	−0.02	−0.89	0.093	300.	
0852+602	−0.12	−1.00			
0856+331	0.00	−0.89	0.34	271.	0.0497
0858+363	+0.19	−0.59	0.197	211.	
0900+734			0.560	215.7	0.040
0902+092			0.296	141.	
0912+536	+0.30	−0.66	1.56	222.	0.0973
0913+442	+0.24	−0.53	0.27	176.	0.034
0913−010	+0.32	−0.48	0.56	277.	
0921+354	+0.20	−0.56	0.14	264.	0.012
0928−713	+0.175	−0.44	0.428	313.	
0934−587	+0.00	−0.92	0.086	172.	

*Adapted from McCook and Sion (1987).

Catalogue of White Dwarf Stars*

Name	R.A.(1950) h m s	Dec.(1950) ° ′	Sp	V (mag)	M_V (mag)
0935- 371.2	09 35 00	- 37 07.0	DQ6	14.30	12.56
0938+299	09 38 31	+29 58.8	DA2	16.08	10.8
0939+071	09 39 21	+07 11.0	DA7	14.91	13.73
0940+068	09 40 16	+06 49.3	DA:2	13.70	10.14
0943+441	09 43 30	+44 08.6	DA4	13.12	11.04
0946+534	09 46 55	+53 29.2	DQ6	15.30	12.72
0950- 572	09 50 00	- 57 12.	DA5	14.94	11.89
0954+247	09 54 59	+24 47.4	DA5	15.09	13.00
0954- 710	09 54 36	- 71 02.	DA5	13.48	11.42
0955+090	09 55 39	+09 01.3	DC8	17.02	13.96
0955+247	09 55 00	+24 47.4	DA6	15.09	12.98
0957- 666	09 57 48	- 66 39.	DA2	14.60	10.64
1002+430	10 02 46	+43 02.6	DA3	16.16	10.65
1003- 023	10 03 19	- 02 19.6	DA3	15.43	11.07
1010+064	10 10 51	+06 27.0	DA1	16.55	9.6
1012+083	10 12 26	+08 21.3	DZ7	16.16	13.84
1013- 559	10 13 12	- 55 55.	DZ9	15.10	14.86
1019+462	10 19 33	+46 16.1	DA7	16.70	13.30
1019+637	10 19 38	+63 42.7	DA7	14.70	13.76
1022+009	10 22 50	+00 59.3	DC9	18.06	14.3
1022+050	10 22 24	+05 01.4	DA4	14.18	11.50
1026- 056	10 26 30	- 05 39.0	DB3	16.94	10.92
1029+329	10 29 16	+32 55.6	DC5	16.07	11.90
1031- 114	10 31 28	- 11 25.5	DA2	12.97	10.77
1033+714	10 33 24	+71 27.0	DK?	16.90	15.4
1036- 204	10 36 32	- 20 25.8	DQP9	16.0	15.67
1039+145	10 39 14	+14 31.6	DQ7	16.52	13.68
1042+593	10 42 54	+59 21.0	DQ8	17.73	13.39
1042- 690	10 42 36	- 69 02.	DA3	13.09	11.03
1045- 017	10 45 58	- 01 45.0	DB4	15.81	11.64
1046+281	10 46 45	+28 10.3	DA4	15.41	12.25
1046- 017	10 46 00	- 01 45.3	DB5	15.81	12.06
1053- 550	10 53 18	- 55 04.	DA5	14.32	11.27
1055- 039	10 55 14	- 03 55.4	DF- DG	16.53	13.81
1055- 072	10 55 05	- 07 15.2	DA6	14.30	13.98

Catalogue of White Dwarf Stars*

Name	B–V (mag)	U–B (mag)	μ ($''$ yr^{-1})	θ ($^\circ$)	π_t^{-1} ($''$)
0935– 371.2	+0.13	– 0.72	0.37	295.	0.046
0938+299	– 0.06	– 0.93			
0939+071	+0.30	– 0.55			0.058
0940+068	– 0.19	– 1.03			
0943+441	+0.07	– 0.54	0.31	359.	0.028
0946+534	+0.13	– 0.70	0.30	264.	0.0446
0950– 572	+0.20	– 0.58	0.180	138.	
0954+247	+0.25	– 0.54			
0954– 710	+0.12	– 0.65	0.174	225.	
0955+090	0.83	0.17	0.34	223.	
0955+247	+0.25	– 0.54	0.38	218.	0.040
0957– 666	– 0.20	– 0.98	0.072	344.	
1002+430	– 0.08	– 0.87			
1003– 023	– 0.10	– 0.90			
1010+064	– 0.25	– 1.17			
1012+083	+0.41	– 0.40	0.36	304.	
1013– 559	+0.68	– 0.12	0.136	298.	
1019+462	+0.33	– 0.63			
1019+637	0.39	– 0.53	0.38	49.	0.0622
1022+009	+0.84	+0.13	1.101	184.6	0.0123
1022+050	+0.19	– 0.51	0.16	282.	
1026– 056	– 0.16				
1029+329	0.00	– 0.88			
1031– 114	– 0.15	– 1.02	0.33	260.	0.030
1033+714	+1.02		1.997	254.2	0.050
1036– 204			0.628	330.3	0.094
1039+145	+0.31	– 0.58	0.31	149.	0.022
1042+593	+0.09		1.770	214.3	0.0123
1042– 690	– 0.04	– 0.84	0.278	267.	
1045– 017	– 0.03	– 0.88			
1046+281	+0.15	– 0.57	0.060		
1046– 017	– 0.03	– 0.88			
1053– 550	+0.095	– 0.62	0.40	305.	
1055– 039	+0.40	– 0.46	0.36	126.	
1055– 072	+0.32	– 0.52	0.816	274.	0.0832

*Adapted from McCook and Sion (1987).

Catalogue of White Dwarf Stars*

Name	R.A.(1950) h m s	Dec.(1950) ° ′	Sp	V (mag)	M_V (mag)
1104+602	11 04 43	+60 14.8	DA3	13.80	11.07
1105- 048	11 05 28	- 04 52.6	DA3	12.92	11.20
1107+265	11 07 19	+26 35.2	DB4	15.89	11.85
1108+207	11 08 23	+20 42.5	DC:9	17.80	15.76
1108+475	11 08 31	+47 34.8	DA5	15.38	12.18
1115- 029	11 15 43	- 02 57.8	DC5	15.30	11.45
1116+026	11 16 38	+02 36.9	DA4	14.57	11.77
1120+439	11 20 11	+43 59.4	DA1	15.81	9.79
1121+145	11 21 39	+14 30.3	DA1	16.63	9.64
1121+216	11 21 38	+21 38.1	DA7	14.24	13.60
1121- 507	11 21 30	- 50 44.	DA5	14.86	11.39
1122+546	11 22 31	+54 36.1	DA4	15.43	11.47
1126+384	11 26 30	+38 25.4	DA3	14.73	10.71
1129- 537	11 29 18	- 53 46.	DA2	15.69	10.25
1133+293	11 33 36	+29 18.1	DA3	15.	11.0
1133- 528	11 33 30	- 52 48.	DA5	15.8	11.78
1138+424	11 38 57	+42 25.3	DA	15.99	10.8
1142- 645	11 42 58	- 64 33.5	DQ6	11.50	13.06
1143+321	11 43 21	+32 06.2	DA4	13.60	11.27
1143+633	11 43 03	+63 22.8	DC9	16.37	14.76
1147+255	11 47 46	+25 35.0	DA5	15.67	12.20
1148+544	11 48 47	+54 28.5	DA5	16.72	12.65
1149+410	11 49 07	+41 03.8	DA4	15.5 pg	11.35
1153- 484	11 53 42	- 48 24.	DA2	12.85	9.75
1154+186	11 54 11	+18 39.0	DZ5	15.54	13.34
1159+803	11 59 15	+80 21.7	DAV5	15.9 pg	11.88
1201+437	12 01 51	+43 47.6	DC6	16.50	12.66
1202+308	12 02 52	+30 51.5	DA2	15.0 pg	10.8
1208+576	12 08 58	+57 41.1	DA9	15.79	14.28
1211+332	12 11 25	+33 13.3	DO2	14.22	0.50
1211+392	12 11 05	+39 17.6	DA3	16.48	10.72
1213+528	12 13 18	+52 48.0	DA4	13.34	14.07
1214+690	12 14 42	+69 05.5	DC8	16.6 pg	14.15
1215+323	12 15 03	+32 21.9	DC7	16.95	13.71
1220+234	12 20 43	+23 24.4	DA2	15.64	10.72

Catalogue of White Dwarf Stars*

Name	B–V (mag)	U–B (mag)	μ ($'' \ yr^{-1}$)	θ (°)	π_t^{-1} ($''$)
1104+602	− 0.02	− 0.81	0.27	225.	0.021
1105− 048	+0.09	− 0.69	0.43	180.	0.0226
1107+265	− 0.07	− 0.99	0.20	148.	
1108+207	+0.85	+0.40	0.816	267.	0.039
1108+475	+0.15	− 0.63			
1115− 029	+0.09	− 0.76	0.563	292.4	0.025
1116+026	+0.19	− 0.59			
1120+439	− 0.20	− 1.04			
1121+145	− 0.22	− 1.18			
1121+216	+0.31	− 0.52	1.07	272.	0.0759
1121− 507	+0.07	− 0.74	0.119	111.	
1122+546	+0.08	− 0.66			
1126+384	− 0.07	− 1.11	0.111	244.	
1129− 537	− 0.12	− 0.93	0.16	275.	
1133+293					
1133− 528	+0.12	− 0.65	0.231	303.	
1138+424	+0.06	− 1.08			
1142− 645	+0.18	− 0.63	2.680	96.	0.2171
1143+321	+0.06	− 0.67	0.28	203.	0.030
1143+633	0.67	0.20	0.42	201.	0.047
1147+255	+0.20	− 0.60	0.30	259.	0.019
1148+544	+0.20	− 0.57	0.34	264.	
1149+410					
1153− 484	− 0.205	− 1.01	0.05	277.	
1154+186	+0.30	− 0.60	0.32	274.	0.038
1159+803			0.28	263.	
1201+437	+0.15	− 0.47			
1202+308					
1208+576	0.55	− 0.22	0.63	131.	0.0504
1211+332	− 0.36	− 1.25	0.11	310.	
1211+392	− 0.07	− 0.82	0.10	310.	
1213+528	+0.53	− 0.48	0.12	206.	0.0259
1214+690			0.47	272.	
1215+323	+0.38	− 0.46	0.242	236.	0.0246
1220+234	− 0.07		0.043		

*Adapted from McCook and Sion (1987).

Catalogue of White Dwarf Stars*

Name	R.A.(1950) h m s	Dec.(1950) ° ′	Sp	V (mag)	M_V (mag)
1223- 659	12 23 48	- 65 56.	DA	13.97	13.56
1230+417	12 30 00	+41 45.8	DA2	15.72	10.93
1232+479	12 32 33	+47 54.2	DA4	14.52	11.12
1236+479	12 36 44	+47 54.3	DA2	15.60	10.59
1236- 495	12 36 06	- 49 33.	DA6	13.96	12.03
1239+454	12 39 15	+45 25.3	DC8	16.44	14.0
1241+651	12 41 14	+65 09.4	DB4	16.0 pg	11.70
1244+149	12 44 45	+14 58.8	DA5	15.86	12.28
1247+550	12 47 56	+55 04.5	DC9	17.80	16.02
1247+553	12 47 52	+55 22.3	DA4	12.31	11.40
1248- 610	12 48 24	- 61 02.	DA2	15.9	10.71
1249+182	12 49 55	+18 13.1	DA3	15.48	10.65
1253+378	12 52 53	+37 48.7	DAO	15.66	8.6
1253+482	12 53 11	+48 13.5	DA3	16.5 pg	11.55
1257+037	12 57 37	+03 45.3	DC9	15.90	14.77
1257+047	12 57 18	+04 47.7	DA2	14.99	10.58
1257+278	12 57 23	+27 50.2	DA6	15.40	12.97
1257- 723	12 57 54	- 72 18.	DA4	15.18	10.82
1258+593	12 58 30	+59 20.2	DA5	15.31	12.06
1300+263	13 00 54	+26 19.0	DC:	18.81	16.5
1302+283	13 02 24	+28 23.5	DA1	15.42	10.46
1302+597	13 02 30	+59 43.5	DAB	14.52	10.71
1304+150	13 04 56	+15 00.5	DB4	17.06	11.66
1307+354	13 07 37	+35 25.5	DAV5	15.33	12.4
1308- 098	13 08 40	- 09 51.1	DA4	15.66	11.46
1309+853	13 09 59	+85 18.6	DG	16.00	15.24
1314+293	13 14 00	+29 22.0	DA1	12.86	8.7
1317+453	13 17 04	+45 20.8	DA4	14.13	11.28
1319+466	13 19 06	+46 38.8	DA3	14.55	11.20
1323- 514	13 23 02	- 51 25.7	DA4	14.60	10.90
1325+279	13 25 29	+27 59.3	DA2	15.88	10.92
1325+581	13 25 49	+58 10.4	DZ7	16.7	13.90
1327- 083	13 27 40	- 08 18.4	DA5	12.30	11.92
1328+307	13 28 42	+30 45.3	DZ7	16.04	13.34
1330+015	13 30 17	+01 32.7	DC8	17.11	14.03

Catalogue of White Dwarf Stars*

Name	B–V (mag)	U–B (mag)	μ ($''$ yr^{-1})	θ ($^\circ$)	π_t^{-1} ($''$)
1223– 659	+0.40	– 0.54	0.186	182.	
1230+417	– 0.04	– 0.82			
1232+479	+0.06	– 0.68			
1236+479	– 0.21	– 1.05			
1236– 495	+0.18	– 0.70	0.556	257.	0.0596
1239+454	+0.43		0.886	230.1	
1241+651					
1244+149	+0.22	– 0.53	0.46	297.	0.0144
1247+550	+1.59		1.286	192.2	0.044
1247+553	+0.03	– 0.93			0.071
1248– 610	– 0.14	– 1.07	0.042	268.	
1249+182	– 0.08	– 0.85			
1253+378	– 0.28	– 1.26	0.063		
1253+482					
1257+037	+0.64	– 0.09	1.00	203.	0.082
1257+047	– 0.08	– 0.91	0.103	212.	
1257+278	+0.28	– 0.58	0.34	273.	
1257– 723	+0.01	– 0.73	0.187	100.	
1258+593	+0.13	– 0.63			
1300+263	+1.38		0.849	275.6	0.035
1302+283	– 0.28	– 1.14	0.013		
1302+597	– 0.13	– 1.08			
1304+150	– 0.05	– 0.99	0.121	300.	
1307+354	+0.18	– 0.59			
1308– 098	+0.04	– 0.84			
1309+853	0.76	0.30	0.34	137.	0.070
1314+293	– 0.10	– 1.14	0.17	244.	0.012
1317+453	+0.03	– 0.56	0.53	289.	
1319+466	0.00	– 0.66	0.27	288.	
1323– 514	0.00	– 0.79	0.50	268.	
1325+279	– 0.04				
1325+581	+0.42	– 0.43	0.49	301.	0.024
1327– 083	+0.08	– 0.61	1.17	249.	0.0618
1328+307	+0.72	+0.78	0.470	258.7	0.0321
1330+015	+0.38	– 0.47	0.246	172.	

*Adapted from McCook and Sion (1987).

Catalogue of White Dwarf Stars*

Name	R.A.(1950) h m s	Dec.(1950) ° ′	Sp	V (mag)	M_V (mag)
1330+036	13 30 48	+03 36.	DA4	15.86	11.27
1333+487	13 33 59	+48 44.0	DB–	14.02	13.18
1334+039	13 34 10	+03 56.7	DZ9	14.70	15.11
1334–160	13 34 18	–16 04.	DA3	15.35	10.79
1334–678	13 34 38	–67 49.4	DA6	15.57	12.63
1337+705	13 37 40	+70 32.4	DA3	12.79	10.75
1340+572	13 40 36	+57 15.	DA4	16.73	11.24
1345+238	13 45 48	+23 49.6	DC9	15.65	15.27
1347+283	13 47 28	+28 19.3	DA7	16.32	11.27
1348–273	13 48 30	–27 20.	DA6	15.0	11.86
1349+144	13 49 28	+14 24.5	DA4	15.34	11.31
1349+552	13 49 27	+55 12.7	DA4	16.0 pg	11.5
1350+656	13 50 47	+65 39.7	DA5	15.5 pg	12.03
1354+340	13 54 54	+34 03.4	DA5	16.16	11.48
1401+523	14 01 55	+52 21.1	DA7	16.0 pg	13.42
1403–010	14 03 46	–01 05.3	DB4	15.90	11.92
1406+590	14 06 54	+59 00.0	DA1	13.38	9.19
1407–475	14 07 30	–47 31.	DA3	14.31	10.84
1408+323	14 08 16	+32 22.7	DA3	13.97	11.04
1411+157	14 11 31	+15 44.3	DA5	16.20	11.67
1413+231	14 13 26	+23 10.6	DA3	16.41	11.04
1415+132	14 15 15	+13 15.6	DA1	15.33	9.6
1415–064	14 15 21	–06 26.2	DC5	16.20	12.26
1418–088	14 18 16	–08 51.5	DA7	15.36	12.39
1419+351	14 19 57	+35 08.6	DB5	16.0 pg	12.21
1421+318	14 21 31	+31 48.5	DA2	15.30	10.46
1422+095	14 22 12	+09 30.9	DA4	14.32	11.99
1425+540	14 25 59	+54 01.6	DBA4	15.04	11.51
1425–811	14 25 24	–81 07.	DA6	13.75	11.97
1426+442	14 26 38	+44 17.2	DZ7	16.94	14.19
1426+613	14 26 09	+61 23.7	DQ8	16.8 pg	13.97
1429+373	14 29 54	+37 19.7	DA1	15.26	10.25
1430+427	14 30 38	+42 43.4	DA2	14.47	10.58
1433+538	14 33 06	+53 48.5	DA–	16.09	11.39
1450+432	14 50 22	+43 13.9	DA1	14.50	10.6

Catalogue of White Dwarf Stars*

Name	B–V (mag)	U–B (mag)	μ ($'' \text{ yr}^{-1}$)	θ (°)	π_t^{-1} ($''$)
1330+036	+0.01	− 0.79	0.17		
1333+487	+0.03	− 0.92	0.126	256.3	0.0295
1334+039	+0.96	+0.37	3.904	253.5	0.124
1334− 160	− 0.06	− 0.83	0.12	246.	
1334− 678	+0.30	− 0.59	0.54	263.	
1337+705	− 0.09	− 0.84	0.43	269.	0.025
1340+572	+0.55	− 0.27			
1345+238	+1.10		1.484	275.6	0.0796
1347+283	+0.33	− 0.59	0.30	278.	
1348− 273	+0.1		0.23	164.	
1349+144	+0.06	− 0.72			
1349+552			0.07	89.	
1350+656			0.28	297.	
1354+340	+0.08	− 0.70	0.14	275.	
1401+523			0.202	261.	
1403− 010	− 0.05	− 0.99	0.243	253.	
1406+590	− 0.28	− 1.07	0.06	297.	
1407− 475	− 0.07	− 0.89	0.107	268.	
1408+323	− 0.01	− 0.76			0.023
1411+157	+0.07	− 0.65	0.31	259.	
1413+231	− 0.10	− 0.95	0.27	283.	
1415+132	− 0.23	− 1.12			
1415− 064	+0.07	− 0.86	0.27	226.	
1418− 088	+0.33	− 0.58	0.56	272.	
1419+351					
1421+318	− 0.12	− 1.00			
1422+095	+0.14	− 0.59	0.25	235.	
1425+540	− 0.07	− 0.98	0.39	294.	0.016
1425− 811	+0.25	− 0.53	0.45	208.	0.044
1426+442	0.52	− 0.56	0.34	135.	
1426+613			0.25	271.	
1429+373	− 0.18	− 1.16			
1430+427	− 0.09	− 0.62			
1433+538	− 0.03	− 0.88	0.129	284.6	0.004
1450+432	− 0.22	− 0.89			

*Adapted from McCook and Sion (1987).

Catalogue of White Dwarf Stars*

Name	R.A.(1950) h m s	Dec.(1950) ° ′	Sp	V (mag)	M_V (mag)
1451+006	14 51 17	+00 37.7	DA3	15.29	10.92
1455+298	14 55 59	+29 49.8	DA6	15.61	13.06
1456+298	14 56 00	+29 49.5	DA6	15.61	13.30
1459+305	14 59 50	+30 34.6	DA3	13.98	10.9
1459+347	14 59 40	+34 43.7	DA2	15.71	10.82
1459+821	14 59 40	+82 08.6	DB4	15.0 pg	11.47
1503−070	15 03 10	−07 03.1	DA7	15.91	13.46
1504−770	15 04 54	−77 00.	DA4	15.17	11.14
1507−105	15 07 46	−10 34.1	DA5	15.42	11.89
1508+637	15 08 44	+63 43.8	DA5	15.0 pg	12.17
1510+566	15 10 34	+56 36.2	DA6	16.24	12.68
1518+636	15 18 59	+63 40.4	DA3	16.63	10.92
1520+447	15 20 34	+44 43.6	DA4	16.54	11.33
1524+566	15 24 30	+56 40.	DC9	16.92	14.82
1524−749	15 24 18	−74 55.	DA3	15.93	10.43
1526+013	15 26 07	+01 23.3	DA2	16.69	10.0
1531+184	15 31 33	+18 24.0	DA5	16.20	12.25
1533−057	15 33 55	−05 42.7	DAP3	15.33	10.79
1534+503	15 34 47	+50 23.8	DA6	15.71	13.00
1537+651	15 37 08	+65 11.3	DA5	14.64	12.45
1538+269	15 38 17	+26 58.2	DA1	13.85	10.5
1544+008	15 44 12	+00 53.	DA1	15.27	10.82
1544−377	15 44 12	−37 45.	DA7	12.80	12.45
1548+405	15 48 49	+40 34.9	DA2	15.89	8.3
1550+183	15 50 12	+18 19.1	DA4	14.83	11.89
1550+626	15 50 25	+62 36.2	DC6	16.5 pg	13.13
1555−089	15 55 22	−08 59.5	DA5	14.80	11.65
1559+369	15 59 33	+36 57.2	DAV5	14.36	11.88
1600+369	16 00 05	+36 55.4	DA4	14.36	11.9
1602+010	16 02 23	+01 03.4	DC9	17.73	15.39
1606+422	16 06 42	+42 13.5	DA5	13.85	11.75
1609+044	16 09 19	+04 27.5	DA2	15.29	10.22
1609+135	16 09 06	+13 30.3	DA6	15.10	13.00
1610+166	16 10 46	+16 39.6	DA4	15.67	11.43
1612−111	16 12 38	−11 11.1	DB3	15.53	11.31

Catalogue of White Dwarf Stars*

Name	B–V (mag)	U–B (mag)	μ ($'' \text{ yr}^{-1}$)	θ (°)	π_t^{-1} ($''$)
1451+006	- 0.14	- 0.96			
1455+298	+0.31	- 0.59	0.67	163.	0.028
1456+298	+0.31	- 0.59			
1459+305	+0.30	- 0.84			
1459+347	- 0.07				
1459+821			0.42	300.	
1503- 070	+0.40	- 0.42			0.0393
1504- 770	+0.04	- 0.91	0.071	54.	
1507- 105	+0.23	- 0.50			
1508+637					
1510+566	+0.23	- 0.60	0.32	215.	
1518+636	- 0.04	- 0.94	0.27	313.	
1520+447	+0.02				
1524+566	+0.69		0.180	202.	
1524- 749	- 0.06	- 0.93	0.438	238.	
1526+013	- 0.13	- 1.22			
1531+184	+0.16	- 0.58			
1533- 057					
1534+503	+0.25	- 0.44	0.191	152.	
1537+651	+0.18	- 0.49			
1538+269	- 0.19	- 0.94			
1544+008	- 0.32	- 1.19			
1544- 377	+0.30	- 0.40	0.48	243.	0.0746
1548+405	- 0.28	- 1.19			
1550+183	+0.11	- 0.84			
1550+626					
1555- 089	+0.09	- 0.63	0.12	198.	
1559+369	+0.17	- 0.56	0.5562	169.1	0.019
1600+369	+0.17	- 0.56			
1602+010	+1.07		0.755	266.6	0.0250
1606+422	+0.06	- 0.54	0.176	308.	0.023
1609+044	- 0.14	- 1.11			
1609+135	+0.23	- 0.65	0.541	181.1	0.0557
1610+166	+0.07	- 0.63			0.0143
1612- 111	- 0.11	- 0.94			

*Adapted from McCook and Sion (1987).

Catalogue of White Dwarf Stars*

Name	R.A.(1950) h m s	Dec.(1950) ° ′	Sp	V (mag)	M_V (mag)
1614−128	16 14 42	− 12 50.	DA4	16.14	12.24
1615−154	16 15 06	− 15 28.	DA2	12.40	10.48
1616−390	16 16 06	− 39 00.0	DA4	14.1	10.58
1616−591	16 16 12	− 59 09.	DA5	15.08	11.91
1620+260	16 20 31	+26 02.2	DA1	15.53	9.0
1620−391	16 20 12	− 39 07.	DA2	11.00	10.77
1620−536	16 20 36	− 53 37.	DA4	15.64	11.00
1623−540	16 23 00	− 54 05.	DA5	15.74	11.86
1624+477	16 24 48	+47 45.2	DA6	16.47	13.00
1625+093	16 25 30	+09 19.2	DC8	16.12	13.84
1626+368	16 26 39	+36 52.4	DZA6	13.84	12.78
1628−873	16 28 48	− 87 19.	DA6	14.58	12.63
1630+618	16 30 23	+61 48.7	DA3	15.47	10.51
1633+433	16 33 25	+43 23.8	DA8	14.82	14.13
1633+572	16 33 28	+57 15.6	DQ8	15.00	14.20
1635+137	16 35 22	+13 46.7	DC7	16.90	13.85
1636+160	16 36 25	+16 00.1	DA5	15.60	12.12
1636+351	16 36 37	+35 06.1	DA1	15.02	9.9
1637+335	16 37 36	+33 31.3	DA6	14.64	12.52
1639+153	16 39 20	+15 18.6	DA7	15.66	13.34
1641+387	16 41 19	+38 46.7	DA6	14.41	13.0
1641+732	16 41 16	+73 16.0	DC6	16.56	12.71
1645+325	16 45 25	+32 33.7	DB2	13.65	10.77
1647+375	16 47 35	+37 33.4	DA3	15.01	11.4
1647+591	16 47 38	+59 08.7	DAV4	12.22	11.81
1655+210	16 55 00	+21 00.1	DA6	16.60	12.68
1655+215	16 55 01	+21 31.7	DA5	14.06	12.44
1659+303	16 59 12	+30 19.9	DA4	14.99	11.99
1659−531	16 59 00	− 53 06.	DA4	13.47	11.36
1705+030	17 05 37	+03 01.6	DZ7	15.19	13.71
1706+332	17 06 58	+33 16.8	DA4	15.92	11.55
1708−147	17 08 30	− 14 45.	DQ6	14.30	12.25
1709+230	17 09 50	+23 04.7	DB4	14.90	11.04
1709−575	17 09 48	− 57 34.	DA2	15.10	10.89
1710+683	17 10 08	+68 23.4	DA8	17.1 pg	13.84

Catalogue of White Dwarf Stars*

Name	B–V (mag)	U–B (mag)	μ ($'' \ yr^{-1}$)	θ (°)	π_t^{-1} ($''$)
1614– 128	+0.03	– 0.68	0.25	193.	
1615– 154	– 0.25	– 1.02	0.25	223.	0.006
1616– 390	– 0.09	– 1.05			
1616– 591	+0.10	– 0.84	0.243	209.	
1620+260	– 0.17	– 1.08	0.08		
1620– 391	– 0.145	– 0.955	0.078	96.	0.0669
1620– 536	– 0.03	– 0.69	0.098	204.	
1623– 540	+0.10	– 0.80	0.066	10.	
1624+477	+0.24	– 0.63	0.30	333.	
1625+093	+0.37	– 0.44	0.476	191.	0.0453
1626+368	+0.17	– 0.65	0.890	326.9	0.0654
1628– 873	+0.22	– 0.63	0.066		
1630+618	– 0.10	– 0.59			
1633+433	+0.44	– 0.43	0.42	137.	0.0672
1633+572	+0.49	+0.36	1.62	319.	0.0683
1635+137	+0.41	– 0.50	0.27	186.	0.0261
1636+160	+0.14	– 0.63			
1636+351	– 0.03	– 1.14			
1637+335	+0.22	– 0.58	0.57	186.	0.032
1639+153	+0.35	– 0.57	0.72	179.	0.006
1641+387	+0.28	– 0.69	0.181	7.	
1641+732	+0.12	– 0.78	0.37	345.	
1645+325	– 0.11	– 1.04			
1647+375	– 0.15	– 0.96			
1647+591	+0.19	– 0.65	0.33	153.	0.0817
1655+210	– 0.21	– 1.18	0.226	216.	
1655+215	+0.25	– 0.55	0.578	176.9	0.0436
1659+303	+0.12	– 0.65			
1659– 531	+0.10	– 0.70	0.134	175.	0.0297
1705+030	+0.46	– 0.25	0.43	181.	0.0577
1706+332	+0.17	– 0.55	0.14	170.	
1708– 147	+0.02	– 0.68	0.36	132.	
1709+230	– 0.09	– 0.98			
1709– 575	+0.03	– 0.76	0.205	207.	
1710+683			0.22	347.	

*Adapted from McCook and Sion (1987).

Catalogue of White Dwarf Stars*

Name	R.A.(1950) h m s	Dec.(1950) ° ′	Sp	V (mag)	M_V (mag)
1712+215	17 12 23	+21 30.7	DC5	16.51	13.45
1713+332	17 13 44	+33 16.4	DA3	14.46	10.68
1713+695	17 13 28	+69 34.9	DA3	13.27	11.12
1714−547	17 14 48	−54 44.	DA7	15.55	12.03
1716+020	17 16 05	+02 00.2	DA6	14.26	11.78
1726−578	17 26 06	−57 53.	DB4	15.27	11.36
1728+560	17 27 58	+56 00.2	DBQ4	16.15	11.35
1733−544	17 33 12	−54 24.	DA8	15.8	14.23
1737+419	17 37 01	+41 54.0	DA3	15.51	10.65
1743−132	17 43 04	−13 17.3	DA7	14.22	12.87
1743−521	17 43 48	−52 06.	DAP6	15.74	12.88
1748+708	17 48 58	+70 52.7	DXP9	14.15	15.25
1750+098	17 50 33	+09 49.0	DC5	15.72	12.07
1756+827	17 56 24	+82 45.2	DA7	14.30	13.51
1809+284	18 09 43	+28 29.0	DA4	15.11	11.25
1811+327.1	18 11 41	+32 47.7	DA7	16.40	13.59
1814+248	18 14 06	+24 53.7	DC7	16.92	13.76
1822+410	18 22 01	+41 02.2	DBA4	14.39	11.28
1824+040	18 24 45	+04 02.0	DA5	13.90	11.54
1826−045	18 26 30	−04 31.	DA6	14.54	12.66
1827−106	18 27 53	−10 39.0	DA5	14.25	11.49
1829+547	18 29 21	+54 45.2	DXP8	15.50	14.13
1831+197	18 31 49	+19 43.7	DC6	16.45	13.18
1834−781	18 34 42	−78 09.	DA3	15.45	11.07
1837−619	18 37 36	−61 56.	DC5	14.9	12.85
1840−111	18 40 12	−11 11.7	DA5	14.18	12.81
1845+019	18 45 05	+01 54.1	DA2	12.96	9.57
1855+338	18 55 40	+33 53.1	DA5	14.64	12.14
1857+119	18 57 29	+11 54.3	DA5	15.52	12.21
1858+393	18 58 36	+39 18.4	DA6	15.63	12.91
1900+705	19 00 40	+70 35.2	DXP5	13.19	12.40
1911+135	19 11 20	+13 31.3	DA4	14.00	11.56
1917+386	19 17 15	+38 38.0	DC?8	14.56	13.82
1917−077	19 17 54	−07 45.	DBZ5	12.30	12.24
1918+110	19 18 14	+11 05.0	DA3	16.23	10.44

Catalogue of White Dwarf Stars*

Name	B–V (mag)	U–B (mag)	μ ($''$ yr^{-1})	θ ($^{\circ}$)	π_t^{-1} ($''$)
1712+215	+0.32	- 0.52	0.32	195.	
1713+332	- 0.11	- 0.90	0.132	148.	
1713+695	+0.07	- 0.70	0.34	190.	0.0366
1714- 547	+0.27	- 0.59	0.164	151.	
1716+020	+0.13	- 0.59	0.52	234.	0.023
1726- 578	- 0.04	- 1.02	0.110	9.	
1728+560	- 0.08		0.27	338.	
1733- 544	+0.46	- 0.45	0.45	193.	0.048
1737+419	- 0.08	- 1.08			
1743- 132	+0.16	- 0.61	0.09	25.	0.054
1743- 521	+0.09	- 0.99	0.112	148.	
1748+708	+0.40	- 0.30	1.66	312.	0.1638
1750+098	+0.10	- 0.78	0.10	174.	
1756+827	+0.35	- 0.52	3.63	337.	0.0635
1809+284	+0.02	- 0.75	0.174	164.	
1811+327.1	+0.28		0.253	226.9	0.0199
1814+248	+0.39		0.28	172.	
1822+410	- 0.19	- 0.96			
1824+040	+0.05	- 0.55	0.41	223.	0.014
1826- 045	+0.24	- 0.56	0.27	191.	0.034
1827- 106	+0.14	- 0.63	0.32	140.	
1829+547	+0.49	- 0.44	0.385	318.3	0.0675
1831+197	+0.26	- 0.58	0.46	174.	0.017
1834- 781	- 0.06	- 0.80	0.332	161.	
1837- 619	+0.11	- 0.69	0.39	226.	0.039
1840- 111	+0.15	- 0.61	0.41	224.	0.053
1845+019	- 0.23	- 1.09	0.110		
1855+338	+0.17	- 0.60	0.40	7.	0.030
1857+119	+0.20	- 0.57	0.27	42.	
1858+393	+0.22	- 0.62	0.28	167.	
1900+705	+0.05	- 0.85	0.536	9.6	0.0746
1911+135	+0.12	- 0.60	0.10	180.	
1917+386	+0.44	- 0.37	0.28	179.	0.0858
1917- 077	+0.05	- 0.84	0.1823	204.9	0.1014
1918+110	- 0.11	- 0.86			

*Adapted from McCook and Sion (1987).

Catalogue of White Dwarf Stars*

Name	R.A.(1950) h m s	Dec.(1950) ° ′	Sp	V (mag)	M_V (mag)
1919+145	19 19 23	+14 34.9	DA5	13.01	11.51
1932– 136	19 32 54	– 13 36.	DA5	15.95	11.54
1935+276	19 35 11	+27 36.5	DAV5	12.98	11.92
1940+374	19 40 24	+37 24.	DB4	14.51	11.51
1943+163	19 43 16	+16 20.4	DA3	13.99	10.88
1950+250	19 50 22	+25 01.5	DAV4	16.0 pg	11.70
1952– 206	19 52 54	– 20 37.0	DA6	15.0	12.38
1953– 011	19 53 56	– 01 10.2	DA6	13.69	13.28
1953– 715	19 53 12	– 71 31.	DA4	15.15	11.20
1959+059	19 59 46	+05 59.3	DA5	16.41	12.18
2002– 110	20 02 46	– 11 05.4	DC9	16.87	15.61
2007– 219	20 07 18	– 21 55.	DA6	14.40	12.50
2007– 303	20 07 54	– 30 22.	DA4	12.18	11.67
2008+510	20 08 14	+51 03.3	DA3	16.0 pg	11.16
2011+065	20 11 29	+06 34.0	DQ7	15.73	13.73
2018– 585	20 18 18	– 58 31.	DC?	15.55	12.44
2025+488	20 25 52	+48 48.1	DA5	16.5 pg	12.21
2027+073	20 27 37	+07 19.8	DC5	16.26	12.41
2028+390	20 28 05	+39 03.4	DA2	13.37	10.42
2029+183	20 29 46	+18 21.0	DA5	16.31	12.25
2032+188	20 32 58	+18 49.1	DA3	15.34	10.86
2032+248	20 32 13	+24 53.8	DA3	11.52	10.96
2034– 532	20 34 30	– 53 16.	DB4	14.46	11.64
2039– 202	20 39 42	– 20 16.	DA3	12.34	10.82
2039– 682	20 39 36	– 68 16.	DA3	13.53	11.38
2043– 635	20 43 00	– 63 31.	DA3	15.52	10.82
2047+372	20 47 10	+37 16.9	DA4	13.40	11.64
2048+263	20 48 13	+26 19.6	DC9	16.3 pg	16.04
2054– 050	20 54 05	– 05 02.0	DC9	16.68	15.59
2058+181	20 58 58	+18 09.1	DA4	15.0	11.34
2058+342	20 58 20	+34 14.5	DB5	15.5 pg	12.35
2059+247	20 59 59	+24 45.7	DC7	16.57	14.37
2059+316	20 59 41	+31 37.0	DQ5	15.04	12.58
2105– 820	21 05 12	– 82 01.	DA6	13.50	12.61
2111+261	21 11 34	+26 09.3	DA6	14.69	13.10

Catalogue of White Dwarf Stars*

Name	B–V (mag)	U–B (mag)	μ ($'' \ yr^{-1}$)	θ (°)	π_t^{-1} ($''$)
1919+145	+0.06	– 0.66			0.0507
1932– 136	+0.05	– 0.76	0.14	189.	
1935+276	+0.17	– 0.56	0.46	85.	0.0556
1940+374	– 0.09	– 0.97	0.20	353.	
1943+163	– 0.03	– 0.81	0.291	187.1	0.023
1950+250			0.20	10.	
1952– 206	+0.18	– 0.52	0.36	165.	
1953– 011	+0.27	– 0.60	0.84	213.	0.0879
1953– 715	– 0.01	– 0.99	0.22	180.	
1959+059	+0.15	– 0.56			
2002– 110	+1.04		1.081	94.9	0.0576
2007– 219	+0.20	– 0.70	0.29	155.	0.042
2007– 303	+0.07	– 0.66	0.44	237.	
2008+510			0.27	204.	
2011+065	+0.36	– 0.39	0.633	202.8	0.0447
2018– 585	+0.29	– 0.56	0.511	144.	
2025+488					
2027+073	+0.10		0.260	135.	
2028+390	– 0.15	– 0.98			
2029+183	+0.16	– 0.59			
2032+188	– 0.04	– 0.80			
2032+248	– 0.07	– 0.87	0.66	217.	0.052
2034– 532	– 0.05	– 0.94	0.20	162.	
2039– 202	– 0.07	– 0.83	0.33	106.	0.0392
2039– 682	+0.11	– 0.78	0.247	138.	0.037
2043– 635	– 0.16	– 1.09	0.044	131.	
2047+372	+0.00		0.22	46.	0.055
2048+263	+0.95	+0.40	0.63	233.	0.0503
2054– 050	+1.13	+0.64	0.82	105.	0.060
2058+181	+0.01	– 0.75			
2058+342					
2059+247	+0.50	– 0.28	0.39	45.	
2059+316	+0.07	– 0.75	0.48	212.	0.028
2105– 820	+0.27	– 0.59	0.366		0.0567
2111+261	+0.25	– 0.54	0.49	159.	0.0318

*Adapted from McCook and Sion (1987).

Catalogue of White Dwarf Stars*

Name	R.A.(1950) h m s	Dec.(1950) ° ′	Sp	V (mag)	M_V (mag)
2111+498	21 11 03	+49 53.7	DA2	13.09	10.55
2115- 560	21 15 48	- 56 03.	DA6	14.28	12.56
2116+675	21 16 30	+67 32.1	DA3	16.0 pg	11.04
2117+539	21 17 22	+53 59.9	DA3	12.33	11.08
2119+581	21 19 18	+58 06.7	DA6	16.0 pg	13.13
2123- 229	21 23 24	- 22 56.8	DA4	15.0	11.39
2124+550	21 24 50	+55 00.4	DA4	14.70	11.43
2126+734	21 26 42	+73 25.8	DA4	12.82	11.16
2129+000	21 29 41	+00 01.8	DB4	14.73	11.66
2134+218	21 34 18	+21 51.1	DA3	14.45	11.19
2136+229	21 36 28	+22 55.7	DA5	15.25	12.20
2136+828	21 36 22	+82 49.8	DA3	13.02	11.05
2139+115	21 39 02	+11 32.8	DA5	15.8	11.73
2139+132	21 39 43	+13 15.0	DA7	16.58	13.45
2140+207	21 40 22	+20 46.6	DQ6	13.24	12.94
2143+353	21 43 16	+35 18.3	DA2	15.5 pg	10.70
2144- 079	21 44 58	- 07 58.1	DB4	14.82	11.39
2147+280	21 47 39	+28 03.0	DB5	14.68	11.85
2148+539	21 48 14	+53 54.5	DA5	16.74	12.51
2149+021	21 49 53	+02 09.4	DA3	12.77	10.93
2149+372	21 49 56	+37 12.1	DA4	15.0 pg	11.51
2150+338	21 50 19	+33 53.6	DA3	15.0 pg	11.04
2151- 015	21 51 30	- 01 31.	DA6	14.41	12.97
2154- 437	21 54 48	- 43 42.	DB3	14.69	10.72
2154- 512	21 54 30	- 51 14.	DQ7	14.68	15.01
2157+161	21 57 10	+16 11.2	DA3	16.2	10.93
2159- 754	21 59 48	- 75 28.	DA5	15.06	12.96
2203- 485	22 03 54	- 48 34.	DA3	15.55	11.17
2207+142	22 07 21	+14 14.9	DA6	15.61	13.28
2215+388	22 15 52	+38 53.8	DZ5	16.20	12.58
2216- 657	22 16 12	- 65 44.	DZ5	14.43	12.29
2222+683	22 22 09	+68 21.9	DB4	15.65	11.69
2232- 575	22 32 00	- 57 31.	DA4	14.96	11.18
2234+527	22 34 36	+52 47.6	DC5	16.6 pg	12.13
2240- 017	22 40 31	- 01 43.4	DA6	16.21	12.75

Catalogue of White Dwarf Stars*

Name	B–V (mag)	U–B (mag)	μ ($''$ yr^{-1})	θ (°)	π_t^{-1} ($''$)
2111+498	- 0.24	- 1.15			
2115- 560	+0.26	- 0.59	0.450	114.	0.045
2116+675					
2117+539	+0.07	- 0.67	0.27	333.	0.056
2119+581					
2123- 229			0.188	143.	
2124+550	+0.14	- 0.68	0.3383	55.7	0.0274
2126+734	+0.02	- 0.65	0.32	170.	0.0491
2129+000	- 0.07	- 0.92	0.39	84.	0.0331
2134+218	- 0.04	- 0.83			
2136+229	+0.14	- 0.50	0.29	70.	0.024
2136+828	- 0.01	- 0.72	0.62	24.	0.0387
2139+115	+0.08	- 0.68	0.24		
2139+132	+0.29	- 0.56	0.35	110.	
2140+207	+0.16	- 0.70	0.677	200.0	0.0799
2143+353					
2144- 079	- 0.09	- 0.95	0.28	116.	0.0154
2147+280	- 0.01	- 0.88	0.28	102.	0.0273
2148+539	+0.25	- 0.49	0.51	67.	
2149+021	0.00	- 0.78	0.33	182.	0.0409
2149+372					
2150+338					
2151- 015	+0.26	- 0.51	0.29	177.	0.051
2154- 437	- 0.01	- 0.90	0.22	144.	
2154- 512	+0.16	- 0.80	0.40	190.	
2157+161	- 0.04	- 0.81			
2159- 754	+0.16	- 0.63	0.504	277.9	
2203- 485	- 0.045	- 1.05	0.117	265.	
2207+142	+0.32	- 0.58	0.362	44.5	0.0401
2215+388	+0.281	- 0.14	0.232	209.	
2216- 657	+0.135	- 0.77	0.66	160.1	
2222+683	- 0.05	- 0.95	0.284	32.	0.0138
2232- 575	+0.105	- 0.815	0.200	96.	
2234+527			0.29	226.	
2240- 017	+0.31	- 0.60	0.231	217.3	0.009

*Adapted from McCook and Sion (1987).

Catalogue of White Dwarf Stars*

Name	R.A.(1950) h m s	Dec.(1950) ° ′	Sp	V (mag)	M_V (mag)
2246+223	22 46 39	+22 20.5	DA5	14.35	12.28
2248+293	22 48 57	+29 23.7	DA9	15.6	13.94
2251- 070	22 51 09	- 07 02.3	DZ9	15.65	16.13
2253- 062	22 53 12	- 06 16.9	DB4	15.5 pg	11.37
2253- 081	22 53 12	- 08 06.0	DA8	16.5	13.95
2254+076	22 54 55	+07 39.8	DQ5	17.24	12.41
2256+249	22 56 22	+24 59.7	DA4	13.68	11.32
2257+162	22 57 18	+16 13.1	DA3	16.14	11.07
2302+457	23 02 25	+45 43.2	DA3	15.5 pg	10.92
2304+013	23 04 05	+01 19.0	DO?	16.02	10.14
2311+552	23 11 50	+55 12.1	DAV5	16. pg	12.03
2311- 068	23 11 51	- 06 49.1	DQ?6	15.42	13.25
2312- 024	23 12 43	- 02 26.3	DZ8	16.29	13.82
2314+471	23 14 24	+47 10.8	DA3	16.0 pg	11.08
2316- 064	23 16 37	- 06 27.9	DC9	18.25	15.52
2316- 173	23 16 57	- 17 21.9	DB3	14.04	11.85
2319+691	23 19 15	+69 09.9	DA2	14.5 pg	10.67
2323+256	23 23 28	+25 36.0	DA9	17.06	14.46
2326+049	23 26 16	+04 58.5	DAV4	13.05	11.77
2329+267	23 29 33	+26 42.1	DA5	15.33	12.53
2329+407	23 29 09	+40 44.9	DA3	13.82	10.88
2333- 049	23 33 20	- 04 58.9	DA6	15.65	12.63
2337- 760	23 37 42	- 76 03.	DA3	13.53	11.29
2341+322	23 41 21	+32 16.2	DA4	12.92	11.70
2342+806	23 42 57	+80 40.2	DA2	14.52	10.50
2345+304	23 45 11	+30 27.4	DA2	16.39	10.7
2347+128	23 47 20	+12 49.5	DA5	15.9 pg	12.03
2347+292	23 47 23	+29 17.7	DA9	15.72	14.41
2352+401	23 52 24	+40 11.1	DQ6	15.13	13.13
2359- 434	23 59 34	- 43 25.6	DA5	13.05	13.47

Catalogue of White Dwarf Stars*

Name	B–V (mag)	U–B (mag)	μ ($''$ yr^{-1})	θ ($^\circ$)	π_t^{-1} ($''$)
2246+223	+0.20	- 0.68	0.506	86.2	0.0530
2248+293	+0.67	- 0.16	1.28		0.0486
2251- 070	+1.88		2.565	106.	0.125
2253- 062					
2253- 081	+0.44	- 0.45	0.547	96.	0.0361
2254+076	- 0.04	- 0.87	0.31	222.	0.003
2256+249	+0.04	- 0.91			
2257+162	- 0.17				
2302+457					
2304+013	- 0.44	- 1.15			
2311+552					
2311- 068	+0.22	- 0.63	0.38	244.7	0.0381
2312- 024	+0.51	- 0.37	0.61	69.	0.0391
2314+471					
2316- 064	+1.24		1.728	201.7	0.0262
2316- 173	- 0.01	- 0.92	0.26	80.	0.037
2319+691			0.129	271.	
2323+256	+0.47	- 0.23	0.29	154.	
2326+049	+0.16	- 0.64	0.56	237.	0.0735
2329+267	+0.19	- 0.63	0.42	87.	0.025
2329+407	+0.03	- 0.72	0.282	110.	
2333- 049	+0.22	- 0.53	0.27	227.	
2337- 760	+0.53	- 0.11	0.26	241.	
2341+322	+0.14	- 0.61	0.27	254.	0.0563
2342+806	- 0.33				
2345+304	- 0.15				
2347+128			0.36	78.	
2347+292	+0.58	- 0.24	0.531	190.8	0.0474
2352+401	+0.15	- 0.90	0.60	157.	0.0390
2359- 434	+0.07	- 0.87	0.90	138.	0.1269

*Adapted from McCook and Sion (1987).

21.3 Cross–Reference Name Index for White Dwarf Stars

Name	WD	Name	WD	Name	WD
Abell 7	0500– 157	BPM 12843	1953– 715	BPM 26734	2018– 585
Abell 15	0625– 253	BPM 13491	2039– 682	BPM 26944	2034– 532
α CMa B	0642– 166	BPM 13537	2043– 635	BPM 27273	2115– 560
α CMi B	0736+053	BPM 14525	2159– 754	BPM 27606	2154– 512
AC +25°68981	2032+248	BPM 14703	2216– 657	BPM 27891	2232– 575
AC +58°2500	0426+588	BPM 15727	2337– 760	BPM 28016	2248– 504
AC +58°43662	1340+572	BPM 16115	0024– 556	BPM 30551	0104– 464
BD +16°516	0347+171	BPM 16274	0047– 524	BPM 31594	0341– 459
BD – 00°4234	1129+004	BPM 16285	0048– 544	BPM 31852	0419– 487
BD – 07°3632	1327– 083	BPM 16501	0126– 532	BPM 33039	0732– 427
BD – 08°5980B	2253– 081	BPM 16571	0138– 559	BPM 36430	1153– 484
BPM 784	1425– 811	BPM 17088	0308– 565	BPM 37093	1236– 495
BPM 890	1628– 873	BPM 17113	0311– 543	BPM 38165	1407– 475
BPM 1266	2105– 820	BPM 17731	0418– 539	BPM 44275	2154– 437
BPM 2819	0255– 705	BPM 18164	0615– 591	BPM 44347	2203– 485
BPM 3116	0342– 673	BPM 18394	0701– 587	BPM 46232	0018– 339
BPM 3523	0446– 789	BPM 18615	0740– 570	BPM 46460	0034– 211
BPM 4225	0624– 756	BPM 18764	0800– 533	BPM 46931	0104– 331
BPM 4729	0752– 676	BPM 19652	0934– 587	BPM 70331	0041– 102
BPM 4834	0806– 661	BPM 19738	0950– 572	BPM 70524	0100– 068
BPM 5102	0848– 730	BPM 19929	1013– 559	BPM 77964	1531– 022
BPM 5109	0850– 617	BPM 20383	1053– 550	BPM 85584	0316+345
BPM 5639	0928– 713	BPM 20912	1121– 507	BPM 88611	1254+223
BPM 6082	0954– 710	BPM 21021	1129– 537	BPM 88763	1304+150
BPM 6114	0957– 666	BPM 21065	1133– 528	BPM 89123	1330+036
BPM 6502	1042– 690	BPM 21970	1323– 514	BPM 91358	1610+166
BPM 7108	1142– 645	BPM 24047	1616– 591	BPM 91679	1641+387
BPM 7543	1223– 659	BPM 24107	1620– 536	BPM 92077	1709+230
BPM 7855	1248– 610	BPM 24150	1623– 540	BPM 92172	1713+332
BPM 7961	1257– 723	BPM 24601	1659– 531	BPM 92960	1809+284
BPM 8394	1334– 678	BPM 24723	1709– 575	BPM 93487	1845+019
BPM 9323	1504– 770	BPM 24754	1714– 547	BPM 94172	1919+145
BPM 9518	1524– 749	BPM 24866	1726– 578	BPM 94484	1936+327
BPM 11593	1834– 781	BPM 24960	1733– 544	BPM 95701	2032+188
BPM 11668	1837– 619	BPM 25114	1743– 521	BPM 96804	2132+367

Cross–Reference Name Index

Name	WD	Name	WD	Name	WD
BPM 97895	2309+105	CBS 114	0954+342	EG 016	0213+427
C 1	1213+528	CC 398	0644+375	EG 017	0214+568
C 2	1606+422	CD – 32°5613	0839– 327	EG 018	0220+222
C 3	2329+407	CD – 37°6571B	1544– 377	EG 019	0227+050
CBS 3	0902+293	CD – 38°10980	1620– 391	EG 020	0232+035
CBS 4	0906+296	CI 20– 13	0011+000	EG 021	0310– 688
CBS 6	0920+306	CI 20– 960	1559+369	EG 022	0236– 273
CBS 9	0924+289	CI 20– 1215	2032+248	EG 023	0332+320
CBS 12	0929+290	CPD – 69°177	0310– 688	EG 024	0341+182
CBS 21	1028+328	CSO 4	0855+329	EG 025	0349+247
CBS 30	0938+299	CSO 5	0903+290	EG 026	0352+096
CBS 37	1108+325	CSO 43	1020+315	EG 028	0401+250
CBS 43	1131+320	CSO 57	1039+302	EG 029	0406+169
CBS 44	1131+315	CSO 64	1057+307	EG 030	0407+179
CBS 46	1134+300	CSO 74	1105+321	EG 031	0410+117
CBS 48	1143+321	CSO 75	1106+316	EG 033	0413– 077
CBS 49	1152+320	CSO 105	1137+311	EG 034	0417+361
CBS 51	1200+310	CSO 154	1235+321	EG 035	0418+153
CBS 52	1202+308	CSO 160	1240+326	EG 036	0421+216
CBS 54	1211+320	CSO 197	0835+340	EG 037	0425+168
CBS 60	1224+309	EG 001	0007+308	EG 038	0429+176
CBS 61	1227+307	EG 002	0011+000	EG 039	0431+135
CBS 64	1246+299	EG 003	0017+136	EG 040	0433+270
CBS 65	1249+296	EG 004	0033+016	EG 041	0435– 088
CBS 68	0810+334	EG 005	0046+051	EG 042	0438+108
CBS 70	0815+376	EG 006	0053– 117	EG 043	0518+333
CBS 77	0835+366	EG 007	0101+048	EG 044	0551+123
CBS 78	0838+375	EG 008	0107+267	EG 045	0552– 041
CBS 79	0839+379	EG 009	0115+159	EG 046	0612+177
CBS 84	0834+358	EG 010	0133– 116	EG 047	0625+100
CBS 86	0847+354	EG 011	0135-052	EG 048	0639+447
CBS 90	0859+337	EG 012	0142+230	EG 049	0642– 166
CBS 93	0901+348	EG 013	0143+216	EG 050	0644+375
CBS 94	0906+341	EG 014	0150+089	EG 051	0706+377
CBS 98	0920+366	EG 015	0205+250	EG 052	0727+482

Cross–Reference Name Index

Name	WD	Name	WD	Name	WD
EG 053	0736+053	EG 089	1215+323	EG 128	1857+119
EG 054	0738− 172	EG 090	1230+417	EG 129	1900+705
EG 055	0749− 383	EG 092	1244+149	EG 130	1911+135
EG 056	0752− 676	EG 093	1253+378	EG 131	1917− 077
EG 057	0752− 146	EG 094	1257+278	EG 132	1932− 136
EG 058	0816+387	EG 095	1257+037	EG 133	1940+374
EG 059	0836+201	EG 097	1302+283	EG 134	1943+163
EG 060	0836+199	EG 098	1314+293	EG 135	1953− 011
EG 061	0837+199	EG 099	1327− 083	EG 137	2007− 219
EG 062	0839− 327	EG 100	1334+039	EG 138	2011+065
EG 063	0845− 188	EG 101	1334− 160	EG 139	2032+248
EG 064	0913+442	EG 102	1337+705	EG 140	2039− 682
EG 065	0921+354	EG 103	1340+572	EG 141	2039− 202
EG 066	0935− 371.2	EG 104	1348− 273	EG 142	2105− 820
EG 067	0943+441	EG 106	1411+157	EG 143	2124+550
EG 068	0943+330	EG 107	1415+132	EG 144	2126+734
EG 069	0954+247	EG 108	1421+318	EG 145	2129+000
EG 069	0955+247	EG 110	1425− 811	EG 146	2130− 047
EG 070	1031− 114	EG 111	1448+077	EG 147	2136+828
EG 072	1039+145	EG 112	1510+566	EG 148	2140+207
EG 073	1046+281	EG 113	1544+008	EG 149	2144− 079
EG 074	1055− 072	EG 114	1544− 377	EG 150	2149+021
EG 075	1100+604	EG 115	1559+369	EG 151	2151− 015
EG 075	1104+602	EG 115	1600+369	EG 152	2154− 437
EG 076	1105− 048	EG 116	1606+422	EG 153	2224− 344
EG 077	1107+265	EG 117	1609+135	EG 154	2240− 017
EG 078	1115− 029	EG 118	1615− 154	EG 155	2246+223
EG 079	1121+216	EG 119	1626+368	EG 156	2254+076
EG 080	1126+384	EG 120	1637+335	EG 159	2326+049
EG 082	1142− 645	EG 122	1708− 147	EG 160	2329+407
EG 083	1143+321	EG 123	1716+020	EG 161	2329+267
EG 084	1147+255	EG 124	1750+098	EG 162	2341+322
EG 085	1154+186	EG 125	1824+040	EG 163	2351− 335
EG 086	1211+332	EG 126	1826− 045	EG 164	2351− 368
EG 087	1213+528	EG 127	1855+338	EG 166	0023− 109

Cross–Reference Name Index

Name	WD	Name	WD	Name	WD
EG 167	0036+312	EG 203	0004+330	EG 238	1641+387
EG 168	0208+396	EG 204	0037+312	EG 239	1645+325
EG 169	0423+120	EG 205	0126+422	EG 240	1737+419
EG 171	0715+125	EG 206	0155+069	EG 241	1809+284
EG 172	0855+604.2	EG 207	0231– 054	EG 242	1822+410
EG 173	1354+340	EG 208	0316+345	EG 243	2028+390
EG 174	1555– 089	EG 209	0453+418	EG 244	2111+498
EG 175	1743– 132	EG 210	0549+158	EG 245	0038+555
EG 176	1821– 131	EG 211	0606+282	EG 246	0038– 226
EG 177	1827– 106	EG 212	0637+477	EG 247	0501+527
EG 178	2253– 081	EG 213	0651– 020	EG 248	0548– 001
EG 179	0106+372	EG 214	0710+216	EG 249	0827+328
EG 180	0426+588	EG 215	0714+458	EG 250	0912+536
EG 181	0654+027	EG 216	0716+404	EG 251	0946+534
EG 182	0856+331	EG 217	0826+455	EG 252	0959+149
EG 183	1012+083	EG 218	0854+404	EG 253	0913– 010
EG 184	1134+300	EG 219	0858+363	EG 255	1211+392
EG 185	1143+321	EG 220	1019+462	EG 257	1406+590
EG 186	1232+479	EG 221	1052+273	EG 258	1633+572
EG 187	1254+223	EG 222	1422+095	EG 259	1635+137
EG 188	1317+453	EG 223	1539– 035	EG 260	1952– 206
EG 189	1408+323	EG 224	1709+230	EG 261	2047+372
EG 190	1503– 070	EG 225	1840+042	EG 262	2059+316
EG 191	1509+322	EG 226	1936+327	EG 263	2258+406
EG 192	1531– 022	EG 227	2134+218	EG 264	2316– 173
EG 193	1542+182	EG 228	2226+061	EG 265	0416+272
EG 194	1612– 111	EG 229	2240– 045	EG 266	0855+604.1
EG 195	1614– 128	EG 230	2253– 062	EG 298	1456+298
EG 196	1639+153	EG 231	2254+126	EG 387	1045– 017
EG 197	1655+215	EG 232	2256+249	ER 8	1310– 472
EG 198	1706+332	EG 233	2309+105	Feige 4	0017+136
EG 199	1756+827	EG 234	1258+593	Feige 7	0041– 102
EG 200	1840– 111	EG 235	1302+597	Feige 17	0155+069
EG 201	1919+145	EG 236	1429+373	Feige 22	0227+050
EG 202	2054– 050	EG 237	1525+433	Feige 24	0232+035

Cross–Reference Name Index

Cross–Reference Name Index

Name	WD	Name	WD	Name	WD
G 111- 049	0756+437	G 135- 029	1411+157	G 158- 078	0023- 109
G 111- 054	0802+386	G 136- 022	1448+077	G 158- 132	0005- 163
G 111- 071	0816+387	G 138- 008	1609+135	G 159- 012	0151+017
G 115- 009	0830+371	G 138- 031	1625+093	G 160- 060	0413- 077
G 116- 016	0913+442	G 138- 047	1635+137	G 161- 068	0941- 068
G 116- 052	0943+441	G 138- 049	1636+057	G 163- 027	1055- 072
G 117- 025	0930+294	G 138- 056	1639+153	G 163- 028	1055- 039
G 117- B11B	0943+330	G 139- 013	1705+030	G 163- 050	1105- 048
G 117- B15A	0921+354	G 140- 002	1736+052	G 165- 007	1328+307
G 119- 047	1056+345	G 140- B 1B	1750+098	G 165- 018	1334+366
G 120- 045	1121+216	G 141- 002	1818+126	G 165- B 5B	1354+340
G 121- 022	1147+255	G 141- 054	1857+119	G 166- 014	1413+231
G 122- 031	1132+470	G 142- 050	1943+163	G 166- 058	1455+298
G 124- 026	1418- 088	G 142- B 2A	1911+135	G 167- 008	1455+298
G 124- 209	1415- 064	G 144- 051	2059+190	G 169- 034	1655+215
G 125- 003	1917+386	G 145- 004	2059+190	G 170- 002	1655+215
G 126- 018	2136+229	G 147- 065	1133+358	G 170- 027	1712+215
G 126- 025	2139+132	G 148- 007	1143+321	G 171- 002	2329+407
G 126- 027	2140+207	G 148- B 4B	1215+323	G 171- 027	2352+401
G 127- 058	2246+223	G 149- 028	1257+278	G 171- 052	0019+423
G 128- 004	2246+223	G 149- 028	1257+278	G 172- 004	0030+444
G 128- 007	2248+293	G 150- 043	1347+283	G 173- 005	0129+458
G 128- 062	2323+256	G 152- B 4B	1555- 089	G 174- 005	0232+525
G 128- 072	2329+267	G 153- 040	1614- 128	G 174- 014	0245+541
G 130- 005	2341+322	G 153- 041	1615- 154	G 175- 034B	0426+588
G 130- 015	2347+292	G 154- 085B	1743- 132	G 175- 046	0440+510
G 130- 015	2347+292	G 155- 015	1821- 131	G 177- 031	1317+453
G 130- 049	0007+308	G 155- 019	1827- 106	G 177- 034	1319+466
G 131- 019	0003+177	G 155- 034	1840- 111	G 180- 023	1559+369
G 132- 012	0036+312	G 156- 064	2253- 081	G 180- 057	1626+368
G 133- 008	0121+401	G 157- 034	2311- 068	G 180- 063	1633+433
G 133- 036	0142+312	G 157- 035	2312- 024	G 180- 065	1637+335
G 133- 072	0208+396	G 157- 082	2333- 049	G 181- B 5B	1706+332
G 134- 008	0208+396	G 158- 039	0009- 058	G 183- 035	1814+248
G 134- 022	0213+427	G 158- 045	0011- 134	G 184- 012	1831+197

Cross–Reference Name Index

Name	WD	Name	WD	Name	WD
G 185- 032	1935+276	G 221- 002	0236+745	G 256- 007	1309+853
G 186- 031	2032+248	G 221- 010	0324+738	G 256- 018	1459+821
G 187- 008	2048+263	G 223- 024	1344+572	G 257- 038	1641+732
G 187- 009	2055+221	G 223- 063	1426+613	G 259- 021	1756+827
G 187- 015	2059+316	G 224- 059	1518+636	G 260- 015	1900+705
G 187- 016	2059+247	G 225- 068	1633+572	G 261- 043	2126+734
G 187- 032	2111+261	G 226- 017	1633+572	G 261- 045	2136+828
G 188- 027	2147+280	G 226- 029	1647+591	G 266- 032	0000- 170
G 189- 040	2248+293	G 227- 005	1728+560	G 266- 135	0032- 175
G 191- 016	0455+553	G 227- 028	1820+609	G 266- 141	0034- 211
G 191- B 2B	0501+527	G 227- 035	1829+547	G 266- 157	0038- 226
G 193- 074	0749+526	G 230- 030	2008+510	G 266- 157	0038- 226
G 193- 078	0751+578	G 231- 040	2117+539	G 267- 018	0000- 345
G 195- 019	0912+536	G 231- 043	2124+550	G 267- 032	0004- 315
G 195- 042	0946+534	G 232- 038	2148+539	G 268- 005	0032- 175
G 197- 004	1104+602	G 233- 019	2234+527	G 268- 012	0034- 211
G 197- 035	1148+544	G 234- 004	0728+642	G 268- 027	0038- 226
G 197- 047	1208+576	G 234- 051	0855+604.1	G 268- 040	0042- 238
G 198- B 6A	1211+392	G 234- 051	0855+604.2	G 268- 074	0053- 237
G 199- 071	1325+581	G 235- 067	1019+637	G 269- 093	0103- 278
G 200- 039	1425+540	G 237- 028	1143+633	G 269- 160	0123- 262
G 200- 042	1426+442	G 237- 056	1214+690	G 270- 048	0041- 102
G 201- 039	1510+566	G 238- 044	1337+705	G 270- 123	0100- 036
G 202- 049	1624+477	G 238- 053	1350+656	G 270- 124	0100- 068
G 205- 052	1858+393	G 240- 047	1710+683	G 271- 047A	0119- 004
G 206- 017	1811+327.1	G 240- 051	1713+695	G 271- 081	0127- 050
G 206- 018	1811+327.2	G 240- 072	1748+708	G 271- 106	0133- 116
G 207- 009	1855+338	G 241- 006	2222+683	G 271- 115	0135- 052
G 208- 017	1917+386	G 241- 046	2307+636	G 272- B 5B	0200- 171
G 210- 036	2047+372	G 242- 063	0038+730	G 272- 152	0203- 181
G 211- 008	2059+316	G 244- 036	0148+641	G 273- 040	2322- 181
G 212- B 1A	2107+427	G 245- 058	0246+734	G 273- 097	2336- 187
G 216- B14B	2258+406	G 250- 026	0648+641	G 273- B 1B	2350- 083
G 217- 037	0009+501	G 252- 027	0900+734	G 274- 039	0125- 236
G 218- 008	0038+555	G 255- 002	1159+803	G 274- 150	0204- 233

Cross–Reference Name Index

Name	WD	Name	WD	Name	WD
GD 002	0004+330	GD 084	0714+458	GD 196	1610+166
GD 008	0037+312	GD 085	0716+404	GD 198	1612- 111
GD 009	0058- 044	GD 086	0730+487	GD 200	1620+260
GD 011	0106+372	GD 089	0743+442	GD 202	1636+160
GD 012	0107+267	GD 090	0816+376	GD 203	1655+210
GD 013	0126+422	GD 091	0826+455	GD 205	1709+230
GD 014	0127+270	GD 095	0843+358	GD 215	1840+042
GD 020	0155+069	GD 096	0846+346	GD 218	1918+110
GD 025	0213+396	GD 098	0854+404	GD 219	1919+145
GD 030	0230+343	GD 099	0858+363	GD 222	1936+327
GD 031	0231- 054	GD 111	1002+430	GD 229	2010+310
GD 038	0259+378	GD 117	1019+462	GD 230	2029+183
GD 040	0300- 013	GD 122	1029+329	GD 231	2032+188
GD 041	0302+027	GD 123	1033+464	GD 232	2058+181
GD 043	0312+223	GD 124	1046- 017	GD 233	2130- 047
GD 045	0316+345	GD 125	1052+273	GD 234	2134+218
GD 047	0339- 035	GD 128	1107+265	GD 235	2139+115
GD 050	0346- 011	GD 129	1108+475	GD 236	2226+061
GD 051	0347- 137	GD 133	1116+026	GD 240	2240- 045
GD 052	0348+339	GD 140	1134+300	GD 243	2253- 062
GD 056	0408- 041	GD 148	1232+479	GD 244	2254+126
GD 060	0416+334	GD 151	1249+182	GD 245	2256+249
GD 061	0435+410	GD 153	1254+223	GD 246	2309+105
GD 064	0453+418	GD 154	1307+354	GD 248	2323+157
GD 066	0517+307	GD 163	1408+323	GD 251	2331+290
GD 067	0518+005	GD 165	1422+095	GD 257	0548+000
GD 069	0532+414	GD 173	1451+006	GD 266	1959+059
GD 070	0546+265	GD 175	1503- 070	GD 267	1257+047
GD 071	0549+158	GD 176	1507- 105	GD 268	1304+150
GD 072	0606+282	GD 178	1509+322	GD 269	1330+036
GD 074	0625+415	GD 185	1531- 022	GD 272	2157+161
GD 077	0637+477	GD 186	1531+184	GD 273	0103+558
GD 078	0648+368	GD 189	1539- 035	GD 275	0115+521
GD 080	0651- 020	GD 190	1542+182	GD 276	0120+475
GD 083	0710+216	GD 194	1550+183	GD 279	0148+467

Cross–Reference Name Index

Name	WD	Name	WD	Name	WD
GD 282	0205+551	GD 372	1802+213	GD 466	1102+748
GD 283	0231+570	GD 375	1809+284	GD 479	1241+651
GD 290	0543+579	GD 378	1822+410	GD 505	1606+677
GD 294	0713+584	GD 385	1950+250	GD 515	1654+637
GD 295	0810+234	GD 387	2003+437	GD 517	1659+622
GD 303	1011+570	GD 390	2025+488	GD 518	1659+662
GD 304	1034+492	GD 391	2028+390	GD 525	1735+610
GD 305	1108+563	GD 392	2058+342	GD 526	1739+804
GD 307	1122+546	GD 393	2058+506	GD 530	1820+709
GD 310	1126+384	GD 394	2111+498	GD 533	1918+725
GD 312	1149+410	GD 395	2132+367	GD 539	1958+675
GD 314	1202+608	GD 396	2143+353	GD 543	2009+622
GD 317	1230+417	GD 397	2149+372	GD 544	2010+613
GD 319	1247+553	GD 398	2150+338	GD 546	2025+554
GD 320	1253+482	GD 401	2215+388	GD 547	2116+675
GD 322	1258+593	GD 402	2216+484	GD 548	2119+581
GD 323	1302+597	GD 404	2302+457	GD 554	2250+746
GD 325	1333+487	GD 405	2314+471	GD 556	2311+552
GD 330	1401+523	GD 406	2328+510	GD 557	2313+682
GD 335	1419+351	GD 408	0002+729	GD 559	2319+691
GD 336	1429+373	GD 411	0033+771	GD 561	2342+806
GD 337	1433+538	GD 419	0134+833	GD 597	0016– 220
GD 340	1508+637	GD 420	0136+768	GD 603	0018– 339
GD 344	1525+433	GD 421	0147+674	GD 617	0028– 274
GD 347	1534+503	GD 422	0159+754	GD 619	0031– 274
GD 348	1537+651	GD 425	0248+601	GD 656	0048– 202
GD 352	1550+626	GD 426	0302+621	GD 659	0050– 332
GD 354	1630+618	GD 427	0314+648	GD 661	0052– 250
GD 356	1639+537	GD 429	0416+701	GD 674	0101– 250
GD 357	1641+387	GD 433	0513+756	GD 681	0105– 340
GD 358	1645+325	GD 435	0517+771	GD 683	0106– 358
GD 360	1713+332	GD 446	0658+624	GD 685	0107– 192
GD 362	1729+371	GD 448	0710+741	GD 687	0107– 342
GD 363	1737+419	GD 457	0811+644	GD 691	0109– 264
GD 370	1755+194	GD 462	1005+642	GD 693	0113– 243

Cross–Reference Name Index

Name	WD	Name	WD	Name	WD
GD 695	0116– 231	GL 836.5	2140+207	GR 295	1425+540
GD 765	0102– 142	GL 837.1	2144– 079	GR 296	1434+289
GD 821	0114– 034	GL 838.4	2149+021	GR 297	1451+006
GD 1383	0137– 229	GL 893.1	2311– 068	GR 298	1455+298
GD 1669	2329– 291	GL 895.2	2326+049	GR 299	1655+210
GH 7– 23	0347+171	GL 905.2B	2341+322	GR 300	2116+675
GH 7– 41	0352+096	GL 1178.0	1344+106	GR 301	2119+581
GH 7– 112	0406+169	GR 267	0041– 102	GR 302	2207+142
GH 7– 191	0421+162	GR 268	0148+641	GR 303	2319+691
GH 7– 233	0425+168	GR 269	0148+467	GR 304	2342+806
GH 7– 255	0429+176	GR 270	0302+027	GR 305	0002+729
GH 8– 7	0401+250	GR 271	1220+234	GR 306	0030+444
GL 35	0046+051	GR 272	1403– 010	GR 307	0123– 262
GL 64	0135– 052	GR 273	1550+183	GR 308	0134+833
GL 91.3	0213+427	GR 274	1620– 391	GR 310	0147+674
GL 151	0341+182	GR 275	1620+260	GR 311	0151+017
GL 159.1	0401+250	GR 276	1616– 390	GR 312	0213+396
GL 166 B	0413– 077	GR 277	1935+276	GR 314	0232+525
GL 169.1B	0426+588	GR 278	2032+188	GR 315	0435+410
GL 246	0644+375	GR 279	2058+181	GR 316	0437+138
GL 275.2B	0727+482	GR 280	2139+115	GR 317	0440+510
GL 339.1	0912+536	GR 281	2150+338	GR 318	0455+553
GL 427	1121+216	GR 282	2157+161	GR 319	0532+414
GL 492	1257+037	GR 283	2248+293	GR 321	0728+642
GL 518	1334+039	GR 284	2307+636	GR 322	0751+578
GL 626.2	1626+368	GR 285	0034– 211	GR 323	0846+346
GL 630.1B	1633+572	GR 286	0212– 231	GR 324	0930+294
GL 742	1900+705	GR 287	0259+378	GR 325	1056+345
GL 754.1A	1917– 077	GR 288	0346– 011	GR 326	1413+231
GL 772	1953– 011	GR 289	0548+000	GR 327	1625+093
GL 794	2032+248	GR 290	0553+053	GR 328	1636+057
GL 812 B	2054– 050	GR 291	0840+262	GR 329	1639+537
GL 828.3	2124+550	GR 292	1241+651	GR 330	1811+327.2
GL 828.5	2126+734	GR 293	1257+047	GR 331	1918+725
GL 836.2	2136+828	GR 294	1330+036	GR 332	2009+622

Cross–Reference Name Index

Name	WD	Name	WD	Name	WD
GR 333	2010+310	GR 368	1647+591	GR 405	2347+128
GR 334	2107+427	GR 369	1710+683	GR 406	0000- 345
GR 335	2323+157	GR 370	1713+695	GR 407	0058- 044
GR 337	0033+771	GR 371	1736+052	GR 408	0101+866
GR 338	0103+558	GR 372	1748+708	GR 409	0112+104
GR 339	0136+768	GR 373	1811+327.1	GR 410	0115+521
GR 340	0533+322	GR 374	1829+547	GR 411	0228+269
GR 341	0543+579	GR 375	1917+386	GR 412	0231+570
GR 342	0648+641	GR 376	2008+510	GR 413	0236+745
GR 343	0743+442	GR 377	2059+190	GR 414	0302+621
GR 344	0749+526	GR 378	2117+539	GR 415	0340- 243
GR 345	0752+365	GR 379	2234+527	GR 416	0341- 248
GR 346	0802+386	GR 380	2323+256	GR 417	0407+197
GR 347	0810+234	GR 381	0009+501	GR 418	0419+266
GR 348	0816+376	GR 383	0038+730	GR 419	0437+093
GR 349	0900+734	GR 384	0300- 013	GR 420	0452+103
GR 350	1019+637	GR 385	0324+738	GR 421	0600+735
GR 351	1108+563	GR 386	1011+570	GR 423	0720+304
GR 352	1122+546	GR 387	1046- 017	GR 424	0725+318
GR 353	1143+633	GR 388	1133+358	GR 425	0737+288
GR 354	1149+410	GR 390	1304+150	GR 426	0747+073.1
GR 355	1159+803	GR 391	1328+307	GR 427	0747+073.2
GR 356	1214+690	GR 392	1401+523	GR 428	0756+437
GR 357	1253+482	GR 393	1459+821	GR 429	0807+484
GR 358	1325+581	GR 394	1950+250	GR 430	0852+630
GR 359	1333+487	GR 395	2003+437	GR 431	0939+071
GR 359	1333+487	GR 396	2025+488	GR 432	1005+642
GR 360	1344+106	GR 397	2028+390	GR 433	1020+315
GR 361	1419+351	GR 398	2058+342	GR 434	1042+593
GR 362	1426+613	GR 399	2111+498	GR 435	1148+544
GR 363	1433+538	GR 400	2116+675	GR 436	1309+853
GR 364	1508+637	GR 401	2149+372	GR 437	1330+015
GR 365	1534+503	GR 402	2150+338	GR 438	1345+238
GR 366	1550+626	GR 403	2302+457	GR 439	1421+028
GR 367	1641+732	GR 404	2311+552	GR 440	1514+033

Cross–Reference Name Index

Name	WD	Name	WD	Name	WD
GR 441	1728+560	GR 475	0248+601	GR 515	0114−034
GR 442	1958+675	GR 476	0257+080	GR 516	0119−004
GR 442	1958+675	GR 477	0314+648	GR 518	0127−050
GR 443	2010+613	GR 478	0352+076	GR 520	0200−171
GR 444	2025+554	GR 479	0357−233	GR 521	0312+223
GR 445	2047+809	GR 480	0409+237	GR 523	0419+500
GR 446	2058+506	GR 481	0416+701	GR 525	0513+756
GR 447	2111+261	GR 482	0423+044	GR 526	0517+771
GR 448	2132+367	GR 483	0518+005	GR 527	0546+265
GR 449	2143+353	GR 484	0644+025	GR 530	0658+712
GR 450	2148+539	GR 485	0657+320	GR 533	0942+236.1
GR 451	2215+388	GR 486	0830+371	GR 534	0942+236.2
GR 452	2216+484	GR 488	1524+438	GR 535	1036−204
GR 453	2251−070	GR 489	1524+566	GR 537	1307+354
GR 454	2301+762	GR 490	1536+000	GR 538	1350+656
GR 455	2313+682	GR 491	1541+003	GR 539	1404+670
GR 456	2314+471	GR 492	1602+010	GR 541	1606+677
GR 457	0003+177	GR 493	1643+807	GR 542	1630+245
GR 458	0013−241	GR 494	1705+030	GR 543	1659+622
GR 459	0019+423	GR 495	1717−014	GR 544	1659+662
GR 460	0021−234	GR 496	1802+213	GR 545	1729+371
GR 461	0040−220	GR 497	1939+662	GR 546	1735+610
GR 462	0102+210.1	GR 498	2002−110	GR 547	1739+804
GR 463	0102+210.2	GR 500	2041+731	GR 548	1755+194
GR 464	0120+475	GR 501	2101+398	GR 549	1820+709
GR 465	0125−236	GR 502	2123−229	GR 550	1959+637
GR 466	0137−229	GR 503	2201−037	GR 552	2242+383
GR 467	0145−174	GR 504	2250+746	GR 553	2311−068
GR 468	0159+754	GR 505	2329−291	GR 554	2312−024
GR 469	0203+207	GR 506	2347+292	GR 555	2316−064
GR 470	0205+551	GR 507	2352+401	GR 556	2322−181
GR 471	0230−144	GR 509	0005−163	GR 557	2336−187
GR 472	0239+109	GR 512	0100−036	GR 558	2350−083
GR 473	0245+541	GR 513	0100−068	GR 559	0009−058
GR 474	0246+734	GR 514	0102−142	GR 560	0032−175

Cross–Reference Name Index

Name	WD	Name	WD	Name	WD
GR 561	0037- 006	GW +69°6829	1713+695	KPD	1949+472
GR 562	0052+226	GW +70°5824	1337+705	KPD	2006+481
GR 563	0107- 192	GW +70°8247	1900+705	KPD	2046+396
GR 564	0126+101	GW +73°8031	2126+734	KPD	2058+499
GR 565	0243- 026	GW +82°3818	2136+828	KPD	2154+408
GR 566	0322- 019	H Per 1166	0214+568	KPD	2236+541
GR 567	0339+523	HD 149499B	1634- 573	KPD	2328+510
GR 568	0339- 035	HE 3	0644+375	KUV	0343- 007
GR 569	0354+556	HL 4	0552- 041	KUV	0352+096
GR 570	0354+463	HL Tau 76	0416+272	KUV	0425+168
GR 571	0408- 041	HZ 2	0410+117	KUV	0429+176
GR 572	0517+307	HZ 4	0352+096	KUV	0437+152
GR 573	1610+166	HZ 7	0431+125	KUV	0510+163
GR 574	1624+477	HZ 9	0429+176	KUV	0802+386
GR 575	1636+160	HZ 10	0407+179	KUV	0802+413
GR 576	1704+481.1	HZ 14	0438+108	KUV	0817+386
GR 577	1704+481.2	HZ 21	1211+332	KUV	0827+410
GR 578	1713+332	HZ 28	1230+417	KUV	0854+404
GR 579	1818+126	HZ 34	1253+378	KUV	0933+383
GR 580	1858+393	HZ 39	1302+283	KUV	1009+416
GR 581	2107- 216	HZ 43A	1314+293	KUV	1126+384
GR 582	2136+229	IC 2391- 1	0839- 528	KUV	1137+423
GR 583	2147+280	K 1- 12	1448+077	KUV	1138+424
GR 584	2211+372	K 1- 16	1821+643	KUV	1148+408
GR 585	2331+290	KPD	0005+511	KUV	1302+283
GR 900	0108+100	KPD	0253+508	KUV	1408+323
GR 901	0109+111	KPD	0416+402	KUV	1428+373
GR 902	0500- 157	KPD	0420+520	KUV	1429+373
GR 903	0625- 253	KPD	0516+365	KUV	1636+351
GR 904	0853+163	KPD	0556+172	KUV	1647+375
GR 905	0956+359	KPD	0558+165	KUV	1648+371
GR 906	2229+139	KPD	0631+107	KUV	2256+249
GR 907	2234+064	KPD	1138+424	KUV	2306+124
GR 908	2246+120	KPD	1845+019	KUV	2306+130
GR 909	2310+175	KPD	1914+094	KUV	2309+105

Cross–Reference Name Index

Name	WD	Name	WD	Name	WD
KUV	2316+123	L 185- 053	0740- 570	L 356- 123	2203- 485
KUV	2328+107	L 188- 086	0934- 587	L 384- 024	0732- 427
L 007- 044	1708- 871	L 189- 036	0950- 572	L 427- 060	2154- 437
L 008- 061	1628- 873	L 190- 021	1013- 559	L 462- 056B	0935- 371.2
L 019- 002	1425- 811	L 202- 179	1616- 591	L 481- 060	1544- 377
L 024- 052	2105- 820	L 203- 131	1709- 575	L 505- 001	0000- 345
L 026- 023	2337- 760	L 204- 118	1726- 578	L 505- 042	2351- 368
L 041- 026	1504- 770	L 210- 114	2014- 575	L 532- 081	0839- 327
L 044- 095	1834- 781	L 210- 160	2018- 585	L 573- 108	2224- 344
L 048- 015	2159- 754	L 212- 019	2115- 560	L 577- 071	2351- 335
L 054- 005	0255- 705	L 214- 057	2232- 575	L 581- 026	0123- 262
L 064- 027	0954- 710	L 219- 048	0047- 524	L 587- 077	0316+345
L 069- 047	1257- 723	L 220- 145	0048- 544	L 587- 077A	0326- 273
L 072- 091	1524- 749	L 227- 140	0311- 543	L 619- 050	1348- 273
L 080- 056	1953- 715	L 242- 083	0800- 533	L 651- 029	0034- 211
L 088- 059	0141- 675	L 250- 052	1053- 550	L 651- 057	0038- 226
L 097- 003	0806- 661	L 251- 024	1121- 507	L 709- 020	1952- 206
L 097- 012	0752- 676	L 252- 073	1133- 528	L 710- 030	2007- 219
L 101- 026	0950- 572	L 252- 112	1129- 537	L 711- 010	2039- 202
L 101- 026	0957- 666	L 257- 047	1323- 514	L 745- 046A	0738- 172
L 101- 080	1042- 690	L 266- 180	1620- 536	L 748- 070	0845- 188
L 104- 002	1223- 659	L 266- 195	1623- 540	L 762- 021	1334- 160
L 106- 073	1334- 678	L 268- 092	1659- 531	L 770- 003	1615- 154
L 116- 079	2039- 682	L 269- 072	1714- 547	L 791- 040	2316- 173
L 119- 034	2216- 657	L 270- 031	1743- 521	L 793- 018	0000- 170
L 139- 026	0850- 617	L 270- 137	1733- 544	L 795- 007	0041- 102
L 145- 141	1142- 645	L 279- 025	2034- 532	L 796- 010	0053- 117
L 147- 011	1248- 610	L 283- 007	2154- 512	L 817- 013	0752- 146
L 158- 053	1837- 619	L 285- 014	2248- 504	L 825- 014	1031- 114
L 162- 160	2043- 635	L 292- 041	0104- 464	L 842- 032	1614- 128
L 170- 027	0024- 556	L 300- 034	0341- 459	L 845- 070	1708- 147
L 172- 015	0138- 559	L 302- 089	0419- 487	L 849- 015	1840- 111
L 175- 034	0308- 565	L 325- 214	1153- 484	L 852- 037	1932- 136
L 182- 061	0615- 591	L 327- 186	1236- 495	L 870- 002	0135- 052
L 184- 075	0701- 587	L 332- 123	1407- 475	L 879- 014	0435- 088

Cross–Reference Name Index

Cross–Reference Name Index

Name	WD	Name	WD	Name	WD
LFT 0142	0135– 052	LHS 0007	0046+051	LHS 1028	0007+308
LFT 0158	0141– 675	LHS 0027	0426+588	LHS 1038	0009+501
LFT 0245	0255– 705	LHS 0032	0552– 041	LHS 1044	0011– 134
LFT 0286	0326– 273	LHS 0034	0752– 676	LHS 1076	0024– 556
LFT 0306	0341+182	LHS 0043	1142– 645	LHS 1126	0038– 226
LFT 0349	0419– 487	LHS 0046	1334+039	LHS 1131	0040– 220
LFT 0486	0642– 166	LHS 0056	1756+827	LHS 1158	0048– 207
LFT 0487	0644+375	LHS 0069	2251– 070	LHS 1219	0112– 018
LFT 0600	0839– 327	LHS 0147	0145– 174	LHS 1227	0115+159
LFT 0753	1055– 072	LHS 0151	0208+396	LHS 1247	0123– 262
LFT 0792	1115– 029	LHS 0153	0213+427	LHS 1270	0135– 052
LFT 0800	1121+216	LHS 0179	0341+182	LHS 1415	0230– 144
LFT 0844	1142– 645	LHS 0194	0435– 088	LHS 1442	0243– 026
LFT 0931	1236– 495	LHS 0212	0553+053	LHS 1446	0245+541
LFT 0960	1257+037	LHS 0230	0727+482	LHS 1474	0255– 705
LFT 1004	1323– 514	LHS 0235	0738– 172	LHS 1547	0322– 019
LFT 1014	1327– 083	LHS 0239	0747+073.1	LHS 1549	0326– 273
LFT 1023	1334+039	LHS 0240	0747+073.2	LHS 1611	0349+495
LFT 1025	1334– 678	LHS 0253	0839– 327	LHS 1617	0357+081
LFT 1146	1448+077	LHS 0282	1022+009	LHS 1636	0407+197
LFT 1242	1559+369	LHS 0285	1033+714	LHS 1660	0419– 487
LFT 1280	1626+368	LHS 0290	1043– 188	LHS 1670	0423+044
LFT 1339	1716+020	LHS 0291	1042+593	LHS 1693	0437+093
LFT 1446	1900+705	LHS 0304	1121+216	LHS 1734	0503– 174
LFT 1503	1953– 011	LHS 0339	1237– 230	LHS 1764	0525+526
LFT 1540	2018– 585	LHS 0342	1247+550	LHS 1822	0600+735
LFT 1554	2032+248	LHS 0354	1327– 083	LHS 1838	0618+067
LFT 1585	2047+809	LHS 0361	1345+238	LHS 1870	0644+375
LFT 1649	2136+828	LHS 0378	1444– 174	LHS 1889	0657+320
LFT 1655	2140+207	LHS 0422	1633+572	LHS 1927	0732– 427
LFT 1679	2159– 754	LHS 0455	1748+708	LHS 1980	0802+387
LFT 1705	2216– 657	LHS 0483	2002– 110	LHS 1986	0807+190
LFT 1749	2253– 081	LHS 0529	2248+293	LHS 2022	0827+328
LFT 1835	2351– 335	LHS 0542	2316– 064	LHS 2086	0855+604.2
LFT 1837	2351– 368	LHS 1008	0000– 345	LHS 2087	0855+604.1

Cross–Reference Name Index

Name	WD	Name	WD	Name	WD
LHS 2101	0900+734	LHS 3501	1953− 011	LP 030− 079	0159+754
LHS 2273	1026+117	LHS 3532	2011+065	LP 030− 203	0236+745
LHS 2293	1036− 204	LHS 3562	2032+248	LP 030− 265	0246+734
LHS 2333	1055− 072	LHS 3589	2048+263	LP 031− 140	0324+738
LHS 2364	1108+207	LHS 3601	2054− 050	LP 033− 276	0600+735
LHS 2392	1115− 029	LHS 3636	2107− 216	LP 034− 137	0658+712
LHS 2478	1153+135	LHS 3703	2140+207	LP 034− 185	0710+741
LHS 2522	1208+576	LHS 3752	2159− 754	LP 036− 115	0900+734
LHS 2559	1224+354	LHS 3779	2211+372	LP 037− 186	1033+714
LHS 2594	1236− 495	LHS 3794	2216− 657	LP 039− 076	1214+690
LHS 2596	1239+454	LHS 3857	2246+223	LP 043− 146	1641+732
LHS 2661	1257+037	LHS 3917	2312− 024	LP 044− 113	1748+708
LHS 2673	1300+263	LHS 4019	2347+292	LP 046− 147	2041+731
LHS 2710	1313− 198	LHS 4040	2351− 335	LP 048− 292	2250+746
LHS 2737	1323− 514	LHS 4041	2351− 368	LP 049− 275	0002+729
LHS 2745	1328+307	LHS 4043	2352+401	LP 052− 029	0147+674
LHS 2769	1334− 678	LHS 5016	0052+226	LP 058− 053	0648+641
LHS 2800	1344+106	LHS 5023	0102+210.1	LP 058− 247	0728+642
LHS 2808	1346+121	LHS 5024	0102+210.2	LP 060− 359	0852+630
LHS 2856	1404+670	LHS 5064	0257+080	LP 062− 147	1019+637
LHS 2951	1434+437	LHS 5212	1214+690	LP 063− 267	1143+633
LHS 2984	1448+077	LP 001− 170B	0101+866	LP 066− 262	1350+656
LHS 3007	1455+298	LP 002− 534	0134+833	LP 066− 433	1404+670
LHS 3088	1533+469	LP 007− 200	1159+803	LP 068− 060	1518+636
LHS 3146	1559+369	LP 007− 226	1309+853	LP 070− 172	1710+683
LHS 3151	1602+010	LP 008− 046	1309+853	LP 070− 238	1653+630
LHS 3163	1609+135	LP 008− 157	1459+821	LP 073− 052	1939+662
LHS 3200	1626+368	LP 009− 175	1643+807	LP 073− 276	1959+637
LHS 3230	1636+057	LP 009− 231	1756+827	LP 077− 024	2319+691
LHS 3231	1643+807	LP 012− 438	0033+771	LP 077− 057	2351+650
LHS 3250	1653+630	LP 013− 249	0136+768	LP 090− 070	0855+604.2
LHS 3254	1655+215	LP 015− 308	0513+756	LP 090− 071	0855+604.1
LHS 3278	1716+020	LP 025− 436	2047+809	LP 092− 065	1011+570
LHS 3384	1821− 131	LP 027− 275	2301+762	LP 093− 021	1042+593
LHS 3424	1900+705	LP 029− 015	0038+730	LP 093− 370	1104+602

Cross–Reference Name Index

Name	WD	Name	WD	Name	WD
LP 094- 268	1128+564	LP 223- 013	1525+433	LP 380- 005	1345+238
LP 096- 040	1302+597	LP 234- 1099B	2101+398	LP 381- 028	1413+231
LP 096- 298	1325+581	LP 234- 1664	2107+427	LP 386- 028	1630+245
LP 098- 057	1426+613	LP 254- 026	0644+375	LP 387- 036	1703+261
LP 101- 016	1633+572	LP 257- 028	0802+386	LP 406- 062	0102+210.1
LP 101- 148	1647+591	LP 266- 055	1224+354	LP 406- 063	0102+210.2
LP 102- 162	1735+610	LP 275- 078	1637+335	LP 414- 101	0406+169
LP 103- 294	1820+609	LP 287- 035	2211+372	LP 414- 106	0407+197
LP 106- 017	2009+622	LP 287- 040	2215+388	LP 414- 120	0310+188
LP 106- 019	2010+613	LP 288- 050	2242+383	LP 414- 120	0410+188
LP 119- 048	0525+526	LP 308- 010	0657+320	LP 415- 046	0421+162
LP 129- 010	1108+563	LP 313- 016	0906+296	LP 424- 015	0807+190
LP 129- 171	1122+546	LP 320- 644	1215+323	LP 434- 097	1154+186
LP 129- 586	1148+544	LP 321- 138	1246+299	LP 439- 356	1411+157
LP 130- 013	1148+544	LP 321- 160	1249+296	LP 443- 069	1550+183
LP 131- 066	1247+550	LP 322- 267	1257+278	LP 444- 031	1610+166
LP 133- 144	1349+552	LP 322- 800	1300+263	LP 458- 064	2140+207
LP 133- 353	1401+523	LP 324- 080	1408+323	LP 460- 003	2217+211
LP 135- 154	1510+566	LP 327- 193	1509+322	LP 467- 027	0108+143
LP 135- 438	1524+566	LP 340- 428	2059+316	LP 469- 055	0155+069
LP 137- 043	1639+537	LP 347- 415	2341+322	LP 475- 070	0423+120
LP 155- 332	0349+495	LP 354- 381	0228+269	LP 475- 242	0437+138
LP 161- 012	0637+477	LP 354- 382	0229+270	LP 475- 247	0437+093
LP 162- 003	0714+458	LP 357- 027	0349+247	LP 475- 249	0437+122
LP 163- 121	0807+484	LP 357- 134	0401+250	LP 476- 267	0452+103
LP 167- 033	1019+462	LP 357- 186	0409+237	LP 486- 048	0902+092
LP 171- 040	1239+454	LP 358- 142	0419+266	LP 487- 021	0913+103
LP 176- 060	1533+469	LP 358- 525	0433+270	LP 488- 019	0937+093
LP 176- 065	1534+503	LP 358- 676	0416+272	LP 493- 078	1153+135
LP 182- 044	1835+507	LP 370- 050	0942+236.1	LP 497- 114	1311+129
LP 196- 042	0213+396	LP 370- 051	0942+236.2	LP 498- 026	1336+123
LP 210- 008	0854+404	LP 374- 004	1108+207	LP 498- 066	1346+121
LP 216- 074	1211+392	LP 375- 051	1147+255	LP 500- 017	1422+095
LP 221- 114	1426+442	LP 378- 099	1254+223	LP 518- 035	2139+132
LP 221- 217	1434+437	LP 378- 537	1304+227	LP 521- 049	2254+126

Cross–Reference Name Index

Name	WD	Name	WD	Name	WD
LP 522– 034	2318+126	LP 702– 007	2316– 064	LP 873– 045	2123– 229
LP 535– 011	0423+044	LP 702– 048	2311– 068	LP 880– 451	0004– 315
LP 543– 032	0747+073.1	LP 704– 001	0011– 134	LP 888– 064	0326– 273
LP 543– 033	0747+073.2	LP 705– 094	0041– 102	LP 891– 012	0443– 275
LP 550– 052	1022+050	LP 706– 065	0053– 117	LP 895– 041	0642– 285
LP 550– 292	1026+023	LP 710– 047	0230– 144	LP 907– 037	1350– 090
LP 551– 021	1048+048	LP 713– 034	0347– 137	LP 916– 027	1542– 275
LP 556– 035	1257+047	LP 734– 006	1159– 098	LP 936– 013	2351– 335
LP 561– 013	1448+077	LP 744– 040	1615– 154	LP 936– 025	2351– 368
LP 575– 016	2027+073	LP 753– 005	1932– 136	LP 936– 069	0000– 345
LP 580– 021	2226+061	LP 754– 016	2002– 110	LTT 0003	0000– 345
LP 581– 035	2253+054	LP 768– 500	0145– 174	LTT 0011	0000– 170
LP 587– 016	0112– 018	LP 777– 001	0503– 174	LTT 0235	0024– 556
LP 587– 044	0119– 004	LP 783– 003	0738– 172	LTT 0329	0034– 211
LP 591– 260	0302+027	LP 790– 029	1036– 204	LTT 0375	0038– 226
LP 610– 010	1022+009	LP 791– 055	1043– 188	LTT 0615	0103– 278
LP 612– 033	1115– 029	LP 797– 033	1313– 198	LTT 0632	0104– 464
LP 615– 183	1237– 028	LP 798– 014	1334– 160	LTT 0784	0123– 262
LP 618– 001	1330+015	LP 801– 009	1444– 174	LTT 0934	0141– 675
LP 624– 027	1602+010	LP 815– 045	2035– 174	LTT 1419	0255– 705
LP 627– 022	1717– 014	LP 822– 050	2316– 173	LTT 1648	0326– 273
LP 634– 001	1953– 011	LP 822– 081	2322– 181	LTT 1908	0413– 077
LP 640– 069	2240– 017	LP 824– 273	0013– 241	LTT 1951	0419– 487
LP 648– 024	0135– 052	LP 825– 559	0034– 211	LTT 2049	0435– 088
LP 650– 250	0231– 054	LP 825– 669	0038– 226	LTT 2511	0615– 591
LP 653– 026	0339– 035	LP 825– 713	0040– 220	LTT 2884	0732– 427
LP 655– 042	0435– 088	LP 825– 786	0042– 238	LTT 2980	0752– 146
LP 658– 002	0552– 041	LP 826– 143	0048– 207	LTT 3059	0806– 661
LP 671– 011	1055– 072	LP 827– 217	0123– 262	LTT 3537	0935– 371.2
LP 672– 001	1105– 048	LP 827– 232	0125– 236	LTT 3870	1031– 114
LP 686– 032	1656– 062	LP 829– 017	0204– 233	LTT 4013	1053– 550
LP 696– 004	2044– 043	LP 849– 059	1107– 257	LTT 4020	1055– 072
LP 699– 030B	2201– 037	LP 853– 015	1237– 230	LTT 4099	1105– 048
LP 700– 073	2240– 045	LP 856– 053	1348– 273	LTT 4281	1115– 029
LP 701– 029	2251– 070	LP 873– 008	2107– 216	LTT 4816	1236– 495

Cross–Reference Name Index

Name	WD	Name	WD	Name	WD
LTT 5178	1323– 514	LTT 9768	2351– 335	LTT 14394	1448+077
LTT 5382	1348– 273	LTT 9774	2351– 368	LTT 14705	1550+183
LTT 5712	1425– 811	LTT 10078	0011+000	LTT 14769	1559+369
LTT 6302	1544– 377	LTT 10199	0033+016	LTT 14906	1626+368
LTT 6494	1614– 128	LTT 10292	0046+051	LTT 14945	1637+335
LTT 6497	1615– 154	LTT 10380	0101+048	LTT 14958	1641+732
LTT 6847	1708– 147	LTT 10425	0107+267	LTT 15125	1716+020
LTT 6859	1709– 575	LTT 10469	0107+267	LTT 15423	1818+126
LTT 6999	1733– 544	LTT 10525	0126+101	LTT 15456	1824+040
LTT 7347	1826– 045	LTT 10618	0143+216	LTT 15569	1855+338
LTT 7406	1837– 619	LTT 10723	0205+250	LTT 15765	1943+163
LTT 7421	1840– 111	LTT 10886	0239+109	LTT 15808	1950+250
LTT 7658	1917– 077	LTT 11239	0341+182	LTT 15921	2011+065
LTT 7873	1952– 206	LTT 11733	0549+158	LTT 16005	2032+248
LTT 7875	1953– 715	LTT 11818	0612+177	LTT 16093	2047+372
LTT 7879	1953– 011	LTT 11917	0644+375	LTT 16151	2059+316
LTT 7983	2007– 219	LTT 12215	0826+455	LTT 16224	2111+261
LTT 7987	2007– 303	LTT 12265	0955+090	LTT 16270	2124+550
LTT 8148	2034– 532	LTT 12351	0856+331	LTT 16294	2129+000
LTT 8189	2039– 202	LTT 12591	0943+441	LTT 16348	2140+207
LTT 8190	2039– 682	LTT 12661	0955+247	LTT 16482	2207+142
LTT 8381	2105– 820	LTT 12680	0959+149	LTT 16738	2253+054
LTT 8452	2115– 560	LTT 12749	1012+083	LTT 16749	2254+076
LTT 8579	2130– 047	LTT 12808	1026+117	LTT 16907	2326+049
LTT 8702	2144– 079	LTT 12854	1039+145	LTT 16922	2329+267
LTT 8747	2151– 015	LTT 13002	1107+265	LTT 16991	2341+322
LTT 8768	2154– 512	LTT 13038	1114+067	LTT 17144	0038+555
LTT 8771	2154– 437	LTT 13087	1121+216	LTT 17876	0548– 001
LTT 8816	2159– 754	LTT 13260	1147+255	LTT 17891	0553+053
LTT 8962	2216– 657	LTT 13724	1254+223	LTT 18061	0827+328
LTT 9031	2224– 344	LTT 13742	1257+278	LTT 18157	1104+602
LTT 9082	2232– 575	LTT 13746	1257+037	LTT 18341	1337+705
LTT 9427	2311– 068	LTT 14141	1408+323	LTT 18455	1713+695
LTT 9491	2316– 173	LTT 14182	1413+231	LTT 18524	2126+734
LTT 9648	2337– 760	LTT 14236	1422+095	LTT 18580	2246+223

Cross–Reference Name Index

Name	WD	Name	WD	Name	WD
M 5- 228	1514+033	PB 2947	1131+320	PG	0107+172
Mira B	0216- 032	PB 3534	1410+317	PG	0107+267
MK 320	2317+268	PB 3990	1332+162	PG	0108+100
MK 362	0145+234	PB 4117	1349+144	PG	0109+111
MK 377	0710+741	PB 4421	1257+032	PG	0111+177
MK 380	0713+745	PB 5280	2308+050	PG	0112+104
MK 392	0856+331	PB 5312	2314+064	PG	0115+159
MK 393	0905+605	PB 5379	2324+060	PG	0125+093
MK 396	0921+354	PB 5462	2333- 002	PG	0126+101
NGC 2287- 2	0644- 207.1	PB 5486	2336+063	PG	0127+270
NGC 2287- 3	0644- 207.3	PB 5529	2343+043	PG	0129+246
NGC 2287- 5	0644- 207.2	PB 5617	2353+026	PG	0132+254
NGC 2422- 1	0734- 143	PB 6089	0037- 006	PG	0134+181
NGC 2451- 1	0738- 382	PB 6250	0101+059	PG	0136+152
NGC 2451- 5	0746- 375	PB 6456	0125+093	PG	0136+251
NGC 2451- 6	0737- 384	PB 6544	0156+015	PG	0143+216
NGC 2477- 1	0749- 383	PG	0000+171	PG	0156+015
NGC 2516- 1	0757- 606.2	PG	0004+061	PG	0156+155
NGC 2516- 2	0757- 606.1	PG	0009+190	PG	0210+168
NGC 2516- 3	0756- 607	PG	0010+280	PG	0216+143
NGC 2516- 5	0757- 603	PG	0014+097	PG	0221+217
NGC 2516- 7	0801- 608	PG	0017+061	PG	0229+270
NGC 2516- 8	0751- 595	PG	0017+136	PG	0232+035
NGC 6752	1906- 600.1	PG	0019+150	PG	0235+064
NGC 6752	1906- 600.2	PG	0026+136	PG	0237+115
NGC 6752	1906- 600.3	PG	0033+016	PG	0237+241
NGC 6752	1906- 600.4	PG	0035+124	PG	0243+155
Omicron Cet B	0216- 032	PG	0037- 006	PG	0258+184
OX +25°6725	0205+250	PG	0046+077	PG	0307+149
Procyon B	0736+053	PG	0048+202	PG	0308+096
PB 0520	1039+412	PG	0052+190	PG	0308+188
PB 0772	1348+442	PG	0059+257	PG	0317+196
PB 0999	1353+409	PG	0101+048	PG	0319+054
PB 1549	1407+425	PG	0101+059	PG	0402+543
PB 1665	1410+425	PG	0102+095	PG	0805+654

Cross–Reference Name Index

Name	WD	Name	WD	Name	WD
PG	0808+595	PG	0915+201	PG	0950+185
PG	0814+569	PG	0915+526	PG	0954+134
PG	0816+297	PG	0916+064	PG	0954+247
PG	0817+386	PG	0920+216	PG	0954+697
PG	0819+363	PG	0920+363	PG	0955+247
PG	0821+632	PG	0920+375	PG	0955−008
PG	0823+316	PG	0921+091	PG	0956+020
PG	0826+455	PG	0921+354	PG	0956+045
PG	0834+500	PG	0922+162	PG	1000+220
PG	0836+237	PG	0922+183	PG	1000−001
PG	0839+231	PG	0924+199	PG	1001+203
PG	0840+262	PG	0928+085	PG	1003−023
PG	0841+603	PG	0929+270	PG	1005+642
PG	0843+516	PG	0933+025	PG	1010+043
PG	0846+249	PG	0933+383	PG	1010+064
PG	0846+557	PG	0933+729	PG	1013+256
PG	0852+602	PG	0934+337	PG	1015+014
PG	0852+658	PG	0937+505	PG	1015+076
PG	0853+163	PG	0938+286	PG	1015+161
PG	0854+404	PG	0938+299	PG	1017+125
PG	0856+331	PG	0938+550	PG	1018+410
PG	0858+363	PG	0939+071	PG	1019+129
PG	0900+142	PG	0939+262	PG	1022+050
PG	0900+554	PG	0940+068	PG	1023+009
PG	0901+140	PG	0941+432	PG	1025+257
PG	0901+597	PG	0943+441	PG	1026+002
PG	0904+391	PG	0945+245	PG	1026+023
PG	0904+511	PG	0946+534	PG	1026+453
PG	0908+171	PG	0947+325	PG	1026−056
PG	0909+271	PG	0947+639	PG	1031+063
PG	0910+621	PG	0948+013	PG	1031+234
PG	0912+536	PG	0949+094	PG	1033+464
PG	0913+103	PG	0949+256	PG	1034+001
PG	0913+204	PG	0950+023	PG	1034+492
PG	0913+442	PG	0950+077	PG	1035+532

Cross–Reference Name Index

Name	WD	Name	WD	Name	WD
PG	1036+085	PG	1115–029	PG	1159–098
PG	1037+512	PG	1116+026	PG	1200+548
PG	1038+290	PG	1119+385	PG	1201+437
PG	1038+633	PG	1120+439	PG	1201–001
PG	1039+412	PG	1121+145	PG	1201–049
PG	1039+747	PG	1121+216	PG	1202+308
PG	1041+580	PG	1122+546	PG	1204+450
PG	1045–017	PG	1123+189	PG	1207–032
PG	1046+281	PG	1124–018	PG	1210+464
PG	1046–017	PG	1125–025	PG	1210+533
PG	1047+694	PG	1126+185	PG	1211+332
PG	1049+103	PG	1126+384	PG	1214+267
PG	1052+273	PG	1128+564	PG	1216+036
PG	1053–092	PG	1129+071	PG	1218+497
PG	1055–072	PG	1129+155	PG	1220+234
PG	1056+345	PG	1129+373	PG	1223+478
PG	1057+719	PG	1132+470	PG	1224+309
PG	1057–059	PG	1133+293	PG	1225–079
PG	1058–129	PG	1133+489	PG	1229–012
PG	1100+604	PG	1134+073	PG	1230+417
PG	1101+242	PG	1134+124	PG	1231+465
PG	1101+364	PG	1134+300	PG	1232+238
PG	1101+384	PG	1138+424	PG	1232+479
PG	1102+748	PG	1141+077	PG	1233+337
PG	1103+384	PG	1141+504	PG	1236+479
PG	1104+602	PG	1143+321	PG	1237–028
PG	1105–048	PG	1144–084	PG	1240+754
PG	1107+265	PG	1145+080	PG	1241+235
PG	1108+325	PG	1145+187	PG	1241–010
PG	1108+475	PG	1147+255	PG	1244+149
PG	1109+244	PG	1149+057	PG	1246+586
PG	1113+413	PG	1149+410	PG	1247+553
PG	1114+223	PG	1149–133	PG	1247+575
PG	1115+158	PG	1158+432	PG	1253+378
PG	1115+166	PG	1159–034	PG	1254+223

Cross–Reference Name Index

Name	WD	Name	WD	Name	WD
PG	1255+426	PG	1335+700	PG	1422+095
PG	1257+032	PG	1337+705	PG	1424+240
PG	1257+047	PG	1339+346	PG	1424+534
PG	1257+278	PG	1342+443	PG	1425+540
PG	1258+593	PG	1344+106	PG	1428+102
PG	1300- 098	PG	1344+509	PG	1428+373
PG	1301+544	PG	1344+572	PG	1429+373
PG	1302+283	PG	1347+253	PG	1430+427
PG	1302+597	PG	1348+442	PG	1431+153
PG	1305+018	PG	1349+144	PG	1433+538
PG	1305- 017	PG	1349+552	PG	1434+289
PG	1307+354	PG	1350+656	PG	1436+526
PG	1308- 098	PG	1350- 090	PG	1437+398
PG	1310+583	PG	1351+489	PG	1439+304
PG	1311+129	PG	1352+004	PG	1443+336
PG	1312+098	PG	1353+409	PG	1444- 096
PG	1314+293	PG	1401+005	PG	1445+152
PG	1314- 067	PG	1403- 010	PG	1446+286
PG	1317+453	PG	1403- 077	PG	1448+077
PG	1319+466	PG	1407+425	PG	1449+168
PG	1320+645	PG	1408+323	PG	1450+432
PG	1322+076	PG	1410+317	PG	1451+006
PG	1324+077	PG	1410+425	PG	1452- 042
PG	1325+167	PG	1411+218	PG	1454+172
PG	1325+279	PG	1412- 049	PG	1455+298
PG	1326- 037	PG	1412- 109	PG	1456+103
PG	1327- 083	PG	1413+015	PG	1456+298
PG	1328+343	PG	1413+231	PG	1457- 086
PG	1330+473	PG	1415+132	PG	1458+171
PG	1332+162	PG	1415+234	PG	1459+219
PG	1333+487	PG	1415- 064	PG	1459+305
PG	1333+497	PG	1418- 005	PG	1459+347
PG	1333+510	PG	1421+318	PG	1459+644
PG	1333+524	PG	1421- 011	PG	1501+032
PG	1335+369	PG	1422+028	PG	1502+351

Cross–Reference Name Index

Name	WD	Name	WD	Name	WD
PG	1507+021	PG	1548+405	PG	1636+351
PG	1507+220	PG	1549– 000	PG	1637+335
PG	1508+548	PG	1550+183	PG	1639+153
PG	1508+637	PG	1553+353	PG	1639+537
PG	1509+322	PG	1554+215	PG	1640+113
PG	1511+009	PG	1554+262	PG	1640+457
PG	1513+442	PG	1559+128	PG	1640+690
PG	1515+668	PG	1559+369	PG	1641+387
PG	1519+383	PG	1600+308	PG	1642+385
PG	1519+500	PG	1600+369	PG	1642+413
PG	1520+447	PG	1601+581	PG	1643+143
PG	1520+525	PG	1603+432	PG	1644+198
PG	1521+310	PG	1605+683	PG	1645+325
PG	1525+257	PG	1608+118	PG	1646+062
PG	1525+422	PG	1608+419	PG	1647+375
PG	1526+013	PG	1609+044	PG	1648+371
PG	1527+090	PG	1609+135	PG	1654+160
PG	1531– 022	PG	1609+631	PG	1654+637
PG	1532+033	PG	1610+166	PG	1655+215
PG	1532+239	PG	1610+239	PG	1657+343
PG	1533– 057	PG	1614+136	PG	1658+440
PG	1534+503	PG	1614+160	PG	1659+303
PG	1535+293	PG	1614+270	PG	1659+442
PG	1537+651	PG	1619+123	PG	1703+319
PG	1538+269	PG	1619+525	PG	1707+427
PG	1539+255	PG	1620+513	PG	1707+475
PG	1539+530	PG	1620+647	PG	1713+332
PG	1539– 035	PG	1622+323	PG	1720+360
PG	1540+680	PG	1626+409	PG	1725+586
PG	1541+650	PG	1630+249	PG	1728+560
PG	1542+182	PG	1632+177	PG	1747+450
PG	1545+244	PG	1632+222	PG	2056+033
PG	1547+015	PG	1633+433	PG	2115+010
PG	1547+057	PG	1633+676	PG	2120+054
PG	1548+149	PG	1635+608	PG	2122+157

Cross–Reference Name Index

Name	WD	Name	WD	Name	WD
PG	2128+112	PG	2341+322	Ross 548	0133– 116
PG	2131+066	PG	2343+043	Ross 627	1121+216
PG	2150+021	PG	2345+304	Ross 640	1626+368
PG	2204+070	PG	2349+286	Ross 808	1559+369
PG	2204+071	PG	2353+026	Ross 813	1641+732
PG	2207+142	PG	2354+126	Rubin 70	0339+523
PG	2220+133	PG	2354+159	Rubin 79	0354+556
PG	2226+061	PG	2357+296	Rubin 80	0354+463
PG	2229+139	PHL 0028	2129+000	RWT 103	0419+500
PG	2234+033	PHL 0145	2144– 079	Sirius B	0642– 166
PG	2234+064	PHL 0380	2240– 045	SA 26– 82	0639+447
PG	2239+081	PHL 0386	2240– 017	SA 29– 130	0943+441
PG	2244+031	PHL 0400	2303+017	SA 51– 59	0734+308
PG	2246+120	PHL 0555	2329– 291	SA 51– 484	0737+288
PG	2246+154	PHL 0560	2322– 181	SA 51– 776	0725+318
PG	2246+223	PHL 0670	0004+061	SA 51– 822A	0720+304
PG	2257+138	PHL 0671	0004– 315	SA 54– 79	1020+315
PG	2257+162	PHL 0790	0017+061	SA 82– 189	1415+132
PG	2303+017	PHL 0802	0019+150	SA 107– 28	1541+003
PG	2303+242	PHL 0810	0026+136	SA 107– 136	1536+000
PG	2306+124	PHL 0861	0048– 202	SB 143	0018– 339
PG	2306+130	PHL 0940	0058– 044	SB 360	0050– 332
PG	2308+050	PHL 0962	0100– 068	SB 485	0109– 264
PG	2309+105	PHL 0965	0101– 250	STN 2051B	0426+588
PG	2310+175	PHL 0972	0102+095	SY For B	0249– 381
PG	2314+064	PHL 0980	0102– 142	TC 142	0016– 220
PG	2322+118	PHL 1003	0109– 264	TC 161	0028– 274
PG	2322+206	PHL 1062	0133– 116	TC 163	0031– 274
PG	2324+060	PHL 1358	0231– 054	TC 201	0109– 264
PG	2326+049	PHL 2856	0016– 220	TC 205	0116– 231
PG	2328+107	PHL 3287	0100– 036	Ton 10	0840+262
PG	2329+267	PHL 6421	0011– 134	Ton 16	0929+290
PG	2331+290	Ross 137	1824+040	Ton 20	0938+286
PG	2333– 002	Ross 193B	2054– 050	Ton 21	0939+262
PG	2336+063	Ross 198	2124+550	Ton 40	1038+290

Cross–Reference Name Index

Name	WD	Name	WD	Name	WD
Ton 53	1101+242	Ton 573	1107+265	Wolf 72	0126+101
Ton 60	1108+325	Ton 610	1220+234	Wolf 82	0143+216
Ton 61	1109+244	Ton 797	1535+293	Wolf 219	0341+182
Ton 75	1202+308	Ton 816	1620+260	Wolf 309	0826+455
Ton 82	1231+465	Ton 953	0846+346	Wolf 457	1257+037
Ton 96	1236+479	Ton 1054	0920+363	Wolf 485A	1327– 083
Ton 170	1347+253	Ton 1061	0924+199	Wolf 489	1334+039
Ton 191	0104– 331	Ton 1080	0934+337	Wolf 672A	1716+020
Ton 197	1421+318	Ton 1145	1000+220	Wolf 1346	2032+248
Ton 210	1434+289	Ton 1150	1001+203	Wolf 1516	0115+159
Ton 214	1446+286	TPS 34	0050– 332	ξ 1 Cet B	0207+083
Ton 229	1521+310	TPS 54	0109– 264	ζ Cap B	2123– 226
Ton 235	1525+257	TS 102	2329– 291	ZZ Ceti	0133– 116
Ton 241	1532+239	TS 147	0013– 241	40 Eri B	0413– 077
Ton 245	1538+269	TS 155	0021– 234	56 Peg B	2304+251
Ton 246	1539+255	TS 217	0125– 236		
Ton 249	1545+244	TS 243	0212– 231		
Ton 252	1600+308	TS 372	0340– 243		
Ton 264	1616– 390	TS 374	0341– 248		
Ton 320	0823+316	TS 392	0357– 233		
Ton 353	0846+249	T 518 Com	1025+257		
Ton 393	0909+271	US 906	0941+432		
Ton 443	0938+299	VB 3	0743– 340		
Ton 458	0947+325	VB 11	2054– 050		
Ton 462	0949+256	VMa 2	0046+051		
Ton 494	1013+256	VR 7	0421+162		
Ton 523	1028+328	VR 16	0425+168		
Ton 527	1031+234	VV 47	0753+535		
Ton 547	1046+281	Wolf 1	0011+000		
Ton 556	1052+273	Wof 28	0046+051		

22
Pulsars

22.1 Basic Data for Pulsars

The radio radiation from pulsars is detected at the Earth as periodic pulses. Both the astonishing regularity and the length of their periods, P, can be attributed to the rotation of a neutron star. Measurements of the phase of a pulsar's waveform at a number of epochs can be used to determine the secular evolution of its period, as well as its celestial coordinates. Such observations indicate that pulsar periods gradually increase with typical derivatives $\dot{P}$ around $10^{-15} s/s$. This lengthening has been attributed to the gradual loss of rotational energy by the spinning neutron star.

A logarithmic P,$\dot{P}$ diagram is shown in the first accompanying diagram together with the death and spin-up lines. The death line defines the locus of P,$\dot{P}$ values for which the pulsar radio emission becomes very weak or turns off. The spin-up line applies to binary pulsar systems in which the period increases as the result of the accretion of matter from its companion. This process will not work for values of $\dot{P}$ above the spin-up line. The next diagram gives the distribution of pulsars in galactic coordinates, together with the location of particularly interesting examples such as the Crab, eclipsing, M 4, M 15, millisecond and Vela pulsars.

The observed and derived parameters for 330 pulsars were given by Manchester and Taylor (1981). Our tables are based upon a computerized and updated version of this compilation provided by Joe Taylor (1989). Here pulsar names are given in the PSR convention, that is, in hours and minutes of right ascension (1950) followed by sign and degree of declination (1950).

The first accompanying table provides the mean values and ranges for many of the pulsar parameters. The heliocentric distances, D, for most of the pulsars were computed from the dispersion measure using the interstellar electron density model of Lyne, Manchester, and Taylor (1985). The radio luminosity, L_r, and the magnetic field strength, B, at the surface of the neutron star were computed using formulae given by Manchester and Taylor (1981). Characteristic ages, T, are computed from the relation $T = P/(2\dot{P})$.

The next table lists the fastest pulsars in order of increasing period, together with the object name, $\dot{P}$, B and the age. Rapidly-rotating, relatively-young pulsars are still found within visible supernovae remnants. Fast, older pulsars, such as those found within globular star clusters, are probably "recycled" old pulsars which have been spun-up through accretion from a binary stellar companion.

The orbital parameters of binary pulsar systems are given in the following table, including the semi-major axis, $a \sin i$, in units of the Sun's radius, $R_\odot$, the mass function $f(m) = (m_2 \sin i)^3/(m_1 + m_2)^2$ in units of the Sun's mass, $M_\odot$, for a pulsar of mass m_1 and its companion of mass m_2, the longitude of the perihelion, ω, in degrees, and $\dot{\omega}$, the rate of periastron advance in degrees per year.

The binary pulsar system PSR 1913 + 16 has been used to test the general theory of relativity (Taylor and Weisberg, 1989). The masses of the pulsar and its companion are determined to be $m_1 = 1.442 \pm 0.003$ and $m_2 = 1.386 \pm 0.003$ times the mass of the Sun, and the orbit is found to be decaying at a rate equal to 1.01 $\pm$ 0.001 times the general relativity prediction for damping by gravitational radiation.

Annual pulsar proper motions are next given in units of milliarcseconds in the sky in the right ascension, R.A., and declination, Dec., directions. They have been combined with the assumed distances to infer peculiar velocities, V, that are often much larger than those of typical stars in the galactic disk.

Our main catalogue of pulsar parameters is given in two parts. The first part contains the celestial coordinates, R.A. and Dec., for the epoch 1950.0, the galactic coordinates, the time-average flux density, S_{400}, of the pulsed signal at 400 MHz in milliJanskys, the pulsar period, P, in seconds, $\dot{P}$ in units of $10^{-15} s/s$, the epoch in Julian days - 2400000 to which P and $\dot{P}$ are referred, and the equivalent width, W_E, and full width at half intensity, W_{50}, of the mean pulse shape in units of milliseconds.

The second part of our main catalogue of pulsar parameters contains the pulsar name, the dispersion measure, DM, the rotation measure, RM, the height, z, above the galactic plane in kiloparsecs, and the logarithms of the radio luminosity, L_r, in $erg\ s^{-1}$, and the age, $P/(2\dot{P})$ in years.

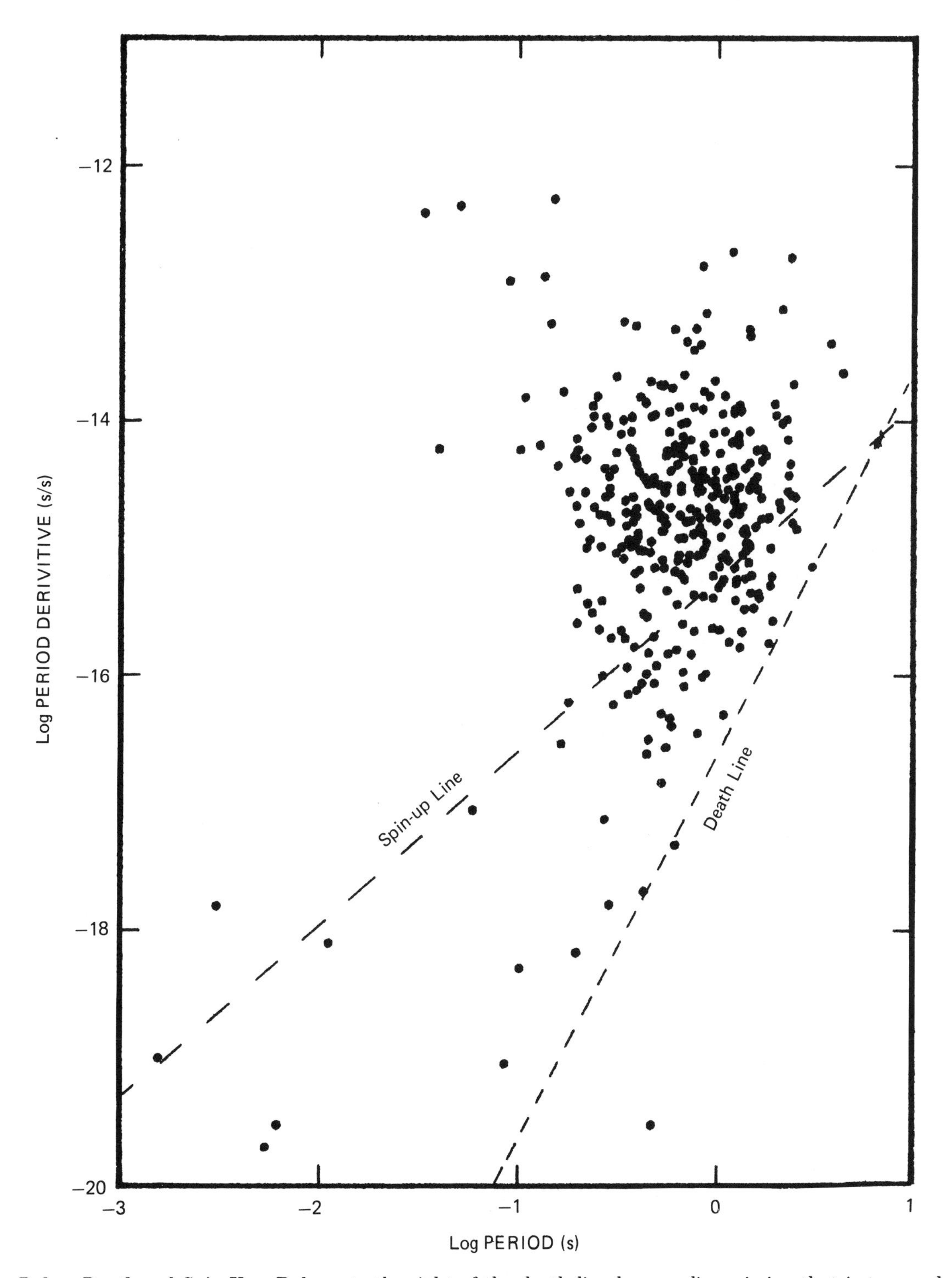

Pulsar Death and Spin-Up. Pulsars to the right of the death line have radio emission that is too weak to be detected; they have essentially turned off their radio beacons. The spin-up line applies to binary pulsar systems in which the period increases as the result of the accretion of matter from a companion; this process does not work above the spin-up line.

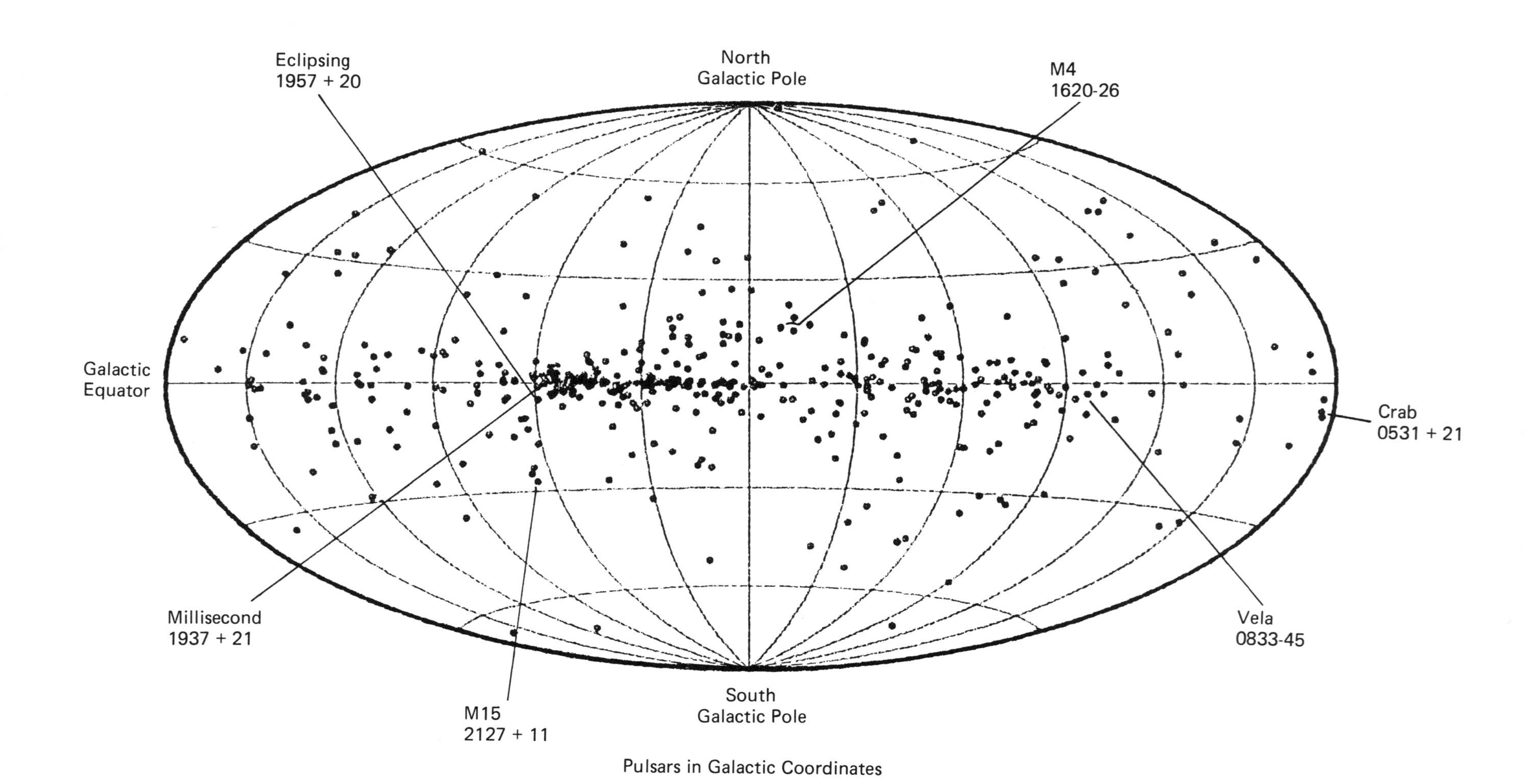

Pulsars in Galactic Coordinates

Pulsar Statistics

Parameters	Mean	Max.	Min.	Disp., σ
Period, P (seconds)	0.79	4.30	0.001	0.58
Period Derivative, $\dot{P}$ ($\times 10^{15}$)	10.47	540.19	-0.34	45.82
Dispersion Measure, DM (cm^{-3} pc)	134.98	1140.0	2.97	147.82
Rotation Measure, RM (m^{-2} rad)	29.79	1000.0	- 611.0	152.76
Equivalent Width, W_E (ms)	28.32	520.0	0.2	39.20
Full Width at Half Intensity, W_{50} (ms)	31.02	740.0	1.0	52.65
Flux Density, S_{400} (mJy)	57.51	5000.0	1.0	279.49
Heliocentric Distance, D (kpc)	3.83	25.0	0.06	3.77
Galactocentric Coordinate, Z (kpc)	- 0.02	1.20	- 2.70	0.44
Log Radio Luminosity, log L_r (erg sec^{-1})	27.55	30.99	24.58	1.07
Log Magnetic Field Strength, log B (Gauss)	12.02	13.33	8.52	0.58
Log Age, T (years)	6.73	9.66	3.09	0.93

The Fastest Pulsars*

PSR	Period (seconds)	Object***	$\dot{P}$ (10^{-15})	Log B (Gauss)	Log Age (years)
1937+21	0.0016	Millisecond Pulsar +	0.0001	8.601	8.392
1957+20	0.0016	Eclipsing Pulsar ++			
1821– 24	0.0030	Glob. Cluster M 28	0.00155	9.343	7.494
0021– 72 A	0.0045	Glob. Cluster 47 Tuc**			
1855+09	0.0054	Binary Pulsar	0.00002	8.520	9.628
0021– 72 B	0.0061	Glob. Cluster 47 Tuc**			
1953+29	0.0061	Binary Pulsar	0.00003	8.638	9.510
1620– 26	0.0111	Glob. Cluster M 4**	0.00080	9.479	8.341
0531+21	0.0333	Crab Nebula SNR	421.288	12.579	3.098
1951+32	0.0395	CTB 80 SNR	5.92	11.690	5.024
0540– 69	0.0503	LMC SNR	479.8383	12.696	3.220
1913+16	0.0590	Binary Pulsar	0.0864	11.915	7.798
1830– 08	0.0853		9	11.948	5.176
0833– 45	0.0893	Vela X SNR	124.371	12.528	4.056
1823– 13	0.1014		51	12.362	4.498
0114+58	0.1014		5.84479	11.892	5.439
0906– 49	0.1067		15.151	12.110	5.048
2127+11	0.1107	Glob. Cluster M 15	– 0.02		
1758– 24	0.1248			12.248	6.644
1356– 60	0.1275		6.3385	11.959	5.503
1800– 21	0.1336		134.154	12.632	4.198
1509– 58	0.1502	G320.4–1.2 SNR	1540.1980	13.182	3.189

*Adapted from Taylor (1989). Accurate coordinates and other data may be found in the table of Pulsar Parameters. Also see Backer et al. (1982), Boriakoff, Buccheri and Fauci (1983); Fruchter, Stinebring and Taylor (1988), Lyne et al. (1987), Lyne et al. (1988), Segelstein et al. (1986), and Wolszczan et al. (1989).

** PSR 0021-71A and B may not be within 47 Tuc, but PSR 0021-71C, with a period of 5.757 milliseconds [Nature 345, 599 (1990)], and ten other pulsars with periods within 1 and 6 milliseconds are most likely within this globular cluster [Nature 352, 219 (1991)].

*** SNR = Supernova Remnant.

+ Since this table was prepared, more than twenty "millisecond" pulsars have been found in twelve globular clusters (see page 261).

++ The bow shock of an emission nebula around the Eclipsing Binary Pulsar PSR 1957+20 provides evidence for the interaction of this pulsar with the surrounding medium (see page 613) [Nature 335, 801 (1988)].

Binary Pulsars*

PSR	Orbital Period (days)	Pulsar Period (seconds)	Orbital Eccentricity	$a \sin i$ ($R_\odot$)	f(m) ($M_\odot$)
0021− 72 A•	0.022272	0.0045	0.33	0.00084	0.000000016
1913+16†	0.3229974726	0.0590	0.617127	1.009	0.1322
1957+20‡	0.3819664	0.0016	0.000	0.0384	0.00005
0655+64	1.028669703	0.1957	0.00000	1.777	0.0712
1831− 00	1.811103	0.5209	0.000	0.3115	0.000123
1855+09	12.3271701	0.0054	0.00002142	3.976	0.0055
2303+46	12.339541	1.0664	0.65838	14.08	0.2463
0021− 72 B•	51	0.0061			
1953+29	117.34911	0.0061	0.000330	13.53	0.0024
1620− 26•	191.4427	0.0111	0.02532	27.92	0.0080
0820+02	1232.47	0.8649	0.011868	69.84	0.0030

•Within a globular cluster 0021 - 72 A, B are in 47 Tuc and 1620 - 26 is in M4.
†This binary pulsar has been used to test general relativistic effects and to provide confirmation of the existence of gravitational radiation (see Taylor and Weisberg 1989). Mass estimates for the pulsar and its companion are 1.442±0.003 ($M_\odot$) and 1.386±0.003 ($M_\odot$) respectively.
‡The eclipsing pulsar.

Binary Pulsars*

PSR	$\dot{P}$ (10^{-15})	log B (Gauss)	ω (deg)	$\dot{\omega}$ (deg/year)	$m_p + m_c$ ($M_\odot$)
1913+16	0.00864	10.359	178.8643	4.2263	2.8
0655+64	0.000677	10.066	180		
1831− 00	0.0143	10.941	70		
1855+09	0.00002	8.520	279.8361		
2303+46	0.5693	11.897	35.040	0.0092	2.3±0.6
1953+29	0.00003	8.638	29.5474		
1620− 26	0.0008	9.479	117.13		
0820+02	0.1039	11.482	332.02	$\leq$0.006	

*Adapted from Taylor (1989) with $R_\odot = 6.599 \times 10^{10}$ cm and $c = 2.9979 \times 10^{10}$ cm s^{-1}. Accurate pulsar periods, and other data may be found in the table of Pulsar Parameters. Also see Boriakoff, Buccheri and Fauci (1983), Damashek et al. (1982), Dewey et al. (1986), Fruchter, Stinebring and Taylor (1988), Hulse and Taylor (1975), Lyne et al. (1988), Stokes, Taylor and Dewey (1985), Segelstein et al. (1986) and Taylor and Dewey (1988).

Pulsar Proper Motions*

PSR	Dist (kpc)	R. A. (mas/yr)	Err	V R. A. (km/s)	Dec. (mas/yr)	Err	V Dec (km/s)	Posn Epoch (JD - 2400000)
0301+19	0.56	5.8	7.2	15.40	−37.2	4.2	-98.74	42326
0329+54	2.30	17.4	0.9	189.69	−11.7	0.9	−127.55	42050
0355+54	1.60	5.3	4.2	40.20	6.2	3.4	47.02	42524
0525+21	2.00	2.2	27.7	20.86	14.7	17.6	139.36	41994
0531+21	2.00	−12	3	−113.80	5	4	47.40	40709
0611+22	3.30	0.0	5.6	0.00	7.5	3.7	117.31	42881
0809+74	0.17	15.1	7.0	12.17	−49.0	5.5	−39.48	42077
0823+26	0.71	61.0	3.0	205.29	−89.6	2.2	−301.54	42717
0834+06	0.43	1.9	4.7	3.87	51.1	3.3	104.15	42385
0943+10	0.56	−37.9	19.1	−100.60	−20.8	11.6	−55.21	41665
0950+08	0.09	14.9	7.9	6.36	30.6	5.3	13.05	41501
1133+16	0.15	−101.5	4.6	−72.17	356.7	3.1	253.61	42364
1237+25	0.33	−106.0	3.5	−165.81	42.1	2.5	65.85	41834
1508+55	0.73	−73.3	3.6	−253.63	−68.0	2.7	−235.29	42058
1541+09	1.30	−12.3	4.0	−75.79	3.2	2.8	19.72	42304
1604–00	0.36	−0.7	14.1	−1.19	−7.2	8.9	−12.29	42307
1642–03	1.30	40.7	16.8	250.79	−24.8	11.2	−152.82	41902
1818–04	1.50	2.8	3.3	19.91	27.1	2.2	192.68	41901
1929+10	0.08	79.1	5.6	29.99	39.2	3.6	14.86	42391
1937+21	5.00	−0.3	0.2	−7.11	−0.5	0.3	−11.85	45303
1944+17	0.43	0.8	5.3	1.63	−9.2	3.7	−18.75	42320
1952+29	0.20	25.4	16.5	24.08	−35.6	9.9	−33.75	42434
2016+28	1.30	2.0	2.3	12.32	1.3	1.5	8.01	41739
2020+28	1.30	−8.5	2.6	−52.38	−13.2	2.0	−81.34	41696
2021+51	0.68	5.6	4.1	18.05	16.6	3.5	53.51	41605
2217+47	1.50	−12.2	8.0	−86.74	−30.0	5.5	−213.30	41744
2303+30	1.90	12.7	8.4	114.38	−32.7	5.5	−294.50	42341

* Adapted from Taylor (1989). Coordinates, periods and other data may be found in the table of Pulsar Parameters.

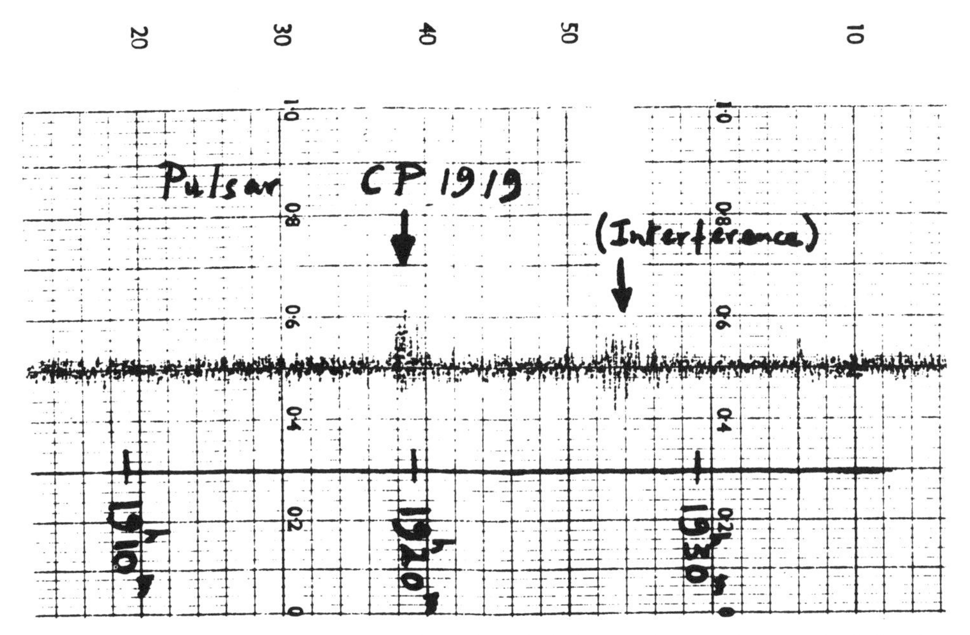

CP 1919. One of the first observations of the pulsar CP 1919, taken on August 6, 1967. It indicates that the pulsating radio signal was then barely discernable from terrestrial interference; but now the pulsed radiation from hundreds of these objects has been reliably identified. It is attributed to the rotation of a "light-house" beam of radiation from a rotating neutron star. (Courtesy of Anthony Hewish.)

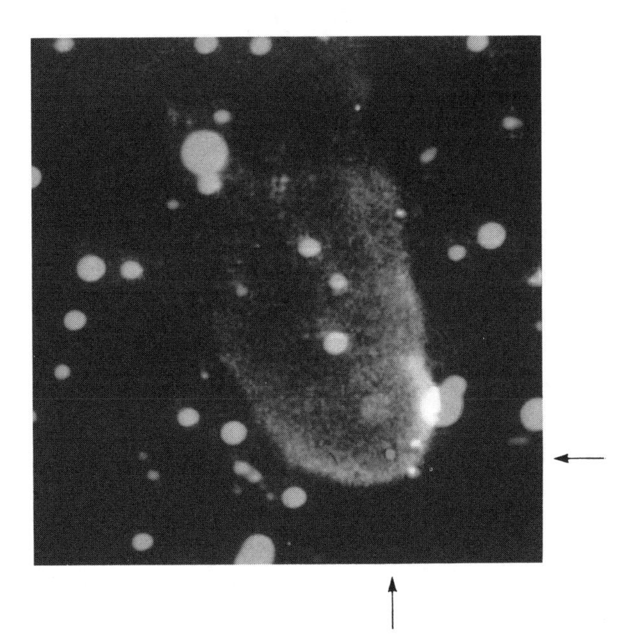

Binary Eclipsing Millisecond Nebula. The optical nebula surrounding the binary, eclipsing millisecond pulsar 1957 + 20 is detected in the Balmer-alpha line of hydrogen. The pulsar is the yellowish star at the intersection of the axes indicated by the arrows. High-energy particles, radiation and magnetic fields ejected from the pulsar push the interstellar medium, producing the brilliant nebula. Its cometary shape is due to the motion of the pulsar toward the lower right. (Photo taken by Jeff Hester and Shrinivas Kulkarni at the Palomar Observatory.)

22.2 Catalogue of Pulsars - Part I

Pulsar Parameters*

PSR	R.A.(1950) h m s	Dec.(1950) ° ′ ″	l °	b °	S_{400} (mJy)	D (kpc)
0011+47	00 11 39.83	+47 29 52.6	116.5	- 14.6	15	1.10
0021- 72A	00 21 53	- 72 21 30	305.9	- 44.9		2.50
0021- 72B	00 21 53	- 72 21 30	305.9	- 44.9		2.50
0031- 07	00 31 36.48	- 07 38 33	110.4	- 69.8	25	0.39
0037+56	00 37 41.04	+56 59 57.0	121.5	- 5.6	3.5	3.40
0042- 73	00 42 30	- 73 30	303.6	- 43.9	2	3.80
0045+33	00 45 50.99	+33 55 47	122.3	- 28.7	4.5	1.60
0052+51	00 52 52.15	+51 01 10.7	123.6	- 11.6	3.6	1.50
0053+47	00 53 33.38	+47 40 01.7	123.8	- 14.9	5.8	0.58
0059+65	00 59 20.7	+65 21 06.1	124.1	+2.8	25	2.10
0105+65	01 05 05.1	+65 52 32.2	124.6	+3.3	15	0.85
0105+68	01 05 05.84	+68 49 52.5	124.5	+6.3	3.7	2.10
0114+58	01 14 28.548	+58 58 50.45	126.3	- 3.5	5	1.50
0136+57	01 36 01.72	+57 59 19.8	129.2	- 4.0	50	2.50
0138+59	01 38 17.3	+59 54 21.8	129.1	- 2.1	55	3.00
0144+59	01 44 20.788	+59 07 06.07	130.1	- 2.7	4	1.10
0148- 06	01 48 52.5	- 06 49 49	160.4	- 65.0	35	0.97
0149- 16	01 49 46.42	- 16 52 39.3	179.3	- 72.5	21	0.44
0153+39	01 53 54.3	+39 34 52	136.4	- 21.3	3	
0154+61	01 54 14.9	+61 57 49	130.6	+0.3	5	0.66
0203- 40	02 03 57.202	- 40 42 21.8	258.6	- 69.6	11	0.47
0226+70	02 26 47.1	+70 13 15.9	131.2	+9.2	3.5	1.70
0254- 53	02 54 24.201	- 53 16 25.3	269.9	- 55.3	17	0.59
0301+19	03 01 42.418	+19 21 12.5	161.1	- 33.3	28	0.56
0320+39	03 20 10.15	+39 34 16	152.2	- 14.3	35	0.89
0329+54	03 29 10.98	+54 24 36.8	145.0	- 1.2	1400	2.30
0331+45	03 31 47.633	+45 45 55.28	150.3	- 8.0	10	1.60
0339+53	03 39 25.9	+53 03 22	147.0	- 1.4	20	2.00
0353+52	03 53 56.292	+52 28 20.5	149.1	- 0.5	12	2.90
0355+54	03 55 00.44	+54 04 42.1	148.2	+0.8	60	1.60
0402+61	04 02 07.93	+61 30 33.0	144.0	+7.0	12	2.40
0403- 76	04 03 15.17	- 76 16 25.8	290.3	- 35.9	19	0.79
0410+69	04 10 39.220	+69 46 39.74	138.9	+13.7	4.3	0.94
0447- 12	04 47 49.52	- 12 53 12.4	211.1	- 32.6	15	1.40
0450+55	04 50 00.234	+55 38 48.5	152.6	+7.5	40	0.41

Pulsar Parameters*

PSR	Period (seconds)	$\dot{P}$ (10^{-15})	Epoch (JD - 2400000)	W_E (ms)	W_{50} (ms)
0011+47	1.2406988730	0.561	46259.92	55	130
0021− 72A	0.004478953				
0021− 72B	0.006127				
0031− 07	0.94295078486	0.4083	40689.67	42	50
0037+56	1.11822446071	2.879	46111.33		
0042− 73	0.926499		46811		
0045+33	1.2170935289	2.354	46104.36		
0052+51	2.1151682248	9.541	46116.33		
0053+47	0.47203562065	3.567	46260.95		
0059+65	1.6791623969	5.947	46110.37	48	28
0105+65	1.28365292622	13.169	43891.931	20	30
0105+68	1.07111809422	0.049	46116.36		
0114+58	0.101437227225	5.84479	46110.37		
0136+57	0.27244563408	10.6867	43889.962	7.5	7.5
0138+59	1.22294826723	0.3904	41793.695	38	36
0144+59	0.196321200154	0.25694	46110.39		
0148− 06	1.4646643399	0.445	43889.971	60	140
0149− 16	0.83274112609	1.300	43890.971	14	14
0153+39	1.811560563	0.18	46110.40		
0154+61	2.351653037	188.99	44079.467	30	35
0203− 40	0.63054980456	1.198	43555.333	11	10
0226+70	1.4668193434	3.116	46110.43		
0254− 53	0.44770843654	0.0311	44018.8788	8	9
0301+19	1.38758366697	1.29613	42325.5	42	57
0320+39	3.0320716577	0.71	43890.036	42	42
0329+54	0.71451866398	2.04959	40621	8.7	5.9
0331+45	0.269200538516	0.00734	46110.46		
0339+53	1.9344746763	13.42	46385.72		40
0353+52	0.197029941508	0.4760	46110.48		
0355+54	0.156381271509	4.3912	46600.0	6.3	5.2
0402+61	0.59457139808	5.575	43891.066	20	20
0403− 76	0.54525231097	1.540	43555.3624	22	22
0410+69	0.390715065014	0.0765	46111.49		
0447− 12	0.4380141010	0.103	43889.099	18	19
0450+55	0.340728715400	2.3581	46382.78	13	8.5

*Adapted from Taylor (1989).

Pulsar Parameters*

PSR	R.A.(1950) h m s	Dec.(1950) ° ′ ″	l °	b °	S_{400} (mJy)	D (kpc)
0450−18	04 50 21.43	−18 04 20.8	217.1	−34.1	55	1.60
0458+46	04 58 21.87	+46 49 46.6	160.4	+3.1	10	1.20
0523+11	05 23 09.52	+11 12 41.0	192.7	−13.2	20	3.20
0525+21	05 25 51.748	+21 58 00.6	183.9	−6.9	93	2.00
0529−66	05 29 30	−66 57	277.0	−32.8		4.90
0531+21	05 31 31.405	+21 58 54.39	184.6	−5.8	800	2.00
0538−75	05 38 18.9	−75 45 38.0	287.2	−30.8	75	0.65
0540+23	05 40 06.938	+23 27 46	184.4	−3.3	30	2.60
0540−69	05 40 33.9	−69 21 23.2	279.7	−31.5		
0559−05	05 59 31.62	−05 27 46.4	212.2	−13.5	10	3.30
0559−57	05 59 59.4	−57 56 52	266.5	−29.3	2.1	1.10
0609+37	06 09 23.931	+37 22 25.5	175.5	+9.1	3	0.86
0611+22	06 11 14.949	+22 26 06.2	188.9	+2.4	25	3.30
0621−04	06 21 51.38	−04 23 10.8	213.8	−8.0	10	2.70
0626+24	06 26 02.0	+24 17 57	188.8	+6.2	15	3.20
0628−28	06 28 51.80	−28 32 33.7	237.0	−16.8	90	1.30
0643+80	06 43 53.51	+80 55 30.7	133.2	+26.8	30	1.20
0655+64	06 55 49.394	+64 22 22.87	151.6	+25.2	40	0.27
0656+14	06 56 57.91	+14 18 37	201.1	+8.3	10	0.40
0727−18	07 27 19.36	−18 30 25.5	233.8	−0.3	13	1.60
0736−40	07 36 50.92	−40 35 45.0	254.2	−9.2	190	2.50
0740−28	07 40 47.7	−28 15 41	243.8	−2.4	195	1.50
0743−53	07 43 50.5	−53 44 01	266.6	−14.3	23	2.40
0751+32	07 51 29.26	+32 39 52	188.2	+26.7	15	1.50
0756−15	07 56 11.34	−15 19 58.1	234.5	+7.2	7	2.30
0808−47	08 08 12.58	−47 45 00.7	263.3	−8.0	46	5.70
0809+74	08 09 02.85	+74 38 12.5	140.0	+31.6	50	0.17
0818−13	08 18 06.01	−13 41 22.7	235.9	+12.6	90	1.50
0818−41	08 18 29.74	−41 05 04.0	258.7	−2.7	65	0.66
0820+02	08 20 34.03	+02 08 54.3	222.0	+21.2	22	0.84
0823+26	08 23 50.459	+26 47 18.98	197.0	+31.7	70	0.71
0826−34	08 26 19.1	−34 07 08	254.0	+2.6	16	0.45
0833−45	08 33 39.27	−45 00 10.2	263.6	−2.8	5000	0.50
0834+06	08 34 26.166	+06 20 43.50	219.7	+26.3	65	0.43
0835−41	08 35 33.276	−41 24 42.0	260.9	−0.3	197	2.40

Pulsar Parameters*

PSR	Period (seconds)	$\dot{P}$ (10^{-15})	Epoch (JD - 2400000)	W_E (ms)	W_{50} (ms)
0450− 18	0.5489353684	5.7494	41535.54	26	28
0458+46	0.63856435537	5.582	46382.79	19	22
0523+11	0.35443756681	0.071	43889.125	12	16
0525+21	3.74549702902	40.0565	41993.5	75	183
0529− 66	0.97571407		44732.2		
0531+21	0.033326323455	421.288	46626.0	1.2	14
0538− 75	1.2458554349	0.57	43555.3699	68	75
0540+23	0.245966105161	15.42869	42391.5	7.9	6.0
0540− 69	0.0502812873	479.8383	45940.8659		
0559− 05	0.39596855140	1.309	43889.15	13	11
0559− 57	2.261364513	2.78	43557.3604	60	60
0609+37	0.297982308320	0.0589	46111.57		
0611+22	0.33492505401	59.630	42880.5	7.5	6.5
0621− 04	1.0390760459	0.846	43889.165	38	55
0626+24	0.47662189633	1.990	43984.899	12	12
0628− 28	1.2444170726	7.107	44123.633	73	66
0643+80	1.21443965739	3.8009	46110.59	24	25
0655+64	0.195670944915	0.000677	44546.9286	10	6.3
0656+14	0.38487474134	55.032	46260.19	13	15
0727− 18	0.51015078278	18.948	43889.211	13	8.5
0736− 40	0.37491871098	1.6140	42554.3342	27.5	29
0740− 28	0.16675244661	16.8317	42554.3026	5.0	4.5
0743− 53	0.2148363514	2.73	43779.7121	34	28
0751+32	1.4423490281	1.074	43890.227	22	10
0756− 15	0.68226433641	1.617	43889.229	11	6.6
0808− 47	0.54719837196	3.085	43556.353	48	38
0809+74	1.29224132384	0.1676	40688.97	45	42
0818− 13	1.23812810723	2.1056	41006.14	24	24
0818− 41	0.5454455279	0.027	43557.3813	100	160
0820+02	0.86487275188	0.1039	43419.3472	25	23
0823+26	0.53065995906	1.7236	42716.5	9.1	7.1
0826− 34	1.848918399	1.00	43435.7358	520	740
0833− 45	0.08928188753	124.371	47148.1349	2.6	2.3
0834+06	1.27376417152	6.79918	41707.5	17	23
0835− 41	0.75162112843	3.546	43557.4118	8.8	8.4

*Adapted from Taylor (1989).

Pulsar Parameters*

PSR	R.A.(1950) h m s	Dec.(1950) ° ′ ″	l °	b °	S_{400} (mJy)	D (kpc)
0839−53	08 39 09.20	−53 21 51.9	270.8	−7.1	19	3.10
0840−48	08 40 29.72	−48 40 32.3	267.2	−4.1	6.2	4.00
0841+80	08 41 24.8	+80 39 59.5	132.7	+31.5	1.5	1.30
0844−35	08 44 07.87	−35 22 38.8	257.2	+4.7	16	0.59
0853−33	08 53 38	−33 19 58	256.8	+7.5	10	0.59
0855−61	08 55 54.02	−61 26 15.8	278.6	−10.4	11	3.60
0901−63	09 01 31.80	−63 13 17.1	280.4	−11.1	4.5	2.80
0903−42	09 03 08.19	−42 34 13.0	265.1	+2.9	8	1.70
0904−74	09 04 28.76	−74 47 40.9	289.7	−18.3	11	1.90
0905−51	09 05 40.59	−51 45 50.8	272.2	−3.0	35	0.86
0906−17	09 06 18.83	−17 27 25.2	246.1	+19.8	10	0.52
0906−49	09 06 54.46	−49 00 54.4	270.3	−1.0		3.20
0909−71	09 09 22.32	−71 59 51.1	287.7	−16.3	6.5	2.00
0917+63	09 17 15.649	+63 06 59.63	151.4	+40.7	4.6	0.45
0919+06	09 19 35.09	+06 51 11.8	225.4	+36.4	40	1.00
0922−52	09 22 30.79	−52 49 47.2	274.7	−1.9	12	2.70
0923−58	09 23 04.9	−58 01 08.2	278.4	−5.6	22	1.00
0932−52	09 32 46.85	−52 36 02.5	275.7	−0.7	18	1.10
0940−55	09 40 37.76	−55 39 10.7	278.6	−2.2	55	4.90
0940+16	09 40 46	+16 45 08	216.6	+45.4	18	0.75
0941−56	09 41 18.65	−56 43 57.4	279.3	−3.0	13	4.90
0942−13	09 42 04.39	−13 40 52.4	249.1	+28.8	30	0.42
0943+10	09 43 27.12	+10 05 45.	225.4	+43.1	15	0.56
0950−38	09 50 11.80	−38 25 02.1	268.7	+12.0	8.5	4.00
0950+08	09 50 30.533	+08 09 45.19	228.9	+43.7	900	0.09
0953−52	09 53 41.780	−52 50 00.9	278.3	+1.2	29	3.50
0957−47	09 57 30.08	−47 55 23.0	275.7	+5.4	16	1.30
0959−54	09 59 51.62	−54 52 39.1	280.2	+0.1	80	0.06
1001−47	10 01 23.82	−47 32 29.0	276.0	+6.1	6	1.60
1010−23	10 10 10	−23 23 02	262.1	+26.4	12	0.97
1014−53	10 14 37.40	−53 30 13.9	281.2	+2.5	3.5	2.00
1015−56	10 15 22.78	−56 06 29.4	282.7	+0.3	15	14.00
1016−16	10 16 15.22	−16 27 06.6	258.3	+32.6	5.5	1.90
1030−58	10 30 15	−58 55	285.9	−1.0	14	15.00
1039−19	10 39 12	−19 26 07	265.6	+33.6	20	1.20

Pulsar Parameters*

PSR	Period (seconds)	$\dot{P}$ (10^{-15})	Epoch (JD - 2400000)	W_E (ms)	W_{50} (ms)
0839−53	0.7206117681	1.651	43557.4191	41	34
0840−48	0.64435405580	9.497	43557.436	15	14
0841+80	1.6022278243	0.442	46111.68		
0844−35	1.1160964496	1.57	43557.4633	24	16
0853−33	1.267532672	6.5	44328	20	20
0855−61	0.9625085573	1.677	43557.4792	25	23
0901−63	0.66031337153	0.107	43557.495	20	18
0903−42	0.9651709987	1.887	43557.5148	28	28
0904−74	0.54955311909	0.463	43555.4259	26	23
0905−51	0.25355550115	1.8335	43555.4202	23	19
0906−17	0.40162534295	0.6709	43889.279	10	9
0906−49	0.10675459250	15.151	46546.7841		
0909−71	1.3628899542	0.333	43557.5301	27	20
0917+63	1.56799290342	3.6081	46110.70		
0919+06	0.43061431165	13.7248	43890.286	12	10
0922−52	0.74629520954	35.477	43555.4605	17	15
0923−58	0.7394995609	4.83	43555.4439	50	42
0932−52	1.4447714964	4.653	43557.5569	28	28
0940−55	0.66436112446	22.739	43555.4767	15	15
0940+16	1.087417727	0.9	44328.5	65	65
0941−56	0.80811632004	39.6171	43557.5728	15	14
0942−13	0.57026410323	0.0462	43890.301	7.0	7.1
0943+10	1.09770363952	3.5291	41664.5	38	34
0950−38	1.3738151214	0.58	43557.5914	40	36
0950+08	0.25306506819	0.22915	41500.5	12.5	10.8
0953−52	0.86211773933	3.517	43555.5182	15	12
0957−47	0.67008576040	0.082	43557.6071	42	98
0959−54	1.4365681929	51.665	43557.6153	20	16
1001−47	0.30707169975	22.0737	43555.5027	12	11
1010−23	2.517944579	1.4	44327.51	38	34
1014−53	0.76958352720	1.926	43558.4826	16	15
1015−56	0.50345881546	3.130	43557.6221	24	22
1016−16	1.8046944181	1.746	46184.55		
1030−58	0.464208010	3	44095.23	60	58
1039−19	1.386367735	1.1	44328.51	42	60

*Adapted from Taylor (1989).

Pulsar Parameters*

PSR	R.A.(1950) h m s	Dec.(1950) ° ′ ″	l °	b °	S_{400} (mJy)	D (kpc)
1039– 55	10 39 59.35	– 55 05 22.3	285.2	+3.0	14	12.00
1044– 57	10 44 19.46	– 57 58 02.4	287.1	+0.7	18	7.10
1054– 62	10 54 28.06	– 62 42 44.7	290.3	– 3.0	45	6.00
1055– 52	10 55 48.59	– 52 10 51.9	286.0	+6.6	80	0.92
1056– 78	10 56 27.9	– 78 58 19.7	297.6	– 17.6	6.5	1.90
1056– 57	10 56 55.25	– 57 26 08.5	288.3	+1.9	19	3.30
1105– 59	11 05 51.1	– 59 30 49	290.2	+0.5	7	2.70
1110– 65	11 10 36.33	– 65 56 44.9	293.2	– 5.2	19	9.30
1110– 69	11 10 54.07	– 69 10 10.9	294.4	– 8.2	13	5.50
1112+50	11 12 48.42	+50 46 35.8	154.4	+60.4	20	0.32
1114– 41	11 14 20.635	– 41 06 20.9	284.5	+18.1	26	1.50
1118– 79	11 18 11.1	– 79 20 04.2	298.7	– 17.5	7	0.95
1119– 54	11 19 01.677	– 54 27 38.5	290.1	+5.9	24	7.60
1133+16	11 33 27.427	+16 07 36.77	241.9	+69.2	340	0.15
1133– 55	11 33 38.92	– 55 08 32.2	292.3	+5.9	23	2.90
1143– 60	11 43 41.70	– 60 14 19.4	295.0	+1.3	17	3.20
1154– 62	11 54 43.72	– 62 08 08.3	296.7	– 0.2	145	7.00
1159– 58	11 59 54.080	– 58 03 51.3	296.5	+3.9	23	5.00
1221– 63	12 21 34.71	– 63 51 16.3	300.0	– 1.4	48	2.70
1222– 63	12 22 54.53	– 63 52 06.7	300.1	– 1.4	11	15.00
1232– 55	12 32 31	– 55 00	300.6	+7.5	4.5	3.50
1236– 68	12 36 57.00	– 68 15 59.6	301.9	– 5.7	6.5	3.30
1237+25	12 37 11.991	+25 10 17.01	252.5	+86.5	160	0.33
1237– 41	12 37 33.93	– 41 08 24.0	300.7	+21.4	3.5	1.60
1240– 64	12 40 18.33	– 64 06 57.6	302.1	– 1.5	110	12.00
1254– 10	12 54 28.07	– 10 10 52.8	305.2	+52.4	5.5	1.10
1256– 67	12 56 08.61	– 67 25 29.4	303.7	– 4.8	4.5	3.20
1302– 64	13 02 10.9	– 64 39 22.7	304.4	– 2.1	29	19.00
1309– 53	13 09 03.10	– 53 46 47.3	306.0	+8.7	15	4.80
1309– 12	13 09 14.47	– 12 12 05.7	310.7	+50.1	3.8	1.30
1309– 55	13 09 50.67	– 55 00 53.2	306.0	+7.5	16	4.80
1317– 53	13 17 49.079	– 53 43 24.0	307.3	+8.6	18	3.50
1322– 66	13 22 33.2	– 66 45 16	306.3	– 4.4	28	7.30
1322+83	13 22 39.2	+83 39 17.5	121.9	+33.7	10	0.48
1323– 63	13 23 09.2	– 63 53 09.3	306.7	– 1.5	18	18.00

Pulsar Parameters*

PSR	Period (seconds)	$\dot{P}$ (10^{-15})	Epoch (JD - 2400000)	W_E (ms)	W_{50} (ms)
1039−55	1.1708590770	6.733	43557.6592	32	32
1044−57	0.36942669620	1.144	43555.5535	20	17
1054−62	0.42244618669	3.570	43555.6074	38	35
1055−52	0.197107608187	5.8335	43555.6172	19	89
1056−78	1.3474021612	1.327	43558.4965	28	26
1056−57	1.18499740291	4.288	43557.6746	21	21
1105−59	1.5165309654	0.34	43558.5147	45	45
1110−65	0.33421283978	0.824	43555.6226	21	19
1110−69	0.82048420411	2.841	43557.6831	22	20
1112+50	1.65643808033	2.4929	41535.795	27	31
1114−41	0.94315457524	7.940	43564.8062	16	15
1118−79	2.2805971963	3.67	43557.7091	32	46
1119−54	0.53578313115	2.7669	43556.7459	19	15
1133+16	1.18791153608	3.73273	41664.5	18.2	27.5
1133−55	0.36470327369	8.2284	43555.6393	18	17
1143−60	0.27337242988	1.7941	43555.6543	16	15
1154−62	0.40052094651	3.930	43555.6697	17	17
1159−58	0.452800508195	2.1265	43555.6852	13	13
1221−63	0.216474814297	4.9547	43555.6933	5.5	5.0
1222−63	0.4196177915	0.946	43558.5929	24	20
1232−55	0.63823381		43558.6099	40	44
1236−68	1.3019081139	11.892	43558.6251	25	24
1237+25	1.38244861210	0.95954	40610.5	25	50
1237−41	0.51224209318	1.739	43564.8443	10	10
1240−64	0.38847933086	4.5003	42710.2311	11	10
1254−10	0.617307560845	0.3625	46184.66		
1256−67	0.6633290494	1.206	43564.9026	15	14
1302−64	0.5716469444	4.032	43555.6989	45	38
1309−53	0.7281542173	0.148	43557.7397	33	33
1309−12	0.447517665849	0.1506	46184.67		
1309−55	0.8492356848	5.706	43556.8405	28	28
1317−53	0.279726553496	9.2589	43556.7969	13	12
1322−66	0.5430085792	5.31	43555.7073	58	60
1322+83	0.67003728616	0.565	46184.70		21
1323−63	0.7926701073	3.103	43558.7263	50	40

*Adapted from Taylor (1989).

Pulsar Parameters*

PSR	R.A.(1950) h m s	Dec.(1950) ° ′ ″	l °	b °	S_{400} (mJy)	D (kpc)
1323– 58	13 23 44.3	– 58 43 56.4	307.5	+3.6	120	9.80
1323– 62	13 23 57.1	– 62 07 10	307.1	+0.2	135	7.90
1325– 43	13 25 09.05	– 43 42 12.8	309.9	+18.4	18	1.50
1325– 49	13 25 31.09	– 49 06 02.7	309.1	+13.1	11	4.30
1336– 64	13 36 27.50	– 64 41 31.4	308.0	– 2.6	9	2.30
1338– 62	13 38 30.	– 62 07	308.8	– 0.1		25.00
1352– 51	13 52 44.32	– 51 39 13.9	313.0	+9.7	12	4.00
1353– 62	13 53 50	– 62 13	310.5	– 0.6		12.00
1356– 60	13 56 26.18	– 60 23 36.2	311.2	+1.1	105	8.80
1358– 63	13 58 10.47	– 63 43 17.2	310.6	– 2.1	34	2.90
1359– 51	13 59 41	– 51 10	314.1	+9.9	10	1.30
1417– 54	14 17 02.74	– 54 02 38.8	315.8	+6.4	9	4.50
1424– 55	14 24 54.68	– 55 17 26.3	316.4	+4.8	35	2.70
1426– 66	14 26 35.30	– 66 09 45.6	312.7	– 5.4	130	2.10
1436– 63	14 36 31.549	– 63 31 55.8	314.6	– 3.4	21	4.00
1449– 64	14 49 25.28	– 64 01 00.3	315.7	– 4.4	230	2.20
1451– 68	14 51 29.060	– 68 31 30.92	313.9	– 8.5	350	0.23
1454– 51	14 54 08.56	– 51 10 52	322.1	+6.7	4	1.10
1503– 51	15 03 04.3	– 51 46 37	323.1	+5.5	5	1.90
1503– 66	15 03 23.16	– 66 29 26.0	315.9	– 7.3	13	4.50
1504– 43	15 04 14.01	– 43 40 33.1	327.3	+12.5	16	1.70
1507– 44	15 07 27.3	– 44 10 47	327.6	+11.7	14	3.00
1508+55	15 08 03.75	+55 42 56.0	91.3	+52.3	125	0.73
1509– 58	15 09 59.06	– 58 56 58.0	320.3	– 1.2		6.70
1510– 48	15 10 44.61	– 48 23 10.5	325.9	+7.8	8.5	1.70
1523– 55	15 23 50.31	– 55 41 42	323.6	+0.6	17	9.30
1524– 39	15 24 42.11	– 39 21 12	333.1	+14.0	11	1.70
1530+27	15 30 04.63	+27 55 56.7	43.5	+54.5	10	0.53
1530– 53	15 30 22.46	– 53 24 16.7	325.7	+1.9	70	0.65
1540– 06	15 40 50.38	– 06 11 17.0	0.6	+36.6	50	0.66
1541– 52	15 41 12.591	– 52 59 22.9	327.3	+1.3	23	0.91
1541+09	15 41 14.354	+09 38 42.81	17.8	+45.8	100	1.30
1550– 54	15 50 05.3	– 54 47 15	327.2	– 0.9	13	5.60
1552– 31	15 52 10	– 31 25 07	342.7	+16.8	12	2.60
1552– 23	15 52 32	– 23 33 10	348.4	+22.5	8	1.90

Pulsar Parameters*

PSR	Period (seconds)	$\dot{P}$ (10^{-15})	Epoch (JD - 2400000)	W_E (ms)	W_{50} (ms)
1323−58	0.4779896901	3.211	43555.7294	90	65
1323−62	0.5299062943	18.89	43555.7163	14	11
1325−43	0.53269880337	3.014	43613.5346	18	18
1325−49	1.4787213559	0.61	43557.8283	30	20
1336−64	0.37862206628	5.0512	43564.9414	14	13
1338−62	0.1932024		45039.5		
1352−51	0.64430109622	2.813	43556.8491	13	12
1353−62	0.455761			25	22
1356−60	0.12750077685	6.3385	43555.7229	34	31
1358−63	0.84278490076	16.905	43555.8221	13	12
1359−51	1.380177		43419.5	25	15
1417−54	0.9357719221	0.237	43557.8701	26	25
1424−55	0.57028987883	2.087	43555.7535	19	18
1426−66	0.78543998083	2.771	43555.7617	17	17
1436−63	0.45960536385	1.1201	43555.7681	14	13
1449−64	0.179483893925	2.74754	43176.6735	5.0	4.5
1451−68	0.263376778654	0.09878	42553.5875	14.0	12.5
1454−51	1.7483010801	5.29	43563.9725	24	14
1503−51	0.840738545	6.37	43613.6802	16	14
1503−66	0.35565490117	1.157	43564.0041	11	11
1504−43	0.28675704522	1.6045	43555.7885	8.2	9.2
1507−44	0.9438712975	0.61	43557.9853	50	50
1508+55	0.73967789896	5.0327	40625.46	13	10
1509−58	0.15023079664	540.1908	45144.2437		
1510−48	0.45483920949	0.925	43564.9908	13	12
1523−55	1.0487048866	11.28	43558.002	42	41
1524−39	2.4175822701	19.07	43558.6872	44	44
1530+27	1.1248355211	0.803	46388.20	27.0	26.5
1530−53	1.3688805090	1.428	43558.6798	26	22
1540−06	0.70906364986	0.8830	43889.545	9.8	7.9
1541−52	0.178553799688	0.0607	43555.7965	10	8
1541+09	0.74844817748	0.43030	42303.5	75	50
1550−54	1.0813282469	15.65	43558.0118	56	52
1552−31	0.5181097514	0.05	44327.51	14	26
1552−23	0.532577402	0.70	44327.53	15	13

*Adapted from Taylor (1989).

Pulsar Parameters*

PSR	R.A.(1950) h m s	Dec.(1950) ° ′ ″	l °	b °	S_{400} (mJy)	D (kpc)
1555– 55	15 55 23.35	– 55 37 09.0	327.2	– 2.0	15	6.30
1556– 44	15 56 11.031	– 44 30 16.9	334.5	+6.4	110	1.90
1556– 57	15 56 14.66	– 57 42 47.4	326.0	– 3.7	20	5.70
1557– 50	15 57 08.76	– 50 35 55.9	330.7	+1.6		7.80
1558– 50	15 58 33.89	– 50 51 44.8	330.7	+1.3	45	2.50
1600– 27	16 00 05	– 27 04 12	347.1	+18.8	20	1.70
1600– 49	16 00 42.037	– 49 01 45.7	332.2	+2.4	44	4.20
1601– 52	16 01 25.38	– 52 49 26	329.7	– 0.5	30	0.80
1604– 00	16 04 37.863	– 00 24 41.67	10.7	+35.5	45	0.36
1607– 13	16 07 55.00	– 13 14 34.	359.4	+26.9	4.6	1.80
1609– 47	16 09 51.08	– 47 06 49.5	334.6	+2.8	17	5.00
1612+07	16 12 15.28	+07 45 01.0	20.6	+38.2	12	0.77
1612– 29	16 12 45	– 29 32 38	347.4	+15.1	5	1.40
1620– 42	16 20 18.10	– 42 49 57	338.9	+4.6	24	9.00
1620– 09	16 20 34	– 09 01 13	5.3	+27.2	5	2.60
1620– 26	16 20 34.14	– 26 24 58.0	351.0	+16.0	15	2.20
1630– 59	16 30 48.44	– 59 48 29.1	327.7	– 8.3	7	4.60
1633+24	16 33 20.179	+24 24 53.52	43.0	+39.9	12	0.88
1641– 45	16 41 10.313	– 45 53 38.7	339.2	– 0.2	375	5.30
1641– 68	16 41 40.67	– 68 26 25.6	321.8	– 14.8	23	1.50
1642– 03	16 42 24.63	– 03 12 31.1	14.1	+26.1	300	1.30
1647– 528	16 47 43.29	– 52 50 45	334.6	– 5.5	12	5.40
1647– 52	16 47 46.67	– 52 17 55.8	335.0	– 5.2	23	5.80
1648– 42	16 48 17	– 42 40	342.5	+0.9	100	14.00
1648– 17	16 48 38.37	– 17 04 19	2.8	+16.9	8	1.00
1649– 23	16 49 57	– 23 58 40	357.3	+12.5	8	2.30
1657– 13	16 57 04.38	– 13 00 42	7.5	+17.6	4.4	2.10
1659– 60	16 59 47.9	– 60 12 43	329.8	– 11.4	23	1.90
1700– 32	17 00 07	– 32 51 22	351.6	+5.2	45	3.40
1700– 18	17 00 55.36	– 18 42 06	3.2	+13.6	10	1.70
1701– 75	17 01 15.7	– 75 35 23	316.7	– 20.2	2.5	1.30
1702– 19	17 02 41	– 19 01 34	3.2	+13.0	25	0.74
1706– 16	17 06 33.16	– 16 37 12	5.8	+13.7	60	0.81
1707– 53	17 07 50.22	– 53 46 38.1	335.7	– 8.5	9	3.60
1709– 15	17 09 03.645	– 15 06 03.3	7.4	+14.0	11	2.00

Pulsar Parameters*

PSR	Period (seconds)	$\dot{P}$ (10^{-15})	Epoch (JD - 2400000)	W_E (ms)	W_{50} (ms)
1555– 55	0.9572424424	20.479	43558.0288	24	19
1556– 44	0.257055723521	1.01955	42553.6183	7	6
1556– 57	0.19445388693	2.1248	43556.7043	17	15
1557– 50	0.192598318908	5.06256	42553.6681	4.0	3.8
1558– 50	0.8642020784	69.572	43558.6729	21	21
1600– 27	0.778311740	2.92	44327.54	19	19
1600– 49	0.327417287977	1.0117	43556.8643	11	10
1601– 52	0.6580131007	0.256	43556.872	45	74
1604– 00	0.42181611020	0.30607	42306.5	11.5	11.5
1607– 13	1.01839267444	0.230	46075.09		
1609– 47	0.38237595278	0.632	43556.8812	17	14
1612+07	1.2068002120	2.357	43890.565	14	12
1612– 29	2.477567322	2.5	44326.5	30	20
1620– 42	0.3645903553	1.022	43556.8895	52	38
1620– 09	1.276444847	3.0	44327.5	20	16
1620– 26	0.011075750806	0.00080	47200.0	1.0	1.0
1630– 59	0.52912125685	1.368	43564.0529	16	13
1633+24	0.490506485353	0.1194	46075.10		
1641– 45	0.45505464292	20.137	43633.7101	9.2	8.2
1641– 68	1.7856112396	1.70	43558.102	78	90
1642– 03	0.38768879135	1.7810	40621.54	4.0	3.7
1647– 528	0.8905339578	2.070	43558.758	32	28
1647– 52	0.63505597647	1.812	43556.9135	19	19
1648– 42	0.8440794	0	44095.5722	340	380
1648– 17	0.97339208101	3.042	43889.591	19	28
1649– 23	1.703740		43329.5	36	36
1657– 13	0.64095800488	0.6210	46075.12		
1659– 60	0.3063230254	0.91	43557.0061	60	85
1700– 32	1.211784624	0.7	42004.6	43	43
1700– 18	0.80434055233	1.730	46382.28	20	20
1701– 75	1.1910239966	1.88	43565.1011	25	25
1702– 19	0.2989858521	4.14	44326.48	8.3	7.1
1706– 16	0.65305047326	6.380	40621.57	11	11
1707– 53	0.89921779930	15.494	43565.087	17	16
1709– 15	0.86880388358	1.104	46075.13		

*Adapted from Taylor (1989).

Pulsar Parameters*

PSR	R.A.(1950) h m s	Dec.(1950) ° ′ ″	l °	b °	S_{400} (mJy)	D (kpc)
1717- 29	17 17 23	- 29 30 09	356.5	+4.2	30	1.20
1717- 16	17 17 31.9	- 16 30 32	7.4	+11.5	7	1.40
1718- 02	17 18 20.95	- 02 09 27.7	20.1	+18.9	24	2.40
1718- 32	17 18 48	- 32 05 04	354.6	+2.5	45	3.70
1719- 37	17 19 35.47	- 37 09 06	350.5	- 0.5	25	2.50
1726- 00	17 26 00.91	- 00 05 24.5	23.0	+18.3	5	1.30
1727- 47	17 27 55.38	- 47 42 21.4	342.6	- 7.7	190	4.10
1729- 41	17 29 17.69	- 41 26 42	348.0	- 4.5	9	6.10
1730- 22	17 30 25	- 22 26 43	4.0	+5.7	20	1.30
1732- 02	17 32 09.38	- 02 10 38	21.9	+15.9	5.7	2.30
1732- 07	17 32 22	- 07 22 57	17.3	+13.3	15	2.60
1735- 32	17 35 51	- 32 09	356.5	- 0.5		1.50
1736- 31	17 36 08	- 31 29	357.1	- 0.2		12.00
1736- 29	17 36 28	- 29 03	359.2	+1.0		3.60
1737- 30	17 37 22	- 30 14 12	358.3	+0.2		3.50
1737+13	17 37 49.22	+13 13 29.4	37.1	+21.7	70	1.80
1737- 39	17 37 49.52	- 39 26 08.6	350.6	- 4.7	35	5.10
1738- 08	17 38 38.69	- 08 39 06	17.0	+11.3	40	2.60
1740- 03	17 40 30	- 03 36 08	21.7	+13.4	8	1.20
1740- 13	17 40 47.38	- 13 50 24	12.7	+8.2	4	3.90
1742- 30	17 42 42	- 30 39 02	358.6	- 1.0	55	2.30
1745- 12	17 45 28.20	- 12 59 56	14.0	+7.7	15	3.40
1745- 56	17 45 31.51	- 56 04 25	336.6	- 14.3	3	2.00
1747- 46	17 47 57.01	- 46 56 39.5	345.0	- 10.2	70	0.66
1749- 28	17 49 49.24	- 28 06 00.2	1.5	- 1.0	1300	1.00
1750- 24	17 50 28	- 24 58	4.3	+0.5		19.00
1753+52	17 53 14.69	+52 01 39.9	79.6	+29.6	2.8	1.30
1753- 24	17 53 46	- 24 36	5.0	+0.1		7.60
1754- 24	17 54 37	- 24 21 40	5.3	+0.0		4.00
1756- 22	17 56 23	- 22 05 33	7.5	+0.8	25	4.50
1757- 23	17 57	- 23 43	6.1	- 0.1	4	6.10
1758- 23	17 58 15.	- 23 05	6.8	- 0.1		23.00
1758- 24	17 58 15.	- 24 53	5.3	- 1.0		6.40
1758- 03	17 58 44.252	- 03 57 54.6	23.6	+9.3	19	4.00
1800- 21	18 00 51.2	- 21 37 12	8.4	+0.1		5.20

Pulsar Parameters*

PSR	Period (seconds)	$\dot{P}$ (10^{-15})	Epoch (JD - 2400000)	W_E (ms)	W_{50} (ms)
1717−29	0.620447861	0.8	42004.5	30	30
1717−16	1.5655994316	5.82	46262.63		
1718−02	0.47771530393	0.087	43889.612	55	60
1718−32	0.4771570774	0.7	42004.6	12	13
1719−37	0.23616867232	10.816	43557.0305	12	11
1726−00	0.38600506360	1.124	46075.14		
1727−47	0.82972364148	163.6716	43494.1836	19	20
1729−41	0.6279806844	12.838	43565.0515	28	22
1730−22	0.871682810	0.0	42004.5	35	25
1732−02	0.8393860273	0.42	46382.30		
1732−07	0.419335		43324.5	12	10
1735−32	0.76850		46627.0		
1736−31	0.529437		46564.0		
1736−29	0.3228806	7.861	46611.0		
1737−30	0.606561385650	466	46612		
1737+13	0.80304971623	1.454	43892.616	22.5	32
1737−39	0.51221006003	1.809	43557.9179	23	20
1738−08	2.0430815106	2.27	43890.622	60	80
1740−03	0.444644291	3.17	44328.49	8.6	8.6
1740−13	0.4053387113	0.481	46183.86		
1742−30	0.3674214929	10.7	42004.6	8	4
1745−12	0.39413270428	1.212	43890.631	12	11
1745−56	1.3323097198	2.12	43558.8546	24	18
1747−46	0.74235203333	1.295	43557.0878	17	17
1749−28	0.56255316830	8.154	40127.69	7.5	7.0
1750−24	0.528331		46611.0		
1753+52	2.3913963093	1.568	46075.17		
1753−24	0.670480		46627.0		
1754−24	0.234094343	13.0	44327.48	26	18
1756−22	0.4609690621	10.8	44327.48	14	12
1757−23	1.03082				
1758−23	0.4157644		45039.5		
1758−24	0.1248315		45039.5		
1758−03	0.92148958467	3.316	46117.04		
1800−21	0.13357862	134.154	46141.0		

*Adapted from Taylor (1989).

Pulsar Parameters*

PSR	R.A.(1950) h m s	Dec.(1950) ° ′ ″	l °	b °	S_{400} (mJy)	D (kpc)
1802+03	18 02 40.02	+03 06 13.8	30.4	+11.7	5	2.80
1804- 12	18 04	- 12 00	17.1	+4.2	4	3.80
1804- 27	18 04 02	- 27 15 40	3.8	- 3.3	25	8.90
1804- 20	18 04 50	- 20 59	9.4	- 0.3		14.00
1804- 08	18 04 53.921	- 08 48 10.2	20.1	+5.6	55	3.70
1806- 21	18 06 06	- 21 10	9.4	- 0.7		9.10
1806- 53	18 06 39.98	- 53 38 45.7	340.3	- 15.9	12	1.60
1809- 173	18 09 15	- 17 19	13.1	+0.5		5.10
1809- 175	18 09 41	- 17 30	13.0	+0.4		16.00
1810+02	18 10 22.28	+02 26 07.8	30.7	+9.7	5	3.50
1811+40	18 11 35.98	+40 12 45.6	67.4	+24.0	8	1.50
1813- 17	18 13 04	- 17 30	13.4	- 0.4		11.00
1813- 26	18 13 28	- 26 50 55	5.2	- 4.9	30	4.20
1813- 36	18 13 43.028	- 36 19 11.3	356.8	- 9.4	22	3.20
1814- 23	18 14	- 23 13	8.5	- 3.3	4	7.20
1815- 14	18 15 15	- 14 22	16.4	+0.7		15.00
1817- 13	18 17 14	- 13 50	17.1	+0.5		19.00
1818- 04	18 18 13.65	- 04 29 05	25.5	+4.7	170	1.50
1819- 14	18 19 54	- 14 04	17.2	- 0.2		11.00
1819- 22	18 19 57	- 22 57 58	9.4	- 4.4	25	3.90
1820- 31	18 20 31	- 31 08 16	2.1	- 8.3	18	1.70
1820- 11	18 20 37	- 11 12	19.8	+1.0		11.00
1821- 19	18 21 02.78	- 19 47 29	12.3	- 3.1	52	6.80
1821+05	18 21 04.01	+05 48 47.4	35.0	+8.9	25	2.30
1821- 24	18 21 27.382	- 24 53 50.95	7.8	- 5.6		3.90
1821- 11	18 21 52	- 11 22	19.8	+0.7		16.00
1822- 14	18 22 20	- 14 50	16.8	- 1.0		9.30
1822+00	18 22 41.60	5 00 02 36.4	30.0	+5.9	23	1.70
1822- 09	18 22 46.2	- 09 37 31	21.4	+1.3	32	0.51
1823- 11	18 23 06	- 11 32	19.8	+0.3		10.00
1823- 13	18 23 19	- 13 37	18.0	- 0.7		5.50
1824- 09	18 24 06	- 09 58	21.3	+0.9		11.00
1826- 17	18 26 48	- 17 52 45	14.6	- 3.4	70	6.70
1828- 10	18 28 20	- 10 58	20.9	- 0.5		4.40
1828- 60	18 28 44.5	- 60 25 19	334.8	- 21.2	5.5	1.20

Pulsar Parameters*

PSR	Period (seconds)	$\dot{P}$ (10^{-15})	Epoch (JD - 2400000)	W_E (ms)	W_{50} (ms)
1802+03	0.21871123480	0.999	46116.06		
1804−12	0.52276				
1804−27	0.827770612	12.25	44326.48	28	24
1804−20	0.91840		46612.0		
1804−08	0.16372736083	0.02868	43890.64	7.7	7.0
1806−21	0.702413		46627.0		
1806−53	0.26104931796	0.383	43558.1388	18	16
1809−173	1.205368		46612.0		
1809−175	0.53835		46612.0		
1810+02	0.79390161052	3.5997	46075.17		
1811+40	0.931087849	2.553	43984.394	18	20
1813−17	0.782308		46612.0		
1813−26	0.592885093	−0.3	42004.1	65	75
1813−36	0.38701692095	2.0465	43558.1261	13	11
1814−23	0.62547				
1815−14	0.2914888		46612.0		
1817−13	0.921460		46627.0		
1818−04	0.59807263930	6.3376	40621.6	9	8
1819−14	0.2147710		46612.0		
1819−22	1.874267731	0.6	42004.2	65	65
1820−31	0.2840528876	2.92	44326.48	6.3	5.9
1820−11	0.279822		46333.0		
1821−19	0.18933213477	5.2378	43556.9232	30	24
1821+05	0.75290644846	0.225	43889.659	13	10
1821−24	0.0030543144932	0.00155	46977.61		
1821−11	0.435758		46612.0		
1822−14	0.2791816		46612.0		
1822+00	0.77894912481	0.8733	46075.18		
1822−09	0.7689585865	52.32	43780.4685	17	14
1823−11	2.093110		46627.0		
1823−13	0.10143785	51	46565.0		
1824−09	0.245753		46612.0		
1826−17	0.3071291968	5.59	42004.6	55	45
1828−10	0.405028		46627.0		
1828−60	1.8894355004	0.27	43558.8657	48	58

*Adapted from Taylor (1989).

Pulsar Parameters*

PSR	R.A.(1950) h m s	Dec.(1950) ° ′ ″	l °	b °	S_{400} (mJy)	D (kpc)
1829− 10	18 29 28	− 10 26	21.5	− 0.5		11.00
1829− 08	18 29 59	− 08 28	23.3	+0.3		6.60
1830− 08	18 30 54	− 08 29	23.4	+0.1		8.50
1831− 03	18 31 04	− 03 40 55	27.7	+2.3	95	7.10
1831− 00	18 31 43.25	− 00 13 13.3	30.8	+3.7	15	2.70
1831− 04	18 31 46.46	− 04 28 59	27.0	+1.7	75	2.20
1832− 06	18 32 32	− 06 46	25.1	+0.5		10.00
1834− 10	18 34 08.3	− 10 10 44	22.3	− 1.4	30	8.80
1834− 06	18 34 28	− 06 56	25.2	+0.0		6.70
1834− 04	18 34 36	− 04 40	27.2	+1.0		5.40
1838− 04	18 38 25	− 04 28	27.8	+0.3		7.10
1839+09	18 39 32.829	+09 09 10.5	40.1	+6.3	18	1.60
1839− 04	18 39 44	− 04 06	28.3	+0.2		4.70
1839+56	18 39 50.3	+56 38 56	86.1	+23.8	110	0.94
1841− 05	18 41 20	− 05 41	27.1	− 0.9		11.00
1841− 04	18 41 49	− 04 35	28.1	− 0.5		2.50
1842− 02	18 42 19	− 02 51	29.7	+0.2		9.10
1842− 04	18 42 36	− 04 35	28.2	− 0.7		7.20
1842+14	18 42 38.52	+14 51 03.6	45.6	+8.1	30	1.30
1844− 04	18 44 45	− 04 05 32	28.9	− 0.9	100	3.80
1845− 19	18 45 20.55	− 19 55 49	14.8	− 8.3	15	0.53
1845− 01	18 45 49	− 01 27 30	31.3	+0.0	60	3.70
1846− 06	18 46 26	− 06 40 26	26.8	− 2.5	25	4.50
1848+13	18 48 17.468	+13 32 24.4	45.0	+6.3	4	1.90
1848+04	18 48 34.39	+04 14 35.6	36.7	+2.0		3.30
1848+12	18 48 54.48	+12 55 59.4	44.5	+5.9	4	2.30
1849+00	18 49 54.3	+00 28 20	33.5	+0.0		19.00
1851− 79	18 51 46.8	− 79 55 51.1	314.3	− 27.1	6.5	1.40
1851− 14	18 51 53.85	− 14 25 21	20.5	− 7.2	10	4.40
1852+10	18 52 06	+10 47	42.9	+4.3	11	8.40
1854+00	18 54 28	+00 53 21	34.4	− 0.8	5	2.30
1855+09	18 55 13.683	+09 39 13.32	42.3	+3.1	15	0.35
1855+02	18 55 19	+02 07	35.6	− 0.4		12.00
1857− 26	18 57 43	− 26 04 54	10.3	− 13.5	120	1.30
1859+03	18 59 01.945	+03 26 45.61	37.2	− 0.6	125	11.00

Pulsar Parameters*

PSR	Period (seconds)	$\dot{P}$ (10^{-15})	Epoch (JD - 2400000)	W_E (ms)	W_{50} (ms)
1829−10	0.330354		46628.0		
1829−08	0.647275		46627.0		
1830−08	0.08528106	9	46141.0		
1831−03	0.686676816	41.5	42003.8	35	25
1831−00	0.52095430816	0.0143	46071.2018		
1831−04	0.29010620279	0.197	44054.225	45	45
1832−06	0.305818		46613.0		
1834−10	0.5627056550	11.775	43892.656	130	100
1834−06	1.90581		46628.0		
1834−04	0.354236		46613.0		
1838−04	0.18614511		46613.0		
1839+09	0.38131888123	1.0916	43889.669	9.5	9.0
1839−04	1.83995	0.51	46613.0		
1839+56	1.6528613	1.7	44326.59	30	23
1841−05	0.2556965		46612.0		
1841−04	0.99102		46627.0		
1842−02	0.507718		46612.0		
1842−04	0.1622506		46333.0		
1842+14	0.37546250760	1.866	43986.411	8.3	8.9
1844−04	0.5977390	51.9	42003.8	32	24
1845−19	4.3081792620	23.31	43556.9421	65	65
1845−01	0.65942849	5.2	42004.7	26	22
1846−06	1.451292579	45.7	42004.7	30	30
1848+13	0.34558145170	1.4926	46260.70		
1848+04	0.2846970	0.0016	45062.261		
1848+12	1.2052994672	11.521	46077.19		
1849+00	2.1803		46627.0		
1851−79	1.2791931935	1.86	43565.1375	40	34
1851−14	1.1465926813	4.171	43894.666	24	18
1852+10	0.572341				
1854+00	0.3569290				
1855+09	0.005362100452367	0.00002	46421.2363	0.6	
1855+02	0.415814		46613.0		
1857−26	0.61220908	0.16	42004.7	43	48
1859+03	0.655445115169	7.4874	42099.5	30	26

*Adapted from Taylor (1989).

Pulsar Parameters*

PSR	R.A.(1950) h m s	Dec.(1950) ° ′ ″	l °	b °	S_{400} (mJy)	D (kpc)
1859+01	18 59 02.665	+01 52 18.6	35.8	− 1.4	10	2.80
1859+07	18 59 20	+07 13	40.6	+1.0		6.70
1900+05	19 00 15.541	+05 52 00.88	39.5	+0.2	20	4.30
1900+06	19 00 20	+06 11 25	39.8	+0.3	20	14.00
1900+01	19 00 57.961	+01 31 09.41	35.7	− 2.0	60	5.00
1900− 06	19 00 59	− 06 36 30	28.5	− 5.7	22	6.40
1901+10	19 01 40	+10 00	43.3	+1.8	2	4.10
1902− 01	19 02 52.72	− 01 01 08	33.7	− 3.5	12	7.20
1903+07	19 03 08	+07 05	40.9	+0.1		8.80
1904+06	19 04 09	+06 37	40.6	− 0.3		12.00
1904+12	19 04 49	+12 42	46.1	+2.4	2	8.40
1905+39	19 05 54.7	+39 57 18.3	70.9	+14.2	25	1.00
1906+09	19 06 35.379	+09 11 21.5	43.2	+0.4	5	6.20
1907+00	19 07 01.586	+00 03 03.60	35.1	− 4.0	8	3.60
1907+02	19 07 07.746	+02 49 56.34	37.6	− 2.7	20	5.40
1907+10	19 07 27.329	+10 57 07.67	44.8	+1.0	55	4.00
1907+03	19 07 39.7	+03 53 30	38.6	− 2.3	30	2.30
1907− 03	19 07 52.30	− 03 14 50.7	32.3	− 5.7	30	6.80
1907+12	19 07 53.949	+12 26 43.00	46.2	+1.6	5	8.50
1910+10	19 10 30	+10 30	44.8	+0.1	2	3.40
1910+20	19 10 34.050	+20 59 26.05	54.1	+5.0	10	2.90
1911+13	19 11 06.406	+13 55 42.36	47.9	+1.6	30	4.20
1911− 04	19 11 15.17	− 04 45 59.7	31.3	− 7.1	120	3.00
1911+09	19 11 30	+09 30	44.0	− 0.6	3	4.00
1911+11	19 11 49.063	+11 16 49.87	45.6	+0.2	5	2.00
1913+167	19 13 04.438	+16 41 50.43	50.6	+2.5	6	1.90
1913+10	19 13 07.60	+10 04 25.3	44.7	− 0.7	20	6.50
1913+16	19 13 12.474	+16 01 08.02	50.0	+2.1	20	5.20
1914+09	19 14 09.513	+09 46 02.72	44.6	− 1.0	15	1.60
1914+13	19 14 39.715	+13 07 25.27	47.6	+0.5	45	6.10
1915+13	19 15 21.573	+13 48 28.67	48.3	+0.6	30	2.40
1915+22	19 15 36	+22 19	55.8	+4.6	3	4.10
1916+14	19 16 07.12	+14 39 21	49.1	+0.9	6	0.76
1917+00	19 17 17.218	+00 16 03.32	36.5	− 6.2	30	3.00
1918+26	19 18 36.341	+26 44 57.87	60.1	+6.0	10	0.80

Pulsar Parameters*

PSR	Period (seconds)	$\dot{P}$ (10^{-15})	Epoch (JD - 2400000)	W_E (ms)	W_{50} (ms)
1859+01	0.28821851116		46116.09		
1859+07	0.643998		46627.0		
1900+05	0.74656970999	12.896	42854.5	26	24
1900+06	0.673501		44485.5		16
1900+01	0.72930163274	4.0322	42345.5	13	11
1900−06	0.4318847189	3.4	42004.3	22	20
1901+10	1.856568				50
1902−01	0.64317549703	3.056	46075.21		
1903+07	0.648039		46613.0		
1904+06	0.2672741		46564.0		
1904+12	0.827096				30
1905+39	1.2357572279	0.53	43890.684	34	50
1906+09	0.83026988471	0.098	42832.5	47	37
1907+00	1.01694545765	5.5151	42646.5	8.7	6.8
1907+02	0.989827728010	2.76434	42415.5	14	11
1907+10	0.283638670836	2.63637	42539.5	5.8	5.1
1907+03	2.330260471	4.53	43984.433	170	320
1907−03	0.50460347381	2.189	43987.428	11	9.8
1907+12	1.44173740007	8.245	42845.5	48	34
1910+10	0.409338				25
1910+20	2.23296353405	10.1819	42491.5	14	9.3
1911+13	0.52147222250	0.8052	42826.5	10	7.2
1911−04	0.82593368968	4.0696	40623.63	7.5	8.2
1911+09	1.241964				45
1911+11	0.60099749634	0.6555	42826.5	18	9
1913+167	1.61623128389	0.408	42839.5	32	38
1913+10	0.40453042033	15.244	42846.5	36	30
1913+16	0.059029995269	0.00864	42320.9332	5.0	8.2
1914+09	0.270252897887	2.51825	42824.5	9.8	9.6
1914+13	0.28184027865	3.6172	42846.5	13	7
1915+13	0.19462634149	7.20286	42301.5	4.5	3.7
1915+22	0.425906				
1916+14	1.1808836647	211.4	42955.5	26	20
1917+00	1.27225573197	7.6764	42425.5	18	12
1918+26	0.785521838082	0.0351	46071.23		

*Adapted from Taylor (1989).

Pulsar Parameters*

PSR	R.A.(1950) h m s	Dec.(1950) ° ′ ″	l °	b °	S_{400} (mJy)	D (kpc)
1918+19	19 18 52.601	+19 43 02.45	53.9	+2.7	30	5.00
1919+14	19 19 06.383	+14 13 32.17	49.1	+0.0	20	2.20
1919+21	19 19 36.163	+21 47 16.30	55.8	+3.5	240	0.33
1919+20	19 19 40	+20 00	54.2	+2.6	1	2.10
1920+20	19 20 08	+20 12 16	54.4	+2.6	6	6.70
1920+21	19 20 43.979	+21 04 52.09	55.3	+2.9	45	7.30
1921+17	19 21 06	+17 00	51.7	+0.9	2	3.70
1922+20	19 22 30	+20 34 06	55.0	+2.3	4	7.00
1923+04	19 23 55.56	+04 25 27.4	41.0	- 5.7	10	3.40
1924+19	19 24 16	+19 20	54.1	+1.4	2	14.00
1924+16	19 24 30.302	+16 42 27.23	51.9	+0.1	12	4.20
1924+14	19 24 39.581	+14 28 48.49	49.9	- 1.0	9	5.80
1925+18	19 25 00	+18 50	53.8	+1.0	3	7.30
1925+22	19 25 00.104	+22 28 49.69	57.0	+2.7	6	6.00
1925+188	19 25 27	+18 50	53.8	+0.9	3	2.40
1926+18	19 26 58	+18 39 50	53.9	+0.5	3	2.80
1927+13	19 27 41.709	+13 09 52.24	49.1	- 2.3	5	6.60
1929+15	19 29 30	+15 30	51.4	- 1.6	3	3.40
1929+10	19 29 51.928	+10 53 03.74	47.4	- 3.9	130	0.08
1929+20	19 29 57.16	+20 14 17.6	55.6	+0.6	50	5.70
1930+22	19 30 12.48	+22 15 19.0	57.4	+1.6	18	6.60
1930+13	19 30 58	+13 00	49.4	- 3.1	2	5.40
1931+24	19 31	+24 15	59.2	+2.4	8	2.70
1933+17	19 33 19	+17 40	53.7	- 1.3	3	6.30
1933+16	19 33 31.860	+16 09 58.32	52.4	- 2.1	260	6.00
1933+15	19 33 44.782	+15 29 52.83	51.9	- 2.5	2	5.30
1935+25	19 35 01.48	+25 37 14.8	60.8	+2.3	13	1.80
1937+24	19 37 08	+24 43	60.3	+1.4	3	2.80
1937+21	19 37 28.746	+21 28 01.46	57.5	- 0.3		5.00
1937- 26	19 37 58	- 26 08 54	13.9	- 21.8	14	1.80
1939+17	19 39 47	+17 40	54.5	- 2.7	3	5.70
1940- 12	19 40 38.13	- 12 44 53	27.3	- 17.2	15	1.00
1941- 17	19 41 12.0	- 17 57 27	22.3	- 19.4	7	2.00
1942+17	19 42 15	+17 50	54.9	- 3.1	1	5.30
1942- 00	19 42 53.82	- 00 48 17.6	38.6	- 12.3	40	2.00

Pulsar Parameters*

PSR	Period (seconds)	$\dot{P}$ (10^{-15})	Epoch (JD - 2400000)	W_E (ms)	W_{50} (ms)
1918+19	0.82103460383	0.8952	42576.5	50	60
1919+14	0.618179730019	5.612	42833.5	24	26
1919+21	1.337301192269	1.34809	40689.45	25	29
1919+20	0.760682				30
1920+20	1.172761				55
1920+21	1.07791915514	8.1899	42546.5	16	11
1921+17	0.547209				25
1922+20	0.237790138	2.09	43957.49	12	16
1923+04	1.0740771471	2.465	43892.689	15	15
1924+19	1.346010				55
1924+16	0.57981189556	18.0035	42834.5	14	13
1924+14	1.32492185957	0.223	42823.5	55	28
1925+18	0.482765				35
1925+22	1.431066414236	0.771	42830.5	44	40
1925+188	0.298312				30
1926+18	1.2204686				26
1927+13	0.76003196831	3.661	42828.5	21	17
1929+15	0.314351				30
1929+10	0.22651715301	1.15675	41703.5	7.2	6.2
1929+20	0.26821490434	4.179	43028.5	17	15
1930+22	0.14442788986	57.78	42675.5	9	8
1930+13	0.928325				35
1931+24	0.81368				
1933+17	0.654408				35
1933+16	0.358736248270	6.00354	42264.5	6.5	8.0
1933+15	0.96733836953	4.039	42830.5	26	22
1935+25	0.20097999418	1.555	46077.23		
1937+24	0.645277				
1937+21	0.001557806448873	0.0001	45303.2940	0.2	
1937–26	0.4028574341	0.96	44326.48	5.8	5.0
1939+17	0.696261				65
1940–12	0.97242816020	1.659	43890.708	12	11
1941–17	0.8411572623	0.980	43892.702	19	19
1942+17	1.996898				45
1942–00	1.04563235954	0.536	46259.74	38	50

*Adapted from Taylor (1989).

Pulsar Parameters*

PSR	R.A.(1950) h m s	Dec.(1950) ° ′ ″	l °	b °	S_{400} (mJy)	D (kpc)
1943+18	19 43 26	+18 28	55.6	− 3.0	3	7.30
1943− 29	19 43 44	− 29 21 02	11.1	− 24.1	10	1.60
1944+22	19 44 16.159	+22 37 34.78	59.3	− 1.1	3	3.90
1944+17	19 44 38.754	+17 58 15.35	55.3	− 3.5	60	0.43
1946− 25	19 46 24	− 25 32 17	15.2	− 23.4	7	0.79
1946+35	19 46 33.950	+35 32 38.29	70.7	+5.0	120	4.60
1949+14	19 49 43	+14 03	52.5	− 6.5	6	2.00
1951+32	19 51 02.47	+32 44 50.2	68.8	+2.8		1.30
1952+29	19 52 21.848	+29 15 22.18	65.9	+0.8	20	0.20
1953+29	19 53 26.731	+29 00 43.73	65.8	+0.4		2.70
1953+50	19 53 57.35	+50 51 53.6	84.8	+11.6	35	1.10
1957+20	19 57 24.993	+20 39 59.82	59.2	− 4.6		0.83
2000+32	20 00 16	+32 08	69.3	+0.8		3.70
2000+40	20 00 59.94	+40 42 26.6	76.6	+5.3	22	4.60
2002+31	20 02 53.701	+31 28 34.48	69.0	+0.0	14	8.00
2003− 08	20 03 34.3	− 08 15 36	34.1	− 20.3	10	0.90
2011+38	20 11 21.56	+38 36 38.0	75.9	+2.5	5	8.60
2016+28	20 16 00.169	+28 30 30.18	68.1	− 4.0	150	1.30
2020+28	20 20 33.298	+28 44 43.16	68.9	− 4.7	250	1.30
2021+51	20 21 25.3	+51 45 07	87.9	+8.4	60	0.68
2022+50	20 22 14.075	+50 27 49.34	86.9	+7.5	6.8	1.00
2025+21	20 25 03	+21 35	63.5	− 9.6	3	3.50
2027+37	20 27 31.21	+37 34 05.3	76.9	− 0.7	10	5.40
2028+22	20 28 28.544	+22 18 13.15	64.6	− 9.8	5	2.60
2034+19	20 34 59.28	+19 32 23.3	63.2	− 12.7	2	1.20
2035+36	20 35 31.80	+36 10 51	76.7	− 2.8	13	2.90
2036+53	20 36 38.7	+53 08 37	90.4	+7.3	4.0	6.20
2043− 04	20 43 22.44	− 04 32 25.0	42.7	− 27.4	18	1.30
2044+15	20 44 19.60	+15 29 30.2	61.1	− 16.8	12	1.40
2045+56	20 45 30.08	+56 57 32.9	94.2	+8.6	4.2	3.70
2045− 16	20 45 46.99	− 16 27 53	30.5	− 33.1	130	0.38
2048− 72	20 48 41.43	− 72 12 02.6	321.9	− 35.0	29	0.64
2053+21	20 53 24.98	+21 57 56.1	67.8	− 14.7	6	1.30
2053+36	20 53 33.24	+36 18 49.6	79.1	− 5.6	20	3.40
2106+44	21 06 31.73	+44 29 37.1	86.9	− 2.0	38	4.60

Pulsar Parameters*

PSR	Period (seconds)	$\dot{P}$ (10^{-15})	Epoch (JD - 2400000)	W_E (ms)	W_{50} (ms)
1943+18	1.068707				35
1943− 29	0.959447364	1.7	44326.48	13	13
1944+22	1.33444985809	0.889	42854.5	30	22
1944+17	0.44061846173	0.02404	41500.5	21	18
1946− 25	0.957615458	3.4	43956.04	9	9
1946+35	0.717306765257	7.05212	42220.5	24	20
1949+14	0.275044				
1951+32	0.03952975206	5.92	46993.66		3
1952+29	0.426676785594	0.00201	42433.5	16	11
1953+29	0.006133166488729	0.00003	46113.1435	0.85	
1953+50	0.51893741261	1.366	43889.727	7.5	6.5
1957+20	0.001607401683650	0.00002	47402.0730		
2000+32	0.6967314		46565.0		
2000+40	0.90506625212	1.744	46105.16		
2002+31	2.111216952302	74.5706	42303.5	19	15
2003− 08	0.5808713142	0.040	43984.47	28	16
2011+38	0.23019082737	8.8548	46075.26		
2016+28	0.557953407280	0.14936	40688.5	14	14
2020+28	0.343400791509	1.89549	41347.69	6.0	13
2021+51	0.52919532782	3.0518	40624.67	11.5	6.7
2022+50	0.372618228135	2.5148	46105.18		
2025+21	0.3981726			17	11
2027+37	1.2168007731	12.299	46071.28		
2028+22	0.63051210273	0.8838	42840.5	22	22
2034+19	2.0743770781	2.04	46105.19		
2035+36	0.6187136536	4.542	46259.78		
2036+53	1.424567909	1.06	46383.44		
2043− 04	1.54693750023	1.476	43890.75	20	20
2044+15	1.1382856067	0.185	43889.754	19	13
2045+56	0.47672876154	11.125	46183.98		
2045− 16	1.96156687985	10.9610	40694.51	42	78
2048− 72	0.34133616161	0.196	43557.1067	29	33
2053+21	0.81518067999	1.339	46075.29		
2053+36	0.22150803351	0.3648	43890.764	13	13
2106+44	0.41487049185	0.0863	42693.056	24	24

*Adapted from Taylor (1989).

Pulsar Parameters*

PSR	R.A.(1950) h m s	Dec.(1950) ° ′ ″	l °	b °	S_{400} (mJy)	D (kpc)
2110+27	21 10 54.214	+27 41 37.99	75.0	- 14.0		0.82
2111+46	21 11 37.77	+46 31 42.3	89.0	- 1.3	190	4.30
2113+14	21 13 51.08	+14 01 47.7	64.5	- 23.4	10	2.10
2122+13	21 22 23.321	+13 54 22.0	65.8	- 25.1	4	1.10
2123- 67	21 23 19.62	- 67 01 30.8	326.4	- 39.8	7	1.30
2127+11	21 27 33.1	11 56 49	65.0	- 27.3		2.20
2148+63	21 48 36.79	+63 15 40.0	104.3	+7.4	25	5.00
2148+52	21 48 51.16	+52 33 45.13	97.5	- 0.9	7.9	4.30
2151- 56	21 51 34.0	- 56 56 10	337.0	- 47.1	2.1	0.50
2152- 31	21 52 18.43	- 31 33 08.6	15.8	- 51.6	11	0.52
2154+40	21 54 57.21	+40 03 26.1	90.5	- 11.3	39	2.60
2210+29	22 10 06.83	+29 18 14.5	86.1	- 21.7	3.6	2.80
2217+47	22 17 45.882	+47 39 48.18	98.4	- 7.6	63	1.50
2224+65	22 24 17.37	+65 20 14.5	108.6	+6.8	20	1.10
2227+61	22 27 56.74	+61 50 12.0	107.2	+3.6	20	4.50
2241+69	22 41 22.99	+69 35 07.3	112.2	+9.7	3	1.40
2255+58	22 55 54.22	+58 53 10.4	108.8	- 0.6	60	4.40
2303+30	23 03 34.098	+30 43 48.61	97.7	- 26.7	25	1.90
2303+46	23 03 39.17	+46 51 31.8	104.9	- 12.0		2.30
2306+55	23 06 02.50	+55 31 19.8	108.7	- 4.2	20	1.50
2310+42	23 10 47.65	+42 36 52.5	104.4	- 16.4	40	0.56
2315+21	23 15 29.02	+21 33 26	95.8	- 36.1	12	0.75
2319+60	23 19 41.40	+60 08 02.3	112.1	- 0.6	70	2.80
2321- 61	23 21 33.7	- 61 10 33	320.4	- 53.2	4	0.58
2323+63	23 22 00.2	+63 00 28	113.3	+2.1	10	7.40
2324+60	23 24 26.8	+60 55 53	112.9	- 0.0	30	3.20
2327- 20	23 27 49.72	- 20 22 04.1	49.4	- 70.2	19	0.29
2334+61	23 34 45.03	+61 34 25.3	114.3	+0.2	9.3	1.50
2351+61	23 51 34.8	+61 39 06.6	116.2	- 0.2	25	2.50

Pulsar Parameters*

PSR	Period (seconds)	$\dot{P}$ (10^{-15})	Epoch (JD - 2400000)	W_E (ms)	W_{50} (ms)
2110+27	1.20285114987	2.6226	46075.29		
2111+46	1.01468444504	0.7195	41005.68	16	12
2113+14	0.44015295481	0.290	43985.516	8.5	8.5
2122+13	0.69405322696	0.7691	46075.30		
2123− 67	0.32577128742	0.226	43557.1151	18	16
2127+11	0.110664734		47159		
2148+63	0.38014034472	0.1681	43890.8	16	18
2148+52	0.332202511939	10.1090	46075.32		
2151− 56	1.373654387	4.23	43557.1304	50	50
2152− 31	1.0300015591	1.235	43557.1471	20	20
2154+40	1.5252633491	3.417	41532.26	58	52
2210+29	1.00459237450	0.4948	46075.33		
2217+47	0.538467394548	2.76421	40623.76	8.5	7.5
2224+65	0.68253370807	9.671	43888.831	28	78
2227+61	0.443053826812	2.2532	46072.36		35
2241+69	1.6644992795	4.821	46075.36		
2255+58	0.36824365392	5.7501	42629.13	19	17
2303+30	1.57588474427	2.89567	42340.5	17	17
2303+46	1.06637107153	0.5693	46107.2919		
2306+55	0.47506759150	0.202	43890.853	20	25
2310+42	0.34943363975	0.1155	43890.861	8.5	9.2
2315+21	1.4446526747	1.05	43986.599	20	20
2319+60	2.2564837049	7.037	41535.32	105	120
2321− 61	2.347485196	2.60	43557.1967	55	65
2323+63	1.4363084791	2.89	43956.49	130	140
2324+60	0.2336517827	0.309	43804.114	15	14
2327− 20	1.6436196560	4.634	43556.3157	19	20
2334+61	0.495239875085	191.9098	46072.40		
2351+61	0.9447768664	16.226	43890.886	20	30

*Adapted from Taylor (1989).

22.3 Catalogue of Pulsars - Part II

Pulsar Parameters*

PSR	DM (cm^{-3} pc)	RM (m^{-2} rad)	z (kpc)	Log L_r (erg sec^{-1})	Log Age (years)
0011+47	30		- 0.27	27.312	7.545
0021- 72A	65		- 1.70		
0021- 72B	65		- 1.70		
0031- 07	10.89	+9.8	- 0.37	26.380	7.563
0037+56	90.6		- 0.33		6.789
0042- 73	101		- 2.70		
0045+33	41		- 0.76		6.913
0052+51	43		- 0.31		6.546
0053+47	18		- 0.15		6.322
0059+65	67		+0.10	27.334	6.651
0105+65	30.15	- 29	+0.05	26.479	6.189
0105+68	59.5		+0.23		8.540
0114+58	48.5		- 0.09		5.439
0136+57	73.7	- 90	- 0.18	28.012	5.606
0138+59	34.80	- 48	- 0.11	28.235	7.696
0144+59	39.3		- 0.05		7.083
0148- 06	25.1	+2	- 0.88	27.567	7.717
0149- 16	11.94	+15	- 0.41	25.897	7.006
0153+39					8.203
0154+61	25.7	- 29	+0.00	25.577	5.295
0203- 40	12.9		- 0.44	25.661	6.921
0226+70	47		+0.27		6.873
0254- 53	15.9		- 0.48	26.140	8.358
0301+19	15.69	- 8.3	- 0.31	26.627	7.229
0320+39	25.8	+58	- 0.22	26.655	7.830
0329+54	26.776	- 63.7	- 0.05	28.858	6.742
0331+45	46.2		- 0.23		8.764
0339+53	69		- 0.05		6.359
0353+52	102.5		- 0.03		6.817
0355+54	57.03	+79	+0.02	27.755	5.751
0402+61	65.7	+9	+0.30	27.442	6.228
0403- 76	21.7		- 0.46	26.754	6.749
0410+69	27.2		+0.22		7.908
0447- 12	35.7	+13	- 0.75	27.168	7.829
0450+55	14.3	+10	+0.05	26.289	6.360

Pulsar Parameters*

PSR	DM (cm^{-3} pc)	RM (m^{-2} rad)	z (kpc)	Log L_r (erg sec^{-1})	Log Age (years)
0450− 18	39.93	+13.8	− 0.88	27.907	6.180
0458+46	41.0	− 43	+0.06	26.776	6.258
0523+11	78.6	+29	− 0.73	28.033	7.898
0525+21	50.955	− 39.6	− 0.24	28.331	6.171
0529− 66	125.		− 2.70		
0531+21	56.791	− 42.3	− 0.20	30.200	3.098
0538− 75	18.33		− 0.33	27.352	7.539
0540+23	77.58	+8.7	− 0.15	27.781	5.402
0540− 69					3.220
0559− 05	81.2	+67	− 0.76	27.547	6.681
0559− 57	30		− 0.55	25.914	7.110
0609+37	26.7		+0.14		7.904
0611+22	96.70	+67.0	+0.13	27.790	4.949
0621− 04	72.0	+42	− 0.38	27.669	7.289
0626+24	84.4	+82	+0.35	27.669	6.579
0628− 28	34.36	+46.19	− 0.36	27.946	6.443
0643+80	32.5	− 32	+0.56	27.043	6.704
0655+64	8.73	− 7	+0.12	26.049	9.661
0656+14	14	+22	+0.06	25.870	5.045
0727− 18	60.9	+53	− 0.01	26.815	5.630
0736− 40	160.8	+13.5	− 0.40	29.035	6.566
0740− 28	73.77	+150.43	− 0.06	28.145	5.196
0743− 53	122.3		− 0.60	28.322	6.096
0751+32	39.4	− 7	+0.69	26.460	7.328
0756− 15	63.7	+55	+0.29	26.627	6.825
0808− 47	228.3		− 0.78	29.081	6.449
0809+74	5.757	− 11.7	+0.09	25.759	8.087
0818− 13	40.99	− 1.2	+0.32	27.655	6.969
0818− 41	111		− 0.03	27.997	8.505
0820+02	23.6	+13	+0.31	26.690	8.120
0823+26	19.4634	+5.9	+0.37	26.746	6.688
0826− 34	47.3	+59	+0.02	27.177	7.467
0833− 45	69.08	+38.17	− 0.02	28.580	4.056
0834+06	12.8550	+23.6	+0.19	26.407	6.472
0835− 41	147.6	+135.8	− 0.01	28.175	6.526

*Adapted from Taylor (1989).

Pulsar Parameters*

PSR	DM (cm^{-3} pc)	RM (m^{-2} rad)	z (kpc)	Log L_r (erg sec^{-1})	Log Age (years)
0839− 53	156		− 0.38	27.998	6.840
0840− 48	197.0		− 0.28	27.394	6.031
0841+80	34		+0.68		7.759
0844− 35	91.9	+144	+0.05	25.975	7.052
0853− 33	87.7	+165	+0.08	25.805	6.490
0855− 61	95		− 0.64	27.594	6.959
0901− 63	76		− 0.54	27.056	7.990
0903− 42	146.3		+0.08	26.886	6.909
0904− 74	51.1		− 0.59	27.286	7.274
0905− 51	104.0		− 0.05	27.361	6.341
0906− 17	15.7	− 36	+0.18	25.850	6.977
0906− 49	192		− 0.06		5.048
0909− 71	54.3		− 0.56	26.653	7.812
0917+63	12.8		+0.30		6.838
0919+06	27.25	+32	+0.61	27.069	5.696
0922− 52	152.9		− 0.09	27.320	5.523
0923− 58	60		− 0.10	27.195	6.385
0932− 52	99.4		− 0.01	26.680	6.692
0940− 55	180.2	− 61.9	− 0.19	28.545	5.665
0940+16	20	+53	+0.53	26.853	7.282
0941− 56	160.8		− 0.25	27.801	5.509
0942− 13	12.6	− 7	+0.20	25.898	8.291
0943+10	15.35	+13.3	+0.38	26.231	6.693
0950− 38	167		+0.84	27.631	7.574
0950+08	2.969	+1.35	+0.06	26.511	7.243
0953− 52	156.9		+0.07	27.755	6.589
0957− 47	92.7		+0.12	27.651	8.112
0959− 54	130.6		+0.00	24.578	5.644
1001− 47	98.1		+0.18	26.835	5.343
1010− 23	26.5	+52	+0.43	26.254	7.455
1014− 53	66.8		+0.08	26.494	6.801
1015− 56	439.1		+0.08	29.163	6.406
1016− 16	48.0		1.00		7.214
1030− 58	419		− 0.26	29.688	6.389
1039− 19	32.1	− 16	+0.67	27.176	7.300

Pulsar Parameters*

PSR	DM (cm^{-3} pc)	RM (m^{-2} rad)	z (kpc)	Log L_r (erg sec^{-1})	Log Age (years)
1039−55	306.4		+0.62	28.799	6.440
1044−57	240.2		+0.09	28.694	6.709
1054−62	323.4		−0.31	29.200	6.273
1055−52	30.1	+47.2	+0.11	28.553	5.729
1056−78	51.7		−0.57	26.724	7.206
1056−57	107.9		+0.11	27.626	6.641
1105−59	105		+0.03	27.265	7.849
1110−65	249.3		−0.85	29.043	6.808
1110−69	148.4		−0.78	28.046	6.660
1112+50	9.16	+3.2	+0.28	25.656	7.022
1114−41	41.4		+0.47	27.046	6.275
1118−79	27.4		−0.29	26.176	6.993
1119−54	204.7		+0.78	28.665	6.487
1133+16	4.8479	+3.9	+0.14	26.348	6.703
1133−55	85.5		+0.30	28.041	5.846
1143−60	112.8		+0.08	28.056	6.383
1154−62	325.2	+508.2	−0.02	29.551	6.208
1159−58	145.8		+0.34	28.293	6.528
1221−63	96.9	−3.6	−0.07	27.985	5.840
1222−63	416.2		−0.36	29.127	6.847
1232−55	100		+0.46	27.662	
1236−68	96		−0.33	27.185	6.239
1237+25	9.296	−0.33	+0.33	26.872	7.358
1237−41	44.1		+0.59	26.325	6.669
1240−64	297.4	+157.8	−0.32	29.682	6.136
1254−10	29.0		+0.87		7.431
1256−67	95		−0.27	27.054	6.940
1302−64	506		−0.70	29.927	6.351
1309−53	133		+0.72	28.260	7.892
1309−12	35.4		+1.00		7.673
1309−55	134.1		+0.62	28.149	6.373
1317−53	97.6		+0.52	28.037	5.680
1322−66	211		−0.56	29.292	6.210
1322+83	13.7		+0.26		7.274
1323−63	505		−0.49	29.556	6.607

*Adapted from Taylor (1989).

Pulsar Parameters*

PSR	DM (cm^{-3} pc)	RM (m^{-2} rad)	z (kpc)	Log L_r (erg sec^{-1})	Log Age (years)
1323- 58	283		+0.61	30.268	6.373
1323- 62	318.4		+0.03	29.315	5.648
1325- 43	42.0		+0.48	27.215	6.447
1325- 49	117.8		+0.97	27.506	7.584
1336- 64	77.3		- 0.10	27.276	6.075
1338- 62	880.		- 0.03		
1352- 51	112.1		+0.67	27.621	6.560
1353- 62	434		- 0.12		
1356- 60	295.0		+0.17	30.369	5.503
1358- 63	98.0		- 0.11	27.675	5.898
1359- 51	39		+0.22	26.333	
1417- 54	129.6		+0.49	27.751	7.796
1424- 55	82.4		+0.22	27.969	6.636
1426- 66	65.3	- 12.0	- 0.20	28.170	6.652
1436- 63	124.2		- 0.24	28.047	6.813
1449- 64	71.07	- 22.3	- 0.17	28.535	6.015
1451- 68	8.60	- 2.0	- 0.04	27.028	7.626
1454- 51	37		+0.13	25.694	6.719
1503- 51	61.0		+0.19	26.572	6.320
1503- 66	129.8		- 0.57	27.984	6.688
1504- 43	48.7		+0.36	27.239	6.452
1507- 44	84		+0.60	27.885	7.389
1508+55	19.599	+0.8	+0.58	27.028	6.367
1509- 58	235.		- 0.14		3.189
1510- 48	51.5		+0.23	26.882	6.892
1523- 55	358		+0.10	28.833	6.168
1524- 39	49		+0.42	26.844	6.303
1530+27	14.606	+54	+0.43	25.892	7.346
1530- 53	24.82	+17.5	+0.02	26.745	7.182
1540- 06	18.5	+4	+0.39	26.458	7.105
1541- 52	35.2		+0.02	27.003	7.668
1541+09	34.99	+21	+0.94	28.132	7.440
1550- 54	210		- 0.09	28.366	6.039
1552- 31	73.3	- 49	+0.75	27.687	8.215
1552- 23	51.4	- 22	+0.71	26.903	7.081

Pulsar Parameters*

PSR	DM (cm^{-3} pc)	RM (m^{-2} rad)	z (kpc)	Log L_r (erg sec^{-1})	Log Age (years)
1555– 55	211		– 0.22	28.149	5.870
1556– 44	58.8	– 2.6	+0.21	28.038	6.601
1556– 57	176.9		– 0.37	28.773	6.161
1557– 50	270	+119	+0.22		5.780
1558– 50	169.5	+71.5	+0.06	27.907	5.294
1600– 27	46.2	– 5	+0.53	27.196	6.626
1600– 49	140.8		+0.18	28.454	6.710
1601– 52	32		– 0.01	27.407	7.610
1604– 00	10.72	+6.5	+0.21	26.267	7.339
1607– 13	48		+0.80		7.846
1609– 47	161.3		+0.25	28.258	6.982
1612+07	21.3	+40	+0.48	25.925	6.909
1612– 29	40	– 30	+0.36	25.964	7.196
1620– 42	295		+0.73	29.381	6.752
1620– 09	70.2	– 85	+1.20	26.685	6.829
1620– 26	62.87		+0.61		8.341
1630– 59	134.9		– 0.67	27.638	6.787
1633+24	23.8	+31	+0.56		7.814
1641– 45	475	– 611	– 0.02	29.350	5.554
1641– 68	43		– 0.39	27.495	7.221
1642– 03	35.665	+15.84	+0.57	27.752	6.538
1647– 528	164		– 0.52	28.111	6.834
1647– 52	179.1		– 0.52	28.440	6.745
1648– 42	525		+0.22	30.987	
1648– 17	30.5	+4	+0.30	26.475	6.705
1649– 23	66.7	– 24	+0.50	27.034	
1657– 13	59.0		+0.64		7.214
1659– 60	54		– 0.37	28.420	6.727
1700– 32	105.4	– 21.7	+0.31	28.343	7.438
1700– 18	48.3	– 29	+0.39	26.918	6.867
1701– 75	37		– 0.46	26.038	7.002
1702– 19	23.1	– 9	+0.17	26.583	6.059
1706– 16	24.88	– 1.3	+0.19	26.894	6.210
1707– 53	106.1		– 0.53	27.396	5.964
1709– 15	58		+0.49		7.096

*Adapted from Taylor (1989).

Pulsar Parameters*

PSR	DM (cm^{-3} pc)	RM (m^{-2} rad)	z (kpc)	Log L_r (erg sec^{-1})	Log Age (years)
1717-29	42.8	+21	+0.09	27.425	7.089
1717-16	42		+0.28		6.630
1718-02	65.5	+17	+0.77	28.296	7.940
1718-32	125.8	+90	+0.16	28.302	7.033
1719-37	99.7		-0.02	27.924	5.539
1726-00	38		+0.42		6.736
1727-47	121.9	-429.1	-0.55	28.959	4.905
1729-41	195		-0.48	28.143	5.889
1730-22	41.5	-9	+0.13	27.032	
1732-02	64		+0.63		7.501
1732-07	73.9	+38	+0.60	27.456	
1735-32	60		-0.01		
1736-31	620		-0.04		
1736-29	139		+0.07		5.813
1737-30	149		+0.01		4.314
1737+13	48.9	+73	+0.66	28.019	6.942
1737-39	158.5		-0.42	28.616	6.652
1738-08	75.4	+124	+0.51	28.105	7.154
1740-03	35		+0.28	26.411	6.347
1740-13	115		+0.55		7.126
1742-30	88.8	+101	-0.04	27.569	5.736
1745-12	100.0	+67	+0.45	27.750	6.712
1745-56	57.5		-0.50	26.295	6.998
1747-46	21.7	+18.1	-0.12	26.919	6.958
1749-28	50.88	+96.0	-0.02	28.281	6.039
1750-24	800		+0.17		
1753+52	35.3		+0.65		7.383
1753-24	370		+0.01		
1754-24	178.2	+153	+0.00		5.455
1756-22	177.3	+6	+0.06	28.196	5.830
1757-23	280		-0.01		
1758-23	140.		-0.02		
1758-24	250.		-0.11		
1758-03	117.6		+0.65		6.644
1800-21	230		+0.01		4.198

Pulsar Parameters*

PSR	DM (cm^{-3} pc)	RM (m^{-2} rad)	z (kpc)	Log L_r (erg sec^{-1})	Log Age (years)
1802+03	79.4		+0.57		6.540
1804– 12	120		+0.28		
1804– 27	313.3	– 47	– 0.51	28.831	6.030
1804– 20	660		– 0.09		
1804– 08	112.8	+166	+0.36	28.583	7.956
1806– 21	380		– 0.11		
1806– 53	44.5		– 0.43	27.329	7.033
1809– 173	210		+0.05		
1809– 175	720		+0.10		
1810+02	101.6		+0.59		6.543
1811+40	41.8	+47	+0.63	26.687	6.762
1813– 17	490		– 0.07		
1813– 26	129.2	+90	– 0.36	28.889	
1813– 36	94.3		– 0.52	27.884	6.477
1814– 23	240		– 0.41		
1815– 14	595		+0.18		
1817– 13	750		+0.17		
1818– 04	84.38	+69.2	+0.12	27.781	6.175
1819– 14	560		– 0.03		
1819– 22	123.2	+140	– 0.30	28.198	7.695
1820– 31	50.5	+95	– 0.24	27.085	6.188
1820– 11	430		+0.20		
1821– 19	224.3	– 303	– 0.37	29.550	5.758
1821+05	67.2	+145	+0.35	27.317	7.724
1821– 24	120.0		– 0.38		7.494
1821– 11	620		+0.19		
1822– 14	360		– 0.17		
1822+00	54.4		+0.18		7.150
1822– 09	19.9	+65.2	+0.01	26.247	5.367
1823– 11	460		+0.06		
1823– 13	219		– 0.06		4.498
1824– 09	410		+0.16		
1826– 17	217.8	+317	– 0.40	29.729	5.940
1828– 10	180		– 0.04		
1828– 60	35		– 0.45	26.493	8.045

*Adapted from Taylor (1989).

Pulsar Parameters*

PSR	DM (cm^{-3} pc)	RM (m^{-2} rad)	z (kpc)	Log L_r (erg sec^{-1})	Log Age (years)
1829−10	440		−0.09		
1829−08	286		+0.03		
1830−08	402		+0.01		5.176
1831−03	235.8	−41	+0.28	29.312	5.419
1831−00	88.3		+0.18		8.761
1831−04	78.8	+100	+0.07	28.809	7.368
1832−06	410		+0.09		
1834−10	318	+1000	−0.22	29.688	5.879
1834−06	310		+0.00		
1834−04	200		+0.10		
1838−04	304		+0.04		
1839+09	49.1	+3	+0.17	27.087	6.743
1839−04	200		+0.01		7.757
1839+56	26.2	−3	+0.38	27.207	7.188
1841−05	400		−0.17		
1841−04	100		−0.02		
1842−02	410		+0.03		
1842−04	280		−0.09		
1842+14	41.2	+121	+0.19	27.176	6.504
1844−04	142.6	+117	−0.06	28.826	5.261
1845−19	18.26	+7	−0.08	25.873	6.467
1845−01	159.1	+580	+0.00	28.509	6.303
1846−06	147.6		−0.19	28.084	5.702
1848+13	59.0		+0.21		6.564
1848+04	112.0		+0.12		9.450
1848+12	71		+0.24		6.219
1849+00	840		+0.00		
1851−79	39		−0.65	26.623	7.037
1851−14	130.1	+103	−0.55	27.548	6.639
1852+10	250		+0.62		
1854+00	90		−0.03		
1855+09	13.2952		+0.02		9.628
1855+02	480		−0.09		
1857−26	37.6	−7.3	−0.30	28.262	7.783
1859+03	402.9	−237.4	−0.12	29.832	6.142

Pulsar Parameters*

PSR	DM (cm^{-3} pc)	RM (m^{-2} rad)	z (kpc)	Log L_r (erg sec^{-1})	Log Age (years)
1859+01	102.1		- 0.07		
1859+07	240		+0.12		
1900+05	179.7	- 113	+0.02	28.151	5.962
1900+06	530		+0.08		
1900+01	246.4	+72.3	- 0.17	28.426	6.457
1900- 06	195.7	+203	- 0.64	28.694	6.304
1901+10	140		+0.13		
1902- 01	225		- 0.45		6.523
1903+07	380		+0.02		
1904+06	465		- 0.06		
1904+12	260		+0.35		
1905+39	30.1	+7	+0.25	27.102	7.568
1906+09	250		+0.04	28.008	8.128
1907+00	112.9	- 40	- 0.25	26.919	6.466
1907+02	172.1		- 0.26	27.883	6.754
1907+10	148.4	+540	+0.07	28.279	6.232
1907+03	78.8	- 127	- 0.09	28.398	6.911
1907- 03	205.5	+152	- 0.67	28.499	6.563
1907+12	274.4		+0.24	28.000	6.443
1910+10	140		+0.01		
1910+20	88.0	+148	+0.25	26.624	6.541
1911+13	144.4	+435	+0.12	27.938	7.011
1911- 04	89.43	+12	- 0.38	28.110	6.507
1911+09	155		- 0.04		
1911+11	80		+0.01	26.535	7.162
1913+167	65.6	+161	+0.08	26.772	7.798
1913+10	246.1	+431	- 0.07	28.875	5.624
1913+16	167		+0.19	28.947	8.034
1914+09	61.40	+100	- 0.03	27.207	6.231
1914+13	236.80	+280	+0.05	28.688	6.091
1915+13	94.8	+233	+0.03	27.601	5.632
1915+22	120		+0.32		
1916+14	30		+0.01	25.844	4.947
1917+00	90.7	+120	- 0.33	27.490	6.419
1918+26	27.3		+0.08		8.550

*Adapted from Taylor (1989).

Pulsar Parameters*

PSR	DM (cm^{-3} pc)	RM (m^{-2} rad)	z (kpc)	Log L_r (erg sec^{-1})	Log Age (years)
1918+19	154.4	+160	+0.23	28.806	7.162
1919+14	91.94		+0.00	27.691	6.242
1919+21	12.4309	− 16.5	+0.02	26.813	7.196
1919+20	70		+0.09		
1920+20	203		+0.31		
1920+21	217.1	+282	+0.38	28.464	6.319
1921+17	135		+0.06		
1922+20	213		+0.28		6.256
1923+04	101.8	0	− 0.34	27.285	6.839
1924+19	420		+0.34		
1924+16	177.00	+320	+0.00	27.755	5.708
1924+14	205		− 0.11	27.882	7.974
1925+18	250		+0.12		
1925+22	180		+0.28	27.846	7.468
1925+188	90		+0.04		
1926+18	109		+0.02		
1927+13	207.3		− 0.27	27.765	6.517
1929+15	120		− 0.09		
1929+10	3.176	− 6.1	+0.00	25.373	6.492
1929+20	211.025	+10	+0.06	29.032	6.007
1930+22	211.26	+173	+0.18	28.706	4.598
1930+13	165		− 0.29		
1931+24	89		+0.11		
1933+17	210		− 0.14		
1933+16	158.53	− 1.9	− 0.22	29.391	5.976
1933+15	165		− 0.23	27.173	6.579
1935+25	62		+0.07		6.311
1937+24	100		+0.07		
1937+21	71.044		− 0.03		8.392
1937− 26	50.0	− 26	− 0.67	26.827	6.823
1939+17	175		− 0.26		
1940− 12	29.1	− 10	− 0.30	26.302	6.968
1941− 17	55.7	− 40	− 0.67	26.877	7.134
1942+17	160		− 0.28		
1942− 00	58.1	− 45	− 0.44	27.975	7.490

Pulsar Parameters*

PSR	DM (cm^{-3} pc)	RM (m^{-2} rad)	z (kpc)	Log L_r (erg sec^{-1})	Log Age (years)
1943+18	215		- 0.38		
1943- 29	43.8		- 0.65	26.606	6.951
1944+22	140		- 0.07	26.954	7.376
1944+17	16.3	- 28.0	- 0.03	26.733	8.463
1946- 25	22.8		- 0.31	25.688	6.650
1946+35	129.05	+116	+0.40	28.919	6.207
1949+14	60		- 0.23		
1951+32	45		+0.06		5.024
1952+29	7.91	- 18	+0.00	25.381	9.527
1953+29	104.58		+0.02		9.510
1953+50	31.8	- 22	+0.21	26.773	6.780
1957+20	29.128		- 0.07		
2000+32	135		+0.05		
2000+40	128		+0.43		6.915
2002+31	234.67	+30	+0.00	27.876	5.652
2003- 08	26	- 52	- 0.31	26.424	8.362
2011+38	238.6		+0.37		5.615
2016+28	14.176	- 34.6	- 0.09	27.875	7.772
2020+28	24.62	- 74.7	- 0.11	28.276	6.458
2021+51	22.580	- 6.5	+0.10	26.623	6.439
2022+50	32.4		+0.13		6.371
2025+21	96.75		- 0.58	27.080	
2027+37	189		- 0.07		6.195
2028+22	71.83		- 0.44	27.132	7.053
2034+19	36		- 0.27		7.207
2035+36	92		- 0.14		6.334
2036+53	158		+0.78		7.328
2043- 04	35.9	- 1	- 0.61	26.677	7.220
2044+15	39.5	- 101	- 0.41	26.509	7.989
2045+56	99		+0.56		5.832
2045- 16	11.51	- 10.00	- 0.21	26.955	6.453
2048- 72	18.1		- 0.37	27.138	7.441
2053+21	35.8		- 0.32		6.984
2053+36	97.5	- 68	- 0.34	28.217	6.983
2106+44	139.6	- 146	- 0.16	28.732	7.882

*Adapted from Taylor (1989).

Pulsar Parameters*

PSR	DM (cm^{-3} pc)	RM (m^{-2} rad)	z (kpc)	Log L_r (erg sec^{-1})	Log Age (years)
2110+27	24.7	− 65	− 0.20		6.861
2111+46	141.50	− 223.7	− 0.09	28.693	7.349
2113+14	56.3	− 25	− 0.84	27.005	7.381
2122+13	30.0		− 0.46		7.155
2123− 67	35		− 0.84	26.840	7.359
2127+11	58		− 1.00		
2148+63	128	− 160	+0.65	28.547	7.554
2148+52	145.7		− 0.07		5.717
2151− 56	14		− 0.37	25.351	6.711
2152− 31	14.4	+21	− 0.41	25.831	7.121
2154+40	71.0	− 44.0	− 0.52	28.040	6.850
2210+29	73.3		− 1.00		7.507
2217+47	43.54	− 35.3	− 0.19	27.341	6.489
2224+65	35.3	− 21	+0.13	27.533	6.049
2227+61	122.6		+0.28		6.494
2241+69	40.7		+0.24		6.738
2255+58	151.1	− 322	− 0.04	28.795	6.006
2303+30	49.9	− 84	− 0.86	27.068	6.936
2303+46	60.9		− 0.47		7.472
2306+55	47.0	− 34	− 0.11	27.420	7.571
2310+42	17.3	+7	− 0.16	26.593	7.681
2315+21	20.5	− 37	− 0.44	26.043	7.338
2319+60	93.8	− 230	− 0.03	28.537	6.706
2321− 61	16		− 0.47	25.649	7.155
2323+63	195	− 102	+0.26	28.795	6.896
2324+60	123	− 221	+0.00	28.335	7.078
2327− 20	8.39	+9.5	− 0.27	25.365	6.750
2334+61	57.6		+0.01		4.612
2351+61	95	− 77	− 0.01	27.769	5.965

*Adapted from Taylor (1989).

23 Candidate Black Holes

23.1 Catalogue of Candidate Black Holes

When a very massive star has used up its thermonuclear fuel, there is no equilibrium state and it is thought to collapse forever into a black hole. This hypothetical object is by itself invisible, for no radiation can escape from its surface, but its presence can be inferred when the invisible black hole is close enough to a large visible companion. Gas is then pulled from the visible star into the black hole; the gas heats up and emits X-rays before disappearing from view. Orbital and spectral analysis of the binary system can lead to estimates of the mass, M_c, of the visible companion, and the mass, M_X, of the invisible object responsible for the X-ray radiation. When the invisible object has a mass greater than three solar masses, or when $M_X \geq 3M_{\odot}$, it is too massive to be a white dwarf or neutron star, and it is in all probability a black hole.

There are four such candidate black holes presently known. The positions and physical parameters for these objects, given in the accompanying tables, are adapted from Bolton (1972) and Gies and Bolton (1986) for Cygnus X-1, McClintock (1986) and McClintock and Remillard (1986) for A 0620-00, Hutchings et al. (1987) for LMC X-1, and Cowley et al. (1983) for LMC X-3. These last two objects are in the Large Magellanic Cloud, LMC, located at a distance of D = 55 *kpc*, while the other two are relatively nearby objects in our Galaxy.

Here we give the name of the X-ray source and its optically visible companion, the right ascension, R.A., and declination, Dec., of the visible star for epoch (1950.0), its galactic latitude and longitude, l and b, apparent visual magnitude, V, and spectral type. These are followed by the maximum X-ray luminosity, L_X max, the source distance, D, the orbital period of the binary system, P, the sinusoidal fit to the optical radial velocity variations, V_o and K, the semi-major axis of the orbit, a sin i, in units of the solar radius, $R_{\odot}$, the optical mass function f(M) $= (M_X \sin i)^3/(M_X + M_c)^2 = P(V_c \sin i)^3/(2\pi G)$, and the two masses M_c and M_X in solar units, $M_{\odot}$.

Candidate Black Holes*

X-Ray Source	Optical Companion	R.A.(1950) h m s	Dec.(1950) ° ′ ″	l °	b °	V mag	Spectral Type	L_X max (erg/sec)
Cygnus X-1	HDE 226868	19 56 28.87	+35 03 55.0	71.3	+ 3.1	8.9	O9.7Iab	2×10^{37}
LMC X-3	WP8	05 38 39.7	− 64 06 34	273.6	−32.1	16.9	B3V	3×10^{38}
LMC X-1	*F.C.*** *of R*148	05 40 05.5	− 64 46 03.6	280.2	−31.5	14	O7	2×10^{38}
A 0620-00	V616 Mon	06 20 11.12	− 00 19 11.0	210.0	− 6.5	18	K5V	2×10^{38}

** *F.C.* means Faint Companion

X-Ray Source	D (kpc)	P (days)	V_0 (km/sec)	K (km/sec)	$a \sin i$ ($R_\odot$)	f(M) ($M_\odot$)	M_c ($M_\odot$)	M_x ($M_\odot$)
Cygnus X-1	2.5	5.5995	−6	76	8.36	0.25	16	≥ 6
LMC X-3	55	1.7049	310	235	7.92	2.3	9	≥ 3
LMC X-1	55	4.2288	221	68	5.77	0.144	20	≥ 4
A 0620-00	1.0	0.3230	−5	457	2.91	3.18	0.7	≥ 3

* Adapted from Bolton (1972), Cowley et al. (1983), Gies and Bolton (1986), Hutchings et al. (1987), McClintock (1986), and McClintock and Remillard (1986).

Part VII
Interacting Binary Systems

24
Cataclysmic Binary Systems

24.1 Basic Data for Cataclysmic Binary Systems

Cataclysmic binary systems consist of a white dwarf star, conventionally called the primary star, and a companion secondary star that is usually a cool main-sequence star (red dwarf) of late spectral type G, K or M. These variable binary star systems have orbital periods lying between 1 and 15 hours. The orbital period and ratio of the masses of the two stars are such that the secondary star fills its Roche lobe, causing gas to flow from the secondary toward the degenerate primary star. In the absence of a magnetic field, this gas forms an accretion disk around the primary star; the gas streams onto the magnetic poles when there is a strong magnetic field (Wade and Ward, 1985).

These interacting binary systems are classified according to the character of their light variations. One class, called novae or classical novae, undergo very luminous outbursts that are visible to the unaided eye or with only modest telescopic aid. They have usually been observed only once. The nova outbursts are thought to be due to the sudden nuclear fusion of hydrogen in the hot dense envelope of the degenerate primary star. Classical novae occur within our Galaxy at the rate of about 73 per year, which makes them about 3,300 times more common than supernovae in our Galaxy (Liller and Mayer, 1987). Each nova event releases about 0.00005 solar masses.

Another class, called dwarf novae, have frequent weaker outbursts with amplitudes between 2 and 6 magnitudes and an interval between outbursts ranging from days to months and even years. The dwarf nova outbursts are attributed to a sudden increase in the brightness of the accretion disk.

The mean absolute visual magnitudes of novae at minimum are around +4.0, while those of dwarf novae are considerably fainter, at +7.5. All novae have orbital periods longer than three hours, while dwarf novae have orbital periods that are both shorter and longer than three hours.

Recurrent novae have been recorded to undergo more than one outburst of smaller amplitude than nova eruptions; the interval between outbursts is 10 to 50 years. Dwarf novae have been divided into three subtypes according to the light curves of their stellar prototype - the U Geminorum, Z Camelopardalis and SU Ursae Majoris subtypes (see Smak, 1984). Nova-like cataclysmic variables include the UX Ursae Majoris stars that resemble novae before or after eruption, or dwarf novae in outburst, but they have not been observed to undergo eruptive outbursts. The magnetic accreting, or AM Herculis, white dwarf stars, have circularly polarized radiation that indicates an intense magnetic field. Some examples of the various types of cataclysmic variables are given in the first accompanying table.

Stellar magnetic fields can be amplified during the collapse to the white dwarf stage, resulting in surface magnetic field strengths of more than ten million Gauss (ten mega-Gauss). Some cataclysmic variables therefore consist of an intensely-magnetized white dwarf primary and a red dwarf secondary that fills it Roche lobe. The magnetic field then controls the flow of the accreting material at large distances from the white dwarf surface, and the accretion disk is partially or completely disrupted.

There are two classes of these magnetic cataclysmic binary systems – the polars, or AM Herculis objects, and the intermediate polars, or DQ Herculis objects. The magnetic field of the polars is strong enough to force the white dwarf to rotate synchronously. The magnetic field therefore always presents the same aspect to the incoming accretion stream, which is funneled onto the white dwarf's surface near a magnetic pole. (In intermediate polars the white dwarf rotation period is shorter than the orbital period, so the spin is not synchronized and a remnant accretion disk may exist.) The next table of polars, adopted from Crooper (1990), gives their names, celestial coordinates, orbital periods, P, the surface magnetic field strengths, B, with a subscript z or c for determination by Zeeman spectroscopy or cyclotron spectral features, respectively, and the distance, D, in parsecs.

The polars AM Her and V834 Cen have been detected at radio wavelengths (Dulk, Bastian and Chanmugan, 1983; Wright et al., 1988), as has a flare from the intermediate polar AE Aquarii (Bastian, Dulk and Chanmugan, 1988). The polar V 1500 Cyg and the intermediate polar GK Per are both classical optical novae (Nova Cyg 1975, Nova Per 1901). The X-ray properties of both polars and intermediate polars have been reviewed by Watson (1987). Many of these objects are listed in the following table of cataclysmic variables with known periods (for example the intermediate polars GK Per, BG CMi, EX Hya, FO Aqr, and AO Psc). Their magnetic field strengths are given in the next table of magnetic fields on accreting white dwarfs, which is followed by a table of X-ray emission from classical novae.

The distances, D, of a few novae can be inferred from the angular rates of expansions and the expansion velocities of their nova shells. These distances are given in the table of nova shells together with estimates for the apparent visual magnitude, m, at maximum, max, the interstellar visual extinction, A, and the absolute visual magnitude, M, at maximum.

Cataclysmic binaries with known orbital periods have been complied by Ritter (1987). Here we subdivide this data into classical novae, dwarf novae, and nova-like variables, each in order of increasing right ascension, R.A., in the epoch 1950; the declination, Dec., for this epoch is also given together with the minimum, min, and maximum, max, apparent visual magnitudes, m, the orbital period, P, the orbital inclination, inc, the spectral classification, Sp, of the secondary, the mass ratio and the respective masses, M, of the primary, 1, and secondary, 2, star in solar masses.

The catalogue of classical novae in our Galaxy is from Hilmar W. Duerbeck's (1987) compilation, as corrected and updated by him in 1989. We did not reproduce the valuable finding charts given in his compilation. The objects in our catalogue are listed in order of increasing right ascension, R.A., with both celestial coordinates for epoch 1950.0. Here the names and dates of discovery are followed by the right ascension, R.A., and declination, Dec., galactic longitude and latitude, l and b, the brightness range, B.R., from observed maximum to minimum magnitude followed by p, v, B, V, j or r denoting the photographic, visual, Johnson B, Johnson V, unfiltered IIIa-J photographic part of the spectrum, or red, the light-curve decay time, t, and the type of nova. For the magnitudes, ([) indicates fainter than, and (]) indicates brighter than. The subscript 3 to the light-curve decay time, t, indicates that it is the time in days for the nova to decrease by 3 magnitudes from maximum brightness. Here we have only listed the classical novae, with designations NA for a fast nova (t less than 100 days); NB for a slow nova (t greater than 100 days) and NC for an extremely slow nova with time scales of decades. The symbol NR denotes a recurrent nova and the designation N denotes a nova whose light curve was too poorly known to designate t. A (:) following a nova designation means that the nova was not confirmed by spectroscopic observations. A (?) denotes only one photographic or a few independent visual observations, and a (??) is a nova based on no more than two visual observations. These somewhat dubious novae have also been marked by (**) after the name.

The value of the maximum absolute visual magnitude, M, can be estimated from $M = -11.75 + 2.5 \log t$, and the distance, D, in parsecs can then be inferred from the relation $M - m = 5 \log D - 5$ (Canterna Schwartz, 1977). Distance estimates for dwarf novae and nova-like systems are not well determined, and distances accurate to 25 to 50% are known only for a handful (Berriman, 1987).

The catalogue of dwarf novae has been compiled from the lists of Bruch, Fischer, and Wilmsen (1987), for the northern hemisphere, and Vogt and Bateson (1982) for the southern hemisphere. Positions accurate to 0.4 arc seconds in both coordinates and minimum, min, and maximum, max, photographic magnitudes, m, are from Bruch et al. (1987) for the northern dwarf novae; the positions and magnitude data for the southern objects are taken from Kholopovu's (1985, 1987) General Catalogue of Variable Stars. In both cases the right ascensions, R.A., and declinations, Dec., are for the epoch 1950.0.

Examples of Cataclysmic Variables*

Object	Orbital period (h)	'Normal' visual (mag)	Visual magnitude at outburst	Spectral class of secondary star	Remarks
Novae					
DQ Herculis	4.6	14.6	1.5	M	Erupted 1934. A slow nova. Deep eclipses every orbital period. Regular 71–s oscillations in brightness.
V1500 Cygni	3.4	>17.0	2.2	?	Erupted 1975. A very fast nova. Fainter than 20th magnitude before outburst.
Dwarf novae					
Z Camelopardalis	7.0	14.5	10.2	K	Standstills near magnitude 11.7.
U Geminorum	4.2	14.5	8.9	M	Eclipses of the bright spot.
VW Hydri	1.8	13.9	9.5	?	Eclipses. Superoutbursts to magnitude 8.5.
Z Chamaeleontis	1.8	15.2	12.4	M	Distinct eclipses of white dwarf and bright spot. Superoutbursts to magnitude 11.9.
Recurrent novae					
T Coronae Borealis	227 days	10.8	3.4	M	Eruptions in 1866 and 1946. The secondary star is a giant star.

Examples of Cataclysmic Variables*

Object	Orbital period (h)	'Normal' visual (mag)	Visual magnitude at outburst	Spectral class of secondary star	Remarks
Ultrashort-period variables					
AM Canum Venaticorum	0.3	14.1			Thought to be two helium-rich white dwarfs orbiting each other. The shortest known binary orbital period.
Other categories					
UX Ursae Majoris	4.7	13.5		K–M?	Deep eclipses. DQ Her-like oscillations sometimes seen. The prototypical 'novalike' star.
AM Herculis	3.1	12.9	15.2	M	Complicated variations of brightness around the orbital cycle. 'High' and 'low' states. Polarised light. The prototypical magnetic variable.
AO Piscium	3.6	13.3		?	A 'DQ Her' star, showing brightness variations at three distinct periods.
VY Sculptoris	?	~13.0	~17.0	?	Prototypical 'anti-dwarf nova'. 'High' states similar to the UX UMa stars, occasional 'low' states similar to quiescent dwarf novae.

*Adapted from Wade and Ward (1985).

Magnetic Fields on Accreting White Dwarfs*

Object	Surface Magnetic Field† (Gauss)	Object	Surface Magnetic Field† (Gauss)	Object	Surface Magnetic Field†† (Gauss)
AE Aqr	$10^5 - 10^6$	EX Hyi	$10^6 - 10^7$	ST LMi	$1 - 5 \times 10^7$
DQ Aqr	$10^5 - 10^6$	SW UMa	$10^6 - 10^7$	VV Pup	$1 - 5 \times 10^7$
TV Col	$10^6 - 10^7$	E1048	$1 - 5 \times 10^7$	E1405	$1 - 5 \times 10^7$
FO Aqr	$10^6 - 10^7$	E2003	$1 - 5 \times 10^7$	MR Ser	$1 - 5 \times 10^7$
BM CMi	$10^6 - 10^7$	H0541	$1 - 5 \times 10^7$	BL Hyi	$1 - 5 \times 10^7$
AO Psc	$10^6 - 10^7$	E1114	$1 - 5 \times 10^7$	AN UMa	$1 - 5 \times 10^7$
V1223 Sgr	$10^6 - 10^7$	EF Eri	$1 - 5 \times 10^7$	AM Her	$1 - 5 \times 10^7$

*Adapted from Schmidt and Liebert (1987).
†Inferred field strength, ††Measured field strength.

X–ray Emission from Classical Novae*

Object	X–ray Intensity (erg cm^{-2} s^{-1})	Assumed Distance† (pc)	X–ray Luminosity (erg s^{-1})
V603 Aql††	7.53×10^{-12}	376	1.28×10^{32}
T Aur	$< 1.84 \times 10^{-13}$	830	$< 1.5 \times 10^{31}$
CP Lac	$< 3.3 \times 10^{-13}$	1340	$< 7.1 \times 10^{31}$
V841 Oph	5.13×10^{-13}	860	4.55×10^{31}
GK Per	4.80×10^{-12}	470	1.27×10^{32}
RR Pic	8.37×10^{-13}	480	2.31×10^{31}
CP Pup	1.62×10^{-12}	700	9.53×10^{31}
V1059 Sag	3.78×10^{-13}	1370	8.51×10^{31}
CK Vul	$< 1.8 \times 10^{-13}$	380	$< 3.1 \times 10^{30}$

*Adapted from Becker and Marshall (1981).
†Distances taken from Mc Laughlin (1945, 1960).
††V603 Aql has a variable X–ray intensity that goes from 7×10^{-12} to 2×10^{-10} erg cm^{-2} s^{-1} and an X–ray luminosity of up to 3×10^{33} erg s^{-1} (Drechsel et al., 1983; Haefner et al., 1988).

Polars – White Dwarf Component With Mega–Gauss Fields*

Name	Other Designation	R.A.(1950.0) h m s	Dec.(1950.0) ° ′ ″	P_{orb} (days)	B_Z (10^6G)	B_C (10^6G)	D (pc)
BL Hyi	H0139–68	01 39 37.5	– 68 08 32	113.6	30		128
EXO023432–5232.3		02 34 32.2	– 52 32 15	114.6			500
EF Eri	2A0311–227	03 12 00.0	– 22 46 49	81.0			$\geq$ 89
EXO032957–2606.9		03 29 57.0	– 26 06 56	228			
EXO033319–2554.2		03 33 20.9	– 25 54 17	126.5		56	250
H0538+608		05 38 15.9	+60 50 03	~ 198		40.8 ± 1.5	
VV Pup		08 12 52.2	– 18 54 02	100.4		31.7	145
1E1048.5+5241		10 48 33.6	+54 20 33	114.5		47 ± 3	
AN UMa		11 01 35.6	+45 19 26	114.8		35.8 ± 1.0	$\geq$ 270
ST LMi	CW1103+254	11 02 58.0	+25 22 42	113.9	19		128
DP Leo	E1114+182	11 14 38.1	+18 14 05	89.8		44.0 ± 1.4	$\geq$ 380
V834 Cen	E1405–451	14 05 58.2	– 45 03 06	101.5	22 ± 2		86
MR Ser	PG1550+191	15 50 33.1	+19 05 18	113.6		24.6 ± 0.6	112
AM Her	3U1809+50	18 14 58.6	+49 50 55	185.6	13		75
QQ Vul	E2003+225	20 03 30.7	+22 31 28	222.5			$\geq$ 400
V1500 Cyg	Nova Cyg 1975	21 09 52.8	+47 56 41	201.0			1200
Grus V1		21 34 45.2	– 43 55 46	108.6			

*Adapted from Crooper (1990).

Nova Shells*

Name	Year	θ ($''$)	$(\Delta\theta/\Delta t)_{exp}$ ($''$ yr^{-1})	V_{exp} (km s^{-1})	A_V (mag)	m_{Vmax} (mag)	M_{Vmax} (mag)	D (pc)	
V603 Aql	1918	(60)	0.95	265	0.2	− 1.1	− 9.15	372	(376)
T Aur	1891	9.5	0.11	655	1.5	+4.2	− 7.9	1318	(830)
V476 Cyg	1920	5.7	0.093	790:	0.6	+2.0	− 9.85	1778	(1590)
V1500 Cyg	1975	(2.6)	0.44	1180	1.6	+1.85	− 10.2	1230	
HR Del	1967	1.8	0.13	520	0.2	+4.8	− 5.05	851	
DQ Her	1934	10.5	0.22	315	0.2	+1.4	− 6.2	302	(230)
V533 Her	1963	1.6	0.09	580	0.2	+3.5	− 7.45	1412	
CP Lac	1936	(11.3)	0.25	295:	0.8	+2.1	− 9.35	1349	(1340)
GK Per	1901	41.5	0.52	1200	0.3	+0.2	− 8.55	490	(470)
RR Pic	1925	11.5	0.20	475	0.2	+1.2	− 7.3	457	(480)
CP Pup	1942	7	0.18	710	0.3	+0.4	− 9.55	851	(700)
FH Ser	1970	2.0	0.18	560	2.8	+4.4	− 7.55	676	

*Adapted from Cohen and Rosenthal (1983) and Lance, Mc Call and Uomoto (1988) for V1500 Cygni. The distances, D, have been computed from the distance modulus $m - M = 5 \log D - 5$ corrected for A_V. Distances in parenthesis are those given by Mc Laughlin (1960).

24.2 Cataclysmic Binaries with Known Periods

Classical Novae*

Name	R.A.(1950) h m s	Dec.(1950) ° ′ ″	m_{min}	m_{max}	Porb (days)	Inc.	Sp	M_1/M_2	M_1 ($M_\odot$)	M_2 ($M_\odot$)
GK Per	03 27 47.2	+43 44 04	10.2	0.2	1.996803	<73	KO/4	3.6	0.9	0.25
T Aur	05 28 46.5	+30 24 36	14.9	4.1	0.204378	57:	–	1.1:	0.68:	0.63:
RR Pic	06 35 09.8	- 62 35 49	12.0	1.2	0.145026	65:	–	2.4:	0.95:	0.40:
BT Mon	06 41 15.8	- 01 58 09	15.4	4.5	0.333814		K5- 7	–	–	–
CP Pup	08 09 52.1	- 35 12 04	15:	0.2	0.06115		–	–	–	–
DQ Her	18 06 05.3	+45 51 01	14.2	1.4	0.193621	70	M3v	1.41	0.62	0.44
V533 Her	18 12 46.4	+41 50 22	14.3	3.0	0.28		–	–	–	–
V603 Aql	18 46 21.4	+00 31 36	11.4	- 1.1	0.138154	17	–	2.3	0.66	0.2
WY Sge	19 30 29.7	+17 38 24	19.3	5.4	0.153634	–	–	–	–	–
HR Del	20 40 04.2	+18 58 51	11.9	3.3	0.214167	41:	–	1.7:	0.9:	0.5
V1500 Cyg	21 09 52.8	+47 56 41	16.0	2.2	0.139613		–	–	–	–
V1668 Cyg	21 40 38.2	+43 48 10	>19	6.0	0.4392		–	–	–	–
DI Lac	22 33 46.5	+52 27 26	14.3	4.6	0.543773		–	–	–	–

*Adapted from Ritter (1987).

Dwarf Novae*

Name	R.A.(1950) h m s	Dec.(1950) ° ′ ″	m_{min}	m_{max}	Porb (days)	Inc.	Sp	M_1/M_2	M_1 ($M_\odot$)	M_2 ($M_\odot$)
WW CET	00 08 51.7	- 11 45 25	15.0	9.3	0.17578	55	–	1.4	0.50	0.35
RX AND	01 01 45.9	+41 01 54	12.6	10.9	0.21154	51	–	2.4	1.14	0.48
AR AND	01 04 06.9	+37 41 32	16.9	11.0	0.0938	–	–	–	–	–
HT CAS	01 07 05.2	+59 48 39	16.4	10.8	0.073647	76.6	–	3.3	0.60	0.20
TY PSC	01 22 50.4	+32 07 35	15.3	12.2	0.071	–	–	–	–	–
WX HYI	02 08 29.4	- 63 32 48	14.7	12.5	0.074813	40	–	5.5	0.9	0.16
VW HYI	04 09 33	- 71 25 27	13.4	9.5	0.074271	60	–	6	0.63	0.11
TU MEN	04 43 32.8	- 76 42 18	>16	11.6	0.1176	65	–	1.7	0.6	0.35
CN ORI	05 49 40.4	- 05 25 41	14.2	11.9	0.1639	67	–	1.7	0.94	0.56
SS AUR	06 09 35.4	+47 45 15	14.5	10.5	0.1828	38	–	2.8	1.08	0.39
CW MON	06 34 22	+00 05	16.3	11.9	0.1762	–	M3v	–	–	–
HL CMA	06 43 03.4	- 16 48 24	13.2	11.7	0.2145	45:	–	2.2:	1.0:	0.45
IR GEM	06 44 25.8	+28 09 43	16.3	11.7	0.0684	–	–	–	–	–
U GEM	07 52 07.8	+22 08 03	14.0	9.1	0.176906	67	M4.5v	2.09	1.18	0.56
YZ CNC	08 07 52.6	+28 17 33	14.1	11.9	0.0864	38	–	2.3	0.39	0.27
SU UMA	08 08 05.5	+62 45 23	14.2	12.2	0.07635	–	–	–	–	–
Z CHA	08 08 49.6	- 76 23 09	15.3	12.4	0.074499	81.7	M0- 5v	6.7	0.54	0.081
Z CAM	08 19 39.9	+73 16 24	13.6	10.5	0.289840	57	K7v	1.41	0.99	0.70
AT CNC	08 25 15	+25 31	15.0	12.7	0.238691	–	–	–	–	–
SW UMA	08 32 58.6	+53 39 04	16.5	10.6	0.05681	45	–	7.1	0.71	0.10
PG0834+488	08 34 48.5	+48 48 37	14.9	–	0.26810	–	–	–	–	–
CU VEL	08 56 44.6	- 41 36 10	15.5	10.7	0.0769	–	–	–	–	–
SY CNC	08 58 14.3	+18 05 44	13.5	11.1	0.380	26	G8v	0.81	0.89	1.10
X LEO	09 48 21	+12 06 36	15.8	12.4	0.1644	–	M2v	–	–	–
OY CAR	10 05 16.8	- 69 59 24	15.3	12.4	0.063121	82.6	–	8.8	0.90	0.100

Dwarf Novae*

Name	R.A.(1950) h m s	Dec.(1950) ° ′ ″	m_{min}	m_{max}	Porb (days)	Inc.	Sp	M_1/M_2	M_1 $(M_\odot)$	M_2 $(M_\odot)$
V346 CEN	11 11 36.9	- 37 24 26	15.3	12.4	0.062501	65:	-	4	0.7:	0.17:
T LEO	11 35 52.9	+03 38 46	15.2	11.0	0.058819	65	-	1.4	0.16	0.11
TW VIR	11 42 47.7	- 04 09 25	15.8	12.1	0.18267	43	M2- 4v	2.30	0.91	0.02
BV CEN	13 28 09.5	- 54 43 06	12.6	10.5	0.610116	62	G5- 8v	0.92	0.83	0.90
EK TRA	15 09 40.9	- 64 54 31	>17	12.1	0.0636	-	-	-	-	-
AH HER	16 42 06.1	+25 20 31	13.9	11.3	0.258116	46	K2- M0v	1.25	0.95	0.76
V2051 OPH	17 05 14.0	- 25 44 38	15.0	13.0	0.062428	80.5	-	3.4	0.44	0.13
BD PAV	18 38 54.7	- 57 34 38	15.4	12.4	0.17930	-	-	-	-	
AY LYR	18 42 40.5	+37 56 52	18.4	13.2	0.0730	-	-	-	-	-
EM CYG	19 36 42.1	+30 23 34	13.3	12.5	0.290909	63	K5v	0.75	0.57	0.
AB DRA	19 51 04.0	+77 36 40	14.5	12.3	0.15198	-	-	-	-	
UU AQL	19 54 35.2	- 09 27 26	16.1	11.0	0.14049	-	-	-	-	
RZ SGE	20 01 02.3	+16 54 23	16.9	12.8	0.067	-	-	-	-	
WZ SGE	20 05 20.6	+17 33 30	14.9	-	0.056688	78:	-	9.5	0.8:	0.
CM DEL	20 22 39.9	+17 08 07	13.4	-	0.162	73:	-	1.3	0.48	0.
VW VUL	20 55 34.0	+25 18 48	13.6	-	0.0731	44	-	1.7	0.24	0.
VY AQR	21 09 28.4	- 09 01 57	17.1	8.0	0.22	-	-	-	-	-
SS CYG	21 40 44.5	+43 21 23	11.4	8.2	0.275130	37	K5v	1.72	1.20	0.
RU PEG	22 11 35.5	+12 27 17	12.7	9.0	0.3746	33	K2- 3v	1.29	1.21	0.
TY PSA	22 46 55.5	- 27 22 46	16	-	0.08063	-	-	-	-	
IP PEG	23 20 38.6	+18 08 33	14.0	12	0.158208	85.	M4v	2.3	5 0.8	0

*Adapted from Ritter (1987).

Nova–like Variables*

Name	R.A.(1950) h m s	Dec.(1950) ° ′ ″	m_{min}	m_{max}	Porb (days)	Inc.	Sp	M_1/M_2	M_1 ($M_\odot$)	M_2 ($M_\odot$)
HV AND	00 38 10	+43 07	15.0	–	0.05599	–	–	–	–	–
BL HYI	01 39 37.5	– 68 08 32	14.3	17.3	0.078914	–	M3– 4v	–	–	–
TT ARI	02 04 19.9	+15 03 27	9.5	14.5	0.137551	23:	–	2.1:	0.8:	0.4:
RW TRI	02 22 41.5	+27 52 21	12.6	–	0.231883	82:	–	0.76	0.44	0.58
EF ERI	03 12 00	– 22 46 47	13.7	16.5	0.056266	70	–	–	–	0.13
TV COL	05 27 34.4	– 32 51 25	13.6	–	0.228600	>60				
V363 AUR	05 30 09.8	+36 57 29	14.2	–	0.321242	70	K0v	1.12	0.86	0.77
HO538+608	05 38 15.9	+60 50 03	14.6	>17B	0.129					
1H0542– 407	05 41 44.5	– 41 03 13	15.7	–	0.258	–	–	–	–	–
KR AUR	06 12 33.8	+28 36 10	11.3	16.9	0.16280	38	–	1.7	0.59	0.35
0623+71	06 23 46.5	+71 06 34	12:	–	0.1390	–	–	–	–	–
BG CMI	07 28 44.4	+10 02 46	14.3	–	0.13479	33	–	2.1	0.8	0.38
VV PUP	08 12 52.2	– 18 54 02	14.5	17.5	0.069747	60:	M4v	4:	>1	>0.2
IX VEL	08 13 49.6	– 49 04 02	9.1	–	0.1220	–	–	–	–	–
AC CNC	08 41 41.7	+13 03 26	13.8	–	0.300478	72	G8– K2v	0.81	0.82	1.02
SW SEX	10 12 37.2	– 02 53 35	14.8	–	0.134938	79		1.8	0.58	0.33
KO VEL	10 13 57.2	– 42 43 12	16.7	–	0.07183	–	–	–	–	–
RW SEX	10 17 27.2	– 08 26 51	10.4	–	0.2451	43	–	1.5	0.8	0.54
PG1030+590	10 30 37.6	+59 02 22	14.9	–	0.1361	–	–	–	–	–
1E1048+542	10 48 33.6	+54 20 33	18.0	–	0.07948	–	–	–	–	–
AN UMA	11 01 35.6	+45 19 26	14.5	16.0	0.079753	–	–	–	–	–
ST LMI	11 02 58.5	+25 22 42	–	15.0	0.079089	–	M5– 6v	–	–	–
QU CAR	11 03 48.9	– 68 21 46	11.1	–	0.454	–	–	–	–	–
DP LEO	11 14 38.1	+18 14 05	17.5	19.5	0.062363	76	–	4:	0.4:	0.1:
DO DRA	11 40 48.9	+71 57 59	10.6	–	0.1658	–	M3v	–	–	–
AM CVN	12 32 28.3	+37 54 15	14.1	–	–	–	–	–	–	–
EX HYA	12 49 42.5	– 28 58 39	13.0	11.7	0.068234	75	M5.5v	4.3	0.57	0.1
GP COM	13 03 15.8	+18 17 05	15.7	–	0.03231	–	–	–	–	–
UX UMA	13 34 42.1	+52 10 04	12.7	–	0.196671	<72	K8– M6v	0.90	0.43	0.4
PG134+082	13 46 25.9	+08 12 27	13.0	–	0.01725	–	–	–	–	–

Nova–like Variables*

Name	R.A.(1950) h m s	Dec.(1950) ° ′ ″	m_{min}	m_{max}	Porb (days)	Inc.	Sp	M_1/M_2	M_1 ($M_\odot$)	M_2 ($M_\odot$)
V834 CEN	14 05 58.2	- 45 03 05	14.2	–	0.070497	38	–	–	–	–
LX SER	15 35 45.0	+19 01 49	14.5		0.158432	75	–	1.14	0.41	0.3
MR SER	15 50 33.1	+19 05 18	14.9	17	0.078873	35:	M5- 6v	–	–	–
V795 HER	17 11 05.7	+33 34 49	12.5	–	0.6157	–	–	–	–	–
V442 OPH	17 29 22.0	- 16 13 15	>14.5	12.6	0.1406	67:	–	1.1	0.34	0.3
V380 OPH	17 47 46.9	+06 06 18	14.5	–	0.16	42	–	1.6	0.58	0.3
V426 OPH	18 05 24.8	+05 51 20	11.5	–	0.250	–	K4v:	–	–	–
AM HER	18 14 58.7	+49 50 55	12.0	15.0	0.128927 :	60:	M4.5v	2.4	0.39:	0.2
V1223 SGR	18 51 49.0	- 31 13 39	12.3	16:	0.140232	21	–	1.3	0.5	0
MV LYR	19 05 44.3	+43 56 21	12.1	17.7	0.1336	–	M5v	–	–	0
V1315 AQL	19 11 35	+12 12 37	14.4	–	0.139690 5	78	–	2.2	0.9	0
V3885 SGR	19 44 12.6	- 42 07 55	9.6	–	0.206 :	<50		1.0	0.8:	0
QQ VUL	20 03 30.7	+22 31 28	14.5	–	0.154522	60	M2- 4v	–	–	
V794 AQL	20 14 56.6	- 03 49 12	13.7	20.2	0.23	39	–	1.7	0.88	0
V SGE	20 18 02.1	+20 56 39	12.2	10.5	0.514198	90	F6- G0v	0.27	0.74	2
AE AQR	20 37 34.3	- 01 02 57	10.9	9.8	0.411654	58	K5v	1.26	0.9	0
V751 CYG	20 50 26.7	+44 08 05	13.2	16:	0.25	–	–	–	–	
FO AQR	22 15 17.2	- 08 35 02	13.5	–	0.16771	–	–	–	–	
PHL 227	22 17 53.8	+01 45 46	13.5	–	0.1356	–	–	–	–	
AO PSC	22 52 43.1	- 03 26 40	13.3	–	0.149626	–	–	–	–	
V425 CAS	23 01 34.9	+53 01 05	14.5	–	0.14964	25	–	2.8	0.86	
VY SCL	23 26 21.4	- 30 03 16	12.9	18.5	0.1662	–	–	–	–	
VZ SCL	23 47 33.8	- 26 39 33	15.6	>18	0.144622 :	90:	–	0.7	0.3:	

*Adapted from Ritter (1987).

24.3 Catalogue of Classical Novae*

Name	R.A.(1950) h m s	Dec.(1950) ° ′ ″	l °	b °
LS And	00 29 28.42	+41 41 39.2	119.104	- 20.786
UZ Tri	01 35 29	+33 17	134.028	- 28.322
V Per	01 58 29.30	+56 29 36.8	132.521	- 4.820
UW Tri	02 42 14.60	+33 18 48.6	148.627	- 23.605
SU Ari**	02 45 43.76	+17 10 26.6	158.795	- 37.206
V400 Per	03 04 12.435	+46 56 09.0	145.663	- 9.644
NSV00856 Tri**	03 15 56.9	+35 17 38	153.944	- 18.325
W Ari**	03 17 41.04	+28 46 52.0	158.278	- 23.484
SV Ari	03 22 11.28	+19 39 22.0	165.417	- 30.090
GK Per	03 27 47.36	+43 44 05.1	150.955	- 10.104
SZ Per**	03 43 54.20	+34 11 15.1	159.458	- 15.729
XX Tau	05 16 31.05	+16 39 57.9	187.104	- 11.653
GR Ori	05 19 00.20	+01 07 15.9	201.172	- 19.310
QZ Aur	05 25 16.96	+33 15 57.4	174.355	- 0.730
T Aur	05 28 46.47	+30 24 36.1	177.143	- 1.698
V529 Ori**	05 55 24	+20 15 12	188.94	- 1.95
KT Mon	06 22 38.43	+05 28 16.3	205.100	- 3.299
RR Pic	06 35 09.85	- 62 35 49.3	272.355	- 25.672
SY Gem**	06 37 23.49	+31 14 37.2	183.637	+11.552
DM Gem	06 41 00.51	+29 59 47.1	185.127	+11.728
BT Mon	06 41 15.81	- 01 58 08.8	213.859	- 2.623
DN Gem	06 51 39.72	+32 12 18.8	184.018	+14.714
NSV03313 Gem**	06 55 51	+17 06 19	198.448	+ 9.248
CG CMa**	07 01 59.55	- 23 41 03.2	235.596	- 7.978
GI Mon	07 24 20.60	- 06 34 23.8	222.930	+ 4.749
HS Pup	07 51 27.41	- 31.30 58.8	247.756	- 2.112
NSV03846 Pup**	07 57 21.24	- 43 41 09.0	258.809	- 7.424
HZ Pup	08 01 20.215	- 28 19 58.9	246.179	+ 1.385
VZ Gem**	08 04 42.65	+30 58 45.9	190.899	+28.924
CP Pup	08 09 52.04	- 35 12 04.4	252.926	- 0.835
DY Pup	08 11 42.57	- 26 24 48.6	245.823	+ 4.361
CQ Vel	08 57 21.06	- 53 08 35.2	272.333	- 4.895
T Pyx	09 02 37.14	- 32 10 47.3	257.207	+ 9.707
NSV04550 Leo**	09 34 15	+15 28 36	217.423	+43.441
U Leo**	10 21 22.7	+14 15 24	226.341	+53.263

Catalogue of Classical Novae*

Name	Year	B.R. (mag)	t_3 (days)	Type
LS And	1971	11.7p–20.5p	8	NA:
UZ Tri	1980	14.2p–21.0:p		N?
V Per	1887	9.2p–18.5p		N
UW Tri	1983	15p–[21p		N:
SU Ari**	1854	9.5v–[14.5v		N?
V400 Per	1974	7.8p–20p	43	NA
NSV00856 Tri**	1853	9.5v– ?		N??
W Ari**	1855	9.5v–[20v		N??
SV Ari	1905	12p–22p		N?
GK Per	1901	0.2v–14.0	13	NA
SZ Per**	1908	9.5v– ?		N??
XX Tau	1927	5.9p– 18.5p	42	NA
GR Ori	1916	11.5p–[21p		N:
QZ Aur	1964	6.0p–18.0p	23	NA
T Aur	1892	4.2p–15.2p	100	NB
V529 Ori**	1678	6v–20.5p?		N??
KT Mon	1942	10.3p–[21	40	NA
RR Pic	1925	1.0v–11.9p	150	NB
SY Gem**	1856	9.2v–[13v		N??
DM Gem	1903	4.8v–16.7p	22	NA
BT Mon	1939	8.5p–15.5		NA
DN Gem	1912	3.5p–15.8p	37	NA
NSV03313 Gem**	1892	7.0v–[14p		N??
CG CMa**	1934	13.7p–15.9p		N?
GI Mon	1918	5.6p–18p	23	NA
HS Pup	1963	8.0p–20.5p	65	NA
NSV03846 Pup**	1673	3v–20p?		N??
HZ Pup	1963	7.7p–17p	70	NA
VZ Gem**	1856	8.7v–[21?		N??
CP Pup	1942	0.5v–15.0v	8	NA
DY Pup	1902	7.0p–20p	160	NB:
CQ Vel	1940	9.0p–21j	50	NA:
T Pyx	1902	6.5p–15.3p	88	NR
NSV04550 Leo**	1612	4v– ?		N??
U Leo**	1855	10.5v–[15v		N??

*Adapted from H.W. Duerbeck, Space Science Reviews 45, 1–212 (1987) and private communication (1989).

**Novae designated by ** are somewhat dubious.

Catalogue of Classical Novae*

Name	R.A.(1950) h m s	Dec.(1950) ° ′ ″	l °	b °
V411 Car	10 29 29.37	- 59 42 59.35	286.232	- 1.726
N Car	10 36 35.04	- 62 52 51.6	288.547	- 4.048
N Car	10 37 59.08	- 62 58 27.3	288.732	- 4.503
CN Vel	11 00 28.30	- 54 06 59.6	287.431	- 5.179
V365 Car	11 01 09.30	- 58 11 17.2	289.168	+ 1.493
RS Car	11 06 01.46	- 61 39 49.2	291.098	- 1.469
MT Cen	11 41 35.88	- 60 16 59.9	294.736	+ 1.234
GQ Mus	11 49 34.99	- 66 55 39.0	297.212	- 4.996
V359 Cen	11 55 42.39	- 41 29 26.0	292.414	+20.014
AP Cru	12 28 28.59	- 64 09 50.8	300.765	- 1.653
V812 Cen	13 10 47.88	- 57 24 52.05	305.943	+ 5.061
HV Vir	13 18 30.17	+02 09 11.5	319.882	+63.784
RR Cha	13 20 55.70	- 82 04 07.4	304.165	- 19.541
AB Boo**	14 04 43.20	+20 58 55.0	16.706	+71.604
T Boo	14 11 40.87	+19 18 02.1	14.274	+69.404
V842 Cen	14 32 13.33	- 57 24 30.6	316.574	+ 2.453
X Cir	14 38 33.64	- 64 58 50.0	314.259	- 4.797
AR Cir	14 44 19.28	- 59 47 56.6	317.039	- 0.372
AI Cir	14 45 03.02	- 68 39 09.2	313.280	- 8.395
IL Nor	15 25 45.56	- 50 24 42.05	326.835	+ 4.810
IM Nor	15 35 42.26	- 52 09 39.8	327.097	+ 2.485
CT Ser	15 43 19.58	+14 31 51.2	24.482	+47.563
T CrB	15 57 24.50	+26 03 38.25	42.373	+48.165
V341 Nor	16 09 50.68	- 53 11 31.95	330.423	- 1.607
NSV07542 Oph**	16 10 51.67	- 05 11 04.45	7.160	+31.415
T Sco	16 14 03.79	- 22 51 09.4	352.675	+19.462
X Ser	16 16 41.32	- 02 22 17.8	10.841	+31.873
U Sco	16 19 37.49	- 17 45 42.9	357.669	+21.869
OY Ara	16 36 55.28	- 52 20 04.0	333.902	- 3.937
RW UMi	16 49 55.77	+77 07 16.2	109.638	+33.152
V840 Oph	16 51 33.89	- 29 32 39.7	353.096	+ 8.729
V841 Oph	16 56 41.885	- 12 48 59.45	7.621	+17.779
V2214 Oph	17 08 50.83	- 29 33 58.0	355.373	+ 5.724
V360 Her	17 14 33.88	+24 30 04.4	46.520	+30.957
V2109 Oph	17 21 12.08	- 24 34 08.05	1.074	+ 6.342

Catalogue of Classical Novae*

Name	Year	B.R. (mag)	t_3 (days)	Type
V411 Car	1953	14.5p–19p		N
N Car	1972	13?–18?		N
N Car	1971	13?–18?		N
CN Vel	1905	10.2p–17p?	> 800	NB
V365 Car	1948	10.1p–21.6j	530	NB
RS Car	1895	7.0p–[22j		N
MT Cen	1931	8.35p–22j		N:
GQ Mus	1983	7.2v–22j	45	NA
V359 Cen	1939	13.8p–21.0j		N:
AP Cru	1935	10.7p–21.7j		N:
V812 Cen	1973	11p–18j		N
HV Vir	1929	11p–19p		N:
RR Cha	1953	7.1p–19.3j	60	NA
AB Boo**	1877	4.5v– ?		N??
T Boo	1860	9.75v–18.5p?		N??
V842 Cen	1986	4.6v–20.3j		N
X Cir	1927	6.5p–[23j	170	NB
AR Cir	1906	10.3p–15p	415	N?
AI Cir	1914	10.9p–23j		N:
IL Nor	1893	7.0p–18j 1	08	
IM Nor	1920	9.0p–22.0j		N:
CT Ser	1948	7.9v–16.6p		N
T CrB	1866	2.0p–11.3p	6.8	NR
V341 Nor	1983	9.4v–[17j		N?
NSV07542 Oph**	1893	11.5p–21j?		N?
T Sco	1860	7v– ?	21	N:
X Ser	1903	8.9p–18.3p		NBS
U Sco	1863	8.8v–19.2v	7	NR
OY Ara	1910	6.0p–17.5p	80	NA
RW UMi	1956	6p–21p	140	NB
V840 Oph	1917	6.5p–20j?	36	NA
V841 Oph	1848	4.2v–13.5v	130	NB
V2214 Oph	1988	8.5p–20.5B	73	NA
V360 Her	1892	6.3– ?		N?
V2109 Oph	1969	10.8r–18j		N

*Adapted from H.W. Duerbeck, Space Science Reviews 45, 1–212 (1987) and private communication (1989).

**Novae designated by ** are somewhat dubious.

Catalogue of Classical Novae*

Name	R.A.(1950) h m s	Dec.(1950) ° ′ ″	l °	b °
BB Oph	17 21 25.90	- 24 45 24.9	0.947	+ 6.194
V902 Sco	17 22 41.20	- 39 01 31.0	349.294	- 2.069
V906 Oph	17 23 28	- 21 50 06	3.654	+ 7.430
V908 Oph	17 24 42	- 27 43	358.893	+ 3.932
V972 Oph	17 31 34.61	- 28 08 38.7	359.374	+ 2.430
N Oph	17 35 00.400	- 03 12 54.5	21.329	+14.799
V728 Sco	17 35 31.22	- 45 27 06.5	345.191	- 7.566
V794 Oph	17 35 47.53	- 22 49 10.1	4.380	+ 4.502
V1012 Oph	17 38 31.92	- 23 22 06.0	4.250	+ 3.677
V721 Sco	17 39 09.235	- 34 39 18.15	354.757	- 2.438
V2024 Oph	17 39 18	- 24 58	2.986	+ 2.681
V553 Oph	17 39 49.425	- 24 50 11.7	3.159	+ 2.649
V2110 Oph**	17 40 31.60	- 22 44 18.5	5.029	+ 3.620
KP Sco	17 40 54.90	- 35 42 07.6	354.058	- 3.297
V719 Sco	17 42 25.07	- 33 59 45.5	355.673	- 2.663
N Sco	17 44 20.75	- 33 10 44.0	356.581	- 2.579
V707 Sco	17 45 03.12	- 36 36 57.0	353.713	- 4.486
V722 Sco	17 45 17	- 34 56 53	355.168	- 3.663
V3888 Sgr	17 45 44.74	- 18 44 41.1	9.084	+ 4.659
V1274 Sgr	17 46 00	- 17 51	9.885	- 5.067
V825 Sco	17 46 35.76	- 33 31 22.3	356.531	- 3.159
V723 Sco	17 46 40.30	- 35 23 02.9	354.940	- 4.132
V3964 Sgr	17 46 47.77	- 17 22 44.3	10.389	- 5.146
V1172 Sgr	17 47 24.69	- 20 39 41.8	7.639	+ 3.335
RS Oph	17 47 31.55	- 06 41 39.8	19.799	+10.372
V697 Sco	17 47 57.11	- 37 24 10.0	353.333	- 5.387
N Sco 89	17 48 33.96	- 32 31 16.2	357.603	- 2.997
V382 Sco	17 48 34.99	- 35 24 22.8	355.122	- 4.477
V720 Sco	17 48 36.94	- 35 22 39.1	355.150	- 4.469
LW Ser	17 48 59.66	- 14 43 08.7	12.959	+ 6.047
V696 Sco	17 49 49.625	- 35 49 38.3	354.887	- 4.909
V2415 Sgr	17 50 00	- 29 33 48	0.303	- 1.745
V744 Sco	17 50 03.93	- 31 12 58.7	358.889	- 2.602
NSV09808 Sco	17 50 27	- 30 44 57	359.333	- 2.435
V4092 Sgr	17 50 31.25	- 29 01 33.9	0.823	- 1.567

Catalogue of Classical Novae*

Name	Year	B.R. (mag)	t_3 (days)	Type
BB Oph	1897	11.3p–19j		N:
V902 Sco	1949	11p–20.0j	200	NB
V906 Oph	1952	8.4v– ?	25	NA
V908 Oph	1954	9– ?		N
V972 Oph	1957	8.0p–17.0j	176	NB
N Oph	1938	15.6p–21v		N:
V728 Sco	1862	5.0v–20j?	<9	N??
V794 Oph	1939	11.7p–18p	220	NB
V1012 Oph	1961	13.8p–22j		N
V721 Sco	1950	10.1p–[18.0j		N
V2024 Oph	1967	11.0v–[18		N
V553 Oph	1940	12.5p–22j	45	NA
V2110 Oph**	1950	12.0p–20j		NB/ZAN
KP Sco	1928	9.4p–21j?	38	NA:
V719 Sco	1950	9.8p–20.5j	24	NA
N Sco	1952	11p–21j?		N
V707 Sco	1922	9.9p–20j	49	NA
V722 Sco	1952	9.5– ?	18	NA
V3888 Sgr	1974	9.0v–16j		N
V1274 Sgr	1954	10.5– ?		N
V825 Sco	1964	12p–19.0j		N
V723 Sco	1952	9.8p–19j?	17	NA:
V3964 Sgr	1975	9.4p–[17p	32	NA
V1172 Sgr	1951	9.0p–18j		N
RS Oph	1901	4.3v–12.5v	23	NR
V697 Sco	1941	10.2p–17j	<15	NA
N Sco 89	1989	11.v–20j		N
V382 Sco	1901	9.5p–22j		N:
V720 Sco	1950	7.5p–21		N
LW Ser	1978	8.3v–21p	50	NA
V696 Sco	1944	7.5p–19.5j	9	NA
V2415 Sgr	1951	13p– ?		N
V744 Sco	1935	13.3p–21p		N:
NSV09808 Sco	1954	13.8p– ?		N
V4092 Sgr	1984	10.5v– ?		N:

*Adapted from H.W. Duerbeck, Space Science Reviews 45, 1–212 (1987) and private communication (1989).

**Novae designated by ** are somewhat dubious.

Catalogue of Classical Novae*

Name	R.A.(1950) h m s	Dec.(1950) ° ′ ″	l °	b °
V711 Sco	17 50 46.70	− 34 20 42.6	356.268	− 4.325
N Sgr	17 51 33	− 28 41 11	1.230	− 1.588
V745 Sco	17 52 01.35	− 33 14 32.5	357.352	− 3.989
V732 Sgr	17 52 59.18	− 27 21 53.2	2.526	− 1.186
MU Ser	17 53 02.42	− 14 00 52.9	14.092	+ 5.569
V960 Sco	17 53 19.01	− 31 49 14.2	358.720	− 3.506
V990 Sgr	17 54 09	− 28 18 48	1.842	− 1.890
V3889 Sgr	17 55 11.57	− 28 21 38.6	1.917	− 2.113
V1275 Sgr	17 55 43.44	− 36 18 29.1	355.069	− 6.182
V4135 Sgr	17 56 28.94	− 32 16 13.4	358.667	− 4.312
V787 Sgr	17 56 49.02	− 30 30 23.1	0.235	− 3.494
V999 Sgr	17 56 57.05	− 27 33 07.5	2.811	− 2.042
V394 CrA	17 56 58.165	− 39 00 29.4	352.836	− 7.715
FL Sgr	17 57 10.74	− 34 36 11.4	356.707	− 5.596
V1944 Sgr	17 57 28.45	− 27 17 12,8	3.099	− 2.010
DZ Ser	17 58 12.72	− 10 33 50.35	17.708	+ 6.169
KY Sgr	17 58 14.05	− 26 24 38.7	3.943	− 1.719
V384 Sco	17 58 21.54	− 35 39 28.0	535.903	− 6.325
V1174 Sgr	17 58 27	− 28 44 26	1.946	− 2.922
V1431 Sgr	17 58 52	− 30 06 12	0.806	− 3.677
V4027 Sgr	17 59 18.87	− 28 45 23.8	2.026	− 3.095
AT Sgr	18 00 23.915	− 26 28 37.1	4.127	− 2.172
N Sgr 1986	18 00 28.65	− 28 00 17.5	2.276	− 3.246
V2104 Oph	18 01 05.37	+11 47 47.3	38.251	+15.928
N Sgr	18 02 00	− 29 54 56	1.305	− 4.175
V1572 Sgr	18 02 22	− 31 38 00	359.840	− 5.084
V1012 Sgr	18 02 59.09	− 31 44 46.05	359.806	− 5.254
V1014 Sgr	18 03 37.28	− 27 26 36.7	3.639	− 3.274
V737 Sgr	18 03 58.335	− 28 45 16.6	2.531	− 3.982
V927 Sgr	18 04 25.45	− 33 21 43.9	358.532	− 6.302
V4121 Sgr	18 04 44.33	− 28 49 54.5	2.545	− 4.166
V630 Sgr	18 05 28.94	− 34 20 52.65	357.768	− 6.967
V1015 Sgr	18 05 45.78	− 32 29 04.1	359.442	− 6.126
V1148 Sgr	18 05 59.5	− 25 59 40	5.164	− 3.028
DQ Her	18 06 05.28	+45 51 02.2	73.153	+26.444

Catalogue of Classical Novae*

Name	Year	B.R. (mag)	t_3 (days)	Type
V711 Sco	1906	9.7p–19j		N:
N Sgr	1963	13– ?		N
V745 Sco	1937	11.2p–21j	12	N:
V732 Sgr	1936	6.5p– ?	74	NA
MU Ser	1983	7.7v–[21p	5	NA
V960 Sco	1985	10.5v–20j		N
V990 Sgr	1936	11.1p– ?	24	NA:
V3889 Sgr	1975	8.4v–21j	14	NA:
V1275 Sgr	1954	7.0p–18j?	>10	N
V4135 Sgr	1987	10.4p–[22.5j	30	NA
V787 Sgr	1937	9.8p–21j?	74	NA:
V999 Sgr	1910	8.0p–17.35B	160	NB
V394 CrA	1949	7.5p–20:j		N
FL Sgr	1924	8.3p–20j?	32	NA:
V1944 Sgr	1960	13p– ?		N
DZ Ser	1960	14.0p–21p		N
KY Sgr	1926	10.6p–20j	60	NA:
V384 Sco	1893	12.3p–18.5j		N:
V1174 Sgr	1952	12.0p– ?		N
V1431 Sgr	1945	17.2p–[19.4		N?
V4027 Sgr	1968	11.0r–[21j		N
AT Sgr	1900	11.0p–19j	35	NA:
N Sgr 1986	1986	10.4v–16j		N?
V2104 Oph	1976	8.8v–20.5p		N
N Sgr	1953	10.5– ?		N
V1572 Sgr	1955	11p– ?		N
V1012 Sgr	1914	8.0p–20j	32	NA:
V1014 Sgr	1901	10.9p–20j	>50	NA:
V737 Sgr	1933	10.3p–19j?	>70	N:
V927 Sgr	1944	8.0p–20j?	<15	NA
V4121 Sgr	1983	9.5v–21j		N
V630 Sgr	1936	4.5v–19j	11	NA
V1015 Sgr	1905	7.1p–21j?	34	NA:
V1148 Sgr	1948	8.0p– ?		N
DQ Her	1934	1.3v–14.5v	94	NA

*Adapted from H.W. Duerbeck, Space Science Reviews 45, 1–212 (1987) and private communication (1989).

**Novae designated by ** are somewhat dubious.

Catalogue of Classical Novae*

Name	R.A.(1950) h m s	Dec.(1950) ° ′ ″	l °	b °
V1175 Sgr	18 11 03	− 31 08 03	1.175	− 6.470
V849 Oph	18 11 47.18	+11 35 46.1	39.233	+13.480
V1583 Sgr	18 12 23.91	− 23 24 18.6	8.138	− 3.052
V533 Her	18 12 46.38	+41 50 22.1	69.188	+24.274
V4074 Sgr**	18 12 51.90	− 30 52 14.85	1.593	− 6.689
FM Sgr	18 14 15.16	− 23 39 36.7	8.117	− 3.547
V1149 Sgr	18 15 20.81	− 28 18 31.4	4.124	− 5.962
V928 Sgr	18 15 49.83	− 28 07 14.9	4.342	− 5.971
V1150 Sgr	18 15 51.15	− 24 06 46.4	7.892	− 4.084
V726 Sgr	18 16 26.22	− 26 54 38.2	5.479	− 5.519
V4065 Sgr	18 16 30	− 24 45	7.399	− 4.515
V1016 Sgr	18 16 52.53	− 25 12 35.1	7.032	− 4.806
V4049 Sgr	18 17 29.05	− 27 57 49.2	4.652	− 6.217
V441 Sgr	18 19 02.68	− 25 30 23.6	7.000	− 5.379
GR Sgr	18 19 52.88	− 25 36 20.6	7.000	− 5.592
V655 CrA	18 21 20.80	− 37 01 23.9	356.859	− 11.057
V1151 Sgr	18 22 25.60	− 20 13 43.7	12.045	− 3.598
V909 Sgr	18 22 32.26	− 35 03 13.0	358.767	− 10.404
BS Sgr	18 23 38.88	− 27 10 10.0	5.997	− 7.062
HS Sgr	18 25 03.50	− 21 36 20.6	11.110	− 4.783
LQ Sgr	18 25 19.93	− 27 57 17.1	5.464	− 7.752
V366 Sct	18 26 54.6	− 12 20 58	19.516	− 0.857
V3890 Sgr**	18 27 39.66	− 24 03 05.8	9.206	− 6.441
FH Ser	18 28 16.24	+02 34 42.8	32.909	+ 5.786
V2572 Sgr	18 28 20.68	− 32 38 08.4	1.514	− 10.422
V1017 Sgr	18 28 53.30	− 29 25 25.7	4.490	− 9.109
V1905 Sgr	18 30 36.97	− 25 22 56.8	8.317	− 7.643
V4077 Sgr	18 31 32.83	− 26 28 27.1	7.426	− 8.319
V1310 Sgr	18 31 48	− 30 06 48	4.146	− 9.980
FV Sct	18 32 02.67	− 12 57 52.5	19.555	− 2.252
V3645 Sgr	18 32 53.15	− 18 44 14.7	14.515	− 5.093
V4021 Sgr	18 35 12.03	− 23 25 26.95	10.555	− 7.696
V949 Sgr	18 37 55	− 28 12 20	6.482	− 10.351
V693 CrA	18 38 33.6	− 37 34 09	357.830	− 14.392
V827 Her	18 41 26.70	+15 16 12.3	45.808	+ 8.594

Catalogue of Classical Novae*

Name	Year	B.R. (mag)	t_3 (days)	Type
V1175 Sgr	1952	7.0– ?		N
V849 Oph	1919	7.3v–17p	175	NB
V1583 Sgr	1928	8.9p–21j	37	NA:
V533 Her	1963	3.0p–15.0p	44	NA
V4074 Sgr**	1965	8.6p–12.3p	120	NB/ZAND
FM Sgr	1926	8.6p–20.5j?	30	NA:
V1149 Sgr	1945	7.4p–21j	< 210	NB
V928 Sgr	1947	8.9p–20.5j	150	NB
V1150 Sgr	1946	13.3p–[22j?	< 600	NB
V726 Sgr	1936	10.8p–19j?	95	NA
V4065 Sgr	1980	9.0v–[18		N?
V1016 Sgr	1899	8.5p–17p	140	NB:
V4049 Sgr	1978	12p–21j		N
V441 Sgr	1930	8.7p– ?	53	NA
GR Sgr	1924	11.4p–16.5		N
V655 CrA	1967	8p–17.6j		N
V1151 Sgr	1947	11.1p–20j	135	NB
V909 Sgr	1941	6.8p–20j	7	NA
BS Sgr	1917	9.2p–17j	700	NB
HS Sgr	1900	11.5p–17:		NB:
LQ Sgr	1897	13.0p–21j		N:
V366 Sct	1961	15.4–[23j		N
V3890 Sgr**	1962	8.4p–17.0		N/ZAND
FH Ser	1970	4.5v–16.2p	62	NA
V2572 Sgr	1969	6.5p–18j	44	NA:
V1017 Sgr	1919	7.2p–14.7B	130	NR
V1905 Sgr	1932	9.1p–19j		N:
V4077 Sgr	1982	8.0v–22j	100	NB
V1310 Sgr	1935	11.7p– ?	390	NB:
FV Sct	1960	12.5p–20p		N
V3645 Sgr	1970	12.6p–18p	300?	NB
V4021 Sgr	1977	8.8v–18j	70	NA
V949 Sgr	1914	15.7p– ?		N:
V693 CrA	1981	7.0v–23j	12	NA
V827 Her	1987	7.4v–17.5p	55	NA

*Adapted from H.W. Duerbeck, Space Science Reviews 45, 1–212 (1987) and private communication (1989).

**Novae designated by ** are somewhat dubious.

Catalogue of Classical Novae*

Name	R.A.(1950) h m s	Dec.(1950) ° ′ ″	l °	b °
V368 Sct	18 42 59.96	− 08 36 13.6	24.669	− 2.629
GL Sct	18 43 06.65	− 06 28 22.4	26.576	− 1.674
V522 Sgr**	18 44 55.60	− 25 25 44.3	9.175	− 10.565
V603 Aql	18 46 21.45	+00 31 36.1	13.164	+ 0.829
N Sct 89	18 46 58.13	− 06 14 45.2	27.219	− 2.421
HR Lyr	18 51 27.64	+29 09 50.0	59.584	+12.470
V373 Sct	18 52 44.18	− 07 46 59.8	26.504	− 4.397
EL Aql	18 53 24.27	− 03 23 15.95	30.497	− 2.535
EU Sct	18 53 34.50	− 04 16 30.4	29.727	− 2.980
NSV11561 Sct**	18 53 50.635	− 08 39 29.4	25.847	− 5.039
V446 Her	18 55 03.03	+13 10 26.6	45.409	+ 4.707
FS Sct	18 55 36.98	− 05 28 11.4	28.895	− 3.978
V1059 Sgr	18 59 01.62	− 13 14 03.7	22.305	− 8.230
V604 Aql	18 59 27.47	− 04 31 07.5	30.182	− 4.395
QV Vul	19 02 32.15	+21 41 39.2	53.858	+ 6.974
V841 Aql	19 05 17.92	+10 24 57.1	44.111	+ 1.212
V363 Sgr	19 08 05.81	− 29 55 01.45	7.629	− 17.072
V1378 Aql	19 14 05.73	+03 37 55.8	39.124	− 3.880
V356 Aql	19 14 41.69	+01 37 56.0	37.419	− 4.944
V1301 Aql	19 15 26.75	+04 41 49.3	40.225	− 3.681
V605 Aql**	19 15 48.67	+01 41 28.8	37.601	− 5.163
V528 Aql	19 16 45.93	+00 32 19.2	36.687	− 5.910
V606 Aql	19 17 50.07	− 00 13 40.7	36.128	− 6.502
V1370 Aql	19 20 50.08	+02 23 34.8	38.813	− 5.947
V1229 Aql	19 22 15.41	+04 08 50.4	40.537	− 5.438
PW Vul	19 24 03.50	+27 15 54.8	61.098	+ 5.197
V368 Aql	19 24 08.97	+07 30 09.0	43.728	− 4.265
NQ Vul	19 27 03.95	+20 21 43.6	55.355	+ 1.290
DO Aql	19 28 45.10	− 06 32 02.4	31.705	− 11.805
WY Sge	19 30 29.70	+17 38 24.5	53.368	− 0.739
EY Aql	19 32 27.23	+14 55 13.7	51.222	− 2.476
SS Sge**	19 36 52.83	+16 35 44.2	53.206	− 2.581
HS Sge	19 37 08.16	+18 00 57.45	54.471	− 1.931
LU Vul	19 43 34.18	+28 28 07.6	64.261	+ 2.022
CK Vul	19 45 34.97	+27 11 10.6	63.381	+ 0.989

Catalogue of Classical Novae*

Name	Year	B.R. (mag)	t_3 (days)	Type
V368 Sct	1970	6.9v–19.0p	30	NA
GL Sct	1915	13.6p– ?		N?
V522 Sgr**	1931	12.9p–17p		N:/UG
V603 Aql	1918	- 1.1v–12.0v	8	NA
N Sct 89	1989	8.5v–[21j		N
HR Lyr	1919	6.5p–15.8p	74	NA
V373 Sct	1975	7.1v–18.5p	85	NA
EL Aql	1927	6.4p–20p	25	NA
EU Sct	1949	8.4p–18p	42	NA
NSV11561 Sct**	1938	16.2p–17.0p		N??
V446 Her	1960	3.0p–18.0	16	NA
FS Sct	1952	10.1p–18p	86	NA:
V1059 Sgr	1898	4.9p–18.1B		NA
V604 Aql	1905	8.2p–21p	25	NA
QV Vul	1987	7.0v–19B		N
V841 Aql	1951	11.5p–17.5p		N
V363 Sgr	1927	8.8p–20j	64	NA
V1378 Aql	1984	10p–21p		N
V356 Aql	1936	7.7p–17.7p	115	NB
V1301 Aql	1975	10.3v–21p	35	NA
V605 Aql**	1919	10.4v–22.5v	1500	NB
V528 Aql	1945	7.0p–18.1p	37	NA
V606 Aql	1899	6.7p–17.3p	65	NA
V1370 Aql	1982	6:–19.5p		N
V1229 Aql	1970	6.7v–19.4p	37	NA
PW Vul	1984	6.4v–17:	97	NA
V368 Aql	1936	6.5p–17.8p	42	NA
NQ Vul	1976	6.0v–18.5p	65	NA
DO Aql	1925	8.7v–16.5p	900	NB
WY Sge	1783	6v–19.5p		N
EY Aql	1926	10.5p–21p	40	NA
SS Sge**	1916	11.8p–16.3p		ZAND/NC
HS Sge	1977	7.0p–20.5p	20	NA
LU Vul	1968	9.5p–[21p	21	NA
CK Vul	1670	2.6v–20.7?		N:

*Adapted from H.W. Duerbeck, Space Science Reviews 45, 1–212 (1987) and private communication (1989).

**Novae designated by ** are somewhat dubious.

Catalogue of Classical Novae*

Name	R.A.(1950) h m s	Dec.(1950) ° ′ ″	l °	b °
LV Vul	19 45 57.37	+27 02 48.4	63.303	+ 0.846
V500 Aql	19 50 02.89	+08 20 58.9	47.608	− 9.462
V465 Cyg	19 50 47.60	+36 26 03.3	71.908	+ 4.764
V1819 Cyg	19 52 45.90	+35 34 20.7	71.372	+ 3.978
V476 Cyg	19 57 09.64	+53 28 54.5	87.368	+12.417
RR Tel**	20 00 20.13	− 55 52 03.2	342.163	− 32.242
PU Vul**	20 19 01.08	+21 24 44.4	62.575	− 8.532
QU Vul	20 24 40.55	+27 40 47.2	68.511	− 6.026
HR Del	20 40 04.11	+18 58 52.2	63.431	− 13.972
V1330 Cyg	20 50 46.46	+35 48 04.4	78.376	− 5.488
V450 Cyg	20 56 48.15	+35 44 46.2	79.128	− 6.458
V1500 Cyg	21 09 52.95	+47 56 40.95	89.823	− 0.073
Q Cyg	21 39 45.38	+42 36 45.6	89.928	− 7.552
V1668 Cyg	21 40 38.06	+43 48 10.1	90.838	− 6.760
IV Cep	22 02 46.83	+53 15 48.0	99.614	− 1.638
CP Lac	22 13 50.45	+55 22 02.9	102.141	− 0.837
DI Lac	22 33 46.51	+52 27 26.1	103.108	− 4.855
DK Lac	22 47 40.455	+53 01 24.45	105.237	− 5.352
OS And	23 09 47.47	+47 12 00.6	106.052	− 12.117
BC Cas	23 48 48.71	+60 01 29.2	115.544	− 1.703

Catalogue of Classical Novae*

Name	Year	B.R. (mag)	t_3 (days)	Type
LV Vul	1968	5.17v–16.9p	37	NA
V500 Aql	1943	6.5p–17.8p	42	NA
V465 Cyg	1948	8.0p–17.0p	140	NB
V1819 Cyg	1986	8.7v–19p	> 100	NB
V476 Cyg	1920	2.0p–17.2B	16.5	NA
RR Tel**	1908	6.8p–16p	2000	NC/ZAND
PU Vul**	1979	9.0p–16p		NC/ZAND
QU Vul	1984	5.6v–19p	40	NA
HR Del	1967	3.5v–12.0v	230	NB
V1330 Cyg	1970	9p–18.1p	18	NA
V450 Cyg	1942	7.8p–16.3p?	108	NB
V1500 Cyg	1975	2.2B–21.5p	3.6	NA
Q Cyg	1876	3.0v–15.6v	11	NA
V1668 Cyg	1978	6.7p–20.0p	23	NA
IV Cep	1971	7.5B–17.1B	37	NA
CP Lac	1936	2.1v–16.6p	10	NA
DI Lac	1910	4.6v–14.9p	43	NA
DK Lac	1950	5.9p–15.5p	32	NA
OS And	1986	6.3v–17.8p	22	N
BC Cas	1931	10.7p–17.4p	75	NA

*Adapted from H.W. Duerbeck, Space Science Reviews 45, 1–212 (1987) and private communication (1989).

**Novae designated by ** are somewhat dubious.

24.4 Catalogue of Dwarf Novae*

Name	R.A.(1950) h m s	Dec.(1950) ° ′ ″	m_{min} (mag)	m_{max} (mag)
FI Cas	00 03 34.085	+55 42 10.30	18.5:	15.0
WW Cet	00 08 51.7	- 11 45 30	16.8	9.3
V513 Cas	00 15 30.160	+66 01 35.00	<17.2	15.5
V452 Cas	00 49 24.795	+53 35 26.75	17.5:	14
RX And	01 01 45.825	+41 01 54.16	13.57	10.90
HT Cas	01 07 05.425	+59 48 38.70	16.55	12.6
FN And	01 09 09.495	+35 01 31.19	17.5	13.5
FO And	01 12 42.085	+37 21 46.02	17.5	13.5
WX Cet	01 14 38	- 18 12 12	18	9.5
UY Phe	01 20 59	- 42 34 24	15.9	14.6
TY Psc	01 22 50.450	+32 07 35.60	16	12.5
KU Cas	01 27 48.500	+57 38 48.20	17.5	13.5
KT Per	01 34 01.700	+50 42 04.80	15.39	10.6
AR And	01 42 06.915	+37 41 33.10	16.95	12.59
UV Per	02 06 39.40	+56 57 10.0	17.5	11.9
WX Hyi	02 08 28	- 63 33 06	14.85	9.6
TZ Per	02 10 18.295	+58 08 52.60	15.6v	12.3v
FS And	02 22 51.120	+37 20 36.35	<17.5	14.5
PU Per	02 39 11.705	+35 28 03.10	<17.5	15
PT Per	02 39 12.10	+56 28 48.0	18	15
PV Per	02 39 47.255	+37 51 24.00	<18	15.5
V368 Per	02 44 28.020	+34 45 58.15	<17.5	15
RU Hor	02 45 01	- 63 47 36	17.5:	13.9
V372 Per	02 53 31.075	+36 56 19.95	<17.5	15
AF Cam	03 28 14.510	+58 37 12.80	17.6	13.4
OV Tau	03 42 57	+29 33	16.5	15.5
MR Per	03 47 10.415	+48 03 25.50	16.5	14.5
FO Per	04 04 48.230	+51 06 52.70	16.2	13.8
XZ Eri	04 09 10	- 15 31 36	<17.5	14.6
VW Hyi	04 09 30	- 71 25 24	14.4	8.4
NS Per	04 14 56.250	+51 00 15.25	<18	15
AH Eri	04 20 23	- 13 28 54	17.5	13.5
TU Men	04 43 30	- 76 41 42	17	11.4
HV Aur	04 44 54	+38 11	<17	15
AQ Eri	05 03 44	- 04 12 00	16.5	12.5

Catalogue of Dwarf Novae*

Name	R.A.(1950) h m s	Dec.(1950) ° ′ ″	m_{min} (mag)	m_{max} (mag)
BI Ori	05 21 17	+00 57 48	16.7	13.2
FS Aur	05 44 39	+28 34 12	16.2	14.4
V421 Tau	05 45 18	+22 41	20	15
CN Ori	05 49 40	− 05 25 42	16.2	11.0
SS Aur	06 09 35.240	+47 45 16.90	15.7	10.3v
V344 Ori	06 12 28	+15 29	17.5:	14.2
CZ Ori	06 13 51	+15 25 18	15.6	11.2
CW Mon	06 34 22	+00 05	16.3	11.9
UV Gem	06 35 49	+18 18 48	18.5	14.7
IR Gem	06 44 31	+28 08	<14.5	10.7
WZ CMa	07 16 48	− 27 02 06	<16.0	14.5
AW Gem	07 19 33	+28 36 06	<17.5	12.9
SV CMi	07 28 27	+06 05 00	16.3	13.0
UY Pup	07 44 10	− 12 49 36	16.8	13.0
BV Pup	07 46 58	− 23 26 24	15.7	13.12
U Gem	07 52 08	+22 08 06	14.9	8.2
BX Pup	07 52 09	− 24 11 42	16.0	13.76
YZ Cnc	08 07 52	+28 17 36	14.6	10.2
SU UMa	08 08 05.500	+62 45 22.45	14.53	12.16
Z Cha	08 08 38	− 76 23 30	16.23	11.5
Z Cam	08 19 39.760	+73 16 22.85	13.6	10.85
AT Cnc	08 25 15	+25 31	15.0	12.7
SW UMa	08 32 58.405	+53 39 04.80	16.50	11.17v
CC Cnc	08 33 24	+21 31 30	17.6	13.1
BB Vel	08 35 12	− 47 12 18	16.0	13.8
CT Hya	08 48 28	+03 19	20:	14.1
AK Cnc	08 52 35	+11 29 36	<17	13
BZ UMa	08 49 52.485	+58 00 04.10	16	10.5
CU Vel	08 56 41	− 41 36 12	15.5:	10.0
SY Cnc	08 58 13	+18 06 06	13.5	11.1
AR Cnc	09 19 07.925	+31 16 03.90	<17.4	15.3
TU Leo	09 27 00	+21 36 42	15.19	11.7
AG Hya	09 48 13	− 23 31 00	<16.5	14.3
X Leo	09 48 21	+12 06 36	15.7	11.1
RU LMi	09 59 11.325	+34 05 30.80	19.5	13.8

***Adapted from Bruch, Fischer and Wilmsen (1987), Vogt and Batesen (1982), and Kholopovu (1985, 1987).**

Catalogue of Dwarf Novae*

Name	R.A.(1950) h m s	Dec.(1950) ° ′ ″	m_{min} (mag)	m_{max} (mag)
CH UMa	10 03 08.870	+67 47 25.95	15.9	10.7
OY Car	10 05 17	− 69 59 24	17	12.2
CP Dra	10 11 25	+73 41 00	20	15.1
CI UMa	10 14 09.000	+72 10 44.70	<17.5	13.75
RX Cha	10 36 15	− 79 47 06	<16.9	14.4
V436 Cen	11 11 35	− 37 24 24	15.2	11.5
V442 Cen	11 22 25	− 35 37 18	<16.5	11.89
RZ Leo	11 34 49	+02 05 30	17.5	11.5
T Leo	11 35 53	+03 38 54	15.71	10.0
TW Vir	11 42 48	− 04 09 24	16.40	11.2
SW Crt	11 49 39	− 24 12 30	17.5	15.2
MU Cen	12 10 18	− 44 11	15.0:	11.8
V373 Cen	12 23 22	− 45 33	15.8	13.5
AL Com	12 29 54	+14 37 18	20.0	13.0
EX Hya	12 49 43	− 28 58 42	13.99	9.6
GO Com	12 54 11	+26 52 54	20	13.1
V485 Cen	12 54 39	− 32 57 00	<16.5	12.9
BM Cha	13 04 05	− 77 39 06	16.63	14.6
NN Cen	13 11 05	− 60 36 30	17.5	13.2
BV Cen	13 28 10	− 54 43 06	13.6	10.7
GY Hya	14 27 42	− 25 39 24	16	14
UZ Boo	14 41 42	+22 13 18	<16.1	11.5
TT Boo	14 55 51.295	+40 55 39.85	<15.6	12.7
EK TrA	15 09 40	− 64 55 06	15.0	10.4
AG Aps	15 24 20	− 75 34 42	<16.0	14.3
BR Lup	15 32 32	− 40 24 30	17.5	13.1
AB Nor	15 45 49	− 42 55 54	<16.0	13.9
HP Nor	16 16 52	− 54 45 12	16.41	12.8
V589 Her	16 19 54	+19 29	<17.5	14.1
IK Nor	16 21 28	− 55 13 18	16.3	12.9
V544 Her	16 35 39	+08 43	20:	14.5
AH Her	16 42 06	+25 20 36	14.7	10.9
V422 Ara	16 54 46	− 61 38 48	16.2	14.2
V601 Sco	17 00 42	− 36 39 18	<16.5	15.0
FQ Sco	17 04 51	− 32 37 42	18.45	12.0

Catalogue of Dwarf Novae*

Name	R.A.(1950) h m s	Dec.(1950) ° ′ ″	m_{min} (mag)	m_{max} (mag)
V2051 Oph	17 05 14	- 25 44 36	17.5	13.0
V2101 Oph	17 06 31	- 27 14 54	16.5	13.2
V499 Ara	17 10 09	- 58 50 24	<16.2	14.8
DT Aps	17 16 04	- 75 07 12	<15.8	14.4
V478 Sco	17 22 38	- 35 29 36	16.69	14.0
AT Ara	17 26 51	- 46 03 36	14.9	11.5
MM Sco	17 27 12	- 42 08 54	18.53	13.0
FV Ara	17 30 19	- 63 00 42	<18	12
BF Ara	17 34 25	- 47 09 00	<16.0	13.6
V735 Sgr	17 56 40	- 29 33 42	16.5	13.5
V551 Sgr	17 57 36	- 34 35 30	20	13.5
UZ Ser	18 08 33	- 14 56 18	16.7	12.0
V1830 Sgr	18 10 42	- 27 42 54	17.5	11.5
V391 Lyr	18 19 31.575	+38 46 13.75	17.0	14.0
DP Pav	18 21 24	- 64 59 24	<16.5	15
YY Tel	18 29 51	- 54 01 18	<17.4	14.4
CH Her	18 32 43	+24 45 30	17:	13.5
BP CrA	18 33 26	- 37 28 18	15.9	13.5
LL Lyr	18 33 31.325	+38 17 34.45	17.1	12.8
AY Lyr	18 42 44.120	+37 56 42.10	18.5B	12.5B
V344 Lyr	18 43 07.265	+43 19 16.10	<20	14.5
CY Lyr	18 50 39	+26 41 30	17.0	13.2
V800 Aql	18 54 10	+10 44 24	15.7	12.8
V415 Lyr	19 06 09.110	+31 19 07.25	<17	14.8
CG Dra	19 06 23.715	+52 53 39.80	17.5	15
V363 Lyr	19 07 17.340	+42 55 38.85	<18	15.5
FO Aql	19 14 05	+00 02 12	17.5	13.6
V1113 Cyg	19 21 30.265	+52 38 08.10	<17	14
V1114 Cyg	19 22 36.630	+28 20 05.30	<17	14.9
DH Aql	19 23 27	- 10 21 24	<17.0	12.5
V868 Cyg	19 27 04.83	+28 48 19.5	<17.79	14.86
V792 Cyg	19 29 08.435	+33 40 40.25	17	14.1
V793 Cyg	19 31 13.07	+33 08 59.5	18.5	14.4
KX Aql	19 31 36	+14 11 06	17.5:	12.5
HN Cyg	19 31 40.070	+28 49 41.45	16.0	15.2

***Adapted from Bruch, Fischer and Wilmsen (1987), Vogt and Batesen (1982), and Kholopovu (1985, 1987).**

Catalogue of Dwarf Novae*

Name	R.A.(1950) h m s	Dec.(1950) ° ′ ″	m_{min} (mag)	m_{max} (mag)
V795 Cyg	19 32 37.875	+31 25 34.25	>17.9:	13.4
V905 Cyg	19 33 07.55	+30 27 47.0	<18.13	16.5
V1141 Aql	19 34 39	+02 29 12	20	14.5
EM Cyg	19 36 42.040	+30 23 33.00	14.20	12.74
FY Vul	19 39 30	+21 38 48	15.33	13.4
V1153 Cyg	19 46 22.06	+34 44 25.7	<19	16.4
V811 Cyg	19 46 31.645	+36 19 03.55	<17.7	12.7
V1449 Cyg	19 47 23.13	+34 03 13.0	<17	15.5
V1006 Cyg	19 47 45.335	+57 01 44.55	17.0	15.4
V813 Cyg	19 47 46.460	+36 37 01.65	<17.8	16.0
V542 Cyg	19 48 13.695	+58 24 21.60	<17.5	13.0
V1050 Aql	19 49 25	+10 42	17.5	14.5
AB Dra	19 51 03.960	+77 36 40.10	14.57	12.28
V1454 Cyg	19 51 46.170	+35 13 51.15	<17	13.9
EY Cyg	19 52 44.130	+32 13 58.45	15.5	11.4
V725 Aql	19 54 21	+10 41 18	16.2	13.7
UU Aql	19 54 36	− 09 27 18	16.8	11.0
AW Sge	19 56 21	+16 33 06	<17.5	13.8
SW Vul	19 57 56	+22 48 00	18.5	14.5
V1028 Cyg	19 59 50.680	+56 48 12.50	18	13.0
V823 Cyg	19 59 57.330	+35 58 58.75	<17.67	15.5
RZ Sge	20 01 02	+16 54 18	17.4	12.2
V1032 Cyg	20 01 35.820	+57 07 51.50	<18.0	15.5
V1363 Cyg	20 04 15.760	+33 33 57.50	<17.6	13.0
WZ Sge	20 05 21	+17 33 30	15.3	7.0
V1310 Cyg	20 07 46.765	+40 51 35.35	<19	15.9
V1316 Cyg	20 10 30.825	+42 36 50.30	<17.5	14.5
V1101 Aql	20 10 44	+15 26 48	14.9	13.8
V794 Aql	20 14 55	− 03 49 06	16.5	14.0
CM Del	20 22 38	+17 08 36	15.3	13.4
V2276 Sgr	20 22 52	− 43 50 36	<16.7	14.3
EZ Del	20 23 03	+15 36 06	16.5	14.5
KK Tel	20 24 54	− 52 28 24	19.7:	13.5
V503 Cyg	20 25 34.760	+43 31 25.75	16.7	13.4
V1390 Cyg	20 26 32.810	+38 53 53.45	<18.1	16.0

Catalogue of Dwarf Novae*

Name	R.A.(1950) h m s	Dec.(1950) ° ′ ″	m_{min} (mag)	m_{max} (mag)
IS Del	20 28 51	+16 15	<17.5	15
TU Ind	20 29 42	− 45 36 36	15.54	12.9
TT Ind	20 29 46	− 56 44 00	<16.5	14.0
IL Vul	20 36 20.520	+22 31 41.60	20:	15
V516 Cyg	20 45 20.750	+41 44 22.30	16.8	13.8
UY Vul	20 53 43.285	+26 29 18.20	<17	14.1
UZ Vul	20 54 33.190	+23 23 01.55	<17	14
VW Vul	20 55 34	+25 18 48	16.27	13.1
AO Oct	20 59 25	− 75 33 18	21	13.5
VY Aqr	21 09 28	− 09 01 54	16.6	8.0
VZ Aqr	21 27 48	− 03 12 30	17.2	11.3
V1081 Cyg	21 32 28.10	+48 58 02.3	<17.5	14.4
V630 Cyg	21 32 59.435	+40 26 54.20	17.2	13.4
V1570 Cyg	21 38 25.595	+47 46 08.60	<16.8	15.0
SS Cyg	21 40 44.735	+43 21 24.45	11.61	8.72
V1404 Cyg	21 55 27.780	+51 57 38.00	<17.7	15.7
PS Lac	22 03 48.565	+51 24 01.95	17.63	15.01
RU Peg	22 11 36	+12 27 18	13.2	9.0
MN Lac	22 21 06.970	+52 25 46.65	<18.0	15.1
MR Lac	22 22 42.250	+50 16 26.75	<17.3	14.97
BS Cep	22 27 27.805	+64 59 11.40	16	14
EG Lac	22 28 16	+54 54	<17.5	16.4
AN Gru	23 05 04	− 47 41 12	16.1	14.9
KZ Cas	23 06 03.020	+56 11 01.65	19	15.2
CC Cep	23 08 25.705	+66 17 14.20	17.2	14.5
LM Cas	23 10 46.58	+56 34 59.2	<19	15.8
EG Aqr	23 22 44	− 08 35 00	18.5	14.0
DX And	23 27 21.710	+43 28 31.93	16.4	10.9

*Adapted from Bruch, Fischer and Wilmsen (1987), Vogt and Batesen (1982), and Kholopovu (1985, 1987).

25
Symbiotic Stars

25.1 Basic Data for Symbiotic Stars

Symbiotic stars are binary star systems in which a cool, red giant star sheds mass on to a more compact, hotter companion. Their optical spectra contain both the absorption features of a late-type giant star (G, K or M) and bright emission lines of hydrogen and helium. Additional absorption lines of hydrogen and helium and/or bright lines of ions may also be present. The hot components may be main sequence stars (with or without accretion) or accreting white dwarf stars. An ionized nebula similar to those seen in planetary nebulae may also be present. Studies of symbiotic stars are related to mass loss from giants, accretion onto compact stars, the formation of planetary nebulae, and the evolution of nova-like eruptions.

Our data relies heavily upon D.A. Allen's (1984) catalogue and the book by S.J. Kenyon (1986), supplemented with papers by Kenyon and Fernandez-Castro (1987), Kenyon and Garcia (1986), Munari et al.(1988) and Seaquist, Taylor and Button (1984). The first small table gives a few measured proper motions in right ascension, $\mu_{R.A.}$, and declination, $\mu_{Dec.}$, and the parallax, π. This is followed by tables of orbital data and a table of those symbiotic stars detected at either X-ray or radio wavelengths, with the X-ray flux, F_x, visual magnitude, V, the giant's spectral type, Sp, and the flux density, S_6, at 6 centimeters wavelength. The main catalogue of symbiotic stars contains their name, right ascension, R.A., and declination, Dec., at epoch 1950.0, the galactic longitudes and latitudes, l and b, for symbiotics that inhabit the Milky Way, and alternative designations. Allen's (1984) article also contains a uniform set of finding charts that we do not reproduce here.

PK Designation

Many of these systems were originally discovered as planetary nebulae, so their alternative designation often refers to lists of planetary nebulae, including the Perek-Kohoutek (PK) designation. Tables of planetary nebulae are given elsewhere in this volume. Also included are V835 Cen = PK 312-2°1, Hen 1092 = PK 319-9°1, KX Tra = PK 326-10°1, V455 Sco = PK 351+3°1, AS 221 = PK 352+3°1, M1-21 = PK 6+7°1, AE Ara = PK 344-8°1, AS 245 = PK 6+2°2, V2416 Sgr = PK 7+1°2, AS 281=PK 3-4°1, V2506 Sgr = PK 3-4°6, V2756 Sgr = PK 2-5°1, HD 319167 = PK 1-6°1 and AS 316 = PK 12-7°1.

Proper Motions and Parallaxes*

Star	$\mu_{R.A.}$ (s yr^{-1})	$\mu_{Dec.}$ ($''$ yr^{-1})	π ($''$)
Z And	+0.0010	+0.000	− 0.021 ± 0.013
EG And	+0.0009	− 0.015	
R Aqr	+0.0020	− 0.021	
UV Aur	− 0.0013	+0.005	
TX CVn	− 0.0013	− 0.014	
T CrB	− 0.0011	+0.020	
CH Cyg	− 0.0014	− 0.022	
AG Dra	− 0.0006	− 0.013	
RW Hya	− 0.0025	+0.014	
AG Peg	− 0.0005	− 0.013	− 0.003 ± 0.009
BL Tel	+0.0024	+0.005	

*Adapted from Kenyon (1986).

Orbital Periods*

Star	P (days)	Star	P (days)	Star	P (days)	Star	P (days)
Z And	756.9	V748 Cen	564.8	V1329 Cyg	959	AG Peg	827.0
EG And	481.2	T CrB	227.5	AG Dra	554	AX Per	681.6
R Aqr	16060::	BF Cyg	757.3:	RW Hya	372.5:	V2601 Sgr	850:
UV Aur	395.2	CH Cyg	5750::	BX Mon	1380:	V2756 Sgr	243:
TX CVn	70.8:	CI Cyg	855.25	SY Mus	627.0	CL Sco	624.7:
				AR Pav	604.6	BL Tel	778.6

Orbital Elements*

Star	P (days)	K_1 (km s^{-1})	K_2 (km s^{-1})	V_0 (km s^{-1})	e	ω	a sin i (10^{11} cm)	f(M) ($M_\odot$)
EG And	481.2	8.6		− 94.8	0.24	112	47	0.018
TX CVn	70.8	9.8		− 6.3	0.56	103	11	0.004
T CrB	227.5	23.3	33.8	− 27.9	0.06	90	73	0.488
CH Cyg	5750	6.8		− 58.4	0.29	201		0.16
CI Cyg	855.25	23	16	+20	0.0		390	1.0
V1329 Cyg	950		63.0	− 34.0	0.17	40	118	23
AR Pav	605	13.0		− 63.6	0.11	127	154	0.140
AG Peg	820	5.1	17–22	− 16.3	0.23	222	82	
BL Tel	778	19.3		+91.8	0.31	91	281	0.496

*Adapted from Kenyon (1986), Kenyon and Garcia (1986), and Munari et al. (1988).

X-ray and Radio Emission*

Star	F_x (10^{-12} erg cm^{-2} s^{-1})	V (mag)	Giant Sp Type	S_6 (mJy)
RX Pup			M	20V
SS 38		13.8	M	13
RW Hya		10	M1.1 III	0.38
V835 Cen	< 0.11		M	20
T CrB	0.09	10.0	M4.1 ± 0.3 III	< 0.88
AG Dra	2.1	11.2	K3	< 0.41
He2–171	< 0.12		M	2.0V
He2–173			M	0.51
He2–176			M7	14
AS 210		11.5	G:	1.78
V455 Sco	< 0.08	15	M	2.36
V2116 Oph	8000V	19	M	< 0.78
RT Ser		13	M5.5 ± 0.4	1.32
SS 96		12	M2	2.57
H1–36			M	40
AS 245		11.0	M6	0.87
V2416 Sgr		13	M	3.52
SS 122		12.0	M7	1.64
H2–38			M8	13
AS 296		10.5	M5	0.63
He2–390		12	M	0.73
V1017 Sgr	0.56	14	G5	
BF Cyg		12	M5 ± 1 III	2.05
CH Cyg		7	M6.5 ± 0.3 III	1.38
HM Sge	0.83	16	> M4	33V
V1016 Cyg	0.08	16	M6	40
RR Tel	0.18	14	M	28
V1329 Cyg		14	M5	2.16
AG Peg		9.4	M3.0 ± 0.4 III	8.15
Z And		10.5	M3.5 ± 1 III	1.16
R Aqr	0.09	5.8	M7 III	6V

*Adapted from Kenyon (1986), Kenyon and Fernandez- Castro (1987), and Seaquist, Taylor and Button (1984).

25.2 Catalogue of Symbiotic Stars*

Star	R.A.(1950) h m s	Dec.(1950) ° ′ ″	l °	b °	Other
EG And	00 41 52.7	+40 24 22	121.5	− 22.17	HD 4174
SMC N60	00 55 30.2	− 74 29 30			Ln 323
SMC Ln 358	00 57 42.3	− 75 21 29			
SMC N73	01 03 18.9	− 76 04 28			Ln 445a
AX Per	01 33 05.7	+54 00 07	129.5	− 8.04	MWC 411
V471 Per	01 55 32.9	+52 39 15	133.1	− 8.64	M1–2
UV Aur	05 18 33.3	+32 27 50	174.2	− 2.35	34842
Sanduleak's	05 46 02.7	− 71 17 13			(LMC)
LMC S63	05 48 52.1	− 67 37 02			HV 12671
BX Mon	07 22 52.9	− 03 29 51	220.0	+5.88	AS 150
Wray 157	08 04 32.2	− 28 23 18	246.6	+1.95	
RX Pup	08 12 28.2	− 41 33 18	258.5	− 3.93	69190
Hen 160	08 23 26.7	− 51 18 46	267.7	− 7.87	Wra 208
AS 201	08 29 36.8	− 27 35 20	249.1	+6.97	Hen 172
He2–38	09 53 03.7	− 57 04 39	280.8	− 2.24	PK 280–2°1
SS 29	11 06 27.3	− 65 31 02	292.6	− 5.00	
SY Mus	11 29 55.0	− 65 08 36	294.8	− 3.81	HD 100336
BI Cru	12 20 40.3	− 62 21 39	299.7	+0.06	Hen 782
TX CVn	12 42 17.9	+37 02 14	115.1	+80.2	
He2–87	12 42 48.3	− 62 44 09	302.3	− 0.14	PK 302–0°1
Hen 828	12 48 01.6	− 57 34 28	302.9	+5.03	SS 37
SS 38	12 48 21.8	− 64 43 40	302.9	− 2.13	
Hen 863	13 04 49.0	− 47 44 22	305.7	+14.78	
St2–22	13 11 22.4	− 58 36 01	305.9	+3.87	
Hen 905	13 27 23.2	− 57 42 50	308.1	+4.51	SS 40
RW Hya	13 31 31.9	− 25 07 29	315.0	+36.49	HD 117970
Hen 916	13 31 59.1	− 64 30 25	307.6	− 2.29	SS 42
V835 Cen	14 10 22.7	− 63 11 45	312.0	− 2.03	He2–106
BD −21°3873	14 13 45.8	− 21 31 56	327.9	+36.95	
He2–127	15 21 10.4	− 51 39 15	325.5	+4.18	PK 325+4°2
Hen 1092	15 42 29.7	− 66 19 58	319.2	− 9.35	He2–134
Hen 1103	15 45 00.3	− 44 09 50	333.2	+7.90	
He2–139	15 50 48.6	− 55 20 48	326.9	− 1.40	
T CrB	15 57 24.5	+26 03 39	42.4	+48.16	HD 143454
AG Dra	16 01 23.2	+66 56 25	10.3	+40.97	

Catalogue of Symbiotic Stars*

Star	R.A.(1950) h m s	Dec.(1950) ° ′ ″	l °	b °	Other
W16-202	16 03 15.0	- 49 18 40	332.8	+1.95	
He2-147	16 09 55.6	- 56 51 56	327.9	- 4.30	PK 327-4°1
UKS-Ce1	16 12 31.2	- 22 04 50	353.0	+20.25	
Wray 1470	16 20 16.2	- 27 33 16	350.1	+15.24	Hen 1187
He2-171	16 30 47.0	- 34 59 12	346.0	+8.55	PK 346+8°1
Hen 1213	16 31 23.0	- 51 36 14	333.9	- 2.81	PK 339+0°1
He2-173	16 32 58.9	- 39 45 40	342.8	+5.01	PK 342+5°1
He2-176	16 37 54.3	- 45 07 22	339.4	+0.74	PK 339+0°1
KX TrA	16 40 00.3	- 62 31 40	326.4	- 10.94	Hen 1242
AS 210	16 48 15.6	- 25 55 24	355.5	+11.55	Hen 1265
HK Sco	16 51 29.8	- 30 18 17	352.5	+8.27	AS 212
CL Sco	16 51 40.3	- 30 32 30	352.3	+8.09	AS 213
V455 Sco	17 04 04.1	- 34 01 18	351.2	+3.89	AS 217
Hen 1341	17 05 42.3	- 17 22 41	5.0	+13.39	SS 75
Hen 1342	17 05 53.1	- 23 19 48	0.1	+9.92	SS 77
AS 221	17 08 56.2	- 32 34 12	353.0	+3.93	H2-4
H2-5	17 12 04.7	- 31 30 39	354.2	+4.02	PK 354+4°2
Th3-7	17 17 51.9	- 29 19 55	356.7	+4.26	PK 356+4°3
Draco C-1	17 19 08.5	+57 53 01			
Th3-17	17 24 21.4	- 29 00 28	357.8	+3.28	PK 357+3°3
Th3-18	17 25 17.1	- 28 36 09	358.2	+3.33	PK 358+3°5
Hen 1410	17 25 54.9	- 29 41 01	357.4	+2.62	Th3-20
V2116 Oph	17 28 57.9	- 24 42 35	1.9	+4.79	GX 1+4
Th3-30	17 30 34.2	- 28 05 20	359.3	+2.65	PK 359+2°1
Th3-31	17 31 05.6	- 29 27 12	358.2	+1.80	PK 358+1°2
M1-21	17 31 20.4	- 19 07 23	7.0	+7.36	He2-247
Pt-1	17 35 46.5	- 23 52 24	3.5	+3.94	
RT Ser	17 37 04.1	- 11 55 04	13.9	+9.97	MWC 265
AE Ara	17 37 19.7	- 47 01 50	344.0	- 8.66	Hen 1451
SS 96	17 38 04.9	- 36 46 17	352.8	- 3.38	
UU Ser	17 39 46.5	- 15 23 23	11.2	+7.62	AS 237
V2110 Oph	17 40 30.8	- 22 44 16	5.0	+3.6	AS 239
AS 239	17 40 31.6	- 22 44 16	5.0	+3.62	Hen 1465
SSM 1	17 40 32.6	- 36 02 07	353.7	- 3.41	
V917 Sco	17 44 41.5	- 36 07 19	354.1	- 4.17	Hen 1481

*Adapted from Allen (1984) and Kenyon (1986).

Catalogue of Symbiotic Stars*

Star	R.A.(1950) h m s	Dec.(1950) ° ′ ″	l °	b °	Other
H1–36	17 46 24.1	– 37 00 36	353.5	– 4.92	
RS Oph	17 47 31.6	– 06 41 40	19.8	+10.37	HD 162214
He2–294	17 48 28.5	– 32 54 10	357.3	– 3.18	PK 357–3°1
AS 245	17 48 59.5	– 22 18 50	6.3	+2.37	Hen 1510
B13–14	17 49 14.1	– 29 45 19	0.1	– 1.70	PK 0–1°4
B13–6	17 49 42.1	– 31 18 41	358.8	– 2.58	
B1 L	17 50 00.7	– 30 17 25	359.7	– 2.12	PK 359–2°1
AS 255	17 53 47.7	– 35 15 18	355.8	– 5.32	Hen 1525
V2416 Sgr	17 54 15.6	– 21 41 10	7.6	+1.44	M3–18
SS 117	17 59 07.5	– 31 59 14	359.2	– 4.66	PK 7+1°2
Ap 1–8	18 01 19.8	– 28 21 48	2.6	– 3.28	PK 2–3°1
SS 122	18 01 33.2	– 27 09 26	3.7	– 2.73	
AS 270	18 02 35.2	– 20 20 52	9.7	+0.42	Hen 1581
H2–38	18 02 51.5	– 28 17 23	2.8	– 3.54	PK 2–3°4
SS 129	18 03 54.0	– 29 36 50	1.8	– 4.39	T 17
V615 Sgr	18 04 17.5	– 36 06 47	356.1	– 7.60	He2–349
Hen 1591	18 04 25.8	– 25 54 10	5.1	– 2.68	
AS 276	18 05 37.3	– 41 13 58	351.6	– 10.24	Hen 1595
Ap 1–9	18 07 19.5	– 28 08 20	3.4	– 4.33	PK 3–4°2
AS 281	18 07 34.6	– 27 58 31	3.6	– 4.30	Ap 1–10
V2506 Sgr	18 07 51.6	– 28 33 21	3.1	– 4.63	Ap 1–11
SS 141	18 08 53.9	– 33 11 29	359.1	– 7.04	
AS 289	18 09 34.7	– 11 40 55	18.1	+3.19	Hen 1627
Y CrA	18 10 47.3	– 42 51 27	350.6	– 11.83	HD 166813
V2756 Sgr	18 11 22.5	– 29 50 19	2.4	– 5.92	AS 293
HD 319167	18 12 11.5	– 30 32 56	1.8	– 6.41	CnMy 17
YY Her	18 12 25.9	+20 58 20	48.1	+17.25	AS 297
He2–374	18 12 30.9	– 21 36 24	9.7	– 2.21	PK 9–2°1
AS 296	18 12 33.0	– 00 19 53	28.5	+7.93	SS 148
AS 295B	18 12 51.9	– 30 52 16	1.6	– 6.69	Hen 1641
AR Pav	18 15 24.6	– 66 06 07	328.5	– 21.61	MWC 600
Hen 1674	18 17 12.4	– 26 24 10	6.0	– 5.43	Wra 1864
He2–390	18 17 51.4	– 26 49 50	5.7	– 5.76	PK 5–5°2
V3804 Sgr	18 18 13.8	– 31 33 31	1.5	– 8.0	Hen 1676
V443 Her	18 20 02.9	+23 25 47	51.2	+16.60	MWC 603

Catalogue of Symbiotic Stars*

Star	R.A.(1950) h m s	Dec.(1950) ° ′ ″	l °	b °	Other
V3811 Sgr	18 20 28.5	− 21 54 45	10.3	− 3.98	He2− 396
CD −28°14567	18 22 16.7	− 28 37 42	4.5	− 7.46	AS 304
V1017 Sgr	18 28 53.0	− 29 26 06	4.5	− 9.1	
V2601 Sgr	18 35 00.7	− 22 44 30	11.2	− 7.35	AS 313
AS 316	18 39 33.4	− 21 20 46	12.9	− 7.67	He2−417
MWC 960	18 44 58.1	− 20 09 12	14.5	− 8.27	Hen 1726
AS 327	18 50 13.3	− 24 26 43	11.1	− 11.23	Hen 1730
FN Sgr	18 50 58.5	− 19 03 27	16.2	− 9.06	AS 329
Pe2−16	18 51 30.9	− 04 42 40	29.1	− 2.72	PK 29−2°1
V919 Sgr	19 00 51.6	− 17 04 24	19.0	− 10.31	AS 337
CM Aql	19 00 57.9	− 03 07 44	31.6	− 4.09	
AS 338	19 01 32.0	+16 21 47	49.0	+4.77	PK 48+4°1
Ap 3−1	19 08 05.4	+02 44 33	37.6	− 2.97	PK 37−2°1
BF Cyg	19 21 55.2	+29 34 34	62.9	+6.70	MWC 315
CH Cyg	19 23 14.2	+50 08 31	81.9	+15.5	
Hen 1761	19 37 20.8	− 68 14 45	327.7	− 29.76	
HM Sge	19 39 41.4	+16 37 33	53.6	− 3.15	
AS 360	19 43 35.7	+18 29 23	55.6	− 3.02	Hen 1771
CI Cyg	19 48 20.6	+35 33 23	70.9	+4.74	MWC 415
V1016 Cyg	19 55 19.8	+39 41 30	75.2	+5.68	AS 373
RR Tel	20 00 20.1	− 55 52 04	342.2	− 32.24	Hen 1811
PU Vul	20 19 01.1	+21 24 43	62.6	− 8.5	
He2−467	20 33 42.6	− 20 01 02	63.4	− 12.15	PK 63−12°1
He2−468	20 39 20.5	+34 34 07	75.9	− 4.44	
V1329 Cyg	20 49 02.6	+35 23 37	77.8	− 5.48	HBV 475
CD −43°14304	20 56 48.6	− 42 50 34	358.6	− 41.10	Hen 1924
V407 Cyg	21 00 24.1	+45 34 41	87.0	− 0.48	AS 453
AG Peg	21 48 36.2	+12 23 27	69.3	− 30.89	HD 207757
Z And	23 31 15.3	+48 32 31	110.0	− 12.09	HD 221650
R Aqr	23 41 14.3	− 15 33 43	66.5	− 70.33	HD 222800

*Adapted from Allen (1984) and Kenyon (1986).

Part VIII
Supernovae Explosions and their Remnants

26
Supernovae

26.1 Basic Data for Supernovae

Supernovae are stars that explode at the end stages of stellar evolution when all of their thermonuclear fuel has been consumed. At maximum, they can briefly outshine all of the stars in a galaxy combined. But they occur infrequently, perhaps once every 50 years or so for a typical galaxy. So the statistical properties of supernovae are best understood by monitoring the numerous galaxies rather than waiting for one to occur in our own Milky Way.

Such statistical studies indicate that there are two main types of supernovae, Type I and Type II, distinguished by the absence or presence of hydrogen lines in their spectra near maximum light. Type II supernovae occur preferentially in spiral galaxies, and usually in the spiral arms. Type I supernovae are the only type that occurs in elliptical galaxies, but supernovae of Types I and II occur with roughly equal rates in galaxies showing spiral structure. Type I has been subdivided into two categories, Ia and Ib, that respectively do or do not have the strong absorption feature at about λ 6150 Å, while Type II has been separated into two classes known as II-P (plateau) or II-L (linear) depending on whether they do or do not show a temporary flattening (plateau) in their blue light curves at about 30 to 80 days after maximum light.

The physical mechanisms for these explosive outbursts also differ according to type. Type II supernovae are attributed to the sudden implosion of the cores of massive supergiant stars with original main sequence masses of $M > 8M_{\odot}$. A central neutron star can be formed together with the explosive ejection of the supergiant envelope at velocities of thousands of kilometers per second. Vast quantities of explosive energy $E \approx 10^{51}$ erg are released (Trimble, 1982). The progenitor stars of Type I supernovae are less massive, with $M \approx 1$ to 8 $M_{\odot}$. Type Ia supernovae probably occur in close binary systems containing a low-mass white dwarf component (Iben and Tutukov, 1984; Kirshner, 1988). Accretion of mass from a companion pushes the white dwarf over the Chandrasekhar limit with subsequent core collapse; the total star can be blown apart in the resulting nuclear detonation (or deflagration). Type Ib supernovae may be due to the core bounce of more massive Wolf-Rayet stars of 4 to 7 $M_{\odot}$ (originally 15 to 25 $M_{\odot}$ on the main sequence; Ensman and Woosley, 1988). The identifying features of supernovae of Types Ia, Ib and II are given in the first accompanying table.

The predicted supernova rates for our own Milky Way are about one per century for each of these three types (Van den Bergh, McClure and Evans, 1987), but only seven or eight have been recorded in our Galaxy during the past millenium. This is in part due to interstellar dust that obscures the light of all but the nearest ones. A supernova occurring in the galactic plane could, for example, only be seen with the naked eye if it was within 10 kiloparsecs. The second table gives the explosion dates, apparent visual magnitudes and estimated distances for the historical supernovae, together with the diameters of their explosive remnants that are seen today; additional data for four of them are contained in the next table, while their positions can be found in the main table of supernova remnants.

On 23 February 1987 the explosion of a blue supergiant star was observed in the Large Magellanic Cloud, producing the first supernova to be visible to the naked eye in 400 years. General review articles about this supernova, called SN 1987A, have been written by Arnett et al. (1989), Murdin (1989), Trimble (1988) and Woosley and Phillips (1988). Our last accompanying table describes the progenitor star, neutrino emission, explosive debris, and radioactivity of SN 1987A.

The neutrinos released during core collapse were detected on 23.316 February, or about 3 hours before the visible explosion. The Japanese Kamiokande II and the American IMB Cherenkov detectors recorded the release of about 2.5×10^{53} *erg* of neutrino energy that briefly provided a luminous output equal to that of all the stars in the visible Universe (Bionta et al., 1987; Hirata et al., 1987).

Visible radiation accompanied the explosive release of about 10^{51} *erg*, with ultraviolet, optical, radio and X-ray emission (respectively Fransson et al., 1989; Filippenko, 1988; Bartel et al., 1988; and Massei et al., 1988). A secondary maximum and exponential decline in the light output was apparently powered by the explosive synthesis of radioactive nickel-56 that decayed into radioactive cobalt-56 (Woosley, 1988) with the emission of gamma-ray spectral lines (Cook et al., 1988; Mahoney, et al., 1988).

Characteristics of Supernova Types*

Characteristic	Type Ia	Type Ib	Type II
Optical Spectrum	No hydrogen λ 6150 Å present	No hydrogen No λ 6150 Å	Hydrogen present
Maximum Absolute Magnitude†	$M_B = -19.69 \pm 0.45$	$M_B \approx -18.0$	$M_B = -18.0 \pm 0.8$
Supernova Rate††	0.3 h^2 SNU	0.4 h^2 SNU	1.1 h^2 SNU
Ejection Velocity	$\geq 10^4$ km s^{-1}	$\geq 10^4$ km s^{-1}	$\leq 10^4$ km s^{-1}
Ejected Mass	$\approx 1\ M_\odot$	$\approx 1\ M_\odot$	$\approx 5\ M_\odot$
Kinetic Energy	$\approx 10^{51}$ erg	$\approx 10^{51}$ erg	$\geq 10^{51}$ erg
Progenitor Star	Lowish mass white dwarf $M \approx 1\ M_\odot$	Massive Wolf–Rayet $M \approx 4$ to $7\ M_\odot$	Massive Supergiant $M \geq 8\ M_\odot$
Explosion Mechanism	Accretion from binary component, nuclear detonation/deflagration		Core collapse and bounce
Remnant Star	None		Neutron

*Adapted from Trimble (1982), Van den Bergh (1988), Van den Bergh, McClure and Evans (1987), Weiler and Sramek (1988).

†For $H_0 = 50$ km s^{-1} Mpc^{-1}, add 5 $\log(H_0/50)$ to get M_B for other values of H_0.

††$h = H_0/100$ km s^{-1} Mpc^{-1}, and one SNU is one supernova per 10^{10} $L_B(\odot)$ per century.

Historical Supernovae*

Explosion Date (AD)	Maximum Apparent Visual Magnitude, V (mag)	Time Visible to Unaided Eye (months)	Galactic Coordinates	Remnant Name	Distance (kpc)	Remnant Diameter (pc)
185	- 8.0	20	G 315.4–02.3	RCW 86	3.	35.0
386**	+1.5	3	G 11.2–00.3		$\geq$ 5.	$\geq$ 6.0
393	0.0	8	G 348.5+00.1	CTB 37A	10.4	24.0
			or G 348.7+00.3	CTB 37B	10.4	24.0
1006	- 9.5	> 24	G 327.6+14.6	PKS 1459–41	1.0	8.8
1054	- 5.0	22	G 184.6–05.8	Crab Nebula, 3C 144	2.0	2.9
1181	0.0	6	G 130.7+03.1	3C 58	2.6	5.3
1572	- 4.0	16	G 120.1+01.4	Tycho, 3C 10	2.3	5.4
1604	- 3.0	12	G 4.5+06.8	Kepler, 3C 358	4.4	3.8

*Adapted from Clark and Stephenson (1977), with distances given by Green (1984) and maximum observable radio diameters from Berkhuijsen (1986). Uncertainties in the distances and sizes can be found in the table of supernova remnants at known distances. Celestial coordinates of the historical supernovae are given in the main table of supernova remnants. See the following table for additional data for the Type I supernovae AD 185, 1006, 1572 and 1604.

**Possible nova.

Historical Type I Supernovae*

Supernova	185	1006	1572	1604
Maximum Diameter, θ_{MAX}	44.8′	32.2′	512″	216″
Minimum Diameter, θ_{MIN}	35.4′	29.3′	452″	188″
Average Diameter, θ_{AV}	40.8′	30.6′	498″	194″
Assumed Distance, D	1.2 kpc	1.4 kpc	2.5 kpc	4.2 kpc
Linear Diameter, 2R	14.4 pc	12..7 pc	6.0 pc	3.9 pc
Maximum Absolute Visual Magnitude, M_V	-19.2 ± 2	-19.8 ± 1	-17.6 ± 0.5	-19.6 ± 0.5
Ambient Density, n_0	0.2 cm^{-3}	0.1 cm^{-3}	0.8 cm^{-3}	5.5 cm^{-3}
Mass Ejected, M_{ej}		0.2 to 0.5 $M_\odot$	0.4 $M_\odot$	
Mass Swept-Up, M_{sw}	6.9 $M_\odot$	2.5 $M_\odot$	2.1 $M_\odot$	4.2 $M_\odot$
Shock Velocity, V_s	1600 km s^{-1}	2500 km s^{-1}	2900 km s^{-1}	2000 km s^{-1}

*Adapted from Strom (1988).

Supernova 1987A (SN 1987A)

Progenitor Star

Name	Sk – 69°202	=	Sanduleak – 69°202
Location	LMC	=	Large Magellanic Cloud
Right Ascension	R.A.(1950.0)	=	05^h 35^m 49.992^s
Declination	Dec.(1950.0)	=	– 69° 17′ 50.08″
Distance	D	=	50 kpc
		=	170,000 light years
Distance Modulus	m – M	=	18.7 mag
Spectral Type	Sp	=	B3 Ia (blue supergiant)
Apparent Visual Magnitude	V	=	12.4 mag
Color Index	B – V	=	+0.04 mag
Visual Absorption	A_V	≈	0.5 mag
Bolometric Correction	B.C.	=	+1.15
Bolometric Absolute Magnitude	M_{bol}	=	– 7.9
Absolute Luminosity	L	=	4.6×10^{38} erg s^{-1}
		≈	10^5 $L_\odot$
Effective Temperature	T_{eff}	=	16,000 K
Radius	R	=	3×10^{12} cm ≈ 50 $R_\odot$
Initial Main Sequence Mass	M	=	20 $M_\odot$
Evolved Helium Core Mass	M_{He}	=	6 $M_\odot$
Evolved Hydrogen Envelope Mass	M_H	≈	10 $M_\odot$

Neutrino Burst

Kamiokande II Detector

Event Start Time	t_0	=	07^h 35^m 35^s UT (±1 min) on 23 Feb 1987
Number of Anti-Neutrinos Detected	N	=	11
Energy Range of Anti-Neutrinos	ΔE	=	7.5 to 36 MeV
Burst Duration	Δt	=	12.5 s

Irvine - Michigan - Brookhaven (IMB) Detector

Event Start Time	t_0	=	07^h 35^m 41.37^s UT (±10 ms) on 23 Feb 1987
Number of Anti-Neutrinos Detected	N	=	8
Energy Range of Anti-Neutrinos	ΔE	=	20 to 40 MeV
Burst Duration	Δt	=	5.6 s

Supernova 1987A (SN 1987A)

Neutrino Burst

Total Anti-Neutrino Energy	E	$\approx$	3×10^{52} erg
Total Neutrino Energy	E	$\approx$	2.5×10^{53} erg
Total Neutrino Luminosity	L	$\approx$	10^{55} erg s^{-1} $\approx 10^{22}$ $L_\odot$
Average Neutrino Temperature	T	=	4 MeV $\approx 10^{10}$ K
Number of Neutrinos Produced	N	=	10^{58} neutrinos
Neutrino Flux at Earth	F	$\approx$	5×10^{10} cm^{-2}
Inferred Neutrino Mass Limit	m	$\leq$	16 eV
Core Collapse Time	Δt	=	0.01 s = 10 ms
Collapsed Remnant Mass	M	=	1.4 $M_\odot$
Collapsed Remnant Radius	R	=	10 km
Gravitational Potential Energy Released	E	=	2.5×10^{53} erg

Visible Explosion*

Optical Discovery	V	=	5.0 mag on 24.122 Feb 1987 (t $\approx$ 19 hours) (Ian Shelton)
Explosive Optical Brightening	V	=	6.36 mag on 23.444 Feb 1987 (t $\approx$ 3 hours)
	M_{bol}	=	− 12.8 mag
	L	=	4.2×10^{40} erg s−1 $\approx 10^{7}$ $L_\odot$
Secondary Visual Maximum	V	=	2.97 mag on 20 May 1987 (t $\approx$ 90 days)
	M_{bol}	=	− 16.2 mag
	L	$\approx$	10^{42} erg s^{-1} $\approx 10^{8}$ $L_\odot$

*Time t from neutrino burst at t_0 = 23.316 Feb 1987

Supernova 1987A (SN 1987A)

Visible Explosion

Maximum Temperature	T_{max}	=	10^6 K
Maximum Expansion Velocity	V_{max}	=	35,000 km s^{-1}
Mean Expansion Velocity	V_{mean}	=	2,850 km s^{-1}
Explosive Kinetic Energy	K.E.	=	10^{51} erg
Element Abundance	N/C	=	$7.8 \pm 4 = 37\ (N/C)_{\odot}$
of Debris	N/O	=	$1.6 \pm 0.8 = 12\ (N/O)_{\odot}$

Radio Burst

Maximum Observed Flux Density	S_{max}	=	140 mJy at 1.4 GHz
Maximum Radio Luminosity	L_r	=	10^{33} erg s^{-1}
Angular Radius Limit	θ_r	$\geq$	0.00125″ at t=5.2 days
Linear Radius Limit	R_r	$\geq$	8.3×10^{14} cm at t = 5.2 days

X-Ray Emission

X-Ray Flux	F_x	=	(1 to 80) $\times 10^{-12}$ erg cm^{-2} s^{-1} (for 6 to 28 keV at t = 100 to 500 days)

Radioactivity

Nickel-56

Amount Synthesized	^{56}Ni	=	0.075 $M_{\odot}$
Half-life	$\tau_{1/2}$	=	6.1 days
Electron Capture	$^{56}Ni_{28} + e^-$	$\rightarrow$	$^{56}Co_{27} + \gamma + \nu_e$

Cobalt-56

Half-life	$\tau_{1/2}$	=	77.12 days
Electron Capture	$^{56}Co_{27} + e^-$	$\rightarrow$	$^{56}Fe_{26} + \gamma + \nu_e$
Gamma Ray Lines	γ	=	847 keV (100% decays)
	γ	=	1238 keV (68% decays)
Peak Gamma Ray Flux	F_{γ}	=	2×10^{-3} photons cm^{-2} s^{-1} (t = months)

27
Supernova Remnants

27.1 Basic Data for Supernova Remnants

Supernovae explosions give rise to expanding debris and shock fronts that interact with the interstellar medium and produce radiation at optical, X-ray and radio wavelengths; these objects are called supernova remnants. A great majority of the radio supernova remnants are shells that appears as bright rings in projection against the sky. These shell-shaped structures are designated by the letter S. The shell emits nonthermal radio radiation with a steep spectrum. It originates from relativistic (high-speed) electrons spiralling about magnetic fields; they are related to the original explosion and/or the interaction of the expanding supernova explosion shock front with the interstellar medium.

A subset of supernova remnants has a filled center at radio wavelengths; they have been designated by the letter F. These objects have also been given the name plerions from the Greek word pleres for full or filled. The central radio emission also has a nonthermal origin but a flat spectrum. Filled center supernova remnants like the Crab Nebula require a continued source of relativistic particles, such as a pulsar, to sustain optical and X-ray emission. A composite class, designated C, combines the properties of both shell and plerionic supernova remnants.

The identifying features of shells, plerions and composites have been reviewed by Weiler (1985); he also provides a detailed catalogue of plerionic and composite supernova remnants. At X-ray wavelengths the shell has a thermal origin and spectrum while the core is nonthermal with a power-law spectrum.

Observations at optical wavelengths can reveal thermal filaments that emit either the Balmer lines of hydrogen or strong lines of ionized oxygen [O III]. This has led to another classification scheme for young supernova remnants. They are divided into the oxygen-rich, plerionic and Balmer-dominated categories with the Cassiopeia A, Crab and Tycho supernova remnants serving as the respective prototypes (Van den Bergh, 1988). Older, more evolved supernova remnants can be added to these three types to give the classification scheme in the first accompanying table (also see Weiler and Sramek, 1988).

The properties of two of these prototypes, Cassiopeia A and the Crab Nebula, are given in the second table. Although Cassiopeia A is the youngest known galactic supernova remnant, it was not detected as a naked-eye supernova explosion and is therefore omitted from our preceding table of historical supernovae. The explosion that gave rise to Cassiopeia A may nevertheless have been detected as a "new star" by Flamsteed in 1680 (Ashworth, 1980). The Crab Nebula supernova remnant is the expanding debris of a supernova explosion that occurred in A.D. 1054. The Crab has almost everything, including a central pulsar, synchrotron radiation, expanding filaments, and detectable emission across most of the electromagnetic spectrum.

The distances of only a few supernova remnants are known with any accuracy. Those with good or reasonable distance estimates are given in the next table together with the surface brightness, Σ_R, and the linear diameter, D_R, at a radio frequency of 1 GHz. The surface brightness for a flux density, S, in Janskys and an angular extent, θ, in arcminutes is $\Sigma_R = 1.5 \times 10^{-19} S/\theta^2$ in units of $W\ m^{-2} Hz^{-1}\ sr^{-1}$.

The shell-type supernova remnants exhibit an approximate $\Sigma - D$ relation given by $\Sigma_R \approx 3.4 \times 10^{-15} D_R^{-3.8}$ at 1 GHz, where D_R is in parsecs, or $D_R \approx 1.7 \times 10^{-4} \Sigma_R^{-0.26}$ parsecs (Milne, 1979, Ilovaisky and Lequex, 1972). The $\Sigma - D$ relation can be used to derive approximate distances d, from $d \approx D_R/\theta$ for shell-type supernova remnants when the surface brightness and angular size are measured and the distance is not known. Green (1984, 1988) has argued that the $\Sigma - D$ relation only provides an upper limit to the diameter, and Berkhuijsen (1986) gives the relation $\Sigma_R \approx 2.5 \times 10^{-14} D_R^{-3.5}$ for the maximum observable radio diameter, D_R, at 1 GHz of supernova remnants with known distances in our Galaxy, the Magellanic Clouds, M 31 and M 33. The plerions generally adhere to a $S\theta^2 - d$ relation (Weiler, 1985).

The exact ages of the five remnants of historical supernovae can be inferred from the observed dates of their explosion. Those of shell-type supernova remnants can be obtained from the approximate relation $D_R \approx 0.9\ t^{2/5}$, where the linear diameter at 1 GHz is in parsecs and the age, t, is in years (Clark and Caswell, 1976; Caswell and Lerche, 1979; Milne, 1987). For filled supernova remnants, or plerions, there is the relation $Sd^2 \approx 7 \times 10^5\ t^{-0.8}$, for a flux density S at 1 GHz in Janskys, a distance, d, in kiloparsecs, and an age, t, in years. These two approximate relations have been used to obtain the ages given in the next table.

The upper limit to the age of an observable radio shell-type supernova remnant is apparently about $t \leq 10,000$ years. The mean expansion velocity probably has an upper limit given by $V \leq 10,000\ km\ s^{-1}$, so the largest observable radio diameter is $D_R \leq t\ V \approx 100\ pc$, for the shells. This, in turn, gives an upper limit to the distance $d \leq 100\, pc/\theta$, where the angular extent θ is given in radians. As the supernova remnants expand to larger ages and sizes, they thin out and slow down, eventually disappearing from view.

The next table lists 37 supernova remnants detected at optical wavelengths. Photographs of 23 of these objects were given by van den Bergh, Marscher and Terzian (1973). Spectral lines emitted by the optical filaments can be used to infer the expansion velocities. Maximum values of the observed radial velocities and proper motions are given in the next table together with the coordinates of the expansion center.

This is followed by tables of supernova remnants with synchrotron X-ray emission, those with compact X-ray objects, and the Einstein X-ray survey of supernova remnants.

The central pulsars of some supernova remnants have been detected at radio wavelengths and are described in the next table. The long radio wavelengths also enable astronomers to see through obscuring interstellar dust and detect supernova remnants throughout our Galaxy. Altogether 155 of these remnants are included in the next table which is adapted from Green (1988). It includes four newly-discovered shell-type supernova remnants (Fürst et al., 1987), and 12 of 32 remnants tentatively identified as new supernova remnants from deep radio continuum surveys (Reich et al., 1988). Here we provide the galactic coordinates of the radio source centroid, its right ascension, R.A., and declination, Dec., for the epoch 1950.0, the angular size in arcminutes, supernova type S, F, or C, the flux density at 1 GHz in Janskys, and the spectral index of the integrated radio emission, α, in the sense $S \propto \nu^{-\alpha}$ for flux density, S, and frequency, ν. It is followed by a cross-reference name index for all supernova remnants with multiple names.

Classification of Supernova Remnants*

Classification	Young Supernova Remnants			Older Evolved Remnants
	Balmer Dominated	Oxygen Rich	Plerionic –Composite	
Prototype	Tycho	Cassiopeia A	Crab	
Radio				
	1. Nonthermal	Nonthermal	Nonthermal	Nonthermal
	2. Full or partial shell form	Full or partial shell form	Extended, filled center form, possible surrounding shell	Partial shell form
	3. Steep spectrum, index $\alpha \leq -0.3$	Steep spectrum, index $\alpha \leq -0.3$	Flat spectrum plerion, index $\alpha \leq -0.3$, plus possible steep spectrum shell, index $\alpha \leq -0.3$	Steep spectrum, index $\alpha \leq -0.3$
	4. Linear polarization	Linear polarization	Linear polarization	Linear polarization
	5. Radial magnetic field when young	Radial magnetic field when young	Regular magnetic field in plerion plus radial or tangential field in shell	Mixed magnetic field direction

Classification of Supernova Remnants*

Classification	Young Supernova Remnants			Older Evolved Remnants
	Balmer Dominated	Oxygen Rich	Plerionic –Composite	
Prototype	Tycho	Cassiopeia A	Crab	
Optical				
	1. Thermal line emitting filaments of ISM material, strong in Balmer lines, weak in [O III] and [S II]	Thermal line emitting filaments of processed material, strong in [O III], high velocity dispersion	Thermal line emitting filaments throughout, plus nonthermal continuum from plerion	Thermal line emitting filaments of ISM material, [S II]/Hα > 0.7
	2. Full or partial shell organization of filaments	Full or partial shell organization of filaments	Filaments throughout source, continuum centrally concentrated	Partial shell organization of filaments

Classification of Supernova Remnants*

Classification	Young Supernova Remnants			Older Evolved Remnants
	Balmer Dominated	Oxygen Rich	Plerionic –Composite	
Prototype	Tycho	Cassiopeia A	Crab	
X–ray	1. Thermal emission	Thermal emission, high luminosity	Nonthermal plerion emission plus possible thermal shell emission	Thermal emission
	2. Both shell and filled forms seen	Both shell and filled forms seen	Extended, centrally concentrated emission plus compacted emitter, plus possible outer shell	Irregular shell form
Gamma–ray	1. None known	None known	Possible emitters	None known
Physical Properties	1. High velocity, collisionless, nonradiative shock in neutral gas	Shock interacting with internal processed material	Pulsar driven plus, for composites, external shock	Slow shock, 50–200 km s^{-1}, emission arises from shocked ISM cloudlets

*Adapted from Van den Bergh (1988) and Weiler and Sramek (1988).

Cassiopeia A*

Cassiopeia A* = G 111.7–02.1 = 3C 461			
Expansion Center	R.A.(1950.0)	=	23^h 21^m $12.0^s \pm 0.1^s$
	Dec.(1950.0)	=	$+58°$ $32'$ $17.9'' \pm 0.8''$
Explosion	Date	=	1681 ± 15 A.D. (no deceleration)
Progenitor Star Mass	M	≈	15 $M_\odot$ (main sequence)
Angular Extent	θ	=	390″= 6.5′
Distance	D	=	2.8 kpc
Linear Radius	R	=	2.65 pc
Maximum Radial Velocity	V_{rmax}	=	7,000 to 8,500 km s^{-1}
Maximum Proper Motion	μ_{max}	=	0.50 to 0.65 ″ yr^{-1}
Maximum Expansion Velocity	V_{exp}	=	10,000 km s^{-1}
Radio Flux Density	S	=	2723 Jy at 1 GHz
Spectral Index	α	=	– 0.856 (valid for 0.4 to 25 GHz)

*Adapted from Ashworth (1980), Baars et al. (1977), Fesen, Becker and Goodrich (1988), Green (1988), Jansen et al. (1988), Kamper and van den Bergh (1976) and Tuffs (1986).

The Crab Nebula*

Crab Nebula* = G 184.6–05.8 = 3C 144 = SN 1054

Expansion Center	R.A.(1950.0)	=	05^h 31^m $32.2^s \pm 0.1^s$
	Dec.(1950.0)	=	$+21°$ $58'$ $50'' \pm 1''$
Explosion	Date	=	1054 A.D.
Progenitor Star Mass	M	≈	9 $M_\odot$ (main sequence)
Ejected Mass	M_{ejc}	=	2 to 3 $M_\odot$
Neutron Star Mass	M_N	=	1.4 $M_\odot$
Pulsar Position	R.A.(1950.0)	=	05^h 31^m 31.405^s
	Dec.(1950.0)	=	$+21°$ $58'$ $54.39''$
Pulsar Period	P	=	0.033326323455 s
Pulsar Period Derivative	dP/dt	=	421.288×10^{-15} s s^{-1}
Angular Extent	θ	=	$4.5'$ x $7.0'$
Distance	D	=	2.0 ± 0.1 kpc
Linear Radius	R	=	2.7 pc x 4.2 pc
Maximum Radial Velocity	V_{rmax}	=	1,450 ± 40 km s^{-1}
Maximum Proper Motion	μ_{max}	=	0.222 ± 0.002 $''$ yr^{-1}
Maximum Expansion Velocity	V_{exp}	=	1,500 km s^{-1}
Radio Flux Density	S	=	1040 Jy at 1 GHz
Spectral Index	α	=	– 0.30 (integrated radio)
X–ray Luminosity	L_X	=	$10^{37.38}$ erg s^{-1}
Total Luminosity	L	=	$10^{38.14}$ erg s^{-1}

*Adapted from Aller and Reynolds (1985), Davidson and Fesen (1985), Kafatos and Henry (1985), Mitton (1979), Nomoto (1985), Trimble (1968, 1973, 1982, 1985), and Woltjer (1958).

Galactic Supernova Remnants at Known Distances*

Galactic Coordinates	Remnant Name	Distance (kpc)	D_R (pc)	$\log \Sigma_R$ (w Hz^{-1} m^{-2} sr^{-1})
Remnants with good distance determinations				
34.7–00.4	W 44	3.0 ± 1.0	26.0	19.409
111.7–02.1	Cas A	2.8 ± 0.6	4.1	16.721
120.1+01.4	Tycho†	2.3 ± 0.3	5.4	18.921
130.7+03.1	3C 58	2.6 ± 0.5	5.3	19.000
184.6–05.8	Crab††	2.0 ± 0.5	2.9	17.244
320.4–01.2	RCW 89	4.2 ± 0.5	37.0	20.013
332.4–00.4	RCW 103	3.3 ± 0.5	9.6	19.377
348.5+00.1	CTB 37A	10.4 ± 3.5	24.0	18.770
348.7+00.3	CTB 37B	10.4 ± 3.5	24.0	19.215
349.7+00.2		18.3 ± 4.6	11.0	18.125
Remnants with reasonable distance estimates				
4.5+06.8	Kepler	4.4	3.8	18.481
18.8+00.3	Kes 67	14.	61.0	19.745
43.3–00.2	W 49B	10.	11.0	18.432
116.9+00.2	CTB 1	3.	32.0	20.959
132.7+01.3	HB 3	3.	61.0	20.721
263.9–03.3	Vela XYZ	0.5	36.0	20.367
315.4–02.3	RCW 86	3.	35.0	20.276
327.6+14.6	SN 1006	1.	8.8	20.569

*Adapted from Green (1984) and Berkhuijsen (1986). D_R is the maximum observable radio diameter, and Σ_R is the radio surface brightness; both are given for a frequency of 1 GHz. To convert brightness into c.g.s. units use 1 w Hz^{-1} m^{-2} sr^{-1} = 10^3 erg s^{-1} Hz^{-1} cm^{-2} sr^{-1}.

†Black and Raymond (1984) showed that interstellar absorption lines in the direction of Tycho's supernova remnant are consistent with a distance of 2.0 to 2.5 kpc.

††Trimble (1973) used twelve lines of evidence for the distance to the Crab Nebula and its pulsar NP 0532 to obtain a distance of 1.930 ± 0.110 kpc.

Ages of Supernova Remnants*

Remnant	Age years	Type
G 5.4− 1.0	2500	Shell
G 6.4− 0.1	6000	Shell
G 7.7− 3.7	2000	Shell
G 18.8+0.3	4000	Shell
G 21.5− 0.9	500?	Filled
G 21.8− 0.6	4000	Shell
G 34.6− 0.5	2500	Shell
G 74.0− 8.6	12000	Shell
G 74.9+1.2	3000	Filled
G 89.0+4.7	7000	Shell
G 93.3+6.9	1500	Shell
G 111.7− 2.1	300	Shell
G 120.1+1.4	500†	Shell
G 127.1+0.5	18000	Shell
G 130.7+3.1	1000†	Filled
G 184.6− 5.8	1000†	Filled
G 189.1+2.9	2500	Shell
G 260.4− 3.4	3000?	Shell
G 263.9− 3.0	11000	Shell
G 291.0− 0.1	4000	Filled
G 296.5+10.0	5000	Shell
G 315.4− 2.3	1800†	Shell
G 316.3− 0.0	700	Shell
G 320.4− 1.2	1600?	Shell
G 326.3− 1.8	4000	Filled
G 327.6+14.0	980†	Shell
G 327.4+0.4	6000	Shell
G 332.4+0.1	5000	Shell

*Adapted from Caswell and Lerche (1979), Milne (1987), and Weiler and Pangia (1980).

†Use the explosion dates given in the table of historical supernovae to obtain the exact ages of these objects at the time of reading this table.

Supernova Remnants Detected at Optical Wavelengths*

Galactic Coordinates	Name	Size (′)	Galactic Coordinates	Name	Size (′)
119.5+10.2	CTA 1	90	327.6+14.6	PKS 1459–41, SN 1006	30
120.1+01.4	Tycho, 3C 10, SN 1572	8	320.4– 01.2	RCW 89, MSH 15–52	30
126.2+01.6		70	326.3– 01.8	MSH 15–56	36
130.7+03.1	3C 58, SN 1181	9 x 5	332.4– 00.4	RCW 103	9
132.7+01.3	HB 3	80	338.1+00.4		12
160.9+02.6	HB 9	140 x 120	4.5+06.8	Kepler 3C 358, SN 1604	3
166.2+02.5	OA 184	90 x 70	6.4– 00.1	W 28	42
166.0+04.3	VRO 42.05.01	55 x 35	5.4– 01.2	Milne 56	35
184.6– 05.8	Crab, 3C 144, SN 1054	7 x 5	39.7– 02.0	W 50, SS 433	120 x 60
180.0– 01.7	S 147**	180	65.3+05.7	S 91 + S 94	310 x 240
189.1+03.0	IC 443, 3C 157	45	53.6– 02.2	3C 400.2, NRAO 611	28
205.5+00.5	Monoceros	220	69.0+02.7	CTB 80	80:
206.9+02.3	PKS 0646+06	60 x 40	78.2+02.1	DR 4, γ Cygni	60
260.4– 03.4	Puppis A, MSH 08–44	60 x 50	74.0– 08.5	Cygnus Loop	230 x 160
263.9– 03.3	Vela XYZ	255	109.1– 01.0	CTB 109	28
290.1– 00.8	MSH 11–61A	15 x 10	111.7– 02.1	Cassiopeia A, 3C 461	5
292.0+01.8	MSH 11–54	12 x 8	116.9+00.2	CTB 1	34
296.5+10.0	PKS 1209–51/52	90 x 65			
315.4– 02.3	RCW 86, MSH 14–63, SN 185	40			

*Adapted from Van den Bergh (1981, 1983) with total angular extents, or sizes, from Green (1988). Here we do not include G 284.2–01.7, G 339–00.4, and G 342.0+00.1 that are listed by Van den Bergh, but do not appear in Green's (1988) catalogue of supernova remnants. Positions and radio flux densities are given in our main catalogue of supernova remnants.

**The designation S 147 is due to the catalogue compiled at Simeiz by Gaze and Shain in 1952; it has nothing to do with the Sharpless catalogue of H II regions.

Expansion of Supernova Remnants*

Galactic Coordinates	Remnant Name	Expansion Center† R.A.(1950.0) h m s	Dec.(1950.0) ° ′ ″	Radial Velocity†† (km s^{-1})	Proper Motion†† (″ yr^{-1})
111.7− 02.1	Cas A, 3C 461	23 21 12.0	+58 32 17.9	7,000 to 8,500	0.50 to 0.65
120.1+01.4	Tycho, 3C 10, SN 1572	00 23 31	+63 51 36	1,800 ± 200	0.20 ± 0.01
130.7+03.1	3C 58, SN 1181	02 01 55	+64 35	1,100 ± 100	0.05 to 0.07
180.0− 01.7	S 147	05 36 00	+27 50	80	
184.6− 05.8	Crab, 3C 144, SN 1054	05 31 32.2	+21 58 50.0	1,450 ± 40	0.220 to 0.225
260.4− 03.4	Puppis A, MSH 08–44	08 20 44.3	− 42 47 48	400 to 1,600	0.09 to 0.22
327.6+14.6	PKS 1459–41, SN 1006	14 59 35	− 41 44	5,000 to 6,500	0.30 ± 0.04

†Accurate positions are for the center of expansion, less accurate ones for the radio source centroid.

††Maximum observed radial velocity or proper motion for optical filaments.

*Adapted from Chevalier, Krishner and Raymond (1980), Dickel et al. (1988), Fesen, Becker and Goodrich (1988), Fesen (1984), Fesen, Kirshner and Becker (1988), Fesen et al. (1988), Kamper and Van Den Bergh (1976), Kirshner and Arnold (1979), Kirshner, Winkler and Chevalier (1987), Long, Blair and Van Den Bergh (1988), Reynolds and Aller (1988), Strom, Goss and Shaver (1982), Trimble (1968) and Winkler et al. (1988).

Pulsars In Supernova Remnants*

Remnant Name	PSR	R.A.(1950) h m s	Dec.(1950) ° ′ ″	Period (seconds)	$\dot{P}$ (10^{-15})	D (kpc)
Crab Nebula	0531+21	05 31 31.405	+21 58 54.39	0.0333	421.3	2.0
LMC 0540–69	0540– 69	05 40 33.9	– 69 21 23.2	0.0503	479.8	55.0
Vela X	0833– 45	08 33 39.27	– 45 00 10.2	0.0893	124.4	0.5
G 320.4–1.2	1509– 58	15 09 59.06	– 58 56 58.0	0.1502	540.2	6.7
W 44	1853+01	18 53 38.45	+01 09 22.7	0.2674	208.4	3.2
CTB 80	1951+32	19 51 02.47	+32 44 50.2	0.3953	5.9	1.3
G 109.1–1.0	2259+58	22 59 03.2	+58 36 30	6.9789	710.	5.0

*Adapted from Morini et al. (1988), Taylor (1989) and Wolszczan (1989). Accurate periods and other data are given in the table of Pulsar Parameters, except for the X–ray pulsar, PSR 2259+58, with P = 6.978725 s.

27.2 X-Ray Radiation from Supernova Remnants

Synchrotron X–Ray Emission*

Galactic Coordinates	Remnant Name	N_H 10^{21} atom cm^{-2}	D (kpc)	log $L_X^{\dagger}$ (erg s^{-1})
21.5− 0.9		20	5	35.45
29.7− 0.2	Kes 75	50	21	36.93
68.8+2.6	CTB 80	6	3	34.68
74.9+1.2	CTB 87	16	12	35.17
111.7− 2.1	Cas A	10	3	< 35.28
130.7+3.1	3C 58	2	2.6	34.13
184.6− 5.8	Crab	3 ± 1	2.0 ± 0.2	37.38
263.8− 1.7	Vela XYZ	0.25 ± 0.1	0.5 ± 0.1	33.77
296.5+10.0	PKS 1209–52	1.4	1.5	32.97
320.3− 1.2	MSH 15–52	9 ± 3	4.2 ± 1.0	35.27
332.4− 0.4	RCW 103	5	3.3	33.76††
332.4− 0.4	RCW 103	5	3.3	< 34.60
279.7− 31.5	0540–69.3	2	55 ± 5	37.02

*Adapted from Seward and Wang (1990).
†Pulsar plus synchrotron nebula
††Unresolved source only.

Compact Objects in Supernova Remnants*

Galactic Coordinates	Remnant Name	R.A.(1950.0) h m s	Dec.(1950.0) ° ′ ″	D (kpc)	$L_X^\dagger$ (10^{35} erg s^{-1})	R (pc)
G 127.1+05		01 25 08.3	+62 50 59	3.8	0.02	28.5
G 130.7+3.1	3C 58	02 01 51.1	+64 35 23	2.6	0.13	2.0
G 184.6− 5.8	Crab	05 31 31.4	+21 58 54	2.0	240.0	1.5
G 279.7− 31.5	in LMC	05 40 33.9	− 69 21 23	55.0	104.7	0.65
G 263.9− 3.3	Vela X	08 33 39.3	− 45 00 10	0.5	0.06	4.5
G 27.4+0.0	KES 73	08 37 45	− 04 59	26.0	15.0	15.6
G 296.5+10.0	PKS 1209–52	12 07 23.5	− 52 09 49	2.0	8.0	12.2
G 320.4− 1.2	MSH 15–52	15 09 59	− 58 56 57	4.2	1.86	16.4
G 326.3− 1.8	MSH 15–56	15 48 25	− 56 03 30	$\geq$ 1.5		$\geq$ 1.8
G 332.4− 0.4	RCW 103	16 13 48.3	− 50 55 05	3.3	0.06	4.5
G 0.9+0.1		17 44 12	− 28 08 20	10.0	$\leq$ 1.4	2.9
G 6.6− 0.2	W 28	17 57 55.7	− 23 19 59	2.4	0.01	17.6
G 29.7− 0.3		18 43 47.6	− 03 01 51	19.0	40.0	1.4
G 34.6− 0.5	W 44	18 53 36	+01 09 00			
G 39.6− 1.8	W 50/SS 433	19 09 21.3	+04 53 54	5.0	1.0	48.8
G 68.8+2.6	CTB 80	19 51 02.4	+32 44 45	3.0	0.48	0.45
G 109.1− 1.0	CTB 109	22 59 02.8	+58 36 33	4.2	0.3	15.1

*Adapted from Angelini et al. (1988), Helfand and Becker (1984, 1987), Matsui, Long and Tuohy (1988), Seward and Wang (1988), and Wolszczan (1989).

†Compact object plus synchrotron nebula, X-ray luminosity in the band 0.2 to 4 keV.

Einstein X–Ray Survey of Supernova Remnants*

Galactic Coordinates	Remnant Name	IPC Rate† (c s^{-1})	HRI Rate† (c s^{-1})	MPC Rate† (c s^{-1})	Type††
4.5+6.8	Kepler	7.3 ± 0.4	$1.73 \pm .05$	$4.50 \pm .05$	S
6.4– 0.1	W 28	3.2 ± 0.4	0.5 ± 0.1	3.47 ± 0.2	F, CO
	1E1757–233	NRS	$0.008 \pm .001$	NRS	
11.2– 0.3		1.00 ± 0.1	0.11 ± 0.008	1.8 ± 0.2	S
21.5– 0.9		$0.49 \pm .05$	$0.029 \pm .003$	4.15 ± 0.1	F
21.8– 0.6	Kes 69	> 0.1	ND	NRB	F?
27.4+0.0	Kes 73	1.07 ± 0.1	$0.078 \pm .008$	4.66 ± 0.1	F?, CO
	1E1838–049	NRS	$0.016 \pm .002$	NRS	NB
29.7– 0.2	Kes 75	$0.22 \pm .03$	$0.0067 \pm .0017$	2.39 ± 0.1	F
31.9+0.0	3C 391	$0.24 \pm .03$	$0.048 \pm .012$	$1.40 \pm .15$	IR
33.6+0.0	Kes 79	$0.44 \pm .05$	ND	1.66 ± 0.2	S?
34.7– 0.4	W 44	3.3 ± 0.3	0.35 ± 0.1	4.13 ± 0.2	F
39.2– 0.3	3C 396	$.06 \pm .01$	ND	0.6 ± 0.2	IR
39.7– 2.0	W 50	1.6 ave.	ND	NRB	IR, CO
	SS 433	1.2 ave.	0.10 ave.	7.0 ave.	NB
41.1– 0.3	3C 397	0.75 ± 0.1	$0.11 \pm .025$	1.76 ± 0.1	IR
43.3– 0.2	W 49B	$0.67 \pm .06$	$0.041 \pm .006$	4.7 ± 0.2	F
49.2– 0.7	W 51	0.70 ± 0.2	INC	$0.89 \pm .06$	IR
53.6– 2.2	3C 400.2	0.80 ± 0.1	$0.25 \pm .05$	$0.20 \pm .06$	F
54.1+0.3		0.016 ± 0.004	ND	NRB	
65.3+5.7	GKP	$> 1.7 \pm 0.3$	ND	$> 0.21 \pm .12$	
	1E1928+313	$0.042 \pm .006$	ND	NRS	
69.0+2.7	CTB 80	$0.025 \pm .05$	$0.025 \pm .004$	$0.81 \pm .04$	F, CO
	1E1951+327	$0.14 \pm .01$	$0.009 \pm .001$	NRS	NI
74.3– 8.5	Cyg Loop	660 ± 60	177 ± 16	14.5 ± 2.2	S
74.9+1.2	CTB 87	$.040 \pm .01$	ND	NRB	F
78.2+2.1	W 66	$> 0.7 \pm .15$	INC	1.3 ± 0.2	
82.2+5.3	W 63	0.4 ± 0.1	ND	0.0 ± 0.15	IR
89.0+4.7	HB 21	> 0.2	ND	$> 0.1 \pm 0.1$	IR
109.1– 1.0	CTB 109	5.2 ± 0.4	ND	3.1 ± 0.3	S, CO
	1E2259+586	1.1 ± 0.1	$0.09 \pm .01$	NRS	NB?
111.7– 2.1	Cas A	61 ± 2	$6.28 \pm .13$	109 ± 1	S
119.5+10.2	CTA 1	$0.75 \pm .15$	ND	0.75 ± 0.15	F
	1E0000+726	$0.025 \pm .004$	ND	NRS	

Einstein X–Ray Survey of Supernova Remnants*

Galactic Coordinates	Remnant Name	IPC Rate† ($c\ s^{-1}$)	HRI Rate† ($c\ s^{-1}$)	MPC Rate† ($c\ s^{-1}$)	Type††
120.1+1.4	Tycho	22.3 ± 1	$3.36 \pm .08$	27.5 ± 0.3	S
130.7+3.1	3C 58	$0.35 \pm .04$	$0.051 \pm .008$	$0.61 \pm .05$	F, CO
	1E0201+645	NRS	$0.0039 \pm .0008$	NRS	
132.7+1.3	HB 3	2.1 ± 0.4	ND	$0.0 \pm .15$	IR
160.4+2.8	HB 9	$> 1.6 \pm 0.5$	ND	$> 0.6 \pm 0.2$	IR, CO
	4C +46.09	$.049 \pm .01$	ND	NRS	
184.6– 5.8	Crab	684 ± 35	120 ± 10	1383 ± 10	F, CO
	PSR 0531+21	NRS	5.2 ± 0.3	NRS	NI
189.1+3.0	IC 443	12.4 ± 0.6	$2.55 \pm .35$	6.5 ± 0.6	IR
260.4– 3.4	Pup A	250 ± 20	79 ± 2	36 ± 4	S
263.5– 2.7	Vela SNR	520 ± 25	ND	54 ± 8	IR, F, CO
	PSR 0833–45	2.15 ± 0.15	$0.69 \pm .03$	7.6 ± 0.1	NI
290.1– 0.8	MSH 11–61A	0.47 ± 0.1	$0.13 \pm .03$	$0.92 \pm .05$	F
291.0– 0.1	MSH 11–62	$0.22 \pm .05$	ND	$0.80 \pm .07$	F
292.0+1.8	MSH 11–54	9.1 ± 1.0	$1.41 \pm .05$	3.72 ± 0.1	IR
296.1– 0.5		3.2 ± 0.3	1.2 ± 0.2	2.3 ± 0.3	S
296.5+10.0	PKS 1209–52	3.6 ± 0.3	ND	0.0 ± 0.2	S, CO
	1E1207–521	0.082 ± 0.08	$0.028 \pm .004$	NRS	NI
315.4– 2.3	RCW 86	8.5 ± 1.0	1.9 ± 0.1	6.0 ± 1.0	S
320.4– 1.2	MSH 15–52	2.40 ± 0.2	$0.18 \pm .05$	9.3 ± 0.6	S, F, CO
	PSR 1509–58	$0.30 \pm .04$	$0.011 \pm .001$	NRS	NI
326.3– 1.8	MSH 15–56	$1.10 \pm .10$	INC	1.5 ± 0.2	IR
327.1– 1.1		$.085 \pm .02$	$0.035 \pm .012$	NRB	IR
327.4+0.4	Kes 27	0.40 ± 0.1	ND	1.18 ± 0.1	F
327.6+14.6	SNR 1006	11.1 ± 1.0	5.6 ± 0.7	3.4 ± 0.2	S
332.4– 0.4	RCW 103	9.3 ± 1.0	$2.17 \pm .08$	4.35 ± 0.1	S, CO
	1E1613–509	NRS	$0.007 \pm .001$	NRS	NI

*Adapted from Seward (1990).

†Order of magnitude conversion factors for the detectors are:

HRI – 1 count $\sim 1 \times 10^{-10}$ erg cm^{-2} in the energy band 0.2–4 keV.
IPC – 1 count $\sim 3 \times 10^{-11}$ erg cm^{-2} in the energy band 0.2–4 keV.
MPC – 1 count $\sim 2 \times 10^{-11}$ erg cm^{-2} in the energy band 1–10 keV.

NRS = not resolved from other parts of SNR
INC = incomplete data, much of SNR not observed
NRB = not resolved from nearby bright source
ND = no data, not observed

††S = shell
F = filled center, plerionic
IR = irregular
CO = central object
NI = neutron star, isolated
NB = neutron star in binary system

27.3 Catalogue of Galactic Supernova Remnants*

Galactic Coordinates l b	R.A.(1950) h m s	Dec.(1950) ° ′	Size (′)	Type	S (Jy)	α
119.5+10.2	00 04 00	+72 30	90	S	36	0.3
120.1+01.4	00 22 30	+63 52	8	S	56	0.61
126.2+01.6	01 18 30	+64 00	70	S?	7	varies
127.1+00.5	01 25 00	+62 55	45	S	13	0.6
130.7+03.1	02 01 55	+64 35	9 x 5	F	33	0.10
132.7+01.3	02 14 00	+62 30	80	S	45	0.6
152.2- 01.2	04 05 30	+48 24	110?	S?	16?	0.7?
160.9+02.6	04 57 00	+46 36	140 x 120	S	110	0.6
166.2+02.5	05 15 30	+41 50	90 x 70	S	11	0.5
166.0+04.3	05 23 00	+42 52	55 x 35	S	7?	0.4?
184.6- 05.8	05 31 30	+21 59	7 x 5	F	1040	0.30
180.0- 01.7	05 36 00	+27 50	180	S	65	varies
179.0+02.6	05 50 30	+31 05	70	S?	7	0.4
192.8- 01.1	06 06 30	+17 20	78	S	20?	0.6?
189.1+03.0	06 14 00	+22 36	45	S	160	0.36
205.5+00.5	06 36 00	+06 30	220	S	160	0.5
211.7- 01.1	06 43 10	+00 24	70?	S?	15?	0.5?
206.9+02.3	06 46 00	+06 30	60 x 40	S?	6	0.5
240.9- 00.9	07 40 30	- 25 06	95?	S?	24?	0.1?
260.4- 03.4	08 20 30	- 42 50	60 x 50	S	130	0.5
263.9- 03.3	08 32 30	- 45 35	255	C	1750	varies
261.9+05.5	09 02 20	- 38 30	40 x 30	S	10?	0.4?
290.1- 00.8	11 01 00	- 60 40	15 x 10	S	42	0.6
291.0- 00.1	11 09 45	- 60 22	10	F	16	0.29
292.0+01.8	11 22 20	- 59 00	12 x 8	C?	15	0.4
293.8+00.6	11 32 40	- 60 37	20	C	5.5?	0.6?
296.1- 00.5	11 48 40	- 62 17	33?	S	8?	0.6?
296.8- 00.3	11 56 00	- 62 18	14	S	9	0.6
296.5+10.0	12 07 00	- 52 10	90 x 65	S	48	0.5
298.5- 00.3	12 10 00	- 62 35	5?	?	5	0.4
298.6- 00.0	12 11 00	- 62 20	12 x 8?	S	4.3	0.3
299.0+00.2	12 15 00	- 62 12	11?	S?	9?	?
302.3+00.7	12 42 55	- 61 52	15	S	5.5	0.4
304.6+00.1	13 02 50	- 62 26	8?	S?	14	0.5
308.7+00.0	13 38 00	- 62 00	17 x 7	F?	12	0.4

*Adapted from Green (1988).

Catalogue of Galactic Supernova Remnants*

Galactic Coordinates l b	R.A.(1950) h m s	Dec.(1950) ° ′	Size (′)	Type	S (Jy)	α
309.2– 00.6	13 43 00	– 62 39	17 x 13	S	7?	0.4?
309.8+00.0	13 47 00	– 61 50	24	S	17	0.5
311.5– 00.3	14 02 00	– 61 44	3?	?	3.7	0.5
312.4– 00.4	14 09 20	– 61 29	36 x 27	S	44?	0.3?
315.4– 00.3	14 32 10	– 60 23	15 x 10	S	8	0.4
316.3– 00.0	14 37 40	– 59 47	20?	S	24	0.5
315.4– 02.3	14 39 00	– 62 17	40	S	49	0.6
327.6+14.6	14 59 35	– 41 44	30	S	19	0.6
330.0+15.0	15 05 00	– 39 30	180?	S	350?	0.5?
320.4– 01.2	15 10 30	– 58 58	30	C	60?	0.4?
321.9– 00.3	15 16 45	– 57 23	30 x 20	S	13	0.3
323.5+00.1	15 24 50	– 56 11	10?	S	3?	0.4?
327.4+00.4	15 44 30	– 53 40	20	S	34	0.6
326.3– 01.8	15 49 00	– 56 00	36	C	145	varies
328.0+00.3	15 49 30	– 53 20	6?	?	2.4?	0.6?
327.1– 01.1	15 50 30	– 55 00	14?	S?	8?	?
328.4+00.2	15 51 40	– 53 08	6	F	16	0.2
330.2+01.0	15 57 20	– 51 26	10	S?	7	0.3
332.0+00.2	16 09 30	– 50 45	10	S	9	0.5
332.4+00.1	16 11 30	– 50 35	15	S	26	0.5
332.4– 00.4	16 13 45	– 50 55	9	S	28	0.5
335.2+00.1	16 24 00	– 48 40	19	S	18	0.5
336.7+00.5	16 28 30	– 47 13	13 x 10	S	6	0.5
337.3+01.0	16 29 00	– 46 30	12	S	16	0.5
337.0– 00.1	16 32 10	– 47 27	14?	S?	17?	0.5?
338.1+00.4	16 34 20	– 46 18	12	S	4.5	0.4
337.8– 00.1	16 35 20	– 46 53	7?	?	18	0.5
337.2– 00.7	16 35 45	– 47 45	4?	?	2.1	0.7
338.3– 00.0	16 37 20	– 46 28	8?	S?	15?	0.7?
338.5+00.1	16 37 30	– 46 13	8?	?	28?	0.3?
340.4+00.4	16 42 55	– 44 34	6	S	6	0.4
340.6+00.3	16 44 05	– 44 29	5	S	5.1	0.4
342.0– 00.2	16 51 15	– 43 48	11 x 7	S?	3.5?	0.4?
341.9– 00.3	16 51 25	– 43 56	6	S	3.2	0.5
344.7– 00.1	17 00 20	– 41 38	8?	S?	3.0	0.5

Catalogue of Galactic Supernova Remnants*

Galactic Coordinates l b	R.A.(1950) h m s	Dec.(1950) ° ′	Size (′)	Type	S (Jy)	α
346.6− 00.2	17 06 50	− 40 07	8	S	10	0.5
348.7+00.3	17 10 30	− 38 08	10	S	26	0.3
348.5+00.1	17 10 40	− 38 29	10	S	72	0.3
349.7+00.2	17 14 35	− 37 23	2.5 x 2	S?	20	0.5
350.1− 00.3	17 17 40	− 37 24	4?	?	5.6	0.7
351.2+00.1	17 19 05	− 36 08	7	S	5.8	0.4
350.0− 01.8	17 23 40	− 38 20	30?	S?	31	0.5
352.7− 00.1	17 24 20	− 35 05	6 x 5	S?	6?	0.6?
4.5+06.8	17 27 42	− 21 27	3	S	19	0.64
357.7+00.3	17 35 20	− 30 42	24	S	10	0.4?
357.7− 00.1	17 37 15	− 30 56	3 x 8?	?	37	0.4
6.4+04.0	17 42 10	− 21 20	31	S	1.3?	0.4?
359.1− 00.5	17 42 20	− 29 56	24	S	14	0.4?
0.0+00.0	17 42 33	− 28 59	3.5 x 2.5	S	100?	0.8?
355.9− 02.5	17 42 35	− 33 42	13	S	8	0.5
0.9+00.1	17 44 12	− 28 08	8	C	18?	varies
5.9+03.1	17 44 20	− 22 15	20	S	3.3?	0.4?
1.9+00.3	17 45 37	− 27 09	1.2	S	0.6	0.7
6.4− 00.1	17 57 30	− 23 25	42	C	310	varies
5.4− 01.2	17 59 00	− 24 55	35	C?	35?	0.2?
9.8+00.6	18 02 10	− 20 14	12	S	3.9	0.5
8.7− 00.1	18 02 35	− 21 25	45	S?	90	0.25
10.0− 00.3	18 05 40	− 20 26	8?	S?	2.9	0.8
11.4− 00.1	18 07 50	− 19 06	8	S?	6	0.5
11.2− 00.3	18 08 30	− 19 26	4	S	22	0.49
12.0− 00.1	18 09 15	− 18 38	5?	?	3.5	0.7
7.7− 03.7	18 14 20	− 24 05	18	S	10	0.32
15.9+00.2	18 16 00	− 15 03	7 x 5	S?	4.5?	0.7?
18.8+00.3	18 21 10	− 12 25	18 x 13	S	27	0.4
16.8− 01.1	18 22 30	− 14 48	30 x 24?	?	2?	?
20.0− 00.2	18 25 20	− 11 37	10	F	10	0.0
18.9− 01.1	18 27 00	− 13 00	33	C?	37	varies
21.8− 00.6	18 30 00	− 10 10	20	S	69	0.5
23.6+00.3	18 30 20	− 08 15	10?	?	8?	0.3
22.7− 00.2	18 30 30	− 09 15	26	S?	33	0.6

*Adapted from Green (1988).

Catalogue of Galactic Supernova Remnants*

Galactic Coordinates l b	R.A.(1950) h m s	Dec.(1950) ° ′	Size (′)	Type	S (Jy)	α
21.5- 00.9	18 30 37	- 10 37	1.2	F	6	0.0
24.7+00.6	18 31 30	- 07 07	30 x 15	C?	20?	0.2?
23.3- 00.3	18 32 00	- 08 50	27	S	70	0.5
24.7- 00.6	18 36 00	- 07 35	15?	S?	8	0.5
27.8+00.6	18 37 06	- 04 28	50 x 30	F	30	varies
27.4+00.0	18 38 40	- 04 59	4	S	6	0.68
30.7+01.0	18 42 10	- 01 35	24 x 18	S?	6	0.4
29.7- 00.3	18 43 48	- 03 02	3	C?	10	0.7
31.9+00.0	18 46 50	- 00 59	5	S	24	0.55
31.5- 00.6	18 48 35	- 01 35	18?	S?	2?	?
32.8- 00.1	18 48 50	- 00 12	17	S?	11?	0.2?
33.6+00.1	18 50 15	+00 37	10	S	22	0.5
33.2- 00.6	18 51 12	- 00 05	18	S	5?	varies
34.7- 00.4	18 53 30	+01 18	35 x 27	S	230	0.30
36.6- 00.7	18 58 05	+02 52	25?	S?	?	?
39.2- 00.3	19 01 40	+05 23	8 x 6	S	18	0.6
32.0- 04.9	19 03 00	- 03 00	60?	S?	22?	0.5?
40.5- 00.5	19 04 45	+06 26	22	S	11	0.5
42.8+00.6	19 04 55	+09 00	24	S	3?	0.5?
41.1- 00.3	19 05 08	+07 03	4.5 x 2.5	S	22	0.48
43.3- 00.2	19 08 44	+09 01	4 x 3	S	38	0.48
39.7- 02.0	19 10 00	+04 50	120 x 60	?	85?	0.7?
45.7- 00.4	19 14 05	+11 04	22	S	4.2?	0.4?
46.8- 00.3	19 15 50	+12 04	17 x 13	S	14	0.42
55.7+03.4	19 19 10	+21 38	23	S	1.4	0.6
49.2- 00.7	19 21 30	+14 00	25?	S?	160?	0.3?
54.1+00.3	19 28 28	+18 46	1.5	F?	0.5	0.1
65.3+05.7	19 31 00	+31 05	310 x 240	S?	52?	0.6?
54.4- 00.3	19 31 10	+18 50	40	S	28	0.5
57.2+00.8	19 32 50	+21 50	12?	S?	1.8?	?
53.6- 02.2	19 36 30	+17 08	28	S	8	0.6
65.7+01.2	19 50 10	+29 18	18	F?	5.1	0.6
69.0+02.7	19 51 30	+32 45	80?	?	120?	varies
73.9+00.9	20 12 20	+36 03	22?	S?	9?	0.3?
74.9+01.2	20 14 10	+37 03	8 x 6	F	9	varies

Catalogue of Galactic Supernova Remnants*

Galactic Coordinates l b	R.A.(1950) h m s	Dec.(1950) ° ′	Size (′)	Type	S (Jy)	α
82.2+05.3	20 17 30	+45 20	95 x 65	S	120?	0.7?
78.2+02.1	20 19 00	+40 15	60	S	340	0.7
89.0+04.7	20 43 30	+50 25	120 x 90	S	220	0.40
74.0− 08.5	20 49 00	+30 30	230 x 160	S	210	varies
93.3+06.9	20 51 00	+55 10	27 x 20	S	9	0.54
84.2− 00.8	20 51 30	+43 16	20 x 16	S	11	0.5
94.0+01.0	21 23 10	+51 40	30 x 25	S	15	0.44
93.7− 00.2	21 27 45	+50 35	80	S	65	0.3
109.1− 01.0	22 59 30	+58 37	28	S	20	0.50
112.0+01.2	23 13 40	+61 30	30?	S?	7?	0.6?
111.7− 02.1	23 21 10	+58 32	5	S	2720	0.77
114.3+00.3	23 34 45	+61 38	90 x 55	S	6?	0.3?
116.5+01.1	23 51 20	+62 58	80 x 60	S	11?	0.8?
117.4+05.0	23 52 30	+67 30	60 x 80?	S?	30?	0.5?
116.9+00.2	23 56 40	+62 10	34	S	9?	0.5?

*Adapted from Green (1988).

27.4 Cross-Reference Name Index for Supernova Remnants

Galactic Coordinates	Names	Galactic Coordinates	Names
119.5+10.2	CTA 1	4.5+06.8	Kepler, SN 1604, 3C 358
120.1+01.4	Tycho, 3C 10, SN 1572	0.0+00.0	Sgr A East
127.1+00.5	R5	6.4- 00.1	W28
130.7+03.1	3C 58, SN 1181	5.4- 01.2	Milne 56
132.7+01.3	HB3	8.7- 00.1	(W30)
160.9+02.6	HB9	7.7- 03.7	1814- 24
166.2+02.5	OA 184	21.8- 00.6	Kes 69
166.0+04.3	VRO 42.05.01	18.8+00.3	Kes 67
184.6- 05.8	Crab Nebula, 3C 144, SN 1054	27.4+00.0	4C- 04.71, Kes 73
180.0- 01.7	S147	23.3- 00.3	W41
192.8- 01.1	PKS 0607+17	31.9+00.0	3C391
189.1+03.0	IC443, 3C 157	29.7- 00.3	Kes 75
205.5+00.5	Monoceros Nebula	32.8- 00.1	Kes 78
206.9+02.3	PKS 0646+06	33.6+00.1	Kes 79, 4C 00.70, HC13
260.4- 03.4	Puppis A, MSH 08- 44	34.7- 00.4	W44, 3C 392
263.9- 03.3	Vela XYZ	39.2- 00.3	3C 396, HC24, NRAO 593
290.1- 00.8	MSH 11- 61A	32.0- 04.9	3C 396.1
291.0- 00.1	(MSH 11- 62)	40.5- 00.5	Flo
292.0+01.8	MSH 11- 54	41.1- 00.3	3C 397
296.8- 00.3	1156- 62	43.3- 00.2	W49B
296.5+10.0	PKS 1209- 51/52	39.7- 02.0	W50, SS 433
304.6+00.1	Kes 17	46.8- 00.3	(HC30)
316.3- 00.0	(MSH 14- 57)	49.2- 00.7	(W51)
315.4- 02.3	RCW 86, MSH 14- 63, SN 185	54.4- 00.3	(HC40)
327.6+14.6	SN1006, PKS 1459- 41	57.2+00.8	(4C21.53)
330.0+15.0	Lupus Loop	53.6- 02.2	3C 400.2, NRAO 611
320.4- 01.2	MSH 15- 52, RCW 89	65.7+01.2	DA 495
327.4+00.4	Kes 27	69.0+02.7	CTB 80
326.3- 01.8	MSH 15- 56	74.9+01.2	CTB 87
328.4+00.2	(MSH 15- 57)	82.2+05.3	W63
332.4+00.1	MSH 16- 51, Kes 32	78.2+02.1	DR4, γ Cygni
332.4- 00.4	RCW 103	89.0+04.7	HB21
337.3+01.0	Kes 40	74.0- 08.5	Cygnus Loop
337.0- 00.1	(CTB 33)	93.3+06.9	DA 530, 4C(T)55.38.1
337.8- 00.1	Kes 41	94.0+01.0	3C 434.1
348.7+00.3	CTB 37B	93.7- 00.2	CTB 104A, DA 551
348.5+00.1	CTB 37A	109.1- 01.0	CTB 109
357.7- 00.1	MSH 17- 39	111.7- 02.1	Cassiopeia A, 3C 461
		116.9+00.2	CTB 1

*Adapted from Green (1988).

Part IX
High Energy Radiation From Stars

28
X-Ray and Gamma Ray Sources

28.1 Compact X-Ray Sources

The ubiquitous nature of cosmic X-ray sources was not fully realized until satellites such as Uhuru, Ariel, HEAO and Einstein surveyed the X-ray sky. For instance, the fourth Uhuru (the Swahili word for freedom) catalogue contains 339 X-ray sources in the photon energy range 2 to 6 keV (1 $keV = 10^3$ $eV = 1.6 \times 10^{-9} erg$), while one HEAO (High Energy Astrophysical Observatory) catalogue contains 842 sources in the 0.25 to 25 keV range. As illustrated in the first accompanying diagram, these sources can be divided into extragalactic and galactic components, but this book is only concerned with the galactic ones.

The most intense X-ray sources have been named after the constellation they appear in, followed by an X for X-ray and a number ordered by decreasing brightness. Famous examples of named galactic sources include Cygnus X-1, Cygnus X-3, Centaurus X-3, Hercules X-1 and Scorpius X-1. The more numerous, fainter sources are often named by a catalogue designation, such as 2A or 4U for the second Ariel or fourth Uhuru catalogues, followed by the source's right ascension in hours and minutes and the sign and degrees of declination.

Apart from the Crab Nebula, the most luminous galactic X-ray sources are interacting binary systems which contain a neutron star or black hole and an optically-visible stellar companion. These binary X-ray systems have X-ray luminosities, L_x, in the range $L_x = 10^{36}$ to $10^{38} erg\ s^{-1}$. Fainter X-ray objects that have been identified have $L_x = 10^{27}$ to $10^{36} ergs^{-1}$. The optical counterparts of these less-luminous stellar X-ray sources include flare stars, Section 13.2, RS CVn and BY Dra stars, Section 13.4, naked T Tauri stars, Section 17.2, hot white dwarfs, Section 21.1, classical novae, Section 24.1, and supernova remnants, Section 27.2. But because the X-ray emission from each of these types of objects is discussed in the designated section, here we will only concern ourselves with the luminous binary stellar systems that consist of a compact primary and a Roche lobe-filling secondary.

In the intense accretion-powered X-ray binaries, the source of energy comes from the fall of matter into the enormous gravity of a neutron star or black hole. Matter from the outer mantle of the companion star spirals slowly into the compact object via an accretion disk. The infalling matter gains kinetic energy through the release of large amounts of gravitational energy; this is converted into heat and emitted as X-ray radiation by the accreting material.

The optical counterparts of bright, accretion -driven X-ray binaries have been reviewed by Bradt and McClintock (1983), Van Pardijis (1983), and Ritter (1987). They are usually identified with either a massive $M \geq 10M_\odot$ early-type star or a low-mass, $M \leq M_\odot$, late-type star.

One type of X-ray binary system contains a massive early-type star as the mass donating component which dominates the optical light output of the binary system. The ratio of X-ray to optical luminosity, L_x/L_o, of these massive X-ray binary systems is usually less than ten. The other class usually has a luminosity ratio, L_x/L_o, that is larger than ten. This suggests that the mass-donating companion is a low-mass, optically faint dwarf star with a mass less than, or comparable to, that of the Sun. The massive, early-type systems also show a strong tendency to be pulsars, and only the low-mass systems exhibit X-ray bursting behavior.

Perhaps the most exotic type of X-ray binaries are those containing candidate black holes. They include both massive and low-mass X-ray binary systems. Cygnus X-1 has a massive, blue supergiant optical companion, while A0620-00 has been optically identified with a K-type dwarf star. Cygnus X-1 is a bright source of X-rays in the 1 to 10 keV range, but it also emits hard X-rays from 50 to 300 keV as well as episodic MeV gamma rays. For convenience, we reproduce our table of candidate black holes (Section 23.1) in the first accompanying table.

The enigmatic low-mass X-ray binary, Cygnus X-3, probably ranks next on the hit parade of galactic powerhouses, followed closely by the singular binary system SS 433. Data for these two objects are given in the next table.

The combination of high X-ray luminosity $L_x = 10^{38} erg\ s^{-1}$ (at 2 to 20 keV with an assumed distance of 8.5 kiloparsecs or more), short orbital period, P = 4.79 hours, giant radio flares with flux densities reaching 20 Jy, and reports of energetic gamma-ray emission make Cygnus X-3 a unique object amongst X-ray binaries [see Bonnet-Bidaud and Chardin (1988) for a review]. Its position has been accurately determined from radio interferometry measurements. The flaring radio emissions have been interpreted as the synchrotron radiation of an expanding double-sided jet, aligned with the rotation axis of the accretion disk and moving at bulk velocities between 0.16 and 0.31 times the velocity of light, c [see Molnar, Reid and Grindlay (1988)]. Cygnus X-3 has been proposed to be one of the most powerful emitters of energetic particles on the basis of possible detections of very high energy gamma rays [see Weekes (1988)].

Discovered as a strong stellar source of hydrogen emission lines, and numbered 433 in the catalogue of such objects by Stephenson and Sanduleak, SS 433 dissipates a total power of $10^{39} erg\ s^{-1}$, which is comparable to the X-ray luminosity of the brightest binary systems. But SS 433 appears to have an overloaded accretion disk that converts accretional energy into the kinetic energy of ejected matter with only weak X-ray emission. Spectroscopic observations of highly Doppler shifted emission lines, of up to 50,000 $km\ s^{-1}$ toward the red and up to 30,000 $km\ s^{-1}$ toward the blue, have been ascribed to the ejection of two narrow, collimated jets of matter in opposite directions from a compact object at velocities of 0.26 c [see Margon (1989)]. The jets are probably ejected along the axis of an accretion disk, whose plane precesses and rotates the jet axis with a 164-day period. Two collimated, colinear, precessing jets that originate from SS 433 have been detected in radio and X-ray observations.

The next table summarizes the radio detections of X-ray stars. It contains the X-ray source, name, right ascension, R.A., and declination, Dec., for the epoch 1950.0, the orbital period of the binary system, the distance, D, in kiloparsecs, and the flux density, S, observed at frequency ν.

The massive X-ray binary systems have easily-identified optical companions and often exhibit eclipses and pulsations of their X-ray radiation. Long orbital periods of 5 to 10 days are typical of these systems, and the pulse periods range from 0.7 to 835 seconds. These X-ray pulsars are thought to be neutron stars undergoing accretion from their giant stellar companions as the result of a stellar wind or by Roche-lobe overflow. The strong magnetic field of the neutron star disrupts the inflowing material, and constrains its fall toward the magnetic poles. As the neutron star rotates, localized spots at the poles sweep by an observer, producing pulses of X-ray radiation.

As indicated in the next table, observations

of the brightest, massive X-ray binaries have led to accurate estimates of their system parameters. The measured pulse arrival times have been used to establish accurate orbits and to measure the masses and radii of the companion stars, as well as the masses of the neutron stars.

A substantial fraction of massive X-ray binary systems exhibit regular X-ray pulsations, and most low-mass systems do not. Our next table contains the X-ray source, name, celestial coordinates, R.A. and Dec., galactic coordinates, l and b, pulse period, P, period derivative, $\dot{P}$, orbital period and X-ray luminosity, L_x, for 21 binary X-ray pulsars. In contrast to radio pulsars which slow down, the X-ray pulsars spin-up with a negative $\dot{P}$ and a pulsation period that decreases as time goes on. The decreasing period is due to torques exerted by matter accreting onto the neutron star and spinning it up.

Sufficient information for many of these systems is now available to allow a determination of the projected semi-major axis of the X-ray star, $a\sin i$, and the mass function $f(M) = M_c^3 \sin^3 i/(M_x + M_c^3)^2$. They are given in the next table together with the name and spectral type of the optical companion. The ratio M_x/M_c, of the mass of the X-ray star, M_x, and its companion, M_c, can be determined when the Doppler velocity curve for the optical companion is known. One can then infer the masses of both the companion and the neutron star.

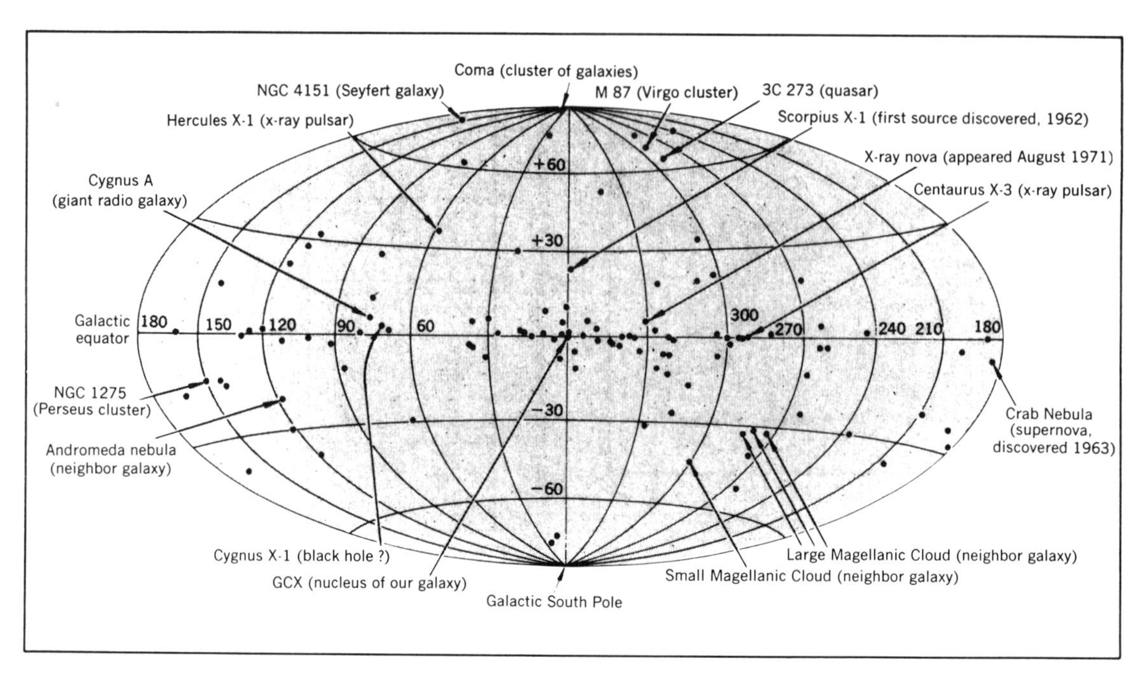

The X-Ray Sky. Bright X-ray sources can be divided into two groups in this distribution in galactic coordinates. One group consists of stellar sources that are found within our Galaxy and tend to lie near the galactic equator; the other group includes extragalactic sources that are usually located far away from this equator. The extragalactic sources include nearby galaxies like the Large Magellanic Cloud, LMC, and the Andromeda galaxy, M 31, Seyfert galaxies, clusters of galaxies, radio galaxies **and** quasars. Their X-ray luminosities, L_x, have to be enormous just to be able to detect radiation coming from vast distances, with $L_x = 10^{41}$ to 10^{45} $erg\ s^1$. Even the brightest galactic sources are less luminous, with $L_x = 10^{36}$ to 10^{38} $erg\ s^{-1}$, but they are still much more luminous than the Sun whose total luminous output at all wavelengths is $L_\odot = 4 \times 10^{33}$ $erg\ s^{-1}$. The source concentration toward the galactic center is called the galactic bulge. The nearest and brightest galactic X-ray sources include candidate black holes like Cygnus X-1, the massive binary X-ray systems that contain X-ray pulsars like Centaurus X-3 and Hercules X-1, low mass X-ray binary systems like Scorpius X-1, and supernova remnants such as the Crab Nebula

Candidate Black Holes*

X-Ray Source	Optical Companion	R.A.(1950) h m s	Dec.(1950) ° ′ ″	l °	b °
Cygnus X-1	HDE 226868	19 56 28.87	+35 03 55.0	71.3	+ 3.1
LMC X–3	WP8	05 38 39.7	– 64 06 34	273.6	–32.1
LMC X–1	F.C.* of R148	05 40 05.5	– 64 46 03.6	280.2	–31.5
A 0620-00	V616 Mon	06 20 11.12	– 00 19 11.0	210.0	– 6.5

*F.C. means Faint Companion

X-Ray Source	V mag	Spectral Type	$L_X$max (erg s^{-1})	D (kpc)	P (days)	V_0 (km s^{-1})
Cygnus X-1	8.9	O9.7Iab	2×10^{37}	2.5	5.5995	–6
LMC X–3	16.9	B3V	3×10^{38}	55	1.7049	310
LMC X–1	14	O7	2×10^{38}	55	4.2288	221
A 0620-00	18	K5V	1.5×10^{38}	1.0	0.3230	–5

X-Ray Source	K (km s^{-1})	a sin i ($R_\odot$)	f(M) ($M_\odot$)	M_C ($M_\odot$)	M_X ($M_\odot$)
Cygnus X-1	76	8.36	0.25	16	≥ 6
LMC X–3	235	7.92	2.3	9	≥ 3
LMC X–1	68	5.77	0.144	20	≥ 4
A 0620-00	457	2.91	3.18	0.7	≥ 3

*Adapted from Bolton (1972), Cowley et al. (1983), Gies and Bolton (1986), Hutchings et al. (1987), Mc Clintock (1986), and Mc Clintock and Remillard (1986).

Cygnus X–3 and SS 433

Cygnus X–3*

Equatorial Coordinates	R.A.(1950.0)	=	20 h 30 m 37.616 s ± 0.001 s
	Dec.(1950.0)	=	+40° 47′ 12.73″ ± 0.01″
Distance	D	=	8.5 kpc (adopted – lower limit)
Orbital Period	P_{orb}	=	0.19968354 ± 0.00000015 days
	P_{orb}	=	4.7924 hours
Orbital Period Derivative	$\dot{P}_{orb}$	=	$(0.90 \pm 0.05) \times 10^{-9}$ s s^{-1}
X–ray Luminosity	L_X	=	$(0.2 - 2) \times 10^{38}$ erg s^{-1}
Radio Luminosity	L_R	=	$10^{33} - 10^{35}$ erg s^{-1}
Radio Size (Flares)	R_F	=	3.1×10^{16} cm
Radio Size (Quiescent)	R_Q	=	2.5×10^{14} cm
Radio Flux Density	S_R	=	0.1 Jy (Quiescent)
	S_R	=	20 Jy (Flares)
Expansion Velocity	V	=	0.13 c (each jet)

*Adapted from Bonnet–Bidaud and Chardin (1988), Molnar, Reid and Grindlay (1984, 1988).

SS 433*

Equatorial Coordinates	R.A.(1950.0)	=	19 h 09 m 21.282 s ± 0.003 s
	Dec.(1950.0)	=	+04° 53′ 54.04″ ± 0.05″
Distance	D	=	5 kpc
Orbital Period	P_{orb}	=	13.087 ± 0.003 days
Orbital Period Derivative	$\dot{P}_{orb}$	≤	9×10^{-5} s s^{-1}
Precession Period	P_{prec}	=	164 days
X–ray Luminosity	L_X	=	1.0×10^{35} erg s^{-1}
X–ray Flux	F_X	=	10^{-10} erg cm^{-2} s^{-1}
Radio Luminosity	L_R	=	10^{32} erg s^{-1}
Radio Flux Density	S_R	=	1 Jy (variable)
Apparent Visual Magnitude	V	=	14.2 mag (variable ± 1.0 mag)
Interstellar Extinction	A_V	=	8 mag
Optical Luminosity	L_0	=	2.6×10^{38} erg s^{-1}
Radio Size	R	=	2.5×10^{17} cm
Velocity Variation (Doppler Shifts)	ΔV	=	± 0.26 c
Expansion Velocity	V	=	0.52 c

*Adapted from Margon (1989).

Radio Emitting X–Ray Stars*

Source	Name	R.A.(1950.0) h m s	Dec.(1950) ° ′ ″	P_{orb} (days)	D (kpc)	S_ν^{**} (Jy)	ν (GHz)
0236+610 B	LSI +61°303	02 36 40.6	+61 00 53.9	26.5	2.3	0.28	5.0
0531+219 P	Crab Pulsar	05 31 31.405	+21 58 54.39		2.0	0.48	0.4
0620– 003 BH,N	A0620–003	06 20 11.12	– 00 19 11.0		1.1	0.3	1.4
0833– 450 P	Vela Pulsar	08 33 39.27	– 45 00 10.2		0.50	0.5	2.8
1455– 314 N	Cen X–4	14 55 19.63	– 31 28 09.0	0.3230	1.5	0.008	4.9
1516– 569 B	Cir X–1	15 16 48.37	– 56 59 11.6	16.6	10.0	4.0	5.0
1617– 155 B	Sco X–1	16 17 04.49	– 15 31 14.8	0.8	0.7	0.26	3.0
1742– 289 N		17 42 26.3	– 28 59 57		1.2	0.48	1.0
1758– 250	GX 5–1	17 58 03.182	– 25 04 40.2		10.0	0.0013	1.4
1811– 171	GX 13+1	18 11 37.688	– 17 10 22.9			0.0018	1.4
1813– 140	GX 17+2	18 13 10.928	– 14 03 14.50		1.4	0.001	1.4
1820– 304 GC	NGC 6624	18 20 27.44	– 30 23 16.1	0.008?	8.5	0.002	1.4
1909+098 B	SS 433	19 09 21.282	+04 53 54.04	13.087	5.0	1.4	2.7
1916– 053		19 16 08.51	– 05 19 50.89			0.0025	1.4
1956+350 BH	Cyg X–1	19 56 28.87	+35 03 55.0	5.5995	2.5	0.02	1.4
2030+407 B	Cyg X–3	20 30 37.616	+40 47 12.73	0.19968	≥ 8.5	20.0	10.0
2142+380 B	Cyg X–2	21 42 36.9	+38 05 28	9.8	8.0	0.006	8.0

†B = X–ray binary, P = Rotation–powered pulsar, BH = Black hole candidate, N = X–ray nova, GC = Globular cluster

*Adapted from Bonnet–Bidaud and Chardin (1988), Bradt and McClintock (1983), Geldzahler (1983), Gregory et al. (1979), Grindlay and Seaquist (1986), Hayes et al. (1986), Hjellming (1986) and Taylor and Gregory (1982). RS CVn stars and dwarf M flare stars also emit quiescent and flaring radiation at both radio and X–ray wavelengths. These objects are catalogued elsewhere in this work.

**Flux density, S_ν, at frequency, ν, usually the maximum value observed when flaring.

Masses of Binary X–Ray Pulsars*

Source	Name	Companion Mass ($M_\odot$)	Companion Radius ($R_\odot$)	Neutron–Star Mass ($M_\odot$)
0115– 737	SMC X–1	17.0	16.5	1.05
0532– 664	LMC X–4	19.0	9.0	1.70
0900– 403	4U0900–40†	23.0	31.0	1.85
1119– 603	Cen X–3	19.0	12.2	1.07
1538– 522	4U1538–52	18.5	16.0	1.87
1656+354	Her X–1	2.35	4.05	1.45

*Adapted from Joss and Rappaport (1984).
†Vela X–1 = 4U0900–40

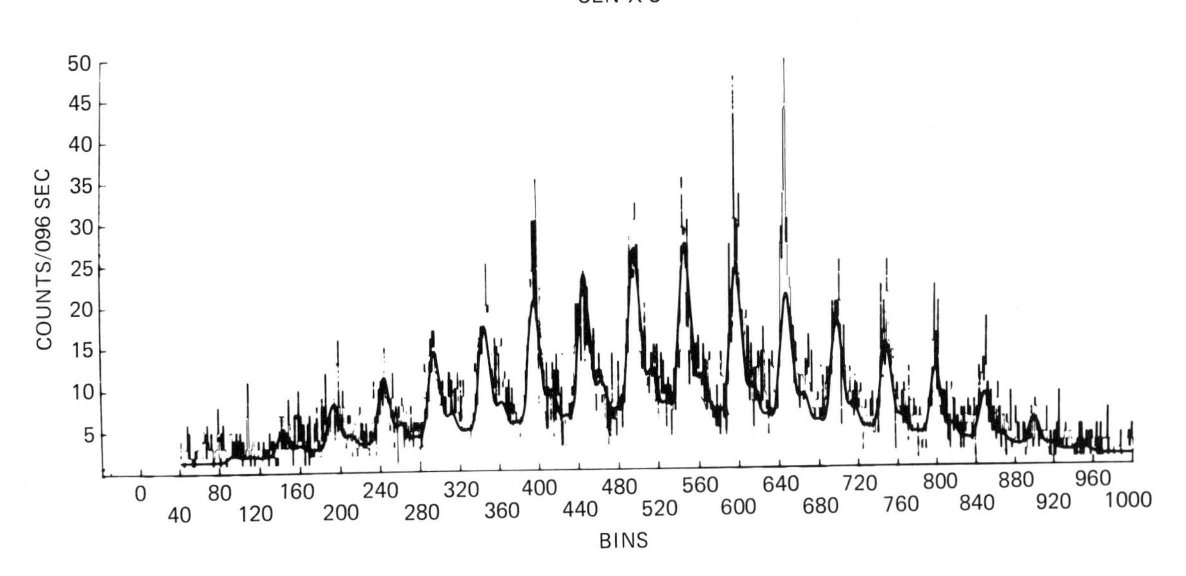

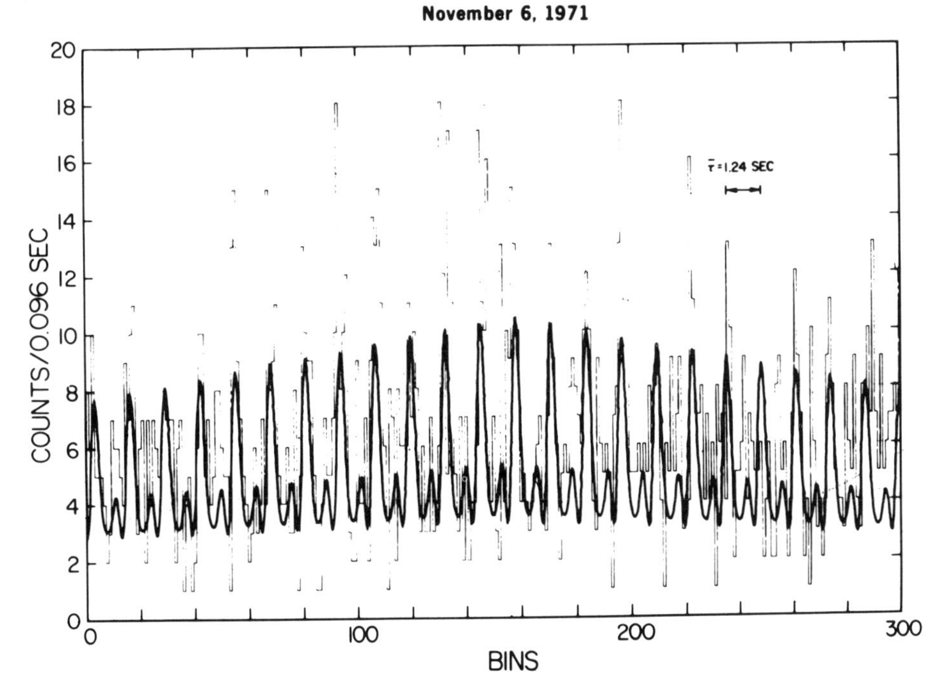

X-Ray Pulsars. Time profiles of the X-ray emission from Centaurus X-3 and Hercules X-1 made in intervals or "bins" of roughly 0.1 second duration. The main pulses of these two X-ray pulsars have respective periods of 4.84 and 1.24 seconds. This periodicity is attributed to the rotation of a neutron star. (UHURU time profiles, courtesy of Harvey Tananbaum, Harvard-Smithsonian Center for Astrophysics.)

Binary X–Ray Pulsars*

Source	Name	R.A.(1950.0) h m s	Dec.(1950) ° ′ ″	l °	b °
0115+634	4U0115+63	01 15 13.8	+63 28 38	125.9	+01.0
0115– 737	SMC X–1	01 15 45.6	– 73 42 22	300.4	– 43.6
0352+309	4U0352+30	03 52 15.13	+30 54 00.8	163.1	– 17.1
0532– 664	LMC X–4	05 32 47.3	– 66 24 13	276.3	– 32.5
0535– 668	A0538–66	05 35 42.5	– 66 52 40	275.8	– 32.2
0535+262	A0535+26	05 35 47.98	+26 17 18.3	181.7	– 02.8
0900– 403	Vela X–1†	09 00 13.18	– 40 21 25.3	263.1	+03.9
1118– 616	A1118–61	11 18 45.21	– 61 38 31.1	292.5	– 00.9
1119– 603	Cen X–3	11 19 01.9	– 60 20 57	292.1	+00.3
1145– 616	1E1145.1–6141	11 45 02.8	– 61 40 33	295.5	– 00.0
1145– 619	2S1145–619	11 45 34.0	– 61 55 45	295.6	– 00.2
1223– 624	GX 301–2	12 23 49.7	– 62 29 37	300.1	– 00.0
1258– 613	GX 304–1	12 58 11.8	– 61 19 58	304.1	+01.2
1417– 624	2S1417–62	14 17 25.47	– 62 28 11.2	313.0	– 01.6
1538– 522	4U1538–52	15 38 38.62	– 52 13 36.9	327.4	+02.1
1553– 542	2S1553–54	15 53 55.6	– 54 16 15	327.9	– 00.9
1627– 673	4U1627–67	16 27 14.65	– 67 21 17.6	321.8	– 13.1
1656+354	Her X–1	16 56 01.68	+35 25 04.8	58.2	+37.5
1657– 415	OAO1653–40	16 57 16.3	– 41 34 37	344.7	+00.8
1728– 247	GX 1+4	17 28 57.89	– 24 42 34.9	1.9	+04.8
2138+56	GS 2138+56	21 38	+56 50		
2259+586	1E2259+586	22 59 03.4	+58 36 38	109.1	– 01.0

*Adapted from Bradt and McClintock (1983), Fahlman, and Gregory (1981), Hutchings, Crampton and Cowley (1981), Johnston et al. (1979), Parmar et al. (1980).

†Vela X–1 = 4U0900–40

Binary X–Ray Pulsars*

Source	Pulse Period P (seconds)	$-\dot{P}/P$ (yr^{-1})	Orbital Period (days)	$L_X^\dagger$ (10^{37} erg s^{-1})
0115+634	3.61	3.2×10^{-5}	24.31	< 3. tr
0115−737	0.714	7.1×10^{-4}	3.892	50.
0352+309	835.	1.8×10^{-4}	580.(?)	4×10^{-4}
0532−664	13.5019	$< 1.2 \times 10^{-3}$	1.408	35.
0535−668††	0.069		16.6515	80. tr
0535+262	104.	3.5×10^{-2}	111.(?)	< 2. tr
0900−403†††	283.	1.7×10^{-4}	8.96443	0.15
1118−616	405.			< 0.5 tr
1119−603	4.84	2.8×10^{-4}	2.087	5.
1145−616	297.7	$< 1.0 \times 10^{-4}$	> 12.	0.3
1145−619	291.7	$< 1.0 \times 10^{-4}$	187.	0.025
1223−624	696.	7.0×10^{-3}	41.4	0.1
1258−613	272.		132.(?)	0.2
1417−624	17.6	9.0×10^{-3}	> 15.	
1538−522	529.	$< 2.0 \times 10^{-3}$	3.730	0.4
1553−542	9.26		30.7	
1627−673	7.68	1.9×10^{-4}	0.0288	3 at 10 kpc
1656+354	1.24	2.9×10^{-6}	1.700	1.
1657−415	38.2	5.4×10^{-3}		
1728−247	122.	2.1×10^{-2}	> 15.	4.
2138+56**	66.2490		23	
2259+586	3.49		< 0.08	

†Typical 2 to 11 keV X–ray luminosities uncertain by at least a factor of two, the tr denotes transient. Also see White, Swank and Holt (1983).

††0535− 668 = A0 0538− 66.

†††0900− 403 = Vela X–1 = 4U0900− 40.

** Koyama et al, Astrophysical Journal (Letters) 366, L19 (1991).

Binary X–Ray Pulsars*

Source	Companion Star Name	Companion Star Sp	$a_X \sin i$ (lt–sec)	f(M) ($M_\odot$)
0115+634	V635 Cas	Be	140.13	5.00
0115– 737	Sk 160	B0I	53.46	10.8
0352+309	X Per	O9.5 III–Ve		
0532– 664	Sk/Ph	O7 III–Ve	26.0	9.4
0535– 668		B7 IIe–B2 IVe		0.027
0535+262	HDE 245770	O9.7 IIIe		
0900– 403	HD 77581	B0.5 Ib	112.70	19.12
1118– 616	Hen 3–640?	O9.5 III–Ve		
1119– 603	V779 Cen	O6.5 IIIe	39.792	15.5
1145– 616		~ B1 I	> 50.	
1145– 619	Hen 715	B0.–1 Ve	> 100.	
1223– 624	WRA 977	B1.5 Ia	367.	31.
1258– 613	MMV	B2 Vne		
1417– 624		Hα	> 25.	
1538– 522	QV Nor	B0 Ib	55.2	13.
1553– 542			165.	5.1
1627– 673	KZ TrA		< 0.04	$< 8 \times 10^{-5}$
1656+354	HZ Her	A9–B	13.1831	0.85
1657– 415				
1728– 247	V2116 Oph	M6 giant	> 60.	
2138+56				
2259+586				

*Adapted from Boynton et al. (1986), Henrichs (1983), Hutchings et al. (1985), Joss and Rappaport (1984), Pietsch et al. (1985), and van der Klis and Bonnet–Bidaud (1984).

28.2 Low-Mass X-Ray Binaries

Low mass X-ray binaries are binary systems with an accreting component, or primary, that is thought to be a neutron star, and a mass-donating, optical secondary that is a low-mass ($M \leq 1M_\odot$) star driven to overflow its Roche lobe. They have a large ratio of X-ray to optical luminosity, L_x/L_o = 100 to 10,000. The optical components are intrinsically faint, and most of their light comes from an accretion disk. This group of low-mass X-ray binaries is thought to include the bright bulge sources, the compact sources in globular clusters, and the X-ray bursters.

The low-mass optical companions could not have evolved into giant stars during the lifetime of our Galaxy, so Roche overflow is thought to be due to orbital decay rather than the evolutionary expansion of the companion. Loss of angular momentum from the binary system, perhaps as the result of gravitational radiation, must drive the overflow.

The minority of low-mass systems are X-ray pulsators, and with a few notable exceptions such as Hercules X-1, they do not exhibit well-defined eclipses. In a number of inclined systems, the X-ray emitting star is hidden from direct view by the accretion disk. The lack of observed X-ray eclipses is also thought to be due to the low probability of the small companion being able to eclipse the neutron star primary. Orbital periods are therefore identified by partial X-ray eclipses, by periodic dips in the X-ray light curves, or by periodic variations in the faint optical counterparts. The orbital periods of many of these systems are of the order of a few hours.

The absence of detectable pulsations in low-mass systems, with the notable exceptions of Hercules X-1 and 4U 1626-67, is attributed to the occultation of the compact neutron star by its accretion disk or to the fact that the magnetic fields of these old stellar systems have decayed with age. If the magnetic fields of the neutron stars are not sufficiently strong to disrupt the inflowing material, the X-ray emission will be from the accretion disk rather than hot spots at the magnetic poles.

The first accompanying table of low-mass X-ray binaries includes the X-ray source, name, right ascension, R.A., and declination, Dec., for epoch 1950.0, the galactic coordinates, l and b, the optical counterpart, the X-ray flux, S_x, the source distance, D, in kiloparsecs, the X-ray luminosity, L_x, and its ratio, L_x/L_o, to the optical luminosity, L_o, the apparent magnitude, V, and the optical period. This information has been extracted from Bradt and McClintock (1983), Paradijs, (1983) and Ritter (1987).

The high X-ray luminosity of $L_x = 10^{37}$ to $10^{38} erg\ s^{-1}$ is at or near the Eddington limit for an accreting neutron star, and well above that for a white dwarf star. The X-ray luminosity of the accreting neutron star increases with an increasing rate of mass accretion, but this rate has an upper limit set when the force derived from the minimum outward radiation pressure, due to electron scattering only, equals the gravitational force. The corresponding maximum luminosity, the Eddington limit, is $L_x = 10^{38} erg\ s^{-1}$. The fact that the observed luminosities are near this limit strongly suggests that the compact objects in low-mass X-ray binaries are neutron stars.

There are also optically unidentified X-ray sources which are generally believed to be low-mass X-ray binaries; they include the bright sources lying close to the galactic center and the compact X-ray sources in the cores of globular clusters. The high extinction to the galactic center precludes the identification of an optical counterpart, and none of the globular cluster X-ray sources have been optically identified with stellar objects. Such identifications would be very difficult in the crowded, centrally condensed cores of globular clusters. Both of these unidentified sources are thought to be associated with old, low-mass stellar companions. For completeness, the next table provides data for the compact X-ray sources found in globular clusters.

Intense X-ray bursts, which have typical rise times of 1 second and decay times of several seconds to minutes, are observed in about half the low-mass X-ray binaries; most of the globular cluster sources also emit X-ray bursts. The commonly observed X-ray bursts are probably due to thermonuclear flashes of freshly accreted material on the surface of the neutron star. Other bursts, such as those from the Rapid Burster MXB1730-335, have been attributed to an instability in the accretion flow onto the neutron star. The satisfactory explanation of X-ray bursts in terms of the thermonuclear flash model also argues that the compact object is a neutron star.

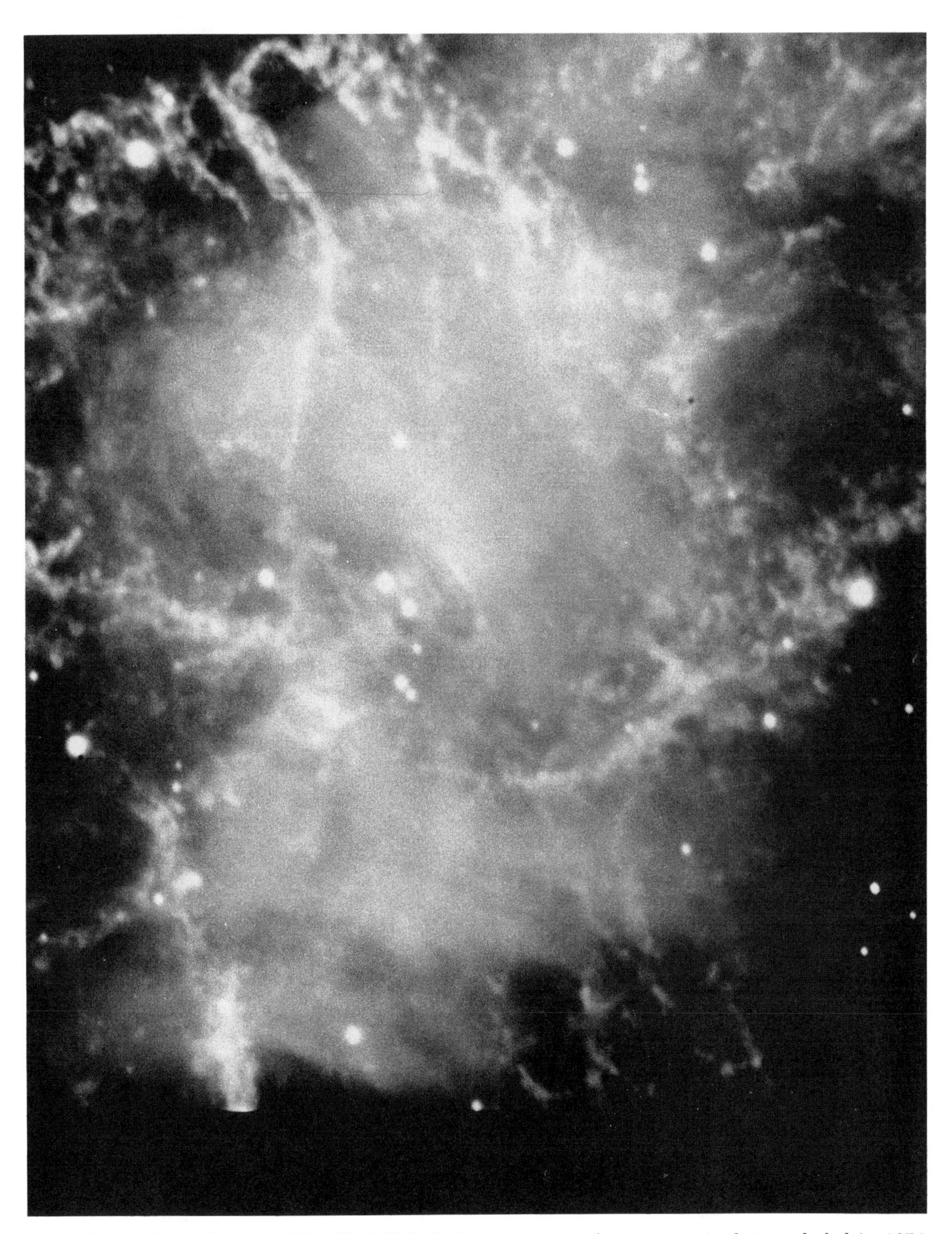

Crab Nebula Central Area. The Crab Nebula is a remnant of a supernova that exploded in 1054. The supernova remnant is now one of the most intense X-ray and gamma ray sources in the sky. A central pulsar powers the diffuse background light shown here. The pulsar is the lower right of the two brighter stars near the center; it has been detected at optical, radio and X-ray wavelengths. (Courtesy of the European Southern Observatory.)

Low–Mass X–Ray Binaries*

Source	Name†	R.A.(1950) h m s	Dec.(1950) ° ′ ″	l °	b °	Optical Counterpart
0521– 720	LMC X–2	05 21 18.0	– 72 00 26	283.1	– 32.7	star
0614+091		06 14 22.84	+09 09 22.3	200.9	– 03.4	V1055 Ori
0620– 003	BH	06 20 11.12	– 00 19 11.0	210.0	– 06.5	V616 Mon
0748– 676		07 48 25.0	– 67 37 32	280.0	– 19.8	
0918– 549		09 18 55.3	– 54 59 39	275.9	– 03.8	star
0921– 630		09 21 25.08	– 63 04 48.0	281.8	– 09.3	V395 Car
1254– 690		12 54 21.00	– 69 01 07.9	303.5	– 06.4	GR Mus
1323– 619		13 23 17.1	– 61 52 35	307.1	+00.3	
1455– 314	Cen X–4	14 55 19.63	– 31 28 09.0	332.2	+23.9	V822 Cen
1524– 617	TrA X–1	15 24 05.8	– 61 42 35	320.3	– 04.4	KY TrA
1543– 624		15 43 34.06	– 62 24 51.5	321.8	– 06.3	
1556– 605		15 56 45.83	– 60 35 52.4	324.1	– 05.9	star
1608– 522	N	16 08 52.2	– 52 17 43	330.9	– 00.9	QX Nor
1617– 155	Sco X–1	16 17 04.49	– 15 31 14.8	359.1	+23.8	V818 Sco
1624– 490		16 24 17.8	– 49 04 46	334.9	– 00.3	
1627– 673		16 27 14.65	– 67 21 17.6	321.8	– 13.1	KZ TrA
1636– 536		16 36 56.41	– 53 39 18.1	332.9	– 04.8	V801 Ara
1656+354	Her X–1	16 56 01.68	+35 25 04.8	58.2	+37.5	HZ Her
1658– 298		16 58 55.44	– 29 52 27.5	353.8	+07.3	V2134 Oph
1659– 487	GX 339–4	16 59 01.96	– 48 43 07.0	338.9	– 04.3	V821 Ara

†BH = black hole candidate, GC = globular cluster, N = nova.

Low–Mass X–Ray Binaries*

Source	S_X (μJy)	D (kpc)	L_X (erg s^{-1})	L_X/L_0	V (mag)	Orb. Per. (days)
0521–720	26	50	3.0×10^{38}	600	18.5	0.266
0614+091	50	~ 6	1.3×10^{37}	1200	18.5	5.2
0620–003	50000	1.1	1.6×10^{38}	200	11.2	0.323014
0748–676					16.9	0.159338
0918–549	10				19.6	
0921–630	2.3	~ 8	4.0×10^{35}	1.6	15.3	9.0115
1254–690	29			200	19.1	0.16389
1323–619						0.1223
1455–314	4200	~ 1.5	1.0×10^{38}	300	12.8	0.62908
1524–617	800			700	17.5	
1543–624	40			1400	~ 20.0	
1556–605	26			2000	19.5	
1608–522	1100			800	> 20.0	
1617–155	19000	~ 0.7	3.0×10^{37}	500	12.2	0.787313
1624–490	88					0.88:
1627–673	20			170	18.5	0.028762
1636–536	260			1700	17.5	0.15889
1656+354	160	~ 5	7.0×10^{36}	14	13.2	
1658–298	40			600	18.3	0.296421
1659–487	400			10	15.5	

*Adapted from Bradt and McClintock (1983), Paradijs (1983), Ritter (1987).

Low–Mass X–Ray Binaries*

Source	Name†	R.A.(1950) h m s	Dec.(1950) ° ′ ″	l °	b °	Optical Counterpart
1702−363	Sco X−2	17 02 23.7	−36 21 23	349.1	+02.7	
1705−250	N	17 05 10.4	−25 01 38	358.0	+09.1	
1728−169	GX 9+9	17 28 50.22	−16 55 31.6	08.5	+09.0	
1728−247	GX 1+4	17 28 57.89	−24 42 34.9	01.9	+04.8	V2116 Oph
1735−444		17 35 19.28	−44 25 20.3	346.1	−07.0	V926 Sco
1755−338		17 55 21.52	−33 48 13.7	357.2	−04.9	
1813−140	GX 17+2	18 13 10.89	−14 03 15.0	16.4	+01.3	
1820−303	GC	18 20 27.74	−30 23 15.1	2.8	−07.9	NGC 6624
1822−371		18 22 22.67	−37 08 03.5	356.9	−11.3	V691 CrA
1822−000		18 22 48.2	−00 02 24	29.9	+05.8	
1837+049	Ser X−1	18 37 29.51	+04 59 19.8	36.1	+04.8	MM Ser
1908+005	Aql X−1	19 08 42.83	+00 30 04.9	35.7	−04.1	V1333 Aql
1916−053		19 16 08.0	−05 19 42	31.4	−08.5	star
1957+115		19 57 02.16	+11 34 15.2	51.3	−09.3	
2030+407	Cyg X−3	20 30 37.62	+40 47 12.6	79.9	+00.7	V1521 Cyg
2127+119	GC	21 27 33.03	+11 56 49.6	65.0	−27.3	NGC 7078
2129+470		21 29 36.24	+47 04 08.4	91.6	−03.0	V1727 Cyg
2142+380	Cyg X−2	21 42 36.9	+38 05 28	87.3	−11.3	V1341 Cyg
2259+586		22 59 03.2	+58 36 31	109.1	−01.0	

†BH = black hole candidate, GC = globular cluster, N = nova.

Low–Mass X–Ray Binaries*

Source	S_X (μJy)	D (kpc)	L_X (erg s^{-1})	L_X/L_0	V (mag)	Orb. Per. (days)
1702–363	400					8.71156:
1705–250	1200			800	15.9	
1728–169	250			400	16.6	0.175
1728–247	125	10?	6.0×10^{37}	12	19.0	
1735–444	200	~ 7	4.0×10^{37}	1100	17.5	0.1939
1755–338	60			1400	19.3	0.1858
1813–140	1060	1.4	5.0×10^{36}	3000	17.5	0.813
1820–303	410	8.5	8.0×10^{37}			0.007928
1822–371	9			20	15.3	0.232110
1822–000	43				19.0	
1837+049	225			1200	19.2	3.40252:
1908+005	1300	1.7	1.0×10^{37}	900	14.8	1.3:
1916–053	20			10000	22.0	0.03465
1957+115	38	~ 7	1.0×10^{37}	800	18.7	0.3887
2030+407	260	10	1.0×10^{38}	60	> 23.0	0.199685
2127+119	50	8	8.0×10^{36}		15.9	0.35575
2129+470	4	~ 2.2	4.0×10^{35}	30	16.6	0.218259
2142+380	480	8	1.2×10^{38}	140	14.7	9.843
2259+586	1	3.6	3.0×10^{34}	> 100	23.5	0.0266:

*Adapted from Bradt and McClintock (1983), Paradijs (1983), Ritter (1987).

Compact X-Ray Sources In Globular Clusters*

Globular Cluster	X-Ray Source	R.A.(1950) h m s	Dec.(1950) ° ′ ″	Position Error (″)
NGC 104, 47 Tuc	1E 0021.8–7221	00 21 51.58	– 72 21 34.6	±1
NGC 1851	4U 0513–40	05 12 27.88	– 40 05 59.7	±1
NGC 1904, M 79	1E 0522.1–2433	05 22 07.7	– 24 33 55	±60
NGC 5139, ω Cen	1E 1323.8–4713	13 23 49.8	– 47 13 13	±30
NGC 5272, M 3	1E 1339.8+2837	13 39 51.4	+28 37 52	±30
NGC 5824	1E 1500.7–3251	15 00 49.4	– 32 52 07	±10
Terzan 2	4U 1722–30	17 24 20.09	– 30 45 39.4	±1
Grindlay 1	4U 1728–34	17 28 39.2	– 33 47 55	±5
Liller 1	MXB 1730–335	17 30 06.63	– 33 21 13.2	±1
NGC 6440	MX 1746–20	17 45 55.0	– 20 21 07	±60
NGC 6441	4U 1746–37	17 46 48.49	– 37 02 17.8	±1
NGC 6541	1E 1804.4–4343	18 04 24.8	– 43 43 36	±60
NGC 6624	4U 1820–30†	18 20 27.84	– 30 23 17.0	±1
NGC 6656, M 22	1E 1833.3–2357	18 33 20.4	– 23 56 56	±20
NGC 6712	4U 1850–08	18 50 21.18	– 08 46 04.4	±1
NGC 7078, M 15	4U 2127+11†	21 27 33.14	+11 56 51.0	±1

Compact X-Ray Sources In Globular Clusters*

X-Ray Source	Assumed Distance (kpc)	X-Ray Luminosity (erg/sec)	X-Ray Source	Assumed Distance (kpc)	X-Ray Luminosity (erg/sec)
1E 0021.8–7221	4.6	34.59	MXB 1730–335	10.0	36.81
4U 0513–40	10.8	36.12	MX 1746–20	3.5	32.82
1E 0522.1–2433	13.3	33.89	4U 1746–37	10.1	36.81
1E 1323.8–4713	5.1	32.93	1E 1804.4–4343	6.8	33.33
1E 1339.8+2837	9.8	33.56	4U 1820–30†	8.7	37.95
1E 1500.7–3251	23.5	34.29	1E 1833.3–2357	3.0	32.39
4U 1722–30	10.5	36.71	4U 1850–08	7.4	36.41
4U 1728–34	10.0	37.61	4U 2127+11†	9.4	36.75

*Adapted from Hertz and Grindlay (1983), Grindlay et al. (1984), Grindlay (1988). The positions are those of the X-ray sources. The optical counterpart of the X-ray source in NGC 7078 is AC 211 (Aurière et al. 1985). For NGC 5139 and NGC 6656 the positions of the cluster centers are given, these two clusters contain multiple X-ray sources and the luminosity of the brightest one is given. †X-ray sources 4U 1820-30 and 4U 2127+12 are periodic, with respective periods of 11.4 minutes and 8.5 hours.

28.3 Gamma Ray Sources

Gamma rays are like X-rays, only more energetic. They are photons with energies that can range from 0.511 million-electron-volts ($1\ MeV = 10^6\ eV = 1.6 \times 10^{-6} erg$), the rest mass energy of the electron, to an extremely high energy of $10^{20} eV$. This broad energy interval has been subdivided into the low (0.51 to 10 MeV), medium (10 to 30 MeV) and high (30 MeV to 10 GeV) energy ranges, as well as the very high energy regime of 10 GeV to 100 TeV (10^{10} to $10^{14} eV$) and ultra high energy range of 100 TeV to 100 PeV (10^{14} to 10^{17} eV).

Spectral lines have been observed in the low-energy gamma-ray range. Nuclear excitation lines are created when solar flare protons collide with iron (at 0.847 MeV), magnesium (at 1.369 MeV), neon (at 1.634 MeV), silicon (at 1.778 MeV), carbon (at 4.439 MeV), and oxygen (at 6.129 MeV). A gamma-ray line is also emitted at 2.223 MeV when solar flare neutrons are captured by atmospheric hydrogen. The electron-positron annihilation line at 0.511 MeV has been detected during solar flares and from a source in the vicinity of the center of the Galaxy (Ramaty and Lingenfelter, 1983).

The galactic annihilation radiation has been attributed to a time-varying, compact source and a constant, extended component. A black hole could be supplying the positrons; some place it in a compact X-ray source 1E 1740.7−2942 located some 45 arcminutes away from the dynamical center of our Galaxy. The extended source could be related to positrons generated during stellar explosions or during cosmic ray interactions with interstellar material. A superhot, electron-positron pair-dominated plasma, such as that found near the event horizon of the black hole candidate Cygnus X-1, could emit broad spectral features near 1 MeV.

A diffuse gamma-ray glow has been produced throughout our Galaxy by exploding stars of aeons past. The radioactive isotope of aluminum, ^{26}Al, with a mean life of about a million years, generates a gamma-ray line at 1.809 MeV that is distributed throughout the galactic plane. Explosive events like novae or supernovae, or the milder stellar winds of massive giant stars, could have produced the diffuse source of 1.809 MeV gamma rays during the past million years of galactic history.

More recently, gamma-ray signatures from the decay of radioactive cobalt, ^{56}Co, have been detected in the expanding debris of supernova SN 1987A in the Large Magellanic Cloud. This isotope decays into iron with a half-life of 79 days, emitting gamma ray lines at 0.847, 1.238 and 2.599 MeV (also see Section 26.1).

Gamma rays in the high energy range around 100 MeV are produced during the interaction of cosmic ray electrons and protons with the highly-structured interstellar medium, creating a diffuse, high-energy source of emission that is irregularly distributed along the galactic plane. The sky at 100 MeV is dominated by this broad band of emission, making it difficult to detect and identify discrete sources within the large field of view of the gamma-ray telescopes. However, the second catalogue of sources detected with the COS-B satellite includes four discrete sources that have been unambiguously detected, and as many as twenty otherwize unidentified point-like sources.

The known sources of high-energy gamma rays are listed in the first accompanying table that includes their right ascension, R.A., and declination, Dec., for the epoch 1950.0, the galactic coordinates l and b, and the flux of high-energy gamma rays. Two of them are the young radio pulsars, the Crab pulsar PSR 0531+21 and the Vela pulsar PSR 0833 − 45, identified by a timing analysis of their gamma-ray pulses. More information about these objects can be found in the catalogue of pulsars (Sections 22.2 and 22.3). These isolated, rapidly-spinning neutron stars with strong magnetic fields are capable of accelerating particles to extremely high energy.

The enigmatic gamma-ray source Geminga (2CG 195+04) emits virtually all of its power at energies above 50 MeV; but it has nevertheless been identified with the X-ray source 1E 0630+178 and a faint 25th magnitude blue object. (In Milanese slang, Geminga means "the source that is not there".) The celestial coordinates for the X-ray counterpart are R.A.(1950.0) = 06^h 30^m 59.15^s and Dec.(1950.0) = 17° 48' 33.0". At 100 MeV energies, the only extragalactic source to be detected so far is the quasar 3C 273.

The second accompanying table lists unidentified sources from the second COS-B catalogue of gamma-ray sources (Bignami and Hermsen 1983). The reality of these sources is still under discussion. Their gamma-ray luminosity is in the range 0.4 to 5 x 10^{36} $ergs^{-1}$ for an assumed distance of 2 to 7 kiloparsecs; but they have not been identified as X-ray sources, implying that their X-ray luminosity is at least a thousand times fainter. Dark molecular cloud complexes, such as the Rho Ophiuchus and the Orion Nebula regions, have apparently been resolved as extended sources that become visible in gamma rays when cosmic rays collide with matter in these clouds.

We have not included the gamma-ray burst sources because of their short transient nature, poor location information, and unknown optical or X-ray counterparts. They were first detected decades ago by the Vela satellites designed to monitor compliance with the ban on nuclear explosions in space. During their brief duration of 0.1 to 10 seconds, these bursts can be the brightest gamma-ray sources in the sky, with luminosities of up to $10^{38} erg\ s^{-1}$ at an energy of 1 MeV if they are at distances of about 1 kiloparsec. The gamma-ray bursters may also emit cyclotron emission and absorption in the 20 to 80 keV range, thereby indicating the presence of an amplified magnetic field as strong as 10^{12} Gauss. They may therefore be associated with neutron stars and explained by an extension and modification of the thermonuclear flash model of X-ray bursts.

Very high energy gamma-ray astronomy in the range of 10 GeV to 100 TeV has been reviewed by Weekes (1988). Although there have been reports of emission at these energies from Cygnus X-3 and other X-ray binaries, the statistical significance of many of the observations is not high and many reported effects await confirmation. Ground-based observations of secondary particles produced by ultra high energy (100 TeV to 100 PeV) gamma rays are also of low statistical significance, and the detection of discrete sources in this energy range has not been established beyond a reasonable doubt (Nagle, Gaisser and Protheroe, 1988). Candidate objects for possible future detections in the very high and ultra high energy ranges include the Crab pulsar, Vela pulsar, Cygnus X-3, Hercules X-1, Vela X-1, Geminga, SS 433, LMC X-4, and 4U 0115+63; these sources are discussed in the previous two sections.

Discrete Gamma Ray Sources*

Name	R.A. h m	Dec. °	l °	b °	Error Radius °	Flux E > 100 MeV (10^{-6} photons cm^{-2} s^{-1})	P	Other Name
2CG 184–05	05 31	+21	184.5	– 5.8	0.4	3.7	33.3 ms	Crab Pulsar*
2CG 195+04	06 33	+19	195.1	+4.5	0.4	4.8		Geminga**
2CG 263–02	08 33	– 45	263.6	– 2.5	0.3	13.2	89.2 ms	Vela Pulsar†
2CG 289+64	12 26	+02	289.3	+64.6	0.8	0.6		Quasar 3C 273††
000+00	17 06	– 29	0.0	+0.0	4.0			Galactic Center***

*Crab Pulsar PSR 0531+21, distance ≈ 2.0 kpc

**Geminga, 1E0630+178, pulsar with period 237.0974 ms.

†Vela Pulsar PSR 0833–45, distance ≈ 0.5 kpc

††Quasar 3C 273, distance ≈ 8.6×10^5 kpc

***A source of positron annihilation radiation at 0.511 MeV has been observed from the galactic center (within the ± 4°observational uncertainty). It is a single, compact ($< 10^{18}$ cm) source which is inherently variable on time scales of six months or less and has a luminosity of at least 6×10^{37} erg s^{-1} (Ramaty, 1983).

Possible Discrete Gamma Ray Sources*

Name	R.A. h m	Dec. °	l °	b °	Error Radius °	Flux E > 100 MeV (10^{-6} photons cm^{-2} s^{-1})
2CG 121+04	00 25	+70	121.0	+4.0	1.0	1.0
2CG 135+01	02 38	+62	135.0	+1.5	1.0	1.0
2CG 218–00	06 57	− 05.1	218.5	− 0.5	1.3	1.0
2CG 235–01	07 29	− 20.4	235.5	− 1.0	1.5	1.0
2CG 284–00	10 22	− 58	284.3	− 0.5	1.0	2.7
2CG 288–00	10 48	− 60	288.3	− 0.7	1.3	1.6
2CG 311+01	14 04	− 63	311.5	− 1.3	1.0	2.1
2CG 333+01	16 17	− 50	333.5	+1.0	1.0	3.8
2CG 353+16**	16 27	− 24.7	353.3	+16.0	1.5	1.1
2CG 342–02	17 05	− 44.5	342.9	− 2.5	1.0	2.0
2CG 356+00	17 33	− 31.7	356.5	+0.3	1.0	2.6
2CG 359–00	17 44	− 29.7	359.5	− 0.7	1.0	1.8
2CG 006–00	18 00	− 23.4	6.7	− 0.5	1.0	2.4
2CG 013+00	18 10	− 16	13.7	+0.6	1.0	1.0
2CG 036+01	18 50	+03.8	36.5	+1.5	1.0	1.9
2CG 054+01	19 23	+19.6	54.2	+1.7	1.0	1.3
2CG 065+00†	19 53	+29	65.7	+0.0	0.8	1.2
2CG 010–31	20 12	− 32	10.5	− 31.5	1.5	1.2
2CG 075+00	20 19	+36	75.0	+0.0	1.0	1.3
2CG 078+01	20 24	+41	78.0	+1.5	1.0	2.5
2CG 095+04	21 15	+55	95.5	+4.2	1.5	1.1

*Adapted from Bignami and Hermsen (1983) and White (1989). The reality of these sources is still under discussion. Another possible gamma ray source is Cygnus X–3 that is described on page 738.

**2CG 353+16 is identified with the Rho Ophiuchi dark cloud and is thought to be a diffuse source of gamma rays. Another diffuse source may exist in the Orion region.

†Radio Pulsar PSR 1953+29, distance ≈ 3.5 kpc, radio periodicity 6.13 ms.

29 Appendix

29.1 Catalogue of Nearby Stars

The following data for stars nearer than 22 parsecs, or trigonometric parallax, π_t, less than 0".045, have been compiled from the catalogues of Gliese (1969), and Gliese and Jahreiss (1979). The decimal number from the Gliese catalogue, often designated G, or the numbers in the thousands from the Gliese and Jahreiss catalogue, designated GJ, are given on the left side of the tables. Numbers in the DM, HD and Giclas catalogues and some star names are given in the original catalogues. Right ascensions (R.A.) and declinations (Dec.) in the epoch 1950.0 are given to 1 s and 6 seconds of arc, respectively. Yearly proper motions, μ, can be used together with precession to obtain positions at another epoch; the R.A. components of μ are in seconds of time per year and the Dec. components are in seconds of arc per year. These components have been computed from the total annual proper motions, μ, and the position angle, θ, of the GJ catalogue using $\mu_{Dec.} = \mu \cos\theta$ and $\mu_{R.A} = \mu \sin\theta/[15\cos(Dec.)]$. The radial velocity, V_r, is given to 1 $km\ s^{-1}$; it is followed by spectral type, Sp, and luminosity class, LC. The apparent visual magnitude, V, trigonometric parallax, π_t, in units of 0.001 seconds of arc, and absolute magnitude, M_V, for the Gliese stars are from Grenon and Rufener (1981), while the color index B − V is from the original G or GJ catalogues. Parallax values for those GJ stars marked with an (*) are spectroscopic or photographic. The G and GJ catalogues also contain U − B and R − I, and the six Geneva colors are listed by Grenon and Rufener (1981). The distance, D, to the star in parsecs can be inferred from $D = \pi_t^{-1}$ when π_t is in seconds of arc, and the stellar space velocity tangent to the line of sight, V_t in $km\ s^{-1}$ can be inferred from the total annual proper motion, μ, and D from $V_T = 4.75 D\mu$.

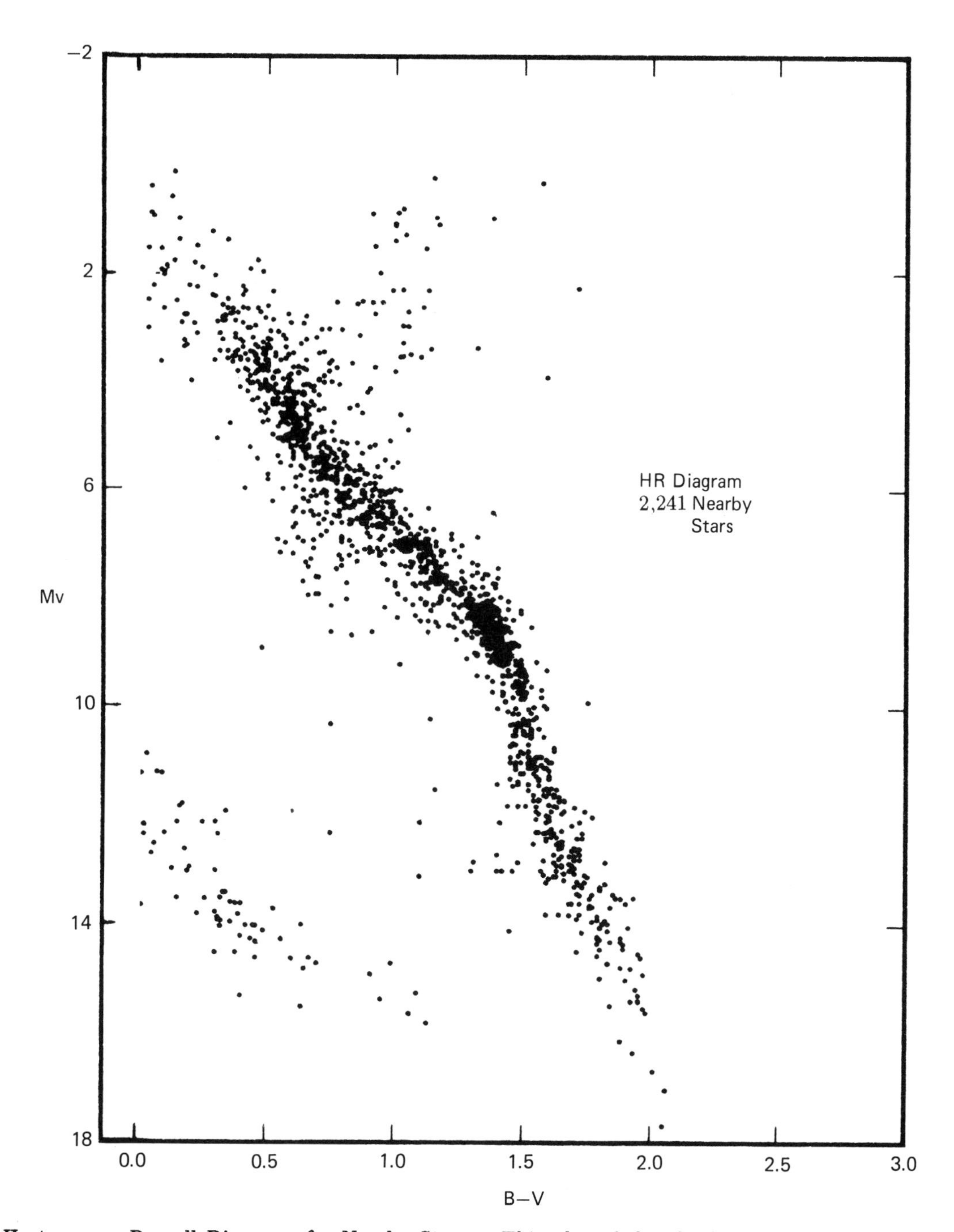

Hertzsprung-Russell Diagram for Nearby Stars. This plot of the absolute visual magnitude, M_V, against color index, B-V, illustrates the heavily populated main sequence running from the hot, luminous upper left to the cooler, less-luminous lower right, as well as the faint white dwarf stars in the lower left. It has been compiled from data for 2,241 nearby stars given in the accompanying tables.

Catalogue of Nearby Stars*

Name	R.A.(1950) h m s	Dec.(1950) ° ′ ″	$\mu_{R.A.}$ (s yr^{-1})	$\mu_{Dec.}$ (″ yr^{-1})	V_r (km s^{-1})
1001	00 02 05	− 40 57 48	+0.0604	− 1.468	− 1
1.0	00 02 28	− 37 36 12	+0.4753	− 2.329	+23
2.0	00 02 32	+45 30 36	+0.0820	− 0.140	− 1
3.0	00 02 48	− 68 06 12	− 0.0270	− 0.540	+41
4.0A	00 03 02	+45 32 12	+0.0835	− 0.127	+3
4.0B	00 03 02	+45 32 06	+0.0835	− 0.189	+2
4.1A	00 03 38	+58 09 30	+0.0331	+0.036	− 12
4.1B	00 03 38	+58 09 30	+0.0331	+0.036	− 16
4.2A	00 03 44	− 49 21 12	+0.0578	− 0.032	+2
4.2B	00 03 44	− 49 21 12	+0.0578	− 0.032	
5.0	00 04 01	+28 44 42	+0.0289	− 0.177	− 8
1002	00 04 13	− 07 47 30	− 0.0558	− 1.864	
1003	00 04 46	+28 58 48	+0.1149	− 1.137	
6.0	00 06 06	+36 21 00	− 0.0084	− 0.141	− 14
7.0	00 06 29	− 27 24 06	+0.0490	+0.130	
8.0	00 06 30	+58 52 24	+0.0679	− 0.177	+12
9.0A	00 06 31	− 54 16 48	+0.0060	+0.022	
9.0B	00 06 31	− 54 16 48	+0.0060	+0.022	
9.1	00 06 53	− 46 01 24	+0.0124	− 0.177	− 9
10.0	00 08 43	− 15 44 30	− 0.0056	− 0.264	+14
10.1	00 09 03	− 59 11 18	+0.0377	− 0.001	− 5
1004	00 09 40	+50 09 12	− 0.0429	− 0.590	
11.0A	00 10 30	+69 02 12	+0.1370	− 0.300	
11.0B	00 10 30	+69 02 12	+0.1370	− 0.300	
1005	00 12 53	− 16 24 18	+0.0435	− 0.604	− 29
12.0	00 13 12	+13 16 24	+0.0430	+0.320	
1006A	00 13 36	+19 35 36	+0.0502	− 0.761	
1006B	00 13 38	+19 35 48	+0.0502	− 0.761	
1007	00 14 22	+04 51 18	− 0.0044	− 0.627	
13.0	00 14 24	− 52 55 54	+0.0323	+0.206	− 9
14.0	00 14 26	+40 40 12	+0.0490	+0.110	+2
14.1	00 14 45	− 44 07 48	+0.0375	− 0.036	+9
15.0A	00 15 31	+43 44 24	+0.2653	+0.404	+13
15.0B	00 15 31	+43 44 24	+0.2653	+0.404	+20
16.0	00 15 42	+09 55 30	+0.0000	− 0.020	

Catalogue of Nearby Stars*

Name	Sp LC	V (mag)	B–V (mag)	π_t (0.001″)	M_V (mag)
1001		12.84	+1.63	95*	12.9
1.0	M4V	8.554	+1.45	224	10.31
2.0	M2VE	9.989	+1.50	92	9.81
3.0	K5	8.45	+1.06	61	7.40
4.0A	K6VE	8.213	+1.44	92	8.03
4.0B	M0.5E	8.97	+1.45		8.84
4.1A	G4V	5.979	+0.64	45	4.24
4.1B	(DG8)	7.20	+0.78		5.61
4.2A	G1IV	5.692	+0.52	49	4.14
4.2B		11.50			9.80
5.0	K0VE	6.100	+0.75	69	5.29
1002	dM	13.74	+1.95	214	15.39
1003		14.18	+1.48	54	12.84
6.0	F5V	6.218	+0.47	46	4.53
7.0	M	12.30		56	11.00
8.0	F2IV	2.282	+0.34	66	1.38
9.0A	G4IV	6.318	+0.74	27	3.80
9.0B		8.60		27	5.80
9.1	K0III	3.891	+1.03	65	2.96
10.0	F6V	4.901	+0.49	66	4.00
10.1	G4V	8.331	+0.62	51	6.87
1004	DA-P	14.51:	+0.45	79*	14.0
11.0A	M6	12.30		59	11.20
11.0B		12.60		59	11.50
1005		11.53	+1.74	117*	11.9
12.0	K	13.80		88	13.50
1006A		12.26	+1.55	65	11.81
1006B		13.22	+1.60	65	12.27
1007		13.9	+1.40	57	12.7
13.0	G2V	6.826	+0.64	68	5.99
14.0	M0.5V	8.968	+1.36	65	8.03
14.1	G5IV-V	7.932	+0.66	45	6.20
15.0A	M2V	8.090	+1.56	287	10.38
15.0B	M4.5VE	11.063	+1.80	287	13.35
16.0	M	10.40		55	9.10

*Adapted from Gliese (1969), Gliese and Jahreiss (1979), and Grenon and Rufener (1981).

Catalogue of Nearby Stars*

Name	R.A.(1950) h m s	Dec.(1950) ° ′ ″	$\mu_{R.A.}$ (s yr^{-1})	$\mu_{Dec.}$ (″ yr^{-1})	V_r (km s^{-1})
16.1	00 16 07	- 08 19 42	+0.0276	- 0.135	- 11
1008	00 16 33	- 10 14 18	- 0.0065	- 0.263	
17.0	00 17 29	- 65 10 06	+0.2714	+1.167	+9
17.1	00 18 52	- 46 00 42	+0.0060	- 0.800	
1009	00 19 27	- 31 41 00	+0.0030	- 0.124	
17.2	00 19 52	- 27 18 18	+0.0191	- 0.533	
17.3	00 20 18	- 12 29 12	+0.0266	+0.065	- 5
1010A	00 20 41	+76 54 42	- 0.2438	+0.043	
1010B	00 20 43	+76 54 48	- 0.2438	+0.043	
1011	00 20 52	+24 01 30	- 0.0179	+0.136	
18.0	00 21 53	- 27 18 18	+0.0505	+0.090	+4
19.0	00 23 09	- 77 32 06	+0.6884	+0.327	+23
20.0	00 23 45	- 43 57 24	+0.0098	+0.035	+11
21.0	00 24 00	+69 52 06	- 0.0310	- 0.140	
1012	00 26 08	- 06 55 48	- 0.0228	- 0.801	+51
1013	00 29 02	- 06 07 54	+0.0228	- 1.046	
22.0A	00 29 20	+66 57 48	+0.2980	- 0.210	+10
22.0B	00 29 20	+66 57 48	+0.2980	- 0.210	
22.1	00 30 22	+41 43 24	+0.0270	+0.170	+3
22.2	00 32 05	- 52 39 00	+0.0246	+0.036	+35
23.0A	00 32 40	- 03 52 06	+0.0274	- 0.021	+9
23.0B	00 32 40	- 03 52 06	+0.0274	- 0.021	
1014	00 33 16	+10 12 24	+0.0746	- 0.423	
24.0A	00 34 14	- 49 24 18	+0.0378	- 0.130	- 3
24.0B	00 33 40	- 49 24 06	+0.0388	- 0.138	+5
25.0A	00 34 47	- 25 02 30	+0.1022	- 0.007	+17
25.0B	00 34 47	- 25 02 30	+0.1022	- 0.007	
26.0	00 36 13	+30 20 30	+0.1200	+0.040	9
27.0	00 36 45	+20 58 54	- 0.0331	- 0.366	- 34
27.1	00 37 34	- 44 31 30	+0.0420	- 0.250	
27.2	00 38 02	- 24 04 24	+0.0469	- 0.328	- 53
28.0	00 38 04	+39 55 18	+0.0301	- 0.666	- 62
29.0	00 38 06	- 59 44 06	+0.1180	+0.454	+2
1015A	00 38 28	+55 33 36	+0.0378	- 0.074	
1015B	00 38 29	+55 33 42	+0.0378	- 0.074	

Catalogue of Nearby Stars*

Name	Sp LC	V (mag)	B–V (mag)	π_t (0.001″)	M_V (mag)
16.1	G5IV-V	6.459	+0.68	52	5.04
1008	K7V	9.94	+1.33	45*	8.2
17.0	G2V	4.237	+0.58	140	4.97
17.1	M1	10.20		46	8.50
1009	dM2	11.16	+1.48	67*	10.3
17.2	K3V	8.317	+0.90	51	6.85
17.3	G2V	6.378	+0.66	48	4.78
1010A		11.34	+1.47	68	10.50
1010B		14		68	13
1011		14.28	+1.59	61	13.21
18.0	K3V	7.904	+0.95	48	6.31
19.0	G2IV	2.827	+0.62	159	3.83
20.0	A7VN	3.948	+0.17	72	3.23
21.0	M0	10.20		60	9.10
1012		12.23	+1.48	81*	11.8
1013		12.75	+1.64	62	11.71
22.0A	M2	10.51	+1.50	96	10.42
22.0B	M4.5	2.40		96	12.30
22.1	M0	11.00		36	8.80
22.2	F6V	5.548	+0.46	49	4.00
23.0A	F8V	5.211	+0.58	62	4.17
23.0B		6.29		63	5.30
1014		15.32	+1.89	64	14.35
24.0A	G3V	6.778	+0.64	58	5.60
24.0B	K0V	8.373	+0.78	58	7.19
25.0A	G5V	5.565	+0.72	74	4.91
25.0B		6.35		74	5.70
26.0	M4	11.02	+1.53	80	10.54
27.0	K0V	5.893	+0.85	96	5.80
27.1	M	2.80P		51	11.30
27.2	G3V	6.125	+0.70	42	4.24
28.0	K2VE	7.370	+0.93	72	6.66
29.0	G1V	5.898	+0.56	57	4.68
1015A		14.02	+1.55	45	12.28
1015B	DC	14.08	+0.02	45*	12.14

*Adapted from Gliese (1969), Gliese and Jahreiss (1979), and Grenon and Rufener (1981).

Catalogue of Nearby Stars*

Name	R.A.(1950) h m s	Dec.(1950) ° ′ ″	$\mu_{R.A.}$ (s yr^{-1})	$\mu_{Dec.}$ (″ yr^{-1})	V_r (km s^{-1})
1016	00 39 06	− 33 33 48	− 0.0305	− 0.257	
1017	00 39 45	− 52 38 30	+0.0724	− 0.336	
1018	00 39 53	− 36 59 30	− 0.0138	+0.070	
29.1	00 40 05	+35 16 24	+0.0200	+0.090	+10
30.0	00 40 52	+33 34 36	− 0.0175	− 0.346	− 34
1019	00 40 56	+28 11 06	− 0.0098	− 1.052	
31.0	00 41 05	− 18 15 36	+0.0163	+0.036	+13
31.1A	00 41 07	− 57 44 12	− 0.0002	+0.016	+2
31.1B	00 41 07	− 57 44 12	− 0.0002	+0.016	
31.2	00 42 07	− 19 13 24	+0.0210	+0.090	+27
31.3	00 42 16	− 22 16 48	− 0.0049	+0.087	+12
31.4	00 42 31	+01 31 12	− 0.0028	− 0.563	+6
31.5	00 42 32	− 65 54 48	+0.0247	− 0.744	+98
1020	00 42 58	− 13 09 06	− 0.0026	− 0.196	− 13
32.0A	00 43 25	− 42 10 54	+0.0255	− 0.070	− 26
32.0B	00 43 25	− 42 10 54	+0.0255	− 0.070	− 23
1021	00 43 25	− 47 49 36	+0.0166	+0.089	− 11
33.0	00 45 45	+05 01 24	+0.0506	− 1.142	− 12
34.0A	00 46 03	+57 33 06	+0.1368	− 0.521	+9
34.0B	00 46 03	+57 33 06	+0.1368	− 0.521	+13
34.1	00 46 21	+16 40 18	− 0.0001	− 0.200	+2
35.0	00 46 31	+05 09 12	+0.0850	− 2.710	+54
36.0	00 46 57	− 23 29 12	+0.0377	+0.112	+5
1022	00 47 18	− 61 17 42	+1.1173	− 0.078	− 2
37.0	00 47 37	− 10 54 48	− 0.0156	− 0.223	+8
38.0	00 48 23	+58 01 30	+0.1900	+0.460	− 19
39.0	00 48 43	+18 28 18	+0.0040	− 0.272	+8
40.0	00 49 04	− 23 10 42	+0.0451	− 0.278	+15
41.0	00 50 04	+60 51 00	− 0.0098	+0.178	+21
42.0	00 50 34	− 30 37 42	+0.0484	+0.041	− 7
1023A	00 51 05	+68 46 36	− 0.0161	− 0.048	
1023B	00 51 07	+68 46 36	− 0.0105	− 0.019	
42.1	00 52 19	+23 49 54	− 0.0143	− 0.176	− 13
43.0	00 53 10	− 52 06 30	+0.0330	+0.360	
1024	00 53 58	+17 11 36	+0.0465	− 0.297	

Catalogue of Nearby Stars*

Name	Sp LC	V (mag)	B–V (mag)	π_t (0.001″)	M_V (mag)
1016		10.60	+1.44	46*	9.0
1017		11		46	9
1018		12.74	+1.56	52*	11.3
29.1	M0E	10.38	+1.38	42	8.50
30.0	K5V	8.747	+1.13	52	7.33
1019		14.53	+1.58	51	13.07
31.0	K1III	2.064	+1.01	58	0.88
31.1A	A0IV	4.352	– 0.01	45	2.62
31.1B		11.50		45	9.80
31.2	M0V	10.724		57	9.50
31.3	F2V	5.206	+0.34	47	3.57
31.4	K2V	8.031	+0.99	43	6.20
31.5	G3V	6.525	+0.65	55	5.23
1020	G0V	6.15	+0.60	45*	4.4
32.0A	K5V	7.882	+1.26	83	7.48
32.0B	K7V	8.90		75	8.30
1021	G5IV	5.80	+0.64	48	4.2
33.0	K2V	5.743	+0.88	143	6.52
34.0A	G0V	3.471	+0.57	171	4.64
34.0B	M0V	7.51	+1.39	170	8.66
34.1	F8V	5.060	+0.50	46	3.37
35.0	G	12.37	+0.56	239	14.26
36.0	K0V	7.138	+0.79	64	6.17
1022		12.18	+1.45	58*	11.0
37.0	F8V	5.172	+0.50	63	4.17
38.0	M2	11.00		62	9.90
39.0	K6V	9.221	+1.21	71	8.48
40.0	(K5)	8.945	+1.29	69	8.14
41.0	F8V	4.789	+0.53	64	3.82
42.0	K3V	7.148	+0.94	80	6.66
1023A	G0	9.9		52	8.5
1023B		10.0		52*	8.6
42.1	(G5)	7.411		49	5.86
43.0	M	13.70		63	12.70
1024		13.72	+1.65	56	12.46

*Adapted from Gliese (1969), Gliese and Jahreiss (1979), and Grenon and Rufener (1981).

Catalogue of Nearby Stars*

Name	R.A.(1950) h m s	Dec.(1950) ° ′ ″	$\mu_{R.A.}$ (s yr^{-1})	$\mu_{Dec.}$ (″ yr^{-1})	V_r (km s^{-1})
44.0	00 54 48	– 02 05 00	– 0.0183	– 0.197	– 45
45.0	00 55 10	– 62 31 00	+0.1510	+0.170	+11
46.0	00 55 58	– 28 07 18	+0.0960	– 0.290	
46.1	00 57 36	+17 55 48	+0.0018	– 0.070	– 41
47.0	00 58 13	+61 06 30	+0.0540	– 0.830	+12
1025	00 58 20	– 04 43 30	+0.0836	+0.455	+8
48.0	00 58 48	+71 25 00	+0.3580	– 0.400	+6
49.0	00 59 27	+62 04 30	+0.1090	+0.110	– 6
50.0	00 59 27	– 10 08 54	– 0.0140	– 0.480	
50.1	00 59 34	+59 01 06	+0.0000	– 0.017	– 25
51.0	01 00 08	+62 05 48	+0.1090	+0.110	
1026A	01 00 32	+19 49 48	+0.0459	+0.045	
1026B	01 00 32	+19 49 48	+0.0459	+0.045	
1027	01 01 14	+04 48 18	+0.0216	+0.252	+61
1028	01 02 21	– 18 23 54	+0.0890	+0.436	
1029	01 02 47	+28 13 36	+0.1439	– 0.166	
52.0	01 03 45	+63 40 12	+0.2290	+0.290	– 2
1030	01 04 03	+15 00 30	– 0.0068	– 0.273	
52.1	01 04 35	– 51 15 30	+0.0567	– 0.058	+9
52.2	01 04 56	+33 56 06	+0.1130	+0.510	
53.0A	01 04 56	+54 40 30	+0.3947	– 1.575	– 97
53.0B	01 04 56	+54 40 30	+0.3947	– 1.575	
53.1A	01 04 56	+22 41 42	+0.0073	– 0.482	+2
53.1B	01 04 56	+22 41 42	+0.0073	– 0.482	
1031	01 05 52	– 29 04 18	+0.0539	– 0.137	
53.2	01 06 01	+16 59 06	– 0.0060	– 0.580	– 36
1032	01 06 47	– 24 57 24	+0.0191	+0.009	
53.3	01 06 55	+35 21 24	+0.0146	– 0.110	
53.4	01 07 28	+41 49 00	– 0.0122	– 0.037	– 11
54.0	01 08 34	– 67 43 06	+0.0680	+0.600	
54.1	01 09 59	– 17 16 00	+0.0820	+0.620	+28
1033	01 11 00	– 23 10 00	+0.0065	+0.003	
54.2A	01 11 53	– 08 11 30	+0.0082	+0.277	+22
54.2B	01 11 51	– 08 10 42	+0.0088	+0.282	+16
54.3	01 12 51	– 65 01 24	+0.0300	– 0.300	

Catalogue of Nearby Stars*

Name	Sp LC	V (mag)	B–V (mag)	π_t (0.001″)	M_V (mag)
44.0	K5V	9.432	+0.83	70	8.66
45.0	K7V	9.496	+1.32	60	8.39
46.0	M	12.50		71	11.80
46.1	(G5)	7.353		50	5.85
47.0	M2.5E	10.70		80	10.20
1025		13.33	+1.73	63*	12.3
48.0	M3.5VE	10.039	+1.47	113	10.30
49.0	M1.5V	9.554	+1.50	109	9.74
50.0	K5V	10.455		42	8.57
50.1	K2V	7.780		52	6.30
51.0	M7E	3.66	+1.68	107	13.81
1026A		12.8		57	11.6
1026B		13.3		57*	12.1
1027	DA	13.96	+0.31	47	12.32
1028		14.4		100*	14.4
1029		14.81	+1.88	79	14.30
52.0	K7V	8.991	+1.30	71	8.25
1030		11.37	+1.47	65	10.4
52.1	K1V	8.832	+0.94	47	7.19
52.2	M4	14.70		50	13.20
53.0A	G5VP	5.163	+0.69	126	5.66
53.0B		0		130	
53.1A	K4V	8.398	+1.12	50	6.89
53.1B	M3V	4.00		52	12.60
1031	dM5	13.42	+1.71	64*	12.4
53.2	K6	11.20		35	8.90
1032	dM2	12.38	+1.53	55*	11.1
53.3	M0IIIE	2.109	+1.57	44	+0.33
53.4	F8V	5.686	+0.60	50	4.20
54.0	K	11.20		113	11.50
54.1	M5E	11.60		267	14.19
1033		14.16	+1.58		
54.2A	F5V	5.136	+0.46	51	3.67
54.2B	G7V	7.835	+0.78	51	6.37
54.3	K0V	9.029	+0.77	48	7.44

*Adapted from Gliese (1969), Gliese and Jahreiss (1979), and Grenon and Rufener (1981).

Catalogue of Nearby Stars*

Name	R.A.(1950) h m s	Dec.(1950) ° ′ ″	$\mu_{R.A.}$ (s yr^{-1})	$\mu_{Dec.}$ (″ yr^{-1})	V_r (km s^{-1})
55.0	01 12 56	− 45 47 54	+0.0636	+0.190	+12
55.1A	01 13 19	− 69 05 06	+0.0734	+0.110	
55.1B	01 13 19	− 69 05 06	+0.0734	+0.110	
1034	01 13 40	+24 04 12	+0.1245	− 0.689	
55.2	01 13 54	+25 04 12	+0.0290	− 0.080	− 30
55.3A	01 14 05	− 69 08 30	+0.0755	+0.110	+9
55.3B	01 14 04	− 69 08 24	+0.0664	+0.113	
1035	01 14 18	+83 53 18	− 0.6042	+0.456	
1036	01 14 57	− 35 58 36	+0.0067	− 0.183	
56.0	01 15 05	− 15 45 36	+0.0188	− 0.472	
1037	01 15 20	+15 55 06	− 0.0016	− 0.650	
56.1	01 15 47	− 13 09 12	+0.0120	− 0.690	
56.2	01 15 49	− 48 24 54	+0.0220	+0.110	
56.3A	01 16 06	− 01 07 36	+0.0288	− 0.264	+13
56.3B	01 16 05	− 01 08 00	+0.0288	− 0.264	
56.4	01 16 43	+79 53 30	+0.1070	− 0.090	− 20
56.5	01 17 50	+76 27 00	− 0.0035	− 0.031	− 23
57.0	01 19 16	− 41 54 42	+0.1120	− 0.450	
57.1A	01 20 33	− 13 13 36	+0.0326	− 0.036	+31
57.1B	01 20 31	− 13 13 06	+0.0310	− 0.040	
57.1C	01 20 31	− 13 13 06	+0.0310	− 0.040	
58.0	01 22 27	− 28 05 42	+0.0239	− 0.282	+87
58.1	01 22 31	+59 58 36	+0.0400	− 0.045	+7
1038	01 22 43	− 33 06 48	+0.0191	+0.144	
1039	01 23 02	− 26 16 00	+0.0163	− 0.515	
58.2	01 26 20	+21 28 00	+0.0331	− 0.190	+44
59.0	01 30 53	− 24 25 54	+0.0215	− 0.152	+2
59.1	01 30 55	+68 41 30	− 0.0680	+0.124	− 30
59.2	01 31 12	− 07 16 48	+0.0118	− 0.077	− 15
59.3	01 31 59	+62 49 36	− 0.0006	− 0.015	− 28
60.0A	01 32 42	− 30 10 00	+0.0104	+0.124	+34
60.0B	01 32 42	− 30 10 00	+0.0104	+0.124	
60.0C	01 32 42	− 30 10 00	+0.0104	+0.124	
61.0	01 33 51	+41 09 24	− 0.0153	− 0.379	− 28
62.0	01 34 49	− 29 38 48	+0.0236	− 0.070	

Catalogue of Nearby Stars*

Name	Sp LC	V (mag)	B–V (mag)	π_t (0.001″)	M_V (mag)
55.0	F8V	4.952	+0.58	69	4.15
55.1A	K2V	7.244	+0.98	37	5.08
55.1B		8.30		45	6.60
1034		15.05	+1.86	49	13.50
55.2	M0V	10.112	+1.36	46	8.40
55.3A	F6V	4.867	+0.47	37	2.71
55.3B		7.29		45	5.60
1035		14.77	+1.79	78	14.23
1036		11.42	+1.53	63*	10.4
56.0	(K3)	9.743		68	8.91
1037	DC	13.83	+0.13	67	12.96
56.1	M	12.20		50	10.70
56.2	K	13.00		47	11.40
56.3A	(K0)	7.999	+0.82	46	6.31
56.3B	(M0)	10.708	+1.40	46	9.02
56.4	K8	9.66	+1.29	45	7.90
56.5	K0	7.11	+0.82	51	5.60
57.0	M1V	10.146		65	9.21
57.1A	G8V	7.831	+0.91	50	6.33
57.1B	K6V	10.274	+1.38	50	8.77
57.1C		13.50		46	11.80
58.0	G3V	8.277	+0.69	73	7.59
58.1	A5V	2.669	+0.13	31	0.13
1038		9.81	+1.45	56*	8.6
1039	DC	15.0	+0.40	53*	13.6
58.2	K2V	7.712	+0.96	35	5.43
59.0	G8V	6.964	+0.77	69	6.16
59.1	G6V	6.528	+0.68	38	4.43
59.2	(DG2)	5.760	+0.65	68	4.90
59.3	K1V	6.727		73	6.10
60.0A	K3V	7.095	+0.92	57	5.87
60.0B		8.00		56	6.70
60.0C		11.30		56	10.00
61.0	F8V	4.105	+0.54	62	3.07
62.0	K0V	8.116	+0.86	41	6.18

*Adapted from Gliese (1969), Gliese and Jahreiss (1979), and Grenon and Rufener (1981).

Catalogue of Nearby Stars*

Name	R.A.(1950) h m s	Dec.(1950) ° ′ ″	$\mu_{R.A.}$ (s yr^{-1})	$\mu_{Dec.}$ (″ yr^{-1})	V_r (km s^{-1})
63.0	01 35 07	+56 58 54	− 0.0280	− 0.420	
64.0	01 35 26	− 05 14 42	+0.0380	− 0.350	
1040	01 35 30	+44 20	+0.0143	+0.112	
65.0A	01 36 25	− 18 12 42	+0.2320	+0.580	+29
65.0B	01 36 25	− 18 12 42	+0.2320	+0.580	+32
65.1	01 36 33	+45 37 42	+0.0219	− 0.222	+6
65.2	01 37 37	− 46 45 42	+0.0020	− 0.080	
66.0A	01 37 54	− 56 26 54	+0.0337	+0.027	+21
66.0B	01 37 54	− 56 26 54	+0.0337	+0.027	
67.0	01 38 44	+42 21 48	+0.0730	− 0.147	+4
67.1	01 39 09	− 83 13 48	+0.0681	+0.129	− 5
67.2	01 39 28	− 45 40 18	+0.0220	+0.010	− 4
68.0	01 39 47	+20 01 36	− 0.0209	− 0.668	− 34
69.0	01 40 12	+63 34 48	− 0.0611	− 0.567	− 45
70.0	01 40 46	+04 04 54	− 0.0290	− 0.760	− 3
71.0	01 41 45	− 16 12 00	− 0.1194	+0.858	− 16
72.0	01 42 12	+19 50 00	− 0.0029	− 0.105	− 44
73.0	01 42 18	+16 06 00	− 0.0490	− 0.390	
74.0	01 43 58	+12 09 48	+0.0024	− 0.079	+19
75.0	01 44 06	+63 36 24	+0.0876	− 0.246	+3
76.0	01 45 20	− 26 59 42	− 0.0077	− 0.276	+25
77.0	01 46 12	− 41 44 42	+0.0354	+0.150	+36
78.0	01 49 19	− 11 02 36	+0.0380	− 0.570	
78.1	01 50 13	+29 20 12	+0.0008	− 0.229	− 13
79.0	01 50 25	− 22 40 54	+0.0598	+0.007	+20
80.0	01 51 52	+20 33 54	+0.0069	− 0.117	− 2
81.0A	01 54 01	− 51 51 24	+0.0733	+0.299	− 6
81.0B	01 54 01	− 51 51 24	+0.0733	+0.299	
81.1	01 54 43	− 10 29 00	− 0.0257	− 0.228	− 7
81.2	01 54 52	− 60 28 24	+0.0586	+0.164	
81.3	01 55 06	− 52 00 48	+0.0386	+0.255	+4
82.0	01 55 54	+58 16 54	+0.0380	− 0.230	
82.1	01 56 10	+32 58 18	+0.0188	− 0.359	− 37
1041A	01 56 36	+03 16 30	+0.0184	+0.049	
1041B	01 56 36	+03 16 30	+0.0184	+0.049	

Catalogue of Nearby Stars*

Name	Sp LC	V (mag)	B–V (mag)	π_t (0.001″)	M_V (mag)
63.0	M4	11.70		87	11.40
64.0	A	12.84	+0.34	66	11.90
1040	dKB	10.74	+1.14	78	10.2
65.0A	M5.5E	12.45		367	15.27
65.0B	M5.5E	12.95		367	15.80
65.1	G5IV	6.634		51	5.10
65.2	M0	10.90		66	10.00
66.0A	K2V	5.753	+0.88	148	6.60
66.0B	K0V	5.869		148	6.72
67.0	G1.5V	4.970	+0.62	87	4.67
67.1	G2V	5.887	+0.61	58	4.70
67.2	G5V	9.288	+0.70	53	7.91
68.0	K1V	5.262	+0.84	134	5.90
69.0	K5V	8.402	+1.22	72	7.69
70.0	M2	10.95	+15.9	98	10.90
71.0	G6VP	3.503	+0.72	277	5.72
72.0	G5IV	6.292	+0.72	27	3.45
73.0	M4	14.40		63	13.40
74.0	K8	8.90	+1.04	42	7.00
75.0	K0V	5.630	+0.81	105	5.74
76.0	K1V	8.963	+0.81	56	7.70
77.0	G4V	7.116	+0.64	64	6.15
78.0	M4	11.20		88	10.90
78.1	F6IV	3.429	+0.48	51	1.97
79.0	(M1E)	8.881		84	8.50
80.0	A5V	2.663	+0.13	66	1.76
81.0A	G5IV	3.747	+0.85	58	2.56
81.0B		10.70		63	9.70
81.1	(G5)	6.419	+0.82	41	4.48
81.2	K5V	8.592		55	7.29
81.3	F8V	6.078	+0.48	54	4.74
82.0	M4E	12.10		82	11.70
82.1	G7V	7.157	+0.78	48	5.60
1041A		10.92	+1.51	72*	10.3:
1041B		13.5:			

*Adapted from Gliese (1969), Gliese and Jahreiss (1979), and Grenon and Rufener (1981).

Catalogue of Nearby Stars*

Name	R.A.(1950) h m s	Dec.(1950) ° ′ ″	$\mu_{R.A.}$ (s yr^{-1})	$\mu_{Dec.}$ (″ yr^{-1})	V_r (km s^{-1})
83.0	01 57 12	− 61 48 48	+0.0380	+0.033	+7
83.1	01 57 28	+12 49 54	+0.0730	− 1.790	
83.2	01 57 33	+02 51 30	+0.0152	− 0.246	− 17
83.3	01 58 23	+61 39 54	− 0.0002	+0.002	− 12
83.4A	02 01 55	− 45 39 12	+0.0334	+0.067	
83.4B	02 01 55	− 45 39 12	+0.0334	+0.067	
84.0	02 02 37	− 17 51 06	+0.0900	− 0.160	− 35
84.1A	02 03 07	− 28 18 48	+0.0250	+0.440	
84.1B	02 03 08	− 28 17 54	+0.0250	+0.440	
84.2	02 03 48	+44 57 12	+0.0260	− 0.440	+62
84.3	02 04 21	+23 13 36	+0.0138	− 0.144	− 14
85.0	02 05 56	− 66 48 42	+0.2950	+0.400	
85.1	02 07 07	+35 12 00	− 0.0270	− 0.590	
1042	02 08 13	+39 41 36	+0.0887	− 0.500	+29
86.0	02 08 25	− 51 04 06	+0.2244	+0.655	+53
86.1	02 08 27	− 28 27 18	+0.0065	+0.002	+40
87.0	02 09 51	+03 22 00	− 0.1180	− 1.880	+7
1043	02 10 00	+28 58 36	+0.0122	+0.003	+6
87.1A	02 10 14	− 02 37 36	+0.0244	+0.064	− 3
87.1B	02 10 12	− 02 37 36	+0.0244	− 0.064	− 4
88.0	02 10 27	− 17 55 24	+0.0330	+0.210	
89.0	02 10 42	− 30 57 30	+0.0015	+0.010	+10
1044	02 10 50	− 21 25 36	+0.0233	+0.057	
90.0	02 11 35	+67 26 36	+0.0932	− 0.304	− 14
91.0	02 11 40	− 32 16 00	+0.0590	− 0.570	
1045	02 12 14	+17 11 30	+0.0242	− 0.477	
91.1A	02 13 16	+06 23 42	− 0.0074	− 0.062	− 2
91.1B	02 13 16	+06 23 42	− 0.0074	− 0.062	
91.2A	02 13 25	− 18 28 12	− 0.0047	− 0.183	+5
91.2B	02 13 25	− 18 28 12	− 0.0047	− 0.183	
91.3	02 13 46	+42 44 36	+0.0770	− 0.640	
92.0	02 14 00	+33 59 48	+0.0928	− 0.237	− 6
92.1	02 14 52	+56 20 00	+0.0409	− 0.231	+2
92.2	02 15 00	+44 02 24	+0.0470	− 0.110	
93.0	02 15 50	− 54 13 30	+0.0540	+0.380	

Catalogue of Nearby Stars*

Name	Sp LC	V (mag)	B–V (mag)	π_t (0.001″)	M_V (mag)
83.0	F3V	2.869	+0.28	47	1.23
83.1	M5V	12.284	+1.80	224	14.04
83.2	G0IV	5.884	+0.62	37	3.72
83.3	K5V	7.512		100	7.40
83.4A	G3V	7.273	+0.70	50	5.77
83.4B		11.50		50	10.00
84.0	M3V	10.169	+1.53	114	10.45
84.1A	K	1.39	+1.40	54	10.00
84.1B	M	2.77	+1.52	54	11.40
84.2	M0	11.10		40	9.00
84.3	K2IIIE	2.024	+1.15	44	+0.24
85.0	M	13.50		66	12.60
85.1	M3	15.00		46	13.30
1042	DA-P	14.53	+0.34	60	13.4
86.0	K0V	6.108	+0.82	89	5.85
86.1	K2V	7.060	+1.02	77	6.50
87.0	M2.5V	10.058	+1.44	100	10.06
1043	F5V	5.29	+0.42	46*	3.6
87.1A	F8V	5.680	+0.56	45	3.95
87.1B	G4V	7.734	+0.68	45	6.00
88.0	M	11.90		48	10.30
89.0	A1V	5.249	- 0.02	63	4.25
1044	K7V	9.88	+1.36	47*	8.2
90.0	K2V	7.074	+0.92	54	5.74
91.0	(M)	10.301		109	10.49
1045		14.39	+1.62	49	12.84
91.1A	K8	9.79	+1.10	35	7.50
91.1B		9.80		35	7.50
91.2A	K6V	7.944		47	6.70
91.2B		9.30		47	7.70
91.3	C	6.23	+0.73	49	14.70
92.0	G0VE	4.872	+0.61	96	4.78
92.1	K2III	8.263		48	6.67
92.2	M4	13.20		46	11.50
93.0	M	12.60		74	12.00

*Adapted from Gliese (1969), Gliese and Jahreiss (1979), and Grenon and Rufener (1981).

Catalogue of Nearby Stars*

Name	R.A.(1950) h m s	Dec.(1950) ° ′ ″	$\mu_{R.A.}$ (s yr^{-1})	$\mu_{Dec.}$ (″ yr^{-1})	V_r (km s^{-1})
94.0	02 16 00	+35 07 42	+0.0570	− 0.430	
95.0	02 16 44	− 26 10 54	− 0.0162	+0.451	+7
1046	02 16 59	− 39 00 54	+0.1193	+0.562	+47
1047AB	02 17 59	+36 39 48	+0.0602	− 0.566	
1047C	02 17 57	+36 39 30	+0.0602	− 0.566	
96.0	02 18 57	+47 39 06	+0.0220	+0.050	− 34
97.0	02 20 15	− 24 02 36	+0.0144	− 0.056	+18
97.1	02 20 51	− 68 53 12	− 0.0078	+0.008	+6
97.2	02 22 52	+59 26 06	+0.0123	− 0.031	+19
98.0A	02 25 09	+04 12 18	+0.0100	+0.215	+8
98.0B	02 25 09	+04 12 18	+0.0100	+0.215	
99.0A	02 25 46	+32 02 06	+0.0350	+0.110	− 13
99.0B	02 25 46	+32 02 06	+0.0350	+0.110	
99.1	02 25 52	+29 42 30	− 0.0053	+0.078	+40
100.0A	02 26 41	− 20 12 18	+0.0462	+0.229	
100.0B	02 26 41	− 20 12 18	+0.0462	+0.229	
100.1	02 27 39	+05 02 36	+0.0064	− 0.019	+30
101.0	02 27 44	+57 09 30	+0.1400	− 0.020	
101.1	02 29 18	− 46 54 24	+0.0010	− 0.300	
102.0	02 30 44	+24 42 54	+0.0030	− 0.680	
103.0	02 32 28	− 44 00 36	+0.0053	− 0.301	+42
104.0	02 33 03	+20 00 18	+0.0170	− 0.140	
105.0A	02 33 20	+06 39 00	+0.1208	+1.461	+24
105.0B	02 33 31	+06 38 00	+0.1208	+1.461	
1048	02 33 45	− 23 44 18	+0.0058	+0.006	
105.1	02 34 54	− 34 47 30	− 0.0015	− 0.266	+10
105.2	02 36 16	− 42 07 12	+0.0107	+0.063	+48
105.3	02 36 56	− 26 31 48	+0.0134	− 0.211	+16
105.4A	02 37 09	− 12 05 00	+0.0096	− 0.234	+15
105.4B	02 37 09	− 12 05 00	+0.0096	− 0.234	
1049	02 37 36	− 58 24 06			
1050	02 37 44	− 34 19 30	+0.0429	− 1.636	− 71
105.5	02 38 07	+00 58 54	+0.0183	+0.234	+79
105.6	02 39 05	+39 59 00	− 0.0014	− 0.184	− 22
106.0	02 40 30	+19 13 06	+0.0299	− 0.012	+27

Catalogue of Nearby Stars*

Name	Sp LC	V (mag)	B–V (mag)	π_t (0.001″)	M_V (mag)
94.0	M4.5	14.50		56	13.20
95.0	G5V	6.322	+0.73	96	6.23
1046		11.62	+1.50	83*	11.2
1047AB		13			
1047C		14.22	+1.70	47	12.58
96.0	M1.5VE	9.413	+1.49	87	9.11
97.0	G1V	5.172	+0.60	67	4.58
97.1	A2V	4.076	+0.03	48	2.48
97.2	K0V	6.980		63	5.90
98.0A	M1.5V	8.710	+1.39	63	7.71
98.0B		9.50		64	8.50
99.0A		10.20	+1.36	40	8.20
99.0B		10.50		40	8.50
99.1	G0V	5.902		49	4.40
100.0A	K4V	8.762	+1.18	40	7.10
100.0B		11.50		40	9.70
100.1	A	2.65	- 0.06	45	10.90
101.0	M5	15.10		68	14.30
101.1	M	12.20		65	11.30
102.0	M6	14.50		133	15.10
103.0	K7VE	8.909	+1.39	88	8.63
104.0	M2	10.80		60	9.70
105.0A	K3V	5.825	+0.97	138	6.52
105.0B	M4.5VE	11.673	+1.61	138	12.37
1048	K2V	8.43	+1.08	52*	7.0
105.1	G5IV	5.763	+0.66	55	4.46
105.2	K0V	7.223	+0.94	59	6.10
105.3	G2VI	8.750	+0.60	48	7.16
105.4A	F5IV-V	4.827	+0.44	69	4.02
105.4B		15.60		69	4.80
1049	M0V	9.65	+1.39	60*	8.6
1050		11.75	+1.52	74*	11.1
105.5	M0VP	9.498	+1.19	41	7.56
105.6	F9V	4.904	+0.59	40	2.91
106.0	K4V	8.214	+1.07	55	6.92

*Adapted from Gliese (1969), Gliese and Jahreiss (1979), and Grenon and Rufener (1981).

Catalogue of Nearby Stars*

Name	R.A.(1950) h m s	Dec.(1950) ° ′ ″	$\mu_{R.A.}$ (s yr^{-1})	$\mu_{Dec.}$ (″ yr^{-1})	V_r (km s^{-1})
106.1A	02 40 42	+03 01 36	– 0.0097	– 0.148	– 5
106.1B	02 40 42	+03 01 36	– 0.0097	– 0.148	– 12
106.1C	02 39 54	+03 09 54	– 0.0110	– 0.120	
107.0A	02 40 46	+49 01 06	+0.0342	– 0.083	+25
107.0B	02 40 46	+49 01 06	+0.0342	– 0.083	+25
108.0	02 40 51	– 51 00 54	+0.0349	+0.229	+16
109.0	02 41 18	+25 19 00	+0.0630	– 0.380	+47
1051	02 41 22	– 09 02 24	+0.0399	– 0.705	+6
110.0	02 42 34	– 66 55 30	1.90	– 0.66	– 20
111.0	02 42 46	– 18 47 00	+0.0229	+0.042	+26
112.0	02 43 20	+25 26 30	+0.0169	– 0.150	+12
112.1	02 43 34	+11 34 06	+0.0183	– 0.207	+11
1052	02 43 58	– 02 39 36	+0.0141	– 0.475	
113.0	02 45 12	+26 51 42	+0.0205	– 0.108	+7
113.1	02 45 42	+30 54 36	+0.0169	– 0.172	+10
114.0	02 47 49	+15 30 36	+0.0235	– 0.395	– 25
114.1A	02 48 31	– 53 21 18	– 0.0130	+0.520	
114.1B	02 48 31	– 53 21 18	– 0.0130	+0.520	
115.0	02 48 52	– 44 17 00	– 0.0015	– 0.285	+30
116.0	02 48 58	+34 11 54	+0.0790	– 0.980	– 46
117.0	02 50 07	– 12 58 18	+0.0266	– 0.172	+19
118.0	02 51 14	– 63 53 30	+0.1470	+0.600	
118.1A	02 52 00	– 36 06 18	+0.0444	– 0.162	+3
118.1B	02 52 00	– 36 06 54	+0.0444	– 0.162	
118.2A	02 52 41	+26 40 24	+0.0197	– 0.185	+32
118.2B	02 52 41	+26 40 24	+0.0197	– 0.185	+27
118.2C	02 52 38	+26 40 18	+0.0197	– 0.185	
119.0A	02 52 50	+55 14 30	+0.0840	– 0.450	
119.0B	02 52 50	+55 14 48	+0.0840	– 0.450	
120.0	02 54 43	+10 35 48	+0.1230	– 0.350	+49
120.1A	02 55 02	– 25 10 06	+0.0012	– 0.016	
120.1B	02 55 02	– 25 10 06	+0.0012	– 0.016	
120.1C	02 55 01	– 25 10 30	– 0.0012	– 0.023	
120.2	02 59 28	+26 24 54	+0.0176	– 0.161	+11
121.0	03 00 11	– 23 49 12	– 0.0107	– 0.051	– 10

Catalogue of Nearby Stars*

Name	Sp LC	V (mag)	B–V (mag)	π_t (0.001″)	M_V (mag)
106.1A	A2V	3.468	+0.09	50	1.96
106.1B	(DF3)	6.20		46	4.50
106.1C	K5	10.20		46	8.50
107.0A	F7V	4.107	+0.49	79	3.60
107.0B	M2E	9.87		79	9.36
108.0	G3IV	5.377	+0.56	70	4.60
109.0	M3.5VE	10.597	+1.55	128	11.13
1051	M2	11.92	+1.46	52	10.5
110.0	F7V	6.240	+0.53	40	4.20
111.0	F6V	4.472	+0.48	73	3.79
112.0	K2IV	7.874		60	6.76
112.1	(K5)	8.6044		55	7.31
1052		15.53	+0.36	48	13.94
113.0	K1VE	7.560	+0.82	56	6.30
113.1	(G9E)	6.791		47	5.15
114.0	K6V	8.921	+1.27	59	7.78
114.1A	K-M	11.50		45	9.80
114.1B		12.00		45	10.00
115.0	F8V	8.148	+0.55	63	7.14
116.0	(S0K6)	9.582		67	8.71
117.0	K0VE	6.071	+0.87	127	6.59
118.0	M	10.70		82	10.30
118.1A	K3V	8.216	+0.93	48	6.62
118.1B	M4	12.50		48	10.90
118.2A	K2V	7.442	+0.93	42	5.56
118.2B	MIV	9.80	+1.40	48	8.20
118.2C	M	3.86	+1.58	48	12.30
119.0A	M1	10.41		62	9.40
119.0B	M3.5	11.60		62	10.60
120.0	M4	12.20		69	11.40
120.1A	(G5)	7.332	+0.87	50	6.60
120.1B		8.20		50	6.70
120.1C	(G5)	7.819	+0.95	50	6.30
120.2	G8V	6.623		55	5.40
121.0	A5V	4.082	+0.16	58	2.90

*Adapted from Gliese (1969), Gliese and Jahreiss (1979), and Grenon and Rufener (1981).

Catalogue of Nearby Stars*

Name	R.A.(1950) h m s	Dec.(1950) ° ′ ″	$\mu_{R.A.}$ (s yr^{-1})	$\mu_{Dec.}$ (″ yr^{-1})	V_r (km s^{-1})
121.1	03 00 30	− 18 21 00	+0.0310	+0.140	
121.2	03 01 09	− 05 51 30	+0.0233	− 0.263	− 20
122.0	03 31 05	+75 51 48	+0.0590	− 0.060	+34
123.0	03 03 50	+01 47 12	+0.0258	− 0.884	− 24
1053	03 05 16	+73 35 48	+0.4289	− 1.055	
124.0	03 05 27	+49 25 24	+0.1295	− 0.078	+50
1054A	03 05 49	− 28 24 24	− 0.0271	− 0.130	
1054B	03 05 47	− 28 25 30	− 0.0271	− 0.130	
125.0	03 06 09	+45 32 54	− 0.0420	− 0.340	+3
1055	03 06 17	+09 50 30	+0.0197	− 0.570	
1056	03 07 13	− 60 21 54	+0.0290	+0.223	
126.0	03 09 33	− 46 42 42	+0.0180	+0.400	
127.0A	03 09 57	− 29 11 00	+0.0254	+0.644	− 20
127.0B	03 09 57	− 29 11 00	+0.0254	+0.644	
127.1A	03 10 04	− 68 47 12	+0.0110	− 0.050	
127.1B	03 10 04	− 68 47 12	+0.0110	− 0.050	
128.0A	03 10 13	− 01 02 29	+0.0129	− 0.063	+18
128.0B	03 10 13	− 01 02 29	+0.0129	− 0.063	
129.0	03 10 29	+18 03 94	+0.0890	− 1.080−	+102
130.0	03 10 30	− 38 17 12	+0.1030	+0.730	
1057	03 10 39	+04 35 12	+0.1147	+0.120	
130.1A	03 12 17	+57 59 18	+0.0630	− 0.310	
130.1B	03 12 17	+57 59 18	+0.0630	− 0.310	
131.0	03 12 35	− 26 38 00	+0.0180	+0.070	+15
132.0	03 13 25	− 45 51 06	− 0.0143	+0.139	+4
133.0	03 13 56	+79 46 54	+0.1450	+0.300	
134.0	03 14 52	+38 04 42	+0.0390	− 0.560	+9
135.0	03 16 30	− 03 01 24	+0.0165	− 0.106	+21
136.0	03 16 41	− 62 46 00	+0.1946	+0.661	+12
137.0	03 16 44	+03 11 18	+0.0178	+0.098	+20
138.0	03 17 07	− 62 41 48	+0.1936	+0.658	+11
138.1A	03 17 54	+08 51 18	+0.0197	− 0.066	+12
138.1B	03 17 59	+08 51 24	+0.0197	− 0.066	
139.0	03 17 56	− 43 15 36	+0.2783	+0.748	+87
1058	03 19 27	+02 46 24	+0.0217	− 0.731	

Catalogue of Nearby Stars*

Name	Sp LC	V (mag)	B–V (mag)	π_t (0.001″)	M_V (mag)
121.1		13.20		64	12.20
121.2	(G5)	8.069	+0.67	46	6.38
122.0	M0	9.76	+1.32	51	8.30
123.0	M0V	9.068	+1.35	65	8.13
1053	dM6	14.66	+1.79	86	14.33
124.0	G0V	4.056	+0.60	86	3.73
1054A	K7V	10.20	+1.40	46*	8.5
1054B		13.10	+1.45		
125.0	M2E	10.10		75	9.50
1055		14.86	+1.71	84	14.48
1056	K5V	9.35	+1.22	49*	7.8
126.0	K4	12.50		76	11.90
127.0A	F8IV	3.878	+0.51	74	3.22
127.0B		6.50		74	5.80
127.1A		11.40	+0.05	52	10.00
127.1B		14.00		52	13.00
128.0A	F8V	5.072	+0.57	50	3.57
128.0B		11.50		55	10.20
129.0		4.30	+1.46	55	13.00
130.0	M5	12.00		76	11.40
1057		13.78	+1.83	116	14.10
130.1A	M2	10.70		47	9.10
130.1B	M2	10.80		47	9.20
131.0	K7V	9.134	+1.23	55	7.84
132.0	G3V	6.760	+0.58	37	4.60
133.0	M2	11.28		91	11.10
134.0	M1.5E	10.28	+1.48	73	9.60
135.0	(G0)	7.048	+0.67	65	6.11
136.0	G2V	5.513	+0.64	89	5.26
137.0	G5VE	4.867	+0.68	106	4.99
138.0	G2V	5.220	+0.60	89	4.97
138.1A	(G2)	8.462	+0.68	49	6.91
138.1B		16.20		27	13.40
139.0	G5V	4.254	+0.71	161	5.29
1058		14.78	+1.76	60	13.67

*Adapted from Gliese (1969), Gliese and Jahreiss (1979), and Grenon and Rufener (1981).

Catalogue of Nearby Stars*

Name	R.A.(1950) h m s	Dec.(1950) ° ′ ″	$\mu_{R.A.}$ (s yr^{-1})	$\mu_{Dec.}$ (″ yr^{-1})	V_r (km s^{-1})
1059	03 19 41	+41 50 18	+0.0252	− 0.663	
140.0A	03 21 09	+23 36 36	+0.0150	− 0.090	
140.0B	03 21 09	+23 36 36	+0.0150	− 0.090	
140.0C	03 21 16	+23 35 54	+0.0150	− 0.090	
140.1A	03 21 43	− 50 10 30	+0.0217	+0.297	− 8
140.1B	03 21 42	− 50 10 36	+0.0217	+0.297	
141.0	03 22 32	− 05 31 42	− 0.0167	− 0.775	− 12
141.1	03 23 33	+28 32 30	+0.0016	− 0.106	+16
141.2	03 24 07	− 30 47 42	+0.0173	+0.203	+20
142.0	03 25 36	− 19 58 54	+0.0362	+0.334	+34
143.0	03 26 08	− 63 40 06	+0.0547	− 0.246	+57
1060A	03 26 45	− 27 29 18	+0.0555	+0.377	
1060B	03 26 45	− 27 29 24			
143.1	03 26 57	− 11 50 42	+0.0040	− 0.280	
143.2A	03 28 30	− 63 06 48	+0.0560	+0.376	+12
143.2B	03 28 36	− 63 07 18	+0.0560	+0.376	
143.3	03 28 59	+14 09 42	+0.0030	− 0.700	
144.0	03 30 34	− 09 37 36	− 0.0662	+0.019	+16
145.0	03 31 17	− 44 52 18	− 0.0300	+0.140	
146.0	03 33 26	− 48 35 18	+0.0395	+0.335	+20
1061	03 34 16	− 44 40 18	+0.0680	− 0.402	− 20
147.0	03 34 19	+00 14 42	− 0.0157	− 0.481	+28
1062	03 35 49	− 11 36 48	+0.0968	− 2.675	
147.1	03 37 49	− 03 22 30	+0.0465	− 0.214	+114
148.0	03 38 34	+03 27 18	− 0.0019	− 0.229	+1
149.0	03 40 36	− 24 37 12	+0.0031	− 0.377	
150.0	03 40 51	− 09 55 54	− 0.0067	+0.746	− 6
150.1A	03 41 02	+16 31 06	+0.0080	− 0.310	
150.1B	03 40 55	+16 30 48	+0.0080	− 0.310	
150.2	03 41 16	+45 52 48	+0.0299	− 0.115	+18
151.0	03 41 41	+18 17 42	+0.0330	− 1.160	
1063	03 42 05	+11 45 52	+0.0210	+0.118	+82
152.0	03 42 18	− 38 26 30	+0.0173	+0.295	+32
153.0A	03 43 04	+68 30 54	+0.0200	+0.292	− 5
153.0B	03 43 05	+68 31 12	+0.0200	+0.292	

Catalogue of Nearby Stars*

Name	Sp LC	V (mag)	B–V (mag)	π_t (0.001″)	M_V (mag)
1059		15.33	+1.89	66	14.43
140.0A	M0	10.64	+1.51	57	9.40
140.0B		12.00		57	10.80
140.0C		13.70		57	12.50
140.1A	K5V	8.578	+1.13	52	7.16
140.1B	K	10.161		52	8.74
141.0	K4V	7.876	+1.16	65	6.94
141.1	K0IV	6.431		50	5.20
141.2	G5V	7.873	+0.72	54	6.53
142.0	K7V	8.380	+1.34	70	7.61
143.0	K5V	8.083	+1.12	65	7.15
1060A	DA	14.0		56	12.7
1060B	dM3	14.1		56	12.8
143.1	M0	10.00		49	8.50
143.2A	F5V	4.733	+0.39	58	3.55
143.2B	(M)	10.681	+1.38	58	9.50
143.3	M3	12.27	+1.55	85	11.90
144.0	K2VE	3.729	+0.88	303	6.14
145.0	M4.5	11.10		84	10.70
146.0	K7V	8.589	+1.32	89	8.34
1061		13.03	+1.90	130*	13.6
147.0	F8V	4.288	+0.57	61	3.21
1062	M2	13.02	+1.65	61	11.94
147.1	F9V	6.678	+0.54	47	5.04
148.0	M0P	9.72	+1.43	57	8.50
149.0	K4	9.17	+1.15	60	8.10
150.0	K0IV	3.544	+0.92	109	3.73
150.1A	M2	10.00		45	8.30
150.1B	M1	10.70		45	9.00
150.2	(K2)	7.720		49	6.17
151.0		5.20	+0.30	72	14.50
1063	dK8	9.15	+1.18	48*	7.6
152.0	K0V	6.979	+0.88	65	6.04
153.0A	M0V	9.305	+1.30	49	7.76
153.0B	(M2.5)	10.697	+1.55	49	9.15

*Adapted from Gliese (1969), Gliese and Jahreiss (1979), and Grenon and Rufener (1981).

Catalogue of Nearby Stars*

Name	R.A.(1950) h m s	Dec.(1950) ° ′ ″	$\mu_{R.A.}$ (s yr^{-1})	$\mu_{Dec.}$ (″ yr^{-1})	V_r (km s^{-1})
153.0C	03 43 05	+68 31 12	+0.0200	+0.292	
154.0	03 43 18	+26 03 48	+0.0250	- 0.210	+32
154.1A	03 43 19	- 28 01 12	+0.0246	+0.162	+33
154.1B	03 43 19	- 28 01 12	+0.0246	+0.162	
154.2	03 43 34	- 64 57 48	+0.0495	+0.078	+51
1064A	03 43 37	+41 17 24	+0.0517	- 1.251	+50
1064B	03 43 37	+41 17 30	+0.0517	- 1.251	+54
155.0	03 44 42	- 23 23 48	- 0.0118	- 0.527	+6
155.1	03 45 23	+02 38 30	- 0.0260	- 0.380	
1064	03 46 45	+43 17 48	+0.0385	- 1.377	
1065	03 48 18	- 06 13 36	- 0.0278	- 1.358	
155.2	03 50 44	+61 01 24	+0.0619	- 0.250	+48
155.3	03 51 31	- 37 11 54	- 0.0310	- 1.080	
156.0	03 52 09	- 06 58 48	- 0.0015	+0.531	+60
156.1A	03 52 55	+53 25 18	+0.0380	- 0.430	- 5
156.1B	03 52 55	+53 25 18	+0.0380	- 0.430	
156.2	03 54 49	+76 01 36	+0.0968	- 0.520	+20
1066	03 54 53	- 41 29 18	+0.0061	+0.110	
157.0A	03 54 57	- 01 18 00	- 0.0125	- 0.158	+6
157.0B	03 54 57	- 01 18 00	- 0.0125	- 0.158	+14
157.1	03 56 49	+25 57 12	+0.0560	- 0.230	+94
157.2	03 58 42	+18 36 12	+0.0200	- 1.210	
158.0	03 59 53	+35 09 18	+0.1420	- 1.346	- 30
159.0	04 00 03	+00 24 12	+0.0097	- 0.248	+18
159.1	04 01 33	+25 00 54	+0.0120	- 0.230	+88
160.0	04 02 22	+21 52 30	+0.0122	- 0.133	+26
160.1A	04 04 14	+37 56 42	+0.0145	- 0.218	+26
160.1B	04 04 14	+37 56 42	+0.0145	- 0.218	
160.2	04 04 23	- 20 58 30	+0.0027	- 0.769	+28
161.0	04 04 23	+69 24 48	+0.0162	- 0.288	- 11
161.1	04 05 16	+37 54 36	+0.0139	- 0.198	+25
161.2	04 05 21	- 40 53 24	- 0.0080	+0.330	
162.0	04 05 23	+33 30 12	+0.0450	+0.140	
162.1	04 06 49	- 64 21 30	+0.0324	+0.328	+28
1067	04 07 26	+70 04 24	- 0.0184	+0.033	

Catalogue of Nearby Stars*

Name	Sp LC	V (mag)	B–V (mag)	π_t (0.001″)	M_V (mag)
153.0C		11.35		52	9.90
154.0	M0V	9.573		84	9.19
154.1A	K5V	8.185	+1.00	36	5.97
154.1B		12.20		47	10.60
154.2	K0IV	3.863	+1.13	49	2.31
1064A	K1V	8.15	+0.78	46	6.48
1064B	K2V	8.74	+0.90	46	7.05
155.0	F3V	4.224	+0.42	57	3.00
155.1	M1	10.50		54	9.20
1064-	dMp	13.90	+1.57	25	10.9
1065		12.79	+1.69	105	12.90
155.2	K0V	7.828	+0.83	49	6.28
155.3	K	12.80		54	11.50
156.0	M0.5V	9.024	+1.36	69	8.22
156.1A	M1.5E	10.86	+1.44	49	9.30
156.1B		13.30		49	11.80
156.2	K4V	8.212	+1.15	50	6.71
1066	K5V	8.94	+1.21	57*	7.7
157.0A	K4V	8.020	+1.12	102	8.06
157.0B	M3E	11.48	+1.52	81	11.00
157.1	M4	12.60		55	11.30
157.2		5.52	+1.83	70	14.70
158.0	K1V	8.500	+0.88	52	7.08
159.0	F6V	5.365	+0.50	55	4.07
159.1		3.80	+0.10	50	12.30
160.0	G5V	5.897	+0.62	69	5.09
160.1A	K2V	7.119	+0.85	29	4.43
160.1B	K2	9.30		45	7.60
160.2	M0V	9.691	+1.23	52	8.27
161.0	K2V	7.696		56	6.44
161.1	F7V	5.494	+0.54	46	3.81
161.2		12.80		47	11.20
162.0	M1	10.20		73	9.50
162.1	G3V	6.361	+0.64	37	4.20
1067	dM0	10.0		48	8.4

*Adapted from Gliese (1969), Gliese and Jahreiss (1979), and Grenon and Rufener (1981).

Catalogue of Nearby Stars*

Name	R.A.(1950) h m s	Dec.(1950) ° ′ ″	$\mu_{R.A.}$ (s yr^{-1})	$\mu_{Dec.}$ (″ yr^{-1})	V_r (km s^{-1})
162.2	04 07 43	- 59 39 18	- 0.0108	- 0.158	
163.0	04 07 56	- 53 30 42	+0.1160	+0.600	
164.0	04 09 09	+52 29 42	- 0.0410	- 0.810	
1068	04 09 17	- 53 42 00	- 0.0877	- 2.397	+27
165.0A	04 09 27	+50 24 12	- 0.0530	- 0.260	
165.0B	04 09 27	+50 24 12	- 0.0530	- 0.260	
165.1	04 11 20	+58 24 00	+0.0223	- 0.214	+24
165.2	04 11 53	+02 53 36	+0.0051	+0.275	+57
1069	04 12 40	- 04 32 30	+0.0065	- 0.101	
166.0A	04 12 58	- 07 43 48	- 0.1501	- 3.420	- 42
166.0B	04 13 04	- 07 44 06	- 0.1469	- 3.440	- 21
166.0C	04 13 04	- 07 44 06	- 0.1469	- 3.440	- 45
167.0	04 14 39	- 53 26 18	+0.0845	+0.419	- 23
167.1	04 14 43	- 51 36 42	+0.0113	+0.189	+25
167.2	04 15 10	- 26 10 00	+0.0430	+0.210	
167.3	04 15 37	- 59 25 18	- 0.0067	- 0.162	+29
168.0	04 17 55	+48 13 06	- 0.0030	+0.030	- 72
1070	04 19 08	+18 54 12	+0.0362	- 0.532	
168.1	04 19 15	+19 21 54	+0.0001	+0.003	
168.2	04 19 18	+19 22 06	+0.0100	- 0.270	
168.3A	04 19 27	- 25 50 42	+0.0032	- 0.049	+23
168.3B	04 19 27	- 25 50 42	+0.0032	- 0.049	
168.3C	04 19 29	- 25 50 06	+0.0032	- 0.049	
169.0	04 26 02	+21 48 42	- 0.0045	+0.183	- 34
169.1A	04 26 47	+58 53 54	+0.1700	- 1.960	
169.1B	04 26 47	+58 53 54	+0.1700	- 1.960	
170.0	04 26 59	+39 44 54	+0.0190	- 0.580	
170.1	04 27 42	+16 05 12	+0.0074	- 0.026	+38
171.0	04 31 59	+55 18 54	+0.0667	- 0.280	+50
1071	04 32 26	- 43 37 36	+0.0045	- 0.087	
171.1A	04 33 03	+16 24 36	+0.0045	- 0.189	+54
171.1B	04 33 03	+16 24 36	+0.0045	- 0.189	
171.2A	04 33 42	+27 02 00	+0.0183	- 0.132	+36
171.2B	04 33 39	+27 03 54	+0.0183	- 0.132	
172.0	04 33 43	+52 48 00	+0.0346	- 0.460	+35

Catalogue of Nearby Stars*

Name	Sp LC	V (mag)	B–V (mag)	π_t (0.001″)	M_V (mag)
162.2	M0V	9.719		50	8.00
163.0		12.30		65	11.40
164.0	M5	13.20		66	12.30
1068	pec	13.58	+1.93	95*	13.5
165.0A	M5	15.00		63	14.00
165.0B		5.00		63	
165.1	(K0)	8.657		51	7.19
165.2	K3V	8.796		50	7.29
1069	K5V	9.39	+1.22	46*	7.7
166.0A	K1V	4.419	+0.82	207	6.00
166.0B	DA	9.270	+0.03	207	10.85
166.0C	M4.5E	1.17	+1.68	205	12.73
167.0	K5V	7.628	+1.13	75	7.00
167.1	F5V	4.247	+0.31	59	3.10
167.2	G	13.40		51	11.90
167.3	K2IV	4.461	+1.06	64	3.49
168.0	M0V	9.619	+1.17	49	8.07
1070		15.22	+1.71	54	13.88
168.1		16.00		69	15.00
168.2		17.30		58	16.00
168.3A	F0IV–V	6.061	+0.35	55	4.76
168.3B		6.80		33	4.40
168.3C		8.257	+0.57	55	6.96
169.0	K7V	8.313	+1.35	90	8.08
169.1A	M4	1.09	+1.64	192	12.51
169.1B		2.44	+0.31	192	13.86
170.0	M7	13.68		99	13.70
170.1	A5.5VN	4.775	+0.17	52	3.35
171.0	K2V	8.345	+0.89	41	6.41
1071	K5V	8.86	+1.11	47*	7.2
171.1A	K5III	0.921	+1.54	49	– 0.63
171.1B	M2	13.50		50	12.00
171.2A	K5VEP	8.145	+1.12	60	7.04
171.2B		5.94	+0.65	60	14.80
172.0	M0V	8.595	+1.41	90	8.37

*Adapted from Gliese (1969), Gliese and Jahreiss (1979), and Grenon and Rufener (1981).

Catalogue of Nearby Stars*

Name	R.A.(1950) h m s	Dec.(1950) ° ′ ″	$\mu_{R.A.}$ (s yr^{-1})	$\mu_{Dec.}$ (″ yr^{-1})	V_r (km s^{-1})
172.1	04 34 27	+46 08 06	+0.0000	− 0.027	− 11
173.0	04 35 21	− 11 08 06	− 0.0150	− 0.200	− 13
173.1A	04 36 58	+09 46 48	− 0.0005	− 0.358	− 22
173.1B	04 36 59	+09 46 06	− 0.0005	− 0.358	
174.0	04 38 22	+20 48 36	− 0.0166	− 0.260	+7
174.1A	04 38 57	− 41 57 30	− 0.0136	− 0.078	− 1
174.1B	04 38 57	− 41 57 30	− 0.0136	− 0.078	
175.0A	04 39 29	− 59 02 30	+0.0087	+0.189	+10
175.0B	04 39 29	− 59 02 30	+0.0087	+0.189	
176.0	04 39 58	+18 52 48	+0.0500	− 1.080	+29
176.1	04 40 17	− 37 14 30	+0.0032	+0.194	+27
176.2	04 40 29	+27 35 54	+0.0047	− 0.239	+21
176.3	04 44 23	− 50 09 36	− 0.0474	− 0.354	+17
177.0	04 45 21	− 17 01 30	+0.0088	+0.171	+17
177.1	04 46 08	− 05 45 24	+0.0201	− 0.241	+79
178.0	04 47 07	+06 52 30	+0.0311	+0.016	+24
1072	04 47 49	+22 02 42	+0.0451	− 0.392	
1073	04 49 02	+40 38 18	+0.1034	− 1.098	
179.0	04 49 24	+06 23 48	+0.0110	− 0.320	
180.0	04 51 35	− 17 50 42	+0.0310	− 0.640	
1074	04 54 50	+50 52 24	+0.0531	− 0.327	
181.0	04 55 00	+49 46 30	+0.0110	− 0.100	− 33
181.1	04 55 14	− 61 13 54	+0.1280	− 0.600	
182.0	04 56 59	+01 42 36	− 0.0030	− 0.150	+39
182.1	04 57 43	+14 18 36	+0.0033	+0.054	− 16
183.0	04 58 20	− 05 48 36	+0.0369	− 1.091	+27
184.0	04 59 17	+53 04 48	+0.1440	− 1.510	+66
185.0A	05 00 20	− 21 19 24	− 0.0114	− 0.264	− 15
185.0B	05 00 20	− 21 19 24	− 0.0114	− 0.264	
186.0	05 01 15	− 23 19 18	+0.0214	+0.142	+124
186.1	05 01 20	− 56 09 42	− 0.0080	+0.622	+3
187.0	05 01 30	− 49 13 18	− 0.0051	+0.026	+21
187.1	05 02 30	− 42 25 48	+0.0160	− 0.140	
187.2A	05 02 45	+51 32 00	− 0.0030	− 0.172	− 7
188.0A	05 04 30	+18 34 48	+0.0378	+0.019	+20

Catalogue of Nearby Stars*

Name	Sp LC	V (mag)	B–V (mag)	π_t (0.001″)	M_V (mag)
172.1	G8V	6.930		53	5.50
173.0	M2V	10.347		94	10.21
173.1A	K5V	9.189		49	7.64
173.1B		15.70		42	13.80
174.0	K3V	7.995	+1.19	82	7.56
174.1A	F2V	4.427	+0.34	44	2.64
174.1B		12.50		50	11.00
175.0A	G5V	6.524	+0.68	64	5.55
175.0B		7.34		43	5.50
176.0	M2.5VE	9.961	+1.54	101	9.98
176.1	F8V	5.022	+0.38	57	3.80
176.2	(K3)	8.011	+0.92	45	6.28
176.3	K0V	7.587	+0.88	46	5.90
177.0	(G0)	5.483	+0.64	78	4.94
177.1	G0V	5.757	+0.65	37	3.60
178.0	F6V	3.191	+0.46	132	3.79
1072		15.23	+1.95	72	14.52
1073	K	13.5:	+1.3:	80	13.0
179.0	M4E	12.00		82	11.60
180.0	M3	12.50		83	12.10
1074	dM0	10.93	+1.46	55	9.6
181.0	M2VE	9.808	+1.42	71	9.06
181.1		13.50		48	11.90
182.0	M1E	9.60		68	8.80
182.1	G2IV-V	6.761		48	5.05
183.0	K3V	6.263	+1.06	109	6.45
184.0	M0.5V	9.981	+1.41	65	9.05
185.0A	(M1)	8.311	+1.41	130	8.88
185.0B		10.50		131	11.10
186.0	M0V	9.282	+1.28	58	8.10
186.1	G5V	7.011	+0.63	51	5.55
187.0	F4V	5.354	+0.42	53	3.98
187.1	K0V	10.019	+0.75	52	8.60
187.2A	F0V	4.975	+0.34	49	3.43
188.0A	G4V	4.924	+0.65	59	3.78

*Adapted from Gliese (1969), Gliese and Jahreiss (1979), and Grenon and Rufener (1981).

Catalogue of Nearby Stars*

Name	R.A.(1950) h m s	Dec.(1950) ° ′ ″	$\mu_{R.A.}$ (s yr^{-1})	$\mu_{Dec.}$ (″ yr^{-1})	V$_r$ (km s^{-1})
188.0B	05 04 30	+18 34 48	+0.0378	+0.019	
189.0	05 04 39	- 57 32 24	- 0.0043	+0.116	- 2
1075	05 04 54	- 57 37 00	- 0.0014	+0.109	
189.1	05 05 05	- 12 33 18	+0.0091	- 0.080	+50
189.2	05 06 15	- 04 31 12	+0.0026	+0.018	+9
190.0	05 06 21	- 18 12 54	+0.0360	- 1.400	
191.0	05 09 41	- 44 59 54	+0.6218	- 5.702	+245
192.0	05 09 44	+19 36 12	+0.0190	+0.240	
193.0	05 12 04	- 15 52 48	+0.0148	- 0.226	+33
194.0A	05 12 59	+45 57 00	+0.0078	- 0.423	+30
194.0B	05 12 59	+45 57 00	+0.0078	- 0.423	
195.0A	05 13 42	+45 47 30	+0.0060	- 0.400	+36
195.0B	05 13 42	+45 47 30	+0.0060	- 0.400	
196.0	05 14 17	+79 10 42	- 0.0285	+0.160	- 10
197.0	05 15 37	+40 03 24	+0.0456	- 0.661	+66
198.0	05 16 37	- 18 10 54	+0.0263	+0.059	+40
199.0A	05 16 40	- 21 26 42	- 0.0090	- 0.040	+25
199.0B	05 16 40	- 21 26 42	- 0.0090	- 0.040	
200.0A	05 16 40	- 03 07 36	+0.0482	+0.127	+86
200.0B	05 16 40	- 03 07 36	+0.0482	+0.127	
200.1	05 16 48	- 53 43 36	+0.0260	- 0.470	
1076	05 19 13	+54 45 30	+0.0080	- 0.394	+49
1077	05 19 38	- 78 19 36	+0.0255	- 1.107	- 17
201.0	05 20 43	+17 16 42	+0.0187	- 0.002	+37
1078	05 20 47	+22 30 12	+0.0169	- 0.299	
202.0	05 21 30	+17 20 18	+0.0172	- 0.008	+36
1079	05 24 22	- 32 32 42	+0.0195	- 0.080	
203.0	05 25 16	+09 36 48	- 0.0130	- 0.800	
1080	05 25 38	+02 56 42			
204.0	05 25 57	- 03 31 42	- 0.0219	- 0.797	- 54
204.1	05 27 03	- 60 27 18	- 0.0205	- 0.103	+13
204.2	05 27 23	- 03 28 24	- 0.0220	- 0.470	
205.0	05 28 55	- 03 41 06	+0.0512	- 2.099	+11
206.0	05 29 30	+09 47 18	- 0.0140	- 0.210	+17
1081	05 29 38	+44 47 12	+0.0047	- 0.356	

Catalogue of Nearby Stars*

Name	Sp LC	V (mag)	B–V (mag)	π_t (0.001″)	M_V (mag)
188.0B		5.66		59	4.50
189.0	F8V	4.694	+0.52	84	4.32
1075	K7V	9.02	+1.40	87*	8.7
189.1	(F8)	5.960	+0.60	49	4.40
189.2	F5V	5.110	+0.45	48	3.52
190.0	M5.5	12.10		91	11.90
191.0	M1VIP	8.855	+1.56	257	10.90
192.0	M5	11.30		70	10.50
193.0	(DG6)	7.437	+0.73	62	6.40
194.0A	G8III	10.092	+0.80	75	– 0.53
194.0B	(F)	10.96		75	0.34
195.0A	M1V	10.167	+1.50	75	9.54
195.0B	M5V	13.70		74	13.00
196.0		5.04	+0.48	53	3.70
197.0	G0IV	4.706	+0.62	67	3.84
198.0	(DG0)	5.961	+0.57	64	4.99
199.0A	(M0)	9.332		68	8.49
199.0B		13.50		62	12.50
200.0A	K3V	7.790	+1.04	72	7.08
200.0B	M2	12.60		74	12.00
200.1	K	3.19	+1.16	45	11.50
1076	dKe	9.46	+1.04	51	8.0
1077		11.91	+1.48	79*	11.4
201.0	K5VE	7.942	+1.09	59	6.80
1078		15.52	+1.83	49	13.97
202.0	F8VE	5.015	+0.53	64	4.05
1079	K2V	7.74	+0.94		
203.0	M5	2.48	+1.64	115	12.78
1080		12.81	+1.46	53	11.43
204.0	K5V	7.656	+1.10	75	7.03
204.1	G5VE	6.978	+0.78	45	5.24
204.2	M5	13.60		54	12.30
205.0	M1.5V	7.978	+1.47	170	9.13
206.0	M4E	11.50	+1.62	70	10.73
1081		12.19	+1.60	65	11.25

*Adapted from Gliese (1969), Gliese and Jahreiss (1979), and Grenon and Rufener (1981).

Catalogue of Nearby Stars*

Name	R.A.(1950) h m s	Dec.(1950) ° ′ ″	$\mu_{R.A.}$ (s yr^{-1})	$\mu_{Dec.}$ (″ yr^{-1})	V$_r$ (km s^{-1})
207.0	05 29 53	+29 21 24	- 0.0270	- 0.160	
207.1	05 31 09	+01 54 48	- 0.0170	- 0.160	
207.2	05 32 43	- 23 29 42	+0.0260	- 0.453	
1082	05 32 48	+41 28 12	- 0.0086	+0.026	
208.0	05 33 44	+11 17 54	- 0.0008	- 0.044	+23
209.0	05 34 04	+20 42 24	- 0.0054	- 0.419	- 13
209.1	05 35 47	- 28 43 06	- 0.0030	+0.051	+36
210.0	05 36 30	- 42 59 42	+0.0092	+0.255	+34
211.0	05 37 17	+53 27 48	+0.0008	- 0.518	+2
1083AB	05 37 21	+24 46 54	+0.0056	- 0.393	
212.0	05 37 27	+53 28 18	+0.0022	- 0.494	- 3
213.0	05 39 14	+12 29 18	+0.1370	- 1.550	+103
214.0	05 39 38	- 15 39 06	+0.0147	- 0.107	+60
215.0	05 41 05	+62 13 42	+0.0330	- 0.800	- 18
216.0A	05 42 23	- 22 27 48	- 0.0212	- 0.373	- 10
216.0B	05 42 21	- 22 26 12	- 0.0225	- 0.355	- 10
217.0	05 42 35	+37 16 24	+0.0403	- 0.508	- 32
217.1	05 44 41	- 14 50 24	- 0.0016	- 0.003	+20
217.2	05 45 06	- 70 10 48	- 0.0724	+1.199	+14
218.0	05 45 53	- 36 20 36	+0.0570	- 0.100	
1084	05 46 00	- 48 32 12	+0.0022	- 0.319	
1085	05 46 06	- 04 06 24	+0.0048	- 0.219	+29
219.0	05 46 06	- 51 05 00	+0.0005	+0.087	+20
1086	05 48 46	- 00 11 18	+0.0067	+0.250	
220.0	05 50 10	+24 15 24	+0.0130	- 0.590	
221.0	05 50 34	- 06 00 00	- 0.0010	- 0.250	+24
222.0	05 51 25	+20 16 06	- 0.0133	- 0.086	- 14
223.0	05 51 52	+02 08 36	+0.0040	- 0.660	- 30
223.1	05 52 11	- 09 24 18	+0.0090	+0.430	
223.2	05 52 39	- 04 08 48	+0.0370	- 2.320	
223.3	05 52 57	- 50 22 42	+0.0080	+0.571	+1
224.0	05 53 12	+13 55 30	+0.0265	- 0.466	- 2
224.1	05 53 43	- 63 06 18	+0.0206	+0.543	+25
1087	05 53 47	+05 22 12	- 0.0305	- 0.935	
225.0	05 54 08	- 14 10 30	- 0.0034	+0.136	- 2

Catalogue of Nearby Stars*

Name	Sp LC	V (mag)	B–V (mag)	π_t (0.001″)	M_V (mag)
207.0	K7	11.00		53	9.60
207.1	M3E	11.68		66	10.80
207.2	K3V	8.811	+0.94	47	7.17
1082	DA	14.75	+0.32	57*	13.5
208.0	M0V	8.796	+1.38	85	8.44
209.0	G4IV–V	7.680	+0.67	55	6.38
209.1	(DF4)	5.262	+0.46	46	3.58
210.0	G0V	7.421	+0.61	60	6.31
211.0	K1VE	6.221	+0.84	86	5.89
1083AB		14.85	+1.88	97	14.78
212.0	M2V	9.725	+1.49	86	9.40
213.0	M5.5	11.60	+1.65	168	12.73
214.0	(G5)	7.376	+0.79	80	6.89
215.0	M0VEP	8.999	+1.40	69	8.19
216.0A	F6V	3.608	+0.47	123	4.06
216.0B	K2V	6.159	+0.94	123	6.61
217.0	K1V	7.354	+0.83	69	6.55
217.1	A3V	3.541	+0.10	46	1.85
217.2	K0V	8.088	+0.76	47	6.45
218.0	M2	11.60		94	11.50
1084	M0V	9.74	+1.38	21	6.4
1085	dG4	5.97	+0.64	61*	4.9
219.0	A3V	3.855	+0.17	60	2.75
1086	DC2P	14.60	+0.46	89*	14.3
220.0	M2	12.50		62	11.50
221.0	M0.5V	9.707	+1.32	60	8.60
222.0	G0V	4.397	+0.59	101	4.42
223.0	K3V	8.840	+1.01	65	7.90
223.1	M0	10.63	+1.14	33	8.20
223.2	K	4.52	+1.06	166	15.62
223.3	K1V	6.526	+0.90	33	4.12
224.0	G5IV	6.630	+0.65	46	4.94
224.1	K3IV	4.657	+1.05	46	2.97
1087	DCP	14.11	+0.60	126	14.61
225.0	F0V	3.708	+0.33	65	2.77

*Adapted from Gliese (1969), Gliese and Jahreiss (1979), and Grenon and Rufener (1981).

Catalogue of Nearby Stars*

Name	R.A.(1950) h m s	Dec.(1950) ° ′ ″	$\mu_{R.A.}$ (s yr^{-1})	$\mu_{Dec.}$ (″ yr^{-1})	V$_r$ (km s^{-1})
225.1	05 58 12	- 37 03 36	+0.0163	+0.186	
225.2A	05 58 28	- 31 02 12	- 0.0320	+0.400	+102
225.2B	05 58 28	- 31 02 12	- 0.0320	+0.400	
225.2C	05 58 28	- 31 02 12	- 0.0320	+0.400	+105
226.0	05 59 42	+82 07 54	+0.0450	- 1.300	- 21
226.1	06 00 47	+26 09 24	+0.0300	- 0.540	
226.2	06 02 37	+67 59 06	- 0.0040	- 0.090	+2
226.3	06 02 48	+35 23 48	- 0.0105	- 0.304	- 12
227.0	06 03 49	+15 33 00	- 0.0079	- 0.109	- 12
228.0A	06 08 09	+10 20 36	+0.0050	- 0.940	+51
228.0B	06 08 09	+10 20 36	+0.0050	- 0.940	
229.0	06 08 28	- 21 50 36	- 0.0094	- 0.725	+4
229.1	06 08 34	+22 14 36	+0.0060	- 0.430	
1088	06 09 25	- 43 24 30	+0.0106	+0.731	+28
230.0	06 10 26	+10 38 42	+0.0062	- 0.285	+3
231.0	06 11 44	- 74 44 12	+0.0310	- 0.211	+35
231.1A	06 14 37	+05 07 00	- 0.0147	+0.158	+13
231.1B	06 14 32	+05 08 06	- 0.0170	+0.160	
231.2	06 16 39	+13 28 12	- 0.0003	- 0.009	+9
231.3	06 16 55	- 06 37 36	- 0.0070	- 0.620	
232.0	06 21 37	+23 28 06	+0.0400	- 0.520	
233.0	06 23 14	+18 47 18	- 0.0084	- 0.183	- 16
234.0A	06 26 51	- 02 46 12	+0.0490	- 0.690	+24
234.0B	06 26 51	- 02 46 12	+0.0490	- 0.690	
235.0A	06 28 16	+52 26 54	+0.0110	+0.080	+3
235.0B	06 28 16	+52 26 54	+0.0110	+0.080	
236.0	06 30 07	- 26 59 30	+0.0290	- 0.240	
1089	06 30 08	- 51 47 24	+0.0101	+0.090	+16
237.0	06 30 22	- 43 29 48	- 0.0250	- 0.020	
238.0	06 33 07	- 58 29 54	- 0.0520	+0.770	
239.0	06 34 19	+17 36 12	- 0.0550	+0.350	- 59
239.1	06 34 30	- 19 12 42	+0.0040	- 0.077	+2
240.0	06 35 55	- 49 59 42	+0.0210	+0.000	
240.1	06 37 44	+79 37 24	- 0.0304	- 0.607	+12
241.0	06 38 12	+24 00 36	+0.0150	- 0.275	- 45

Catalogue of Nearby Stars*

Name	Sp LC	V (mag)	B–V (mag)	π_t (0.001″)	M_V (mag)
225.1	(G0)	8.572	+0.64	47	6.93
225.2A	K5	9.00		50	7.50
225.2B		9.80		50	8.30
225.2C	K5	8.60		50	7.10
226.0	M2V	10.465		104	10.55
226.1	M5	15.10		49	13.60
226.2	K8	9.75	+1.25	43	7.90
226.3	G0V	6.132	+0.60	46	4.45
227.0	K0V	6.787	+0.81	65	5.85
228.0A	M3	10.57	+1.46	104	10.65
228.0B		12.60		104	12.70
229.0	M1V	8.149	+1.50	174	9.35
229.1		13.80		52	12.40
1088		12.32	+1.59	56*	11.1
230.0	G8V	6.463	+0.66	53	5.20
231.0	G5V	5.091	+0.72	115	5.39
231.1A	G0.5V	5.708	+0.61	50	4.20
231.1B		3.42	+1.41	55	12.10
231.2	K1V	7.646	+1.13		
231.3	M5	14.00		69	13.20
232.0	M6	13.30		117	13.60
233.0	K3VE	6.758	+0.94	63	5.75
234.0A	M7	11.07	+1.74	252	13.08
234.0B		14.0		252	16.0
235.0A	M0V	9.715	+1.26	35	8.10
235.0B		10.50		35	8.20
236.0		12.70		68	11.90
1089	F8V	5.59	+0.54	53*	4.2
237.0		11.30		59	10.20
238.0		12.91		72	12.20
239.0	M1V	9.641	+1.50	103	9.71
239.1	K1IV	3.961	+1.05	56	2.70
240.0	(K0)	9.606		70	8.83
240.1	F8V	5.435	+0.50	47	3.80
241.0	K6V	8.078	+1.02	72	7.36

*Adapted from Gliese (1969), Gliese and Jahreiss (1979), and Grenon and Rufener (1981).

Catalogue of Nearby Stars*

Name	R.A.(1950) h m s	Dec.(1950) ° ′ ″	$\mu_{R.A.}$ (s yr^{-1})	$\mu_{Dec.}$ (″ yr^{-1})	V_r (km s^{-1})
242.0	06 42 29	+12 57 06	- 0.0079	- 0.194	+25
243.0	06 42 52	- 27 17 36	- 0.0013	+0.306	- 14
244.0A	06 42 57	- 16 38 48	- 0.0379	- 1.211	- 8
244.0B	06 42 57	- 16 38 48	- 0.0379	- 1.211	
1090	06 43 08	+23 25 30	- 0.0006	+0.049	
245.0	06 43 08	+43 37 48	- 0.0002	+0.164	- 24
245.1	06 43 31	- 31 44 06	- 0.0175	- 0.323	+32
246.0	06 44 15	+37 35 06	- 0.0180	- 0.930	+80
1091	06 45 27	+37 12 00	- 0.0116	- 0.017	
247.0	06 45 27	+60 23 12	+0.0321	+0.402	- 49
1092	06 45 41	+37 11 42	+0.0186	- 1.584	
248.0	06 47 41	- 61 53 12	- 0.0101	+0.266	+21
249.0	06 47 49	+47 26 12	- 0.0250	- 0.690	+22
249.1	06 48 30	- 46 33 36	- 0.0008	+0.373	+20
250.0A	06 49 52	- 05 06 42	- 0.0365	- 0.004	- 10
250.0B	06 49 52	- 05 07 42	- 0.0365	- 0.004	
251.0	06 51 35	+33 20 18	- 0.0590	- 0.420	+36
251.1	06 52 10	+12 13 42	- 0.0030	- 0.340	+30
252.0	06 52 14	+25 26 24	- 0.0030	+0.020	- 11
253.0	06 53 51	- 55 11 30	- 0.0046	- 0.189	+40
254.0	06 53 52	+30 49 36	+0.0070	- 0.230	- 14
255.0A	06 55 30	- 35 26 24	- 0.0036	+0.006	+10
255.0B	06 55 30	- 35 26 24	- 0.0036	+0.006	
256.0	06 56 07	- 12 55 18	+0.0058	- 0.083	+2
257.0A	06 56 21	- 44 13 18	- 0.1050	- 0.120	
257.0B	06 56 21	- 44 13 18	- 0.1050	- 0.120	
1093	06 56 29	+19 25 48	+0.0630	- 0.891	
257.1	06 57 49	+48 27 24	+0.0554	- 0.439	- 23
258.0	06 59 07	+68 21 42	+0.0600	+0.070	
259.0	06 59 11	- 25 52 36	+0.0149	+0.038	+13
260.0	06 59 25	- 61 16 06	- 0.0237	+0.252	+21
261.0	06 59 29	- 06 22 36	- 0.0050	- 0.820	
1094	07 00 17	- 06 43 18	- 0.0135	- 0.298	
262.0	07 00 20	+29 25 24	+0.0121	- 0.827	+22
263.0	07 01 56	- 10 25 18	- 0.0080	- 0.790	

Catalogue of Nearby Stars*

Name	Sp LC	V (mag)	B–V (mag)	π_t (0.001″)	M_V (mag)
242.0	F5III	3.343	+0.43	52	1.92
243.0	G2V	6.435	+0.54	61	5.36
244.0A	A1V	– 1.44		376	1.44
244.0B		8.68		377	11.56
1090	G8V	6.5		63*	5.5:
245.0	G0V	5.258	+0.56	68	4.42
245.1	(DF6)	5.914	+0.49	50	4.41
246.0		12.10	– 0.08	60	11.00
1091		13.8	+1.10	46	12.1
247.0	M0VP	8.601	+1.20	56	7.34
1092		13.77	+1.66	76	13.17
248.0	A5V	3.264	+0.21	51	1.80
249.0	K6V	8.962	+1.24	75	8.34
249.1	F5V	5.130	+0.45	47	3.49
250.0A	K6	6.60	+1.05	104	6.68
250.0B	M2	10.11	+1.50	104	10.20
251.0	M4V	10.027	+1.60	168	11.15
251.1	M1VE	10.569		42	8.69
252.0	G0V	5.776	+0.58	55	4.48
253.0	G7V	8.153	+0.79	70	7.38
254.0	(K6E)	9.708	+1.36	73	9.02
255.0A	F8IV–V	6.213	+0.46	69	5.41
255.0B		7.10		40	5.10
256.0	(K8)	9.159	+1.13	27	6.32
257.0A	M4.5	11.47	+1.51	115	11.80
257.0B	M4.5	11.70		115	12.00
1093		14.83	+1.95	130	15.40
257.1	K3V	7.975	+0.99	33	5.57
258.0	M5	12.80		57	11.60
259.0	K0VE	6.705	+0.90	68	5.87
260.0	K0IV–VE	6.81	+0.81	62	5.80
261.0		15.95		70	15.20
1094	K5V	8.38	+1.08	51*	6.9
262.0	G2V	5.941	+0.60	58	4.76
263.0	M5	11.30		60	10.20

*Adapted from Gliese (1969), Gliese and Jahreiss (1979), and Grenon and Rufener (1981).

Catalogue of Nearby Stars*

Name	R.A.(1950) h m s	Dec.(1950) ° ′ ″	$\mu_{R.A.}$ (s yr^{-1})	$\mu_{Dec.}$ (″ yr^{-1})	V_r (km s^{-1})
264.0	07 02 18	− 43 29 30	− 0.0105	+0.406	+88
264.1A	07 02 25	− 43 32 18	− 0.0106	+0.385	+86
264.1B	07 02 25	− 43 32 18	− 0.0106	+0.385	+90
265.0A	07 02 35	+27 32 54	− 0.0040	− 0.100	− 42
265.0B	07 02 35	+27 32 54	− 0.0040	− 0.100	
266.0	07 04 32	+03 31 48	+0.0000	− 0.280	
267.0	07 05 47	− 09 53 18	− 0.0136	+0.024	+26
268.0	07 06 39	+38 37 30	− 0.0430	− 0.950	+39
268.1	07 11 08	− 46 40 30	− 0.0136	+0.100	− 1
1095	07 12 08	+47 19 54	+0.0032	− 0.187	+85
268.2	07 12 42	− 63 15 54	− 0.0430	+0.590	
1096	07 13 03	+33 14 48	− 0.0093	− 0.435	
268.3	07 13 14	+27 14 00	− 0.0020	− 0.220	
268.4	07 14 19	− 40 56 36	− 0.0234	+0.226	+16
268.5	07 15 11	− 13 54 36	− 0.0110	− 0.220	
269.0A	07 16 03	− 46 53 42	− 0.0022	+0.590	+59
269.0B	07 16 03	− 46 53 42	− 0.0022	+0.590	
270.0	07 16 15	+32 55 42	+0.0330	− 0.360	− 61
271.0A	07 17 08	+22 04 36	− 0.0018	− 0.014	+3
271.0B	07 17 08	+22 04 36	− 0.0018	− 0.014	+2
272.0	07 19 37	+46 11 18	− 0.0130	− 0.250	
273.0	07 24 43	+05 22 42	+0.0400	− 3.720	+26
273.1	07 25 49	+32 05 42	+0.0124	+0.177	+1
274.0A	07 25 54	+31 53 06	+0.0118	+0.171	− 6
274.0B	07 25 54	+31 53 06	+0.0118	+0.171	
275.0	07 26 11	− 51 18 00	− 0.0335	+0.003	+3
1097	07 26 14	− 03 11 12	+0.0304	− 0.822	
275.1	07 26 52	+68 43 42	− 0.0330	− 0.120	+4
275.2A	07 27 02	+48 19 24	− 0.0260	− 1.310	
275.2B	07 27 06	+48 17 48	− 0.0260	− 1.310	
276.0	07 28 18	+14 43 30	+0.0040	− 0.300	+68
277.0A	07 28 40	+36 19 48	− 0.0285	− 0.274	+1
1098	07 28 49	+64 16 06	+0.0110	− 0.250	
277.0B	07 28 39	+36 20 24	− 0.0285	− 0.274	− 2
277.1	07 29 57	+63 03 06	− 0.0720	− 0.120	

Catalogue of Nearby Stars*

Name	Sp LC	V (mag)	B–V (mag)	π_t (0.001″)	M_V (mag)
264.0	K5V	8.697	+1.18	50	7.19
264.1A	G3V	5.556	+0.65	50	4.05
264.1B	K0V	6.853	+0.80	50	5.35
265.0A	M0V	10.137	+1.32	60	9.03
265.0B		14.63	+1.10	50	13.10
266.0	M0	10.00		55	8.70
267.0	(K8)	8.874	+0.96	44	7.09
268.0	M4.5VE	11.536	+1.71	165	12.62
268.1	F0V	4.482	+0.32	46	2.80
1095	G0V	5.64	+0.58	37	3.5
268.2	(K5)	9.074	+1.26	51	7.61
1096		14.48	+1.76	68	13.64
268.3	M0V	10.841		46	8.80
268.4	G3V	9.082	+0.66	46	7.40
268.5	K5	12.70		45	11.00
269.0A	K2V	6.703	+0.99	83	6.30
269.0B		7.90		83	7.50
270.0	M1.5V	10.048	+1.36	51	8.59
271.0A	F0IV	3.545	+0.34	61	2.47
271.0B	K6V	8.16		61	7.10
272.0	M2	10.53	+1.59	80	10.00
273.0	M5V	9.845	+1.56	262	11.94
273.1	K2V	7.728	+0.95	55	6.43
274.0A	F0V	4.170	+0.32	55	2.87
274.0B		12.50		54	11.20
275.0	G5IV–V	6.717	+0.70	57	5.50
1097		11.48	+1.45	92	11.30
275.1	M0	10.79	+1.59	50	9.30
275.2A	M5	13.52	+1.69	104	13.60
275.2B		14.62	+0.99	104	14.70
276.0	K8V	8.957	+1.10	45	7.22
277.0A	M3.5VE	10.529	+1.46	82	10.10
1098	DC–K	16.38:	+0.91	51*	14.9
277.0B	M4.5E	11.76	+1.59	88	11.48
277.1	M0	10.80		48	9.20

*Adapted from Gliese (1969), Gliese and Jahreiss (1979), and Grenon and Rufener (1981).

Catalogue of Nearby Stars*

Name	R.A.(1950) h m s	Dec.(1950) ° ′ ″	$\mu_{R.A.}$ (s yr^{-1})	$\mu_{Dec.}$ (″ yr^{-1})	V_r (km s^{-1})
278.0A	07 31 25	+32 00 00	- 0.0134	- 0.102	+3
278.0B	07 31 25	+32 00 00	- 0.0134	- 0.102	
278.0C	07 31 26	+31 58 48	- 0.0154	- 0.108	+2
1099	07 31 43	+01 06 12	+0.0022	- 0.629	
279.0	07 31 55	- 22 11 12	- 0.0033	+0.040	+61
280.0A	07 36 41	+05 21 18	- 0.0473	- 1.029	- 3
280.0B	07 36 41	+05 21 18	- 0.0473	- 1.029	
281.0	07 36 48	+02 18 12	- 0.0080	- 0.250	+10
282.0A	07 37 29	- 03 28 42	+0.0048	- 0.290	- 21
282.0B	07 37 33	- 03 29 00	+0.0048	- 0.290	- 11
283.0A	07 38 02	- 17 17 24	+0.0780	- 0.570	+11
283.0B	07 38 02	- 17 17 24	+0.0780	- 0.570	
284.0	07 41 24	- 45 02 42	- 0.0065	- 0.559	+28
285.0	07 42 04	+03 40 48	- 0.0260	- 0.470	+18
285.1	07 42 11	+70 19 54	- 0.0184	- 0.144	- 24
286.0	07 42 16	+28 08 54	- 0.0474	- 0.051	+4
287.0	07 42 25	+02 15 42	+0.0050	- 0.200	- 21
288.0A	07 43 43	- 34 04 24	- 0.0236	+1.662	+102
288.0B	07 43 51	- 33 49 12	- 0.0270	+1.590	
1100	07 44 46	- 13 48 36	+0.0083	- 0.485	
1101	07 44 56	+83 31 18	- 0.1618	- 0.589	
289.0	07 45 15	+20 30 24	+0.0940	- 0.980	
1102A	07 47 33	+07 20 42	+0.0146	- 1.767	
1102B	07 47 32	+07 20 54	+0.0146	- 1.767	
290.0	07 48 07	+80 23 42	- 0.1899	+0.076	- 8
1103A	07 49 20	+00 08 06	+0.0169	- 0.738	+52
1103B	07 49 21	+00 08 12	+0.0169	- 0.738	
291.0A	07 49 27	- 13 45 48	- 0.0046	- 0.344	- 18
291.0B	07 49 27	- 13 45 48	- 0.0046	- 0.344	
291.1	07 49 42	- 50 10 42	- 0.0140	+0.155	
1104	07 50 22	+30 45 42	+0.0547	- 1.839-	+240
292.0A	07 50 24	- 34 34 42	- 0.0163	+0.240	+28
292.0B	07 50 24	- 34 34 42	- 0.0163	+0.240	
292.1	07 51 59	+19 22 30	+0.0075	- 0.465	- 19
292.2	07 52 03	- 01 16 48	- 0.0178	- 0.060	+96

Catalogue of Nearby Stars*

Name	Sp LC	V (mag)	B–V (mag)	π_t (0.001″)	M_V (mag)
278.0A	A1V	1.566	+0.04	75	0.88
278.0B	AM	12.85		69	2.04
278.0C	M0.5VE	9.07	+1.49	69	8.26
1099		11.93	+1.47	69	11.12
279.0	F7IV	4.437	+0.51	53	3.06
280.0A	F5V	0.373	+0.42	285	2.65
280.0B	DF			285	13.00
281.0	M0V	9.631	+1.38	76	9.03
282.0A	K2V	7.187	+0.96	73	6.50
282.0B	K5V	8.937	+1.33	73	8.25
283.0A		3.00	+0.30	142	13.76
283.0B		17.60		142	18.40
284.0	G5IV	5.042	+0.77	55	3.74
285.0		1.20	+1.59	165	12.29
285.1	G6V	7.056	+0.76	45	5.32
286.0	K0IIIE	1.161	+1.00	97	1.09
287.0		10.20		50	8.70
288.0A	G0VI	5.372	+0.58	63	4.37
288.0B		18.30		63	17.30
1100	M1	12		48	10
1101		13.09	+1.69	81	12.63
289.0		11.40		74	10.70
1102A	DC	16.8		58	15.7
1102B	DC	17.1		58	16.0
290.0	G8V	6.553	+0.73	70	5.78
1103A		13.30	+1.68	115	13.60
1103B		16		115	16
291.0A	G1V	5.152	+0.57	62	4.11
291.0B		16.17	+0.65	69	5.36
291.1		10.90		47	9.30
1104	G2VI	8.34	+0.61	45	6.61
292.0A	F5V	5.020	+0.44	73	4.34
292.0B		8.10		73	7.40
292.1	K6V	7.773	+0.95	36	6.20
292.2	G8V	7.438	+0.73	52	6.02

*Adapted from Gliese (1969), Gliese and Jahreiss (1979), and Grenon and Rufener (1981).

Catalogue of Nearby Stars*

Name	R.A.(1950) h m s	Dec.(1950) ° ′ ″	$\mu_{R.A.}$ (s yr^{-1})	$\mu_{Dec.}$ (″ yr^{-1})	V_r (km s^{-1})
293.0	07 53 20	− 67 38 00	+0.2540	− 1.450	
1105	07 54 47	+41 26 54	+0.0192	− 0.666	
293.1A	07 55 25	+00 40 42	− 0.0106	− 0.006	− 4
293.1B	07 55 25	+00 40 42	− 0.0106	− 0.006	
1106	07 55 54	− 33 49 12	+0.0085	+0.291	
293.2	07 55 57	− 25 29 12	+0.0269	− 0.266	− 5
294.0A	07 56 52	− 60 10 06	+0.0692	+0.117	+14
294.0B	07 57 00	− 60 09 48	+0.0692	+0.117	
294.0C	07 57 00	− 60 09 48	+0.0692	+0.117	
295.0	07 57 27	+29 22 00	− 0.0114	− 1.167	+12
295.1	07 57 28	+13 56 12	− 0.0060	− 0.110	+30
296.0	07 58 15	− 39 53 30	+0.0450	− 0.690	+6
1107	08 02 29	+34 13 24	+0.0140	− 0.257	− 5
296.1	08 04 58	− 29 15 12	+0.0269	− 0.369	− 18
296.2	08 05 11	+69 52 12	+0.0309	+0.105	− 6
1108A	08 05 48	+32 58	− 0.0034	− 0.185	+4
1108B	08 05 48	+32 58	+0.0000	+0.000	
297.1	08 08 11	− 61 09 00	− 0.0219	− 0.283	+25
297.2A	08 08 21	− 13 39 06	− 0.0170	+0.052	+38
297.2B	08 08 15	− 13 39 54	− 0.0170	+0.052	
298.0	08 08 42	− 52 49 42	− 0.0590	+0.610	
299.0	08 09 11	+08 59 42	+0.0800	− 5.100	− 35
300.0	08 10 29	− 21 23 30	+0.0050	− 0.730	
301.0A	08 10 50	− 13 45 30	− 0.0160	− 0.470	+12
301.0B	08 10 50	− 13 45 30	− 0.0160	− 0.470	
301.1	08 15 05	+30 46 06	− 0.0228	− 0.822	+12
302.0	08 16 01	− 12 27 42	+0.0181	− 0.983	+30
1109	08 16 41	− 36 30 12	− 0.0089	+0.090	+5
303.0	08 17 02	+27 22 54	− 0.0012	− 0.382	+32
304.0	08 19 41	− 39 32 54	− 0.0177	+0.244	+48
305.0	08 19 51	− 76 45 42	+0.0328	+0.108	− 14
305.1	08 20 34	+22 00 54	+0.0217	− 0.239	− 19
306.0	08 22 05	− 03 35 18	− 0.0143	− 0.028	+72
306.1	08 23 46	− 29 45 42	+0.0113	− 0.304	+16
307.0	08 24 05	+22 07 18	+0.0010	+0.030	

Catalogue of Nearby Stars*

Name	Sp LC	V (mag)	B–V (mag)	π_t (0.001″)	M_V (mag)
293.0		14.34		173	15.50
1105		12.04	+1.63	124	12.51
293.1A	(K2)	8.064	+1.04	63	7.00
293.1B		13.30		63	12.30
1106	K5V	8.84	+1.16	52*	7.43
293.2	(K0)	8.414	+1.05	51	6.95
294.0A	G2V	5.586	+0.57	63	4.58
294.0B		9.87	+1.35	61	8.80
294.0C		13.50		61	12.40
295.0	G8V	6.977	+0.71	58	5.79
295.1	(K5E)	10.349	+1.28	56	9.09
296.0	K7V	9.660	+1.33	57	8.44
1107	dM0p	10.15	+1.37	57	8.9
296.1	G4V	6.784	+0.60	53	5.41
296.2	F8V	6.551		46	4.86
1108A	dM0.5e	10.1		35	7.8
1108B	dM3e	11.4		35	9.1
297.1	F5V	4.743	+0.43	55	3.44
297.2A	(G0)	5.517	+0.49	49	3.97
297.2B		11.50		46	9.80
298.0		13.30		39	11.20
299.0		2.77	+1.77	151	13.66
300.0		13.80		171	15.00
301.0A	M0V	9.382	+1.38	57	8.16
301.0B		10.70		58	9.50
301.1	K4V	8.817	+1.14	46	7.13
302.0	G7.5V	5.948	+0.76	79	5.44
1109	A5V	4.44	+0.22	53	3.1
303.0	F6V	5.148	+0.47	63	4.14
304.0	G6IV–V	7.158	+0.71	60	6.05
305.0	F6IV	4.054	+0.40	52	2.63
305.1	M0VE	9.493	+1.18	51	8.03
306.0	(DF2)	5.598	+0.46	55	4.30
306.1	(G5)	7.799	+0.68	54	6.46
307.0		10.40		52	9.00

*Adapted from Gliese (1969), Gliese and Jahreiss (1979), and Grenon and Rufener (1981).

Catalogue of Nearby Stars*

Name	R.A.(1950) h m s	Dec.(1950) ° ′ ″	$\mu_{R.A.}$ (s yr^{-1})	$\mu_{Dec.}$ (″ yr^{-1})	V$_r$ (km s^{-1})
307.1	08 24 08	+45 49 24	− 0.0025	− 0.354	− 34
308.0A	08 25 15	+35 11 12	− 0.0840	− 0.380	
308.0B	08 25 15	+35 11 12	− 0.0840	− 0.380	
1110	08 25 20	+20 18 54	+0.0067	− 0.385	
308.1	08 25 39	+61 54 12	+0.0680	− 0.710	
1111	08 26 53	+26 57 12	− 0.0838	− 0.596	
308.2	08 27 07	− 01 34 00	+0.0260	− 0.890	
1112	08 27 32	+32 52 24	− 0.0113	− 0.531	− 18
308.3	08 30 27	− 50 01 18	− 0.0210	+0.220	
309.0	08 30 54	− 31 20 24	− 0.0873	+0.757	+16
310.0	08 31 55	+67 28 06	− 0.1830	+0.000	+14
310.1A	08 33 12	+06 47 42	− 0.0087	− 0.146	+25
310.1B	08 33 12	+06 47 54	− 0.0087	− 0.146	+27
311.0	08 34 47	+65 11 42	− 0.0039	+0.086	− 12
312.0	08 35 31	− 39 58 18	− 0.0274	+0.029	− 2
313.0	08 36 24	− 13 04 48	− 0.0045	+0.049	+27
314.0A	08 36 57	− 22 29 24	− 0.0174	+0.432	+44
314.0B	08 36 57	− 22 29 24	− 0.0174	+0.432	
315.0	08 37 07	+11 42 24	− 0.0068	− 0.512	− 13
1113	08 37 15	+43 17 48	− 0.0184	− 0.310	+56
316.0	08 37 32	− 06 17 42	+0.0074	− 0.126	− 6
316.1	08 38 10	+18 35 00	− 0.0550	− 0.450	
317.0	08 38 49	− 23 17 18	− 0.0310	+0.800	
318.0	08 39 36	− 32 46 54	− 0.0840	+1.310	+40
318.1	08 39 58	+44 40 36	− 0.0430	− 0.200	
319.0A	08 40 02	+09 44 42	+0.0150	− 0.620	+21
319.0B	08 40 02	+09 44 42	+0.0150	− 0.620	
319.0C	08 40 09	+09 44 30	+0.0160	− 0.620	
319.1A	08 40 22	− 42 44 54	− 0.0267	− 0.086	
319.1B	08 40 22	− 42 45 36	− 0.0267	− 0.086	
320.0	08 41 26	− 38 42 24	− 0.0256	+0.330	+13
321.0	08 41 53	+41 51 48	− 0.0254	− 0.646	− 26
321.1	08 42 37	− 42 28 00	− 0.0017	+0.016	− 2
321.2	08 42 58	− 42 26 54	− 0.0005	− 0.280	+5
321.3A	08 43 19	− 54 31 30	+0.0025	− 0.079	+2

Catalogue of Nearby Stars*

Name	Sp LC	V (mag)	B–V (mag)	π_t (0.001″)	M_V (mag)
307.1	G4V	6.313	+0.63	47	4.67
308.0A		11.10		52	9.70
308.0B		11.20		52	9.80
1110		13.08	+1.48	45	11.35
308.1		10.00		46	8.30
1111		14.81	+2.06	278	17.03
308.2		13.00		49	11.40
1112	DAs	15.71	+0.32	46	14.02
308.3		11.60		54	10.30
309.0	K0V	6.385	+0.79	86	6.06
310.0	M1V	9.287	+1.42	72	8.57
310.1A	F8V	5.967	+0.52	48	4.37
310.1B	G5V	7.246	+0.71	48	5.65
311.0	G0V	5.632	+0.62	72	4.92
312.0	(G0)	6.526	+0.60	59	5.38
313.0	K5V	9.671	+1.12	36	7.45
314.0A	G5V	5.044	+0.72	62	4.01
314.0B	F6V	16.80		62	5.80
315.0	K1V	7.595	+0.82	51	6.13
1113	dKB	9.31	+1.08	62	8.3
316.0		9.80		52	8.40
316.1		19.20		139	19.90
317.0		13.20		116	13.50
318.0		2.05		113	12.30
318.1		15.50		48	13.90
319.0A	M1V	9.631	+1.37	66	8.73
319.0B		13.10		64	12.10
319.0C		11.83	+1.55	64	10.90
319.1A		8.10	+0.90	49	6.50
319.1B		14.30		49	12.70
320.0	K1V	6.564	+0.93	96	6.48
321.0	K3V	8.513	+0.94	53	7.13
321.1	(G5)	4.055	+0.87	49	2.51
321.2	G5V	7.209	+0.74	47	5.57
321.3A	A0V	1.939	+0.04	49	0.39

*Adapted from Gliese (1969), Gliese and Jahreiss (1979), and Grenon and Rufener (1981).

Catalogue of Nearby Stars*

Name	R.A.(1950) h m s	Dec.(1950) ° ′ ″	$\mu_{R.A.}$ (s yr^{-1})	$\mu_{Dec.}$ (″ yr^{-1})	V_r (km s^{-1})
321.3B	08 43 19	- 54 31 30	+0.0025	- 0.079	
321.3C	08 43 26	- 54 30 54	+0.0025	- 0.079	
321.3D	08 43 26	- 54 30 54	+0.0025	- 0.079	
322.0	08 47 34	+66 19 06	+0.0148	+0.116	- 21
323.0A	08 48 02	+08 03 06	- 0.0040	+0.000	- 10
323.0B	08 48 02	+08 03 06	- 0.0040	+0.000	
1114	08 48 57	+18 18 54	- 0.0623	- 0.062	
324.0A	08 49 37	+28 31 24	- 0.0366	- 0.239	+27
324.0B	08 49 42	+28 30 30	- 0.0366	- 0.239	
1115	08 50 40	+35 25 00	- 0.0164	- 0.287	+45
325.0A	08 50 44	+70 59 24	- 0.2744	- 0.352	+44
325.0B	08 50 44	+70 59 24	- 0.2744	- 0.352	+47
326.0A	08 51 42	- 12 55 36	+0.0250	- 0.500	
326.0B	08 51 42	- 12 55 36	+0.0250	- 0.500	
327.0	08 51 50	- 05 14 36	- 0.0282	+0.026	+29
328.0	08 52 32	+01 45 06	+0.0060	- 1.040	+3
329.0	08 52 47	- 24 12 18	- 0.0260	+0.135	+67
330.0	08 54 21	+11 50 42	- 0.0020	- 0.360	
1116A	08 55 27	+19 57 24	- 0.0630	- 0.047	
1116B	08 55 27	+19 57 30	- 0.0630	- 0.047	
330.1	08 55 44	+20 44 36	+0.0477	- 0.163	- 47
331.0A	08 55 48	+48 14 24	- 0.0443	- 0.235	+12
331.0B	08 55 48	+48 14 24	- 0.0443	- 0.235	+15
331.0C	08 55 48	+48 14 24	- 0.0443	- 0.235	
1117	08 56 11	+33 08 48	- 0.0257	+0.000	
1118	08 56 58	- 31 01 00	+0.0536	- 0.739	+50
1119	08 57 11	+46 47 18	- 0.0462	- 0.528	
332.0A	08 57 24	+41 58 54	- 0.0394	- 0.253	+27
332.0B	08 57 24	+41 58 54	- 0.0394	- 0.253	
333.0	08 57 52	- 47 15 00	- 0.0490	+0.660	
333.1	08 58 11	- 58 53 30	- 0.0227	+0.277	+11
333.2A	08 58 11	+05 26 36	- 0.0170	- 0.300	+7
333.2B	08 58 13	+05 26 24	- 0.0170	- 0.300	- 11
1120A	08 58 31	+15 28 00	- 0.0065	- 0.327	- 13
1120B	08 58 31	+15 27 54	- 0.0065	- 0.327	

Catalogue of Nearby Stars*

Name	Sp LC	V (mag)	B–V (mag)	π_t (0.001″)	M_V (mag)
321.3B		15.10		50	3.60
321.3C		11.00		50	9.50
321.3D		13.50		50	12.00
322.0	M1V	9.237	+1.34	51	7.77
323.0A	M0VP	9.080	+1.34	58	8.60
323.0B		9.90		58	8.70
1114		11.5		58	10.3
324.0A	G8V	5.960	+0.87	74	5.31
324.0B		13.15	+1.65	74	12.50
1115	dM0	9.26	+1.15	39	7.2
325.0A	M1V	8.037	+1.39	89	7.78
325.0B		8.90		91	8.70
326.0A		12.40		82	12.00
326.0B		12.70		82	12.30
327.0	G3V	6.010	+0.67	85	5.66
328.0	M1V	9.982		64	9.01
329.0	(K5)	8.652	+1.00	66	7.75
330.0	M5V	10.613	+1.53	38	8.51
1116A		14.06	+1.84	192	15.48
1116B		14.92	+1.93	192	16.34
330.1	K5V	9.237	+1.11	36	7.02
331.0A	A7IV	3.125	+0.19	66	2.22
331.0B	M1V	10.80		66	9.90
331.0C		11.10		66	10.20
1117	C2P	15.18	+0.01	49	13.63
1118		13.80	+1.64	56*	12.5
1119		13.32	+1.72	97	13.25
332.0A	F5V	3.951	+0.37	70	3.18
332.0B		15.98	+0.65	74	5.33
333.0		14.40		74	13.70
333.1	F3V	5.164	+0.42	53	3.79
333.2A		2.38	+1.45	47	10.70
333.2B		2.70	+1.47	47	11.00
1120A	dM0	9.43	+1.30	51	8.0
1120B		9.49		51	8.0

*Adapted from Gliese (1969), Gliese and Jahreiss (1979), and Grenon and Rufener (1981).

Catalogue of Nearby Stars*

Name	R.A.(1950) h m s	Dec.(1950) ° ′ ″	$\mu_{R.A.}$ (s yr^{-1})	$\mu_{Dec.}$ (″ yr^{-1})	V$_r$ (km s^{-1})
333.3	09 01 40	− 66 11 48	+0.0007	− 0.102	+5
334.0	09 04 20	− 08 36 30	− 0.0206	+0.220	+32
334.1	09 05 22	+73 36 54	− 0.0380	− 0.270	
334.2	09 05 47	+34 05 12	− 0.0150	− 0.124	+27
335.0A	09 06 01	+67 20 24	− 0.0034	− 0.077	− 2
335.0B	09 06 01	+67 20 24	− 0.0034	− 0.077	
1121	09 06 14	+40 18 42	− 0.0486	− 0.575	
336.0	09 06 29	+33 01 54	− 0.0260	− 0.600	
336.1	09 08 11	+46 49 24	− 0.0380	− 0.020	
336.2A	09 09 03	− 45 06 12	− 0.0180	+0.030	
336.2B	09 09 03	− 45 06 12	− 0.0180	+0.030	
337.0A	09 09 34	+15 11 54	− 0.0361	+0.242	+45
337.0B	09 09 34	+15 11 54	− 0.0361	+0.242	
337.1	09 10 25	+61 37 54	+0.0001	− 0.034	− 14
338.0A	09 10 59	+52 54 06	− 0.1737	− 0.587	+11
338.0B	09 11 01	+52 54 12	− 0.1712	− 0.679	+10
338.1A	09 11 57	+77 27 12	− 0.3260	− 0.040	− 4
338.1B	09 11 57	+77 27 12	− 0.3260	− 0.040	
339.0	09 12 17	+04 39 00	− 0.0081	+0.030	+19
339.1	09 12 29	+53 38 54	− 0.1170	− 1.160	
339.2	09 12 40	− 69 30 42	− 0.0288	+0.102	− 5
339.3	09 13 45	− 37 12 12	+0.0017	− 0.013	+6
340.0A	09 14 56	+28 46 42	+0.0052	− 0.505	− 18
340.0B	09 14 56	+28 46 42	+0.0052	− 0.505	
1122A	09 16 12	+38 44 00	− 0.0205	+0.004	
1122B	09 16 12	+38 44 06	− 0.0205	+0.004	
340.1A	09 16 36	− 68 28 42	− 0.0185	− 0.031	+32
340.1B	09 16 36	− 68 28 42	− 0.0185	− 0.031	
340.2	09 16 54	+01 06 36	− 0.0091	− 0.107	+41
1123	09 17 32	− 77 37 00	+0.2098	− 0.777	
340.3	09 18 16	− 05 32 24	− 0.0237	− 0.119	+41
1124	09 19 17	+40 25 12	− 0.0304	− 0.360	− 43
341.0	09 20 24	− 60 04 12	− 0.1140	+0.180	
341.1	09 21 06	+80 48 12	− 0.0028	− 0.458	− 21
342.0	09 21 35	+76 09 06	− 0.1007	− 0.103	− 6

Catalogue of Nearby Stars*

Name	Sp LC	V (mag)	B–V (mag)	π_t (0.001″)	M_V (mag)
333.3	A5III	4.002	+0.14	50	2.50
334.0	M0V	9.504	+1.42	84	9.13
334.1	K5V	10.052		49	8.50
334.2	G0VE	5.963	+0.57	46	4.30
335.0A	F7IV–V	4.791	+0.49	54	3.45
335.0B		8.44		52	7.00
1121		14.53	+1.82	46	12.84
336.0	M2V	9.958		39	7.91
336.1		10.90		48	9.30
336.2A		9.76	+0.81	45	8.00
336.2B		10.30	+0.91	45	8.60
337.0A	G9V	6.503	+0.73	64	5.53
337.0B		7.25		64	6.30
337.1	G0IV	5.189	+0.58	46	3.50
338.0A	M0VE	7.666	+1.38	163	8.73
338.0B	M0VE	7.767	+1.34	163	8.83
338.1A	K5V	10.068	+1.38	42	8.18
338.1B		11.00		41	9.10
339.0	K5V	7.963		45	6.23
339.1		13.79	+0.30	70	13.00
339.2	A1IV	1.673	+0.01	17	– 2.17
339.3	F3IV–V	4.632	+0.45	63	3.63
340.0A	K3V	7.187	+1.02	58	6.00
340.0B		8.00		58	6.80
1122A		14.50	+1.70	49	12.95
1122B		14.60	+1.70	49	13.05
340.1A	F5V	5.399	+0.42	57	4.18
340.1B		16.13		40	4.10
340.2	K0V	8.113		48	6.52
1123		13.10	+1.64	140*	13.8
340.3	(K8)	9.085	+1.17	46	7.40
1124	K2V	7.62	+0.98	47	6.0
341.0		10.90		104	11.00
341.1	K5V	9.281	+1.23	58	8.10
342.0	K5VE	9.028	+1.19	58	7.85

*Adapted from Gliese (1969), Gliese and Jahreiss (1979), and Grenon and Rufener (1981).

Catalogue of Nearby Stars*

Name	R.A.(1950) h m s	Dec.(1950) ° ′ ″	$\mu_{R.A.}$ (s yr^{-1})	$\mu_{Dec.}$ (″ yr^{-1})	V$_r$ (km s^{-1})
343.0	09 22 49	+18 53 30	− 0.0340	− 0.330	
343.1	09 24 20	+39 43 30	+0.0120	− 0.130	− 21
344.0A	09 25 18	− 05 51 06	− 0.0153	− 0.079	+54
344.0B	09 25 18	− 05 51 06	− 0.0153	− 0.079	
345.0	09 25 51	− 80 19 18	+0.0780	+1.240	− 11
346.0	09 26 24	− 09 02 48	+0.0020	− 0.050	
347.0A	09 26 25	− 07 08 30	− 0.0090	− 0.700	
347.0B	09 26 28	− 07 08 30	− 0.0090	− 0.700	
348.0A	09 26 37	− 02 33 00	+0.0084	− 0.018	+10
348.0B	09 26 37	− 02 31 54	+0.0096	− 0.019	+22
349.0	09 27 19	+05 52 24	− 0.0345	+0.089	+27
350.0A	09 28 02	+27 11 00	− 0.0100	+0.140	
350.0B	09 28 05	+27 12 18	− 0.0100	+0.140	
1125	09 28 12	+00 33 00	− 0.0391	− 0.529	− 40
1126	09 28 20	− 31 53 12	− 0.0058	+0.322	
351.0A	09 28 44	− 40 14 48	− 0.0165	+0.069	+9
351.0B	09 28 44	− 40 14 48	− 0.0165	+0.069	
351.1	09 28 44	+20 30 48	+0.0000	− 0.800	
352.0A	09 28 53	− 13 16 06	+0.0510	+0.030	+8
352.0B	09 28 53	− 13 16 06	+0.0510	+0.030	
353.0	09 28 54	+36 32 54	− 0.0180	− 0.520	+22
354.0A	09 29 31	+51 54 24	− 0.1029	− 0.540	+15
354.0B	09 29 31	+51 54 24	− 0.1029	− 0.540	
354.1	09 29 50	+27 12 48	− 0.0107	− 0.239	+14
355.0	09 30 01	− 10 57 48	− 0.0184	+0.026	+14
355.1	09 30 06	+70 03 06	− 0.0124	+0.076	− 27
355.2	09 30 54	− 20 53 36	− 0.0020	+0.010	+13
356.0A	09 32 40	+36 02 12	− 0.0582	− 0.248	+13
356.0B	09 32 40	+36 02 12	− 0.0582	− 0.248	
357.0	09 33 43	− 21 25 24	+0.0110	− 1.090	
1127	09 34 22	+22 55 18	− 0.0096	− 0.176	− 32
358.0	09 37 49	− 40 50 42	− 0.0470	+0.370	
359.0	09 38 11	+22 15 30	+0.0360	− 0.420	
360.0	09 38 22	+70 15 54	− 0.1310	− 0.270	
361.0	09 38 30	+13 26 24	− 0.0480	− 0.110	+19

Catalogue of Nearby Stars*

Name	Sp LC	V (mag)	B–V (mag)	π_t (0.001″)	M_V (mag)
343.0		14.30		61	13.20
343.1	K8VE	9.831	+1.32	45	8.10
344.0A	G2V	5.381	+0.65	37	3.22
344.0B		16.65		50	5.10
345.0	(SDG)	10.116	+0.48	57	8.90
346.0	M0V	10.508	+1.41	54	9.10
347.0A		12.00		64	11.00
347.0B		16.40		64	15.40
348.0A	F6IV	4.597	+0.45	71	3.85
348.0B	K0V	7.207	+0.87	71	6.46
349.0	K3VE	7.211	+1.00	88	6.93
350.0A		10.60		50	9.10
350.0B		14.60		50	13.10
1125		11.72	+1.60	81*	11.3
1126	K3V	8.38	+0.98	54*	7.0
351.0A	F2III	3.581	+0.36	65	2.65
351.0B		14.65		65	3.70
351.1		13.10		50	11.60
352.0A		10.81	+1.53	110	11.02
352.0B		10.80		110	11.00
353.0	M2V	10.198	+1.50	66	9.30
354.0A	F6IV	3.182	+0.46	52	1.76
354.0B		13.80		59	12.60
354.1	K0V	7.030	+0.77	54	5.69
355.0	(K0)	7.603	+0.91	80	7.12
355.1	G4III-IV	4.575	+0.77	39	2.53
355.2	K0IV	5.007	+1.02	51	3.54
356.0A	G8IV-VE	5.408	+0.77	108	5.58
356.0B		13.00		109	13.20
357.0		12.70		122	13.10
1127	dM0	9.50	+1.29	50*	8.0
358.0		12.40		96	12.30
359.0		15.70		92	15.50
36 0.0	M3V	10.564		85	10.21
361.0		10.40	+1.50	75	9.80

*Adapted from Gliese (1969), Gliese and Jahreiss (1979), and Grenon and Rufener (1981).

Catalogue of Nearby Stars*

Name	R.A.(1950) h m s	Dec.(1950) ° ′ ″	$\mu_{R.A.}$ (s yr^{-1})	$\mu_{Dec.}$ (″ yr^{-1})	V_r (km s^{-1})
362.0	09 38 39	+70 16 18	- 0.1310	- 0.270	
363.0	09 38 57	+56 13 12	- 0.0930	- 0.490	
364.0	09 39 59	- 23 41 24	- 0.0291	+0.255	+34
365.0	09 40 17	+42 55 54	+0.0040	- 0.828	- 12
366.0	09 41 42	+76 17 18	+0.0240	- 0.980	- 3
1128	09 41 57	- 68 40 18	- 0.0107	1.118	- 21
366.1A	09 41 58	- 27 32 24	- 0.0037	+0.030	+24
366.1B	09 41 58	- 27 32 24	- 0.0037	+0.030	
1129	09 42 32	- 17 58 48	- 0.1109	- 0.170	+9
367.0	09 42 37	- 45 32 18	- 0.0420	- 0.590	+60
368.0	09 45 22	+46 15 18	+0.0216	- 0.095	+5
368.1A	09 47 02	- 52 23 06	- 0.0288	+0.242	+4
368.1B	09 47 02	- 52 23 06	- 0.0288	+0.242	
369.0	09 48 40	- 12 04 30	+0.0726	- 1.440	+61
370.0	09 49 05	- 43 15 42	+0.0402	- 0.446	- 9
371.0	09 49 37	+03 27 24	- 0.0292	+0.019	+19
372.0	09 50 41	- 03 26 54	- 0.0060	- 0.470	
1130A	09 51 17	- 31 30 54	- 0.0126	- 0.135	
1130B	09 51 17	- 31 30 54	- 0.0126	- 0.135	
373.0	09 52 29	+63 02 00	- 0.0480	- 0.600	+10
374.0	09 54 09	- 40 32 54	- 0.0120	- 0.060	- 12
375.0	09 56 34	- 46 10 42	+0.0470	- 0.490	
375.1	09 57 02	- 49 45 48	- 0.0150	+0.030	
375.2	09 57 36	+27 30 24	+0.0000	+0.100	
376.0	09 58 08	+32 10 12	- 0.0414	- 0.433	+56
377.0	09 59 01	- 30 09 30	- 0.0830	+0.670	
378.0	09 59 14	+48 21 06	- 0.0630	- 1.410	- 4
378.1	09 59 23	+44 49 18	- 0.0261	- 0.064	+20
378.2	10 04 22	+03 12 36	- 0.0050	- 0.090	- 20
378.3	10 04 29	+35 29 24	+0.0041	+0.001	- 18
379.0A	10 06 24	+75 23 00	+0.0620	+0.269	- 49
379.0B	10 06 24	+75 23 00	+0.0620	+0.269	
379.1A	10 07 22	- 35 36 42	- 0.0360	+0.002	+41
379.1B	10 07 22	- 35 36 42	- 0.0360	+0.002	

Catalogue of Nearby Stars*

Name	Sp LC	V (mag)	B–V (mag)	π_t (0.001″)	M_V (mag)
362.0	M4V	11.205		85	10.85
363.0		14.20		71	13.50
364.0	G0V	4.921	+0.53	79	4.41
365.0	K5V	8.136	+1.15	69	7.33
366.0		10.70		65	9.80
1128		12.78	+1.73	125*	13.3
366.1A	F7V	4.781	+0.50	51	3.32
366.1B		15.54		48	4.00
1129		12.60	+1.59	91	12.4
367.0		10.60		109	10.80
368.0	G0.5V	5.089	+0.62	66	4.19
368.1A	K1V	7.920	+0.90	45	6.19
368.1B		12.00		45	10.30
369.0		10.00	+1.42	81	9.54
370.0	K5V	7.660	+1.17	99	7.64
371.0	M0VP	8.876	+1.23	44	7.09
372.0	M0V	10.544		76	9.80
1130A	M0V	10.21	+1.38	48*	8.6
1130B		14.42	+1.70	48*	12.8
373.0	M1.5V	8.956	+1.43	84	8.58
374.0	K4V	8.983	+0.96	68	8.15
375.0		11.60		75	11.00
375.1		12.30		46	10.60
375.2		10.90		46	9.20
376.0	G2V	5.387	+0.66	54	4.05
377.0		10.70		74	10.00
378.0	M2V	10.085	+1.37	65	9.15
378.1	K8V	9.088	+1.07	53	7.71
1131		14.33	+1.75	57	13.11
378.2	M0VE	9.934	+1.38	47	8.29
378.3	A7V	4.483	+0.18	45	2.75
379.0A	K6VE	9.501	+1.40	56	8.24
379.0B		10.30		51	8.80
379.1A	F9V	6.150	+0.60	48	4.56
379.1B		10.90		48	9.30

*Adapted from Gliese (1969), Gliese and Jahreiss (1979), and Grenon and Rufener (1981).

Catalogue of Nearby Stars*

Name	R.A.(1950) h m s	Dec.(1950) ° ′ ″	$\mu_{R.A.}$ (s yr^{-1})	$\mu_{Dec.}$ (″ yr^{-1})	V_r (km s^{-1})
379.2	10 07 40	– 36 30 48	– 0.0265	+0.369	+111
380.0	10 08 19	+49 42 30	– 0.1404	– 0.509	– 26
381.0	10 09 31	– 02 25 42	+0.0350	– 0.600	
382.0	10 09 46	– 03 29 42	– 0.0105	– 0.232	+12
383.0	10 09 47	– 18 22 12	– 0.0370	+0.000	
383.1	10 10 44	+52 45 54	+0.0103	– 0.747	– 25
384.0A	10 10 56	– 47 13 48	– 0.0227	+0.143	
384.0B	10 10 56	– 47 13 48	– 0.0227	+0.143	
385.0	10 10 59	– 84 52 24	+0.0030	– 0.010	
385.1	10 11 12	– 32 47 06	– 0.0289	+0.056	+42
1132	10 12 54	– 46 54 42	– 0.1031	+0.427	– 1
386.0	10 14 20	– 11 42 12	– 0.0280	– 0.580	
387.0A	10 14 30	+23 21 30	– 0.0296	– 0.105	+38
387.0B	10 14 30	+23 21 30	– 0.0296	– 0.105	
388.0	10 16 54	+20 07 18	– 0.0346	– 0.049	+11
388.1	10 17 01	+19 43 30	– 0.0165	– 0.220	+6
388.2	10 18 25	– 15 13 54	– 0.0158	+0.275	+79
1133	10 19 37	+63 42 36	+0.0375	+0.216	
389.0A	10 20 37	– 59 55 06	+0.0490	– 0.440	
389.0B	10 20 37	– 59 54 54	+0.0490	– 0.440	
389.1	10 21 45	– 10 08 54	+0.0200	– 0.270	
390.0	10 22 44	– 09 58 36	– 0.0480	+0.100	
391.0	10 23 24	– 73 46 36	– 0.0044	– 0.031	– 5
392.0A	10 24 59	+49 03 12	+0.0081	– 0.888	– 7
392.0B	10 24 59	+49 03 12	+0.0081	– 0.888	
392.1	10 25 10	+82 48 54	– 0.0445	+0.026	+7
393.0	10 26 23	+01 06 24	– 0.0410	– 0.740	+12
394.0	10 27 14	+56 15 24	– 0.0202	– 0.039	+4
395.0	10 27 26	+56 14 18	– 0.0214	– 0.035	+9
396.0	10 28 15	+84 39 24	+0.0150	+0.050	– 1
397.0	10 28 27	+45 47 30	– 0.0550	– 0.580	+23
397.1	10 28 30	+57 22 12	– 0.0060	+0.150	– 2
397.2	10 29 25	– 53 27 42	– 0.0470	+0.201	+20
398.0	10 33 28	+05 22 42	– 0.0430	+0.140	+21
398.1	10 34 03	– 11 57 42	+0.0178	– 0.680	– 6

Catalogue of Nearby Stars*

Name	Sp LC	V (mag)	B–V (mag)	π_t (0.001″)	M_V (mag)
379.2	G3V–VI	8.073	+0.60	46	6.39
380.0	K7VE	6.581	+1.36	220	8.29
381.0		10.80		116	11.12
382.0	M2V	9.246	+1.47	117	9.59
383.0	M0V	9.913		71	9.17
383.1	M0V	9.536	+1.10	49	7.99
384.0A	G8V	8.170	+0.79	66	7.27
384.0B		10.80		40	8.80
385.0		10.23		57	9.00
385.1	G1V	6.375	+0.60	54	5.04
1132		13.52	+1.73	60*	12.4
386.0	(K)	10.961		107	11.11
387.0A	F8V	5.821	+0.50	60	4.71
387.0B	M1V	11.40		60	10.30
388.0	M4.5VE	9.420	+1.54	205	10.98
388.1	F6IV	4.785	+0.45	51	3.32
388.2	F8V	7.179	+0.55	46	5.49
1133	DA	14.70	+0.39	65*	13.8
389.0A		11.60		61	10.50
389.0B		13.50		61	12.40
389.1	K8V	9.977	+1.23	51	8.51
390.0	M0V	10.157		69	9.35
391.0	F2III	3.992	+0.36	85	3.64
392.0A	G1V	6.435	+0.60	51	4.97
392.0B		12.50		52	11.10
392.1	F5IV	5.255	+0.37	45	3.52
393.0	M2.5V	9.659	+1.52	130	10.23
394.0	K7VE	8.637	+1.36	82	8.21
395.0	F8V	4.837	+0.52	82	4.41
396.0	K0V	7.271		65	6.34
397.0	K7V	8.861	+1.33	65	7.93
397.1	M0VE	9.613		48	8.02
397.2	F6V	4.893	+0.50	49	3.34
398.0		2.61	+1.56	66	11.70
398.1	F8V	5.722	+0.52	44	3.94

*Adapted from Gliese (1969), Gliese and Jahreiss (1979), and Grenon and Rufener (1981).

Catalogue of Nearby Stars*

Name	R.A.(1950) h m s	Dec.(1950) ° ′ ″	$\mu_{R.A.}$ (s yr^{-1})	$\mu_{Dec.}$ (″ yr^{-1})	V$_r$ (km s^{-1})
398.2	10 36 41	+43 21 48	+0.0030	- 0.020	+17
399.0	10 37 12	- 06 39 42	- 0.0460	- 0.110	
1134	10 38 53	+37 52 36	- 0.1228	- 0.363	
1135	10 38 54	- 36 38 00	+0.0164	- 0.184	
1136A	10 39 35	- 36 22 12	- 0.0092	+0.067	
1136B	10 39 35	- 36 22 12	- 0.0092	+0.067	
1137	10 42 02	- 33 18 42	- 0.0061	- 0.163	+42
400.0A	10 42 30	+38 46 24	- 0.0031	+0.134	- 7
400.0B	10 42 30	+38 46 24	- 0.0031	+0.134	
401.0	10 43 19	- 18 50 30	- 0.1300	- 0.600	
1138	10 47 00	+35 49 30	- 0.0551	- 1.032	
402.0	10 48 19	+07 05 06	- 0.0570	- 0.830	+4
402.1	10 49 29	+00 06 30	+0.0040	- 0.270	
403.0	10 49 30	+14 15 42	- 0.0770	+0.200	
1139	10 50 40	+76 19 54	- 0.1305	+0.124	- 24
403.1	10 51 03	- 19 52 06	+0.0053	- 0.244	- 5
404.0	10 51 12	- 44 08 42	- 0.0243	+0.002	+7
404.1	10 51 27	- 58 35 12	+0.0088	+0.027	+8
405.0	10 52 53	+56 18 00	- 0.0710	+0.100	
406.0	10 54 06	+07 19 12	- 0.2590	- 2.700	+13
406.1	10 54 21	+69 51 48	- 0.1211	+0.113	+7
1140	10 55 07	- 07 15 24	- 0.0542	+0.071	
407.0	10 56 40	+40 41 54	- 0.0281	+0.052	+13
408.0	10 57 25	+23 06 18	- 0.0300	- 0.280	+29
409.0	10 59 29	- 17 41 12	- 0.0010	- 0.020	
1141A	10 59 38	+16 47 24	+0.0065	- 0.177	
1141B	10 59 38	+16 47 24	+0.0065	- 0.177	
410.0	10 59 57	+22 14 12	+0.0110	- 0.040	- 20
411.0	11 00 37	+36 18 18	- 0.0468	- 4.743	- 84
412.0A	11 03 00	+43 47 00	- 0.4095	+0.939	+65
412.0B	11 03 02	+43 46 42	- 0.4095	+0.939	
412.1	11 04 29	- 62 09 12	- 0.0056	+0.009	- 2
412.2	11 05 31	- 29 54 06	- 0.0396	- 0.143	+13
1142A	11 05 34	- 04 57 12	- 0.0025	- 0.428	
1142B	11 05 28	- 04 52 54	- 0.0025	- 0.428	+15

Catalogue of Nearby Stars*

Name	Sp LC	V (mag)	B–V (mag)	π_t (0.001″)	M_V (mag)
398.2		1.12	- 0.30	60	10.00
399.0		12.69		103	12.80
1134		12.96	+1.68	96	12.87
1135	M0V	9.97	+1.48	79*	9.5
1136A	K7V	10.19	+1.46	59*	8.8
1136B		11.67	+1.52	54*	10.3
1137	K2V	8.30	+0.94	52*	6.9
400.0A		9.30	+1.41	85	8.95
400.0B		12.20		85	11.80
401.0		12.40		94	12.30
1138		12.99	+1.66	103	13.05
402.0		1.66		142	12.42
402.1	(K8)	10.187		47	8.55
403.0		13.90		85	13.50
1139	dKB	9.60	+1.10	59	8.4
403.1	F6V	5.227	+0.46	52	3.81
404.0	F8IV–V	8.078	+0.54	58	6.90
404.1	K0III–IV	3.793	+0.95	56	2.53
405.0		12.30		63	11.30
406.0		13.53	+2.01	426	16.68
406.1	M0.5V	10.264	+1.37	41	8.33
1140	DA	14.31	+0.32	83	13.91
407.0	G0V	5.044	+0.61	74	4.39
408.0	M3V	10.054	+1.54	151	10.95
409.0		12.28		56	11.00
1141A	dM2	11.0		68	10.2
1141B	dM	11.2		68	10.4
410.0	M2VE	9.581	+1.47	90	9.35
411.0	M2VE	7.488	+1.51	397	10.48
412.0A	M2VE	8.798	+1.55	188	10.17
412.0B		14.53		186	15.88
412.1	G8III	4.614	+1.03	54	3.28
412.2	G2V	6.516	+0.60	48	4.92
1142A	dM6	12.55	+1.52	45*	10.8
1142B	DA	12.92	+0.09	46*	11.2

*Adapted from Gliese (1969), Gliese and Jahreiss (1979), and Grenon and Rufener (1981).

Catalogue of Nearby Stars*

Name	R.A.(1950) h m s	Dec.(1950) ° ′ ″	$\mu_{R.A.}$ (s yr^{-1})	$\mu_{Dec.}$ (″ yr^{-1})	V_r (km s^{-1})
412.3	11 05 43	− 27 59 48	− 0.0374	− 0.026	
413.0	11 05 53	+16 02 36	+0.0090	− 0.360	+106
413.1	11 07 07	− 24 19 18	− 0.0560	− 0.510	
414.0A	11 08 20	+30 43 12	+0.0447	− 0.208	− 14
414.0B	11 08 18	+30 43 12	+0.0447	− 0.208	− 26
414.1A	11 08 35	+43 41 42	− 0.0590	− 0.430	− 19
414.1B	11 08 35	+43 41 42	− 0.0590	− 0.430	
415.0	11 08 43	− 10 41 12	− 0.0590	+0.660	+40
416.0	11 09 01	− 14 42 42	+0.0493	− 0.586	− 6
416.1	11 09 12	− 22 33 06	+0.0000	− 0.102	+6
417.0	11 09 49	+36 05 18	− 0.0225	− 0.175	− 3
418.0	11 10 39	+04 45 18	− 0.0205	− 0.034	+19
419.0	11 11 27	+20 47 54	+0.0102	− 0.135	− 21
420.0A	11 11 57	+73 44 48	− 0.0955	+0.105	+8
420.0B	11 11 57	+73 44 48	− 0.0955	+0.105	
1143A	11 12 21	− 22 49 36	− 0.0187	− 0.356	
1143B	11 12 21	− 22 49 36	− 0.0187	− 0.356	
421.0A	11 12 50	− 17 51 42	+0.0130	− 0.740	+6
421.0B	11 12 51	− 17 51 36	+0.0130	− 0.740	+17
421.0C	11 12 46	− 17 50 36	+0.0130	− 0.740	
421.1A	11 13 12	+53 02 42	+0.0177	+0.052	− 41
421.1B	11 13 12	+53 02 54	+0.0165	+0.041	
1144	11 13 57	− 14 25 00	− 0.0135	− 0.100	
422.0	11 14 03	− 57 17 30	− 0.3040	+1.150	
1145	11 14 40	− 27 32 24	+0.0124	− 0.095	
423.0A	11 15 31	+31 48 36	− 0.0339	− 0.591	− 16
423.0B	11 15 31	+31 48 36	− 0.0339	− 0.591	
423.1	11 15 47	− 04 47 30	+0.0528	− 0.150	+10
424.0	11 17 29	+66 07 00	− 0.4853	+0.148	+51
425.0A	11 18 57	− 20 10 42	+0.0133	− 0.139	+5
425.0B	11 18 57	− 20 10 42	+0.0133	− 0.139	
1146	11 19 06	+06 26 06	− 0.0515	− 1.573	
426.0A	11 19 12	+18 27 54	− 0.0104	− 0.104	− 4
426.0B	11 19 12	+18 27 54	− 0.0104	− 0.104	
426.1A	11 21 19	+10 48 18	+0.0113	− 0.079	− 10

Catalogue of Nearby Stars*

Name	Sp LC	V (mag)	B–V (mag)	π_t (0.001″)	M_V (mag)
412.3		9.50		49	8.00
413.0		9.77	+1.12	35	7.50
413.1		12.10		101	12.10
414.0A		8.38	+1.35	81	7.90
414.0B	(DM2)	9.968	+1.48	88	9.69
414.1A		1.40		64	10.40
414.1B		11.50		64	10.50
415.0	K5V	9.210	+1.08	51	7.75
416.0	K7V	9.037	+1.17	54	7.70
416.1	A2III–IV	4.465	+0.03	51	3.00
417.0	G0V	6.416	+0.60	52	5.00
418.0	K5V	8.772	+1.18	54	7.40
419.0	A4V	2.582	+0.12	40	0.59
420.0A	K5V	7.599	+1.04	79	7.09
420.0B	(M2)	11.40		74	10.70
1143A	K5V	8.98	+1.14	54*	7.6
1143B		13.5		54*	12
421.0A	M0.5V	10.021	+1.31	62	8.98
421.0B		10.05	+1.32	62	9.00
421.0C		15.00		62	14.00
421.1A	F6V	6.538	+0.43	54	5.20
421.1B		8.023	+0.60	54	6.68
1144	K7V	10.00	+1.40	50*	8.5
422.0		12.80		77	12.20
1145	M0V	9.79	+1.40	56*	8.5
423.0A	G0VE	3.791	+0.59	127	4.31
423.0B	GOVE	14.81		130	5.38
423.1		7.30	+0.73	46	5.60
424.0		9.32	+1.42	119	9.70
425.0A	(K5)	8.602		100	8.70
425.0B		11.00		100	11.00
1146	K–M	14.4		54	13.1
426.0A	K0V	7.907		99	7.89
426.0B	K6V	10.76		59	9.60
426.1A	F2IV	3.960	+0.41	47	2.32

*Adapted from Gliese (1969), Gliese and Jahreiss (1979), and Grenon and Rufener (1981).

Catalogue of Nearby Stars*

Name	R.A.(1950) h m s	Dec.(1950) ° ′ ″	$\mu_{R.A.}$ (s yr^{-1})	$\mu_{Dec.}$ (″ yr^{-1})	V_r (km s^{-1})
426.1B	11 21 19	+10 48 18	+0.0113	− 0.079	
427.0	11 21 39	+21 38 06	− 0.0740	+0.020	+59
428.0A	11 22 29	− 61 22 24	− 0.0677	+0.061	+5
428.0B	11 22 29	− 61 22 24	− 0.0677	+0.061	+5
429.0A	11 24 13	+03 17 06	− 0.0483	+0.177	− 3
429.0B	11 24 14	+03 16 42	− 0.0484	+0.169	+2
429.1	11 25 09	− 25 35 06	− 0.0045	+0.036	− 15
429.2	11 25 26	− 08 53 42	+0.0350	− 0.820	
429.3	11 27 28	− 51 23 18	− 0.0365	+0.085	+7
429.4	11 28 18	− 56 51 30	− 0.0624	+0.025	− 3
430.0	11 28 22	+63 26 00	+0.0090	+0.010	− 6
430.1	11 29 08	+22 56 30	− 0.0430	− 0.010	− 8
431.0	11 29 23	− 40 46 18	− 0.0590	+0.240	
431.1A	11 29 32	+61 21 36	+0.0000	− 0.073	− 46
431.1B	11 29 32	+61 21 36	+0.0000	− 0.073	
431.2	11 30 07	+49 46 18	− 0.0027	− 0.023	− 3
432.0A	11 32 03	− 32 34 00	− 0.0536	+0.821	− 23
432.0B	11 32 04	− 32 34 06	− 0.0550	+0.790	
433.0	11 32 58	− 32 15 06	− 0.0058	− 0.828	
433.1	11 34 28	+30 04 36	− 0.0120	− 0.020	
1147	11 36 03	− 41 05 36	− 0.0845	+0.084	+28
433.2A	11 36 07	+45 23 06	− 0.0566	+0.021	− 18
433.2B	11 36 06	+45 23 06	− 0.0598	− 0.009	− 14
434.0	11 38 25	+34 29 00	− 0.0010	− 0.388	− 5
435.0	11 38 37	− 44 07 54	− 0.0627	+0.217	+13
1148	11 39 09	+43 01 48	− 0.0557	− 0.108	
435.1	11 39 14	+05 25 30	+0.0130	− 0.430	+19
436.0	11 39 31	+26 59 48	+0.0640	− 0.770	+10
437.0	11 40 01	− 74 57 00	+0.0110	+0.000	
438.0	11 40 49	− 51 33 18	+0.0730	− 0.540	
438.1	11 41 50	+48 47 36	− 0.0029	+0.005	
439.0	11 42 04	+31 14 30	− 0.0032	− 0.379	+28
440.0	11 42 58	− 64 33 30	+0.4130	− 0.330	
1149	11 43 03	+63 22 54	− 0.0217	− 0.362	
441.0	11 43 08	+72 22 18	+0.0011	+0.041	− 24

Catalogue of Nearby Stars*

Name	Sp LC	V (mag)	B–V (mag)	π_t (0.001″)	M_V (mag)
426.1B		16.90		42	5.00
427.0		4.24	+0.31	84	13.90
428.0A	K7V	7.232	+1.26	90	7.00
428.0B		8.59		90	8.36
429.0A	K0IV	6.490	+0.80	59	5.34
429.0B	K2V	7.565	+1.00	59	6.42
429.1	(K0)	6.749	+1.08		0.0
429.2		13.40		49	11.80
429.3	(F2)	7.380	+0.41	52	5.96
429.4		8.33	+1.06	50	6.80
430.0		9.00	+1.10	46	7.30
430.1		10.50		56	9.20
431.0		12.82		85	12.50
431.1A	F6V	5.466	+0.50	44	3.68
431.1B		17.10		44	5.30
431.2	K3V	8.062		87	8.00
432.0A	K0V	5.956	+0.81	105	6.06
432.0B		15.00		105	15.00
433.0	M2V	9.790	+1.49	91	9.59
433.1		12.50	- 0.06	46	10.80
1147		13.78	+1.72	45*	12.1
433.2A	G0V	6.303	+0.57	49	4.75
433.2B	K2V	8.40	+0.96	45	6.70
434.0	G8VE	5.318	+0.74	112	5.56
435.0	K5V	7.782	+1.07	81	7.32
1148	M5	12		97	12
435.1	(DK8)	9.569	+1.24	46	7.88
436.0	M3.5V	10.662	+1.52	111	10.89
437.0	G1V	6.477	+0.54	39	4.40
438.0		11.30		120	11.70
438.1	K4III	7.098		100	7.80
439.0	K8V	8.996	+1.13	46	7.30
440.0		11.44	+0.19	206	13.01
1149	BC–G	16.37	+0.67	45*	14.6
441.0		9.02	+1.17	50	7.50

*Adapted from Gliese (1969), Gliese and Jahreiss (1979), and Grenon and Rufener (1981).

Catalogue of Nearby Stars*

Name	R.A.(1950) h m s	Dec.(1950) ° ′ ″	$\mu_{R.A.}$ (s yr^{-1})	$\mu_{Dec.}$ (″ yr^{-1})	V_r (km s^{-1})
442.0A	11 44 08	- 40 13 42	- 0.1341	+0.396	+15
442.0B	11 44 09	- 40 13 24	- 0.1340	+0.360	
443.0	11 44 08	- 13 43 36	+0.0520	- 0.780	
443.1	11 44 21	+27 18 00	+0.0004	- 0.009	+8
1150	11 44 23	+51 15 24	+0.0085	- 0.089	+2
444.0A	11 44 32	- 11 32 42	- 0.0138	- 0.055	+21
444.0B	11 44 32	- 11 32 42	- 0.0138	- 0.055	
445.0	11 44 35	+78 57 42	+0.2610	+0.480-	+119
446.0	11 44 45	- 30 00 18	- 0.0214	- 0.245	+12
447.0	11 45 09	+01 06 00	+0.0430	- 1.220	- 13
447.1A	11 46 04	+14 33 42	- 0.0074	+0.002	+6
447.1B	11 46 04	+14 33 42	- 0.0074	+0.002	
448.0	11 46 31	+14 51 06	- 0.0343	- 0.119	- 1
449.0	11 48 05	+02 02 48	+0.0494	- 0.275	+5
1151	11 48 29	+48 40 06	- 0.1556	- 0.964	
450.0	11 48 33	+35 32 48	- 0.0220	+0.260	+4
1152	11 49 02	+35 28 06	- 0.0098	- 0.010	
451.0A	11 50 06	+38 04 42	+0.3387	- 5.806	- 98
451.0B	11 50 06	+38 04 42	+0.3387	- 5.806	
452.0	11 50 43	- 07 05 18	- 0.0100	- 0.520	
452.1	11 51 34	+10 05 42	+0.0070	- 0.770	
452.2A	11 51 55	- 37 28 18	- 0.0263	+0.056	
452.2B	11 51 55	- 37 28 18	- 0.0263	+0.056	
452.3A	11 51 59	+19 41 24	- 0.0320	- 0.016	+8
452.3B	11 52 02	+19 42 24	- 0.0319	- 0.028	+4
452.4	11 52 22	+29 01 12	+0.0160	- 0.300	- 7
452.5A	11 52 29	- 55 48 54	+0.0241	- 0.205	+51
452.5B	11 52 29	- 55 48 54	+0.0241	- 0.205	
453.0	11 55 27	- 27 25 12	- 0.0808	- 0.628	+50
11 56 3	11 56 36	- 20 04 12	+0.0109	- 0.423	
454.0	11 58 10	- 10 09 42	+0.0082	- 0.482	
454.1	11 58 35	- 01 27 24	- 0.0310	+0.200	
454.2A	11 59 13	- 34 22 18	- 0.0160	+0.012	
454.2B	11 59 13	- 34 22 18	- 0.0160	+0.012	
454.3	11 59 47	+43 25 06	- 0.0015	- 0.023	+4

Catalogue of Nearby Stars*

Name	Sp LC	V (mag)	B–V (mag)	π_t (0.001″)	M_V (mag)
442.0A	G5V	4.894	+0.66	98	4.85
442.0B		15.00		98	15.00
443.0		13.40		83	13.00
443.1	K3V	9.227		72	8.00
1150	dM0p	9.62	+1.26	45*	7.9
444.0A		9.00	+1.12	48	7.40
444.0B		14.20		48	12.60
445.0	(SDM4)	10.829		192	12.25
446.0	G5V	6.479	+0.68	63	5.48
447.0	M4.5V	11.128	+1.76	299	13.51
447.1A	A7V	5.881		52	4.46
447.1B		10.10		29	7.40
448.0	A3V	2.136	+0.08	76	1.54
449.0	F8V	3.603	+0.55	100	3.60
1151		13.26	+1.84	121	13.67
450.0		9.80		95	9.70
1152	G7IV	9.64	+0.78	45	
451.0A	G8VI	6.435	+0.75	108	6.60
451.0B		12.00		113	12.00
452.0		11.99		83	11.60
452.1		14.20		95	14.10
452.2A	(DF8)	6.435	+0.52	51	4.97
452.2B		8.10		34	5.80
452.3A	G6V	8.213	+0.71	53	6.83
452.3B		8.43	+0.76	30	5.80
452.4		10.20		47	8.60
452.5A	G3V	6.705	+0.64	46	5.02
452.5B		17.70		39	5.60
453.0	K5V	6.977	+1.16	102	7.02
11 56 3	K5V	7.94	+0.98	53	6.6
454.0	K0IV	5.559	+0.76	80	5.07
454.1		12.00		48	10.40
454.2A	(G0)	6.894	+0.59	46	5.21
454.2B		8.10		46	6.40
454.3	K0V	8.331		60	7.30

*Adapted from Gliese (1969), Gliese and Jahreiss (1979), and Grenon and Rufener (1981).

Catalogue of Nearby Stars*

Name	R.A.(1950) h m s	Dec.(1950) ° ′ ″	$\mu_{R.A.}$ (s yr^{-1})	$\mu_{Dec.}$ (″ yr^{-1})	V_r (km s^{-1})
455.0	11 59 48	+28 52 00	− 0.0590	− 0.010	
455.1	12 03 17	− 18 35 24	− 0.0020	− 0.310	
455.2	12 04 15	− 64 20 06	+0.0052	− 0.039	+9
455.3	12 05 50	− 24 27 00	+0.0062	− 0.045	+4
456.0	12 05 52	+00 12 12	− 0.0630	− 0.070	+31
456.1A	12 07 20	− 45 55 42	− 0.0363	− 0.081	− 4
456.1B	12 07 23	− 45 54 48	− 0.0363	− 0.081	
457.0	12 09 36	+59 12 18	+0.0130	+0.010	− 15
458.0A	12 09 50	+54 45 42	+0.0300	+0.080	− 14
458.0B	12 09 51	+54 45 54	+0.0300	+0.080	
458.1	12 09 57	− 02 48 42	− 0.0404	+0.401	+11
1154	12 11 46	+00 54 12	− 0.0625	− 0.287	
458.2	12 12 41	+49 00 42	− 0.0270	− 0.020	
459.0	12 12 58	+57 18 36	+0.0126	+0.004	− 12
459.1	12 13 16	+52 47 30	− 0.0060	− 0.110	0
459.2	12 14 14	+44 40 48	− 0.0033	− 0.010	− 53
1155A	12 14 20	+03 14 36	− 0.0430	+0.274	
1155B	12 14 20	+03 14 36	− 0.0430	+0.274	
1156	12 16 32	+11 24 00	− 0.0864	+0.224	
459.3	12 16 56	+28 39 30	− 0.0500	+0.060	− 25
460.0	12 17 47	+26 16 42	− 0.0108	+0.015	+10
461.0	12 17 52	+00 51 42	+0.0060	− 0.020	− 16
462.0	12 19 25	+42 25 06	+0.0200	− 0.520	+12
1157	12 20 26	− 46 20 42	− 0.0705	− 0.325	+42
463.0	12 20 45	+64 18 12	− 0.0930	+0.390	+60
464.0	12 21 21	+12 51 36	+0.0040	− 0.170	+10
464.1	12 22 01	+31 33 24	− 0.0157	+0.021	− 6
465.0	12 22 13	− 17 56 00	+0.0800	− 2.250	+58
466.0	12 23 26	+08 20 24	− 0.0080	− 0.090	− 5
467.0A	12 25 48	− 71 12 12	− 0.0870	+1.090	
467.0B	12 25 48	− 71 12 12	− 0.0870	+1.090	
468.0	12 25 55	− 18 01 06	+0.0122	− 0.202	+3
469.0	12 26 27	+08 42 24	− 0.0420	− 0.300	
1158	12 26 49	− 55 42 48	− 0.1165	− 0.770	− 27
1159A	12 26 58	+53 49 12	− 0.1383	+0.107	

Catalogue of Nearby Stars*

Name	Sp LC	V (mag)	B–V (mag)	π_t (0.001″)	M_V (mag)
455.0		15.00		59	13.80
455.1	K7V	9.957		46	8.40
455.2	F0III	4.156	+0.35	52	2.74
455.3	F2IV	4.025	+0.32	69	3.22
456.0		1.30	+1.38	48	9.70
456.1A	K4V	8.436	+1.13	49	6.89
456.1B		12.20		59	11.10
457.0	M0V	10.074	+1.28	45	8.34
458.0A	M1.5V	9.763	+1.42	76	9.17
458.0B		15.30		77	14.70
458.1		7.43	+0.71	40	5.40
1154		13.42	+1.59	119	13.80
458.2		10.40		47	8.80
459.0	A3V	3.309	+0.08	53	1.93
459.1		3.34	+0.53	120	13.70
459.2	K4V	9.135		83	8.20
1155A	dM3	13.28	+1.62	46	11.59
1155B	DA	15.32	+0.38	46	13.6
1156	dMe	13.79	+1.83	153	14.71
459.3	M2V	10.601	+1.42	49	9.05
460.0	F0IV	6.145	+0.30	60	5.04
461.0	M2VE	9.984		60	9.10
462.0	M0VE	9.362	+1.35	70	8.59
1157		13.62	+1.62	79*	13.1
463.0		1.60		75	11.00
464.0		10.40		53	9.00
464.1		8.90			
465.0		1.70		110	11.90
466.0		9.90		55	8.60
467.0A		13.62		61	12.50
467.0B		15.50		61	14.40
468.0		9.19	+1.23	53	7.80
469.0		12.00		78	11.50
1158		13.27	+1.62	78*	12.7
1159A		14.19	+1.54	46	12.5

*Adapted from Gliese (1969), Gliese and Jahreiss (1979), and Grenon and Rufener (1981).

Catalogue of Nearby Stars*

Name	R.A.(1950) h m s	Dec.(1950) ° ′ ″	$\mu_{R.A.}$ (s yr^{-1})	$\mu_{Dec.}$ (″ yr^{-1})	V_r (km s^{-1})
1159B	12 26 58	+53 49 30	- 0.1383	+0.107	
469.1	12 27 10	- 03 03 00	- 0.0225	- 0.582	- 4
469.2A	12 27 30	- 13 07 00	- 0.0173	- 0.047	
469.2B	12 27 30	- 13 07 00	- 0.0173	- 0.047	
470.0	12 28 23	- 56 50 00	+0.0035	- 0.267	+21
471.0	12 28 46	+09 05 36	- 0.0432	- 0.525	+21
1160	12 28 57	+55 23 42	+0.0117	- 0.002	- 5
471.1	12 29 17	+33 45 48	- 0.0014	- 0.012	- 1
471.2	12 29 29	- 15 55 12	- 0.0298	- 0.066	- 4
472.0	12 30 39	- 68 28 30	- 0.0975	- 0.305	
473.0A	12 30 51	+09 17 36	- 0.1180	+0.260	- 5
473.0B	12 30 51	+09 17 36	- 0.1180	+0.260	
474.0	12 31 20	+33 39 36	+0.0000	- 0.013	- 43
475.0	12 31 22	+41 37 42	- 0.0631	+0.288	+7
476.0	12 32 30	+10 06 30	- 0.0320	- 0.330	
1161A	12 32 54	- 34 36 18	- 0.0180	- 0.113	
1161B	12 32 58	- 34 37 42	- 0.0180	- 0.113	
477.0	12 33 15	- 45 39 12	- 0.0070	- 0.710	
477.1	12 33 30	+46 03 18	- 0.0014	+0.013	- 1
478.0	12 33 45	- 76 40 42	- 0.2420	- 0.070	
479.0	12 35 11	- 51 43 36	- 0.1100	+0.040	
479.1	12 35 44	+79 29 24	- 0.0455	+0.006	- 18
1162	12 36 15	- 04 02 48	- 0.0473	- 0.216	
480.0	12 36 25	+11 58 24	- 0.0780	- 0.240	+8
480.1	12 38 10	- 43 18 00	- 0.0710	+0.700	
480.2	12 38 27	+34 22 42	+0.0024	- 0.013	+1
481.0	12 38 35	+15 39 24	+0.0073	- 0.419	+23
482.0A	12 39 07	- 01 10 30	- 0.0379	+0.008	- 20
482.0B	12 39 07	- 01 10 30	- 0.0379	+0.008	- 20
1163	12 41 09	+25 22 48	- 0.0255	- 0.055	
483.0	12 41 59	+52 02 06	- 0.0431	- 0.193	+9
484.0	12 42 38	+39 33 00	- 0.0309	+0.135	+81
484.1	12 44 42	+32 50 30	+0.0031	- 0.028	
485.0	12 44 54	+31 29 24	+0.0040	- 0.250	- 2
486.0	12 45 29	+10 01 54	- 0.0640	- 0.440	+5

Catalogue of Nearby Stars*

Name	Sp LC	V (mag)	B–V (mag)	π_t (0.001″)	M_V (mag)
1159B		19.7			
469.1	G8V	9.053	+0.70	51	7.59
469.2A	(G0)	6.378	+0.58	48	4.78
469.2B		10.20		40	8.20
470.0	M4III	1.622	+1.59		
471.0	M1V	9.669	+1.45	69	8.86
1160	K2V	8.10	+0.94	46*	6.4
471.1	K4V	9.139		77	8.20
471.2	F0IV	4.289	+0.37	52	2.87
472.0	K0V	7.116	+0.85	76	6.52
473.0A	M5.5V	13.16	+1.80	231	14.98
473.0B		13.40		231	15.20
474.0	K0III	6.259	+1.05	53	4.88
475.0	G0V	4.276	+0.59	109	4.46
476.0		11.40		70	10.60
1161A	K5V	7.91	+1.04	65*	7.0
1161B		11.91	+1.60	65*	11.0
477.0		11.50		76	10.90
477.1		7.13		164	8.20
478.0		12.80		55	11.50
479.0		10.65	+1.47	121	11.06
479.1	G2VE	7.008	+0.59	51	5.55
1162		13.52	+1.60	51	12.06
480.0		1.30		82	10.90
480.1		12.20		48	10.60
480.2	K4V	7.781		128	8.20
481.0	K8V	7.935	+1.12	78	7.40
482.0A	F0V	2.754	+0.36	99	2.73
482.0B	F0V	3.50		99	3.48
1163	dM4	12.96	+1.63	51*	11.5
483.0	K3VE	7.018	+0.94	63	6.01
484.0	G0V	5.961	+0.55	65	5.03
484.1	K3V	7.701		91	8.00
485.0		9.84	+1.29	49	8.30
486.0		1.40	+1.56	114	11.68

*Adapted from Gliese (1969), Gliese and Jahreiss (1979), and Grenon and Rufener (1981).

Catalogue of Nearby Stars*

Name	R.A.(1950) h m s	Dec.(1950) ° ′ ″	$\mu_{R.A.}$ (s yr^{-1})	$\mu_{Dec.}$ (″ yr^{-1})	V$_r$ (km s^{-1})
1164A	12 45 32	− 24 32 12	− 0.0235	+0.163	
1164B	12 45 31	− 24 32 00	− 0.0235	+0.163	
1165	12 45 54	− 15 26 54	+0.0053	+0.046	
486.1	12 46 21	+25 06 48	− 0.0249	− 0.114	− 8
487.0	12 47 04	+66 23 00	− 0.0760	− 0.100	− 18
488.0	12 48 10	+00 29 24	− 0.0029	− 0.395	+3
1166A	12 48 57	+22 22 30	− 0.0127	+0.071	
1166B	12 49 02	+22 23 18	− 0.0127	+0.071	
488.1	12 50 40	− 39 54 24	+0.0055	− 0.026	− 2
488.2	12 52 22	− 06 03 54	− 0.0170	− 0.140	
489.0	12 55 07	− 14 11 36	− 0.0239	+0.028	+5
490.0A	12 55 19	+35 29 48	− 0.0210	− 0.150	− 10
490.0B	12 55 18	+35 29 36	− 0.0210	− 0.150	
491.0A	12 56 28	− 09 34 00	− 0.0554	+0.185	− 4
491.0B	12 56 28	− 09 34 00	− 0.0554	+0.185	
492.0	12 57 38	+03 45 30	− 0.0290	− 0.920	
493.0	12 57 45	− 02 26 06	− 0.0535	+0.006	− 11
493.1	12 58 05	+05 57 06	− 0.0640	+0.280	− 40
494.0	12 58 19	+12 38 42	− 0.0460	+0.000	− 10
495.0	12 59 27	− 01 48 48	+0.0020	− 0.490	+156
496.0A	13 01 05	− 20 18 54	+0.0099	+0.013	+34
496.0B	13 01 05	− 20 18 54	+0.0099	+0.013	+28
496.1	13 02 04	− 52 09 48	− 0.0850	− 0.810	+38
497.0A	13 02 41	+56 10 12	− 0.0190	+0.040	
497.0B	13 02 41	+56 10 12	− 0.0190	+0.040	
498.0	13 03 29	+49 44 12	+0.0052	+0.148	− 6
499.0A	13 03 49	+20 59 42	− 0.0039	+0.061	− 5
499.0B	13 03 49	+20 59 42	− 0.0039	+0.061	
499.1	13 03 57	+22 53 00	+0.0018	− 0.053	− 5
500.0	13 07 00	− 21 55 18	+0.0110	− 0.347	− 7
1167A	13 07 13	+29 15 12	− 0.0258	− 0.195	
1167B	13 07 19	+29 18 06	− 0.0258	− 0.195	
501.0A	13 07 33	+17 47 36	− 0.0302	+0.131	− 18
501.0B	13 07 33	+17 47 36	− 0.0302	+0.131	
501.1	13 08 18	+36 12 00	− 0.0048	+0.023	− 15

Catalogue of Nearby Stars*

Name	Sp LC	V (mag)	B–V (mag)	π_t (0.001″)	M_V (mag)
1164A	K5V	9.02	+1.12	54*	7.7
1164B		10.04	+1.38	49*	8.5
1165	K2V	7.94	+0.96	55*	6.6
486.1	G7V	6.291	+0.70	34	3.95
487.0		10.90		116	11.20
488.0	M0.5V	8.499	+1.40	91	8.29
1166A	dM4	12.99	+1.60	46*	11.3
1166B	dM	14.33	+1.72	46*	12.6
488.1	A7III	4.263	+0.21	54	2.92
488.2	K8V	10.414	+1.34	46	8.73
489.0	(K2)	9.109	+1.12	62	8.07
490.0A		10.60	+1.42	57	9.40
490.0B		13.16	+1.61	57	11.90
491.0A	K0V	7.559	+0.79	64	6.59
491.0B		12.50		51	11.00
492.0		15.90	+0.64	82	15.50
493.0	K7V	9.757	+1.16	54	8.42
493.1		3.34	+1.73	90	0.0
494.0		9.79	+1.44	82	9.40
495.0		2.78	+0.75	79	12.30
496.0A	F8V	5.569	+0.56	39	4.20
496.0B		6.40		39	4.30
496.1	K9V	9.048	+1.36	50	7.54
497.0A		10.80		46	9.10
497.0B		11.60		46	9.90
498.0		9.30	+1.17	42	7.40
499.0A	M0V	9.422	+1.29	59	8.28
499.0B		14.00		57	12.80
499.1	M5III	5.596	+1.59	46	3.91
500.0	(G5)	7.351	+0.73	76	6.75
1167A	dM5	14.18	+1.72	54*	12.8
1167B		16		54*	15
501.0A	F5V	4.334	+0.45	53	2.96
501.0B	F5V	15.08		53	3.71
501.1		8.00			

*Adapted from Gliese (1969), Gliese and Jahreiss (1979), and Grenon and Rufener (1981).

Catalogue of Nearby Stars*

Name	R.A.(1950) h m s	Dec.(1950) ° ′ ″	$\mu_{R.A.}$ (s yr^{-1})	$\mu_{Dec.}$ (″ yr^{-1})	V_r (km s^{-1})
501.2	13 09 15	− 37 32 18	− 0.0324	+0.039	− 15
502.0	13 09 32	+28 07 54	− 0.0605	+0.879	+6
1168	13 10 41	+20 27 12	− 0.0428	+0.150	
503.0	13 11 08	− 58 50 12	− 0.0336	− 0.161	− 65
503.1	13 11 30	− 19 40 06	− 0.0088	+0.162	− 45
503.2	13 11 34	+56 58 24	+0.0131	− 0.033	− 9
503.3	13 13 16	− 19 40 42	+0.0215	− 0.121	+34
1169	13 14 14	+28 08 00	− 0.0549	+0.222	
504.0	13 14 18	+09 41 06	− 0.0227	+0.187	− 26
505.0A	13 14 22	+17 17 00	+0.0440	− 0.270	+6
505.0B	13 14 22	+17 17 00	+0.0440	− 0.270	+9
1170	13 15 41	+36 34 00	− 0.0064	− 0.290	+23
506.0	13 15 47	− 18 02 00	− 0.0754	− 1.072	− 8
506.1	13 16 24	− 02 48 30	− 0.0410	− 0.150	+126
506.2	13 16 45	+85 00 54	− 0.1031	+0.019	+11
507.0A	13 17 14	+35 23 00	+0.0329	− 0.787	− 6
507.0B	13 17 14	+35 23 00	+0.0329	− 0.787	
507.1	13 17 22	+33 36 36	− 0.0230	− 0.110	− 4
508.0A	13 17 36	+48 02 24	+0.0132	− 0.031	+8
508.0B	13 17 36	+48 02 24	+0.0132	− 0.031	
508.1	13 17 47	− 36 26 54	− 0.0282	− 0.089	
508.2	13 18 38	+34 32 42	+0.0410	− 0.300	− 23
508.3	13 21 07	− 13 46 48	− 0.0420	− 0.380	
509.0A	13 21 14	+29 29 42	− 0.0341	+0.254	− 38
509.0B	13 21 14	+29 29 42	− 0.0341	+0.254	
509.1	13 21 26	+58 10 00	+0.0141	− 0.023	− 7
510.0	13 23 04	− 28 06 48	− 0.0370	− 0.110	
511.0	13 23 56	− 24 02 00	− 0.0274	− 0.097	− 11
511.1	13 24 17	+63 31 00	− 0.0589	+0.214	− 31
511.2	13 24 47	− 15 42 54	− 0.0085	+0.019	− 14
512.0A	13 25 46	− 02 05 36	+0.0110	− 0.480	− 17
512.0B	13 25 46	− 02 05 36	+0.0110	− 0.480	
512.1	13 25 59	+14 02 42	− 0.0164	− 0.581	+5
513.0	13 26 52	+11 42 54	+0.0230	− 1.200	+40
514.0	13 27 27	+10 39 00	+0.0730	− 1.040	+15

Catalogue of Nearby Stars*

Name	Sp LC	V (mag)	B–V (mag)	π_t (0.001″)	M_V (mag)
501.2	(G5)	4.844	+0.70	53	3.47
502.0	G0V	4.249	+0.58	120	4.64
1168	dM	13.02	+1.58	50*	11.5
503.0	F8V	4.909	+0.48	52	3.49
503.1	(G6)	5.324		8	
503.2	G1V	6.814	+0.60	50	5.31
503.3	K1IV	5.220	+1.03	46	3.53
1169		13.26	+1.65	63	12.26
504.0	F8V	5.201	+0.58	76	4.60
505.0A		6.58	+0.92	87	6.28
505.0B		9.60		87	9.30
1170	dM2	11.1		44	9.3
506.0	G6V	4.740	+0.71	120	5.14
506.1		10.82	+1.02	47	9.20
506.2	F7V	7.264	+0.50	49	5.71
507.0A		9.52	+1.47	99	9.50
507.0B		12.09	+1.59	99	12.10
507.1		10.60		50	9.10
508.0A		8.96	+1.48	117	9.30
508.0B		9.80		117	10.10
508.1	A2V	2.748	+0.03	57	1.53
508.2		10.71	+1.36	41	8.80
508.3		12.60		46	10.90
509.0A		9.56	+1.33	60	8.50
509.0B		9.80		60	8.70
509.1	M0V	9.722	+1.26		
510.0		12.40		57	11.20
511.0	K3V	8.747	+0.93	32	6.27
511.1	G6V	6.505	+0.74	27	3.66
511.2	K1III	4.764	+1.10	57	3.54
512.0A		1.24	+1.52	84	10.90
512.0B		14.20		84	13.80
512.1	G2.5V	4.969	+0.71	41	3.03
513.0		3.20P		57	12.00
514.0	M1V	9.035	+1.49	129	9.59

*Adapted from Gliese (1969), Gliese and Jahreiss (1979), and Grenon and Rufener (1981).

Catalogue of Nearby Stars*

Name	R.A.(1950) h m s	Dec.(1950) ° ′ ″	$\mu_{R.A.}$ (s yr^{-1})	$\mu_{Dec.}$ (″ yr^{-1})	V_r (km s^{-1})
514.1	13 27 29	− 08 26 36	− 0.0740	− 0.490	
515.0	13 27 40	− 08 18 36	− 0.0760	− 0.430	
1171	13 28 08	+19 26 00	− 0.0352	− 1.298	
516.0A	13 30 18	+17 04 12	+0.0200	− 0.220	
516.0B	13 30 18	+17 04 12	+0.0200	− 0.220	+8
1172	13 31 49	+04 55 30	+0.0097	− 0.122	+14
517.0	13 32 07	− 08 05 06	− 0.0186	− 0.098	− 20
1173	13 32 51	− 00 08 12	+0.0036	+0.203	+10
518.0	13 34 13	+03 57 00	− 0.2470	− 1.130	
518.1	13 34 29	+08 01 36	− 0.0530	− 0.350	
518.2A	13 34 55	+30 20 18	− 0.0108	+0.026	+3
518.2B	13 34 57	+30 20 24	− 0.0108	+0.026	
519.0	13 35 13	+35 58 24	+0.0279	− 0.048	− 9
520.0A	13 35 49	+48 23 36	− 0.0220	− 0.120	− 27
520.0B	13 35 49	+48 23 36	− 0.0220	− 0.120	
520.0C	13 35 38	+48 23 36	− 0.0220	− 0.120	
521.0	13 37 20	+46 26 00	− 0.0040	+0.382	− 40
521.1	13 37 33	− 03 56 24	− 0.0257	+0.495	+33
1174	13 38 08	+44 01 18	− 0.1016	+0.314	
1175	13 38 10	− 34 12 36	+0.0159	− 0.154	
521.2A	13 38 24	+50 46 18	− 0.0143	+0.056	− 10
521.2B	13 38 26	+50 46 06	− 0.0143	+0.056	
522.0	13 39 22	+00 07 42	− 0.0104	− 0.419	+46
1176	13 39 53	− 01 25 54	− 0.0188	− 0.103	
523.0	13 41 13	+39 30 06	+0.0065	− 0.086	− 1
524.0	13 42 00	− 53 51 00	− 0.0280	− 0.360	
524.1	13 42 30	− 04 22 06	− 0.0110	− 0.060	+12
525.0	13 42 39	+18 03 42	+0.0304	− 1.845	+25
525.1	13 42 50	− 32 47 30	− 0.0365	− 0.150	− 22
526.0	13 43 12	+15 09 42	+0.1228	− 1.454	+15
1177A	13 44 49	− 32 10 48	+0.0000	+0.000	
1177B	13 44 49	− 32 10 48	+0.0000	+0.000	
527.0A	13 44 53	+17 42 18	− 0.0338	+0.034	− 16
527.0B	13 44 53	+17 42 18	− 0.0338	+0.034	
1178	13 44 59	+10 36 42	− 0.0598	− 0.124	

Catalogue of Nearby Stars*

Name	Sp LC	V (mag)	B–V (mag)	π_t (0.001″)	M_V (mag)
514.1		14.34	+1.63	57	13.10
515.0	DA	12.290	+0.07	60	11.18
1171		14.74	+1.79	68	13.90
516.0A		12.00	+1.53	62	11.00
516.0B		2.30		62	11.30
1172	dM0.5	9.97	+1.38	35	7.7
517.0	(K5)	9.252	+1.18	55	7.95
1173	K7V	10.28	+1.44	52*	8.9
518.0		14.71	+0.95	135	15.36
518.1	(M0)	9.981	+1.01	47	8.34
518.2A	G8V	9.265	+0.64	48	7.67
518.2B		10.502		48	8.91
519.0	M1VE	9.048	+1.40	103	9.11
520.0A		10.10	+1.38	54	8.80
520.0B		11.00		54	9.70
520.0C		16.00		54	14.70
521.0		10.00		106	10.10
521.1		9.60	+1.39	69	8.80
1174	dM	12.75	+1.66	62	11.71
1175	K2V	6.98	+0.86	69*	6.2
521.2A	F8V	6.325	+0.54	50	4.82
521.2B		10.49	+1.37	42	8.60
522.0		9.77	+1.30	60	8.70
1176	K7V	9.28	+1.19	46*	7.6
523.0		9.28	+1.10	39	7.20
524.0		14.30		86	14.00
524.1		10.20		48	8.60
525.0		9.83	+1.42	97	9.76
525.1	F2III	4.224	+0.38	50	2.72
526.0	M4VE	8.467	+1.43	192	9.88
1177A	K5V	8.94	+1.32	72*	8.2
1177B		9.12	+1.36	72*	8.4
527.0A	F7V	4.495	+0.48	57	3.27
527.0B	(DM2)	10.70		62	9.70
1178	DAwk	15.09	+0.36	50	13.58

*Adapted from Gliese (1969), Gliese and Jahreiss (1979), and Grenon and Rufener (1981).

Catalogue of Nearby Stars*

Name	R.A.(1950) h m s	Dec.(1950) ° ′ ″	$\mu_{R.A.}$ (s yr^{-1})	$\mu_{Dec.}$ (″ yr^{-1})	V_r (km s^{-1})
1179A	13 45 58	+23 51 36	− 0.1072	+0.155	
1179B	13 45 48	+23 49 36	− 0.1072	+0.155	
528.0A	13 46 47	+27 13 42	− 0.0328	− 0.092	− 20
528.0B	13 46 47	+27 13 42	− 0.0328	− 0.092	− 21
529.0	13 47 05	− 21 51 24	− 0.1261	− 0.502	− 35
529.1	13 48 08	+61 44 18	+0.0088	− 0.102	− 11
530.0	13 48 35	− 24 08 24	− 0.0419	− 0.305	+2
531.0	13 49 27	− 50 40 30	− 0.0641	− 0.055	− 25
1180	13 49 48	+26 55 00			
532.0	13 50 01	+50 11 54	+0.0440	− 0.141	− 41
532.1	13 50 57	− 35 04 06	− 0.0221	− 0.079	+7
533.0	13 51 02	+13 11 48	− 0.0120	− 0.680	+6
533.1	13 51 29	+65 52 30	− 0.0910	− 0.120	
1181AB	13 52 12	− 28 50 42	− 0.0199	− 0.100	
534.0	13 52 18	+18 38 54	− 0.0045	− 0.363	
534.1	13 53 14	− 54 27 24	− 0.0041	− 0.219	+5
534.2	13 54 01	+79 05 42	− 0.0930	+0.110	
534.3	13 55 43	− 33 45 12	− 0.0367	− 0.292	+59
535.0	13 57 00	+23 06 42	− 0.0124	+0.006	− 53
536.0	13 58 31	− 02 25 18	− 0.0530	+0.610	
536.1A	13 59 34	+15 44 06	+0.0080	− 0.020	+6
536.1B	13 59 34	+15 44 06	+0.0080	− 0.020	
537.0A	14 00 32	+46 34 54	+0.0580	− 0.050	− 31
537.0B	14 00 32	+46 34 54	+0.0580	− 0.050	
538.0	14 01 05	+11 01 48	+0.0055	− 0.311	− 17
538.1	14 03 31	− 26 26 30	+0.0032	− 0.144	+27
539.0	14 03 44	− 36 07 30	− 0.0429	− 0.523	+1
539.1	14 03 48	− 74 36 54	− 0.0600	+0.170	− 22
539.2	14 06 34	− 30 41 30	− 0.0360	− 0.230	
540.0	14 09 12	+80 50 24	+0.0780	− 0.565	+17
540.1	14 09 54	− 27 01 36	− 0.0009	− 0.038	+27
540.2	14 10 26	− 11 47 12	− 0.0450	− 0.440	
540.3	14 12 28	− 44 46 00	+0.0127	− 0.146	+3
1182	14 13 04	+04 54 00	− 0.0502	− 0.750	+4
541.0	14 13 23	+19 26 30	− 0.0776	− 1.999	− 5

Catalogue of Nearby Stars*

Name	Sp LC	V (mag)	B–V (mag)	π_t (0.001″)	M_V (mag)
1179A	dM4:	15.29	+1.97		14.91
1179B	DC	15.63	+1.09	84	15.25
528.0A	K4V	7.044	+1.12	87	6.74
528.0B	K6V	8.03		86	7.70
529.0	(K5)	8.153	+1.27	80	7.67
529.1	G3V	5.979	+0.99		
530.0	(G0)	6.429	+0.69	62	5.39
531.0	K1V	7.374	+0.89	76	6.78
1180		10.83	+0.75	77	10.3
532.0	M0VP	8.862	+1.33	75	8.30
532.1	K1IV–V	6.176	+1.02	48	4.60
533.0		9.81	+1.37	60	8.70
533.1		11.40		47	9.80
1181AB	K7V	9.57	+1.45	91*	9.4
534.0	G0IV	2.706	+0.58	102	2.75
534.1	G6V	6.010	+0.78	80	5.50
534.2		10.60		44	8.80
534.3	G5V	8.135	+0.70	46	6.45
535.0		9.06	+1.16	49	7.50
536.0	(K5)	9.701		91	9.50
536.1A		10.80		32	8.30
536.1B		11.00		32	8.50
537.0A	M3VE	9.179	+1.48	86	8.85
537.0B	(DM3E)	9.95		84	9.60
538.0	G8V	6.290	+0.74	61	5.22
538.1	K2III	3.278	+1.12	45	1.54
539.0	K0III	2.093	+1.00	64	1.12
539.1	G1V	6.019	+0.58	40	4.03
539.2		12.70		49	11.20
540.0		10.30		65	9.30
540.1	(K0)	5.075	+1.14	46	3.39
540.2		13.50		72	12.80
540.3	G1V	6.310	+0.60	48	4.70
1182		14.31	+1.72	72	13.60
541.0	K1IIIP	- 0.043	+1.23	91	- 0.25

*Adapted from Gliese (1969), Gliese and Jahreiss (1979), and Grenon and Rufener (1981).

Catalogue of Nearby Stars*

Name	R.A.(1950) h m s	Dec.(1950) ° ′ ″	$\mu_{R.A.}$ (s yr^{-1})	$\mu_{Dec.}$ (″ yr^{-1})	V_r (km s^{-1})
541.1	14 15 21	- 07 18 30	+0.0168	- 0.245	- 14
541.2	14 15 29	+45 40 30	+0.0056	+0.001	
542.0	14 15 30	- 59 08 18	- 0.0606	- 0.829	- 15
542.1A	14 16 10	- 25 35 24	- 0.0274	+0.347	- 21
542.1B	14 16 10	- 25 35 24	- 0.0274	+0.347	0
542.2	14 16 20	- 06 22 06	+0.0000	- 0.410	+14
543.0	14 16 36	- 07 03 48	- 0.0740	- 0.800	
544.0A	14 17 00	- 04 55 12	- 0.0441	- 0.144	- 8
544.0B	14 17 00	- 04 55 12	- 0.0441	- 0.144	
545.0	14 17 29	- 09 22 48	- 0.0440	- 0.920	
545.1	14 18 20	- 40 09 54	- 0.0470	- 0.110	
546.0	14 19 48	+29 51 42	- 0.0508	- 0.312	- 37
547.0	14 20 42	+01 28 30	+0.0148	- 0.481	- 18
548.0A	14 23 24	+23 51 24	+0.0574	- 1.128	+14
548.0B	14 23 27	+23 51 36	+0.0576	- 1.122	+6
549.0A	14 23 30	+52 04 54	- 0.0262	- 0.400	- 11
549.0B	14 23 29	+52 03 42	- 0.0262	- 0.400	
549.0C	14 23 45	+53 33 12	- 0.0350	- 0.450	
550.0	14 24 10	- 51 42 36	- 0.0318	+0.023	+12
1183A	14 25 23	- 00 09 12	- 0.0254	+0.081	
1183B	14 25 24	- 00 09 06	- 0.0254	+0.081	
550.1	14 25 31	+24 03 48	- 0.0380	+0.080	- 59
550.2A	14 25 37	- 02 00 18	- 0.0094	- 0.005	- 10
550.2B	14 25 37	- 02 00 18	- 0.0094	- 0.005	- 8
550.3	14 26 03	- 46 14 18	- 0.0040	- 0.200	
551.0	14 26 19	- 62 28 06	- 0.5440	+0.790	- 16
552.0	14 27 11	+15 44 12	- 0.0720	+1.350	+19
552.1	14 27 30	- 53 52 36	- 0.0300	- 0.180	
553.0	14 28 12	- 08 25 18	- 0.0851	- 0.226	- 27
553.1	14 28 20	- 12 04 06	- 0.0250	- 0.370	
554.0	14 28 42	+35 40 18	- 0.0397	+0.190	- 12
1184	14 29 48	+11 34 12	+0.0064	+0.232	- 39
555.0	14 31 35	- 12 18 36	- 0.0210	+0.620	
556.0	14 31 51	+53 07 24	- 0.0233	+0.240	+12
557.0	14 32 30	+29 57 42	+0.0145	+0.129	

Catalogue of Nearby Stars*

Name	Sp LC	V (mag)	B–V (mag)	π_t (0.001″)	M_V (mag)
541.1	G8V	6.469	+0.73	62	5.40
541.2		10.00		46	8.30
542.0	K3V	6.662	+1.02	104	6.75
542.1A		5.86	+0.50	45	4.10
542.1B		3.30		45	11.60
542.2	K7V	9.106	+1.30	44	7.32
543.0		14.50		75	13.90
544.0A	K1V	7.593	+0.84	53	6.21
544.0B	(M6)	14.50		53	13.10
545.0		13.02		92	12.80
545.1	K5V	9.002		45	7.27
546.0	K8V	8.566	+1.26	70	7.79
547.0	G1V	6.274	+0.63	61	5.20
548.0A	M1V	9.762	+1.41	68	8.92
548.0B	M2V	10.010	+1.44	68	9.17
549.0A	F7V	4.049	+0.50	68	3.21
549.0B		11.10		68	10.30
549.0C		13.50		68	12.70
550.0	G5V	7.830	+0.70	57	6.61
1183A		13.96	+1.65	62	12.92
1183B		14.04	+1.68	62	13.00
550.1		10.91	+1.27	34	8.60
550.2A	G2IV	4.807	+0.74	44	3.02
550.2B		9.00		45	7.30
550.3		12.00		47	10.40
551.0	M5VE	11.094	+1.97	772	15.53
552.0		10.68	+1.47	73	10.00
552.1		13.10		54	11.80
553.0	K7V	9.380	+1.41	52	7.96
553.1		13.40		83	13.00
554.0	K3V	8.705	+1.13	66	7.80
1184	dM0p	9.69	+1.21	45*	8.0
555.0		11.36	+1.63	160	12.38
556.0	K3V	7.251	+0.99	69	6.45
557.0	F2V	4.466	+0.37	63	3.46

*Adapted from Gliese (1969), Gliese and Jahreiss (1979), and Grenon and Rufener (1981).

Catalogue of Nearby Stars*

Name	R.A.(1950) h m s	Dec.(1950) ° ′ ″	$\mu_{R.A.}$ (s yr^{-1})	$\mu_{Dec.}$ (″ yr^{-1})	V_r (km s^{-1})
558.0	14 32 55	+33 57 42	− 0.0580	+0.260	− 52
558.1	14 33 32	− 67 42 42	− 0.0620	− 0.283	− 30
559.0A	14 36 11	− 60 37 48	− 0.4904	+0.712	− 22
559.0B	14 36 11	− 60 37 48	− 0.4904	+0.712	
559.1	14 37 56	+64 30 24	− 0.0236	− 0.007	− 30
560.0A	14 38 26	− 64 45 30	− 0.0291	− 0.238	+7
560.0B	14 38 26	− 64 45 30	− 0.0291	− 0.238	+7
561.0	14 41 09	+26 57 42	− 0.0226	− 0.006	− 80
561.1A	14 43 06	− 25 13 54	− 0.0111	− 0.107	− 13
561.1B	14 43 07	− 25 14 00	− 0.0103	− 0.083	− 20
562.0	14 44 03	+16 43 06	− 0.0087	− 0.923	+46
563.0	14 44 23	− 12 31 42	− 0.0320	− 0.190	
1185	14 45 21	− 02 57 30	− 0.0347	+0.391	
563.1	14 46 03	+38 40 36	− 0.0050	+0.110	− 15
563.2A	14 46 42	− 25 53 48	− 0.0890	− 0.190	
563.2B	14 46 40	− 25 54 00	− 0.0890	− 0.190	
563.3	14 47 55	+07 01 18	− 0.0406	− 0.049	− 31
563.4	14 47 55	− 15 47 24	− 0.0070	− 0.072	− 23
564.0	14 48 02	+24 07 00	+0.0110	+0.028	− 1
564.1	14 48 06	− 15 50 06	− 0.0075	− 0.071	− 10
565.0	14 48 50	− 24 05 36	− 0.0676	− 0.422	− 65
566.0A	14 49 05	+19 18 24	+0.0096	− 0.103	+4
566.0B	14 49 05	+19 18 24	+0.0096	− 0.103	+5
567.0	14 51 07	+19 21 12	− 0.0319	+0.213	− 34
1186	14 51 13	+11 47 00	+0.0061	− 0.734	
568.0A	14 51 41	+23 45 30	− 0.0510	+0.190	
568.0B	14 51 41	+23 45 30	− 0.0510	+0.190	
569.0	14 52 08	+16 18 18	+0.0210	− 0.130	+5
569.1	14 53 46	+53 52 30	− 0.1107	+0.472	− 15
569.2	14 53 55	+17 56 54	− 0.0130	+0.060	
570.0A	14 54 32	− 21 11 30	+0.0745	− 1.740	+26
570.0B	14 54 31	− 21 11 18	+0.0696	− 1.668	+26
570.1	14 54 42	− 48 39 30	− 0.0017	− 0.315	+36
570.2	14 55 30	+31 36 42	− 0.0580	− 1.190	+24
571.0	14 56 14	− 43 53 30	− 0.0250	− 0.270	− 2

Catalogue of Nearby Stars*

Name	Sp LC	V (mag)	B–V (mag)	π_t (0.001″)	M_V (mag)
558.0	M0V	9.553	+1.28	51	8.09
558.1	F7V	6.038	+0.50	49	4.49
559.0A	G2V	- 0.297	+0.68	750	4.08
559.0B		11.33	+0.88	743	5.69
559.1	G0VE	7.527	+0.61	48	5.93
560.0A	F0VP	3.190	+0.24	55	1.89
560.0B		8.47	+1.15	58	7.30
561.0	(K0)	9.657		53	8.28
561.1A	F0III	4.932		47	3.29
561.1B	F9V	7.14		33	4.70
562.0	K5V	9.089	+1.26	73	8.41
563.0		13.50		62	12.50
1185		13.28	+1.63	51	11.82
563.1	M2V	9.744	+1.31	15	8.10
563.2A		13.00		45	11.30
563.2B		13.20		45	11.50
563.3	K2V	9.065	+0.97	52	7.64
563.4	F5IV	5.135	+0.41	53	3.76
564.0	G2V	5.862	+0.56	70	5.09
564.1	A3IV	2.749	+0.15	53	1.37
565.0	(K2)	7.797	+1.00	69	6.99
566.0A	G8VE	4.544	+0.72	152	5.45
566.0B	K4VE	6.84	+1.00	148	7.69
567.0	K1V	6.019	+0.84	84	5.64
1186		15.29	+1.82	53	13.91
568.0A		11.60		96	11.50
568.0B		12.20		96	12.10
569.0		10.20	+1.48	96	10.10
569.1	K1V	7.774	+0.79	46	6.09
569.2		10.90		48	9.30
570.0A	K5V	5.760	+1.10	180	7.04
570.0B	M2VE	8.078	+1.50	180	9.35
570.1	G4V	6.339	+0.71	50	4.83
570.2		1.08	+1.32	47	9.40
571.0		10.16	+1.31	45	8.40

*Adapted from Gliese (1969), Gliese and Jahreiss (1979), and Grenon and Rufener (1981).

Catalogue of Nearby Stars*

Name	R.A.(1950) h m s	Dec.(1950) ° ′ ″	$\mu_{R.A.}$ (s yr^{-1})	$\mu_{Dec.}$ (″ yr^{-1})	V_r (km s^{-1})
1187	14 56 29	+56 51 48	+0.0328	− 0.635	
571.1	14 58 01	− 10 55 54	+0.0004	− 0.469	+14
572.0	14 59 09	+45 37 06	+0.0236	+0.342	− 11
573.0	14 59 09	+16 04 06	+0.0058	− 0.255	+4
574.0	14 59 42	− 46 05 54	− 0.0280	+0.000	− 65
1188	15 00 57	+03 58 06	− 0.0600	+0.702−	+128
574.1	15 01 08	− 25 05 12	− 0.0054	− 0.047	− 4
575.0A	15 02 08	+47 50 54	− 0.0409	+0.032	− 25
575.0B	15 02 08	+47 50 54	− 0.0409	+0.032	
575.1	15 02 15	+29 40 24	+0.0180	− 0.180	
576.0	15 02 27	+05 50 18	− 0.0390	− 0.480	− 68
577.0	15 04 57	+64 14 12	− 0.0214	+0.088	− 6
578.0	15 05 06	+25 03 48	+0.0136	− 0.171	− 7
579.0	15 05 16	+25 07 12	− 0.0644	+0.494	− 65
579.1	15 06 31	+13 25 24	− 0.0039	+0.061	− 49
1189	15 06 54	+24 12 12	− 0.0359	+0.169	− 51
579.2A	15 07 28	− 16 08 30	− 0.0711	− 3.530	+292
579.2B	15 07 28	− 16 13 30	− 0.0708	− 3.534	+306
579.3	15 10 14	+00 58 30	− 0.0163	− 0.312	− 60
579.4	15 10 33	− 25 07 18	− 0.0293	− 0.073	+4
580.0A	15 11 20	− 01 09 30	− 0.0856	− 0.491	− 68
580.0B	15 11 20	− 01 09 30	− 0.0856	− 0.491	
1190	15 11 25	− 03 36 54			− 106
580.1	15 13 35	− 58 37 00	− 0.0122	− 0.140	+9
580.2	15 14 03	+67 32 12	+0.0368	− 0.393	− 47
581.0	15 16 50	− 07 32 24	− 0.0830	− 0.090	− 30
582.0	15 18 25	− 48 08 06	− 0.1615	− 0.271	− 69
582.1A	15 19 08	− 47 44 24	− 0.0362	− 0.272	− 22
582.1B	15 19 08	− 47 44 24	− 0.0362	− 0.272	
583.0	15 19 27	− 04 35 54	− 0.0199	+0.008	− 15
1191	15 19 54	− 10 28 48	− 0.0039	− 0.202	
1192	15 20 11	+01 36 06	− 0.0237	− 0.380	− 30
584.0A	15 21 08	+30 28 00	+0.0104	− 0.191	− 7
584.0B	15 21 08	+30 28 00	+0.0104	− 0.191	
585.0	15 21 35	+17 39 36	− 0.0260	− 1.230	

Catalogue of Nearby Stars*

Name	Sp LC	V (mag)	B–V (mag)	π_t (0.001″)	M_V (mag)
1187		15.52	+1.95	90	15.29
571.1	K7V	9.467	+1.38	51	8.00
572.0	M0V	9.169	+1.43	69	8.36
573.0		9.13	+1.04	37	7.00
574.0		9.87	+1.18	42	8.00
1188		12.09	+1.45	54	10.7
574.1	M3III	3.305	+1.71	62	2.27
575.0A	G2V	4.820	+0.65	84	4.44
575.0B	G2V	15.85		84	5.47
575.1		14.20		47	12.60
576.0	K5V	9.811	+1.30	57	8.59
577.0	G5VE	8.436	+0.68	53	7.06
578.0	F5V	4.927	+0.43	62	3.89
579.0	M0.5V	10.045	+1.36	75	9.42
579.1	(DG6)	6.162	+0.98		
1189	dKB	9.30	+1.07	57	8.1
579.2A	K0VI	9.054	+0.79	38	6.95
579.2B	K2VI	9.399	+0.85	38	7.30
579.3	(K0)	9.260		56	8.00
579.4	(G5)	6.444	+0.70	32	3.97
580.0A	G8V	6.579	+0.77	63	5.58
580.0B		17.34		64	6.37
1190	K5V	9.83	+1.13	52	8.4
580.1	A3V	4.061	+0.09	52	2.64
580.2	F8V	5.169	+0.53	47	3.53
581.0	M5V	10.546	+1.61	154	11.48
582.0	G2V	5.647	+0.65	65	4.71
582.1A		8.30	+0.69	36	6.10
582.1B		8.60		36	6.40
583.0	M1V	9.445	+1.30	59	8.30
1191	K2V	7.99	+0.95	51*	6.53
1192	K3V	8.30	+0.99	41	6.4
584.0A	G2V	5.001	+0.58	61	3.93
584.0B	G2V	15.86		60	4.75
585.0		15.00		91	14.80

*Adapted from Gliese (1969), Gliese and Jahreiss (1979), and Grenon and Rufener (1981).

Catalogue of Nearby Stars*

Name	R.A.(1950) h m s	Dec.(1950) ° ′ ″	$\mu_{R.A.}$ (s yr^{-1})	$\mu_{Dec.}$ (″ yr^{-1})	V$_r$ (km s^{-1})
585.1	15 25 12	+02 46 18	− 0.0030	− 0.010	
586.0A	15 25 27	− 09 10 12	+0.0046	− 0.357	+2
586.0B	15 25 30	− 09 10 48	+0.0062	− 0.367	+7
586.0C	15 25 03	− 08 51 00	+0.0050	− 0.350	
587.0	15 25 44	− 49 46 48	− 0.0242	− 0.090	− 43
587.1	15 25 53	+25 57 48	+0.0080	− 0.060	
588.0	15 28 58	− 41 05 36	− 0.1030	− 1.020	
588.1	15 32 03	+38 04 54	+0.0020	− 0.080	
1193	15 32 13	+14 26 18	− 0.0465	− 0.143	
589.0A	15 33 08	+17 52 54	− 0.0860	− 0.130	
589.0B	15 33 08	+17 52 54	− 0.0860	− 0.130	
590.0	15 33 50	− 37 43 00	− 0.0280	− 0.820	
591.0	15 34 10	+39 59 42	− 0.0394	+0.056	− 72
592.0	15 34 13	− 13 57 30	− 0.0320	− 0.640	
593.0A	15 34 15	+39 58 00	− 0.0391	+0.040	− 70
593.0B	15 34 15	+39 58 00	− 0.0391	+0.040	
593.1	15 37 19	− 23 39 24	− 0.0013	− 0.018	− 22
594.0	15 37 45	− 44 29 48	− 0.0167	− 0.263	− 6
1194A	15 38 16	+43 39 30	+0.1088	− 0.339	
1194B	15 38 17	+43 39 24	+0.1088	− 0.339	
595.0	15 39 20	− 19 18 36	− 0.1420	− 0.980	
1195	15 40 22	− 10 46 18	− 0.0841	− 0.356	− 171
596.0	15 40 55	+26 26 12	+0.0080	− 0.130	
596.1A	15 41 31	+02 40 24	− 0.0055	− 0.153	+14
596.1B	15 41 31	+02 40 24	− 0.0055	− 0.153	
596.2	15 41 48	+06 34 54	+0.0091	+0.044	+3
597.0	15 42 15	+76 09 36	+0.2460	− 0.880	
598.0	15 44 01	+07 30 30	− 0.0151	− 0.068	− 66
599.0A	15 44 14	− 37 45 36	− 0.0364	− 0.218	− 5
599.0B	15 44 14	− 37 45 36	− 0.0364	− 0.218	+51
599.1	15 47 22	− 50 32 12	+0.0270	− 0.050	
600.0	15 49 46	+11 01 36	− 0.0176	− 0.244	+1
601.0	15 50 43	− 63 16 42	− 0.0278	− 0.396	
602.0	15 50 57	+42 35 24	+0.0396	+0.628	− 55
1196	15 51 38	− 25 51 36	− 0.0168	+0.106	

Catalogue of Nearby Stars*

Name	Sp LC	V (mag)	B–V (mag)	π_t (0.001″)	M_V (mag)
585.1		10.15	+1.40	47	8.50
586.0A	(K1)	6.893	+0.81	61	5.82
586.0B	K2V	7.570	+0.90	61	6.50
586.0C		15.35	+1.84	61	14.30
587.0	G5V	7.674	+0.78	57	6.45
587.1		11.07	+1.61		
588.0		10.10		169	11.20
588.1		11.10		46	9.40
1193		13.83	+1.56	49	12.28
589.0A		12.00		93	11.80
589.0B		15.20		93	15.00
590.0		13.80		82	13.40
591.0	K3V	7.650	+0.95	47	6.01
592.0		13.80		90	13.60
593.0A	K2V	6.802	+0.92	47	5.16
593.0B		17.60		54	6.30
593.1	(K0)	4.966	+1.32	48	3.37
594.0	F5IV–V	4.634	+0.41	59	3.49
1194A		12.48	+1.70	74	11.83
1194B		13.8			13.2
595.0		11.70		111	11.90
1195	F3VI	7.20	+0.50	52	5.8
596.0		10.80		56	9.50
596.1A	G5V	5.859	+0.68	46	4.17
596.1B		12.00		46	10.30
596.2	K2III	2.649	+1.17	49	1.10
597.0		11.60		65	10.70
598.0	G0V	4.435	+0.60	94	4.30
599.0A	(G0)	6.000	+0.72	74	5.35
599.0B		2.80	+0.30	73	12.10
599.1		11.30		51	9.80
600.0	M0V	9.360	+1.40	41	7.42
601.0	F2IV	2.823	+0.29	83	2.42
602.0	F9V	4.615	+0.57	56	3.36
1196	K5V	9.28	+1.24	48*	7.7

*Adapted from Gliese (1969), Gliese and Jahreiss (1979), and Grenon and Rufener (1981).

Catalogue of Nearby Stars*

Name	R.A.(1950) h m s	Dec.(1950) ° ′ ″	$\mu_{R.A.}$ (s yr^{-1})	$\mu_{Dec.}$ (″ yr^{-1})	V$_r$ (km s^{-1})
603.0	15 54 08	+15 49 24	+0.0212	− 1.285	+7
604.0	15 54 16	− 42 28 42	− 0.0224	− 0.202	+36
604.1	15 56 52	− 45 18 36	− 0.0143	− 0.125	− 24
605.0	15 56 55	+59 24 36	− 0.0430	+0.220	
606.0	15 57 11	− 08 06 48	+0.0150	− 0.030	
606.1A	15 58 21	− 84 05 48	− 0.2071	− 0.007	− 10
606.1B	15 58 21	− 84 05 48	− 0.2071	− 0.007	
606.2	15 59 08	+33 27 12	− 0.0159	− 0.768	+18
607.0	15 59 45	+30 19 00	− 0.0250	+0.140	
608.0	15 59 53	+61 48 00	− 0.0633	+0.026	− 37
609.0	16 00 43	+20 44 36	− 0.0690	− 1.240	
609.1	16 00 57	+58 41 54	− 0.0417	+0.336	− 8
609.2	16 01 59	+25 22 54	− 0.0389	+0.672	− 44
610.0	16 02 44	− 20 18 36	+0.0228	− 0.341	+34
611.0	16 03 13	+39 17 24	− 0.0493	+0.055	− 58
611.1	16 03 51	− 70 55 54	− 0.0410	− 0.390	+26
611.2	16 03 58	+80 45 48	− 0.0111	+0.016	
611.3	16 04 18	+08 31 12	− 0.0340	+0.040	− 48
612.0	16 04 42	+38 46 24	+0.0203	− 0.541	+24
612.1	16 04 48	+34 45 54	+0.0220	− 0.580	+8
1197	16 05 09	+26 58 36	+0.0274	− 0.437	
1198	16 05 31	− 10 17 18	− 0.0237	− 1.304	
613.0	16 05 41	− 56 19 06	− 0.0152	+0.324	+38
614.0	16 08 47	+43 57 00	+0.0114	− 0.304	− 6
1199	16 09 06	+13 30 24	− 0.0006	− 0.540	+131
615.0	16 09 47	− 57 25 30	− 0.1080	− 1.400	+11
615.1A	16 10 58	+13 39 36	+0.0122	− 0.412	+18
615.1B	16 10 58	+13 39 36	+0.0122	− 0.412	+21
1200	16 12 27	+19 13 18	− 0.1411	+0.352	
615.2A	16 12 48	+33 59 00	− 0.0221	− 0.080	− 11
615.2B	16 12 48	+33 59 00	− 0.0221	− 0.080	− 17
615.2C	16 12 03	+33 53 54	− 0.0230	− 0.070	
616.0	16 12 54	− 08 14 18	+0.0154	− 0.505	+11
616.1A	16 13 04	− 53 34 18	− 0.0074	− 0.095	− 19
616.1B	16 13 04	− 53 34 18	− 0.0074	− 0.095	

Catalogue of Nearby Stars*

Name	Sp LC	V (mag)	B–V (mag)	π_t (0.001″)	M_V (mag)
603.0	F6V	3.875	+0.48	80	3.39
604.0	K5V	8.069	+1.12	60	6.96
604.1	G5IV	7.531	+0.75	45	5.80
605.0	M0V	10.344		50	9.10
606.0	M0V	10.475		58	9.00
606.1A	K0V	7.713	+0.80	49	6.16
606.1B		12.80		44	11.00
606.2	G0IV	5.402	+0.60	41	3.47
607.0		14.10		63	13.10
608.0		9.98	+1.28	41	8.00
609.0		14.20		99	14.20
609.1	F8IV–V	4.021	+0.52	46	2.33
609.2	G8V	7.08	+0.77	53	5.74
610.0	K3IV–V	7.393	+0.97	58	6.21
611.0	G8V	6.672	+0.73	77	6.10
611.1	G8V	7.248	+0.73	47	5.61
611.2		17.45		46	5.80
611.3		1.60		52	10.20
612.0	K3V	8.579	+0.96	57	7.36
612.1	M0V	10.425	+1.24	46	8.74
1197		13.33	+1.64	46	11.64
1198	M2	14.51	+1.4	49	13.0
613.0	K3V	7.106	+0.85	76	6.51
614.0	K0V	6.610	+0.87	63	5.61
1199	DA	15.09	+0.23	55	13.79
615.0	K0V	7.521	+0.82	66	6.62
615.1A		7.39	+0.77	48	5.80
615.1B		7.48		48	5.90
1200		12.90	+1.54	55	11.60
615.2A	F8V	5.198	+0.50	48	3.60
615.2B	G1V	6.72		45	5.00
615.2C	(M)	12.227		48	10.63
616.0	G2V	5.489	+0.65	62	4.45
616.1A	G5V	6.392	+0.80	49	5.30
616.1B		7.60		49	6.00

*Adapted from Gliese (1969), Gliese and Jahreiss (1979), and Grenon and Rufener (1981).

Catalogue of Nearby Stars*

Name	R.A.(1950) h m s	Dec.(1950) ° ′ ″	$\mu_{R.A.}$ (s yr^{-1})	$\mu_{Dec.}$ (″ yr^{-1})	V$_r$ (km s^{-1})
616.2	16 15 59	+55 23 48	+0.0120	− 0.520	− 30
617.0A	16 16 37	+67 21 30	− 0.0899	+0.086	− 14
617.0B	16 16 39	+67 22 36	− 0.0899	+0.086	− 24
618.0A	16 16 47	− 37 25 24	− 0.0590	+1.000	
618.0B	16 16 47	− 37 25 24	− 0.0590	+1.000	
618.1	16 17 48	− 04 08 54	− 0.0270	− 0.010	
618.2	16 17 50	+51 52 24	− 0.0200	+0.220	
618.3	16 17 54	+21 15 06	− 0.0013	− 0.052	− 25
618.4	16 19 01	− 48 31 54	− 0.0590	− 0.460	
619.0	16 19 12	+41 04 36	− 0.0027	+0.126	+5
620.0	16 20 07	− 24 35 06	− 0.0250	− 0.680	
620.1	16 20 38	− 39 04 42	+0.0067	− 0.008	+10
620.2	16 20 45	− 46 36 30	− 0.0500	− 0.740	
621.0	16 21 32	− 13 31 30	− 0.0162	− 0.213	+9
622.0	16 22 17	− 21 49 06	− 0.0430	− 0.310	
623.0	16 22 39	+48 28 24	+0.1160	− 0.460	− 29
624.0	16 23 04	− 69 58 30	+0.0404	+0.101	+8
624.1A	16 23 18	+61 37 36	− 0.0033	+0.059	− 14
624.1B	16 23 18	+61 37 36	− 0.0033	+0.059	
625.0	16 24 14	+54 25 06	+0.0510	− 0.150	
1201	16 25 30	+09 19 18	− 0.0066	− 0.460	
626.0	16 25 32	+07 25 12	− 0.0178	− 0.270	− 32
626.1	16 25 43	− 78 47 18	− 0.0404	− 0.069	+5
626.2	16 26 40	+36 52 12	− 0.0400	+0.720	+24
627.0A	16 26 41	+18 31 06	− 0.0228	+0.397	− 36
627.0B	16 26 41	+18 31 06	− 0.0228	+0.397	
628.0	16 27 31	− 12 32 18	− 0.0050	− 1.180	− 13
629.0	16 28 08	− 38 54 06	− 0.0356	− 0.330	− 59
1202	16 29 22	+17 40 54	− 0.0231	− 0.816	
629.1	16 30 11	− 12 29 00	− 0.0230	− 0.210	
1203	16 30 28	+12 43 06	− 0.0525	− 0.135	
629.2A	16 32 05	− 04 07 00	− 0.0110	− 0.780−	+163
629.2B	16 32 51	− 03 51 18	− 0.0080	− 0.720	
629.3	16 32 50	− 49 11 00	− 0.0050	− 0.590	
630.0	16 33 17	+33 24 18	− 0.0190	− 0.100	

Catalogue of Nearby Stars*

Name	Sp LC	V (mag)	B–V (mag)	π_t (0.001″)	M_V (mag)
616.2	M1.5VE	9.972	+1.46	48	8.38
617.0A	M1V	8.607	+1.41	89	8.35
617.0B	M3V	10.736	+1.50	89	10.48
618.0A		10.60		131	11.20
618.0B		16.00		131	16.60
618.1		10.66	+1.44	49	9.10
618.2	M0VP	9.854		46	8.50
618.3		6.04	+0.93		0.03
618.4		11.80		49	10.20
619.0	M0VP	8.991	+1.30	72	8.28
620.0		10.40		70	9.60
620.1	(G0)	5.361	+0.64	55	4.06
620.2		13.90		53	12.50
621.0	K4V	8.380	+0.95	62	7.34
622.0		10.30		68	9.50
623.0	M3V	10.277	+1.48	132	10.88
624.0	G0V	4.904	+0.55	93	4.75
624.1A	G8III	2.736	+0.91	43	0.90
624.1B	K1V	8.80		46	7.10
625.0	M2V	10.109		161	11.14
1201	DC	16.13	+0.38	47	14.5
626.0		8.88	+1.22	57	7.70
626.1	K0IV	3.873	+0.91	54	2.53
626.2		3.86	+0.18	56	12.60
627.0A	K2V	6.976	+0.86	58	5.79
627.0B		7.75		53	6.40
628.0	M5V	10.085	+1.60	248	12.06
629.0	(G5)	7.218	+0.87	57	6.00
1202		12.76	+1.56	54	11.42
629.1		10.90		48	9.30
1203		12.13	+1.47	51	10.67
629.2A	(G5)	9.583	+0.75	51	8.12
629.2B		15.00		23	11.80
629.3		14.00		52	12.60
630.0		11.03	+1.40	40	9.00

*Adapted from Gliese (1969), Gliese and Jahreiss (1979), and Grenon and Rufener (1981).

Catalogue of Nearby Stars*

Name	R.A.(1950) h m s	Dec.(1950) ° ′ ″	$\mu_{R.A.}$ (s yr^{-1})	$\mu_{Dec.}$ (″ yr^{-1})	V_r (km s^{-1})
630.1A	16 33 29	+57 14 48	− 0.1330	+1.200	
630.1B	16 33 31	+57 15 12	− 0.1330	+1.200	
1204	16 33 43	+08 54 54	− 0.0363	− 0.154	
631.0	16 33 44	− 02 13 12	+0.0303	− 0.315	− 15
632.0	16 34 28	+79 53 42	+0.0306	− 0.080	− 14
632.1	16 34 52	+31 12 12	+0.0271	− 0.445	− 9
632.2A	16 35 48	+76 04 54	+0.0320	+0.140	
632.2B	16 35 47	+76 04 54	+0.0280	+0.130	
632.3	16 36 13	+05 31 24	− 0.0120	− 0.010	
633.0	16 37 30	− 45 54 00	+0.0340	− 0.390	
634.0	16 37 56	− 43 53 00	− 0.0360	− 0.470	
634.1	16 38 31	− 02 45 18	− 0.0017	− 0.439	− 42
635.0A	16 39 24	+31 41 30	− 0.0369	+0.394	− 70
635.0B	16 39 24	+31 41 30	− 0.0369	+0.394	
1205	16 39 49	+53 47 00	− 0.0143	− 0.204	
635.1	16 40 41	+79 00 48	− 0.0091	+0.036	− 20
636.0	16 41 11	+39 01 00	+0.0028	− 0.083	+8
637.0	16 42 30	− 72 52 48	− 0.1160	− 0.460	
637.1	16 42 44	+68 11 18	− 0.0519	+0.426	+6
638.0	16 43 15	+33 35 42	− 0.0037	+0.375	− 30
638.1	16 45 16	− 47 37 54	− 0.0111	− 0.030	+29
639.0	16 46 50	+37 06 18	− 0.0060	− 0.381	+3
639.1	16 46 55	− 34 12 18	− 0.0492	− 0.255	− 2
639.2	16 47 00	− 64 21 12	− 0.0800	− 0.110	
1206	16 47 38	+59 08 48	+0.0000	+0.320	
640.0	16 47 54	+18 59 12	− 0.0022	− 0.088	− 3
641.0	16 50 27	+00 04 30	− 0.0481	− 1.505	+41
641.1	16 50 28	− 20 20 00	− 0.0036	− 0.031	− 17
642.0	16 51 54	+11 59 30	− 0.0390	+0.370	− 61
643.0	16 52 45	− 08 13 54	− 0.0540	− 0.880	+22
644.0A	16 52 48	− 08 14 42	− 0.0537	− 0.875	+19
644.0B	16 52 48	− 08 14 42	− 0.0537	− 0.875	
644.0C	16 52 55	− 08 18 12	− 0.0540	− 0.860	
645.0	16 53 24	− 36 58 42	− 0.0120	− 0.450	
1207	16 54 26	− 04 16 00	+0.0341	− 0.370	

Catalogue of Nearby Stars*

Name	Sp LC	V (mag)	B–V (mag)	π_t (0.001″)	M_V (mag)
630.1A		12.90	+1.60	66	12.00
630.1B		15.00	+0.49	66	14.10
1204		13.79	+1.65	66	12.89
631.0	K0V	5.776	+0.81	90	5.55
632.0	G3V	7.032	+0.61	56	5.77
632.1		9.49	+1.20	45	7.80
632.2A		10.00	+1.17	51	8.50
632.2B		13.00		51	11.50
632.3		12.00		46	10.00
633.0		14.40		105	14.50
634.0		10.80		57	9.60
634.1		7.24	+0.64	38	5.10
635.0A	G0IV	2.819	+0.64	105	2.92
635.0B	K0V	15.49	+0.75	104	5.57
1205	DC	15.06	+0.33	46*	13.43
635.1		6.20			
636.0	G8III	3.488	+0.92	40	1.50
637.0		13.00		68	12.20
637.1	K1V	7.564	+0.77	49	6.01
638.0	K7V	8.081	+1.37	99	8.06
638.1		7.38	+0.91	56	6.10
639.0	K2V	8.381	+0.82	51	6.92
639.1	K2III	2.310	+1.16	54	0.97
639.2		12.40		52	11.00
1206	DA	12.24	+0.16	82	11.81
640.0	K5V	8.856	+1.02	35	6.58
641.0	G6V	6.616	+0.75	67	5.75
641.1	(G5)	5.855		58	4.60
642.0		1.10		46	9.40
643.0		1.70	+1.70	161	12.73
644.0A		9.76	+1.62	161	10.79
644.0B		9.80		161	10.80
644.0C		16.66	+2.05	161	17.69
645.0		13.50		58	12.30
1207	M5	12.33	+1.59	105	12.44

*Adapted from Gliese (1969), Gliese and Jahreiss (1979), and Grenon and Rufener (1981).

Catalogue of Nearby Stars*

Name	R.A.(1950) h m s	Dec.(1950) ° ′ ″	$\mu_{R.A.}$ (s yr^{-1})	$\mu_{Dec.}$ (″ yr^{-1})	V_r (km s^{-1})
1208	16 55 01	+21 31 48	+0.0022	- 0.579	+51
646.0A	16 55 26	- 39 29 12	+0.0236	+0.187	- 42
646.0B	16 55 26	- 39 29 12	+0.0236	+0.187	
647.0	16 55 28	+13 22 00	+0.0030	+0.090	
648.0	16 55 45	+65 12 42	+0.0371	+0.046	- 23
649.0	16 56 07	+25 49 36	- 0.0080	- 0.530	+6
649.1A	16 56 30	+47 26 18	- 0.0172	+0.261	- 6
649.1B	16 56 30	+47 26 18	- 0.0172	+0.261	
649.1C	16 56 19	+47 26 00	- 0.0150	+0.260	
650.0	16 58 22	- 13 29 24	- 0.0016	- 0.323	- 98
651.0	17 01 12	+47 08 24	+0.0117	+0.846	- 46
652.0	17 01 19	- 28 30 36	+0.0069	- 0.261	+15
652.1	17 01 58	+64 40 12	- 0.0089	+0.025	- 25
1209	17 02 08	+17 00 54	+0.0083	- 1.134	
653.0	17 02 27	- 04 59 00	- 0.0622	- 1.127	+34
654.0	17 02 37	- 05 00 42	- 0.0620	- 1.120	+31
654.1	17 02 44	+00 46 30	- 0.0006	- 0.339	- 18
654.2	17 04 00	+15 17 48	- 0.0072	- 0.084	- 25
654.3	17 04 36	- 41 39 18	- 0.0188	- 0.322	- 19
654.4	17 04 50	+88 41 48	+0.0071	+0.026	
655.0	17 05 01	+21 37 06	- 0.0370	+0.000	
1210	17 05 17	+07 26 18	- 0.0349	- 0.391	
1211	17 05 37	+03 01 42	- 0.0005	- 0.400	
656.0	17 05 40	- 60 40 24	+0.0129	+0.605	+9
656.1A	17 07 30	- 15 39 54	+0.0026	+0.095	- 1
656.1B	17 07 30	- 15 39 54	+0.0026	+0.095	
657.0	17 08 34	- 43 10 30	+0.0021	- 0.285	- 27
658.0	17 08 34	+24 35 24	- 0.0163	+0.223	- 56
659.0A	17 09 08	+54 33 24	+0.0101	- 0.116	+9
659.0B	17 09 10	+54 33 06	+0.0120	- 0.110	- 1
660.0A	17 09 18	- 01 47 18	- 0.0300	- 0.320	
660.0B	17 09 18	- 01 47 18	- 0.0300	- 0.320	
660.1	17 10 11	- 05 03 24	+0.0120	- 0.660	
661.0A	17 10 40	+45 44 48	+0.0240	- 1.570	- 21
661.0B	17 10 40	+45 44 48	+0.0240	- 1.570	

Catalogue of Nearby Stars*

Name	Sp LC	V (mag)	B–V (mag)	π_t (0.001″)	M_V (mag)
1208	DA	14.06	+0.25	44	12.1
646.0A	K5V	8.322	+1.16	74	7.67
646.0B		10.40		66	9.50
647.0		10.60		50	9.10
648.0	F6V	4.883	+0.48	59	3.74
649.0		9.72	+1.50	95	9.60
649.1A		7.79	+0.99	57	6.60
649.1B		11.10		57	9.90
649.1C		7.90	+1.00	57	6.70
650.0	(G3)	7.113	+0.56	59	5.97
651.0	G8V	6.773	+0.73	62	5.73
652.0	G8IV–V	6.567	+0.84	72	5.85
652.1	G5V	6.108		30	3.49
1209		12.26	+1.55	58	11.08
653.0	K5V	7.726	+1.16	92	7.54
654.0	M4V	10.051	+1.43	92	9.87
654.1	F9V	6.010	+0.58	49	4.46
654.2		7.05		59	5.90
654.3	K5V	8.285	+1.06	54	6.95
654.4	K0V	8.311		48	6.72
655.0		11.59		102	11.60
1210		14.02	+1.88	79	13.51
1211	DF–G	15.19	+0.46	58	14.01
656.0	K0V	7.380	+0.89	65	6.44
656.1A	A2.5V	2.398	+0.05	51	+0.94
656.1B		13.45		49	1.90
657.0	F0IV–N	3.314	+0.40	61	2.24
658.0	K2V	8.339	+0.87	35	6.06
659.0A		8.83	+1.16	49	7.30
659.0B		9.34	+1.26	49	7.80
660.0A		11.98	+1.60	81	11.50
660.0B		12.20		81	11.70
660.1		13.00		49	11.40
661.0A	M4V	9.364	+1.49	157	10.34
661.0B	M4V	10.33		155	11.28

*Adapted from Gliese (1969), Gliese and Jahreiss (1979), and Grenon and Rufener (1981).

Catalogue of Nearby Stars*

Name	R.A.(1950) h m s	Dec.(1950) ° ′ ″	$\mu_{R.A.}$ (s yr^{-1})	$\mu_{Dec.}$ (″ yr^{-1})	V_r (km s^{-1})
1212	17 10 59	− 08 21 24	− 0.0281	− 0.432	− 15
1213	17 11 26	+42 23 42	− 0.0915	− 0.369	+6
662.0A	17 12 11	− 38 32 00	− 0.0157	− 0.412	− 55
662.0B	17 12 11	− 38 32 00	− 0.0157	− 0.412	
663.0A	17 12 16	− 26 31 48	− 0.0369	− 1.132	− 1
663.0B	17 12 16	− 26 31 54	− 0.0345	− 1.141	
1214	17 12 49	+05 01 42	+0.0371	− 0.709	
664.0	17 13 09	− 26 28 36	− 0.0359	− 1.124	− 1
665.0A	17 14 39	− 69 59 30	− 0.0073	− 0.201	− 5
665.0B	17 14 39	− 69 59 30	− 0.0073	− 0.201	
665.1	17 15 04	− 24 01 12	+0.0072	− 0.071	− 15
666.0A	17 15 15	− 46 35 06	+0.0952	+0.218	+24
666.0B	17 15 15	− 46 35 06	+0.0952	+0.218	+21
1215	17 15 25	+11 43 42	− 0.0231	− 0.420	
667.0A	17 15 33	− 34 56 12	+0.0951	− 0.172	+1
667.0B	17 15 33	− 34 56 12	+0.0951	− 0.172	+2
667.0C	17 15 34	− 34 56 30	+0.0951	− 0.172	
667.1	17 15 40	− 75 17 42	− 0.2510	− 0.214	+59
668.0A	17 16 25	− 11 04 18	+0.0030	− 0.100	
668.0B	17 16 25	− 11 04 18	+0.0030	− 0.100	
668.1	17 17 19	− 05 51 54	+0.0024	− 0.182	− 33
669.0A	17 17 54	+26 32 48	− 0.0160	+0.390	− 28
669.0B	17 17 53	+26 32 48	− 0.0160	+0.390	
670.0A	17 18 00	− 21 03 42	+0.0165	− 0.209	− 9
670.0B	17 18 00	− 21 03 42	+0.0165	− 0.209	
671.0	17 18 17	+41 46 30	+0.0260	− 0.810	
672.0	17 18 47	+32 31 54	+0.0100	− 1.042	− 79
1216	17 19 26	+49 19 12	+0.0562	− 1.178	
1217	17 19 54	− 14 55 00	− 0.0067	− 0.208	− 56
672.1	17 21 00	− 32 13 00	− 0.0130	− 0.600	
673.0	17 23 16	+02 10 12	− 0.0391	− 1.193	− 24
673.1	17 23 19	− 24 07 54	+0.0000	− 0.116	− 37
1218	17 23 33	− 62 24 06	− 0.0440	− 0.942	+28
674.0	17 24 53	− 46 50 36	+0.0560	− 0.890	
675.0	17 25 09	+67 20 54	− 0.0921	+0.002	− 40

Catalogue of Nearby Stars*

Name	Sp LC	V (mag)	B–V (mag)	π_t (0.001″)	M_V (mag)
1212	dM1	12.05	+1.61	62	11.0
1213	dM1	10.07	+1.28	40	8.1
662.0A	G3IV	5.934	+0.58	43	4.10
662.0B		6.80			
663.0A	K0V	4.331	+0.86	182	5.63
663.0B		5.09		184	6.41
1214		14.67	+1.73	78	14.13
664.0	K5VE	6.323	+1.16	182	7.62
665.0A	G2IV–V	6.524	+0.60	33	4.20
665.0B		9.40		33	7.00
665.1	(G5)	6.583		48	5.00
666.0A	G8V	5.461	+0.80	132	6.06
666.0B		8.69		131	9.28
1215		15.12	+1.80	73	14.44
667.0A	K3V	5.888	+1.04	140	6.62
667.0B	K5V	7.20		140	7.90
667.0C	M2V	10.431		140	11.16
667.1	G2V	6.977	+0.60	46	5.29
668.0A		10.40		55	9.10
668.0B		12.50		55	11.20
668.1	G9V	6.312	+0.85	81	5.90
669.0A		11.36	+1.55	95	11.25
669.0B		12.92	+1.58	95	12.81
670.0A	F2V	4.368	+0.40	60	3.26
670.0B	(DK3)	8.50		60	7.40
671.0		11.47		82	11.00
672.0	G2V	5.386	+0.62	73	4.70
1216		14.40	+1.70	60	13.29
1217	K5V	10.84	+1.41	45	9.1
672.1		12.80		49	11.20
673.0	K7V	7.527	+1.36	130	8.10
673.1	A9V	4.134	+0.28	45	2.40
1218		12.73	+1.51		
674.0	M4V	9.367	+1.53	216	11.04
675.0	K0V	6.437	+0.76	75	5.81

*Adapted from Gliese (1969), Gliese and Jahreiss (1979), and Grenon and Rufener (1981).

Catalogue of Nearby Stars*

Name	R.A.(1950) h m s	Dec.(1950) ° ′ ″	$\mu_{R.A.}$ (s yr^{-1})	$\mu_{Dec.}$ (″ yr^{-1})	V_r (km s^{-1})
1219	17 25 27	+14 31 48	− 0.0774	− 0.322	
676.0A	17 26 15	− 51 35 42	− 0.0270	− 0.160	
676.0B	17 26 20	− 51 35 36	− 0.0270	− 0.160	
677.0A	17 27 24	+29 26 00	− 0.0152	− 0.283	− 9
677.0B	17 27 24	+29 26 00	− 0.0152	− 0.283	
678.0A	17 27 49	− 01 01 24	− 0.0080	− 0.170	− 77
678.0B	17 27 49	− 01 01 24	− 0.0080	− 0.170	
678.1A	17 27 55	+05 35 24	+0.0024	− 0.261	− 11
678.1B	17 27 55	+05 35 24	+0.0024	− 0.261	
679.0	17 30 15	+34 18 18	− 0.0191	+0.050	− 52
680.0	17 31 24	− 48 39 18	+0.0050	+0.470	
681.0	17 32 37	+12 35 42	+0.0080	− 0.227	+12
682.0	17 33 28	− 44 16 36	− 0.0650	− 0.920	
683.0	17 33 59	− 54 28 12	− 0.0041	− 0.148	− 3
683.1	17 34 18	− 42 32 00	+0.0162	− 0.346	+4
683.2A	17 34 26	− 37 49 48	− 0.0017	− 0.124	+3
683.2B	17 34 26	− 37 49 48	− 0.0017	− 0.124	
684.0A	17 34 28	+61 54 48	+0.0352	− 0.510	− 13
684.0B	17 34 28	+61 54 48	+0.0352	− 0.510	
685.0	17 35 02	+61 43 06	+0.0340	− 0.520	− 11
686.0	17 35 39	+18 36 24	+0.0660	+0.990	− 8
686.1A	17 35 44	+22 59 06	− 0.0120	− 0.130	+5
686.1B	17 35 44	+22 59 06	− 0.0120	− 0.130	+1
686.2	17 36 32	− 49 23 12	+0.0081	− 0.175	+4
687.0	17 36 42	+68 23 06	− 0.0658	− 1.258	− 22
688.0	17 36 48	+03 35 00	− 0.0120	− 0.100	+19
1220	17 36 50	+82 06 48	− 0.1382	+0.494	
689.0	17 37 02	+71 54 24	+0.0162	− 0.051	− 24
690.0A	17 39 18	+71 21 12	− 0.0287	+0.319	− 1
690.0B	17 39 18	+71 21 12	− 0.0287	+0.319	
690.1	17 39 39	− 16 36 30	− 0.0080	− 0.690	
691.0	17 40 10	− 51 48 36	− 0.0014	− 0.193	− 11
692.0	17 40 26	− 21 39 36	− 0.0068	− 0.046	+11
692.1	17 41 09	+21 38 24	− 0.0099	− 0.642	+20
693.0	17 42 24	− 57 16 54	− 0.1330	− 1.340	

Catalogue of Nearby Stars*

Name	Sp LC	V (mag)	B–V (mag)	π_t (0.001″)	M_V (mag)
1219		13.70	+1.77	46	12.01
676.0A		10.50		85	10.10
676.0B		14.80		85	14.40
677.0A	K8V	8.975	+1.14	55	7.68
677.0B		9.91		46	8.20
678.0A	G9IV–V	5.312	+0.72	52	3.89
678.0B		6.10		52	4.68
678.1A		9.32	+1.47	102	9.36
678.1B		14.00		102	14.00
679.0	G5V	6.543	+0.64	54	5.20
680.0		10.50		83	10.10
681.0	A5III	2.104	+0.15	60	0.99
682.0		11.20		213	12.80
683.0	A7V	5.235	+0.20	56	3.98
683.1	G5V	7.141	+0.65	45	5.41
683.2A	G8V	6.684	+0.77	53	5.40
683.2B		9.10		53	7.70
684.0A	G0V	5.215	+0.56	69	4.41
684.0B	(DM0.5)	8.06	+1.10	67	7.19
685.0	M1VE	9.992		69	9.19
686.0	M1V	9.621	+1.53	124	10.09
686.1A		10.04	+1.35	49	8.50
686.1B		10.23	+1.37	49	8.70
686.2	F2V	4.741	+0.40	43	2.91
687.0	M3.5V	9.156	+1.50	216	10.83
688.0	K3V	6.526	+0.96	81	6.07
1220		14.18	+1.77	71	13.44
689.0	K4V	8.586	+1.10	47	6.95
690.0A		9.20	+1.10	51	7.70
690.0B		13.00		51	11.50
690.1		14.00		56	12.70
691.0	G5V	5.113	+0.70	76	4.52
692.0	F5V	4.863	+0.47	60	3.75
692.1	K0V	7.509	+0.77	47	5.87
693.0		12.90		170	14.00

*Adapted from Gliese (1969), Gliese and Jahreiss (1979), and Grenon and Rufener (1981).

Catalogue of Nearby Stars*

Name	R.A.(1950) h m s	Dec.(1950) ° ′ ″	$\mu_{R.A.}$ (s yr^{-1})	$\mu_{Dec.}$ (″ yr^{-1})	V_r (km s^{-1})
694.0	17 42 25	+43 24 24	− 0.0010	− 0.620	− 23
694.1A	17 42 49	+72 10 24	+0.0033	− 0.267	− 10
694.1B	17 42 51	+72 10 54	+0.0041	− 0.278	− 10
694.2	17 44 11	+46 52 24	− 0.0020	− 0.020	+18
695.0A	17 44 30	+27 44 54	− 0.0233	− 0.748	− 16
695.0B	17 44 28	+27 44 42	− 0.0233	− 0.748	− 16
695.0C	17 44 28	+27 44 42	− 0.0233	− 0.748	
695.1	17 44 39	− 33 59 42	− 0.0181	− 0.554	+2
696.0	17 47 53	− 06 02 06	− 0.0010	− 0.140	− 21
1221	17 48 58	+70 52 24	− 0.2555	1.096	
697.0	17 51 22	+21 20 00	− 0.0059	+0.065	− 12
1222	17 51 55	+07 23 30	− 0.0405	− 0.294	
697.1	17 52 56	+03 45 42	− 0.0060	+0.040	+3
698.0	17 53 34	+18 30 24	− 0.0014	− 0.053	− 28
698.1	17 54 59	− 51 37 06	+0.0080	− 0.230	
699.0	17 55 23	+04 33 18	− 0.0500	+10.310−	−108
699.1	17 57 00	+82 44 00	− 0.7150	3.330−	+154
699.2	17 57 50	− 03 41 18	+0.0096	− 0.046	− 43
700.0A	18 00 18	− 59 12 42	− 0.0265	− 0.060	+3
700.0B	18 00 18	− 59 12 42	− 0.0265	− 0.060	
700.1A	18 00 21	− 08 10 54	+0.0017	− 0.039	− 40
700.1B	18 00 21	− 08 10 54	+0.0017	− 0.039	
700.1C	18 00 27	− 08 11 54	+0.0017	− 0.039	
700.2	18 00 29	+26 19 12	+0.0286	− 0.593	+23
1223	18 01 03	+37 31 48	+0.0136	− 1.149	
701.0	18 02 28	− 03 01 54	+0.0368	− 0.333	+34
702.0A	18 02 56	+02 30 36	+0.0172	− 1.094	− 7
702.0B	18 02 56	+02 30 36	+0.0172	− 1.094	
702.1	18 03 01	− 36 01 30	+0.0091	+0.014	+13
702.2	18 03 09	+04 39 24	− 0.0011	− 0.306−	+124
1224	18 04 42	− 15 58 00	− 0.0564	− 0.340	
703.0	18 05 04	+15 56 30	− 0.0042	− 0.202−	+104
704.0A	18 05 08	+30 33 12	− 0.0075	+0.071	+1
704.0B	18 05 08	+30 33 12	− 0.0075	+0.071	
705.0	18 05 10	+15 56 54	+0.0006	+0.010	− 4

Catalogue of Nearby Stars*

Name	Sp LC	V (mag)	B–V (mag)	π_t (0.001″)	M_V (mag)
694.0	M3V	10.477	+1.53	96	10.39
694.1A		4.58	+0.43	47	3.00
694.1B	F8V	5.784	+0.52	47	4.14
694.2	M1.5VE	10.717		42	8.83
695.0A	G5IV	3.444	+0.75	129	4.00
695.0B		10.33	+1.49	124	10.80
695.0C		10.79		124	11.26
695.1	(G5)	7.181	+0.71	51	5.72
696.0		10.10		77	9.50
1221		14.15	+0.40	169	15.30
697.0	K5V	8.526	+0.95	46	6.84
1222		13.11	+1.55	61	12.04
697.1		10.03	+1.46	57	8.80
698.0		9.22	+1.18	48	7.60
698.1		10.60		47	9.00
699.0	M4VI	9.520	+1.74	547	13.21
699.1	F3V	4.605	+0.33	45	2.87
699.2		4.62	+0.39	49	4.10
700.0A	G4IV	6.860	+0.63	20	3.30
700.0B		12.60		20	9.10
700.1A	(F2)	4.747	+0.39	59	3.60
700.1B	(F2)	15.94		45	4.20
700.1C		9.40		45	7.70
700.2	K0V	6.998	+0.80	49	5.45
1223		14.89	+1.79	86	14.56
701.0	(M2)	9.355	+1.52	136	10.02
702.0A	K0VE	4.003	+0.86	203	5.54
702.0B	K5VE	16.00		195	7.45
702.1	G5V	5.949	+0.62	59	4.80
702.2	G2V	6.798	+0.62	45	5.06
1224		13.64	+1.80	133	14.26
703.0	G6V	8.642	+0.66	72	7.93
704.0A	F7V	5.066	+0.52	59	3.92
704.0B	K5V	8.44		61	7.40
705.0	K2V	8.145	+1.32	55	6.85

*Adapted from Gliese (1969), Gliese and Jahreiss (1979), and Grenon and Rufener (1981).

Catalogue of Nearby Stars*

Name	R.A.(1950) h m s	Dec.(1950) ° ′ ″	$\mu_{R.A.}$ (s yr^{-1})	$\mu_{Dec.}$ (″ yr^{-1})	V_r (km s^{-1})
705.1	18 05 47	- 62 00 54	- 0.0110	+0.225	+30
705.2	18 06 48	+52 47 24	- 0.0470	- 0.070	
706.0	18 07 58	+38 27 12	- 0.0264	- 0.471	- 19
707.0	18 08 44	- 43 27 06	+0.0127	- 0.433	- 2
708.0	18 13 07	+18 28 54	+0.0040	+0.090	+8
708.1	18 13 36	+64 22 48	+0.0526	+0.032	- 35
708.2	18 13 45	+13 54 06	+0.0070	- 0.500	+11
708.3	18 13 48	+01 30 48	- 0.0280	- 0.630	
708.4	18 14 06	+45 11 36	- 0.0070	- 0.109	- 64
709.0	18 15 05	+45 32 00	- 0.0020	+0.320	
710.0	18 17 15	- 01 57 42	- 0.0006	- 0.022	- 23
1225	18 17 33	+68 32 48	- 0.1287	- 1.590	
710.1	18 17 48	- 29 51 06	+0.0030	- 0.029	- 20
1226	18 18 24	- 01 03 48	- 0.0338	- 0.954	+50
711.0	18 18 43	- 02 54 48	- 0.0370	- 0.697	+9
712.0	18 19 44	+06 18 54	- 0.0820	+0.170	
713.0	18 21 57	+72 42 42	+0.1169	- 0.356	+32
1227	18 22 05	+62 02 18	- 0.1325	- 1.238	
713.1	18 24 53	- 25 27 06	- 0.0032	- 0.185	- 43
714.0	18 25 51	- 58 18 06	- 0.0050	- 0.430	
715.0	18 27 16	- 01 51 00	+0.0116	- 0.207	- 53
716.0	18 28 23	- 18 56 30	- 0.0099	- 0.195	- 46
1228	18 29 22	+54 45 00	- 0.0278	+0.268	
1229	18 29 52	+13 41 48	+0.0081	+0.122	
717.0	18 30 42	- 11 40 18	- 0.0219	- 0.247	- 84
718.0	18 31 12	+22 16 54	- 0.0131	- 0.470	+37
719.0	18 32 45	+51 41 00	+0.0197	- 0.311	- 27
720.0A	18 33 50	+45 41 48	+0.0450	+0.340	- 25
720.0B	18 33 59	+45 42 54	+0.0450	+0.340	
720.1	18 33 54	- 28 33 18	+0.0020	- 0.135	+25
721.0	18 35 15	+38 44 12	+0.0171	+0.285	- 14
722.0	18 35 54	- 21 05 42	- 0.0057	- 0.152	+36
722.1	18 36 28	+42 37 12	+0.0260	+0.069	+32
723.0	18 37 32	- 10 30 18	- 0.0090	- 0.540	
724.0	18 38 08	- 13 25 06	- 0.0050	- 0.680	

Catalogue of Nearby Stars*

Name	Sp LC	V (mag)	B–V (mag)	π_t (0.001″)	M_V (mag)
705.1	G0V	5.460	+0.58	52	4.04
705.2		12.50		52	11.10
706.0	K2V	6.391	+0.87	92	6.21
707.0	K7VE	8.365	+1.32	67	7.50
708.0	M1V	10.069		62	9.03
708.1	F5V	5.003	+0.40	47	3.36
708.2		10.10		50	8.60
708.3		12.51	+1.62	66	11.60
708.4	G0V	6.295	+0.65	46	4.61
709.0	M0V	10.288		38	8.19
710.0	K7V	9.647		71	8.90
1225		15.38	+1.88	59	14.23
710.1	K3III	2.712	+1.38	45	+0.98
1226		12.71	+1.66	68*	11.9
711.0	K0III–IV	3.252	+0.94	56	1.99
712.0		13.80		64	12.80
713.0	F7V	3.539	+0.49	130	4.11
1227		13.40	+1.79	122	13.83
713.1	K1III	2.841	+1.04	49	1.29
714.0		11.90		70	11.10
715.0	(K5)	8.031	+1.10	54	6.69
716.0	K3V	6.803	+0.84	78	6.26
1228	DIP	15.58:	+0.46:	50*	14.6
1229	G8V	7.2		46*	5.5
717.0	M0V	9.989	+1.28	54	8.65
718.0	K4V	8.910	+1.12	59	7.76
719.0	K3V	8.198		64	7.23
720.0A	M2V	9.859	+1.42	82	9.40
720.0B		14.90		82	14.50
720.1	K1V	6.770	+0.98	80	6.30
721.0	A0V	10.059		124	0.53
722.0	(DG4)	5.839	+0.66	73	5.16
722.1	K0VE	8.321	+0.82	44	6.54
723.0		11.40		66	10.50
724.0		10.70		55	9.40

*Adapted from Gliese (1969), Gliese and Jahreiss (1979), and Grenon and Rufener (1981).

Catalogue of Nearby Stars*

Name	R.A.(1950) h m s	Dec.(1950) ° ′ ″	$\mu_{R.A.}$ (s yr^{-1})	$\mu_{Dec.}$ (″ yr^{-1})	V_r (km s^{-1})
1230A	18 39 04	+24 44 12	+0.0379	+0.063	
1230B	18 39 04	+24 44 18	+0.0379	+0.063	
724.1	18 39 48	− 50 12 48	− 0.0073	− 0.169	− 32
724.2	18 40 10	− 77 55 24	− 0.0030	+0.183	+15
725.0A	18 42 12	+59 33 18	− 0.1751	+1.869	
725.0B	18 42 13	+59 33 00	− 0.1794	+1.818	+10
725.1	18 43 05	+43 46 48	− 0.0119	+0.026	
725.2	18 43 31	+20 29 48	− 0.0008	− 0.334	+24
725.3	18 44 39	− 50 45 24	− 0.0019	− 0.189	− 6
726.0	18 44 50	− 03 41 30	− 0.0082	− 0.247	+18
727.0	18 46 07	+10 41 42	+0.0096	− 0.439	− 18
728.0	18 46 40	+17 23 12	− 0.0292	− 0.401	− 18
729.0	18 46 45	− 23 53 30	+0.0510	− 0.170	− 4
730.0	18 47 31	+03 02 06	− 0.0100	− 0.430	+4
731.0	18 49 38	+16 31 48	− 0.0180	− 0.460	− 11
732.0A	18 50 30	− 38 40 00	+0.0260	− 0.950	
732.0B	18 50 30	− 38 40 00	+0.0260	− 0.950	
732.1	18 50 28	+52 54 36	− 0.0054	+0.272	+2
733.0	18 51 31	− 22 38 30	+0.0091	− 0.325	
734.0A	18 52 33	+10 54 36	+0.0080	− 0.010	− 17
734.0B	18 52 33	+10 54 36	+0.0080	− 0.010	
735.0	18 53 03	+08 20 18	+0.0070	− 0.080	− 3
736.0	18 53 12	+04 12 06	− 0.0007	− 0.083	+18
737.0A	18 53 21	− 56 03 06	+0.0019	− 0.463	− 14
737.0B	18 53 21	− 56 03 06	+0.0019	− 0.463	− 16
738.0A	18 55 09	+32 50 12	+0.0133	− 0.157	− 47
738.0B	18 55 09	+32 50 12	+0.0133	− 0.157	
739.0	18 55 20	− 48 20 12	+0.0140	− 0.480	
740.0	18 55 34	+05 51 24	− 0.0139	− 1.218	+15
740.1	19 00 10	+00 47 00	− 0.0025	− 0.125	− 12
741.0	19 00 29	− 13 38 06	− 0.0390	− 0.490	
742.0	19 00 39	+70 35 06	+0.0190	+0.520	
743.0	19 03 00	+22 59 48	+0.0162	+0.235	
743.1A	19 03 02	− 37 08 12	+0.0079	− 0.273	− 52
743.1B	19 03 02	− 37 08 12	+0.0079	− 0.273	− 53

Catalogue of Nearby Stars*

Name	Sp LC	V (mag)	B–V (mag)	π_t (0.001″)	M_V (mag)
1230A	K–M	12.4	+1.69	131	13.0
1230B		14.4:		131	15.0
724.1	G5V	9.292	+0.69	53	7.91
724.2	G2V	6.447	+0.60	45	4.70
725.0A	M4V	8.938	+1.54	283	11.20
725.0B	M5V	9.708	+1.59	285	11.98
725.1	K0V	7.522		49	5.97
725.2	F6V	4.188	+0.46	49	2.64
725.3	G3V	8.660	+0.65	54	7.32
726.0	(M0P)	8.790	+1.28	72	8.08
727.0	K4V	7.946	+1.08	60	6.84
728.0	M1V	9.202	+1.29	59	8.06
729.0		10.60		345	13.30
730.0		10.72	+1.49	66	9.83
731.0	M2V	10.170	+1.57	79	9.66
732.0A		12.00		75	11.40
732.0B		16.00		75	15.00
732.1		5.50	+0.84		
733.0		10.70		56	9.40
734.0A		9.52	+1.37	56	8.26
734.0B		12.00		56	10.70
735.0		10.07	+1.75	92	9.90
736.0	G8V	8.008	+0.90	67	7.14
737.0A	K5VE	8.832	+1.44	74	8.18
737.0B	K7VE	9.80		63	8.80
738.0A	F9.5V	5.202	+0.59	54	3.86
738.0B		7.74		54	6.40
739.0		11.00		83	10.60
740.0		9.23	+1.45	85	8.87
740.1	(DG5)	8.355		54	7.02
741.0		14.60		71	13.90
742.0	DA	13.162	+0.05	80	12.68
743.0	G9V	8.532	+0.82	54	7.19
743.1A	F8V	4.191	+0.52	54	2.85
743.1B	(F8)	15.01		58	3.80

*Adapted from Gliese (1969), Gliese and Jahreiss (1979), and Grenon and Rufener (1981).

Catalogue of Nearby Stars*

Name	R.A.(1950) h m s	Dec.(1950) ° ′ ″	$\mu_{R.A.}$ (s yr^{-1})	$\mu_{Dec.}$ (″ yr^{-1})	V_r (km s^{-1})
743.2	19 03 18	+25 50 42	- 0.0021	- 0.035	- 74
744.0	19 03 30	- 37 53 00	- 0.0157	- 0.350	+58
745.0A	19 04 58	+20 48 48	- 0.0350	- 0.330	+34
745.0B	19 05 05	+20 48 06	- 0.0350	- 0.330	+35
746.0	19 05 43	+16 46 36	+0.0040	- 0.306	+14
747.0A	19 05 45	+32 27 00	+0.0960	+1.090	- 31
747.0B	19 05 45	+32 27 00	+0.0960	+1.090	
1231	19 06 15	+26 30 36	- 0.0151	- 0.501	
747.1	19 06 29	- 14 49 24	+0.0110	- 0.480	
747.2	19 07 04	+33 58 48	- 0.0055	+0.063	+13
747.3	19 07 10	- 47 14 00	- 0.0050	- 0.630	+2
1232	19 07 40	+17 35 30	- 0.0424	- 0.457	
747.4	19 08 19	- 55 57 00	- 0.0330	- 0.180	
748.0	19 09 38	+02 48 36	+0.1180	- 0.450	- 40
748.1	19 11 01	+76 28 42	+0.0127	- 0.121	- 4
1233	19 11 16	+57 34 48	+0.0262	+0.397	
748.2A	19 12 26	+02 04 18	+0.0230	+0.400	
748.2B	19 12 26	+02 04 18	+0.0230	+0.400	
749.0	19 12 30	- 24 16 00	+0.0085	- 0.095	- 26
750.0A	19 13 03	- 45 58 24	+0.0200	- 0.420	- 37
750.0B	19 13 03	- 45 58 24	+0.0200	- 0.420	
751.0	19 13 14	+24 48 18	+0.0180	+0.210	- 52
752.0A	19 14 29	+05 05 48	- 0.0390	- 1.350	+33
752.0B	19 14 32	+05 04 42	- 0.0310	- 1.420	
753.0	19 15 27	+88 13 48	- 0.1181	+0.157	- 10
754.0	19 17 07	- 45 36 42	+0.0580	- 2.870	
1234	19 17 15	+38 38 00	+0.0022	- 0.249	
754.1A	19 17 53	- 07 45 36	- 0.0040	- 0.190	
754.1B	19 17 51	- 07 45 18	- 0.0040	- 0.190	
754.2	19 17 53	+37 14 24	- 0.0059	- 0.186	+2
755.0	19 18 12	- 35 04 36	+0.0086	- 0.098	- 10
755.1	19 19 23	+14 34 54	+0.0000	- 0.090	+35
1235	19 19 32	+20 47 30	- 0.0675	- 1.459	
1236	19 19 39	+06 57 12	- 0.0498	- 0.394	
756.0	19 19 49	+28 34 00	+0.0660	+0.250	

Catalogue of Nearby Stars*

Name	Sp LC	V (mag)	B–V (mag)	π_t (0.001″)	M_V (mag)
743.2	K2V	7.193	+1.03	75	6.60
744.0	G5IV	6.144	+0.72	62	5.11
745.0A	(DM2)	10.769	+1.57	114	11.05
745.0B	(DM2)	10.770	+1.57	114	11.05
746.0	G5V	6.079	+0.71	59	4.93
747.0A		11.77		122	12.20
747.0B		12.07		122	12.50
1231		15.19	+1.91	59	14.04
747.1		13.12		48	11.50
747.2	K6V	9.428	+1.25	36	7.21
747.3	K7V	9.337	+1.31	49	7.79
1232		13.54	+1.85	95	13.43
747.4		13.60		54	12.30
748.0		11.17		98	11.13
748.1	F2V	5.118	+0.31	46	3.43
1233	G8V	7.03	+0.79	41	5.1
748.2A		11.50		46	9.80
748.2B		12.00		46	10.30
749.0	(DF9)	6.239	+0.54	40	4.30
750.0A	K9V	9.384	+1.42	70	8.61
750.0B		10.10		70	9.30
751.0	M0V	9.722		12	5.12
752.0A	M3V	9.128	+1.50	170	10.28
752.0B		7.38	+2.12	173	18.57
753.0		9.96	+1.06	29	7.30
754.0		13.70		175	14.90
1234	DC	14.56	+0.44	86	14.25
754.1A		12.38	+0.02	104	12.32
754.1B		12.75	+1.31	104	12.83
754.2	G8V	6.300	+0.67	41	4.36
755.0	G5V	6.483	+0.63	46	4.80
755.1		3.01	+0.06	79	12.50
1235		13.42	+1.71	100	13.42
1236		12.32	+1.69	93	12.16
756.0		11.50		55	10.20

*Adapted from Gliese (1969), Gliese and Jahreiss (1979), and Grenon and Rufener (1981).

Catalogue of Nearby Stars*

Name	R.A.(1950) h m s	Dec.(1950) ° ′ ″	$\mu_{R.A.}$ (s yr^{-1})	$\mu_{Dec.}$ (″ yr^{-1})	V$_r$ (km s^{-1})
756.1	19 20 14	− 66 34 18	− 0.0385	+0.220	+38
756.2	19 20 50	+07 25 42	+0.0002	− 0.003	
757.0	19 21 36	− 22 09 18	− 0.0160	− 0.450	
758.0	19 21 41	+33 07 18	+0.0066	+0.168	− 20
759.0	19 22 35	+11 50 12	+0.0492	+0.640	+100
760.0	19 22 59	+03 00 48	+0.0169	+0.083	− 30
1237	19 25 02	+49 21 12	+0.0462	+0.696	− 66
1238	19 25 40	+75 26 42	+0.0994	+0.579	
761.0A	19 25 56	+12 26 00	− 0.0042	− 0.050	− 16
761.0B	19 25 56	+12 26 00	− 0.0042	− 0.050	
761.1	19 27 26	+31 30 36	− 0.0008	− 0.415	− 12
1239	19 29 14	− 35 33 42	− 0.0239	+0.155	
1240	19 29 20	− 11 23 00	+0.0155	+0.028	
761.2	19 30 04	+00 28 12	+0.0160	+0.070	− 39
762.0	19 30 20	− 62 57 00	− 0.0550	− 0.330	
762.1	19 30 18	+58 29 00	− 0.0668	− 0.392	+11
762.2	19 31 16	+21 44 00	− 0.0011	− 0.197	+2
763.0	19 32 09	+04 28 06	+0.0336	+0.308	− 52
764.0	19 32 28	+69 34 36	+0.1098	− 1.744	+27
764.1A	19 34 02	− 10 33 06	− 0.0188	− 0.270	+68
764.1B	19 34 02	− 10 33 06	− 0.0188	− 0.270	
764.2	19 34 45	− 14 24 48	− 0.0075	− 0.142	− 21
765.0A	19 35 06	+50 06 18	− 0.0025	+0.256	− 28
765.0B	19 35 06	+50 06 18	− 0.0025	+0.256	
1241	19 35 10	+27 36 24	+0.0331	+0.023	
1242	19 39 24	+03 02 30	− 0.0178	− 0.481	
765.1A	19 40 29	+50 24 30	− 0.0162	− 0.151	− 26
765.1B	19 40 32	+50 24 00	− 0.0146	− 0.160	− 28
765.2	19 40 40	+76 18 12	+0.0409	+0.138	− 9
765.3	19 42 16	+57 53 48	+0.0166	− 0.059	− 22
765.4A	19 43 39	+33 29 06	+0.0014	− 0.431	+6
765.4B	19 43 39	+33 29 06	+0.0014	− 0.431	
766.0A	19 43 43	+27 01 12	− 0.0020	− 1.250	
766.0B	19 43 43	+27 01 12	− 0.0020	− 1.250	
767.0A	19 44 26	+31 53 54	+0.0369	− 0.409	

Catalogue of Nearby Stars*

Name	Sp LC	V (mag)	B–V (mag)	π_t (0.001″)	M_V (mag)
756.1	K5V	8.386	+1.04	39	6.34
756.2	K5V	8.136		72	7.40
757.0		10.70		58	9.50
758.0	K0V	6.355	+0.81	57	5.13
759.0	G8IV	5.164	+0.78	60	4.05
760.0	F0IV	3.357	+0.32	70	2.58
1237	K3V	8.03	+0.93	43	6.2
1238		15.36	+1.94	92	15.18
761.0A		9.20	+1.10	44	7.40
761.0B		14.20		44	12.40
761.1	(DG5)	6.958	+0.72	49	5.41
1239	K5V	8.71	+1.11	50*	7.2
1240	K2V	7.57	+0.92	58*	6.4
761.2	M0V	10.428	+1.42	48	8.83
762.0		13.80		58	12.60
762.1	K1V	6.615	+0.87	39	4.57
762.2	G5V	6.892	+0.70	46	5.20
763.0	M1V	9.343	+1.39	78	8.80
764.0	K0V	4.673	+0.80	175	5.89
764.1A	K2V	8.375	+1.01	48	6.78
764.1B	(DK5)	10.20		43	8.40
764.2	(DF6)	5.451	+0.50	45	3.72
765.0A	F4V	4.475	+0.39	56	3.22
765.0B		13.00		56	11.70
1241	DA	12.98	+0.17	57	11.76
1242		12.88	+1.58	44	11.10
765.1A	G3V	5.973	+0.64	36	3.75
765.1B	G4V	6.20	+0.66	47	4.60
765.2	K0V	8.066	+0.88	50	6.56
765.3	F8V	6.301	+0.55	45	4.57
765.4A	K3V	7.670	+0.98	47	6.03
765.4B		8.50		47	6.90
766.0A		12.60		101	12.60
766.0B		13.60		101	13.60
767.0A		10.28	+1.48	77	9.72

*Adapted from Gliese (1969), Gliese and Jahreiss (1979), and Grenon and Rufener (1981).

Catalogue of Nearby Stars*

Name	R.A.(1950) h m s	Dec.(1950) ° ′ ″	$\mu_{R.A.}$ (s yr^{-1})	$\mu_{Dec.}$ (″ yr^{-1})	V_r (km s^{-1})
767.0B	19 44 26	+31 53 54	+0.0369	− 0.409	− 6
767.1A	19 44 32	+33 36 36	+0.0014	− 0.445	+5
767.1B	19 44 34	+33 36 48	+0.0021	− 0.438	+1
768.0	19 48 21	+08 44 06	+0.0362	+0.387	− 26
768.1A	19 48 38	+10 17 24	+0.0162	− 0.138	− 1
768.1B	19 48 38	+10 17 24	+0.0162	− 0.138	0
1243	19 49 37	+46 21 00	+0.0168	+0.268	
769.0	19 50 22	− 47 55 36	− 0.0130	− 1.050	
1244	19 50 37	− 72 29 30	− 0.0046	− 0.299	
770.0	19 51 18	− 24 04 00	− 0.0097	− 0.412	− 7
1245A	19 52 16	+44 17 30	+0.0414	− 0.591	
1245B	19 52 17	+44 17 30	+0.0414	− 0.591	
770.1	19 52 47	− 26 26 00	+0.0155	+0.083	− 21
771.0A	19 52 51	+06 16 48	+0.0029	− 0.479	− 40
771.0B	19 52 51	+06 16 48	+0.0029	− 0.479	
1246	19 53 26	− 31 28 12	+0.0319	+0.029	
772.0	19 53 56	− 01 10 00	− 0.0300	− 0.690	+81
773.0	19 54 33	− 12 41 42	− 0.0056	− 0.509	− 2
1247	19 54 35	− 55 04 18	+0.0143	− 0.132	
773.1	19 55 07	− 15 37 30	+0.0010	− 0.098	− 2
773.2	19 55 13	+29 41 06	+0.0073	+0.247	− 23
773.3	19 57 04	− 10 05 24	− 0.0186	− 0.398	+23
773.4	19 57 07	− 33 50 18	+0.0108	− 0.305	− 10
773.5	19 58 26	− 22 52 36	+0.0012	− 0.008	+8
773.6	19 58 50	− 50 11 12	+0.0375	− 0.368	+19
774.0A	19 59 20	− 65 43 42	+0.0140	− 0.850	
774.0B	19 59 20	− 65 43 42	+0.0140	− 0.850	
1248	20 01 24	+05 51 54	− 0.0327	− 0.780	
775.0	20 00 17	+03 11 00	− 0.0062	+0.115	− 28
775.1	20 00 17	+15 27 36	− 0.0108	− 0.587	+12
776.0	20 00 34	− 67 27 12	+0.1460	− 0.684	− 12
776.1	20 01 20	− 45 48 00	− 0.0560	+0.030	
777.0A	20 01 34	+29 45 42	+0.0521	− 0.527	− 43
777.0B	20 01 23	+29 43 54	+0.0520	− 0.530	
778.0	20 01 47	+23 12 42	− 0.0734	− 0.914	− 3

Catalogue of Nearby Stars*

Name	Sp LC	V (mag)	B–V (mag)	π_t (0.001″)	M_V (mag)
767.0B		1.30		77	10.70
767.1A	F5V	4.997	+0.46	47	3.36
767.1B	K6V	8.593	+1.04	47	6.95
768.0	A7V	0.780	+0.22	197	2.25
768.1A	F8V	5.129	+0.56	40	3.14
768.1B		3.10		53	11.70
1243		12.83	+1.64	85	12.48
769.0		12.50		35	10.20
1244	K5V	8.48	+1.07	48*	6.9
770.0	K2V	6.160	+1.01	73	5.48
1245A		13.41	+1.90	211	15.01
1245B		13.99	+1.98	211	15.61
770.1	(DG5)	4.686	+0.75	61	3.61
771.0A	G8IV	3.736	+0.86	76	3.14
771.0B	M3V	11.40		70	10.60
1246	K4V	8.3		59*	7.2
772.0		3.69	+0.26	92	13.51
773.0	(K5)	9.289	+1.30	45	7.55
1247	K3V	8.60	+1.08	47*	7.0
773.1	A2IV	4.989	+0.05	28	2.22
773.2	K0VE	7.895	+0.80	47	6.26
773.3	G0V	5.875	+0.58	40	3.89
773.4	F8V	5.653	+0.49	48	4.06
773.5	(DG7)	5.994		85	5.60
773.6	K5V	8.669	+1.13	48	7.08
774.0A		13.10		75	12.50
774.0B		14.40		75	13.80
1248		12.08	+1.58	80	11.60
775.0	K4VE	7.464	+1.12	71	6.90
775.1	G8V	7.154	+0.72	49	5.60
776.0	G2V	6.073	+0.65	52	4.65
776.1		13.60		47	12.00
777.0A	G6IV	5.745	+0.72	56	4.49
777.0B		14.40	+1.57	53	13.00
778.0	K1V	7.284	+0.82	51	5.82

*Adapted from Gliese (1969), Gliese and Jahreiss (1979), and Grenon and Rufener (1981).

Catalogue of Nearby Stars*

Name	R.A.(1950) h m s	Dec.(1950) ° ′ ″	$\mu_{R.A.}$ (s yr^{-1})	$\mu_{Dec.}$ (″ yr^{-1})	V_r (km s^{-1})
779.0	20 01 51	+16 56 00	- 0.0277	- 0.411	+4
779.1	20 02 04	+25 38 54	- 0.0057	- 0.034	- 7
1249	20 03 20	+38 20 00	+0.0219	+0.109	- 24
780.0	20 03 50	- 66 18 42	+0.1988	- 1.141	- 22
781.0	20 03 55	+54 18 12	- 0.1410	- 0.900	
781.1A	20 05 00	- 31 54 00	+0.0230	- 0.730	
781.1B	20 05 00	- 31 54 00	+0.0230	- 0.730	
1250	20 06 21	+33 09 00	+0.0266	+0.372	
781.2	20 06 48	- 14 26 00	+0.0076	- 0.048	- 14
781.3	20 07 30	- 21 55 00	+0.0090	- 0.260	+35
782.0	20 07 26	- 20 38 12	- 0.0313	- 0.356	+19
783.0A	20 07 55	- 36 13 42	+0.0371	- 1.568–	+130
783.0B	20 07 55	- 36 13 42	+0.0371	- 1.568	
783.1A	20 08 34	+43 47 42	+0.0004	+0.083	- 40
783.1B	20 08 34	+43 47 42	+0.0004	+0.083	
783.2A	20 08 50	+16 02 00	- 0.0287	+0.404	- 51
783.2B	20 08 57	+16 01 48	- 0.0320	+0.420	
784.0	20 10 19	- 45 18 48	+0.0731	- 0.133	- 30
784.1	20 11 30	+13 14 12	+0.0290	+0.000	
784.2A	20 11 32	+06 32 30	- 0.0190	- 0.600	
784.2B	20 11 29	+06 34 00	- 0.0190	- 0.600	
785.0	20 12 10	- 27 11 00	+0.0932	- 0.178	- 54
786.0	20 12 24	+77 04 48	+0.0270	+0.511	- 2
786.1	20 14 11	+42 49 30	+0.0010	- 0.060	
787.0	20 16 07	- 46 35 06	- 0.0358	- 0.090	+30
788.0	20 17 02	+66 41 36	+0.0784	+0.299	- 5
788.1	20 18 35	- 58 26 24	+0.0950	- 0.360	
788.2	20 18 38	- 06 35 24	+0.0130	- 0.490	
788.3	20 19 04	- 42 12 30	+0.0032	- 0.091	- 10
789.0	20 20 32	+14 23 24	+0.0051	+0.001	+2
1251	20 21 33	- 76 49 54	+0.1895	- 1.274	+122
1252	20 23 46	- 56 35 42	+0.0503	- 1.210	+6
790.0	20 24 38	- 31 01 36	- 0.0007	- 0.521	- 1
791.0	20 24 40	- 27 54 06	- 0.0150	- 0.870	
1253	20 24 58	+58 24 00	+0.0328	+0.553	

Catalogue of Nearby Stars*

Name	Sp LC	V (mag)	B–V (mag)	π_t (0.001″)	M_V (mag)
779.0	G1V	5.783	+0.61	58	4.60
779.1	K3V	7.808	+0.91	51	6.40
1249	G5V	6.16	+0.64	44	4.4
780.0	G8V	3.535	+0.76	175	4.75
781.0	M3VE	12.015		60	10.91
781.1A		12.90		51	11.40
781.1B		13.30		51	11.80
1250		14.88	+1.82	47	13.24
781.2	K5V	9.743	+1.12	47	8.10
781.3		14.20		58	13.00
782.0	K7V	8.893	+1.29	63	7.89
783.0A	K3V	5.302	+0.87	177	6.54
783.0B		11.50		177	12.70
783.1A	G5V	7.414	+0.63	46	5.73
783.1B		8.42	+0.70	30	5.80
783.2A	K1V	7.337	+0.85	36	5.12
783.2B		13.94	+1.62	47	12.30
784.0	M0V	7.961	+1.41	164	9.04
784.1		12.50		71	11.80
784.2A		13.20	+1.54	50	11.70
784.2B		15.67	+0.40	50	14.20
785.0	K0V	5.728	+0.88	120	6.12
786.0	M0V	8.866	+1.36	64	7.90
786.1	M0VP	9.985		46	8.70
787.0	K5V	8.716	+1.16	63	7.71
788.0	G3V	5.909	+0.58	67	5.04
788.1		12.30		51	10.80
788.2		13.10		49	11.60
788.3	A0V	5.558		30	2.94
789.0	F8V	6.169		40	4.18
1251		13.81	+1.72	98*	13.8
1252		12.23	+1.45	55*	10.9
790.0	G5V	6.608	+0.73	57	5.39
791.0		11.60		79	11.10
1253		14.04	+1.79	108	14.21

*Adapted from Gliese (1969), Gliese and Jahreiss (1979), and Grenon and Rufener (1981).

Catalogue of Nearby Stars*

Name	R.A.(1950) h m s	Dec.(1950) ° ′ ″	$\mu_{R.A.}$ (s yr^{-1})	$\mu_{Dec.}$ (″ yr^{-1})	V_r (km s^{-1})
791.1A	20 26 01	− 17 58 48	− 0.0011	− 0.020	+18
791.1B	20 26 01	− 17 58 48	− 0.0011	− 0.020	
791.2	20 27 21	+09 31 12	+0.0470	+0.100	
791.3	20 29 33	+33 36 18	+0.0127	− 0.004	− 25
792.0	20 29 34	+38 22 54	+0.0180	+0.720	
792.1A	20 29 40	− 10 01 30	+0.0206	+0.103	+9
792.1B	20 29 40	− 10 01 30	+0.0206	+0.103	0
793.0	20 29 50	+65 16 36	+0.0710	+0.280	+24
793.1	20 31 06	+41 43 12	− 0.0147	+0.450	− 10
794.0	20 32 14	+24 54 00	− 0.0300	− 0.550	+65
1254	20 32 40	+61 34 00			
794.1	20 34 04	− 47 28 00	+0.0055	+0.070	− 1
794.2	20 35 55	− 60 43 06	+0.0421	− 0.569	− 32
794.3	20 36 47	+38 27 42	+0.0157	− 0.187	− 21
795.0A	20 37 05	+04 47 36	+0.0576	+0.087	− 41
795.0B	20 37 05	+04 47 36	+0.0576	+0.087	
796.0	20 37 13	− 23 57 30	+0.0360	+0.464	− 44
1255AB	20 38 03	+75 25 00	+0.0885	+0.557	
1255C	20 38 03	+75 25 00	+0.0885	+0.557	
1256	20 38 10	+15 18 42	+0.0911	+0.672	
797.0A	20 38 29	+19 45 12	+0.0087	+0.306	− 37
797.0B	20 38 29	+19 43 06	+0.0080	+0.320	
798.0	20 38 37	− 52 51 48	+0.0040	− 1.060	− 43
799.0A	20 38 44	− 32 36 36	+0.0216	− 0.331	− 4
799.0B	20 38 44	− 32 36 36	+0.0216	− 0.331	− 3
1257	20 38 45	− 22 29 24	+0.0452	− 0.545	
799.1	20 39 41	− 20 15 18	+0.0220	− 0.090	
800.0A	20 40 04	− 19 05 12	+0.0460	− 0.890	+5
800.0B	20 40 04	− 19 05 12	+0.0460	− 0.890	
1258	20 41 03	+35 19 30	− 0.0181	− 0.579	+42
801.0	20 41 07	− 21 32 00	+0.0060	− 0.280	
802.0	20 41 53	+55 08 48	+0.0950	+1.740	− 23
803.0	20 42 04	− 31 31 06	+0.0212	− 0.348	+1
804.0	20 42 06	+19 34 30	+0.0010	− 0.570	+9
805.0	20 43 08	− 25 27 06	− 0.0036	− 0.156	+26

Catalogue of Nearby Stars*

Name	Sp LC	V (mag)	B–V (mag)	π_t (0.001″)	M_V (mag)
791.1A	F2IV	4.750	+0.37	47	3.11
791.1B		10.00		39	8.00
791.2		13.06	+1.71	106	13.20
791.3	K8V	9.240	+1.13	49	7.69
792.0		15.30		66	14.40
792.1A	G2.5IV	5.657	+0.69	32	3.18
792.1B		1.30		48	9.70
793.0		10.43	+1.59	127	10.95
793.1	G9V	7.068	+0.79	51	5.61
794.0		1.53	- 0.06	73	10.85
1254		12.51	+1.52	63	11.51
794.1	K0III	3.136	+1.00	45	1.40
794.2	F8V	5.108	+0.53	47	3.47
794.3	G2V	6.764	+0.61	48	5.17
795.0A	K5V	7.860	+1.22	61	6.79
795.0B		9.30		62	8.30
796.0	G8V	6.364	+0.72	70	5.59
1255AB	K0V	7.3	+0.85	47	
1255C		10.3			
1256	dM4–5	13.41	+1.72	102	13.45
797.0A	G5V	6.441	+0.62	49	4.89
797.0B		13.40		48	11.80
798.0	K7V	8.838	+1.31	85	8.49
799.0A	M4.5E	10.241	+1.58	114	10.53
799.0B	M4.5E	10.90		113	11.20
1257	K0	9.1		48	7.5
799.1		12.00		100	12.00
800.0A	M2V	10.734		67	9.86
800.0B		14.00		67	13.10
1258	dM3	11.5		40	9.5
801.0	M0V	9.862		50	8.70
802.0		5.10		73	14.40
803.0	M0VE	8.701	+1.44	108	8.87
804.0		10.20	+1.43	59	9.00
805.0	F5V	4.117	+0.43	96	4.03

*Adapted from Gliese (1969), Gliese and Jahreiss (1979), and Grenon and Rufener (1981).

Catalogue of Nearby Stars*

Name	R.A.(1950) h m s	Dec.(1950) ° ′ ″	$\mu_{R.A.}$ (s yr^{-1})	$\mu_{Dec.}$ (″ yr^{-1})	V_r (km s^{-1})
806.0	20 43 18	+44 18 42	+0.0400	+0.290	– 15
806.1A	20 44 11	+33 46 54	+0.0285	+0.329	– 11
806.1B	20 44 05	+33 46 48	+0.0285	+0.329	
807.0	20 44 16	+61 38 36	+0.0128	+0.822	– 87
808.0	20 44 30	– 79 29 00	+0.2450	– 0.990	
808.1	20 45 06	– 44 10 18	+0.0172	– 0.104	– 15
808.2	20 48 04	+29 11 54	+0.0027	– 0.041	– 14
808.3	20 50 22	– 11 45 48	+0.0032	+0.047	– 1
1255–A	20 52 04	+74 34 54	+0.0987	+0.565	– 30
809.0	20 52 18	+61 58 30	– 0.0010	– 0.760	– 15
1259	20 52 42	+12 58 48	+0.0375	+0.384	– 32
810.0A	20 52 48	– 14 13 12	+0.0970	– 0.430	
810.0B	20 52 47	– 14 15 00	+0.0970	– 0.430	
811.0	20 53 49	– 26 29 18	+0.0070	– 0.063	– 16
811.1	20 54 04	– 10 37 36	– 0.0050	– 1.140	+51
812.0A	20 54 07	– 05 02 12	+0.0530	– 0.230	– 32
812.0B	20 54 06	– 05 02 00	+0.0530	– 0.350	
812.1	20 54 22	– 44 18 30	– 0.0490	– 0.956	– 10
813.0	20 55 09	+22 10 24	+0.0600	– 0.150	
814.0	20 56 59	+00 52 06	+0.0145	+0.116	– 20
815.0A	20 58 09	+39 52 42	+0.0540	– 0.260	– 57
815.0B	20 58 09	+39 52 42	+0.0540	– 0.260	
816.0	20 59 20	– 06 30 36	– 0.0150	– 0.460	
816.1A	21 00 53	+45 41 06	+0.0352	+0.149	– 12
816.1B	21 00 53	+45 41 06	+0.0352	+0.149	
816.2A	21 01 34	– 20 03 12	– 0.0025	– 0.035	+24
816.2B	21 01 34	– 20 03 12	– 0.0025	– 0.035	
817.0	21 02 09	– 17 07 48	– 0.0690	– 2.070	
818.0	21 02 51	+06 52 36	+0.0054	– 0.561	– 63
818.1A	21 04 11	– 73 22 18	+0.1020	– 0.334	– 10
818.1B	21 04 11	– 73 22 18	+0.1020	– 0.334	
818.1C	21 04 11	– 73 22 18	+0.1020	– 0.334	
819.0A	21 04 24	– 14 07 24	+0.0267	– 0.027	– 34
819.0B	21 04 24	– 14 07 24	+0.0267	– 0.027	
820.0A	21 04 40	+38 30 00	+0.3523	3.185	– 64

Catalogue of Nearby Stars*

Name	Sp LC	V (mag)	B–V (mag)	π_t (0.001″)	M_V (mag)
806.0	M3V	10.799		86	10.47
806.1A	K0IIIE	2.501	+1.03	46	0.81
806.1B		13.40	+1.66	46	11.70
807.0	K0IVE	3.423	+0.92	73	2.74
808.0		13.20		62	12.20
808.1	F1V	5.086	+0.36	49	3.54
808.2	K5V	8.330	+1.06	55	7.10
808.3	(DG1)	6.371	+0.65	46	4.70
1255–A	G8V	7.76	+0.72	45	6.0
809.0	M2VE	8.550	+1.49	134	9.19
1259	K3V	8.81	+1.05	45	7.1
810.0A		14.50		80	14.00
810.0B		15.70		80	15.20
811.0	F8V	5.686	+0.50	39	3.64
811.1		1.40		56	10.10
812.0A		1.87	+1.49	66	11.00
812.0B		16.68	+1.13	66	15.80
812.1	G0V	6.504	+0.59	54	5.17
813.0		11.97		78	11.40
814.0	G4VE	8.598	+0.67	54	7.26
815.0A	M3VE	10.169	+1.49	70	9.39
815.0B		12.30		80	11.80
816.0		11.00		82	10.60
816.1A	K2.5V	7.666	+0.97	47	6.03
816.1B		13.00		52	11.60
816.2A	A3M	4.830	+0.18	50	3.32
816.2B		16.50		50	5.00
817.0		12.40		55	11.10
818.0	K7V	8.306	+1.22	52	6.89
818.1A	G3IV	5.668	+0.59	46	3.98
818.1B		16.50		46	4.80
818.1C		13.50		46	11.80
819.0A	K1VE	7.102	+0.90	69	6.30
819.0B		10.20		66	9.30
820.0A	K5VE	5.229	+1.17	294	7.57

*Adapted from Gliese (1969), Gliese and Jahreiss (1979), and Grenon and Rufener (1981).

Catalogue of Nearby Stars*

Name	R.A.(1950) h m s	Dec.(1950) ° ′ ″	$\mu_{R.A.}$ (s yr^{-1})	$\mu_{Dec.}$ (″ yr^{-1})	V_r (km s^{-1})
820.0B	21 04 40	+38 30 00	+0.3523	3.185	
820.1	21 05 20	- 82 01 00	+0.0400	- 0.360	
821.0	21 06 30	- 13 28 42	+0.0490	- 1.960	
821.1	21 10 40	+15 36 12	- 0.0050	- 0.210	
1260	21 11 34	+26 09 24	+0.0072	- 0.388	
822.0A	21 12 03	+09 48 12	+0.0031	- 0.301	- 15
822.0B	21 12 03	+09 48 12	+0.0031	- 0.301	
822.1A	21 12 48	+37 49 54	+0.0134	+0.437	- 21
822.1B	21 12 48	+37 49 54	+0.0134	+0.437	
822.1C	21 12 47	+37 48 24	+0.0140	+0.470	
822.2	21 13 45	+25 13 30	- 0.0046	- 0.032	- 29
823.0	21 13 57	+62 37 42	+0.0158	+0.245	
824.0	21 14 05	+09 11 06	+0.0110	- 0.123	- 14
825.0	21 14 20	- 39 03 42	- 0.2805	- 1.152	+21
825.1	21 14 47	- 61 33 24	+0.0671	- 0.431	- 20
825.2	21 15 11	- 43 32 42	+0.0209	+0.032	+11
825.3	21 15 28	+00 02 48	+0.0301	- 0.179	- 26
825.4A	21 16 52	- 26 33 36	- 0.0402	- 0.352	- 30
825.4B	21 16 52	- 26 33 36	- 0.0402	- 0.352	+2
1261	21 17 23	+53 59 48	- 0.0134	+0.243	
826.0	21 17 23	+62 22 24	+0.0215	+0.052	- 10
826.1	21 17 25	- 20 03 18	- 0.0125	- 0.724	+20
1262	21 20 19	- 68 26 42	+0.0267	+0.163	+12
826.2	21 21 30	- 46 55 00	+0.0710	- 0.090	
827.0	21 22 20	- 65 35 36	+0.0154	+0.797	- 29
827.1	21 23 18	- 56 20 42	+0.0818	+0.132	- 41
828.0A	21 23 45	- 45 01 42	+0.0234	+0.181	+8
828.0B	21 24 00	- 45 01 00	+0.0220	+0.160	
828.1	21 24 11	+03 31 12	- 0.0050	- 0.060	+2
828.2	21 24 38	- 07 03 24	- 0.0010	- 0.420	
828.3	21 24 50	+55 00 18	+0.0300	+0.190	
828.4	21 25 45	+10 23 30	+0.0087	+0.033	
828.5	21 26 43	+73 25 54	+0.0110	- 0.310	+8
829.0	21 27 12	+17 25 06	+0.0700	+0.380	+1
830.0	21 27 16	- 12 43 36	+0.0699	- 0.265	- 81

Catalogue of Nearby Stars*

Name	Sp LC	V (mag)	B–V (mag)	π_t (0.001″)	M_V (mag)
820.0B	K7VE	6.069	+1.37	294	8.41
820.1		13.50	+0.27		
821.0		10.70		93	10.50
821.1		16.30		51	14.80
1260	DA	14.8		52*	13.4
822.0A	F8V	4.500	+0.50	55	3.20
822.0B	F8V	15.25		55	3.95
822.1A	F0IV	3.738	+0.40	54	2.40
822.1B		6.42		50	4.90
822.1C		12.00	+1.53	50	10.50
822.2	G5IV	6.988	+0.89	55	5.70
823.0	(G5)	9.546		66	8.64
824.0		7.95	+1.02	64	7.00
825.0	M0VE	6.676	+1.38	260	8.75
825.1	G5V	6.585	+0.69	47	4.95
825.2	G5V	6.739	+0.65	46	5.05
825.3	(K2)	8.168	+0.99	46	6.48
825.4A	G5V	6.559	+0.74	54	5.22
825.4B		9.50		54	8.20
1261	DA	12.31:	– 0.10:	55*	11.2
826.0	A7IV–V	2.458	+0.22	64	1.49
826.1	(K5)	9.076	+1.35	49	7.53
1262	G5V	6.97	+0.73	45*	5.2
826.2		13.90		53	12.50
827.0	F6V	4.231	+0.49	117	4.57
827.1	K3V	8.657	+0.93	48	7.06
828.0A	K0V	7.481	+0.91	68	6.64
828.0B		14.10		60	13.00
828.1		10.57	+1.36	46	8.90
828.2		11.08	+1.49	53	9.70
828.3		14.66	+0.15	58	13.50
828.4	K0V	8.770		51	7.31
828.5		2.88	+0.01	45	11.20
829.0	M4VE	10.423		148	11.27
830.0	M0V	9.075	+1.29	61	8.00

*Adapted from Gliese (1969), Gliese and Jahreiss (1979), and Grenon and Rufener (1981).

Catalogue of Nearby Stars*

Name	R.A.(1950) h m s	Dec.(1950) ° ′ ″	$\mu_{R.A.}$ (s yr^{-1})	$\mu_{Dec.}$ (″ yr^{-1})	V_r (km s^{-1})
831.0	21 28 34	– 10 00 36	+0.0790	– 0.060	
832.0	21 30 14	– 49 13 12	– 0.0060	– 0.810	+8
833.0	21 33 17	– 51 04 06	+0.0451	– 0.221	+28
834.0A	21 34 38	+39 14 00	– 0.0190	– 0.160	
834.0B	21 34 38	+39 14 00	– 0.0190	– 0.160	
835.0	21 35 45	+27 29 54	+0.0320	– 0.030	– 13
835.1	21 36 00	– 77 36 48	+0.0176	– 0.239	+34
836.0	21 36 06	– 24 22 30	– 0.0730	– 0.680	
836.1	21 36 15	– 27 31 54	+0.0282	– 0.079	– 18
836.2	21 36 22	+82 49 36	+0.1650	+0.560	
836.3	21 39 02	– 41 21 00	+0.0179	– 0.282	+58
836.4	21 39 25	– 12 23 00	– 0.0040	– 0.690	
836.5	21 40 22	+20 46 42	– 0.0170	– 0.650	
836.6A	21 41 54	+28 31 00	+0.0200	– 0.238	+18
836.6B	21 41 54	+28 31 00	+0.0200	– 0.238	+16
836.7	21 42 07	+14 32 36	+0.0183	– 0.089	– 19
836.8	21 42 30	+41 22 00	– 0.0040	+0.060	– 21
836.9A	21 43 02	– 57 55 18	+0.0100	– 0.926	– 7
836.9B	21 43 02	– 57 55 18	+0.0100	– 0.926	
1263	21 44 04	– 00 24 00	+0.0503	– 0.509	+39
837.0	21 44 17	– 16 21 18	+0.0182	– 0.294	
1264	21 44 34	– 72 19 54	+0.0733	– 0.301	
837.1	21 44 58	– 07 58 06	+0.0210	– 0.160	
838.0	21 45 01	– 47 31 54	+0.0162	– 0.295	– 7
838.1A	21 46 40	+05 29 24	+0.0370	– 0.025	– 11
838.1B	21 47 04	+05 25 00	+0.0360	– 0.010	
838.2	21 49 05	+00 36 54	+0.0213	– 0.064	– 28
838.3A	21 49 52	+42 06 48	– 0.0148	– 0.322	– 27
838.3B	21 49 52	+42 06 48	– 0.0148	– 0.322	
838.4	21 49 53	+02 09 30	– 0.0010	– 0.330	
838.5	21 50 34	– 13 47 18	+0.0212	+0.014	– 22
838.6	21 52 00	– 47 14 00	– 0.0330	– 0.370	
839.0	21 51 55	+41 32 48	+0.0350	– 0.360	– 35
840.0	21 52 32	+32 05 42	+0.0160	– 0.238	– 16
841.0A	21 53 35	– 51 14 24	– 0.0060	– 0.390	

Catalogue of Nearby Stars*

Name	Sp LC	V (mag)	B–V (mag)	π_t (0.001″)	M_V (mag)
831.0		11.95	+1.64	134	12.59
832.0	M1V	8.648	+1.46	214	10.30
833.0	K2V	7.132	+0.89	74	6.48
834.0A	M0V	10.192	+1.53	77	9.60
834.0B		12.20		77	11.60
835.0	M0VE	9.901	+1.49	72	9.19
835.1	M0III	3.770	+0.99	51	2.31
836.0		14.60		75	14.00
836.1	G4IV–V	6.746	+0.62	45	5.01
836.2		13.02	- 0.02	47	11.40
836.3	K5V	8.813	+1.04	52	7.39
836.4		14.30		46	12.60
836.5		13.23	+0.17		
836.6A	F4V	4.509	+0.49	46	2.82
836.6B	F3V	6.20		46	4.50
836.7	G0VE	5.980	+0.58	66	5.08
836.8		9.63	+1.34	50	8.10
836.9A	K7V	8.778	+1.32	52	7.36
836.9B		9.60		55	8.30
1263		12.70	+1.61	80	12.22
837.0	AM	2.849	+0.29	69	2.04
1264	M2Ve	9.80	+1.46	100*	9.8
837.1		14.80	- 0.06	52	13.40
838.0	G2V	5.557	+0.60	73	4.87
838.1A	K3V	8.667	+1.04	44	6.88
838.1B		14.80	+1.63	44	13.00
838.2	K1V	8.612	+0.83	48	7.02
838.3A	G5V	7.852	+0.78	47	6.21
838.3B		12.30		46	10.60
838.4		12.77			0.0
838.5	F0V	5.063	+0.38	47	3.42
838.6		12.10		89	11.90
839.0	M2V	10.339	+1.40	50	8.83
840.0	K0VE	7.748	+0.92	44	5.97
841.0A		10.40		69	9.60

*Adapted from Gliese (1969), Gliese and Jahreiss (1979), and Grenon and Rufener (1981).

Catalogue of Nearby Stars*

Name	R.A.(1950) h m s	Dec.(1950) ° ′ ″	$\mu_{R.A.}$ (s yr^{-1})	$\mu_{Dec.}$ (″ yr^{-1})	V$_r$ (km s^{-1})
841.0B	21 53 35	− 51 14 24	− 0.0060	− 0.390	
841.1	21 54 04	+21 00 06	+0.0001	+0.020	+2
842.0	21 55 56	− 59 59 24	+0.1150	− 0.090	
842.1	21 56 19	− 04 36 36	− 0.0003	− 0.252	− 44
842.2	21 57 58	+75 20 54	+0.0590	+0.030	− 11
843.0	21 59 12	− 19 43 30	+0.0650	+0.030	
844.0	21 59 24	+16 13 24	+0.0270	+0.150	
845.0	21 59 33	− 56 59 36	+0.4822	− 2.555	− 40
846.0	21 59 39	+01 09 42	− 0.0320	− 0.270	+20
847.0A	21 59 55	− 70 09 54	+0.1180	− 0.050	
847.0B	21 59 55	− 70 09 54	+0.1180	− 0.050	
847.1	22 03 38	− 45 38 00	+0.0357	− 0.289	− 23
848.0	22 04 41	+25 06 00	+0.0220	+0.028	− 5
848.1A	22 04 55	− 51 27 42	+0.0390	− 0.140	
848.1B	22 04 34	− 51 21 42	+0.0360	− 0.130	
848.1C	22 04 33	− 51 21 54	+0.0360	− 0.130	
848.2	22 05 05	− 47 12 12	+0.0128	− 0.149	+11
848.3A	22 06 40	− 08 09 42	+0.0020	− 0.680	
848.3B	22 06 40	− 08 09 42	+0.0020	− 0.680	
848.4	22 06 52	− 07 47 18	+0.0057	− 0.444	− 24
849.0	22 07 00	− 04 53 12	+0.0730	− 0.020	− 12
849.1	22 07 13	− 32 47 42	+0.0344	+0.015	− 15
850.0	22 09 01	+36 00 48	+0.0028	− 0.234	− 20
851.0	22 09 05	+18 10 36	+0.0230	+0.210	− 42
851.1	22 09 53	+31 19 12	− 0.0270	− 0.390	+16
1265	22 10 56	− 17 55 48	+0.0578	− 0.333	+24
851.2	22 11 36	− 41 37 12	+0.0502	− 0.788	− 19
851.3	22 11 56	− 16 03 42	− 0.0003	− 0.356	+12
851.4	22 12 13	+56 52 18	+0.0140	+0.040	
851.5	22 12 14	+27 35 54	− 0.0190	+0.550	+17
852.0A	22 14 42	− 09 03 00	− 0.0320	− 0.280	+54
852.0B	22 14 42	− 09 03 00	− 0.0320	− 0.280	
852.1	22 14 53	+15 06 36	+0.0150	− 0.060	
853.0A	22 15 00	− 53 52 06	+0.0485	− 0.664	− 14
853.0B	22 15 00	− 53 52 06	+0.0485	− 0.664	

Catalogue of Nearby Stars*

Name	Sp LC	V (mag)	B–V (mag)	π_t (0.001″)	M_V (mag)
841.0B		12.50		69	11.70
841.1	K5V	6.386			
842.0		11.40		79	10.90
842.1	(DK2)	6.217	+1.00	33	3.81
842.2		10.00		50	8.50
843.0		13.30		79	12.80
844.0	M4V	10.696	+1.58	71	9.95
845.0	K5V	4.698	+1.05	290	7.01
846.0		9.17	+1.47	102	9.20
847.0A		10.50		47	8.90
847.0B		14.10		47	12.50
847.1	K2V	8.428	+0.85	55	7.13
848.0	F5V	3.774	+0.44	76	3.18
848.1A		11.20		48	9.60
848.1B		14.00		48	12.40
848.1C		14.30		48	12.70
848.2	B5V	1.759	− 0.13	57	0.54
848.3A		13.90		47	12.30
848.3B		15.50		47	13.90
848.4	(G0)	6.548	+0.75	47	4.91
849.0		10.42	+1.52	112	10.67
849.1	F6V	4.920	+0.49	52	3.50
850.0	K0V	7.235	+0.82	51	5.77
851.0	M2V	10.238		78	9.70
851.1		10.00		47	8.40
1265		13.60	+1.72	65*	12.7
851.2	G5V	6.206	+0.65	46	4.52
851.3	G8V	6.534	+0.89	34	4.19
851.4	K0V	9.846		48	8.25
851.5	M0.5VE	10.359	+1.49	37	8.20
852.0A		13.50		103	13.60
852.0B		14.50		103	14.60
852.1		15.00		53	13.60
853.0A	G1V	5.362	+0.60	82	4.93
853.0B		10.00		82	9.60

*Adapted from Gliese (1969), Gliese and Jahreiss (1979), and Grenon and Rufener (1981).

Catalogue of Nearby Stars*

Name	R.A.(1950) h m s	Dec.(1950) ° ′ ″	$\mu_{R.A.}$ (s yr^{-1})	$\mu_{Dec.}$ (″ yr^{-1})	V$_r$ (km s^{-1})
1266	22 15 04	+70 41 340	+0.1805	- 0.031	
854.0	22 15 08	+68 05 24	+0.0657	- 0.002	- 7
1267	22 19 04	- 54 49 00	- 0.0179	+0.268	
855.0	22 20 13	- 57 28 06	+0.0740	- 0.320	
855.1A	22 20 22	- 72 30 00	+0.2888	- 0.689	+25
855.1B	22 20 22	- 72 30 00	+0.2888	- 0.689	
856.0A	22 21 12	+32 12 30	+0.0190	- 0.210	
856.0B	22 21 12	+32 12 30	+0.0190	- 0.210	
857.0	22 21 38	- 58 02 48	+0.0184	- 0.339	+8
857.1A	22 22 24	+22 17 54	- 0.0143	- 0.068	- 1
857.1B	22 22 24	+22 17 54	- 0.0143	- 0.068	
1268	22 22 56	+51 44 48	+0.0324	+0.399	
858.0	22 23 18	- 21 05 12	- 0.0340	- 0.610	
859.0A	22 23 52	- 16 59 48	+0.0154	+0.000	- 6
859.0B	22 23 51	- 16 59 48	+0.0181	- 0.010	- 3
1269A	22 25 39	+11 59 36	+0.0143	+0.004	
1269B	22 25 39	+11 59 36	+0.0143	+0.004	
860.0A	22 26 13	+57 26 48	- 0.0970	- 0.350	- 26
860.0B	22 26 13	+57 26 48	- 0.0970	- 0.350	
861.0	22 26 16	+05 34 06	+0.0320	- 1.610-	+157
862.0	22 26 25	- 30 15 48	+0.0165	- 0.805	+7
1270	22 27 32	+41 13 06	+0.1067	+0.462	
862.1	22 28 41	- 06 48 36	+0.0114	- 0.102	- 14
863.0	22 30 31	+09 07 06	+0.0350	+0.160	+9
863.1A	22 30 50	+53 32 06	+0.1650	+0.000	- 2
863.1B	22 30 50	+53 32 06	+0.1650	+0.000	
863.1C	22 30 50	+53 32 06	+0.1650	+0.000	
863.2	22 31 58	- 20 57 54	+0.0158	- 0.142	- 2
863.3	22 32 57	- 54 52 00	+0.0418	- 0.255	+67
864.0	22 33 35	- 01 05 36	+0.0040	- 0.590	+21
865.0	22 34 57	- 65 38 12	+0.1330	- 0.170	
866.0	22 35 45	- 15 35 36	+0.1620	+2.270	- 60
867.0A	22 36 01	- 20 52 48	+0.0325	- 0.056	- 9
867.0B	22 36 01	- 20 52 48	+0.0325	- 0.056	
867.1A	22 36 55	- 12 52 24	+0.0160	- 0.152	- 11

Catalogue of Nearby Stars*

Name	Sp LC	V (mag)	B–V (mag)	π_t (0.001″)	M_V (mag)
1266		12.12	+1.47	46	10.43
854.0		9.00		50	7.50
1267	M0V	9.06	+1.35	72*	8.4
855.0		10.73	+1.51	57	9.50
855.1A	G0IV	5.265	+0.65	32	2.79
855.1B		6.10		45	4.40
856.0A		11.39	+1.57	48	9.80
856.0B		11.60		48	10.00
857.0	G4V	5.313	+0.67	57	4.09
857.1A	K7VE	8.810	+1.19	46	7.12
857.1B		12.30		53	10.90
1268		14.94	+1.81	64	13.97
858.0		13.70		25	10.70
859.0A	(DG1)	5.548		59	4.40
859.0B	(DG1)	6.44		54	5.10
1269A	K0V	7.15	+0.90	41	5.2
1269B		10.2		41	9.0
860.0A	M2V	9.606	+1.62	254	11.63
860.0B	M6VE	11.30	+1.80	253	13.30
861.0		4.19	+1.42	57	13.00
862.0	K5V	7.630	+1.10	76	7.03
1270		13.25	+1.65	73	12.57
862.1		6.14	+0.56	46	4.40
863.0		10.37	+1.50	63	9.40
863.1A	(DM1)	10.097	+1.34	43	8.26
863.1B		10.80		39	8.80
863.1C		14.00		39	12.00
863.2	F3V	5.186	+0.44	45	3.45
863.3	G5V	7.557	+0.67	45	5.82
864.0	(K5)	9.976	+1.42	66	9.07
865.0		13.00		87	12.70
866.0	MV	12.18	+1.96	305	14.60
867.0A	(M2E)	9.062	+1.49	109	9.25
867.0B		11.45	+1.62	120	11.80
867.1A	G9E	7.737	+0.78	47	6.10

*Adapted from Gliese (1969), Gliese and Jahreiss (1979), and Grenon and Rufener (1981).

Catalogue of Nearby Stars*

Name	R.A.(1950) h m s	Dec.(1950) ° ′ ″	$\mu_{R.A.}$ (s yr^{-1})	$\mu_{Dec.}$ (″ yr^{-1})	V_r (km s^{-1})
867.1B	22 36 55	− 12 52 24	+0.0160	− 0.152	− 7
867.1C	22 37 02	− 12 50 54	+0.0150	− 0.170	
868.0	22 37 55	− 29 56 06	+0.0308	− 0.030	+1
869.0	22 38 06	− 32 15 06	+0.0294	+0.038	− 21
870.0	22 38 18	+42 45 12	+0.0080	− 0.130	− 35
871.0A	22 39 39	− 47 28 06	+0.0008	− 0.321	+17
871.0B	22 39 39	− 47 28 06	+0.0008	− 0.321	
1271	22 40 08	+17 24 00	+0.0779	+0.544	
871.1A	22 42 30	− 33 31 00	+0.0160	− 0.100	
871.1B	22 42 30	− 33 31 00	+0.0160	− 0.100	
1272A	22 42 58	+10 55 54	+0.0008	− 0.170	+5
1272B	22 42 58	+10 55 54	+0.0008	− 0.170	
871.2	22 44 11	+49 56 48	+0.0262	+0.030	− 51
872.0A	22 44 12	+11 55 00	+0.0159	− 0.493	− 5
872.0B	22 44 12	+11 55 00	+0.0159	− 0.493	
873.0	22 44 40	+44 04 36	− 0.0640	− 0.460	− 2
874.0	22 45 25	− 37 03 00	+0.0620	− 0.280	
1273	22 46 39	+22 20 30	+0.0367	+0.036	+94
875.0	22 47 43	− 07 21 24	− 0.0080	+0.120	− 8
1274	22 48 14	+34 35 12	+0.0677	+0.304	
1275	22 48 56	+29 23 42	+0.0956	+0.154	
875.1	22 49 30	+31 29 24	+0.0400	− 0.060	
876.0	22 50 35	− 14 31 12	+0.0650	− 0.630	+9
1276	22 51 09	− 07 02 18	+0.1659	− 0.708	
877.0	22 52 12	− 75 42 42	− 0.2790	− 1.000	
878.0	22 52 26	+60 43 42	− 0.0920	− 0.100	
878.1A	22 53 12	− 08 05 18	+0.0380	− 0.058	− 24
878.1B	22 53 11	− 08 06 06	+0.0400	− 0.010	
1277	22 53 25	− 60 18 18	− 0.0691	− 0.927	+68
879.0	22 53 37	− 31 49 48	+0.0256	− 0.158	+10
880.0	22 54 10	+16 17 24	− 0.0717	− 0.269	− 21
881.0	22 54 54	− 29 53 18	+0.0258	− 0.161	+6
882.0	22 55 00	+20 30 00	+0.0145	+0.063	− 31
883.0	22 57 16	− 11 38 54	+0.0150	− 0.030	− 20
884.0	22 57 38	− 22 47 36	− 0.0658	+0.065	+18

Catalogue of Nearby Stars*

Name	Sp LC	V (mag)	B–V (mag)	π_t (0.001″)	M_V (mag)
867.1B		8.55		34	6.20
867.1C		14.65	+1.61	34	12.30
868.0	K5VE	7.824	+1.13	70	7.05
869.0	G8V	7.391	+0.80	61	6.32
870.0		9.81	+1.11	31	7.20
871.0A	G1V	5.973	+0.58	66	5.07
871.0B		10.00		52	8.60
1271		11.7		51	10.2
871.1A		13.00		60	11.90
871.1B		14.40		60	13.30
1272A	dK6	9.98	+1.13	52	8.6
1272B		12		52	11
871.2	K0V	7.867		47	6.23
872.0A	F6IV–V	4.190	+0.50	48	2.60
872.0B	M1V	11.70		48	10.10
873.0	M4.5VE	10.048	+1.60	201	11.56
874.0		13.40		49	11.90
1273	DA	14.35	+0.20	52	12.93
875.0	(M1E)	9.863		67	8.99
1274		11.72	+1.54	50	10.21
1275	DR	15.54	+0.64	49	13.99
875.1		11.80		57	10.60
876.0	M5V	10.177	+1.60	209	11.78
1276	DKp	15.65	+1.88	125	16.13
877.0		11.69		122	12.10
878.0		14.20		72	13.50
878.1A	G6V	7.995	+0.64	48	6.40
878.1B		16.50	+0.42	32	14.00
1277		14.08	+1.45	100*	14.1
879.0	K4V	6.471	+1.10	129	7.02
880.0	M2.5VE	8.692	+1.51	144	9.48
881.0	A3V	1.157	+0.09	149	2.02
882.0	G2.5IV	5.458	+0.67	73	4.77
883.0		10.80		55	9.50
884.0	K7V	7.869	+1.39	130	8.44

*Adapted from Gliese (1969), Gliese and Jahreiss (1979), and Grenon and Rufener (1981).

Catalogue of Nearby Stars*

Name	R.A.(1950) h m s	Dec.(1950) ° ′ ″	$\mu_{R.A.}$ (s yr^{-1})	$\mu_{Dec.}$ (″ yr^{-1})	V_r (km s^{-1})
885.0A	22 57 45	- 26 34 42	+0.0080	- 0.160	+12
885.0B	22 57 45	- 26 34 42	+0.0080	- 0.160	
886.0	22 59 15	- 04 06 54	+0.0267	- 0.217	- 46
886.1A	23 00 18	+42 29 18	+0.0048	+0.000	+2
886.1B	23 00 18	+42 29 18	+0.0048	+0.000	
886.2	23 00 44	- 35 01 12	+0.0058	+0.085	- 7
1278	23 02 32	+66 29 42	+0.0541	- 0.099	
887.0	23 02 39	- 36 08 30	+0.5594	+1.310	+10
888.0A	23 03 43	+42 03 42	+0.0050	- 0.260	
888.0B	23 03 43	+42 03 42	+0.0050	- 0.260	
889.0	23 04 26	- 23 25 36	+0.0114	- 0.302	+17
889.1	23 05 32	+03 03 18	+0.0270	+0.310	
890.0	23 05 41	- 15 40 48	+0.0040	+0.010	+2
1279	23 06 33	- 68 00 06	- 0.0574	- 0.201	
1280	23 07 08	- 69 06 54	+0.0164	+0.328	
891.0	23 07 32	- 26 12 12	+0.0530	+0.020	
1281	23 08 03	- 19 28 42	+0.0035	- 1.419	+55
891.1	23 10 15	+49 08 00	+0.0092	+0.102	+12
892.0	23 10 52	+56 53 30	+0.2530	+0.299	- 18
893.0	23 11 16	+39 08 54	+0.0210	- 0.340	- 32
893.1	23 11 51	- 06 49 06	- 0.0240	- 0.170	
893.2A	23 13 16	- 09 21 36	+0.0249	- 0.012	- 26
893.2B	23 13 14	- 09 21 06	+0.0257	- 0.006	- 25
893.2C	23 13 14	- 09 21 06	+0.0257	- 0.006	
1282	23 13 58	- 62 16 24	+0.0239	- 0.030	
893.3	23 14 00	- 41 03 36	+0.0000	- 0.260	
893.4	23 14 11	+19 21 00	- 0.0150	- 0.190	
894.0	23 14 51	- 42 27 54	+0.0270	- 0.050	
894.1	23 15 56	+46 00 48	+0.0340	+0.120	
1283	23 16 10	- 60 47 36	- 0.0116	- 0.029	
894.2A	23 16 29	- 13 43 54	+0.0202	- 0.098	+11
894.2B	23 16 29	- 13 43 36	+0.0210	- 0.088	+10
894.3	23 17 24	- 05 26 24	+0.0002	- 0.003	+25
894.4	23 19 11	+43 49 12	+0.0590	+0.225	+2
894.5	23 20 27	- 11 02 30	+0.0299	+0.258	+36

Catalogue of Nearby Stars*

Name	Sp LC	V (mag)	B–V (mag)	π_t (0.001″)	M_V (mag)
885.0A	(M0E)	9.598		46	8.60
885.0B		10.50		46	8.80
886.0	K4V	7.483		58	6.30
886.1A	A3VN	5.121	+0.08	50	3.62
886.1B		8.80			
886.2	F0IV	5.102	+0.29	50	3.60
1278	dM1	9.89	+1.40	34	7.6
887.0	M2V	7.352	+1.46	279	9.58
888.0A		11.40		40	9.40
888.0B		12.00		40	10.00
889.0	(M0)	9.576	+1.34	50	8.20
889.1		10.90		47	9.20
890.0	M2.5V	10.905		50	9.10
1279	K5V	8.39	+1.20	74*	7.7
1280	K4V	8.78	+1.02	45*	7.0
891.0		12.40		69	11.60
1281		12.47	+1.49	56*	11.2
891.1	A8III	4.526	+0.30	47	2.89
892.0	K3V	5.558	+1.00	146	6.38
893.0	(SDF6)	10.963	+0.52	14	6.20
893.1		15.50		76	14.90
893.2A	K0III	4.256	+1.11	47	2.62
893.2B	K3V	9.172	+1.05	47	7.53
893.2C		10.10		33	7.70
1282	F8V	5.65	+0.51	49*	4.1
893.3		13.20		49	11.70
893.4		11.20		48	9.60
894.0		10.20		62	9.20
894.1	M1V	10.919		48	9.33
1283	K4V	8.97	+1.15	51*	7.5
894.2A	G5IV–V	5.076	+0.79	39	3.03
894.2B		7.60	+0.91	48	6.00
894.3		11.50	– 0.30	60	10.40
894.4	K1V	7.364	+0.80	55	6.07
894.5	K2V	7.793	+0.89	60	6.68

*Adapted from Gliese (1969), Gliese and Jahreiss (1979), and Grenon and Rufener (1981).

Catalogue of Nearby Stars*

Name	R.A.(1950) h m s	Dec.(1950) ° ′ ″	$\mu_{R.A.}$ (s yr^{-1})	$\mu_{Dec.}$ (″ yr^{-1})	V$_r$ (km s^{-1})
895.0	23 22 14	+57 35 00	− 0.0030	− 0.310	− 5
895.1	23 22 55	− 45 53 06	+0.0450	− 0.040	
895.2	23 26 16	+04 58 30	− 0.0310	− 0.310	+34
895.3	23 26 49	− 47 18 36	+0.0220	− 0.070	− 4
1284	23 27 35	− 20 39 48	+0.0220	− 0.185	
895.4	23 28 56	+58 53 18	+0.1413	+0.114	− 26
1285	23 29 09	− 03 01 42	+0.0088	− 0.123	
896.0A	23 29 20	+19 39 42	+0.0390	− 0.020	− 1
896.0B	23 29 20	+19 39 42	+0.0390	− 0.020	− 4
897.0A	23 30 09	− 17 01 30	+0.0220	− 0.230	
897.0B	23 30 09	− 17 01 30	+0.0220	− 0.230	
898.0	23 30 12	− 17 07 06	+0.0218	− 0.234	+1
899.0	23 31 33	+00 05 06	− 0.0700	− 0.910	
900.0	23 32 26	+01 19 42	+0.0210	+0.060	− 4
1286	23 32 34	− 02 39 18	+0.0523	− 0.841	
900.1	23 32 52	− 47 13 06	+0.0209	− 0.303	− 5
901.0	23 34 14	+00 53 24	+0.0820	− 0.090	
1287	23 35 21	− 69 22 30	+0.1700	− 0.127−	+117
902.0	23 36 41	− 72 59 18	+0.0313	− 0.731	+67
902.1	23 37 13	− 33 01 00	+0.0103	− 0.283	− 8
903.0	23 37 17	+77 21 12	− 0.0198	+0.156	− 42
904.0	23 37 23	+05 21 18	+0.0251	− 0.432	+5
904.1A	23 38 19	+20 05 18	+0.0171	+0.092	
904.1B	23 38 19	+20 05 18	+0.0171	+0.092	
905.0	23 39 26	+43 55 12	+0.0100	− 1.600	− 81
905.1	23 40 20	+00 39 06	− 0.0030	− 0.330	
1288	23 40 23	+30 32 54	− 0.0261	− 0.283	
1289	23 40 33	+36 15 42	+0.0767	− 0.147	
905.2A	23 41 23	+32 19 00	− 0.0200	− 0.060	
905.2B	23 41 21	+32 16 12	− 0.0200	− 0.060	+17
1290	23 41 48	+21 19 18	+0.0331	+0.082	
906.0	23 43 18	+35 58 36	+0.0270	+0.050	− 8
907.0	23 45 31	+48 44 12	+0.0650	+0.030	
907.1	23 45 50	− 13 15 54	+0.0140	+0.050	− 4
908.0	23 46 36	+02 08 12	+0.0656	− 0.962	− 65

Catalogue of Nearby Stars*

Name	Sp LC	V (mag)	B–V (mag)	π_t (0.001″)	M_V (mag)
895.0		10.05	+1.51	85	9.70
895.1		13.50		45	11.80
895.2		13.10	+0.20		
895.3	K7V	10.175	+1.32	49	8.63
1284	M2Ve	11.16	+1.51	74*	10.5:
895.4	K0V	6.762	+0.83	24	3.66
1285	dM4e	14.2		59*	13.0
896.0A		10.38	+1.56	155	11.33
896.0B		12.40		155	13.40
897.0A		11.00		77	10.40
897.0B		11.40		77	10.80
898.0	(K5E)	8.594	+1.25	81	8.14
899.0		11.10		80	10.60
900.0	M1VE	9.591	+1.35	58	8.41
1286		14.68	+1.92	139	15.40
900.1	G8V	8.538	+0.79	46	6.85
901.0		14.20		34	11.90
1287		13.44	+1.44	48*	11.8
902.0	K3V	7.096	+0.99	101	7.12
902.1	K1V	7.180	+0.81	43	5.35
903.0	K1III–IV	3.237	+1.03	65	2.30
904.0	F7V	4.125	+0.51	71	3.38
904.1A	K2V	8.281		45	6.55
904.1B		11.70		45	10.00
905.0	6V	12.29	+1.92	318	14.80
905.1		16.50		52	15.10
1288		14.36	+1.77	82	13.93
1289		12.67	+1.60	124	13.14
905.2A		11.68	+1.55	69	10.90
905.2B		12.90	+0.15	69	12.10
1290		13.31	+1.59	46	11.62
906.0	M0V	9.922		55	8.62
907.0		12.10		77	11.50
907.1	K8V	9.593		51	8.20
908.0	M2VE	8.991	+1.48	180	10.27

*Adapted from Gliese (1969), Gliese and Jahreiss (1979), and Grenon and Rufener (1981).

Catalogue of Nearby Stars*

Name	R.A.(1950) h m s	Dec.(1950) ° ′ ″	$\mu_{R.A.}$ (s yr^{-1})	$\mu_{Dec.}$ (″ yr^{-1})	V_r (km s^{-1})
1291	23 47 39	− 29 40 48	+0.0145	+0.023	
908.1	23 47 55	+30 04 30	+0.0199	+0.000	− 5
908.2	23 48 22	+19 40 00	+0.0620	− 0.140	
909.0A	23 49 57	+75 15 54	+0.0824	+0.056	+1
909.0B	23 49 57	+75 15 54	+0.0824	+0.056	
909.1	23 50 05	− 06 16 18	+0.0290	− 0.030	
910.0	23 50 36	+28 44 24	− 0.0110	+0.050	
911.0	23 52 12	− 22 03 18	− 0.0040	+0.160	+18
912.0	23 53 07	− 06 24 54	− 0.0320	− 0.370	
1292	23 55 07	+23 02 24	+0.0735	− 1.050	
913.0	23 56 07	+46 27 00	+0.0548	− 0.027	+4
1293	23 58 51	− 17 13 24	+0.0187	− 0.225	+8
914.0A	23 59 33	+26 49 00	+0.0628	− 0.985	− 37
914.0B	23 59 33	+26 49 00	+0.0628	− 0.985	
1294A	23 59 36	− 68 33 00	+0.0398	− 0.260	
1294B	23 59 36	− 68 33 00	+0.0398	− 0.260	

Catalogue of Nearby Stars*

Name	Sp LC	V (mag)	B–V (mag)	π_t (0.001″)	M_V (mag)
1291	K2V	7.94	+0.84	45*	6.2
908.1	K8V	9.349	+1.26	49	7.80
908.2		17.00		50	15.50
909.0A	K3V	6.369	+0.98	93	6.21
909.0B	(DM2)	11.70		93	11.50
909.1	(K0)	9.541		45	7.81
910.0	M0VE	9.755	+1.39	60	8.60
911.0	M0V	10.780		34	8.44
912.0		11.14	+1.46	63	10.10
1292		12.0	+1.4	76	11.4
913.0	M0V	9.620	+1.44	60	8.51
1293	dM1.5	10.9		45*	9.2
914.0A	G3V	5.758	+0.67	84	5.38
914.0B		11.00		84	10.60
1294A	M0V	9.66	+1.39	66*	8.8:
1294B		10.6		66*	10:

*Adapted from Gliese (1969), Gliese and Jahreiss (1979), and Grenon and Rufener (1981).

References

Abbott, D.C., Beiging, J.H., Churchwell, E., Torres, A.V.: Radio emission from galactic Wolf-Rayet stars and the structure of Wolf-Rayet winds. Astrophysical Journal 303, 239-261 (1986).

Abbott, D.C., Conti, P.S.: Wolf-Rayet Stars. Annual Review of Astronomy and Astrophysics 25, 113-150 (1987).

Acker, A., Ochsenbein, F., Stenholm, B., Tylenda, R., Marcout, J., Schohn, C. The Strasbourg - ESO catalogue of galactic planetary nebulae. Private communication (1989).

Acker, A., Stenholm, B., Tylenda, R: The absolute Hα fluxes for southern planetary nebulae. Astronomy and Astrophysics Supplement 77, 487-495 (1989).

Alcaino, G: Basic morphological data for galactic globular clusters. Vistas in Astronomy 23, 1-43 (1979).

Alcaino, G., Liller, W.: BVRI CCD photometry of the globular cluster NGC 6362. Astronomical Journal 91, 303-311 (1986).

Alcaino, G., Liller, W.: The ages of globular clusters derived from BVRI CCD photometry. In: *The Harlow Shapley symposium on globular cluster systems in galaxies.* (eds. J.E. Grindlay and A.G. Davis Philip). Boston: Kluwer Academic Publishers, 1988.

Alcaino, G., Liller, W., Alvarado, F.: BVRI CCD photometry of the globular cluster M 4 (NGC 6121). Astrophysical Journal 330, 569-578 (1988).

Allen, C.W.: *Astrophysical Quantities Third Edition.* London: The Athlone Press, (1976).

Allen, D.A.: A catalogue of symbiotic stars. Proceedings of the Australian Astronomical Society 5, 369-377 (1984).

Aller, H.D., Reynolds, S.P.: Evolution of the centimeter flux of 3C 58 and The Crab Nebula. In: *The Crab Nebula and related supernova remnants.* (eds. M.C. Kafatos, R.B.C. Henry). New York: Cambridge University Press, 1985).

Aller, L.J., Keyes, C.D.: A spectroscopic survey of 51 planetary nebulae. Astrophysical Journal Supplement 65, 405-428 (1987).

Anada, M.P., Sjorgren, W.L., Phillips, R.N., Bills, B.G.: A low-order gravity field of Venus and dynamical implications. Journal of Geophysical Research 85, 8303-8318 (1980).

Anders, E., Grevesse, N.: Abundances of the elements - meteoritic and solar. Geochimica et Cosmochimica 53, 197-214 (1989).

Anderson, J.D., et al.: Radio science with Voyager 2 at Uranus - Results on masses and densities of the planet and five principal satellites. Journal of Geophysical Research 92, 14877-14883 (1987).

Angel, J.R.P.: Magnetic white dwarfs. Annual Review of Astronomy and Astrophysics 16, 487-519 (1978).

Angel, J.R.P., Borra, E.F., Landstreet, J.D.: The magnetic fields of white dwarfs. Astrophysical Journal Supplement 45, 457-474 (1981).

Angelini, L. et al.: An EXOSAT observation of CTB 80. Astrophysical Journal (Letters) 330, L43-L46 (1988).

Armanoff, T.E.: Color-magnitude diagrams for six metal-rich, low latitude globular clusters. Astronomical Journal 96, 588-634 (1988).

Armanoff, T.E., Zinn, R.: Integrated-light spectroscopy of globular clusters at the infrared Ca II line. Astronomical Journal 96, 92-104 (1988).

Arnett, W.D., Bahcall, J.N., Kirshner, R.P., Woosley, S.E.: Supernova 1987A. Annual Review of Astronomy and Astrophysics 27, 629-700 (1989).

Aschenbach, B.: Galactic supernova remnants. Space Science Reviews 40, 4447-465 (1985).

Ashworth, W.B.: A probable Flamsteed observation of the Cassiopeia A supernova. Journal of the History of Astronomy 11, 1-9 (1980).

Aumann, H.H.: IRAS observations of matter around nearby stars. Publications of the Astronomical Society of the Pacific 97, 885-891 (1985).

Auriere, M., Le Fevre, O., Terzan, A: A possible optical counterpart for the M 15 globular cluster X-ray source. Astronomy and Astrophysics 138, 415-420 (1984).

Avduevski, V.S., et al.: Soft landing of Venera 7 on the Venus surface. Journal of Atmospheric Science 28, 263-269 (1971).

Baars, J.W.M., et al.: The absolute spectrum of Cas A; an accurate flux density scale and a set of secondary calibrators. Astronomy and Astrophysics 61, 99-106 (1977).

Babcock, H.W.: A catalog of magnetic stars. Astrophysical Journal Supplement 3, 141-148 (1958).

Backer, D.C., Fomalont, E.B., Goss, W.M.: Accurate timing and interferometer positions for the millisecond pulsar 1937 + 21 and the binary pulsar 1913 + 16. Astronomical Journal 90, 2275-2280 (1985).

Backer, D.C., Kulkarni, S.R., Heiles, C., Davis, M.M., Goss, W.M.: A millisecond pulsar. Nature 300, 615-618 (1982).

Bahcall, J.N., Davis, R., Wolfenstein, L.: Solar neutrinos: a field in transition. Nature 334, 487-493 (1988).

Bahcall, J.N. et al.: Standard solar models and the uncertainties in predicted capture rates of solar neutrinos. Reviews of Modern Physics 554, 767-799 (1982).

Bahcall, J.N., Ulrich, R.K.: Solar models, neutrino experiments and helioseismology. Reviews of Modern Physics 60, 297-372 (1988).

Barnes, T.G., Evans, D.S.: Stellar angular diameters and visual surface brightness I. Late spectral types. Monthly Notices of the Royal Astronomical Society 174, 489-502 (1976).

Barnes, T.G., Evans, D.S., Moffet, T.J.: Stellar angular diameters and visual surface brightness III. An improved definition of the relationship. Monthly Notices of the Royal Astronomical Society 183, 285-304 (1978).

Barnes, T.G., Evans, D.S., Parsons, S.B.: Stellar angular diameters and visual surface brightness II. Early and intermediate spectral types. Monthly Notices of the Royal Astronomical Society 174, 503-512 (1976).

Bartel, N. et al.: Radio observations of SN 1987A. In: *Supernova 1987A in the Large Magellanic Cloud* (eds. M. Kafatos, A.G. Michalitsianos). New York: Cambridge University Press 1988.

Bastian, T.S., Dulk, G.A., Chanmugam, G.: Radio flares from AE Aquarii - a low-power analog to Cygnus X-3. Astrophysical Journal 324, 431-440 (1988).

Baugher, J.F.: The space-age solar system. New York: John Wiley, 1988.

Becker, R.H., Helfand, D.J.: The dual nature of the supernova remnant

G351.2+0.1. Astronomical Journal 95, 883-886 (1988).

Becker, R.H., Marshall, F.E.: An X-ray survey of the supernova remnant G351.2+0.1. Astronomical Journal 95, 883-886 (1988).

Beech, M.: The Wolf-Rayet mass luminosity relation. Astrophysics and Space Science 139, 149-153 (1987).

Behannon, K.W., et al.: The magnetotail of Uranus. Journal of Geophysical Research 92, 15354-15366 (1987).

Bender, D.F.: Osculating elements of the asteroids. In: *Asteroids* (ed. T. Gehrels). Tucson: University of Arizona Press, 1979.

Bergh, S. Van Den: A study of reflection nebulae. Astronomical Journal 71, 990-998 (1966).

Bergh, S. Van Den: Supernova remnants and their relationship to W50 and S 433. Vistas in Astronomy 25, 109-118 (1981).

Bergh, S. Van Den: A catalogue of galactic supernova remnants. In: *Supernova remnants and their X-ray emission.* (eds. J. Danziger, P. Gorenstein). New York, D. Reidel, 1983.

Bergh, S. Van Den: Classification of supernovae and their remnants. Astrophysical Journal 327, 156-163 (1988).

Bergh, S. Van Den: The large-scale distribution of galactic supernova rremnants. Publications of the Astronomical Society of the Pacific 100, 205-206 (1988).

Bergh, S. Van Den: Summary of the George Mason University SN 1987A workshop. In: *Supernova 1987A in the Large Magellanic Cloud* (eds. M. Kafatos, A.G. Michalitsianos). New York: Cambridge University Press, 1988.

Bergh, S. Van Den, Herbst, W.: Catalogue of southern stars embedded in N nebulosity. Astronomical Journal 80, 208-210 (1975).

Bergh, S. Van Den, Marscher, A.P., Terzian, Y.: An optical atlas of galactic supernova remnants. Astrophysical Journal Supplement 26, No. 227, 19-36 (1973).

Bergh, S. Van Den, McClure, R.D., Evans, R.: The supernova rate in the Shapley-Ames galaxies. Astrophysical Journal 323, 44-53 (1987).

Berkhuijsen, E.M.: Properties of supernova remnants at known distances I. Surface brightness and radio spectral index. Astronomy and Astrophysics 166, 257-270 (1986).

Berkhuijsen, E.M.: Effect of ambient density on statistical properties of SNRS: Σ-D relation, N-D relation, energy content. In: *Supernova remnants and the interstellar medium.* (ed. R.S. Roger, T.L. Landecker. New York: Cambridge University Press, 1988.

Berriman, G.: A compilation of distances to cataclysmic variable stars. Astronomy and Astrophysics Supplement 68, 41-50 (1987).

Bertaud, C., Floquet, M.: New catalogue of A stars with peculiar spectra (Ap) and with metallic lines (Am). Astronomy and Astrophysics Supplement 16, 71-153 (1974).

Bignami, G.F., Hermsen, W.: Galactic gamma-ray sources. Annual Review of Astronomy and Astrophysics 21, 67-108 (1983).

Bills, B.G, Ferrari, A.J.: A harmonic analysis of lunar gravity. Journal of Geophysical Research 85, 1013-1025 (1980).

Bionta, R.M. et al.: Observation of a neutrino burst in coincidence with Supernova 1987A in the Large Magellanic Cloud. Physical Review Letters 58, 1494-1496 (1987).

Bixby, J.E., Van Flandern, T.C.: The diameter of Neptune. Astronomical Journal 74, 1220-1222 (1969).

Blaauw, A.: The O associations in the solar neighborhood. Annual Reviews of Astronomy and Astrophysics 2, 213-246 (1964).

Black, J.H., Raymond, J.C.: The distribution of interstellar matter toward Tycho's supernova remnant and its relation to distance estimates. Astronomical Journal 89, 411-416 (1984).

Blackwell, S.R., Purton, C.R.: Optical positions for northern stellar planetary nebulae. Astronomy and Astrophysics Supplement 46, 181-183 (1981).

Böhm-Vitense, E.: The effective temperature scale. Annual Review of Astronomy and Astrophysics 19, 295-318 (1981).

Böhm-Vitense, E.: *Introduction to stellar astrophysics 1. Basic stellar observations and data.* New York: Cambridge University Press, 1989.

Bolton, C.T.: Dimensions of the binary system HDE 226868: Cygnus X-1. Nature Physical Science 240, 124-127 (1972). reproduced in *A Source Book in Astronomy and Astrophysics, 1900-1975.* (eds. K.R. Lang and O. Gingerich). Cambridge, MA: Harvard University Press, 1979.

Bonnet-Bidaud, J.-M., Chardin, G.: Cygnus X-3, a critical review. Physics Reports 170, 325-404 (1988).

Bopp, B.W., Espenak, F.: Survey of the BY Draconis syndrome among dMe stars. Astronomical Journal 82, 916-924 (1977).

Boriakoff, V., Buccheri, R., Fauci, F.: Discovery of a 6.1-ms binary pulsar PSR 1953 + 29. Nature 304, 417-419 (1983).

Borra, E.F.: Magnetic stars. Annual Reviews of Astronomy and Astrophysics 20, 191-220 (1982).

Borra, E.F., Landstreet, J.D.: The magnetic fields of the Ap stars. Astrophysical Journal Supplement 42, 421-445 (1980).

Bowell, E., Gehrels, T., Zellner, B.: Magnitudes, colors, types and adopted diameters of the asteroids. In: *Asteroids* (ed. T. Gehrels). Tucson: University of Arizona Press, 1979.

Boynton, P.E. et al.: Vela X-1 pulse timing I. Determination of the neutron star orbit. Astrophysical Journal 307, 545-563 (1986).

Bradt, H.V.D., McClintock, J.E.: The optical counterparts of compact galactic X-ray sources. Annual Reviews of Astronomy and Astrophysics 21, 13-66 (1983).

Bridge, H.S., et al.: Plasma observations near Jupiter - initial results from Voyager 1. Science 204, 987-991 (1979).

Broadfoot, A.L., et al.: Extreme ultraviolet observations from Voyager 1 encounter with Jupiter. Science 204, 979-982 (1979).

Broadfoot, A.L. et al.: Ultraviolet spectrometer observations of Neptune and Triton. Science 246, 1459-1465 (1989).

Bruch, A., Fischer F.-J., Wilmsen, U.: An atlas and catalogue of northern dwarf novae. Astronomy and Astrophysics Supplement 70, 481-516 (1987).

Bruch, A., Sanders, W.L.: The absolute masses of 72 galactic clusters and 12 OB associations. Astronomy and Astrophysics 121, 237-240 (1983).

Burns, J.A. (ed.): *Planetary Satellites.* Tucson: University of Arizona Press, 1977.

Burns, J.A., Showalter, M.R., Morfill, G.E.: The ethereal rings of Jupiter and Saturn. In: *Planetary Rings* (eds. R. Greenberg and A. Brahic).Tucson: University of Arizona Press, 1984.

Burrows, A.: A statistical derivation of an upper limit to the electron neutrino mass from the SN 1987A neutrino data. Astrophysical Journal Letters 328, L51-L54 (1988).

Burrows, A.: Peering into the abyss - the neutrinos from SN 1987A. In: *Supernova 1987A in the Large Magellanic Cloud* (eds. M. Kafatos, A.G. Michalitsianos). New York: Cambridge University Press, 1988.

Cabrit, S., Goldsmith, P.F., Snell, R.L.: Identification of RNO 43 and B335 as two highly collimated bipolar flows oriented in the plane of the sky. Astrophysical Journal 334, 196-208 (1988).

Caillault, J.-P., Drake, S., Florkowski, D.: A VLA survey of BY Draconis Variables. Astronomical Journal 95, 887-893 (1988).

Calder, N.: *Timescale - an atlas of the fourth dimension.* Viking Press: New York, 1983.

Cameron, A.G.W.: Elemental and nuclidic abundances in the solar system. In: *Essays in nuclear astrophysics* (eds. C.A. Barnes, D.D. Clayton and D.N. Schramm). New York: Cambridge University Press, 1982.

Campbell, J.K., Synnott, S.P.: Gravity field of the Jovian system from Pioneer and Voyager tracking data. Journal of Geophysical Research 90, 364-372 (1985).

Canterna, R., Schwartz, R.D.: Photometry of LMC nova 1977b. Astrophysical Journal (Letters) 216, L91-L94 (1977).

Carney, B.W.: The ages and distances of eight globular clusters. Astrophysical Journal Supplement 42, 481-500 (1980).

Carney, B.W., Latham, D.W., Jones, R.V., Beck, J.A.: The distances and ages of globular clusters. In: *Calibration of stellar ages* (ed. A.G. Phillip Davis) Schenectady, N.Y.: Van Vleck Observatory Contribution No. 7, 1988.

Carr, M.H.: *The surface of Mars.* New Haven: Yale University Press, 1981.

Caswell, J.L.: Radio studies of supernova remnants - patterns and statistics. In: *Supernova remnants and the interstellar medium.* (eds. R.S. Roger, T.L. Landecker). New York: Cambridge University Press, 1988.

Caswell, J.L.: The frequency of supernovae in our Galaxy, estimated from supernova remnants detected at 178 MHz. Astronomy and Astrophysics 7, 59-64 (1970).

Caswell, J.L. et al.: Neutral hydrogen absorption measurements yielding kinematic distances for 42 continuum sources in the galactic plane. Astronomy and Astrophysics 45, 239-258 (1975).

Caswell, J.L., Lerche, I.: Galactic supernova remnants - dependence of radio brightness on galactic height and its implications. Monthly Notices of the Royal Astronomical Society 187, 201-216 (1979).

Catalano, F.A., Renson, P.: Catalogue des périodes observées pour des étoiles Ap. Astronomy and Astrophysics Supplement 55, 371-392 (1984).

Catalano, F.A., Renson, P.: Premier supplement au catalogue des périodes observées pour des étoiles Ap. Astronomy and Astrophysics Supplement 72, 1-17 (1988).

Chamberlain,J.W., Hunten, D.M.: *Theory of planetary atmospheres, second edition.* New York: Academic Press, 1987.

Chapman, C.R., Williams, J.G., Hartmann, W.K.: The asteroids. Annual Reviews of Astronomy and Astrophysics 16, 33-75 (1978).

Charles, P.A., Jones, D.C., Naylor, T: An emission-line object in the core of M 5. Nature 323, 417-419 (1986).

Chevalier, R.A., Kirshner, R.P., Raymond, J.C.: The optical emission from a fast shock wave with application to supernova remnants. Astrophysical Journal 235, 186-195 (1980).

Christensen, E.J., Balmino, G.: Development and analysis of a twelfth degree and order gravity model for Mars. Journal of Geophysical Research 84, 7943-7953 (1979).

Chu, Y.-H., Treffers, R.R., Kwitter, K.B.: Galactic ring nebulae associated with Wolf-Rayet stars VIII. Summary and atlas. Astrophysical Journal Supplement 53, 937-944 (1983).

Clark,D.H., Caswell, J.L.: A study of galactic supernova remnants, based on Molonglo-Parkes observational data. Monthly Notices of the Royal Astronomical Society 174, 267-305 (1976).

Clark, D.H., Stephenson, F.R.: *The historical supernovae.* New York: Pergamon Press, 1977.

Code, A.D., Davis, J., Bless, R.C., Hanbury Brown R.: Empirical effective temperatures and bolometric corrections for early-type stars. Astrophysical Journal 203, 417-434 (1976).

Cohen, E.R., Taylor, B.N.: The 1986 adjustment of the fundamental constants. Reviews of Modern Physics 59, 1121-1148 (1987).

Cohen, J.G., Rosenthal, A.J.: Nova shells. Astrophysical Journal 268, 689-697 (1983).

Cohen, M. et al.: The peculiar object HD 44179, the red rectangle. Astrophysical Journal 196, 179-189 (1975).

Cohen, M. Kuhi, L.V.: Observational studies of pre-main-sequence evolution. Astrophysical Journal Supplement 41, 743-843 (1979).

Connerney, J.E.P., Acuna, M.H., Ness, N.F.: The magnetic field of Uranus. Journal of Geophysical Research 92, 15329-15336 (1987).

Conrath, B. et al.: The helium abundance of Uranus from Voyager measurements. Journal of Geophysical Research 92, 15003-15010 (1987).

Conrath, B. et al.: Infrared observations of the Neptunian system. Science 246, 1454-1459 (1989).

Cook, W.R. et al.: An imaging observation of SN 1987A at gamma-ray energies. Astrophysical Journal (Letters) 334, L87-L90 (1988).

Cowley, A.P., Crampton, D., Hutchings, J.B., Remillard, R., Penfold, J.E.: Discovery of a massive unseen star in LMC X-3. Astrophysical Journal 272, 118-122 (1983).

Crampton, D., Cowley, A.P., Humphreys, R.M.: Spectroscopic observations of CRL 2688. Astrophysical Journal (Letters) 198, L135-137 (1975).

Crampton, D., Georgelin, Y.M.: The distribution of optical H II regions in our Galaxy. Astronomy and Astrophysics 40, 317-321 (1975).

Cropper, M.: The Polars. Space Science Reviews 54, 195-295 (1990).

Cunningham, C.J.: *Introduction to asteroids - the next frontier.* Richmond, Virginia: Willmann-Bell, 1988.

Cuzzi, J.N., et al.: Saturn rings - properties and processes. In: *Planetary rings* (eds. R. Greenberg and A. Brahic). Tucson: University of Arizona Press: 1984.

Da Costa, G.S., Seitzer, P.: NGC 6366 - the most metal-rich halo globular cluster? Astronomical Journal 97, 405-413 (1989).

Dachs, J., Isserstedt, J.: The dipole nebula IC 2220, a southern reflection nebula around a variable red giant HD 65750. Astronomy and Astrophysics 23, 241-245 (1973).

Dachs, J., Isserstedt, J., Rahe, J.: On the photometric variations of the red giant HD 65750 and of the surrounding reflection nebula IC 2220. Astronomy and Astrophysics 63, 353-362 (1978).

Damashek, M., Backus, P.R., Taylor, J.H., Burkhardt, R.K.: Northern hemisphere pulsar survey: a third radio pulsar in a binary system. Astrophysical Journal (Letters) 253, L57-L60 (1972).

Danks, A.C., Dennefeld, M., Wamsteker, W., Shaver, P.A.: Near infrared spectroscopy and infrared photometry of a new WC9 star. Astronomy and Astrophysics 118, 301-305 (1983).

Davidson, K., Fesen, R.A.: Recent developments concerning the Crab Nebula. Annual Review of Astronomy and Astrophysics 23, 119-146 (1985).

Davies, M.E., Colvin, T.R., Katayama, F.Y., Thomas, P.C.: The control networks of the satellites of Uranus. Icarus 71, 137-147 (1987).

Davies, M.E., et al.: Report of the IAU/IAG/COSPAR working group on cartographic coordinates and rotational elements of the planets and satellites 1985, 1982. Celestial Mechanics 39, 103-113 (1986), 29, 309-321 (1983).

Degewij, J., Van Houten, C.A.: Distant asteroids and outer Jovian satellites. In: *Asteroids* (ed. T. Gehrels). Tucson: University of Arizona Press, 1979.

De Jager, C., Nieuwenhuijzen, H., Van Der Hulst, K.A.: Mass loss rates in the Hertzsprung-Russell Diagram. Astronomy and Astrophysics Supplement 72, 259-289 (1988).

Delaporte, E.: *Delimitation scientifique des constellations.* Cambridge, England: Cambridge University Press, 1930.

Delsemme, A.H.: Chemical composition of cometary nuclei. In: *Comets* (ed. L.L. Wilkening). Tucson: University of Arizona Press, 1982.

Delsemme, A.H.: The nature of the cometary nucleus. Publications of the Astronomical Society of the Pacific 97, 861-870 (1985).

Delsemme, A.H.: The chemistry of comets. Philosophical Transactions of the Royal Society (London) A325, 509-523 (1988).

Dermott, S.F., Gold, T., Sinclair, A.T.: The rings of Uranus - nature and origin. Astronomical Journal 84, 1225-1234 (1979).

Desch, M.D., Kaiser, M.L.: Voyager measurement of the rotation period of Saturn's magnetic field. Geophysical Research Letters 8, 253-256 (1981).

Dewey, R.J., Maguire, C.M., Rawley, L.A., Stokes, G.H., Taylor, J.H.: Binary pulsar with a very small mass function. Nature 322, 712-714 (1986).

Dewey, R.J., Taylor, J.H., Maguire, C.M., Stokes, G.H.: Period derivatives and improved parameters for 66 pulsars. Astrophysical Journal 332, 762-769 (1988).

Dickel, J.R. et al.: The evolution of radio emission from Kepler's supernova remnant. Astrophysical Journal 330, 254-263 (1988).

Didelon, P.: Catalog of magnetic field measurements. Astronomy and Astrophysics Supplement 53, 119-137 (1983).

Didelon, P.: Stellar magnetic fields. Astronomy and Astrophysics Supplement 55, 69-75 (1984).

Dinerstein, H.L.: Classical novae detected in the IRAS survey. Astronomical Journal 92, 1381-1386 (1986).

Dolez, N., Vauclair, G., Chevreton, M.: Identification of the gravity modes in the newly discovered ZZ Ceti variable GD 66. Astronomy and Astrophysics Letters 21, L23-L26 (1983).

Doom, C., De Greve, J.P., De Loore, C.: Stellar evolution in the upper Hertzsprung-Russell diagram. Astrophysical Journal 303, 136-145 (1986).

Drake, S.A., Simon, T., Linsky, J.L.: A survey of the radio continuum of RS Canum Venaticorum and related active binary systems. Astrophysical Journal Supplement 71, 905-930 (1989).

Drechsel, H. et al.: The X-ray emission of the old nova V603 Aquilae (1918). Astronomy and Astrophysics 126, 357-362 (1983).

Drummond, J.D.: Earth orbit approaching comets and their theoretical meteor radiants. Icarus 47, 500-517 (1981).

Drummond, J.D.: A test of comet and meteor shower associations. Icarus 45, 545-553 (1981).

Duerbeck, H.W.: Light curve types, absolute magnitudes, and physical properties of galactic novae. Publications of the Astronomical Society of the Pacific 93, 165-175 (1981).

Duerbeck, H.W.: A reference catalogue and atlas of galactic novae. Space Science Reviews 45, 1-212 (1987).

Dulk, G.A.: Radio emission from the Sun and stars. Annual Review of Astronomy and Astrophysics 23, 169-224 (1985).

Dulk, G.A., Bastian, T.S., Chanmugam, G.: Radio emission from AM Herculis - the quiescent component and an outburst. Astrophysical Journal 273, 249-254 (1983).

Duncan, M., Quinn, T., Tremaine, S.: The formation and extent of the solar system comet cloud. Astronomical Journal 94, 1330-1338 (1987).

Dyal, P., Parkin, C.W., Daily, W.D.: Magnetism and the interior of the Moon. Review of Geophysics and Space Physics 12, 23-70 (1974).

Eggen, O.J., Greenstein, J.L.: Spectra, colors, luminosities and motions of white dwarfs. Astrophysical Journal 141, 83-108 (1965).

Eggen, O.J., Greenstein, J.L.: Observations of proper-motion stars II, III. Astrophysical Journal 142, 925-933 (1165), 150, 927-942 (1967).

Elliot, J.L.: Stellar occultation studies of the solar system. Annual Review of Astronomy and Astrophysics 17, 445-475 (1979).

Ensman, L.M., Woosley, S.E.: Explosions in Wolf-Rayet stars and Type Ib supernovae. Astrophysical Journal 333, 754-776 (1988).

Eshleman, V.R.: Pluto's atmosphere - models based on refraction, inversion and vapor pressure equilibrium. Icarus 80, 439-443 (1989).

Eshleman, V.R., et al.: Radio science with Voyager 1 at Jupiter - preliminary profiles of the atmosphere and ionosphere. Science 204, 976-978 (1979).

Eshleman, V.R., et al.: Radio science with Voyager at Jupiter - initial Voyager 2 results and a Voyager 1 measure of the Io torus. Science 206, 959-962 (1979).

Esposito, P.B., Anderson, J.D., NG, A.T.Y.: Experimental determination of Mercury's mass and oblateness. Space Research 17, 639-644 (1977).

Eubanks, T.M., Steppe, J.A., Dickey, J.O., Rosen, R.D., Salstein, D.A.: Causes of rapid motions of the Earth's pole. Nature 334, 115-119 (1988).

Everhart, E., Marsden, B.G.: New, original and future cometary orbits. Astronomical Journal 88, 135-137 (1983).

Fahlman, G.G., Gregory, P.C.: An X-ray pulsar in SNR G109.1-1.0. Nature 293, 202-205 (1981).

Fazio, C.G., et al: The effective temperature of Uranus. Astrophysical Journal 209, 633-637 (1976).

Fekel, F.C., Moffett, T.J., Henry, G.W.: A survey of chromospherically active stars. Astrophysical Journal Supplement 60, 551-576 (1986).

Feldman, W.C., Asbridge, J.R., Bame, S.J., Gosling, J.T.: Plasma and magnetic fields from the Sun. In: *The solar output and its variations* (ed. O.R. White) Boulder: Colorado Associated University Press, 1977.

Fernandez, J.A.: Dynamical aspects of the origin of comets. Astronomical Journal 87, 1318-1332 (1982).

Fesen, R.A.: The nature of the filaments northeast of the supernova remnant IC 443. Astrophysical Journal 281, 658-664 (1984).

Fesen, R.A., Becker, R.H., Goodrich, R.W.: Highest velocity ejecta of Cassiopeia A. Astrophysical Journal (Letters) 329, L89-L92 (1988).

Fesen, R.A. et al.: High-velocity ultraviolet iron, silicon, oxygen and sulfer absorption features associated with the remnant of SN 1006. Astrophysical Journal 327, 164-177 (1988).

Fesen, R.A., Kirshner, R.P., Becker, R.H.: 3C 58's filamentary radial velocities, line intensities, and proper motions. In: *Supernova remnants and the interstellar medium.* (eds. R.S. Roger, T.L. Landecker). New York: Cambridge University Press, 1988.

Filippenko, A.V.: On the nature and apparent uniqueness of SN 1987A. In: *Supernova 1987A in the Large Magellanic Cloud* (eds. M. Kafatos, A.G. Michalitsianos). New York: Cambridge University Press, 1988.

Fontaine, G. et al.: On the statistics of ZZ Ceti stars. Astrophysical Journal 258, 651-658 (1982).

Fontaine, G., Wesemael, F.: G 238-53, a new pulsating white dwarf. Astronomical Journal 89, 1728-1731 (1984).

Fransson, C. et al.: Narrow ultraviolet emission lines from SN 1987A - evidence for CNO processing in the progenitor. Astrophysical Journal 336, 429-441 (1989).

French, R.G., et al.: Uranian ring orbits from earth-based and Voyager occultation observations, Icarus 73, 349-378 (1988).

Friel, E.D., Heasley, J.N., Christian, C.A.: CCD photometry of the globular cluster NGC 4147. Publications of the Astronomical Society of the Pacific 99, 1248-1260 (1987).

Fruchter, A.S., Stinebring, D.R., Taylor, J.H.: A millisecond pulsar in an eclipsing binary. Nature 333, 237-239 (1988).

Furst, E. et al.: Multifrequency radio continuum observations of extended galactic objects II. Astronomy and Astrophysics Supplement 69, 403-413 (1987).

Gary, D.E.: Quiescent stellar radio emission. In: *Radio stars* (eds. R.M. Hjellming, D.M. Gibson) Boston: D. Reidel, 1985.

Gatewood, G.D., Gatewood, C.V.: A study of Sirius. Astrophysical Journal 225, 191-197 (1978).

Gautier, D., et al.: The helium abundance of Jupiter from Voyager. Journal of Geophysical Research 86, 8713-8720 (1981).

Gautier, D., Owen, T.: Cosmological implications of elemental and isotopic abundances in atmospheres of the giant planets. Nature 304, 691-694 (1983).

Gehrels, T. (ed.): *Jupiter*. Tucson: University of Arizona Press, 1976.

Gehrels, T. (ed.): *Asteroids*. Tucson: University of Arizona Press, 1979.

Geldzahler, B.J.: Radio frequency observations of galactic X-ray sources. Astrophysical Journal (Letters) 264, L49-L53 (1983).

Georgelin, Y.M., Georgelin, Y.P.: The spiral structure of our Galaxy determined from H II regions. Astronomy and Astrophysics 49, 57-79 (1976).

Georgelin, Y.M., Georgelin, Y.P., Roux, S.: Observations de nouvelles régions H II galactiques et d'étoiles excitatrices. Astronomy and Astrophysics 25, 337-350 (1973).

Georgelin, Y.P.: Comparison des vitesses radiales de régions H II obtenues en optique et en radio. Astronomy and Astrophysics 7, 322-329 (1970).

Georgelin, Y.P.,Georgelin, Y.M.: Régions H II de l'hemishere austral. Astronomy and Astrophysics Supplement 3, 1-49 (1970).

Georgelin, Y.P., Georgelin, Y.M.: Vitesses radiales et distances des régions H II galactiques. Astronomy and Astrophysics 6, 349-363 (1970).

Gies, D.R., Bolton, C.T.: The optical spectrum of HDE 226868 = Cygnus X-l. II. Spectrophotometry and mass estimates. Astrophysical Journal 304, 371-388 (1986).

Gliese, W.: Catalogue of nearby stars. *Veroffentlichungen des astronomischen rechen-instituts Heidelberg nr. 22*. Karlsruhe: Verlag G. Braun, 1969.

Gliese, W.: The nearest stars. In: *Landolt-Bornstein new series, volume 2b, astronomy and astrophysics - stars and star clusters* (eds. K. Schaifers, H.H. Voigt). New York: Springer Verlag, 1982.

Gliese, W., Jahreiss, H.: Nearby star data published 1969-1978. Astronomy and Astrophysics Supplement 38, 423-448 (1979).

Gliese, W., Jahreiss, H.: The third catalogue of nearby stars with special emphasis on wide binaries. Astrophysics and Space Science 142, 49-56 (1988).

Goettel, K.A., Shields, J.A., Decker, D.A.: Density constraints on the composition of Venus. Proceedings Lunar Science Conference 12B, 1507-1516 (1981).

Goldreich, P.: The dynamics of planetary rings. Annual Review of Astronomy and Astrophysics 20, 249-283 (1982).

Goldsmith, P.F.: Molecular clouds - an overview. In: *Interstellar Processes* (eds. D.J. Hollenbach, H.A. Thronson, Jr.). Boston: D.Reidel, 1987.

Gradie, J.C., Chapman, C.R., Williams, J.G.: Families of minor planets. In: *Asteroids* (ed. T. Gehrels). Tucson: University of Arizona Press, 1979.

Graham, A., Bevan, A.W.R., Hutchinson, R.: *Catalogue of meteorites.* Tucson: University of Arizona Press, 1985.

Grauer, A.D. et al.: PG 1456 + 103 - a new pulsating DB white dwarf. Astronomical Journal 95, 879-882 (1988).

Green, D.A.: Statistical studies of supernova remnants. Monthly Notices of the Royal Astronomical Society 209, 449-478 (1984).

Green, D.A.: A revised reference catalogue of galactic supernova remnants. Astrophysics and Space Science 148, 3-74 (1988).

Green, D.A.: Statistical studies of SNRs - selection effects. In: *Supernova shells and their birth events. Lecture notes in physics 316.* (ed. W. Kundt). New York: Springer Verlag, 1988.

Green, D.A.: The bulk radio expansion of Cassiopeia A. In: *Supernova remnants and the interstellar medium.* (eds. R.S. Roger, T.L. Landecker). New York: Cambridge University Press, 1988.

Greenstein, J.L.: White dwarfs in wide binaries I. Physical properties. Astronomical Journal 92, 859-866 (1986).

Greenstein, J.L.: White dwarfs in wide binaries II. Double degenerates and composite spectra. Astronomical Journal 92, 867-877 (1986).

Greenstein, J.L., Boksenberg, A., Carswell, R., Shortridge, K.: The rotation and gravitational redshift of white dwarfs. Astrophysical Journal 212, 189-197 (1977).

Greenstein, J.L., Oke, J.B.: An interpretation of the spectrum of the red rectangle. Publications of the Astronomical Society of the Pacific 89, 131-138 (1977).

Greenstein, J.L., Oke, J.B., Shipman, H.L.: Effective temperature, radius and gravitational redshift of Sirius B. Astrophysical Journal 169, 563-566 (1971).

Greenstein, J.L., Trimble, V.L.: The Einstein redshift in white dwarfs. Astrophysical Journal 149, 283-298 (1967).

Gregory, P.C. et al.: The radio, optical, X-ray (?), γ-ray (?)star LS I +61 303. Astrophysical Journal 84, 1030-1036 (1979).

Grenon, M., Rufener, F.: The colours, magnitudes and parallaxes of the nearby stars. Astronomy and Astrophysics Supplement 46, 25-40 (1981).

Grieve, R.A.F.: The impact cratering rate in recent time. Journal of Geophysical Research Supplement 89, B403-B408 (1984).

Grieve, R.A.F., Robertson, P.B.: The terrestrial cratering record I. Current status of observations. II. The crater production rate. Icarus 38, 212-229, 230-242 (1979).

Grieve, R.A.F., Dence, M.R.: The terrestrial cratering rate II. The crater production rate. Icarus 38, 230-242 (1979).

Grindlay, J.E.: X-ray binaries in globular clusters. In: *The Harlow Shapley symposium on globular cluster systems in galaxies.* (eds. J.E. Grindlay and A.G. Davis Philip) Boston: Kluwer Academic Publishers, 1988.

Grindlay, J.E., Bailyn, C.D.: Birth of millisecond pulsars in globular clusters. Nature 336, 48-50 (1988).

Grindlay, J.E., Hertz, P., Steiner, J.E., Murray, S.S., Lightman, A.P.: Determination of the mass of globular cluster X-ray sources. Astrophysical Journal (Letters) 282, L13-L16 (1984).

Grindlay, J.E., Seaquist, E.R.: Radio observations of galactic bulge and globular cluster X-ray sources. Astrophysical Journal 310, 172-175 (1986).

Gum, C.S.: A survey of southern H II regions. Memoirs of the Royal Astronomical Society 67, 155-177 (1955).

Gurzadyan, G.A.: *Flare Stars.* New York: Pergamon Press, 1980.

Haefner, R., Pietsch, W., Metz, K.: X-ray and optical observations of the old nova V 603 Aql. Astronomy and Astrophysics 200, 75-78 (1988).

Hall, D.S.: The RS Canum Venaticorum binaries. In: *Solar phenomena in stars and stellar systems* (eds. R.M. Bonnet, A.K. Dupree) Boston: D. Reidel, 1981. Also in: *Multiply periodic variable stars, IAU Colloquium 29* (ed. W.S. Fitch) Boston: D. Reidel, 1976.

Hamada, T., Salpeter, E.E.: Models for zero-temperature stars. Astrophysical Journal 134, 683-695 (1961).

Hanbury-Brown, R., Davis, J., Allen, L.R.: The angular diameter of 32 stars. Monthly Notices of the Royal Astronomical Society 167, 121-136 (1974).

Hanel, R.A. et al.: Infrared observations of the Jovian system from Voyager 1. Science 204, 972-976 (1979).

Hanel, R.A. et al.: Infrared observations of the Jovian system from Voyager 2. Science 206, 952-956 (1979).

Hanel, R.A., et al.: Albedo, internal heat, and energy balance of Jupiter, preliminary results from the Voyager infrared investigation. Journal of Geophysical Research 86, 8705-8712 (1981).

Hanel, R.A., et al.: Infrared observations of the Saturnian system from Voyager 2. Science 215, 544-548 (1982).

Hanel, R.A., et al.: Albedo, internal heat flux, and energy balance of Saturn. Icarus 53, 262-285 (1983).

Hanel, R.A., et al.: Infrared observations of the Uranian system. Science 233, 70-74 (1986).

Harrington, R.S., Christy, J.W.: The satellite of Pluto III. Astronomical Journal 86, 442-443 (1981).

Harris, D.L.: The stellar temperature scale and bolometric corrections. In: *Basic astronomical data - Stars and stellar systems III.* (ed. K. Strand). Chicago: University of Chicago Press, 1963.

Harris, D.L., Strand, K.A., Worley, C.E.: Empirical data on stellar masses, luminosities and radii. In: *Basic astronomical data - Stars and stellar systems III.* (ed.K. Strand). Chicago: University of Chicago Press, 1963.

Harris, W.E., Racine, R.: Globular clusters in galaxies. Annual Review of Astronomy and Astrophysics 17, 241-274 (1979).

Harris, W.E: The galactic distance scale - globular clusters. In: *Star Clusters* (ed. J.E. Hesser). Boston: Reidel, 1980.

Havlen, R.J.: OB star distribution in Puppis. Astronomy and Astrophysics 17, 413-424 (1972).

Hayes, R.F. et al.: A radio nebula associated with Circinus X-1. Nature 324, 233-235 (1986).

Hearn, D.R. et al.: MX 1313+29 - a compact source of very low energy X-rays in Coma Berenices. Astrophysical Journal (Letters) 203, L21-L24 (1976).

Heasley, J.N., Janes, K.A., Christian, C.A.: CCD and photographic photometry of NGC 1904 (M 79). Astronomical Journal 91, 1108-1116 (1986).

Heintz, W.D.: Astrometric study of four visual binaries. Astronomical Journal 79, 819-825 (1974).

Heise, J.: X-ray emission from isolated white dwarfs. Space Science Reviews 40, 79-90 (1985).

Heise, J. et al.: The high-resolution soft X-ray spectrum of HZ 43 II. Bolometric X-ray flux, effective temperature and radius. Astrophysical Journal 334, 958-970 (1988).

Helfand, D.J.: Bang- the supernova of 1987. Physics Today 40, 25-32 (1987)-August.

Helfand, D.J., Becker, R.H.: Observations of stellar remnants from recent supernovae. Nature 307, 215-221 (1984).

Helfand, D.J., Becker, R.H.: GO.9+0.1 and the emerging class of composite supernova remnants. Astrophysical Journal 314, 203-214 (1987).

Henderson-Sellers, A.: *Origin and evolution of planetary atmospheres.* Bristol: Adam Hilger, 1983.

Henrichs, H.F.: Spinup and spindown of accreting neutron stars. In: *Accretion-driven stellar X-ray sources.* (eds. W.H.G. Lewin and E.P.J. Van Den Heuvel). New York, Cambridge University Press, 1983.

Herbig, G.H.: The spectrum and structure of Minkowski's footprint, M1-92. Astrophysical Journal 200, 1-5 (1975).

Herbig, G.H., Kameswara Rao, N.: Second catalog of emission-line stars of the Orion population. Astrophysical Journal 174, 401-423 (1972).

Herbst, W.: R associations I. UBV photometry and MK spectroscopy of stars in southern reflection nebulae. Astronomical Journal 80, 212-226 (1975a).

Herbst, W.: R associations III. local optical spiral structure. Astronomical Journal 80, 503-509 (1975b)

Herbst, W.: R associations IV. Vela R2, a young stellar group. Astronomical Journal 80, 683-688 (1975c).

Herbst, W., Racine, R.: R associations V. Monoceros R2. Astronomical Journal 81, 840-844 (1976).

Hertz, P., Grindlay, J.E.: An X-ray survey of globular clusters and their X-ray luminosity function. Astrophysical Journal 275, 105-119 (1983).

Hesser, J.E., Harris, W.E., Venberg, D.A., Allwright, J.W.B., Shott, P., Stetson, P.B.: A CCD color-magnitude diagram of 47 Tucanae (NGC 104). Publications of the Astronomical Society of the Pacific 99, 739-808 (1987).

Hidayat, B., Supelli, K., Hucht, K.A. Van Der: The galactic distribution of Wolf-Rayet stars. In: *Wolf-Rayet stars - observations, physics, evolution. IAU symposium no. 99.* (eds. C.W.H. De Loore, A.J. Willis). Boston: D. Reidel, 1982.

Hippel, T. Von, Burnell, S.J. Bell, Williams, P.M.: A working catalogue of Herbig - Haro objects. Astronomy and Astrophysics Supplement 74, 431-442 (1988).

Hirano, N., Kemeya, O., Nakayama, M., Takadubo,K.: Bipolar outflow in B335. Astrophysical Journal (Letters) 327, L69-L72 (1988).

Hirata, K. et al.: Observation of a neutrino burst from the supernova SN 1987A. Physical Review Letters 58, 1490-1493 (1987).

Hirshfeld, A., Sinnott, R.W.: *Catalogue 2000.0. Volume 2. Double Stars, Variable Stars and Nonstellar Objects.* New York: Cambridge University Press, 1985.

Hjellming, R.M.: Radio emission from X-ray binaries and the proto-type jets of SS 433. In: *The physics of accretion onto compact objects.* (eds. K.O. Mason, M.G. Watson, N.E. White). New York: Springer-Verlag. 1986.

Hoffleit, D.: *Catalogue of bright stars, fourth revised edition.* Yale University Observatory and National Space Science Data Center, 1982.

Hoffmeister, C.: Meteor streams. In: *Landolt-Bornstein Group VI, Volume 1, Astronomy and Astrophysics* (ed. H.H. Voigt). New York: Springer-Verlag, 1965.

Holberg, J.B., Wesemael, F.: Analysis of Lyman alpha profiles in hot DA white dwarfs. Memorie della Societa Astronomica Italiana. Nuovo Serie 58, 67-73 (1987).

Holberg, J.B., Wesemal, F., Hubeny, I.: The far-ultraviolet energy distribution of Sirius B from Voyager 2. Astrophysical Journal 280, 679-687 (1984).

Howard, T., et al.: Mercury - results on mass, radius, ionosphere, and atmosphere from Mariner 10 dual frequency radio signal. Science 185, 179-180 (1974).

Hubbard, W.B., et al.: Interior structure of Saturn from Pioneer 11 gravity data. Journal of Geophysical Research 85, 5909-5916 (1980).

Hubbard, W.B., et al.: Results from observations of the 15 June 1983 occultation of the Neptune system. Astronomical Journal 90, 655-667 (1985).

Hubbard, W.B. et al.: Occultation evidence for an atmosphere on Pluto. Nature 336, 452-454 (1988).

Hucht, K.A. Van Der, Conti P.S., Lundstrom, I., Stenholm, B.: The sixth catalogue of galactic Wolf-Rayet stars, their past and present. Space Science Reviews 28, 227-306 (1981).

Hulse, R.A., Taylor, J.H.: Discovery of a pulsar in a binary system. Astrophysical Journal (Letters) 195, L51-L53 (1975).

Humphreys, R.M.: Studies of luminous stars in nearby galaxies I. Supergiants and O stars in the Milky Way. Astrophysical Journal Supplement 38, 309-350, 1978.

Humphreys, R.M., Nichols, M., Massey, P.: On the initial masses and evolutionary origins of Wolf-Rayet stars. Astronomical Journal 90, 101-108 (1985).

Hutchings, J.B., Crampton, D., Cowley, A.P.: The X-ray pulsars 2S 1145-619 and E 1145.1-6141: Optical identifications and a nearby supernova remnant. Astronomical Journal 86, 871-874 (1981).

Hutchings, J.B., Crampton, D., Cowley, A.P.: A spectrographic orbit for LMC X-1: Another massive X-ray source? Astrophysical Journal (Letters) 275, L43-L47 (1983).

Hutchings, J.B., Crampton, D., Cowley, A.P., Bianchi, L., Thompson, I.B.: Optical and uv spectroscopy of the black hole candidate LMC X-l. Astronomical Journal 94, 340-344 (1987).

Hutchings, J.B. et al.: The optical orbit of the X-ray pulsar binary 0535-668 (=A0538-66). Publications of the Astronomical Society of the Pacific 97, 418-422 (1985).

Iben, I., Tutukov, A.V.: Supernovae of Type I as end products of the evolution of binaries with components of moderate initial mass. Astrophysical Journal Supplement 54, 335-372 (1974).

Ilovaisky, S.A., Lequeux, J.: A study of galactic supernova remnants I. Distances, radio luminosity function and galactic distribution. Astronomy and Astrophysics 18, 169-185 (1972).

Ingersoll, A.P., et al.: Results of the infrared radiometer experiment on Pioneers 10 and 11. In: *Jupiter* (ed. T. Gehreis). Tucson: University of Arizona Press, 1976.

Ip, W.-H., Axford, W.I.: Atmospheres of the planets and satellites. In: *Landolt-Bornstein new series, volume 2a, astronomy and astrophysics.* (eds. K. Schaifers and H.H. Voigt). New York: Springer-Verlag, 1981.

Jackson, P.D., Kundu, M.R., White, S.M.: Quiescent and flaring radio emission from the flare stars AD Leonis, EQ Pegasi, UV Ceti, Wolf 630, YY Geminorum, and YZ Canis Minoris. Astronomy and Astrophysics 210, 284-294 (1989).

Jahreiss, H.: White dwarfs in the third catalogue of nearby stars. Memoire della Societa Astronomica Italiana 58, 53-62 (1987).

Janes, K.A.: Private communication (1989).

Janes, K.A., Adler, D.: Open clusters and galactic structure. Astrophysical Journal Supplement 49, 425-446 (1982).

Janes, K.A., Tilley, C., Lynga, G.: Properties of the open cluster system. Astronomical Journal 95, 771-784 (1988).

Jansen, F. et al.: EXOSAT observations of the Cassiopeia A supernova remnant. Astrophysical Journal 331, 949-957 (1988).

Jeffreys, H.: The variation of latitude. Monthly Notices of the Royal Astronomical Society 141, 255-268 (1968).

Jensen, K.A. et al.: EXOSAT observations of V471 Tauri. Astrophysical Journal (Letters) 309, L27-L31 (1986).

Jessberger, E.K.: Meteors. In: *Landolt-Bornstein, Group VI, Astronomy, astrophysics and space research. Volume 2a* (eds. K. Schaifers, H.H. Voigt). New York: Springer-Verlag, 1981.

Jessberger, E.K., Christoforldis, A., Kissel, J.: Aspects of the major elemental composition of Halley's dust. Nature 332, 691-695 (1988).

Johnson, H.M.: Hubble's variable nebula NGC 2261. Astronomical Journal 71, 224-227 (1966).

Johnson, T.V., Brown R.H., Pollack, J.B.: Uranus satellites - density and composition. Journal of Geophysical Research 92, 14884-14894 (1987).

Johnston, M.D. et al.: Location of the recurrent (LMC?) X-ray transient AO538-66 with the HEAO 1 scanning modulation collimator. Astrophysical Journal (Letters) 230, L11-L14 (1979).

Jordan, J.F., Lorell, J.: Mariner 9 - an instrument of dynamical science. Icarus 25, 146-165 (1975).

Joss, P.C., Rappaport, S.A.: Neutron stars in interacting binary systems. Annual Review of Astronomy and Astrophysics 22, 537-592 (1984).

Kafatos, M.C., Henry, R.C.B.: *The Crab Nebula and related supernova remnants.* New York: Cambridge University Press, 1985.

Kafatos, M.C., Michalitsianos, A.G.: *Supernova 1987A in the Large Magellanic Cloud.* New York: Cambridge University Press, 1988.

Kahn, S.M. et al.: Photospheric soft X-ray emission from hot DA white dwarfs. Astrophysical Journal 278, 255-265 (1984).

Kamper, K., Van Den Bergh, S.: Optical studies of Cassiopeia A. A definitive study of proper motions. Astrophysical Journal Supplement 32, 351-366 (1976).

Kamper, K., Van Den Bergh, S.: A definitive study of proper motions in Cassiopeia A. Publications of the Astronomical Society of the Pacific 88, 587 (1976).

Kaula, W.M.: *Theory of satellite goedesy.* Waltham, Mass.: Blaisdell, 1966.

Kazes, I., Caswell, J.L.: Distance estimates for the SNR G74.9+1.2 and a nearby point source. Astronomy and Astrophysics 58, 449-450 (1977).

Keenan, P.C.: The MK classification and its calibration. In: *Calibration of fundamental stellar quantities.* (eds. D.S. Hayes, L.E. Pasinetti, A.G. Davis Phillip). Boston: D. Reidel, 1985.

Keenan, P.C.: Spectral types and their uses. Publications of the Astronomical Society of the Pacific 99, 713-723 (1987).

Keihm, S.J., Langseth, M.G.: Lunar thermal regime at 300 km. Proceedings Eighth Lunar Science Conference 1, 371-398 (1977).

Kemp, J.C., Swedlund, J.B., Landstreet, J.D., Angel, J.R.P.: Discovery of circularly polarized light from a white dwarf star. Astrophysical Journal (Letters) 161, L77-L79 (1970). Reproduced in: *Source book in astronomy and astrophysics 1900-1975.* (eds. K.R. Lang, O. Gingerich). Cambridge, Mass.: Harvard University Press, 1979.

Kenyon, S.J.: *The symbiotic stars.* New York: Cambridge University Press, 1986.

Kenyon, S.J., Fernandez-Castro, T.: The cool components of symbiotic stars I. Optical spectral types. Astronomical Journal 93, 938-949 (1987).

Kenyon, S.J., Garcia, M.R.: Spectroscopic orbits for symbiotic stars I. The recurrent nova T Coronae Borealis. Astronomical Journal 91, 125-131 (1986).

Kepler, S.O.: The ZZ Ceti star GD 385 revisited. Astrophysical Journal 278, 754-760 (1984).

Kepler, S.O. et al.: The pulsation periods of the pulsating white dwarf G117-B15A. Astrophysical Journal 254, 676-682 (1982).

Kepler, S.O., Robinson, E.L., Nather, R.E.: The light curve of the ZZ Ceti star G 226-29. Astrophysical Journal 271, 744-753 (1983).

Kholopov, P.N.: A revised list of T-associations and their members. Soviet Astronomy A.J. 3, 291-300 (1959).

Kholopov, P.N.: *General Catalogue of Variable Stars, I, II, III.* Moscow: Nauka Publishing House, 1985, 1985, 1987.

Kieffer, H.H.: Soil and surface temperatures at the Viking landing sites. Science 194, 1344-1346 (1976).

Kirshner, R.P.: Supernova remnants and their Supernovae. In: *Supernova remnants and the interstellar medium.* (eds. R.S. Roger, T.L. Landecker), New York: Cambridge University Press, 1988.

Kirshner, R.P., Arnold, C.N.: Kinematics of the supernova remnant S 147. Astrophysical Journal 229, 147-152 (1979).

Kirshner, R.P., Winkler, P.F., Chevalier, R.A.: High-velocity emission in young supernova remnants SN 1006 and SN 1572. Astrophysical Journal (Letters) 315, L135-L139 (1987).

Klassen, K.P.: Mercury's rotation axis and period. Icarus 28, 469-478 (1976).

Klinglesmith, D.A., Hollis, J.M.: A Palomar observatory sky survey of selected molecular clouds. Astrophysical Journal Supplement 64, 127-218 (1987).

Koester, D.: Gravitational redshift in white dwarfs. Memorie della Societa Italiana, Nuovo Serie 58, 45-47 (1987).

Koester, D., Schulz, H., Wiedemann, V.: Atmospheric parameters and mass distribution of DA white dwarfs. Astronomy and Astrophysics 76, 262-275 (1979).

Krimigis, S.M.,et al.: Low-energy charged particle environment at Jupiter - a first look. Science 204, 998-1003 (1979).

Krimigis, S.M., et al.: Low-energy hot plasma and particles in Saturn's magnetosphere. Science 215, 571-577 (1982).

Krimigis, S.M. et al.: Hot plasma and energetic particles in Neptune's magnetosphere. Science 246, 1483-1489 (1989).

Kronk, G.W.: *Comets - a descriptive catalogue.* Hillside, New Jersey: Enslow Publishers, 1984.

Kronk, G.W.: *Meteor showers - a descriptive catalogue.* Hillside, New Jersey: Enslow Publishers, 1988.

Kundu, M.R., Jackson, P.D., White, S.M. Melozzi, M.: Microwave observations of the flare stars UV Ceti, AT Microscopii, and AU Microscopii. Astrophysical Journal 312, 822-829 (1987).

Kunkel, W.E.: Solar neighborhood flare stars - a review. In: *Variable stars and stellar evolution. IAU Symposium 67* (eds. V.E. Sherwood and L. Plaut). Boston: D. Reidel, 1975.

Lacy, C.H.: Radii of nearby stars. An application of the Barnes-Evans relation. Astrophysical Journal Supplement 34, 479-492 (1977).

Lada, C.J.: Cold outflows, energetic winds, and enigmatic jets around young stellar objects. Annual Review of Astronomy and Astrophysics 23, 267-317 (1985).

Lambeck, K.: *The Earth's variable rotation - geophysical causes and consequences.* New York: Cambridge University Press, 1980.

Lance, C.M., Mc Call, M.L., Uomoto, A.K.: Portrait of a novel nova - V1500 Cygni. Astrophysical Journal Supplement 66, 151-182 (1988).

Landecker, T.L., Roger, R.S., Higgs, L.,A.: Atomic hydrogen in a field of Cygnus X containing the supernova remnant G78.2+2.1. Astronomy and Astrophysics Supplement 39, 133-151 (1980).

Landstreet, J.D.: The measurement of magnetic fields in stars. Astronomical Journal 85, 611-620 (1980).

Lane, A.L., et al.: Photometry from Voyager 1 - initial results from the Uranian atmosphere, satellites and rings. Science 233, 65-70 (1986).

Lang, K.R.: The uncertainties of space and time. Vistas in Astronomy 28, 277-288 (1985).

Lang, K.R.: *Astrophysical Formulae.* New York: Springer-Verlag, 1986.

Lang, K.R.: Flare stars at radio wavelengths. In: *IAU Symposium No. 137 Flare Stars.* Boston: Kluwer Academic Publishers, 1990.

Lanzerotti, L.J., Krimigis, S.M.: Comparative magnetospheres. Physics Today 38, 24-34 (1985)- November.

Latter, W.B., Schmidt, G.D., Green, R.F.: The rotationally modulated Zeeman spectrum at nearly a billion Gauss of the white dwarf PG 1031+234. Astrophysical Journal 320, 308-314 (1987).

Lattimer, J.M., Yahil, A.: Neutrino energetics of SN 1987A. In: *Supernova 1987A in the Large Magellanic Cloud* (eds. M. Kafatos, A.G. Michalitsianos). New York: Cambridge University Press, 1988.

Ledoux, P., Renson, P.: Magnetic stars. Annual Review of Astronomy and Astrophysics 4, 293-351 (1966).

Levin, A.: Areas of the constellations. British Astronomical Association Handbook (1935).

Levreault, R.M.: Molecular outflows and mass loss in pre-main sequence stars. Astrophysical Journal 330, 897-910 (1988).

Liebert, J.: White dwarf stars. Annual Review of Astronomy and Astrophysics 18, 363-398 (1980).

Liebert, J., Angel, J.R.P., Landstreet, J.D.: The detection of an Hα Zeeman pattern in the cool magnetic white dwarf G99-47. Astrophysical Journal (Letters) 202, L139-L143 (1974).

Liebert, J., Dahn, C.C., Monet, D.G.: The luminosity function of white dwarfs. Astrophysical Journal 332, 891-909 (1988).

Liebert, J. et al.: Two new color-selected magnetic DA white dwarfs. Publications of the Astronomical Society of the Pacific 97, 158-164 (1985).

Liebert, J., Stockman, H.S.: The AM Herculis magnetic variables. In: *Cataclysmic variables and low-mass X-ray binaries.* (eds. D.Q. Lamb, J. Patterson). Boston: D. Reidel, 1985.

Liller, W., Mayer, B.: The rate of nova production in the Galaxy. Publications of the Astronomical Society of the Pacific 99, 606-609 (1987).

Lindal, G.F. et al.: Viking radio occultation measurements of the atmosphere and topography of Mars: data acquired during one Martian year of tracking. Journal of Geophysical Research 84, 8443 (1979).

Lindal, G.F., et al.: The atmosphere of Jupiter - an analysis of the Voyager radio occultation measurements. Journal of Geophysical Research 86, 8721-8727 (1981).

Lindal, G.F. et al.: The atmosphere of Titan: an analysis of the Voyager 1 radio occultation measurements. Icarus 53, 348-363 (1983).

Lindal, G.F., et al.: The atmosphere of Saturn: an analysis of the Voyager radio occultation measurements. Astronomical Journal 90, 1136-1146 (1985).

Lindal, G.F., et al.: The atmosphere of Uranus - results of radio occultation measurements with Voyager 2. Journal of Geophysical Research 92, 14937-15001 (1987).

Lindal, G.F., et al.: The atmosphere of Neptune: results of radio occultation measurements with the Voyager 2 spacecraft. Geophysical Research Letters (in press 1990).

Little, L.T., Bergman, P., Cunningham, C.T., Heaton, B.D., Knee, L.B.G., MacDonald, G.H., Richards, P.J., Toriseva, M.: IRAS 20188–3928 - molecular cloud with a very dense bipolar outflow. Astronomy and Astrophysics 205, 129-134 (1988).

Lizano, S., Heiles, C., Rodriguez, L.F., Koo, B.-C., Shu, F.H., Hasegawa, T., Hayashi, S., Mirable, I.F.: Neutral stellar winds that drive bipolar outflows in low-mass protostars. Astrophysical Journal 328, 763-776 (1988).

Long, K.S., Blair, W.P., Van Den Bergh, S.: An improved proper motion measurement of SN 1006. Astrophysical Journal 333, 749-753 (1988).

Lowenstein, R.F., Harper, D.A., Moseley, H.: The effective temperature of Neptune. Astrophysical Journal (Letters) 218, L145-L146 (1977).

Lundstrom, I., Stenholm, B.: Wolf-Rayet stars in open clusters and associations. Astronomy and Astrophysics Supplement 58, 163-192 (1984).

Lynds, B.T.: Catalogue of Bright Nebulae. Astrophysical Journal Supplement 12, No. 105, 163-185 (1965).

Lyne, A.G., Biggs, J.D., Brinklow, A., Ashworth, M., Mc Kenna, J.: Discovery of a binary millisecond pulsar in the globular cluster M 4. Nature 332, 45-47 (1988). Also see Nature 336, 226-227 (1988).

Lyne, A.G., Brinklow, A., Middleditch, J., Kulkarni, S.R., Backer, D.C., Clifton, T.R.: The discovery of a millisecond pulsar in the globular cluster M 28. Nature 328, 399-401 (1987).

Lyne, A.G., Manchester, R.N., Taylor, J.H.: The galactic population of pulsars. Monthly Notices of the Royal Astronomical Society 213, 613-639 (1985).

Lynga, G.: *Catalogue of open cluster data.* Sweden: Lund Observatory, 1987.

Lynga, G., Hannsson, G.: Nebulae of the southern Milky Way, an atlas. Astronomy and Astrophysics Supplement 6, 327-414 (1972).

Lynga, G., Palous, J.: The local kinematics of open star clusters. Astronomy and Astrophysics 188, 35-38 (1987).

Maciel, W.J.: A catalogue of distances of planetary nebulae. Astronomy and Astrophysics Supplement 55, 253-258 (1984).

Maddalena, R.J. et al.: The system of molecular clouds in Orion and Monoceros. Astrophysical Journal 303, 375-391 (1986).

Maeder, A.: Changes of surface chemistry for standard massive star evolution - cartography in the H-R diagram. Astronomy and Astrophysics 173, 247-262 (1987).

Mahoney, W.A. et al.: Gamma-ray observations of 56 Co in SN 1987A. Astrophysical Journal (Letters) 334, L81-L85 (1988).

Manchester, R.N., Taylor, J.H.: Observed and derived parameters for 330 pulsars. Astronomical Journal 86, 1953-1973 (1981).

Margon, B.: Observations of SS 433. Annual Review of Astronomy and Astrophysics 22, 507-536 (1984).

Margon, B., Bolte, M., Anderson, S.F.: An X-ray selected white dwarf of intermediate luminosity. Astronomical Journal 93, 1229-1235 (1987).

Margon, B. et al.: An intense extreme-ultraviolet source in Cetus. Astrophysical Journal (Letters) 210, L79-L82 (1976).

Margulis, M., Lada, C.J., Snell, R.L.: Molecular outflows in the Monoceros OB1 molecular cloud. Astrophysical Journal 333, 316-331 (1988).

Marsden, B.G.: The sungrazing comet group I. Astronomical Journal 72, 1170-1178 (1967).

Marsden, B.G.: *Catalog of cometary orbits.* International Astronomical Union, 1986.

Marsden, B.G.: The sungrazing comet group II. Astronomical Journal 98, 2306-2321 (1989).

Marsden, B.G., Roemer, E.: Basic information and references. In: *Comets* (ed. L.L. Wilkening). Tuscon: University of Arizona Press, 1982.

Marsden, B.G., Sekanina, Z., Everhart, E.: New osculating orbits for 110 comets and analysis of original orbits for 200 comets. Astronomical Journal 83, 64-71 (1978).

Massai, K. et al.: Circumstellar matter of SN 1987A and soft X-ray emission. Nature 335, 804-806 (1988).

Massey, P.: The masses of Wolf-Rayet stars. Astrophysical Journal 246, 153-160 (1981).

Massey, P.: Wolf-Rayet stars with massive companions. In: *Wolf-Rayet stars observations, physics, evolution. IAU symposium no. 99.* (eds C.W.H. De Loore, A.J. Willis). Boston: D. Reidel, 1982.

Masson, P.: Comparative geology of the satellites of the giant planets. Space Science Reviews 38, 281-324 (1984).

Mathieu, R.D., Benson, P.J., Fuller, G.A., Myers, P.C., Schild, R.E.: L43 an example of interaction between molecular outflows and dense cores. Astrophysical Journal 330, 385-398 (1988).

Matsui, Y., Long, K.S., Tuohy, I.R.: X-ray imaging observation of the supernova remnant PKS 1209-52 and its central compact X-ray source. Astrophysical Journal 329, 838-848 (1988).

McAlister, H.A.: High angular resolution measurements of stellar properties. Annual Review of Astronomy and Astrophysics 23, 59-87 (1985).

McClintock, J.E.: Black holes in X-ray binaries. In: *The physics of accretion onto compact objects.* (eds. K.O. Mason, M.G. Watson, N.E. White) New York:Springer-Verlag, 1986.

McClintock, J.E., Remillard, R.A.: The black hole binary A0620-00. Astrophysical Journal 308, 110-122 (1986).

McClure, R.D., Vandenberg, D.A., Bell, R.A., Hesser, J.E. and Stetson, P.B.: CCD photometry of the globular cluster M 68 (NGC 4590). Astronomical Journal 93, 1144-1165 (1987).

McCluskey, G.E., Kondo, Y.: On the mass-luminosity relation. Astrophysics and Space Science 17, 134-149 (1972).

McCook, G.P., Sion, E.M.: A catalog of spectroscopically identified white dwarfs. Astrophysical Journal Supplement 65, 603-671 (1987).

McGraw, J.T.: White dwarf pulsations - a review. Space Science Reviews 27, 601-611 (1980).

McGraw, J.T. et al.: Two new variable white dwarfs G 185-32 and G 191-16. Astrophysical Journal 250, 349-354 (1981).

McKinnon, W.B., Mueller, S.: Pluto's structure and composition suggest origin in the solar, not a planetary, nebula. Nature 335, 240-243 (1988).

McLaughlin, D.B.: The space distribution of novae. Astronomical Journal 51, 136-143 (1945).

McLaughlin, D.B.: The spectra of novae. In: *Stellar Atmospheres* (ed. J.L.

Greenstein). Chicago: University of Chicago Press, 1960.

Meisel, D.D., Morris, C.S.: Comet head photometry - past, present and future. In: *Comets* (ed. L.L. Wilkening). Tucson: University of Arizona Press, 1982.

Mermilliod, J.C.: Comparative studies of young open clusters III. Empirical isochronous curves and zero age main sequence. Astronomy and Astrophysics 97, 235-244 (1981).

Metheringham, S.J.,Wood, P.R., Faulkner, H.J.: A study of some southern planetary nebulae. Astrophysical Journal 334, 862-874 (1988).

Mewe, R. et al.: Detection of X-ray emission from stellar coronae with ANS. Astrophysical Journal (Letters) 202, L67-L91 (1975).

Mikolajewska, J., Friedjung, M., Kenyon, S.J.: *The symbiotic phenomenon. Proceedings of IAU colloquium 103.* Boston: Kluwer, 1988.

Millis, R.L., et al.: The diameter of Juno from its occultation of AG+01022. Astronomical Journal 86, 306-313 (1981).

Millis, R.L., et al.: The diameter of 88 Thisbe from its occultation of SAO 187124. Astronomical Journal 88, 229-235 (1983).

Milne, D.K.: Improved optical positions for 153 planetary nebulae. Astronomical Journal 78, 239-242 (1973).

Milne, D.K.: Optical positions for planetary nebulae II. Astronomical Journal 81, 753-758 (1976).

Milne, D.K.: A new catalogue of galactic SNRs corrected for distance from the galactic plane. Australian Journal of Physics 32, 83-92 (1979).

Milne, D.K.: An atlas of supernova remnant magnetic fields. Australian Journal of Physics 40, 7-23 (1987).

Mitchell, G.F., Allen, M., Beer, R., Dekany, R., Huntress, W., Maillard, J.-P.: The detection of high-velocity outflows from M8E-IR. Astrophysical Journal (Letters) 327, L17-L21 (1988). The detection of a discrete outflow from the young stellar object GL 490. Astronomy and Astrophysics 201, L16-L18 (1988).

Mitton, S.: *The Crab Nebula.* Boston: Faber and Faber, 1979.

Moffett, T.J., Barnes, T.G., Evans, D.S.: Applicability of the visual surface brightness relation to white dwarfs. Astronomical Journal 83, 820-827 (1978)

Molnar, L.A., Reid, M.J., Grindlay, J.E.: Low-level radio flares from Cygnus X-3. Nature 310, 662-665 (1984).

Molnar, L.A., Reid, M.J., Grindlay, J.E.: VLBI observations of expansion in Cygnus X-3. Astrophysical Journal 331, 494-508 (1988).

Moneti, A., Forrest, W.J., Pipher, J.L., Woodward, C.E.: High spatial resolution of L1551-IRS5 - direct observations of its circumstellar envelope. Astrophysical Journal 327, 870-875 (1988).

Morabito, L.A., et al.: Discovery of currently active extraterrestrial volcanism. Science 204, 972 (1979).

Morgan, W.W., Keenan, P.C.: Spectral classification. Annual Review of Astronomy and Astrophysics 11, 29-50 (1973).

Moriarity - Schieven, G.H., Snell, R.L.: High-resolution images of the L1551 molecular outflow II. structure and kinematics. Astrophysical Journal 332, 364-378 (1988).

Morini, M. et al.: EXOSAT observations of the supernova remnant G109.1-1.0 and the X-ray pulsar lE 2259+586. Astrophysical Journal 333, 777-787 (1988).

Morris, D.H., Mutel, R.L.: Radio emission from RS CVn binaries III. A VLA survey of 103 systems. Astronomical Journal 95, 204-214 (1988).

Morrison, D.: The satellites of Jupiter and Saturn. Annual Review of Astronomy and Astrophysics 20, 469-495 (1982).

Morrison, D., Cruikshank, D.P., Burns, J.A.: Introducing the satellites. In: *Planetary Satellites* (ed. J.A. Burns). Tucson: University of Arizona Press, 1977.

Morrison, D., Owen, T.: *The planetary system*. New York: Addison-Wesley, 1988.

Munari, U. et al.: The spectroscopic orbit of the symbiotic star EG Andromedae. Astronomy and Astrophysics 198, 173-178 (1988).

Mundy, L.G., Adelman, G.A.: Circumstellar material associated with GL 490. Astrophysical Journal 329, 907-913 (1988).

Munk, W.H., MacDonald, G.J.F.: *The rotation of the Earth*. New York: Cambridge University Press, 1975.

Murdin, P.: The LMC supernova in order of magnitude. Quarterly Journal of the Royal Astronomical Society 30, 419-429 (1989).

Nagle, D.E., Gaisser, T.K.: Extensive air showers associated with discrete astrophysical sources. Annual Review of Nuclear Science 38, 609-657 (1988).

Ness, N.F., et al.: Magnetic field studies at Jupiter by Voyager 1 - preliminary results. Science 204, 982-987 (1979).

Ness, N.F., et al.: Magnetic field studies at Jupiter by Voyager 2 - preliminary results. Science 206, 966-971 (1979).

Ness, N.F., et al.: Magnetic field studies by Voyager 2 - preliminary results at Saturn. Science 215, 558-563 (1982).

Ness, N.F., et al.: Magnetic fields at Uranus. Science 233, 85-89 (1986).

Ness, N.F. et al.: Magnetic fields at Neptune. Science 246, 1473-1477 (1989).

Ney, E.P. et al.: Studies of the infrared source CRL 2688. Astrophysical Journal (Letters) 198, L129-L134 (1975).

Nomoto, K.: Evolutionary models of the Crab Nebula's progenitor. In: *The Crab Nebula and related supernova remnants*. (eds. M.C. Kafatos, R.C.B. Henry). New York: Cambridge University Press, 1985.

Null, G.W.: Gravity field of Jupiter and its satellites from Pioneer 10 and Pioneer 11 tracking data. Astronomical Journal 81, 1153-1161 (1976).

Null, G.W., et al.: Saturn gravity results obtained from Pioneer 11 tracking data and Earth-based Saturn satellite data. Astronomical Journal 86, 456-468 (1981).

Ochsenbein, F., Halbawachs, J.L.: A list of stars with large expected angular diameters. Astronomy and Astrophysics Supplement 47, 523-531 (1982).

Ockert, M.E., et al.: Uranian ring photometry - results from Voyager 2. Journal of Geophysical Research 92, 14969-14978 (1987).

O'Donoghue, D.E., Warner, B.: Observations and analysis of the low amplitude ZZ Ceti variable L19-2. Monthly Notices of the Royal Astronomical Society 200, 563-583 (1982).

Oort, J.H.: The structure of the cloud of comets surrounding the solar system and a hypothesis concerning its origin. Bulletin of the Astronomical Institutes of the Netherlands 11, 91-98 (1950). In: *Source book in astronomy and astrophysics 1900-1975* (eds. K.R. Lang, O.Gingerich) Cambridge, Mass: Harvard University Press, 1979).

Osterbrock, D.E.: *Astrophysics of gaseous nebulae*. San Francisco: W.H. Freeman and Co., 1974.

Owen, W.M., Synnott, S.P.: Orbits of the ten small satellites of Uranus. Astronomical Journal 93, 1268-1271 (1987).

Paerels, F.B.S. et al.: The high-resolution photospheric soft X-ray spectrum of Sirius B. Astrophysical Journal 329, 849-862 (1988).

Pallavicini, R., Tagliaferri, G., Stella, L.: X-ray emission from solar neighborhood flare stars: a comprehensive survey of EXOSAT results. Astronomy and Astrophysics 228, 403-425 (1990).

Pandey, A.K., Bhatt, B.C., Mahra, H.S.: Age distribution of open clusters as a function of their linear diameter and age-dependence of cluster masses. Astrophysics and Space Science 129, 293-302 (1987).

Paradijs, J. Van: Optical observations of compact galactic X-ray sources In: *Accretion-driven stellar X-ray sources.* (eds. W.H.G. Lewin and E.P.J. Van Den Heuvel). New York: Cambridge University Press, 1983.

Parmar, A.N. et al.: X-ray observations of the OAO 1053-40 field. Monthly Notices of the Royal Astronomical Society 193, 49p-53p (1980).

Patterson, J.: Rapid oscillations in cataclysmic variables VI. periodicities in erupting dwarf novae. Astrophysical Journal Supplement 45, 517-539 (1981)

Peale, S.J.: The gravitational fields of the major planets. Space Science Reviews 14, 412-423 (1973).

Perek, L., Kohoutek, L.: *Catalogue of galactic planetary nebulae.* Prague: Academia Publishing House: Czechoslovak Academy of Sciences, 1967.

Perkins, H.G., King, D.J., Scarrott, S.M.: The Toby Jug Nebula (IC 2220) - a bipolar and biconical nebula. Monthly Notices of the Royal Astronomical Society 196, 403-408 (1981).

Petersen, C.J.: Ages of globular clusters. Publications of the Astronomical Society of the Pacific 99, 1153-1160 (1987).

Petersen, C.J., Reed, B.C.: Structural parameters and luminosities of globular clusters. Publications of the Astronomical Society of the Pacific 99, 20-26 (1987).

Peterson, R.C.: Radial velocities of remote globular clusters - stalking the missing mass. Astrophysical Journal 297, 309-313 (1985).

Petre, R., Shipman, H.L., Canizares, C.R.: An X-ray survey of hot white dwarf stars - evidence for a n(He)/n(H) versus T effective correlation. Astrophysical Journal 304, 356-364 (1986).

Pettengill, G., et al.: Pioneer Venus radar results - altimetry and surface properties. Journal of Geophysical Research 85, 8261-8270 (1980).

Pettersen, B.R.: Discovery of flare activity on - the dM5e star Gliese 268; BD 66 34 = Gliese 22A; the dM4e star Gliese 82; the red dwarf system Gliese 490 AB, BD + 36 2322; and G 141-29. Astronomy and Astrophysics 41, 87-90 (1975); 41, 113 (1975); Suppl. 30, 113-117 (1977); 97, 199-200 (1981), and Publications of the Astronomical Society of the Pacific 92, 188-190 (1980).

Pettersen, B.R.: Physical parameters of solar neighborhood flare stars. Astronomy and Astrophysics 82, 53-60 (1980).

Pettersen, B.R.: Global and photospheric physical parameters of active dwarf stars. In: *Activity in red-dwarf stars. IAU Colloquium No. 71.* (eds. P.B. Byrne, M. Rodono). Boston: D. Reidel, 1983.

Pietsch, W. et al.: LMC X-4, 13.5 s pulsations and an X-ray flare observed by EXOSAT. Space Science Reviews 40, 371-377 (1985).

Pilcher, F.: Circumstances of minor planet discovery. In: *Asteroids* (ed. T. Gehrels). Tucson: University of Arizona Press, 1979.

Pollack, J.B.: Origin and history of the outer planets. Annual Review of Astronomy and Astrophysics 22, 389-424 (1984).

Pollock, A.M.T.: The EINSTEIN view of the Wolf-Rayet stars. Astrophysical Journal 320, 283-295 (1987).

Popper, D.M.: Stellar masses. Annual Review of Astronomy and Astrophysics 18, 115-164 (1980).

Porco, C.C., Goldreich, P.: Shepherding the Uranian rings I. Kinematics. Astronomical Journal 93, 724-729 (1987).

Pottasch, S.R.: *Planetary Nebulae.* Boston: D. Reidel, 1984.

Pound, M.W., Janes, K.A., Heasley, J.N.: CCD photometry of the globular cluster NGC 288. Astronomical Journal 94, 1185-1201 (1987).

Racine, R.: Stars in reflection nebulae. Astronomical Journal 73, 233-245 (1968).

Rahe, J.: Comets. In: *Landolt-Bornstein Group VI, Volume 2a* (eds. K. Schaifers, H.H. Voigt). New York: Springer-Verlag, 1981.

Ramaty, R., Lingenfelter, R.E.: Gamma-ray line astronomy. Space Science Reviews 36, 305-317 (1983).

Reich, W. et al.: New supernova remnants from deep radio continuum surveys. In: *Supernova remnants and the interstellar medium.* (eds. T.S. Roger, T.L.Landecker). New York: Cambridge University Press, 1988.

Reich, W., Furst, E.: Statistical properties of recently new identified supernova remnants. In: *Supernova shells and their birth events. Lecture notes in physics 316.* (ed. W. Kundt). New York: Springer-Verlag, 1988.

Reifenstein, E.C. et al.: A survey of H 109α recombination line emission in galactic H II regions of the northern sky. Astronomy and Astrophysics 4, 357-377 (1970).

Reynolds, S.P., Aller, H.D.: Radio observations of the crab-like supernova remnant 3C 58 I. total intensity observations. Astrophysical Journal 327, 845-852 (1988).

Richer, H.B., Fahlman, G.G., Vandenberg, D.A.: Deep CCD photometry in globular cluster M 30 (NGC 7099). Astrophysical Journal 329, 187-207 (1988).

Ritter, H.: Catalogue of cataclysmic binaries, low-mass X-ray binaries and related objects. Astronomy and Astrophysics Supplement 70, 335-367 (1987).

Robinson, E.L.: The observational properties of the ZZ Ceti stars. In: *IAU Colloquium No. 53. White dwarfs and variable degenerate stars.* (eds. H.M. Van Horn, V. Weidmann) New York: University of Rochester, 1979.

Robiinson, E.L., Clemens,J.C., Hine, B.P.: Detection of the optical counterpart of the 555 second X-ray pulsations from V471 Tauri. Astrophysical Journal (Letters) 331, L29-L32 (1988).

Robinson, E.L. et al.: PG 1654+160 - a new pulsating DB white dwarf. Astrophysical Journal (Letters) 279, L15-L17 (1984).

Robinson, E.L., Kepler, S.O.: ZZ Ceti stars and the rate of evolution of white dwarfs. Space Science Reviews 27, 613-620 (1980).

Robinson, G.J., Reay, N.K., Atherton, P.D.: Measurements of expansion velocities in planetary nebulae. Monthly Notices of the Royal Astronomical Society 199, 649-657 (1982).

Roger, R.S., Landecker, T.L.: *Supernova remnants and the interstellar medium.* New York: Cambridge University Press, 1988.

Rogers, A.W., Campbell, C.T., Whiteoak, J.B.: A catalogue of Hα emission nebulae in the southern Milky Way. Monthly Notices of the Royal Astronomical Society 121, 103-110 (1960).

Ruiz, M.T., Maza, J.: ESO 439-162, a common proper motion binary formed by a magnetic DQ and a DC type white dwarf. Astrophysical Journal (Letters) 335, L15-L17 (1988).

Ruprecht, J.: Tables of OB associations. In: *Transactions of the International Astronomical Union XIIB* (ed. J.-C. Pecker). New York: Academic Press, 1966. pp. 336-356.

Russell, C.T., Elphic, R.C., Slavin, J.A.: Limits on the possible intrinsic magnetic field of Venus. Journal of Geophysical Research 85, 8319-8322 (1980).

Sabbadin, F.: A catalogue of expansion velocities in planetary nebulae. Astronomy and Astrophysics Supplement 58, 273-285 (1984).

Sabbadin, F., Hamzaouglu, E.: Internal motions in planetary nebulae. Astronomy and Astrophysics 110, 105-110 (1982).

Sabbadin, F., Strafella, F., Bianchini, A.: Internal motions of thirty-two genuine planetary nebulae and a misclassified object. Astronomy and Astrophysics Supplement 65, 259-265 (1986).

Saffer, R.A., Liebert, J., Olszewski, W.: Discovery of a close detached binary DA white dwarf system. Astrophysical Journal 334, 947-957 (1988).

Sandage, A.: The Osterhoff period groups and the age of globular clusters III. The age of the globular cluster system. Astrophysical Journal 252, 553-573 (1982).

Sandage, A.: On the age of M 92 and M 15. Astronomical Journal 88, 1159-1165 (1983).

Sandel, B.R., et al.: Extreme ultraviolet observations from the Voyager 2 encounter with Saturn. Science 215, 548-553 (1982).

Sandell, G., Magnani, L., Lada, E.A.: Detection of shock-enhanced CH emission in bipolar outflows. Astrophysical Journal 329, 920-926 (1988).

Sandie, W.G. et al.: High-resolution observations of gamma-ray line emission from SN 1987A. Astrophysical Journal (Letters) 334, L91-L94 (1988).

Sarajedini, A., King, C.R.: Evidence for an age spread among galactic globular clusters. Astronomical Journal 98, 1624-1633 (1989).

Sargent, A.I., Beckwith, S.: Kinematics of the circumstellar gas of HL Tauri and R. Monocerotis. Astrophysical Journal 323, 294 - 305 (1987).

Sargent, A.I., Beckwith, S., Keene, J., Masson, C.: Small-scale structure of the circumstellar gas around L1551-IRS5. Astrophysical Journal **333**, 936-942 (1988).

Scarf, F.L., et al.: Plasma wave observations at Saturn. Science 215, 587-594 (1982).

Schaeffer, O.A.: Lunar chronology as determined from the radiometric ages of returned lunar samples. Philosophical Transactions of the Royal Society (London) A285, 137-143 (1977).

Scheffler, H.: Reflection nebulae. In: *Landolt-Bornstein 2c. astronomy and astrophysics: interstellar matter, galaxy, universe.* (eds. K. Schaifers, H.H. Voigt) New York: Springer-Verlag, 1982).

Scheffler, H.: H II regions. In: *Landolt-Bornstein. Group VI, Volume 2c. astronomy and astrophysics, interstellar matter, galaxy, universe.* (eds. K. Schaifers, H.H. Voigt). New York: Springer-Verlag, 1982.

Scheffler, H., Elsasser, H.: *Physics of the Galaxy and interstellar matter.* New York: Springer-Verlag, 1987.

Schmidt, G.D.: The magnetic white dwarfs. Memorie della Societa Astronomica Italiana, Nuovo Serie 58, 77-81 (1987).

Schmidt, G.D.: Magnetic fields in white dwarfs. In: *White dwarfs, proceedings of IAU colloquium no. 114* (ed. G. Wegner). New York: Springer-Verlag, 1988.

Schmidt, G.D.: Private communication (1990).

Schmidt, G.D. et al.: The new magnetic white dwarf PG 1031+234 - polarization and field structure at more than 500 million Gauss. Astrophysical Journal 309, 218-229 (1986).

Schmidt, G.D., Liebert, J.: The period and magnetic field distributions of cataclysmic variables: implications for their evolution. Astrophysics and Space Science 131, 549-555 (1987).

Schmidt-Kaler, T.H.: Physical parameters of the stars. In: *Landolt-Bornstein New Series, Volume 2b, as-*

tronomy and astrophysics - stars and star clusters (eds. K. Schaifers, H.H. Voigt). New York: Springer-Verlag, 1982.

Schneider, S.E., Terzian, Y., Purgathofer, A., Perinotto, M.: Radial velocities of planetary nebulae. Astrophyical Journal Supplement 52, 399-423 (1983).

Schubert, J., Matson, D.L.: Masses and densities of asteroids. In: *Asteroids* (ed. T. Gehrels). Tucson: University of Arizona Press, 1979.

Seaquist, E.R., Taylor, A.R., Button, S.: A radio survey of symbiotic stars. Astrophysical Journal 284, 202-210 (1984).

Seargent, D.A.: *Comets - vagabonds of space.* New York: Doubleday and Company 1982.

Segelstein, D.J., Rawley, L.A., Stinebring, D.R., Fruchter, A.S., Taylor, J.H.: New millisecond pulsar in a binary system. Nature 322, 714-717 (1986).

Seidelmann, P.K., Doggett, L.E., Deluccia, M.R.: Mean elements of the principal planets. Astronomical Journal 79, 57-60 (1974).

Sekanina, Z.: Statistical model of meteor streams IV. A study of radio streams from the synoptic year. Icarus 27, 265-321 (1976).

Sekanina, Z.: The problem of split comets in review. In: *Comets* (ed. L.L. Wilkening). Tucson: University of Arizona Press, 1982.

Seward, F.D.: The Crab-like supernova remnants. Space Science Reviews 49, 385-423 (1989).

Seward, F.D.: Einstein observations of galactic supernova remnants. Astrophysical Journal Supplement 73, 781-819 (1990).

Seward, F.D., Wang, Z-R.: Pulsars, X-ray synchrotron nebulae, and guest stars. Astrophysical Journal 332, 199-205 (1988).

Shao, M. et al: Initial stellar diameter measurements with the Mark III interferometer. Astrophysical Journal 327, 905-910 (1988).

Shapiro, I.I., Campbell, D.B., DeCampli, W.M.: Nonresonance rotation of Venus? Astrophysical Journal (Letters) 230, L123-L126 (1979).

Sharpless, S.: A catalogue of H II regions. Astrophysical Journal Supplement 4, No. 41, 257-279 (1959).

Shaw, R.A., Kaler, J.B.: Apparent magnitudes of luminous planetary nebula nuclei I. Method and application. Astrophysical Journal 295, 537-546 (1985).

Shawl, S.J., White, R.E.: Accurate optical positions for the centers of galactic globular clusters. Astronomical Journal 91, 312-316 (1986).

Shipman, H.L.: Masses and radii of white dwarfs. Astrophysical Journal 177, 723-743 (1972).

Shipman, H.L.: Masses and radii of white-dwarf stars III. Results for 110 hydrogen-rich and 28 helium-rich stars. Astrophysical Journal 228, 240-256 (1979).

Shipman, H.L.: Hot DA white dwarfs seen with Einstein, an ultrahot H and He-poor white dwarf, and Zanstra temperatures of planetary nebulae nuclei. Memorie della Societa Astronomica Italiana, Nuovo Serie 58, 49-52 (1987).

Shipman, H.L., Sass, C.A.: Masses and radii of white-dwarf stars IV. The two-color diagram. Astrophysical Journal 235, 177-185 (1980).

Shoemaker, E.M., Williams, J.G., Helin, E.F., Wolfe, R.F.: Earth-crossing asteroids- orbital classes, collision rates with Earth and origin. In: *Asteroids* (ed. T. Gehrels). Tucson: University of Arizona Press, 1979.

Shore, S.N., Adelman, S.J.: An application of the Barnes-Evans relation to the peculiar A stars. Astronomical Journal 84, 559-561 (1979).

Sion, E.M., et al.: A proposed new white

dwarf spectral classification system. Astrophysical Journal 269, 253-257 (1983).

Slavin, J.A., Holzer, R.E.: The solar wind interaction with Mars revisited. Journal of Geophysical Research, 87, 10285 - 10296 (1982).

Smak, J.: Outbursts of dwarf novae. Publications of the American Astronomical Society of the Pacific 96, 5-18 (1984).

Smith, B.A., et al.: The Jupiter system through the eyes of Voyager 1. Science 204, 951-971 (1979).

Smith, B.A., et al.: The Galilean satellites and Jupiter - Voyager 2 imaging science results. Science 206, 927-950 (1979).

Smith, B.A., et al.: Encounter with Saturn - Voyager 1 imaging science results. Science 212, 163-191 (1981).

Smith, B.A., et al.: A new look at the Saturn system - the Voyager 2 images. Science 215, 504-537 (1982).

Smith, B.A., et al.: Voyager 2 in the Uranian system - imaging science results. Science 233, 43-64 (1986).

Smith, B.A., et al.: Voyager 2 at Neptune - imaging science results. Science 246, 1422-1450 (1989).

Smith, E.J., Davis, L., Jr., Jones, D.E.: Jupiter's magnetic field and magnetosphere. In: *Jupiter* (ed. T. Gehrels). Tucson: University of Arizona Press, 1976.

Smith, E.J., et al.: Saturn's magnetosphere and its interaction with the solar wind. Journal of Geophysical Research 85, 5655-5674 (1980).

Smith, L.F.: A revised spectral classification system and a new catalogue for galactic Wolf-Rayet stars. Monthly Notices of the Royal Astronomical Society 138, 109-121 (1968).

Standish, E.M., JR.: Orientation of the JPL ephemerides, DE200/LE200, to the dynamical equinox of J2000. Astronomy and Astrophysics 114, 297-302 (1982).

Stauffer, J.R.: Dynamical mass determinations for the white dwarf components of HZ 9 and Case 1. Astronomical Journal 94, 996-1002 (1987).

Stauffer, J.R., Hartmann, L.W.: Chromospheric activity, kinematics, and metallicities of nearby M dwarfs. Astrophysical Journal Supplement 61, 531-568 (1986).

Stephenson, F.R., Lieske, J.H.: Changes in the Earth's rate of rotation between A.D. 1672 and 1806 as deduced from solar eclipse timings. Astronomy and Astrophysics 200, 218-224 (1988).

Stephenson, F.R., Morrison, L.V.: Long-term changes in the rotation of the Earth - 700 B.C. to A.D. 1980. Philosophical Transactions of the Royal Society (London) A 313, 47-70 (1984).

Stern, S.A.: Constraints on Pluto's density and composition. Icarus 73, 269-278 (1988).

Stetson, P.B., Harris, W.E.: CCD photometry of the globular cluster M 92 (NGC 6341). Astronomical Journal 96, 909-975 (1988).

Stift, M.J.: The radii of the Ap stars. Astronomy and Astrophysics 34, 153-155 (1974).

Stift, M.J.: On the period-radius relation for magnetic stars. Astronomy and Astrophysics 50, 125-126 (1976).

Stockman, H.S., Schmidt, G.D., Lamb, D.Q.: V1500 Cygni - discovery of a magnetic nova. Astrophysical Journal 332, 282-286 (1988).

Stokes, G.H., Taylor, J.H., Dewey, R.J.: A new binary pulsar in a highly eccentric orbit. Astrophysical Journal (Letters) 294, L21-L24 (1985).

Stone, E.C.: The Voyager 2 encounter with Uranus. Journal of Geophysical Research 92, 14873-14876 (1987).

Stone, E.C., Lane, A.L.: Voyager 1 encounter with the Jovian system. Science 204, 945-948 (1979).

Stone, E.C., Lane, A.L.: Voyager 2 encounter with the Jovian system. Science 206, 925-929 (1979).

Stone, E.C., Miner, E.D.: Voyager 1 encounter with the Saturnian system. Science 212, 159-163 (1981).

Stone, E.C., Miner, E.D.: Voyager 2 encounter with the Saturnian system. Science 215, 499-504 (1982).

Stone, E.C., Miner, E.D.: The Voyager 2 encounter with the Uranian system. Science 233, 39-43 (1986).

Stone, E.C., Miner, E.D.: The Voyager 2 encounter with the Neptunian system. Science 246, 1417-1422 (1989).

Stothers, R.B.: Structure of Oort's comet cloud inferred from terrestrial impact craters. Observatory 108, 1-9 (1988).

Stover, R.J., et al.: Period stability of the pulsating white dwarf R548 = ZZ Ceti. Astrophysical Journal 240, 865-870 (1980).

Strassmeier, K.G., Hall, D.S., Zeilik, M., Nelson, E., Eker, Z., Fekel, F.C.: A catalogue of chromospherically active binary stars. Astronomy and Astrophysics Supplement 72, 291-345 (1988).

Strom, K.M.: Circumstellar material associated with solar-type pre-main-sequence stars - a possible constraint on the timescale for planet building. Astronomical Journal 97, 1451-1470 (1989).

Strom, R.G.: Distances to the remnants of historical Type I supernovae. Monthly Notices of the Royal Astronomical Society 230, 331-344 (1988).

Strom, R.G., Goss, W.M., Shaver, P.A.: Expansion of the supernova remnant 3C 10 (Tycho) and its implications for models of young remnants. Monthly Notices of the Royal Astronomical Society 200, 473-487 (1982).

Synnott, S.P., et al.: Orbits of the small satellites of Saturn. Science 212, 191-192 (1981).

Tammann, G.A.: On the frequency of supernovae as a function of the integral properties of intermediate and late type spiral galaxies. Astronomy and Astrophysics 8, 458-475 (1970).

Taylor, A.R., et al.: Radio images of the expanding ejecta of nova QU Vulpeculae 1984. Nature 335, 235-238 (1988).

Taylor, A.R., Gregory, P.C.: Periodic radio emission from LS I +61 303. Astrophysical Journal 255, 210- (1982).

Taylor, J.H.: Observed and derived parameters of pulsars, compiled by A.G. Lyne, R.N. Manchester and J.H. Taylor (1989).

Taylor, J.H., Dewey, R.J.: Improved parameters for four binary pulsars. Astrophysical Journal 332, 770-776 (1988).

Taylor, J.H., Fowler, L.A., Mc Colloch, P.M.: Measurements of general relativistic effects in the binary pulsar PSR 1913 + 16, Nature 277, 437-440 (1979).

Taylor, J.H., Weisberg, J.M.: Further experimental tests of relativistic gravity using the binary pulsar PSR 1913 + 16. Astrophysical Journal 345, 434-450 (1989).

Taylor, K.N.R., Scharrott, S.M.: The boomerang nebula - a highly polarized bipolar. Monthly Notices of the Royal Astronomical Society 193, 321-327 (1980).

Taylor, R.J.: The Sun as a star. Quarterly Journal of the Royal Astronomical Society 30, 125-161 (1989).

Tedesco, E.F.: Lightcurve parameters of asteroids. In: *Asteroids* (ed. T. Gehrels). Tucson: University of Arizona Press, 1979.

Tholen, D.J. et al.: Improved orbital and physical parameters for the Pluto - Charon system. Science 237, 512-514 (1987).

Tinsley, B.M.: What stars become supernovae? Publications of the Astronomical Society of the Pacific 87, 837-848 (1975).

Trimble, V.: Motions and structure of the filamentary envelope of the Crab

Nebula. Astronomical Journal 73, 535-547 (1968).

Trimble, V.: The distance to the Crab Nebula and NP 0532. Publications of the Astronomical Society of the Pacific 85, 579-585 (1973).

Trimble, V.: Supernovae. Part I - the events. Part II. the aftermath. Reviews of Modern Physics 54, 1183-1224 (1982), 55, 511-563 (1983).

Trimble, V.: The Crab Nebula and related supernova remnants - an overview. In: *The Crab Nebula and related supernova remnants* (eds. M.C. Kafatos, R.C.B. Henry). New York: Cambridge University Press, 1985.

Trimble, V.: 1987A - the greatest supernova since Kepler. Reviews of Modern Physics 60, 859-871 (1988).

Trimble, V., Greenstein, J.L.: The Einstein redshift in white dwarfs. Astrophysical Journal 177, 441-452 (1972).

Tuffs, R.J.: Secular changes within Cassiopeia A at 5 GHz. Monthly Notices of the Royal Astronomical Society 219, 13-38 (1986).

Turck-Chieze, S., Cahen, S., Casse, M., Doom, C.: Revisiting the standard solar model. Astrophysical Journal 335, 415-424 (1988).

Tylenda, R., Acker, A., Gleizes, F., Stenholm, B.: Magnitudes of central stars of southern planetary nebulae. Astronomy and Astrophysics Supplement 77, 39-44 (1989).

Tyler G.L., et al.: Radio science investigations of the Saturnian system with Voyager 1. Science 212, 201-206 (1981).

Tyler, G.L., et al.: Radio science with Voyager 2 at Saturn - atmosphere and ionosphere and the masses of Mimas, Tethys and Iapetus. Science 215, 553-558 (1982).

Tyler, G.L., et al.: Voyager 2 radio science observations of the Uranian system atmosphere, rings and satellites. Science 233, 79-85 (1986).

Tyler, G.L., et al.,: Voyager radio science observations of Neptune and Triton. Science 246, 1466-1473 (1989).

Tyson, J.A., Boeshaar, P.C.: Early optical spectroscopy of supernova 1987A. Publications of the Astronomical Society of the Pacific 99, 905-914 (1987).

Underhill, A.B.: The riddle of the Wolf-Rayet stars. Publications of the Astronomical Society of the Pacific 98, 897-913 (1986).

Ungerechts, H., Thaddeus, P.: A CO survey of the dark nebulae in Perseus, Taurus and Auriga. Astrophysical Journal Supplement 63, 6645-660 (1987).

Unsold, A.: Solar corona, interplanetary plasma and prominences. Astronomy and Astrophysics 4, 220-228 (1970).

U.S. Standard Atmosphere. National Oceanic and Atmospheric Administration NOAA-S/T76-1562., Washington, D.C.: U.S. Government Printing Office 1976.

Van De Kamp, P.: The nearby stars. Annual Review of Astronomy and Astrophysics 9, 127-146 (1971).

Vandenberg, D.A.: Ages of galactic globular clusters. In:*Harlow Shapley symposium on globular cluster systems in galaxies* (eds. J.E. Grindlay and A.G. Davis Philip) Boston: Kluwer Academic Publishers, 1988.

Van Den Bergh, S.: see Bergh, S. Van Den.

Van Der Klis, M., Bonnet-Bidaud, J.M.: The orbital parameters and the X-ray pulsation of Vela X-1 (4U 0900-40). Astronomy and Astrophysics 135, 155-170 (1984).

Vauclair, G., Dolez, N., Chevreton, M.: G255-2 - a new ZZ Ceti variable star. Astronomy and Astrophysics Letters 103, L17-L19 (1981).

Veverka, J., Burns, J.A.: The moons of Mars. Annual Review of Earth and Planetary Science 8, 527-558 (1980).

Veverka, J. et al.: Satellites of Uranus - disk integrated photometry from Voyager imaging observations. Journal of Geophysical Research 92, 14895-14904 (1987).

Veverka, J., Thomas, P., Synnott, S.: The inner satellites of Jupiter. Vistas in Astronomy 25, 245-262 (1982).

Voigt, G.-H., Behannon, K.W., Ness, N.F.: Magnetic field and current structures in the magnetosphere of Uranus. Journal of Geophysical Research 92, 15337-15346 (1987).

Vogt, N., Bateson, F.M.: An atlas of southern and equatorial dwarf novae. Astronomy and Astrophysics Supplement 48, 383-407 (1982).

Vogt, N., Moffat, A.F.J.: A comparison of photometric and kinematic distances for southern H II regions. Astronomy and Astrophysics 45, 405-409 (1975).

Vreux, J.Y., Dennefeld, Y., Andrillat, Y.: A catalogue of near infrared spectra of southern galactic Wolf-Rayet stars. Astronomy and Astrophysics Supplement 54, 437-458 (1983).

Wade, R.A., Ward, M.J.: Cataclysmic variables - observational overview. In:*Interacting binary stars* (eds. J.E. Pringle, R.A. Wade). New York: Cambridge University Press, 1985.

Walker C.K., Lada, C.J., Young, E.T., Martulis, M.: An unusual outflow around IRAS 16293-2422. Astrophysical Journal 332, 335-345 (1988).

Walter, F.W., Brown, A., Mathieu, R.D., Myers, P.C., Vrba, F.J.: X-ray sources in regions of star formation III. Naked T Tauri stars associated with the Taurus-Auriga complex. Astronomical Journal 96, 297-325 (1988).

Warner, B.: Intermediate polars. In: *Cataclysmic variables and low-mass X-ray binaries.* (eds. D.Q. Lamb, J. Patterson). Boston: D. Reidel, 1985.

Warner, B.: Absolute magnitudes of cataclysmic variables. Monthly Notices of the Royal Astronomical Society 227, 23-73 (1987).

Warren, W.H., Hesser, J.: A photometric study of the Orion OBl association I. Observational data. III. Subgroup analyses. Astrophysical Journal Supplement 34, 115-206 (1977); 36, 497-572 (1978).

Warwick, J.W., et al.: Energetic charged particles in Saturn's magnetosphere - Voyager 2 results. Science 215, 577-587 (1982).

Warwick, J.W., et al.: Voyager 2 radio observations of Uranus. Science 233, 102-106 (1986).

Warwick, J.W. et al.: Voyager planetary radio astronomy at Neptune. Science 246,498-1502 (1989).

Wasserman, L.H., et al.: The diameter of Pallas from its occultation of SAO 85009. Astronomical Journal 84, 259-268 (1979).

Wasson, J.T.: *Meteorites - Classification and Properties.* New York: Springer-Verlag, 1974.

Wasson, J.T.: *Meteorites.* New York: W.H. Freeman, 1985.

Watson, M.G.: X-ray properties of magnetic cataclysmic variable systems. In: *The Physics of Accretion onto Compact Objects. Lecture Notes in Physics 266.* (eds. K.O. Mason, M.G. Watson, N.E. White). New York: Springer-Verlag, 1986.

Webbink, R.F.: Structure parameters of galactic globular clusters. In: *Dynamics of star clusters.* Boston: Kluwer Academic Publishers, 1985.

Weekes, T.C.: Very high energy gamma-ray astronomy. Physics Reports 160, 1-121 (1988).

Wegner, G.: A spectroscopic survey of southern hemisphere white dwarfs III. Binaries containing a white dwarf. Monthly Notices of the Royal Astronomical Society 165, 271-294 (1973).

Wegner, G.: A new gravitational redshift for the white dwarf o2 Eri B. Astronomical Journal 84, 650-654 (1979), 85, 1255-1261 (1980).

Wegner, G.: On the kinematics and ages of wide binaries containing a white dwarf. Astronomical Journal 86, 264-270 (1981).

Wegner, G., Glass, I.S.: A new bipolar nebula in Centaurus. Monthly Notices of the Royal Astronomical Society 188, 327-330 (1979).

Weidemann, V., Koester, D.: Mass distribution parameters of white dwarfs and atmospheric parameters of ZZ Ceti stars. Astronomy and Astrophysics 132, 195-202 (1984).

Weiler, K.W.: A catalogue of plerionic and composite supernova remnants. In: *The Crab Nebula and related supernova remnants* (eds. M.C. Kafatos, R.B.C. Henry)New York: Cambridge University Press, 1985.

Weiler, K.W.: Plerions, regular and delux. In: *The Crab Nebula and related supernova remnants.* (eds. M.C. Kafatos, R.B.C. Henry). New York, Cambridge University Press, 1985.

Weiler, K.W., Panagia, N.: Vela X and the evolution of plerions. Astronomy and Astrophysics 90, 269-282 (1980).

Weiler, K.W., Sramek, R.A.: Supernovae and their remnants. Annual Review of Astronomy and Astrophysics 26, 295-341 (1988).

Wesselink, A.J., Paranya, K., De Vorkin, K.: Catalogue of stellar dimensions. Astronomy and Astrophysics 7, 257-289 (1972).

Westbrook, W.E., et al.: Observations of an isolated compact infrared source in Perseus. Astrophysical Journal 202, 407-414 (1975).

Whang, Y.C.: Magnetospheric magnetic field of Mercury. Journal of Geophysical Research 82, 1024-1030 (1977).

Whipple, F.L.: The cometary nucleus - current concepts. Astronomy and Astrophysics 187, 852-858 (1987).

White, N.E., Swank, J.H., Holt, S.S.: Accretion powered X-ray pulsars. Astrophysical Journal 270, 711-734 (1983).

White, N.M., Feierman, B.H.: A catalogue of stellar angular diameters measured by lunar occultation. Astronomical Journal 94, 751-770 (1987).

White, S.M., Jackson, P.D., Kundu, M.R.: A VLA survey of nearby flare stars. Astrophysical Journal Supplement 71, 895-904 (1989).

White, S.R.: Gamma-ray astronomy. In: *Encyclopedia of astronomy and astrophysics.* (eds. R.A. Myers, S.N. Shore). New York: Academic Press, 1989.

Wickramsinghe, D.T., Cooper, M.: Spectropolarimetry of the magnetic white dwarf PG 1015+014 - evidence for a 100-MG field. Monthly Notices of the Royal Astronomical Society 235, 1451-1465 (1988).

Wickramasinghe, D.T., Ferrario, L.: A centered dipole model for the high magnetic white dwarf GRW +70 8247. Astrophysical Journal 327, 222-233 (1988).

Williams, J.G.: Proper orbital elements and family memberships of the asteroids. In: *Asteroids* (ed. T. Gehrels). Tucson: University of Arizona Press, 1979.

Wilson, T.L. et al.: A survey of H 109α recombination line emission in galactic H II regions of the southern sky. Astronomy and Astrophysics 6, 364-384 (1970).

Winget, D.E.: Progress and problems in the study of the pulsating white dwarfs. Highlights of Astronomy 7, 221-228 (1986).

Winget, D.E., et al.: Photometric observations of GD 358 - DB white dwarfs do pulsate. Astrophysical Journal (Letters) 262, L11-L15 (1982).

Winget, D.E., et al.: PG 1654+160 - a new pulsating DB white dwarf. Astrophysical Journal (Letters) 279, L15-L17 (1984).

Winget, D.E., Nather, R.E., Hill, J.A.: Two extreme examples of compact pulsators PG 1115+158 and PG 1351+489. Astrophysical Journal 316, 305-310 (1987).

Winkler, P.F., et al.: Kinematics of oxygen-rich filaments in Puppis A. In: *Supernova remnants and the interstellar medium* (eds. R.S. Roger,

T.L. Landecker). New York: Cambridge University Press, 1988.

Witt, A.N., Schild, R.E.: CCD surface photometry of bright reflection nebulae. Astrophysical Journal Supplement 62, 839-852 (1986).

Wolszczan, A.: Pulsar in supernova remnant W44. Private communication (1989).

Wolszczan, A., Kulkarni, S.R., Middleditch, J., Backer, D.C., Fruchter, A.S. Dewey, R.J.: A 110-ms pulsar, with negative period derivative, in the globular cluster M 15. Nature 337, 531-553 (1989).

Woltjer, L.: The Crab Nebula. Bulletin of the Astronomical Institutes of the Netherlands 14, 39-80 (1958).

Woltjer, L.: Supernova remnants. Annual Review of Astronomy and Astrophysics 10, 129-158 (1972).

Woosley, S.E.: SN 1987A - after the peak. Astrophysical Journal 330, 218-253 (1988).

Woosley, S.E., Phillips, M.M.: Supernova 1987A! Science 240, 750-759 (1988).

Worden, S.P., Stein, M.K.: Angular diameter of the asteroids Vesta and Pallas determined from speckle observations. Astronomical Journal 84, 140-142 (1979).

Wright, A.E. et al.: Detection of the AM Her type cataclysmic variable V834 Cen at radio wavelengths. Monthly Notices of the Royal Astronomical Society 231, 319-324 (1988).

Wyckoff, S. Wehinger, P.A.: Identifications and astrometric positions of old novae. Publications of the Astronomical Society of the Pacific 30, 557-562 (1978).

Yung, Y.L., Allen, M., Pinto, J.P.: Photochemistry of the atmosphere of Titan: comparison between model and observation. Astrophysical Journal Supplement 55, 465-506 (1984).

Zebker, R.A., Tyler, G.L.: Thickness of Saturn's rings inferred from Voyager 1 observations of microwave scatter. Science 223, 396-398 (1984).

Zellner, B., Tholen, D.J., Tedesco, E.F.: The eight-color asteroid survey - results for 589 minor planets. Icarus 61, 355-416 (1985).

Zellner, B., Thirunagari, A., Bender, D.: The large-scale structure of the asteroid belt. Icarus 62, 505-511 (1985).

Zinn, R.: The globular cluster system of the Galaxy IV. The halo and disk subsystems. Astrophysical Journal 293, 424-444 (1985).

Zinn, R.: Private communication (1989).

Zinn, R., West, M.J.: The globular cluster system of the Galaxy III. Measurements of radial velocity and metallicity for 60 clusters and a compilation of metallicities for 121 clusters. Astrophysical Journal Supplement 55, 45-66 (1984).

Index